PRINCIPLES OF ELECTROMAGNETICS

KNOW THE AUTHORS

S C Mahapatra

Sarat Chandra Mahapatra is former professor of electrical engineering, University College of Engineering (UCE) Burla, which is affiliated to Sambalpur University (currently known as Veer Surendra Sai University of Technology), Orissa. He has served the Department of Electrical Engineering at UCE, Burla, from the year 1959, when he had joined as a lecturer, till 1994, when he retired. Dr Mahapatra completed BSc (Engg) in Electrical Engineering in 1958 and MSc (Engg) in Electrical Machine Design in 1964 from Banaras Hindu University (currently known as IT-BHU). He obtained a doctoral degree in Electrical Engineering from Sambalpur University in 1974. He has supervised one PhD thesis in Electrical Engineering and published several papers on starting of single-phase induction motors with a single stator winding. At UCE, Burla, Dr Mahapatra has taught a number of undergraduate level courses like Field Theory, Theory of Electrical Machines, Electrical Machine Design, Basic Electrical Engineering, Linear Control Systems; and postgraduate level courses like Synchronous Machines. As a UGC Fellow, he has taught courses like Network Analysis and Control Systems to first semester postgraduate students of Electronic Sciences at Berhampur University, Orissa.

S Mahapatra

Sudipta Mahapatra is currently Associate Professor in the Department of Electronics and Electrical Communication Engineering (E&ECE) at the Indian Institute of Technology (IIT) Kharagpur, West Bengal. Before joining IIT Kharagpur in 2002, he served the Department of Computer Science and Engineering at National Institute of Technology, Rourkela (formerly Regional Engineering College, Rourkela) in various capacities from April 1993. He completed his BSc (Engg) in Electronics and Telecommunication Engineering from UCE, Burla, in the year 1990. He obtained his MTech and PhD degrees in Computer Engineering from the E&ECE Department at IIT Kharagpur, in 1992 and 1997, respectively. Prof. Mahapatra has taught a number of undergraduate and postgraduate level courses such as Microprocessors, Advanced Microprocessors, Computer Organisation, Computer Architecture, Advanced Operating System, Artificial Neural Networks, and Computer Communications and Networking. He has to his credit a number of publications in peer reviewed international journals and conferences.

PRINCIPLES OF ELECTROMAGNETICS

S C Mahapatra

Formerly Professor
Department of Electrical Engineering
University College of Engineering, Burla
Sambalpur University
Orissa

S Mahapatra

Associate Professor
Department of Electronics and Electrical Communication Engineering
Indian Institute of Technology Kharagpur
West Bengal

Tata McGraw Hill Education Private Limited

NEW DELHI

McGraw-Hill Offices

New Delhi New York St Louis San Francisco Auckland Bogotá Caracas
Kuala Lumpur Lisbon London Madrid Mexico City Milan Montreal
San Juan Santiago Singapore Sydney Tokyo Toronto

Tata McGraw-Hill

Published by the Tata McGraw Hill Education Private Limited,
7 West Patel Nagar, New Delhi 110 008.

Principles of Electromagnetics

78-1-25-900315-8
-25-900315-9

Vice President and Managing Director—McGraw-Hill Education, Asia-Pacific Region: *Ajay Shukla*

Head—Higher Education Publishing and Marketing: *Vibha Mahajan*
Manager: Sponsoring—SEM & Tech Ed.: *Shalini Jha*
Assoc. Sponsoring Editor: *Suman Sen*
Editorial Researcher: *Koyel Ghosh*
Sr Copy Editor: *Nimisha Kapoor*
Sr Production Manager: *Satinder S Baveja*
Production Executive: *Anuj Kr Shriwastava*

Dy Marketing Manager—SEM & Tech Ed.: *Biju Ganesan*
Sr Product Specialist—SEM & Tech Ed.: *John Mathews*

General Manager—Production: *Rajender P Ghansela*
Asst General Manager—Production: *B L Dogra*

Typeset at BeSpoke Integrated Solutions, Puducherry, India 605 008 and printed at Pashupati Printers Pvt. Ltd., 1/429/16, Gali No. 1, Friends Colony, Industrial Area, G.T. Road, Shahdara, Delhi - 110 095
Cover Printer: A. P. OFFSET

RAXACRCCDXCCL

Dedicated to the readers

CONTENTS

PREFACE

Electromagnetics is the study of principles, theory, analysis, computation, and application of electric and magnetic fields, their coexistence, and propagation in unbounded and bounded regions. This is a subject that students in general find difficult to comprehend, especially at the junior undergraduate level. The material in this book, covered in ten chapters, is suitable for a one-semester course, taught in the third or fourth semester, to undergraduate level electrical engineering, electrical and electronics engineering, and electronics and communications engineering students.

Principles of many electrical and electronic equipments and devices are based on the laws of electric and magnetic fields. Some of these are transformers, rotating and linear electromagnetic apparatuses, electromagnetic instruments and relays, electromagnetic levitation, lifting electromagnets, electrostatic precipitators, electronic display systems, radio direction and range (RADAR), particle accelerators, and resonators. Some other important applications of electromagnetic field laws are in terrestrial, satellite and space communications.

Study of electromagnetics will help students better understand electromagnetic devices, and propose innovative solutions to problems such as electromagnetic interference, and electromagnetic compatibility, encountered in various fields like power systems, wireless communication, and biomedical engineering. A design engineer is required to have the basic knowledge of electromagnetics to understand the design principles involved in connecting cell phones, laptop computers, and personal digital assistants to the Internet through wireless access points using RF signals. Another potential area of application of electromagnetics is the use of RF signals in high performance computer systems for intercomponent data transfer. This book therefore aims to assist students in a holistic study of this subject.

What makes this book unique?

This book presents the subject in a simple and systematic manner. It adheres to precise explanations of topics to help students grasp concepts better. A number of two-dimensional and three-dimensional illustrations have been included for easy understanding. Mathematical steps have been explained stepwise; leaving the final step, such as differentiation or integration, for the students to work out. Each chapter starts with an introduction, followed by theory and applications. Most of the sections in a chapter start with an introduction. A dedicated chapter on 'Solution of Boundary-Value Problems', for both static electric and magnetic fields has been included.

Every chapter contains a number of solved examples to acquaint the students with problem solving techniques. Summaries and important formulae are given after the main text in all the chapters to aid in quick revision. Questions, problems, and multiple-choice questions given at the end of each chapter will help students in evaluating their understanding of the chapter's content. Questions have been especially graded with either one, two or three stars to indicate challenge levels. Answers for selected questions have also been provided. Multiple-choice questions with two answers have been marked with a star.

Book Overview

The subject has been discussed in a comprehensive manner wherein a chapter each has been dedicated to an important topic. Contents of the book follow a logical flow and are described ahead briefly.

Electric and magnetic field quantities are vectors in three-dimensional space. These are represented by the use of suitable coordinate systems. Moreover, fields varying sinusoidally with time are vectors as well as phasors. For proper understanding of the subject, students should have adequate knowledge of at least three basic coordinate systems, fundamentals of vector analysis, and representation of sinusoidally time-varying quantities. Brief descriptions of these topics are given in **Chapter 1**.

Chapter 2 covers laws and theorems of electrostatic fields and their applications. Some of the important topics available in this chapter are vector forms of Coulomb's law and electric field intensity, divergence of electric flux density and divergence theorem, potential gradient, relationship between electric susceptibility and relative permittivity, boundary conditions, energy stored in electric field, and calculation of capacitance for various types of capacitors and transmission lines. The study of electrostatics will help students understand the principles of equipments and devices based on electrostatic laws. Insulation coordination of high voltage equipments requires calculation of potential gradient in the insulating medium. Calculation of capacitance of transmission lines is necessary to develop their circuit models.

An important relation between electric field intensity and conduction current density, known as point form of Ohm's law, is derived in **Chapter 3**. This law, also valid for time-varying fields, is an extension of Maxwell's equation. The chapter also contains Kirchhoff's voltage law in terms of field quantities, continuity equation for direct currents, Joule's law, calculation of insulation resistance, and a brief description of current flow in semiconductors.

Chapter 4 starts with the statement of scalar and vector forms of Biot–Savart law. The concepts of magnetic flux density and magnetic flux are introduced next. Electromagnetic force on a current-carrying conductor and electromagnetic torque on a current loop are discussed in some detail because of their practical importance. Some other topics covered in the chapter include Ampere's magnetic circuital law, curl of magnetic field intensity, Stokes' theorem, relation between magnetic susceptibility and relative permeability, boundary conditions, and energy stored in the magnetic field. The magnetic circuital law is applied for magnetic circuit calculations, which in turn is utilised for calculation of internal inductance, self-inductance, and mutual inductance from flux linkage point of view. Knowledge of magnetic circuit calculations is necessary for predetermination of magnetising current of electromagnetic equipments. Development of circuit models of transmission lines, transformers, etc., requires calculation of inductance.

Three techniques for solving boundary-value problems are discussed in **Chapter 5**. These are closed-form solution of one-dimensional and two-dimensional Laplace's equation in rectangular and cylindrical coordinates, the image method, and the technique of free-hand field mapping. The first two methods are useful for solving only a limited class of problems with boundaries of regular shapes. The graphical technique however can be used for problems with any boundary shape. Results obtained from field maps are not accurate. However, they may be used to verify whether results from analytical methods are correct or not.

Time-varying electric and magnetic fields, designated as electromagnetic fields, are described in **Chapter 6**. The chapter covers Faraday's law of electromagnetic induction, Maxwell's equations, Poynting vector, and their applications. The continuity equation, eddy-current loss in a thin iron sheet, displacement current, distribution of current in a round conductor are some other topics included in this chapter. Faraday's law is used for calculation of induced voltage in conductors and conducting coils both by speed and transformer actions. Maxwell's equations are applied to derive the Poynting vector, which is used for power calculations. The theories of propagation of electromagnetic waves given in Chapters 7 and 9 are developed using Maxwell's equations.

In **Chapter 7** we use all the important Maxwell's equations for derivation of wave equations. The chapter contains solution for one-dimensional wave equation in unbounded regions, properties and parameters of uniform plane waves, wave reflection at discontinuities, power calculations using the Poynting vector, and standing wave ratio. The chapter is concluded with descriptions of critical angle of incidence and the Brewster angle.

The analysis of two-conductor uniform long transmission lines used for transmission of electric power and signals is outlined in **Chapter 8**. The chapter contains derivation of transmission line equations from the circuit model of the line. Solutions of the wave equation, properties and parameters of the line, reflection and transmission of voltage and current at discontinuities, power calculations, standing wave ratio, matching of transmission lines with load are some of the topics covered in this chapter. Two-conductor transmission lines are widely used as links between generators (transmitters) and loads (receivers).

Though two conductors (one may be earth) are generally required for transmission of electric power, electric energy at high frequencies can also be transmitted through a hollow metallic cylinder – the waveguide. **Chapter 9** describes wave propagation in a parallel-plate waveguide and through a rectangular waveguide. Solutions of relevant wave equations, various properties and formulae of parameters of the waveguides are included in this chapter. We learn in this chapter that electromagnetic waves in bounded regions can travel in different modes, and wave propagation is possible only above a certain frequency, called the cut-off frequency. The cut-off frequency depends on the distance between the plates of a parallel-plate waveguide, and on the side lengths of the cross section of a rectangular waveguide.

Chapter 10 describes the principle of launching electromagnetic waves into free space with the help of antennas. We learn in this chapter about various radiation properties of antennas from the analysis of a Hertzian dipole and a half-wave dipole. It is shown that the image of a monopole above a conducting plane or earth is also a monopole, and the actual and the image monopoles combine to form a dipole antenna. A brief study of radiation from antenna arrays is given in this chapter. Working principles of the Yagi–Uda antenna and receiving antennas are also described.

Equations, expressions, and formulae are numbered sectionwise. Equation number, section number, and chapter number are mentioned wherever a reference is made to an earlier equation. For example, (3.11-7) refers to Formula 7 of Section 11 in Chapter 3. Whereas (9) refers to Equation 9 of the same section.

Acknowledgements

We would like to thank the reviewers mentioned below for their valuable suggestions and comments on the manuscript.

Navneet Gupta	*Birla Institute of Technology and Science, Pilani, Rajasthan*
Abhijit Sarkar	*Jaipur Engineering College and Research Centre, Jaipur, Rajasthan*
D S Kushwaha	*Kanpur Institute of Technology, Kanpur, Uttar Pradesh*
Debparna Sengupta	*Techno India College of Technology, Kolkata*
Sanghmitra Chatterjee	*Camellia Institute of Technology, Kolkata*
S K Behera	*National Institute of Technology Rourkela, Orissa*
P D Kasyap	*North Eastern Regional Institute of Science and Technology, Nirjuli, Arunachal Pradesh*
Ashok Khedkar	*Cummins College of Engineering for Women, Pune, Maharashtra*
Sarosh Dastoor	*Sarvajanik College of Engineering and Technology, Surat, Gujarat*

K T Mathew	*Viswajyoti College of Engineering and Technology, Vazhakulam, Kerala*
P Dananjayan	*Pondicherry Engineering College, Pondicherry*
K Lakshmi	*K S Rangasamy College of Technology, Tiruchengode, Tamil Nadu*
P Chidambara Nathan	*National Institute of Technology Trichy, Tamil Nadu*
K R Subramanian (Retd.)	*Kumaraguru College of Technology, Coimbatore, Tamil Nadu*
M Karthikeyan	*Paaval Engineering College, Namakkal, Tamil Nadu*
T V Prasad	*Narayan Engineering College, Nellore, Andhra Pradesh*
B L Prakash	*Vignan Institute of Information Technology, Vishakhapatnam, Andhra Pradesh*

We are also grateful to Dr R K Mishra and Dr T Sahu, professors in the Postgraduate Department of Electronic Sciences, Berhampur University, and Dr L K Mahapatra, a mathematician and a senior officer in industries department, Government of Orissa, for their sincere help. We are also grateful to Mr A K Khanda, programmer at the Computer Centre, Berhampur University, for his timely help whenever required.

We thank our family members for their active support and encouragement, which has been of great help in completing this project.

S C Mahapatra
S Mahapatra

Publisher's Note

Have something to tell us? Write to us with your comments at *tmh.elefeedback@gmail.com*. Please mention the title and author's name as the subject.

EARLY INSIGHTS

Chapter Five

SOLUTION OF BOUNDARY-VALUE PROBLEMS

5.1 Introduction

In Chapter 2, we applied Coulomb's law and Gauss's law for calculation of electric field intensity and electrostatic potential of specified charge distributions in unbounded regions. And in chapter four, we applied Biot–Savart law and Ampere's magnetic circuit law to find the magnetic field intensity and vector magnetic potential of some specific current distributions in unbounded regions. Application of Coulomb's law and Biot–Savart law involves evolution of definite integrals. Thus, these methods are useful only if a function can be integrated. Moreover, Gauss's law and Ampere's magnetic circuit law are useful for calculation of fields of only highly symmetrical charge and current distributions.

There are many electric- and magnetic-field problems involving boundary surfaces on which a field quantity is specified. These are called boundary-value problems. Out of several techniques available for solving such problems, we will discuss only three methods. These are (1) Closed-form solution of Laplace's and Poisson's equations for potential functions; (2) The image method; (3) Graphical field mapping.

Laplace's and Poisson's equation are second order partial differential equations of a field quantity. In electric and magnetic fields, the quantity generally is a potential function. The equations, in general, are three-dimensional as the potential is a function of all the three coordinate variables. We will solve problems which can be treated as one-dimensional and two-dimensional problems. We will restrict our study to the solution of two-dimensional problems having rectangular and cylindrical symmetry. It may be noted that closed-form solutions satisfying the appropriate boundary conditions can be obtained only for a limited set of problems.

Dedicated chapter on Solution of Boundary-Value Problems for both static electric and magnetic fields

Introduction to every chapter containing relevance of subject to practical applications

Chapter Two

STATIC ELECTRIC FIELDS

2.1 Introduction

If an electrically neutral body is supplied with electronic charge, it gets negatively charged, and if electrons are removed from the body, it becomes positively charged. There are several methods of charging a body such as by mechanical means, by the use of a battery, by ionization, etc. A charged body may be static (stationary) or may be in motion, and the charge itself may be time invariant or may be a function of time.

When two charged bodies are separated over a finite distance, each body experiences an electric force. The electric force is a function of magnitude of charges, relative displacement between the charges and the dielectric property of the medium in which the charges are located. If one charge is kept fixed and the other charge is moved to different positions, in general, both magnitude and direction of the force on the charge will change. The graphical plot showing the direction of the force at various points will define the force field of the fixed charge. Further, if locations of the charges in space are described using a coordinate system, the electric force can be expressed in terms of the coordinate variables in vector form. The theory and application of the electric field are based on this force. Some other important quantities used in the development of electric field theory are the electric field intensity, scalar electric potential and potential gradient.

The main purpose of analyzing electric field problems is to calculate the force between charged bodies, potential gradient in a dielectric (insulation) medium, capacitance of a system of conductors and energy stored in a capacitor. Knowledge of potential gradient in an insulating medium (including air) is necessary to know whether it is within the breakdown strength of the material or not. Capacitance is used as a circuit parameter to develop circuit models of transmission lines, electronic devices, etc. Capacitors are also used for improvement

Some of the equipments and devices using electric force on charged particles are the cathode ray tubes, electrostatic precipitators, xerographic machines, and electrostatic separators. Cathode ray tubes are used as a display device in oscilloscopes, television receivers and computer monitors. Electrostatic precipitators are installed to remove ash and dust particles from the exhaust gas of coal-fired power plants, cement factories and chemical plants to name a few. Xerographic machines are used for dry-copying of documents. The formation of thunder clouds and the lightning phenomenon are due to static charges. The formation of corona on the surface of conductors of high voltage overhead transmission lines is explained by the use of electric field laws.

Mathematical fundamentals explained using detailed solutions

EXAMPLE 5.5

The angles of two concentric cones are 10° and 20°. The slant length of each cone is 5.0 cm. The relative permittivity of the dielectric material and the potential difference between the cones are 4.6 and 75.0 V respectively. Calculate the charge contained in each cone and the capacitance.

The problem is illustrated in Fig. 5.5(*a*). The given data are

$$\alpha_1 = 10°,\ \alpha_2 = 20°,\ a = 5.0\times10^{-2}\ \text{m},\ \varepsilon_r = 4.6,\ V_0 = 75.0\ \text{V}$$

The electric flux density is normal to the cone surfaces. Hence, the surface charge density on each cone surface is equal to the flux density at the surface. Referring to formula (43*b*), the surface charge density on the outside surface of the inner cone,

$$\rho_{S1} = D_\theta(\theta=\alpha_1) = \varepsilon E_\theta(\theta=\alpha_1) = \frac{\varepsilon V_0}{\ln[\tan(\alpha_2/2)] - \ln[\tan(\alpha_1/2)]}\ \frac{1}{r\sin\alpha_1}$$

The differential surface area in θ direction, $dS_\theta = r\sin\alpha_1\, dr\, d\phi$

Charge on the outside surface of the inner cone,

$$Q_1 = \int_S \rho_{S1}\, dS_\theta = \int_0^a\int_0^{2\pi} \frac{\varepsilon V_0\, dr d\phi}{\ln[\tan(\alpha_2/2)] - \ln[\tan(\alpha_1/2)]}$$

$$= \frac{2\pi\varepsilon a V_0}{\ln[\tan(\alpha_2/2)] - \ln[\tan(\alpha_1/2)]}$$

Substituting the data,

$$Q_1 = \frac{2\pi(8.85\times10^{-12})(4.6)(5.0\times10^{-2})(75.0)}{\ln(\tan 10°) - \ln(\tan 5°)} = 1.368\times10^{-9}\ \text{C}$$

As the slant length a of the outer cone is same as that of the inner cone, charge on the inside surface of the outer cone

$Q_2 = Q_1 = 1.368\times10^{-9}$

QUESTIONS

7.1 Mention the assumptions under which the one-dimensional wave equation $\partial^{\hat{}}\ _x\,\partial z^2 = (\mu\varepsilon)(\partial^2 E_x/\partial t^2)$ for the electric field intensity E_x is obtained.

7.2 Show that $Ce^{jk(t-\sqrt{\mu\varepsilon}z)}$, where k is a real quantity, is a solution of the appropriate wave equation.

7.3 The magnetic field intensity in x direction varies with z, and also varies harmonically with time at ω rad/s. Write the one-dimensional wave equation of the field in a perfect dielectric medium. Show that the function $C\sin(\omega t - \sqrt{\mu\varepsilon}z)$ is a solution of the wave equation.

7.4 Explain with necessary illustrations that the function $C\cos(\omega t - \beta z)$, where $\beta\ \ \omega\ \ \mu\varepsilon$, is a travelling wave that moves at a velocity $v = 1/\sqrt{\mu\varepsilon}$ in the positive z direction.

7.5 The parameter β and the velocity v of a uniform plane wave are called *phase* constant and *phase* velocity. What is the significance of the adjective 'phase' in each case?

ANSWERS TO SELECTED QUESTIONS

7.1 Mention the assumptions under which the one-dimensional wave equation $\partial^{\hat{}}\ _x\,\partial z^2 = (\mu\varepsilon)(\partial^2 E_x/\partial t^2)$ for the electric field intensity E_x is obtained.

Answer: The wave equation is obtained under the following assumptions.

1. The region of the field has no free charges and no losses.
2. The region is homogeneous and linear.
3. The electric field does not vary with the space variables x and y.

Comprehensive pedagogy comprising answers to selected questions

Graded Problems that indicate challenge levels

PROBLEMS

*6.1 A circular coil of 25 turns and mean radius 2.5 cm is on the $z = 0$ plane of cylindrical coordinates in a uniform magnetic field. The flux density is given by $\mathbf{B} = (0.06\mathbf{u}_y + 0.08\mathbf{u}_z)\sin 5000t$ T. Find the value of root-mean-square (rms) voltage induced in the coil.

**6.2 A planar coil of 20 turns has an area of 30 cm². The unit vector normal to the plane of the coil is given by $0.6\mathbf{u}_x + 0.8\mathbf{u}_y$. The coil is linked by an alternating magnetic field of flux density $\mathbf{B} = (0.4\mathbf{u}_y + 0.5\mathbf{u}_z)\cos(100\pi)t$ T. Find the voltage induced in the coil.

*6.3 Derive the integral form of Faraday's law considering a conducting loop of one turn.

**6.4 A circular coil of N turns and mean radius 2.5 cm is on the $z = 0$ plane of cylindrical coordinates. The coil is linked by an alternating magnetic field of flux density $\mathbf{B} = 0.08\sin 2500t\,\mathbf{u}_z$ T. Calculate the electric field intensity along the wire of the coil using the integral form of Faraday's law.

*6.5 A coil of N turns is carrying a time-varying current. Obtain the induced voltage in the coil due to its self-flux in terms of the self-inductance.

**6.6 The number of turns of two magnetically coupled coils 1 and 2 are N_1 and N_2 and currents in them are I_1 and I_2. Obtain the voltage induced in Coil 2 due the current in Coil 1 in terms of the mutual inductance.

***6.7 A search coil of 25 turns has an effective area of 2.4 cm². It is placed coaxially at the centre of an air-core solenoid having a uniformly and closely distributed winding of 2000 turns. The length of the solenoid is 25 cm. Find the rms value of induced voltage in the search coil if the current in the solenoid I t A.

*6.8 A conductor oriented along y direction has a length of 15 cm. It moves in a magnetic field with a velocity of 25 m/s in x direction. The magnetic flux density $\mathbf{B}$ $\mathbf{u}_z$ T. Find the value of voltage induced in the conductor.

Chapter One

INTRODUCTORY TOPICS

1.1 Introduction

Some of the electromagnetic quantities such as force between charged bodies, force between current-carrying conductors, electric and magnetic flux densities, current density, etc., are vector quantities. Electrostatic potential, electric and magnetic flux, electric and magnetic field energy, etc., are scalar quantities. These quantities are expressed mathematically using suitable coordinate systems. Therefore, it is necessary that a student has good knowledge of coordinate systems and vector analysis for proper understanding of the subject of electromagnetics. To fulfil this requirement, a major part of this chapter is devoted to coordinate systems and vector analysis. Three types of coordinate systems, fundamentals of vector algebra and vector calculus, and topics involving both coordinate systems and vectors are described.

Electric and magnetic fields may be static (time-invariant) or they may be functions of time. Field quantities varying sinusoidally with time can be described mathematically in the same way as sinusoidal voltage and current. Results of circuit analysis, given in Section 1.12, can be used for analysis of sinusoidally time-changing electromagnetic fields replacing voltage and current by appropriate field quantities.

Product solution of two-dimensional Laplace's equation in boundary-value problems with plane boundaries may contain infinite sine or cosine series. To obtain the complete solution of a problem, the general solution is matched by the Fourier series of the function at the boundary. The Fourier series of a square wave and a triangular wave, derived in Example 1.13, are used in Chapter 5.

Bessel's equation arises from the product solution of two-dimensional Laplace's equation in some types of problem with cylindrical boundaries. The differential equation for current density in round conductors carrying sinusoidally time-varying current is a Bessel's equation. The solution of Bessel's equation, which is an infinite series, is called Bessel function. Bessel function of zero order, given in Section 1.14, is used in Chapters 5 and 6.

It is quite likely that students taking this course have studied the topics in this chapter earlier. Even then, they should read this chapter and solve some simple problems as it will then be easier to understand the main subject.

1.2 Coordinate Systems

A real function can be described algebraically using a coordinate system. The choice of a coordinate system to be used depends on the physical nature of the function. Three orthogonal coordinate systems are described in this section. These are (1) the rectangular, or the Cartesian, coordinate system; (2) the circular cylindrical, or simply the cylindrical, coordinate system; and (3) the spherical coordinate system.

The Rectangular Coordinate System

The rectangular coordinate system is defined by three mutually perpendicular plane surfaces S_1, S_2 and S_3, as shown in Fig. 1.1(a). The surfaces are called coordinate planes. The common point of intersection of the planes is the *origin*. The line of intersection of two planes is a coordinate axis. Thus, there are three coordinate axes which are perpendicular to each other, and they meet at the origin. The coordinate axes are denoted by the symbols x, y, and z. The positive sense of an axis is in the direction of the arrow viewed from the origin. The coordinate planes are designated as x-y (or $z = 0$) plane, y-z (or $x = 0$) plane and z-x (or $y = 0$) plane.

The coordinate system may be right-handed or left-handed. In the right-handed system, the z axis is along the direction of advance of a right-handed screw when it is rotated from the x axis towards the y axis. By interchanging the directions of the x and y axes of the right-handed system we will obtain the left-handed system. Figure 1.1(*a*) shows the right-handed system and Fig.1.1(*b*) shows the left-handed system. We will use only the right-handed system.

Since the intersection of two surfaces is a line and the intersection of the line and a third surface is a point, the location of a point in space is described by the distances of three mutually perpendicular planes from the coordinate planes.

The distances give the coordinates of the point. For example, when we say a point P_1 has coordinates x_1, y_1, and z_1, as indicated in Fig. 1.1(*c*), it means that the point is at the intersection of $x = x_1$, $y = y_1$ and $z = z_1$ planes. The points $P_1(x, y)$ and $P_1(x)$, shown in the figure, are on the x-y plane and on the x axis respectively. Distance between two points P_1 and P_2, shown in Fig. 1.1(*c*), is given by

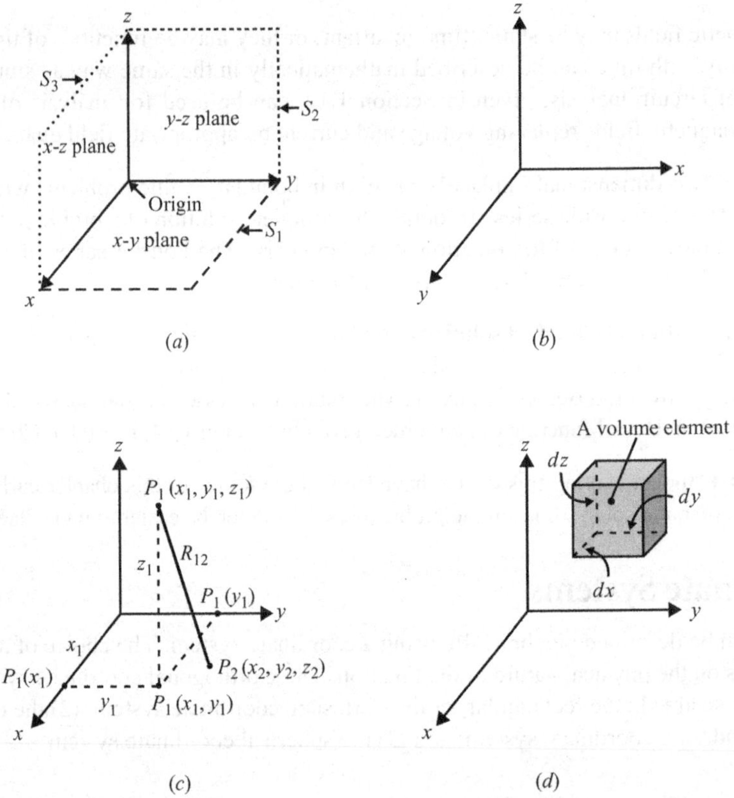

Fig. 1.1 The rectangular coordinate system: (*a*) Right-handed system, (*b*) Left-handed system, (*c*) Coordinates of points, and (*d*) A volume element with side lengths *dx*, *dy* and *dz*.

$$R_{12} = \sqrt{(x_2 - x_1)^2 + (y_2 - y_1)^2 + (z_2 - z_1)^2} \tag{1}$$

If $x_1 = -1.5$ m, $y_1 = 2.6$ m, $z_1 = 3.8$ m, $x_2 = 2.7$ m, $y_2 = 1.6$ m, $z_2 = -1.7$ m, then distance between P_1 and P_2,

$$R_{12} = \sqrt{(2.7 + 1.5)^2 + (1.6 - 2.6)^2 + (-1.7 - 3.8)^2} = 6.992 \text{ m}$$

It will be necessary to use differential elements of length, surface area and volume to formulate field equations. The differential length along x direction is the distance between two planes located at distances x and $x + dx$ from the y-z plane. Thus,

the differential length along x direction is dx. Similarly, the differential lengths along y and z directions are dy and dz respectively. A volume element with side lengths dx, dy and dz is shown in Fig. 1.1(d). The differential areas of its top and bottom surfaces is $dxdy$, of the front and back surfaces is $dydz$ and of the two side surfaces is $dzdx$. The volume of the element, $dv = dxdydz$.

The Cylindrical Coordinate System

A point in space in cylindrical coordinate system is described by the point of intersection of a circular cylindrical surface S_1 and two plane surfaces S_2 and S_3,

as shown in Fig. 1.2(a). These surfaces are specified with reference to the rectangular coordinate system. The axis of the cylinder is along the z axis and its radius is denoted by r.[1] The plane surface S_2 containing the z axis is specified by the angle it makes with the y-z plane. The angle is generally denoted by ϕ. The plane surface S_3 is parallel to, and at a height z from, the x-y plane. We see that the coordinate surfaces are mutually perpendicular to one another. The coordinate directions at a point P are perpendicular to the respective coordinate planes as indicated in the figure. Thus, the coordinate directions are perpendicular to each other.

The intersection of planes S_1 and S_2 is a straight line parallel to the z axis and the intersection of this line and surface S_3 defines a point P in space, as shown in Fig. 1.2(b). The coordinates of the point are given by the radius r, the angle ϕ and the height z. The z coordinate is common to both rectangular and cylindrical coordinate systems.

If the coordinates of two points are given in cylindrical coordinate system, we cannot calculate the distance between the points by direct use of the coordinate variables because ϕ is an angle.

To calculate the distance, we have to obtain the coordinate of the points with reference to rectangular coordinates. To find the relationships between the variable of the two coordinate systems, let us consider a point P with coordinates r, ϕ and z, as shown in Fig. 1.2(b). Also, let point P_{xy} be the projection of P on the x-y plane. As the length of OP_{xy} is equal to r and it is at an angle ϕ with the x axis, the lengths of projections of OP_{xy} along x and y directions are given by $r\cos\phi$ and $r\sin\phi$ respectively. Moreover, z has the same meaning in both the coordinate systems. Thus, the coordinates of P with reference to rectangular coordinate system are as follows:

$$x = r\cos\phi \tag{2a}$$

$$y = r\sin\phi \tag{2b}$$

$$z = z \tag{2c}$$

We can calculate the distance between two points by the use of formula (1), after transforming the variables using relations (2). For example, distance between points $P_1(r_1, \phi_1, z_1)$ and $P_2(r_2, \phi_2, z_2)$,

$$R_{12} = \sqrt{(r_2\cos\phi_2 - r_1\cos\phi_1)^2 + (r_2\sin\phi_2 - r_1\sin\phi_1)^2 + (z_2 - z_1)^2}$$

[1] Some authors use the symbol ρ.

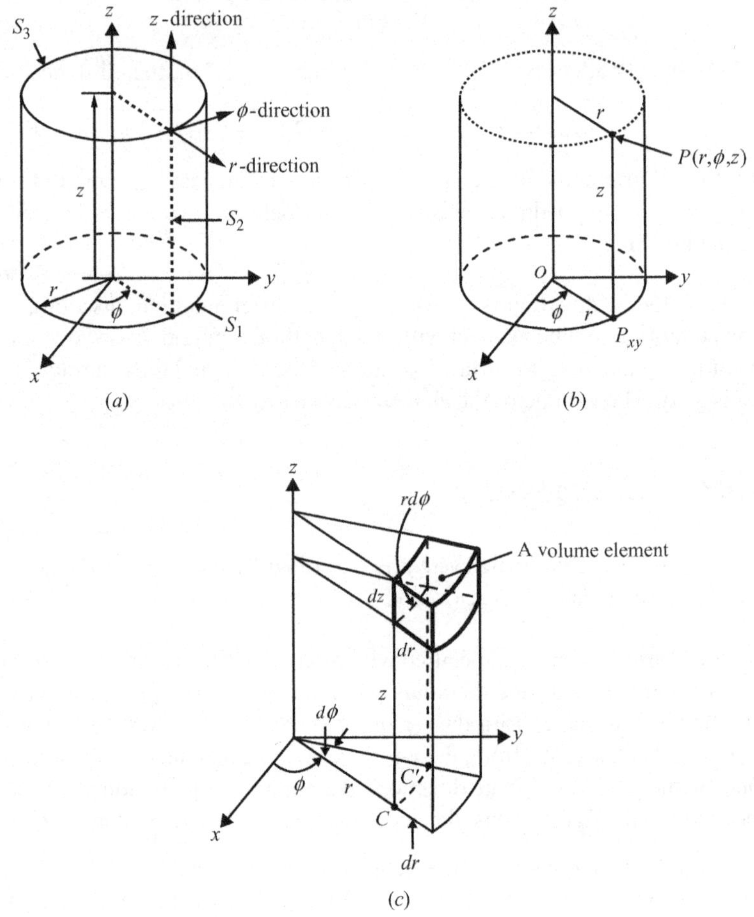

Fig. 1.2 The cylindrical coordinate system: (*a*) Coordinate planes, (*b*) Coordinates of a point, and (c) A volume element having side lengths dr, $rd\phi$ and dz.

Differential Lengths

The differential length dr along r direction is the radial distance between two concentric cylindrical surfaces of radii r and $(r + dr)$. The differential length in ϕ direction, however, is not $d\phi$ because ϕ is an angle. To find the differential length along this direction, let us consider a radial line OC at an angle ϕ with the x axis, as shown in Fig. 1.2(*c*). If the line is rotated by a differential angle $d\phi$, then it will be displaced along ϕ direction to the position OC'. The arc length $CC' = rd\phi$ is the differential length along ϕ direction. As the length is infinitesimally small, we can assume it to be a straight line. The differential length in z direction is dz. An element of volume having side lengths dr, $rd\phi$ and dz is shown in at Fig. 1.2(*c*). The surface areas of the element are $rd\phi dz$, $drdz$ and $rdrd\phi$, in r, ϕ and z direction respectively. The volume of the element, which is equal to the product of its three side lengths, is $dv = rdrd\phi dz$.

The Spherical Coordinate System

In the spherical coordinate system the three coordinate surfaces are those of a sphere, of a cone and of a plane.

These surfaces, indicated by S_1, S_2, and S_3 in Fig. 1.3(*a*), are defined with reference to rectangular coordinates. The spherical surface S_1, which has its centre at the origin of rectangular coordinates, is described by its radius r. Surface S_2 is that of a right circular cone with its vertex at the centre of the sphere and its axis along the z axis. The cone is defined by its semivertical angle θ. The angle is called *polar angle* or *co-latitude*. The plane surface S_3 is at an angle ϕ with the x-z plane. The angle ϕ is common to both cylindrical and spherical coordinates. The three surfaces form an orthogonal set. The coordinate directions are perpendicular to the coordinate surfaces as indicated in the figure. Thus, the coordinate directions are perpendicular to each other.

The intersection between a sphere of radius r and a cone of angle θ is a circle. The radius of the circle is equal to $r\sin\theta$. A plane at an angle ϕ with respect to the x-z plane intersects this circle at P, as shown in Fig. 1.3(*a*). The coordinates of P are therefore r, θ, and ϕ.

To calculate distance between two points, it is necessary to express the spherical-coordinate variables in terms of rectangular-coordinate variables as ϕ and θ are angles. The relationships between the two sets of variables are obtained as follows. In Fig. 1.3(*b*), the coordinates of P are r, θ and ϕ, and the projection of P on the x-y plane is P_{xy}. The length of $OP_{xy} = r\sin\theta$, as the angle between OP and OP_{xy} is $90° - \theta$. Moreover, as the angle between the x axis and OP_{xy} is ϕ, the lengths of projections of OP_{xy} along x and y directions are given by $r\sin\theta\cos\phi$ and $r\sin\theta\sin\phi$ respectively. Also, the component of OP in z direction is $r\cos\theta$. Accordingly, the coordinates of P with reference to rectangular coordinate system are

$$x = r\sin\theta\cos\phi \qquad (3a)$$

$$y = r\sin\theta\sin\phi \qquad (3b)$$

$$z = r\cos\theta \qquad (3c)$$

We can calculate the distance between two points $P_1(r_1, \theta_1, \phi_1)$ and $P_2(r_2, \theta_2, \phi_2)$ by the use of formula (1) after transforming the variables using relations (3).

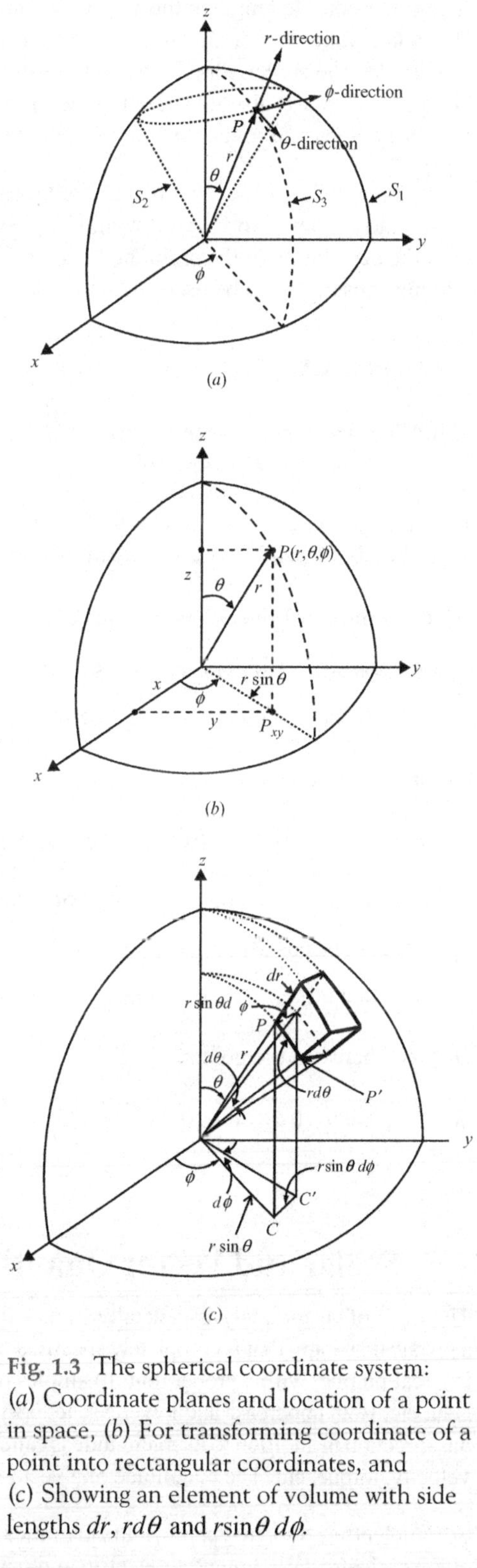

Fig. 1.3 The spherical coordinate system: (*a*) Coordinate planes and location of a point in space, (*b*) For transforming coordinate of a point into rectangular coordinates, and (*c*) Showing an element of volume with side lengths dr, $rd\theta$ and $r\sin\theta\, d\phi$.

Differential Lengths

The differential length in r direction is dr. The element of length in θ direction is equal to the arc length PP' shown in Fig. 1.3(*c*). The radius of the arc is r and it subtends an angle $d\theta$ at the origin. The differential length in θ direction is equal to the arc length $rd\theta$. The element of length in ϕ

direction is equal to length of the arc CC' on the x-y plane (or parallel to the x-y plane), as indicated in the figure. The radius of the arc is $r\sin\theta$ and it makes an angle $d\phi$ at the origin. Thus, the differential length in ϕ direction is $r\sin\theta\, d\phi$. The arc lengths can be considered as straight lines. An element of volume having side lengths dr, $rd\theta$ and $r\sin\theta\, d\phi$ is also illustrated in the figure. The surface areas of the element are $r^2\sin\theta\, d\theta\, d\phi$, $r\sin\theta\, drd\phi$ and $rdrd\phi$ along r, θ and ϕ direction respectively. The volume of the element, $dv = r^2 \sin\theta\, dr\, d\theta d\phi$.

It may be noted that the symbol r is used to denote the radial distance in cylindrical as well as in spherical coordinate systems. However, it would be possible to know for which coordinate system the symbol is meant from the description of the problem. If in the solution of a problem both the coordinate systems are employed, suitable subscripts can be used.

EXAMPLE 1.1

Find the distance between points, (a) P_1(1.2 m, 30°, 1.6 m) and P_2(0.7 m, 45°, 0.9 m), (b) P_3(0.6 m, 30°, 20°) and P_4(0.9 m, 45°, 50°).

The units of coordinates of the points show that the locations of P_1 and P_2 are described in cylindrical coordinates and of P_3 and P_4 are described in spherical coordinates.

(a) Coordinates of the points in terms of rectangular-coordinate variables are given by

$x_1 = 1.2 \cos 30° = 1.039$ m, $y_1 = 1.2 \sin 30° = 0.6$ m, $z_1 = 1.6$ m

$x_2 = 0.7 \cos 45° = 0.495$ m, $y_2 = 0.7 \sin 45° = 0.495$ m, $z_2 = 0.9$ m

Distance between the points,

$$R_{12} = \sqrt{(0.495 - 1.039)^2 + (0.495 - 0.6)^2 + (0.9 - 1.6)^2} = 0.893 \text{ m}$$

(b) Coordinates of the points in terms of rectangular-coordinate variables are given by

$x_3 = 0.6 \sin 30° \cos 20° = 0.282$ m, $y_3 = 0.6 \sin 30° \sin 20° = 0.103$ m, $z_3 = 0.6 \cos 30° = 0.52$ m

$x_4 = 0.9 \sin 45° \cos 50° = 0.409$ m, $y_4 = 0.9 \sin 45° \sin 50° = 0.488$ m, $z_4 = 0.9 \cos 45° = 0.636$ m

Distance between the points,

$$R_{34} = \sqrt{(0.409 - 0.282)^2 + (0.488 - 0.103)^2 + (0.636 - 0.52)^2} = 0.422 \text{ m}$$

1.3 Scalar and Vector Quantities

The topic of vector analysis generally starts with the definition of scalar and vector quantities. A geometrical or a physical quantity that is completely specified by its magnitude alone is called a scalar quantity. A scalar quantity is a real number with a proper unit. Examples of scalar quantities are mass, length, time, volume, work, energy, heat, etc. Both uppercase and lowercase letters in *italic* will be used to denote scalar quantities. A quantity which has direction in addition to its magnitude is called a vector. Examples of vector quantities are force, displacement, velocity, torque, etc. The magnitude of a vector quantity is a real positive quantity with a proper unit.

Some electric- and magnetic-field quantities are scalar functions and some are vector functions. Electric charge, electrostatic potential, electric and magnetic fluxes, electric current, etc., are scalars. Whereas, force between electric charges and between current-carrying conductors, electric field intensity, magnetic-field intensity, current density, etc., are vectors. As vector functions are described by scalar quantities, it is possible to study electric and magnetic fields by the use of scalar functions. However, the use of vector analysis has the following advantages.

1. The field equations can be written in a compact form. Hence, time and space needed to write an equation are reduced.
2. The equations contain all the information. Therefore, physical interpretation of the equations is made easy.
3. Equations of scalar quantities can easily be written from the vector form of the equations.
4. Some theorems and laws of vector analysis find application in electromagnetic theory.

1.4 Representation of Vectors

A vector is represented geometrically by a directed line segment having an initial point and an end point. The direction of a vector is indicated by an arrow mark. Vector quantities will be denoted by uppercase as well as lowercase bold letters. Representations of two vectors **A** and **B** are shown in Fig. 1.4. The length of the line is generally arbitrary. However for geometrical solution of a problem, the line segments have to be drawn to scale.

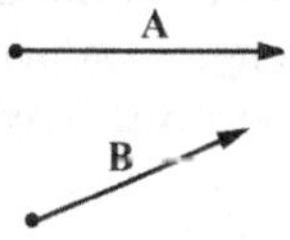

Fig. 1.4 Representation of vectors by directed line segments

The sum and difference of two vectors is also a vector. If **C** is the sum of two vectors **A** and **B** and **D** is the difference between them, then we write

$$\mathbf{C} = \mathbf{A} + \mathbf{B}$$

$$\mathbf{D} = \mathbf{A} - \mathbf{B}$$

The geometric methods of finding the sum and difference of the vectors are shown in Fig. 1.5.

A vector is also represented by showing its magnitude and direction explicitly. The magnitude is denoted by an alphabetic symbol with or without the modulus sign.

The *direction* of the vector is represented by a *unit vector*. The magnitude of a unit vector is equal to one unit and its direction is same as the direction of the original vector.

Fig. 1.5 (*a*) Addition of two vectors and (*b*) Subtraction of two vectors.

Various types of symbols are used to denote a unit vector. We will denote it by **u** with a subscript to indicate its direction. For example, the vector shown in Fig. 1.6 is written as

$$\mathbf{A} = |\mathbf{A}|\mathbf{u}_A = A\mathbf{u}_A = \mathbf{u}_A A$$

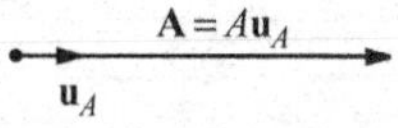

Fig. 1.6 A vector represented by its magnitude and a unit vector

The unit vectors at a point in rectangular coordinate system, which are denoted by $\mathbf{u}_x$, $\mathbf{u}_y$ and $\mathbf{u}_z$, are parallel to the *x*, *y* and *z* axes respectively. Their directions do not vary with the coordinate variables. In cylindrical coordinate system, the unit vectors $\mathbf{u}_r$, $\mathbf{u}_\varphi$ and $\mathbf{u}_z$ at any point are perpendicular to their respective coor dinate surfaces.

The directions of $\mathbf{u}_r$, $\mathbf{u}_\phi$ for different values of ϕ are different. This can be verified by drawing the unit vectors at two different values of ϕ. Therefore, these vectors cannot be treated as constants while differentiating or integrating with respect ϕ.

The direction of $\mathbf{u}_z$ does not change with any of the coordinate variables. The unit vectors in spherical coordinate system, which are denoted by $\mathbf{u}_r$, $\mathbf{u}_\theta$ and $\mathbf{u}_\phi$, are perpendicular to the coordinate surfaces at any point.

The directions of $\mathbf{u}_r$ and $\mathbf{u}_\theta$ vary with θ and ϕ. The direction of $\mathbf{u}_\phi$ varies with ϕ.

The directions of all the unit vectors in all the coordinate systems are positive towards the increasing values of coordinates variables at the point considered. Each set of unit vectors, shown in Fig. 1.7, is a right-handed orthogonal system.

An important method of representing a vector quantity is in terms of its components along the coordinate directions. The value of the scalar component of a vector along a coordinate direction is equal to the product of the magnitude of the vector and cosine of the angle of the vector with the coordinate direction. The angle measured from the coordinate direction towards the vector in the counter-clockwise sense is taken as positive. Thus, the projection of a vector along a coordinate direction is its scalar value along that direction. If a vector **A** makes angles α, β and γ with the x, y and z axes of rectangular coordinates respectively, the scalar components of the vector along the axes are

$$A_x = A\cos\alpha,\ A_y = A\cos\beta,\ A_z = A\cos\gamma$$

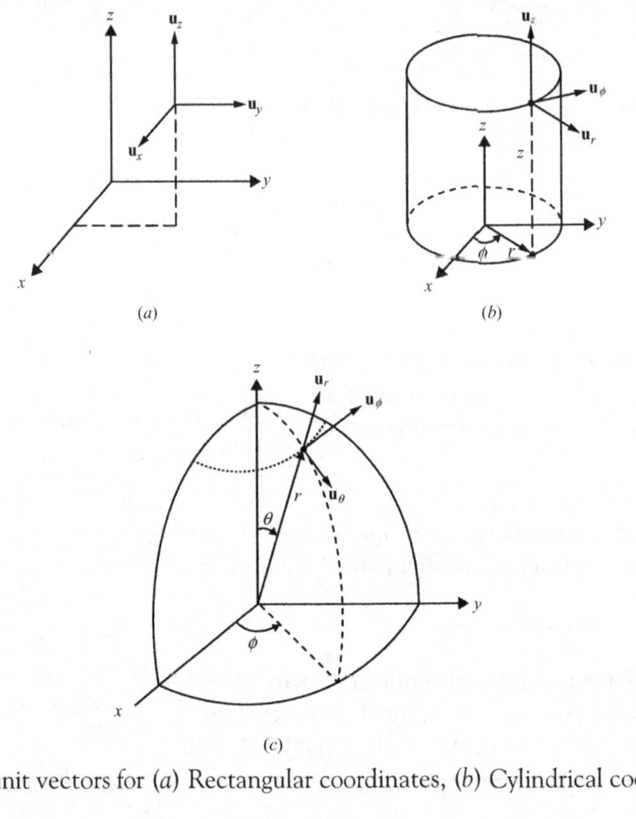

Fig. 1.7 Directions of unit vectors for (*a*) Rectangular coordinates, (*b*) Cylindrical coordinates, and (*c*) Spherical coordinates.

The quantities cos α, cos β and cos γ are called direction cosines of the vector. The vector components of **A** along the coordinate directions are $A_x\mathbf{u}_x$, $A_y\mathbf{u}_y$ and $A_z\mathbf{u}_z$, as indicated in Fig. 1.8. Conversely, the sum of the vector components is equal to the original vector **A**. That is,

$$\mathbf{A} = A_x\mathbf{u}_x + A_y\mathbf{u}_y + A_z\mathbf{u}_z$$

The magnitude of **A** in terms of its scalar components is given by

$$A = \sqrt{A_x^2 + A_y^2 + A_z^2}$$

In cylindrical coordinates,

$$\mathbf{A} = A_r\mathbf{u}_r + A_\phi\mathbf{u}_\phi + A_z\mathbf{u}_z$$

$$A = \sqrt{A_r^2 + A_\phi^2 + A_z^2}$$

In spherical coordinates,

$$\mathbf{A} = A_r\mathbf{u}_r + A_\theta\mathbf{u}_\theta + A_\phi\mathbf{u}_\phi$$

$$A = \sqrt{A_r^2 + A_\theta^2 + A_\phi^2}$$

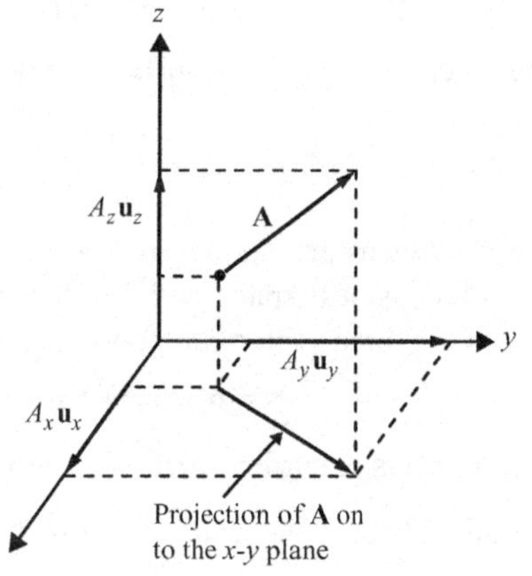

Fig. 1.8 Showing components of a vector **A**

Now, let the components of another vector **B** be denoted by B_x, B_y and B_z in rectangular coordinates. If **C** is equal to the vector sum of **A** and **B**, and C_x, C_y and C_z are the components **C**, then

$$\mathbf{C} = \mathbf{A} + \mathbf{B}$$

Or,

$$C_x\mathbf{u}_x + C_y\mathbf{u}_y + C_z\mathbf{u}_z = (A_x\mathbf{u}_x + A_y\mathbf{u}_y + A_z\mathbf{u}_z) + (B_x\mathbf{u}_x + B_y\mathbf{u}_y + B_z\mathbf{u}_z)$$

Hence,

$$C_x = A_x + B_x,\ C_y = A_y + B_y,\ C_z = A_z + B_z$$

Vector Representations of Differential Lengths and Differential Surfaces

A directed line segment of differential length $d\ell$ is written as

$$d\boldsymbol{\ell} = \mathbf{u}_\ell d\ell$$

The direction of the unit vector $\mathbf{u}_\ell$ is along $d\boldsymbol{\ell}$. Accordingly, elements of length vectors in the coordinate directions of rectangular coordinates are $\mathbf{u}_x dx$, $\mathbf{u}_y dy$ and $\mathbf{u}_z dz$. If a differential length $d\boldsymbol{\ell}$ has components in all the coordinate directions, then

$$d\boldsymbol{\ell} = \mathbf{u}_x dx + \mathbf{u}_y dy + \mathbf{u}_z dz$$

In the same way, the vector forms of elements of lengths in cylindrical and spherical coordinate systems can be written.

In cylindrical coordinates, $d\boldsymbol{\ell} = \mathbf{u}_r dr + \mathbf{u}_\phi(r d\phi) + \mathbf{u}_z dz$

In spherical coordinates, $d\boldsymbol{\ell} = \mathbf{u}_r dr + \mathbf{u}_\theta(r d\theta) + \mathbf{u}_\phi(r \sin\theta\, d\phi)$

A differential surface element is considered as a vector quantity. The vector is described by its area dS and a unit vector normal to its surface. Denoting the normal unit vector by $\mathbf{u}_n$, the vector form of a surface element is written as

$$d\mathbf{S} = \mathbf{u}_n dS$$

For example, the area of a surface element on or parallel to the *y-z* plane of rectangular coordinates is equal to $dydz$, and the normal to the surface is parallel to the x axis. Therefore, denoting the differential surface vector by $d\mathbf{S}_x$, we have

$$d\mathbf{S}_x = \mathbf{u}_x (dydz)$$

The other components of surface elements are

$$d\mathbf{S}_y = \mathbf{u}_y (dxdz)$$
$$d\mathbf{S}_z = \mathbf{u}_z (dxdy)$$

These surfaces are shown in Fig. 1.9. A differential surface vector in three-dimensional space with reference to rectangular coordinates is the sum of the three vector components. Thus

$$d\mathbf{S} = \mathbf{u}_x (dydz) + \mathbf{u}_y (dzdx) + \mathbf{u}_z (dxdy)$$

The elements of surface vector in the other two coordinate systems are

Cylindrical coordinates: $d\mathbf{S} = \mathbf{u}_r (rd\phi dz) + \mathbf{u}_\phi (drdz) + \mathbf{u}_z (rdrd\phi)$

Spherical coordinates: $d\mathbf{S} = \mathbf{u}_r (r^2 \sin\theta d\theta d\phi) + \mathbf{u}_\theta (r\sin\theta dr\, d\phi)$
$+ \mathbf{u}_\phi (r\, dr\, d\theta)$

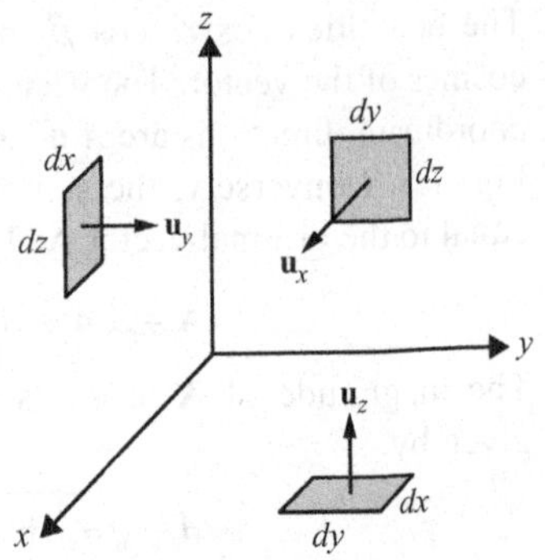

Fig. 1.9 Showing differential vector surfaces along the coordinate directions

The Position Vector

If one end of a directed line segment is located at the origin and the other end at a point $P(x, y, z)$, as shown in Fig. 1.10(a), the directed line is known as the position vector at P. A position vector is generally denoted by $\mathbf{r}$. Referring to the figure, the scalar components of the vector along the coordinate axes are $OP_x = x$, $OP_y = y$, $OP_z = z$. Accordingly, the vector in terms of its components is given by

$$\mathbf{r} = x\mathbf{u}_x + y\mathbf{u}_y + z\mathbf{u}_z$$

The vector equation of a directed line can be obtained by the use of position vectors. In Fig. 1.10(b), two ends of a directed line $\mathbf{R}_{12}$ are at $P_1(x_1, y_1, z_1)$ and $P_2(x_2, y_2, z_2)$. The equations of position vectors at these points are

$$\mathbf{r}_1 = x_1\mathbf{u}_x + y_1\mathbf{u}_y + z_1\mathbf{u}_z$$
$$\mathbf{r}_2 = x_2\mathbf{u}_x + y_2\mathbf{u}_y + z_2\mathbf{u}_z$$

Referring to the figure we find that the sum of vectors $\mathbf{r}_1$ and $\mathbf{R}_{12}$ is equal to $\mathbf{r}_2$. Hence,

$$\mathbf{R}_{12} = \mathbf{r}_2 - \mathbf{r}_1 = (x_2 - x_1)\mathbf{u}_x + (y_2 - y_1)\mathbf{u}_y + (z_2 - z_1)\mathbf{u}_z$$

The length of the line,

$$R_{12} = \sqrt{(x_2 - x_1)^2 + (y_2 - y_1)^2 + (z_2 - z_1)^2}$$

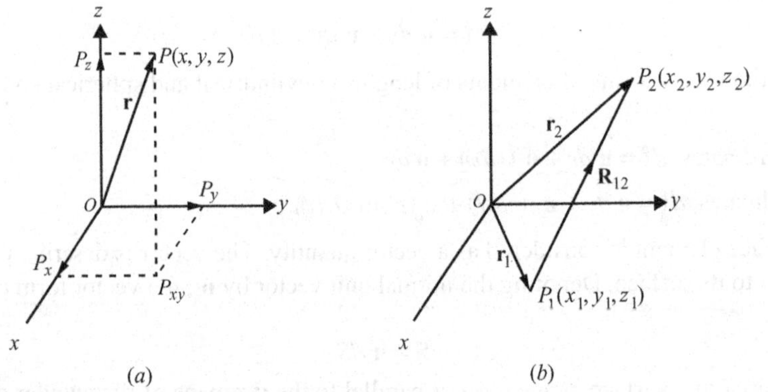

Fig. 1.10 (a) A position vector $\mathbf{r}$ and its components along the axes and (b) Describing a vector $\mathbf{R}_{12}$ in space in terms of two position vectors $\mathbf{r}_1$ and $\mathbf{r}_2$.

The unit vector directed from P_1 to P_2,

$$\mathbf{u}_{12} = \frac{\mathbf{R}_{12}}{R_{12}} = \frac{1}{R_{12}}\left[(x_2 - x_1)\mathbf{u}_x + (y_2 - y_1)\mathbf{u}_y + (z_2 - z_1)\mathbf{u}_z\right]$$

EXAMPLE 1.2

A vector is given by $\mathbf{F} = xyz^2\mathbf{u}_x + y^2z\mathbf{u}_y + \sqrt{x^2+1}\mathbf{u}_z$. Find the vector at $x = 1, y = 2, z = 3$.

$$\mathbf{F} \quad (1)(2)(3^2)\mathbf{u}_x + (2^2)(3)\mathbf{u}_y + \sqrt{1^2+1}\mathbf{u}_z = 18\mathbf{u}_x + 12\mathbf{u}_y + \sqrt{2}\mathbf{u}_z$$

EXAMPLE 1.3

Two points are located at P_1(1.4 m, 0.9 m, –0.6 m) and P_2(–0.25 m, 1.7 m, 1.5 m). Find the length vector directed from P_1 to P_2, and the unit vector directed from P_2 to P_1.

The position vectors at P_1 and P_2 are given by

$\mathbf{r}_1 = 1.4\mathbf{u}_x + 0.9\mathbf{u}_y - 0.6\mathbf{u}_z$

$\mathbf{r}_2 = -0.25\mathbf{u}_x + 1.7\mathbf{u}_y + 1.5\mathbf{u}_z$

The vector directed from P_1 to P_2,

$$\mathbf{R}_{12} = \mathbf{r}_2 - \mathbf{r}_1 = (-0.25 - 1.4)\mathbf{u}_x + (1.7 - 0.9)\mathbf{u}_y + (1.5 + 0.6)\mathbf{u}_z$$
$$= -1.65\mathbf{u}_x + 0.8\mathbf{u}_y + 2.1\mathbf{u}_z$$

The length of the line,

$$R_{12} = \sqrt{1.65^2 + 0.8^2 + 2.1^2} = 2.788 \text{ m}$$

The unit vector directed from P_2 to P_1,

$$\mathbf{u}_{21} = \frac{\mathbf{R}_{21}}{R_{21}}$$

As $\mathbf{R}_{21} = -\mathbf{R}_{12}$ and $R_{21} = R_{12}$, we have

$$\mathbf{u}_{21} = \frac{1.65\mathbf{u}_x - 0.8\mathbf{u}_y - 2.1\mathbf{u}_z}{2.788} = 0.592\mathbf{u}_x - 0.287\mathbf{u}_y - 0.753\mathbf{u}_z$$

EXAMPLE 1.4

Two points in cylindrical coordinates are defined by P_1(2.0 m, 30°, 3.0 m) and P_2(3.0 m, 45°, 4.0 m). Find the unit vector directed from P_2 to P_1.

$x_1 = 2 \cos 30° = 1.732$ m; $y_1 = 2 \sin 30° = 1$ m; $z_1 = 3$ m

$x_2 = 3 \cos 45° = 2.121$ m; $y_2 = 3 \cos 45° = 2.121$ m; $z_2 = 4$ m

$$\mathbf{R}_{21} = (1.732 - 2.121)\mathbf{u}_x + (1.0 - 2.121)\mathbf{u}_y + (3.0 - 4.0)\mathbf{u}_z$$
$$= -0.389\mathbf{u}_x - 1.121\mathbf{u}_y - 1.0\mathbf{u}_z \text{ m}$$

$$R_{21} = \sqrt{0.389^2 + 1.121^2 + 1} = 1.552\,\text{m}$$

$$\mathbf{u}_{21} = \frac{1}{1.552}(-0.389\mathbf{u}_x - 1.121\mathbf{u}_y - 1.0\mathbf{u}_z) = -0.251\mathbf{u}_x - 0.722\mathbf{u}_y - 0.644\mathbf{u}_z$$

EXAMPLE 1.5

Two points in spherical coordinates are given by $P_1(3.0\text{ m}, 0°, 30°)$ and $P_2(2.0\text{ m}, 60°, 0°)$. Find the unit vector directed from P_1 to P_2.

$x_1 = 3 \sin 0 \cos 30° = 0$ m; $y_1 = 3 \sin 0 \sin 30° = 0$ m; $z_1 = 3 \cos 0° = 3$ m

$x_2 = 2 \sin 60° \cos 0° = 1.732$ m; $y_2 = 2 \sin 60° \sin 0° = 0$ m; $z_2 = 2 \cos 60° = 1$ m

$\mathbf{R}_{12} = (1.732 - 0)\mathbf{u}_x + (1.0 - 3.0)\mathbf{u}_z = 1.732\,\mathbf{u}_x - 2.0\mathbf{u}_z$ m

$$R_{21} = \sqrt{1.732^2 + 2^2} = 2.646\text{ m}$$

$$\mathbf{u}_{12} = \frac{1.732\mathbf{u}_x - 2\mathbf{u}_z}{2.646} = 0.658\mathbf{u}_x - 0.756\mathbf{u}_z$$

1.5 Multiplication of a Vector by a Scalar

When a vector is multiplied by a scalar it results in a vector quantity. If we denote a vector by **A**, a scalar (a real quantity) by ϕ and the product of the two quantities by **B**, then

$$\mathbf{B} = \phi\mathbf{A}$$

The direction of **B** is same as that of **A** if ϕ is positive, and opposite to that of **A** if ϕ is negative.

By writing **A** and **B** in terms of their vector components in rectangular coordinates, we get

$$B_x\mathbf{u}_x + B_y\mathbf{u}_y + B_z\mathbf{u}_z = \phi(A_x\mathbf{u}_x + A_y\mathbf{u}_y + A_z\mathbf{u}_z)$$

The above equality is satisfied if each component of lhs is equal to the corresponding component of rhs. Therefore,

$$B_x = \phi A_x,\ B_y = \phi A_y,\ B_z = \phi A_z$$

The magnitude of **B** is

$$|\mathbf{B}| = \phi\sqrt{A_x^2 + A_y^2 + A_z^2} = \phi A$$

Also,

$$(\phi_1 + \phi_2)\mathbf{A} = \phi_1\mathbf{A} + \phi_2\mathbf{A};$$

$$\phi(\mathbf{A} + \mathbf{C}) = \phi\mathbf{A} + \phi\mathbf{C}.$$

1.6 Scalar Product and Vector Product of Two Vectors

The *scalar product* of two vectors **A** and **B** is a scalar quantity. It is equal to the products of the magnitude of **A**, the magnitude of **B** and cosine of the angle α between **A** and **B.** The scalar product is written by putting a *dot* between **A** and **B**. Accordingly,

$$\mathbf{A} \cdot \mathbf{B} = AB \cos \alpha$$

The scalar product is also called *dot* product or *inner* product.

Since $\cos \alpha = \cos(-\alpha)$,

$$\mathbf{A} \cdot \mathbf{B} = \mathbf{B} \cdot \mathbf{A}$$

Also,

$$(\mathbf{A} + \mathbf{B}) \cdot \mathbf{C} = \mathbf{A} \cdot \mathbf{C} + \mathbf{B} \cdot \mathbf{C}$$

The dot product of two vectors **A** and **B** can be obtained in terms of their components as follows. In rectangular coordinates, let

$$\mathbf{A} = A_x\mathbf{u}_x + A_y\mathbf{u}_y + A_z\mathbf{u}_z \quad (1)$$

$$\mathbf{B} = B_x\mathbf{u}_x + B_y\mathbf{u}_y + B_z\mathbf{u}_z \quad (2)$$

Then

$$\begin{aligned}\mathbf{A} \cdot \mathbf{B} &= (A_x\mathbf{u}_x + A_y\mathbf{u}_y + A_z\mathbf{u}_z) \cdot (B_x\mathbf{u}_x + B_y\mathbf{u}_y + B_z\mathbf{u}_z)\\ &= A_x[(\mathbf{u}_x \cdot \mathbf{u}_x)B_x + (\mathbf{u}_x \cdot \mathbf{u}_y)B_y + (\mathbf{u}_x \cdot \mathbf{u}_z)B_z]\\ &\quad + A_y[(\mathbf{u}_y \cdot \mathbf{u}_x)B_x + (\mathbf{u}_y \cdot \mathbf{u}_y)B_y + (\mathbf{u}_y \cdot \mathbf{u}_z)B_z]\\ &\quad + A_z[(\mathbf{u}_z \cdot \mathbf{u}_x)B_x + (\mathbf{u}_z \cdot \mathbf{u}_y)B_y + (\mathbf{u}_z \cdot \mathbf{u}_z)B_z]\end{aligned}$$

The dot product of two unit vectors having the same direction is equal to unity, and the dot product of two unit vectors that are perpendicular to each other is equal to zero. Using these properties in the above relation, we get

$$\mathbf{A} \cdot \mathbf{B} = A_xB_x + A_yB_y + A_zB_z$$

In cylindrical coordinates, $\mathbf{A} \cdot \mathbf{B} = A_rB_r + A_\phi B_\phi + A_zB_z$

In spherical coordinates, $\mathbf{A} \cdot \mathbf{B} = A_rB_r + A_\theta B_\theta + A_\phi B_\phi$

The *vector product* of two vectors **A** and **B** is a vector quantity, say, **C**. The magnitude of **C** is equal to the products of the magnitude of **A**, the magnitude of **B**, and the sine of angle α between **A** and **B**. The direction of **C** is perpendicular to the plane containing **A** and **B**. The positive direction is taken along the advance of a right-handed screw when it is rotated from **A** towards **B**. This is illustrated in Fig. 1.11. The vector product is written by putting a *cross* sign between **A** and **B**. Accordingly,

$$\mathbf{C} = \mathbf{A} \times \mathbf{B} = (AB \sin \alpha)\mathbf{u}_n = |\mathbf{A} \times \mathbf{B}|\ \mathbf{u}_n$$

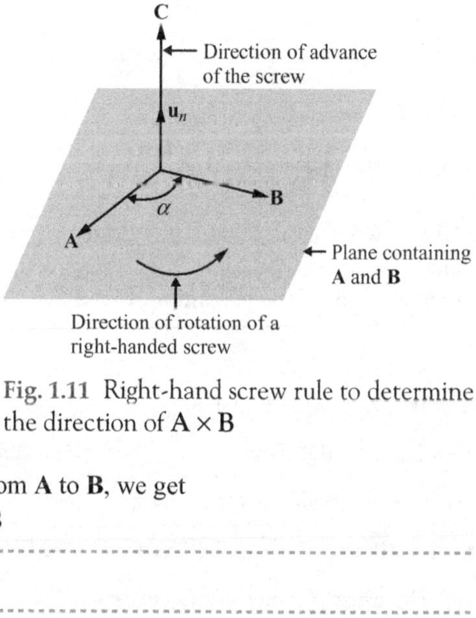

Fig. 1.11 Right-hand screw rule to determine the direction of $\mathbf{A} \times \mathbf{B}$

where $\mathbf{u}_n$ is a unit vector normal to the plane containing **A** and **B**. Since the angle from **B** to **A** is negative of the angle from **A** to **B**, we get

$$\mathbf{B} \times \mathbf{A} = -\mathbf{A} \times \mathbf{B}$$

The vector product is also called *cross* product.

The cross product of **A** and **B** defined in (1) and (2) is

$$\begin{aligned}\mathbf{A} \times \mathbf{B} &= (A_x\mathbf{u}_x + A_y\mathbf{u}_y + A_z\mathbf{u}_z) \times (B_x\mathbf{u}_x + B_y\mathbf{u}_y + B_z\mathbf{u}_z)\\ &= A_x[(\mathbf{u}_x \times \mathbf{u}_x)B_x + (\mathbf{u}_x \times \mathbf{u}_y)B_y + (\mathbf{u}_x \times \mathbf{u}_z)B_z]\\ &\quad + A_y[(\mathbf{u}_y \times \mathbf{u}_x)B_x + (\mathbf{u}_y \times \mathbf{u}_y)B_y + (\mathbf{u}_y \times \mathbf{u}_z)B_z]\\ &\quad + A_z[(\mathbf{u}_z \times \mathbf{u}_x)B_x + (\mathbf{u}_z \times \mathbf{u}_y)B_y + (\mathbf{u}_z \times \mathbf{u}_z)B_z] \quad (3)\end{aligned}$$

The expression contains nine cross products of unit vectors. The cross product of two unit vectors having the same direction is equal to zero. The cross product of two unit vectors that are perpendicular to each other is equal to either plus or minus of the unit vector in the third direction. Using these properties in (3), we get

$$\mathbf{u}_x \times \mathbf{u}_x = 0,\ \mathbf{u}_y \times \mathbf{u}_y = 0,\ \mathbf{u}_z \times \mathbf{u}_z = 0$$

$$\mathbf{u}_x \times \mathbf{u}_y = \mathbf{u}_z,\ \mathbf{u}_y \times \mathbf{u}_z = \mathbf{u}_x,\ \mathbf{u}_z \times \mathbf{u}_x = \mathbf{u}_y$$

$$\mathbf{u}_y \times \mathbf{u}_x = -\mathbf{u}_z,\ \mathbf{u}_z \times \mathbf{u}_y = -\mathbf{u}_x,\ \mathbf{u}_x \times \mathbf{u}_z = -\mathbf{u}_y$$

Substituting these in (3), we get the cross product of the vectors in rectangular coordinates.

$$\mathbf{A} \times \mathbf{B} = \mathbf{u}_x(A_yB_z - A_zB_y) + \mathbf{u}_y(A_zB_x - A_xB_z) + \mathbf{u}_z(A_xB_y - A_yB_x)$$

The cross product can also be written in determinant form.

In rectangular coordinates,

$$\mathbf{A}\times\mathbf{B} = \begin{vmatrix} \mathbf{u}_x & \mathbf{u}_y & \mathbf{u}_z \\ A_x & A_y & A_z \\ B_x & B_y & B_z \end{vmatrix}$$

In cylindrical coordinates,

$$\mathbf{A}\times\mathbf{B} = \begin{vmatrix} \mathbf{u}_r & \mathbf{u}_\phi & \mathbf{u}_z \\ A_r & A_\phi & A_z \\ B_r & B_\phi & B_z \end{vmatrix}$$

In spherical coordinates,

$$\mathbf{A}\times\mathbf{B} = \begin{vmatrix} \mathbf{u}_r & \mathbf{u}_\theta & \mathbf{u}_\phi \\ A_r & A_\theta & A_\phi \\ B_r & B_\theta & B_\phi \end{vmatrix}$$

EXAMPLE 1.6

Given two vectors $\mathbf{E} = 0.5\mathbf{u}_x - 2.2\mathbf{u}_y + 1.6\mathbf{u}_z$ and $\mathbf{F} = 2.6\mathbf{u}_x + 0.8\mathbf{u}_y - 2.7\mathbf{u}_z$. Find (a) $\mathbf{E}\cdot\mathbf{F}$, (b) $\mathbf{E}\times\mathbf{F}$, (c) angle between **E** and **F**, (d) the scalar component of **F** along **E**, and (e) the unit vector which is perpendicular to the plane containing **E** and **F**.

The given data are

$E_x = 0.5,\ E_y = -2.2,\ E_z = 1.6$

$F_x = 2.6,\ F_y = 0.8,\ F_z = -2.7$

(a) The dot product of the vectors,

$\mathbf{E}\cdot\mathbf{F} = (0.5)(2.6) + (-2.2)(0.8) + (1.6)(-2.7) = -4.78$

(b) The cross product of the vectors,

$$\mathbf{E} \times \mathbf{F} = \begin{vmatrix} \mathbf{u}_x & \mathbf{u}_y & \mathbf{u}_z \\ 0.5 & -2.2 & 1.6 \\ 2.6 & 0.8 & -2.7 \end{vmatrix} = 4.66\mathbf{u}_x + 5.51\mathbf{u}_y + 6.12\mathbf{u}_z$$

(c) The magnitudes of the vectors are

$E = \sqrt{0.5^2 + 2.2^2 + 1.6^2} = 2.766,$

$F = \sqrt{2.6^2 + 0.8^2 + 2.7^2} = 3.833$

If α is angle between **E** and **F**, then

$$\cos\alpha = \frac{\mathbf{E}\cdot\mathbf{F}}{EF} = \frac{-4.78}{(2.766)(3.833)} = -0.451$$

$$\alpha = 116.8°$$

(d) The scalar component of **F** along **E** = F cos $\alpha = -1.728$

(e) The unit vector that is perpendicular to the plane containing **E** and **F**,

$$\mathbf{u}_n = \frac{\mathbf{E}\times\mathbf{F}}{EF\sin\alpha} = \frac{4.66\mathbf{u}_x + 5.51\mathbf{u}_y + 6.12\mathbf{u}_z}{(2.766)(3.833)(0.893)} = 0.492\mathbf{u}_x + 0.582\mathbf{u}_y + 0.646\mathbf{u}_z$$

EXAMPLE 1.7

Given two vectors $\mathbf{E} = 5.2\mathbf{u}_r + 6.5\mathbf{u}_z$ and $\mathbf{F} = 8.3\boldsymbol{u}_r + 12.8\mathbf{u}_\phi - 3.0\mathbf{u}_z$. Show that $|\mathbf{E}\times\mathbf{F}| = EF\sin\alpha$

The vectors are described in cylindrical coordinates. The given data are

$E_r = 5.2,\ E_\phi = 0,\ E_z = 6.5$

$F_r = 8.3,\ F_\phi = 12.8,\ F_z = -3.0$

The magnitudes of the vectors are

$$E = \sqrt{5.2^2 + 0 + 6.5^2} = 8.32,$$

$$F = \sqrt{8.3^2 + 12.8^2 + 9.0} = 15.55$$

The dot product of the vectors,

$$\mathbf{E}\cdot\mathbf{F} = (5.2)(8.3) + 0 + (6.5)(-3.0) = 23.66$$

If α is angle between the two vectors,

$$\cos\alpha = 23.6/(8.32)(15.55) = 0.183$$

and $\sin\alpha = 0.983$

$$EF\sin\alpha = (8.32)(15.55)(0.983) = 127.18$$

The cross product of the vectors,

$$\mathbf{E}\quad\mathbf{F} = \begin{vmatrix} \mathbf{u}_r & \mathbf{u}_\phi & \mathbf{u}_z \\ 5.2 & 0 & 6.5 \\ 8.3 & 12.8 & -3.0 \end{vmatrix} = -83.20\mathbf{u}_r + 69.55\mathbf{u}_\phi + 66.56\mathbf{u}_z$$

The magnitude of the cross product,

$$\mathbf{E}\quad\mathbf{F}| = \sqrt{83.20^2 + 69.55^2 + 66.56^2} = 127.18 = EF\sin\alpha$$

EXAMPLE 1.8

Three vectors are given by $\mathbf{A} = 3\mathbf{u}_x + 4\mathbf{u}_y$, $\mathbf{B} = 4\mathbf{u}_y + 5\mathbf{u}_z$, $\mathbf{C} = 5\mathbf{u}_x + 6\mathbf{u}_z$. Find (a) $\mathbf{A}\cdot(\mathbf{B}\times\mathbf{C})$, (b) $\mathbf{A}\times\mathbf{B}\times\mathbf{C}$.

$$\mathbf{B}\quad\mathbf{C} = \begin{matrix} \mathbf{u}_x & \mathbf{u}_y & \mathbf{u}_z \\ 0 & 4 & 5 \\ 5 & 0 & 6 \end{matrix} = \text{[illegible]}\mathbf{u}_x\ \text{[illegible]}\mathbf{u}_y - 20\mathbf{u}_z$$

$$\mathbf{A}\cdot(\mathbf{B}\times\mathbf{C}) = (3\mathbf{u}_x + 4\mathbf{u}_y)\cdot(24\mathbf{u}_x + 25\mathbf{u}_y - 20\mathbf{u}_z) = 172$$

$$\mathbf{A}\times(\mathbf{B}\times\mathbf{C}) = \begin{vmatrix} \mathbf{u}_x & \mathbf{u}_y & \mathbf{u}_z \\ 3 & 4 & 0 \\ 24 & 25 & -20 \end{vmatrix} = -80\mathbf{u}_x + 60\mathbf{u}_y - 21\mathbf{u}_z$$

1.7 Scalar and Vector Fields

A physical quantity, which is a scalar function of position, is a real number. The quantity may or may not be a function of time. The mathematical representation or graphical plot of a scalar function at various points in a region or domain is the *scalar field* of the function in that region. The region may be one-dimensional (a line), two-dimensional (a surface), or three-dimensional (a volume). As an example, let us consider a heated plate. The temperature at each point on the surface of the plate will have a specific value. A graphical representation of temperatures at several points is the temperature field in the region of the plate surface. By joining all the points of equal temperature by a continuous and smooth curve, we will get a contour of constant temperature. When a number of such contours for temperatures, say T_1, T_2, T_3,..., are plotted, the set of contours is a map of constant-temperature field. Since temperature is a scalar quantity, the field is a scalar field. Some other examples of scalar fields are distribution of air-density in a region of earth's atmosphere, contours of equal height from the base of a three-dimensional object, potential distribution in a region of electric charges, etc.

If a physical quantity is a vector function of position, description of its magnitude and direction in a region is the *vector field* of the function in that region. The vector quantity may or may not be a function of time. Velocity distribution of fluid flow in a pipe of non-uniform cross section and distribution of velocity at various points of a rotating body are vector fields. Vector fields are generally plotted showing only the directions of the vector function at various points in a region. If a curve is drawn so that the directions of a vector function are tangential to the curve at all the points, the curve is called a *field line* or a *streamline*. A set of such curves is the *field map*. As an example, let us consider the vector function defined by the equation,

$$\mathbf{F} = x\mathbf{u}_x + y\mathbf{u}_y$$

It is a two-dimensional function and its domain is the x-y plane. At each point on the x-y plane, $\mathbf{F}$ has a specific magnitude and direction. When the directions at various points on the x-y plane are drawn, a pattern will emerge, as shown in Fig. 1.12. The field map of $\mathbf{F}$ consists of radial lines.

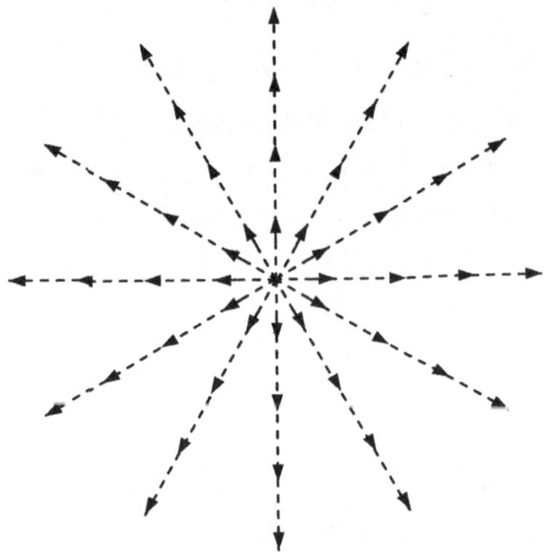

Fig. 1.12 Illustration of a vector field

1.8 Partial Derivative of Vector Functions

The partial derivative of a vector quantity, which is a function of more than one variable, is defined in the same way as for scalar functions. For example, let a vector function in terms of its components in rectangular coordinates be given by

$$\mathbf{A} = y^2\mathbf{u}_x - yx\mathbf{u}_y + az\mathbf{u}_z$$

where a is a constant quantity.

Since the unit vectors do not vary with the coordinate variables, we have

$$\frac{\partial \mathbf{A}}{\partial x} = -y\mathbf{u}_y; \quad \frac{\partial \mathbf{A}}{\partial y} = 2y\mathbf{u}_x - x\mathbf{u}_y; \quad \frac{\partial \mathbf{A}}{\partial z} = a\mathbf{u}_z$$

Partial Derivatives of Unit Vectors in Cylindrical Coordinates

As the directions of unit vectors $\mathbf{u}_r$ and $\mathbf{u}_\phi$ in cylindrical coordinates vary with ϕ, the partial derivatives of these vectors with ϕ are not zero. The procedure for obtaining the partial derivatives of the unit vectors by the use of a graphical method is explained below. Referring to Fig. 1.13(*a*), the radial line *ob* represents a unit vector in r direction at an angle ϕ. The line *oc* represents a unit vector along r direction at angle $\phi + d\phi$. The two vectors are denoted by $\mathbf{u}_r(\phi)$ and $\mathbf{u}_r(\phi + d\phi)$. The directed line from *b* to *c* gives the change of $\mathbf{u}_r$ when ϕ changes by a differential amount $d\phi$. The length of the line is equal to $d\phi$ as $r = 1$ and it is in ϕ direction. Thus, denoting the differential change of $\mathbf{u}_r$ by $d\mathbf{u}_r$, we have

$$d\mathbf{u}_r = \mathbf{u}_r(\phi + d\phi) - \mathbf{u}_r(\phi) = d\mathbf{u}_r = (d\phi)\mathbf{u}_\phi \qquad (1)$$

The differential $d\mathbf{u}_r$ is equal to the rate of change of $\mathbf{u}_r$ with ϕ multiplied by $d\phi$. Accordingly,

$$d\mathbf{u}_r = \frac{\partial \mathbf{u}_r}{\partial \phi} d\phi \qquad (2)$$

From (1) and (2) we get

$$\frac{\partial \mathbf{u}_r}{\partial \phi} = \mathbf{u}_\phi$$

Thus, the partial derivative of $\mathbf{u}_r$ with respect to ϕ is equal to a unit vector along ϕ direction.

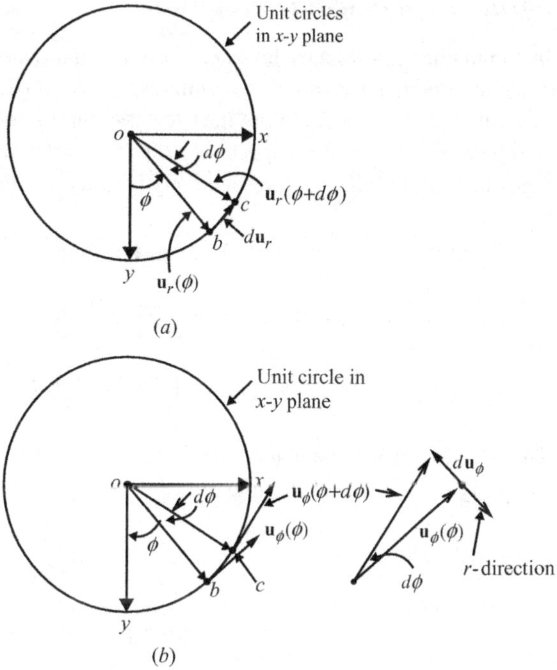

Fig. 1.13 Geometrical methods for determining differential change of (*a*) $\mathbf{u}_r$ with ϕ and (*b*) $\mathbf{u}_\phi$ with ϕ in cylindrical coordinates.

In Fig. 1.13(*b*) and the inset, the ϕ-directed unit vectors at *b* and *c* are shown as $\mathbf{u}_\phi(\phi)$ and $\mathbf{u}_\phi(\phi + d\phi)$ respectively. The vector $d\mathbf{u}_\phi$ is the change of $\mathbf{u}_\phi$ when ϕ changes by a differential amount $d\phi$. The magnitude of $d\mathbf{u}_\phi$ is equal to $d\phi$ and it is along negative r direction. Thus,

$$d\mathbf{u}_\phi = d\phi\,(-\mathbf{u}_r)$$

As

$$d\mathbf{u}_\phi = \frac{\partial \mathbf{u}_\phi}{\partial \phi} d\phi,$$

we have

$$\frac{\partial \mathbf{u}_\phi}{\partial \phi} = -\mathbf{u}_r$$

It may be verified that all other partial derivatives of unit vectors of cylindrical coordinates are zero.

Partial Derivatives of Unit Vectors in Spherical Coordinates

In spherical coordinates, the directions of both $\mathbf{u}_r$ and $\mathbf{u}_\theta$ vary with θ and also with ϕ. The direction of $\mathbf{u}_\phi$ varies with ϕ only. We can obtain the partial derivatives of the unit vectors using graphical techniques similar to those used for cylindrical coordinates. The partial derivatives of unit vectors are given below.

$$\frac{\partial \mathbf{u}_r}{\partial \theta} = \mathbf{u}_\theta; \quad \frac{\partial \mathbf{u}_r}{\partial \phi} = \mathbf{u}_\phi \sin\theta$$

$$\frac{\partial \mathbf{u}_\theta}{\partial \theta} = -\mathbf{u}_r; \quad \frac{\partial \mathbf{u}_\theta}{\partial \phi} = \mathbf{u}_\phi \cos\theta$$

$$\frac{\partial \mathbf{u}_\phi}{\partial \phi} = -\mathbf{u}_r \sin\theta - \mathbf{u}_\theta \cos\theta$$

All other partial derivatives of unit vectors of spherical coordinates are zero.

Gradient of a Scalar Function

In rectangular coordinates, let us consider a scalar quantity V that is a function of all the three variables x, y and z. The first partial derivative of V with respect to x is $\partial V/\partial x$. If V is a single valued function, $\partial V/\partial x$ gives the rate of change of V with x at a point in space assuming y and z as constants. Similarly, $\partial V/\partial y$ and $\partial V/\partial z$ give rate of change of V with y and z respectively. The vector function having $\partial V/\partial x$, $\partial V/\partial y$ and $\partial V/\partial z$ as its scalar components in x, y and z direction respectively is called gradient of V or 'grad V' in abbreviated form. Accordingly,

$$\text{Gradient } V = \text{grad } V = \mathbf{u}_x \frac{\partial V}{\partial x} + \mathbf{u}_y \frac{\partial V}{\partial y} + \mathbf{u}_z \frac{\partial V}{\partial z}$$

Since V is common to all the terms, we can write

$$\text{grad } V = \left(\mathbf{u}_x \frac{\partial}{\partial x} + \mathbf{u}_y \frac{\partial}{\partial y} + \mathbf{u}_z \frac{\partial}{\partial z} \right) V \tag{3}$$

The function inside the bracket of (3) is a differential vector operator. It is called *del* or *nabla* and is denoted by the symbol ∇. In rectangular coordinates,

$$\nabla = \left(\mathbf{u}_x \frac{\partial}{\partial x} + \mathbf{u}_y \frac{\partial}{\partial y} + \mathbf{u}_z \frac{\partial}{\partial z} \right) \tag{4}$$

Thus,

$$\text{grad } V = \nabla V \tag{5}$$

The three components of (3) give the rate of change of V with x, y, z respectively. As grad V is the vector sum of all the components, it gives the maximum rate of change of V at a point in space.

Unit Vector Normal to a Surface

Let us consider that the equation $S(x, y, z) = c$, where c is a constant, represents a continuous and single-valued surface in a scalar field. Then, it can be shown that the gradient of S is along the normal direction to the surface at any point $P(x, y, z)$, and its sense is in the direction of increasing S. The magnitude of ∇S gives the maximum rate of increase of S at P. The unit vector along the normal direction at P is given by

$$\mathbf{u}_n = \pm \frac{\nabla S}{|\nabla S|}$$

where $|\nabla S|$ is the magnitude of ∇S.

Divergence of a Vector Function

Let us consider a vector function $\mathbf{D}$ with D_x, D_y and D_z as components in rectangular coordinates. Thus,

$$\mathbf{D} = \mathbf{u}_x D_x + \mathbf{u}_y D_y + \mathbf{u}_z D_z$$

where, D_x, D_y and D_z are, in general, functions of all the three variables x, y and z. The divergence of $\mathbf{D}$ is defined as the sum of $\partial D_x/\partial x$, $\partial D_y/\partial y$ and $\partial D_z/\partial z$. Thus,

$$\text{div } \mathbf{D} = \frac{\partial D_x}{\partial x} + \frac{\partial D_y}{\partial y} + \frac{\partial D_z}{\partial z} \tag{6}$$

where 'div **D**' is the abbreviation of divergence of **D**.

It can be shown that the divergence of a vector is equal to the dot product of ∇ and **D**.

$$\nabla \cdot \mathbf{D} = \left(\mathbf{u}_x \frac{\partial}{\partial x} + \mathbf{u}_y \frac{\partial}{\partial y} + \mathbf{u}_z \frac{\partial}{\partial z} \right) \cdot (\mathbf{u}_x D_x + \mathbf{u}_y D_y + \mathbf{u}_z D_z)$$

$$= \frac{\partial D_x}{\partial x} + \frac{\partial D_y}{\partial y} + \frac{\partial D_z}{\partial z} \tag{7}$$

As the right-hand sides of (6) and (7) are identical,

$$\text{div } \mathbf{D} = \nabla \cdot \mathbf{D}$$

The divergence of a vector function is a scalar function. A vector field of zero divergence is called a *solenoidal* field.

Curl of a Vector Function

In rectangular coordinates, let us consider a vector function,

$$\mathbf{H} = \mathbf{u}_x H_x + \mathbf{u}_y H_y + \mathbf{u}_z H_z$$

where the components H_x, H_y and H_z are functions of all the three variables x, y and z. The curl of **H**, written in the abbreviated form as 'curl **H**', is defined by the following operation.

$$\text{curl } \mathbf{H} = \mathbf{u}_x \left(\frac{\partial H_z}{\partial y} - \frac{\partial H_y}{\partial z} \right) + \mathbf{u}_y \left(\frac{\partial H_x}{\partial z} - \frac{\partial H_z}{\partial x} \right) + \mathbf{u}_z \left(\frac{\partial H_y}{\partial x} - \frac{\partial H_x}{\partial y} \right) \tag{8}$$

It can be shown that curl **H** is equal to the cross product of the vector operator ∇, defined in (4), and **H**. That is,

$$\text{curl } \mathbf{H} = \nabla \times \mathbf{H}$$

Equation (8) shows that the curl of a vector function is another vector function. A vector field with zero curl is said to be an *irrotational* field.

EXAMPLE 1.9

Given a scalar function, $V = 5x^2y + 3xz$. Find the gradient of V at a point $P(1.5, 2.7, 0.8)$.

The gradient of V,

$\nabla V = (10xy + 3z)\mathbf{u}_x + 5x^2\mathbf{u}_y + 3x\mathbf{u}_z$

Substituting $x = 1.5$, $y = 2.7$, $z = 0.8$,

∇V at P $= [10(1.5)(2.7) + 3(0.8)]\mathbf{u}_x + 5(1.5)^2\mathbf{u}_y + 3(1.5)\mathbf{u}_z$

$= 42.9\mathbf{u}_x + 11.25\mathbf{u}_y + 4.5\mathbf{u}_z$

EXAMPLE 1.10

Find the gradient of the scalar function $\phi = x^2 + y^2 + z^2$ at $(1, -2, 4)$.

$$\nabla_{,} = \frac{\partial \phi}{\partial x}\mathbf{u}_x + \frac{\partial \phi}{\partial y}\mathbf{u}_y + \frac{\partial \phi}{\partial z}\mathbf{u}_z = \frac{x\mathbf{u}_x + y\mathbf{u}_y + z\mathbf{u}_z}{\sqrt{x^2 + y^2 + z^2}}$$

Substituting $x = 1, y = -2, z = 4$,

$$\nabla_{,} = \frac{\mathbf{u}_x - 2\mathbf{u}_y + 4\mathbf{u}_z}{\sqrt{21}}$$

EXAMPLE 1.11

A scalar function is given by $\phi = \ln(x^2 + y^2 + z^2)$. Find $\nabla\phi$ at (2, –1, 3).

$$\nabla_{,} = \frac{\partial \phi}{\partial x}\mathbf{u}_x + \frac{\partial \phi}{\partial y}\mathbf{u}_y + \frac{\partial \phi}{\partial z}\mathbf{u}_z = \frac{2(x\mathbf{u}_x + y\mathbf{u}_y + z\mathbf{u}_z)}{x^2 + y^2 + z^2}$$

Substituting $x = 2, y = -1, z = 3$,

$$\nabla_{,} = \frac{(2\mathbf{u}_x - \mathbf{u}_y + 3\mathbf{u}_z)}{7}$$

EXAMPLE 1.12

A vector field is described by $\mathbf{F} = 500\mathbf{u}_x + 750\mathbf{u}_y$. A plane surface in the region of the field is defined by $2x + 4y + 6z = 12$. Find the vector components of **F** that are normal and tangential to the surface.

Denoting the surface by S, we have

$S = 2x + 4y + 6z$

The gradient of S,

$\nabla S = 2\mathbf{u}_x + 4\mathbf{u}_y + 6\mathbf{u}_z$

The magnitude of the gradient of S,

$$\nabla \;\; = \sqrt{2^{`} + 4^2 + 6^2} = \sqrt{56}$$

The unit vector normal to the surface,

$$\mathbf{u}_n \;\; \frac{\nabla S}{\nabla S} = \frac{1}{\sqrt{56}}(2\mathbf{u}_x + 4\mathbf{u}_y + 6\mathbf{u}_z)$$

The scalar component of the vector field normal to the surface,

$$F_n = \mathbf{F} \cdot \mathbf{u}_n = \frac{1}{\sqrt{56}}\left(500\mathbf{u}_x + 750\mathbf{u}_y\right)\cdot\left(2\mathbf{u}_x + 4\mathbf{u}_y + 6\mathbf{u}_z\right) = \frac{4000}{\sqrt{56}}$$

The vector component of the vector field normal to the surface,

$$\mathbf{F}_n \;\; F_n \cdot \mathbf{u}_n = \frac{(4000)(2\mathbf{u}_x + 4\mathbf{u}_y + 6\mathbf{u}_z)}{56} = 142\ 9\mathbf{u}_x + 285\ 7\mathbf{u}_y + 428.6\mathbf{u}_z$$

The vector component of the vector field tangential to the surface,

$\mathbf{F}_t = \mathbf{F} - \mathbf{F}_n = 357.1\mathbf{u}_x + 464.3\mathbf{u}_y - 428.6\ \mathbf{u}_z$

EXAMPLE 1.13

Given a vector function, $\mathbf{F} = 2x^{1/2}y\mathbf{u}_x + xy^2\mathbf{u}_y + (1/z)\mathbf{u}_z$. Find divergence and curl of **F** at $P(0.5, 0.8, 0.2)$.

The divergence of **F**,

$$\nabla\cdot\mathbf{F} = \frac{\partial}{\partial x}\left(2x^{1/2}y\right) + \frac{\partial}{\partial y}\left(xy^2\right) + \frac{\partial}{\partial z}(1/z) = \frac{y}{\sqrt{x}} + 2xy - \frac{1}{z^2}$$

Substituting $x = 0.5$, $y = 0.8$ and $z = 0.2$,

$\nabla\cdot\mathbf{F}$ at $P = 26.931$

The curl of **F**,

$$\nabla\times\mathbf{F} = \begin{vmatrix} \mathbf{u}_x & \mathbf{u}_y & \mathbf{u}_z \\ \dfrac{\partial}{\partial x} & \dfrac{\partial}{\partial y} & \dfrac{\partial}{\partial z} \\ 2x^{1/2}y & xy^2 & 1/z \end{vmatrix} = \left(y^2 - 2\sqrt{x}\right)\mathbf{u}_z$$

Substituting $x = 0.5$, $y = 0.8$ and $z = 0.2$,

$\nabla\times\mathbf{F}$ at $P = -\,0.774\,\mathbf{u}_z$

EXAMPLE 1.14

The equation to a surface is given by $S(x, y, z) = xy - 4z^2 = 0$. What is the angle between unit normal vectors to the surface at $P_1(8, 2, 2)$ and $P_2(6, 6, -3)$?

Unit vector normal to a surface, $\mathbf{u}_n = \dfrac{\nabla S}{|\nabla S|}$

$\nabla S = y\mathbf{u}_x + x\mathbf{u}_y - 8z\mathbf{u}_z$;

At P_1, $\nabla S_1 = 2\mathbf{u}_x + 8\mathbf{u}_y - 16\mathbf{u}_z$; $|\nabla S_1| = 18$

$$\mathbf{u}_{n1} = \frac{2\mathbf{u}_x + 8\mathbf{u}_y - 16\mathbf{u}_z}{18}$$

At P_2, $\nabla S_2 = 6\mathbf{u}_x + 6\mathbf{u}_y + 24\mathbf{u}_z$; $|\nabla S_2| = 25.45$

$$\mathbf{u}_{n2} = \frac{6\mathbf{u}_x + 6\mathbf{u}_y + 24\mathbf{u}_z}{25.4}$$

Let α be the angle between the unit vectors. Then,

$$\cos\alpha = \mathbf{u}_{n1}\cdot\mathbf{u}_{n2} = \frac{(2\mathbf{u}_x + 8\mathbf{u}_y - 16\mathbf{u}_z)\cdot(6\mathbf{u}_x + 6\mathbf{u}_y + 24\mathbf{u}_z)}{(18)(25.4)} = \frac{-324}{459.72} = -\,0.705$$

$\alpha = 134.5°$

EXAMPLE 1.15

Two surfaces are defined by $S_1 = x^2 + y^2 - z^2 = 7$ and $S_2 = x^2 - y + z^2 = 3$. Find the angle between the surfaces at $P(2, 2, -1)$.

The angle between two surfaces at a point common to them is the angle between the normal to the surtaces at the point.

$\nabla S_1 = 2(x\mathbf{u}_x + y\mathbf{u}_y - z\mathbf{u}_z)$

At P, $\nabla S_1 = 4\mathbf{u}_x + 4\mathbf{u}_y + 2\mathbf{u}_z$ and $|\nabla S_1| = 6$

$\nabla S_2 = 2x\mathbf{u}_x - \mathbf{u}_y + 2z\mathbf{u}_z$

At P, $\nabla S_2 = 4\mathbf{u}_x - \mathbf{u}_y - 2\mathbf{u}_z$ and $|\nabla S_2| = \sqrt{21}$

Let α be the angle between the unit vectors. Then,

$$\cos\alpha = \mathbf{u}_{n1} \cdot \mathbf{u}_{n2} = \frac{\nabla S_1 \cdot \nabla S_2}{|\nabla S_1||\nabla S_2|} = \frac{(4\mathbf{u}_x + 4\mathbf{u}_y + 2\mathbf{u}_z)\cdot(4\mathbf{u}_x - \mathbf{u}_y - 2\mathbf{u}_z)}{(6)(\sqrt{21})} = \frac{8}{6\sqrt{21}}$$

$\alpha = 73.08°$

EXAMPLE 1.16

Two vector functions are given by $\mathbf{A} = 2x\mathbf{u}_y + 3y^2\mathbf{u}_z$ and $\mathbf{B} = z\mathbf{u}_x + 3y\mathbf{u}_z$. Find $\nabla \cdot (\mathbf{A} \times \mathbf{B})$ at (3, 2, 1).

$$\mathbf{A} \times \mathbf{B} = \begin{vmatrix} \mathbf{u}_x & \mathbf{u}_y & \mathbf{u}_z \\ 0 & 2x & 3y^2 \\ z & 0 & 3y \end{vmatrix} = 6xy\mathbf{u}_x + 3y^2 z\mathbf{u}_y - 2xz\mathbf{u}_z$$

$$\nabla \cdot (\mathbf{A} \times \mathbf{B}) = \left(\mathbf{u}_x \frac{\partial}{\partial x} + \mathbf{u}_y \frac{\partial}{\partial y} + \mathbf{u}_z \frac{\partial}{\partial z}\right) \cdot (6xy\mathbf{u}_x + 3y^2 z\mathbf{u}_y - 2xz\mathbf{u}_z)$$

$$= 6y + 6yz - 2x = 18 \text{ at } (3, 2, 1).$$

EXAMPLE 1.17

A vector is defined by $\mathbf{A} = xy^2\mathbf{u}_x + yz^2\mathbf{u}_y + zx^2\mathbf{u}_z$. Find $\nabla \times \nabla \times \mathbf{A}$.

$$\nabla \times \mathbf{A} = \begin{vmatrix} \mathbf{u}_x & \mathbf{u}_y & \mathbf{u}_z \\ \frac{\partial}{\partial x} & \frac{\partial}{\partial y} & \frac{\partial}{\partial z} \\ xy^2 & yz^2 & zx^2 \end{vmatrix} = -2(yz\mathbf{u}_x + zx\mathbf{u}_y + xy\mathbf{u}_z)$$

$$\nabla \times \nabla \times \mathbf{A} = (-2)\begin{vmatrix} \mathbf{u}_x & \mathbf{u}_y & \mathbf{u}_z \\ \frac{\partial}{\partial x} & \frac{\partial}{\partial y} & \frac{\partial}{\partial z} \\ yz & zx & xy \end{vmatrix} = 0$$

EXAMPLE 1.18

A scalar function is given by $\phi = x^3 + y^3 + xyz$. Find $\nabla \cdot \nabla\phi$.

$$\nabla \cdot \nabla\phi = \nabla^2\phi = \left(\frac{\partial^2}{\partial x^2} + \frac{\partial^2}{\partial y^2} + \frac{\partial^2}{\partial z^2}\right)(x^3 + y^3 + xyz)$$

$$= \frac{\partial^2}{\partial x^2}(x^3 + y^3 + xyz) + \frac{\partial^2}{\partial y^2}(x^3 + y^3 + xyz) + \frac{\partial^2}{\partial z^2}(x^3 + y^3 + xyz) = 6x + 6y$$

1.9 Vector Identities

Some useful vector identities are listed below. The identities can easily be proved by assuming that the vector **A** and **B** and the scalar ϕ are functions of rectangular-coordinate variables.

1. $\nabla \times (\nabla \phi) = 0$
2. $\nabla \cdot (\nabla \times \mathbf{A}) = 0$
3. $\nabla \cdot (\phi \mathbf{A}) = \mathbf{A} \cdot (\nabla \phi) + \phi (\nabla \cdot \mathbf{A})$
4. $\nabla \cdot (\nabla \phi) = \nabla^2 \phi; \quad \nabla^2 = \frac{\partial^2}{\partial x^2} + \frac{\partial^2}{\partial y^2} + \frac{\partial^2}{\partial z^2}$
5. $\nabla \times \nabla \times \mathbf{A} = \nabla(\nabla \cdot \mathbf{A}) - \nabla^2 \mathbf{A}$
6. $\nabla \times (\phi \mathbf{A}) = (\nabla \phi) \times \mathbf{A} + \phi (\nabla \times \mathbf{A})$
7. $\nabla \cdot (\mathbf{A} \times \mathbf{B}) = \mathbf{B} \cdot (\nabla \times \mathbf{A}) - \mathbf{A} \cdot (\nabla \times \mathbf{B})$

1.10 Line, Surface, and Volume Integrals

In the case of a definite integral,

$$\int_a^b F(x)dx$$

where the integrand $F(x)$ is a real function of a single variable x. The function is integrated along x direction between limits $x = a$ and $x = b$. Whereas in the case of a line integral, a real function of more than one variable is integrated along a continuous curve from an initial point to an end point. The curve may also be a closed one. The function may be a scalar or a vector function of real variables. In the case of a vector function, its component along a specified path is integrated. The line integral is a definite integral.

The principle of evaluating the line integral of a vector function along a path is as follows. Let us consider a vector function **F** in any coordinate plane. Let us take a continuous curve of any shape, denoted by C in Fig. 1.14, in the plane of the vector. The positive direction of the curve is indicated by an arrow, and its initial and final positions are marked by a and b respectively. If we divide the curve into a large number of very small segments, each segment may be considered as a straight line. Let there be n such segments of lengths $\Delta\ell_1, \Delta\ell_2, \cdots \Delta\ell_k, \cdots \Delta\ell_n$. If the average value of the vector function over the kth segment is $\mathbf{F}_k$, its scalar component along the positive direction of the segment is $F_k \cos \alpha_k$, where α_k is the angle between $\mathbf{F}_k$ and tangent to the kth segment, as shown in the figure. The product of $F_k \cos \alpha_k$ and $\Delta\ell_k$ is equal to the dot product of $\mathbf{F}_k$ and the vector length $\Delta\ell_k$. That is,

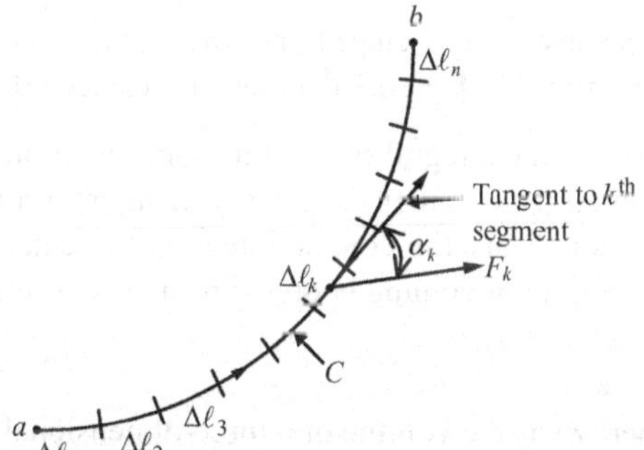

Fig. 1.14 Pertaining to line integration of a vector

$$F_k \Delta\ell_k \cos\alpha_k = \mathbf{F}_k \cdot \Delta\boldsymbol{\ell}_k$$

The sum of such dot products of all the segments between a and b is

$$\sum_{k=1}^{k=n} \mathbf{F}_k \cdot \Delta\boldsymbol{\ell}_k$$

In the limit n approaching infinity, the above summation is called line integral of **F** along the curve C. The curve is the path of integration. The line integral of **F** along the path shown in Fig. 1.14 is

$$\int_a^b \mathbf{F} \cdot d\boldsymbol{\ell}$$

If the line integral is over a closed path, it is written by putting a small circle on the integral sign. Thus, $\oint \mathbf{F} \cdot d\ell$ denotes the line integral of $\mathbf{F}$ over a closed path.

In order to understand the process of integrating a vector function $\mathbf{F}$ over a surface, let us consider a continuous open surface S shown in Fig. 1.15. The surface may be considered as consisting of a large number of very small surface elements with areas $\Delta S_1, \Delta S_2, \ldots, \Delta S_k, \ldots, \Delta S_n$. Let the average of $\mathbf{F}$ over the kth surface element be $\mathbf{F}_k$. The scalar component of $\mathbf{F}_k$ along normal to the surface element is equal to $F_k \cos \alpha_k$, where α_k is the angle between $\mathbf{F}_k$ and the unit vector $\mathbf{u}_n$ normal to the surface element as shown in the figure. The product of $F_k \cos \alpha_k$ and ΔS_k can be expressed as the dot product of the vectors $\mathbf{F}_k$ and $\Delta \mathbf{S}_k$.

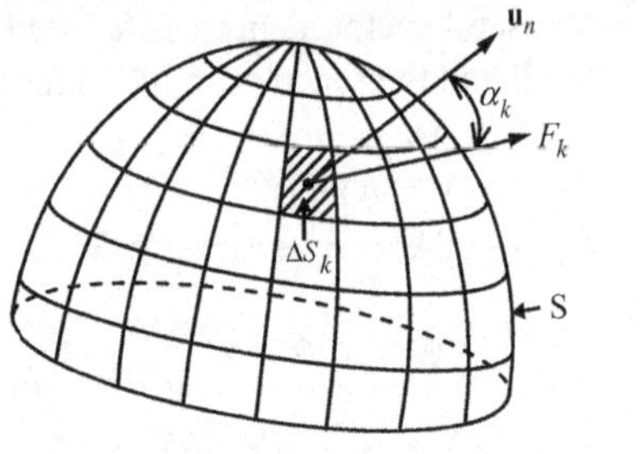

Fig. 1.15 Pertaining to surface integration of a vector

$$F_k \Delta S_k \cos \alpha = \mathbf{F} \cdot (\Delta S_k \mathbf{u}_n) = \mathbf{F}_k \cdot \Delta \mathbf{S}_k$$

The sum of such dot products over all the n surface elements is given by

$$\sum_{k=1}^{k=n} \mathbf{F}_k \cdot \Delta \mathbf{S}_k$$

The above summation, as n approaches infinity, is the surface integral of $\mathbf{F}$ over the surface. The surface integral is a definite double integral which is denoted using either two integral signs or a single integral sign with a subscript s. We will use the latter notation. Accordingly,

$$\int_S \mathbf{F} \cdot d\mathbf{S}$$

denotes the surface integral of a vector function $\mathbf{F}$ over an open surface.

If the surface is a closed one, the surface integral is denoted by putting a small circle on the integral sign. Accordingly, $\oint_S \mathbf{F} \cdot d\mathbf{S}$ denotes the surface integral of $\mathbf{F}$ over a closed surface.

The volume integral of a scalar point function in three-dimensional space is a triple integral. The integral is denoted using either three integral signs or a single integral sign with a subscript v. We will use the latter notation. Thus, the volume integral of a scalar function ρ, which is a continuous function of space variables in a region of volume bounded by a closed surface, is written as

$$\int_v \rho \, dv,$$

where dv is the volume of a three-dimensional element.

EXAMPLE 1.19

Find the line integral of the vector function $\mathbf{F} = (x^2 + 2y)\mathbf{u}_x + (3x - y^2)\mathbf{u}_y$ along the straight line defined by $y = 3x$ from (0, 0) to (1, 3).

An element of length vector in rectangular coordinates,
$d\ell = \mathbf{u}_x dx + \mathbf{u}_y dy + \mathbf{u}_z dz$
The dot product of $\mathbf{F}$ and $d\ell$, $\mathbf{F} \cdot d\ell = (x^2 + 2y)dx + (3x - y^2)dy$

Substituting y by $3x$ in the first term and x by $y/3$ in the second term, we have

$\mathbf{F} \cdot d\ell = (x^2 + 6x)dx + (y - y^2)dy$

The line integral, $\int \mathbf{F} \cdot \quad = \quad + \quad + \int_0^3 (\; - \;) \quad = -7.17$

EXAMPLE 1.20

Integrate the vector function,

$$\mathbf{F} = \frac{x}{(x^2+y^2)^{3/2}}\mathbf{u}_x + \frac{y}{(x^2+y^2)^{3/2}}\mathbf{u}_y,$$

along the path described by the equation $y^2 = 4x$ from $P_1(1, 2)$ to $P_2(4, 4)$.

An element of length vector in rectangular coordinates,

$d\ell = \mathbf{u}_x dx + \mathbf{u}_y dy + \mathbf{u}_z dz$

The dot product of **F** and $d\ell$, $\mathbf{F}\cdot d\ell = \dfrac{x}{(x^2+y^2)^{3/2}}dx + \dfrac{y}{(x^2+y^2)^{3/2}}dy$ (1)

The path of integration is defined by the equation,

$y^2 = 4x$ (2)

Or, $ydy = 2dx$ (3)

Substituting (2) and (3) in (1), we get

$$\mathbf{F}\cdot d\ell = \frac{x+2}{(x^2+4x)^{3/2}}dx$$

Thus, $\displaystyle\int_{P_1}^{P_2}\mathbf{F}\cdot d\ell = \int_1^4 \frac{x+2}{(x^2+4x)^{3/2}}dx = 0.274$

The integral can be evaluated by substituting $x^2 + 4x = u$.

EXAMPLE 1.21

A circular cylinder has its base on the *x-y* plane with its axis along the *z* axis. The radius and height of the cylinder are 20 cm and 50 cm respectively. By using the formula for the differential area $d\mathbf{S} = (rd\phi\, dz)\mathbf{u}_r + (drdz)\mathbf{u}_\phi + (rdrd\phi)\mathbf{u}_z$, find the surface area of the cylinder.

A differential area on the surface of the cylinder is in *r* direction. Thus,

$dS = rd\phi\, dz$

At $r = 0.2$ m, $dS = 0.2d\phi\, dz$

Surface area, $S = \displaystyle\int_{\phi=0}^{2\pi}\int_{z=0}^{0.5} 0.2d\phi dz = 0.2\pi\ \text{m}^2$

EXAMPLE 1.22

The radius of a sphere is 25 cm. By using the formula for the differential area $d\mathbf{S} = (r^2\sin\theta d\theta d\phi)\mathbf{u}_r + (r\sin\theta drd\phi)\,\mathbf{u}_\theta + (rdrd\phi)\,\mathbf{u}_\phi$, find surface area of the lower hemisphere.

A differential area on the surface of the sphere is in *r* direction. Thus,

$dS = r^2\sin\theta d\theta d\phi$

At $r = 0.25$ m, $dS = 0.0625\sin\theta d\theta d\phi$

Surface area, $S \quad \displaystyle\int_{\theta=\pi/2}^{\pi}\int_{\phi=0}^{2\pi} \qquad d\theta d\phi = 0.125\pi\ \text{m}^2$

EXAMPLE 1.23

Given a vector function $\mathbf{F} = x^2y\mathbf{u}_x + xy^2\mathbf{u}_y$. Integrate $\nabla \times \mathbf{F}$ over the rectangular surface shown in Fig. 1.16.

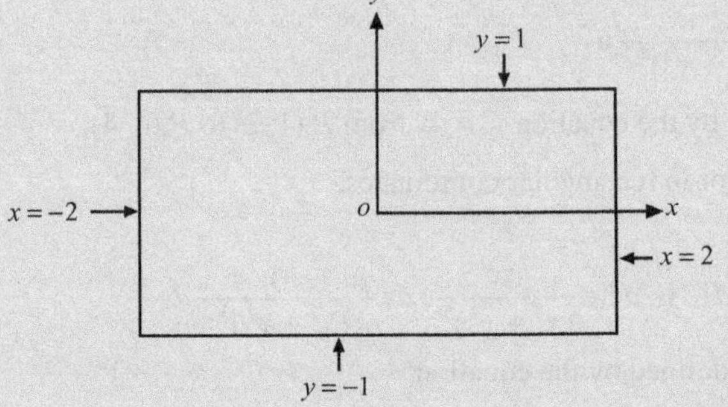

Fig. 1.16 For Example 1.23

The curl of the vector function,

$$\nabla \times \mathbf{F} = \begin{vmatrix} \mathbf{u}_x & \mathbf{u}_y & \mathbf{u}_z \\ \dfrac{\partial}{\partial x} & \dfrac{\partial}{\partial y} & \dfrac{\partial}{\partial z} \\ x^2y & xy^2 & 0 \end{vmatrix} = (y^2 - x^2)\mathbf{u}_z$$

An element of surface vector is given by

$$d\mathbf{S} = \mathbf{u}_x dydz + \mathbf{u}_y dzdx + \mathbf{u}_z dxdy$$

Thus, $(\nabla \times \mathbf{F}) \cdot d\mathbf{S} = (y^2 - x^2)dxdy$

We find from Fig. 1.16 that the lower and upper limits of x are −2.0 and +2.0, and those of y are −1.0 and +1.0. Thus,

$$\int_S \overline{\quad}\ )\cdot d\mathbf{S} = \int_{-1}^{+1}\int_{-2}^{+2}(y^2 - x^2)dxdy = \ 8.0$$

1.11 Coordinate Transformation of Vectors

It is desirable to use a suitable coordinate system at the outset for solving a field problem involving vectors. For example, it is easier to solve a problem having cylindrical symmetry by using cylindrical coordinates. A proper choice of a coordinate system simplifies the analysis, and results in a solution that is compact in form and easy for physical interpretation. However, it is occasionally necessary to convert a vector function from one coordinate system to another system. The principles of transformation of a vector function between cylindrical and rectangular coordinates, and between spherical and rectangular coordinates are described in the following.

Transformations between Cylindrical and Rectangular Coordinates

Let a vector **F** in terms of its components in cylindrical coordinates be given by

$$\mathbf{F} = F_r\mathbf{u}_r + F_\phi\mathbf{u}_\phi + F_z\mathbf{u}_z \tag{1}$$

where in general each component is a function of all the three variables r, ϕ and z. We will transform the vector to rectangular coordinates.

Let the *same* vector in terms of its components in rectangular coordinates is

$$\mathbf{F} = F_x\mathbf{u}_x + F_y\mathbf{u}_y + F_z\mathbf{u}_z \tag{2}$$

In order to transform the vector, we obtain at first the rectangular-coordinate components F_x, F_y and F_z of **F** in terms of F_r, F_ϕ and F_z and then change the cylindrical-coordinate variables to rectangular-coordinate variables. We know that the component of a vector in any direction is given by the dot product of the vector and a unit vector in that direction. Accordingly, F_x is equal to the dot product of **F** described by (1) and the unit vector $\mathbf{u}_x$.

$$\begin{aligned} F_x &= (F_r\mathbf{u}_r + F_\phi\mathbf{u}_\phi + F_z\mathbf{u}_z) \cdot \mathbf{u}_x \\ &= F_r(\mathbf{u}_r \cdot \mathbf{u}_x) + F_\phi(\mathbf{u}_\phi \cdot \mathbf{u}_x) + F_z(\mathbf{u}_z \cdot \mathbf{u}_x) \end{aligned} \tag{3}$$

We see from Fig. 1.17 that the angle between $\mathbf{u}_r$ and $\mathbf{u}_x$ is ϕ, the angle between $\mathbf{u}_\phi$ and $\mathbf{u}_x$ is $90° + \phi$ and the angle between $\mathbf{u}_z$ and $\mathbf{u}_x$ is 90°. Therefore, $\mathbf{u}_r \cdot \mathbf{u}_x = \cos\phi$, $\mathbf{u}_\phi \cdot \mathbf{u}_x = -\sin\phi$ and $\mathbf{u}_z \cdot \mathbf{u}_x = 0$. Substituting these results in (3), we get

$$F_x = F_r \cos\phi - F_\phi \sin\phi$$

The component, $F_y = (F_r\mathbf{u}_r + F_\phi\mathbf{u}_\phi + F_z\mathbf{u}_z) \cdot \mathbf{u}_y = F_r(\mathbf{u}_r \cdot \mathbf{u}_y) + F_\phi(\mathbf{u}_\phi \cdot \mathbf{u}_y) + F_z(\mathbf{u}_z \cdot \mathbf{u}_y)$

The angle between $\mathbf{u}_r$ and $\mathbf{u}_y$ is $90° - \phi$, between $\mathbf{u}_\phi$ and $\mathbf{u}_y$ is ϕ and between $\mathbf{u}_z$ and $\mathbf{u}_y$ is 90°. Thus,

$$F_y = F_r \sin\phi + F_\phi \cos\phi$$

As the z direction is common to both the systems,

$$F_z = F_z$$

Since F_r, F_ϕ and F_z are functions of the cylindrical-coordinate variables, these have to be expressed in terms of rectangular-coordinate variables to complete the transformation. Relations connecting the two sets of variables are given in (1.2-2) [Equation 2 of Section 1.2].

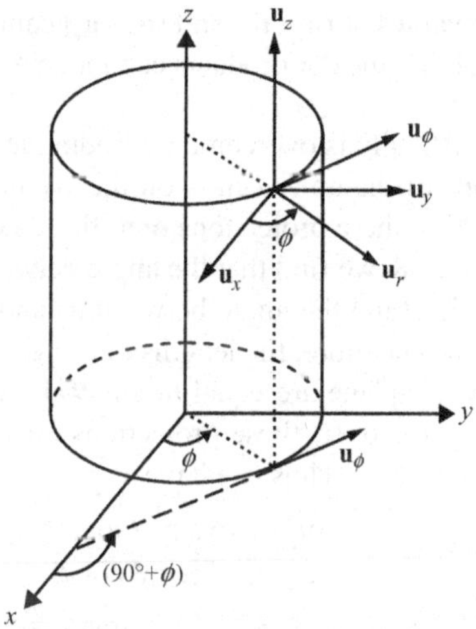

Fig. 1.17 Showing unit vectors of rectangular and cylindrical coordinates

We use the same procedure as described above to transform a vector in rectangular coordinates to cylindrical coordinates. It can be shown that a vector

$$\mathbf{G} = G_x\mathbf{u}_x + G_y\mathbf{u}_y + G_z\mathbf{u}_z$$

in cylindrical coordinates when transformed to rectangular coordinates will have the following scalar components.

$$G_r = G_x \cos\phi + G_y \sin\phi$$

$$G_\phi = -G_x \sin\phi + G_y \cos\phi$$

$$G_z = G_z$$

To complete the transformation, the components have to be expressed in terms of spherical-coordinate variables using the following formulae which can be derived from (1.2-2)

$$r = \sqrt{x^2 + y^2}$$
$$\phi = \tan^{-1}(y/x)$$
$$z = z$$

Transformations between Spherical and Rectangular Coordinates

In spherical coordinates, let a vector function be given by

$$\mathbf{E} = E_r\mathbf{u}_r + E_\theta\mathbf{u}_\theta + E_\phi\mathbf{u}_\phi$$

where E_r, E_θ and E_ϕ are in general functions of all the three variables r, θ and ϕ.

In order to transform the vector to rectangular coordinates, we have to find its components in x, y and z directions, and then express these components in terms of x, y and z. The component of **E** along x direction is

$$E_x = \mathbf{E} \cdot \mathbf{u}_x = E_r(\mathbf{u}_r \cdot \mathbf{u}_x) + E_\theta(\mathbf{u}_\theta \cdot \mathbf{u}_x) + E_\phi(\mathbf{u}_\phi \cdot \mathbf{u}_x) \quad (4)$$

The values of $(\mathbf{u}_r \cdot \mathbf{u}_x)$ and $(\mathbf{u}_\theta \cdot \mathbf{u}_x)$ cannot be found directly because the angles between the unit vectors involving the dot products cannot be expressed as simple functions of θ and ϕ.

This difficulty is overcome by finding at first the projections of the unit vectors on the x-y plane and then projecting these projections onto the x axis. Referring to Fig. 1.18, we find that the angle between $\mathbf{u}_r$ and the z axis is θ and the angle between $\mathbf{u}_\theta$ and the z axis is $90° + \theta$. Therefore, the lengths of projections $\mathbf{u}_r$ and $\mathbf{u}_\theta$ on the x-y plane are equal to $\sin\theta$ and $\cos\theta$ respectively. Moreover, these projections are at an angle ϕ with the x axis. Thus, we have

$$\mathbf{u}_r \cdot \mathbf{u}_x = \sin\theta\cos\phi$$
$$\mathbf{u}_\theta \cdot \mathbf{u}_x = \cos\theta\cos\phi$$

As the unit vector $\mathbf{u}_\phi$ is at angle $90° + \phi$ with the x axis, the dot product of $\mathbf{u}_\phi$ and $\mathbf{u}_x$ is

$$\mathbf{u}_\phi \cdot \mathbf{u}_x = -\sin\phi$$

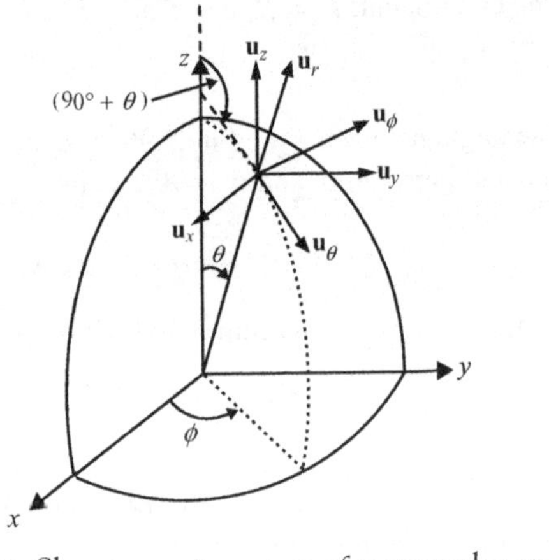

Fig. 1.18 Showing unit vectors of rectangular and spherical coordinates

Substituting the expressions of the dot products in (4), we get

$$E_x = E_r\sin\theta\cos\phi + E_\theta\cos\theta\cos\phi - E_\phi\sin\phi$$

The components of **E** along y and z directions can similarly be calculated. These are given by

$$E_y = E_r\sin\theta\sin\phi + E_\theta\cos\theta\sin\phi + E_\phi\cos\phi$$
$$E_z = E_r\cos\theta - E_\theta\sin\theta$$

To complete the process of transformation, we have to express the components in terms of the rectangular-coordinate variables. The relations connecting the two sets of variables are given by (1.2-3).

Let us next consider a vector **H** in rectangular coordinates defined in terms of its components.

$$\mathbf{H} = H_x\mathbf{u}_x + H_y\mathbf{u}_y + H_z\mathbf{u}_z,$$

We can transform the vector into spherical coordinate system using the procedure similar to that described above. The scalar components of **H** in spherical coordinates are

$$H_r = H_x \sin\theta\cos\phi + H_y \sin\theta\sin\phi + H_z \cos\theta$$
$$H_\theta = H_x \cos\theta\cos\phi + H_y \cos\theta\sin\phi - H_z \sin\theta$$
$$H_\phi = -H_x \sin\phi + H_y \cos\phi$$

To complete the process of conversion, the components will have to be transformed to spherical-coordinate variables using the following formulae. These can be derived from (1.2-3)

$$r = \sqrt{x^2 + y^2 + z^2}$$

$$\theta = \cos^{-1}\frac{z}{\sqrt{x^2 + y^2 + z^2}}$$

$$\phi = \tan^{-1}(y/x)$$

EXAMPLE 1.24

A vector in cylindrical coordinates is given by $\mathbf{F} = \frac{\mathbf{u}_r}{r}$. Transform the vector into rectangular coordinates.

The vector has only one component $F_r = 1/r$ in r direction. Let the vector in rectangular coordinates be $\mathbf{F} = F_x\mathbf{u}_x + F_y\mathbf{u}_y + F_z\mathbf{u}_z$

Then, $F_x = F_r(\mathbf{u}_r \cdot \mathbf{u}_x) = \frac{\cos\phi}{r} = \frac{x}{x^2 + y^2}$

$$F_y = F_r(\mathbf{u}_r \cdot \mathbf{u}_y) = \frac{\sin\phi}{r} = \frac{y}{x^2 + y^2}$$

$$F_z = 0$$

The vector in rectangular coordinates is given by

$$\mathbf{F} \quad \frac{1}{x^2 + y^2}(x\mathbf{u}_x + y\mathbf{u}_y)$$

EXAMPLE 1.25

The vector $\mathbf{F} = r\sin\phi\ \mathbf{u}_\phi$ is given in cylindrical coordinates. Transform the vector into rectangular coordinates.

The vector has only one component $F_\phi = r\sin\phi$ in ϕ direction. Let the vector in rectangular coordinates be $\mathbf{F} = F_x\mathbf{u}_x + F_y\mathbf{u}_y + F_z\mathbf{u}_z$

Then, $F_x = F_r(\mathbf{u}_\phi \cdot \mathbf{u}_x) = r\sin\phi(-\sin\phi) = -\frac{y^2}{\sqrt{x^2 + y^2}}$

$$F_y = F_r(\mathbf{u}_\phi \cdot \mathbf{u}_y) = r\sin\phi(\cos\phi) = \frac{xy}{\sqrt{x^2 + y^2}}$$

$$F_z = 0$$

The vector in rectangular coordinate is given by

$$\mathbf{F} \quad \frac{1}{x^2 + y^2}\text{[illegible]}\,\mathbf{u}_x + xy\mathbf{u}_y)$$

EXAMPLE 1.26

Convert the vector $\boldsymbol{E} = \frac{\boldsymbol{u}_r}{r^2}$ in spherical coordinates into rectangular coordinates.

The vector has only one component $E_r = \frac{1}{r^2}$ in r direction. Let the vector in rectangular coordinates be

$\mathbf{E} = E_x\mathbf{u}_x + E_y\mathbf{u}_y + E_z\mathbf{u}_z$

We have $E_x = E_r(\mathbf{u}_r \cdot \mathbf{u}_x) = \frac{1}{r^2}(\sin\theta\cos\phi) = \frac{x}{(x^2+y^2+z^2)^{3/2}}$

$$E_y \quad E_r(\mathbf{u}_r \cdot \mathbf{u}_y) = \frac{1}{r^2}(\sin\theta\sin\phi) = \frac{y}{(x^2+y^2+z^2)^{3/2}}$$

$$E_z \quad E_r(\mathbf{u}_r \cdot \mathbf{u}_z) = \frac{\cos\theta}{r^2} = \frac{z}{(x^2+y^2+z^2)^{3/2}}$$

The vector in rectangular coordinate is given by

$$\mathbf{E} \quad \frac{1}{(\ ^2+\ ^2+z^2)^{3/2}}(x\mathbf{u}_x + y\mathbf{u}_y + z\mathbf{u}_z)$$

1.12 Sinusoidally Time-Varying Quantities

A scalar or a vector quantity may be time-invariant or it may be a function of time. Quantities that are either sine or cosine function of time are called sinusoidally time-varying quantities. For example, a voltage represented by

$$V(t) = V_0 \cos \omega t \tag{1}$$

is a co-sinusoidal quantity. And a current given by

$$I(t) = I_0 \sin \omega t \tag{2}$$

is a sinusoidal quantity. These are also called *alternating* quantities. In (1) and (2) $V(t)$ and $I(t)$ denote instantaneous values, V_0 and I_0 are the maximum values or amplitudes of the functions, and ω is the angular or radian frequency. The units of voltage and current are *volt* (V) and *ampere* (A) and that of ω is *radians per second* (rad/s). The graphs of variation of voltage and current with time, shown in Fig. 1.19, are the familiar cosine and sine waves. The waves repeat after every $\omega t = 2\pi$ radian. The number of such repetitions or cycles per second is called frequency of the waves. The frequency is denoted by f and its unit is hertz (Hz). The duration of one cycle is equal to *one period*. It is generally denoted by T. Therefore, $T = 1/f$. As ωt changes by 2π radian in one period, $\omega = 2\pi/T = 2\pi f$. Each cycle consists of a positive half-cycle and a negative half-cycle. Both the half-cycles have the same amplitude and identical shapes.

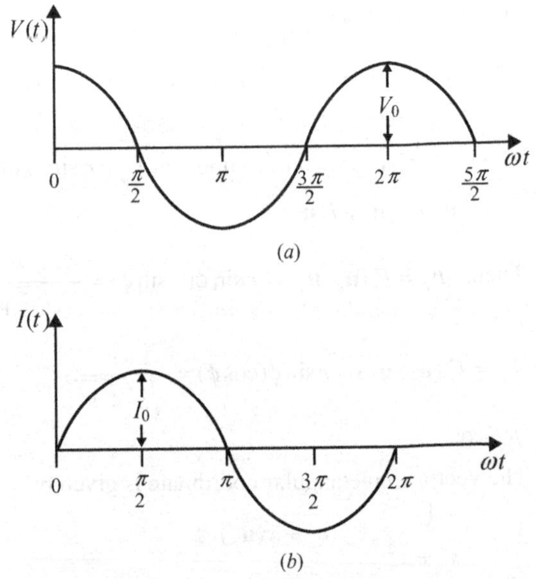

Fig. 1.19 (*a*) A cosine wave and (*b*) A sine wave.

Next, let us consider the functions,

$$V_1(t) = V_{10} \cos (\omega t + \psi_1) \tag{3}$$

and

$$V_2(t) = V_{20} \cos (\omega t - \psi_2) \tag{4}$$

Note that $V_1(t)$ and $V_2(t)$ have the same frequency as that of $V(t)$. Graphs of functions defined by (1), (3) and (4) are shown in Fig. 1.20. The symbols ψ_1 and ψ_2 denote phase angles of $V_1(t)$ and $V_2(t)$ respectively. If we take the time-angle at which $V(t)$ is zero (or maximum) as reference, the corresponding zero of $V_1(t)$ occurs earlier by a time-angle ψ_1 and that of $V_2(t)$ occurs later by a time-angle ψ_2. Thus, $V_1(t)$ leads $V(t)$ by an angle ψ_1 and $V_2(t)$ lags $V(t)$ by an angle ψ_2. As $\cos \omega t = \sin(\omega t + \pi/2)$ a cosine function leads a sine function by $\pi/2$ radian. The phase angles are specified either in radian or in degree.

Representation of Sinusoidal Functions

Any of the equations (1) to (4) is the basic form of representing a sinusoidal quantity in terms of its amplitude, frequency and phase angle. This form gives the value of the quantity at any instant. The waveforms of Fig. 1.19 and 1.20 also provide the same information.

Some other useful ways of representing a sinusoidal quantity are the phasor, the complex and the polar forms. The basis of these forms is the Euler's formula (or identity) of the complex exponential function $e^{j\omega t}$. As per the formula,

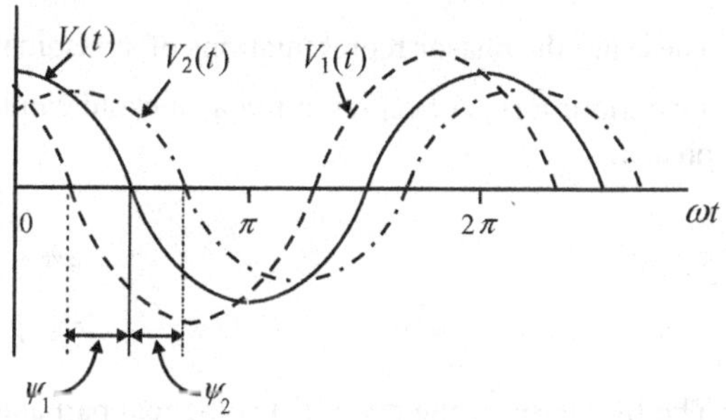

Fig. 1.20 Showing time-phase difference between sinusoidal waves

$$e^{j\omega t} = \cos \omega t + j \sin \omega t$$

where

$$j = \sqrt{-1}$$

$$\cos \omega t = \text{Re } e^{j\omega t} \tag{5}$$

$$\sin \omega t = \text{Im } e^{j\omega t} \tag{6}$$

The notations Re and Im are the abbreviations of '*real part of*' and '*imaginary part of*' respectively. Thus using (5) and (6) in (1) and (2), we have

$$V(t) = \text{Re } V_0 \, e^{j\omega t}$$

and

$$I(t) = \text{Im } I_0 \, e^{j\omega t} = \text{Re } I_0 \, e^{j(\omega t - \pi/2)}$$

Similarly (3) and (4) can be written as

$$V_1(t) = \text{Re} V_{10} e^{j(\omega t + \psi_1)}$$

$$V_2(t) = \text{Re} V_{20} e^{j(\omega t - \psi_2)}$$

These are the complex exponential forms of writing sinusoidally time-changing quantities. One advantage of using the complex exponential function $e^{j\omega t}$ instead of sin ωt or cos ωt is that the form of a function does not change when differentiated or integrated with respect to time.

To proceed further, let us write the last four equations as in the following.

$$V(t) = V_0 \cos \omega t = \text{Re } (V_0 \, e^{j0}) e^{j\omega t}$$

$$I(t) = I_0 \sin \omega t = \text{Re } (I_0 \, e^{-j\pi/2}) e^{j\omega t}$$

$$V_1(t) = V_{10} \quad t + \psi_1) \quad \text{Re}(V_{10} e^{j\psi_1}) e^{j \quad t}$$

$$V_2(t) = V_{20} \quad t - \psi_1) \quad \text{Re}(V_{20} e^{-j\psi_2}) e^{j \quad t}$$

A study of the above equations shows that the factor $e^{j\omega t}$ appears in all the functions. In the mathematical analysis of a problem, it is not necessary to show this factor in all the steps. The presence of $e^{j\omega t}$ in a function is assumed. If all the quantities in a study have the same frequency, it is sufficient to use only the functions shown within the brackets. At the end of an analysis, the final expression is multiplied by $e^{j\omega t}$ and either its real part or the imaginary part is used to express the result in real time. The functions inside the brackets are called *phasors*.

We will denote a phasor by putting a *bar* over the symbol of the quantity. For example,

$$\bar{V}_{10} = V_{10} e^{j\psi_1}$$

and

$$\bar{V}_{20} = V_{20} e^{-j\psi_2}$$

These are the phasor representations of sinusoidally time-varying quantities.

Two more forms of representing a sinusoid can be derived from the phasor form. Let us consider a voltage phasor,

$$\bar{V}_0 = V_0 e^{j\psi}$$

Since

$$e^{j\omega} = \cos\psi + j\sin\psi,$$

$$\bar{V}_0 = V_0(\cos\psi + j\sin\psi) \tag{7}$$

The first term on the rhs of (7) is the real part and the second term is the imaginary part of $\bar{V}_0$. Denoting the real part by V_{0a} and the imaginary part by V_{0i}, we have

$$\bar{V}_0 = V_{0a} + jV_{0i} \tag{8}$$

where $V_{0a} = V_0 \cos\psi$ and $V_{0i} = V_0 \sin\psi$.

The form of (8) suggests that a phasor can be represented in a complex plane, as shown in Fig. 1.21. Equation (8) and Fig. 1.21 are the complex form of representing a sinusoidally time-varying quantity.

We see that to write a sinusoidal quantity in phasor and complex forms only its amplitude and phase angle are needed. Thus, a simpler way of describing a sinusoidal quantity is by its amplitude and phase angle as given below:

$$\bar{V}_0 = V_0 \angle \pm\psi$$

This is known as the polar form of representing a sinusoid. This form is advantageous where multiplication and division of phasors are involved. To illustrate these operations, let

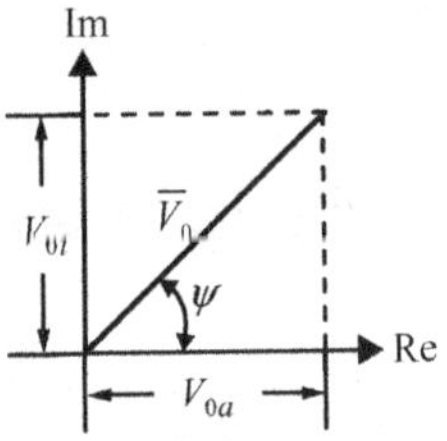

Fig. 1.21 Representation of a sinusoidal function in the complex plane

$$\bar{V}_0 = V_0 e^{\pm j\psi_1}$$

and

$$\bar{I}_0 = I_0 e^{\pm j\psi_2}$$

The product of the two phasors, $\bar{V}_0\bar{I}_0 = V_0 I_0 e^{j[\pm\psi_1 + (\pm\psi_2)]} = V_0 I_0 \angle[\pm\psi_1 + (\pm\psi_2)]$

The ratio of the two phasors, $\dfrac{\bar{V}_0}{\bar{I}_0} = \dfrac{V_0}{I_0} e^{j[\pm\psi_1 - (\pm\psi_2)]} = \dfrac{V_0}{I_0} \angle[\pm\psi_1 - (\pm\psi_2)]$

Thus, the magnitude of the product of two phasors is equal to the product of their magnitudes whereas the phase angle of their product is equal to the algebraic sum of their phase angles. Moreover, the magnitude of the ratio of two phasors is equal to the ratio of their magnitudes and the phase angle is equal to the phase angle of the numerator phasor minus the phase angle of the denominator-phasor.

Root-Mean-Square Value of a Sinusoidal Quantity

In order to define the root-mean-square (rms) value of a function, consider that a sinusoidal current

$$I(t) = I_0 \sin \omega t$$

flows through a resistor of R ohm. Then, electric power dissipated in the resistor at any instant,

$$P(t) = [I(t)]^2 R = I_0^{\,2} R \sin^2 \omega t$$

The average power over a complete cycle,

$$P_{av} = \frac{1}{2\pi}\int_0^{2\pi} I_0^{\,2} R \sin^2 \omega t \, d(\omega t) = \frac{I_0^{\,2} R}{2}$$

Next, consider that a time-invariant current I flows through the same resistor instead of the sinusoidal current. Power dissipated in the resistor is equal to $I^2 R$. If we assume that $I^2 R$ is equal to the average power dissipated by the sinusoidal current $I_0 \sin \omega t$, then

$$I^2 R = \frac{I_0^{\,2} R}{2}$$

Or,

$$I = \frac{I_0}{\sqrt{2}}$$

The constant current I, which produces the same power in the resistor as the average power due to an alternating current, is called *root-mean-square (rms) or effective value* of the time-varying current. The nomenclature is derived from the mathematical procedure involved in deriving it.

The rms value of a sinusoidal quantity is equal to $1/\sqrt{2}$ times of its amplitude. The rms value of any periodic function can be determined using the above procedure. Sinusoidal quantities can also be represented using rms value instead of amplitudes as given below:

Instantaneous form: $I(t) = \sqrt{2} I \sin(\omega t \pm \psi)$

$V(t) = \sqrt{2} V \cos(\omega t \pm \psi)$

Phasor form: $\bar{V} = V e^{\pm j\psi}$

Complex form: $\bar{V} = V_a \pm j V_i$

Polar form: $\bar{V} = V \angle \pm \psi$

The Complex Impedance

The complex impedance $\bar{Z}$ of an electric load is defined as the ratio of its complex voltage and complex current. To calculate the impedance, let us assume that the instantaneous voltage across the electric load shown in Fig. 1.22 is

$$V(t) = V_0 \cos \omega t \tag{9}$$

The complex and polar forms of the voltage are

$$V_v \quad V_0 + j0 \tag{10}$$

$$V_v \quad V_0 \angle 0$$

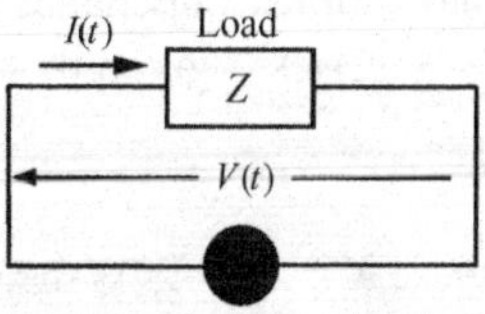

Fig. 1.22 A load connected to an ac source

For a purely resistive load of resistance R ohm, instantaneous current through the load

$$I(t) = \frac{V(t)}{R} = \frac{V_0}{R}\cos\omega t = I_0 \cos\omega t \tag{11}$$

where, $I_0 = V_0/R$. We find from equations (9) and (11) that these are in time-phase. The complex form of load current is

$$\overline{I}_0 = I_0 + j0 \tag{12}$$

The complex impedance is found by dividing (10) by (12).

$$\overline{Z} = \frac{V_0 + j0}{I_0 + j0} = R + j0 = R\angle 0$$

If the load is an ideal (loss-free) inductor of inductance L henry (H),

$$I(t) = \frac{1}{L}\int V(t)dt = \frac{V_0}{\omega L}\sin\omega t = \frac{V_0}{X_L}\cos(\omega t - 90°) = I_0\cos(\omega t - 90°) \tag{13}$$

The constant of integration is assumed to be zero as we are considering alternating quantities under steady state (not transient) condition. In the above expressions, $X_L = \omega L$ is reactance of the inductor and $I_0 = V_0/X_L$ is amplitude of the current. The unit of reactance is ohm. We find from (9) and (13) that the current in a pure inductor lags the voltage across it by 90°. The complex form of current is

$$\overline{I}_0 = 0 - jI_0$$

The complex impedance of the inductor,

$$\overline{Z} = \frac{V_0 + j0}{0 - jI_0} = 0 + jX_L = jX_L = X_L\angle 90°$$

If the load is an ideal (loss-free) capacitor of capacitance C farad (F),

$$I(t) = C\frac{dV(t)}{dt} = -\omega C V_0 \sin\omega t = \frac{V_0}{X_c}\cos(\omega t + 90°) = I_0\cos(\omega t + 90°)$$

where $X_c = (1/\omega c)$ is reactance of the capacitor in ohm and $I_0 = (V_0/X_c)$ is amplitude of the current. We find that current in an ideal capacitor leads the voltage across it by 90°. The complex form of current is given by

$$\overline{I}_0 = 0 + jI_0$$

The complex impedance,

$$\overline{Z} = \frac{\overline{V}}{\overline{I}} = \frac{V_0 + j0}{0 + jI_0} = 0 - jX_c = -jX_c = X_c\angle -90°$$

The complex impedance of an electric load consisting of a resistance R, an inductance X_L and a capacitance X_C in series is given by

$$\overline{Z} = R + j(X_L - X_C) = R + jX \tag{14}$$

where $X = X_L - X_C$ is the effective reactance of the circuit. If X_L is greater than X_C, then the load is inductive and if X_C is greater than X_L, the load is capacitive. The angle $\psi = \tan^{-1}(X/R)$ is called impedance angle.

The resistance and reactance of the load in terms of the load angle are given by

$$R = Z \cos \psi,$$
$$X = Z \sin \psi$$

The magnitude of the impedance $Z = \sqrt{R^2 + X^2}$

The impedance in polar form is represented as

$$\bar{Z} = Z\angle\psi$$

Current $\bar{I}_0$ taken by a load is equal to the ratio of (10) and (14).

$$\bar{I}_0 = \frac{\bar{V}_0}{\bar{Z}} = \frac{V_0 + j0}{R + jX} = \frac{V_0(R - jX)}{Z^2} = I_0(\cos\psi - j\sin\psi) = I_{0a} - jI_{0i}$$

where $I_{0a} = (V_0/Z)\cos\psi = I_0\cos\psi$ is the real part, and $I_{0i} = (V_0/Z)\sin\psi = I_0\sin\psi$ is the imaginary part of the current phasor respectively.

The current in the polar form is

$$\bar{I}_0 = \frac{V_0\angle 0}{Z\angle +\psi} = I_0\angle -\psi \text{ where } I_0 = \frac{V_0}{Z}$$

If the resistance and reactance of an electric load is known, we can calculate values of the circuit impedance Z and the impedance angle ψ. Taking the voltage across the circuit as reference, the equation for instantaneous current is given by

$$I(t) = I_0\cos(\omega t - \psi)$$

We see that the phase angle between voltage and current is equal to the impedance angle ψ. The current lags the voltage for an inductive load and the current leads the voltage for a capacitive load. The impedance angle lies between 0 and 90° if the load is inductive, and between 0 and –90° if the load is capacitive.

The rms value of current, $I = \frac{\text{rms value of voltage}}{Z} = \frac{V}{Z}$

Instantaneous, Average and Complex Powers

Instantaneous power input to a load is equal to the product of its voltage and current at a given instant. If the instantaneous voltage is described by the equation

$$V(t) = V_0 \cos(\omega t + \psi_1) \tag{15}$$

and the current at the same instant is given by

$$I(t) = I_0 \cos(\omega t + \psi_2), \tag{16}$$

then instantaneous power,

$$P(t) = V(t)I(t) = V_0 I_0 \cos(\omega t + \psi_1)\cos(\omega t + \psi_2)$$

$$= \frac{V_0 I_0}{2}[\cos\psi + \cos(2\omega t + \psi_1 + \psi_2)] \tag{17}$$

where $\psi = \psi_2 - \psi_1$ is the phase difference between current and voltage.

The instantaneous power given by (17) has two parts. One part is independent of time and the other part varies with time at twice the frequency. The average power is equal to the average value of instantaneous power over one cycle. Thus the average power,

$$P_{av} = \frac{V_0 I_0}{2}\left\{\frac{1}{2\pi}\int_0^{2\pi}[\cos\psi + \cos(2\omega t + \psi_1 + \psi_2)]d(\omega t)\right\}$$

$$= \frac{V_0 I_0}{2}\cos\psi = \frac{V_0}{\sqrt{2}}\frac{I_0}{\sqrt{2}}\cos\psi = VI\cos\psi \qquad (18)$$

The average power therefore is equal to the product of rms voltage, rms current and the cosine of phase-angle between the voltage and current. The power is also called *active* or *effective* or *real* power. The electric unit of power is *watt* (W). The product of V and I, that is VI, is called apparent power and its unit is *volt-ampere* (VA). The factor cos ψ is called power factor of the load.

Power factor, $\cos\psi = \dfrac{2P_{av}}{V_0 I_0} = \dfrac{P_{av}}{VI}$

Substituting $V = IZ$ and cos ψ = R/Z in (18), we get the average power in terms of the current and the resistance.

$$P_{av} = I^2 R$$

Thus, the active power is the power absorbed by the load.

In an analogous way, the reactive or quadrature power,

$$P_X = \frac{1}{2}V_0 I_0 \sin\psi = VI\sin\psi = I^2 X \qquad (19)$$

The unit of reactive power is *volt-ampere reactive* (var).

It may be noted that I cos ψ of (18) is the component of current that is in phase with the voltage, and I sin ψ of (19) is the component of current that is in quadrature with the voltage. Thus, the general principle is that the average power is equal to the product of in-phase components of rms voltage and rms current, and the reactive power is given by the product of rms voltage and rms current which are out of phase by 90°. We will use this principle to define complex power.

The complex forms of voltage and current defined by (15) and (16) are

$$\bar{V}_0 = V_0(\cos\psi_1 + j\sin\psi_1) = V_{0a} + jV_{0i} \qquad (20)$$

and

$$\bar{I}_0 = I_0(\cos\psi_2 + j\sin\psi_2) = I_{0a} + jI_{0i} \qquad (21)$$

In (20) V_{0a} and V_{0i} are the real and imaginary parts of the voltage, and in (21) I_{0a} and I_{0i} are the real and imaginary parts of the current. These are represented in the complex plane of Fig. 1.23. Since V_{0a} and I_{0a} are in phase and V_{0i} and I_{0i} are also in phase, the average power,

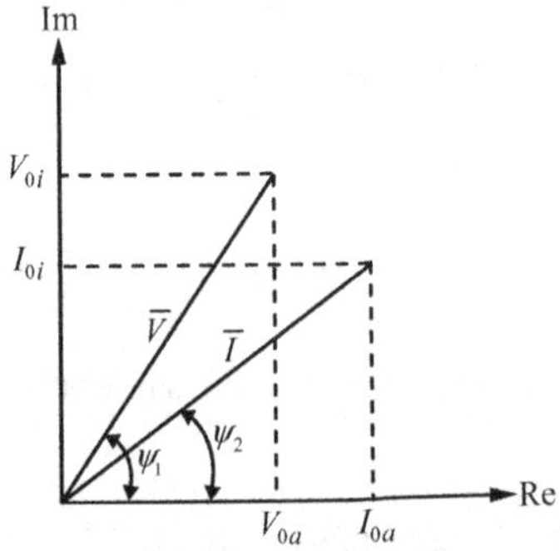

Fig. 1.23 Showing real and imaginary parts of a voltage phasor and a current phasor

$$P_{av} = \frac{1}{2}(V_{0a}I_{0a} + V_{0i}I_{0i}) \qquad (22)$$

Moreover, as V_{0i} and I_{0a} are out of phase by 90°, $(V_{0i}I_{0a}/2)$ gives one part of the reactive power. For the same reason $(V_{0a}I_{0i}/2)$ gives a second part of reactive power. Note that in the first part the current lags the voltage, and hence this part of the reactive power is inductive. The second part of the reactive power is capacitive, as in this case the current leads the voltage. If we assume the inductive power to be positive, then the capacitive power would be negative. Thus the net reactive power,

$$P_X = \frac{1}{2}(V_{0i}I_{0a} - V_{0a}I_{0i}) \qquad (23)$$

We can also obtain the above results mathematically by the use of the complex forms of voltage and current given in (20) and (21) as follows. The conjugate of current is

$$\overline{I}_0^{\,*} = I_{0a} - jI_{0i}$$

Then,
$$\frac{1}{2}\overline{V}_0\overline{I}_0^{\,*} = \frac{1}{2}(V_{0a} + jV_{0i})(I_{0a} - jI_{0i})$$

$$= \frac{1}{2}[(V_{0a}I_{0a} + V_{0i}I_{0i}) + j(V_{0i}I_{0a} - V_{0a}I_{0i})] \tag{24}$$

We see that the rhs of (24) has a real part and an imaginary part. The real part is same as the average power given by (22) and the imaginary part is same as the reactive power given by (23). Thus, $\overline{V}_0\overline{I}_0^{\,*}/2$ gives the complex power. Denoting the complex power by $\overline{P}$, we have

$$\overline{P} = \frac{1}{2}\overline{V}_0\overline{I}_0^{\,*} \tag{25}$$

The average or active power,
$$P_{av} = \text{Re}\,\frac{1}{2}\overline{V}_0\overline{I}_0^{\,*} \tag{26}$$

The reactive power,
$$P_X = \text{Im}\,\frac{1}{2}\overline{V}_0\overline{I}_0^{\,*} \tag{27}$$

In (26) Re means 'Real part of' and in (27) Im means 'Imaginary part of' the complex power. The reactive power is inductive if the imaginary part of (24) is positive, and it is capacitive if the imaginary part is negative. It can be shown that the real part of $(\overline{I}_0\overline{V}_0^{\,*}/2)$ is also equal to P_{av}. However, its imaginary part is negative of the imaginary part of (24).

EXAMPLE 1.27

The voltage and current of a given load are

$V(t) = 120 \sin(1000t + 60°)$ V

$I(t) = 1.5 \cos(1000t - 45°)$ A

(a) Find the phase difference between voltage and current and their root-mean-square values. (b) Represent the voltage and current in phasor, complex and polar forms. (c) Calculate the load impedance, resistance, reactance, inductance (or capacitance), power factor, active power and reactive power using (18) and (19). (d) Calculate the complex power using (25).

(a) In order to find the phase difference between voltage and current, we have to write the voltage equation in the cosine form.

$V(t) = 120 \sin(1000t + 60°) = 120 \cos(1000t - 30°)$ V

$I(t) = 1.5 \cos(1000t - 45°)$ A

The phase angle of the current with respect to the voltage = – 45°–(–30) = –15°. The current lags the voltage by 15°.

Root-mean-square values of voltage and current are

$V \quad V_0/\ 2 = 120/\ 2 = 84.85$ V

$I \quad I_0/\ 2 = 1.5 \quad 2 = .06$ A

(b) Phasor forms: $\bar{V}_0 = 120 e^{-j(\pi/6)}$ V

$\bar{I}_0 = 1.5 e^{-j(\pi/4)}$ A

Complex forms: $\bar{V}_0 = 120(\cos 30° - j\sin 30°) = 103.9 - j60.0$ V

$\bar{I}_0 = 1.5(\cos 45° - j\sin 45°) = 1.06(1-j)$ A

Polar forms: $\bar{V}_0 = 120\angle -30°$ V

$\bar{I}_0 = 1.5\angle -45°$ A

(c) $Z = \dfrac{V_0}{I_0} = \dfrac{120\angle -30°}{1.5\angle -45°} = 80\angle 15° = 80(\cos 15° + j\sin 15°) = 77.27 + j20.71\ \Omega$

$R \quad 77.27\ \Omega,\ X = 20.71\ \Omega$ (Inductive)

Inductance, $L = X/\omega = 20.71/1000 = 20.71$ mH

Power factor = cos ψ = cos 15° = 0.966 (lagging)

Active power, $P_{av} = \dfrac{1}{2} V_0 I_0 \cos\psi = \dfrac{1}{2}(120)(1.5)(0.966) = 86.93$ W

Reactive power, $P_X = \dfrac{1}{2} V_0 I_0 \sin\psi = \dfrac{1}{2}(120)(1.5)(0.259) = 23.29$ var

(d) Complex power, $\bar{P} = \dfrac{1}{2}\bar{V}_0 \bar{I}_0^* = \dfrac{1}{2}(103.9 - j60.0)(1.06 + j1.06) = 86.88 + j23.27$

The real part is the average power in watts, and the imaginary part is the reactive power in vars (inductive).

1.13 Fourier Series

Consider that a function $f(x)$ has the following properties.

1. It varies periodically with x. That is, $f(x + T) = f(x)$, where T is the period.
2. It is a single-valued function. That is, it has only one value for a given value of the variable.
3. It is a continuous function of the variable. It may however have finite number of discontinuities. Such a periodic function is shown in Fig. 1.24. The wave is continuous in the range $0 < x < \pi$ and $\pi < x < 2\pi$ and it is discontinuous at $x = \pi$ and 2π.

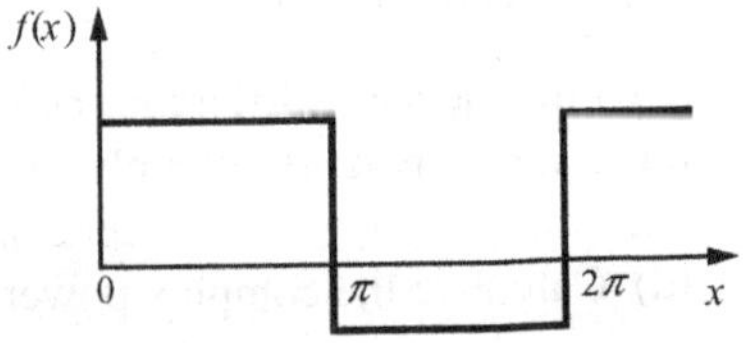

Fig. 1.24 Illustration of a periodic function

Any function $f(x)$ with a period 2π and satisfying the above mentioned properties can be represented by a trigonometric series as given below.

$$f(x) = a_0 + a_1\cos x + a_2\cos 2x + a_3\cos 3x + \cdots + b_1\sin x + b_2\sin 2x + b_3\sin 3x + \cdots$$

$$= a_0 + \sum_{n=1}^{\infty}(a_n\cos nx + b_n\sin nx) \qquad (1)$$

where the constant a_0, a_n and b_n are real numbers. The trigonometric series is called Fourier series. The constants of the series are calculated from the following formulae.

$$a \quad \frac{1}{2\pi}\int_{-\pi}^{\pi} f(x,dx$$

$$a_n = \frac{1}{\pi}\int_{-\pi}^{\pi} f(x)\cos nx\, dx$$

$$b_n = \frac{1}{\pi}\int_{-\pi}^{\pi} f(x)\sin nx\, dx$$

Over a period 2π, the constant a_0 is the average value of $f(x)$ whereas the coefficients a_n and b_n are twice the average values of $f(x)$ cos nx and $f(x)$ sin nx respectively.

If a function $f(x)$ is defined over an interval of length $-\ell$ to ℓ instead of $-\pi$ to π, its Fourier series is obtained by replacing x in (1) by $\pi x/\ell$. Accordingly,

$$f(x) = a_0 + \sum_{n=1}^{\infty}\left(a_n \cos\frac{n\pi x}{\ell} + b_n \sin\frac{n\pi x}{\ell}\right) \tag{2}$$

The constants of (2) are given by

$$a_0 = \frac{1}{2\ell}\int_{-\ell}^{+\ell} f(x)dx$$

$$a_n = \frac{1}{\ell}\int_{-\ell}^{+\ell} f(x)\cos\frac{n\pi x}{\ell}dx$$

$$b_n = \frac{1}{\ell}\int_{-\ell}^{+\ell} f(x)\sin\frac{n\pi x}{\ell}dx$$

If $f(x)$ is an even function of x, i.e., if $f(-x) = f(x)$, then $b_n = 0$ and

$$a_n = \frac{2}{\ell}\int_{0}^{\ell} f(x)\cos\frac{n\pi x}{\ell}dx$$

If $f(x)$ is an odd function of x $[f(-x) = -f(x)]$, then $a_n = 0$ and

$$b_n = \frac{2}{\ell}\int_{0}^{\ell} f(x)\sin\frac{n\pi x}{\ell}dx$$

For an alternating function with identical positive and negative half-waves $[f(x+\pi) = -f(x)]$, the constant $a_0 = 0$.

EXAMPLE 1.28

We will obtain the Fourier series of, (a) a rectangular waveform, and (b) of a triangular waveform, illustrated in Fig. 1.25.

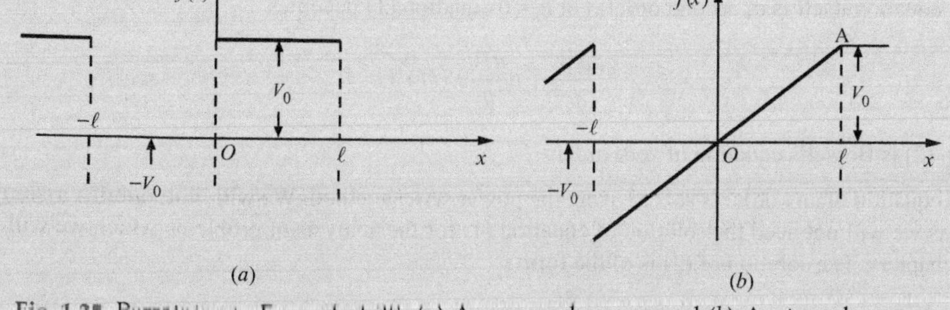

Fig. 1.25 Pertaining to Example 1.28: (*a*) A rectangular wave and (*b*) A triangular wave.

As both the waves are odd functions of x and have identical positive and negative half-cycles, it is necessary to calculate coefficient of the sine series only over the interval $x = 0$ to $x = \ell$.

(a) For the rectangular waveform, $f(x) = V_0$ in the interval $x = 0$ to $x = \ell$.

$$b_n \quad \frac{2}{\ell}\int_0^{\ell} V_0 \sin\frac{n\pi x}{\ell}\,dx = \frac{2V_0}{n\pi}\left[-\cos\frac{n\pi x}{\ell}\right]_0^{\ell} = \frac{2V_0}{n\pi}(1-\cos n\pi)$$

We see that $b_n = 0$ for even values of n, and $b_n = 2$ for odd values of n. Thus, the Fourier series of the function is

$$f\ x) = \frac{4V_0}{\pi}\sum_n \frac{1}{n}\sin\frac{n\pi x}{\ell}, \quad n = 1, 3, 5, \ldots \tag{3}$$

(b) The equation of the triangular waveform from $x = 0$ to $x = \ell$ is that of the straight line OA. Its equation is

$$f\ x) = \frac{V_0 x}{\ell}$$

The coefficient of the nth term of the sine series,

$$b_n = \frac{2}{\ell}\int_0^{\ell}\frac{V_0 x}{\ell}\sin\frac{n\pi x}{\ell}\,dx$$

$$= \frac{2V_0}{\ell^2}\left[-\frac{\ell}{n\pi}x\cos\frac{n\pi x}{\ell} + \left(\frac{\ell}{n\pi}\right)^2\sin\frac{n\pi x}{\ell}\right]_0^{\ell} = -\frac{2V_0}{n\pi}\cos n\pi$$

Hence, the Fourier series of the function is given by

$$f\ x) = \frac{2V_0}{\pi}\left(\sin\frac{\pi x}{\ell} - \frac{1}{2}\sin\frac{2\pi x}{\ell} + \frac{1}{3}\sin\frac{3\pi x}{\ell} - \cdots\right) \tag{4}$$

1.14 Bessel Functions

Bessel functions are solutions of specific types of second order ordinary differential equations. The general form of the equation with V as dependent variable and r as independent variable is

$$r^2\frac{d^2V}{dr^2} + r\frac{dV}{dr} + (kr)^2V - n^2V = 0 \tag{1}$$

where k^2 is positive and n is any number. Equation (1) is called Bessel's equation of order n though the differential equation itself is of second order. For $n = 0$ equation (1) becomes

$$r^2\frac{d^2V}{dr^2} + r\frac{dV}{dr} + (kr)^2V = 0 \tag{2}$$

Equation (2) is Bessel's equation of zero order.

Bessel's equation of any order is solved using the power series method. We will limit our discussion to equation (2) as we will not need the solution of equation (1) for the analysis of problems which we will consider in later chapters. The solution of (2) is of the form

$$V \quad a_0 + a_1 r + a_2 r^2 + a_3 r^3 + a_4 r^4 + a_5 r^5 + a_6 r^6 + - - - \tag{3}$$

The coefficients of the series are found by substituting V, dV/dr and d^2V/dr^2 in (2) and then equating the coefficients of like powers of r to zero. Substituting (3) into the third term of (2), we get

$$k^2r^2V = k^2(a_0r^2 + a_1r^3 + a_2r^4 + a_3r^5 + a_4r^6 + \cdots) \tag{4}$$

Substituting dV/dr and d^2V/dr^2 into the second and first terms of (2), respectively, we get

$$r\frac{dV}{dr} = a_1r + 2a_2r^2 + 3a_3r^3 + 4a_4r^4 + 5a_5r^5 + 6a_6r^6 + \cdots \tag{5}$$

$$r^2\frac{d^2V}{dr^2} = 2a_2r^2 + 6a_3r^3 + 12a_4r^4 + 20a_5r^5 + 30a_6r^6 + \cdots \tag{6}$$

Since the sum of the left-hand sides of (4), (5) and (6) is zero, the sum of their right-hand sides must be equal to zero. Taking the sum and then collecting the terms of like powers of r, we have

$$\begin{aligned} &a_1r + (k^2a_0 + 4a_2)r^2 + (k^2a_1 + 9a_3)r^3 + (k^2a_2 + 16a_4)r^4 \\ &+(k^2a_3 + 25a_5)r^5 + (k^2a_4 + 36)r^6 + \cdots = 0 \end{aligned} \tag{7}$$

Equation (7) will be satisfied if the coefficient of each power of r is equal to zero. This exercise will yield the following results:

$$a_1r = 0 \Rightarrow a_1 = 0$$

$$4a_2 + k^2a_0 = 0 \Rightarrow a_2 = -a_0\frac{k^2}{2^2(1!)^2}; \quad a_0 \text{ is an arbitrary constant.}$$

$$9a_3 + k^2a_1^2 = 0 \Rightarrow a_3 = 0$$

$$16a_4 + k^2a_2 = 0 \Rightarrow a_4 = a_0\frac{k^4}{2^4(2!)^2}$$

$$25a_5 + k^2a_3 = 0 \Rightarrow a_5 = 0$$

$$36a_6 + k^2a_4 = 0 \Rightarrow a_6 = -a_0\frac{k^6}{2^6(3!)^2}$$

We find that the coefficients of odd powers of r of (3) are zero. So, the solution of the differential equation will have terms with even powers of r only. The coefficient of r^{2m}, where m is a positive integer including zero, is given by

$$a_{2m} = a_0\frac{(-1)^m(k/2)^{2m}}{(m!)^2} \tag{8}$$

Replacing a_0 of (8) by a different constant A, the solution of (2) is given by

$$V = A\sum_{m=0}^{\infty}\frac{(-1)^m(kr/2)^{2m}}{(m!)^2} = AJ_0(kr) \tag{9}$$

The function, $J_0(kr) = \sum_{m=0}^{\infty}\frac{(-1)^m(kr/2)^{2m}}{(m!)^2}$

is called Bessel function of first kind and zero order.

As equation (2) is a second-order differential equation, there must be a second independent solution of the equation. The second solution is called Bessel function of second kind and zero order.

SUMMARY

We learnt the basics of three common types of coordinate systems, vector analysis, sinusoidally time-varying functions, Fourier series, and Bessel function. Other mathematical tools are described in later chapters as and when necessary. We found that unlike the unit vectors in rectangular coordinates, the directions of some of the unit vectors in cylindrical and spherical coordinates vary with one or two coordinate variables. Though only the mathematical definitions of divergence, gradient, and curl operations are available, their applications to electric and magnetic fields, and interpretations of results are described in later chapters. The vector operations in cylindrical and spherical coordinates are given in Chapters 2 and 4. The vector identities, line, surface, and volume integrals will find frequent use in the text. Moreover, we will utilise the coordinate transformation technique for calculation of vector magnetic potential of a current loop in Chapter 4.

Representation of sinusoidal quantities and power calculation will be used for the study of time-varying fields. We will apply Fourier series and Bessel function for the solution of boundary-value problems. Bessel function will also be used to study the skin-effect phenomenon in cylindrical conductors.

IMPORTANT FORMULAE

Relation between coordinate variables:

Rectangular and cylindrical coordinates,

$x = r \cos \phi$

$y = r \sin \phi$

$z = z$

Rectangular and cylindrical coordinates,

$x = r \sin \theta \cos \phi$

$y = r \sin \theta \sin \phi$

$z = r \cos \theta$

Differential lengths:

Rectangular coordinates, $d\boldsymbol{\ell} = \mathbf{u}_x dx + \mathbf{u}_y dy + \mathbf{u}_z dz$

Cylindrical coordinates, $d\boldsymbol{\ell} = dr\mathbf{u}_r + rd\phi\mathbf{u}\phi + \mathbf{u}_z dz$

Spherical coordinates, $d\boldsymbol{\ell} = \mathbf{u}_r dr + \mathbf{u}_\theta(rd\theta) + \mathbf{u}_\phi(r sin\ \theta d\phi)$

Differential surfaces:

Rectangular coordinates, $d\mathbf{S} = \mathbf{u}_x(dydz) + \mathbf{u}_y(dzdx) + \mathbf{u}_z(dxdy)$

Cylindrical coordinates, $d\mathbf{S} = \mathbf{u}_r(rd\phi dz) + \mathbf{u}_\phi(drdz) + \mathbf{u}_z(rdrd\phi)$

Spherical coordinates, $d\mathbf{S} = \mathbf{u}_r(r^2 \sin\theta d\theta d\phi) + \mathbf{u}_\theta(r\sin\theta drd\phi) + \mathbf{u}_\phi(rdrd\theta)$

Differential volumes:

Rectangular coordinates, $dv = dxdydz$

Cylindrical coordinates, $rdrd\phi dz$

Spherical coordinates, $r^2\sin\theta drd\theta d\phi$

Position vector: $\mathbf{r} = x\mathbf{u}_x + y\mathbf{u}_y + z\mathbf{u}_z$

Dot product of two vectors:

In rectangular coordinates, $\mathbf{A}\cdot\mathbf{B} = A_xB_x + A_yB_y + A_zB_z$

In cylindrical coordinates, $\mathbf{A}\cdot\mathbf{B} = A_rB_r + A_\phi B_\phi + A_zB_z$

In spherical coordinates, $\mathbf{A}\cdot\mathbf{B} = A_rB_r + A_\theta B_\theta + A_\phi B_\phi$

Cross product of two vectors:

In rectangular coordinates,

$$\mathbf{A}\times\mathbf{B} = \begin{vmatrix} \mathbf{u}_x & \mathbf{u}_y & \mathbf{u}_z \\ A_x & A_y & A_z \\ B_x & B_y & B_z \end{vmatrix}$$

In cylindrical coordinates,

$$\mathbf{A}\times\mathbf{B} = \begin{vmatrix} \mathbf{u}_r & \mathbf{u}_\phi & \mathbf{u}_z \\ A_r & A_\phi & A_z \\ B_r & B_\phi & B_z \end{vmatrix}$$

In spherical coordinates,

$$\mathbf{A}\times\mathbf{B} = \begin{vmatrix} \mathbf{u}_r & \mathbf{u}_\theta & \mathbf{u}_\phi \\ A_r & A_\theta & A_\phi \\ B_r & B_\theta & B_\phi \end{vmatrix}$$

Gradient of a scalar function in rectangular coordinates:

$$\nabla V = \mathbf{u}_x\frac{\partial V}{\partial x} + \mathbf{u}_y\frac{\partial V}{\partial y} + \mathbf{u}_z\frac{\partial V}{\partial z}$$

Divergence of a vector function in rectangular coordinates:

$$\nabla\cdot\mathbf{D} = \frac{\partial D_x}{\partial x} + \frac{\partial D_y}{\partial y} + \frac{\partial D_z}{\partial z}$$

Curl of a vector function in rectangular coordinates:

$$\nabla\times\mathbf{H} = \begin{vmatrix} \mathbf{u}_x & \mathbf{u}_y & \mathbf{u}_z \\ \dfrac{\partial}{\partial x} & \dfrac{\partial}{\partial y} & \dfrac{\partial}{\partial z} \\ H_x & H_y & H_z \end{vmatrix}$$

Vector identities:

$$\nabla\times\nabla\times\mathbf{A} = \nabla(\nabla\cdot \mathrm{A}) - \nabla^2\mathbf{A}$$

$$\nabla\cdot(\mathbf{A}\times\mathbf{B}) = \mathbf{B}\cdot(\nabla\times\mathbf{A}) - \mathbf{A}\cdot(\nabla\times\mathbf{B})$$

Dot product of unit vectors:

In Rectangular and Cylindrical Coordinate systems

	$\cdot\mathbf{u}_r$	$\cdot\mathbf{u}_\phi$	$\cdot\mathbf{u}_z$
$\mathbf{u}_x$	$\cos\phi$	$-\sin\phi$	0
$\mathbf{u}_y$	$\sin\phi$	$\cos\phi$	0
$\mathbf{u}_z$	0	0	1

In Rectangular and Spherical Coordinate systems

	$\cdot\mathbf{u}_r$	$\cdot\mathbf{u}_\theta$	$\cdot\mathbf{u}_\phi$
$\mathbf{u}_x$	$\sin\theta\cos\phi$	$\cos\theta\cos\phi$	$-\sin\phi$
$\mathbf{u}_y$	$\sin\theta\sin\phi$	$\cos\theta\sin\phi$	$\cos\phi$
$\mathbf{u}_z$	$\cos\theta$	$-\sin\theta$	0

Complex impedance:

$$\bar{Z} = R + j(X_L - X_C)$$

Instantaneous power:

$$P(t) = \frac{V_0 I_0}{2}\left[\cos\psi + \cos(2\omega t + \psi_1 + \psi_2)\right]$$

Complex power:

$$\bar{P} = \frac{1}{2}\bar{V}_0 \bar{I}_0^*$$

MULTIPLE-CHOICE QUESTIONS

(starred questions have two answers)

1.1 A point in cylindrical coordinates is defined by (1.4 m, 25°, 0.75 m). Location of the point in rectangular coordinates is
(a) (1.4 m, 0.592 m, 0.75 m)
(b) (1.269 m, 0.592 m, 0.75 m)
(c) (1.4 m, 0.592 m, 0.75 m)
(d) (1.4 m, 0.592 m, 0.375 m)

1.2 A point in spherical coordinates is described by (0.95 m, 35°, 20°). Location of the point in rectangular coordinates is
(a) (0.512 m, 0.186 m, 0.778 m)
(b) (0.186 m, 0.512 m, 0.778 m)
(c) (0.512 m, 0.778 m, 0.186 m)
(d) (0.186 m, 0.778 m, 0.512 m)

1.3 A position vector is given by $\mathbf{r} = 3\mathbf{u}_x + 4\mathbf{u}_y + 5\mathbf{u}_z$. The unit vector in the direction of the position vector is

(a) $0.424\mathbf{u}_x + 0.707\mathbf{u}_y + 0.566\mathbf{u}_z$
(b) $0.566\mathbf{u}_x + 0.707\mathbf{u}_y + 0.424\mathbf{u}_z$
(c) $0.707\mathbf{u}_x + 0.424\mathbf{u}_y + 0.566\mathbf{u}_z$
(d) $0.424\mathbf{u}_x + 0.566\mathbf{u}_y + 0.707\mathbf{u}_z$

1.4 A vector is given by $\mathbf{A} = 6\mathbf{u}_x + p\mathbf{u}_y + 8\mathbf{u}_z$. If the unit vector along $\mathbf{A}$ is $\mathbf{u}_A = 0.424\mathbf{u}_x + 0.707\mathbf{u}_y + 0.566\mathbf{u}_z$, what should be the value of p?

(a) 6 (b) 8 (c) 10 (d) 12

1.5 Two vectors are given by $\mathbf{A} = 3\mathbf{u}_x + 4\mathbf{u}_y$ and $\mathbf{B} = 5\mathbf{u}_x + 12\mathbf{u}_y$. The angle between the vectors is nearest to

(a) 12° (b) 14° (c) 16° (d) 18°

1.6 Two vectors are described by $\mathbf{A} = 2\mathbf{u}_x + 3\mathbf{u}_y + 6\mathbf{u}_z$ and $\mathbf{B} = p\mathbf{u}_x + q\mathbf{u}_y + 9\mathbf{u}_z$. For the vectors to be parallel, the values of p and q should be

(a) $p = 3, q = 3$ (b) $p = 3, q = 9$ (c) $p = 4.5, q = 3$ (d) $p = 3, q = 4.5$

*1.7 The angle between two vectors $\mathbf{A} = 3\mathbf{u}_x + 4\mathbf{u}_y$ and $\mathbf{B} = p\mathbf{u}_x + q\mathbf{u}_y$ is 15°. What is the ratio of p to q?

(a) 1.732 (b) −1.732 (c) 0.577 (d) −0.577

*1.8 Two vectors are given by $\mathbf{A} = 3\mathbf{u}_x + 2\mathbf{u}_z$ and $\mathbf{B} = 3\mathbf{u}_y + 4\mathbf{u}_z$. The unit vector normal to the plane containing $\mathbf{A}$ and $\mathbf{B}$ is given by

(a) $-0.371\mathbf{u}_x - 0.743\mathbf{u}_y + 0.557\mathbf{u}_z$
(b) $-0.741\mathbf{u}_x - 0.577\mathbf{u}_y + 0.371\mathbf{u}_z$
(c) $0.741\mathbf{u}_x - 0.577\mathbf{u}_y + 0.371\mathbf{u}_z$
(d) $0.371\mathbf{u}_x + 0.743\mathbf{u}_y - 0.5571\mathbf{u}_z$

1.9 A scalar function is given by $\phi = xyz$. The magnitude of gradϕ at (1, −1, 1) is equal to

(a) $2\sqrt{3}$ (b) $\sqrt{3}/2$ (c) $1/\sqrt{3}$ (d) $\sqrt{3}$

1.10 If $\boldsymbol{r}$ is the position vector at (x, y, z), the magnitude of $\nabla \cdot \mathbf{r}$ is equal to

(a) 3 (b) $\sqrt{3}$ (c) $1/\sqrt{3}$ (d) 1/3

1.11 The divergence of $\mathbf{F} = xy\,(\mathbf{u}_x + \mathbf{u}_y)$ at (1, 3) is equal to

(a) 1 (b) 4 (c) 3 (d) 2

1.12 A vector in cylindrical coordinates is given by $(\sin\phi/r)\,\mathbf{u}_r$. The vector in rectangular coordinates is

(a) $\dfrac{x^2\mathbf{u}_x + y^2\mathbf{u}_y}{(x^2+y^2)^{3/2}}$ (b) $\dfrac{xy\mathbf{u}_x + y^2\mathbf{u}_y}{(x^2+y^2)^{3/2}}$ (c) $\dfrac{y^2\mathbf{u}_x + x^2\mathbf{u}_y}{(x^2+y^2)^{3/2}}$ (d) $\dfrac{x^2\mathbf{u}_x + xy\mathbf{u}_y}{(x^2+y^2)^{3/2}}$

1.13 A vector in spherical coordinates is given by $\mathbf{u}_r/r^2$. The vector in rectangular coordinate is

(a) $\dfrac{\mathbf{u}_x + \mathbf{u}_y + \mathbf{u}_z}{(x^2+y^2+z^2)^{3/2}}$ (b) $\dfrac{x^2\mathbf{u}_x + y^2\mathbf{u}_y + z^2\mathbf{u}_z}{(x^2+y^2+z^2)^{3/2}}$ (c) $\dfrac{x\mathbf{u}_x + y\mathbf{u}_y + z\mathbf{u}_z}{(x^2+y^2+z^2)^{3/2}}$ (d) $\dfrac{xy\mathbf{u}_x + yz\mathbf{u}_y + zx\mathbf{u}_z}{(x^2+y^2+z^2)^{3/2}}$

1.14 The line integral of $\mathbf{F} = xy\,(\mathbf{u}_x + \mathbf{u}_y)$ along the straight line $y = x$ from (0, 0) to (2, 2) is

(a) 16/3 (b) 8 (c) 16 (d) 4

1.15 The instantaneous voltage and current in an electric circuit are given by $V(t) = 5\sin(\omega t - 30°)$ V and $I(t) = -0.2\cos\omega t$ A. The average power supplied to the circuit is
(a) 0.5 W (b) $\sqrt{3}/2$ W (c) 0.25 W (d) $\sqrt{3}/4$ W

QUESTIONS

1.1 Name the coordinate planes of rectangular coordinates.

1.2 In rectangular coordinates, the coordinates of a point are x, y and z. What is the normal distance of the point from the x-y plane?

1.3 Name the plane on which $x = 0$.

1.4 The coordinate of two points are x_1, y_1, z_1 and x_2, y_2, z_2 respectively. What is the distance between the points?

1.5 In rectangular coordinates, locate the point $P(x, -y, z)$ by tracing the path along or parallel to the coordinate axes, starting from the origin.

1.6 Describe the cylindrical coordinate system. Show that the coordinate directions are mutually perpendicular to each other.

1.7 Describe the spherical coordinate system. Show that the coordinate directions are mutually perpendicular to each other.

1.8 The r and ϕ variable appearing in cylindrical and spherical coordinate systems have the same meaning or different meanings in the two systems? Explain.

1.9 Explain why the distance between two points specified in spherical coordinates cannot be calculated by direct use of the coordinates of the points.

How is this distance calculated?

1.10 Sketch a length vector in rectangular coordinates and show its components in the coordinate directions.

1.11 Express the vector A as sum of its component in rectangular, cylindrical, and spherical coordinates.

1.12 Write the differential length vector and the differential surface vector in rectangular, cylindrical, and spherical coordinates in terms of the components. Give also the formula for the differential volume in the three coordinate systems.

1.13 Define the cross product of two vectors A and B. Explain how the angle between the vectors can be calculated from the definition of the cross product.

1.14 Explain what is meant by a scalar field. Does the value of the field at a point depend on the choice of the coordinate system?

1.15 State what is meant by a vector field. Does the value of the field at a point depend on the choice of the coordinate system?

1.16 Explain with sketches that the unit vectors $\mathbf{u}_r$ and $\mathbf{u}_\phi$ in cylindrical coordinates vary with ϕ.

1.17 Explain the properties of the gradient of a scalar function.

1.18 Describe the procedure for evaluating the line integral of a vector function from an initial point to a final point along a given path.

1.19 For the vector function $\mathbf{A}(x, y, z)$ show that $\nabla \cdot \nabla \times \mathbf{A}(x, y, z) = 0$

1.20 Two sinusoidal functions are given by

$$V(t) = V_0 \cos(\omega t - \tfrac{\pi}{4})$$
$$I(t) = V_0 \sin(\omega t - \tfrac{\pi}{12})$$

What is the phase difference between the two quantities, and which is the leading quantity?

1.21 A sinusoidal quantity is given by

$$\Phi(t) = \Phi_0 \sin(\omega t + \psi)$$

Write the complex exponential, phasor, and polar forms of the quantity?

1.22 Define the rms value of a sinusoidally time-varying current. Explain its significance.

1.23 If W, V, and I are the measured values of power, rms value of voltage, and rms value of current respectively of an electric load, then the ratio $W/(VI)$ is defined as the power factor of the load. Is this definition applicable to only sinusoidal variation of voltage and current with time or for any type of variation? Explain.

1.24 Explain with sketches the meanings of single-valued and multi-valued periodic functions.

ANSWERS TO SELECTED QUESTIONS

1.6 Describe the cylindrical coordinate system. Show that the coordinate directions are mutually perpendicular to each other.

Answer: In cylindrical coordinate system, the three surfaces are those of a circular cylinder and two planes. These surfaces are specified with reference to the rectangular coordinate system. The axis of the cylinder is along the z axis and its radius is denoted by r. One of the plane surfaces containing the z axis is specified by the angle it makes with the y-z plane. The angle is generally denoted by ϕ. The second plane surface is parallel to, and at a height z, from the x-y plane. The three coordinate directions are r, ϕ and z.

The r direction is normal to the surface of the cylinder, hence it is perpendicular to the axis of the cylinder. As the axis of the cylinder is along the z axis, r and z directions are perpendicular to each other. At a point on the cylindrical surface, the r direction is perpendicular to and the ϕ direction is tangential to the cylindrical surface. Hence, they are perpendicular to each other. Moreover, the ϕ direction is parallel to the x-y plane; hence it is perpendicular to the z direction.

1.8 The r and ϕ variable appearing in cylindrical and spherical coordinate systems have the same meaning or different meanings in the two systems? Explain.

Answer: The variables in cylindrical coordinate system are r, ϕ and z, and those of spherical coordinates are r, θ and ϕ. The r direction in cylindrical coordinates is perpendicular to the z axis of rectangular coordinates whereas the r direction in spherical coordinates is at an angle θ with the z axis. Thus, the r-variables have different meaning. In both the systems, the angle ϕ is the angle between the y-z plane and a plane surface containing the z axis. Thus, the variable ϕ has the same meaning in both the systems.

1.9 Explain why the distance between two points specified in spherical coordinates cannot be calculated by direct use of the coordinates of the points.

How is the distance calculated?

Answer: Both θ and ϕ variables of spherical coordinates are angles and radian is their unit. As the unit of distance between two points is that of length, the variables cannot be used directly for calculation of distance.

To calculate the distance, it is necessary to express the spherical coordinates of the point in terms of rectangular coordinate variables.

1.13 Define the cross product of two vectors **A** and **B**. Explain how the angle between the vectors can be calculated from the definition of the cross product.

Answer: The cross product of two vectors **A** and **B** is a vector quantity, say, **C**. The magnitude of **C** is equal to the products of the magnitude of **A**, the magnitude of **B**, and the sine of angle α between **A** and **B**. The direction of **C** is perpendicular to the plane containing **A** and **B**. The positive direction is taken along the advance of a right-handed screw when it is rotated from **A** towards **B**.

As per the definition of the cross product,

$$|\mathbf{A} \times \mathbf{B}| = |\mathbf{A}|\,|\mathbf{B}| \sin \alpha$$

Hence,

$$\sin \alpha = \frac{|\mathbf{A} \times \mathbf{B}|}{|\mathbf{A}||\mathbf{B}|}$$

1.14 Explain what is meant by a scalar field. Does the value of the field at a point depend on the choice of the coordinate system?

Answer: A physical quantity, which is a scalar function of position, is a real number. The quantity may or may not be a function of time. The mathematical representation or graphical plot of a scalar function at various points in a region or domain is the scalar field of the function in that region. The region may be one-dimensional (a line), two-dimensional (a surface), or three-dimensional (a volume). As an example, let us consider a heated plate. The temperature at each point on the surface of the plate will have a specific value. A graphical representation of temperatures at several points is the temperature field in the region of the plate surface.

Value of the field at a point is independent of the choice of a coordinate system.

1.17 Explain the properties of the gradient of a scalar function.

Answer: The gradient of a scalar quantity which is a function of x, y and z in general is a vector quantity. It is given by

$$\nabla V = \left(\mathbf{u}_x \frac{\partial}{\partial x} + \mathbf{u}_y \frac{\partial}{\partial y} + \mathbf{u}_z \frac{\partial}{\partial z} \right) V = \mathbf{u}_x \frac{\partial V}{\partial x} + \mathbf{u}_y \frac{\partial V}{\partial y} + \mathbf{u}_z \frac{\partial V}{\partial z}$$

The first, second and third terms give the rate of change of V with x, y and z respectively. As ∇V is the vector sum of all the three components, it gives the maximum rate of change of V with distance at a point in space.

1.20 Two sinusoidal functions are given by

$$V(t) = V_0 \text{ [illegible] } t - \tfrac{\pi}{4})$$

$$I(t) = V_0 \text{ [illegible] } t - \tfrac{\pi}{12})$$

What is the phase difference between the two quantities, and which is the leading quantity?

Answer: As $V(t)$ is a cosine function of time and $I(t)$ is a sine function of time, we have to express either $V(t)$ as a sine function or $I(t)$ as a cosine function of time. By expressing $I(t)$ as a cosine function, we get

$$I(t) = I_0 \cos(\omega t - \tfrac{\pi}{12} - \tfrac{\pi}{2}) = I_0 \cos(\omega t - \tfrac{7\pi}{12})$$

Thus, the phase difference between the quantities is $(7\pi/12) - (\pi/4) = \pi/3$. $V(t)$ is the leading quantity.

PROBLEMS

*1.1 Show the location of points P_1(1.0 m, 2.0 m, 3.0m); P_2(−1.0 m, 2.0 m, 3.0m); P_3(1.0 m, −2.0 m, 3.0m); P_4(1.0 m, 2.0 m, −3.0m) in rectangular coordinates. Calculate the distances of the points from the origin.

*1.2 Show the location of the point P(0.5 m, 60°, 0.2 m) in cylindrical coordinates. Calculate the distances of P from the origin.

*1.3 Show the location of the point P(0.4 m, 30°, 45°) in spherical coordinates. Calculate the distances of P from the origin.

**1.4 A point in rectangular coordinates is defined by (2 m, −3 m, −1 m). Find coordinates of the point in, (a) cylindrical coordinate and (b) in spherical coordinates.

**1.5 Derive relations between the cylindrical coordinate and spherical coordinate variables. For the radial directions, use subscripts c and s respectively for cylindrical and spherical coordinates.

**1.6 A point in cylindrical coordinates is defined by (3. m, 30^0, 2 m). Find coordinates of the point in spherical coordinates.

**1.7 Find the distance between two points located at (a) (1.0 m, 2.4 m, −0.8 m) and (−2.6 m, 1.5 m, 1.5 m) in rectangular coordinates (b) (1.2 m, 45°, 2.0 m) and (0.5 m, 60°, 0.2 m) in cylindrical coordinates, (c) (0.4 m, 30°, 45°) and (1.0 m, 75°, 15°) in spherical coordinates.

**1.8 Find the length vector from the point P_2(0.75 m, 1.3 m, −0.6 m) to the point P_1(1.25 m, −0.7 m, 0.4 m). Find also the unit vector directed from P_1 to P_2.

**1.9 A point is located at (2.0 m, π/3 rad, 0.8 m) in cylindrical coordinates. Obtain in rectangular coordinates the unit vector directed from the origin to the point.

***1.10 The location of two points are given by P_1(1 m, 2 m, 3 m) and P_2(2 m, 3 m, 5 m). Find the length vector from the origin to the mid-point of the line joining P_1 and P_2.

***1.11 The location of two points are given by P_1(1.5 m, 45°, 2 m) and P_2(5 m, 60°, 30°) in cylindrical and spherical coordinates, respectively. Find the length vector from P_1 to P_2.

**1.12 A point is located at P(5 m, 90°, 60°) in spherical coordinates. Find the unit vector directed from the origin to P in rectangular coordinates.

***1.13 The equation $y = 2 - x$ is that of a straight line in the x-y plane. Obtain the unit normal vector directed from the given line to the point (3, 2).

**1.14 Find the dot and cross products of the vectors $\mathbf{A} = 3\mathbf{u}_x + 8\mathbf{u}_y$; $\mathbf{B} = 5\mathbf{u}_x + 10\mathbf{u}_y$. What is the angle between the vectors?

***1.15 Given the vectors $\mathbf{A} = 4\mathbf{u}_x + 5\mathbf{u}_y$; $\mathbf{B} = 6\mathbf{u}_x + B_y\mathbf{u}_y + 8\mathbf{u}_z$. What should be the value of B_y so that the angle between **A** and **B** is 60°?

***1.16 Vector **A** is directed from (2 m, 3 m, 4 m) to (5 m, 7 m, 9 m), and vector **B** is directed from (0, 1 m, 2 m) to (3 m, 5 m, 7 m). Find the angle between the vectors.

*1.17 Find the gradient of the scalar function $\phi = x^2y + y^2z + xz^2$ at (–1, 2, –3).

**1.18 The equation $2x + 3y + 4z = c$, where c is a constant, is that of a plane surface. Find the unit vector that is normal to the given plane and directed away from the origin.

**1.19 A surface is defined by $x^2 + y^2 - z^2 = 21$. Find the unit vector normal to the surface at (4, 3, 2).

***1.20 For the surface defined in Problem 1.19, find the angle between unit vectors normal to the surface at (4, 3, 2) and $\left(\sqrt{15}, 3, -\sqrt{3}\right)$.

***1.21 A vector function is described by $\mathbf{A} = \dfrac{1}{\sqrt{x^2 + y^2 + z^2}}(x\mathbf{u}_x + y\mathbf{u}_y + z\mathbf{u}_z)$. Find the divergence of the vector function at (2, –1, –2).

**1.22 A vector function is given by $\mathbf{A} = x^2y\mathbf{u}_x + y^2z\mathbf{u}_y + xz^2\mathbf{u}_z$. Find the curl of the vector at (1, 2, 4)

*1.23 For the scalar function $\phi(x, y, z)$, show that $\nabla \times \nabla\phi = 0$.

***1.24 Prove the validity of the vector identity $\nabla \times \nabla \times \mathbf{A} = \nabla(\nabla \cdot \mathbf{A}) - \nabla^2\mathbf{A}$ for the vector function $\mathbf{A} = A(x, y, z)\mathbf{u}_x + A(x, y, z)\mathbf{u}_y + A(x, y, z)\mathbf{u}_z$.

***1.25 A vector field in rectangular coordinates is given by, $\mathbf{E} = \dfrac{\mathbf{u}_x}{x} + \dfrac{\mathbf{u}_z}{z}$. Express the vector in cylindrical coordinates at (4, 45°, 3).

***1.26 A vector is directed from (2, 30°, 4) to (3, 60°, 6) in cylindrical coordinates. Find the angle between the vector and the y axis.

***1.27 A vector function in cylindrical coordinates is given by

$$\mathbf{F} = \frac{\cos\phi}{r}\mathbf{u}_r + \frac{\sin\phi}{r}\mathbf{u}_\phi + \frac{\tan\phi}{z}\mathbf{u}_z.$$

Calculate the components of the vector in rectangular coordinates.

***1.28 A vector function is given by $\mathbf{E} = \dfrac{8\mathbf{u}_x}{x^2 + y^2}$. Find the line integral of **E** along a straight path from (3, 5) to (6, 10).

**1.29 On the surface of a sphere, find the area in the region if $r = 6$, $30° \le \theta \le 60°$, $0 \le \phi \le 30°$.

**1.30 The amplitude of sinusoidal voltage and current of an electric circuit are 15 V and 2 A respectively. The current lags the voltage by $\pi/6$ rad. Calculate (a) the rms values of voltage and current, (b) the power factor, (c) the average power, and (d) the complex impedance of the circuit.

Find also the instantaneous power and complex power taking the voltage as reference. Identify the components of the complex power. Assume ω to be the angular frequency of the voltage.

***1.31 The complex forms of voltage and current amplitudes of an electric load are

$$\bar{V}_0 = 200 + j40 \text{ V}$$

$$\bar{I}_{\_} = 10 - j2 \text{ A}$$

at 50 Hz. Calculate the complex impedance. Give the values of resistance and reactance of the load. Find the inductance and average power. Explain why the load is inductive.

Chapter Two

STATIC ELECTRIC FIELDS

2.1 Introduction

If an electrically neutral body is supplied with electronic charge, it gets negatively charged, and if electrons are removed from the body, it becomes positively charged. There are several methods of charging a body such as by mechanical means, by the use of a battery, by ionization, etc. A charged body may be static (stationary) or may be in motion, and the charge itself may be time invariant or may be a function of time.

When two charged bodies are separated over a finite distance, each body experiences an electric force. The electric force is a function of magnitude of charges, relative displacement between the charges and the dielectric property of the medium in which the charges are located. If one charge is kept fixed and the other charge is moved to different positions, in general, both magnitude and direction of the force on the charge will change. The graphical plot showing the direction of the force at various points will define the force field of the fixed charge. Further, if locations of the charges in space are described using a coordinate system, the electric force can be expressed in terms of the coordinate variables in vector form. The theory and application of the electric field are based on this force. Some other important quantities used in the development of electric field theory are the electric field intensity, scalar electric potential and potential gradient.

The main purpose of analysing electric field problems is to calculate the force between charged bodies, potential gradient in a dielectric (insulation) medium, capacitance of a system of conductors and energy stored in a capacitor. Knowledge of potential gradient in an insulating medium (including air) is necessary to know whether it is within the breakdown strength of the material or not. Capacitance is used as a circuit parameter to develop circuit models of transmission lines, electronic devices, etc. Capacitors are also used for improvement of power factor in power systems and for starting single-phase induction motors. Energy can be stored for a long time in a good capacitor.

Some of the equipments and devices using electric force on charged particles are the cathode ray tubes, electrostatic precipitators, xerographic machines, and electrostatic separators. Cathode ray tubes are used as a display device in oscilloscopes, television receivers and computer monitors. Electrostatic precipitators are installed to remove ash and dust particles from the exhaust gas of coal-fired power plants, cement factories and chemical plants to name a few. Xerographic machines are used for dry-copying of documents. The formation of thunder clouds and the lightning phenomenon are due to static charges. The formation of corona on the surface of conductors of high voltage overhead transmission lines is explained by the use of electric field laws.

This chapter starts with Coulomb's law of force between two charged particles. The subject is developed from this fundamental law. Some important, though well known, laws and theorems described in this chapter are applicable for time-varying fields without or with some modification.

2.2 Forces between Two Point Charges: Coulomb's Law

When two charged bodies are separated over a finite distance then each body experiences an electric force.

The force has the following properties:

1. The force between like charges (both the charges are either positive or negative) is repulsive, and the force between unlike charges (one is positive and the other is negative) is attractive. That is, if the charges are free, like charges will move away from each other, and unlike charges will move towards each other. The directions of electric forces between two very small charges Q_1 and Q_2 are shown in Fig. 2.1.
2. The force between two very small charged bodies acts along the line joining the charge centres.

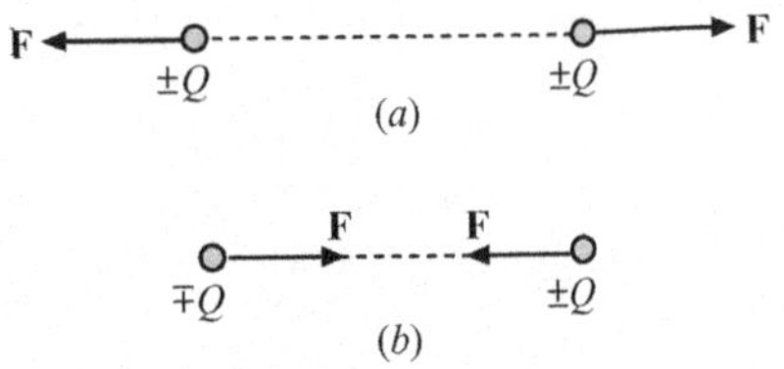

Fig. 2.1 Shows directions of forces between (a) two point charges of same sign and (b) two point charges of opposite sign.

3. If sizes of the charged bodies are small compared to the distance between them, the force is directly proportional to the product of magnitude of charges and inversely proportional to the square of the distance R. This law was formulated by Coulomb from the results of his experiments on charged bodies. Thus, treating each charged body as a point charge (charge that is concentrated at a point is termed as a point charge), the force F between the charges is

$$F \propto \frac{Q_1 Q_2}{R^2}$$

Or

$$F = K\frac{Q_1 Q_2}{R^2} \tag{1}$$

In (1) K is the constant of proportionality. In the International System,

$$K = \frac{1}{4\pi\varepsilon} \tag{2}$$

where ε is permittivity of the medium in which the charges are located. By substituting (2) in (1), we get

$$F = \frac{Q_1 Q_2}{4\pi\varepsilon R^2} \tag{3}$$

This is known as Coulomb's law. In the international system of units, the unit of force is newton (N), the unit of electric charge is coulomb (C) and the unit of distance is metre (m). Accordingly, the unit of permittivity is $C^2/N\text{-}m^2$. The practical unit of permittivity is farad per metre (F/m). The value of free-space permittivity,

$$\varepsilon_0 = 8.854 \times 10^{-12} \cong \frac{10^{-9}}{36\pi}$$

The permittivity of air (Coulomb performed experiments in air) at normal temperature and pressure is very nearly equal to the free-space value. The discrepancy is about 0.05 per cent.

Our study of electric fields will be based on the assumption that the dielectric medium (other than air) is homogeneous (dielectric property is same throughout the medium), linear (permittivity is not a function of electric force) and isotropic (the value of permittivity is same in all directions within the medium) These assumptions greatly simplify the analysis of electric field problems and make the superposition of fields due to several charges possible. The permittivity can be taken as a constant.

The ratio between the permittivity ε and permittivity of free space is called relative permittivity and it is denoted by ε_r. The relative permittivity of different materials is different.

Relative permittivity, $\varepsilon_r = \varepsilon/\varepsilon_0$

Or $$\varepsilon = \varepsilon_0 \varepsilon_r \tag{4}$$

The relative permittivity is a dimensionless quantity. It depends on the atomic structure of the material. The relative permittivity of materials used in electrical and electronic apparatuses generally lies between 2 and 6.

By substituting (4) in (3), we get

$$F = \frac{Q_1 Q_2}{4\pi\varepsilon_0\varepsilon_r R^2} \cong \frac{(9\times10^9)Q_1Q_2}{4\pi\varepsilon_r R^2} \tag{5}$$

Equation (5) gives the magnitude of force experienced by each of the point charges. As mentioned earlier and shown in Fig. 2.1, the force is along the line joining the centres of the two charges. If the charges are of the same sign, the force on one charge is directed away from the other charge. And if the charges are of opposite sign, the force on one charge is directed towards the other charge.

In order to express the electric force between two charges in vector form, let us assume that charges Q_1 and Q_2 are located at $P_1(x_1, y_1, z_1)$ and $P_2(x_2, y_2, z_2)$ in a dielectric medium of permittivity ε. If both the charges are of the same sign, the vector form of the force on charge Q_2 is given by

$$\mathbf{F}_2 = \frac{Q_1Q_2}{4\pi\varepsilon R_{12}^{\ 2}}\mathbf{u}_{12}$$

where $\mathbf{u}_{12}$ is the unit vector directed from P_1 towards P_2, and R_{12} is distance between centres of the charges.

As $$\mathbf{u}_{12} = \frac{\mathbf{R}_{12}}{R_{12}}$$

$$\mathbf{F}_2 = \frac{Q_1Q_2\mathbf{R}_{12}}{4\pi\varepsilon(R_{12})^3} \tag{6}$$

The displacement vector $\mathbf{R}_{12}$ in terms of coordinate variables and unit vectors along the coordinate directions is

$$\mathbf{R}_{12} = (x_2 - x_1)\mathbf{u}_x + (y_2 - y_1)\mathbf{u}_y + (z_2 - z_1)\mathbf{u}_z,$$

and distance between the charges is

$$R_{12} = \sqrt{(x_2 - x_1)^2 + (y_2 - y_1)^2 + (z_2 - z_1)^2}$$

The force on charge Q_1 is

$$\mathbf{F}_1 = \frac{Q_1Q_2\mathbf{R}_{21}}{4\pi\varepsilon(R_{21})^3} = -\mathbf{F}_2,$$

as $\mathbf{R}_{21} = -\mathbf{R}_{12}$ and $R_{21} = R_{12}$

If a region of constant permittivity has more than two charges, the electric force on any charge is equal to the vector sum of forces due to other charges.

Now let us consider that a point charge is located at the origin of a coordinate system.

> If a test charge of very small value is moved to various positions, the direction of force at each position of the test charge would be along the line joining the two charges. A plot of directions of the force at various points in a region will give the force field of Q. The force field of a point charge has spherical symmetry. The force-field of a charge distribution is a continuous function of space variables in a given medium.

To get an idea about the magnitude of the force, let us calculate the force between two point charges each of one coulomb separated by one metre in air. By substituting $Q_1 = Q_2 = 1$, $R = 1$ and $\varepsilon_r = 1$ in (5), we will get $F = 9 \times 10^9$ N, which is very large. This is because one coulomb itself is a large amount of charge. Therefore, the value of charge is generally given in micro-coulomb (10^{-6} C).

EXAMPLE 2.1

Point charges $Q_1 = 50\ \mu$C, $Q_2 = 40\ \mu$C and $Q_3 = -90\ \mu$C are located at P_1(2 m, 0, 0), P_2(−2 m, 3 m, 0) and P_3(2 m, 2, 0) respectively in free space. Find the electric force on Q_3.

Force on Q_3, $$\mathbf{F}_3 = \mathbf{F}_{13} + \mathbf{F}_{23} = \frac{1}{4\pi\varepsilon_0}\left(\frac{Q_1Q_3\mathbf{R}_{13}}{R_{13}^{\ 3}} + \frac{Q_2Q_3\mathbf{R}_{23}}{R_{23}^{\ 3}}\right)$$

$$\frac{1}{4\pi\varepsilon_0} = 9 \times 10^9$$

$$\mathbf{R}_{13} = (2-2)\mathbf{u}_x + (2-0)\mathbf{u}_y = 2\mathbf{u}_y \text{ m}$$

$$R_{13}^{\ 3} = 8\text{m}^3$$

$$\mathbf{R}_{23} = (2+2)\mathbf{u}_x + (2-3)\mathbf{u}_y - 4\mathbf{u}_x - \mathbf{u}_y \text{ m}$$

$$R_{23} = \sqrt{17} \text{ m},\ R_{23}^{\ 3} = 70.09 \text{ m}^3$$

$$\mathbf{F}_3 = 9 \times 10^9 \left[\frac{(50 \times 10^{-6})(-90 \times 10^{-6})(2\mathbf{u}_y)}{8} + \frac{(40 \times 10^{-6})(-90 \times 10^{-6})(4\mathbf{u}_x - \mathbf{u}_y)}{70.09}\right]$$

$$= -1.849\mathbf{u}_x - 9.962\mathbf{u}_y \text{ N}$$

EXAMPLE 2.2

Two point charges, $Q_1 = 1.6\ \mu$C and $Q_2 = -2.2\ \mu$C, are located at P_1(0.5 m, −0.8 m, 1.2 m) and P_2(−0.7 m, 1.5 m, 0.7 m) respectively, in a medium of relative permittivity 2 · 6. Calculate the electric force on Q_1.

The force on charge Q_1,

$$\mathbf{F}_1 \quad \frac{Q_1Q_2}{\varepsilon_r (R_{21})^3} \mathbf{R}_{21} \text{N}$$

$$\mathbf{R}_{21} = (0.5 + 0.7)\mathbf{u}_x + (-0.8 - 1.5)\mathbf{u}_y + (1.2 - 0.7)\mathbf{u}_z$$

$$= 1.2\ \mathbf{u}_x - 2.3\ \mathbf{u}_y + 0.5\ \mathbf{u}_z \text{ m}$$

$$R_{21} = \sqrt{1.2^2 + (-2.3)^2 + 0.5^2} = 2\cdot 642 \text{ m}$$

$$\mathbf{F}_1 \frac{(9\times 10^9)(1\cdot 6\times 10^{-6})(-2\cdot 2\times 10^{-6})}{(2\cdot 6)(2\cdot 642)^3}(1\cdot 2\mathbf{u}_x - 2\cdot 3\mathbf{u}_y + 0\cdot 5\mathbf{u}_z)$$

$$= (-0.79\mathbf{u}_x + 1.52\mathbf{u}_y - 0.33\mathbf{u}_z)10^3 \text{ N}$$

2.3 The Electric Field Intensity

Let us consider that a small positive test charge ΔQ is brought into the neighbourhood of another charged body (or a system of charged bodies). The electric force $\Delta\mathbf{F}$ on the test charge is directed away from a positive charge distribution and towards a negative charge distribution, as shown in Fig. 2.2.

The electric field intensity (or electric field strength) is the ratio between $\Delta\mathbf{F}$ and ΔQ in the limit ΔQ tending to zero.

Thus, electric field intensity at a point is equal to electric force per unit positive charge. The unit of electric field intensity is *newton per coulomb*. Its practical unit is *volt per metre* (V/m). It is a vector quantity, and we will denote it by the symbol **E**. Thus mathematically,

$$\mathbf{E} = \lim_{\Delta Q \to 0} \frac{\Delta\mathbf{F}}{\Delta Q} \text{ V/m}$$

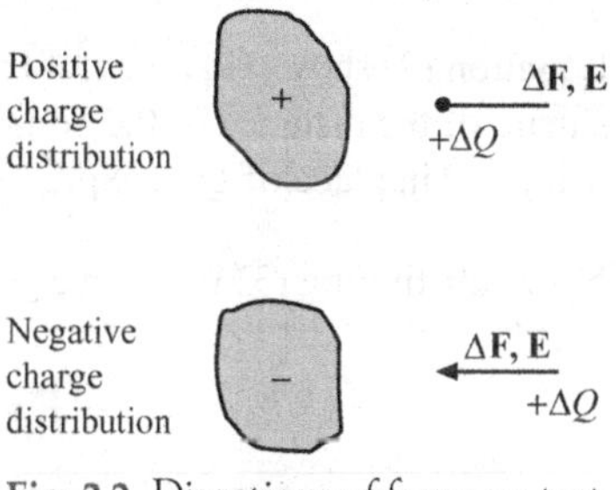

Fig. 2.2 Directions of force on test charge ΔQ and field intensity due to positive and negative charge distributions

The field intensity has the following properties:

1. It is not a force. It is equal to force per unit positive charge, as defined earlier.
2. Its direction is same as the direction of force on a test charge. Its magnitude is independent of the value of the test charge.
3. The field intensity of a positive charge distribution is directed away from the charge and that of a negative charge distribution is directed towards it, as shown in Fig. 2.2.
4. The magnitude of field intensity is inversely proportional to permittivity of the medium.
5. It is a continuous function in a homogenous medium.
6. The force on a point charge of $\pm Q$ coulombs is

$$\mathbf{F} = (\pm Q)\mathbf{E}$$

where **E** is field intensity at the location of the charge.

In defining the electric field intensity we have tacitly assumed that the field of the test charge has negligible effect on the original field distribution.

2.4 Field Intensity of a Point Charge

The simplest example of a charge distribution is a single point charge. The formula for the field intensity of a point charge is useful in the calculation of fields of linear, surface and volume charge distributions.

To calculate the field intensity of a point charge, let us consider that a positive point charge Q and a very small positive test charge ΔQ are located at $P'(x', y', z')$ and $P(x, y, z)$ respectively, as shown in Fig. 2.3. The force exerted by the point charge on the test charge,

$$\Delta \mathbf{F} = \frac{Q\Delta Q}{4\pi\varepsilon R^2}\mathbf{u}_R \tag{1}$$

The unit vector $\mathbf{u}_R$ is directed from P' to P. The distance R between the charges in terms of the coordinate variables is given by

$$R = \sqrt{(x-x')^2 + (y-y')^2 + (z-z')^2}$$

The electric field intensity at P is

$$\mathbf{E} = \lim_{\Delta Q \to 0} \frac{\Delta \mathbf{F}}{\Delta Q} = \frac{Q}{4\pi\varepsilon R^2}\mathbf{u}_R = \frac{Q\mathbf{R}}{4\pi\varepsilon R^3} \tag{2}$$

$$\mathbf{R} = (x-x')\mathbf{u}_x + (y-y')\mathbf{u}_y + (z-z')\mathbf{u}_z \tag{3}$$

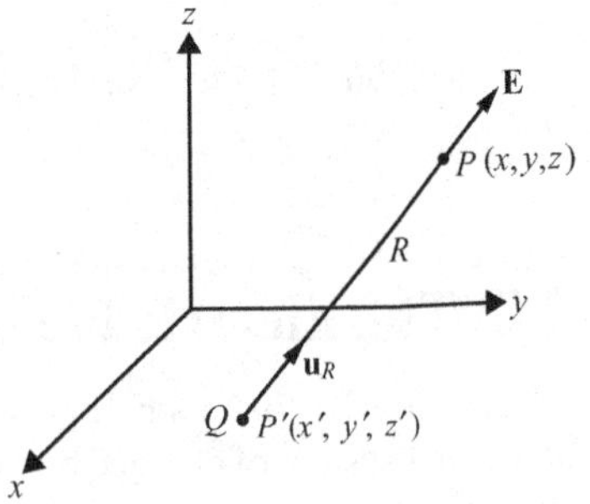

Fig. 2.3 Direction of field intensity at P due to a point charge Q located at P'

Equation (2) shows that the field intensity at a point in the field of a point charge is inversely proportional to square of the distance of the point from the charge. The field intensity of a negative point charge is found by using $-Q$ in place of Q in expression (2).

Now substituting (3) in (2), we get

$$\mathbf{E} = \frac{Q}{4\pi\varepsilon R^3}[(x-x')\mathbf{u}_x + (y-y')\mathbf{u}_y + (z-z')\mathbf{u}_z]$$

Thus, the components of **E** along x, y and z directions are given by

$$E_x = \frac{Q(x-x')}{4\pi\varepsilon R^3} \tag{4}$$

$$E_y = \frac{Q(y-y')}{4\pi\varepsilon R^3} \tag{5}$$

$$E_z = \frac{Q(z-z')}{4\pi\varepsilon R^3} \tag{6}$$

If the position of Q and the point at which the field is to be calculated are given in cylindrical or spherical coordinates, the problem is solved by transforming the coordinates of the given points to rectangular coordinates. After obtaining the components of the field intensity in rectangular coordinates using (4), (5) and (6), they are then transformed back to the original coordinate system *at the given point.*

The variation of field intensity with distance of a point charge is given in Fig. 2.4.

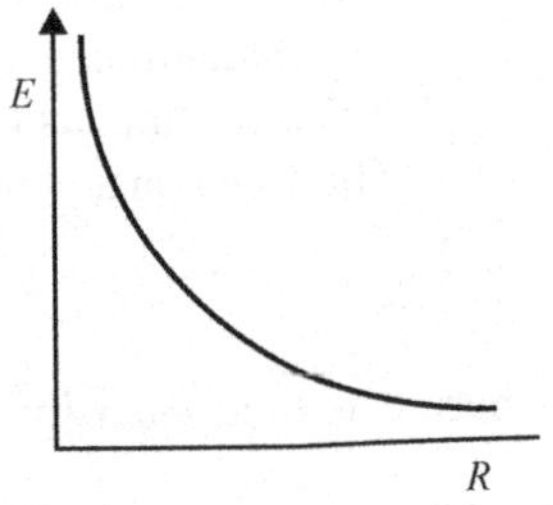

Fig. 2.4 Variation of field intensity E with distance R from a point charge

EXAMPLE 2.3

A point charge of 15 μC is located at (3 m, 4 m, –5 m) in air. Find the electric field intensity at the origin.

Field intensity, $\mathbf{E} = \dfrac{Q\mathbf{R}}{4\pi\varepsilon_0 R^3}$

$\mathbf{R} = (0-3)\mathbf{u}_x + (0-4)\mathbf{u}_y + (0+5)\mathbf{u}_z = -3\mathbf{u}_x - 4\,\mathbf{u}_y + 5\,\mathbf{u}_z$ m

$R^3 = 353.55$ m^3

$$\mathbf{E} = \frac{(9\times10^9)(15\times10^{-6})(-3\mathbf{u}_x - 4\mathbf{u}_y + 4\mathbf{u}_z)}{353.55}$$

$$= -1145.5\mathbf{u}_x - 1527.4\mathbf{u}_y + 1909.6\mathbf{u}_z \text{ V/m}$$

EXAMPLE 2.4

A point charge of –0.08 μC is positioned at P'(2.5 m, 30°, 4 m) in air. Find the field intensity at P(5 m, 45°, –3.5 m).

The locations of the points are given in cylindrical coordinates. These have to be transformed to rectangular coordinates in order to find the displacement between the points. The coordinates of the points in rectangular coordinates are

$x' = 2.5 \cos 30° = 2.165$ m, $y' = 2.5 \sin 30° = 1.25$ m, $z' = 4.0$ m

$x = 5.0 \cos 45° = 3.536$ m, $y = 5.0 \sin 45° = 3.536$ m, $z = -3.5$ m

The displacement of P from P',

$\mathbf{R} = 1.371\mathbf{u}_x + 2.286\mathbf{u}_y - 7.5\,\mathbf{u}_z$ m

$R = \sqrt{1.371^2 + 2.286^2 + 7.5^2} = 7.96$ m

The field intensity at P in rectangular coordinates is given by

$$\mathbf{E}\ \frac{(9\times10^9)(-0\cdot08\times10^{-6})}{(7\cdot96)^3}(1\cdot371\mathbf{u}_x + 2\cdot286\mathbf{u}_y - 7\cdot5\mathbf{u}_z)$$

$$= -1.957\mathbf{u}_x - 3.2643\mathbf{u}_y + 10.707\mathbf{u}_z \text{ V/m}$$

$E_x = -1.957$ V/m, $E_y = -3.263$ V/m, $E_z = 10.707$ V/m

The components of field intensity in cylindrical coordinates at the required point are

$E_r = E_x \cos\phi + E_y \sin\phi = -1.957 \cos 45° - 3.264 \sin 45° = -3.692$ V/m

$E_\phi = -E_x \sin\phi + E_y \sin\phi = 1.957 \sin 45° - 3.264 \cos 45° = -0.925$ V/m

$E_z\,(rect.) = E_z(cyl.) = 10.707$ V/m

Thus, the field intensity at P in cylindrical coordinates is

$\mathbf{E} = -3.692\mathbf{u}_r - 0.925\mathbf{u}_\phi + 10.707\mathbf{u}_z$ V/m

2.5 Field Intensity of Several Point Charges

If several point charges are present in a single dielectric medium of constant permittivity, the field intensity at any point in the region (other than the points at which charges are located) is equal to the vector sum of the fields of all the charges at the given point. Consider that point charges Q_1, Q_2, Q_3, $\ldots Q_n$ are located at $P_1(x_1, y_1, z_1)$, $P_2(x_2, y_2, z_2)$, $P_3(x_3, y_3, z_3)$, $\ldots$ $P_n(x_n, y_n, z_n)$ respectively in a medium of permittivity ε. The total field intensity due to all the charges at $P(x, y, z)$ is

$$\mathbf{E} = \mathbf{E}_1 + \mathbf{E}_2 + \mathbf{E}_3 + \cdots + \mathbf{E}_n$$

The field intensity at P due to Q_n is

$$\mathbf{E}_n = \frac{Q_n \mathbf{R}_n}{4\pi\varepsilon R_n^3}$$

where $\mathbf{R}_n$ is the displacement of P from P_n.

The field intensity at P due to all the charges,

$$\mathbf{E}_n = \sum_n \frac{Q_n \mathbf{R}_n}{4\pi\varepsilon R_n^3}, \quad n = 1, 2, 3, \ldots$$

The component of $\mathbf{E}$ along a coordinate axis is equal to the sum of components of field intensities of all the charges along that axis. Accordingly,

$$E_x = \sum_n \frac{Q_n(x - x_n)}{4\pi\varepsilon R_n^3}, \quad n = 1, 2, 3, \ldots$$

$$E_y = \sum_n \frac{Q_n(y - y_n)}{4\pi\varepsilon R_n^3}, \quad n = 1, 2, 3, \ldots$$

$$E_z = \sum_n \frac{Q_n(z - z_n)}{4\pi\varepsilon R_n^3}, \quad n = 1, 2, 3, \ldots$$

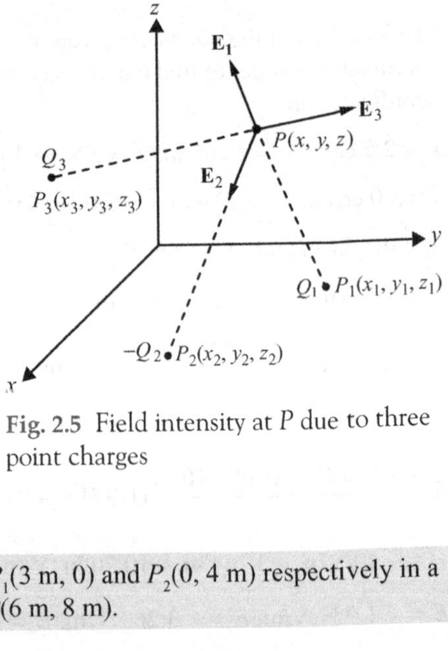

Fig. 2.5 Field intensity at P due to three point charges

For numerical calculations, the specified signs (positive or negative) of charges have to be used. Figure 2.5 shows field intensity at P due to three point charges.

EXAMPLE 2.5

> Two point charges $Q_1 = 8\ \mu\text{C}$ and $Q_2 = -12\ \mu\text{C}$ are located at $P_1(3\text{ m}, 0)$ and $P_2(0, 4\text{ m})$ respectively in a medium of relative permittivity 2.5. Find the field intensity at $P(6\text{ m}, 8\text{ m})$.

Total field intensity at P, $\mathbf{E} = \mathbf{E}_1 + \mathbf{E}_2$

$$\mathbf{E}_1 = \frac{Q_1 \mathbf{R}_1}{4\pi\varepsilon R_1^3}$$

$$\mathbf{R}_1 = (6 - 3)\mathbf{u}_x + (8 - 0)\ \mathbf{u}_y = 3\mathbf{u}_x + 8\mathbf{u}_y\ \text{m}$$

$$R_1^3 = 623.71\ \text{m}^3$$

$$\mathbf{E}_1 \quad \frac{(9 \quad 10^9 \quad 8\times10^{-6})}{(2\ 5)(623.71)} (\ \mathbf{u} \quad 8\mathbf{u}\ ) \quad [illegible]\ \mathbf{u}_x + 369.4\mathbf{u}_y\ \text{V/m}$$

$$\mathbf{E}_2 = \frac{Q_2\mathbf{R}_2}{4\pi\varepsilon R_2^{\,3}}$$

$\mathbf{R}_2 = (6-0)\mathbf{u}_x + (8-4)\,\mathbf{u}_y = 6\mathbf{u}_x + 4\,\mathbf{u}_y$

$R_2^{\,3} = 374.98\ \text{m}^3$

$$\mathbf{E}_2 = \frac{(9\times10^9)(-12\times10^{-6})}{(2.5)(374.98)}(6\mathbf{u}_x + 4\mathbf{u}_y) = -691.2\mathbf{u}_x - 460.8\mathbf{u}_y\ \text{V/m}$$

$\mathbf{E} = -\,552.7\mathbf{u}_x - 91.4\mathbf{u}_y\ \text{V/m}$

EXAMPLE 2.6

Two point charges $Q_1 = 5.0$ nC and $Q_2 = -3.0$ nC are located at $P_1(-1.0$ m, 0.8 m, 2.0 m) and $P_2(2.0$ m, 1.0 m, -1.0 m) respectively in free space. Find the electric field intensity at $P(0.5$ m, $-$ 0.7 m, 1.0 m).

The displacements of P from P_1 and P_2 are

$\mathbf{R}_1 = (0.5 + 1.0)\mathbf{u}_x + (-\,0.7 - 0.8)\,\mathbf{u}_y + (1.0 - 2.0)\,\mathbf{u}_z$

$\quad = 1.5\mathbf{u}_x - 1.5\mathbf{u}_y - 1.0\mathbf{u}_z$ m

$\mathbf{R}_2 = (0.5 - 2.0)\mathbf{u}_x + (-\,0.7 - 1.0)\,\mathbf{u}_y + (1.0 + 1.0)\,\mathbf{u}_z$

$\quad = -1.5\mathbf{u}_x - 1.7\mathbf{u}_y + 2.0\mathbf{u}_z$ m

Also $\qquad R_1^3 = 12.899\ \text{m}^3,\quad R_2^{\,3} = 27.632\ \text{m}^3$

The field intensity due to the charges are given by

$$\mathbf{E}_1 = \frac{(9\times10^9)(5\times10^{-9})}{12\cdot899}(1.5\mathbf{u}_x - 1.5\mathbf{u}_y - 1.0\mathbf{u}_z)$$

$\quad = 5.233\mathbf{u}_x - 5.233\ \mathbf{u}_y - 3.489\ \mathbf{u}_z$ V/m

$$\mathbf{E}_2 = \frac{(9\times10^9)(-3\times10^{-9})}{12\cdot899}(-1.5\mathbf{u}_x - 1.7\mathbf{u}_y + 2.0\mathbf{u}_z)$$

$\quad = 1.466\mathbf{u}_x + 1.661\mathbf{u}_y - 1.954\mathbf{u}_z$ V/m

Total field intensity at P,

$\mathbf{E} = \mathbf{E}_1 + \mathbf{E}_2 = 6.699\mathbf{u}_x - 3.572\mathbf{u}_y - 5.443\mathbf{u}_z$ V/m

2.6 Field Intensity of Continuous Charge Distributions

The procedure for calculating the electric field intensity of a continuous charge distribution consists of obtaining an expression for the field intensity of a differential charge element, treating it as a point charge, and then integrating the scalar components of the field to get the total field. The vector expression of the differential field cannot be integrated directly as the direction of the field depends on the position of the charge element. In this section, we will calculate the field intensities of a line charge, of an infinite sheet of charge, and of a volume charge distribution.

Field Intensity of a Line Charge

Continuous distribution of charge in the form of a thin line is called a line charge. It is specified by the charge per unit length which is called the linear charge density. Accordingly, the unit of linear charge density is C/m. We shall denote the charge density by ρ_ℓ.

We will calculate the field intensity of a line charge of finite length assuming uniform distribution of charge along the line. The field of a line charge of infinite length will be obtained treating it as a special case.

Let us take a line charge AB of length ℓ along the z axis of cylindrical coordinates with its mid-point at the origin, as illustrated in Fig. 2.6(*a*). With this orientation of the line charge, the field intensity is independent of the ϕ coordinate for any r and z. Therefore, we can choose the location of the point at which we wish to calculate the field at $\phi = 0$. The calculations are further simplified by taking the point on the $z = 0$ plane. Thus, the point is at $P(r, 0, 0)$ as indicated in the figure.

Consider a differential length dz' of the charge at $P_1(0, 0, z')$. The charge contained in the differential length is

$$dQ = \rho_\ell \, dz' ,$$

which may be treated as a point charge. The direction of field intensity at P due to this charge is along the line joining P_1 and P. The field, denoted by $d\mathbf{E}_1$, is shown in the figure. We see that the field has components along r and z directions. Therefore to find the field intensity of the charge distribution, we have to evaluate two definite integrals. However, as charge distribution is symmetrical about the $z = 0$ plane, an easier approach is to consider a second charge element at $(0, 0, -z')$ and add the field of this charge to the field of the differential charge at P_1. The second field intensity $d\mathbf{E}_2$, which acts along the line joining P_2 and P, as shown in the figure, also has components along r and z directions. Since the two differential charges are symmetrically located above and below the $z = 0$ plane, the z components of $d\mathbf{E}_1$ and $d\mathbf{E}_2$ have equal magnitudes but they

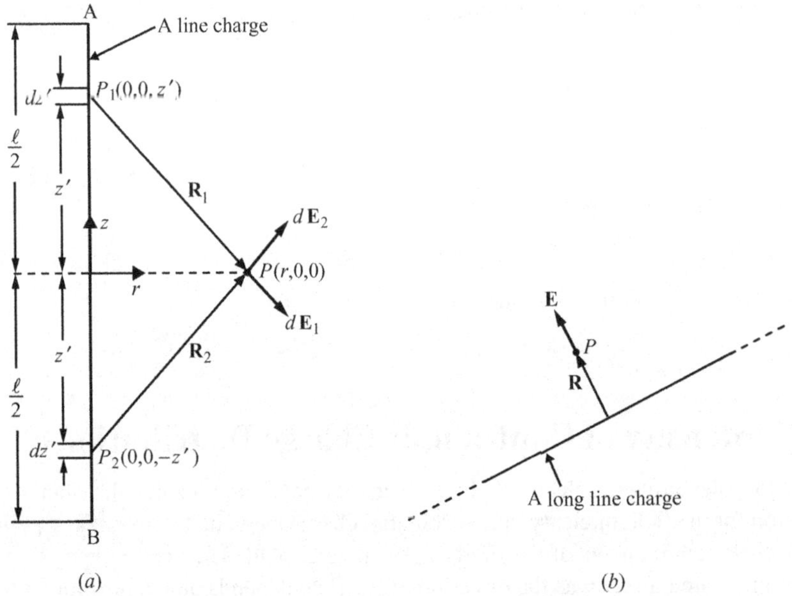

Fig. 2.6 (*a*) Showing field intensities on $z = 0$ plane due to charges contained in two differential lengths at P_1 and P_2 on a line charge, and (*b*) Direction of field intensity of an arbitrarily oriented long line charge.

are in opposite directions, and their r components have equal magnitudes and are in the same directions. As the resultant of the two differential fields has only one scalar component along r direction, it is necessary to evaluate only one definite integral to get the total field.

Referring to formula (2.4-2) that gives the field intensity of a point charge, and Fig. 2.6(a), the field intensity at P due to the differential charge at P_1 is given by

$$d\mathbf{E}_1 = \frac{\rho_\ell \mathbf{R}_1}{4\pi\varepsilon R_1^3} dz'$$

$$\mathbf{R}_1 = r\mathbf{u}_r - z'\mathbf{u}_z \tag{1}$$

and

$$R_1 = \sqrt{r^2 + (z')^2}$$

Similarly, the field intensity at P due to the charge contained in a differential length dz' at $P_2(0, 0, -z')$ is

$$d\mathbf{E}_2 = \frac{\rho_\ell dz}{4\pi\varepsilon R_2^3}\mathbf{R}_2 \tag{2}$$

where

$$\mathbf{R}_2 = r\mathbf{u}_r + z'\mathbf{u}_z$$

and

$$R_2 = \sqrt{r^2 + (z')^2}$$

The sum of the fields given by (1) and (2),

$$d\mathbf{E} = \frac{\rho_\ell r dz'}{2\pi\varepsilon[r^2 + (z')^2]^{3/2}}\mathbf{u}_r$$

Since the unit vector $\mathbf{u}_r$ is not a function of z', total field intensity at P is given by

$$\mathbf{E} = \int_0^{\ell/2} d\mathbf{E} = \frac{\ell\rho_\ell}{2\pi\varepsilon r\sqrt{4r^2 + \ell^2}}\mathbf{u}_r \tag{3}$$

Note that r is perpendicular distance from the mid-point of the line charge to P and $\mathbf{u}_r$ is the unit vector along r.

For a line charge of large length tending to infinity $(\ell \gg r)$, the field intensity at a distance r from the line charge is

$$\mathbf{E} = \frac{\rho_\ell}{2\pi\varepsilon r}\mathbf{u}_r$$

Thus, the field intensity at any point due to a long line charge is in the radial direction. It is *inversely proportional to the distance* r of the point from the line charge. Note that the field intensity of a point charge is inversely proportional to square of the distance.

For any arbitrary orientation of a long line charge as shown in Fig. 2.6(b), the field intensity at a point P that is located at a perpendicular distance R from the line is

$$\mathbf{E} = \frac{\rho_\ell \mathbf{R}}{2\pi\varepsilon R^2} \tag{4}$$

Field Intensity of an Infinite Sheet of Charge

The distribution of charge in the form of a thin plane sheet is called a sheet of charge. The charge distribution is specified by charge per unit area called surface charge density ρ_S. The unit of surface charge density is C/m^2.

The surface charge density in the case of non-uniform charge distribution is defined as

$$\rho_S = \lim_{\Delta S \to 0} \frac{\Delta Q}{\Delta S}$$

where ΔQ is charge contained in an infinitesimally small area ΔS.

Let us now consider an infinite sheet of charge with uniform charge distribution of density ρ_S, as shown in Fig. 2.7. In order to calculate the field intensity at a point P, which is at height h above a point O on the sheet, consider a differential area dS on the sheet centred at P_1. The point P_1 is at a distance R and the line OP_1 makes an angle α with a reference line, as shown in the figure. The differential area is given by

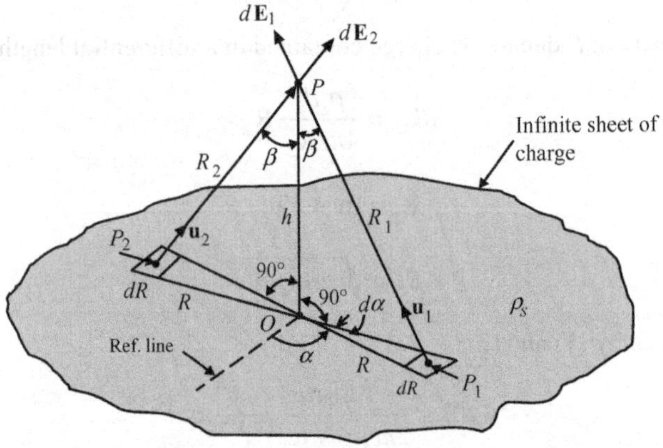

Fig. 2.7 Fields due to charges contained in two diametrically opposite surface elements

$$dS = RdRd\alpha$$

Charge contained in the differential area is

$$dQ = \rho_S RdRd\alpha,$$

which may be assumed to be a point charge. Therefore, field intensity at P due to dQ is

$$d\mathbf{E}_1 = \frac{\rho_S RdRd\alpha}{4\pi\varepsilon R_1^2}\mathbf{u}_1 = \frac{\rho_S RdRd\alpha}{4\pi\varepsilon(h^2 + R^2)}\mathbf{u}_1$$

The unit vector $\mathbf{u}_1$ is directed from P_1 to P.

Similarly, the field intensity of charge in a second differential area dS at P_2, which is diametrically opposite to P_1, as shown in the figure, is

$$d\mathbf{E}_2 = \frac{\rho_S RdRd\alpha}{4\pi\varepsilon R_2^2}\mathbf{u}_2 = \frac{\rho_S RdRd\alpha}{4\pi\varepsilon(h^2 + R^2)}\mathbf{u}_2$$

where $\mathbf{u}_2$ is a unit vector directed from P_2 to P.

Since P_1 and P_2 are symmetrically located with respect to O, the components of $d\mathbf{E}_1$ and $d\mathbf{E}_2$ parallel to the sheet cancel out and their components normal to the sheet add up. The normal component of each field,

$$dE_{n1} = dE_{n2} = \frac{\rho_S RdRd\alpha}{4\pi\varepsilon(h^2 + R^2)}\cos\beta$$

where

$$\cos\beta = \frac{h}{\sqrt{h^2 + R^2}}$$

The sum of the two normal components is

$$dE_n = 2(dE_{n1}) = \frac{\rho_S\, hR\, dR\, d\alpha}{2\pi\varepsilon(h^2 + R^2)^{3/2}} \tag{5}$$

The field intensity of the charge distribution is found by evaluating the double integral of the expression in (5). The lower and upper limits of α are 0 and π, and those of R are 0 and ∞. Thus,

$$E_n = \int_{R=0}^{\infty}\int_{\alpha=0}^{\pi} \frac{hR\rho_S dR d\alpha}{2\pi\varepsilon(h + R^2)^{3/2}} = \frac{\rho_S}{2\varepsilon} \tag{6}$$

Formula (6) shows that the direction of field intensity is perpendicular to the sheet and it has the same magnitude at any height above the sheet. The field intensity below the sheet has the same magnitude as given by (6). However, its direction is opposite to the direction of the field above the charge distribution.

Field Intensity of a Volume Charge Distribution

Continuous distribution of charge in a region of volume is specified by charge per unit volume which is called volume charge density. Its unit is coulomb per metre^3 (C/m^3), and it is generally denoted by the symbol ρ.

If the charge distribution is not uniform, the volume charge density,

$$\rho = \lim_{\Delta v \to 0} \frac{\Delta Q}{\Delta v}$$

where ΔQ is charge contained in a volume element Δv.

Referring to Fig. 2.8, let dv' be a differential volume centred at P' (x', y', z') within the charge distribution and $P(x, y, z)$ is the point at which the field intensity is required. If charge density at P' is $\rho\,(x', y', z')$, charge contained in the differential volume,

$$dQ = \rho\,(x', y', z')dv'$$

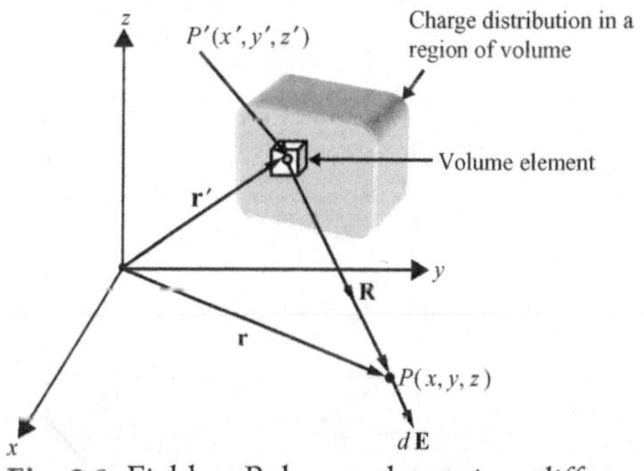

Fig. 2.8 Field at P due to charge in a differential volume element

The displacement of P from P' is

$$\mathbf{R} = \mathbf{r} - \mathbf{r}' = (x - x')\,\mathbf{u}_x + (y - y')\,\mathbf{u}_y + (z - z')\,\mathbf{u}_z$$

Thus, the field intensity at P due to the differential charge,

$$d\mathbf{E} = \frac{\rho(x', y', z')dv'}{4\pi R^3}\mathbf{R} \tag{7}$$

The total field intensity is found by integrating (7) over the region of charge.

$$\mathbf{E} = \int_v \frac{\rho(\ ', y', \ ')\mathbf{R}}{4\pi R^3} dv',$$

It may be noted that it is not easy to evaluate the integral even for uniform distribution of charge as it is a triple integral and the integrand is a vector function.

EXAMPLE 2.7

A line charge of linear charge density 5 nC/m and length 2 m is along the x axis with its mid-point at the origin in air. Find the electric field intensity at $P(0, 3\text{ m}, 5\text{ m})$.

Electric field intensity at P, $\mathbf{E} = \dfrac{\ell \rho_\ell}{2\pi\varepsilon_0 r\sqrt{r^2 + (\ell/2)^2}}\mathbf{u}_r$

$\ell = 2$, $2\pi\varepsilon_0 = 18\times 10^9$ F/m

$\mathbf{r} = (3-0)\mathbf{u}_y + (5-0)\mathbf{u}_z = 3\mathbf{u}_y + 5\,\mathbf{u}_z$ m, $r = \sqrt{3^2+5^2} = 5.83$ m

$\mathbf{u}_r = \mathbf{r}/r = 0.515\mathbf{u}_y + 0.858\mathbf{u}_z$

$$\mathbf{E} \quad \frac{(\quad \times 10^9)(2)(5\times 10^{-9})}{(\quad .83)\sqrt{5.83^2+1}}(0.515\mathbf{u}_y + 0.858\mathbf{u}_z) = 2.69\mathbf{u}_y + 4.47\mathbf{u}_z \text{ V/m}$$

EXAMPLE 2.8

A long line charge having linear charge density of 25 nC/m is parallel to the z axis, and intersects the x-y plane at $x = 0.7$ m, and $y = 1.6$ m, as shown in Fig. 2.9. Calculate the field intensity at $P(0.2\text{ m}, 0.4\text{ m}, 0.8\text{ m})$ in free space.

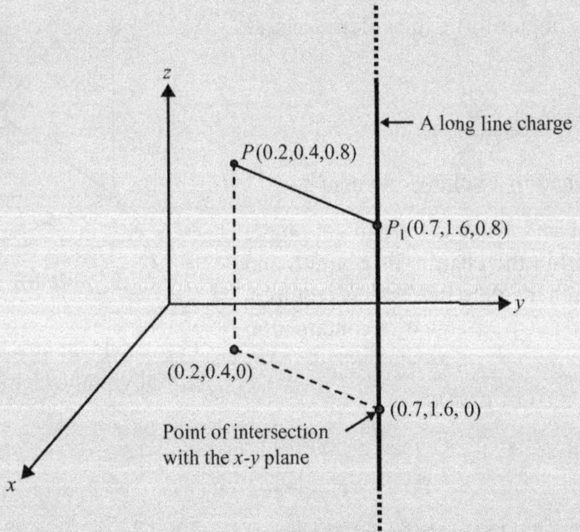

Fig. 2.9 Illustration of a line charge described in Example 2.8. The line PP_1 is perpendicular to the line charge.

In order to calculate the field intensity, we have to know the perpendicular distance between P and the line charge. Referring to the figure, let the line PP_1 be perpendicular to the line charge. As the charge distribution is parallel to the z axis, the distance of P_1 above the x-y plane would be same as that of P. Therefore, the coordinates of P_1 are (0.7 m, 1.6 m, 0.8 m). Hence,

$\mathbf{R} = (0.2-0.7)\mathbf{u}_x + (0.4-1.6)\,\mathbf{u}_y + (0.8-0.8)\,\mathbf{u}_z = -0.5\mathbf{u}_x - 1.2\mathbf{u}_y$ m

$R = 1.3$ m

Using formula (4),

$$\mathbf{E} \quad \frac{\rho_\ell}{2\pi\varepsilon_0 R^2}\mathbf{R} = \frac{25\times10^{-9}}{2\pi(10^{-9}/36\pi)(1.3)^2}(-0.5\mathbf{u}_x - 1.2\mathbf{u}_y) = -133\cdot1\mathbf{u}_x - 319\cdot5\mathbf{u}_y \text{ V/m}$$

EXAMPLE 2.9

Two oppositely charged lines are parallel to the z axis of cylindrical coordinates. The line with positive charge is at $P_1(0.5 \text{ m}, 60°)$ and the line with negative charge is at $P_2(0.8 \text{ m}, 150°)$. Find the field intensity at $P(1.5 \text{ m}, 90°)$ in free space. The charge density in each line is 0.075 pC/m.

Since distances are involved in the calculation of field intensity, the coordinates of the given points have to be transformed to the rectangular coordinate system. It may be noted that all the points are on the $z = 0$ plane. The coordinates of P, P_1 and P_2 in rectangular coordinates are given by

$x = 1.5 \cos 90° = 0$, $y = 1.5 \sin 90° = 1.5$ m

$x_1 = 0.5 \cos 60° = 0\ 25$ m, $y_1 = 0.5 \sin 60° = 0.433$ m

$x_2 = 0.8 \cos 150° = -\ 0\ 693$ m, $y_2 = 0.8 \sin 150° = 0.4$ m

The vector distance of P from P_1 is

$\mathbf{R}_1 = -\ 0.25\ \mathbf{u}_x + 1.067\ \mathbf{u}_y$ m

$R_1^2 = 1\cdot201 \text{ m}^2$

The displacement of P from P_2 is

$\mathbf{R}_2 = 0.693\ \mathbf{u}_x + 1.1\mathbf{u}_y$ m

$R_2^2 = 1\cdot69 \text{ m}^2$

The total field intensity is equal to the vector sum of the fields of two line charges. Thus, using formula (4), we get

$$\mathbf{E} = \frac{0.075\times10^{-12}}{2\pi(10^{-9}/36\pi)}\left(\frac{-0\cdot25\mathbf{u}_x + 1\cdot067\mathbf{u}_y}{1\cdot201} - \frac{0\cdot693\mathbf{u}_x + 1\cdot1\mathbf{u}_y}{1\cdot69}\right)$$

$= -\ 0.834\mathbf{u}_x + 0.320\mathbf{u}_y$ mV/m

The components of field intensity are

$E_x = -\ 0.834$ mV/m, $E_y = 0.320$ mV/m, $E_z = 0$

The r and ϕ components of the field strength in cylindrical coordinates at (1.5 m, 90°) are

$E_r = -0.834 \cos 90° + 0.320 \sin 90° = 0.320$ mV/m

$E_\phi = -0.834\ (-\sin 90°) + 0.32 \cos 90° = 0.834$ mV/m

$E_z = 0$

Thus, the field intensity at P in cylindrical coordinates is given by

$\mathbf{E} = 0.320\ \mathbf{u}_r + 0.0834\ \mathbf{u}_\phi$ mV/m

EXAMPLE 2.10

Two large sheets of charge with densities ρ_{S1} and $-\rho_{S2}$ are located at $x = 0$ and $x = a$. Find field intensities in all the regions.

The directions of fields due to the charge distributions are as shown in Fig. 2.10.

(1) In the region $x < 0$

$$\mathbf{E} \quad \mathbf{E}_2 - \mathbf{E}_1 = \frac{\rho_{S2} - \rho_{S1}}{2\varepsilon}\mathbf{u}_x \text{ V/m}$$

(2) In the region $0 < x < a$

$$\mathbf{E} \quad \mathbf{E}_2 + \mathbf{E}_1 = \frac{\rho_{S2} + \rho_{S1}}{2\varepsilon}\mathbf{u}_x \text{ V/m}$$

(3) In the region $x > a$

$$\mathbf{E} \quad \mathbf{E}_1 - \mathbf{E}_2 = \frac{\rho_{S1} - \rho_{S2}}{2\varepsilon}\mathbf{u}_x \text{ V/m}$$

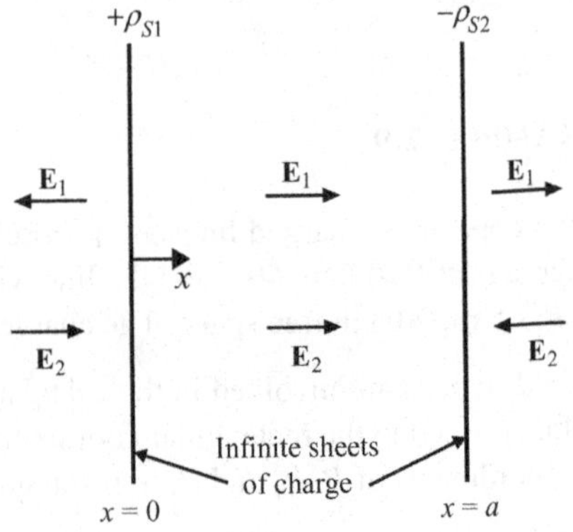

Fig. 2.10 Showing directions of field intensities produced by two parallel sheets of charge

2.7 Stream Lines of Electric Field Intensity

The electric field intensity being a vector quantity has magnitude as well as direction. For some simple charge distributions, it is possible to get an idea about the variation of magnitude and direction of the field from their mathematical expressions. For example, the field of a point charge is in the radial direction, and its magnitude at a point is inversely proportional to square of the radial distance of the point from the charge. The magnitude and direction of the field of a uniformly charged line can also be visualized from its mathematical expression. But, variation of the field of some other charge configurations (e.g., of two point charges) cannot be visualized easily from their mathematical expressions. The problem can be solved to some extent by mapping the field.

The field map consists of lines, called *field lines* or *stream lines*, having the following properties:

1. The stream lines start from positive charges and end on the negative charges. They are either straight lines or smooth curves.
2. The tangent at any point to the lines gives the direction of field intensity.
3. In a single medium, the field lines are continuous curves. However, they are not continuous at the boundary surface between two media.
4. Field maps are generally limited to two-dimensional fields.
5. A good map of the field provides information regarding relative magnitude of field intensity in a region. If the stream lines converge along a certain direction, then it shows that the magnitude of the field intensity increases in that direction.

Maps of fields of some simple charge configuration are given in Fig. 2.11.

2.8 Equation of Field Lines

Let us assume that the curve shown in Fig. 2.12 represents an electric field line in the x-y plane. At any point $P(x, y)$, let E_x and E_y be the field intensities in x and y directions respectively. The slope of the curve at P is

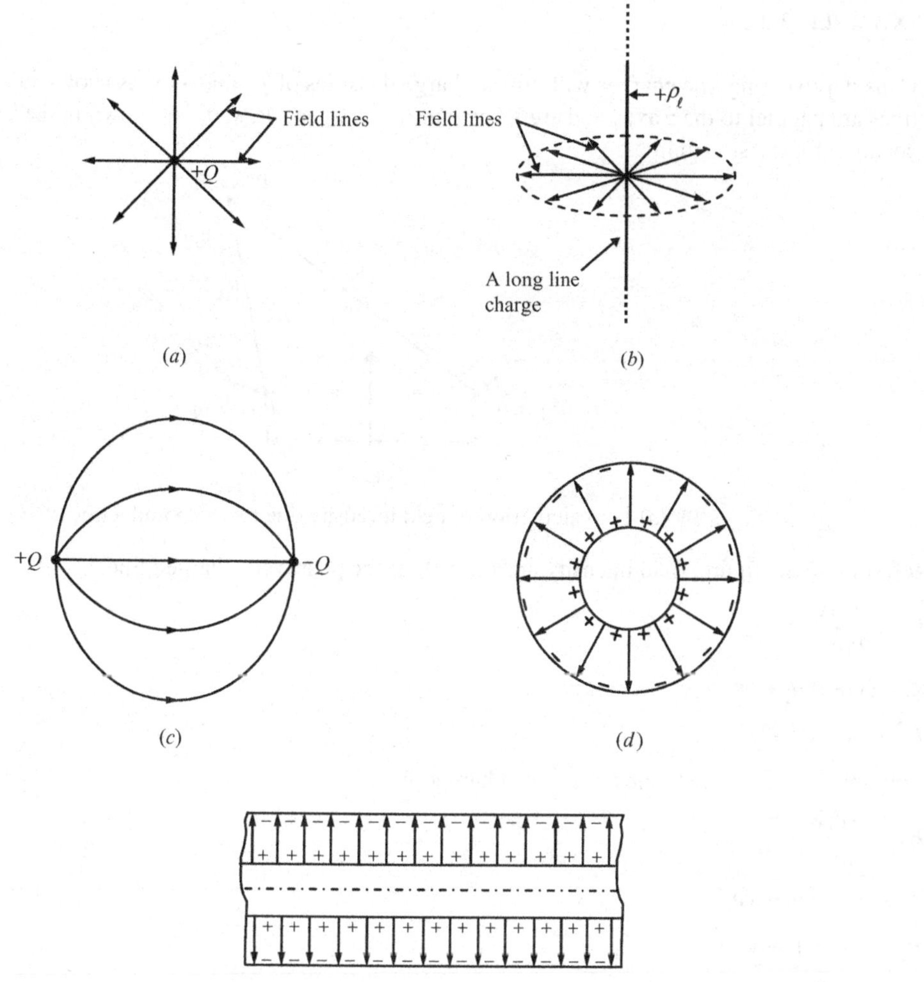

Fig. 2.11 Showing stream lines: (*a*) of a point charge, (*b*) of a line charge, (*c*) between two point charges, (*d*) between concentric spherical shells with uniform surface charges, and (*e*) between concentric cylinders with uniform surface charges.

$$\frac{dy}{dx} = \frac{E_y}{E_x} \qquad (1)$$

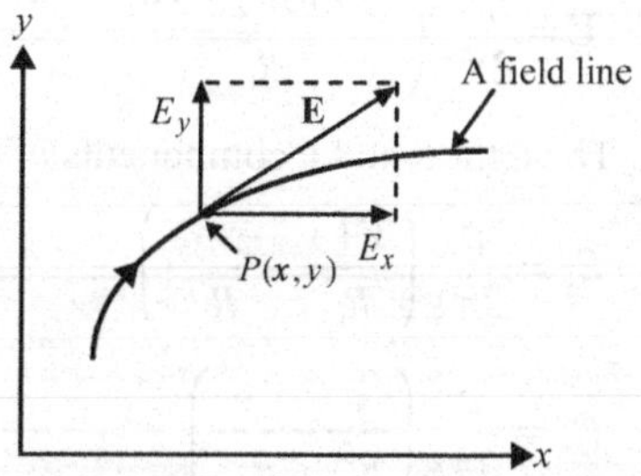

Fig. 2.12 For deriving equation of field lines

Assuming that the fields vary with x and y, we may write

$$E_x = f_1(x, y)$$

and

$$E_y = f_2(x, y)$$

Substituting in (1)

$$\frac{dy}{dx} = \frac{f_1(x, y)}{f_2(x, y)}$$

The above equation is a first order differential equation. Its solution is the equation of field lines. However, except for some simple charge distribution, a direct solution of the equation cannot be obtained easily.

EXAMPLE 2.11

Consider two long line charges with linear charge densities of ρ_ℓ and $-\rho_\ell$ as shown in Fig. 2.13(*a*). The lines are parallel to the *z* axis, and are located at $P_1(-s, 0)$ and $P_2(s, 0)$, as shown in the figure. Obtain the equation for the stream lines.

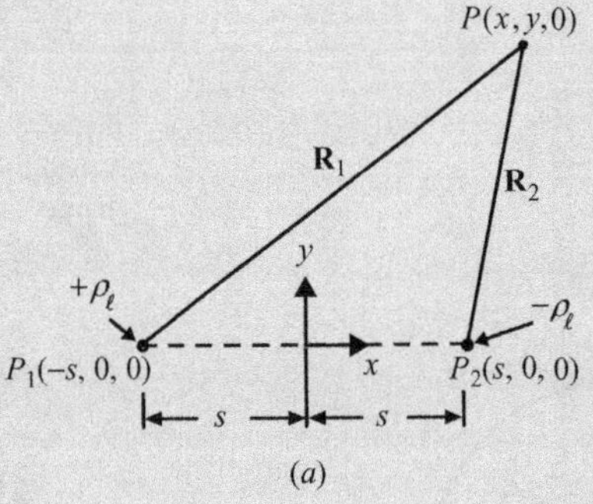

Fig. 2.13 (*a*) For calculation of field intensity due to two parallel line charges.

Referring to the figure, field intensity at $P(x, y, 0)$ of the positively charged line,

$$\mathbf{E}_1 = \frac{\rho_\ell \mathbf{R}_1}{2\pi\varepsilon R_1^2};$$

$$\mathbf{R}_1 = (x + s)\,\mathbf{u}_x + y\mathbf{u}_y, \tag{2}$$

$$R_1^2 = (x + s)^2 + y^2 \tag{3}$$

Field intensity at P due to the negatively charged line,

$$\mathbf{E}_2 = \frac{-\rho_\ell \mathbf{R}_2}{2\pi\varepsilon R_2^2};$$

$$\mathbf{R}_2 = (x - s)\,\mathbf{u}_x + y\mathbf{u}_y, \tag{4}$$

$$R_2^2 = (x - s)^2 + y^2 \tag{5}$$

The total field intensity at P,

$$\mathbf{E} = \mathbf{E}_1 + \mathbf{E}_2 = \frac{\rho_\ell}{2\pi\varepsilon}\left(\frac{\mathbf{R}_1}{R_1^2} - \frac{\mathbf{R}_2}{R_2^2}\right) \tag{6}$$

Substituting in (6) the expressions of $\mathbf{R}_1$ and $\mathbf{R}_2$ given by (2) and (4), we get

$$\mathbf{E} = \frac{\rho_\ell}{2\pi\varepsilon}\left(\frac{(x+s)\mathbf{u}_x + y\mathbf{u}_y}{R_1^2} - \frac{(x-s)\mathbf{u}_x + y\mathbf{u}_y}{R_2^2}\right)$$

Thus, the *x* and *y* components of the field are

$$E_x = \frac{\rho_\ell}{2\pi\varepsilon}\left(\frac{x+s}{R_1^2} - \frac{x-s}{R_2^2}\right)$$

$$E_y = \frac{\rho_\ell}{2\pi\varepsilon}\left(\frac{y}{R_1^2} - \frac{y}{R_2^2}\right)$$

Using the expressions of R_1^2 and R_2^2 given by (3) and (5), it can be shown that the *x* and *y* components of the field are

$$E_x = \frac{2s(y^2 - x^2 + s^2)}{R_1^2 R_2^2}$$

$$E_y = -\frac{4sxy}{R_1^{\,2} R_2^{\,2}}$$

Substituting the expressions of E_x and E_y in (1), we get

$$\frac{dy}{dx} = \frac{2xy}{x^2 - y^2 - s^2}$$

Or $\quad 2xydx + (y^2 - x^2 + s^2)dy = 0$ (7)

Equation (7) is not an *exact* differential equation. It is made exact by multiplication by an integrating factor $1/y^2$. Thus, multiplying (7) by $1/y^2$ we have

$$\frac{2x}{y}dx + (1 - \frac{x^2}{y^2} + \frac{s^2}{y^2})dy = 0 \qquad (8)$$

Let $f(x, y) = c_1$ be a solution of (8). Then the total differential of $f(x, y)$,

$$df(x, y) = \frac{\partial f(x, y)}{\partial x}dx + \frac{\partial f(x, y)}{\partial y}dy = 0 \qquad (9)$$

Comparison between (8) and (9) yields

$$\frac{\partial f(x, y)}{\partial x} = \frac{2x}{y} \qquad (10)$$

and $$\frac{\partial f(x, y)}{\partial y} = 1 - \frac{x^2}{y^2} + \frac{s^2}{y^2} \qquad (11)$$

Integrating (10) with respect to x assuming y to be constant, we get

$$f\ x, y) = \frac{x^2}{y} + k(y) \qquad (12)$$

where $k(y)$ is a function of y but not of x.

To determine $k(y)$, we at first find the partial derivative of (12) with respect to y and then substitute the result in (11) to get $\partial[k(y)]/\partial y$. Accordingly,

$$\frac{\partial[\ (y)]}{\partial y} = 1 + \frac{s^2}{y^2}$$

Thus, $$k(y) = \int\left(1 + \frac{s^2}{y^2}\right)dy = y - \frac{s^2}{y} + c_2$$

By substituting the above expression in (12), we will get the solution of (8). The solution is

$$x^2 + y^2 - s^2 = \pm 2cy$$

Or $$x^2 + (y \mp c)^2 = c^2 + s^2 \qquad (13)$$

where $$\pm 2c = c_1 - c_2$$

Equation (13) represents a family of circles with ( ) as the parameter. The centres of the circles are at (, $\mp c$) and their radii are $(c^2 + s^2)^{1/2}$. A field map is shown in Fig. 2.13(*b*).

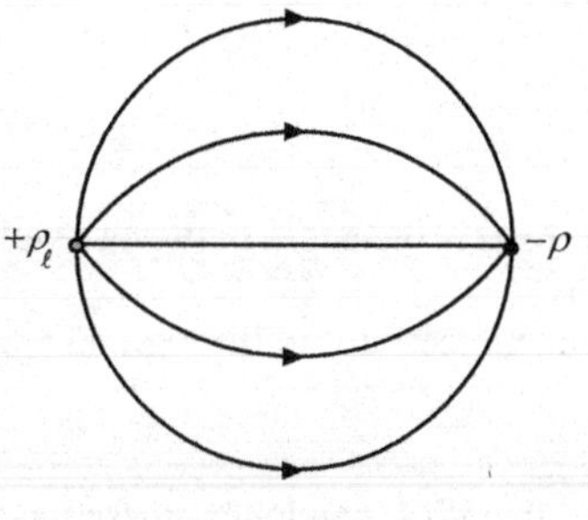

(*b*)

Fig. 2.13 (*b*) Stream lines of the charges.

2.9 Electric Flux Density and Electric Flux

The magnitude of electric field intensity depends on the magnitude and distribution of charges, and the location of the point at which the field is to be found. It is also inversely proportional to permittivity of the medium. The concept of field lines is helpful in the drawing of field maps. These two concepts are, however, not adequate for solving field problems involving more than one dielectric medium. In these problems it is convenient to deal with quantities independent of permittivity of the medium. Two such quantities are the electric flux density and the electric flux.

The electric flux density is a vector quantity. In an isotropic medium, the direction of flux density is same as that of field intensity. Its magnitude is ε times of the field intensity. It is denoted by **D**.

Thus, the vector form of flux density,

$$\mathbf{D} = \varepsilon \mathbf{E} = \varepsilon_0 \varepsilon_r \mathbf{E} \tag{1}$$

Accordingly, flux density of a point charge of Q coulombs located at the origin of spherical coordinates is

$$\mathbf{D} = \frac{Q}{4\pi r^2}\mathbf{u}_r \tag{2}$$

The flux density due to a long filamentary line charge with linear charge density ρ_ℓ C/m and positioned along the z axis of cylindrical coordinates is

$$\mathbf{D} = \frac{\rho_\ell}{2\pi r}\mathbf{u}_r \tag{3}$$

It is clear from (2) and (3) that the unit of flux density is C/m^2.

The electric flux is found by integrating the normal component of flux density over a given surface. The electric flux is a scalar quantity and its unit is *coulomb*. We will denote the electric flux by the symbol Ψ.

In Fig. 2.14, S represents an open surface in the region of an electric field, and dS is a differential area on S centred at P. The unit vector $\mathbf{u}_n$ is normal to dS at P, and **D** is the average flux density vector over the differential surface. The vector form of the differential area is

$$d\mathbf{S} = dS\mathbf{u}_n$$

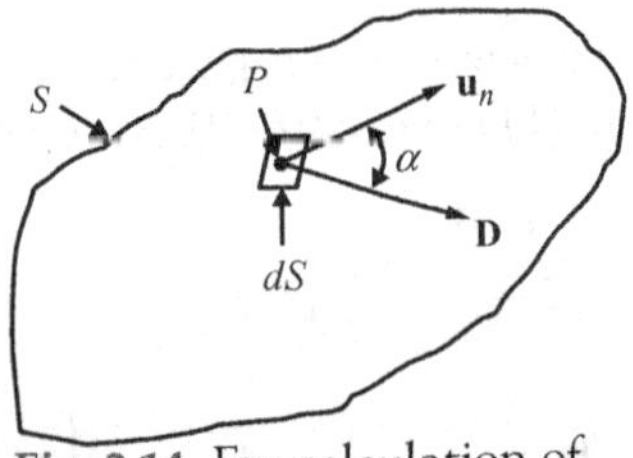

Fig. 2.14 For calculation of flux through surface S

The component of flux density along $\mathbf{u}_n$ is

$$D_n = D\cos\alpha$$

where α is the angle between **D** and $\mathbf{u}_n$.

Hence, electric flux through dS is given by

$$d\Psi = D\cos\alpha\, dS \tag{4}$$

The lhs of (4) can be expressed as the dot product of **D** and $d\mathbf{S}$. Thus,

$$d\Psi = \mathbf{D}\cdot d\mathbf{S}$$

The electric flux through the surface is obtained by integrating $d\Psi$ over the surface. A single integral sign with a symbol S at its bottom is used to mean surface integration.

$$\Psi = \int_S \mathbf{D} \cdot d\,\mathbf{S} \tag{5a}$$

If the surface is a closed one, then

$$\Psi = \oint_S \mathbf{D} \cdot d\,\mathbf{S} \tag{5b}$$

where the small circle on the integral sign indicates that the integration has to be evaluated over a closed surface.

Though the definitions of electric flux density and electric flux given by (1) and (5) are purely mathematical, one may acquire some physical meaning with their repeated use.

As a simple example, let us calculate flux due to a positive point charge of Q coulombs. If the charge is at the centre of a sphere of radius r, flux density on the surface of the sphere is

$$\mathbf{D} = \varepsilon \mathbf{E} = \frac{Q}{4\pi r^2}\mathbf{u}_r$$

Since the magnitude of flux density is constant over the surface and it is in the radial direction, flux out of the spherical surface is given by

$$= \frac{Q}{4\pi r^2}(4\pi r^2) = Q$$

That is, the magnitude of total flux due to a point charge is equal to the charge magnitude in SI unit. Further, flux through each solid angle of the sphere is $Q/4\pi$ (C/steradian).

Because of historic reasons, the electric flux density is also called *displacement density* and the electric flux is called *displacement flux*. These names are derived from the findings of Faraday, and his experiments with two concentric conducting spheres. He found that when the inner sphere is given a charge of $+Q$, the charge induced on the inside surface of the outer sphere is $-Q$. The same result is obtained for any size of the outer sphere, and for any dielectric medium between the spheres. He hypothesized that some sort of *displacement* or *flow* takes place from the inner sphere to the outer sphere through the medium. This *flow* or *flux* or *displacement* is equal to the charge of the inner sphere in SI unit. The displacement density or electric flux density is equal to flux per unit area.

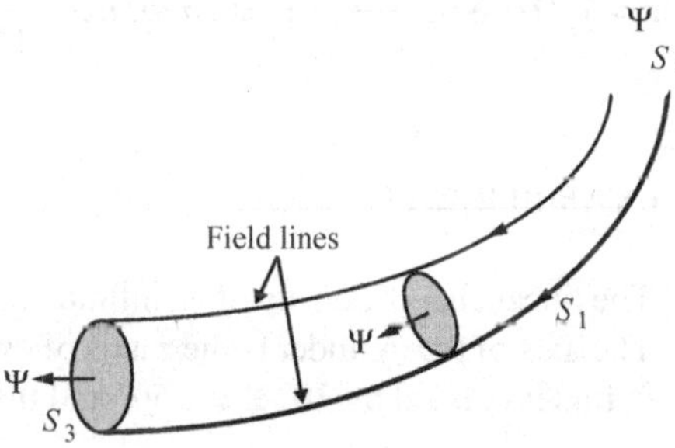

Fig. 2.15 A flux tube

Tubes of Electric Flux

In an electric field, imagine a region in the form of a tube having an outer surface S_1 and end surfaces S_2 and S_3, as shown in Fig. 2.15. The shape of S_1 is determined from the consideration that the field intensity is tangential everywhere on it so that flux through this surface is zero. Thus,

> if the region does not enclose any charge, flux throughout the tube would be same. Therefore, flux entering through one end-surface, say S_2, is equal to flux coming out of the other end-surface S_3. Such a region is called a flux tube. The concept of flux tubes is useful in drawing field maps.

Figure 2.16 shows flux tubes of two equal and opposite point charges.

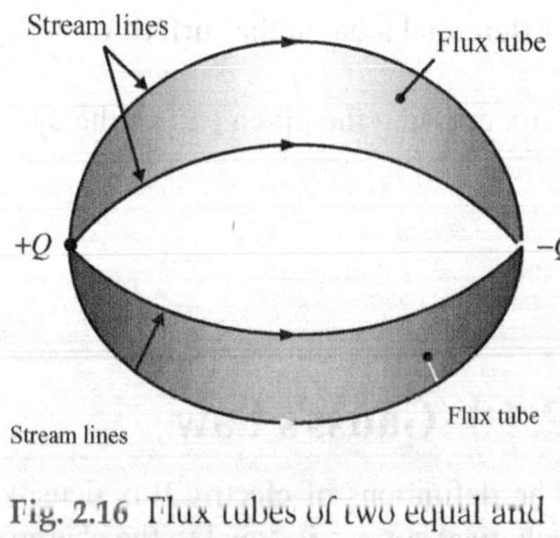

Fig. 2.16 Flux tubes of two equal and opposite point charges

EXAMPLE 2.12

A point charge of 2.0 nC is at the centre of a sphere, as shown in Fig. 2.17. Find the flux crossing a part of the sphere defined by $30° \le \phi \le 60°$ and $0 \le \theta \le 90°$.

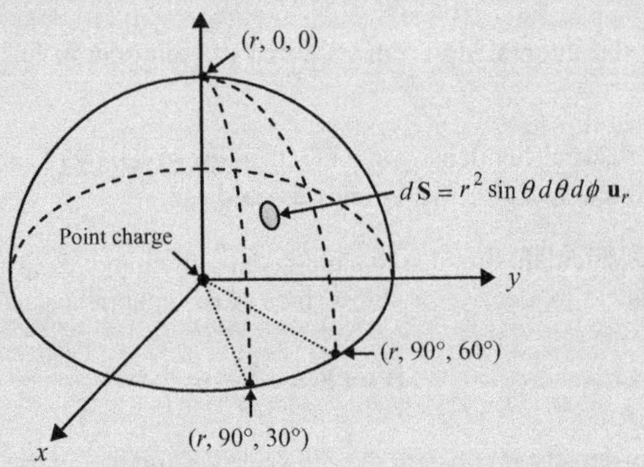

Fig. 2.17 Pertaining to Example 2.12

Flux density on the surface of a sphere of radius r, $\mathbf{D} = \dfrac{2\times10^{-9}}{4\pi r^2}\mathbf{u}_r$ C

A differential area in r direction,

$d\mathbf{S} = r^2 \sin\theta d\theta d\phi \mathbf{u}_r$ m²

Flux through the surface is given by

$$\Psi = \int_S \mathbf{D}\cdot \mathbf{S} = \frac{10^{-9}}{2\pi}\int_{\frac{\pi}{6}}^{\frac{\pi}{3}}\int_0^{\frac{\pi}{2}} \sin\theta\, d\theta\, d\phi = \frac{1}{12}\text{nC}$$

EXAMPLE 2.13

The linear charge density of an infinite line charge located along the axis of a cylinder of radius r is 8 nC/m. The axis of the cylinder is the z axis of cylindrical coordinates. Find the electric flux crossing a part of the cylinder defined by $30° \le \phi \le 60°$ and $0 \le z \le 3.6$ m.

Flux density on the surface of the cylinder, $\mathbf{D}_r = \dfrac{\rho_\ell}{2\pi r}\mathbf{u}_r$

Differential area on the surface of the cylinder, $d\mathbf{S} = rd\phi dz\mathbf{u}_r$

Flux crossing the given part of the cylinder, $\Psi = \int_S \mathbf{D}\cdot d\,\mathbf{S} = \int_{\phi=\pi/6}^{\pi/3}\int_0^{3.6}\dfrac{\rho_\ell d\phi dz}{2\pi}$

$$= \left(\frac{8\times10^{-9}}{2\pi}\right)(3.6)\left(\frac{\pi}{6}\right) = 2.4 \text{ nC}$$

2.10 Gauss's Law

The definitions of electric flux density and flux when applied to a point charge show that flux crossing a spherical surface is equal to the charge enclosed by the surface. The relation is obtained for one point charge located at the centre of the sphere.

Gauss's law is a general statement regarding the relationship between electric flux crossing a closed surface and the charge enclosed by it. The law does not put any restriction on the sign or distribution of charges, and shape of the surface enclosing the charges. The law states that '*the total flux crossing any closed surface is equal to the algebraic sum of charges enclosed by the surface*'.

If the algebraic sum of all the charges enclosed is positive then flux would be *out* of the surface, and if the sum is negative then flux would be *into* the surface. Gauss's law states

Electric flux *out* of a closed surface = charge enclosed by the surface. That is,

$$\Psi = Q \tag{1}$$

Relation (1) is the mathematical statement of Gauss's law.

In order to prove Gauss's law, let us take a positive point charge at the origin of spherical coordinates in a region of constant permittivity. Let the charge be enclosed by a closed surface, as shown in Fig. 2.18. Electric flux density at a point $P(r, \theta, \phi)$ on the surface is

$$\mathbf{D} = \frac{Q}{4\pi r^2}\mathbf{u}_r \tag{2}$$

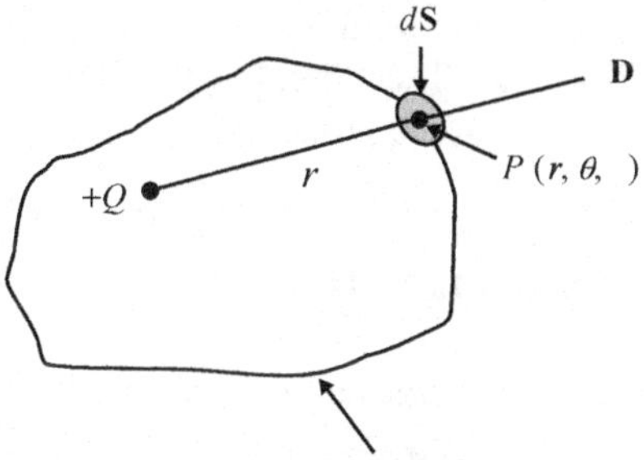

Fig. 2.18 Pertaining to proof of Gauss's law

A differential vector surface in terms of its components at P is

$$d\mathbf{S} = r^2 \sin\theta\, d\theta\, d\phi\, \mathbf{u}_r + r\sin\theta\, dr d\phi\, \mathbf{u}_\theta + r dr d\theta\, \mathbf{u}_\phi \tag{3}$$

Flux through this differential surface,

$$d\Psi = \mathbf{D}\cdot d\mathbf{S} \tag{4}$$

By substituting (2) and (3) in (4), we get

$$d\Psi = \frac{Q}{4\pi}\sin\theta\, d\theta\, d\phi$$

Flux out of the closed surface,

$$\Psi = \frac{Q}{4\pi}\int_{\theta=0}^{\pi}\int_{\phi=0}^{2\pi}\sin\theta\, d\theta\, d\phi = Q,$$

which proves Gauss's law.

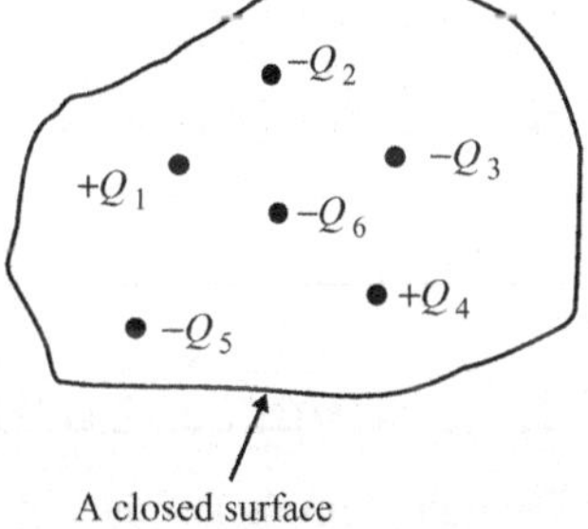

Fig. 2.19 A closed surface enclosing several point charges

If a closed surface encloses a number of point charges, both positive and negative, as shown in Fig. 2.19, total flux through the surface is equal to the algebraic sum of all the charges.

It may be noted that Gauss's law is proved starting from Coulomb's law; no other experimental result is used. This law finds applications in the calculation of electric field intensity of symmetrical charge distributions for deriving the divergence of electric flux density and for deriving boundary relations in static electric field as well as in time varying fields.

EXAMPLE 2.14

Three point charges $Q_1 = 0.12\ \mu$C, $Q_2 = -0.28\ \mu$C, and $Q_3 = 0.08\ \mu$C are located at $P_1(0.8$ m, 0, 0), $P_2(0, 0.8$ m, 0) and $P_3(0.6$ m, 0.8 m, 0), respectively in rectangular coordinates. Find flux out of spherical surfaces having radii (a) 0.6 m, (b) 0.85 m, and (c) 1.2 m. All the spheres are centred at the origin.

(a) The spherical surface of 0.6 m radius does not enclose any charge. Hence, no flux comes out of this surface.
(b) The spherical surface of 0.8 m radius encloses both Q_1 and Q_2. Flux *out* of this surface = – 0.16 μC.
(c) The spherical surface of 1.2 m radius encloses all the three charges. Flux *out* of this surface = – 0.08 μC.

EXAMPLE 2.15

A point charge of – 0.3 μC is located at $x = 0.5$ m, and a line charge on the y axis has its two ends at $y = -0.8$ m and $y = 0.6$ m. The charge density of the line charge is 0.5 μC/m. Find electric flux out of a spherical surface of radius 0.75 m with its centre at the origin.

The spherical surface encloses the point charge and the part of the line charge between $y = -0.75$ m and $y = 0.6$ m. Therefore, length of the line charge that lies within the spherical surface is equal to 1.35 m. Thus, flux *out* of the surface = – 0.3 + (0.5 × 1.35) = 0.375 μC.

EXAMPLE 2.16

A line charge on the x axis has its one end at the origin and the other end at 2.4 m. The charge density of the line is defined by $\rho_\ell = 2e^{-0.4x}$ μC/m. Find the flux out of a closed surface enclosing the line charge.

As per Gauss's law, flux out of the closed surface is equal to the charge enclosed by the surface.

Charge in a differential length $d\ell$, $dQ = \rho_\ell d\ell = (2\times10^{-6})e^{-0.4x}dx$

Charge enclosed, $Q = (2\times10^{-6})\int_0^{2.4} e^{-0.4x}dx = -(0.8\times10^{-6})(e^{-0.4x})_0^{2.4} = 0.494\ \mu\text{C}$

Flux out of the closed surface = 0.494 μC

2.11 Integral Form of Gauss's Law

Though Gauss's law has been proved for discrete distribution of charges, it can also be applied for linear, surface and volume distributions of charges. A general form of Gauss's law is its integral form which can be used for continuous charge distributions. To obtain the integral form of Gauss's law, let us consider a region of volume where charge is continuously distributed, as shown in Fig. 2.20. If ρ is the volume charge density at any point within the volume, charge in a differential volume dv,

$$dQ = \rho\, dv$$

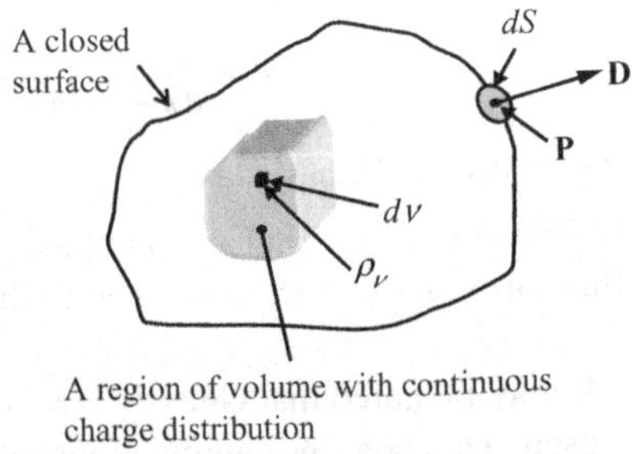

Fig. 2.20 For derivation of integral form of Gauss's law

Total charge in the region is,

$$Q \int_v \rho dv \tag{1}$$

Referring to Fig. 2.20, P is a point on a closed surface enclosing the charged region and $d\mathbf{S}$ is a differential area centred at P. If $\mathbf{D}$ is flux density at P due to the volume charge, flux through the differential area is,

$$d\Psi = \mathbf{D} \cdot d\mathbf{S} \tag{2}$$

Flux crossing the surface is found by integrating (2) over the closed surface.

$$\Psi = \oint_S \mathbf{D} \cdot d\mathbf{S} \tag{3}$$

By Gauss's law, (1) and (3) must be equal. Thus,

$$\oint_S \mathbf{D} \cdot d\mathbf{S} = \int_v \rho dv \tag{4}$$

Equation (4) is the integral form of Gauss's law.

If the closed surface encloses a line charge of finite length with density ρ_ℓ C/m, charge enclosed is found by integrating ρ_ℓ over the line. Hence,

$$\oint_S \mathbf{D} \cdot d\mathbf{S} = \int \rho_\ell d\ell \tag{5}$$

If the closed surface encloses a surface charge spread over a source S' of finite area with density $\rho_{S'}$ C/m^2, charge enclosed is found by integrating $\rho_{S'}$ over the surface. Hence,

$$\oint_S \mathbf{D} \cdot d\mathbf{S} = \int_{S'} \rho_{S'} dS' \tag{6}$$

We find from relations (4), (5) and (6) that the total flux crossing a closed surface is obtained by integrating the normal component of flux density over the surface, and to find the charge enclosed by the surface we have to integrate the charge density over the region within the surface.

2.12 Calculation of Field Intensity using Gauss's Law

One useful application of Gauss's law is to calculate the electric field intensity of charge distributions possessing some kind of symmetry (rectangular, cylindrical and spherical). For these cases the field strength can be obtained more easily using Gauss's law than by the use of Coulomb's law. This is because, the fields of symmetrical charge-distributions are also symmetrical. Hence, the surface integral of (2.11-4) can be evaluated easily.

In order to apply Gauss's law to a problem, it is necessary to choose a suitable closed surface satisfying the following requirements:

1. The surface contains the point at which the field intensity is to be found.
2. Flux density is normal as well as constant over some parts of the surface. Hence, flux crossing these parts of the surface is simply equal to the product of flux density and surface area.
3. Flux density on other parts of the closed surface is either zero or tangential so that flux crossing these parts is zero. A closed surface which fulfils the above requirements is called a Gaussian surface.

To use Gauss's law, it is also necessary to evaluate the volume integral of (2.11-4). This is easily done as the charge is uniformly and symmetrically distributed.

We will now calculate field intensities due to some simple charge distributions.

Spherical Distribution of Charge

If an isolated metallic sphere (hollow or solid) is given a positive charge of Q coulombs, the charge gets distributed uniformly on the surface of the sphere with a surface charge density given by

$$\rho_S = \frac{Q}{4\pi a^2} \text{ C/m}^2 \tag{1}$$

where a is radius of the sphere having its centre at the origin of spherical coordinates, as shown in Fig. 2.21(a). We will calculate the field intensity at a radial distance r outside the sphere ($r > a$).

A little consideration will show that the field intensity is in r direction and it is independent of θ and ϕ because of spherical symmetry. Therefore, the Gaussian surface is that of a sphere of radius r concentric with the charged sphere, as indicated in the figure. If D_r is the flux density on this surface, then flux out of the surface is

$$\Psi = 4\pi r^2 D_r = 4\pi \varepsilon r^2 E_r,$$

which according to Gauss's law is equal to the charge enclosed. Therefore,

$$4\pi\varepsilon r^2 E_r = Q$$

Or

$$E_r = \frac{Q}{4\pi\varepsilon r^2}, \quad r > a \tag{2}$$

This shows that the field intensity outside the sphere is same as that of a point charge of Q coulombs located at the centre of the charged sphere.

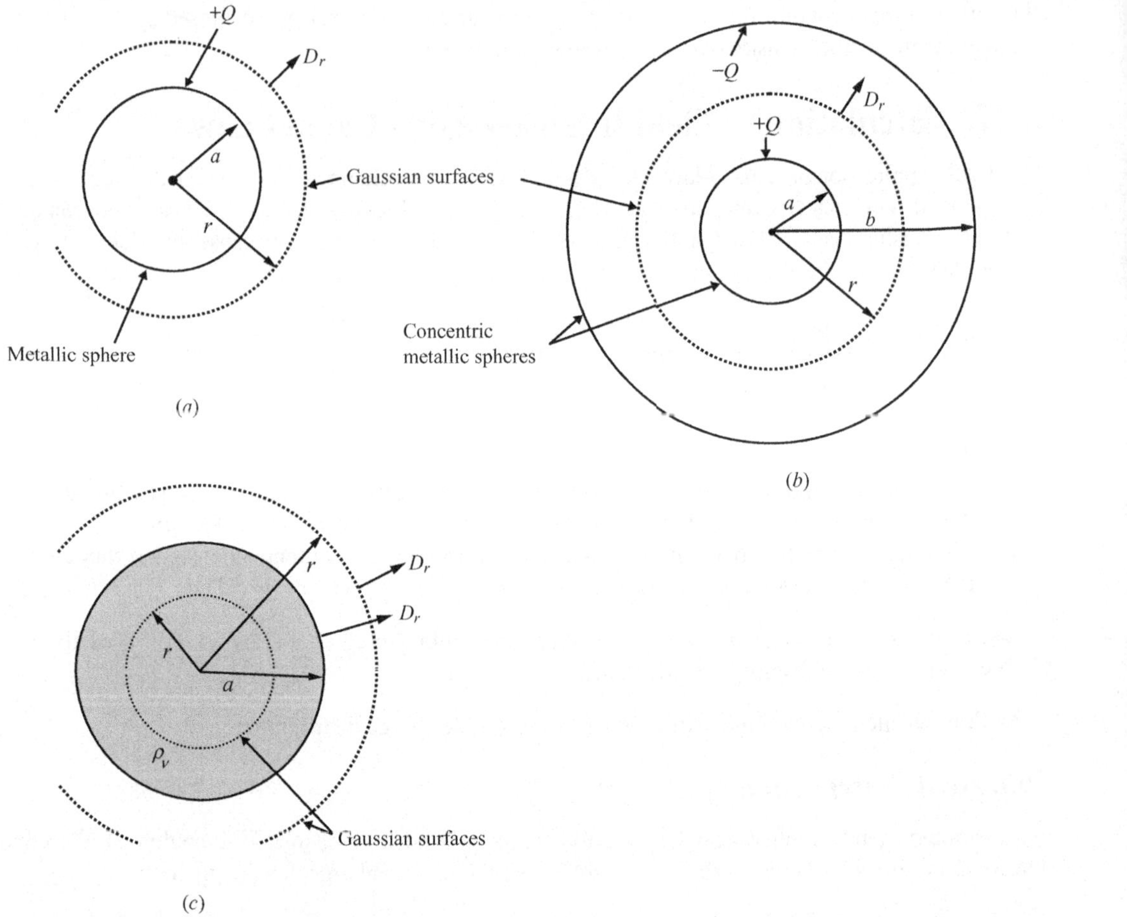

Fig. 2.21 Illustrations of spherical distribution of charges: (a) On the surface of an isolated metallic sphere, (b) On the surfaces of two concentric metallic spheres, and (c) Charge distribution over a spherical volume.

We can express the field intensity in terms of the charge density by eliminating Q from (1) and (2). Thus,

$$E_r = \frac{\rho_S a^2}{\varepsilon r^2}, \quad r > a$$

The field inside the sphere ($r < a$) is zero since there is no charge in the region.

Next, let us consider two equally and oppositely charged concentric metallic spheres, as shown in Fig. 2.21(b). The value of charge in each sphere is taken as Q. In this case there is no field inside the inner sphere ($r < a$) and outside the outer sphere ($r > b$). The field intensity in the region between the spheres is obtained following the same procedure as for the isolated sphere. It is given by

$$E_r = \frac{Q}{4\pi\varepsilon r^2}, \quad b > r > a$$

Lastly, let us assume that charges are uniformly distributed throughout a spherical volume with a total charge Q. The volume charge density is

$$\rho = \frac{3Q}{4\pi a^3}$$

where a is radius of the spherical charge distribution, as shown in Fig. 2.21(c).

Within the sphere ($r < a$), let the field intensity at a radial distance r from the charge centre be E_r. Then, flux out of a spherical Gaussian surface of radius r is

$$\Psi = 4\pi\varepsilon r^2 E_r \tag{3}$$

$$\text{Charge enclosed by the surface } = \frac{4}{3}\pi r^3 \rho = \frac{Qr^3}{a^3} \tag{4}$$

By Gauss's law, (3) and (4) should be equal. Hence, we get

$$E_r = \frac{Qr}{4\pi\varepsilon a^3}, \quad r < a$$

Thus, the field intensity is proportional to r within the spherical charge.

It can easily be shown that the field intensity at a point outside the sphere,

$$E_r = \frac{Q}{4\pi\varepsilon r^2}, \quad r > a.$$

Cylindrical Distribution of Charge

A special case of cylindrical distribution of charge is a uniformly charged thin line of infinite length, as shown in Fig. 2.22(a). The line charge with density ρ_ℓ C/m is taken along the z axis of cylindrical coordinates. We will calculate the field at a point P located at a radial distance r from the line charge.

It is not difficult to reason out that the flux density is along r direction and it does not vary along ϕ and z directions. Therefore, the Gaussian surface is that of a cylinder of radius r and of any length ℓ with its axis along the line charge, as shown in the figure. The cylinder has three surfaces. These are the end surfaces S_1 and S_2, and the curved surface S_3. As the field is tangential to the surfaces S_1 and S_2, flux through these surfaces is zero. However, the flux density D_r is normal to surface S_3 and it is constant over this surface. Therefore, flux out of this surface is equal to the product of D_r and area of S_3. Denoting the flux by Ψ, we have

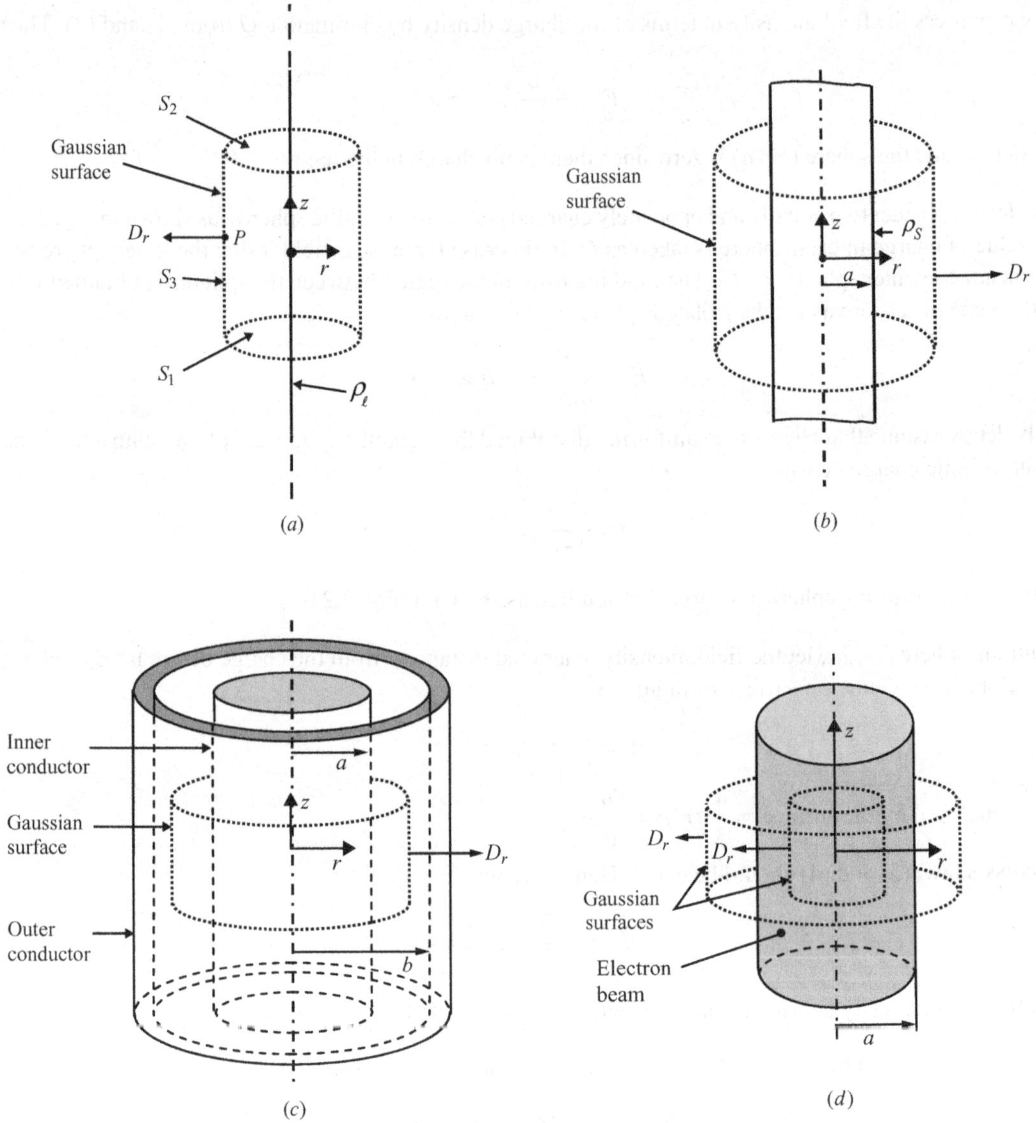

Fig. 2.22 Illustrations of cylindrical distribution of charges: (*a*) A long line charge, (*b*) Charge distribution on the surface of a long cylindrical conductor, (*c*) A coaxial transmission line, and (*d*) A cylindrical electron beam.

$$\Psi = 2\pi r \ell D_r = 2\pi \varepsilon r \ell E_r$$

where E_r is the field intensity at P.

Charge enclosed by the Gaussian surface,

$$Q = \ell \rho_\ell$$

By Gauss's law,

$$\Psi = Q$$

Hence,

$$E_r = \frac{\rho_\ell}{2\pi\varepsilon r} \tag{5}$$

The result is the same as that obtained in Section 2.6. We can use the formula without much error if the length of the line charge is large compared to the distance of the point at which the field is required $(\ell \gg r)$.

We will next calculate the field intensity due to charge distribution on the surface of a cylindrical conductor. We assume that the conductor of radius a is of infinite length, and the charge distribution is uniform with density ρ_s C/m^2. The conductor axis is along the z axis of cylindrical coordinates, as shown in Fig. 2.22(*b*). It is required to find the field intensity at a point outside the conductor and at a radial distance r from its axis $(r > a)$. In this case also the field is in r direction and it does not vary with ϕ and z. Accordingly, the Gaussian surface is that of a cylinder of radius r and of any length ℓ, as shown in the figure.

Flux out of the Gaussian surface is

$$\Psi = 2\pi \varepsilon r \ell E_r, \quad r > a$$

where E_r is field intensity at any point on the curved part of the Gaussian surface.

Charge enclosed by the Gaussian surface,

$$Q = 2\pi a \ell \rho_S$$

By Gauss's law,

$$\Psi = Q$$

Hence

$$E_r = \frac{a\rho_S}{\varepsilon r}, \quad r > a$$

As charge per unit length of the conductor $\rho_\ell = 2\pi a \rho_S$, the field intensity in terms of charge per unit length is given by

$$E_r = \frac{\rho_\ell}{2\pi\varepsilon r}, \quad r > a$$

This result is same as that given by (5). The field intensity inside the cylinder $(r < a)$ is zero.

Our next example of cylindrical charge distribution is a coaxial transmission line used in electrical power systems and communication systems. It consists of an inner conductor and an outer conductor separated by one or more layers of insulation. If the conductors are connected to a dc source, they get equally and oppositely charged. We will consider a coaxial transmission line with a single layer of insulation between the conductors. The outside radius of the inner conductor and the inside radius of the outer conductor are taken as a and b respectively. The axis of the transmission line is along the z axis of cylindrical coordinates, as shown in Fig. 2.22(*c*). We assume that the transmission line is long and equal and opposite charges per unit length are uniformly distributed on the outside surface of the inner conductor and on the inside surface of the outer conductor. Let the charge per unit length be ρ_ℓ.

With the above assumptions, the field intensity within the inner conductor $(r < a)$ is zero as there is no charge in this region and the field outside the outer conductor $(b > r)$ is also zero because the total charge enclosed by a Gaussian surface is zero. The field in the region between the conductors $(b > r > a)$ can be found by the use of Gauss's law.

Because of cylindrical symmetry, the field intensity at any point at a radial distance r between the conductors is in the radial direction and it is a function of r only. Thus, the Gaussian surface is that of a cylinder of radius r and of any length ℓ.

Flux out of the Gaussian surface is given by

$$\Psi = 2\pi \varepsilon r \; E_r, \quad b > \;\; > a$$

Charge enclosed by the Gaussian surface,

$$Q = \ell \rho_\ell$$

By Gauss's law,

$$\Psi = Q$$

Hence $E_r = \dfrac{\rho_\ell}{2\pi\varepsilon r}, \quad b > r > a$

Finally, we will calculate the field intensity due to a cylindrical beam of electrons of radius a in vacuum, as illustrated in Fig. 2.22(d). Since the charges are in motion, it may seem that this is not an example of static electric field. However, if we assume that the electrons travel at constant velocity v (Greek alphabet Nu) and are distributed uniformly throughout the beam, the number of electrons in any length of the beam and at a given instant of time remains constant. That is, the electrons are in a quasi-static state. The field produced by such a beam is stationary. If we further assume that the field is radial and it does not vary with ϕ and z, the field intensity inside as well as outside the beam can be calculated applying Gauss's law. We will calculate the fields in terms of volume charge density and charge per unit length of the beam, and also in terms of current carried by it.

If the volume charge density is ρ C/m^3, then charge per unit length of the beam,

$$\rho_\ell = (\pi a^2)\rho \text{ C/m} \tag{6}$$

The magnitude of unidirectional (direct) current I_{dc} carried by the beam is given by

$$I_{dc} = \rho(\pi a^2) v \text{ A} \tag{7}$$

In order to find the field at any point inside the beam, we equate the flux out of a cylindrical Gaussian surface containing the point with the charge enclosed by the surface. If D_r is flux density at a point on the Gaussian surface of radius r ($r < a$) and ℓ is its length, flux out of the surface

$$\Psi = 2\pi r \ell D_r = 2\pi r \ell \varepsilon_0 E_r \tag{8}$$

Charge enclosed by the surface, $Q = \rho(\pi r^2 \ell)$ (9)

Equating (8) with (9), we get

$$E_r = \frac{\rho r}{2\varepsilon_0}, \quad r < a \tag{10}$$

By eliminating ρ from (6) and (10), we will obtain the field intensity in terms of charge per unit length. And by eliminating ρ from (7) and (10), we will get the field in terms of current carried by the beam. Thus,

$$E_r = \frac{\rho_\ell r}{2\pi\varepsilon_0 a^2} = \frac{I_{dc} r}{2\pi\varepsilon_0 a^2 v}, \quad r < a$$

Using the same procedure, it can be shown that the field intensity outside the beam,

$$E_r = \frac{\rho a^2}{2\varepsilon_0 r} = \frac{\rho_\ell}{2\pi\varepsilon_0 r} = \frac{I_{dc}}{2\pi\varepsilon_0 r v}, \quad r > a$$

Distribution of Charge on a Plane Surface

Consider that charge is uniformly distributed on a large plane surface with charge density of ρ_S C/m^2. We will calculate the field intensity at a normal distance h above or below the sheet of charge using Gauss's law.

Symmetrical considerations show that if a positive test charge is placed above or below the sheet, force and hence field intensity will be normal to the charged surface. Further, the field is also constant on a plane parallel to the sheet of charge. These properties of the field suggests that the Gaussian surface in this case is that of a rectangular box with its top and bottom surfaces parallel to the sheet of charge, as illustrated in Fig. 2.23(*a*). The Gaussian surface consists of six surfaces – four side surfaces, a top surface and a bottom surface. As the field intensity is parallel to the side surfaces, flux through these surfaces is zero. Let us take the flux density, which is normal to top and bottom surfaces, as D_n. If the area of the top as well as the bottom surface is S, flux out of the two surfaces,

$$\Psi = 2SD_n = 2S\varepsilon E_n$$

Charge enclosed by the closed surface,

$$Q = S\rho_S$$

Since by Gauss's law $\Psi = Q$, the field intensity,

$$E_n = \frac{\rho_S}{2\varepsilon}$$

Gauss's law can also be used to find the field intensity between two sheets of charges. Let us consider two large parallel sheets of charge separated by a dielectric medium with surface charge densities of $+\rho_S$ and $-\rho_S$, as shown in Fig. 2.23(*b*). The field intensity due to the two charge distributions cancel out outside the sheets and add up between the sheets. We may assume that the magnitude of the field between the sheets is constant and its direction is normal to the sheets. Accordingly, if we choose a Gaussian surface as shown in the figure, flux densities at all the surfaces except at the top surface may be assumed to be zero. By taking the flux density on the top surface as D_n, and equating the flux out of this surface to the charge enclosed by the Gaussian surface, we will get

$$D_n = \rho_S$$

Or

$$E_n = \frac{\rho_S}{\varepsilon}$$

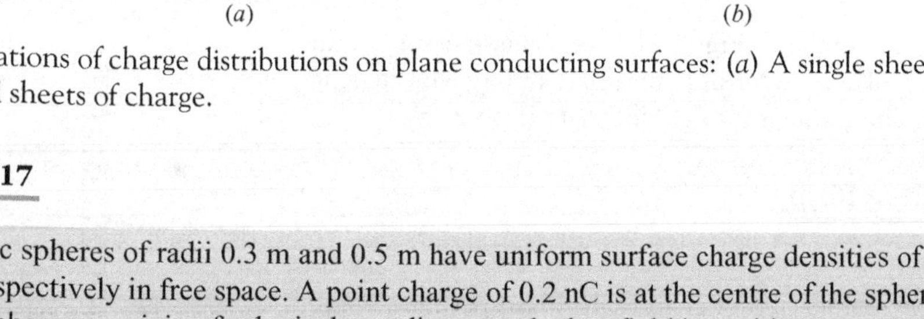

Fig. 2.23 Illustrations of charge distributions on plane conducting surfaces: (*a*) A single sheet of charge and (*b*) Two parallel sheets of charge.

EXAMPLE 2.17

Two concentric spheres of radii 0.3 m and 0.5 m have uniform surface charge densities of 1.0 nC/m^2 and – 0.3 nC/m^2 respectively in free space. A point charge of 0.2 nC is at the centre of the spheres. Taking the centre of the spheres as origin of spherical coordinates, calculate field intensities at $r = 0.25$ m, 0.4 m and 1.0 m.

The example is illustrated in Fig. 2.24. The Gaussian surface in any region is that of a sphere of appropriate radius centered at the origin. The formula for field intensity is given by

$$E_r = \frac{Q}{4\pi\varepsilon_0 r^2} = \frac{(9\times10^9)Q}{r^2}$$

where Q is the net charge within a sphere.

A Gaussian surface at $r = 0.25$ m encloses only the point charge. Field intensity at a point on this surface,

$$E_r = \frac{(\times 0^9)(0.2\times10^{-9})}{0.25^2} = 28.8 \text{ V/m}$$

A Gaussian surface at $r = 0.4$ m encloses the point charge and the charge on the inner sphere.

Charge contained on the surface of the inner sphere

$= 4\pi \times 0.3^2 \times 1.0 \times 10^{-9} = 1.131 \times 10^{-9}$C

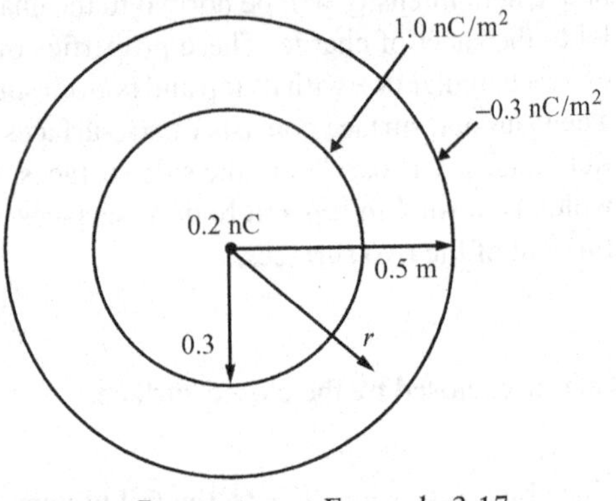

Fig. 2.24 Pertaining to Example 2.17

Total charge enclosed $= (0.2 + 1.131) \times 10^{-9} = 1.331 \times 10^{-9}$ C. Field intensity at a point on this surface,

$$E_r = \frac{(\times 0^9)(1.331\times10^{-9})}{0.4^2} = 74.87 \text{ V/m}$$

A Gaussian surface at $r = 1.0$ m encloses all the charges.

Charge contained on the surface of the outer sphere
$= 4\pi \times 0.5^2 \times (-0.3 \times 10^{-9}) = -0.942 \times 10^{-9}$C

Net charge enclosed $= (0.2 + 1.131 - 0.942) \times 10^{-9} = 0.389 \times 10^{-9}$ C. Field intensity at a point on this surface,

$$E_r = \frac{(\times 0^9)(0.389\times10^{-9})}{1.0^2} = 3.50 \text{ V/m}$$

EXAMPLE 2.18

Charges are uniformly distributed in the region between two hollow concentric cylinders with inner and outer radii of 0.10 m and 0.12 m respectively in free space. The volume charge density in the region is 0.4 μC/m³. There is also a line charge with linear charge density of –0.5 nC/m along the common axis of the cylinders. Calculate the field intensity at $r = 5.0$ cm, 11.0 cm and 15.0 cm.

The example is illustrated in Fig. 2.25. The Gaussian surface in any region is that of a cylinder of appropriate radius with its axis along the z axis. The formula for field intensity is given by

$$E_r = \frac{\rho_\ell}{2\pi\varepsilon_0 r^2} = \frac{(18\times10^9)\rho_\ell}{r^2}$$

where ρ_ℓ is the algebraic sum of charges enclosed per unit length of a Gaussian surface.

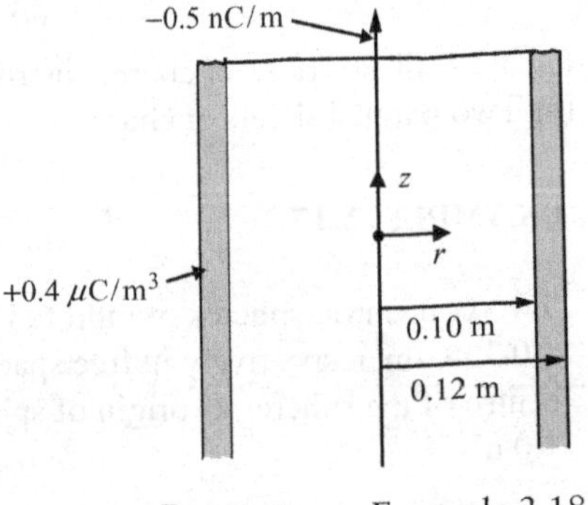

Fig. 2.25 Pertaining to Example 2.18

A cylindrical Gaussian surface at $r = 5.0$ cm encloses only the line charge.

$$E_r = \frac{(18\times10^9)(-0.5\times10^{-9})}{0.05} = -180.0 \text{ V/m}.$$

A cylindrical Gaussian surface at r = 11.0 cm encloses the line charge as well as the charge contained between r = 10.0 cm and r = 11.0 cm.

Charge per unit axial length in the region between r = 10.0 cm and r = 11.0 cm

$= \pi(0.11^2 - 0.10^2)(0.4 \times 10^{-6}) = 2.639 \times 10^{-9}$ C/m

Field intensity at r = 11.0 cm,

$$E_r = \frac{(18\times10^9)[(2.639-0.5)\times10^{-9}]}{0.11} = 350.0 \text{ V/m}.$$

A cylindrical Gaussian surface at r = 15.0 cm encloses all the charges.

Charge per unit axial length of volume charge distribution between the given cylindrical surfaces

$=[\pi(0.12^2 - 0.10^2)](0.4 \times 10^{-6}) = 5.529 \times 10^{-9}$ C/m.

Field intensity at r = 15.0 cm,

$$E_r = \frac{(18\times10^9)[(5.529-0.5)\times10^{-9}]}{0.15} = 603.5 \text{ V/m}$$

EXAMPLE 2.19

Two large sheets of charge with uniform charge distribution are located at $x = 0$ and $x = a$ in free space. The surface charge densities in sheets 1 and 2 are $\rho_{S1} = 6.0$ μC/m and $\rho_{S2} = 3.0$ μC/m respectively. Find the electric flux density in all the regions.

In any region, magnitude of flux density due to sheet 1, $D_1 = \rho_{S1}/2 = 3.0$ μC/m

In any region, magnitude of flux density due to sheet, $D_2 = \rho_{S2}/2 = 1.5$ μC/m

(a) Between the sheets D_1 is in positive x direction and D_2 is in negative x direction. Thus, in this region, $\mathbf{D} = (D_1 - D_2)\mathbf{u}_x = 1.5$ μC
(b) For $x > 0$, the flux density due to both the charge distributions are in positive x direction. Thus, in this region, $\mathbf{D} = (D_1 + D_2)\mathbf{u}_x = 4.5$ μC
(c) For $x < 0$, the flux density due to both the charge distributions are in negative x direction. Thus, in this region, $\mathbf{D} = -(D_1 + D_2)\mathbf{u}_x = -4.5$ μC

EXAMPLE 2.20

Charge is uniformly distributed on the plane surface $S = 3x + 5y + 6z = 30$ m located in air. The surface charge density is 1.5 nC/m². Find the field intensity.

The magnitude of the field intensity on both sides of the surface is equal to $\rho_S/2\varepsilon_0$. The field is along the normal direction to the surface. Thus,

$$\mathbf{E} = (\rho_S/2\varepsilon_0)\mathbf{u}_n$$

$$\frac{\rho_S}{2\varepsilon_0} = \frac{15\times10^{-9}}{2(8.85\times10^{-12})} = 84.75 \text{ V/m}$$

$$\mathbf{u}_n = \frac{\nabla S}{|\nabla S|} = \pm\frac{3\mathbf{u}_x + 5\mathbf{u}_y + 6\mathbf{u}_z}{8.367} = \pm(0.359\mathbf{u}_x + 0.598\mathbf{u}_y + 0.717\mathbf{u}_z)$$

$$\mathbf{E} = \pm(84.75)(0.359\mathbf{u}_x + 0.598\mathbf{u}_y + 0.717\mathbf{u}_z)$$

$$= \pm(30.42\mathbf{u}_x + 50.68\mathbf{u}_y + 60.77\mathbf{u}_z \text{ V/m}$$

The plane intersects the x axis at 10 m, the y axis at 6 m and the z axis at 5 m. Therefore, the plus sign before the bracket is for the field on the side of the plane away from the origin, and the minus sign is on the side of the plane containing the origin.

2.13 Divergence of Electric Flux Density: Point Form of Gauss's Law

In the last section we learnt that the field intensity of symmetrical charge distributions can be found by the use of Gauss's law. In the case of symmetrical charge distribution, it is possible to conceive a suitable Gaussian surface containing the point at which the field is required. If the field is not uniform, it is not possible to find the field intensity using Gauss's law, as it is not possible to define a closed surface fulfilling the requirements of a Gaussian surface. However if we choose a region of very small volume, we can calculate the electric flux out of a closed surface enclosing this volume in terms of the space rate of change of flux density at the geometric centre of this volume. Moreover, the total charge contained in the region is nearly equal to the average volume charge density multiplied by the volume of the region.

By equating the charge enclosed to the flux out and allowing the volume to approach to zero, we will obtain an important partial differential equation, known as *point form* of Gauss's law.

We will derive the relation in rectangular coordinate first and then extend it to cylindrical and spherical coordinates.

In an electric field, consider an infinitesimal rectangular box with sides Δx, Δy, Δz and centred at $P(x, y, z)$ in rectangular coordinates, as shown in Fig. 2.26. The volume of the box is

$$\Delta v = \Delta x\,\Delta y\Delta z$$

Let the electric flux density at P in terms of its components along the coordinate axes be given by

$$\mathbf{D} = D_x\mathbf{u}_x + D_y\mathbf{u}_y + D_z\mathbf{u}_z$$

where D_x, D_y and D_z are the component of $\mathbf{D}$. In general, each is a function of all the three coordinate variables x, y and z.

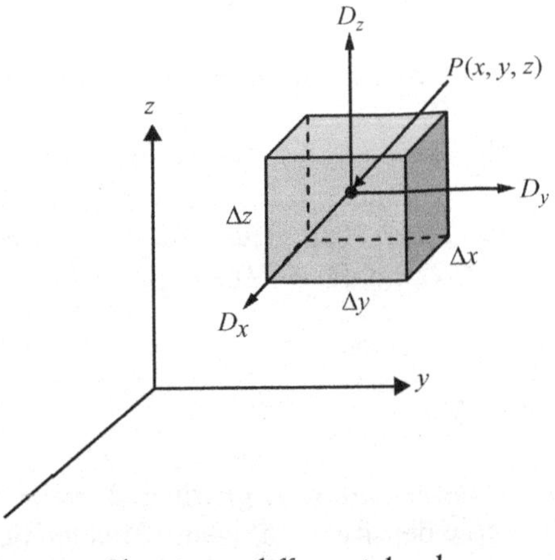

Fig. 2.26 Showing a differential volume element and components of flux density in rectangular coordinates

With the directions of flux densities as indicated in the figure, it is obvious that flux enters through the back, left and bottom surfaces and emerges from the front, right and top surfaces of the box. If we assume that flux leaving a surface is positive and entering into a surface is negative, flux *out* of two opposite surfaces is equal to the difference between the two.

Let us calculate at first flux *out* of the region in the x direction. The flux *out* is equal to the flux emerging from the front surface minus the flux entering through the back surface. The two flux components are obtained in terms of D_x as follows. As the sectional area of the box at P is $\Delta y \Delta z$, differential flux over this section is equal to the product of D_x and the sectional area. Denoting the differential flux by $\Delta\Psi'_x$, we have

$$\Delta\Psi'_x = D_x \Delta y \Delta z,$$

where D_x is the average flux density over the area. If we assume that the rate of change of flux is constant and positive, *change* of flux from the surface at P to the front surface is equal to rate of change of flux with x multiplied by$\Delta x / 2$. It is given by

$$\frac{\partial(D_x \Delta y \Delta z)}{\partial x} \frac{\Delta x}{2}$$

Thus, flux out of the front surface,

$$\Delta\Psi''_x = \Delta\Psi'_x + \frac{\partial(D_x \Delta y \Delta z)}{\partial x} \frac{\Delta x}{2}$$

And flux entering through the back surface,

$$\Delta\Psi'''_x = \Delta\Psi'_x - \frac{\partial(D_x \Delta y \Delta z)}{\partial x} \frac{\Delta x}{2}$$

Therefore, net flux *out* of the region in x direction,

$$\Delta\Psi_x = \Delta\Psi''_x - \Delta\Psi'''_x = \frac{\partial(D_x \Delta y \Delta z)}{\partial x} \Delta x$$

As Δy and Δz are independent of x these can be brought out of the differential sign. Thus,

$$\Delta\Psi_x = \frac{\partial D_x}{\partial x} \Delta x \Delta y \Delta z = \frac{\partial D_x}{\partial x} \Delta v$$

Proceeding in the same way as above, we can obtain flux *out* of the region along y and z directions. These are given by

$$\Delta\Psi_y = \frac{\partial D_y}{\partial y} \Delta v$$

$$\Delta\Psi_z = \frac{\partial D_z}{\partial z} \Delta v$$

Thus, total flux out of the region,

$$\Delta\Psi = \Delta\Psi_x + \Delta\Psi_y + \Delta\Psi_z = \left(\frac{\partial D_x}{\partial x} + \frac{\partial D_y}{\partial y} + \frac{\partial D_z}{\partial z} \right) \Delta v \tag{1}$$

Dividing both sides of (1) by Δv and then taking the limit, we get

$$\lim_{\Delta v \to 0} \frac{\Delta\Psi}{\Delta v} = \frac{\partial D_x}{\partial x} + \frac{\partial D_y}{\partial y} + \frac{\partial D_z}{\partial z} \tag{2}$$

Referring to relations (6) and (7) of Section 1.8, we find that the right side of (2) is divergence of the flux density vector **D**. Thus in rectangular coordinates,

$$\text{div } \mathbf{D} = \nabla \cdot \mathbf{D} = \frac{\partial D_x}{\partial x} + \frac{\partial D_y}{\partial y} + \frac{\partial D_z}{\partial z} \tag{3}$$

Divergence of **D** in cylindrical and spherical coordinates may be obtained by following the same procedure as for rectangular coordinates. In cylindrical coordinates, the side lengths of a volume element are Δr, $r\Delta\phi$ and Δz and hence the volume of the element,

$$\Delta v = r\Delta r\Delta\phi\Delta z$$

Differential flux out of the region in the three coordinate directions are

$$\Delta\Psi_r = \frac{\partial(D_r r\Delta\phi\Delta z)}{\partial r}\Delta r$$

$$\Delta\Psi_\phi = \frac{\partial(D_\phi \Delta r\Delta z)}{r\partial\phi}(r\Delta\phi)$$

$$\Delta\Psi_z = \frac{\partial(D_z r\Delta r\Delta\phi)}{\partial z}\Delta z$$

Total flux out of the region,

$$\Delta\Psi = \frac{\partial(D_r r\Delta\phi\Delta z)}{\partial r}\Delta r + \frac{\partial(D_\phi \Delta r\Delta z)}{r\partial\phi}(r\Delta\phi) + \frac{\partial(D_z r\Delta r\Delta\phi)}{\partial z}\Delta z$$

$$= \left[\frac{1}{r}\frac{\partial(rD_r)}{\partial r} + \frac{1}{r}\frac{\partial D_\phi}{\partial\phi} + \frac{\partial D_z}{\partial z}\right]\Delta v$$

By dividing both sides by Δv and taking the limit we will get the divergence of the flux density vector.

Thus, in cylindrical coordinates,

$$\nabla \cdot \mathbf{D} = \frac{1}{r}\frac{\partial(rD_r)}{\partial r} + \frac{1}{r}\frac{\partial D_\phi}{\partial\phi} + \frac{\partial D_z}{\partial z} \tag{4}$$

In spherical coordinates, the side lengths of a volume element are Δr, $r\Delta\theta$ and $r \sin\theta\Delta\phi$. The volume of the element,

$$\Delta v = r^2 \sin\theta\Delta r\Delta\theta\Delta\phi$$

Differential flux out of the region in the three coordinate directions are

$$\Delta\Psi_r = \frac{\partial(D_r r^2 \sin\theta\Delta\theta\Delta\phi)}{\partial r}\Delta r$$

$$\Delta\Psi_\theta = \frac{\partial(D_\theta r \sin\theta\Delta r\Delta\phi)}{r\partial\theta}(r\Delta\theta)$$

$$\Delta\Psi_\phi = \frac{\partial(D_\phi r\Delta r\Delta\theta)}{(r\sin\theta)\partial\phi}(r\sin\theta\Delta\phi)$$

Total flux out of the region,

$$\Delta\Psi = \frac{\partial(D_r r^2 \sin\theta\,\Delta\theta\,\Delta\phi)}{\partial r}\Delta r + \frac{\partial(D_\theta r \sin\theta\,\Delta r\,\Delta\phi)}{r\,\partial\theta}(\;\Delta\theta) + \frac{\partial(D_\phi\, r\Delta r\Delta\theta)}{r\sin\theta\;\;\phi}(\;\sin\theta\Delta\phi)$$

$$= \left[\frac{1}{r^2}\frac{\partial(r^2 D_r)}{\partial r} + \frac{1}{r\sin\theta}\frac{\partial(\sin\theta D_\theta)}{\partial\theta} + \frac{1}{r\sin\theta}\frac{\partial(D_\phi)}{\partial\phi}\right]\Delta v$$

Thus, in spherical coordinate system divergence of the flux density vector,

$$\nabla \cdot \mathbf{D} = \frac{1}{r^2}\frac{\partial (r^2 D_r)}{\partial r} + \frac{1}{r\sin\theta}\frac{\partial(\sin\theta\, D_\theta)}{\partial\theta} + \frac{1}{r\sin\theta}\frac{\partial D_\phi}{\partial\phi} \tag{5}$$

It may be noted that the differential vector operator ∇ has a clear definition only in rectangular coordinates, as given in Section 1.8. The operator ∇ has no clear-cut definition in other coordinate systems. It is treated as a symbol only.

The physical interpretation of divergence of electric flux density is as follows. It is equal to the net outward flux per unit volume of a closed surface S enclosing an infinitesimal volume Δv in the limit Δv tending to zero.

As flux out of a closed surface is equal to the surface integral of $\mathbf{D} \cdot d\mathbf{S}$, from (2) and (3) we have

$$\nabla \cdot \mathbf{D} = \lim_{\Delta v \to 0} \frac{\oint_S \mathbf{D}\cdot d\mathbf{S}}{\Delta v} \tag{6}$$

The numerator of rhs of (6) is equal to the charge contained in the infinitesimal volume, as per Gauss's law. If charge is continuously distributed with an average volume charge density ρ, charge in the region is $\rho\Delta v$. Thus,

$$\nabla \cdot \mathbf{D} = \lim_{\Delta v \to 0} \frac{\rho\,\Delta v}{\Delta v} = \rho \tag{7}$$

Thus, the divergence of electric flux density is equal to the volume charge density. Equation (7) is a first order partial differential equation relating $\mathbf{D}$ with ρ at a point. It is known as *differential form* or *point form* of Gauss's law. In a charge-free region $\rho = 0$, hence

$$\nabla \cdot \mathbf{D} = 0$$

EXAMPLE 2.21

Charge is uniformly distributed over the volume of a sphere of radius a. The centre of the sphere is at the origin and the volume charge density is ρ. The electric flux density at a point within the sphere and in spherical coordinates is given by

$$\mathbf{D} \quad \frac{\rho r}{3}\mathbf{u}_r, \; r < a$$

Show that $\nabla \cdot \mathbf{D} = \rho$.

The flux density has only one component in r direction. Thus, in spherical coordinates,

$$\nabla \cdot \mathbf{D} = \frac{1}{r^2}\frac{\partial}{\partial r}(r^2 D_r) = \frac{1}{r^2}\frac{\partial}{\partial r}(\frac{\rho r^3}{3}) = \rho$$

EXAMPLE 2.22

A spherical charge of 0.3 m radius with uniform density of 2.1 μC/m^3 is located in free space. Using the point form of Gauss's law, find field intensity in the region of the charge. Also, calculate the maximum value of field intensity.

Let us take the centre of the charge distribution at the origin of spherical coordinates. Accordingly, the field is a function of r only. Therefore,

$$\nabla \cdot \mathbf{D} - \frac{1}{r^2}\frac{\partial (r^2 D_r)}{\partial r} = \rho \text{ C/m}^3$$

Or $\quad \frac{1}{r^2}\frac{\partial(r^2 D_r)}{\partial r} = 2.1\times10^{-6}$

Multiplying both sides by r^2 and then integrating we get

$$r^2 D_r = 2.1\times10^{-6}\int r^2 dr = 0.7\times10^{-6} r^3$$

$\Rightarrow D_r = (0.7 \times 10^{-6})r$

$E_r = D_r / \varepsilon_0 = (79.17 \times 10^3)r$ V/m, $\quad 0.3 \geq r \geq 0$

The maximum value of field intensity occurs at the surface of the charge distribution. It is equal to 23.75×10^3 V/m.

2.14 The Divergence Theorem

The divergence theorem is a specific relation between a surface integral and a volume integral involving a vector function. We will establish the theorem for the electric flux density vector using the integral and differential forms of Gauss's law.

The integral form of Gauss's law is

$$\oint_S \mathbf{D}\cdot d\mathbf{S} = \int_v \rho dv \qquad (1)$$

And the differential form of Gauss's law is

$$\nabla\cdot\mathbf{D} = \rho \qquad (2)$$

By eliminating ρ from (1) and (2) we get

$$\oint_S \mathbf{D}\cdot d\mathbf{S} = \int_v (\nabla\cdot\mathbf{D})dv \qquad (3)$$

The differential vector $d\mathbf{S} = \mathbf{u}_n dS$, where $\mathbf{u}_n$ is a unit vector in the *outward* normal direction to the differential area dS. Equation (3) is known as *divergence theorem or Gauss's theorem*. This theorem is useful if the surface integral of a vector is to be replaced by a volume integral and vice versa. The theorem may be stated as follows. The integral of the normal component of **D** over a closed surface is equal to the integral of divergence of **D** over the volume enclosed by the closed surface. The theorem is applicable to any vector field.

EXAMPLE 2.23

A uniformly charged solid sphere of radius a has its centre at the origin of the spherical coordinate system. The electric flux density at a radius $r < a$ is given by

$\mathbf{D} = \frac{Qr}{4\pi a^3}\mathbf{u}_r$ where Q is total charge in the sphere.

Prove the validity of the divergence theorem for the flux density.

The divergence theorem is given by

$$\oint_S \mathbf{D}\cdot d\mathbf{S} = \int_v \nabla\cdot\mathbf{D}dv$$

Flux density on surface of the sphere,

$$\mathbf{D} = \frac{Q}{4\pi a^2}\mathbf{u}_r = D_r\mathbf{u}_r$$

Differential area on surface of the sphere, $d\mathbf{S} = a^2 \sin\theta d\theta d\phi$

$$\oint_S \mathbf{D}\cdot d\mathbf{S} = \frac{Q}{4\pi}\int_{\theta=0}^{\pi}\int_{\phi=0}^{2\pi} \sin\theta d\theta d\phi = Q$$

In spherical coordinates, $\nabla\cdot\mathbf{D} = \frac{1}{r^2}\frac{\partial}{\partial r}(r^2 D_r) = \frac{1}{r^2}\frac{\partial}{\partial r}\left(\frac{Qr^3}{4\pi a^3}\right) = \frac{3Q}{4\pi a^3}$

Differential volume at radius r, $dv = r^2 \sin\theta dr d\theta d\phi$

$$\int \nabla\cdot\mathbf{D}dv = \frac{3Q}{4\pi a^3}\int_{r=0}^{a}\int_{\theta=0}^{\pi}\int_{\phi=0}^{2\pi} r^2 \sin\theta dr d\theta d\phi = Q$$

Thus, the two sides of divergence theorem are equal.

EXAMPLE 2.24

Evaluate both sides of divergence theorem in the cylindrical region defined by $r = 3.0$ m and $1.0\text{ m} \le z \le 6.0$ m if electric flux density in the region,

$$\mathbf{D} = ae^{-r}\mathbf{u}_r + \frac{b}{z}\mathbf{u}_z$$

where a and b are constants.

The region is bounded by three surfaces S_1, S_2 and S_3, as indicated in Fig. 2.27. The differential area, $d\mathbf{S} = dS_1\mathbf{u}_r + dS_2\mathbf{u}_z - dS_3\mathbf{u}_z$

The components of flux density in the region are

$D_r = ae^{-r}$, $D_\phi = 0$, $D_z = b/z$

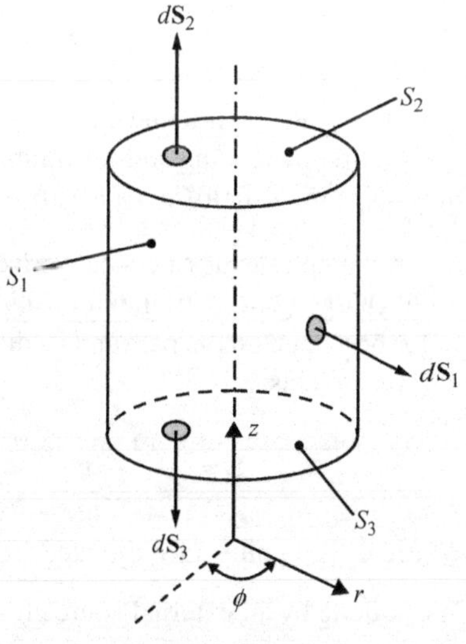

Fig. 2.27 Pertaining to Example 2.24

The left side of the divergence theorem given by (3) is

$$\oint \mathbf{D}\cdot d\mathbf{S} = \int_{S1} D_r dS_1\Big|_{r=3} + \int_{S2} D_z dS_2\Big|_{z=6} - \int_{S3} D_z dS_3\Big|_{z=1}$$

$$= \int_1^6\int_0^{2\pi} 3ae^{-3}\, d\phi\, dz - \int_0^{2\pi}\int_0^3 \frac{5b}{6} r\, dr\, d\phi$$

$$= 30\,\pi a e^{-3} - 7.5\,\pi b \qquad (4)$$

$dv = r dr d\phi dz$

$$\nabla\cdot\mathbf{D} = \frac{1}{r}\frac{\partial(rD_r)}{\partial r} + \frac{\partial D_z}{\partial z} = \frac{1}{r}\frac{\partial(are^{-r})}{\partial r} + \frac{\partial(b/z)}{\partial z} = \frac{ae^{-r}}{r} - ae^{-r} - \frac{b}{z^2}$$

The right side of (3) is

$$\int_v (\quad) dv = \int_1^6\int_0^{2\pi}\int_0^3\left(\frac{ae^{-r}}{r} - ae^{-r} - \frac{b}{z^2}\right) r\, dr\, d\phi\, dz$$

$$= 30\,\pi a e^{-3} - 7.5\,\pi b = \text{The left-hand side of (4)}$$

2.15 Work and Energy

If a charged body is placed in an electric field, a force is exerted on it. Work is done when the charge is displaced from one position to another position by an external source. If the displacement of the charge is opposite to the electric force, work is done by the external source. The energy so expended is converted into field energy and stored in the system. In a static field the only way energy can be stored is in the form of potential energy. The increase in energy stored is equal to the work done provided there is no loss of energy due to friction and other causes. This is analogous to the displacement of a mass in a direction opposite to the gravitational force. If the displacement of the charge is in the same direction as the electric force, work is done by the field. Consequently, field-energy decreases. Further, if the charged body is moved at right angles to the electric force no work is done and hence the field-energy does not change.

Consider that a positive charge Q is moved from an initial position P_1 to a final position P_2 along an arbitrary path in an electric field, as shown in Fig. 2.28. At any point P on this path, let the field intensity be $\mathbf{E}$. Then, electric force on the charge at this point is

$$\mathbf{F} = Q\mathbf{E}$$

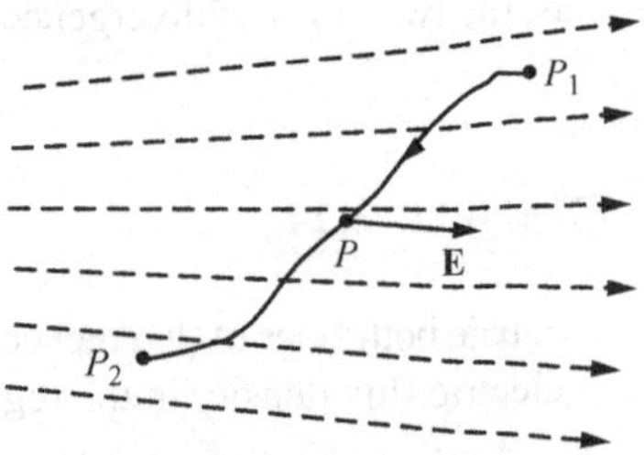

Fig. 2.28 Showing an arbitrary path P_1P_2 along which a charged body is moved in an electric field

The charge will move in the direction of the electric force if it is free. In order to keep the charge stationary, an equal and opposite force has to be applied by an external source. Thus, the applied force,

$$\mathbf{F}' = -Q\mathbf{E}$$

Work done in moving the charge by a differential vector length $d\boldsymbol{\ell}$ that is tangential to the path at P, and assuming $\mathbf{F}'$ is constant over this length, is

$$dW = -Q(\mathbf{E}\cdot d\boldsymbol{\ell})$$

Work done by the external source in moving the charge along the path from P_1 to P_2,

$$W = -\int_{P_1}^{P_2} Q(\mathbf{E}\cdot d\boldsymbol{\ell}) \tag{1}$$

The above integral is a line integral and to evaluate it both $\mathbf{E}$ and $d\boldsymbol{\ell}$ have to be expressed in terms of space variables. After integrating and substituting for the limits, a positive result will mean that energy is transferred from the source to the field, and a negative result will mean that energy is supplied by the field to move the charge.

As an example, let us calculate work done in displacing a positive point charge Q' in the field of another positive point charge Q from an initial position P_1 to a final position P_2 along an arbitrary path, as shown in Fig. 2.29. Let P be a point on the path and its distance from Q is R. Then, field intensity at P due to Q is

$$\mathbf{E} = \frac{Q}{4\pi\varepsilon R^2}\mathbf{u}_R$$

where $\mathbf{u}_R$ is a unit vector directed from Q towards P.

Work done by an external source to move Q' from P_1 to P_2 is

$$W \quad Q'\int \cdot d = -\int_{P_1}^{P_2} \frac{QQ'}{4\pi\varepsilon R^2}(\mathbf{u}_R\cdot d\boldsymbol{\ell})$$

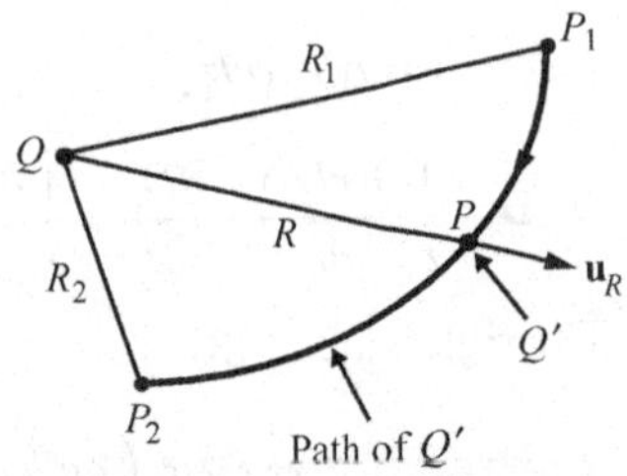

Fig. 2.29 For calculation of work done in displacing a point charge in the field of another point charge

As $\mathbf{u}_R \cdot d\boldsymbol{\ell}$ is equal to the differential length dR along R, the limits of integration would be R_1 and R_2, as shown in the figure.

$$W = -\int_{R_1}^{R_2} \frac{QQ'}{4\pi\varepsilon R^2} dR = \frac{QQ'}{4\pi\varepsilon}\left(\frac{1}{R_2} - \frac{1}{R_1}\right) \quad (2)$$

Note that R_1 and R_2 are the initial and final distances between Q' and Q.

Work done in moving a point charge in the field of another point charge has the following properties:

1. Work done depends only on the initial and final positions of Q' in the field of Q. It does not depend on the shape of path taken by the charge.
2. Work done is inversely proportional to permittivity of the medium.
3. If a charge is moved over a closed path then $R_2 = R_1$, hence net work done is zero. This is because energy expended in moving a charge around a closed path becomes equal to the energy supplied by the field.

As the energy is conserved, the static electric field is a *conservative field*. Mathematically, it means that the integral appearing in (1) becomes a line integral of **E** around a closed path and its value is equal to zero. That is,

$$\oint \mathbf{E} \cdot d\ell = 0 \text{ (Conservative property)} \quad (3)$$

The above properties are satisfied for any distribution of static charges in an ideal dielectric medium.

EXAMPLE 2.25

A point charge of –2.8 μC is located at P(1.3 m, – 0.9 m, 1.75 m). Another point charge of 0.75 μC is moved from P_1(–2.0 m, 1.2 m, 0.8 m) to P_2(1.1 m, –0.75 m, –1.5 m) in the field of the first charge. Find the work done if relative permittivity of the medium is 2.7

Let $Q_1 = -2.8\ \mu C$ and $Q_2 = 0.75\ \mu C$

Initial distance between the charges, $R_1 = \sqrt{[(-2.0-1.3)^2 + (1.2+0.9)^2 + (0.8-1.75)^2}$

$= 4.03$ m

Final distance between the charges, $R_2 = \sqrt{[(1.1-1.3)^2 + (-0.75+0.9)^2 + (-1.5-1.75)^2}$

$= 3.27$ m

$$W = \frac{Q_1Q_2}{4\pi\varepsilon}\left(\frac{1}{R_2} - \frac{1}{R_1}\right) = \frac{(-2.8\times10^{-6})(0.75\times10^{-6})}{4\pi(8.854\times10^{-12}\times2.7)}\left(\frac{1}{3.26} - \frac{1}{4.03}\right) = -4.1\times10^{-4}\text{ J}$$

The negative sign shows that work is done by the field.

EXAMPLE 2.26

A long line charge located in air has its axis along z direction of cylindrical coordinates. The linear charge density of the line is 0.5 μC/m. Find the work done in moving a point charge of 0.8 μC from $r = 5$ m to $r = 2$ m.

Field intensity of the line charge, $\mathbf{E} = \dfrac{\rho_\ell \mathbf{u}_r}{2\pi\varepsilon_0 r}$

Differential length, $d\ell = dr\mathbf{u}_r + rd\phi\mathbf{u}_\phi + z\mathbf{u}_z$

Differential work done, $dW = -\,Q(\mathbf{E}\cdot d\boldsymbol{\ell})$

$$= -\frac{Q\rho_\ell}{2\pi\varepsilon_0 r} = (18\times10^9)(8\times10^{-6})(0.5\times10^{-6})\frac{dr}{r}$$

$$= -(\quad\times\quad^{-3})\frac{dr}{r}$$

Work done, $W = -(72\times10^{-3})\int_5^2 \frac{dr}{r} = -(72\times10^{-3}) = 65.97$ mJ

2.16 Potential Difference and Absolute Potential

The energy expended in displacing a charge in the electric field of other charges is stored in the field as potential energy. This suggests that an electric field can also be described by an energy function. One such function is the energy required to shift a *unit positive* charge from one point to another point in a static electric field. The function is called *potential difference* between the points. The second function, which is a special case of the first function, is the *absolute potential* at a point in an electric field.

The electrostatic potential is a *scalar quantity* and hence the potential field is a scalar field. The unit of the potential function is joules per coulomb. Its practical unit is volt (V). We will denote the electrostatic potential by the symbol V.

Potential Difference

Let us imagine two points P_1 and P_2 in the field of static electric charges. If a positive test charge ΔQ is moved from P_1 to P_2 along any arbitrary path, then work done as per equation (2.15-1) is

$$\Delta W = -\Delta Q\int_{P_1}^{P_2}\mathbf{E}\cdot d\boldsymbol{\ell} \tag{1}$$

By dividing both sides of (1) by ΔQ and then taking the limit we get

$$\lim_{\Delta Q\to 0}\frac{\Delta W}{\Delta Q} = -\int_{P_1}^{P_2}\mathbf{E}\cdot d\boldsymbol{\ell}$$

The lhs, which gives work done per unit positive charge, is called potential difference between points P_2 and P_1. Denoting the potential difference by V_{21}, we have

$$V_{21} = -\int_{P_1}^{P_2}\mathbf{E}\cdot d\boldsymbol{\ell} \tag{2}$$

Thus, the potential difference between two points in a static electric field is equal to the negative of the line integral of **E** between the points. Note that the first subscript of V_{21} corresponds to the upper limit (final position of the test charge) and the second subscript corresponds to the lower limit (initial position of the test charge) of the integral. As both ΔW and ΔQ are scalar quantities, the potential difference is a *scalar function*. Moreover, as work done does not depend on the shape of path of integration, the potential difference is a *single-valued function.*

Absolute Potential

After evaluating the line integral (2) and substituting the limits, the upper limit will yield one value and the lower limit will yield another value. Let us assume that the values for the upper and lower limits are V_2 and V_1 respectively. Accordingly, the potential difference between P_2 and P_1 is

$$V_{21} = V_2 - V_1$$

Now suppose that for some location of P_1, the value of the line integral for the lower limit comes out to be zero. Then,

$$V_{21} = V_2 - 0 = V_2$$

Under this condition, V_2 is termed as the *absolute potential* or *electrostatic potential* or *scalar electric potential* or simply *potential* at P_2. In general, potential at a point P in an electric field is the work done in bringing a unit positive charge from a point (or from a surface) of zero potential to P. The zero-potential point is regarded as a reference point. Earth and earthed conductors are assumed to be at zero potential.

Potential due to Point Charges

We will at first define the potential of a single point charge and then use the result to obtain potential of several point charges. In Fig. 2.30, Q is a point charge and P_1P_2 is an arbitrary path in the field of the charge. The distances of P_1 and P_2 from the charge are R_1 and R_2 respectively. A point P on the path is at a distance R from the charge. The field intensity at P is,

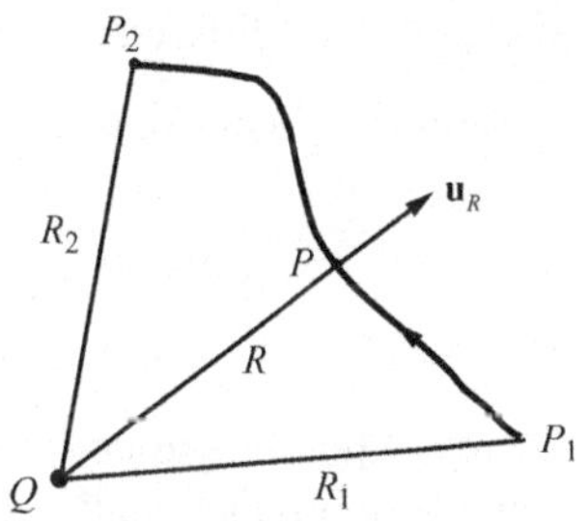

Fig. 2.30 For calculation of potential difference between two points in the field of a point charge

$$\mathbf{E} = \frac{Q}{4\pi\varepsilon R^2}\mathbf{u}_R$$

where $\mathbf{u}_R$ is a unit vector directed from the point charge to P.

Potential difference between points P_2 and P_1,

$$V_{21} = -\int_{R_1}^{R_2} \mathbf{E}\cdot d\boldsymbol{\ell}$$

$$= -\int_{R_1}^{R_2} \frac{Q}{4\pi\varepsilon R^2}(\mathbf{u}_R \cdot d\boldsymbol{\ell})$$

$$= -\int_{R_1}^{R_2} \frac{Q}{4\pi\varepsilon R^2}dR = \frac{Q}{4\pi\varepsilon}\left(\frac{1}{R_2} - \frac{1}{R_1}\right) = V_2 - V_1$$

where $\qquad V_2 = \dfrac{Q}{4\pi\varepsilon R_2}$ and $V_1 = \dfrac{Q}{4\pi\varepsilon R_1}$

Now, imagine that P_1 is located at infinity. Then V_1 becomes zero and V_2 gives the absolute potential at P_2. Thus potential at a point, which is located at a distance R from a point charge Q, is given by

$$V = \int_{\infty}^{R} \frac{Q}{4\pi\varepsilon R^2}dR = \frac{Q}{4\pi\varepsilon R} \tag{3}$$

The electrostatic potential give by (3) is inversely proportional to the distance of the point from the charge. It is also inversely proportional to the permittivity of the medium.

If a linear, homogenous and isotropic dielectric region has n point charges, potential at a given point due to each charge is at first calculated using (3). Then, the potentials due to all the charges are added algebraically to get the total potential at the point. If R_k is distance of P from the kth charge, total potential at the point is

$$V = \sum_{k=1}^{n} \frac{Q_k}{4\pi\varepsilon R_k} \tag{4}$$

Potential Difference in the Field of a Line Charge of Infinite Length

Figure 2.31 illustrates a line charge with uniform charge distribution ρ_ℓ C/m. The two ends P_1 and P_2 of a path are at perpendicular distances R_1 and R_2 from the line charge. A point P on the path is at a perpendicular distance R. We will calculate potential difference between points P_2 and P_1.

Electric field intensity at point P is given by

$$\mathbf{E} = \frac{\rho_\ell}{2\pi\varepsilon R}\mathbf{u}_R \tag{5}$$

where $\mathbf{u}_R$ is unit vector directed from the line towards P as shown.

Potential difference between P_2 and P_1,

$$V_{21} = -\int_{P_1}^{P_2} \mathbf{E}\cdot d\boldsymbol{\ell} \tag{6}$$

where $d\boldsymbol{\ell}$ is the differential vector length tangential to the path at P.

Substituting (5) in (6), and since $\mathbf{u}_r \cdot d\boldsymbol{\ell} = dR$, we get

$$V_{21} = -\int_{R_1}^{R_2} \frac{\rho_\ell\, dR}{2\pi\varepsilon R} = \frac{\rho_\ell}{2\pi\varepsilon}\ln\frac{R_1}{R_2} \tag{7}$$

It is not possible to define potential at a point for a single line charge of infinite length as potential is nowhere zero because of infinite nature of the charge distribution. We can use the above formula without much error if the length of the line charge is much greater than R_2 and R_1.

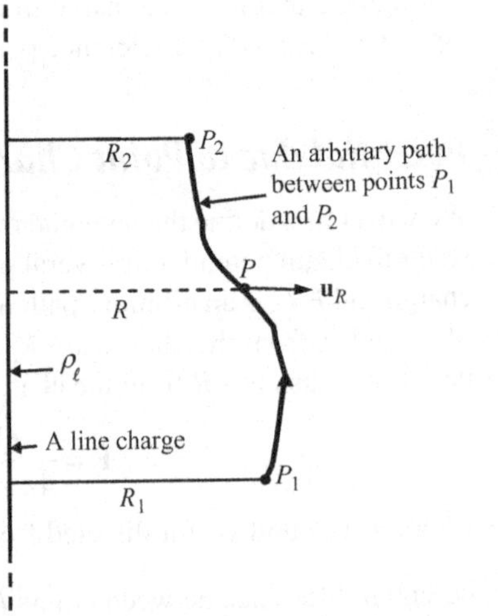

Fig. 2.31 For calculation of potential difference in the field of a line charge

Equipotential Surfaces

The electrostatic potential is a scalar function of space variables and the potential field is a scalar field. If points having the same value of potential are joined in a two dimensional field, we will get a curve of constant potential. Such a curve is called an equipotential line. In three dimensions, the map of points having same value of potential is an equipotential surface.

Equipotential surfaces possess the following properties:

1. These are continuous surfaces. The shape of the surface depends on the distribution of charges. For a point charge, equipotential surfaces are concentric spheres centred at the charge. Figure 2.32 shows three equipotential lines of a point charge on a plane containing the charge.
2. Equipotential surfaces of different potentials do not intersect as the electrostatic potential is a single-valued function.
3. Work done in moving a charge from one point to another point along an equipotential surface is zero. Therefore, the tangential component of electric field intensity on the surface is zero. Accordingly, the direction of electric field intensity can only be normal to an equipotential surface.
4. Work done in moving a charge from any point on an equipotential surface to any point on another equipotential surface is independent of the shape of path taken by the charge.

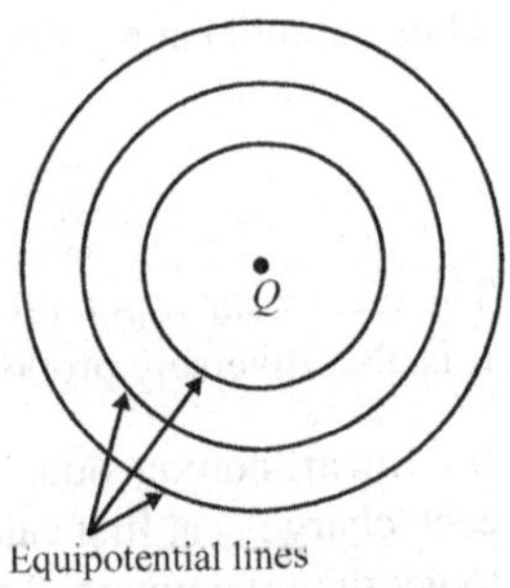

Fig. 2.32 Equipotential lines of a point charge on a plane containing the charge

These properties of equipotential surfaces are useful for drawing field maps.

EXAMPLE 2.27

Electric field intensity in a perfect dielectric medium is described by,

$\mathbf{E} = 4y\mathbf{u}_x + 4x\mathbf{u}_y$.

Find the potential difference between $P_2(-1.0$ m, 4.0 m, $0)$ and $P_1(1.0$ m, 2.0 m, $0)$.

We will evaluate the line integral along two different paths 1 and 2, as shown in Fig. 2.33. We will at first use the straight-line path between P_1 and P_2. The general equation of the line is given by

$$y = mx + c \tag{8}$$

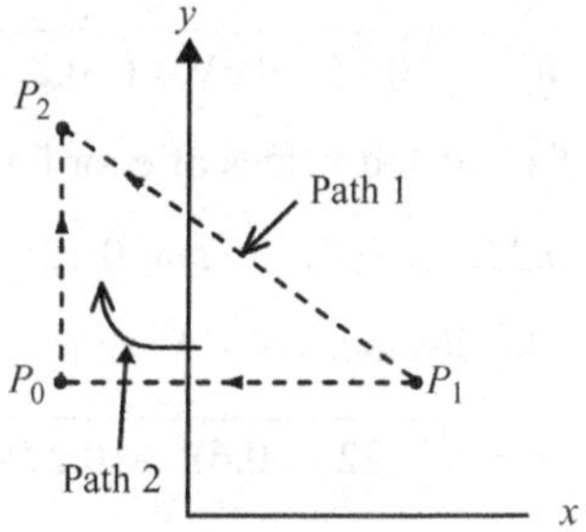

Fig. 2.33 Pertaining to Example 2.27

The slope of the line,

$$m = \frac{2.0 - 4.0}{1.0 - (-1.0)} = -1$$

Thus $y = c - x$

The value of c which can be found by using the values of x_1, y_1 (or x_2, y_2) in the above equation is 3.0. Substituting this value in (8), we have

$$y = 3 - x, \tag{9}$$

$$dy = -\,dx \tag{10}$$

A differential length in rectangular coordinates is given by

$d\ell - dx\mathbf{u}_x \mid dy\mathbf{u}_y$

The dot product of $\mathbf{E}$ and $d\ell$ is

$$\mathbf{E} \quad \ell = (4y\mathbf{u}_x + 4x\mathbf{u}_y)\cdot(dx\mathbf{u}_x + dy\mathbf{u}_y) = 4y\,dx + 4x\,dy$$

Substituting (9) and (10) in the above relation, we get

$$\mathbf{E}\cdot d\ell = (12 - 8x)dx$$

Potential difference between P_1 and P_2

$$V_{21} = -\int_{P_1}^{P_2} \mathbf{E}\cdot d\ell$$

$$= -\int_1^{-1} (12 - 8x)dx = 24 \text{ V}$$

We next choose a path consisting of two straight-line segments; one is P_1P_0 parallel to the x axis and the other is P_0P_2 parallel to the y axis.

For the segment P_1P_0; $y = 2$, $dy = 0$; $d\ell = dx\mathbf{u}_x$; $\mathbf{E}\cdot d\ell = 8\,dx$

For the segment P_0P_2; $x = -1$, $dx = 0$; $d\ell = du\mathbf{u}_y$; $\mathbf{E}\cdot d\ell = -4dy$

Therefore, $\quad V_{21} \quad \int_1^{-1} dx \quad \int_2^{4} dy = 24$ V

We note that the potential difference has the same value though the paths are different.

EXAMPLE 2.28

A point charge Q of 0.03 μC is located at P_0(–0.6 m, 0.2 m, 0.35 m) in a medium of relative permittivity 2.5. Calculate potential difference between P_1(0.3 m, 0.75 m, –0.25 m) and P_2(0.15 m, –0.2 m, 0.5 m) in the field of the charge. Calculate also the potential at mid-point of the straight line joining P_1 and P_2.

The example is illustrated in Fig. 2.34.

The distances of P_1 and P_2 from P_0 are

$$R_1 = \sqrt{(0.3+0.6)^2 + (0.75-0.2)^2 + (-0.25-0.35)^2} = 1.213 \text{ m}$$

$$R_2 = \sqrt{(0.15+0.6)^2 + (-0.2-0.2)^2 + (0.5-0.35)^2} = 0.863 \text{ m}$$

Calculated values of coordinates of the mid-point P are

0.225 m, 0.275 m and 0.125 m

The distances of P from P_0,

$$R = \sqrt{(0.225+0.6)^2 + (0.275-0.2)^2 + (0.125-0.35)^2} = 0.858 \text{ m}$$

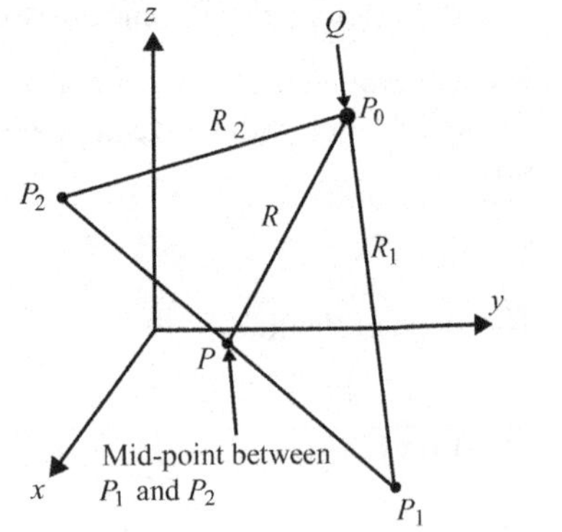

Fig. 2.34 Illustration for Example 2.28

Potential difference between P_1 and P_2,

$$V_{12} = \frac{Q}{4\pi\varepsilon}\left(\frac{1}{R_1} - \frac{1}{R_2}\right) = \frac{(9\times10^9)(0.03\times10^{-6})}{2.5}\left(\frac{1}{1.213} - \frac{1}{0.863}\right) = -36.1 \text{ V}$$

$$\text{Potential at } P,\ V = \frac{Q}{4\pi\varepsilon R} = \frac{(9\times10^9)(0.03\times10^{-6})}{(2.5)(0.858)} = 125.9 \text{ V}$$

EXAMPLE 2.29

Three point charges $Q_1 = 1.2$ μC, $Q_2 = -0.75$ μC and $Q_3 = 0.4$ μC are located at P_1(–0.8 m, 0.6 m, 0), P_2(0.6 m, 0.7 m, –0.2 m) and P_3(1.2 m, –0.5 m, 0.7 m) respectively. Find the potential at P(0.2 m, 0.4 m, 1.2 m). The relative permittivity of the medium is 2.7.

Calculated values of distances of the charges from P are

$R_1 = 1.575$ m, $R_2 = 1.487$ m, $R_3 = 1.435$ m

Potential at P due to the charges is given by

$$V = \frac{(9\times10^9)(10^{-6})}{2.7}\left(\frac{1.2}{1.575} - \frac{0.75}{1.487} + \frac{0.4}{1.435}\right) = 178.7 \text{ V}$$

EXAMPLE 2.30

A long line charge with uniform charge density $\rho_\ell = 1.5$ nC per metre is along the y axis. Find the potential difference between points P_2(1.0 m, 2.0 m, 3.0 m) and P_1(2.0 m, –1.0 m, 4.0 m). The relative permittivity of the medium, $\varepsilon_r = 1.8$

The example is illustrated in Fig. 2.35. The projections of P_1 and P_2 on to the line charge are at (0, –1.0 m, 0) and (0, 2.0 m, 0) respectively. Thus, perpendicular distances of P_1 and P_2 from the line charge are given by

$$R_1 = \sqrt{2^2 + 0 + 4^2} = \sqrt{20} \text{ m}$$

$$R_2 = \sqrt{1^2 + 0 + 3^2} = \sqrt{10} \text{ m},$$

Potential difference between P_2 and P_1,

$$V_{21} = \frac{\rho_\ell}{2\pi\varepsilon} \ln\frac{R_1}{R_2} = \frac{18\times10^9 \times 1.5\times10^{-9}}{1.8} \ln\frac{\sqrt{20}}{\sqrt{10}} = 5.2 \text{ V}$$

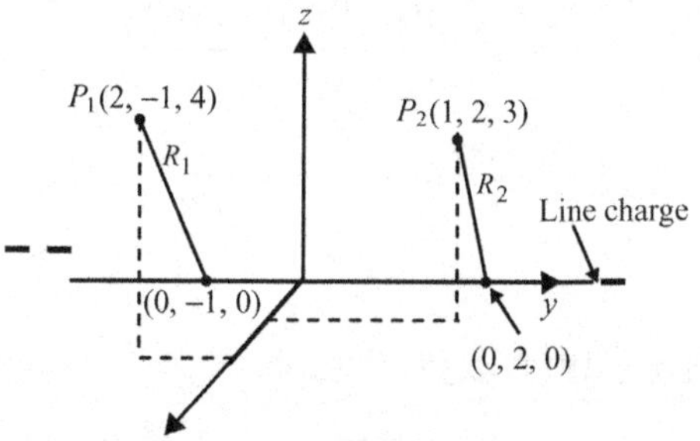

Fig. 2.35 Illustration for Example 2.30

EXAMPLE 2.31

A uniformly charged wire with $\rho_\ell = 2.0$ nC per metre is on the x axis and extends from $x = 0.5$ m to $x = 2.5$ m in air. Calculate the absolute potential at P(1.0 m, 2.0 m, 0).

The example is illustrated in Fig. 2.36. Let us consider a differential length dx' at P' on the line at a distance x' from the origin. Coordinates of P' are $x = x'$, $y = z = 0$. Distance R between P and P' is given by

$$R \quad \sqrt{(1-x')^2 + 2^2}$$

Potential at P due to the charge contained in the differential length,

$$dV = \frac{\rho_\ell\, dx'}{4\pi\varepsilon_0 R} = \frac{(9\times10^9)(2\times10^{-9})}{\sqrt{(1-x')^2 + 2^2}} dx'$$

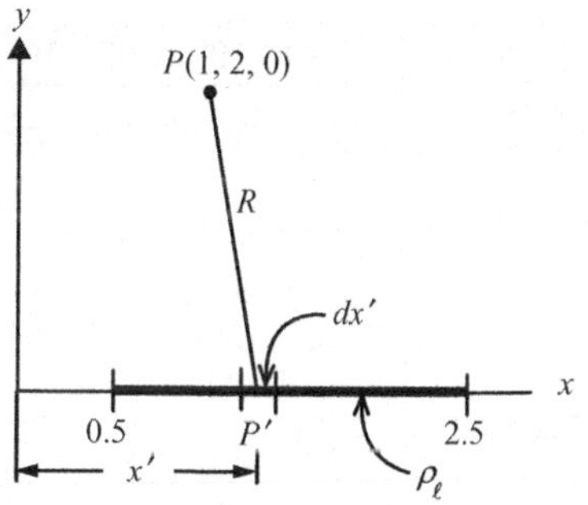

Fig. 2.36 Illustration for Example 2.31

$$V = \int_{0.5}^{2.5} \frac{18}{\sqrt{(1-x')^2 + 2^2}} dx' = -18\ln\left[(1-x') + \sqrt{(1-x')^2 + 4}\right]_{0.5}^{2.5} = 16.93 \text{ V}$$

EXAMPLE 2.32

The radii of the inner and outer conductors of a coaxial transmission line are 2.0 cm and 2.5 cm respectively. The inner conductor has uniformly distributed charge of linear charge density $\rho_\ell = 16.0$ nC/m. The relative permittivity of the dielectric between the conductors $\varepsilon_r = 5.6$. Find the potential difference between the inner and outer conductors.

Application of Gauss's law gives,

$$E_r = \frac{\rho_\ell}{2\pi\varepsilon_0\varepsilon_r r}, \quad < r < 2.5 \text{ cm}$$

Differential length in r direction, d dr

$$dV = -E_r dr = -\frac{(18\times10^9)(16\times10^{-9})}{5.6r}dr = -\frac{51.43}{r}dr$$

$$V = -\int_{0.025}^{0.02}\frac{51.43}{r}dr = 51.43\ln\left(\frac{2.5}{2}\right) = 11.476 \text{ V}$$

EXAMPLE 2.33

The radii of the inner and outer conductors of two concentric spheres are 8.0 cm and 9.0 cm respectively. The inner conductor has uniformly distributed charge of surface charge density $\rho_S = 24$ nC/m². The relative permittivity of the dielectric between the conductors is $\varepsilon_r = 3.8$. Find the potential difference between the inner and outer conductors.

Application of Gauss's law gives,

$$E_r = \frac{Q}{4\pi\varepsilon_0\varepsilon_r r^2}, \quad 8.0 \text{ cm} < r < 9.0 \text{ cm}$$

$Q = 4\pi(0.08)^2(24 \times 10^{-9}) = 1.93 \times 10^{-9}$ C

Differential length in r direction, $d\ell = dr$

$$dV = -E_r dr = -\frac{(9\times10^9)(1.93\times10^{-9})}{3.8r^2} = -\frac{4.571}{r^2}$$

$$V = -4.571\int_{0.09}^{0.08}\frac{dr}{r^2} = 4.571\left(\frac{1}{0.08} - \frac{1}{0.09}\right) = 6.35 \text{ V}$$

EXAMPLE 2.34

Electric charge is uniformly distributed over the volume of a sphere located in air. The radius of the sphere is 5 cm and the volume charge density is 2 μC/m³. Find the potential V at a radial distance 16 cm.

As the point at which potential is to be determined is outside the sphere, a Gaussian spherical surface concentric to the spherical charge will enclose the total charge Q in the sphere. Thus, taking the centre of the sphere at the origin, application of Gauss's law at a radial distance r gives

$$E_r = \frac{Q}{4\pi\varepsilon_0 r^2}$$

Since the potential at infinity is zero,

$$V = -\int_{\infty}^{r} E_r dr = -\int_{\infty}^{r}\frac{Q dr}{4\pi\varepsilon_0 r^2} = \frac{Q}{4\pi\varepsilon_0 r}$$

$$Q = \frac{4\pi}{3}(0.05^3)(2\times10^{-6}) = 1.047\times10^{-9} \text{ C}$$

$r = 0.16$ m

$$V = \frac{(9\times10^9)(1.047\times10^{-9})}{0.16} = 58.9 \text{ V}$$

2.17 Relation between Electric Field Intensity and Electric Potential in Differential Form: The Potential Gradient

The relation between the electric field intensity and the electric potential in integral form, derived in the last section, is useful for calculating the potential difference between two points from the field intensity function. In this section we will derive the relation between the two field quantities in differential form. This relation is useful for finding the electric field intensity from the potential function. We will find in this and other chapters that the differential relation between **E** and V has wide applications. We will at first define *gradient* of electrostatic potential V and then derive it in rectangular, cylindrical and spherical coordinate systems.

To define gradient of V, let us consider two equipotential surfaces S_1 and S_2, as shown in Fig. 2.37. The potential of S_1 is taken as V and that of S_2 as $V + \Delta V$ so that the potential difference between the two surfaces is ΔV.

If the distance between these equipotential surfaces is small, we may assume that the field lines between them are straight, as shown in the figure. Moreover, as the direction of field intensity is perpendicular to an equipotential surface; the field lines start from and end on the surfaces normally.

A few field lines are shown in the figure.

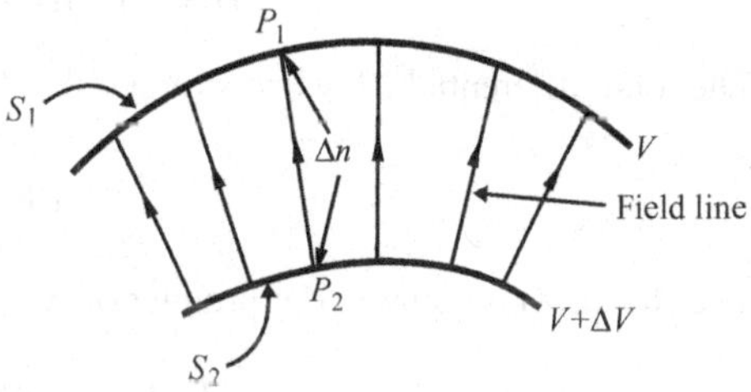

Fig. 2.37 Shows two equipotential surfaces S_1 and S_2 and field lines between them

The potential difference between S_2 and S_1 is equal to negative of the line integral of **E** along any arbitrary path from a point on the lower potential surface S_1 to a point on the higher potential surface S_2. However, if we choose specifically the path of a field line, such as P_1 to P_2, then the potential difference is simply equal to the negative of the product of average value of field intensity over the path and length of the path. Thus, denoting the average value of the field intensity by E_{av} and length of the path by Δn, we have

$$\Delta V = -E_{av}\Delta n \tag{1}$$

Dividing both sides of (1) by Δn, we get

$$\frac{\Delta V}{\Delta n} = -E_{av} \tag{2}$$

The above relation gives a method of computing the field intensity from a plot of equipotential surfaces. In the limit Δn approaching zero, the ratio $\Delta V/\Delta n$ of (2) will become the derivative dV/dn and the average field intensity will be the field intensity at a point. Thus mathematically,

$$\lim_{\Delta n \to 0} \frac{\Delta V}{\Delta n} = \frac{dV}{dn} = -E \tag{3}$$

Relation (3) shows that the derivative dV/dn is equal in magnitude but opposite in direction to the field intensity. As electric field intensity at a point is normal to an equipotential surface passing through the point, dV/dn is also normal to the surface. The positive direction of dV/dn is towards the increasing potential. As a vector quantity, it is called *potential gradient* or *gradient* of V. Thus, if $\mathbf{u}_n$ is a unit normal vector directed from a lower to a higher potential (from P_1 to P_2 in Fig. 2.37), then we have

$$\text{Potential gradient} = \frac{dV}{dn}\mathbf{u}_n \quad E\mathbf{u}_n = -\mathbf{E}$$

The potential gradient gives the maximum rate of change of potential with distance as Δn in (2) is the shortest length between two consecutive equipotential surfaces.

To derive the gradient of V in various coordinate systems, let us consider two points P_2 and P_1 which are separated by a differential displacement $d\ell$ in an electric field, as shown in Fig. 2.38. Also, let the potential difference between P_2 and P_1 be dV and the field intensity at P_2 be **E**. If we assume that both magnitude and direction of the field intensity is constant over the differential length, then the line integral of **E** from P_1 to P_2 is simply equal to $-\mathbf{E}\cdot d\ell$. Hence,

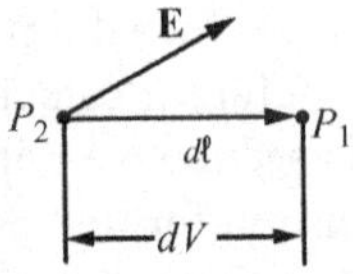

Fig. 2.38 For derivation of relation between **E** and V

$$dV = -\mathbf{E}\cdot d\ell \qquad (4)$$

the vectors **E** and $d\ell$ in terms of their components in rectangular coordinates are

$$\mathbf{E} = E_x\mathbf{u}_x + E_y\mathbf{u}_y + E_z\mathbf{u}_z \qquad (5)$$

$$d\ell = dx\mathbf{u}_x + dy\mathbf{u}_y + dz\mathbf{u}_z \qquad (6)$$

Substituting (5) and (6) in (4), we get

$$dV = -(E_x\mathbf{u}_x + E_y\mathbf{u}_y + E_z\mathbf{u}_z)\cdot(dx\mathbf{u}_x + dy\mathbf{u}_y + dz\mathbf{u}_z) \qquad (7)$$

The total differential dV when expressed in terms of its partial derivatives is given by

$$dV = \frac{\partial V}{\partial x}dx + \frac{\partial V}{\partial y} + \frac{\partial V}{\partial z}dz \qquad (8)$$

The rhs of (8) is equal to dot product of two vector functions as given below:

$$dV = \left(\frac{\partial V}{\partial x}\mathbf{u}_x + \frac{\partial V}{\partial y}\mathbf{u}_y + \frac{\partial V}{\partial z}\mathbf{u}_z\right)\cdot(dx\,\mathbf{u}_x + dy\,\mathbf{u}_y + dz\,\mathbf{u}_z) \qquad (9)$$

Comparison between (7) and (9) yields

$$\frac{\partial V}{\partial x}\mathbf{u}_x + \frac{\partial V}{\partial y}\mathbf{u}_y + \frac{\partial V}{\partial z}\mathbf{u}_z = -(E_x\mathbf{u}_x + E_y\mathbf{u}_y + E_z\mathbf{u}_z) \qquad (10)$$

As two vector functions can be equal only if each component of one is equal to the corresponding component of the other, the above relation gives

$$\frac{\partial V}{\partial x} = -E_x,\ \frac{\partial V}{\partial y} = -E_y,\ \frac{\partial V}{\partial z} = -E_z$$

The physical meaning of these relations is, at any point in an electric field, the rate of change of potential along a coordinate direction is equal to the negative of field intensity along that direction.

Using (5) in (10), we get

$$\left(\frac{\partial V}{\partial x}\mathbf{u}_x + \frac{\partial V}{\partial y}\mathbf{u}_y + \frac{\partial V}{\partial z}\mathbf{u}_z\right) = -\mathbf{E} \qquad (11)$$

This is the differential relation between **E** and V in rectangular coordinates. The vector function within the brackets on lhs of (11) is the gradient of V. The gradient of V in short form is written as grad V or ∇V. Thus,

$$\left(\frac{\partial V}{\partial x}\mathbf{u}_x + \frac{\partial V}{\partial y}\mathbf{u}_y + \frac{\partial V}{\partial z}\mathbf{u}_z\right) = \text{grad}\,V = \nabla V = -\mathbf{E} \qquad (12)$$

In cylindrical coordinates the field intensity and differential vector length in terms of their components are given by

$$\mathbf{E} = E_r\mathbf{u}_r + E_\phi\mathbf{u}_\phi + E_z\mathbf{u}_z \tag{13}$$

and

$$d\ell = dr\mathbf{u}_r + rd\phi\mathbf{u}_\phi + dz\mathbf{u}_z \tag{14}$$

Substituting (13) and (14) in (4),

$$dV = -(E_r\mathbf{u}_r + E_\phi\mathbf{u}_\phi + E_z\mathbf{u}_z)\cdot(dr\,\mathbf{u}_r + r\,d\phi\,\mathbf{u}_\phi + dz\mathbf{u}_z) \tag{15}$$

The total differential dV in terms of partial derivatives is given by

$$dV = \frac{\partial V}{\partial r}dr + \frac{\partial V}{\partial \phi}d\phi + \frac{\partial V}{\partial z}dz$$

$$= \left(\frac{\partial V}{\partial r}\mathbf{u}_r + \frac{1}{r}\frac{\partial V}{\partial \phi}\mathbf{u}_\phi + \frac{\partial V}{\partial z}\mathbf{u}_z\right)\cdot(dr\,\mathbf{u}_r + rd\phi\,\mathbf{u}_\phi + dz\,\mathbf{u}_z) \tag{16}$$

Comparison between (15) and (16) yields,

$$\frac{\partial V}{\partial r}\mathbf{u}_r + \frac{1}{r}\frac{\partial V}{\partial \phi}\mathbf{u}_\phi + \frac{\partial V}{\partial z}\mathbf{u}_z = -(E_r\mathbf{u}_r + E_\phi\mathbf{u}_\phi + E_z\mathbf{u}_z) \tag{17}$$

Hence

$$\frac{\partial V}{\partial r} = -E_r,\ \frac{1}{r}\frac{\partial V}{\partial \phi} = -E_\phi,\ \frac{\partial V}{\partial z} = -E_z$$

Using (13) in (17), we get

$$\left(\frac{\partial V}{\partial r}\mathbf{u}_r + \frac{1}{r}\frac{\partial V}{\partial \phi}\mathbf{u}_\phi + \frac{\partial V}{\partial z}\mathbf{u}_z\right) = -\mathbf{E} \tag{18}$$

The function within the bracket of (18) is the gradient of V in cylindrical coordinates. Thus, we have

$$\frac{\partial V}{\partial r}\mathbf{u}_r + \frac{1}{r}\frac{\partial V}{\partial \phi}\mathbf{u}_\phi + \frac{\partial V}{\partial z}\mathbf{u}_z = \text{grad } V = \nabla V = -\mathbf{E} \tag{19}$$

In spherical coordinates,

$$\mathbf{E} = E_r\mathbf{u}_r + E_\theta\mathbf{u}_\theta + E_\phi\mathbf{u}_\phi \tag{20}$$

$$d\ell = dr\mathbf{u}_r + rd\theta\mathbf{u}_\theta + r\sin\theta\, d\phi\mathbf{u}_\phi \tag{21}$$

Substituting (20) and (21) in (4), we get

$$dV = -(E_r\mathbf{u}_r + E_\theta\mathbf{u}_\theta + E_\phi\mathbf{u}_\phi)\cdot(dr\,\mathbf{u}_r + r\,d\theta\,\mathbf{u}_\theta + r\sin\theta\,d\phi\mathbf{u}_\phi) \tag{22}$$

The total differential dV in terms of its partial derivatives is given by

$$dV = \frac{\partial V}{\partial r}dr + \frac{\partial V}{\partial \theta}d\theta + \frac{\partial V}{\partial \phi}d\phi$$

$$= \left(\frac{\partial V}{\partial r}\mathbf{u}_r + \frac{1}{r}\frac{\partial V}{\partial \theta}\mathbf{u}_\theta + \frac{1}{r\sin\theta}\frac{\partial V}{\partial \phi}\mathbf{u}_\phi\right)\cdot(dr\,\mathbf{u}_r + r\,d\theta\,\mathbf{u}_\theta + r\sin\theta\,d\phi\,\mathbf{u}_\phi) \tag{23}$$

From (22) and (23) we get

$$\frac{\partial V}{\partial r}\mathbf{u}_r + \frac{1}{r}\frac{\partial V}{\partial \theta}\mathbf{u}_\theta + \frac{1}{r\sin\theta}\frac{\partial V}{\partial \phi}\mathbf{u}_\phi = -(E_r\mathbf{u}_r + E_\theta\mathbf{u}_\theta + E_\phi\mathbf{u}_\phi) \tag{24}$$

Hence $\quad \dfrac{\partial V}{\partial r} = -E_r, \; \dfrac{1}{r}\dfrac{\partial V}{\partial \theta} = -E_\theta, \; \dfrac{1}{r\sin\theta}\dfrac{\partial V}{\partial \phi} = -E_\phi$

Using (20) in (24), we get

$$\left(\frac{\partial V}{\partial r}\mathbf{u}_r + \frac{1}{r}\frac{\partial V}{\partial \theta}\mathbf{u}_\theta + \frac{1}{r\sin\theta}\frac{\partial V}{\partial \phi}\mathbf{u}_\phi\right) = -\mathbf{E}$$

Thus in spherical coordinates,

$$\frac{\partial V}{\partial r}\mathbf{u}_r + \frac{1}{r}\frac{\partial V}{\partial \theta}\mathbf{u}_\theta + \frac{1}{r\sin\theta}\frac{\partial V}{\partial \phi}\mathbf{u}_\phi = \text{grad } V = \nabla V = -\mathbf{E} \tag{25}$$

Equations (12), (19) and (25) are the relations between the electric field intensity and electric potential in the three coordinate systems in differential form. If the potential is a known function of space variables, we can calculate the field intensity using these equations.

As the electric field intensity is normal to an equipotential surface, grad V is also normal to the surface over which V is constant. But its sense is opposite to that of the electric field intensity.

EXAMPLE 2.35

The potential function in an electric field is given by $V = e^{-z}/(xy)$ V. Find the electric field intensity at (1.0 m, 2.0 m, 2.5 m).

$$\mathbf{E} = -\frac{\partial}{\partial x}\left(\frac{e^{-z}}{xy}\right)\mathbf{u}_x - \frac{\partial}{\partial y}\left(\frac{e^{-z}}{xy}\right)\mathbf{u}_y - \frac{\partial}{\partial z}\left(\frac{e^{-z}}{xy}\right)\mathbf{u}_z$$

$$= \frac{e^{-z}}{x^2 y}\mathbf{u}_x + \frac{e^{-z}}{xy^2}\mathbf{u}_y + \frac{e^{-z}}{xy}\mathbf{u}_z$$

At (1.0 m, 2.0 m, 2.5 m), $\mathbf{E} = e^{-2.5}(0.5\mathbf{u}_x + 0.25\mathbf{u}_y + 0.5\mathbf{u}_z)$

$= 0.0410\mathbf{u}_x + 0.0205\mathbf{u}_y + 0.0410\mathbf{u}_z$ V/m

EXAMPLE 2.36

Given: $V = 10\dfrac{\sin\theta\cos\phi}{r^2}$ V, find the field intensity at (2.5 m, –60°, 45°)

The potential is given in spherical coordinates,

$$\mathbf{E} = -\left(\mathbf{u}_r\frac{\partial}{\partial r} + \frac{\mathbf{u}_\theta}{r}\frac{\partial}{\partial \theta} + \frac{\mathbf{u}_\phi}{r\sin\theta}\frac{\partial}{\partial \phi}\right)\left(10\frac{\sin\theta\cos\phi}{r^2}\right)$$

$$= 10\left(\frac{\sin\theta\cos\phi}{r}\mathbf{u}_r - \frac{\cos\theta\cos\phi}{r^3}\mathbf{u}_\theta + \frac{\sin\phi}{r^3}\mathbf{u}_\phi\right)$$

Substituting $r = 2.5$ m, $\theta = -60°$, $\phi = 45°$, we get

$\mathbf{E} = -2.45\mathbf{u}_r - 0.23\mathbf{u}_\theta + 0.45\mathbf{u}_\phi$ V/m

2.18 Calculation of Field Intensity from the Potential Function

Two methods for calculating the electric field intensity of a charge distribution have been described in earlier sections. One method is by the use of Coulomb's law which requires integration of a vector function. The other method is by application of Gauss's law which is suitable if the charge distributions have some symmetry. The relation between field intensity and the potential function in differential form provides an alternative method for finding the field intensity from the potential function.

This method is advantageous for some types of field distribution as the potential function is a scalar quantity and the integration involved for calculating the potential is easier.

To calculate the field intensity from the potential function, we have to calculate at first the potential at a point due to a charge distribution. The field intensity is then found from the relation $\mathbf{E} = -\nabla V$.

We will consider some simple charge distributions to illustrate the technique.

Fields on the Axis of a Circular Loop of Charge

Figure 2.39 shows a circular loop of charge of radius a. The charge is uniformly distributed around the loop with linear charge density ρ_ℓ C/m. The loop is taken on $z = 0$ plane of cylindrical coordinates with its centre at the origin. We will calculate the field intensity at a point P on the axis of the loop, as shown in the figure.

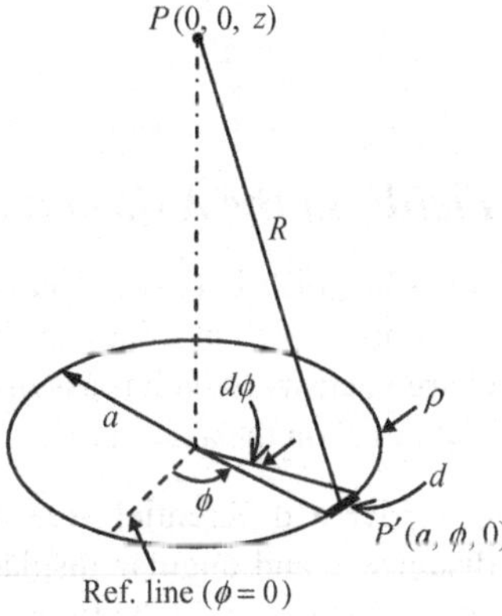

Fig. 2.39 For calculation of fields on the axis of a circular loop of charge

Let us consider a differential length $d\ell$ at $P'(a, \phi, 0)$, as shown in the figure. The differential length subtends an angle $d\phi$ at the centre of the loop.

Charge contained in the differential length, $dQ = a\rho_\ell\, d\phi$

Distance between P' and P, $R = \sqrt{a^2 + z^2}$

Potential at $P(0, 0, z)$ due to dQ is given by

$$dV = \frac{dQ}{4\pi\varepsilon R} = \frac{a\rho_\ell\, d\phi}{4\pi\varepsilon\sqrt{a^2 + z^2}}$$

Hence,
$$V = \int_{\phi}^{2\pi+\phi} dV = \frac{a\rho_\ell}{2\varepsilon\sqrt{a^2 + z^2}} = \frac{Q}{4\pi\varepsilon\sqrt{a^2 + z^2}} \quad (1)$$

where Q is total charge contained in the ring.

We see that the potential is a function of z only.

The field intensity at the point considered is given by

$$\mathbf{E} = -\frac{\partial V}{\partial z}\mathbf{u}_z = \frac{a\rho_\ell z\,\mathbf{u}_z}{2\varepsilon(a^2 + z^2)^{3/2}} = \frac{Qz\,\mathbf{u}_z}{4\pi\varepsilon(a^2 + z^2)^{3/2}} \quad (2)$$

The field intensity is zero at $z = 0$. If $z \gg a$, then

$$\mathbf{E} = \frac{Q}{4\pi\varepsilon z^2}\mathbf{u}_z$$

which is same as the field intensity of a point charge.

EXAMPLE 2.37

A circular ring of radius 0.5 m has a linear charge density of 0.02 μC/m. Find the potential and field intensity on the axis and at a height 4.0 m from the plane of the ring. The relative permittivity of the medium is 1.3.

Given: $a = 0.5$ m, $\rho_\ell = 0.02\ \mu$C/m, $z = 4.0$ m, $\varepsilon_r = 1.3$

$$\frac{a\rho_\ell}{2\varepsilon} = \frac{0\ 5(0.02\times10^{-6})}{2\ 1\ 3\times10^{-9}/36\pi)} = 434.99$$

Potential at the given point, $V = \dfrac{434.99}{\sqrt{0.5^2+4^2}} = 107.91$ V

Electric field intensity at the given point, $\mathbf{E} = \dfrac{434.99\times4}{(0.5^2+4^2)^{3/2}}\mathbf{u}_z = 26.56\mathbf{u}_z$ V/m

Fields on the Axis of a Circular Disc of Charge

A surface distribution of charge in the form of a circular disc is shown in Fig. 2.40. The disc is taken on the $z = 0$ plane with its centre at the origin in cylindrical coordinates. The radius of the disc is a and the surface charge density, which is assumed to be uniform, is ρ_S C/m². We will calculate the potential and field intensity at point P on the axis of disc, as shown in the figure.

Consider a differential area at a point P' that is at a radial distance r and angular displacement ϕ from a reference line ($\phi = 0$), as shown in the figure. The differential area,

$$d\mathbf{S} = r dr d\phi$$

Charge contained in the differential area,

$$dQ = \rho_S\, r dr d\phi$$

Distance between P' and P,

$$R = \sqrt{r^2 + z^2}$$

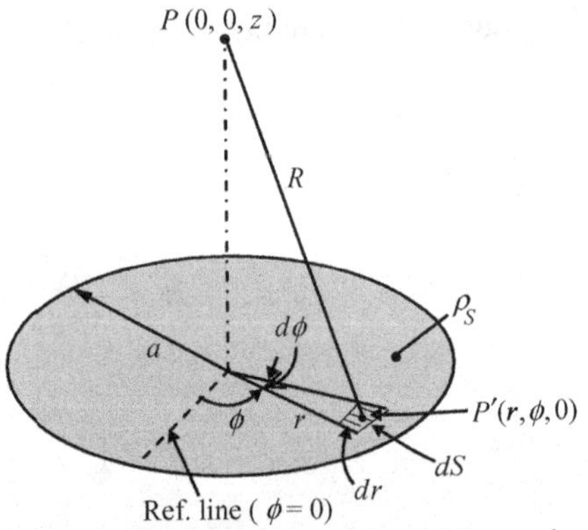

Fig. 2.40 For calculation of fields on the axis of a circular disc of charge

Treating the differential charge dQ as a point charge, potential at P is

$$dV = \frac{dQ}{4\pi\varepsilon R} = \frac{\rho_S\, r dr d\phi}{4\pi\varepsilon\sqrt{z^2+r^2}}$$

Thus, potential at P due to the charge distribution is,

$$V \int_{\phi}^{2\pi+\phi}\int_0^a dV = \frac{\rho_S}{2\varepsilon}\left(a \quad z - z\right) \qquad (3)$$

As V is a function of z only, the field intensity,

$$\mathbf{E} = -\frac{\partial V}{\partial z}\mathbf{u}_z = \frac{\rho_S}{2\varepsilon}\left(1 - \frac{z}{\sqrt{a^2+z^2}}\right)\mathbf{u}_z \tag{4}$$

The magnitude of the field intensity at $z = 0$ is $\rho_s / 2\varepsilon$, and it is zero at $z = \infty$.

EXAMPLE 2.38

A total charge of 15 nC is uniformly distributed in the form of a circular disc of radius 1.2 m in air. Find the potential and field intensity on the axis of the disc at a height of 3.75 m from the plane of the disc.

Given: $a = 1.2$ m, $z = 3.75$ m, $\varepsilon = \varepsilon_0$.

$$\rho_S = \frac{15\times10^{-9}}{\pi(1.2)^2} = 3.315\times10^{-9}\ \text{C/m}^2$$

$$\frac{\rho_S}{2\varepsilon_0} = \frac{3.315\times10^{-9}}{2(10^{-9}/36\pi)} = 187.46$$

Using the above values in formulae (3) and (4), the potential and electric field intensity at the given point are given by

$$V = 187.46\times\left(\sqrt{1.2^2+3.75^2} - 3.75\right) = 35.12\ \text{V}$$

$$\mathbf{E} = (187.46)\left(1 - \frac{3.75}{\sqrt{1.2^2+3.75^2}}\right)\mathbf{u}_z = 8.92\mathbf{u}_z\ \text{V/m}$$

Fields of a Line Charge of Finite Length

A line charge of length ℓ having uniform linear charge density of ρ_ℓ C/m is shown in Fig. 2.41. It is taken on the x axis of rectangular coordinates with its mid point at the origin. We will calculate the potential and field intensity due to the line charge at a point $P(x, y)$.

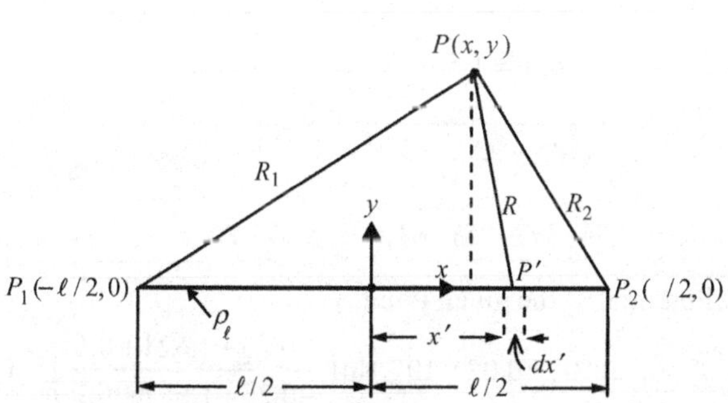

Fig. 2.41 For calculation of potential of a line charge of finite length

Let us consider a differential length dx' of the line charge at P' which is at a distance x' from the origin. Charge contained in the differential length,

$$dQ = \rho_\ell dx'$$

Distance R between P' and P is given by

$$R = \sqrt{(x-x')^2 + y^2}$$

Potential at P due to the differential charge dQ, assuming it to be a point charge, is

$$dV = \frac{dQ}{4\pi\varepsilon R} = \frac{\rho_\ell dx'}{4\pi\varepsilon R}$$

Total potential at P is found by integrating dV from $x = -\ell/2$ to $x = \ell/2$. Thus,

$$V = \int_{-\ell/2}^{+\ell/2} dV = \frac{\rho_\ell}{4\pi\varepsilon}\ln\left(\frac{R_1 + x + \ell/2}{R_2 + x - \ell/2}\right) \tag{5}$$

where $R_1 = \sqrt{(x+\ell/2)^2 + y^2}$ and $R_2 = \sqrt{(x-\ell/2)^2 + y^2}$

It may be noted that R_1 and R_2 are the distances of two ends of the line from P, as indicated in the figure. Further, the potential function of (5) is a function of both x and y.

The electric field intensity,

$$\mathbf{E} = -\nabla V = -\frac{\partial V}{\partial x}\mathbf{u}_x - \frac{\partial V}{\partial y}\mathbf{u}_y$$

The x and y components of field intensities are given by

$$E_x = -\frac{\partial V}{\partial x} = -\frac{\rho_\ell}{4\pi\varepsilon}\left(\frac{1}{R_1} - \frac{1}{R_2}\right) \tag{6}$$

$$E_y = -\frac{\partial V}{\partial y} = \frac{\rho_\ell}{4\pi\varepsilon y}\left(\frac{x+\ell/2}{R_1} - \frac{x-\ell/2}{R_2}\right) \tag{7}$$

EXAMPLE 2.39

Two ends of a uniformly charged line are at (–0.5 m, 0) and (0.5 m, 0) on the x-y plane in free space. Find the potential and field intensity at (–0.25 m, 1.5 m) if the total charge of the line is 0.02 μC.

The configuration of the line charge with respect to rectangular coordinates is as shown in Fig. 2.41. The formula for potential is given by (5), and the formulae for x and y components of field intensities are given by (6) and (7).

Length of the line, $\ell = 1.0$ m

$x = -0.25$ m, $y = 1.5$ m, $\rho_\ell = 0.02$ μC/m.

$$R_1 = \sqrt{(-0\ 25+0.5)^2 + 1.5^2} = 1.521\,\text{m}$$

$$R_2 = \sqrt{(-0\ 25-0.5)^2 + 1.5^2} = 1.677\,\text{m}$$

Potential at the given point,

$$V = (\ \times\ 0^9 \times 0.02\times10^{-6})\ln\left(\frac{-0.25+1.521+0.5}{-0.25+1.677-0.5}\right) = 116.5\,\text{V}$$

The components of the field intensities are

$$E_x = -(9\times10^9)(0.02\times10^{-6})\left(\frac{1}{1.521} - \frac{1}{1.677}\right) = -11.01\ \text{V/m}$$

$$E_y = -\frac{(9\times10^9)(0.02\times10^{-6})}{1.5}\left[\frac{(-0.25+0.5)}{1.521} - \frac{(-0.25-0.25)}{1.521}\right] = 110\ 08\ \text{V/m}$$

The total field intensity,

$\mathbf{E} = -11.01\mathbf{u}_x + 110.08\mathbf{u}_y$ V/m

Far-Fields of an Electric Dipole

A system of two equal and opposite point charges separated by a small distance is said to form an *electric dipole*. An electric dipole is illustrated in Fig. 2.42(*a*) where Q is magnitude of each charge and s is distance between the charges.

A dipole is specified by the *dipole moment,* which is a vector quantity. The magnitude of the dipole moment is equal to the product of Q and s and its direction is from the negative charge towards the positive charge. It is generally denoted by the symbol **p** and its unit is *coulomb-metre*. The magnitude of the dipole moment,

$$p = Qs$$

and its vector form is given by

$$\mathbf{p} = Q\mathbf{s}$$

The direction of **s** is from the negative charge towards the positive charge of the dipole.

We will derive equations for potential and field intensity at a point *assuming the distance of the point from the dipole to be very large compared to the distance between the charges*. With this assumption, field equations can be derived easily using spherical coordinates. Solution of the problem is further simplified by selecting the origin of the coordinate system at the mid-point of the dipole and aligning the dipole along $\theta = 0$ axis. With the dipole oriented as described above, the potential at any point $P(r, \theta)$ is a function of r and θ only.

Referring to the figure, potential at P due to the charges is given by

$$V = \frac{Q}{4\pi\varepsilon}\left(\frac{1}{R_1} - \frac{1}{R_2}\right) \tag{8}$$

Since point P is far away from the dipole, we can assume that angles θ_1 and θ_2 are each equal to θ. Accordingly,

$$R_1 = r - \frac{S}{2}\cos\theta$$

and

$$R_2 = r + \frac{S}{2}\cos\theta$$

Substituting the above relations in (8) and then simplifying, we get

$$V = \frac{Q\,s\cos\theta}{4\pi\varepsilon(r^2 - 0.25s^2\cos^2\theta)} = \frac{p\cos\theta}{4\pi\varepsilon(r^2 - 0.25s^2\cos^2\theta)}$$

As $s \ll r$, the second term within the bracket of the denominator can be neglected. Accordingly,

$$V = \frac{p\cos\theta}{4\pi\varepsilon r^2} = \frac{\mathbf{p}\cdot\mathbf{u}_r}{4\pi\varepsilon r^2} \tag{9}$$

where $\mathbf{u}_r$ is a unit vector in the direction of OP.

We find from (9) that the potential at a point far away from of a dipole is inversely proportional to r^2. It is directly proportional to the dipole moment and cos θ. Whereas, the potential of a single point charge is inversely proportional to r and it is not a function of θ.

For a specific value of the potential function, say V_0, we have from (9)

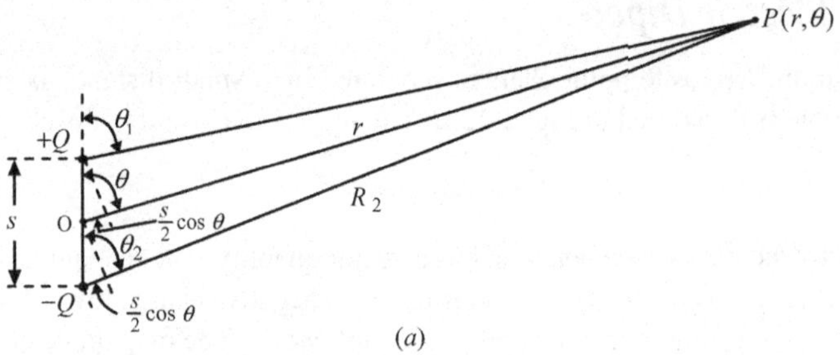

(a)

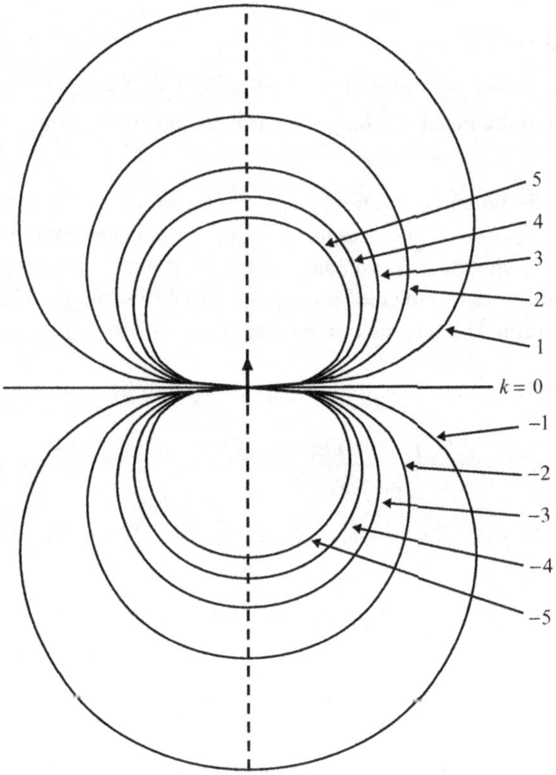

(b)

Fig. 2.42 (a) For calculation of far fields of an electric dipole and (b) Equipotential lines of the dipole.

$$\cos\theta = \left(\frac{4\pi\varepsilon V_0}{p}\right) r^2$$

In a medium of constant permittivity and for a given value of the dipole moment, the function within the bracket of the above relation is a constant. Denoting the constant by k, we get

$$\cos\,\theta = kr^2$$

The graphical plots of this relation for several values of k, both positive and negative, on a vertical plane containing the dipole are shown in Fig. 2.42(b). Each curve represents an equipotential line. The value of θ lies between 0 and $\pi/2$ for positive values of k, and between $\pi/2$ and π for negative values of k. Rotation of the equipotential lines about the axis of the dipole will result in equipotential surfaces.

The field intensity is found from the relation $\mathbf{E} = -\nabla V$ in spherical coordinates. By substituting the potential function in the gradient relation given in (25) of the preceding section we will get

$$E_r = -\frac{\partial V}{\partial r} = \frac{p\cos\theta}{2\pi\varepsilon r^3} \tag{10}$$

$$E_\theta = -\frac{1}{r}\frac{\partial V}{\partial \theta} = \frac{p\sin\theta}{4\pi\varepsilon r^3} \tag{11}$$

$$E_\phi = 0$$

The total field intensity is

$$\mathbf{E} = \frac{p}{4\pi\varepsilon r^3}(2\cos\theta\,\mathbf{u}_r + \sin\theta\,\mathbf{u}_\theta) \tag{12}$$

Equation (12) shows that the field intensity at a point far away from a dipole is inversely proportional to r^3 and it is directly proportional to the dipole moment. It also depends on θ. However, the field intensity of a point charge is inversely proportional to r^2 and it is independent of θ. Thus, the values of both potential and field intensity of a dipole decrease more rapidly with distance than those of a point charge.

EXAMPLE 2.40

The positive and negative point charges of an electric dipole are located in air at $z = 8$ cm and $z = -8$ cm respectively. The value of each charge is 60 μC. What is the dipole moment? Find (a) the potential, (b) the electric field intensity at $r = 50$ m and $\theta = 45°$.

Dipole moment, $p = sQ = (0.16)(60 \times 10^{-6}) = 9.6 \times 10^{-6}$ C-m

(a) $V = \dfrac{p\cos\theta}{4\pi\varepsilon_0 r^2} = (9\times10^9)(9.6\times10^{-6})(0.707)/(50^2) = 24.43$ V

(b) $E_r = \dfrac{p\cos\theta}{2\pi\varepsilon_0 r^3} = (18\times10^9)(9.6\times10^{-6})(0.707)/(50^3) = 0.977 = 0.977$ V/m

$E_\theta = \dfrac{p\sin\theta}{4\pi\varepsilon_0 r^3} = (9\times10^9)(9.6\times10^{-6})(0.707)/(50^3) = 0.489 = 0.489$ V/m

$\mathbf{E} = 0.977\mathbf{u}_r + 0.489\mathbf{u}_\theta$ V/m

Two Parallel Line Charges of Finite Length

Let us consider two parallel line charges A and B having linear charge densities ρ_ℓ and $-\rho_\ell$, as illustrated in Fig. 2.43. Each line is of length ℓ. The lines, which are oriented along z direction, are located at $x = -s/2$ and at $x = s/2$. Thus, spacing between the lines is equal to s. The y axis is normally into the plane of the paper.

The potential due to the charge distribution at any point in space is a function of all the three coordinate variables. However, on the x-y plane it is a function of x and y only. We will calculate the potential and field intensity on this plane at a point $P(x, y, 0)$.

Let us consider a differential length dz at $P_A(-s/2, 0, z)$ on the positive line charge A, as indicated in the figure. Distance between P and P_A is

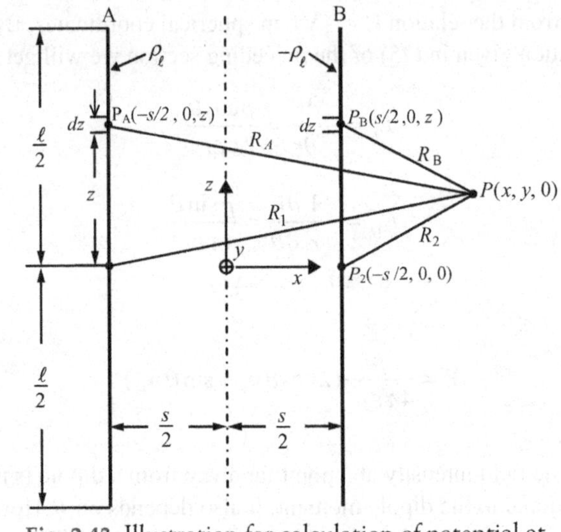

Fig. 2.43 Illustration for calculation of potential at a point due to two parallel line charges of equal and opposite linear charge densities

$$R_A = \sqrt{(x+\tfrac{s}{2})^2 + y^2 + z^2} = \sqrt{R_1^2 + z^2}$$

where $R_1 = \sqrt{(x+\tfrac{s}{2})^2 + y^2}$ is the distance of P from the mid-point of line charge A.

Charge contained in the differential length,

$$dQ = \rho_\ell \, dz$$

Potential at P due to the differential charge,

$$dV_A = \frac{\rho_\ell \, dz}{4\pi\varepsilon R_A}$$

Similarly, potential at P due to charge in a differential length dz at $P_B(s/2, 0, z)$ on the negative line charge B is

$$dV_B = -\frac{\rho_\ell \, dz}{4\pi\varepsilon R_B}$$

$$R_B = \sqrt{(x-\tfrac{s}{2})^2 + y^2 + z^2} = \sqrt{R_2^2 + z^2}$$

where $R_2 = \sqrt{(x-\tfrac{s}{2})^2 + y^2}$ is the distance of P from the mid point of line charge B.

Potential due to the two differential charges,

$$dV = dV_A + dV_B = \frac{\rho_\ell}{4\pi\varepsilon}\left(\frac{1}{R_A} - \frac{1}{R_B}\right)dz$$

Thus total potential,

$$V = \frac{\rho_\ell}{4\pi\varepsilon}\int_{-\ell/2}^{\ell/2}\left(\frac{1}{R_A} - \frac{1}{R_B}\right)dz = \frac{\rho_\ell}{4\pi\varepsilon}\int_{-\ell/2}^{\ell/2}\left(\frac{1}{R_1^2 + z^2} - \frac{1}{R_2^2 + z^2}\right)dz$$

$$= \frac{\rho_\ell}{4\pi\varepsilon}\left[\ln\frac{z+\sqrt{R_1^2+z^2}}{z+\sqrt{R_2^2+z^2}}\right]_{-\ell/2}^{\ell/2} = \frac{\rho_\ell}{2\pi\varepsilon}\ln\frac{\left(\ell+\sqrt{\ell^2+4R_1^2}\right)R_2}{\left(\ell+\sqrt{\ell^2+4R_2^2}\right)R_1} \tag{13}$$

The potential is zero on $x = 0$ plane. It is positive for $x < 0$ and negative for $x > 0$.

The field intensity,

$$\mathbf{E} = -\nabla V = -\frac{\partial V}{\partial x}\mathbf{u}_x - \frac{\partial V}{\partial y}\mathbf{u}_y = E_x\mathbf{u}_x + E_y\mathbf{u}_y$$

$$E_x = -\frac{\partial V}{\partial x} = \frac{\ell\rho_\ell}{4\pi\varepsilon}\left[\frac{x+s/2}{R_1^2\sqrt{(\ell/2)^2+R_1^2}} - \frac{x-s/2}{R_2^2\sqrt{(\ell/2)^2+R_2^2}}\right]$$

$$E_y = -\frac{\partial V}{\partial y} = \frac{\ell\rho_\ell}{4\pi\varepsilon}\left[\frac{y}{R_1^2\sqrt{(\ell/2)^2+R_1^2}} - \frac{y}{R_2^2\sqrt{(\ell/2)^2+R_2^2}}\right]$$

Two Parallel Line Charges of Infinite Length

Two parallel line charges each of infinite length with linear charge densities ρ_ℓ and $-\rho_\ell$ are shown schematically in Fig. 2.44(*a*). The line charges are along *z* direction. The line having positive charge is at $P_1(-s/2, 0, 0)$ and the line with negative charge is at $P_2(s/2, 0, 0)$. We will at first calculate the potential at a point $P(x, y, 0)$ and then show that the equipotential lines on the *x*-*y* plane are nonconcentric circles.

We can obtain the potential at *P* either from (13) of this section by substituting $\ell = \infty$ or by using the formula for potential difference between two points in the field of a line charge of infinite length given in Section 2.16. We will use the latter method. As per the method, potential difference between two points *P* and *P′* located at radial distances *R* and *R′* respectively from a line charge having linear charge density of ρ_ℓ is

$$V = \frac{\rho_\ell}{2\pi\varepsilon}\ln\frac{R'}{R}$$

Thus, potential difference between points *P* and *O* due to the positively charged line of Fig. 2.44(*a*),

$$V_{PO}^{+} = \frac{\rho_\ell}{2\pi\varepsilon}\ln\frac{s/2}{R_1} \tag{14}$$

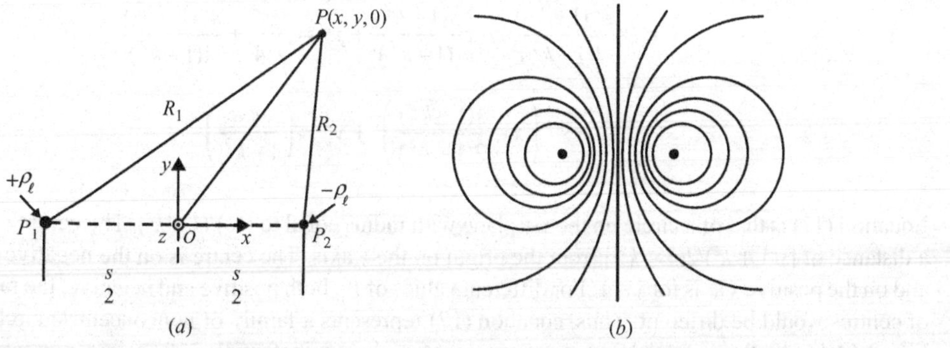

Fig. 2.44 (*a*) Shows two infinitely long line charges of equal and opposite linear charge densities. The charges are parallel to *z* axis which is normal to the paper. And (*b*) Equipotential lines on the *x y* plane.

$$R_1 = \sqrt{(x + \tfrac{s}{2})^2 + y^2}$$

As O is located at midway between the line charges of equal and opposite charge densities, the absolute potential at O is zero. Therefore, V^{+}_{PO} is the absolute potential at P.

Similarly, absolute potential at P caused by the negatively charged line is

$$V^{-}_{\text{PO}} = \frac{-\rho_\ell}{2\pi\varepsilon} \ln \frac{s/2}{R_1} \tag{15}$$

$$R_2 = \sqrt{(x - \tfrac{s}{2})^2 + y^2}$$

The total potential at P is equal to the algebraic sum of potentials given by (14) and (15). Denoting the potential by V, we have

$$V = \frac{\rho_\ell}{2\pi\varepsilon} \ln \left(\frac{s/2}{R_1} - \frac{s/2}{R_2} \right) = \frac{\rho_\ell}{2\pi\varepsilon} \ln \frac{R_2}{R_1} \tag{16}$$

Equipotential Surfaces Let the potential of an equipotential surface be V_1. Then, replacing V by V_1 in (16), we have

$$V_1 = \frac{\rho_\ell}{2\pi\varepsilon} \ln \frac{R_2}{R_1}$$

Or

$$\frac{R_2}{R_1} = \exp\left(\frac{2\pi\varepsilon V_1}{\rho_\ell} \right)$$

The rhs of the above equation has a constant value for a given value of V_1. Denoting the constant by k, we get

$$\frac{R_2}{R_1} = k$$

Or

$$\sqrt{(x - \tfrac{s}{2})^2 + y^2} = k\sqrt{(x + \tfrac{s}{2})^2 + y^2}$$

By squaring both sides and then rearranging, we get

$$x^2 - x\frac{s(1+k^2)}{1-k^2} + y^2 = -\frac{s^2}{4}$$

By adding $[s^2(1+k^2)^2]/[4(1-k^2)^2]$ to both sides and then simplifying the right-hand side, we get

$$x^2 + \frac{s^2(1+k^2)^2}{4(1-k^2)^2} - x\frac{s(1+k^2)}{(1-k^2)} + y^2 = -\frac{s^2}{4} + \frac{s^2(1+k^2)}{4(1-k^2)}$$

$$\Rightarrow \left(x - \frac{s(1+k^2)}{2(1-k^2)} \right)^2 + y^2 = \left(\frac{ks}{1-k^2} \right)^2 \tag{17}$$

Equation (17) is that of a circle on the x-y plane with radius equal to $(ks)/(1-k^2)$. The centre of the circle is at a distance of $[s(1+k^2)/2(1-k^2)]$ from the origin on the x axis. The centre is on the negative x axis for $k > 1$ and on the positive x axis for $k < 1$. For different values of V_1, both positive and negative, the radii and location of centres would be different. Thus, equation (17) represents a family of nonconcentric circles, as shown at Fig. 2.44(b). As the potential is not a function of z, the equipotential surfaces are nonconcentric circular cylinders enclosing the line charges. The $x = 0$ plane may be considered as a cylinder of infinite radius.

The field intensity at P is given by $\mathbf{E} = -\nabla V = -\dfrac{\partial V}{\partial x}\mathbf{u}_x - \dfrac{\partial V}{\partial y}\mathbf{u}_y$

The x and y components of field intensities are

$$E_x = -\frac{\partial V}{\partial x} = \frac{\rho_\ell}{2\pi\varepsilon}\left[\frac{x+s/2}{R_1^2} - \frac{x-s/2}{R_2^2}\right]$$

$$E_y = -\frac{\partial V}{\partial y} = \frac{\rho_\ell}{2\pi\varepsilon}\left[\frac{y}{R_1^2} - \frac{y}{R_2^2}\right]$$

These are same as those obtained in the Example 2.7 where the distance between the line charges has been taken as $2s$.

The total electric field is,

$$\mathbf{E} = \frac{\rho_\ell}{2\pi\varepsilon}\left[\frac{x+s/2}{R_1^2} - \frac{x-s/2}{R_2^2}\right]\mathbf{u}_x + \frac{\rho_\ell}{2\pi\varepsilon}\left[\frac{y}{R_1^2} - \frac{y}{R_2^2}\right]\mathbf{u}_y \qquad (18)$$

The field lines are circles and these are orthogonal to the equipotential lines.

EXAMPLE 2.41

Two equally and oppositely charged lines are along z direction in free space. The distance between the lines is 0.4 m and the linear charge density is 5.0 nC/m. Taking the mid-point between the lines as origin of rectangular coordinates, calculate the potential and the potential gradient at P(0.1 m, 0.1 m, 0).

The arrangement of the lines is same as that given in Fig. 2.44(a). We can calculate V and $\mathbf{E}$ by substituting $x = -0.1$ m, $y = 0.1$ m, $s = 0.4$ m, $\rho_\ell = 5.0\times10^{-9}$ C/m in formulae (15) and (17).

$$\frac{\rho_\ell}{2\pi\varepsilon_0} = \frac{5.0\times10^{-9}}{2\pi(10^{-9}/36\pi)} = 90.0 \text{ V}$$

$$R_1^2 = (x+s/2)^2 + y^2 = (-0.1+0.2)^2 + (0.1)^2 = 0.02 \text{ m}^2$$

$$R_2^2 = (x-s/2)^2 + y^2 = (-0.1-0.2)^2 + (0.1)^2 = 0.1 \text{ m}^2$$

Substituting the results in (16),

$$V = (90)\ln\sqrt{\frac{0.1}{0.02}} = 72.42 \text{ V}$$

Substituting the results in (18), the two components of field intensities are

$$E_x = (90)\left(\frac{-0.1+0.2}{0.02} - \frac{-0.1-0.2}{0.1}\right) = 720 \text{ V/m}$$

$$E_y = (90)\left(\frac{0.1}{0.02} - \frac{0.1}{0.1}\right) = 360 \text{ V/m}$$

Magnifude of potential gradient, $E = ([illegible])^2 + ([illegible])^2 = 804.$[illegible] /m

2.19 Energy Stored in the Field of Point Charges

When a point charge is brought from infinity and placed in the field of another point charge, work is done. The work done is directly proportional to the product of charge magnitudes and it is inversely proportional to distance between the charges.

The work done can also be expressed in terms of the potential due to one charge at the location of the other charge. We will calculate the work done in placing several point charges in a perfect dielectric region in this section.

The energy expended in doing the work is stored in the electric field as *potential energy*.

Suppose that a point charge Q_1 is brought from infinity and placed at a point P_1 in a region where no electric field is present initially, as shown in Fig. 2.45. No work will be done in placing the charge because it will not be acted upon by any electric force. Work done,

$$W_1 = 0$$

Next, let a second point charge Q_2 be brought from infinity and placed at P_2 which is at a distance R_{21} from Q_1. Then by substituting $R_1 = \infty$, $R_2 = R_{21}$, $Q = Q_1$ and $Q' = Q_2$ in formula (2) of Section 2.15, work done in placing Q_2 in the field of Q_1 is

$$W_2 = \frac{Q_2 Q_1}{4\pi\varepsilon R_{21}}$$

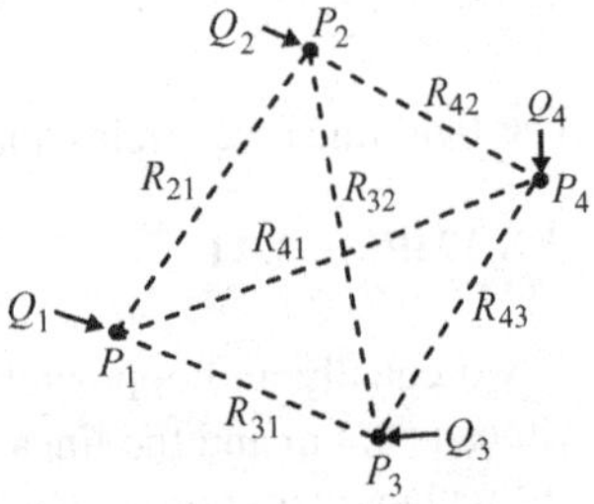

Fig. 2.45 For calculation of energy in the field of several point charges

If a third point charge Q_3 is brought from infinity and placed at P_3 in the fields of Q_1 and Q_2, then work done is given by

$$W_3 = \frac{Q_3}{4\pi\varepsilon}\left(\frac{Q_1}{R_{31}} + \frac{Q_2}{R_{32}}\right)$$

Similarly, work done in moving a fourth point charge Q_4 from infinity to P_4 is

$$W_4 = \frac{Q_4}{4\pi\varepsilon}\left(\frac{Q_1}{R_{41}} + \frac{Q_2}{R_{42}} + \frac{Q_3}{R_{43}}\right)$$

Total work done in positioning the four charges is given by

$$W = \frac{1}{4\pi\varepsilon}\left(\frac{Q_2Q_1}{R_{21}} + \frac{Q_3Q_1}{R_{31}} + \frac{Q_3Q_2}{R_{32}} + \frac{Q_4Q_1}{R_{41}} + \frac{Q_4Q_2}{R_{42}} + \frac{Q_4Q_3}{R_{43}}\right) \quad (1)$$

Since energy expended in positioning the charges is converted to potential energy, equation (1) gives energy stored in the field of the charges. The above procedure may be employed to calculate energy stored in the field of any number of charges. However, the formula in the present form cannot be written as a summation series. To obtain an expression in the form of a summation series, we will use the equalities $Q_i Q_j = Q_j Q_i$ and $R_{ij} = R_{ji}$. Accordingly,

$$\frac{Q_iQ_j}{R_{ij}} = \frac{Q_jQ_i}{R_{ji}}, \quad i \neq j$$

Using these identities in (1), we get

$$W = \frac{1}{4\pi\varepsilon}\left(\frac{Q_1Q_2}{R_{12}} + \frac{Q_1Q_3}{R_{13}} + \frac{Q_2Q_3}{R_{23}} + \frac{Q_1Q_4}{R_{14}} + \frac{Q_2Q_4}{R_{24}} + \frac{Q_3Q_4}{R_{34}}\right) \quad (2)$$

Adding (1) and (2) and then rearranging the terms it can be shown that

$$2W = Q_1\left[\frac{1}{4\pi\varepsilon}\left(\frac{Q_2}{R_{12}}+\frac{Q_3}{R_{13}}+\frac{Q_4}{R_{14}}\right)\right]+Q_2\left[\frac{1}{4\pi\varepsilon}\left(\frac{Q_1}{R_{21}}+\frac{Q_3}{R_{23}}+\frac{Q_4}{R_{24}}\right)\right]$$

$$+Q_3\left[\frac{1}{4\pi\varepsilon}\left(\frac{Q_1}{R_{31}}+\frac{Q_2}{R_{32}}+\frac{Q_4}{R_{34}}\right)\right]+Q_4\left[\frac{1}{4\pi\varepsilon}\left(\frac{Q_1}{R_{41}}+\frac{Q_2}{R_{42}}+\frac{Q_3}{R_{43}}\right)\right]$$

We find that the quantities appearing within the square brackets are the sum of potentials of three charges at the location of the remaining charge. For example, the function inside the square bracket of the first term is the sum of potentials of charges Q_2, Q_3 and Q_4 at the location of Q_1. Denoting the potentials by V_1, V_2, V_3 and V_4, energy stored in terms of the potentials is given by

$$W = \frac{1}{2}(Q_1V_1 + Q_2V_2 + Q_3V_3 + Q_4V_4) \tag{3}$$

The right-hand side of (3) is in the form of a regular series, which may be extended for any number of N charges. Thus, energy stored in the field of N number of discrete point charges is

$$W = \frac{1}{2}\sum_{i=1}^{N} Q_iV_i \tag{4}$$

where V_i is potential at the location of the ith charge due to all other charges, given by

$$V_i = \frac{1}{4\pi\varepsilon}\sum_{j=1}^{N}\frac{Q_j}{R_{ij}}, \quad j \neq i \tag{5}$$

Substituting (5) in (4), we get

$$W = \frac{1}{8\pi\varepsilon}\sum_{i=1}^{N} Q_i \sum_{j=1}^{N}\frac{Q_j}{R_{ij}}, \quad j \neq i \tag{6}$$

We can calculate the energy in an electric field of a system of finite number of point charges using either (1) or (2) or (4) or (6).

EXAMPLE 2.42

Two point charges $Q_1 = 4\ \mu C$ and $Q_2 = 6\ \mu C$ are located at P_1(2 m, 4 m, 0) and P_2(5 m, 8 m, 0) respectively in air. Find the work done in placing a third point charge $Q_3 = 3.5\ \mu$C at P_3(3 m, 6 m, 2 m).

Work done, $W = \dfrac{Q_1Q_3}{4\pi\varepsilon_0 R_{13}} + \dfrac{Q_2Q_3}{4\pi\varepsilon_0 R_{23}}$

where R_{13} is the distance between P_1 and P_3, and R_{23} is the distance between P_2 and P_3.

$$R_{13} = \sqrt{(3-2)^2 + (6-4)^2 + 2^2} = 3.0 \text{ m}$$

$$R_{23} = \sqrt{(3-5)^2 + (6-8)^2 + 2^2} = 3.464 \text{ m}$$

$$W = (9 \times 10^9)(3.5 \times 10^{-6})(10^{-6})\left(\frac{4}{3} + \frac{6}{3.464}\right) = 96.55 \text{ mJ}$$

EXAMPLE 2.43

Three point charges $Q_1 = 0.2\ \mu\text{C}$, $Q_2 = -0.15\ \mu\text{C}$ and $Q_3 = 0.35\ \mu\text{C}$ are located at $P_1(1.5\text{ m}, 0.75\text{ m}, 0.4\text{ m})$, $P_2(0.75\text{ m}, 0.5\text{ m}, 0.3\text{ m})$ and $P_3(0.1\text{ m}, 0.8\text{ m}, 0.6\text{m})$ respectively in free space. Compute energy stored in the field of the charges.

We will use (1) instead of (4) or (6).

$$\text{Energy stored, } W = W_1 + W_2 + W_3 = \frac{1}{4\pi\varepsilon_0}\left(\frac{Q_2Q_1}{R_{21}} + \frac{Q_3Q_1}{R_{31}} + \frac{Q_3Q_2}{R_{32}}\right)$$

where R_{21}, R_{31} and R_{32} are distances between charges as per the subscripts. These are given by

$$R_{21} = \sqrt{(0.75-1.5)^2 + (0.5-0.75)^2 + (0.3-0.4)^2} = 0.797\text{ m},$$

It can similarly be shown that $R_{31} = 1.415$ m, $R_{32} = 0.776$ m

$$W = (\ \times\ ^{9})\left(-\frac{0.15\times 0.2}{0.797} + \frac{0.35\times 0.2}{1.415} - \frac{0.35\times 0.15}{0.776}\right)\times 10^{-12} = -5.024\times 10^{-4}\text{ J}$$

2.20 Energy in the Field of a Continuous Charge Distribution: Energy Density

An expression for energy stored in the field of continuous charge distribution over a volume is obtained by treating the charge distribution as a continuous assemblage of point charges. Accordingly, summation (4) of the preceding section giving energy stored in the field of finite number of point charges will become a volume integral. Moreover by using the differential form of Gauss's law and the differential relation between **E** and V, we can express the integrand of the volume integral in terms of electric flux density and field intensity. This will result in the definition of energy density in an electrostatic field.

Let us consider a region of volume where charge is distributed continuously. Let dv' be a differential volume at any point in the region. If ρ is the volume charge density at this point then charge contained in dv',

$$dQ = \rho dv'$$

The differential charge dQ may be treated as a point charge. Substituting $\rho dv'$ and V in places of Q_i and V_i respectively in (4), and replacing the summation sign by the volume integral sign, we obtain the work done in building a continuous charge distribution over the volume. Thus,

$$W = \frac{1}{2}\int_{v'} V(\rho dv') \tag{1}$$

The energy expended in doing the work is stored in the electric field of the charge distribution.

The next step is to express the energy function in terms of field quantities **D** and **E** using the differential form of Gauss's law, and the differential relation between V and **E**. The differential form of Gauss's law is

$$\nabla \cdot \mathbf{D} = \rho \tag{2}$$

The differential relation between V and **E** is

$$\nabla V = -\mathbf{E} \tag{3}$$

By eliminating ρ from (1) and (2), we get

$$W = \frac{1}{2}\int_{v'} V(\nabla \cdot \mathbf{D})\, dv' \tag{4}$$

The volume integral of (4) is over the region containing the charge. However, since $\nabla \cdot \mathbf{D} = 0$ in a charge-free region, the rhs of (4) will not alter even if the integration is evaluated over a volume covering the entire space. Hence we can write,

$$W = \frac{1}{2}\int_{v} V(\nabla \cdot \mathbf{D})\, dv \tag{5}$$

where the symbol v under the integral sign means that the integration has to be performed over the entire space including the region containing the charge.

By vector identity (3) given in Section 1.9, we have

$$V(\nabla \cdot \mathbf{D}) = \nabla \cdot (V\mathbf{D}) - \mathbf{D}(\nabla V) \tag{6}$$

Replacing the integrand of (5) by the rhs of (6) and then using (3), we get

$$W = \frac{1}{2}\int_{v} \nabla \cdot (V\mathbf{D}) dv + \frac{1}{2}\int_{v} (\mathbf{D} \cdot \mathbf{E})\, dv \tag{7}$$

By transforming the first integral of (7) into a surface integral utilising the divergence theorem, we get

$$W = \frac{1}{2}\oint_{S} (V\mathbf{D}) \cdot d\mathbf{S} + \frac{1}{2}\int_{v} (\mathbf{D} \cdot \mathbf{E}) dv \tag{8}$$

Since the surface integral is to be evaluated on a surface enclosing the entire space, we may assume that the closed surface is in the form of a sphere of very large radius r approaching infinity. From this surface, any charge distributed over a finite region would appear like a point charge. It means that V and $\mathbf{D}$ are proportional to r^{-1} and r^{-2} respectively. Further, $d\mathbf{S}$ is proportional to r^2. Therefore, the surface integral of (8) is inversely proportional r and in the limit r approaching infinity the integral will become zero. Accordingly, energy stored in the field of a volume charge distribution is

$$W = \int_{v} \left(\frac{1}{2}\mathbf{D} \cdot \mathbf{E}\right) dv \tag{9}$$

The unit of the function inside the bracket of (9) is joules per unit volume. It is called *energy density*. Thus, in an electric field,

$$\text{Energy density} = \frac{\mathbf{D} \cdot \mathbf{E}}{2}\ \text{J/m}^3 \tag{10}$$

For a linear, homogenous and isotropic medium $D = \varepsilon E$. Substituting this relation in (10), we get

$$\text{Energy density} = \frac{\varepsilon E^2}{2} = \frac{D^2}{2\varepsilon}\ \text{J/m}^3 \tag{11}$$

And

$$W = \frac{1}{2}\int_{v} \varepsilon E^2 dv = \frac{1}{2\varepsilon}\int_{v} D^2 dv \tag{12}$$

We may use formula (12) to calculate energy stored in the field of a continuous charge distribution.

EXAMPLE 2.44

The radius of the inner conductor of a coaxial transmission line $a = 2.0$ cm and that of the outer conductor $b = 2.5$ cm. The inner conductor has uniformly distributed charge of linear density $\rho_\ell = 16.0$ nC/m. The relative permittivity of the dielectric between the conductors $\varepsilon_r = 5.6$. (a) Find the expression for the energy density at $a < r < b$ (b) Using the expression of energy density, calculate energy stored in the dielectric region per unit length of the line.

(a) Electric field intensity at r, $E_r = \dfrac{\rho_\ell}{2\pi\varepsilon r}$, $a < r < b$

$$\text{Energy density} = \frac{\varepsilon E_r^{\,2}}{2} = \frac{\rho_\ell^{\,2}}{8\pi^2\varepsilon r^2}$$

(b) Differential volume, $dv = r dr d\phi dz$

Energy stored in the differential volume,

$$dW = (\text{energy density})(dv) = \frac{\rho_\ell^{\,2}}{8\pi^2\varepsilon r}(dr d\phi dz)$$

$$\text{Energy stored per unit length} = \frac{\rho_\ell^{\,2}}{8\pi^2\varepsilon}\int_{r=a}^{r=b}\int_{\phi=0}^{2\pi}\int_{z=0}^{1}\frac{dr d\phi dz}{r} = \frac{\rho_\ell^{\,2}}{4\pi\varepsilon_0\varepsilon_r}\ln\frac{b}{a}$$

Substituting the data, energy stored per unit length

$$= \frac{(\times\ [\text{illegible}])(16\times10^{-9})^2}{5.6}\ln\frac{2.5}{2} = 5.737\ \text{nJ/m}$$

EXAMPLE 2.45

Two concentric conducting spherical shells are separated by a dielectric medium of relative permittivity $\varepsilon_r = 3.8$. The radii of the inner and outer spheres are $a = 8.0$ cm and $b = 9.0$ cm respectively. The inner sphere has on its outside surface a uniformly distributed charge of surface density $\rho_S = 24$ nC/m². (a) Find the expression for the energy density at $a < r < b$. (b) Using the expression of energy density, calculate energy stored in the dielectric region.

(a) Total charge on the surface of the inner sphere, $Q = 4\pi a^2 \rho_S$

$$\text{Electric field intensity at } r,\ E_r = \frac{Q}{4\pi\varepsilon r^2} = \frac{a^2\rho_S}{\varepsilon r^2},\quad a < r < b$$

$$\text{Energy density} = \frac{\varepsilon E_r^{\,2}}{2} = \frac{a^4\rho_S^{\,2}}{2\varepsilon r^4}$$

(b) Differential volume, $dv = r^2 \sin\theta\, dr d\theta d\phi$

Energy stored in the differential volume, $dW = (\text{energy density})(dv)$

$$= \frac{a^4\rho_S^{\,2}}{2\varepsilon r^4}(r^2\sin\theta dr d\theta d\phi) = \frac{a^4\rho_S^{\,2}}{2\varepsilon r^2}(\sin\theta dr d\theta d\phi)$$

$$\text{Total energy stored, } W = \frac{a^4\rho_S^{\,2}}{2\varepsilon}\int_{r=a}^{b}\int_{\theta=0}^{\pi}\int_{\phi=0}^{2\pi}\frac{\sin\theta dr d\theta d\phi}{r^2}$$

$$= \frac{2\pi a^4\rho_S^{\,2}}{\varepsilon_0\varepsilon_r}\left(\frac{1}{a} - \frac{1}{b}\right)$$

$$\text{Substituting the data, } W = \frac{2(\pi)(0.08)^4(24\times10^{-9})^2(12.5 - 11.11)}{(8.85\times10^{-12})(3.8)} = 6.106\ \text{nJ}$$

EXAMPLE 2.46

A total charge of Q coulombs is uniformly distributed in the form of a solid sphere with radius a, located in free space. The centre of the sphere is at the origin of spherical coordinates. Calculate the stored energy in the field of the charge distribution.

We will calculate the stored energy using both (1) and (12).

The volume charge density,

$$\rho = \frac{3Q}{4\pi a^3}$$

Charge contained within a radius r within the charge distribution,

$$Q' = \frac{Qr^3}{a^3}, \quad r \le a$$

Field intensity at this radius,

$$E'_r = \frac{Q'}{4\pi\varepsilon_0 r^2} = \frac{Qr}{4\pi\varepsilon_0 a^3}, \quad r \le a$$

Field intensity at a radial distance r outside the spherical charge,

$$E_r = \frac{Q}{4\pi\varepsilon_0 r^2}, \quad r \ge a$$

Potential at a radial distance r inside the charge,

$$V = -\int_\infty^a E_r dr - \int_a^r E' dr$$

$$= -\int_\infty^a \frac{Q}{4\pi\varepsilon_0 r^2} dr - \int_a^r \frac{Qr}{4\pi\varepsilon_0 a^3} dr = \frac{Q}{8\pi\varepsilon_0 a}\left(3 - \frac{r^2}{a^2}\right)$$

In spherical coordinates, a differential volume = $r^2 \sin\theta \, dr d\theta d\phi$

By substituting the expressions of ρ and V in (1) and then integrating over the spherical volume of charge,

$$W = \frac{3Q^2}{64\pi^2 a^4 \varepsilon_0} \int_{r=0}^{a} \int_{\theta=0}^{\pi} \int_{\phi=0}^{2\pi} \left(3 - \frac{r^2}{a^2}\right) r^2 \sin\theta \, dr d\theta d\phi = \frac{3Q^2}{20\pi a \varepsilon_0} \text{ J}$$

As per formula (12), we have to consider the field within as well as the field outside the charge distribution.

$$W \; \frac{\varepsilon_0}{2}\left[\int_{v'} (E'_r)^2 \, dv' + \int_v (E_r)^2 \, dv\right]$$

By substituting the expressions of E_r and E'_r and then integrating,

$$W = \frac{Q^2}{32\pi^2 \varepsilon_0}\left[\int_{\theta=0}^{\pi} \int_{\phi=0}^{2\pi} (\sin\theta \, d\theta d\phi)\left(\int_{r=0}^{a} \frac{r^4}{a^6} dr + \int_{r=a}^{\infty} \frac{dr}{r^2}\right)\right]$$

$$= \frac{3Q^2}{20\pi\varepsilon_0 a} \text{ J}$$

We find that the energy stored in the field of a uniformly charged sphere is inversely proportional to the radius of the sphere. Hence, energy required to create a charge of infinitesimal radius would be infinitely large.

2.21 Good Conductors in Static Electric Field: Boundary Conditions

We know that metals contain large number of free electrons or conduction electrons. These electrons move freely in the crystal lattice as these are not attached to any particular atom. In this section, we will briefly study the response of good conductors when subjected to an external static electric field.

Fields Inside a Good Conductor

Let a plane conductor that has no initial charge be placed in a uniform static electric field in free space, as shown in Fig. 2.46. Every free electron of the conductor will experience an electric force equal to $-eE$, where e is value of charge of an electron and E is the intensity of the applied field. The electrons will move in the opposite direction of the applied field until they reach the bottom surface of the conductor and then get distributed on the conductor surface in the form of a surface charge. The positive charges, which are generated in the conductor after the migration of the electrons, move to the top surface and get distributed over a thin layer. The charges cannot move beyond the conducting surfaces because the medium surrounding it is an insulator. The surface charges produce an induced electric field within the conductor. As the induced field is directed from the top surface having positive charge towards the bottom surface, it is opposite to the applied field.

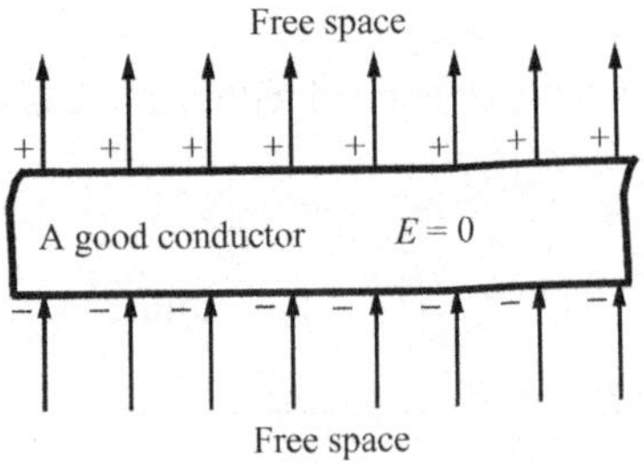

Fig. 2.46 A flat conductor in static electric field

> When the magnitude of the induced field becomes equal to that of the applied field, the net field intensity inside the conductor will become zero.

Consequently, no further movement of charges will take place. That is, static condition will be reached after a brief transient period during which charges move to the surfaces. External field lines terminate on the bottom surface and emerge from the top surface, as shown in the figure.

> Since field intensity inside a good conductor is zero, the potential gradient, which is numerically equal to the field intensity, is also zero. Accordingly, the potential has the same value everywhere inside a good conductor.

Fields on the Surface of a Good Conductor: Boundary Conditions

In a static electric field there is no movement of charges even on the surface of a conductor. It means that the charges are not acted upon by any electric force along its surface. Hence,

> the tangential component of the field intensity on the surface of a good conductor is zero. It means that the electric field intensity as well as flux density can only be perpendicular to the surface of a good conductor. It also means that the surface of a conductor is an equipotential surface.

We can establish these properties mathematically by the use of two integral equations derived earlier. The equations are

$$\oint_S \mathbf{E}\cdot d\boldsymbol{\ell} = 0. \quad \text{(Conservative property of static electric fields)}$$

$$\oint_S \mathbf{D}\cdot d\mathbf{S} = \text{charge enclosed}. \quad \text{(Integral form of Gauss's law)}$$

In order to apply the first of the integral equations, let us take a very small rectangular path *pqrsp*, which is partly inside the conductor and partly in the dielectric medium, as shown in Fig. 2.47(*a*). The sides *pq* and *rs*, each of very small length $\Delta\ell$, are parallel to the conductor surface. The height of the rectangle above the conductor is Δh. By equating the line integral of **E** around this path to zero, we get

$$\int_p^q \mathbf{E}\cdot d\boldsymbol{\ell} + \int_q^r \mathbf{E}\cdot d\boldsymbol{\ell} + \int_r^s \mathbf{E}\cdot d\boldsymbol{\ell} + \int_s^p \mathbf{E}\cdot d\boldsymbol{\ell} = 0 \tag{1}$$

The first integral in (1) is zero because the field intensity inside the conductor is zero. Moreover, as our aim is to find the field on the surface of the conductor, we have to make Δh extremely small. Under this condition, we may assume that each of the second as well as the fourth integral is zero. Thus, we have to evaluate only the third integral and then equate it to zero. If the field intensity that is tangential to the conductor surface is E_t, then the third integral is simply equal to the product of E_t and $\Delta\ell$, as $\Delta\ell$ is very small. Thus, we have

$$\lim_{\Delta h \to 0} \int_r^s \mathbf{E}\cdot d\boldsymbol{\ell} = E_t\,\Delta\ell = 0.$$

$$\Rightarrow \quad E_t = 0$$

That is, the field intensity that is tangential to the surface of a conductor is zero, and hence the field intensity and the flux density can only be normal to the surface of a good conductor. It also means that no work is done in moving a test charge from one point to another point on a conductor surface. Thus, the surface of a good conductor is an equipotential surface.

Next, let us consider a very small closed surface in the form of a cylinder enclosing the boundary surface, as shown in Fig. 2.47(*b*). The cylinder has two end surfaces S_1 and S_2, and the curved surface S_3. As per Gauss's law, the electric flux out of these surfaces is equal to the charge enclosed. Using the integral form of the law, we have

$$\int_{S1} \mathbf{D}\cdot d\mathbf{S} + \int_{S2} \mathbf{D}\cdot d\mathbf{S} + \int_{S3} \mathbf{D}\cdot d\mathbf{S} = \text{charge enclosed} \tag{2}$$

The first integral in (2) is zero as the field intensity inside the conductor is zero. Moreover in the limit $\Delta h \to 0$, the area of S_3 will become negligibly small and hence the third integral may be taken as zero. To evaluate the second integral, let D_n be the flux density normal to the surface S_2 and ΔS be the area of the surface element. Then,

$$\int_{S2} \mathbf{D}\cdot d\mathbf{S} = D_n\,\Delta S$$

If ρ_S is average surface charge density on the conductor surface, electric flux within the closed cylindrical surface is equal to ρ_S multiplied by ΔS. Equating the flux out of surface S_2 with the charge enclosed, we get

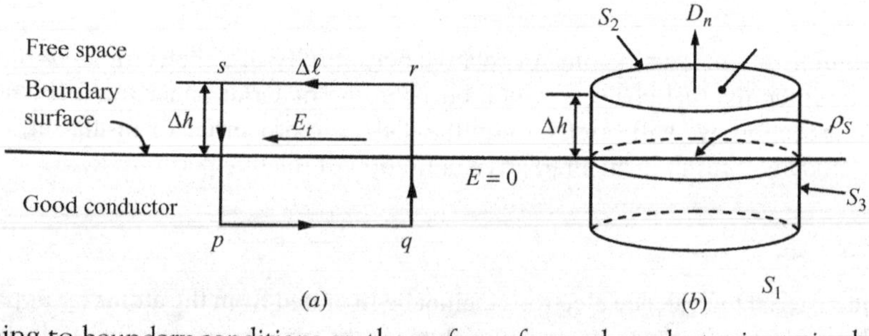

Fig. 2.47 Pertaining to boundary conditions on the surface of a good conductor in static electric field. (*a*) A closed path and (*b*) A Gaussian surface enclosing the boundary.

$$D_n \Delta S = \rho_S \Delta S$$

Or
$$D_n = \rho_S \tag{3a}$$

$$\Rightarrow \quad E_n = \frac{\rho_S}{\varepsilon} \tag{3b}$$

Summarizing the results, we have

1. The tangential component of field intensity on the surface of a good conductor is zero.

That is,
$$\mathbf{E} \times \mathbf{u}_n = 0,$$

where $\mathbf{u}_n$ is a unit vector normal to the conductor surface.

2. The normal component of flux density on the surface of a good conductor is equal to the surface charge density.

That is,
$$\mathbf{D} \cdot \mathbf{u}_n = \rho_S$$

$$\Rightarrow \quad \mathbf{E} \cdot \mathbf{u}_n = \frac{\rho_S}{\varepsilon}$$

3. The conductor surface is an equipotential surface.

These are the boundary conditions on the surface of a good conductor in a static electric field.

2.22 Dielectric Materials in Static Electric Field

Materials in which the valence band is full and the conduction band is empty are bad conductors of electricity. These materials are used to insulate two conductors at different potentials, and hence are called insulating materials or insulators. Some common insulating materials are glass, synthetic enamel, ceramics, asbestos, mica, silk, vinyl, polyethylene, transformer oil, etc. Insulating materials when used between conductors of a capacitor to augment its energy storage capacity are generally termed as dielectric materials or dielectrics. The electric properties of an insulating material in the presence of an external electric field are known as its dielectric properties. An important dielectric property used in electrical engineering is the relative permittivity of the material. It is defined as the ratio of permittivity of the material to that of free space.

One advantage of identifying a dielectric material by its relative permittivity is that it can easily be determined experimentally.

However, the definition and measured value of relative permittivity of a dielectric material do not give any information regarding the internal changes which occur in the material under the influence of an external electric field. In this section, we will briefly study the dielectric properties of insulating materials in static electric field. The study will result in an alternative definition of relative permittivity.

Polarisation

In a pure dielectric material the valence electrons cannot be liberated from the atoms by application of normal electric force. These electronic charges are called bound charges. However, internal changes take place inside the material if it is subjected to an externally applied electric field. When an external field is applied,

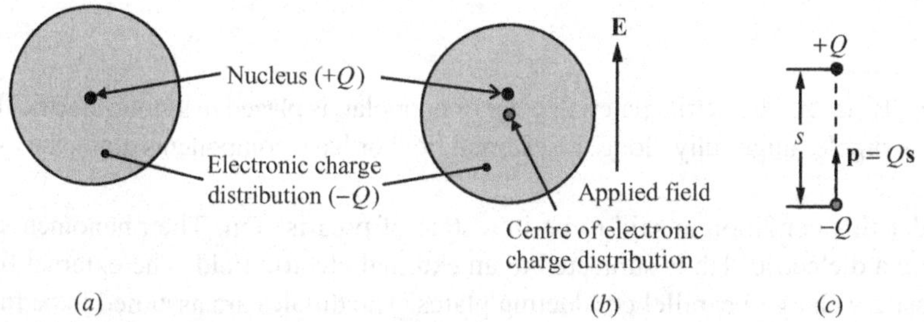

Fig. 2.48 (*a*) Schematic diagram of a neutral, nonpolar atom (*b*) The atom in a static electric field and (*c*) The equivalent dipole of (*b*).

the charges in an atom (or molecule) of the materials get displaced from their neutral positions, thus forming an electric dipole. The phenomenon can be explained by taking a simple model of an atom. The model consists of a nucleus, having a total charge of $+Q$, surrounded by a revolving electronic cloud in the shape of a sphere and having a total charge of $-Q$. In materials like diamond and phosphor, the centres of electronic charge distribution of the atoms coincide with those of the positive charges, as illustrated for one atom in Fig. 2.48(*a*). These materials are called nonpolar materials. In materials like solid nitrobenzene, H_2O and NaCl, a permanent shift between charge centres of electrons and nuclei exists. This is illustrated in Fig. 2.49(*a*) for a single atom. These materials are known as polar materials.

If a nonpolar dielectric material is subjected to an external electric field, the negative and positive charges of the atoms are acted upon by electric forces. Consequently, the centres of assumed spherical distribution of the electrons in the atoms get shifted slightly. The direction of the shift is opposite to that of the applied field. The shift causes internal Coulomb forces between the positive and negative charges. The direction of the Coulomb force is opposite to the electric force due to the applied field. At certain distance between the charge centres, magnitude of the two forces become equal and an equilibrium condition is reached. The situation is shown in Fig. 2.48(*b*). Thus, we see that electric dipoles are created when a nonpolar dielectric material is acted upon by an external field. One such dipole is shown in Fig. 2.48(*c*). The dipole moment of a dipole is equal to the product of the charge Q and the distance s between the opposite charges. The direction of the dipole moment is from the negative charge to the positive charge.

Since a permanent shift exists between positive and negative charge-centres in the atoms of polar dielectric materials, dipoles are present in it even in the absence of an applied field. However, the net dipole moment even over a very small volume in these materials is zero because of random orientation of the dipole moments. When such a material is placed in a static electric field, torque is exerted on its dipoles due to the electric force $Q\mathbf{E}$. This is illustrated for a single dipole in Fig. 2.49(*b*). The direction of the torque is such that the dipoles get oriented along the applied field.

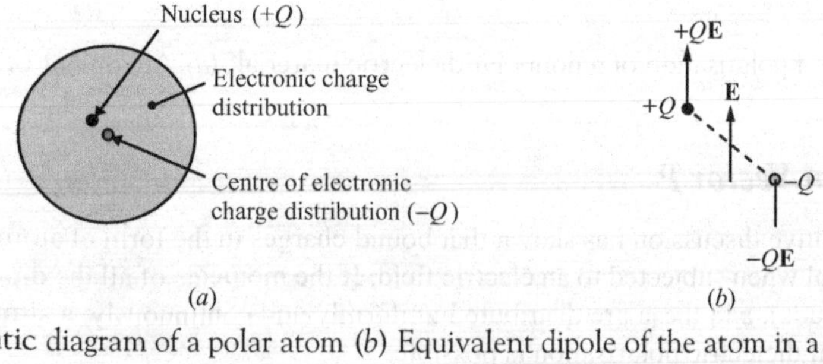

Fig. 2.49 (*a*) Schematic diagram of a polar atom (*b*) Equivalent dipole of the atom in a static electric field

Thus,

when an electrically neutral dielectric material, polar or nonpolar, is placed in a static electric field, dipoles are induced in it. The dipoles align fully along the external field or have components along the external field.

The material under this condition is said to be in a state of polarisation. The phenomenon is illustrated in Fig. 2.50(*a*) where a dielectric slab is subjected to an external electric field. The external field is due to two equally and oppositely charged parallel conducting plates. The dipoles are assumed to be fully aligned along the external field. The positive and negative charges inside the dielectric slab, shown with brackets in the figure, cancel out. This results in a layer of positive bound-charges on the dielectric surface near the negatively charged plate and a layer of negative bound-charges on the dielectric surface near the positively charged plate, as shown at Fig. 2.50(*b*). The value of the bound positive surface-charges must be equal to that of the bound negative surface-charges so that electrical neutrality of the material is maintained. Thus,

a polarised dielectric substance may be considered as equivalent to two layers of *bound charges in free space*. The direction of the electric field due to the bound charges is opposite to that of the applied field. Hence, the electric field intensity within the material is equal to the applied field minus the field due to the bound charges.

It is clear that energy has to be supplied by an external source to polarise a dielectric material. Assuming that no energy is lost, energy supplied by the external source is stored in the material. When the external source is removed, the material is depolarised and the atoms return to their pre-polarised state in most of the dielectric material. Consequently, energy stored in the material is released.

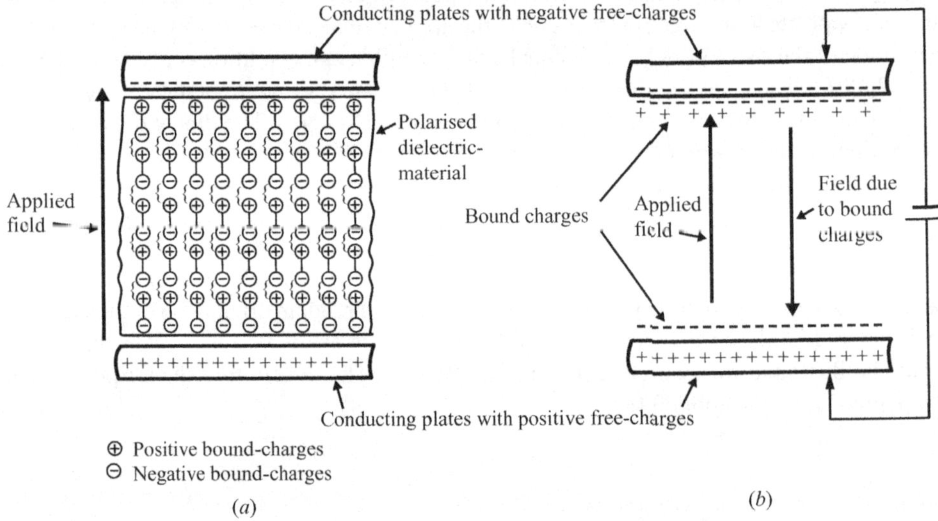

Fig. 2.50 Illustration of polarisation of a nonpolar dielectric material: (*a*) Alignment of dipoles and (*b*) The Equivalent of (*a*).

The Polarisation Vector P

The foregoing qualitative discussion has shown that bound charges in the form of atomic dipoles are formed in a dielectric material when subjected to an electric field. If the moments of all the dipoles are same (both in direction and magnitude), and these are distributed uniformly and continuously, a simple quantitative study of the phenomenon of dielectric polarisation is possible.

Let us assume that there are n polarised dipoles per unit volume in a dielectric material which is acted upon by a uniform external electric field. Then in an elementary volume Δv, there are $n\Delta v$ polarised dipoles each of moment equal to $Q\mathbf{s}$. The vector sum of all the dipole-moments within the volume Δv is given by

$$\Delta \mathbf{p} = \mathbf{p}_1 + \mathbf{p}_2 + \mathbf{p}_3 + \cdots \mathbf{p}_{(n\Delta v)}$$

In a uniform applied electric field all the dipole moments would be in the same direction. Hence, the resultant dipole moment is

$$\Delta \mathbf{p} = (n\Delta v)\, Q\mathbf{s}$$

Dividing both sides by Δv and allowing Δv to tend to zero, we get

$$\lim_{\Delta v \to 0} \frac{\Delta \mathbf{p}}{\Delta v} = nQ\mathbf{s}$$

The left-hand side is equal to the dipole moment per unit volume, and it is a vector quantity.

The dipole moment per unit volume is called *polarisation* and it is generally denoted by **P**.

Thus the polarisation of a uniformly polarised material,

$$\mathbf{P} = nQ\mathbf{s} \tag{1}$$

The unit of **P** is coulomb per square metre which is same as that of the electric flux density.

Bound Charge Density

In Section 2.21, we obtained the relation between the surface density of free charges and the normal component of electric flux density on the surface of a good conductor ($D_n = \rho_S$). We can derive a similar relation between surface density of bound charges and polarisation. To derive the relation, let us consider a uniformly polarised dielectric slab, as shown in Fig. 2.50(a). It is clear from the figure that bound charges at the top and bottom dielectric surfaces are due to the topmost and bottommost dipole layers only. The number of these dipoles over an infinitesimal surface area ΔS and depth s (distance between charge centres of a dipole) is equal to $ns\Delta S$. The total bound charge ΔQ_{Sb} in the infinitesimal volume is

$$\Delta Q_{Sb} = (ns\Delta S)Q \tag{2}$$

Dividing both sides by ΔS, in the limit $\Delta S \to 0$, we have

$$\lim_{\Delta S \to 0} \frac{\Delta Q_{Sb}}{\Delta S} = nsQ$$

The left-hand side is the surface charge density of bound charges and the right-hand side is the magnitude of polarisation. Denoting the surface charge density by ρ_{Sb}, we get

$$\rho_{Sb} = P \tag{3}$$

If the polarisation is at an angle α to the surface then the value of bound charges on it would be cos α times that given by (2). Hence,

$$\rho_{Sb} = P\cos\alpha$$

Or

$$\rho_{Sb} = \mathbf{P} \cdot \mathbf{u}_n$$

where $\mathbf{u}_n$ is a unit vector normal to ΔS.

Electric Susceptibility and Relative Permittivity

Many dielectric materials used in engineering industry are isotropic and linear within practical ranges of applied field intensity. For these materials, the distance between charge-centres and hence the moments of dipoles are nearly proportional to the field intensity within the material. Therefore,

we may assume that the polarisation **P** is proportional to field intensity **E** in the material.

That is,

$$P \propto E$$

The constant of proportionality is taken as ε_0 times another constant, which is generally denoted by χ_e (chi with subscript *e*).

Accordingly,

$$\mathbf{P} = \varepsilon_0 \chi_e \mathbf{E} \tag{4}$$

The quantity χ_e is called electric susceptibility of a material. It is a nondimensional quantity and its value is different for different dielectric materials.

We have seen that polarisation P has the same unit as that of electric flux density D. Moreover, the degree of polarisation depends on the field within the dielectric material. The field intensity within the material is equal to the vector sum of the applied field and the field due to the bound charges. Therefore, it is possible to obtain a relation between field intensity and flux density inside the material. We shall derive the relation by considering the field between two large conducting plates which are separated by a small distance. Let us first assume that the plates are situated in free space and have equal and opposite charges distributed uniformly on their surfaces. The surface charge density on the lower and upper plates are taken as $+\rho_{S0}$ and $-\rho_{S0}$ respectively. With the above assumptions, field intensity between the plates can be considered to be uniform and normal to the plate surfaces. As per formula (2.21-3), the flux density and field intensity between the plates are given by

$$D_0 = \rho_{S0}, \tag{5}$$

Or

$$E_0 = \frac{D_0}{\varepsilon_0} = \frac{\rho_{S0}}{\varepsilon_0} \tag{6}$$

Now, without altering the charge densities on the plates let the space between the plates be completely filled with a linear, homogeneous and isotropic dielectric material. The material will be polarised. The polarised material is equivalent two layers of bound charges distributed uniformly near the plate surfaces, with the material removed. The surface charge densities are ρ_{Sb} and $-\rho_{Sb}$. The induced field intensity due to the bound charges,

$$E_i = \frac{\rho_{Sb}}{\varepsilon_0} \tag{7}$$

Since $\rho_{Sb} = P$ as per (3), and $P = \varepsilon_0 \chi_e E$ as per (4),

$$\rho_{Sb} = \varepsilon_0 \chi_e E \tag{8}$$

where E is field intensity inside the material. Substituting (8) in (7), we get

$$E_i = \chi_e E \tag{9}$$

As the direction of E_i is opposite to that of the applied field E_0, the resultant field intensity inside the material is

$$E = E_0 - E_i \tag{10}$$

Using (6) and (9) in (10), we have

$$E = \frac{1}{\varepsilon_0}(D_0 - \varepsilon_0 \chi_e E)$$

$$\Rightarrow \; E = \frac{D_0}{\varepsilon_0(1+\chi_e)} = \frac{E_0}{1+\chi_e}$$

In the above relation E is the field intensity in the dielectric material and E_0 is value of the field intensity without the material. Thus, the field intensity between the plates will get reduced if a dielectric material is inserted between the plates.

Next, let us assume that the plates are in free space and they are connected to a constant voltage source as shown in Fig. 2.50(*b*). The source will establish an electric field between the plates. The field intensity is equal to the potential difference between the plates divided by the distance between the plates. If E_0 is field intensity, the value of surface charge density on each plate,

$$\rho_{S0} = D_0 = \varepsilon_0 E_0 \tag{11}$$

Now, without disconnecting the source and without altering the distance between the plates let the space between the plates be filled with a perfect dielectric material. The material will be polarised. The polarised material is equivalent to two layers of uniformly distributed bound charges in free space. The field due to the bound charges would tend to reduce the value of the applied field. But it will not change because the applied voltage and the distance between the plates are held constant. Consequently, additional free-charges will flow from the source to the plates. The values of the additional free-charges are equal to the bound charges. Hence, the total surface charge density on each plate would be

$$\rho_S = \rho_{S0} + \rho_{Sb}$$

It can be shown by application of Gauss's law that flux density D at any point between the plates is equal to the charge density ρ_S. That is,

$$D = \rho_{S0} + \rho_{Sb} \tag{12}$$

By substituting (8) and (11) in (12), we have

$$D = \varepsilon_0 E_0 + \varepsilon_0 \chi_e E \tag{13}$$

As it has been explained earlier, the field intensity has the same value without or with a dielectric material between the plates. Thus replacing E_0 by E in (13), we get

$$D = \varepsilon_0(1 + \chi_e)E \tag{14}$$

In Section (2.9) we defined the electric flux density D as ε times the field intensity E. That is,

$$D = \varepsilon E = \varepsilon_0 \varepsilon_r E \tag{15}$$

where ε_r is relative permittivity of a dielectric material.

Comparison between (14) and (15) yields,

$$\varepsilon_r = (1 + \chi_e) \tag{16}$$

Formula (16) gives the relationship between the relative permittivity and electric susceptibility of a dielectric material. The relative permittivity of free space is unity. The relative permittivity of air and of many good conductors may also be taken as unity.

Boundary Conditions

In problems involving more than one medium with common boundary surfaces, solutions of field equations for different regions are obtained first. These solutions are then matched at the interfaces using boundary conditions to obtain complete solutions. The relationships between the tangential components of field intensities and between normal components of flux densities on two sides of a common boundary surface of two regions having different permittivities are called boundary conditions. Another boundary condition is the relation between potentials on two sides of a boundary surface. To derive the relation between tangential components of field intensities, we use the conservative property of static electric field given by,

$$\oint \mathbf{E} \cdot d\boldsymbol{\ell} = 0 \tag{17}$$

The relation between normal components of flux densities is obtained using the integral form of Gauss's law, namely,

$$\oint_S \mathbf{D} \cdot d\mathbf{S} = \text{Charge enclosed} \tag{18}$$

The relation between potentials is found from the consideration that the line integral of **E** across the interface is the potential difference between two sides of the boundary.

In order to derive the boundary conditions, let us consider two perfect dielectric regions 1 and 2 with a common boundary *AB*, as shown in Fig. 2.51. The permittivities of the regions are ε_1 and ε_2. To apply equation (17), we choose a very small rectangular path *pqrsp* enclosing the boundary surface, as shown in (*a*) of the figure. One half of the path is in Region 1 and the other half is in Region 2. The length of sides of the rectangle perpendicular to the interface is Δh and the length of sides parallel to the interface is $\Delta \ell$. The line integral of **E** around the path is

$$\oint \mathbf{E} \cdot d\boldsymbol{\ell} = \int_p^q \mathbf{E} \cdot d\boldsymbol{\ell} + \int_q^r \mathbf{E} \cdot d\boldsymbol{\ell} + \int_r^s \mathbf{E} \cdot d\boldsymbol{\ell} + \int_s^p \mathbf{E} \cdot d\boldsymbol{\ell} \tag{19}$$

As our aim is to find the relationship at the boundary, Δh should be made infinitesimally small. Under this condition the second and fourth terms of (19) would be zero assuming that the field intensities along *qr* and *sp* are finite. To evaluate the other two integrals, let the field intensity tangential to the boundary surface in Region 1 be E_{t1} and the field intensity tangential to the boundary surface in Region 2 be E_{t2}. As $\Delta \ell$ is very small the value of the first integral is simply equal to the product of E_{t1} and $\Delta \ell$. Similarly, the value of the third integral is equal to the negative of the product of E_{t2} and $\Delta \ell$. It is negative because the direction of E_{t2} is opposite to the direction of integration around the path. Hence, we have

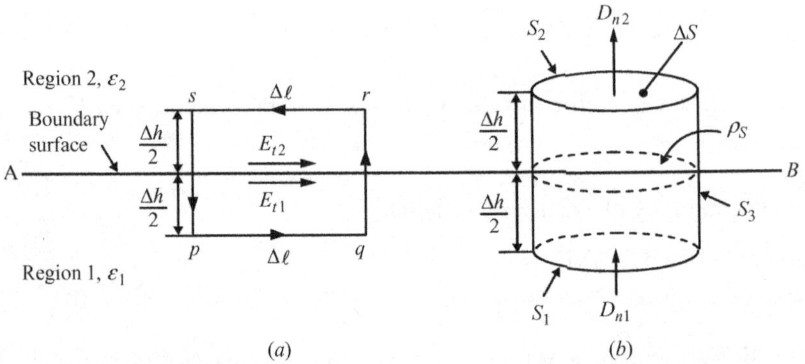

Fig. 2.51 (*a*) A closed path, and (*b*) a Gaussian surface enclosing the common boundary surface of two dielectric regions.

$$\lim_{\Delta h \to 0} \oint \mathbf{E} \cdot d\ell = \Delta\ell(E_{t1} - E_{t2}) = 0$$

Or

$$E_{t2} = E_{t1} \Rightarrow \frac{D_{t2}}{\varepsilon_2} = \frac{D_{t1}}{\varepsilon_1} \tag{20}$$

Thus, the tangential components of field intensity on two sides of the boundary are equal. In other words, the tangential field intensity is continuous across the boundary surface. However, the tangential components of flux densities are not equal.

To apply equation (18) we choose a Gaussian surface in the form of a very small cylinder enclosing the boundary surface, as shown at Fig. 2.51(*b*). One half of the cylinder is in Region 1 and the other half in Region 2. The axis of the cylinder is perpendicular to the interface. The cylinder has three surfaces, viz., two end surfaces S_1 and S_2, each of area ΔS, and the curved surface S_3. The surface integral of **D** over the closed surface,

$$\oint_S \mathbf{D} \cdot d\mathbf{S} = \int_{S1} \mathbf{D} \cdot d\mathbf{S} + \int_{S2} \mathbf{D} \cdot d\mathbf{S} + \int_{S3} \mathbf{D} \cdot d\mathbf{S} \tag{21}$$

In the limit Δh approaching zero, flux out of the curved surface represented by the third term on the rhs of (21) can be neglected, provided the flux density normal to the surface is finite. The first term gives flux entering into surface S_1 and the second term gives flux out of surface S_2. Assuming that flux entering the region as negative and flux out as positive, flux *out* of the closed surface is equal to $\Delta S(D_{n2} - D_{n1})$. Since perfect dielectric materials have no free charges and the bound charges are included in defining the respective permittivities, charge enclosed by the Gaussian surface is zero. Hence,

$$\Delta S(D_{n2} - D_{n1}) = 0$$

Or

$$D_{n2} = D_{n1} \tag{22a}$$

$$\Rightarrow \varepsilon_2 E_{n2} = \varepsilon_1 E_{n1} \tag{22b}$$

However, if the there is a thin sheet of free charge at the boundary with surface density ρ_S, charge enclosed by the Gaussian surface is equal to $\rho_S \Delta S$. Therefore,

$$(D_{n2} - D_{n1})\,\Delta S = \rho_S \Delta S$$

Or

$$D_{n2} - D_{n1} = \rho_S \tag{23}$$

Relations (22) show that the normal components of flux density on two sides of the interface of two perfect dielectric materials are equal. However, the relation between normal components of field intensity depends on the relative permittivities. Moreover, from relation (23) we find that if the interface has a sheet of free charge the difference between normal components of flux density is equal to the surface charge density.

The potential difference between two sides of the interface is equal to negative of the line integral of normal component of field intensity between two points across the boundary surface. As the distance between two points is very small, the value of the integral can be taken as zero. Therefore,

$$V_2 = V_1 \tag{24}$$

Equations (20), (23) and (24) constitute the boundary conditions or boundary relations.

We will now use these equations to derive relation between magnitudes of field intensities on two sides of the interface, and between angles subtended by the fields with the normal to the interface. Referring to Fig. 2.52, E_1 and E_2 are the magnitudes of field intensities on two sides of the interface and α_1 and α_2 are the angles which the fields make with normal to the boundary surface. The tangential and normal components of the fields are,

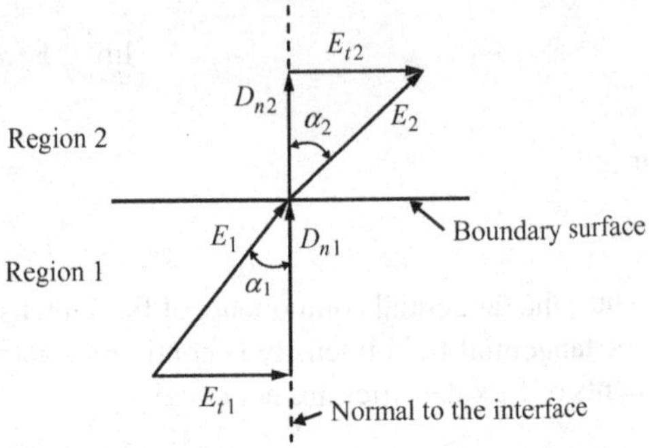

Fig. 2.52 For deriving relations between fields on two sides of common boundary surface of two dielectric regions

$$E_{t1} = E_1 \sin \alpha_1,\ E_{t2} = E_2 \sin \alpha_2$$

$$E_{n1} = E_1 \cos \alpha_1,\ E_{n2} = E_2 \cos \alpha_2$$

Using boundary conditions (20) and (23), we get

$$E_1 \sin \alpha_1 = E_2 \sin \alpha_2 \tag{25}$$

and

$$\varepsilon_1 E_1 \cos \alpha_1 = \varepsilon_2 E_2 \cos \alpha_2 \tag{26}$$

By dividing (25) by (26) we will get the following relation:

$$\frac{\tan \alpha_1}{\tan \alpha_2} = \frac{\varepsilon_1}{\varepsilon_2} = \frac{\varepsilon_{r1}}{\varepsilon_{r2}} \tag{27}$$

Formulae (25) to (27) are useful for calculating the fields on one side of the boundary if the fields on the other side are known. This is illustrated by the following example.

EXAMPLE 2.47

The permittivity of a loss-free homogeneous dielectric material is $\varepsilon = 44.25 \times 10^{-12}$ F/m. The electric field intensity in the material is 2.5 kV/m. Find (a) the electric susceptibility, (b) polarisation, and (c) moment of a dipole if there are 9×10^9 dipoles per unit volume.

(a) Electric susceptibility, $\chi_e = \varepsilon_r - 1 = \dfrac{\varepsilon}{\varepsilon_0} - 1 = \dfrac{44.25 \times 10^{-12}}{8.85 \times 10^{-12}} - 1 = 4$

(b) Polarisation, $P = \varepsilon_0 \chi_e E = (8.85 \times 10^{-12})(4)(2.5 \times 10^3) = 88.5 \times 10^{-9}\ \text{C/m}^2$

(c) Dipole moment per unit volume = $P = 88.5 \times 10^{-9}$ C/m²

Moment of a dipole, $p = \dfrac{88.5 \times 10^{-9}}{9 \times 10^9} = 9.88 \times 10^{-18}$ C-m

EXAMPLE 2.48

Two dielectric regions 1 and 2 of relative permittivities $\varepsilon_{r1} = 3.5$ and $\varepsilon_{r2} = 4.6$ respectively have a common boundary surface. The magnitude of electric field intensity at the boundary of Region 1 is 350 V/m, and it is at an angle 30° with normal to the interface. Find at the boundary of Region 2, (a) the angle of the resultant field intensity with the normal and (b) the magnitude of the resultant field intensity.

(a) If α_1 and α_2 are the angles subtended by the fields with the normal to the interface, then

$\tan \alpha_2 = (\varepsilon_{r2} / \varepsilon_{r1})\tan \alpha_1 = (4.6/3.5)\tan 30° = 0.7588$

$\alpha_2 = 37.2°$

(b) As per the boundary condition, $E_{t2} = E_{t1} = E_1 \sin \alpha_1 = 350 \sin 30° = 175$ V/m

Magnitude of the electric field at the boundary in Region 2,

$E_2 = E_{t2}/\sin \alpha_2 = 175/\sin 37.2° = 248.1$ V/m

EXAMPLE 2.49

The common boundary surface between two dielectric regions 1 and 2 is at the $x = 0$ plane. The relative permittivity of region 1 ($x < 0$) is 2.4 and that of Region 2 ($x > 0$) is 3.8. The electric field intensity at the boundary of Region 2 is $\mathbf{E}_2 = 200(2\mathbf{u}_x + \mathbf{u}_y)$ V/m. Find the field intensity in Region 1.

The y-component of $\mathbf{E}_2$ is tangential to the interface. Thus,

$E_{y1} = E_{y2} = 200$ V/m

The x component of $\mathbf{E}_2$ is normal to the interface. Thus,

$$E_{x1} = \frac{\varepsilon_{r2} E_{x2}}{\varepsilon_{r1}} = \frac{3.8}{2.4}(400) = 633.3 \text{ V/m}$$

The field intensity at the boundary in Region 1,

$\mathbf{E}_1 = 633.3\mathbf{u}_x + 200\mathbf{u}_y$ V/m

EXAMPLE 2.50

The $y = 0$ plane is the common boundary surface between two dielectric regions. The relative permittivity of Region 1 ($y < 0$) is 5.3 and that of Region 2 ($y > 0$) is 1.8. The boundary has a surface charge with density 2.5 nC/m². The electric field intensity in Region 1 is $\mathbf{E}_1 = 450\mathbf{u}_x + 600\mathbf{u}_y$ V/m. Find the field intensity in Region 2.

The x components of the fields are tangential, and the y components of the fields are normal, to the interface. Thus,

$E_{x2} = E_{x1} = 450$ V/m

$D_{y2} - D_{y1} = \rho_S$

Or $\quad \varepsilon_{r2} E_{y2} - \varepsilon_{r1} E_{y1} = \rho_S / \varepsilon_0$

$1\ 8E_{y2} - (5.3)(600) = (2.5 \times 10^{-9}) / (8.85 \times 10^{-12}) = 282.48$

$E_y = 1923.6$ V/m

Field intensity at the boundary of Region 2, $\mathbf{E}_2 = 450.0\mathbf{u}_x + 1923.6\mathbf{u}_y$ V/m

EXAMPLE 2.51

The common boundary surface of two dielectric regions is on the x-y plane. The relative permittivity of Region 1 ($z < 0$) and that of Region 2 ($z > 0$) are 1.1 and 6.0 respectively. The field intensity at the boundary in Region 1 is

$\mathbf{E}_1 = 0.5\mathbf{u}_x - \mathbf{1.2u}_y + 1.5\mathbf{u}_z$ V/m

Calculate the magnitudes of field intensities on two sides of the boundary and the angles which they make with normal to the interface.

As the interface is on the x-y plane, the x- and y-components of the field intensity E_1 at the boundary in Region 1 are tangential and its z component is normal to the interface. Similarly, the x and y components of the field intensity E_2 at the boundary in Region 2 are tangential and its z component is normal to the interface. Thus, as per the boundary conditions, the components of field intensity E_2 at the boundary in Region 2 are

$$E_{x2} = E_{x1} = 0.5 \text{ V/m}$$

$$E_{y2} = E_{y1} = -1.2 \text{ V/m}$$

$$E_{z2} = \frac{\varepsilon_1 E_{z1}}{\varepsilon_2} = \frac{(1.1)(1.5)}{6.0} = 0.275 \text{ V/m}$$

The magnitudes of the field intensities are

$$E_1 = \sqrt{0.5^2 + 1.2^2 + 1.5^2} = 1.985 \text{ V/m}$$

$$E_2 = \sqrt{0.5^2 + 1.2^2 + 0.275^2} = 1.329 \text{ V/m}$$

If E_1 and E_2 make angles α_1 and α_2 with the normal to the interface,

$$\cos\alpha_1 = \frac{E_{z1}}{E_1} = \frac{1.5}{1.985} = 0.7557$$

$$\alpha_1 = 40.91°$$

$$\cos\alpha_2 = \frac{E_{z2}}{E_2} = \frac{0.275}{1.329} = 0.2069$$

$$\alpha_2 = 78.06°$$

EXAMPLE 2.52

The plane interface between two dielectric regions is described by $S = 3x + 5y + 6z = 30$. The relative permittivity of Region 1, that is towards the origin from the interface, is $\varepsilon_{r1} = 4.2$, and the relative permittivity of Region 2, that is away from the origin from the interface, is $\varepsilon_{r2} = 2.6$. The electric field intensity at the interface in Region 1 is $\mathbf{E}_1 = \mathbf{u}_x + 2\mathbf{u}_y + 3\mathbf{u}_z$. Find the electric field intensity at the interface in Region 2.

Magnitude of field intensity normal to the interface in Region 1, $E_{n1} = \mathbf{E}_1 \cdot \mathbf{u}_n$

$$\mathbf{u}_n = \frac{\nabla S}{|\nabla S|} = \frac{3\mathbf{u}_x + 5\mathbf{u}_y + 6\mathbf{u}_z}{8.367} = 0.359\mathbf{u}_x + 0.598\mathbf{u}_y + 0.717\mathbf{u}_z$$

Hence, $E_{n1} = 3.706$

Field intensity vector normal to the interface in Region 1,

$$\mathbf{E}_{n1} = E_{n1}\mathbf{u}_n = (3.706)(0.359\mathbf{u}_x + 0.598\mathbf{u}_y + 0.717\mathbf{u}_z)$$

$$= 1.33\mathbf{u}_x + 2.22\mathbf{u}_y + 2.66\mathbf{u}_z \text{ V/m}$$

Field intensity vector tangential to the interface in Region 1,

$$\mathbf{E}_{t1} = \mathbf{E}_1 - \mathbf{E}_{n1} = -0.33\mathbf{u}_x - 0.22\mathbf{u}_y + 0.34\mathbf{u}_z$$

As per the boundary conditions,

$$\mathbf{E}_{t2} = \mathbf{E}_{t1} = -0.33\mathbf{u}_x - 0.22\mathbf{u}_y + 0.34\mathbf{u}_z \text{ V/m}$$

$$\mathbf{E}_{n2} = \frac{\varepsilon_{r1}\mathbf{E}_{n1}}{\varepsilon_{r2}} = \frac{4.2}{2.6}(1.33\mathbf{u}_x + 2.22\mathbf{u}_y + 2.66\mathbf{u}_z) \text{ V/m}$$

$$= 2.15\mathbf{u}_x + 3.59\mathbf{u}_y + 4.30\mathbf{u}_z$$

Field intensity at the interface in Region 2,

$$\mathbf{E}_2 = \mathbf{E}_{n1} + \mathbf{E}_{t1} = 1.82\mathbf{u}_x + 1.37\mathbf{u}_y + 4.64\mathbf{u}_z \text{ V/m}$$

2.23 The Dielectric Strength

If the potential gradient at any point inside a dielectric (insulating) material exceeds a limiting value, some electrons are liberated from its molecules and act as conduction electrons. Consequently, the material looses its insulating property. The phenomenon is called dielectric breakdown. The dielectric strength of an insulating material is the maximum potential gradient which it can withstand without breakdown. The dielectric strengths of different insulating materials are different.

In a non-uniform field (field lines are convergent or divergent), as in the case of an overhead transmission line, a glow called visual corona is formed at the conductor surface if the potential gradient exceeds a critical value. Further increase of potential gradient causes the air to breakdown. With uniform field (field lines are parallel), as between two parallel plates, breakdown occurs without formation of corona.

The dielectric strength of an insulating material decreases with increase of temperature and ageing. Impurities and moisture also reduce the dielectric strength of an insulator.

The potential gradient near sharp, convex bends of electrodes is several times higher than the value at its straight parts. Therefore to reduce local breakdown of insulation, which may lead to complete breakdown, sharp bends in cables and windings of high voltage equipments are avoided. For the same reason, conductors with rough surfaces are not used.

2.24 Capacitance and Capacitor

Let us consider a system of two conductors A and B of any shape separated by a dielectric medium, as shown in Fig. 2.53(a). Consider that conductor A has a positive charge of Q coulombs and conductor B has an equal amount of negative charge. The charges are distributed on the conductor surfaces in such a way so as to make the surface of each conductor an equipotential one. If potential of A is taken as V_A and that of B as $-V_B$, potential difference between the two plates is,

$$V = V_A + V_B$$

It is found that if the dielectric medium in which the conductors are located is linear and homogeneous, the ratio between charge Q and potential difference V between them has a constant value. The ratio is called *capacitance*, and a system of two conductors separated by a dielectric medium, including air, is called a

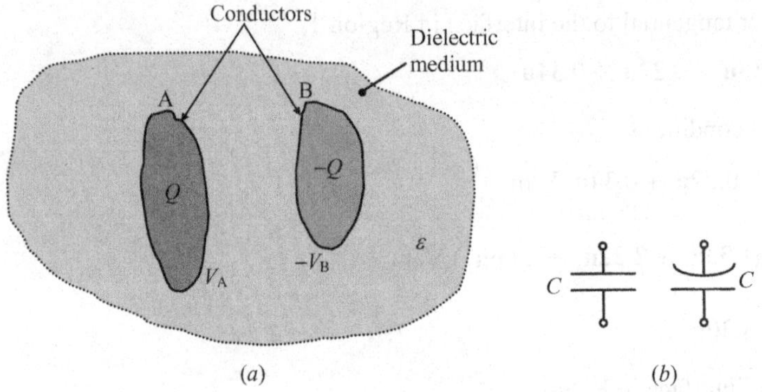

Fig. 2.53 (*a*) A two-conductor capacitor and (*b*) Symbols for denoting a capacitor.

capacitor. The capacitance is denoted by the symbol *C*, and its unit is *farad* (F). Thus capacitance of a capacitor consisting of two conductors is

$$C = \frac{Q}{V} \tag{1}$$

That is, a capacitor with potential difference of one volt between conductors and charge of one coulomb in each conductor has a capacitance of one farad (F). Since a capacitor of one farad is of very large size, capacitors are specified in microfarad and picofarad.

The total charge on the positively charged conductor is equal to the surface integral of its surface charge density ρ_S. Further, as per the boundary condition, ρ_S is equal to the normal component of flux density on a conductor surface. Therefore,

$$Q = \oint_S \rho_S dS = \oint_S (\varepsilon \mathbf{E}) \cdot d\mathbf{S} \tag{2}$$

The potential difference between the conductors is,

$$V = -\int_B^A \mathbf{E} \cdot d\ell \tag{3}$$

Substitution of (2) and (3) into (1) gives

$$C = \left[\oint_S \varepsilon \mathbf{E} \cdot d\mathbf{S}\right] \Big/ \left[-\int_B^A \mathbf{E} \cdot d\ell\right] \tag{4}$$

Equation (4) shows that the capacitance depends only on the surface area and distance between the conductors. It also depends on the permittivity of the medium. It dose not depend on *Q* or *V*. The maximum voltage of a capacitor is limited by the breakdown strength of the dielectric medium. Hence, capacitors are specified by their capacitance and safe operating voltage.

Fig. 2.53(*b*) gives symbols used to represent a capacitor.

Energy Stored in a Capacitor

Energy has to be spent in charging a capacitor. The energy expended in charging a capacitor is stored in it. To calculate the energy stored, consider that the capacitor shown in Fig. 2.53 is charged in small steps, starting from zero state, until the value of charge is *Q* and potential difference is *V*. Since *V* is proportional to *Q*, the relation between the two quantities during the charging period can be represented by a straight line *OP*, as shown in Fig. 2.54. At some stage of charging let the charge be Q' and the corresponding voltage be V'. At this stage if the charge is increased by a differential amount dQ' and the mean value of potential difference is V' during this period, increase in the energy stored is equal to the hatched area shown in the figure. Thus differential increase in energy stored,

$$dW = V'dQ' = \frac{Q'}{C}dQ'$$

Total energy stored in the capacitor,

$$W = \int_0^Q dW = \frac{Q^2}{2C} = \frac{CV^2}{2} = \frac{QV}{2} \tag{5}$$

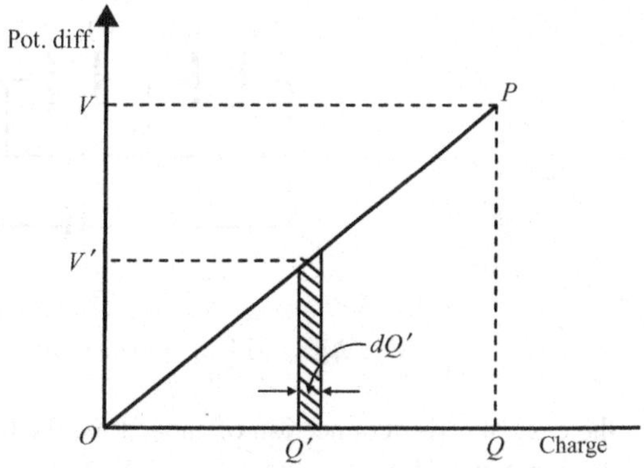

Fig. 2.54 For calculation of energy stored in a two-conductor capacitor

Under ideal conditions energy can be stored in a capacitor for a long time. The energy can be utilised as and when needed. Capacitor is an important component in almost all electronic and power-electronic circuits. Capacitors are also used for improvement of power factor in power systems and industries and for starting of single-phase induction motors.

Combination of Capacitors

It is advantageous to use a bank of capacitors connected either in series or in parallel or in series-parallel combination instead of a large single unit. For a given combination, it is necessary to calculate the equivalent capacitance across its terminals. The criterion for determining the equivalent capacitance is that the terminal conditions of the equivalent capacitor are same as those of the actual capacitor bank. The terminal conditions are (1) the terminal voltages are same, and (2) the charges on the plates connected to the terminals must be equal. If these conditions are fulfilled, the capacitance of the equivalent capacitor and energy stored in it would be same as those of the capacitor bank it represents.

Capacitors in Series Fig. 2.55(*a*) shows three capacitors with capacitances C_1, C_2 and C_3 connected in series. All the capacitors have the same value of charge Q, as same current flows for the same time period in all of them. Voltages across the capacitors are given by

$$V_1 = \frac{Q}{C_1}, \quad V_2 = \frac{Q}{C_2}, \quad V_3 = \frac{Q}{C_3},$$

The terminal voltage,

$$V = V_1 + V_2 + V_3 = Q\left(\frac{1}{C_1} + \frac{1}{C_2} + \frac{1}{C_3}\right) \tag{6}$$

Let us assume that the series combination of the capacitors is equivalent to a single capacitor having the same terminal conditions, as shown in Fig. 2.55(*b*). Thus taking the capacitance of the equivalent capacitor as C, we have

$$V = \frac{Q}{C} \tag{7}$$

Comparison between (6) and (7) yields,

$$\frac{1}{C} = \frac{1}{C_1} + \frac{1}{C_2} + \frac{1}{C_3} \tag{8}$$

Equation (8) can be extended to any number of capacitors connected in series.

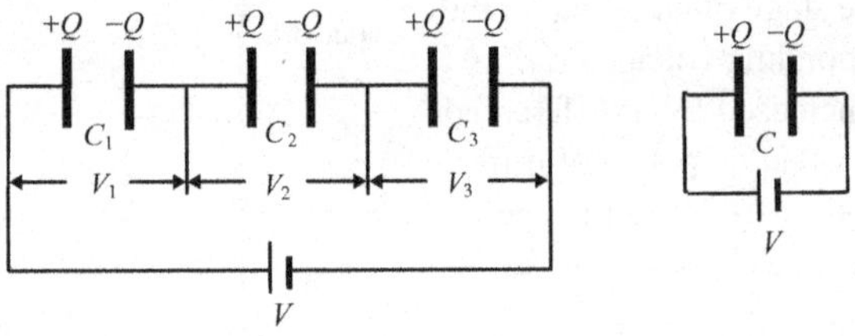

Fig. 2.55 (*a*) Capacitors connected in series and (*b*) Equivalent of (*a*).

In the case of series connection of capacitors, the total capacitance of the combination is less than that of any of its single unit. However, the voltage rating of a unit is less than that of the combination. It can be shown that energy stored in the equivalent capacitor is same as the sum of energies of all the series-connected capacitors.

Capacitors in Parallel We will now obtain the equivalent capacitance of three capacitors with capacitances C_1, C_2 and C_3, connected in parallel, as shown in Fig. 2.56(*a*). In this case the voltage across all the capacitors has the same value. However, the charges are different. The charges in the three capacitors are

$$Q_1 = VC_1;\ Q_2 = VC_2;\ Q_3 = VC_3$$

Total charge of the capacitors,

$$Q = Q_1 + Q_2 + Q_3 = V(C_1 + C_2 + C_3) \tag{9}$$

Let us assume that the parallel combination of the capacitors is equivalent to a single capacitor having the same terminal conditions, as shown in Fig. 2.56(*b*). The capacitance of the equivalent capacitor,

$$V = \frac{Q}{C} \tag{10}$$

Comparison between (9) and (10) yields,

$$C = C_1 + C_2 + C_3 \tag{11}$$

Thus, the value of the equivalent capacitance is equal to the sum of capacitances of all the capacitors. We can extend equation (11) for any number of capacitors connected in parallel. It can be shown that the energy stored in the equivalent capacitor is same as the sum of energies of all the parallel-connected capacitors.

If a capacitor bank has series-parallel grouping, we can calculate the total capacitance using (8) and (11).

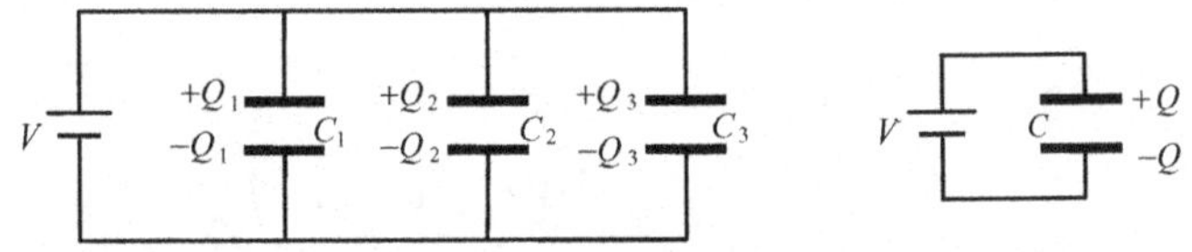

Fig. 2.56 (*a*) Capacitors connected in parallel and (*b*) Equivalent of (*a*).

EXAMPLE **2.53**

A capacitor of capacitance 120 μF is connected to a variable voltage source. If the capacitor voltage is increased from 1500 V to 2000 V, by how much is the energy stored in the capacitor increased?

Energy stored, $W = \dfrac{CV^2}{2}$

Energy stored at 1500 volt, $W_1 = \dfrac{(120\times10^{-6})(1500)^2}{2} = 135\ \text{J}$

Energy stored at 2000 volt, $W_2 = \dfrac{(120\times10^{-6})(2000)^2}{2} = 240\ \text{J}$

Increase in energy stored $= W_2 - W_1 = 105\ \text{J}$

EXAMPLE 2.54

Capacitances of four capacitors are given by $C_1 = 12\ \mu\text{F}$, $C_2 = 15\ \mu\text{F}$, $C_3 = 20\ \mu\text{F}$, and $C_4 = 25\ \mu\text{F}$. If the series combination of C_1 and C_3 is connected in parallel with the series combination of C_2 and C_4, what is the capacitance of the series-parallel combination?

Capacitance of series combination of C_1 and C_3, $C_{se1} = \dfrac{C_1C_3}{C_1+C_3} = \dfrac{240\times10^{-12}}{32\times10^{-6}} = 7.5\ \mu\text{F}$

Capacitance of series combination of C_2 and C_4, $C_{se2} = \dfrac{375\times10^{-12}}{40\times10^{-6}} = 9.375\ \mu\text{F}$

Capacitance of parallel combination of C_{se1} and C_{se2}

$= C_{se1} + C_{se2} = 7.5 + 9.375 = 16.875\ \mu\text{F}$

EXAMPLE 2.55

A capacitor bank of twelve capacitors, each of 1.2 μF, has four identical rows connected in parallel. Calculate (a) energy stored in the bank if the terminal voltage is 12 V, (b) energy stored when the capacitors are arranged in three parallel rows and the voltage across the terminals is 12 V, and (c) What should be the voltage needed for storing the same amount of energy as in (a) if the capacitors are arranged in three parallel rows?

(a) Each row has three capacitors in series.

The capacitance of each row $= \dfrac{1.2\times10^{-6}}{3} = 0.4\times10^{-6}\ \mu\text{F}$

Capacitance of the four parallel rows $= 4(0.4\times10^{-6}) = 1.6\ \mu\text{F}$

Energy stored $= \dfrac{1}{2}CV^2 = \dfrac{1}{2}(1.6\times10^{-6})(12^2) = 115.2\ \mu\text{J}$

(b) Capacitance of the bank $= \dfrac{3(1.2\times10^{-6})}{4} = 0.9\ \mu\text{F}$

Energy stored $= \dfrac{1}{2}(0.9\times10^{-6})(12^2) = 64.8\ \mu\text{J}$

(c) Let V' be the voltage required to store the same amount of energy with three parallel rows as in (a). Then,

$\dfrac{1}{2}(0.9\times10^{-6})(V')^2$ [illegible] $\times10^{-6}$

$V' = 16.0\ \text{V}$

2.25 The Parallel-Plate Capacitor

In its simplest form a parallel-plate capacitor consists of two parallel conducting-plates as shown in Fig. 2.57(*a*). The positively charged plate is at $x = 0$ and negatively charged plate is at $x = s$. The plates are separated by a single homogeneous dielectric medium. A study of this simple device will provide some useful information. The study is carried out assuming that the linear dimensions of the plates are large compared to the distance between them and equal and opposite charges are uniformly distributed on the inside surfaces of the plates. Accordingly, the surface charge density ρ_S is constant. The field lines, which are normal to the plates, are straight, parallel and are spaced uniformly, as shown at Fig. 2.57(*b*). Application of Gauss's law will show that the field intensity between the plates is constant. In practice neither the surface charge density nor the fields are uniform. Further, electric fields do exist outside the plates because of fringing at the plate edges as shown at Fig. 2.57(*c*).

Let the area of each plate be A, and the field intensity between the plates be E. Then, charge on each plate,

$$Q = A\rho_S,$$

Potential difference between the plates,

$$V = -\int_{-plate}^{+plate} E dx = -\int_{s}^{0} E dx = sE \tag{1}$$

As per the boundary condition,

$$D = \rho_S$$

Or

$$E = \frac{\rho_S}{\varepsilon} \tag{2}$$

Substituting (2) in (1), we have

$$V = \frac{Qs}{A\varepsilon}$$

The capacitance of a capacitor with two parallel plates,

$$C = \frac{Q}{V} = \frac{A\varepsilon}{s} \tag{3}$$

We find that the capacitance is not a function of the field quantities. It is directly proportional to the plate area and the permittivity, and inversely proportional to distance between the plates. If fringing at the edges is taken into account, the capacitance would be higher than that given by (3).

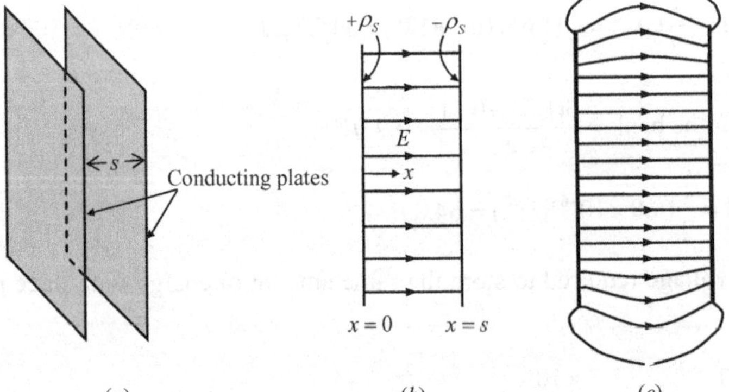

Fig. 2.57 (*a*) A parallel-plate capacitor, (*b*) Assumed uniform field distribution, and (c) Shows fringing of field lines at the plate edges.

Multiple-Plate Capacitor

The capacitance can be increased by reducing the distance between the plates, by using a dielectric material of higher permittivity and/or by increasing the plate area. The distance between the plates is limited from consideration of breakdown strength of the dielectric material which is chosen from economic and other considerations. This also nearly fixes the value of relative permittivity. Further, a capacitor with two large plates is not convenient to use. Hence capacitors are constructed with multiple plates. A multiple-plate capacitor consists of two sets of plates. All the plates of a set are electrically connected, as shown in Fig. 2.58. The tuning capacitor used in radio receivers is a multiple-plate capacitor that consists of a fixed set and a movable set of plates. If the capacitor has n plates and spacing between consecutive positive and negative plates as s, it is equivalent to $(n - 1)$ capacitors in parallel. Therefore, capacitance of the unit is

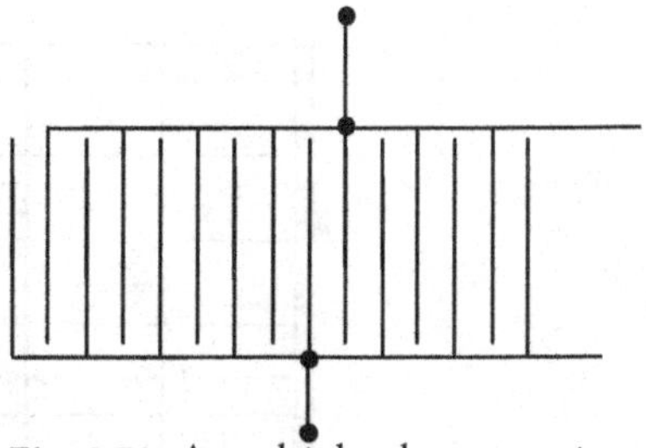

Fig. 2.58 A multiple-plate capacitor

$$C = \frac{(n-1)\,A\varepsilon}{s}$$

where A is the effective plate area. The effective plate area of a tuning capacitor is the area of that portion of a movable plate covered by the neighbouring fixed plates.

Next let us consider two identical parallel plate capacitors, one with air as medium and the other with a dielectric medium of relative permittivity ε_r. Then

$$\frac{C_{dielectric}}{C_{air}} = \varepsilon_r$$

The above relation gives a method of determining the relative permittivity of a dielectric material experimentally.

Finally, we will calculate the energy stored in the capacitor. The formula for energy density in an electric field where D and E are in the same direction is

$$W = \int_v \frac{DE}{2}\,dv$$

As both D and E are constants and volume of the region between the plates is equal to As, we have

$$W = \frac{1}{2}(DE)(As) = \frac{1}{2}(\rho_S A)(Es) = \frac{1}{2}QV$$

Capacitors with Multiple Dielectric Layers

A parallel-plate capacitor may also have more than one layer of dielectric between plates, as shown in Fig. 2.59(*a*) and Fig. 2.59(*b*). Another form of parallel-plate capacitor, shown in Fig. 2.59(*c*), has an additional conducting plate between the main plates. The interfaces in Fig. 2.59(*a*) and Fig. 2.59(*c*) are parallel to the plates, and in Fig. 2.59(*b*) the interfaces are normal to the plates. We will calculate the capacitance of the capacitors with the same assumptions as stated earlier.

Interface Parallel to the Plates Fig. 2.59(*a*) shows a parallel-plate capacitor having three dielectric layers with interfaces parallel to the plates. We find from the figure that the field lines are normal to the interfaces. Since the normal components of flux density at the interfaces are continuous, the flux density D in all the regions must have the same value. Moreover, as the value of flux density at a conducting surface is equal to the surface charge density, we have

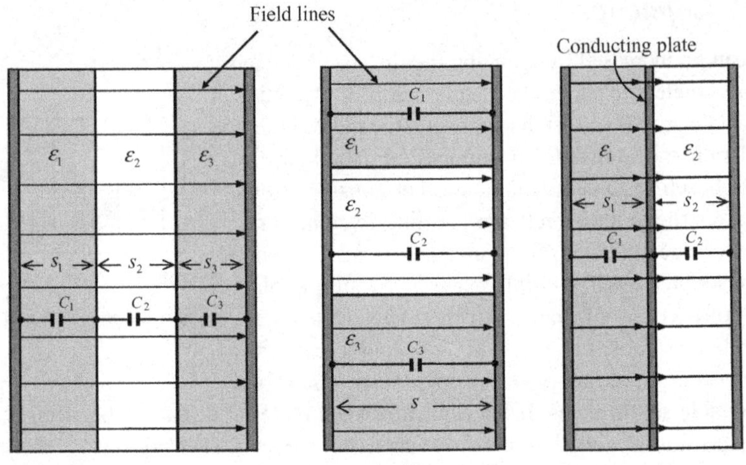

Fig. 2.59 Illustrations of composite parallel-plate capacitors: (*a*) Dielectric interfaces parallel to the plates, (*b*) Dielectric interfaces normal to the plates, and (c) With a conducting plate between the dielectrics.

$$D = \rho_S$$

The field intensities in the three regions are given by

$$E_1 = \frac{\rho_S}{\varepsilon_1};\ E_2 = \frac{\rho_S}{\varepsilon_2};\ E_3 = \frac{\rho_S}{\varepsilon_3}$$

Potential difference between the plates, $V = E_1 s_1 + E_2 s_2 + E_3 s_3 = \rho_S \left(\frac{s_1}{\varepsilon_1} + \frac{s_2}{\varepsilon_2} + \frac{s_3}{\varepsilon_3} \right)$

Value of charge on each plate, $Q = A\rho_S$

Capacitance,
$$C = \frac{Q}{V} = \frac{A}{\frac{s_1}{\varepsilon_1} + \frac{s_2}{\varepsilon_2} + \frac{s_3}{\varepsilon_3}} \tag{4}$$

As the field intensity between the plates is normal to the boundary surfaces, these are equipotential surfaces. Hence, the electrical characteristics of the capacitor will not change even if very thin conducting plates are inserted at the interfaces. Accordingly, each region can be treated as a parallel-plate capacitor, as indicated in the figure.

The capacitances of the three regions are given by

$$C_1 = \frac{A\varepsilon_1}{s_1};\ C_2 = \frac{A\varepsilon_2}{s_2};\ C_3 = \frac{A\varepsilon_3}{s_3}$$

The three capacitors are in series. The total capacitance of the combination when calculated using formula (2.24-8) will be same as that given by (4).

Interface Normal to the Plates A parallel-plate capacitor having three dielectric regions with interfaces normal to the plates is shown in Fig. 2.59(*b*). As the field intensity E is tangential to the interfaces, it has the same value in all the regions. However, the flux density and hence the surface charge density on the plates in the three regions are different. The flux densities are given by

$$D_1 = \rho_{S1} = \varepsilon_1 E;\ D_2 = \rho_{S2} = \varepsilon_2 E;\ D_3 = \rho_{S3} = \varepsilon_3 E$$

Potential difference between the plates, $V = sE$

Charge on each plate, $Q = A_1\rho_{S1} + A_2\rho_{S2} + A_3\rho_{S3} = E(A_1\varepsilon_1 + A_2\varepsilon_2 + A_3\varepsilon_3)$

where A_1, A_2 and A_3 are areas of the three regions.

Capacitance, $C = \dfrac{Q}{V} = \dfrac{1}{s}(A_1\varepsilon_1 + A_2\varepsilon_2 + A_3\varepsilon_3)$ (5)

We can also calculate the capacitance by considering each region as a parallel-plate capacitor, as indicated in the figure. The capacitances of the regions are

$$C_1 = \frac{A_1\varepsilon_1}{s},\ C_2 = \frac{A_2\varepsilon_2}{s},\ C_3 = \frac{A_3\varepsilon_3}{s},$$

Since the capacitors are in parallel, the total capacitance is equal to the sum of the three capacitances.

$$C\ \ C_1 + C_2 + C_3 = \frac{1}{s}(A_1\varepsilon_1 + A_2\varepsilon_2 + A_3\varepsilon_3),\ \text{which is same as (5).}$$

Capacitor with an Intermediate Plate The parallel-plate capacitor shown at Fig. 2.59(*c*) has a third conducting plate placed in between and parallel to the outer plates. The intermediate plate has equal and opposite charges on its two surfaces, and field lines terminate on one surface and emanate from the other surface of the plate. It may be seen from the figure that the capacitor unit is equivalent to two parallel-plate capacitors in series. The capacitances of the two regions are

$$C_1 = \frac{A\varepsilon_1}{s_1};\ C_2 = \frac{A\varepsilon_2}{s_2}$$

The total capacitance, $C = \dfrac{C_1 + C_2}{C_1 C_2} = \dfrac{A\varepsilon_1\varepsilon_2}{s_1\varepsilon_2 + s_2\varepsilon_1}$

EXAMPLE 2.56

The area of each plate of a parallel-plate capacitor with air as dielectric is 0.8 m^2. The distance between the plates is 5 mm. (a) Find the energy stored in the capacitor if it is connected to a dc source of 500 V. (b) Find the energy stored if the space between the plates is filled with a dielectric of relative permittivity 4.5 after disconnecting the capacitor from the source.

Capacitance, $\dfrac{A}{\varepsilon s}$

Energy stored, $W = \dfrac{CV^2}{2} = \dfrac{Q^2}{2C}$

(a) Capacitance, $C_1 = \dfrac{(0.8)(8.85\times10^{-12})}{0.005} = 1.416\times10^{-9}$ F

Energy stored, $W = \dfrac{(\ .416\times10^{-9})(500)^2}{2} = 177\ \mu\text{J}$

(b) Charge in each plate, $Q = C_1V = (1.416\times10^{-9})(500) = 708\times10^{-9}$ C

$$\text{Capacitance, } C_2 = \frac{(0.8)(8.85\times10^{-12})(4.5)}{0.005} = 6.372\times10^{-9}\text{ F}$$

$$\text{Energy stored, } W = \frac{(708\times10^{-9})^2}{2(6.372\times10^{-9})} = 39.33\ \mu\text{J}$$

EXAMPLE 2.57

A parallel-plate capacitor has two dielectric layers with their interface parallel to the plates. The relative permittivity and thickness of Layer 1 are $\varepsilon_{r1} = 2.0$ and $s_1 = 10$ mm respectively and those of Layer 2 are $\varepsilon_{r2} = 5.0$ and $s_2 = 1.0$ mm respectively. The potential difference between the plates $V = 11.0$ kV and area of each plate is 1.5 m². Calculate C, W, ρ_S, E_1 and E_2.

$$\text{Capacitance of Region } 1 = C_1 = \frac{A\varepsilon_1}{s_1} = \frac{(1.5)(2\varepsilon_0)}{0.01} = 300\varepsilon_0\text{ F}$$

$$\text{Capacitance of Region } 2 = C_2 = \frac{A\varepsilon_2}{s_2} = \frac{(1.5)(5\varepsilon_0)}{0.001} = 7500\varepsilon_0\text{ F}$$

As the two capacitances are in series, the total capacitance,

$$C = \frac{C_1C_2}{C_1\ \ C_2} = \frac{300\times7500}{300+7500}\varepsilon_0 = 2.554\text{ nF}$$

$$W\ \ \frac{1}{2}\ \ V^2 = \frac{1}{2}(2.554\times10^{-9})(11\times10^3)^2 = 0.1545\text{ J}$$

$$\rho_{\_}\ \ \frac{Q}{A}\ \ \frac{CV}{A} = \frac{(2.554\times10^{-9})(11\times10^3)}{1.5} = 18.73\ \mu\text{C/m}^2$$

$$E_1 = \frac{D}{\varepsilon_1} = \frac{\rho_S}{\varepsilon_1} = \frac{18.73\times10^{-6}}{2(8.854\times10^{-12})} = 1057.7\text{ kV/m}$$

$$E_2 = \frac{D}{\varepsilon_2} = \frac{\rho_S}{\varepsilon_2} = \frac{18.73\times10^{-6}}{5(8.854\times10^{-12})} = 423.1\text{ kV/m}$$

EXAMPLE 2.58

The distance between two large conducting plates is 12 mm. (a) With air as dielectric, find the working voltage if it has to be 60% of the breakdown voltage. (b) A bakelite plate 2 mm thick and of the same size as the plates is inserted between the plates. Find the working voltage if it has to be 60% of the breakdown voltage.

The dielectric strength of air is 3000 V/mm. The dielectric strength and permittivity of bakelite are 20000 V/mm and 4.3 respectively.

(a) Breakdown voltage is equal the product of the dielectric strength of air and distance between the plates.

Breakdown voltage = (3000)(12) = 36000 V

Working voltage = (0.6)(36000) = 21600 V

(b) The space between the plates has two layers of dielectrics. Layer 1 is air of 10 mm thickness, and Layer 2 is bakelite of 2 mm thickness. The interface is parallel to the plates. We have

$D_1 = D_2$

Or $E_1 = \varepsilon_{r2} E_2 = 4.3E_2$

Potential difference between the plates,

$V = s_1 E_1 + s_2 E_2 = 0.01E_1 + (0.002)(E_1/4.3) = (10.465 \times 10^{-3})E_1$

Also, $V = (0.01)(4.3\ E_2) + (0.002E_2) = (45 \times 10^{-3})E_2$

The dielectric strength of air is less than that of bakelite. Hence, the breakdown voltage is limited by the dielectric strength of air.

Breakdown voltage = $(10.465 \times 10^{-3})(3 \times 10^6) = 31395$ V.

Working voltage = $(0.6)(31395) = 18837$ V

This value is less than that obtained in (a)

EXAMPLE 2.59

A parallel-plate capacitor has two dielectric layers 1 and 2 with the interface normal to the plates. The spacing between the plates is 5 mm, and the relative permittivities of layers 1 and 2 are 2.5 and 4.5 respectively. The plate area in the region of dielectric 1 is 20 cm² and the plate area in the region of dielectric 2 is 30 cm². Find the capacitance.

Capacitance, $C = \dfrac{A\varepsilon}{d}$

Capacitance of Region 1, $C_1 = \dfrac{(20\times10^{-4})(8.85\times10^{-12})(2.5)}{5\times10^{-3}} = 88.5\times10^{-13}$

Capacitance of Region 2, $C_2 = \dfrac{(30\times10^{-4})(8.85\times10^{-12})(4.5)}{5\times10^{-3}} = 239\times10^{-13}$

As C_1 and C_2 are in parallel, total capacitance, $C = C_1 + C_2 = 32.75$ pF

2.26 Capacitance of Two Concentric Spheres

Let us consider two concentric metallic spherical shells separated by a dielectric material of permittivity ε, as shown in Fig. 2.60. The radius of the inner sphere is a and the radius of the outer sphere is b. The inner surfaces of the spheres have uniformly distributed charges of $+Q$ and $-Q$. The field intensity in the region between the spheres, which can be found by the use of Gauss's law is

$$E_r = \frac{Q}{4\pi\varepsilon r^2}, \quad b > r > a \qquad (1)$$

The field intensity elsewhere is zero.

Potential difference between the spheres,

$$V = \int_b^a E_r\, dr = -\int_b^a \frac{Q\,dr}{4\pi\varepsilon r^2} = \frac{Q}{4\pi\varepsilon}\left(\frac{b-a}{ab}\right) \qquad (2)$$

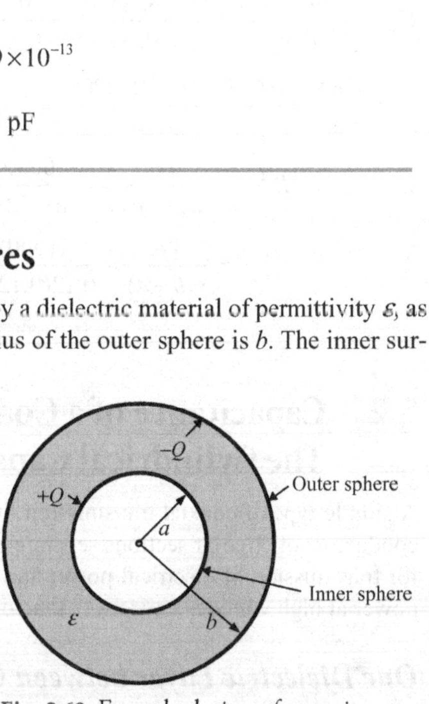

Fig. 2.60 For calculation of capacitance of two concentric conducting spheres

By eliminating Q from (1) and (2) we will obtain E_r in terms of V.

$$E_r = \frac{Vab}{(b-a)r^2}$$

Thus for a given potential difference between the spheres, E_r is inversely proportional to r^2. It is maximum at $r = a$, and minimum at $r = b$.

Capacitance of the concentric spheres,

$$C = \frac{Q}{V} = \frac{4\pi\varepsilon ab}{b-a} \tag{3}$$

The capacitance of a single sphere is calculated by assuming that the outer sphere is of infinite radius. Thus, substituting b equal to infinity in (3), the capacitance of an isolated sphere,

$$C = 4\pi\varepsilon a$$

EXAMPLE 2.60

A capacitor consisting of two concentric spherical shells of radii 10 cm and 12 cm are separated by a dielectric material of relative permittivity 2.5. Find the capacitance and energy stored in the capacitor if the potential difference between the spheres is 1000 V. Also, find the minimum and maximum values of potential gradient.

Radius of the inner sphere, $a = 0.1$ m

Radius of the outer sphere, $b = 0.12$ m

Potential difference between the spheres, $V = 1000$ V

$\varepsilon_r = 2.5$

$$C = \frac{4 \ldots ab}{b \quad a} = \frac{4\pi(2.5\times 8.854\times 10^{-12})(0.10)(0.12)}{(0.12-0.10)} = 166.9 \text{ pF}$$

$$W = \frac{CV^2}{2} = \frac{(166.9\times 10^{-12})(1000)^2}{2} = 83.45\ \mu\text{J}$$

$$E_r \ldots \quad E_r\big|_{r=a} = \frac{Vb}{a(b-a)} = \frac{(1000)(0.12)}{0.1(0.12-0.10)} = 60.0 \text{ kV/m}$$

$$E_r \ldots \quad E_r\big|_{r=b} = \frac{Va}{b(b-a)} = \frac{(1000)(0.1)}{0.12(0.12-0.10)} = 41.67 \text{ kV/m}$$

2.27 Capacitance of a Coaxial Transmission Lines: The Cylindrical Capacitor

A simple type of coaxial transmission line, also called coaxial cable, consists of two cylindrical concentric conductors of circular sections separated by a dielectric medium. Coaxial transmission lines are widely used for transmission of electrical power and electronic signals. Coaxial cables used for transmission of electric power at high voltages have more than one layer of insulation in the radial direction.

One Dielectric Layer between Conductors

Figure 2.61 illustrates a coaxial cable with one dielectric layer of permittivity ε between the conductors. The axis of the cable is along z direction of cylindrical coordinates. The radius of the inner solid conductor and

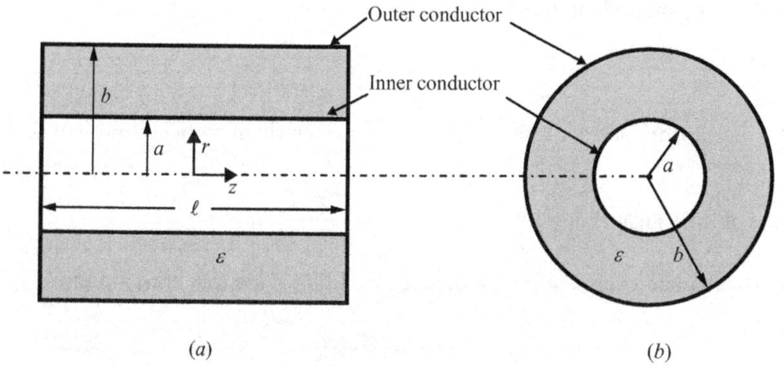

Fig. 2.61 Part of a long coaxial transmission line having one dielectric layer between conductors: (*a*) Section through the axis and (*b*) Section normal to the axis.

inside radius of the outer conductor are taken as *a* and *b* respectively. We assume that the cable is straight and its length is large compared to the radial distance between the conductors. We also assume that the conductors have equal and opposite charges per unit length distributed uniformly on the outside surface of the inner conductor and on the inside surface of the outer conductor. Accordingly, the field intensity is in *r* direction and it does not vary with ϕ and *z*.

Assuming the linear charge density of the inner conductor to be ρ_ℓ, field intensity at a point between the conductors is,

$$E_r = \frac{\rho_\ell}{2\pi\varepsilon r}, \quad b > r > a \tag{1}$$

Potential difference between the conductors,

$$V = -\int_b^a E_r dr = \frac{\rho_\ell}{2\pi\varepsilon}\ln\frac{b}{a} \tag{2}$$

Equation (1) gives the field intensity in terms of ρ_ℓ. We can express the field intensity in terms of V by eliminating ρ_ℓ from (1) and (2). Thus,

$$E_r = \frac{V}{r\ln(b/a)} \tag{3}$$

Equation (3) shows that the potential gradient, which is numerically equal to the field intensity, is inversely proportional to *r*. It has maximum value at $r = a$, and minimum value at $r = b$. The maximum and minimum values of the field intensities are

$$E_r(\text{max.}) = \frac{V}{a\ln(b/a)}, \quad E_r(\text{min.}) = \frac{V}{b\ln(b/a)}$$

For a given value of *b*, there exists a definite relation between *b* and *a* so that the maximum potential gradient is a minimum. The relation is obtained by differentiating E_r(max.) with respect *a*, and then equating the result to zero.

$$\frac{d}{da}[E_r(\text{max.})] = \frac{\ln(b/a) - 1}{[a\ln(b/a)]^2} = 0$$

Or,

$$\ln(b/a) = 1$$

$$\Rightarrow \quad b/a = e$$

That is, if the ratio between *b* and *a* is equal to 2.718, the maximum potential gradient will be minimum. The value is equal to V/a.

Capacitance of one metre length of the transmission line,

$$C = \frac{\rho_\ell}{V} = \frac{2\pi\varepsilon}{\ln(b/a)} \tag{4}$$

This is also the capacitance per unit length of two concentric metallic cylindrical shells separated by a single dielectric layer.

Capacitance of the transmission line of length $\ell = \dfrac{2\pi\varepsilon\ell}{\ln(b/a)}$

We will next calculate energy stored in the electric field of the transmission line.

$$\text{Energy stored} = \int_v \frac{\varepsilon E_r^2}{2} dv$$

The differential volume in cylindrical coordinates, $dv = r dr d\phi dz$

As field intensity inside conductors is zero, the integration need be evaluated over the dielectric region only. Thus, energy stored per unit length of the line,

$$W = \int_a^b \int_0^{2\pi} \int_z^{z+1} \frac{\varepsilon E_r^2}{2} r dr d\phi dz$$

Substituting the expression of the field intensity given by (3) and then evaluating the integral, we have

$$W = \frac{\pi\varepsilon V^2}{\ln(b/a)} = \frac{CV^2}{2}$$

where C is capacitance of the line per unit length.

Two Dielectric Layers between Conductors

> If the distance between the inner and outer conductors is large, as in the case of high voltage cables, two or more layers of insulation in the radial direction are used.

The choice of insulating material of a layer depends on the maximum potential gradient in that region. Accordingly, the dielectric strength of the insulation adjacent to the inner conductor should be of maximum value, and that of the insulation adjacent to the outer conductor is of minimum value. Such a grading of insulation results in an economical design of the cable. Figure 2.62(a) shows a coaxial cable having two dielectric layers with permittivities ε_1 and ε_2. The radii of inner and outer conductors are taken as a and b respectively. The common boundary surface of the dielectrics, which is concentric to the conductors, is at radius c. The capacitance of the cable will be obtained under the same assumption as for the cable with one dielectric layer.

The field intensities in the two dielectric regions are

$$E_{r1} = \frac{\rho_\ell}{2\pi\varepsilon_1 r}, \quad c > r > a \tag{5}$$

$$E_{r2} = \frac{\rho_\ell}{2\pi\varepsilon_2 r}, \quad b > r > c \tag{6}$$

Potential difference between the conductors,

$$V = -\int_b^c E_{r2}\, dr - \int_c^a E_{r1}\, dr = \frac{\rho_\ell}{2\pi}\left(\frac{1}{\varepsilon_1}\ln\frac{c}{a} + \frac{1}{\varepsilon_2}\ln\frac{b}{c}\right) \tag{7}$$

By eliminating ρ_ℓ from (5) and (7), we will obtain the field intensity in Region 1 in terms of V. It is given by

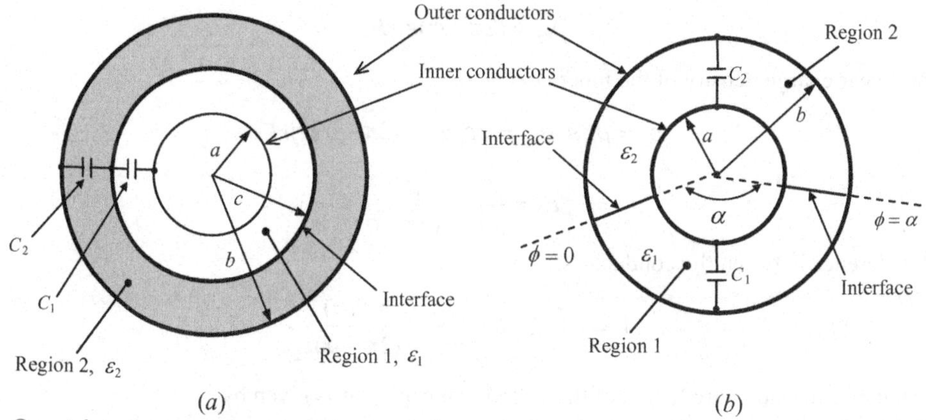

Fig. 2.62 Coaxial transmission lines having two dielectric regions: (*a*) Interface is concentric with the inner conductor and (*b*) Interfaces are normal to conductors.

$$E_{r1} = \frac{V\varepsilon_2}{r[\varepsilon_2 \ln(c/a) + \varepsilon_1 \ln(b/c)]} \tag{8}$$

And eliminating ρ_ℓ from (6) and (7), we have

$$E_{r2} = \frac{V\varepsilon_1}{r[\varepsilon_2 \ln(c/a) + \varepsilon_1 \ln(b/c)]} \tag{9}$$

$$\text{Capacitance per unit length, } C = \frac{\rho_\ell}{V} = \frac{2\pi\varepsilon_1\varepsilon_2}{\varepsilon_2 \ln(c/a) + \varepsilon_1 \ln(b/c)} \tag{10}$$

Formula (10) can be extended for any number of dielectric layers between the conductors.

It can be shown that the interface at radius *c* of Fig. 2.62(*a*) is an equipotential surface. Accordingly, each region can be considered to be a cylindrical capacitor as indicated by C_1 and C_2 in the figure. The capacitance of the transmission line is equal to the series combination of C_1 and C_2.

In the cylindrical capacitor shown in Fig. 2.62(*b*), one dielectric of permittivity ε_1 extends from $\phi = 0$ to $\phi = \alpha$ and a second dielectric of permittivity ε_2 extends from $\phi = \alpha$ to $\phi = 2\pi$. The common boundary surfaces are at $\phi = 0$ and $\phi = \alpha$. The radii of inner and outer conductors are taken as *a* and *b* respectively. If the length of the cylinders is large compared to the distance between the conductors, the field intensity in the dielectric regions would be radial and independent of ϕ and *z*. It follows that the field intensities at both the interfaces are tangential. Since the tangential field intensities on two sides of a boundary surface must be equal, the field intensity in both the dielectric regions has the same value at a given radial distance. Thus,

$$E_{r1} = E_{r2} = E_r \qquad \text{(Say)}$$

The next step is to find the relation between field intensity and linear charge density of the portion of the inner conductor in each region. We will obtain the relation using Gauss's law. To apply the law, let us take a cylindrical surface of radius r ($b > r > a$) that is concentric with the inner conductor. If D_{r1} is flux density at the surface segment in Region 1, flux out of this surface per unit axial length is equal to $\alpha r D_{r1}$. As per the law, this flux is equal to charge per unit length in the portion of the inner conductor of Region 1. Taking the charge per unit length as $\rho_{\ell 1}$, we have

$$\rho_{\ell 1} = \alpha r D_{r1} = \alpha\varepsilon_1 r E_r$$

Similarly, charge per unit length of the capacitor in the portion of the inner conductor of Region 2,

$$\rho_{\ell 2} = (2\pi - \alpha)\varepsilon_2 r E_r$$

Thus, total linear charge density of the inner conductor,

$$\rho_\ell = \rho_{\ell 1} + \rho_{\ell 2} = \alpha\varepsilon_1 r E_r + (2\pi - \alpha)\varepsilon_2 r E_r$$

$$\Rightarrow \quad E_r = \frac{\rho_\ell}{[\alpha\varepsilon_1 + (2\pi - \alpha)\varepsilon_2] r}$$

Potential difference between the conductors,

$$V = -\int_b^a E_r \, dr = \frac{\rho_\ell \ln(b/a)}{\alpha\varepsilon_1 + (2\pi - \alpha)\varepsilon_2}$$

Thus, capacitance of one metre length of the cylindrical capacitor is given by

$$C = \frac{\rho_\ell}{V} = \frac{\alpha\varepsilon_1 + (2\pi - \alpha)\varepsilon_2}{\ln(b/a)} = \frac{\alpha\varepsilon_1}{\ln(b/a)} + \frac{(2\pi - \alpha)\varepsilon_2}{\ln(b/a)} = C_1 + C_2$$

where

$$C_1 = \frac{\alpha\varepsilon_1}{\ln(b/a)}$$

is the capacitance per unit length of the part of the line in Region 1 and

$$C_2 = \frac{(2\pi - \alpha)\varepsilon_2}{\ln(b/a)}$$

is the capacitance per unit length of the part of the line in Region 2. The two capacitances are in parallel as shown in Fig. 2.62(*b*).

EXAMPLE 2.61

A coaxial transmission line operating at 11.0 kV has an inner conductor of radius 10.0 mm and an outer conductor of inside radius 30.0 mm. The relative permittivity of insulation between the conductors is 4.5. Calculate the maximum and minimum potential gradient in the insulation, and capacitance per unit length of the line.

Radius of the inner conductor, $a = 0.01$ m

Inside radius of outer conductor, $b = 0.03$ m

Permittivity of insulation, $\varepsilon = 4.5\varepsilon_0$ F/m

Operating voltage, $V = 11000$ V

The potential gradient is numerically equal to the field intensity. It is given by

$$E_r = \frac{V}{r \quad b/a)}$$

$$E_r \qquad E_r(r = a) = \frac{11000}{(0.01)(\ln 3)} = 1.001 \text{ MV/m}$$

$$E_r \qquad E_r(r = b) = \frac{11000}{(0.03)(\ln 3)} = 0.334 \text{ MV/m}$$

Capacitance per unit length of the line, $C = \dfrac{2\pi\varepsilon}{\ln(b \quad a)} = \dfrac{2 \quad 4\ 5\varepsilon_0)}{\ln 3} = 227.6 \quad /\text{m}$

EXAMPLE 2.62

The coaxial transmission line of the previous example has two layers of insulation with their interface concentric to the inner conductor. The interface is at radius 20.0 mm. The relative permittivities of inner and outer layers of insulation are 6.0 and 4.5 respectively. Calculate (a) maximum and minimum field intensities in each layer of insulation, (b) potential at the interface, and (c) capacitance of each layer and the total capacitance.

Radius of the inner conductor, $a = 0.01$ m

Inside radius of outer conductor, $b = 0.03$ m

Radius of the interface, $c = 0.02$ m

Permittivity of inner layer of insulation, $\varepsilon_1 = 6.0\varepsilon_0$ F/m

Permittivity of outer layer of insulation, $\varepsilon = 4.5\varepsilon_0$ F/m

Operating voltage, $V = 11000$ V

(a) Field intensity between the inner conductor and the interface,

$$E_{r1} = \frac{V\varepsilon_2}{r[\varepsilon_2 \ln(c/a) + \varepsilon_1 \ln(b/c)]}$$

$$- \frac{4.5 \times 11000}{r[4.5(\ln 2) + 6.0(\ln 1.5)]} = \frac{8920}{r}$$

$$E_{r1}(\text{max.}) = E_{r1}(r = 0.01) = \frac{8920}{0.01} = 0.892 \text{ MV/m}$$

$$E_{r1}(\text{min.}) = E_{r1}(r = 0.02) = \frac{8920}{0.02} = 0.446 \text{ MV/m}$$

Field intensity between the outer conductor and the interface,

$$E_{r2} = \frac{V\varepsilon_1}{r[\varepsilon_2 \ln(c/a) + \varepsilon_1 \ln(b/c)]}$$

$$E_{r2}(\text{max.}) = E_{r2}(r = 0.02) = \frac{6.0 \times 11000}{(0.02)[4.5(\ln 2) + 6.0(\ln 1.5)]} = 0.594 \text{ MV/m}$$

$$E_{r2}(\text{min.}) = E_{r2}(r = 0.03) = \frac{6.0 \times 11000}{(0.03)[4.5(\ln 2) + 6.0(\ln 1.5)]} = 0.396 \text{ MV/m}$$

(b) Potential difference between the inner conductor and the interface

$$= -\int_{0.02}^{0.01} E_{r1} dr = -\int_{0.02}^{0.01} \frac{8920}{r} = [8920 \ln r]_{0.01}^{0.02} = 6182.9 \text{ V}$$

(c) Capacitance per unit length of Layer 1, $C_1 = \dfrac{2\pi\varepsilon_1}{\ln(c/a)} = \dfrac{2\pi(8.85 \times 10^{-12})(6)}{\ln 2}$

$$= 481.3 \text{ pF/m}$$

Capacitance per unit length of Layer, $C_2 = \dfrac{2\pi\varepsilon_2}{\ln(b\ c)} = \dfrac{2\ \ 8.85 \times 10^{-12})(4\ 5)}{\ln 1\ 5}$

$$= 617.1 \text{ pF/m}$$

As the two capacitors are in series, total capacitance,

$$C = \frac{C_1 C_2}{C_1 + C_2} = \frac{(481.3)(617.1)}{481.3 + 617.1} = 270.4 \text{ pF/m}$$

2.28 Capacitance of a Transmission Line with Two Parallel Cylindrical Conductors

Parallel-wire transmission lines are used for transmission of electrical energy and electronic signals. Calculation of line capacitance is necessary to obtain a mathematical model of a transmission line and also to represent it by an equivalent circuit. We will find the capacitance of a transmission line consisting of two parallel cylindrical conductors of infinite length, as shown in Fig. 2.63(*a*). The radius of each conductor is *a*. The conductors have uniformly distributed equal and opposite charges with linear charge density ρ_ℓ C/m. The centre of the positively charged conductor is at $(-h/2, 0)$ and that of the negatively charged conductor is at $(h/2, 0)$ in the *x*-*y* plane. In order to calculate the capacitance, it is necessary to obtain potential difference between the lines in terms of the linear charge density.

A direct solution of the problem is not easy because of the following reasons.

1. The cylindrical surfaces of the conductors are equipotential surfaces. A potential function must satisfy these conditions.
2. Charge distribution along the circumferential direction is not uniform because of induction action which is known as proximity effect.

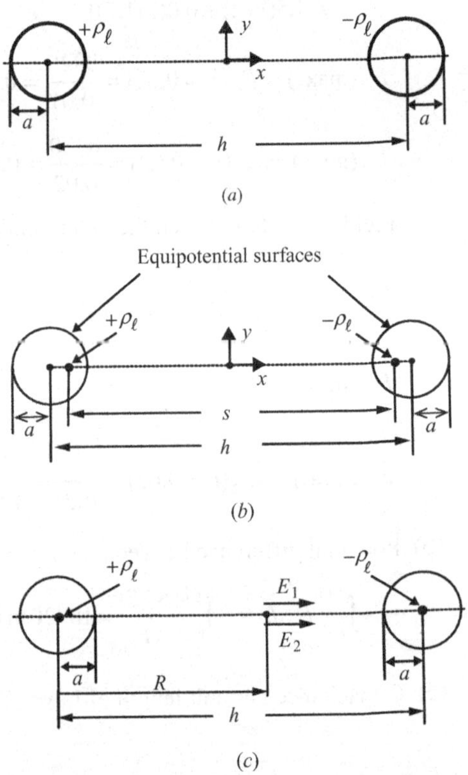

Fig. 2.63 (*a*) A Transmission line with two parallel cylindrical conductors, (*b*) Equivalent of (*a*), and (c) Approximate model of the transmission line.

However, an indirect solution of the problem can be obtained using the results of Section 2.18, where it is shown that equipotential surfaces due to two filamentary line-charges, located at $(-s/2, 0)$ and $(s/2, 0)$, are circular cylinders. As the cylindrical surfaces of the charged conductors shown in Fig. 2.63(*a*) are also equipotential surfaces, these can be replaced by two charge-free parallel cylindrical conductors having the same radius and spacing as in the original transmission line, and two equal and opposite line charges having the same linear charge density as in the original conductors and located at $(-s/2, 0)$ and $(s/2, 0)$. The equivalent model is illustrated in Fig. 2.63(*b*).

Referring to Fig. 2.44 and equation (2.18-16), we find that centres of the equipotential surfaces are at $[s(1 + k^2)]/[2(1 - k^2)]$ and the radii of these surfaces are $(ks)/(1 - k^2)$. Next, referring to the equivalent model of the transmission line under consideration, we see that the centres of the conductors are at (/2,), and radius of each conductor is *a*. Thus by analogy, we have

$$h = \frac{s(1+k^2)}{1-k^2}, \qquad (1)$$

$$a = \frac{sk}{1-k^2}, \tag{2}$$

and

$$k = \exp\left(\frac{2\pi\varepsilon V}{\rho_\ell}\right) \tag{3}$$

where V is potential of either the positively or the negatively charged conductor.

Eliminating s from (1) and (2), we get

$$ak^2 - kh + a = 0 \tag{4}$$

If k_1 and k_2 are the roots of (4), then

$$k_1 = \frac{h + \sqrt{h^2 - 4a^2}}{2a} \tag{5}$$

and

$$k_2 = \frac{h - \sqrt{h^2 - 4a^2}}{2a} \tag{6}$$

By substituting the rhs of (5) in place of k in (3), we will obtain the potential of the positively charged conductor. Denoting it by V^+, we have

$$V^+ = \frac{\rho_\ell}{2\pi\varepsilon} \ln \frac{h + \sqrt{h^2 - 4a^2}}{2a} \tag{7}$$

And substituting the rhs of (6) in place of k in (3), we will obtain the potential of the negatively charged conductor. Denoting it by V^-, we have

$$V^- = \frac{\rho_\ell}{2\pi\varepsilon} \ln \frac{h - \sqrt{h^2 - 4a^2}}{2a} \tag{8}$$

Potential difference between the conductors,

$$V = V^+ - V^- = \frac{\rho_\ell}{2\pi\varepsilon} \ln\left(\frac{h + \sqrt{h^2 - 4a^2}}{h - \sqrt{h^2 - 4a^2}}\right) = \frac{\rho_\ell}{\pi\varepsilon} \ln \frac{h + \sqrt{h^2 - 4a^2}}{2a}$$

Capacitance of a two-conductor transmission line per metre length,

$$C = \frac{\rho_\ell}{V} = \frac{\pi\varepsilon}{\ln\left[\left(h + \sqrt{h^2 - 4a^2}\right)/2a\right]} \tag{9}$$

An easier, though approximate, method of calculating the capacitance of the transmission line is by assuming that the charges are concentrated along the axes of the conductors, as shown in Fig. 2.63(c). Let us take a point P on the line joining the centres of the conductors. It is at a distance R from the centre of the conductor with positive charge. At P, the magnitudes of field intensities E_1 and E_2 due the positive and negative charges respectively are

$$E_1 = \frac{\rho_\ell}{2\pi\varepsilon R}$$

$$E_2 = \frac{\rho_\ell}{2\pi\varepsilon(h-R)}$$

As both the fields are in the same direction, as indicated in the figure, the total field intensity,

$$E_K \quad E_1 + E_2 = \frac{\rho_\ell}{2\pi\varepsilon}\left(\frac{1}{R} + \frac{1}{h-R}\right)$$

The potential difference between the conductors is

$$V = -\int_{h-a}^{a} E_R \, dR = -\frac{\rho_\ell}{2\pi\varepsilon}\int_{h-a}^{a}\left(\frac{1}{R}+\frac{1}{h-R}\right)dR$$

$$= \frac{\rho_\ell}{2\pi\varepsilon}\left[\ln R - \ln(h-R)\right]_{h-a}^{h} = \frac{\rho_\ell}{\pi\varepsilon}\ln\frac{h-a}{h}$$

Capacitance per unit length of the line, $C = \dfrac{\pi\varepsilon}{\ln[(h-a)/a]}$ (10)

If h a, both (9) and (10) give

$$C \approx \frac{\pi\varepsilon}{\ln(h/a)} \text{ F/m} \qquad (11)$$

EXAMPLE 2.63

A long transmission line situated in air consists of two parallel cylindrical conductors. The radius of each conductor is a and the distance between the conductors is h. Calculate capacitance of the line per unit length for $h = 20a$ and $h = 30a$, using, (a) the exact formula (9), (b) the approximate formulae (10), and (c) the approximate formulae (11).

(a) For $h = 20a$, $C = \dfrac{\pi\varepsilon_0}{\ln[(20a+\sqrt{400a^2-4a^2})/(2a)]} = 9.28 \text{ pF/m}$

For $h = 30a$, $C = \dfrac{\pi\varepsilon_0}{\ln[(30a+\sqrt{900a^2-4a^2})/(2a)]} = 8.17 \text{ pF/m}$

(b) For $h = 20a$, $C \approx \dfrac{\pi\varepsilon_0}{\ln[(20a-a)/a]} = 9.43 \text{ pF/m}$

For $h = 30a$, $C \approx \dfrac{\pi\varepsilon_0}{\ln[(30a-a)/a]} = 8.25 \text{ pF/m}$

(c) For $h = 20a$, $C \approx \dfrac{\pi\varepsilon_0}{\ln 20} = 9.27 \text{ pF/m}$

For $h = 30a$, $C \approx \dfrac{\pi\varepsilon_0}{\ln 30} = 8.12$

We note that the approximate values are nearly equal to those obtained using the exact formula. We find that the capacitance decreases with the increase of h/a.

2.29 Relation between Electrostatic Force and Energy

A system of charged conductors is associated with two physical properties. These are (1) each charged conductor is acted upon by electric forces due to charges on other conductors, and (2) energy is stored in the electric field in the form of potential energy. If any conductor is displaced from its original position, mechanical work has to be done. At the same time energy stored in the field, i.e., the field energy, changes.

We will derive the relation between field energy and electric force using the principle of virtual work considering a parallel-plate capacitor.

Figure 2.64(*a*) shows a parallel-plate capacitor consisting of two stationary plates *A* and *B*. The plates are connected to an electric source of constant voltage *V*. The plates have charges $+Q$ and $-Q$ as indicated in the figure. Each plate is acted upon by an electric force *F* due to the charge on the other plate. Imagine that plate *B* is virtually displaced towards plate *A* by a differential distance *dx* without disconnecting the source. Virtual mechanical work done in displacing the plate,

$$dW_m = Fdx \qquad (1)$$

As the displacement of the plate and electric force are in the same direction, work will be done by the field. The decrease in distance between the plates will increase the capacitance of the capacitor. Consequently, charges on the plates will increase as *V* is held constant. If the charge increases by a differential amount *dQ*, charge on each plate would be equal to $Q + dQ$. The additional charge, which is supplied by the source, will increase the energy stored in the capacitor. The differential increase of energy,

$$dW_f = \frac{1}{2}V(Q+dQ) - \frac{1}{2}VQ = \frac{1}{2}VdQ \qquad (2)$$

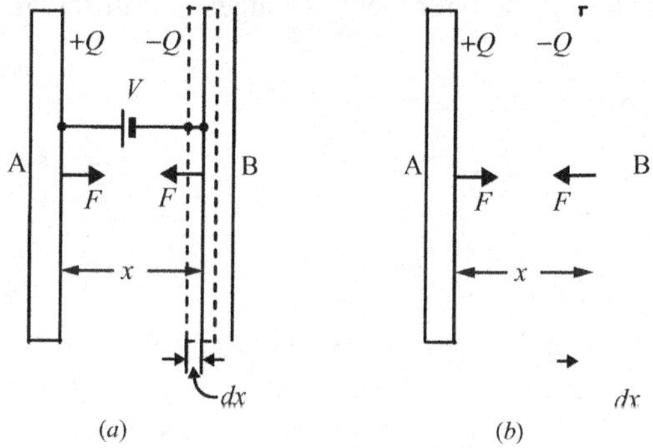

Fig. 2.64 (*a*) Parallel-plate capacitor with constant potential difference and (*b*) Parallel-plate capacitor with fixed charges.

As the source is at a constant voltage, energy drawn from the source during displacement of the plate is,

$$dW_s = V\,dQ = 2dW_f \text{ [Using relation (2)]} \qquad (3)$$

This must be equal to the sum of the mechanical work dW_m and change in the field energy dW_f. Accordingly,

$$dW_s = dW_m + dW_f \qquad (4)$$

By equating the right-hand sides of (3) and (4) and then using (1), we will get

$$dW_f = Fdx \qquad (5)$$

The total differential dW_f is equal to the partial derivative of W_f with respect to *x* multiplied by *dx*.

$$dW_f = \frac{\partial W_x}{\partial x}dx \qquad (6)$$

From (5) and (6) we have

$$F = \frac{\partial W_f}{\partial x} \qquad (7)$$

Energy stored in the field of a parallel-plate capacitor connected to a source of constant voltage is given by

$$W_f = \frac{CV^2}{2} = \frac{\varepsilon AV^2}{2x} \qquad (8)$$

where *A* is surface area of each plate and ε is permittivity of the medium between the plates. By substituting (8) into (7) and then evaluating the partial derivative we will get the force on each plate. Force on each plate,

$$F = \frac{\partial W_f}{\partial x} = -\frac{\varepsilon A V^2}{2x^2} \quad (9)$$

Next, let us consider that the plates have constant charges $+Q$ and $-Q$ and they are not connected to any electric source, as shown in Fig. 2.64(*b*). If plate B is given a virtual displacement dx towards plate A, mechanical work done by the field,

$$dW_m = Fdx$$

The decrease in distance between the plates will increase the capacitance of the capacitor and hence the energy stored in its field will decrease. Since no other source of energy is involved, the sum of the mechanical work done and change in field energy must be equal to zero as per the principle of conservation of energy. Thus, if the field energy changes by a differential amount dW_f,

$$dW_m + dW_f = 0$$

$$\Rightarrow \quad Fdx = -dW_f = -\frac{\partial W_f}{\partial x}dx$$

Or,

$$F = -\frac{\partial W_f}{\partial x} \quad (10)$$

The field energy in terms of charge on each plate and capacitance of the capacitor is,

$$W_f = \frac{Q^2}{2C} = \frac{Q^2 x}{2\varepsilon A} \quad (11)$$

By substituting (11) into (10) and then evaluating the partial derivative, we will get

$$F = -\frac{Q^2}{2\varepsilon A} \quad (12)$$

Now, let us assume that distribution of charge on the plate surfaces is uniform with a surface density ρ_S. Then,

$$Q = A\rho_S$$

By substituting this relation in (12), we will get the electrostatic force on the plates in terms of surface charge density. It is given by

$$F = \frac{A\rho_S^{\,2}}{2\varepsilon}$$

$$\text{Force per unit area} = \frac{\rho_S^{\,2}}{2\varepsilon}\ \text{N/m}^2 \quad (13)$$

As the surface charge density is defined at a point, force per unit area on a charged conductor of any shape is also given by (13).

EXAMPLE 2.64

The area of each plate of a parallel-plate capacitor is 25.0 cm^2. The plates are separated by a dielectric layer of 2.0 mm thick and relative permittivity 4.0. Calculate force on each plate of the capacitor if the potential difference between the plates 1200 V.

Area of each plate, $A = 0.0025$ m^2

Distance between the plates, $x = 0.002$ m

Relative permittivity, $\varepsilon_r = 4.0$

Potential difference, $V = 1200$ V

Force on each plate, $F = \dfrac{\varepsilon_r \varepsilon_0 A V^2}{2x}$

$$= \frac{(\ .0)(8.854\times10^{-12})(0.0025)(1200)^2}{2(0.002)} = 31.9\times10^{-6}\ \text{N}$$

2.30 Poisson's and Laplace's Equations

The relationship between flux density **D** and volume charge density ρ is given by the point form of Gauss's law. Further, gradient of the potential function V is equal to the negative of electric field intensity **E**. Both the relations are partial differential equations of first order. From these relations, a second-order partial differential equation for the potential function can be obtained. The second-order equation is useful in solving boundary value problems.

The point form of Gauss's law is, $\nabla \cdot \mathbf{D} = \rho$ (1)

where ρ is the volume charge density.

Also,

$$\nabla V = -\mathbf{E} \tag{2}$$

Taking divergence of both sides of (2), we get

$$\nabla \cdot \nabla V = \nabla \cdot (-\mathbf{E}) = -\nabla \cdot (\mathbf{D}/\varepsilon) \tag{3}$$

Assuming that the permittivity ε is not a function of the space variables, equation (3) becomes,

$$\nabla \cdot \nabla V = -\frac{1}{\varepsilon}(\nabla \cdot \mathbf{D}) \tag{4}$$

Using (1) in (4), we have

$$\nabla \cdot \nabla V = -\frac{\rho}{\varepsilon} \tag{5}$$

The operator $\nabla \cdot \nabla$ is generally written as ∇^2 and it is known as the Laplacian operator. Thus, equation (5) is written as

$$\nabla^2 V = -\frac{\rho}{\varepsilon} \tag{6}$$

Equation (6) is called Poisson's equation for the scalar potential function V. In a charge-free region the charge density is zero. Accordingly, substituting $\rho = 0$ in (6), we get

$$\nabla^2 V = 0 \tag{7}$$

Equation (7) is called Laplace's equation of the potential function V. The equations in a given coordinate system can be obtained by finding at first the gradient of V and then performing the divergence operation. The divergence of a vector function in rectangular, cylindrical and spherical coordinates are given by equations (3), (4) and (5) in Section 2.13, whereas the gradient of a scalar function in the three coordinate systems are given by (12), (19) and (25) in Section 2.17. Using these equations, we can derive the Laplacian of V for each coordinate system. These are given below.

Rectangular: $$\nabla^{\text{`}} = \frac{\partial^2 V}{\partial x^2} + \frac{\partial^2 V}{\partial y^2} + \frac{\partial^2 V}{\partial z^2} \tag{8}$$

Cylindrical: $$\nabla^{\text{`}} = \frac{1}{r}\frac{\partial}{\partial r}\left(r\frac{\partial}{\partial r}\right) + \frac{1}{r^2}\left(\frac{\partial^2 V}{\partial \phi^2}\right) + \frac{\partial^2 V}{\partial z^2} \tag{9}$$

$$\text{Spherical: } \nabla^2 V = \frac{1}{r^2}\frac{\partial}{\partial r}\left(r^2\frac{\partial V}{\partial r^2}\right) + \frac{1}{r^2 \sin\theta}\frac{\partial}{\partial \theta}\left(\sin\theta\frac{\partial V}{\partial \theta}\right) + \frac{1}{r^2\sin^2\theta}\left(\frac{\partial^2 V}{\partial \phi^2}\right) \tag{10}$$

Poisson's and Laplace's equations are second order partial differential equations. These may be one-dimensional, two-dimensional or three-dimensional depending on whether V is a function of one variable, two variables or three variables. The general solutions of these equations contain constants. The constants are determined from the known or given boundary conditions.

SUMMARY

From the vector form of Coulomb's law we defined the force field due to electric charges. And from the electric force we developed the concept of electric field intensity and electric flux density, which are vectors. We proved Gauss's law and learnt that the macroscopic form of the law is useful for calculation of electric field intensity only of some symmetrical charge distributions. We derived the divergence of flux density and found that it is equal to the electric flux out of a closed surface per unit volume. This interpretation led to the formulation of the differential and the integral forms of Gauss's law, which are parts of Maxwell's equations. We combined the two forms of Gauss's law to establish the divergence theorem for the static electric field. The divergence theorem is useful for transforming a volume integral into a surface integral and vice versa.

We learnt that energy in an electric field is stored in the form of potential energy and that an electric field is a conservative field. From the definition of gradient of a scalar function, the relation between electrostatic potential and electric field intensity, that is, grad $V = -\mathbf{E}$, was derived. We described various properties of an equipotential surface and learnt that electric field intensity is normal to an equipotential surface, and that the surface of a good conductor located in an electric field is an equipotential surface. These properties are useful for calculation of capacitances and for drawing field maps.

We derived the boundary conditions and learnt that the tangential electric field intensity and the normal electric flux density are continuous across the interface of two dielectric regions.

We applied the laws of static electric field for calculations of potential gradient, field energy, energy density, and capacitances for various symmetrical conductor configurations. Though capacitance is defined as the ratio between charge on a conductor and potential difference between the conductors, the formulae for the capacitances show that the capacitance depends on the dimensions of the conductors, their spacing, and permittivity of the dielectric medium.

IMPORTANT FORMULAE

Force between two point charges:

$$\mathbf{F}_1 = \frac{Q_1 Q_2 \mathbf{R}_{21}}{4\pi\varepsilon (R_{21})^3} = -\mathbf{F}_2$$

Field intensity of a point charge:

$$\mathbf{E} = \frac{Q}{4\pi\varepsilon R^3}[(x-x')\mathbf{u}_x + (y-y')\mathbf{u}_y + (z-z')\mathbf{u}_z]$$

Field intensity due to several point charges:

$$\mathbf{E}_n = \sum_n \frac{Q_n \mathbf{R}_n}{4\pi\varepsilon R_n^{\ 3}}, \quad n = 1, 2, 3, \ldots$$

Field intensity of a line charge of finite length:

$$\mathbf{E} \quad \frac{\ell \rho_\ell}{2\pi\varepsilon r\sqrt{4r^2+\ell^2}}\mathbf{u}_r$$

Note that r is perpendicular distance from the mid-point of the line charge to the field point and $\mathbf{u}_r$ is the unit vector along r.

Field intensity of a line charge of infinite length:

$$\mathbf{E} \quad \frac{\rho_\ell}{2\pi\varepsilon r}\mathbf{u}_r$$

Field intensity of an infinite sheet of charge:

$$\mathbf{E} = \frac{\rho_S}{2\varepsilon}$$

Gauss's law

$$\Psi = Q$$

Integral form of Gauss's law:

For volume distribution of charge, $\oint_S \mathbf{D}\cdot d\mathbf{S} = \int_v \rho dv$

For linear charge distribution, $\oint_S \mathbf{D}\cdot d\mathbf{S} - \int \rho_\ell \, d\ell$

For surface charge distribution, $\oint_S \mathbf{D}\cdot d\mathbf{S} = \int_{S'} \rho_{S'} \, dS'$

Divergence of electric flux density:

In rectangular coordinates, $\nabla\cdot\mathbf{D} = \dfrac{\partial D_x}{\partial x} + \dfrac{\partial D_y}{\partial y} + \dfrac{\partial D_z}{\partial z}$

In cylindrical coordinates, $\nabla\cdot\mathbf{D} = \dfrac{1}{r}\dfrac{\partial (rD_r)}{\partial r} + \dfrac{1}{r}\dfrac{\partial D_\phi}{\partial \phi} + \dfrac{\partial D_z}{\partial z}$

In spherical coordinates, $\nabla\cdot\mathbf{D} = \dfrac{1}{r^2}\dfrac{\partial (r^2 D_r)}{\partial r} + \dfrac{1}{r\sin\theta}\dfrac{\partial(\sin\theta\, D_\theta)}{\partial\theta} + \dfrac{1}{r\sin\theta}\dfrac{\partial D_\phi}{\partial\phi}$

Divergence theorem, $\oint_S \mathbf{D}\cdot d\mathbf{S} = \int_v (\nabla\cdot\mathbf{D})\, dv$

Energy stored in the field of two point charges, $W = \dfrac{QQ'}{4\pi\varepsilon}\left(\dfrac{1}{R_2} - \dfrac{1}{R_1}\right)$

Potential difference between points P_2 *and* P_1, $V_{21} = -\int_{P_1}^{P_2} \mathbf{E}\cdot d\ell$

Potential due to a point charge, $V = \dfrac{Q}{4\pi\varepsilon R}$

Relation between field intensity and potential, grad $V = \nabla V = -\mathbf{E}$

Field on the axis of a circular loop of charge, $V = \dfrac{Q}{4\pi\varepsilon \; a^2 \cdot z^2}$, $\mathbf{E} = \dfrac{Q}{4\pi\varepsilon z^2}\mathbf{u}_z$

where Q is total charge on the loop.

Fields on the axis of a circular disc of charge, $V = \frac{\rho_S}{2\varepsilon}(\sqrt{a^2 + z^2} - z)$

$$\mathbf{E} \quad \frac{\rho_S}{2\varepsilon}\left(\quad \frac{z}{\sqrt{a^2 + z^2}}\right)\mathbf{u}_z$$

Fields of an electric dipole, $V = \frac{p\cos\theta}{4\pi\varepsilon r^2}$, $\mathbf{E} = \frac{p}{4\pi\varepsilon r^3}(2\cos\theta\,\mathbf{u}_r + \sin\theta\,\mathbf{u}_\theta)$

Energy stored in the field of N number of discrete point charges,

$$W \quad \frac{1}{2}\sum_{i=1}^{N} Q_i V_i, \quad W = \frac{1}{8\pi\varepsilon}\sum_{i=1}^{N} Q_i \sum_{j=1}^{N} \frac{Q_j}{R_{ij}} \quad j \neq i$$

Energy density $= \frac{\varepsilon E^2}{2} = \frac{D^2}{2\varepsilon}$ J/m^3

Relation between dielectric polarisation and electric field intensity, $\mathbf{P} = \varepsilon_0 \chi_e \mathbf{E}$

Relation between relative permittivity and electric susceptibility, $\varepsilon_r = 1 + \chi_e$

Boundary relation, $E_{t1} = E_{t2}$; $D_{n1} = D_{n2}$; $\tan\alpha_1/\tan\alpha_2 = \varepsilon_1/\varepsilon_2$

Capacitors in series, $\frac{1}{C} = \frac{1}{C_1} + \frac{1}{C_2} + \frac{1}{C_3} + \cdots$

Capacitors in parallel, $C = C_1 + C_2 + C_3 + \cdots$

Capacitance of a parallel-plate capacitor, $C = \frac{A\varepsilon}{s}$

Capacitance of multi-plate capacitor, $C = \frac{(n-1)A\varepsilon}{s}$

Relative permittivity, $\frac{C_{dielectric}}{C_{air}} = \varepsilon_r$

Capacitance of two concentric spheres, $C = \frac{4\pi\varepsilon ab}{b-a}$

Capacitance per unit length of coaxial transmission line, $C = \frac{2\pi\varepsilon}{\ln(b/a)}$

Field intensity in terms of potential difference:

Two concentric spheres, $E_r = \frac{Vab}{(b-a)r^2} \quad b > r > a$

Coaxial transmission line, $E_r = \frac{V}{r\ln(b/a)} \quad b > r > a$

Capacitance of a two-conductor transmission line per metre length,

$$C = \frac{\pi\varepsilon}{\ln\left[\left(h + \sqrt{h^2 - 4a^2}\right)/2a\right]} \approx \frac{\pi\varepsilon}{\ln(h/a)}$$

Electrostatic force per unit area $= \frac{\rho_S^2}{2\varepsilon}$

MULTIPLE-CHOICE QUESTIONS

(Starred questions have two answers)

2.1 A dielectric medium is said to be linear if
(a) the permittivity is not a function of the electric force
(b) the relative permittivity is same in all the directions
(c) the permittivity is a linear function of electric force
(d) the permittivity is same everywhere

2.2 Two point charges $Q_1 = +Q$ and $Q_2 = -Q$ are located at (3 m, 0, 0) and (0, 4 m, 0) respectively in rectangular coordinates. The electric force on charge Q_2 is along the unit vector
(a) $0.6\mathbf{u}_x + 0.8\,\mathbf{u}_y$
(b) $-0.6\mathbf{u}_x - 0.8\,\mathbf{u}_y$
(c) $0.6\mathbf{u}_x - 0.8\,\mathbf{u}_y$
(d) $0.8\mathbf{u}_x - 0.6\,\mathbf{u}_y$

*2.3 The electric field intensity
(a) is equal to the force on a positive test charge in an electric field
(b) has the same direction as the force on a positive test charge in an electric field
(c) directly proportional to the permittivity of the medium
(d) is a continuous function in a homogeneous medium

*2.4 In an electric field, which of the following statements are correct?
(a) The tangent at a point on a field line gives the direction of electric field intensity.
(b) Both the magnitude and direction of the field intensity can be found at a point on a field line.
(c) In a charge-free region, the flux in a flux tube has the same value throughout.
(d) In a charge-free region, both tangential and normal components of field intensity can exist on the surface of a flux tube.

2.5 The region between two large sheets of charges with surface charge densities ρ_S and $-\rho_S/2$ is filled with a perfect dielectric material. The electric field intensity between the plates is E. The relative permittivity of the dielectric material is give by

(a) $\dfrac{\rho_S}{\varepsilon_0 E}$ (b) $\dfrac{\varepsilon_0 E}{\rho_S}$ (c) $\dfrac{\rho_S}{2\varepsilon_0 E}$ (d) $\dfrac{3\rho_S}{4\varepsilon_0 E}$

2.6 Two concentric spherical shells with uniform charge distributions are centred at the origin. The radii of the outer and inner shells are a and b respectively. The total charge of the inner and outer spheres are $-Q$ and $Q/2$, respectively. The electric flux density at a radial distance $r > b$ is

(a) $\dfrac{Q}{4\pi r^2}\mathbf{u}_r$ (b) $-\dfrac{Q}{8\pi r^2}\mathbf{u}_r$ (c) $-\dfrac{Q}{4\pi r^2}\mathbf{u}_r$ (d) $\dfrac{Q}{8\pi r^2}\mathbf{u}_r$

2.7 Two line charges with linear charge densities ρ_ℓ and $-\rho_\ell$ are located in air at $x = -a/2$ and $x = +a/2$ respectively. The field intensity at $x = 0$ is

(a) $\dfrac{2\rho_\ell}{\pi\varepsilon_0 a}\mathbf{u}_x$ (b) $-\dfrac{2\rho_\ell}{\pi\varepsilon_0 a}\mathbf{u}_x$ (c) zero (d) $\dfrac{\rho_\ell}{\pi\varepsilon_0 a}\mathbf{u}_x$

2.8 Charge is uniformly distributed in two coaxial cylinders with their axes along the z direction of cylindrical coordinates. The radii of the inner and outer cylinders are a and b respectively. The charge per unit length of the inner cylinder is $+\rho_\ell$ and that of the outer cylinder is $-3\rho_\ell/4$. The electric flux density at $r > b$ is given by
(a) $\rho_\ell/4\pi r$ (b) $\rho_\ell/2\pi r$ (c) $\rho_\ell/8\pi r$ (d) $7\rho_\ell/8\pi r$

2.9 In an electric field the flux density at (x, y) is given by $\mathbf{D} = \frac{x}{x^2+y^2}\mathbf{u}_x + \frac{y}{x^2+y^2}\mathbf{u}_y$. The divergence of $\mathbf{D}$ is
(a) $\frac{xy}{(x^2+y^2)^2}$ (b) $\frac{x+y}{x^2+y^2}$ (c) zero (d) $\frac{2}{x^2+y^2}$

2.10 The volume charge density in a region is defined by $re^{-r}\sin^2\theta$ in spherical coordinates. The divergence of electric flux density at $r = 0.5$, $\theta = 60°$ is
(a) zero (b) 0.263 (c) 0.618 (d) 0.228

2.11 The electric flux density in a static electric field is $x^2\mathbf{u}_x + xy\mathbf{u}_y + z^2\mathbf{u}_z$ C/m². The volume charge density at $x = 1.0$ m, $y = 2.0$ m, $z = 3$ m is equal to
(a) 12.0 C/m³ (b) 10.0 C/m³ (c) 8.0 C/m³ (d) 9.0 C/m³

*2.12 In a region of electric charge distribution, which of the following statements are true?
(a) The divergence of electric field intensity is equal to the surface charge density.
(b) The divergence of electric flux density is equal to the volume charge density.
(c) The divergence theorem for electric flux density **D** gives relation between volume integral and surface integral of **D**.
(d) The divergence theorem for electric flux density **D** gives relation between surface integral and line integral of **D**.

*2.13 Work done in displacing a point charge in an electric field
(a) depends only on the initial and final positions of the charge
(b) depends on the shape of the path taken by the charge from the initial to the final position
(c) is independent of the permittivity of the medium
(d) is zero if the initial and final positions are same irrespective of the shape of the path taken by the charge

*2.14 Two point charges $+Q$ and $-2Q$ are located at $x = -a$ and $x = a$ respectively. The potential is zero at
(a) $x = -a/6$ (b) infinity (c) $x = a/3$ (d) $x = -a/3$

*2.15 Which of the following statements are true for a static electric field?
(a) Equipotential surfaces that are at different potentials do not intersect.
(b) Work done in moving a test charge from one point to another point on an equipotential surface is zero.
(c) The field intensity normal to an equipotential surface is zero.
(d) Equipotential surfaces may or may not be continuous surfaces.

2.16 The static electric field is a conservative field because
(a) divergence of electric field intensity is zero
(b) line integral of electric field intensity around a closed path is zero
(c) the gradient of the potential function is zero
(d) no work is done in moving a test charge from one point to another point on an equipotential surface

2.17 Potential at a point in an electric field is given by $V = a/\sqrt{b^2 + x^2}$. The electric field intensity at $x \gg b$ is

(a) $(a/x^2)\mathbf{u}_x$ (b) $(a/x)\mathbf{u}_x$ (c) $(a/\sqrt{x})\mathbf{u}_x$ (d) $(a/x^{3/2})\mathbf{u}_x$

2.18 Potential at a point (x, y) in an electric field is given by $V = a/\sqrt{x^2 + y^2}$. The electric field intensity at (x, y) is

(a) $\dfrac{a}{(x^2 + y^2)^{3/2}}(x\mathbf{u}_x + y\mathbf{u}_y)$ (b) $\dfrac{a}{(x^2 + y^2)^{3/2}}(\mathbf{u}_x + \mathbf{u}_y)$

(c) $\dfrac{a}{\sqrt{x^2 + y^2}}(x\mathbf{u}_x + y\mathbf{u}_y)$ (d) $\dfrac{a}{\sqrt{x^2 + y^2}}(\mathbf{u}_x + \mathbf{u}_y)$

2.19 The potential of an electric dipole at a large but finite distance r from the mid-point of a dipole
(a) is inversely proportional to r^3
(b) is inversely proportional to r
(c) is inversely proportional to r^2
(d) is independent of r

*2.20 Energy stored in the field of two point charges is
(a) inversely proportional to square of distance between the charges
(b) inversely proportional to distance between the charges
(c) inversely proportional to permittivity of the medium
(d) directly proportional to permittivity of the medium

2.21 The electric flux density in a dielectric medium of permittivity ε is given by a/r^2 in spherical coordinates. Field energy contained in a differential volume is

(a) $\dfrac{a^2}{\varepsilon r^2} dr d\theta d\phi$ (b) $\dfrac{a^2}{2\varepsilon r^2} dr d\theta d\phi$

(c) $\dfrac{a^2}{2\varepsilon r^4} \sin\theta dr d\theta d\phi$ (d) $\dfrac{a^2}{2\varepsilon r^2} \sin\theta dr d\theta d\phi$

*2.22 In the case of a good conductor located in an electric field,
(a) electric field intensity is tangential to the conductor surface
(b) electric field intensity inside the conductor has a finite value
(c) electric field intensity is normal to the conductor surface
(d) the conductor surface is an equipotential surface

*2.23 If **E**, **P**, χ_e and ε_r are electric field intensity, polarisation, electric susceptibility and relative permittivity of a dielectric material, then
(a) $\mathbf{P} = (\varepsilon_0/\chi_e)\mathbf{E}$ (b) $\mathbf{P} = \varepsilon_0\chi_e\mathbf{E}$
(c) $\varepsilon_r = \chi_e + 1$ (d) $\varepsilon_r = \chi_e - 1$

*2.24 Which of the following statements pertaining to the boundary conditions between two perfect dielectric materials are correct?
(a) Both tangential and normal components of electric field intensities are continuous across the boundary surface.
(b) Tangential components of electric field intensities are equal on both sides of the interface.

(c) Normal component of flux density is continuous across the boundary surface.
(d) Both tangential and normal components of flux densities are continuous across the boundary surface.

*2.25 The dielectric strength of a dielectric material decreases,
(a) with increase in temperature of the material
(b) with decrease in temperature of the material
(c) with increase in moisture in the material
(d) with decrease in moisture in the material

2.26 Two identical parallel circuits are formed with four capacitors, each of 50 μF. The capacitance of the combination is
(a) 25 μF (b) 200 μF (c) 100 μF (d) 50 μF

2.27 Two capacitors with capacitances C and $2C$ are connected in series. The terminal voltage and energy stored of the series combination are V and W_{se} respectively. Next, the capacitors are connected in parallel. The terminal voltage and energy stored of the parallel combination are $V/2$ and W_{pa} respectively. The ratio W_{se}/W_{pa} is
(a) 16/9 (b) 8/9 (c) 4/9 (d) 2/9

*2.28 Consider that a parallel-plate capacitor with air as dielectric is connected to a source of constant voltage. If the distance between the plates is decreased without disconnecting the supply, then
(a) surface charge density on the plates will decrease
(b) energy stored in the capacitor will increase
(c) capacitance of the capacitor will decrease
(d) flux density between the plates will increase

*2.29 A parallel-plate air capacitor is connected to a voltage source for a short time and then disconnected. If the space between the plates is now filled with a perfect dielectric material, then
(a) potential difference between the plates will decrease
(b) flux density between the plates will not change
(c) energy stored in capacitor will increase
(d) field intensity between the plates will not change

2.30 A parallel-plate capacitor has two dielectric layers with the interface parallel to the plates. The thickness and relative permittivity of Layer 1 are s and 2.0 respectively and those of Layer 2 are $1.5s$ and 4 respectively. If C_1 and C_2 are capacitances of the two layers, then C_2/C_1 is
(a) 4/3 (b) 1.0 (c) 3/4 (d) 8/3

2.31 An isolated conducting sphere of radius 50 cm is located in air. The sphere is charged with a potential of 10000 V. The surface charge density in C/m^2 on surface of the sphere is nearest to
(a) 17.7×10^{-8} (b) 1.77×10^{-8} (c) 8.85×10^{-8} (d) 88.5×10^{-8}

2.32 The radii of the inner as well as the outer conductors of two coaxial transmission lines 1 and 2 are the same. However, the permittivity and operating voltage of Line 2 are twice those of Line 1. The ratio between energy stored per unit length of lines 1 and 2 is
(a) 1/2 (b) 1/4 (c) 1/8 (d) 1/16

QUESTIONS

2.1 On what assumptions is Coulomb's law based?

2.2 The study of electric fields is generally based on the assumptions that the dielectric medium is homogeneous, linear, and isotropic. What properties should a dielectric medium have so that it can be treated as a homogeneous, linear, and isotropic medium? Why are these assumptions made?

2.3 What is the approximate value of permittivity of air at normal temperature and pressure?

2.4 Two point charges $2Q$ and $-Q$ are separated by a finite distance. Is there a point at a finite distance from the charges where the electric field intensity is zero? If so, identify the point by means of a sketch. What is the relation between the distances of the point from the charges?

2.5 Two positive point charges have a total charge Q. For a given distance between the charges, show that the force between the charges is a maximum if the value of each charge is $Q/2$.

2.6 Except for the units, electric field intensity has the same properties as the electric force. The electric force is a physical and measurable quantity. What about electric field intensity?

2.7 Why is the direction of electric field intensity away from a positive charge distribution and towards a negative charge distribution?

2.8 The electric field intensity at the mid-point on the line joining two point charges is zero. Is there any other point where the field intensity is also zero?

2.9 A horizontal line AB has its mid-point at O, and line OP is perpendicular to AB. Two equal positive point charges are located at the ends of the horizontal line. What are the directions of field intensities due to the charges at P if (a) P is above the horizontal line and (b) P is below the horizontal line?

2.10 Describe linear, surface and volume charge distributions. State the units used to describe these charge distributions. How are the charge densities for non-uniform charge distributions defined?

2.11 An infinite line charge with uniform charge distribution is oriented along the z axis of cylindrical coordinates. What is the direction of electric field intensity?

2.12 What is the direction of electric field intensity on the axis of a circular ring of charge?

2.13 Describe the properties of electric field lines. Is there a limit to the number of field lines between two opposite charge distributions?

2.14 Under what assumptions is the electric field intensity of a long line charge calculated using Gauss's law?

2.15 Under what assumptions is the electric field intensity of a large sheet of charge calculated using Gauss's law?

2.16 Write the integral form of Gauss's law and explain the meaning of each side.

2.17 Can Gauss's law be applied to calculate the electric field intensity of (a) a cylindrical distribution of charge, and (b) a spherical charge distribution if the volume charge distribution in each case varies with the radius only?

2.18 Is it possible to calculate approximately the field intensity at any point just above a circular disc of uniform charge distribution with finite radius using Gauss's law? If so, what assumption is to be made and what should be the shape of the special Gaussian surface?

2.19 What is the unit of $\nabla \cdot \mathbf{D}$?

2.20 A long line charge is on the axis of cylindrical coordinates. Show that the divergence of flux density due to the line charge is zero for $r > 0$.

2.21 A point charge is situated at the origin of spherical coordinates. Show that the divergence of flux density due to the point charge is zero for $r > 0$.

2.22 If a negative point charge is brought from infinity and placed in the field of some other point charges, how will it affect the energy stored in the field?

2.23 Show that the field intensity of a line charge satisfies the conservative property.

2.24 The negative of line integral of $\mathbf{E}$ from a point P_1 to a point P_2 is found to be positive. Which point is at a higher potential?

2.25 What is the work done in moving a test charge, positive or negative, from one point to another point on an equipotential surface?

2.26 The electric field intensity is perpendicular to an equipotential surface. Explain why.

2.27 Explain with examples that there can be points in an electric field where (a) the field is zero but the potential is not zero, (b) the potential is zero but the field intensity is not zero.

2.28 In an electric field two equipotential surfaces cannot intersect. Explain why.

2.29 A good conductor is located in a static electric field. Why (a) No field exists inside the conductor? (b) Charges are only on its surface? (c) Potential on its surface is same everywhere?

2.30 What is the potential at a point due to an infinite line charge?

2.31 The electric field intensity due to an infinite line charge can be found by the use of Coulomb's law and Gauss's law. Is there a third method available? If so, explain. No derivation is required.

2.32 What is the relation between relative permittivity and electric susceptibility?

2.33 What is the value of dielectric strength of air at normal temperature and pressure?

2.34 State in words the boundary conditions at the interface between two dielectric media.

2.35 Does the capacitance between two conductors depend on the value of potential difference between the conductors and/or magnitude of charge in a conductor?

2.36 The plate area, distance between the plates, and voltage between the plates of two parallel plate capacitors 1 and 2 are same. However, the relative permittivity of the dielectric between plates of capacitor 2 is twice that of capacitor 1. What are the values of field intensity, flux density, capacitance, and stored energy of capacitor 2 in terms of the corresponding values of capacitor 1?

2.37 Answer question 2.36 if both the capacitors have the same plate charges.

2.38 The relative permittivity of a dielectric material can be determined experimentally. Explain how.

2.39 A parallel-plate capacitor has two dielectric layers with the interface parallel to the plates. Will the value of capacitance change if a very thin conducting plate is inserted at the interface?

2.40 The distance and dielectric material between the plates of two parallel plate capacitors 1 and 2 are same. However, the plate area of capacitor 2 is twice that of 1. The series combination of the capacitors is connected to a source with arrangements to vary the voltage. If the voltage is increased until breakdown occurs, which capacitor will rupture first? Note that after one capacitor ruptures, the other is also likely to rupture.

2.41 In which region of a coaxial cable does the potential gradient has the maximum value?

2.42 What for multilayer of insulation between the conductors of a high voltage coaxial transmission line is preferred?

2.43 What assumptions have to be made to calculate approximately the capacitance between two parallel circular cylindrical conductors?

ANSWERS TO SELECTED QUESTIONS

2.2 The study of electric fields is generally based on the assumptions that the dielectric medium is homogeneous, linear and isotropic. What properties should a dielectric medium have so that it can be treated as a homogeneous, linear and isotropic medium? Why are these assumptions made?

Answer: A dielectric medium is said to be homogenous if its dielectric properties are same everywhere. It is linear if the permittivity is not a function of electric force. It is isotropic if its permeability is same in all directions. These assumptions greatly simplify the analysis of electric field problems and make possible for superposition of fields due to several charges. The permittivity can also be taken as constant.

2.7 Why is the direction of electric field intensity away from a positive charge distribution and towards a negative charge distribution?

Answer: The definition of electric field intensity is based on the electric force on a positive test charge. The force on the test charge due to a positive charge distribution is repulsive and the force due to a negative charge distribution is attractive. As the direction of field intensity is same as the direction of force on the test charge, the field intensity is away from a positive charge distribution and towards a negative charge distribution.

2.10 Describe linear, surface and volume charge distributions. State the units used to describe these charge distributions. How are the charge densities for non-uniform charge distributions defined?

Answer: Distribution of charge along a filamentary line is a line charge. For uniform charge distribution, charge per unit length has a constant value. Charge per unit length is called linear charge density and its unit is C/m.

Distribution of charge in the form of a thin sheet is called surface charge distribution. For uniform charge distribution, charge per unit area has a constant value. Charge per unit area is called surface charge density and its unit is C/m^2.

Distribution of charge in a region of volume is called volume charge distribution. For uniform charge distribution charge per unit volume has a constant value. Charge per unit volume is called volume charge density and its unit is C/m^3.

If charge distribution, say over a region of volume is not uniform, then the ratio $\Delta Q/\Delta v$ in the limit Δv tending to zero is defined as volume charge density; ΔQ is charge in a volume element Δv.

2.18 Is it possible to calculate approximate field intensity at any point just above a circular disc of uniform charge distribution with finite radius using Gauss's law? If so, what assumption is to be made and what should be the shape of the special Gaussian surface?

Answer: The field distribution of a circular disc of charge, even with a uniform charge distribution, is not uniform. However just above the disc, the field is nearly uniform and direction of the field intensity is normal to the plane of the disc. Hence, the field intensity just above or below the charge distribution can be found by equating the flux *out* from and charge enclosed by a cylindrical Gaussian surface of very small height.

2.22 If a negative point charge is brought from infinity and placed in the field of some other point charges, how will it affect the energy stored in the field?

Answer: Consider that a region has only positive point charges. When a negative point charge is brought from infinity and placed in the field of the existing charges, work done would be negative and hence energy stored in the field will be reduced. However if the field is due to negative charges only, energy stored in the field will increase. If a region has both positive and negative charges, the field energy will either increase or decrease depending on the charge magnitudes, their signs and distances between the charges.

2.29 A good conductor is located in a static electric field. Why (a) No field intensity exists inside the conductor? (b) Charges are only on its surface? (c) Potential on its surface is same everywhere?

Answer: (a) When a good conductor is placed in an electric field, the free electrons in it are subjected to electric forces and all the electrons move to the conductor surface. As no free charge can remain inside the conductor, the field below the conductor surface is zero. (b) The charges cannot move beyond the surface if the region surrounding the conductor is free space (or a perfect dielectric medium) except when the field intensity is sufficiently large. (c) As per the boundary conditions, the field intensity is normal to the surface of a good conductor.

2.34 State in words the boundary conditions at the interface between two dielectric media.

Answer: At the common boundary of two perfect dielectric regions, (1) the tangential electric field intensity is continuous across the boundary and (2) The normal components of electric flux densities on two sides of the boundary are equal. (3) If the boundary has a sheet of charge, the difference between the flux densities on two sides of the boundary is equal to the charge density.

From (1) and (2) more relations can be derived.

2.35 Does the capacitance between two conductors depend on the value of potential difference between the conductors and/or magnitude of charge in a conductor?

Answer: The capacitance between two conductors is defined as the ratio of charge of one conductor and the potential difference between the conductors. As the charge and potential difference are interdependent, one quantity can be expressed in terms of the other quantity. This results in cancellation of either the charge or the potential difference in the ratio. Therefore, the capacitance does not depend either on the charge or the potential difference.

2.36 The plate area, distance between the plates, and voltage between the plates of two parallel plate capacitors 1 and 2 are same. However, the relative permittivity of the dielectric between plates of capacitor 2 is twice that of capacitor 1. What are the values of field intensity, flux density, capacitance, and stored energy of capacitor 2 in terms of the corresponding values of capacitor 1?

Answer: As the spacing and voltage between the plates of the capacitors are same, hence $E_2 = E_1$. However, $D_2 = 2D_1$ because $\varepsilon_{r2} = 2\varepsilon_{r1}$. The capacitance C_2 of capacitor 2 is twice the capacitance C_1 of capacitor 1 because though the spacing and area of the plates of both the capacitors are same, the relative permittivity of capacitor 2 is twice that of capacitor 1. Energy stored in a capacitor is given by $CV^2/2$. Since $C_2 = 2C_1$ and $V_2 = V_1$, energy stored in capacitor 2 is twice that of capacitor 1.

2.39 A parallel-plate capacitor has two layers of dielectric materials with the interface parallel to the plates. Will the value of capacitance change if a very thin conducting plate is inserted at the interface?

Answer: Assuming the field intensities in the dielectric regions to be constant; and as the interface is parallel to the plates, the interface is an equipotential surface. The surface of a good conductor in a static electric field is also an equipotential surface. Hence if a conducting plate is inserted at the interface, the field distribution will not change. Moreover, because the plate is thin the thickness of the dielectrics will be nearly same as before inserting the plate. Hence, the value of capacitance will be nearly same even after inserting the plate.

2.40 The distance and dielectric material between the plates of two parallel plate capacitors 1 and 2 are same. However, the plate area of capacitor 2 is twice that of 1. The series combination of the capacitors is connected to a source with arrangements to vary the voltage. If the voltage is increased until breakdown occurs, which capacitor will rupture first?

Answer: The capacitance of a parallel plate capacitor is proportional to the plate area. Thus, the capacitance of capacitor 2 is twice that of capacitor 1. When the capacitors are connected in series, the plates of the capacitors have the same charge magnitude but voltages across the capacitors are different. As the voltages are inversely proportional to their capacitances, the voltage across capacitance 1 is twice that of capacitor 2. Hence, capacitor 1 will rupture first.

2.42 What for multiple layers of insulation between the conductors of a high voltage coaxial transmission line is preferred?

Answer: The potential gradient is numerically equal to the electric field intensity of a coaxial transmission line. With a single dielectric layer between conductors, $E_r = \rho_\ell / [r \ln(b/a)]$ where a and b are radii of inner and outer conductors. The maximum and minimum values of potential gradient occur at $r = a$ and $r = b$. Both the values depend on the ratio b/a. To limit the maximum value of potential gradient, the value of b, and hence the radius of the line, has to be large for a given value of a. The difference between the maximum and minimum potential gradient will also be large. However, if two or more dielectric layers with proper grading of thickness and permittivity are used, the potential gradient between the conductors becomes less non-uniform and hence the radius of the line would be less. Moreover, the cost of insulation is reduced.

PROBLEMS

*2.1 A point charge of 60 μC is located on the x axis at a distance of 0.5 m from the origin in free space. A second point charge of –30 μC is located on the y axis at a distance of –0.3 m from the origin. Find the electric force on the 60 μC charge. Show the location of the charges on the x-y plane and direction of the force.

*2.2 In addition to the two charges described in Problem 2.1, a third point charge of 50 μC is situated on the z axis at a distance of 0.8 m from the origin. Find the force on the 50 μC charge.

*2.3 Two point charges $Q_1 = 40\ \mu$C and $Q_2 = 50\ \mu$C are located at P_1(1.2 m, –0.9 m) and P_2(–0.7 m, 1.5 m) respectively in free space. Show locations of the charges on the x-y plane. Calculate the magnitude of the force on Q_1, and the angle of the force vector with the x axis.

**2.4 Charges of 150 nC and –150 nC are situated at (1.3 m, –0.7 m, 1.0 m) and (–0.7 m, 2.3 m, 2.0 m) respectively in free space. Calculate the force on the negative charge. Express the force by its magnitude and unit vector.

**2.5 Two point charges $Q_1 = 25\ \mu C$ and $Q_2 = -20\ \mu C$ are positioned on the x-y plane in free space. The force on charge Q_1 is given by $\mathbf{F}_1 = 0.3\mathbf{u}_x - 0.4\ \mathbf{u}_y$ N. The coordinates of location of Q_1 are $x_1 = -0.75$ m and $y_1 = 1.5$ m. Find the position of Q_2.

**2.6 In free space and in cylindrical coordinates, a point charge of –30 μC is situated at (0.6 m, 135°, –0.9 m). A second point charge 200 nC is situated at (1.3 m, 0°, 0.5 m). Calculate the electric force on the second charge. Give your answer in rectangular coordinates.

**2.7 A point charge of 500 nC is located in free space at (0.5 m, 30°, 0°) in spherical coordinates. A second point charge of –500 nC is at (1.0 m, 0°, ϕ). Calculate the electric force on the positive charge. Give your answer in rectangular coordinates.

**2.8 Two point charges of 30×10^{-10} C and 45×10^{-10} C are located at $(x_1, y_1) = (1$ m, 2 m) and $(x_2, y_2) =$ (2 m, 4.8 m) respectively in air. Find the electric field intensity at the mid-point on the line joining the charges.

**2.9 Point charges Q and $4Q$ are located at (1 m, 2 m, 3 m) and (2 m, 4 m, 8 m), respectively. Find the position of zero field intensity.

*2.10 A line charge of infinite length is oriented along the z axis of cylindrical coordinates. The line carries a uniformly distributed charge of ρ_ℓ C/m. Find the electric field intensity vector at (r, ϕ, z).

**2.11 A positive point charge of Q coulombs is enclosed by a closed surface of an arbitrary shape. Find the electric flux out of the closed surface, without using Gauss's law.

*2.12 A line charge of infinite length with uniform charge distribution lies along the z axis of cylindrical coordinates in air. The linear charge density is 250 pC/m. Find the electric field intensity at (5 m, ϕ, z).

*2.13 A line charge of infinite length has a uniform charge distribution with linear charge density of 120 pC/m. The line is situated along the z axis of rectangular coordinates, in air. Find the electric field intensity at (a) $x = 0$ m, $y = 5$ m, $z = 0$, and (b) $x = 3$ m, $y = 4$ m, $z = 0$.

*2.14 A line charge of infinite length with linear charge density 2.5×10^{-10} C/m is oriented along the z axis of rectangular coordinates, in air. Find the electric field intensity at (2 m, 3 m, 4 m).

**2.15 A line charge of infinite length and carrying a charge of 50 pC/m is oriented parallel to the y axis, in air. The line charge intersects the x-z plane at $x = 1$ m and $z = 2$ m. Calculate the electric field intensity at $(x, y, z) = (2$ m, y, 6 m).

***2.16 A line charge of uniform linear charge density ρ_ℓ C/m is on the x axis. One end of the line is at $x = 0$ and the end of the line is at $x = \ell$. Find the electric field intensity vector at (0, y, 0).

*2.17 A line charge of length 10 m has a uniformly distributed charge of density 0.8 nC/m. It is situated in air along the x axis with its mid-point at the origin. Calculate the electric field intensity vector at $x = 0$, $y = 0$, $z = 6$ m.

**2.18 A straight line is described by the equation $y_1 = 2x_1 + 3$. The line carries a uniformly distributed charge with a linear charge density of 5 nC/m. Find the field intensity at (1.6 m, 2.5 m, 0).

**2.19 A semi-infinite line charge in air is on the positive x axis. One end of the line is at the origin. The line has a uniformly distributed charge with density ρ_ℓ C m. Find the electric field intensity at $x = a$, $y = b$. If $\rho_\ell = 0.4$ nC/ , $a = 2$ m, $b = 3$ m, what is the electric field intensity?

**2.20 Two long line charges carrying uniformly distributed charges of 6 nC/m and –6 nC/m are parallel to the z axis in air. The lines with positive and negative charges are at $y = -25$ cm and $y = 25$ cm respectively. Find the electric field intensity vector at $y = -10$ cm.

**2.21 Two line charges with equal and opposite linear charge densities are located in air, parallel to the z axis. The line with positive charge is at $x = -0.5$ m, $y = 0$, and the line with negative charge is at $x = 0.5$ m, $y = 0$. The value of charge density is 0.4 μC/m. Find the electric field intensity at the points $P(0, 0, 2\text{ m})$ and $P'(0.6\text{ m}, 0.8\text{ m}, 0)$.

*2.22 A large sheet of charge has a uniformly distributed charge of density ρ_S C/m². Find the electric field intensity vector at a height h. Use Gauss's law.

**2.23 A large sheet of charge with surface charge density of 5 nC/m² lies on the x-y plane. A second sheet of charge that is parallel to the first sheet is at $z = 10$ cm. The surface charge density of the second sheet is –7.5 nC/m². Both the charge distributions are located in air. Calculate the field intensity due to the surface charges at (a) $z < 0$, (b) $0 < z < 10$ cm, and (c) $z > 10$ cm.

**2.24 A thin circular disc with radius a has a uniform charge density of ρ_S C/m². The centre of the disc is at the origin and its axis is along z direction of cylindrical coordinates. Find the field intensity vector at $(0, 0, z)$.

***2.25 A plane surface is described by the equation $2x + 3y + 5z = 30$. The surface, situated in air, has a uniformly distributed charge of density 400 pC/m². Find the field intensity vector due to the charge distribution.

**2.26 The electric field intensity on the x-y plane is given by $\mathbf{E} = 2y^2\mathbf{u}_x + 3xy\mathbf{u}_y$ V/m. Obtain the general equation for the field lines.

*2.27 A line charge of length 1.2 m is on the x axis with its mid-point at the origin. The charge density is $\rho_\ell = 2e^{-x}$ nC/m. Calculate the total charge in the line.

***2.28 The electric field intensity in a dielectric region of relative permittivity 2.8 is given by $\mathbf{E} = (x^2 + yz)\mathbf{u}_x$ kV/m. (a) Find the electric flux density vector at $x = 5$ m. (b) Calculate the electric flux crossing a flat surface that is normal to the x axis for $x = 5$ m, $0 \le y \le 3$ m, -1 m $\le z \le 2$ m

**2.29 Charge is distributed on the x-y plane with density $\rho_S = 3x^2 + 4xy$ μC/m². How much charge is contained in the region -1 m $\le x \le 2$ m, $0 \le y \le 1$ m?

***2.30 The charge density of a circular disc of charge with radius 8 cm is given by $15\,re^{-r}$ C/m². Calculate the electric flux crossing a closed surface of arbitrary shape enclosing the disc completely.

***2.31 Charge is distributed over a solid cylinder. The radius and height of the cylinder are a and h respectively. The volume charge density is given by $\rho = \rho_0 e^{-r} \sin^2\phi$. Find the flux out of a closed surface enclosing the cylinder.

**2.32 Charge distribution in spherical coordinates is defined by the volume charge density $\rho = 25\, r \sin(\phi/4)$ C/m³. How much charge is contained in a sphere with radius 10 cm and centred at the origin.

**2.33 The volume charge density in cylindrical coordinates is given by $\rho = 2e^{-r}$ C/m³. How much charge is enclosed in the region of a cylinder with radius 3 cm and height 10 cm? The axis of the cylinder is along the z axis.

**2.34 A line charge of length $\ell = 2\ 0$ m is on the positive x axis with one end at the origin. The linear charge density is given by $150\cos(x/\ )$ pC/m. Find the flux out of a spherical surface with radius 0.5 m and centre at $x = 1.0$ m on the x axis.

***2.35 Charge is uniformly distributed on the outside surface of a long hollow cylindrical conductor situated in air with its axis along the z axis of cylindrical coordinates. The radius of the conductor is 0.8 cm and the surface charge density is 75×10^{-8} C/m^2. Find electric field intensities at (a) $r = 0.4$ cm, and (b) $r = 2.0$ cm. (c) Now consider that a line charge with uniform charge density is on the z axis. What should be its linear charge density so that it will produce the same field intensity as the surface charge at $r = 2.0$ cm?

**2.36 Charge is uniformly distributed in a cylindrical region of volume located in air. The radius of the cylinder is 10 mm and it has charge per unit length of 25 nC/m. Find the volume charge density. Find also the electric field intensities at radial distances of 15 mm, and 5 mm assuming the field intensity to be radial.

**2.37 The volume charge density in a cylindrical region in air is given by $250\sqrt{r}$ μC/m^3. The radius of the cylindrical charge is 0.8 cm. Find electric field intensities at radial distances of 0.6 cm and 1.6 cm. Use Gauss's law.

**2.38 A cylindrical charge of large length and 4 cm radius with volume charge density of $50e^{2r}$ μC/m^3 is situated in air. Calculate, using Gauss's law, the electric field intensity at $r = 3$ cm and $r = 8$ cm.

*2.39 Charge is uniformly distributed on the outside surface of a hollow spherical conductor of radius $r = 12$ cm. The conductor, situated in air, has a surface charge density of 9 μC/m^2. Calculate electric field intensities at $r = 10$ cm and $r = 20$ cm. Now consider that a point charge is located at the centre of the spherical conductor. What should be the value of the point charge so that it will produce the same field intensity as the spherical charge at $r = 20$ cm?

**2.40 A total charge of 30 nC is uniformly distributed over the volume of a sphere with radius $r = 5$ cm in a dielectric medium of relative permittivity 4.2. Find the electric field intensity at $r = 3$ cm and $r = 15$ cm.

**2.41 Charge is distributed in the form of a sphere with volume charge density $40r$ μC/m^3. The spherical charge, located in air, has a radius $r = 8$ cm. Calculate the electric field intensity at $r = 5$ cm and $r = 20$ cm.

***2.42 Charge is distributed in the form of a sphere of radius 5.0 cm in air. The volume charge density is $50e^{-r}$ μC/m^3. Calculate the electric field intensity at $r = 4$ cm and $r = 10$ cm. Use Gauss's law.

**2.43 The flux density vector in an electric field is given by $\mathbf{D} = x^2y\mathbf{u}_x + y^2z\mathbf{u}_y + xz^2\mathbf{u}_z$ μC/m^2. Find the volume charge density at (2 m, 4 m, 6 m).

**2.44 The flux density vector in cylindrical coordinates is described by $\mathbf{D} = r^2 \sin\phi\, \mathbf{u}_r + z\cos\phi\, \mathbf{u}_\phi + rz^2\mathbf{u}_z$ μC/m^2. Find the volume charge density at (0.5 m, 30°, 1.2 m).

***2.45 Derive the divergence of flux density vector $\mathbf{D}$ in cylindrical coordinates. All the components of $\mathbf{D}$ are functions of r, ϕ and z.

***2.46 The flux density vector in spherical coordinates is given by $\mathbf{D} = 5r^2 \sin\theta\, \mathbf{u}_r$. Establish the validity of the divergence theorem over the region of a sphere of radius r_0 centred at the origin.

**2.47 The electric field intensity on the x-y plane is given by $8\mathbf{u}_x + 8\mathbf{u}_y$ V/m. Find the work done in displacing a charge of 2 μC, (a) along the x axis from $x = 0$ to $x = 3$ m, (b) along the y axis from $y = 2$ m to $y = 6$ m, and (c) along the straight line from the origin to the point at (3 m, 6 m).

**2.48 A vector field on the x-y plane is described by $\mathbf{F} = 25y\mathbf{u}_x + 15x\mathbf{u}_y$. Find the line integral of $\mathbf{F}$ from $P_1(-2, 2)$ to $P_2(3, 5)$, (a) along the straight line joining the points, and (b) along the line segments from $P_1(-2, 2)$ to $P_0(3, 2)$ and from $P_0(3, 2)$ to $P_2(3, 5)$. (c) Is the field conservative?

**2.49 A long line charge with uniform charge distribution is oriented along z direction of cylindrical coordinates. The linear charge density of the line is ρ_ℓ C/m. Obtain the potential difference between P_2 (r_2, ϕ, z) and P_1 (r_1, ϕ, z).

**2.50 Find the work done in shifting a point charge of –20 μC from P_1(–2 m, 2 m) to P_2(3 m, 5 m) on the x-y plane in an electric field $\mathbf{E} = 25y\mathbf{u}_x + 25x\mathbf{u}_y$ V/m. Examine whether the field is conservative or not.

**2.51 A point charge of 50 μC is moved from P_1(1 m, 2 m, 4 m) to P_2(2 m, 5 m, 6 m) in an electric field described by $4y\mathbf{u}_x + 4x\mathbf{u}_y - 10\mathbf{u}_z$ V/m. Find the work done.

*2.52 Three point charges $Q_1 = 2.5\ \mu$C, $Q_2 = 3.3\ \mu$C, $Q_3 = 4.6\ \mu$C are located at P_1(3 m, 0, 0), P_2(0, 2.8 m, 0), P_3(0, 0, 1.5 m) in free space. Find the potential at P(–1.8 m, 2 m, 3.5 m) in the field of the charges.

*2.53 A point charge of 12 μC is situated at the origin of rectangular coordinates in air. Calculate the potential at P_1(2.5 m, –1.2 m, 3.2 m) and P_2(–1.8 m, –2 m, 4.5 m). What is the potential difference between points P_2 and P_1?

**2.54 A point charge of 750 nC is located at the origin of spherical coordinates in vacuum. What is the potential at P_2(2 m, 30°, 0) relative to the potential at P_1(5 m, 0, 45°)?

**2.55 A line charge with linear charge density of 50 nC/m is on the z axis of cylindrical coordinates in air. What is the potential difference between points P_1 at (0.4 m, 60°, 0.5m) and P_2 at (1.6 m, 0, 2 m)?

**2.56 In an electric field prove that $\mathbf{E}(x, y, z) = -\nabla V(x, y, z)$ where the symbols have their usual meanings.

2.57 Using the data in Problem 2.21, calculate the potentials at the given points.

***2.58 Electric charges are uniformly distributed over the volume of a sphere. The radius of the sphere and volume charge density are a and ρ respectively. Derive expressions for the electrostatic potential in regions, (a) $r > a$, (b) $r < a$.

**2.59 The electrostatic potential in free space is given by $V = 5xy^2 + 10yz^2$ V. Determine the potential and electric field intensity at (x, y, z) = (1.6 m, 2.2 m, 2.5 m).

**2.60 A line charge of length ℓ and uniform linear charge density ρ_ℓ C/m is oriented along the x axis with its mid-point at the origin. Calculate the electrostatic potential at $P(x, y, 0)$

**2.61 The electrostatic potential in free space and in cylindrical coordinates is described by $r \cos \phi - r \sin \phi + z^2$ V. Find V and $\mathbf{E}$ at $r = 0.7$ m, $\phi = 40°$, $z = 1.4$ m.

***2.62 A circular loop of charge with uniform charge distribution and of radius a is centred at the origin of cylindrical coordinates. The loop axis is along the positive z direction. (a) Find the electrostatic potential at $P(0, 0, z)$. (b) Calculate the electric field intensity at P using the potential function.

**2.63 Given: $V = r^2 \sin \theta \cos \phi$ V in free space and in spherical coordinates. Calculate the electric field intensity at $r = 1.0$ m, $\theta = -45°$, $\phi = 120°$.

**2.64 The far-zone electrostatic potential of an electric dipole at a point P (r, θ, 0) in spherical coordinates is given by

$$V = \frac{p\cos\theta}{4\pi\varepsilon r^2}$$

where p is the dipole moment. Calculate from the potential function the field intensity vector at P.

**2.65 An electric dipole consists of two opposite point charges, each of 1.5 nC and separated by 10 cm. The dipole, located in air, is centred at the origin of spherical coordinates. (a) What is the value of the dipole moment? (b) Calculate the potential and electric field intensity at $r = 5.0$ m, $\theta = 60°$, $\phi = 0$. (c) What is the value of maximum potential at $r = 7.0$ m?

**2.66 The negative and positive charges of an electric dipole are located at $(x, y, z) = (0, 0, -4 \text{ cm})$ and $(0, 0, 4 \text{ cm})$ in air. The value of each charge is 10 nC. Find in spherical coordinates the potential and electric field intensity at $x = 5$ m, $y = 0$, $z = 8$ m.

***2.67 Calculate V and $\mathbf{E}$ at $(x, y, z) = (3 \text{ m}, 4 \text{ m}, 8 \text{ m})$ due to the dipole described in Problem 2.66.

**2.68 Three point charges $Q_1 = 5\ \mu$C, $Q_2 = 3\ \mu$C, $Q_3 = 2\ \mu$C are located at P_1(1 m, 2 m, 3 m), P_2(–1 m, 3 m, 4 m), P_3(0 m, 4 m, 6 m) in free space. Find energy stored in the field of the charges.

**2.69 The potential function in a static electric field in free space is given by $50r^2$ V in cylindrical coordinates. Find the field energy in a region bounded by $1.0 \text{ m} \le r \le 2$ m; $0 \le \phi \le 2\pi$; $0 \le z \le 1.5$ m.

***2.70 Charge is uniformly distributed over a cylindrical region with volume charge density ρ C/m^3 in air. The radius of the cylindrical charge is a metre. (a) Calculate the energy stored per unit length in the region of charge distribution. Assume that the field intensity is in the radial direction only. (b) If $a = 5$ mm and $\rho = 500\ \mu$C/m^3, what is the value of energy stored per unit length in the region of charge distribution?

***2.71 Charge is uniformly distributed on the surface of a spherical conductor located in air with its centre at the origin of spherical coordinates. The radius of the conductor is a, and the surface charge density is ρ_S C/m^2. Derive an expression for the energy stored in the field of the charge distribution. If $a = 10$ cm, $\rho_S = 0.5\ \mu$C/m^2, what is the value of the field energy?

**2.72 Charge is uniformly distributed on the surface of a spherical conductor of radius 0.1 m. The potential at a distance r from the centre of the sphere is $500/r$ ($r > 0.1$ m). Calculate the value of surface charge density on the conductor surface, in free space.

**2.73 The plane interface between a good conductor and free space in an electric field satisfies the equation $S(x, y, z) = 4x + 12y + 3z = 60$. The electric field intensity at the interface is $500\mathbf{u}_x + 1200\mathbf{u}_y$ V/m. Find the charge density on the conductor surface.

**2.74 Assuming the polarisation vector $\mathbf{P}$ of a perfect dielectric material to be proportional to electric field intensity $\mathbf{E}$ in it, obtain the relation between the relative permittivity and electric susceptibility of the material.

**2.75 Two parallel plates located in air are charged by the application of a dc voltage of 24 V. The distance between the plates is 2 mm. After disconnecting the source, the space between the plates is filled with a dielectric material. The value of potential difference between the plates is now 8 V. Find (a) the relative permittivity, (b) electric susceptibility, and (c) electric polarisation of the material.

*2.76 The relative permittivity of a dielectric material is 3.8. What is the polarisation of the material if the electric field intensity in the material is 7500 V/m?

**2.77 Two dielectric regions have a common boundary surface. The relative permittivity of Region 1 is ε_{r1} and that of Region 2 is ε_{r2}. The magnitude of electric field intensity in Region 1 is E_1 and it is at an angle α_1 with the normal to the boundary surface. Find the magnitude of the field intensity in Region 2.

**2.78 The common boundary between two dielectric regions 1 and 2 is a plane surface. The relative permittivity of the regions 1 and 2 are 5.4 and 3.8 respectively. The electric flux density at the interface in Region 1 is 1.2 μC/m^2, and its angle with the boundary surface is 60°. Obtain the value of flux density and its angle with the normal to the interface in Region 2.

**2.79 The common boundary surface of two dielectric media 1 and 2 is on the $x = 0$ plane. The relative permittivity of Medium 1 ($x < 0$) is 3.0 and that of Medium 2 ($x > 0$) is 4.5. The electric flux density in Medium 2 is $\mathbf{D}_2 = 0.2\mathbf{u}_x + 0.3\mathbf{u}_y + 0.6\mathbf{u}_z\ \mu\text{C/m}^2$. Find the energy stored per unit volume in each medium.

***2.80 The common boundary surface of two dielectric regions is perpendicular to the *x-y* plane. It intersects the *x* axis at $x = 4.0$ m and the *y* axis at $y = 3.0$ m. The relative permittivity of the region containing the origin (Region 1) is 4.0, and the relative permittivity of the other region is 5.0. The electric field intensity in Region 1 is given by $400\mathbf{u}_x + 300\mathbf{u}_y$ V/m. Find the field intensity in Region 2.

*2.81 The voltage across the series combination of two capacitors with capacitances C_1 and C_2 is V. Find the voltage across each capacitor.

**2.82 A parallel-plate capacitor has two dielectric layers 1 and 2 with the interface parallel to the plates. The relative permittivity of Layer 1 is 3.2 and that of Layer 2 is 4.5. The plates are connected to a constant voltage source. What is the ratio of the potential gradient in Region 2 to that in Region 1?

**2.83 The region between the plates of a parallel-plate capacitor is filled with two dielectrics with the interface normal to the plates. The relative permittivities of the dielectrics in regions 1 and 2 are 2.8 and 3.2 respectively. The plates are connected to a dc source of constant voltage. What is the ratio of charge densities on the inside surfaces of the two regions?

**2.84 A parallel-plate capacitor has two dielectric layers with the interface parallel to the plates. The relative permittivity and thickness of the dielectric near the positively charged plate are 3 and 5 mm, and those of the dielectric near the other plate are 4 and 3.5 mm. The potential difference between the plates is 12 V. What is the potential difference between the positive plate and the interface?

*2.85 A variable capacitor has 10 fixed plates and equal number of movable plates. The area of each plate is 4 cm^2. The space between fixed and movable plates has an insulating material of thickness 0.5 mm and an air gap of 0.4 mm. The relative permittivity of the insulating material is 2.0. Calculate the capacitance of the multiple-plate capacitor when the overlap between the fixed and movable plates is 60 per cent.

***2.86 A parallel-plate capacitor has two dielectric layers 1 and 2 with the interface parallel to the plates. The thickness, relative permittivity and dielectric strength of Layer 1 are 4 mm, 6.0 and 20.0 MV/m respectively. The values for Layer 2 are 2.0 mm, 5 and 30.0 MV/m. Area of each plate is 0.2 m^2. (a) Calculate the capacitance of each layer and the total capacitance. (b) What should be the operating voltage if it is 50 percent of the breakdown voltage? (c) How much energy is stored in the capacitor at the operating voltage?

***2.87 Using the relevant data in Problem 2.86, find the operating voltage for the following cases. (a) The space between the plates is filled with dielectric 1 only. (b) The space between the plates has a layer of dielectric 1 of thickness 4 mm and an air gap of 2 mm. (c) The medium between the plates is air only.

***2.88 The radii of the inner and outer conductors of a coaxial transmission line are a and b respectively. The region between the conductors has two dielectric layers with the interface concentric with the conductors. The radius of the interface is c. If the potential difference between the conductors is V, find the electric field intensity in the dielectric regions.

**2.89 A coaxial transmission line has three concentric conductors. The middle conductor, which is very thin, acts as an additional shield. The radii of the inner, middle and outer conductors are 1.5 mm, 3.5 mm and 5.0 mm respectively. The relative permittivities of the inner and outer dielectric layers are 3.5 and 2.5 respectively. The potential difference between the inner conductor and outer conductor is 24 V. (a) Calculate the capacitance of each region, and the total capacitance per unit length of the line. (b) Find the potential difference between the outer and the middle conductors.

**2.90 The space between two concentric spherical conducting shells has two dielectric layers with the interface concentric to the inner conductor. The radii of the inner and outer spheres are 5 cm and 10 cm respectively. The interface is at a radial distance of 7.5 cm. The relative permittivity of the dielectric near the inner sphere is 4, and that of the other dielectric is 3.0. Find the capacitance.

**2.91 The distance between the plates of a parallel-plate capacitor is 5.0 mm, in air. The area of each plate is 25 cm^2. The plates are connected to a dc source of constant voltage of 200 V. If the distance between the plates is increased by 2.0 mm without disconnecting the source, (a) what is the magnitude of average force needed for increasing the distance? And (b) how much energy is received by the source?

***2.92 Charge is uniformly distributed over the volume of a sphere of radius a. The volume charge density is ρ C/m^3. Show that the potential at a point within the charge distribution ($r < a$) satisfies Poisson's equation, and the potential at a point outside the charge distribution ($r > a$) satisfies Laplace's equation.

Chapter Three

THE STEADY CONDUCTION-CURRENT

3.1 Introduction

The ability of a material to conduct electricity depends on the concentration of free electrons or charge-carriers in it and the mobility of these carriers. If the energy band is empty, it cannot contribute anything to the process of conduction. Further, if the energy band is completely filled, then also it cannot conduct electricity because no room is available for an electron to jump from one state to another. In other words, a material can conduct electricity only if the energy band is partially filled.

The energy bands in metals either overlap or are partially filled. Therefore, electrons can move from one energy level to another energy level under the influence of an external electric field resulting in flow of current. As the concentration of *free* or *conduction* electrons in metals is quite high, all metals and their alloys are good conductors of electricity. These materials are called *conductors*.

In materials like diamond, glass, and mica, the valence band is full and the conduction band is empty. Hence, no free electrons (negative charge-carriers) or *holes* (positive charge-carriers) are available. Moreover, the forbidden energy gap between the two bands is quite large. For example, the forbidden energy gap of glass is of the order of ten electron-volts. Therefore with increase in temperature or with application of normal electric field, bound electrons in the valence band cannot move to the conduction band. These materials, which are bad conductors of electricity, are called *insulators*.

Though the valence bands of materials like germanium and silicon are full and their conduction bands are empty at 0 K, the widths of forbidden energy gaps are smaller than those of insulators. The forbidden energy gap of silicon is of the order of 1.1 electron-volt and that of germanium is 0.7 electron-volt. At 0 K, the electric properties of these materials are similar to those of insulators. However with an increase in temperature, say to the room temperature, some electrons move to the conduction band. This results in the creation of holes in the valence band. Though the concentration of charge carriers in them is not as high as in metals, these materials also conduct electricity under the influence of an applied electric field even at room temperature. The electric properties of these materials lie between those of conductors and insulators. These materials are called *semiconductors*.

Formulae for conductivities of conductors and semiconductors, the point form of Ohm's law, boundary conditions at the interface of two regions of different conductivities are derived in this chapter. Calculation of insulation resistance of a coaxial transmission line is also given.

3.2 Current Flow in Metals

An important property of a metal is that each of its atoms has at least one valence electron that is free to move in the crystal lattice. These electrons are called *free* or *conduction* electrons. The free electrons do not remain attached to any particular atom. The electrons are in constant motion due to thermal agitation. During movement, the electrons collide with comparatively heavy ions. The average distance between collisions is called *mean free path*, and average time between collisions is called *relaxation time*.

The velocities of the electrons change with each collision. As the movement of electrons in the crystal lattice due to thermal agitation is quite random, their average velocity is zero. Consequently, there is no net flow of current in the absence of an external electric field. Another important property of metals is that their energy bands either overlap or are only partially filled, as mentioned earlier. Hence, only a small amount of external energy is required to move the electrons from one energy level to another.

Now let us examine what will happen if a steady electric field of intensity E is applied to a homogeneous metal. Each electron of charge e of the metal will experience an electric force $\mathbf{F} = -e\mathbf{E}$, as shown in Fig. 3.1. The force will cause the electrons to accelerate. The acceleration of an electron is,

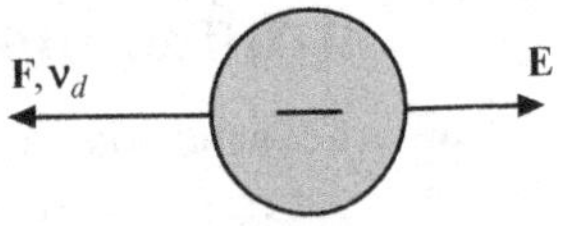

Fig. 3.1 Direction of force and drift velocity of a *free* electron in an electric field

$$\mathbf{a} = \frac{\mathbf{F}}{m}$$

where $m = 9.11 \times 10^{-31}$ kg is the mass of an electron.

The electrons in metals (also in liquids and gases) while moving collide and interact with other ions. The electrons loose energy at each collision as the collisions are not elastic. Therefore, except for a brief transient period after the application of the field, the electrons attain a *drift velocity* $\mathbf{v}_d$ (the symbol $\mathbf{v}$ is the lowercase bold Greek alphabet 'Nu') in a direction opposite to the applied field, as indicated in Fig. 3.1. The drift velocity is given by

$$\mathbf{v}_d = -\frac{e\tau}{m}\mathbf{E} = -\mu_e \mathbf{E} \tag{1}$$

In the above equation τ is the *relaxation time* and $\mu_e = e\tau/m$ is the *mobility* of free electrons. The unit of mobility is metre-square per volt per second ($m^2V^{-1}s^{-1}$).

The relaxation time and hence the mobility of different materials are different. The relaxation time of copper at 0°C is equal to 2.7×10^{-7} second, and the value of its mobility is 0.0032.

Thus, if a metal is subjected to an external electric field, a large number of its free electrons will attain a drift velocity given by (1). This phenomenon is known as conduction of electricity in metals. Since the direction of current flow is opposite to that of the electron drift, current flows in the same direction as the applied field. A steady direct current will flow if the applied field is unidirectional and time-invariant. The applied field should be from an active source. An example of such a source is a battery. Moreover, flow of conduction current is possible only if the conducting path between a source and an electric load is continuous and a closed one. Field due to static charges cannot sustain continuous flow of current.

3.3 Current, Current Density and Conductivity

Let us assume that the conductor shown in Fig. 3.2(a) is homogeneous and it is of uniform sectional area S. If it has n number of free electrons per unit volume, free electronic-charge per unit volume (or charge concentration) is $\rho_e = -ne$. The magnitude of current flow through any section, by definition, is equal to the total charge in coulomb drifting through the section in one second. Further, the direction of current flow is opposite to that of the drift velocity of electrons. Therefore, flow of conduction current through the section is given by

$$I = S(ne)\mathbf{v}_d = S\rho_e\mathbf{v}_d \tag{1}$$

The unit of current is ampere (A).

The distribution of current over a cross-section in a conductor of uniform sectional area is also uniform. The current distribution is described by a quantity called *current density*. It is equal to current per unit sectional area of the conductor. The unit of current density is ampere per metre square (A/m^2), and it is denoted by J. Dividing both sides of (1) by S, we have

$$J = \frac{I}{S} = ne\mathbf{v}_d = \rho_e \mathbf{v}_d \tag{2}$$

Substituting the scalar value of the drift velocity given by formula (3.2-1) in (2), we get

$$J = \rho_e \mu_e E = \sigma E \tag{3}$$

where

$$\sigma = \rho_e \mu_e = \frac{ne^2\tau}{m} \tag{4}$$

The parameter σ is called *electrical conductivity* of a conductor. It is equal to the product of mobility and volume charge density of free electrons. Its unit is siemen per metre (S/m). Current density is a vector quantity. Its magnitude is given by (3) and its direction is that of the applied field intensity E.

In the conductivity formula given by (4) all the quantities except relaxation time are independent of temperature. Relaxation time however depends on temperature, and it decreases when the temperature of the conductor increases. Therefore, the conductivity of a metallic conductor decreases with an increase in its temperature.

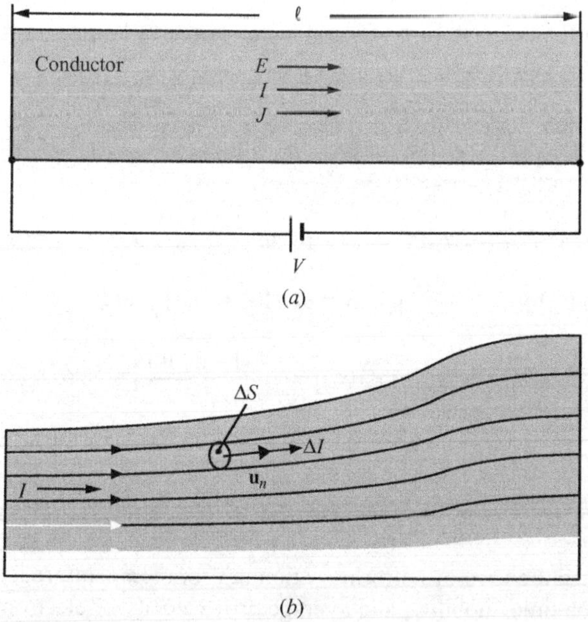

Fig. 3.2 (*a*) A conductor of uniform section connected to a dc source and (*b*) Current streamlines in a conductor of varying cross-sectional area.

The distribution of current in a conductor with varying cross-sectional area is not uniform, as shown in Fig. 3.2(*b*). Hence, the current density is different at different points over a cross-section of the conductor. The current density in the case of non-uniform distribution of current is defined at a point. To define it, let us assume that current through a surface element of area ΔS is ΔI and the direction of current is along the normal to ΔS. The normal direction is indicated by the unit vector $\mathbf{u}_n$. The current density at a point is given by the ratio of ΔI to ΔS in the limit ΔS approaching zero. That is,

$$J = \lim_{\Delta S \to 0} \frac{\Delta I}{\Delta S} \tag{5}$$

Relation (5) defines the current density when the current distribution is not uniform.

If the direction of ΔI is at an angle α with normal to the surface element, then the component of ΔS along ΔI is equal to $\Delta S \cos\alpha$. Hence,

$$J = \lim_{\Delta S \to 0} \frac{\Delta I}{\Delta S \cos\alpha}$$

$$\Rightarrow \qquad \Delta I = J \Delta S \cos\alpha = \mathbf{J} \cdot \Delta \mathbf{S}$$

Current through a cross-section of the conductor is,

$$I = \int_S \mathbf{J} \cdot d\mathbf{S} \tag{6}$$

We find from relation (6) that the current flow over any section of the conductor has the same value and electric current is a scalar quantity. The direction and distribution of the current over any section of a conductor depends on the shape of the conductor.

As **J** **S** is a scalar function, electric current is a scalar quantity.

EXAMPLE 3.1

The current density over the cross-section of a conductor of radius 5 mm is $\mathbf{J} = (4\times10^6)e^{-200r}\mathbf{u}_z$ A/m² in cylindrical coordinates. Calculate current carried by the conductor.

Differential area in *z* direction, $dS_z = r dr d\phi$

Current density in *z* direction, $J_z = (4\times10^6)e^{-200r}$

Current in a differential area, $dI = J_z dS_z = [(4\times10^6)e^{-200r}]\ r dr d\phi$

Current carried by the conductor, $I = \int_{r=0}^{0.005}\int_{\phi=0}^{2\pi} dI = (2\pi)(4\times10^6)\int_0^{0.005} re^{-200r}dr$

$$= (25.13\times10^6)\left[\frac{e^{-200r}(1-200r)}{200^2}\right]_0^{0.005} = 628.3 \text{ A}$$

EXAMPLE 3.2

The conductivity of a metallic wire of uniform section at room temperature is 6.17×10^7 S/m. Calculate current density, relaxation time, mobility, and average drift velocity of electrons if the potential gradient along the wire is 0.11 V/m. Assume that there are 5 8 × 10^{28} conduction electrons per m³.

$$J = \sigma E = 6.17 \times 10^7 \times 0{\cdot}11 = 6{\cdot}787 \times 10^6 \ \text{A/m}^2$$

Relaxation time, $\tau = \dfrac{\sigma m}{ne^2} = \dfrac{(6{\cdot}17\times10^7)\times(9{\cdot}11\times10^{-31})}{(5{\cdot}8\times10^{28})\times(1{\cdot}6\times10^{-19})^2} = 3{\cdot}786\times10^{-14}\ \text{s}$

Mobility, $\mu_e = \dfrac{e\tau}{m} = \dfrac{(1{\cdot}6\times10^{-19})\times(3{\cdot}786\times10^{-14})}{9{\cdot}11\times10^{-31}} = 6{\cdot}645\times10^{-3}\ \text{m}^2\text{V}^{-1}\text{s}^{-1}$

Drift velocity, $v_d = \mu_e E = (6{\cdot}645\times10^{-3})\times(0{\cdot}11) = 7{\cdot}31\times10^{-4}\ \text{m/s}$

3.4 Point-Form of Ohm's Law

Ohm's law states that the voltage difference between two ends of a metallic conductor is proportional to the current through it at a given temperature. For steady current, the constant of proportionality is called *resistance*.

Resistance is denoted by R, and its unit is ohm (Ω). The mathematical form of Ohm's law, therefore, is

$$V = IR \tag{1}$$

$$\Rightarrow \quad I = \frac{V}{R} = VG$$

where I is current through the conductor and V is voltage difference between the ends of the conductor, as illustrated in Fig. 3.2(a).

The quantity $G = 1/R$ is called *conductance* of the material. The unit of conductance is siemen (S).

The resistance of a conductor of uniform section is directly proportional to its length ℓ, and inversely proportional to its sectional area S. The constant of proportionality is called electrical *resistivity*. Denoting the resistivity by ρ_Ω,

$$R = \frac{\rho_\Omega \ell}{S} \tag{2}$$

The unit of resistivity is ohm-metre (Ω-m). The resistivity of homogeneous metallic conductors is almost constant over a wide range of current; provided other physical properties like temperature, tension, etc., remain constant. So, the V-I characteristic of a metallic conductor is almost linear. Further, resistivity is also independent of direction of current flow. Hence, metallic conductors are assumed to be linear and isotropic.

Now, by eliminating R from (1) and (2), we get

$$\frac{I}{S} = \frac{V}{\rho_\Omega \ell}$$

The left-hand side of the above relation is current density and V/ℓ on the right side is the electric field intensity. Thus, we have

$$J = \frac{E}{\rho_\Omega} \tag{3}$$

By comparing the above relation with relation (3) of the preceding section we find that the resistivity is reciprocal of the conductivity. That is,

$$\sigma = \frac{1}{\rho_\Omega} \tag{4}$$

The relation between current density and electric field intensity in terms of the conductivity is,

$$J = \sigma E$$

This is the *point form* of Ohm's law. The current density, as stated in the last section, is a vector quantity with its direction along the field intensity. Thus, the vector form of the law is

$$\mathbf{J} = \sigma \mathbf{E} \tag{5}$$

As described in the preceding section, the conductivity of a given metal decreases when its temperature increases. Because resistivity is reciprocal of conductivity, the resistivity of a metal increases with an increase in temperature. Above a reference temperature, the resistivity of a metallic conductor varies nearly linearly with temperature. The reference temperature is different for different materials. The resistivity ρ_Ω of a metal at any temperature t °C may approximately be expressed in terms of its resistivity $\rho_{\Omega 0}$ at 0 °C by the following equation:

$$\rho_\Omega = \rho_{\Omega 0}(1 + \alpha_0 t), \tag{6}$$

where α_0 is called temperature coefficient of resistivity at zero degree centigrade. The coefficient for copper and aluminium are approximately equal to 0.00426 and 0.00380 per degree centigrade at zero degree centigrade. The conductivities of some materials become very high at temperatures tending to 0 kelvin (K). These materials are called super conductors.

It follows from (2) and (6) that the resistance R at any temperature t is given by

$$R = R_0(1 + \alpha_0 t) \tag{7}$$

where R_0 is the resistance at zero degree centigrade. If R_1 and R_2 are the resistances at temperatures t_1 and t_2 then,

$$\frac{R_1}{R_2} = \frac{1 + \alpha_0 t_1}{1 + \alpha_0 t_2} \tag{8}$$

EXAMPLE 3.3

The internal and external radii of a conductor are 5 mm and 7 mm respectively and its length is 25 m. The current in the conductor is 300 A and the voltage difference between its ends is 3.0 V. Find resistivity of the conductor.

Area of cross-section of the conductor, $S = \pi(7^2 - 5^2) \times 10^{-6} = 75.4 \times 10^{-6}$ m^2

Conductor resistance, $R = \frac{3.0}{300} = 0.01\ \Omega$

Conductivity, $\sigma = \frac{\ell}{RS} = \frac{25}{(0.01 \times 75.4 \times 10^{-6})} =$ [illegible] $\times 10^7$ S/m

EXAMPLE 3.4

The resistivity of a copper wire at 293 K (note that it is not correct to write 293 °K) is $1.724\times10^{-8}\ \Omega$ -m. Calculate its resistivity at 353 K.

Let ρ_1 be the resistivity at temperature t_1 °C, and ρ_2 be the resistivity at temperature t_2 °C. Then,

$$\frac{\rho_2}{\rho_1} = \frac{1+\alpha_0 t_2}{1+\alpha_0 t_1}$$

The temperature coefficient of resistivity, $\alpha_0 = 0.00426$ per °C at zero degree centigrade.

$$t_1 = 293\ \text{K} - 273\ \text{K} = 20\ °\text{C}$$

$$t_2 = 353\ \text{K} - 273\ \text{K} = 80\ °\text{C}$$

$$\rho_2 = (1.724\times10^{-8})\frac{1+(0.00426)(80)}{1+(0.00426)(20)} = 2.13\times10^{-8}\ \Omega\text{-m}$$

3.5 Kirchhoff's Voltage Law

In the preceding sections we learnt that an external field is needed for continuous flow of current and the field should be due to an active source. Moreover, direct current can flow only when the path, consisting of passive elements (resistances) and active elements (sources), is a closed one. Such a closed path is known as an electric circuit.

Kirchhoff's voltage law states that the line integral of electric field intensity around a circuit is zero. That is,

$$\oint \mathbf{E}\cdot d\ell = 0 \tag{1}$$

The electric field intensity in (1) may consist of three components viz., (i) Field intensity within the active sources, (ii) Field intensity due to current flow in the passive part of a circuit; this field intensity is given by the point form of Ohm's law and (iii) Field due to static charges. Since the line integral of field intensity due to static charges over a closed path is zero, it is necessary to consider only the first two components. Thus, denoting the field intensity within the source by $\mathbf{E}_a$, the mathematical form of Kirchhoff's voltage law is,

$$\oint \mathbf{E}_a \cdot d\ell + \oint (\mathbf{J}/\sigma)\cdot d\ell = 0 \tag{2}$$

Assuming that $\mathbf{J}$ and $d\ell$ are taken in the same sense, we have

$$\left(\frac{\mathbf{J}}{\sigma}\right)\cdot d\ell = \frac{J d\ell}{\sigma} = I\left(\frac{d\ell}{\sigma S}\right) = I dR \tag{3}$$

where $dR = d\ell/\sigma S$ is resistance of the differential length $d\ell$. By substituting (3) in (2), we get

$$\oint \mathbf{E}_a \cdot d\ell + \oint I dR = 0 \tag{4}$$

Equation (4) when applied to the electric circuit shown in Fig. 3.3 gives

$$\int_q^{p} \mathbf{E}_a \cdot d\ell + I\int dR = 0 \tag{5}$$

The first term of (5) is the voltage *rise* from the negative to the positive terminal of the source. It is equal to the electromotive force (emf) of an ideal source. However, a practical source has its own resistance called *internal-resistance,* which is generally shown outside the source and combined with the resistance of passive part of the circuit. The second term is voltage *drop* in the total resistance of the circuit. The total resistance is equal to the sum of internal resistance of the source R_i, resistance of the connecting wires R_c and load resistance R_R. Moreover, if the voltage rise is taken as positive then voltage drop would be negative and vice versa. Thus denoting the source emf by $\mathcal{E}$, Kirchhoff's voltage law gives,

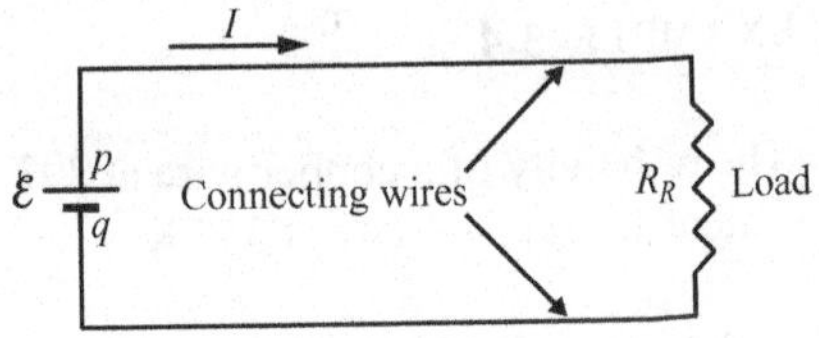

Fig. 3.3 A simple electric circuit

$$\mathcal{E} = I(R_i + R_c + R_R) \tag{6}$$

Voltage across the terminals of the source, $V_s = \mathcal{E} - IR_i = I(R_c + R_R)$

Voltage across the load terminals, $V_R = IR_R$

Voltage *drop* or *ohmic drop* in the connecting wires $= IR_c$

If there are two or more sources in a circuit, then $\mathcal{E}$ in (6) is the net emf of all the sources. Assuming $\mathcal{E}_1$ is greater than $\mathcal{E}_2$ in Fig. 3.4, the net emf around the circuit is equal to $\mathcal{E}_1$ minus $\mathcal{E}_2$ and the direction of current would be as shown in the figure. Hence, by Kirchhoff's voltage law,

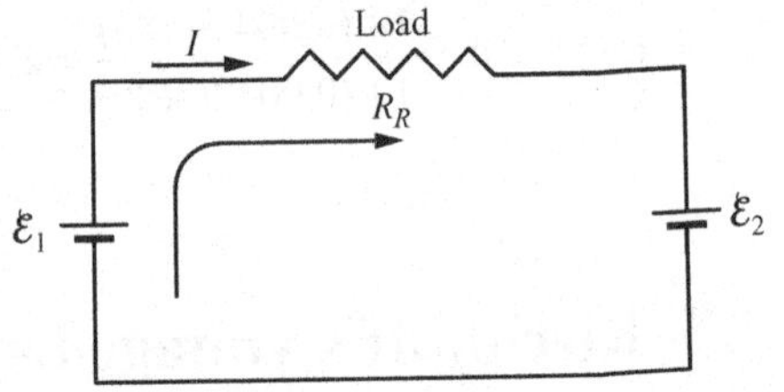

Fig. 3.4 An electric circuit with two sources

$$\mathcal{E}_1 - \mathcal{E}_2 = I(R_{i1} + R_{i2} + R_c + R_R)$$

where R_{i1} and R_{i2} are the internal resistances of the sources, R_c is resistance of connecting wires and R_R is the load resistance.

EXAMPLE 3.5

A load of resistance of 6 Ω is connected to a dc source. The total length of the connecting wires is 12 m, radius 1 mm and conductivity 3.8×10^7 S/m. The terminal voltages of the source without and with the load are 12.3 V and 12.0 V respectively. Find the internal resistance of the source.

Resistance of the connecting wires, $R_c = \dfrac{\ell}{\sigma S} = \dfrac{12}{(3.8\times 10^7)(\pi \times 0.001)^2} = 0.101\ \Omega$

Sum of load resistance and wire resistance = 6.101 Ω

Current supplied by the source, $I = \dfrac{12}{6.101} = 1.967$ A

Internal resistance of the source, $R_i = \dfrac{12.3-12}{1.967} = 0.1525\ \Omega$

3.6 Kirchhoff's Current Law: Divergence of Current Density J

As flow of steady current through a closed conducting path is continuous, current I flowing into a region is equal to that flowing out of it. This is the continuity condition for direct current.

Let us consider a region of volume enclosed by a closed surface through which a steady current I flows. This is illustrated in Fig. 3.5. Let $\mathbf{J}$ be the current density at any point on the surface. Then, current crossing the differential area $d\mathbf{S}$ is given by $\mathbf{J}\cdot d\mathbf{S}$. Since the net current through the closed surface must be zero, the surface integral of current density over the closed surface must also be zero. Thus,

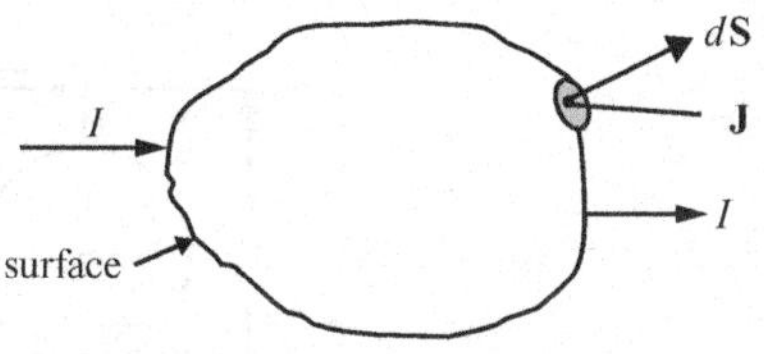

Fig. 3.5 Illustration of a closed surface carrying current

$$\oint_S \mathbf{J}\cdot d\mathbf{S} = 0 \qquad (1)$$

A specific case arises when several currents meet at a point. The meeting point is called a *junction* or a *node*. In Fig. 3.6, currents I_1, I_3 and I_4 flow into the junction whereas I_2 and I_5 flow out of it. Assuming current flowing into the junction as positive and flowing out of the junction as negative, we get

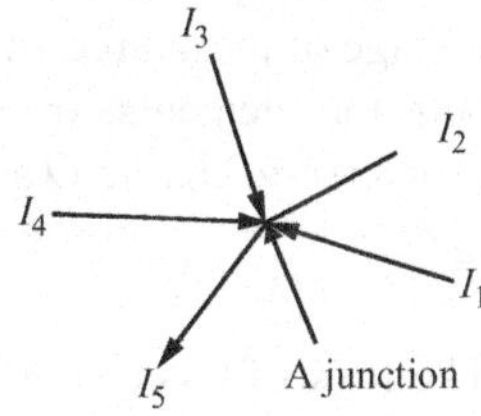

Fig. 3.6 A junction with several currents

$$I_1 - I_2 + I_3 + I_4 - I_5 = 0$$

That is, the algebraic sum of all the currents at a junction is equal to zero. In general, if there are n number of currents meeting at a junction,

$$\sum_n I_n = 0 \qquad (2)$$

Equations (1) and (2) are known as Kirchhoff's current law.

We can find the divergence of current density by using the same procedure as used for deriving the divergence of electric flux density in Section 2.13. The divergence of electric flux density in a static electric field is equal to the net outward flux per unit volume of a closed surface S enclosing the volume. The mathematical form of this statement is given by (6) of the above mentioned section. We define the divergence of current density in the same way.

$$\nabla\cdot\mathbf{J} = \lim_{\Delta v\to 0}\frac{\oint_S \mathbf{J}\cdot d\mathbf{S}}{\Delta v}$$

By (1) the numerator of rhs in the above equation is zero. Hence,

$$\nabla\cdot\mathbf{J} = 0 \qquad (3)$$

That is,

the divergence of steady (time-invariant) current density is equal to zero. Partial differential equation (3) is the *continuity equation* for steady current.

3.7 Electrical Energy and Power: Joule's Law

Let us consider that an electric device (load) is connected to an electric source by means of a transmission system. Then, power will flow from source to the load. However, all the power supplied by the source will not be received by the load because of power loss in the transmission system in the form of heat. The power associated with a circuit element (source, load, transmission system, etc.) is given by the product of its voltage and current.

In order to arrive at the above mentioned relation for a circuit element, let us consider the system shown in Fig. 3.7, where a single source is connected to a load through a two-conductor transmission line. The terminal

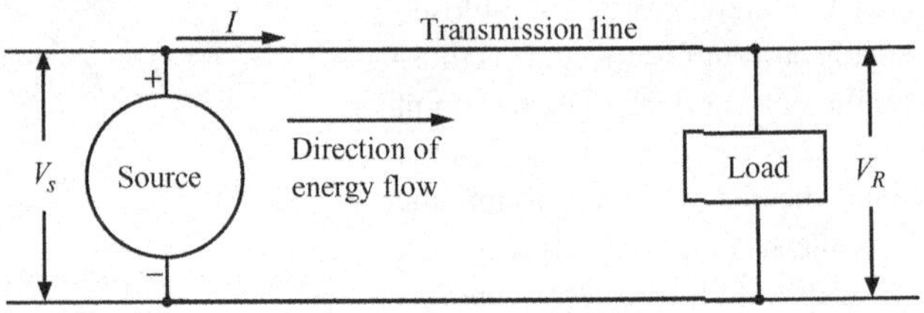

Fig. 3.7 Showing a load connected to a dc source through transmission lines

voltage of the source and voltage across the load are taken as V_s and V_R respectively. Current I is common to both the elements. In Section 3.3, we have defined electric current as the flow of charge per unit time. Conversely, charge Q associated with current I over a period of t second is,

$$Q = It$$

The flow of this charge within the source is at a constant potential difference V_s. Hence, energy supplied by the source,

$$W_s = V_s Q = V_s It$$

Power supplied by the source,

$$P_s = \frac{W_s}{t} = V_s I \tag{1}$$

Thus, power supplied by the source is equal to the product of its terminal voltage and current. Similarly, power received by the load,

$$P_R = V_R I \tag{2}$$

In general, power associated with any circuit element is given by the product of voltage across its terminals and the current flowing through it.

The difference between power supplied by the source given by (1) and power received by the load given by (2) is the loss of power in the transmission line. Denoting the power loss by P_Ω, we have

$$P_\Omega = P_s - P_R = (V_s - V_R)I$$

If R_c is resistance of the transmission line, then $V_s - V_R = IR_c$. Therefore, power loss in the lines,

$$P_\Omega = I^2 R_c \tag{3}$$

Loss of energy over a period of t second, assuming I to be constant during this period, is

$$W_\Omega = I^2 R_c t \tag{4}$$

Power loss given by (3) is known as *ohmic* loss, and expression (4) is called *Joule's* law.

We can also obtain power loss in terms of current density J. Let the length, sectional area, and conductivity of a conductor be ℓ, S, and σ respectively. Let current carried by the conductor be I; then, current density in the conductor is given by

$$J = \frac{I}{S}$$

And
$$R_c = \frac{\ell}{\sigma S}$$

Substituting these relations in (3), we get

$$P_\Omega = \frac{J^2}{\sigma}(S\ell) = \frac{J^2}{\sigma} \times \text{ (Volume of the conductor)}$$

Or power loss per unit volume $= \dfrac{J^2}{\sigma}$ W/m^3

That is, ohmic loss per unit volume of a conductor is proportional to the square of current density. The value of current density for a particular application is chosen from the consideration of ohmic losses and cost of conducting materials. Several other factors are also taken into account.

EXAMPLE 3.6

A 250 m hollow conductor has internal radius of 4 mm and external radius of 6 mm. Power loss in the conductor is 100 W. The resistivity of the conductor is $0.175\times10^{-7}\ \Omega\text{-m}$. Calculate current in the conductor.

Cross-sectional area, $S = \pi(6^2 - 4^2)10^{-6} = 62.83\times10^{-6}\ \text{m}^2$

Conductor resistance, $R_c = \dfrac{\ell\rho_\Omega}{S} = \dfrac{(250)(0.175\times10^{-7})}{62.83\times10^{-6}} = 0.0696\ \Omega$

Power loss, $P_\Omega = I^2 R_c = 100$

Conductor current, $I = \sqrt{\dfrac{100}{0.0696}} = 37.9$ A

EXAMPLE 3.7

A metallic conductor of rectangular section is in the form of a circular bow with its centre at the origin. It extends from $\phi = \phi_1$ to $\phi = \phi_2$ in cylindrical coordinate system, as shown in Fig 3.8. Its inner and outer radii are a and b respectively, and thickness in z direction is c. Calculate resistance of the conductor and power loss.

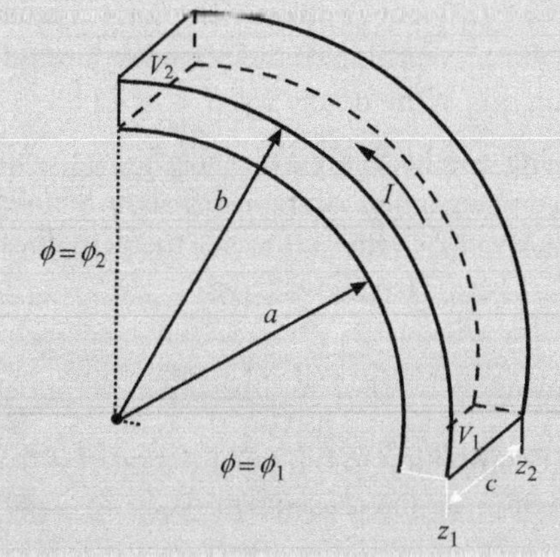

Fig. 3.8 Illustration for Example 3.7

To calculate resistance of a conductor of any regular shape, as described in the example, we first obtain the current density in terms of the voltage drop between the conductor ends. By integrating the current density over a section of the conductor we will get current in terms of the voltage drop. The ratio between the voltage drop and current gives the conductor resistance.

For the conductor illustrated in Fig. 3.8, we assume that V_1 is greater than V_2 and current flows in ϕ direction in cylindrical coordinates. Thus, the current density is also in ϕ direction. Moreover, it varies with r only. If we denote the current density by J_ϕ, the voltage drop over a differential length $d\ell_\phi$ at a radial distance r and subtending an angle $d\phi$ at the origin is given by

$$dV = \frac{J_\phi}{\sigma} d\ell_\phi = \frac{J_\phi}{\sigma}(r\, d\phi)$$

Voltage drop between the conductor ends is given by

$$V_1 - V_2 = \int_{\phi_1}^{\phi_2} \frac{r J_\phi}{\sigma} d\phi = \frac{J_\phi(\phi_2 - \phi_1) r}{\sigma}$$

$$\Rightarrow \quad J_\phi = \frac{\sigma (V_1 - V_2)}{(\phi_2 - \phi_1) r}$$

We see that the current density is inversely proportional to r.

Current, $I = \int_S J_\phi dS_\phi = \int_a^b \int_{z_1}^{z_2} \frac{\sigma(V_1 - V_2)}{(\phi_2 - \phi_1) r} dr\, dz = \frac{c\sigma (V_1 - V_2)}{\phi_2 - \phi_1} \ln \frac{b}{a}$

Resistance of the conductor, $R = \frac{V_1 - V_2}{I} = \frac{\phi_2 - \phi_1}{c\sigma \ln(b/a)}$

Power loss, $P_\Omega = I^2 R = (V_1 - V_2) I = = \frac{c\sigma (V_1 - V_2)^2}{\phi_2 - \phi_1} \ln \frac{b}{a}$

It may be noted that ϕ_1 and ϕ_2 are in radians.

The problem can also be solved using Laplace's equation as a boundary value problem (Problem 5.8).

3.8 Boundary Conditions

When two current-carrying regions of different conductivities have a common boundary, the tangential and normal components of current density on one side of the boundary are related to the corresponding components of current density on the other side of the boundary. We will obtain these relations.

Figure 3.9 shows two regions having conductivities σ_1 and σ_2. The common boundary surface is a plane surface. To find the relations between tangential components of current densities we apply Kirchhoff's voltage law to a closed path *abcda* shown in (*a*) in the figure. As the path does not contain any active source, the first term on the left side of (3.5-2) is zero. Hence, we have

$$\oint \frac{1}{\sigma}(\mathbf{J} \cdot d\boldsymbol{\ell}) = 0 \tag{1}$$

In the limit Δh tending to zero, the contribution to the line integral of J from b to c and from d to a would be negligibly small. Thus, application of (1) to the closed path yields,

$$\left(\frac{J_{t1}}{\sigma_1} - \frac{J_{t2}}{\sigma_2}\right)\Delta\ell = 0$$

$$\Rightarrow \quad \frac{J_{t1}}{\sigma_1} = \frac{J_{t2}}{\sigma_2} \tag{2}$$

Or

$$E_{t1} = E_{t2} \tag{3}$$

where E_{t1} and E_{t2} are field intensities tangential to the two sides of the boundary. Equation (3) shows that the tangential field intensity is continuous across the boundary. However, the tangential current densities are not equal; they are related through their conductivities as in (2).

The relation between the normal components of current densities on two sides of the boundary is obtained by using the continuity property of current flow. To derive this relation, we choose a closed cylindrical surface enclosing the interface as shown in (*b*) in Fig. 3.9. The axis of the cylinder is normal to the interface whereas the top and bottom surfaces are parallel to the interface. As per the continuity condition, current flowing into the cylinder must be equal to the current flowing out of it. If the height of the cylinder is made very small so that it is only enough to enclose the boundary, current through the curved surface may be neglected. Therefore, current flowing into the bottom surface of the cylinder is equal to current flowing out of its top surface. Thus, taking J_{n1} and J_{n2} as the average current densities at the top and bottom surfaces respectively, and assuming the current flowing out of the region to be positive, and current flowing into the region to be negative, we get

$$(J_{n2} - J_{n1})\Delta S = 0$$

$$\Rightarrow \quad J_{n2} = J_{n1} \tag{4}$$

Or

$$\sigma_2 E_{n2} = \sigma_1 E_{n1} \tag{5}$$

where E_{n1} and E_{n2} are field intensities normal to the two sides of the boundary. We find that the normal component of the current density is continuous across the boundary surface whereas the normal components of the field intensities are related through their conductivities.

Equations (2) to (5) are called boundary conditions at the interface between two media of different conductivities. These relations show that the total current densities on two sides of the boundary are not of the same magnitude.

Further, the angles which they make with the normal to the boundary are also not equal, as shown in Fig. 3.9(c). Using the geometry of the figure and the boundary conditions, we will get the following relations:

$$J_{n1} = J_1 \cos\alpha_1, \quad J_{t1} = J_1 \sin\alpha_1;$$

$$J_{n2} = J_2 \cos\alpha_2, \quad J_{t2} = J_2 \sin\alpha_2;$$

$$\frac{\tan\alpha_1}{\tan\alpha_2} = \frac{\sigma_1}{\sigma_2}$$

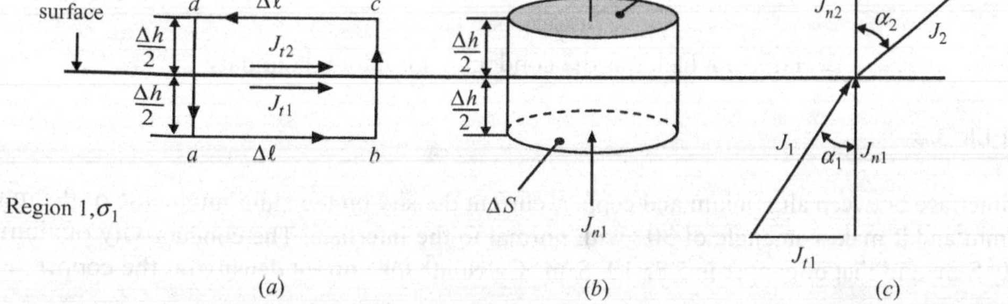

Fig. 3.9 Pertaining to boundary conditions

Now, let us assume that both the regions shown in the figure are good conductors. No modification of the boundary conditions is needed as the conductivities are finite. A special case however arises if one region is a good conductor and the other is a perfect insulator. Let us assume that medium 1 is a good conductor and medium 2 is a perfect insulator. Then, σ_1 has a finite value and σ_2 is zero. Hence, current density in region 2 for finite value of field intensity is zero. Accordingly,

$$J_{n2} = \sigma_2 E_{n2} = 0$$

Moreover, as per boundary relation (4), the normal component of current density in region 1 must also be zero.

$$J_{n1} = 0$$

$$\Rightarrow \quad E_{n1} = (J_{n1}/\sigma_1) = 0$$

Thus, on the conductor-side of the boundary, the current density is totally tangential. The field intensity, which is also totally tangential, is given by

$$E_{t1} = J_{t1}/\sigma_1$$

On the insulator side of the boundary, the tangential component of field intensity is equal to that on the conductor side as per boundary relation (3). That is, $E_{t2} = E_{t1}$. Field intensity along normal direction may exist due to static fields. Tangential component of current density J_{t2} is zero as $\sigma_2 = 0$.

As an example, let us consider a two-conductor transmission line connected to a direct-voltage source of voltage V at one end and a load resistance R_R at the other end, as shown in Fig. 3.10. The conductivity of insulating material separating the conductors is zero. As found earlier, only the tangential component of field intensity can exist on the conductor side of the boundary. This is equal to voltage drop in the resistance per unit length of each conductor. However, both tangential and normal components of field intensities exist on insulation side of the boundary. The tangential component of field intensity is equal to that of the conductor side of the interface. The normal component of field intensity, which is a static field, is due to the applied voltage. The total field intensity is equal to the vector sum of the two components, as shown in the figure.

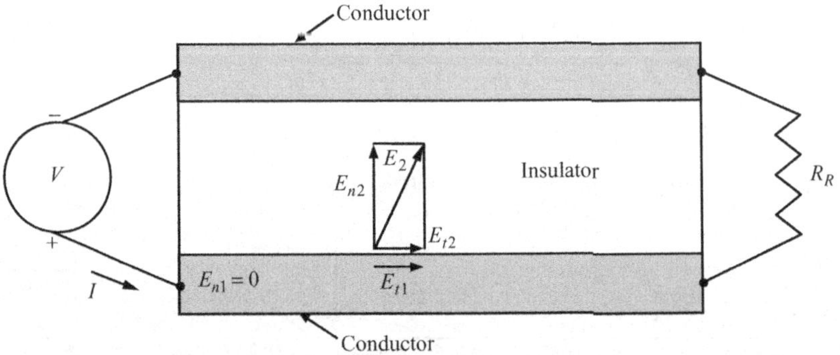

Fig. 3.10 Fields at the conductor-insulation boundary

EXAMPLE 3.8

At the interface between aluminium and copper, current density on the aluminium side of the interface is 2 0 A/mm^2 and it makes an angle of 30° with normal to the interface. The conductivity of aluminium is $3\ 5 \times 10^7$ S m and that of copper is $5\ 8 \times 10^7$ S/m. Calculate the current density on the copper side of the interface.

Using subscripts 1 and 2 for aluminium and copper, the given data are

$J_1 = 2\cdot0\,\text{A/mm}^2, \sigma_1 = 3\cdot5\times10^7\,\text{S/m}, \alpha_1 = 30°, \sigma_2 = 5\cdot8\times10^7\ \text{S/m}$

We have, $\tan\alpha_2 = \dfrac{\sigma_2\tan\alpha_1}{\sigma_1} = \dfrac{5\cdot8}{3\cdot5\times\sqrt{3}} = 0\cdot957$

$\Rightarrow \quad \alpha_2 = 43\cdot73°$

$J_{n2} = J_{n1} = J_1\cos\alpha_1 = \sqrt{3}\ \text{A/mm}^2$

$J_2 = J_{n2}\sec\alpha_2 = 2\cdot397\ \text{A/mm}^2$

EXAMPLE 3.9

The common boundary surface of two conducting media 1 and 2 is at $z = 0$ plane. The conductivities of the two media are $\sigma_1 = 3.8\times10^7$ S/m and $\sigma_2 = 5.8\times10^7$ S/m. Current density in medium 2 is $\mathbf{J}_2 = 3\mathbf{u}_x + 4\mathbf{u}_z$ A/cm². Find the magnitudes of current densities in the two media at the boundary and the angles subtended by the current densities with normal to the boundary.

The z-components of current densities are normal to the boundary surface. Hence,

$J_z = J_{z2} = 4\ \text{A/cm}^2$

The x-components of current densities are tangential to the boundary surface. Hence,

$\dfrac{J_{x1}}{J_{x2}} = \dfrac{\sigma_1}{\sigma_2}$

Or $\quad J_{x1} = (3.8/5.8)(3) = 1.966\ \text{A/cm}^2$

Magnitude of current density in medium 1, $J_1 = \sqrt{1.966^2 + 4^2} = 4.457\ \text{A/cm}^2$

Magnitude of current density in medium 2, $J_2 = \sqrt{3^2 + 4^2} = 5.0\ \text{A/cm}^2$

Angle of J_1 with normal to the boundary surface,

$\alpha_1 = \tan^{-1}(J_{x1}/J_{z1}) = \tan^{-1}(1.966/4) = 26.17°$

Angle of J_2 with normal to the boundary surface,

$\alpha_2 = \tan^{-1}(J_{x2}/J_{z2}) = \tan^{-1}(3/4) = 36.87°$

3.9 Laplace's Equation

In the region of a conductor carrying steady current, let us consider two points which are separated by a differential length $d\ell$. Let the voltage drop between the points be dV and field intensity be $\mathbf{E}$. Then,

$$dV = \mathbf{E}\cdot d\ell$$

Starting from this relation and following the procedure described in Section 2.17, it can be shown that

$$\nabla \quad = \mathbf{E} \tag{1}$$

Taking divergence of both sides of (1), we have

$$\nabla\cdot\nabla \quad = \nabla^{\hat{}} \quad = \nabla\cdot\mathbf{E} \tag{2}$$

If conductivity σ of the conductor is assumed to be constant, then

$$\nabla \cdot \mathbf{E} = \nabla \cdot \left(\frac{\mathbf{J}}{\sigma}\right) = \frac{1}{\sigma}(\nabla \cdot \mathbf{J})$$

As the divergence of current density is zero, the divergence of field intensity will also be zero if the conductivity has a finite value. Accordingly,

$$\nabla \cdot \mathbf{E} = 0 \tag{3}$$

Substituting (3) in (2), we get

$$\nabla^2 V = 0 \tag{4}$$

This is the Laplace's equation for the voltage function V within a conductor carrying steady current.

Equation (4) is similar to Laplace's equation for electrostatic potential in a charge-free dielectric region.

3.10 Insulation Resistance and Conductance

Dielectric materials used as insulation between conductors of transmission lines are not perfect, i.e., the materials have finite conductivities. Hence, conduction current flows from one conductor to the other conductor through the insulation under working condition. This current may be treated as *leakage* current. We will calculate the conduction current, insulation resistance and conductance between the conductors of a coaxial transmission line in Fig. 2.61. We assume that the potential difference between the conductors is V and conductivity of the insulating material is σ_d.

As described in the preceding section, the field intensity in the insulating material has a normal component due to voltage difference between the conductors and a tangential component that is equal to voltage drop per unit length in the conductor. The voltage drop per unit length in a good conductor is very small and hence we may assume the field intensity within the insulating material of a coaxial transmission line to be totally radial. The radial field intensity in terms of potential difference V between the conductors, which has been derived in Section 2.27, is given by

$$E_r = \frac{V}{r\ln(b/a)}, \quad b \geq r \geq a$$

The current density, which is also along the radial direction, is

$$J_r = \sigma_d E_r = \frac{\sigma_d V}{r\ln(b/a)}, \quad b \geq r \geq a$$

We see that the current density is inversely proportional to r. The maximum value of the current density occurs at $r = a$. The leakage current through the insulating material I_{le} at radius r is equal to the current density multiplied by the surface area of the insulating material at this radius. Considering a line of length ℓ, the surface area at radius r is equal to $2\pi r\ell$. Thus the leakage current,

$$I_{le} = J_r(2\pi r\ell) = \frac{2\pi\sigma_d \ell V}{\ln(b/a)}$$

$$\text{Insulation resistance} = \frac{V}{I_{le}} = \frac{\ln(b/a)}{2\pi\sigma_d \ell} \tag{1}$$

We note that the insulation resistance between the conductors is *inversely* proportional to length of the transmission line whereas the resistance of the conductors is *directly* proportional to the line length.

Insulation resistance per unit length of the line is obtained by substituting $\ell = 1$ in (1). Denoting the resistance by R, we have

$$R = \frac{\ln(b/a)}{2\pi\sigma_d} \ \Omega/\text{m}$$

Conductance of the insulating medium (also called *shunt* conductance) per unit length of the line,

$$G = \frac{1}{R} = \frac{2\pi\sigma_d}{\ln(b/a)} \ \text{S/m} \tag{2}$$

This expression has the same form as the capacitance per unit length of a coaxial transmission line which has been derived in Section 2.27. The capacitance per unit length of the coaxial line under study is

$$C = \frac{2\pi\varepsilon}{\ln(b/a)} \ \text{F/m} \tag{3}$$

Dividing the two sides of (2) by the corresponding sides of (3), we get

$$\frac{G}{C} = \frac{\sigma_d}{\varepsilon}$$

Or

$$G = \frac{\sigma_d C}{\varepsilon} \tag{4}$$

Thus, the conductance of the medium between the conductors of a coaxial line is related to the capacitance of the line through the conductivity and permittivity of the medium. As the relation depends only on the electric properties of the material, we can use it to obtain the conductance of any two-conductor transmission line from its capacitance. For example, the approximate formula for the capacitance per unit length of a transmission line with two parallel conductors derived in Section 2.28 is

$$C = \frac{\pi\varepsilon}{\ln(h/a)} \ \text{F/m}$$

By substituting this expression in (4) we will obtain the formula for shunt conductance per unit length of the line. It is given by

$$G = \frac{\pi\sigma_d}{\ln(h/a)} \ \text{S/m}$$

where h is the distance between centres of the conductors and a is radius of each conductor.

EXAMPLE 3.10

The radii of inner and outer conductors of a coaxial transmission line are 2.5 mm and 4.0 mm respectively. The conductivity of insulating material between the conductors is 7.5×10^{-14} S/m. Calculate the insulation resistance for a line length of 100 m.

$$\text{Insulation resistance} = \frac{\ln(b/a)}{2\pi\sigma_d \ell} = \frac{\ln(4.0/2.5)}{2\pi(7.5\times10^{-14})(100)} = 9.97\times10^{9} \ \Omega$$

3.11 Current Flow in Semiconductors

Materials classified as semiconductors have conductivities between those of metals and insulators. Out of many semi-conducting materials, silicon and germanium have numerous applications in engineering. Both silicon and germanium are tetravalent and are of crystalline structure in solid state. The four valence electrons of each atom of a crystal are associated in the formation of electron pairs or covalent bonds with valence electrons of four

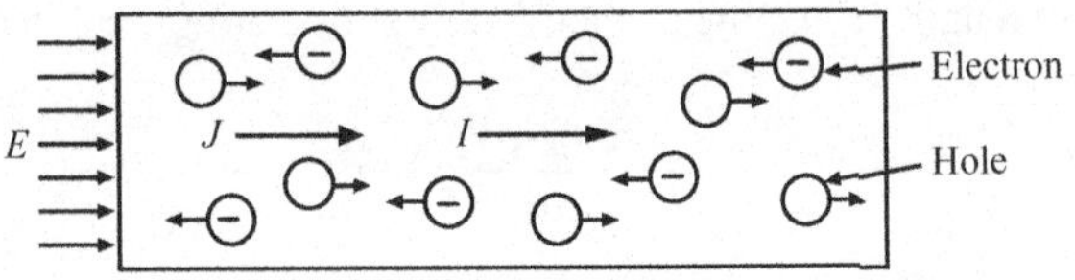

Fig. 3.11 An intrinsic semiconductor in an external electric field E. The electrons move in a direction opposite to E and the holes move in the same direction as E. The current densities due to both charge carriers are in the direction of E.

neighbouring atoms. At very low temperature tending to 0 K, the electron pairs are bound to the nuclei of the atoms. Therefore, in the bonding state the valence band is full and no *free* electrons are available for conduction of electricity. The conductivity of the materials at this temperature is practically zero as that of an insulator.

With an increase in the temperature, say to the room temperature, the thermal energy acquired by the valence electrons enables some of these electrons to get dislodged from the covalent bonds and move to the empty conduction band. Consequently, some covalent bonds become incomplete because of shortage of electrons. The incomplete bonds are called *holes*. A *hole* behaves like a positive charge-carrier having the same magnitude of charge as that of an electron. The life of an electron-hole pair is very short. However, the formation of the pairs is a continuous process. Hence, electrons and *holes* are always available in the material. In pure germanium and silicon, the number of holes in a unit volume (hole-concentration) is equal to the number of electrons per unit volume (electron-concentration). That is, the density of positive charge carriers is equal to that of the negative charge carriers. Materials having this property are called intrinsic semiconductors. When an electric field is applied to an intrinsic semiconductor, the electrons drift in a direction opposite to the applied field and holes drift in the same direction as the applied field, as illustrated in Fig. 3.11. The direction of current density due to both the charge carriers is same as that of the external field.

It may be noted that conduction of electricity in a semiconductor is due to both electrons and holes.

We see that a semiconductor has both positive and negative charge carriers. Some other examples where both positive and negative charge carriers exist are plasma, electrolytes and earth's ionosphere.

In engineering applications, doped (impure) semiconductors are mostly used. If pure germanium or silicon is doped with a very small quantity of a pentavalent impurity, like phosphorus, antimony, arsenic, etc., the four valence electrons of the doping substance form covalent bonds with four atoms of the parent material. The fifth valence electron of the doping substance is comparatively *free* and is available for conduction of electricity. The resulting material is an impure or extrinsic semiconductor. It has excess of negative charge carriers and it is called n-type semiconductor. The pentavalent material, which donates one electron per atom, is called a *donor*. Donor concentrations of one part to 10^6 part of silicon increase the conductivity of the semiconductor by about 10^6 times.

If pure germanium or silicon is doped with a trivalent material, like aluminium, boron, indium, etc., the three valence electrons of an atom of the doping substance form covalent bonds with three neighbouring atoms of the parent material. However, it cannot form a covalent bond with its fourth neighbouring atom as it is short of one electron. This bond which is short of one electron is a *hole*. The resulting impure material has excess of positive charge carriers. It is called p-type semiconductor. As the trivalent substances can accept electrons from the parent material, these are called *acceptors*.

The concentration of electrons is much more than that of holes in n-type semiconductors. Hence, electrons are the majority carriers and holes are the minority carriers in n-type semiconductors. In p-type semiconductors it is just the opposite. When an external electric field is applied to a doped semiconductor, its *free* electrons drift in the opposite direction and the holes drift in the same direction as that of the applied field. Since the two charge carriers are of opposite sign, the current densities due to the two charge carriers add up. The current density is in the same direction as that of the applied field.

The relation between the current density and the applied field can be obtained using the same procedure as for metals. However, in the case of semiconductors we have to consider current flow due to the *free* electrons as well as due to the *holes*.

By referring to formula (3.2-1), which gives drift velocity of *free* electrons in a metal, we can write expressions for the drift velocities of electrons and holes of a semiconductor. These are given by

$$\mathbf{v}_{de} = -\mu_e \mathbf{E}, \quad (1)$$

$$\mathbf{v}_{dh} = \mu_h \mathbf{E} \quad (2)$$

The symbols used in (1) and (2) have the same meaning as defined in section (3.2). Subscripts or additional subscripts e and h in the main symbols correspond to electrons and holes respectively. If there are n_e free electrons and n_h holes per cubic metre of a semiconductor, then the values of volume charge-densities (charge concentrations) are

$$\rho_e = en_e$$

and

$$\rho_h = en_h$$

Total current density, as per formula (3.3-3), is given by

$$\mathbf{J} = (\rho_e\mu_e + \rho_h\mu_h)\mathbf{E} \quad (3)$$

$$= \sigma \mathbf{E}$$

where $\sigma = \rho_e\mu_e + \rho_h\mu_h$ is the conductivity.

In n-type semiconductors $\rho_e \gg \rho_h$ whereas in p-type of semiconductors $\rho_h \gg \rho_e$. Therefore, conductivity and current flow are mainly governed by electrons in n-type semiconductors and by *holes* in p-type semiconductors. Thus, neglecting the second term of (3) the current density for n-type semiconductors is given by

$$\mathbf{J} = (\rho_e \mu_e)\mathbf{E} = \sigma_e \mathbf{E} \quad (4)$$

Neglecting the first term of (3), the current density for p-type semiconductors is given by

$$\mathbf{J} = (\rho_h \mu_h)\mathbf{E} = \sigma_h \mathbf{E} \quad (5)$$

Solving (1) and (4) simultaneously, the drift velocity of electrons in terms of the current density and the charge density is obtained as

$$\mathbf{v}_{de} = -\frac{\mathbf{J}}{\rho_e} \quad (6)$$

Similarly,

$$\mathbf{v}_{dh} = \frac{\mathbf{J}}{\rho_h} \quad (7)$$

The concentration of charge carriers in semiconductors increases substantially with an increase in temperature. Therefore, the conductivity also increases. But, as described earlier, the conductivity of a metallic conductor decreases when its temperature increases.

EXAMPLE 3.11

A semiconductor piece has 2.2×10^{18} electrons/m^3 and 4.8×10^{16} holes/m^3. The mobility of electrons and holes is 0 2 m^{2}V^{-1}s^{-1} and 0.025 m^{2}V^{-1}s^{-1} respectively. The piece has a rectangular section of width 1.5 mm and thickness 0.8 mm. Its length is 12 mm. Calculate the resistance between the end surfaces of the piece.

Given data are

$$n_e = 2.2 \times 10^{18} /\text{m}^3;\; n_h = 4.8 \times 10^{16} /\text{m}^3;$$

$$\mu_e = 0.2 \text{ m}^2/(\text{V-s});\; \mu_h = 0.025 \text{ m}^2/(\text{V-s})$$

Thus, $\rho_e = en_e = (1.6 \times 10^{-19})(2.2 \times 10^{18}) = 0.352 \text{ C/m}^3$

$$\rho_h = en_h = (1.6 \times 10^{-19})(4.8 \times 10^{16}) = 7.68 \times 10^{-3} \text{ C/m}^3$$

$$\sigma = \rho_e\mu_e + \rho_h\mu_h = (0.352 \times 0.2) + (7.68 \times 10^{-3})(0.025) = 0.0706 \text{ S/m}$$

Resistance between the end surfaces

$$\frac{\ell}{\sigma S} = \frac{12.0 \times 10^{-3}}{(0.0706)(1.5 \times 10^{-3})(0.8 \times 10^{-3})} = 141.6 \text{ k}\Omega$$

SUMMARY

From the brief description of the mechanism of current flow, we learnt that current in metals is due to free or conducting electrons whereas current flow in semiconductors is due to both free electrons and holes. We also learned that the conductivity of a metal is equal to the product of the concentration of free electrons and their mobility. The conductivity of a semiconductor, however, is equal to the sum of two conductivities; one is equal to the product of concentration and mobility of free electrons, and the other is equal to the product of concentration and mobility of holes.

Starting from the macroscopic form of Ohm's law we obtained the important relation between electric field intensity and current density, known as the point form of Ohm's law. We also derived the continuity equation from Kirchhoff's current law, and formulated Kirchhoff's voltage law in terms of electric field intensity, current density, and conductivity, thereby establishing the relationship between field theory and circuit theory. We found from the expression for insulation resistance, derived taking the example of a coaxial transmission line, that the insulation resistance is *inversely* proportional to the line length. We also found that the formula for conductance of the insulating medium can be obtained by substituting σ in place of ε in the capacitance formula. We learnt that divergence of current density is equal to zero, and we used this property and Kirchhoff's voltage law for the derivation of boundary conditions. The boundary conditions are analogous to the boundary conditions of static electric field as is evident from the following table. We however note that E for steady current intensity is defined by the point form of Ohm's law, and E in the case of static electric field is due to stationary charges.

Boundary conditions for conduction current	Boundary conditions in static electric field
$E_{t1} = E_{t2}$	$E_{t1} = E_{t2}$
$\frac{J_{t1}}{\sigma_1} = \frac{J_{t2}}{\sigma_2}$	$\frac{D_{t1}}{\varepsilon_1} = \frac{D_{t2}}{\varepsilon_2}$
$J_{n1} = J_{n2}$	$D_{n1} = D_{n2}$
$\sigma_1 E_{n1} = \sigma_2 E_{n2}$	$\varepsilon_1 E_{n1} = \varepsilon_2 E_{n2}$
$\frac{\tan\alpha_1}{\tan\alpha_2} = \frac{\sigma_1}{\sigma_2}$	$\frac{\tan\alpha_1}{\tan\alpha_2} = \frac{\varepsilon_1}{\varepsilon_2}$

IMPORTANT FORMULAE

Drift velocity: $\mathbf{v}_d = -\mu_e \mathbf{E}$

Conductivity, $\sigma = \rho_e \mu_e = \dfrac{ne^2\tau}{m}$

Current: $I = \int_S \mathbf{J} \cdot d\mathbf{S}$

Resistance of good conductor of uniform section: $R = \dfrac{\rho_\Omega \ell}{S}$

Resistance of a good conductor at temperature t °C: $R = R_0(1+\alpha_0 t)$

Point form of Ohm's law: $\mathbf{J} = \sigma \mathbf{E}$

Power loss per unit volume in a good conductor $= \dfrac{J^2}{\sigma}$

Boundary conditions: $E_{t1} = E_{t2}$; $J_{n2} = J_{n1}$

Laplace's equation: $\nabla^2 V = 0$

Insulation resistance $= \dfrac{\ln(b/a)}{2\pi\sigma_d \ell}$

Ratio between conductance and capacitance: $\dfrac{G}{C} = \dfrac{\sigma_d}{\varepsilon}$

Conductance of two-conductor transmission line per unit length: $G = \dfrac{\pi\sigma_d}{\ln(h/a)}$

Drift velocity of electron in a semiconductor: $\mathbf{v}_{de} = -\mu_e \mathbf{E}$

Drift velocity of hole in a semiconductor: $\mathbf{v}_{dh} = \mu_h \mathbf{E}$

Conductivity of semiconductors: $\sigma = \rho_e \mu_e + \rho_h \mu_h$

MULTIPLE-CHOICE QUESTIONS

(Starred questions have two answers)

3.1 The drift velocity of a free electron in a conductor is directly proportional to the electric field intensity. The constant of proportionality is
(a) electronic-mass (b) acceleration (c) mobility (d) volume charge density

3.2 The conductivity of a metal is equal to the product of
(a) volume charge density of free electrons and drift velocity
(b) volume charge density of free electrons and mobility
(c) drift velocity and mobility
(d) drift velocity, charge concentration and mobility

3.3 The SI unit of electric conductivity is
(a) ohm/metre (b) metre/ohm (c) siemen (d) siemen/metre

3.4 The conductance of a good conductor of uniform sectional area S, length ℓ and resistivity ρ_Ω is given by
(a) $S/\rho_\Omega \ell$

(b) $\rho_\Omega \ell / S$
(c) $\ell / \rho_\Omega S$
(d) $\rho_\Omega S / \ell$

3.5 Joule's law gives
(a) energy dissipated in a resistor
(b) power dissipated in a resistor
(c) resistance of a conductor in terms of resistivity
(d) relation between current density and electric field intensity in a conductor

3.6 If current density in a conductor is J and its conductivity is σ, then power dissipated per unit volume of the conductor is
(a) σJ (b) $\sigma^2 J$ (c) σJ^2 (d) J^2/σ [Answer: (d)]

*3.7 At the boundary surface between a good conductor and a perfect insulator
(a) the normal component of current density in the conductor has a finite value
(b) the normal component of electric field intensity in the insulation is zero
(c) the current density in the conductor is tangential to the boundary
(d) the tangential component of electric field intensity in the insulation has a finite value

*3.8 Which of the following statements are correct?
(a) Current flow in p-type semiconductors is due to holes only and current in n-type semiconductors is due to electrons only.
(b) Current flow in p-type as well n-type semiconductors is due to flow of both holes and electrons.
(c) The resistivity of a semiconductor increases with increase in temperature.
(d) The conductivity of a semiconductor increases with an increase in temperature.

QUESTIONS

3.1 What are the quantities on which the mobility of a conductor depends? What is the unit of mobility?

3.2 Discuss why current density is a vector quantity whereas current is a scalar quantity.

3.3 The resistivity of a metal is a function of temperature. It has a constant part and a variable part. The variable part is proportional to the product of temperature and resistivity at 0 °C. What is the constant of proportionality called and what is its unit?

3.4 It has been observed that the filament of an incandescent lamp generally burns out at the time of switching it on. What could be the likely reason?

3.5 State the point form of Ohm's law. What is the analogous relation in the static electric field?

3.6 The earlier unit of electric conductivity was mho/m. What is its present unit?

3.7 Describe Joule's law taking the example of a conductor of uniform cross-sectional area. Express the law in terms of current density.

3.8 Give some examples of active and passive elements in an electric circuit.

3.9 As per Kirchhoff's voltage law, the sum of the emfs of the active sources around a dc electric circuit is equal to the voltage drop in the resistances. Is the sum of the emfs equal to the sum of the magnitudes of the emfs or to the algebraic sum of the emfs?

3.10 Discuss the boundary conditions at the boundary between a good conductor and a perfect insulator.

3.11 The Laplacian of the potential function in a region of charge in a static electric field is not zero (Poisson's equation) whereas the Laplacian of the potential function in the region of steady current flow is zero. Explain why.

3.12 Semiconductors have both positive and negative charge carriers. Give some other examples where both types of charge carriers exist.

3.13 Why are the drift velocities of positive and negative charge carriers in a semiconductor in opposite directions whereas the direction of current flow due to the two charge carriers is same?

3.14 Why does the resistivity of good conductors increase with increase in temperature, while the resistivity of semiconductors decrease with increase in temperature?

ANSWERS TO SELECTED QUESTIONS

3.3 The resistivity of a metal is a function of temperature. It has a constant part and a variable part. The variable part is proportional to the product of temperature and resistivity at 0 °C. What is the constant of proportionality called and what is its unit?

Answer: The resistivity of a metallic conductor at temperature t is approximately given by $\rho_\Omega = \rho_{\Omega 0}(1+\alpha_c t)$ where $\rho_{\Omega 0}$ resistivity of the conductor at 0 °C and ρ_Ω is the resistivity at t °C. The second term gives the increase in the resistivity from temperature 0 °C to t °C. This increase is approximately proportional to t. The constant of proportionality is called temperature coefficient of resistivity. In the formula, α_c is the value at 0 °C; hence it is expressed as per degree centigrade at 0 °C.

3.4 It has been observed that the filament of an incandescent lamp generally burns out at the time of switching it on. What could be the likely reason?

Answer: One likely reason is as follows. The resistivity of a tungsten filament at the working temperature (about 2500 °C) is nearly 20 times higher than its cold value. Since the filament is designed to have proper value at the working temperature, the current at the instant of switching is nearly 20 times of the value under working condition. Moreover, the size of the filament gets reduced in course of use. Because of high current and weakness of the filament it may burn out at the instant of switching.

3.7 Describe Joule's law taking the example of a conductor of uniform cross-sectional area. Express the law in terms of the current density.

Answer: Energy dissipated W in a conductor carrying direct current, found from experiment, is given by I^2Rt where I is current in the conductor (assumed to be constant), R is resistance of the conductor, and t is time. This is known as Joule's law. For a conductor of length ℓ, uniform cross-sectional area S and conductivity σ, resistance of the conductor $R = \ell/\sigma S$. Moreover, $I^2 = J^2S^4$. Hence, $W = (J^2S^2)(\ell/\sigma S)(t) = (J^2/\sigma)(t)(\text{volume})$.

3.14 Why does the resistivity of good conductors increase with increase in temperature, while the resistivity of semiconductors decrease with increase in temperature?

Answer: The resistivity of a good conductor is inversely proportional to the relaxation time, and the relaxation time decreases with increase in temperature. Hence, the resistivity of a good conductor increases with increase in temperature.

In semiconductors, charge carriers per unit volume or charge concentrations increase with increase in temperature. Therefore, the conductivity increases with increase in temperature. Thus the resistivity, which is reciprocal of the conductivity, decreases with increase in temperature.

PROBLEMS

*3.1 The resistance of a metallic conductor of arbitrary length is 0.4 Ω at temperature t_1, and 0.5 Ω at temperature t_2. The temperature coefficient of resistivity of the metal is 0.0038 per degree centigrade at zero °C. Find the relation between t_1 and t_2.

**3.2 The resistance of tungsten filament of an incandescent lamp is 40 Ω at 30 °C, and 520 Ω at the working temperature of 2500 °C. Find the temperature coefficient of resistivity for tungsten.

**3.3 A heating element made of nichrome (nichrome is an alloy of nickel, chromium and iron) has a resistance of 60 Ω at a temperature of 27 °C, and 65 Ω when hot. The temperature coefficient of resistivity for nichrome is 1.1×10^{-4} per degree centigrade at 0 °C. Calculate the working temperature of the element.

***3.4 The length of a tapered conductor of circular section is 2.0 m. The minimum and maximum values of diameter at the two ends are 10.0 mm and 12.0 mm respectively. The resistivity of the material of the conductor is 3.8×10^{-8} Ω-m. Find the resistance of the conductor.

**3.5 A solid cylindrical conductor carries current along the axial direction. The radius of the conductor is a, and the current density at a radial distance r is given by $\mathbf{J}_z = J_{z0}\,e^{-(r/a)}\mathbf{u}_z$ A/mm². If $J_{z0} = 6$ A/mm² and $a = 5$ mm, find the current carried by the conductor.

***3.6 A hollow cylindrical conductor carries current in the axial direction. The inner and outer radii of the cylinder are 4.0 mm and 6.0 mm respectively. The current density varies linearly over the cross-section of the conductor from 5.0 A/mm² at the inner radius to 4.0 A/mm² at the outer radius. Find the current carried by the conductor.

*3.7 Prove that the divergence of steady current density is zero.

**3.8 Two round conductors 1 and 2, each 50 m in length and cross-sectional area of 10 mm², are joined end-to-end. The conductivities of the conductors 1 and 2 are 3.6×10^7S/m and 6.0×10^7S/m respectively. The voltage difference between the ends of the joined conductors is 10 V. Find power losses in conductors 1 and 2.

*3.9 Show that power loss per unit volume in a good conductor is equal to J^2/σ where J is current density and σ is conductivity.

**3.10 Derive the boundary relation at the common boundary between two good conductors of different conductivities carrying steady current.

**3.11 The common boundary surface between two regions is on the $x = 0$ plane. The conductivity of region 1 ($x < 0$) is 5.8×10^7S/m, and the conductivity of region 2 ($x > 0$) is 3.8×10^7S/m. The current density in region 2 is $\mathbf{J}_2 = 5\mathbf{u}_x + 4\mathbf{u}_y + 2\mathbf{u}_z$ A/mm². Find the current density in region 1.

***3.12 The common boundary surface between two conducting regions is a plane surface defined by the equation $6x+3y+5z=12$. The conductivity of region 1 containing the origin is 1.9×10^7S/m, and the conductivity of region 2 is 3.8×10^7S/m. The current density in region 1 is $\mathbf{J}_1 = 4.5\mathbf{u}_x + 3.0\mathbf{u}_y + 5.0\mathbf{u}_z$ A/mm². Find the current density in region 2.

**3.13 Obtain the expression for resistance of insulation between the conductors of a coaxial transmission line.

Chapter Four

MAGNETIC FIELDS DUE TO STATIONARY CURRENTS

4.1 Introduction

We know that when a magnetic needle is brought near a permanent magnet it experiences a force which is called *magnetic force*. The lines of force or the magnetic field lines when plotted close on themselves. Such a plot is called a magnetic field map. We also know that when a magnetic needle is brought near a conductor carrying electric current it experiences a force. Therefore, a current-carrying conductor also produces a magnetic field. The field lines close on themselves around the conductor, as shown in Fig. 4.1. The direction of field lines is given by the right-hand rule. As per the rule, if the conductor is held in the right hand with the thumb pointing in the direction of current flow, then other fingers encircling the conductor give the direction of field lines.

It is an experimental fact that when two conductors carrying electric current are placed at a finite distance apart, magnetic force is exerted on each of the conductors. The force on one conductor can be due to the magnetic field of the other conductor. Alternatively, we may also say that the force is due to interaction between the magnetic fields produced by the two currents. The forces are attractive if the currents are in the same direction and repulsive for opposite current directions. If two current-carrying conductors are straight and parallel, magnitude of force per unit length of either conductor has the same value. The force is a function of current magnitudes, spacing between the conductors and the magnetic property of the medium in which the conductors are located.

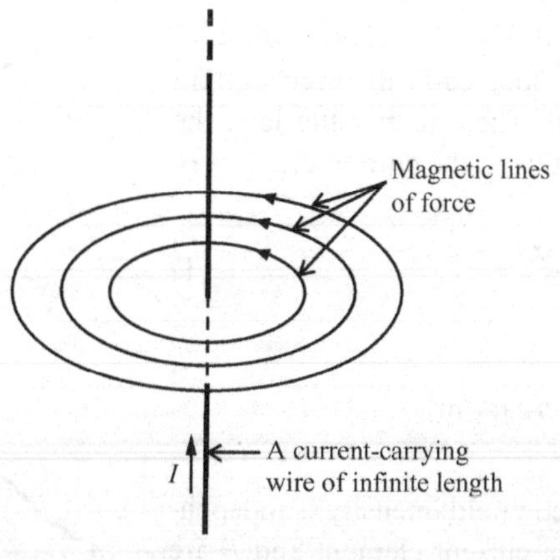

Fig. 4.1 Showing magnetic lines of force due to a straight, current-carrying wire

The magnetic field produced by conductors carrying direct current is called static magnetic field (also called *steady magnetic field* or *stationary magnetic field*). The subject is developed either from Ampere's law of force or from Biot–Savart law. Ampere's law of force gives force between two current loops. From this law, a field quantity called magnetic flux density due to a current loop is defined. Further development of the subject is based on the field concept. This is in line with the study of static electric field which starts from Coulomb's law of force between two point charges. Biot–Savart law defines a field quantity which is called magnetic field intensity (also called *magnetic field strength* or *magnetizing force*) at any point in space due to a current element. As the mathematical form of Biot–Savart law is easier to understand than Ampere's force law, we will start the study of static magnetic field from this law. After developing the subject based on the field concept, the formula for force between two current loops is derived.

Our study of magnetic fields will be based on the assumption that the magnetic medium is homogeneous, linear and isotropic. These assumptions greatly simplify the analysis of magnetic field problems and make possible the superposition of fields due to several currents. Under these assumptions, the permeability of the medium can be taken as constant.

Some of the applications of magnetic fields due to steady currents are lifting magnets, excitation of synchronous generators, direct current motors, magnetic deflection of electron beams, and in studying the properties of semiconductors using *Hall Effect.*

The laws and formulae of static magnetic fields are also applicable for ac fields, at least at power frequency.

4.2 Magnetic Field Intensity: The Biot–Savart Law

When electric current flows through a conductor it produces a magnetic field. The magnetic field is described by a quantity called magnetic field intensity, and it is denoted by H. Biot–Savart law gives the magnetic field intensity of a current element. A current element is defined as the product of current I and a differential-length vector $d\ell$ of a conductor carrying the current. Thus, the current element $Id\ell$ is a vector quantity.

According to Biot–Savart law, the field intensity of a current element at a point P at a distance R from the element is directly proportional to $Id\ell$ and inversely proportional to R^2. It is also proportional to the sine of the angle α between the vectors $d\ell$ and $\mathbf{R}$. The constant of proportionality is unity in SI units.

This is explained with reference to Fig. 4.2 where AB represents a part of a stationary current loop carrying direct current I, and $d\ell$ is a differential length. Then, as per the law, the magnitude of field intensity at P due to the current element is

$$dH = \frac{I d\ell \sin\alpha}{4\pi R^2} \tag{1}$$

The unit of H is *ampere per metre* (A/m).

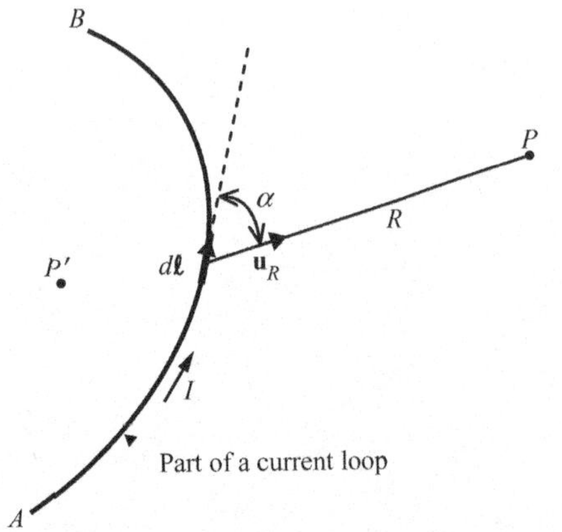

Fig. 4.2 For explanation of Biot–Savart law

It is clear from (1) that the magnetic field intensity is independent of the medium in which the current element and P are located. The field intensity is a vector point function in space and its direction is along normal to the plane containing $d\ell$

and **R**. Its sense is in the direction of progress of a right-handed screw when it is rotated from $d\ell$ towards **R** through the smaller angle α. Accordingly, the direction of the field is normal and into the plane of the paper at P, and out of the plane of the paper at P'. The above definition of field intensity suggests that its vector form can be written in terms of the cross product of $d\ell$ and a unit vector $\mathbf{u}_R$ directed from the current element towards P, as shown in the figure. Thus, the vector form of Biot–Savart law is

$$d\mathbf{H} = \frac{I}{4\pi R^2}(d\ell \times \mathbf{u}_R) = \frac{I}{4\pi R^3}(d\ell \times \mathbf{R}) \tag{2}$$

The field intensity due to the current from A to B is obtained by integrating (2) along the path. Thus,

$$\mathbf{H}_{\mathrm{AB}} = \int_{\mathrm{A}}^{\mathrm{B}} \frac{I}{4\pi R^3}(d\ell \times \mathbf{R}) \tag{3}$$

The field intensity **H** due to a current loop is found by integrating (2) around the loop. Accordingly,

$$\mathbf{H} = \oint \frac{I}{4\pi R^3}(d\ell \times \mathbf{R}) \tag{4}$$

We will now obtain the magnetic field intensity of a current element in rectangular coordinates. Let us assume that the current element is centred at the origin and the point at which we wish to calculate the field is located at (x, y, z). Then,

$$d\ell = dx\,\mathbf{u}_x + dy\,\mathbf{u}_y + dz\,\mathbf{u}_z$$

$$\mathbf{R} = x\mathbf{u}_x + y\mathbf{u}_y + z\mathbf{u}_z$$

$$R^3 = (x^2 + y^2 + z^2)^{3/2}$$

By substituting these expressions in (2), we get

$$d\mathbf{H} = \frac{I}{4\pi R^3}(dx\,\mathbf{u}_x + dy\,\mathbf{u}_y + dz\,\mathbf{u}_z) \times (x\mathbf{u}_x + y\mathbf{u}_y + z\mathbf{u}_z)$$

$$= \frac{I}{4\pi R^3}[(z\,dy - y\,dz)\mathbf{u}_x + (x\,dz - z\,dx)\mathbf{u}_y + (y\,dx - x\,dy)\mathbf{u}_z)] \tag{5}$$

EXAMPLE 4.1

A current element of $0.05\mathbf{u}_x$ A-m is centred at $P'(0, 0, 1.0\text{ m})$ in rectangular coordinates. Find the magnetic field intensity at $P(2\text{ m}, 3\text{ m}, 1.2\text{ m})$.

The magnetic field intensity, $d\mathbf{H} = \dfrac{I}{4\pi R^3}(d\ell \times \mathbf{R})$

The current element, $Id\ell = 0.05\mathbf{u}_x$

The vector distance from P to P',

$$\mathbf{R} \quad (2-0)\mathbf{u}_x + (3-0)\mathbf{u}_y + (-1-1.2)\mathbf{u}_z = 2\mathbf{u}_x + 3\mathbf{u}_y - 2.2\mathbf{u}_z$$

$$R^3 = (4+9+4.84)^{3/2} = 75.35$$

$$d\mathbf{H} = \frac{(\ [illegible].05\mathbf{u}_x [illegible] \mathbf{u}_x + 3\mathbf{u}_y [illegible] \mathbf{u}_z)}{(\ [illegible])(\ [illegible].35)} = ([illegible]\mathbf{u}_y [illegible] \mathbf{u}_z) \times 10^{-4}\ \text{A/m}$$

EXAMPLE 4.2

A current element $Id\ell = 0.01\mathbf{u}_z$ A-m is oriented along z direction of cylindrical coordinates with its mid-point at the origin. Find the magnetic field intensity at (3.6 m, 56 °, 5 m)

Magnetic field intensity, $d\mathbf{H} = \dfrac{I}{4\pi R^3}(d\ell \times \mathbf{R})$

$Id\ell = 0\ 01\mathbf{u}_z$ A-m

$x = \qquad 56° = 2.01$ m; $y = 3.6\sin 56° = 2.98$ m; $z = 5.0$ m

$\mathbf{R} \qquad \mathbf{u}_x + 2.98\mathbf{u}_y + 5\mathbf{u}_z,\ R^3 = (2.01^2 + 2.98^2 + 5^2)^{3/2} = 233.5\ \text{m}^3$

$I\,d\ell \times \bar{\ } = (0.01\mathbf{u}_z)(2.01\mathbf{u}_x + 2.98\mathbf{u}_y + 5\mathbf{u}_z) = 0.0201\mathbf{u}_y - 0.0298\mathbf{u}_x\ \text{m}^2$

$$dH = \frac{0.0201\mathbf{u}_y - 0.0298\mathbf{u}_x}{(4\pi)(233.5)} = -10.156\mathbf{u}_x + 6.850\mathbf{u}_y\ \mu\text{A/m A/m}$$

4.3 Magnetic Flux Density and Magnetic Flux

It is found that magnetic force between current-carrying conductors depends on the medium in which they are located. As magnetic field intensity is not a function of magnetic properties of the medium, another vector quantity which is a function of the magnetic characteristic of the medium is introduced. The quantity is called magnetic flux density. It is a vector quantity and it is denoted by the symbol **B**. It is defined as

$$\mathbf{B} = \mu\mathbf{H} \tag{1}$$

where μ is called permeability of the medium.

> We see that the direction of flux density is same as that of the field intensity and its magnitude is μ times H. The SI unit of flux density is *tesla* (T), and that of permeability is *henry per metre* (H/m). The permeability of free space
>
> $$\mu_0 = 4\pi \times 10^{-7}\ \text{H/m}.$$

The ratio of permeability of a material to that of free space is called *relative permeability*. Thus relative permeability,

$$\mu_r = \frac{\mu}{\mu_0}$$

$$\Rightarrow \qquad \mu = \mu_0\mu_r \tag{2}$$

The relative permeability is a pure number. Its value depends on the magnetic properties of the material. The relative permeability of most materials, except ferromagnetic materials, is nearly unity. The relative permeability of air is also taken as unity.

Substituting (2) in (1), we get

$$\mathbf{B} \quad \mu_0\mu_r \mathbf{H}\ \text{T} \tag{3}$$

The flux density due to a current element, as per Biot–Savart law, is

$$d\mathbf{B} = \frac{\mu I}{4\pi R^3}(d\ell \times \mathbf{R}) \quad (4)$$

It may be noted that the electric flux density is independent of the dielectric properties of a medium whereas the magnetic flux density is a function of the magnetic properties of a medium.

The *magnetic flux* is found by integrating the normal component of flux density over a given surface.

The magnetic flux is a scalar quantity and we will denote it by the symbol Φ. The unit of magnetic flux is *weber* (Wb).

In Fig. 4.3, S represents an open surface in the region of a magnetic field, and dS is a differential area on S centred at P. The unit vector $\mathbf{u}_n$ is normal to dS, and $\mathbf{B}$ is the flux density vector at P. The vector form of the differential area,

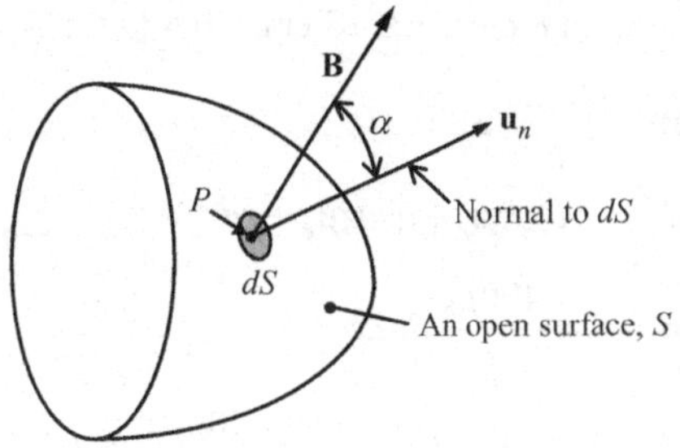

Fig. 4.3 For calculation of flux passing through an open surface located in a magnetic field

$$d\mathbf{S} = dS\,\mathbf{u}_n$$

The component of flux density along $\mathbf{u}_n$ is

$$B_n = B\cos\alpha$$

where α is angle between $\mathbf{B}$ and $\mathbf{u}_n$ as indicated in the figure.

Magnetic flux through the differential surface,

$$d\Phi = B_n\,dS = B\cos\alpha\,dS \quad (5)$$

As the right side of (5) is equal to the dot product of $\mathbf{B}$ and $d\mathbf{S}$, we have

$$d\Phi = \mathbf{B}\cdot d\mathbf{S}$$

The magnetic flux through the surface is obtained by integrating $d\Phi$ over the surface. Thus,

$$\Phi = \int_S \mathbf{B}\cdot d\mathbf{S} \text{ Wb} \quad (6)$$

As the unit of magnetic flux is weber, the old unit of magnetic flux density was *weber per square metre* (Wb/m²).

EXAMPLE 4.3

The flux distribution in a region is given by $\mathbf{B} = \dfrac{\mu I r}{2\pi a^2}\mathbf{u}_\phi$ in cylindrical coordinates. Calculate the flux crossing the region $0 \le r \le a$ m, $0 \le z \le 1$ m.

A differential flux crossing the region, $d\Phi = \mathbf{B}\cdot d\mathbf{S}$

In cylindrical coordinates, $d\mathbf{S} = (r\,d\phi dz)\mathbf{u}_r + (drdz)\mathbf{u}_\phi + (r\,drd\phi)\mathbf{u}_z$

$$d\Phi = \frac{\mu I}{2\pi a^2}(r\,dr\,dz)$$

$$\Phi = \frac{\mu I}{2\pi a^2}\int_0^a\int_0^1 r\,dr\,dz = \frac{\mu I}{4\pi}$$

If $\mu \quad \mu_0$, $I = 15$ A, $\Phi = 15\times10^{-7}$ Wb

EXAMPLE 4.4

The magnetic field intensity in cylindrical coordinates is given by $\mathbf{H} = (400/r)\mathbf{u}_\phi$. Find the flux in the region 8 cm $\le r \le$ 12 cm, -0.5 m $\le z \le$ 2.0 m in air.

$$\mathbf{B} \quad \frac{400\mu_0}{r} \mathbf{u}_\phi$$

$$d \quad rd\,,dz\mathbf{u}_r + drdz\mathbf{u}_\phi + rdrd\phi\mathbf{u}_z$$

$$\mathbf{B} \quad \mathbf{S} = \frac{400\mu_0}{r} dr\,dz$$

$$\Phi = \int_S \mathbf{B} \quad \mathbf{S} = \int_{r=0.08}^{0.12}\int_{z=-0.5}^{2.0} \frac{400\mu_0}{r} drdz$$

$$= (\quad)(4\pi\times10^{-7})\ln(12/8) = 0.5095 \text{ mWb}$$

4.4 Calculation of Flux Density of Some Simple Current Distributions

We will calculate the magnetic flux density of some simple current distributions using Biot–Savart law.

Circular Loop of Current

Figure 4.4 shows a circular loop of wire centred at the origin O of rectangular coordinate system. The radius of the loop and current carried by it are a and I respectively. A general solution of the problem of calculating flux density **B** at a point away from the axis is not easy. So we will calculate **B** at a point P on the axis and at a height h from the plane of the loop. The vector form of Biot–Savart law given by (4.3-4) will be used to calculate the flux density.

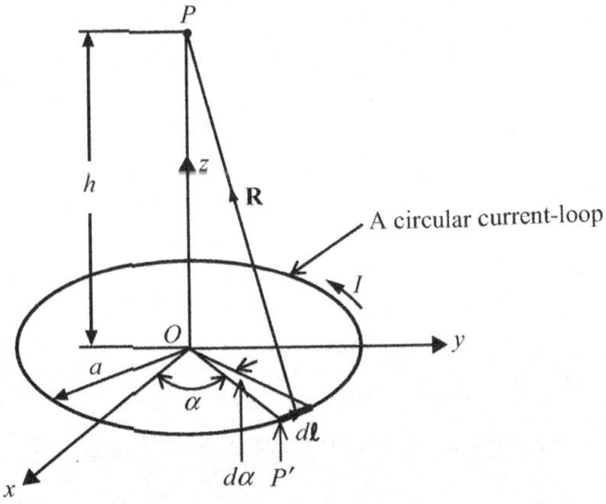

Fig. 4.4 For calculation of flux density on the axis of a circular current loop

Let us take a differential length $d\ell$ at a point P' on the loop, and let the angle between the x axis and the radial line OP' be α. The coordinates of P are (0, 0, h) and the coordinates of P' are ($x = a\cos\alpha, y = a\sin\alpha, z = 0$). Thus, the displacement of P from P' is

$$\mathbf{R} \quad \alpha\mathbf{u}_x - a\sin\alpha\mathbf{u}_y + h\mathbf{u}_z$$

The distance of P from P',

$$R \quad a^2 + h^2 \tag{1}$$

The differential length vector,

$$d\ell = \quad _x + dy\mathbf{u}_y = (-a \quad _x + a \quad \mathbf{u}_y)d\alpha$$

Hence,
$$d\ell \times \mathbf{R} = a\,h(\cos\alpha\,\mathbf{u}_x + \sin\alpha\,\mathbf{u}_y)\,d\alpha + a^2 d\alpha\,\mathbf{u}_z \tag{2}$$

The formula for flux density due to a current loop is

$$\mathbf{B} = \oint \frac{\mu I}{4\pi R^3}\,(d\ell \times \mathbf{R}) \tag{3}$$

By substituting (1) and (2) in (3) and then integrating from $\alpha = 0$ to $\alpha = 2\pi$ we will obtain the flux density at P. Noting that the integral of $\sin\alpha$ and $\cos\alpha$ from 0 to 2π is zero, the flux density is given by

$$\mathbf{B} = \frac{\mu I a^2}{2(a^2 + h^2)^{3/2}}\mathbf{u}_z \tag{4}$$

If the loop has N turns, flux density on the axis is given by

$$\mathbf{B} = \frac{\mu N I a^2}{2(a^2 + h^2)^{3/2}}\mathbf{u}_z$$

At the centre of the loop the flux density,

$$\mathbf{B}(h = 0) = \frac{\mu N I}{2a}\mathbf{u}_z$$

The flux density on the axis of a thin circular coil consisting of several turns of wire is obtained by taking a as the mean radius of the coil.

Short and Long Solenoids

A solenoid is a coil of several turns uniformly wound on a cylindrical insulating former. Solenoids are generally of circular section.

A short solenoid of length ℓ and radius a is shown in Fig. 4.5(a). The solenoid has n turns per metre of its length and it carries a current I. We shall calculate the magnetic flux density at a point P on the axis of the solenoid. The distance of the point from the mid-point O is taken as x.

Let us consider a differential length dx' at a distance x' from O, as shown in the figure. The number of turns contained in the differential length is equal to ndx'. The flux density at P due to the ampere-turns (product of current and number of turns) contained in the differential length may be obtained by substituting $nIdx$ in place of I and $(x - x')$ in place of h in formula for the circular current loop given by (4). The direction of flux density is along the axis of the solenoid. Thus denoting the flux density by $dB_a(x)$, we get

$$dB_a(x) = \frac{\mu_0 I n a^2 dx'}{2[a^2 + (x - x')^2]^{3/2}} \tag{5}$$

Integrating (5) over the length of the solenoid gives the magnetic flux density on the axis of the solenoid.

$$B_a(x) = \int_{-\ell/2}^{\ell/2} \frac{\mu_0 I n a^2 dx'}{2[a^2 + (x - x')^2]^{3/2}}$$

$$= \frac{\mu_0 n I}{2}\left[\frac{x + \ell/2}{\sqrt{a^2 + (x + \ell/2)^2}} - \frac{x - \ell/2}{\sqrt{a^2 + (x - \ell/2)^2}}\right] \tag{6}$$

The flux density at $x = 0$ is

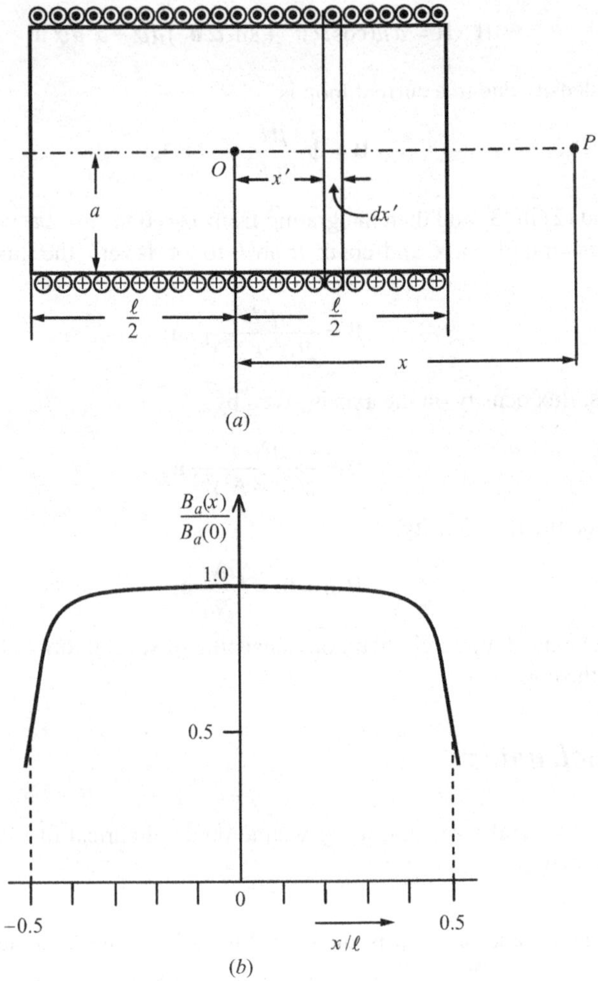

Fig. 4.5 (*a*) Section through the axis of a short solenoid and (*b*) Variation of flux density along its axis.

$$B_a(0) = \frac{\mu_0 n I \ell}{2\sqrt{a^2 + (\ell/2)^2}} = \frac{\mu_0 N I}{2\sqrt{a^2 + (\ell/2)^2}} \tag{7}$$

where $N \quad n\ell$, total number of turns in the solenoid.

If the length of a solenoid is large compared to its radius, i.e., $\ell \gg a$, then flux densities at the centre and at the ends of the solenoid are approximately given by

$$B_a(0) \approx \mu_0 n I, \ \ell \gg a$$

$$B_a(\ell/2) \approx \frac{\mu_0 n I}{2}, \ \ell \gg a$$

Figure 4.5(*b*) gives a plot of the ratio $B_a\ x)/B_a(0)$ of a solenoid for which $a = 0\ 05\ell$. It shows that the field is practically constant over the mid-region; it decreases rapidly towards the ends.

For a solenoid of infinite length the flux density on the axis is

$$B_a = \mu_0 nI$$

$$\Rightarrow \qquad H_a = nI \qquad (8)$$

Two Identical Coaxial Coils: The Helmholtz Pair

Figure 4.6(*a*) shows two identical circular coils *A* and *B* each of mean radius *a*, and *N* number of turns. The coils are coaxial and carry current *I* in the same direction. The distance between the mid-planes of the coils is *s*. The flux density at any point on the common axis of the coils can be found by utilising formula (4). If we take a point *P* on the axis at a distance *x* from the mid-point between the coils, the flux density at *P* due to coil *A* is obtained by replacing *I* by *NI* and *h* by $(s/2)+x$ in the formula. To get the flux density of coil *B* we have to replace *I* by *NI* and *h* by $(s/2)-x$ in the same formula. As the fields are in the same direction, the total field at *P* is

$$B_a(x) = \frac{\mu_0 NIa^2}{2}\left\{\frac{1}{[a^2+(\frac{s}{2}+x)^2]^{3/2}}+\frac{1}{[a^2+(\frac{s}{2}-x)^2]^{3/2}}\right\}$$

It is found that the flux density on the axis between the coils is nearly constant if $s = a$. The field for this condition is

$$B_a(x)\big|_{s=a} = \frac{\mu_0 NIa^2}{2}\left\{\frac{1}{\left[a^2+\left(\frac{a}{2}+x\right)^2\right]^{3/2}}+\frac{1}{\left[a^2+\left(\frac{a}{2}-x\right)^2\right]^{3/2}}\right\}$$

$$\text{At } x=0,\ B_a(0)\big|_{s=a} = \frac{8}{5\sqrt{5}}\frac{\mu_0 NI}{a}$$

Figure 4.6(*b*) shows a plot of the ratio $B_a(x)/B_a(0)$ between the coils for $s = a$. We find that the field on the axis is practically uniform. The coil pair is known as Helmholtz pair.

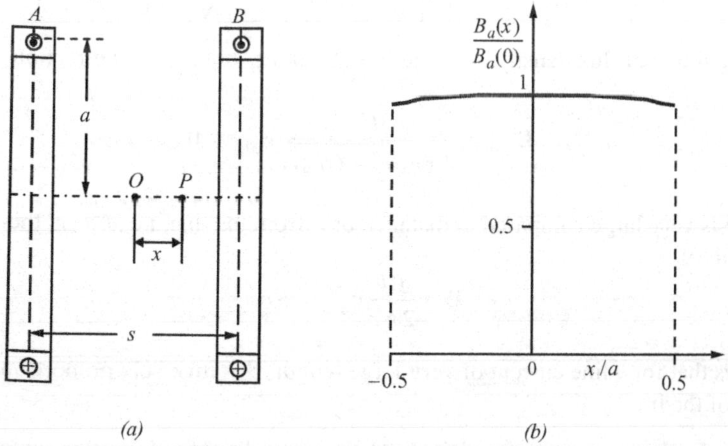

Fig. 4.6 (*a*) For calculation of flux density of two identical flat circular coils and (*b*) Variation of flux density along axis of the coils for $s = a$.

A Line Current

A line current may be considered as one of the two conductors of a transmission line supplying electrical power. We assume that the conductor is straight and its cross-sectional area is negligibly small. We will use the vector form of Biot–Savart law to calculate the flux density at a point P outside the conductor. The problem can be solved using either rectangular or cylindrical coordinates. We will use cylindrical coordinates.

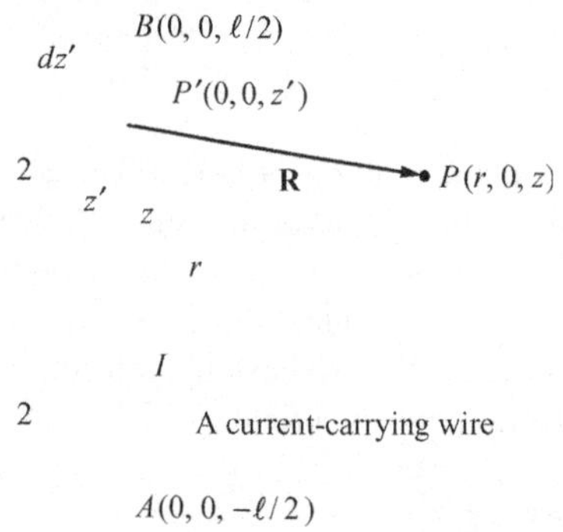

Fig. 4.7 For calculation of flux density due to a line current aligned along the z axis of cylindrical coordinates

AB in Fig. 4.7 represents a straight wire of length ℓ carrying current I. It is taken along the z axis of cylindrical coordinate system with the origin at the mid-point of the wire. The coordinates of the top and bottom ends are $(0, 0, \ell/2)$ and $(0, 0, -\ell/2)$ respectively. Since the flux density is independent of the variable ϕ, we can take P at $(r, 0, z)$.

Now let us consider a point P' located at $(0,0,z')$ on the line. A differential length vector at P' is

$$d\boldsymbol{\ell}' = dz'\mathbf{u}_z$$

The vector length from P' to P is

$$\mathbf{R} = r\mathbf{u}_r + (z - z')\mathbf{u}_z$$

The flux density at P due to the current element,

$$d\mathbf{B} = \frac{\mu I}{4\pi R^3}(d\boldsymbol{\ell}' \times \mathbf{R}) = \frac{\mu I}{4\pi R^3}[dz'\mathbf{u}_z \times \{r\mathbf{u}_r + (z - z')\mathbf{u}_z\}]$$

$$= \frac{\mu I r\, dz'}{4\pi R^3}\mathbf{u}_\phi$$

where

$$R = \sqrt{r^2 + (z - z')^2}$$

Since the unit vector $\mathbf{u}_\varphi$ is not a function of z', the flux density at P is

$$\mathbf{B} = \mathbf{u}_\phi \int_{-\ell/2}^{\ell/2} \frac{\mu I r}{4\pi R^3} dz' = \frac{\mu I}{4\pi r}\left[\frac{z + \ell/2}{\sqrt{r^2 + (z + \ell/2)^2}} - \frac{z - \ell/2}{\sqrt{r^2 + (z - \ell/2)^2}}\right]\mathbf{u}_\phi$$

We see that the magnitude of flux density is a function of r and z, and it is in ϕ direction. The flux density on $z = 0$ plane,

$$\mathbf{B}\Big|_{z=0} = \frac{\mu I \ell}{4\pi r\sqrt{r^2 + (\ell/2)^2}}\mathbf{u}_\phi,\ r > 0 \qquad (9)$$

If length of the line is very large compared to distance of P from the line, i.e., $\ell \gg r$, then $\sqrt{r^2 + (\ell/2)^2} \approx \ell/2$. Hence we have from (9),

$$\mathbf{B} = \frac{\mu I}{2\pi r}\mathbf{u}_\phi,\ \ell \gg r \qquad (10)$$

Formula (10) shows that for a line current of very large length, B is inversely proportional to r and it is not a function of length of the line.

An easier method of calculating the magnetic field intensity due to a long line current is by the use of Ampere's magnetic circuital law described in Section 4.6.

A Rectangular Current Loop

Our final example is a rectangular loop with side lengths $2a$ and $2b$, and carrying current I, as illustrated in Fig. 4.8. We will calculate the flux density on the axis of the loop using the vector form of Biot–Savart law using rectangular coordinates. The loop is taken on the x-y plane with its sides parallel to x and y axes. The centre of the loop is at the origin. Point P at which we wish to find the field is taken at a height h from the plane of the loop. Accordingly, the coordinates of P are (0, 0, h).

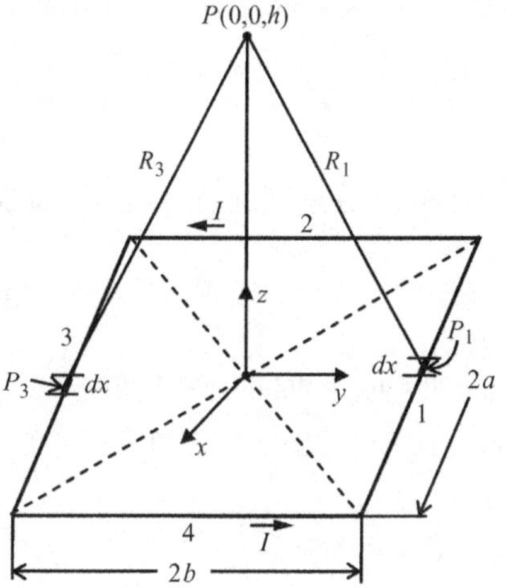

Fig. 4.8 A rectangular current loop on the x-y plane

Now let us take a differential length on side 1 at P_1. The coordinates of P_1 are (x, b, 0). A differential current element vector at P_1 is

$$I d\boldsymbol{\ell} = -I\, dx\, \mathbf{u}_x,$$

The minus sign is used because I and dx are in opposite directions.

Referring to the figure, displacement of P from P_1 is

$$\mathbf{R}_1 = -x\mathbf{u}_x - b\mathbf{u}_y + h\mathbf{u}_z$$

Flux density at P due to the current element,

$$d\mathbf{B}_1 = -\frac{\mu I}{4\pi R_1^3}(d\boldsymbol{\ell} \times \mathbf{R}_1) = -\frac{\mu I}{4\pi R_1^3}[dx\mathbf{u}_x \times (-x\mathbf{u}_x - b\mathbf{u}_y + h\mathbf{u}_z)]$$

$$= \frac{\mu I\, dx}{4\pi R_1^3}(h\mathbf{u}_y + b\mathbf{u}_z)$$

where $\qquad R_1^3 = (x^2 + b^2 + h^2)^{3/2}$

It can similarly be shown that the flux density at P due to the current element at P_3 (x, $-b$, 0) on side 3 is given by

$$d\mathbf{B}_3 = \frac{\mu I\, dx}{4\pi R_3^3}(-h\mathbf{u}_y + b\mathbf{u}_z)$$

where $\qquad R_3 = R_1$

The sum of $d\mathbf{B}_1$ and $d\mathbf{B}_3$, which we will denote by $d\mathbf{B}_{(1+3)}$, is

$$d\mathbf{B}_{(1+3)} = \frac{\mu b I\, dx}{2\pi R_1^3}\mathbf{u}_z$$

Thus, the total field due to currents in sides 1 and 3,

$$\mathbf{B}_{(1+3)} = \mathbf{u}_z \int_{-a}^{a} \frac{\mu b I}{2\pi R_1^3} dx = \frac{\mu a b I}{\pi(b^2 + h^2)\sqrt{a^2 + b^2 + h^2}}\mathbf{u}_z \tag{11}$$

Similarly, it can be shown that flux density at P due to currents in sides 2 and 4,

$$\mathbf{B}_{(2+4)} = \frac{\mu a b I}{\pi(a^2 + h^2)\sqrt{a^2 + b^2 + h^2}}\mathbf{u}_z \tag{12}$$

The flux density **B** on the axis and at a height h from the plane of the loop is equal to the sum of (11) and (12). Thus,

$$\mathbf{B} = \frac{\mu a b I}{\pi\sqrt{a^2 + b^2 + h^2}}\left(\frac{1}{a^2 + h^2} + \frac{1}{b^2 + h^2}\right)\mathbf{u}_z \tag{13a}$$

If the loop has N turns, the flux density is given by

$$\mathbf{B} = \frac{\mu NabI}{\pi\sqrt{a^2+b^2+h^2}}\left(\frac{1}{a^2+h^2}+\frac{1}{b^2+h^2}\right)\mathbf{u}_z \qquad (13b)$$

By substituting $h = 0$ in (13) we will obtain the flux density at the centre of the loop. It is given by

$$\mathbf{B}(h=0) = \frac{\mu NI\sqrt{a^2+b^2}}{\pi ab}\mathbf{u}_z \qquad (14)$$

The fields are along the axial direction.

It may be noted that the side lengths of the rectangular loop are $2a$ and $2b$.

EXAMPLE 4.5

A circular loop consists of 25 turns of very fine wire. The average radius of the loop is 20 cm and it carries a current of 1.6 A. Find the magnetic field intensity at the centre of the loop.

Magnetic field intensity at the centre which is along the axial direction,

$$H_a = \frac{NI}{2a}$$

Substituting N =25; I = 1.6 A; a = 0.2 m,

$$H_a = \frac{(\quad)(1.6)}{2(\quad .2)} = 100 \text{ A/m}$$

EXAMPLE 4.6

The mean radius of a circular coil of 50 turns of fine wire is 8.0 cm. It carries a current of 3.0 A. The coil is located on the y-z plane in air. Find the magnetic field intensity vector at P(20 cm, 0, 0).

Number of turns, $N = 50$

Radius of the coil, $a = 0.08$ m

Current, $I = 3.0$ A

Distance of P from the centre of the coil, $h = 0.2$ m

$$\text{Flux density, } \mathbf{B} = \frac{\mu_0(NI)a^2}{2(a^2+h^2)^{3/2}}\mathbf{u}_x = \frac{(4\pi\times10^{-7})(50\times3)(0.08)^2}{2(0.08^2+0.2^2)^{3/2}}\mathbf{u}_x = 603.49\times10^{-7}\,\mathbf{u}_x \text{ T}$$

EXAMPLE 4.7

An air-core solenoid of circular section has 260 turns. Its radius and length are 8.0 cm and 30.0 cm respectively. Calculate the magnetic flux density at the mid-point on the axis of the solenoid when its current is 0.75 A.

Referring to formula (7), flux density at the required point is

$$B = \frac{\mu_0 NI}{2\sqrt{a^2 + (\ell/2)^2}}$$

By substituting $N = 260$; $I = 0.75$ A; $a = 0.08$ m; $\ell = 0.30$ m,

$$B = \frac{(4\pi \times 10^{-7})(260)(0.75)}{2\sqrt{(0.08)^2 + (0.15)^2}} = 7.207 \times 10^{-4} \text{ T}$$

The flux density is along the axis of the solenoid.

EXAMPLE 4.8

A straight conductor with length 1.2 m and carrying 1.5 A current is located in air. Calculate the magnetic flux density vector at a perpendicular distance of 0.8 m from the mid-point of the line. Assume that the wire is along the z axis of cylindrical coordinates.

Current in the conductor, $I = 1.5$ A

Length of the conductor, $\ell = 1.2$ m

Perpendicular distance, $r = 0.8$ m

We have from (9),

$$\mathbf{B} \quad \frac{\mu_0 I \ell}{4\pi r\sqrt{r^2 + (\ell/2)^2}}\mathbf{u}_\phi = \frac{(4\pi \times 10^{-7})(1.5)(1.2)}{4\pi(0.8)\sqrt{(0.8)^2 + (0.6)^2}}\mathbf{u}_\phi = 2.25 \times 10^{-7}\mathbf{u}_\phi \text{ T}$$

EXAMPLE 4.9

A rectangular coil of side lengths of 16 cm and 12 cm has 20 turns and carries a current of 5.0 A. Find the magnetic field intensity vector at the centre of the loop if the coil is positioned on the x-y plane.

The magnetic field intensity of the coil is in z direction. For a coil of N turns we have from (14),

$$\mathbf{H}(\ = 0) = \frac{NI\sqrt{a^2 + b^2}}{\pi ab}\mathbf{u}_z$$

Given: $a = 0.08$ m; $b = 0.06$ m; $N = 20$; $I = 5.0$ A

$$\mathbf{H}(\ = 0) = \frac{(20 \times 5)\sqrt{0.08^2 + 0.06^2}}{\pi(0.08)0.06)}\mathbf{u}_z = 663.1\mathbf{u}_z \text{ A/m}$$

EXAMPLE 4.10

A rectangular coil of average side lengths 32 cm and 24 cm has 15 turns. It is located in air and carries a current of 2.0 A. Find the magnetic flux density on the axis at a height 20 cm from the plane of the loop.

Substituting $a = 0.16$ m; $b = 0.12$ m; $h = 0.2$ m; $N = 15$; $I = 2.0$ A in formula (13*b*)

$$B_a = \frac{(4 \ \times 10^{-7})(15)(0.16)(0.12)(2.0)}{\pi\sqrt{0.16^2 + 0.12^2 + 0.2^2}}\left(\frac{1}{0.16^2 + 0.2^2} + \frac{1}{0.12^2 + 0.2^2}\right)$$

$$= 273.95 \times 10^{-7} \text{ T}$$

4.5 Electromagnetic Force and Torque

When a current-carrying conductor (or coil) is placed in an external magnetic field, the conductor experiences a force which is called electromagnetic force. The magnitude of the force depends on the size and shape of the conductor, the value of current in the conductor, magnitude of magnetic flux density, and orientation of the current with respect to the direction of the flux density. The torque due to the force is a function of magnitude of the force, length of the torque arm and angle between the force and the torque arm.

The principle of many electromagnetic devices such as direct and alternating current motors, lifting magnets, measuring instruments, relays, electric trains using the principle of magnetic levitation, etc., are based on electromagnetic force and torque.

Force on a Current-Carrying Conductor

Consider that a straight conductor of uniform sectional area and length ℓ carries a current I. The current-carrying conductor when treated as a length vector, denoted by the bold notation $\boldsymbol{\ell}$, has its magnitude equal to ℓ, and its direction is same as the direction of current. When the current-carrying conductor is placed in a uniform magnetic field of flux density B, as illustrated in Fig. 4.9(a), the magnitude of electromagnetic force F on the conductor is found to be proportional to the products of I, ℓ, B, and $\sin\alpha$ where α is the angle between the length vector $\boldsymbol{\ell}$ and the flux density vector **B**. Thus,

$$F \propto I\ell B \sin\alpha$$

In SI unit the constant of proportionality is unity. Hence,

$$F = I\ell B \sin\alpha \tag{1}$$

The direction of F is normal to the plane containing $\boldsymbol{\ell}$ and **B**. Its sense is along the advance of a right-handed screw when it is rotated from $\boldsymbol{\ell}$ towards **B** through the smaller angle between them. Accordingly, if $\boldsymbol{\ell}$ and **B** are in the plane of the paper with directions as shown in Fig. 4.9(a), the force is normally *into* the paper. The magnitude of force given by (1) and its direction as described show that the force can be written in terms of cross product of $\boldsymbol{\ell}$ and **B**. Thus, the vector form of (1) is

$$\mathrm{F} = I(\boldsymbol{\ell} \times \mathbf{B}) \tag{2}$$

To generalise (2), let us consider a current-carrying wire of any shape in a non-uniform magnetic field. The wire may be thought of as part of a closed loop. Let us take a differential vector length $d\boldsymbol{\ell}$ at any point P on the wire. If **B** is flux density at P, differential force on the current element,

$$d\mathbf{F} = I(d\boldsymbol{\ell} \times \mathbf{B}) = -I(\mathbf{B} \times d\boldsymbol{\ell}) \tag{3}$$

Total force on the conductor is obtained by integrating (3) from P_1 to P_2.

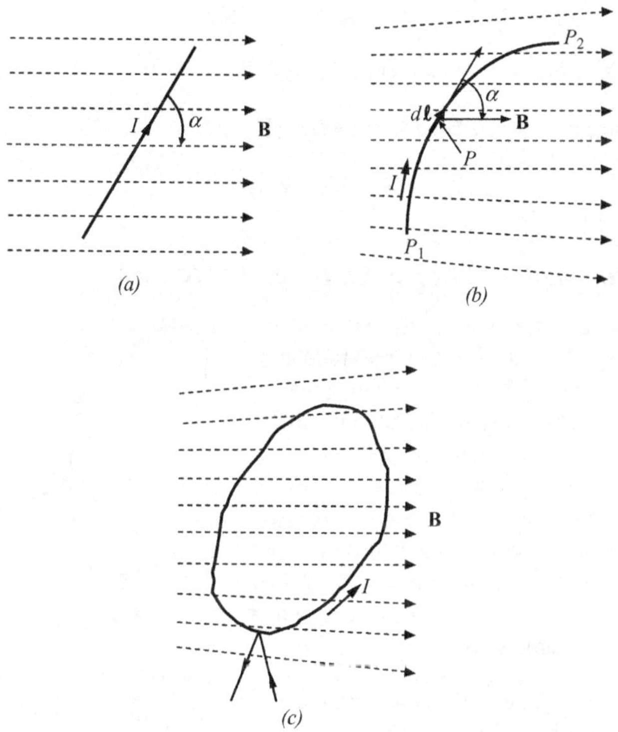

Fig. 4.9 (*a*) A straight current-carrying wire in a uniform magnetic field, (*b*) Current-carrying wire of arbitrary shape in a non-uniform magnetic field, and (c) Current-carrying loop in a magnetic field.

$$\mathbf{F} = I\int_{P_1}^{P_2} (d\ell \times \mathbf{B})$$

Force on a current loop, as shown in Fig. 4.9(*c*), is found by integrating (3) around the loop.

$$\mathbf{F} = I\oint_c (d\ell \times \mathbf{B}) \tag{4a}$$

It may be noted that both **B** and $d\ell$ are, in general, functions of the space variables.

Force on a Current Sheet

When current flows in the form of a very thin sheet, its distribution is defined by current per unit width where the width is perpendicular to the direction of flow of current. The current per unit width is called surface current density. It is expressed in amperes per metre and we will denote it by the letter K. The current density may be regarded as a vector quantity having the same direction as the direction of current flow.

Let us consider a thin rectangular current sheet of length ℓ and width w located in a stationary magnetic field. The magnetic field may or may not be uniform. We will first find the force on the current in a surface element and then integrate the result over the surface of the sheet to get the total force. The differential width and length of the surface element are taken as dw and $d\ell$ respectively.

Formula (3) gives the magnetic force on a linear current element. In the present case, a current element is given by the product of current through the differential width and the differential length. The current through the differential width is equal to the product of the surface current density and the differential width. Thus,

$$I\,d\ell = (\mathbf{K}\,dw)\,d\ell = \mathbf{K}\,dS$$

where $dS = d\ell\,dw$ is area of the surface element.

Force on the current element, $\quad d\mathbf{F} = (\mathbf{K}\,dS)\times\mathbf{B} = (\mathbf{K}\times\mathbf{B})\,dS$

Total force on the strip, $$\mathbf{F} = \int_S (\mathbf{K}\times\mathbf{B})\,dS \qquad (4b)$$

Force between Two Current Loops: Ampere's Force Law

Equation (3) gives magnetic force on a current element linked by an external magnetic field. The external field may be due to a permanent magnet or it may be due to an external current. When two current loops are oriented in such a way that the field produced by one loop links with the other then each loop will be acted upon by an electromagnetic force. We will derive expressions for the electromagnetic force between two filamentary loops separated by a finite distance in a homogeneous medium of permeability μ. The loops denoted by C_1 and C_2 in Fig. 4.10 carry currents I_1 I_2 respectively in the same sense.

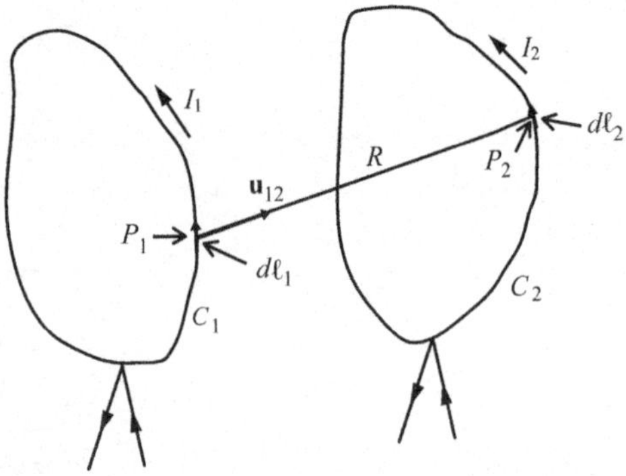

Fig. 4.10 For calculation of force between two current-carrying loops

In order to calculate force experienced by one loop due to the magnetic field of the other loop, we at first find the force on a current element of a loop and then integrate it around the loop. Let us take a current element $I_1 d\ell_1$ at point P_1 on loop C_1 and a current-element $I_2 d\ell_2$ at point P_2 on loop C_2. The distance between P_1 and P_2 is taken as R. Let the flux density at P_2 due to $I_1 d\ell_1$ be $d\mathbf{B}_1$. Then according to Biot–Savart law,

$$d\mathbf{B}_1 = \frac{\mu I_1}{4\pi R^2}(d\ell_1 \times \mathbf{u}_{12})$$

where $\mathbf{u}_{12}$ is a unit vector directed from P_1 to P_2. Hence, flux density at P_2 due to the total current in C_1 is given by

$$\mathbf{B}_1 = \oint_{C_1} \frac{\mu I_1}{4\pi R^2}(d\ell_1 \times \mathbf{u}_{12})$$

The flux density will cause a force $d\mathbf{F}_2$ on the current element $I_2 d\ell_2$. The force as per formula (3) is

$$d\mathbf{F}_2 = I_2(d\ell_2 \times \mathbf{B}_1)$$

$$= \frac{\mu I_1 I_2}{4\pi}\left[d\ell_2 \times \oint_{C_1} \frac{1}{4\pi R^2}(d\ell_1 \times \mathbf{u}_{12})\right] \qquad (5)$$

Total force on loop C_2 is obtained by integrating (5) around the loop. Thus force on the loop,

$$\mathbf{F}_2 = \frac{\mu I_1 I_2}{4\pi}\oint_{C_2} d\ell_2 \times \left[\oint_{C_1} \frac{1}{4\pi R^2}(d\ell_1 \times \mathbf{u}_{12})\right] = \frac{\mu I_1 I_2}{4\pi}\oint_{C_2}\oint_{C_1} \frac{1}{4\pi R^2}[d\ell_2 \times (d\ell_1 \times \mathbf{u}_{12})] \qquad (6)$$

The force $\mathbf{F}_1$ on loop C_1 is obtained by interchanging the subscripts 1 and 2 in the expression of (6). Thus,

$$\mathbf{F}_1 = \frac{\mu I_1 I_2}{4\pi}\oint_{C_1}\oint_{C_2} \frac{1}{4\pi R^2}[d\ell_1 \times (d\ell_2 \times \mathbf{u}_{21})] \qquad (7)$$

Equation (6) or (7) is known as *Ampere's force law*. The electromagnetic force is proportional to the product of current magnitudes and permeability of the medium. It also depends on size and shape of the loops and distance between them.

We will know some other properties of the force between current-carrying conductors from the following example.

Force between Two Parallel Current-Carrying Wires

We will use formula (6) to calculate the electromagnetic force between two parallel current-carrying wires, as shown in Fig. 4.11(*a*). Wires 1 and 2 each of length ℓ are separated by a distance s, and carry currents I_1 and I_2 in opposite directions. The wires are in the x-y plane of rectangular coordinates and they are symmetrically placed with respect to the origin.

Let us take differential lengths $d\boldsymbol{\ell}_1$ and $d\boldsymbol{\ell}_2$ at $P_1(-s/2, y_1)$ on Wire 1 and at $P_2\,(s/2, y_2)$ on Wire 2 respectively. The differential lengths are

$$d\boldsymbol{\ell}_1 = dy_1\,\mathbf{u}_y$$

$$d\boldsymbol{\ell}_2 = dy_2\,\mathbf{u}_y$$

The displacement of P_2 from P_1 is

$$\mathbf{R} = s\mathbf{u}_x + (y_2 - y_1)\mathbf{u}_y$$

Distance between P_1 and P_2,

$$R = \sqrt{s^2 + (y_2 - y_1)^2}$$

The unit vector directed from P_1 to P_2 is given by

$$\mathbf{u}_{12} = \frac{\mathbf{R}}{R} = \frac{s\mathbf{u}_x + (y_2 - y_1)\mathbf{u}_y}{\sqrt{s^2 + (y_2 - y_1)^2}}$$

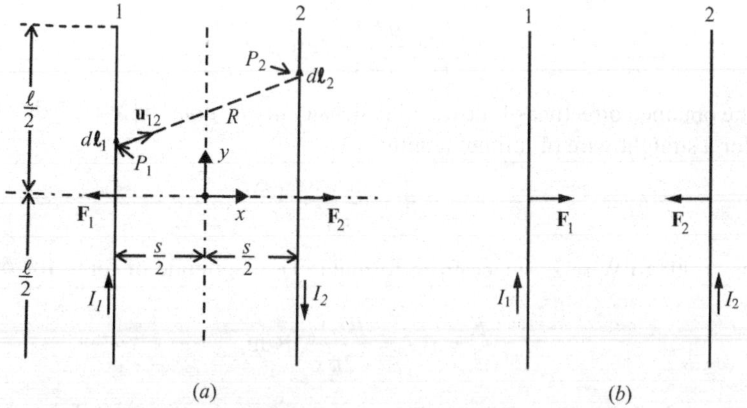

Fig. 4.11 Direction of forces on two straight, parallel current-carrying wires separated by a finite distance (*a*) Currents in opposite directions and (*b*) Currents in same direction.

By substituting the expressions of $d\ell_1$, $d\ell_2$, $\mathbf{u}_{12}$ and R in (6) we will get the force $\mathbf{F}_2$ on Wire 2.

$$\mathbf{F}_2 = \frac{\mu I_1(-I_2)}{4\pi}\int_{-\ell/2}^{+\ell/2}\int_{-\ell/2}^{+\ell/2}\frac{(dy_2\,\mathbf{u}_y)\times\{(dy_1\,\mathbf{u}_y)\times[s\mathbf{u}_x+(y_2-y_1)\mathbf{u}_y]\}}{[s^2+(y_2-y_1)^2]^{3/2}}$$

$$= \frac{\mu I_1 I_2}{4\pi}\int_{-\ell/2}^{+\ell/2}\int_{-\ell/2}^{+\ell/2}\frac{s\,dy_2\,dy_1\,\mathbf{u}_x}{[s^2+(y_2-y_1)^2]^{3/2}}$$

$$= \frac{\mu I_1 I_2 \mathbf{u}_x}{4\pi s}\int_{-\ell/2}^{+\ell/2} dy_2\left[-\frac{y_2-y_1}{\sqrt{s^2+(y_2-y_1)^2}}\,\mathbf{u}_z\right]_{-\ell/2}^{+\ell/2}$$

$$= \frac{\mu I_1 I_2\,\mathbf{u}_x}{4\pi s}\int_{-\ell/2}^{+\ell/2}\left[\frac{y_2+\ell/2}{\sqrt{s^2+(y_2+\ell/2)^2}}-\frac{y_2-\ell/2}{\sqrt{s^2+(y_2-\ell/2)^2}}\right]dy_2$$

$$= \frac{\mu I_1 I_2}{4\pi s}\mathbf{u}_x\left[\sqrt{s^2+(y_2+\ell/2)^2}-\sqrt{s^2+(y_2-\ell/2)^2}\right]_{-\ell/2}^{+\ell/2}$$

$$= \frac{\mu I_1 I_2}{2\pi s}\left[\sqrt{s^2+\ell^2}-s\right]\mathbf{u}_x \tag{8}$$

Similarly, it can be shown that force on Wire 1,

$$\mathbf{F}_1 = -\frac{\mu I_1 I_2}{2\pi s}\left(\sqrt{s^2+\ell^2}-s\right)\mathbf{u}_x \tag{9}$$

Note that the unit vector from P_2 to P_1 is $-\mathbf{u}_{12}$.

We find from formulae (8) and (9) that the forces on the conductors have equal magnitude, but their directions are opposite to each other. The force on Conductor 1 is in negative x direction and it is in positive x direction on Conductor 2, as indicated in Fig. 4.11(a). As the forces are directed away from each other,

the forces between the conductors carrying current in opposite directions are repulsive. If the currents are in same sense, then the forces are directed towards each other, and hence attractive,

as shown in Fig. 4.11(b).

For long parallel currents separated by a small distance ($\ell \gg s$), the lines may be treated as of infinite length. Using this condition in (8), magnitude of force per unit length of each wire,

$$\frac{F}{\ell} = \frac{\mu I_1 I_2}{2\pi s}\ \text{N/m} \tag{10}$$

Formula (10) can be obtained directly as follows. Flux density *at any point* on Wire 2 due to current I_1, as per formula (4.4-10) for a straight wire of infinite length, is

$$B_1 = \frac{\mu I_1}{2\pi s}$$

The flux density is at 90° to Wire 2. Hence from formula (1), magnitude of force for one metre length of Wire 2 is given by

$$\frac{F}{\ell} = B_1 I_2 = \frac{\mu I_1 I_2}{2\pi s}\ \text{N/m}$$

This is also the force per unit length on Wire 1. The magnetic force between two long, parallel, current-carrying conductors is directly proportional to the product of their currents. It is inversely proportional to the perpendicular distance between the conductors. As mentioned earlier, the forces are attractive for the same current direction and repulsive for opposite current directions.

Force and Torque on a Current Loop in a Uniform Magnetic Field

Figure 4.12(*a*) shows a circular current loop of radius *a* located in an external magnetic field. The loop is in the *x-y* plane with its centre at the origin. The current *I* in the loop is in anticlockwise sense. We will obtain the electromagnetic force and torque exerted on the loop assuming that the external magnetic field is uniform.

Let us take a current element $Id\ell$ at point $P(x, y)$ on the loop. The force on the element,

$$d\mathbf{F} = -I(\mathbf{B}\times d\ell)$$

Total force on the loop,

$$\mathbf{F} = -I\oint(\mathbf{B}\times d\ell)$$

As **B** is constant, it can be brought out of the integral sign.

$$\mathbf{F} = -I(\mathbf{B}\times\oint d\ell) \quad (11)$$

The right-hand side of (11) is zero because the net displacement around a closed path is zero. Thus,

the net force on a current loop in a uniform magnetic field is zero. However, force is exerted on every element of the loop.

Next we will derive an expression for the torque on the current-loop assuming, as before, that the external magnetic field is uniform. The magnitude of torque or turning moment *T* of a force *F* acting at a distance *R* (the torque arm) from a given axis is proportional to *R*, *F* and $\sin\alpha$ where α is the angle between **R** and **F**, as indicated in Fig. 4.12(*b*). In SI unit the constant of proportionality is *one*. The unit of torque is newton-metre (N-m). Thus the torque,

$$T = RF\sin\alpha \text{ N-m}$$

Torque is a vector quantity. Its direction is along the axial advance of a right-handed screw when rotated from **R** towards **F** through the smaller angle between them. So the torque vector **T** can be written as the cross product of **R** and **F**.

$$\mathbf{T} = \mathbf{R}\times\mathbf{F}$$

From the definition of torque it is clear that we have to at first select a reference point or axis of the torque arm. For the current loop under consideration we will take the centre of the loop as the reference point. Thus, at point $P(x, y)$ on the loop the torque arm is the vector length directed from the centre of the loop to *P*. That is,

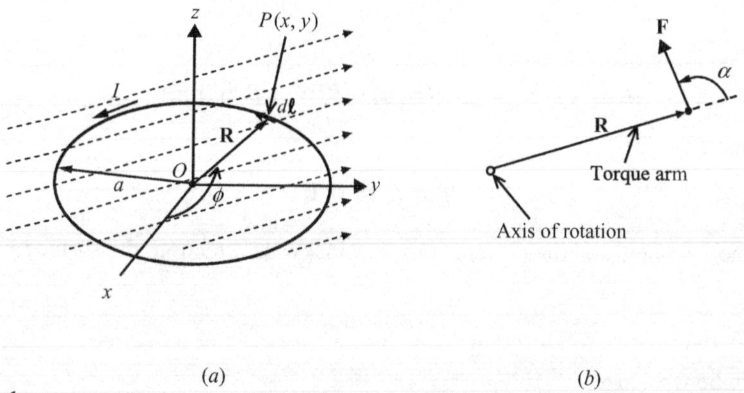

Fig. 4.12 (*a*) A circular current loop in a uniform magnetic field and (*b*) For defining turning moment of a force.

$$\mathbf{R} = x\mathbf{u}_x + y\mathbf{u}_y$$

A differential vector length at P is given by

$$d\ell = dx\,\mathbf{u}_x + dy\,\mathbf{u}_y$$

In the region of the loop, let the flux density,

$$\mathbf{B} = B_x\,\mathbf{u}_x + B_y\,\mathbf{u}_y + B_z\,\mathbf{u}_z$$

Magnetic force on the current element,

$$d\mathbf{F} = (I d\ell) \times \mathbf{B} \tag{12}$$

Substituting the expressions of $d\ell$ and $\mathbf{B}$ in (12) and then calculating the cross product, we get

$$d\mathbf{F} = I[B_z(dy\mathbf{u}_x - dx\mathbf{u}_y) + (B_y dx - B_x dy)\mathbf{u}_z]$$

Torque experienced by the current element

$$d\mathbf{T} = \mathbf{R} \times d\mathbf{F} \tag{13}$$

By substituting the expressions of $\mathbf{R}$ and $d\mathbf{F}$ in (13) and then evaluating the cross product, we will get

$$d\mathbf{T} = [(B_y\,ydx - B_x\,ydy)\mathbf{u}_x - (B_y\,xdx - B_x\,xdy)\mathbf{u}_y - B_z(xdx + ydy)\mathbf{u}_z]$$

The total torque is obtained by integrating the above expression over the closed loop. As integration of xdx and ydy over a closed path is always zero, hence, we have

$$\mathbf{T} = I[(B_x\,\mathbf{u}_y)\int_0^{2\pi} x\,dy + (B_y\,\mathbf{u}_x)\int_0^{2\pi} y\,dx] \tag{14}$$

Flux densities B_x and B_y are written outside the integral sign as these are constants for a uniform field. Taking ϕ as the angle between the x axis and OP, as shown in Fig. 4.12(a),

$$x = a\cos\phi \text{ and } y = a\sin\phi$$

Using these relations and their derivatives in (14), we get

$$\mathbf{T} = I\left[(B_x\,\mathbf{u}_y)\int_0^{2\pi} a^2\cos^2\phi\,d\phi - (B_y\,\mathbf{u}_x)\int_0^{2\pi} a^2\sin^2\phi\,d\phi\right]$$

$$= IS(B_x\,\mathbf{u}_y - B_y\,\mathbf{u}_x) \tag{15}$$

where $S = \pi a^2$ is area of the loop.

It can be shown that

$$B_x\,\mathbf{u}_y - B_y\,\mathbf{u}_x = \mathbf{u}_z \times (B_x\,\mathbf{u}_x + B_y\,\mathbf{u}_y + B_z\,\mathbf{u}_z) = \mathbf{u}_z \times \mathbf{B}$$

Substituting this in (15),

$$\mathbf{T} = IS(\mathbf{u}_z \times \mathbf{B}) \tag{16}$$

As the loop is in the x-y plane and the current is in anticlockwise direction, the surface area of the loop as a vector quantity is

$$\mathbf{S} \quad \mathbf{u}_z$$

Therefore, (16) can be written as

$$\mathbf{T} \quad \mathbf{S} \times \mathbf{B} \tag{17}$$

If the current loop is free and if the electromagnetic torque is strong enough to overcome inertia of the rotating parts, the coil will rotate. It will rotate in the direction so as to align the field produced by it along the external field. At this position the torque on the coil is zero.

For example, if B_y in (15) is taken as zero then the torque-direction (axis of rotation) would be along the positive y axis. Hence, the loop shown in Fig. 4.12(a) will rotate until the normal to the plane of the loop is oriented along x direction. This is another way of determining the direction of the torque.

It may be mentioned that though a circular loop has been considered for the sake of mathematical simplicity, the equations obtained and observations made regarding electromagnetic force and torque are valid for current loops of any shape in a uniform magnetic field. Moreover,

the torque does not depend on the position of the loop relative to the origin of the coordinate system. If the loop (coil) has more than one turn, the torque given by (17) has to be multiplied by the number of turns.

EXAMPLE 4.11

A current element $Id\boldsymbol{\ell} = 8\mathbf{u}_x + 6\mathbf{u}_y$ A-mm is in a magnetic field of flux density $\mathbf{B} = 5\mathbf{u}_x + 7\mathbf{u}_z$ mT. Find the magnetic force on the current element.

Force on the current element, $d\mathbf{F} = (Id\boldsymbol{\ell}) \times \mathbf{B}$

Substituting the given values, $d\mathbf{F} = [(8\mathbf{u}_x + 6\mathbf{u}_y)\times 10^{-3}]\times[(5\mathbf{u}_x + 7\mathbf{u}_z)\times 10^{-3}]$

$= 42\mathbf{u}_x - 56\mathbf{u}_y - 30\mathbf{u}_z\ \mu\text{N}$

EXAMPLE 4.12

A straight conductor of length $\ell = 2.0$ m carries a current of 15.0 A in positive z direction. It is located in a uniform magnetic field of flux density $\mathbf{B} = 0.3\mathbf{u}_x + 0.4\mathbf{u}_y$ T. Calculate the electromagnetic force on the conductor.

Current, $I = 15.0$ A

Vector length of the conductor, $\boldsymbol{\ell} = 2.0\mathbf{u}_z$ A.

Electromagnetic force, $\mathbf{F} = I(\boldsymbol{\ell} \times \mathbf{B}) = (15.0)[(2.0\mathbf{u}_z)\times(0.3\mathbf{u}_x + 0.4\mathbf{u}_y)]$

$= -12.0\mathbf{u}_x + 9.0\mathbf{u}_y$ N

EXAMPLE 4.13

A thin current sheet has its length and width along z and y directions respectively. It carries a y-directed current of 125 A over a width of one metre in z direction. The flux density on the plane of the strip is $0.4\mathbf{u}_y$ T. Calculate the magnetic force on the strip bounded by $-4.0 \text{ cm} \leq y \leq 4.0$ cm and $0 \leq z \leq 1.5$ m.

Surface current density, $\mathbf{K}$ $\mathbf{u}_z$ A/m

Magnetic flux density, $\mathbf{B}$ $\mathbf{u}_y$ T

Force on the strip, $\mathbf{F}$ $\int_S (\mathbf{K}\times\mathbf{B})dS = \int_{z=0}^{1.5}\int_{y=-0.04}^{0.04} \times(.4\mathbf{u}_y) = .0\mathbf{u}_x$ N

EXAMPLE 4.14

A single-turn, rectangular current loop of length 0.25 m and width 0.16 m carries a current of 0.75 A in counter-clockwise direction when viewed from the normal to the plane of the loop. Calculate the electromagnetic torque on the loop when it is on the *y-z* plane in a magnetic field of flux density, $\mathbf{B} = 0.5\mathbf{u}_x + 0.7\mathbf{u}_y$ T.

Area of the loop,

$S = (\ .25)(0.16) = 0.04\ \text{m}^2$

As the loop is on the *y-z* plane, the direction of the area (normal to the loop) is in *x* direction. Thus the vector area,

$\mathbf{S} \quad \mathbf{u}_x\ \text{m}^2$

Electromagnetic torque,

$\mathbf{T} =$ [illegible] $\times \mathbf{B}) = (0.75)[0.04\mathbf{u}_x \times (0.5\mathbf{u}_x + 0.7\mathbf{u}_y)] = 0.021\mathbf{u}_z$ N-m

EXAMPLE 4.15

A straight conductor of length $\ell = 2.0$ m lies along *z* direction. It carries a current $I = 12.0$ A in the positive *z* direction. The flux density in the region is defined by the equation, $\mathbf{B} = 0.16x\mathbf{u}_y$ T. Calculate the work done by an external source to move the conductor by a distance of 2.5 m from the origin in *x* direction. Assume that the conductor remains parallel to the *z* axis during its displacement.

Electromagnetic force on the conductor,

$\mathbf{F} = I(\boldsymbol{\ell} \times \mathbf{B}) = (12.0)(2.0\mathbf{u}_z \times 0.16x\mathbf{u}_y) = -3.84x\mathbf{u}_x$ N

External force, $\mathbf{F}_e = -\mathbf{F} = 3.84x\mathbf{u}_x$ N

Work done by the external source, $W_e = \int_0^{2.5} (3.84x\mathbf{u}_x)\cdot(dx\mathbf{u}_x) = 12.0$ J

EXAMPLE 4.16

A single-turn rectangular coil of side lengths $a = 25$ cm and $b = 15$ cm is on the *x-y* plane with its centre at the origin, as illustrated in Fig. 4.13(*a*). It carries a current of $I = 20$ A in a counter clockwise direction. The coil is situated in a uniform magnetic field of flux density given by $\mathbf{B} = 0.8\mathbf{u}_x + 0.7\mathbf{u}_y$ T. Find (a) the electromagnetic force and torque on each side of the coil, and (b) total torque on the coil, and torques about the *x* and *y* axes.

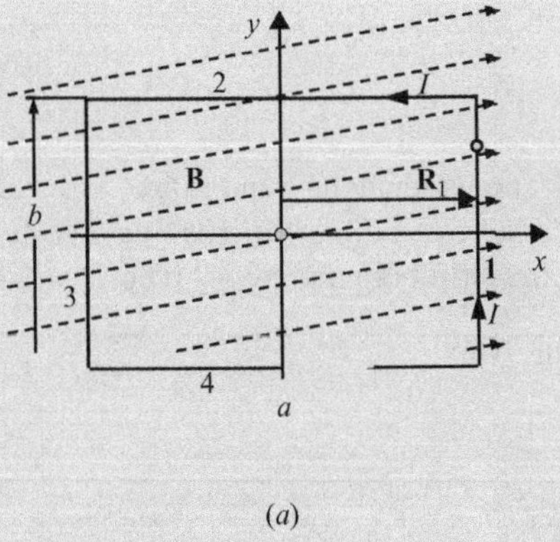

(*a*)

Fig. 4.13. (*a*) Illustration for Example 4.16.

(a) Force on side 1, $\mathbf{F}_1 = I(\boldsymbol{\ell}_1 \times \mathbf{B})$ $(20)(0.15\mathbf{u}_y)\times(0.8\mathbf{u}_x + 0.7\mathbf{u}_y) = -2.4\mathbf{u}_z$ N

Torque on Side 1, $\mathbf{T}_1 = \mathbf{R}_1 \times \mathbf{F}_1 = \left(\frac{0.25}{2}\mathbf{u}_x\right)\times(-2.4\mathbf{u}_z) = 0.3\mathbf{u}_y$ N-m

Force on Side 3, $\mathbf{F}_3 = I(\boldsymbol{\ell}_3 \times \mathbf{B})$ $(20)(-0.15\mathbf{u}_y)\times(0.8\mathbf{u}_x + 0.7\mathbf{u}_y) = 2.4\mathbf{u}_z$ N

Torque on Side 3, $\mathbf{T}_3 = \mathbf{R}_3 \times \mathbf{F}_3 = \left(-\frac{0.25}{2}\mathbf{u}_x\right)\times(2.4\mathbf{u}_z) = 0.3\mathbf{u}_y$ N-m

Force on Side 2, $\mathbf{F}_2 = I(\boldsymbol{\ell}_2 \times \mathbf{B})$ $(20)(-0.25\mathbf{u}_x)\times(0.8\mathbf{u}_x + 0.7\mathbf{u}_y) = -3.5\mathbf{u}_z$ N

Torque on Side 2, $\mathbf{T}_2 = \mathbf{R}_2 \times \mathbf{F}_2 = \left(\frac{0.15}{2}\mathbf{u}_x\right)\times(-3.5\mathbf{u}_z) = -0.263\mathbf{u}_x$ N-m

Force on Side 4, $\mathbf{F}_4 = I(\boldsymbol{\ell}_4 \times \mathbf{B})$ $(20)(0.25\mathbf{u}_x)\times(0.8\mathbf{u}_x + 0.7\mathbf{u}_y) = 3.5\mathbf{u}_z$ N

Torque on Side 4, $\mathbf{T}_4 = \mathbf{R}_4 \times \mathbf{F}_4 = \left(-\frac{0.15}{2}\mathbf{u}_x\right)\times(3.5\mathbf{u}_z) = -0.263\mathbf{u}_x$ N-m

(b) Total torque, $\mathbf{T} = \mathbf{T}_1 + \mathbf{T}_2 + \mathbf{T}_3 + \mathbf{T}_4 = 0.6\mathbf{u}_y - 0.526\mathbf{u}_x$ N-m

Torque about the x axis $= \mathbf{T}_2 + \mathbf{T}_4 = -0.526\mathbf{u}_x$ N-m

Torque about the y axis $= \mathbf{T}_1 + \mathbf{T}_3 - 0.6\mathbf{u}_y$ N m

Alternatively;

Torque on the loop, $\mathbf{T} = I(\mathbf{S}\times\mathbf{B})$

$\mathbf{S} = (0.25)(0.15)\mathbf{u}_z = 0.0375\mathbf{u}_z$ m^2

$\mathbf{T} = (20)(0.0375\mathbf{u}_z)\times(0.8\mathbf{u}_x + 0.7\mathbf{u}_y) = 0.6\mathbf{u}_y - 0.526\mathbf{u}_x$ N-m

The x and y components of the total torque give the torque about the x and y axes respectively.

EXAMPLE 4.17

A semicircular loop of radius 0.25 m, and carrying current 6.0 A in clockwise direction is situated in a uniform magnetic field. It is on the x-y plane with its centre at the origin and the diameter along the x axis, as shown in Fig. 4.13(b). Calculate the force on the circular part and on the diameter of the loop. Also, calculate torque on the loop. The flux density is given by $\mathbf{B} = 0.4\mathbf{u}_x + 0.3\mathbf{u}_y + 0.5\mathbf{u}_z$ T.

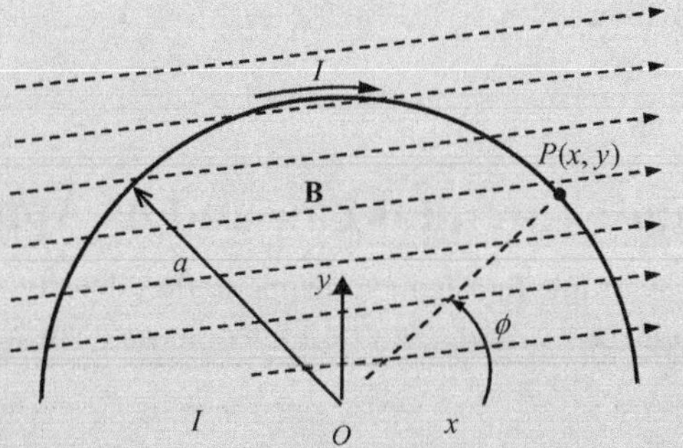

(b)

Fig. 4.13. (b) Illustration for Example 4.17

For the semicircular part of the loop, it is convenient to calculate the force using notations and then use the given data. At a point $P(x, y)$ on the circular part of the loop, a differential current element is given by

$$Id\ell \quad I(dx\,\mathbf{u}_x + dy\,\mathbf{u}_y)$$

The minus sign is used as I and $d\ell$ are in opposite sense.

$$x = a\cos\phi,\ dx = -a\sin\phi\, d\phi$$

$$y = a\sin\phi,\ dy = a\cos\phi\, d\phi$$

Hence, $\quad Id\ell = aI(\sin\phi\,\mathbf{u}_x - \cos\phi\,\mathbf{u}_y)d\phi$

Let $\quad \mathbf{B} = B_x\,\mathbf{u}_x + B_y\,\mathbf{u}_y + B_z\,\mathbf{u}_z$

Force on the current element,

$$d\mathbf{F}_1 = Id\ell \times \mathbf{B} = -\mathbf{B} \times Id\ell$$

As the flux density is independent of the space variables, force on the circular part of the loop,

$$\mathbf{F}_1 \quad \mathbf{B} \times \left[aI\int_0^{\pi} (\sin\phi\,\mathbf{u}_x - \cos\phi\,\mathbf{u}_y)d\phi\right]$$

$$= -aI(B_x\,\mathbf{u}_x + B_y\,\mathbf{u}_y + B_z\,\mathbf{u}_z) \times (2\mathbf{u}_x)$$

$$= 2aI(B_y\,\mathbf{u}_z - B_z\,\mathbf{u}_y)$$

Substituting the given data,

$$\mathbf{F}_1 \quad)(6.0)(0.3\mathbf{u}_z - 0.5\mathbf{u}_y) = 0.9\mathbf{u}_z - 1.5\mathbf{u}_y \text{ N}$$

Force on the straight part of the loop,

$$\mathbf{F}_2 = (I\ell) \times (\mathbf{B}) = (-6.0)(0.5\mathbf{u}_x) \times (0.4\mathbf{u}_x + 0.3\mathbf{u}_y + 0.5\mathbf{u}_z) = -0.9\mathbf{u}_z + 1.5\mathbf{u}_y \text{ N}$$

Total force on the loop, which is equal to the sum of $\mathbf{F}_1$ and $\mathbf{F}_2$, is zero.

Torque on the loop,

$$\mathbf{T} \quad \mathbf{S} \times \mathbf{B}) = -(6.0)\left(\frac{\pi(0.25)^2}{2}\mathbf{u}_z\right) \times (0.4\mathbf{u}_x + 0.3\mathbf{u}_y + 0.5\mathbf{u}_z)$$

$$= 0.177\mathbf{u}_x - 0.236\mathbf{u}_y \text{ N-m}$$

4.6 Ampere's Magnetic Circuital Law and its Applications

In Section 4.4 we used Biot–Savart law to calculate magnetic flux density of some simple current distributions. Another law known as *Ampere's circuital law* or *Ampere's work law* also gives magnetic field intensity due to current distributions.

According to the law, the line integral of magnetic field intensity H taken *once* around a closed path is equal to the current enclosed.

The chosen path may or may not be along a field line. If it is not along a field line, then as per the definition of line integral, the integration has to be evaluated by using the component of H tangential to the path. Thus, the mathematical form of Ampere's magnetic circuital law is

$$\oint \mathbf{H} \cdot d\ell = I \tag{1}$$

where I is current enclosed by the chosen path

The meaning of *current enclosed* is explained with the help of Fig. 4.14. We see that at (*a*) closed paths C_1 and C_2 enclose current I completely. Hence, the line integral of the field intensity along either of the paths will give the same result I. However, the line integral of field intensity around path C_3 is zero as it does not enclose any current. Further, as the closed path at (*b*) encloses several currents the line integral of **H** around this path is equal to the algebraic sum of all the currents.

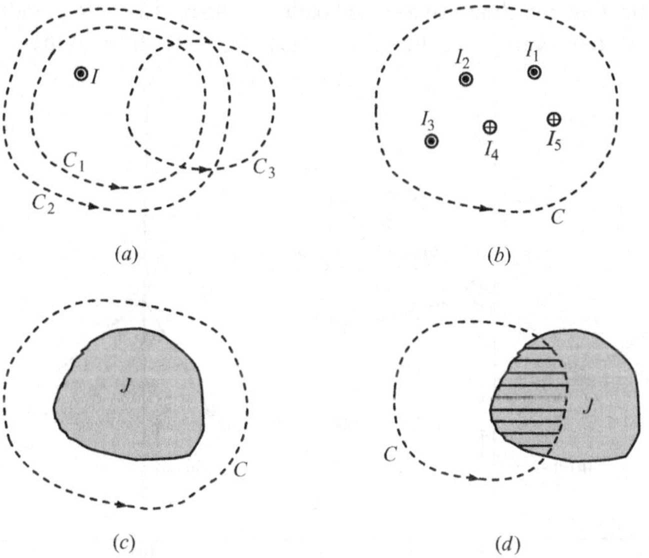

Fig. 4.14 Showing current distributions and closed paths for explanation of Ampere's circuital law

We will now consider an open surface in which current is distributed continuously. The current distribution is defined by current per unit area, i.e., the current density **J**. Therefore, equation (1) may be written as

$$\oint \mathbf{H} \cdot d\ell = \int_S \mathbf{J} \cdot d\mathbf{S} \tag{2}$$

The surface integral of the above relation has to be evaluated over the area of the current-carrying surface enclosed by the chosen closed path. For example, the line integral of **H** around the path shown at (*c*) in Fig. 4.14 is equal to the entire current, whereas at (*d*) the current enclosed is equal to that enclosed within the hatched region only. Relation (2) may be considered as the *integral form* of Ampere's magnetic circuital law.

Before the law is applied to a specific problem, it is necessary to know the nature of field variation so that a suitable closed path can be chosen. A suitable closed path is one along which the field is either tangential or normal so that the line integral of the field intensity is easily evaluated.

The circuital law is useful for calculating field intensity of current distributions having a high degree of symmetry. It is also used for deriving the curl of magnetic field intensity, for deriving the boundary conditions in static as well as in time varying fields, and for magnetic circuit calculations.

Below, we demonstrate the use of Ampere's magnetic circuital law for calculating the magnetic field intensity due to some symmetrical current distributions.

A Long Filamentary Line Current

The magnetic field intensity due to a long line carrying current I has been calculated by the use of Biot–Savart law in Section 4.4. The same result can be obtained more easily by application of the circuital law if we take the line along the z axis of cylindrical coordinates, as shown in Fig. 4.15(a).

Because of symmetry, the field intensity is along ϕ direction and it is a function of r only.

Further, the field lines are concentric circles centred at the origin. Hence to calculate the field intensity at a radial distance r from the line, the choice of a closed path obviously is a circle of radius r as illustrated in the figure. If H_ϕ is the field intensity in ϕ direction along this path, vector form of the field intensity is

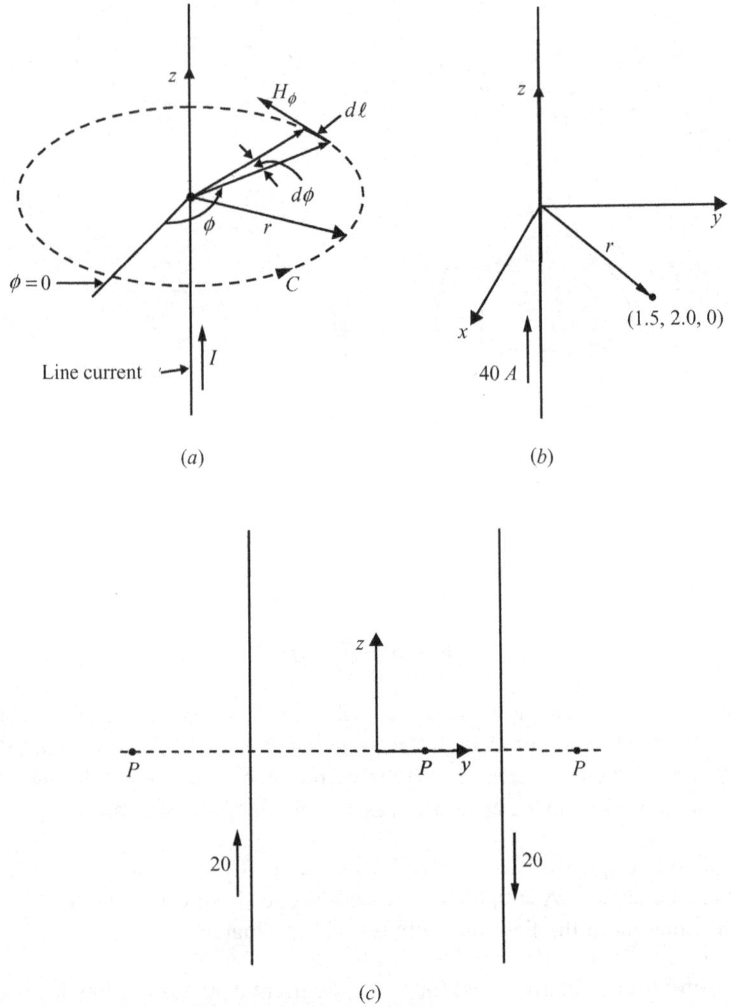

Fig. 4.15 (a) A line current and a circular path C centred at the line current, (b) Illustration for Example 4.18, and (c) Illustration for Example 4.19

$$\mathbf{H} = H_\phi \mathbf{u}_\phi$$

A differential length at radius r and subtending an angle $d\phi$ at the origin,

$$d\ell = r d\phi \mathbf{u}_\phi$$

Thus,
$$\oint \mathbf{H} \cdot d\ell = \oint H_\phi r\, d\phi = 2\pi r H_\phi$$

This must be equal to the current enclosed as per the circuital law. Accordingly,

$$2\pi r H_\phi = I$$

Or
$$H_\phi = \frac{I}{2\pi r}.$$

And
$$\mathbf{H} = \frac{I}{2\pi r}\mathbf{u}_\phi \tag{3a}$$

As $\mathbf{u}_\phi = \mathbf{u}_z \times \mathbf{u}_r$, we can write

$$\mathbf{H} = \frac{(I\mathbf{u}_z)\times\mathbf{u}_r}{2\pi r} = \frac{(I\mathbf{u}_z)\times\mathbf{r}}{2\pi r^2} \tag{3b}$$

If the line current is not along z direction, then $\mathbf{r}$ of cylindrical coordinates has to be replaced by $R\mathbf{u}_n$ where R is the perpendicular distance from the line current to the given point $P(x, y, z)$ at which the field intensity is to be calculated, and $\mathbf{u}_n$ is the unit vector normal to the line current. Moreover, $\mathbf{u}_z$ has to be replaced by a unit vector in the direction of current flow. For example, if the line current is in positive x direction, the magnetic field intensity at $P(x, y, z)$ is given by

$$\mathbf{H} = \frac{(I\mathbf{u}_x)\times(R\mathbf{u}_n)}{2\pi R^2} = \frac{(I\mathbf{u}_x)\times\mathbf{R}}{2\pi R^2}$$

$$\mathbf{R} = (x-x)\mathbf{u}_x + (y-0)\mathbf{u}_y + (z-0)\mathbf{u}_z = y\mathbf{u}_y + z\mathbf{u}_z$$

$$R^2 = y^2 + z^2$$

In general,

$$\mathbf{H} = \frac{I(\mathbf{u}_I \times \mathbf{R})}{2\pi R^2} \tag{3c}$$

where $\mathbf{u}_I$ is a unit vector in the direction of current flow, and $\mathbf{R}$ is the normal displacement vector from the line current to the point where the field intensity is to be calculated.

Uniform Distribution of Current in a Conductor of Circular Section

Consider a solid conductor having circular section of radius a and carrying a current I distributed uniformly over its section. A cross section of the conductor is shown in Fig. 4.16(a). The direction of current is normally out of the paper. The problem is to determine the field intensity outside and inside the conductor assuming that the conductor is long and straight.

If the axis of the conductor is oriented along the z axis of cylindrical coordinates, then the field intensity will have only one component H_ϕ and it will be a function of r only.

Thus, the field intensity outside the conductor at a radial distance r is same as that given by (3). That is,

$$\mathbf{H} = \frac{I}{2\pi r}\mathbf{u}_\phi, \ r \geq a \tag{4}$$

The field intensity inside the conductor is calculated as follows. Current density,

$$J = \frac{I}{\pi a^2}$$

Current within a circle of radius r inside the conductor,

$$I(r) = J\pi r^2 = \frac{Ir^2}{a^2}, \ r \leq a \tag{5}$$

Line integral of the field intensity around a circle of radius r within the conductor is given by

$$\oint \mathbf{H} \cdot d\ell = 2\pi r H_\phi, \ r \leq a \tag{6}$$

By equating (6) with (5), we get

$$2\pi r H_\phi = I(r) = \frac{Ir^2}{a^2}$$

$$\Rightarrow \quad H_\phi = \frac{Ir}{2\pi a^2}, \ 0 \leq r \leq a \tag{7}$$

and

$$\mathbf{H} = \frac{Ir}{2\pi a^2}\mathbf{u}_\phi, \ 0 \leq r \leq a \tag{8}$$

The variation of H with r is shown in Fig. 4.16(b). It may be noted that H has maximum value on the conductor surface.

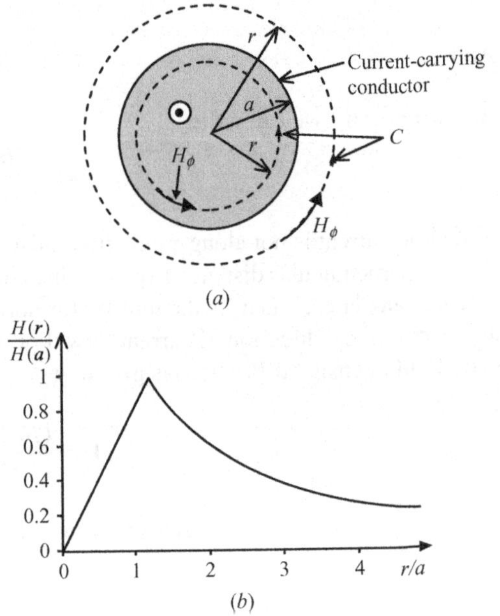

Fig. 4.16 (a) Showing cross section of a cylindrical conductor carrying uniformly distributed current and concentric circular paths C along which H is integrated and (b) Variation of H with r.

A Coaxial Transmission Line

As described in Section 2.12, a coaxial transmission line consists of an outer conductor concentric to and insulated from the inner conductor. Coaxial lines are generally of circular section as illustrated in Fig. 4.17. The radius of the solid inner conductor is taken as a, and the inside and outside radii of the outer conductor are taken as b and c respectively. A coaxial transmission line when connected to an electric source at one end and a load at the other end will carry equal but oppositely directed currents. If we take the current in the outer conductor as $+I$ (into the plane of the page) the current in the inner conductor has to be taken as $-I$ (out of the plane of the page).

We will calculate the magnetic field of the currents assuming that the transmission line is straight and long.

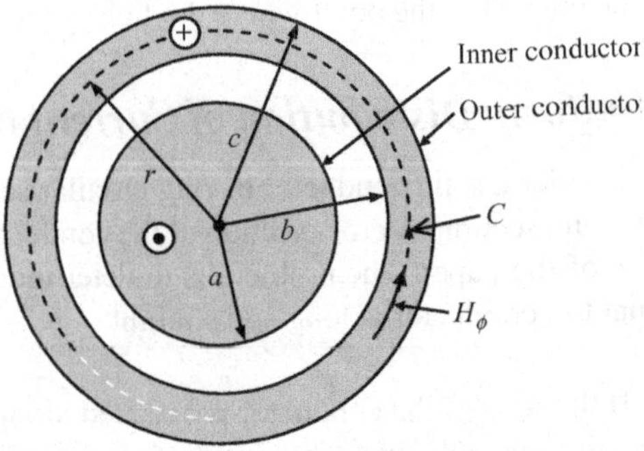

Fig. 4.17 Cross section of a coaxial transmission line

Its axis is taken along the z axis of cylindrical coordinates. Therefore, as in the previous examples, the field intensity at any point within the transmission line has only one component in ϕ direction, and it is a function of r only. The field intensities between the conductors and inside the inner conductor are same as those given by (4) and (8) respectively. Thus,

$$\mathbf{H} = \frac{I}{2\pi r}\mathbf{u}_\phi, \quad b \geq r \geq a$$

$$\mathbf{H} = \frac{Ir}{2\pi a^2}\mathbf{u}_\phi, \quad a \geq r \geq 0$$

Now referring to the figure, net current enclosed by a closed circular path within the outer conductor is equal to the current of the inner conductor minus that part of the outer-conductor current enclosed by the path.

Current density of the outer conductor, $J = \dfrac{I}{\pi(c^2 - b^2)}$

Fraction of outer conductor current enclosed by the path at a radial distance r,

$$I(r) = J[\pi(r^2 - b^2)] = \frac{I(r^2 - b^2)}{c^2 - b^2}, \quad c \geq r \geq b$$

Total current enclosed

$$= I - I(r) = \frac{I(c^2 - r^2)}{(c^2 - b^2)} \tag{9}$$

Thus the field intensity inside the outer conductor,

$$\mathbf{H}_\phi = \frac{I(c^2 - r^2)}{2\pi r(c^2 - b^2)}\mathbf{u}_\phi, \quad c \geq r \geq b \tag{10}$$

The field outside the transmission line is zero as total current enclosed by a closed path enclosing both the conductors is zero.

Field inside a Toroidal Coil

A toroidal coil wound on an annular core is illustrated in Fig. 4.18. The core section may be circular, rectangular or square. The inner and outer radii of the toroid are a and b respectively. We assume that the coil has N turns of very fine wire wound uniformly and closely around the core. Let the current in the coil be I.

Under these assumptions, the direction of the magnetic field intensity at a given radius R inside the core is circumferential and its magnitude is constant. The field outside the core is zero.

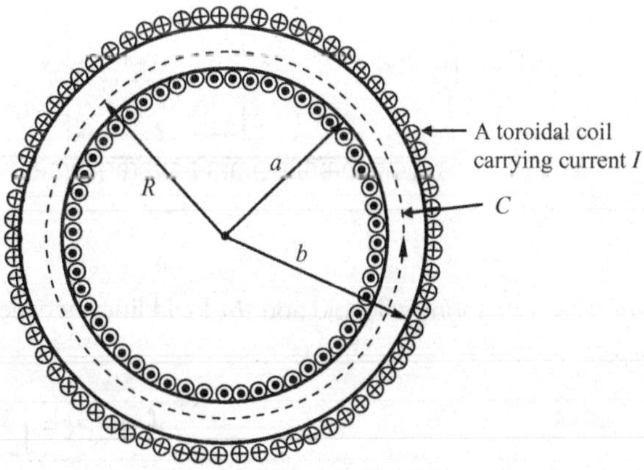

Fig. 4.18 For calculation of H inside a uniformly distributed toroidal coil

The choice of a closed path therefore is a circle of radius R within the core as shown in the figure. The line integral of **H** along this path,

$$\oint \mathbf{H} \cdot d\ell = 2\pi RH, \; b \geq R \geq a$$

Current enclosed by the path is equal to the product of N and I. Hence, application of the circuital law gives,

$$H = \frac{NI}{2\pi R}, \quad b \geq R \geq a$$

The field intensity has maximum value at $R = a$ and minimum value at $R = b$. The field intensity at the mean radius,

$$H_{av} = \frac{NI}{\pi(a+b)}$$

Field inside a Long Solenoid

Consider that the solenoid illustrated in Fig. 4.19(a) is infinitely long, has n turns per metre of its length and it carries a current I. Let us assume that the coil is of very fine wire wound closely and uniformly on a former. In order to find the field of the solenoid by the use of the circuital law we have to choose a suitable closed path so that the line integral of $\mathbf{H}$ can be evaluated easily. And, to choose a path it is necessary to know the nature of the field inside as well as outside the solenoid.

Referring to Fig. 4.19(b) we see that field lines between the conductors, which carry currents in the same direction, are in opposite sense. For closely spaced conductors, the components of the fields normal to the solenoid axis cancel out. It means that the direction of field intensity everywhere inside the solenoid is parallel to its axis. Moreover, as the solenoid is of infinite length, the field intensity has the same value everywhere. So far as the field outside the solenoid is concerned we may assume it to be zero. The above mentioned properties of the field enable us to choose a rectangular path $abcd$, as shown in the figure. The length of the side ab is ℓ and that of side bc is arbitrary. As the field intensity along bc, cd and da is zero, and the current enclosed by the path is equal to $nI\ell$, application of the circuital law yields,

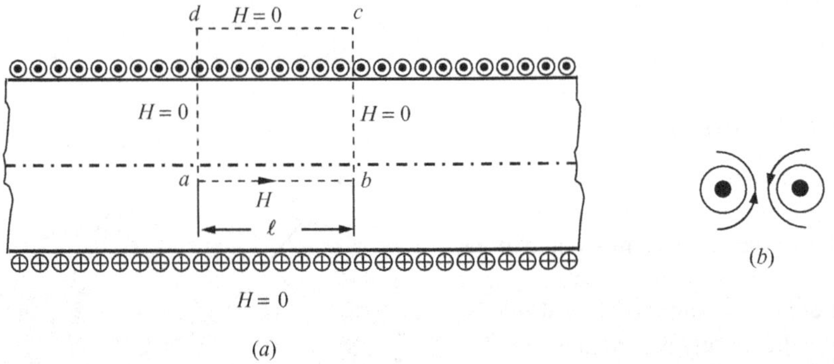

Fig. 4.19 (a) A long solenoid and (b) Field lines between two conductors carrying current in the same direction.

$$\oint \mathbf{H} \cdot d\ell = \int_a^b \mathbf{H} \cdot d\ell = nI\ell$$

Taking the magnetic field intensity along ab as H_a, we have

$$H_a \quad nI\ell$$

$$\Rightarrow \qquad H_a \quad nI$$

This result agrees with formula (4.4-8).

A long solenoid having a uniformly distributed coil with closely spaced turns may be considered as a cylindrical current sheet of uniform surface current density in the circumferential direction. The magnitude of surface current density is equal to nI.

Field due to a Current Sheet

As has been described in Section 4.5, current flow in the form of a thin sheet is described by current per unit width called the surface current density, and denoted by K.

The current density may be considered as a vector quantity having the same direction as the direction of current flow.

We will calculate the field intensity of a plane current sheet assuming its length and width to be infinitely large.

Consider that a current sheet is on the x-y plane with a uniform density of K_x in positive x direction. This is illustrated in Fig. 4.20(a). The vector form of the current density is given by

$$\mathbf{K} = K_x \mathbf{u}_x$$

In order to get an idea about the nature of the field distribution, let us imagine that the sheet is made up of an infinite number of current strips each of differential width dy and parallel to the x axis. The current in each strip is equal to $K_x dy$. Two strips A and B which are symmetrically located on two sides of the origin O are shown separately in Fig. 4.20(b) for clarity. At a point P on the mid-plane between the strips and at a height h above the sheet, let H_A and H_B be the field intensities of currents at A and B respectively. The two fields have the same magnitude and they are at the same angle with respect to y direction. Hence, the resultant of the fields is in negative y direction as shown. Since the resultant fields of all such pairs of current strips have also the same direction, it follows that the field intensity due to the entire current sheet above its surface is in negative y direction. In the same way it can be shown that the field intensity below the sheet is in positive y direction. Also, the magnitude of the field below the sheet and at the same perpendicular distance h from it is equal to that above the sheet. Moreover as the length and width of the sheet are infinitely large, the fields above as well as below the sheet do not vary with x and y. The field intensity in z direction (normal to the sheet) is zero. Thus, the choice of a closed path for application of the circuital law is a rectangle $abcd$, as shown in Fig. 4.20(a). The width (in y direction) of the rectangle is w and its height above and below the sheet is h. The current enclosed by the rectangle is given by wK_x. Now let us assume that the field intensity along ab and cd is H_y with directions as indicated in the figure. Taking the line integral of **H** around $abcda$ and equating it to the current enclosed by the path, we get

$$\oint \mathbf{H} \cdot d\ell = \int_a^b (-H_y(-dy) + \int_c^d H_y dy = wK_x$$

Hence,

$$2H_y w = K_x w$$

Or

$$H_y = \frac{K_x}{2} \tag{11}$$

We see from (11) that

the field intensity is independent of h and it has the same value above as well as below the sheet.

It is in negative y direction above the sheet ($z > 0$), and it is in positive y direction below the sheet ($z < 0$).

Above the sheet ($z > 0$), $\mathbf{H} = -\dfrac{K_x \mathbf{u}_y}{2} = \dfrac{\mathbf{K} \times \mathbf{u}_z}{2}$

Below the sheet ($z < 0$), $\mathbf{H} = \dfrac{K_x \mathbf{u}_y}{2} = \dfrac{\mathbf{K} \times (-\mathbf{u}_z)}{2}$

In general,

$$\mathbf{H} = \frac{\mathbf{K} \times \mathbf{u}_n}{2} \tag{12}$$

The unit vector $\mathbf{u}_n$ in (12) is normal to the current sheet and is directed from the sheet towards the region under consideration.

Next, let us consider two parallel current sheets 1 and 2, as depicted in Fig. 4.20(c). The surface current densities of the sheets have equal magnitude but they are in opposite direction. The field intensities of the sheets are H_y and H_{y2} with directions as indicated in the figure. The fields above Sheet 1 and below Sheet 2 are in opposite direction, and they are in the same direction in the region between the sheets. As the two fields have

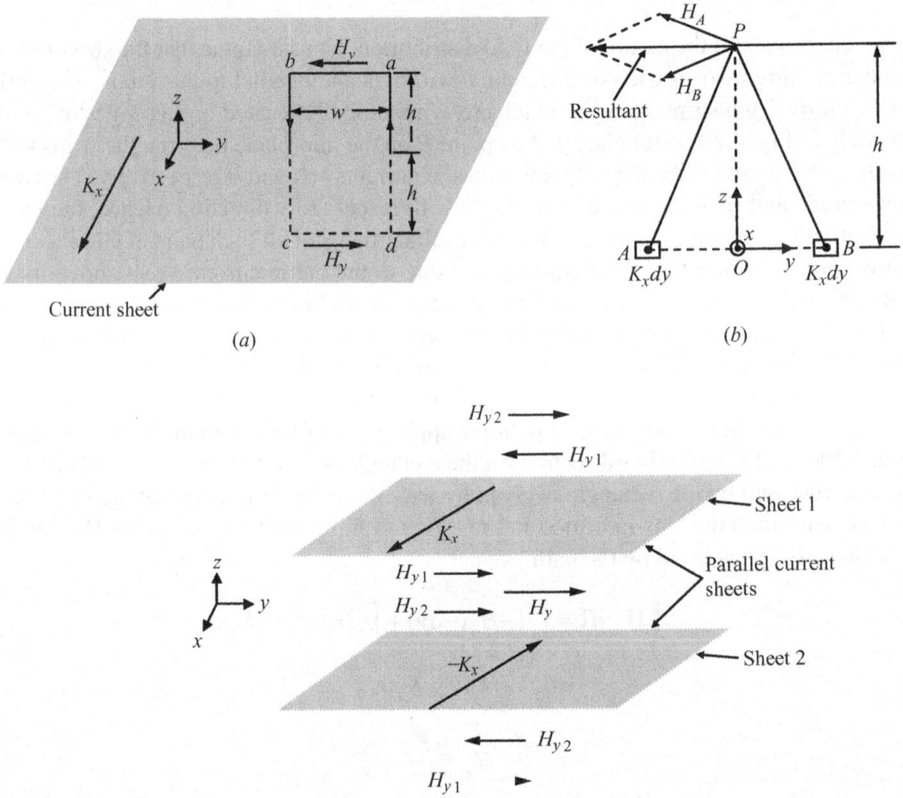

Fig. 4.20 (a) An infinite current sheet in the x-y plane, (b) Field intensities of two current strips and their resultant field, and (c) Field between two parallel current sheets.

equal magnitude, the total field between the sheets is twice that of one sheet and it is zero elsewhere. Thus using formula (11), the magnitude of magnetic field intensity between the sheets,

$$H_y = H_{y1} + H_{y2} = K_x \tag{13}$$

The direction of the field can be determined by the use of the right-hand rule.

EXAMPLE 4.18

A long line current of 40 A is oriented along the z axis of rectangular coordinates. Calculate the magnetic field intensity at $x = 1.5$ m, $y = 2.0$ m and $z = 0$.

The problem is illustrated in Fig. 4.15(b). We will use (3b) to solve the problem.

$\mathbf{r}\ \ 1.5\mathbf{u}_x + 2.0\mathbf{u}_y;\ r^2 = 1.5^2 + 2.0^2 = 6.25$

$$\mathbf{H} = \frac{(40\mathbf{u}_z)\times(1.5\mathbf{u}_x + 2.0\mathbf{u}_y)}{2\pi(6.25)} = -2.037\mathbf{u}_x + 1.528\mathbf{u}_y \ \text{ A/m}$$

EXAMPLE 4.19

Two long line currents 1 and 2, which are parallel to the z axis of rectangular coordinates, are located at $y = -0.25$ m and $y = 0.25$ m respectively. Line 1 has 20 A in positive z direction and Line 2 also has 20 A, but in negative z direction. Find the magnitude of magnetic field intensity at P(0, 0.1 m, 0).

The problem is illustrated in Fig. 4.15(c). Using the right-hand rule to determine the direction of field intensity, we find that the fields due to both the currents at P are into the plane of the paper. Thus, the magnitude of the total field is equal to the sum of the fields of the two line currents.

Magnetic field intensity at a point due to a line current $= \dfrac{I}{2\pi R}$ where R is the perpendicular distance of the point from the line.

Distance of P from Line 1 = 0.35 m

Distance of P from Line 2 = 0.15 m

Total field intensity at $P = \dfrac{20}{2\pi}\left(\dfrac{1}{0.35} + \dfrac{1}{0.15}\right) = 30.32$ A/m

EXAMPLE 4.20

The current density in a solid long cylindrical conductor of radius 2.5 mm is 2.4 A/mm^2. Calculate magnetic field intensity at radial distances of (a) 1.25 mm, (b) 10.0 cm. Assume that the conductor is oriented along the z axis of cylindrical coordinates.

Radius of the conductor, $a = 2.5$ mm

Current density $J = 2.4$ A/mm^2, assumed to be in positive z direction

Current $I = J(\pi a^2) = (2.4)(\pi \times 2.5^2) = 47.12$ A

(a) $\mathbf{H} = \dfrac{Ir\mathbf{u}_\phi}{2\pi a^2} = \dfrac{(47.12)(1.25\times10^{-3})}{2\pi(2.5\times10^{-3})^2}\mathbf{u}_\phi = 1874.8\mathbf{u}_\phi$ A/m

(b) $\mathbf{H} = \dfrac{I\mathbf{u}_\phi}{2\pi r} = \dfrac{47.12}{2\pi\times(0.1)}\mathbf{u}_\phi = 75.12\mathbf{u}_\phi$ A/m

EXAMPLE 4.21

The radius of the inner solid conductor of a coaxial transmission line $a = 5$ mm. The internal and external radii of the outer conductor are $b = 7$ mm and $c = 8.6$ mm respectively. The transmission line supplies 120 kW at 1500 V. Find the magnitude of magnetic field intensity at (a) $r = 3$ mm, (b) $r = 6$ mm and (c) $r = 7.5$ mm.

Current carried by the line, $I = \dfrac{120\times10^3}{1500} = 80$ A

(a) The point is inside the inner conductor. Field intensity,

$$H = \frac{Ir}{2\pi a^2} = \frac{(80)(0.003)}{2\pi(0.005)^2} = 1527.9 \text{ A/m}$$

(b) The point is between the two conductors. Field intensity,

$$H = \frac{I}{2\pi r} = \frac{(80)}{2\pi(0.006)} = 2122.1 \text{ A/m}$$

(c) The point is inside the outer conductor. Field intensity,

$$H = \frac{I(c^2 - r^2)}{2\pi r(c^2 - b^2)} = \frac{(80)[(8.6^2 - 7.5^2)\times10^{-6}]}{2\pi(7.5\times10^{-3})[(8.6^2 - 7^2)\times10^{-6}]} = 1204.6 \text{ A/m}$$

EXAMPLE 4.22

The internal and external radii of the core of a toroid are 40 cm and 45 cm respectively. It carries a uniformly distributed winding of 750 turns. The winding current is 1.8 A. Find (a) the maximum and minimum values of magnetic field intensities in the core and (b) Calculate the average value of the field intensity and compare the result with that of the field intensity at the mean radius.

Magnetic field intensity in the core, $H = \dfrac{NI}{2\pi R}$, $0.4 \text{ m} \le R \le 0.45$ m

$N = 750$; $I = 1.8$ A; $NI = (750)(1.8) = 1350$ A (or amp-turn)

(a) Maximum value of field intensity $= \dfrac{1350}{(2\pi)(0.40)} = 537.15$ A/m

Minimum value of field intensity $= \dfrac{1350}{(2\pi)(0.45)} = 477.46$ A/m

(b)Average value of field intensity,

$$H_{avl} = \frac{1}{(0.45 - 0.4)}\int_{0.4}^{0.45}\frac{1350}{2\pi R}\,dR = 4297.2\ln\frac{45}{40} = 506.14 \text{ A/m}$$

Field intensity at the mean radius, $H_{av2} = \dfrac{1350}{(2\pi)(0.425)} = 505.55 \text{ A/m}$

This is nearly same as the average value.

EXAMPLE 4.23

A current sheet with density $\mathbf{K} = 12.5\mathbf{u}_x$ A/m is located on $z = 0$ plane, and another current sheet with density $\mathbf{K} = -12.5\mathbf{u}_x$ A/m is located on $z = 5.0$ m plane. Find magnetic field intensity in all the regions.

Assuming that the current sheets are of very large linear dimensions, the magnetic field due to the currents is zero in all the regions, except between the sheets. The unit normal vector above the sheet located at $z = 0$ plane is given by $\mathbf{u}_n = \mathbf{u}_z$, and the unit normal vector below the sheet located at $z = 5.0$ m plane is given by $\mathbf{u}_n = -\mathbf{u}_z$. Using formula (12), the field intensity between the sheets,

$$\mathbf{H} = \frac{(12.5\mathbf{u}_x)\times\mathbf{u}_z}{2} + \frac{(-12.5\mathbf{u}_x)\times(-\mathbf{u}_z)}{2} = -12.5\mathbf{u}_y \text{ A/m}$$

The field intensity in rest of the regions is zero.

4.7 Curl of H

In a region of current flow, consider an infinitesimally small, plane, rectangular surface *abcd* on or parallel to the *y-z* plane, as shown in Fig. 4.21. The side lengths of the rectangle are Δy and Δz. The surface element, which is in x direction, has an area

$$\Delta S_x = \Delta y \Delta z$$

If current through the surface is ΔI_x, then it is equal to the line integral of $\mathbf{H}$ around the path *abcda*. Thus,

$$\oint \mathbf{H} \cdot d\ell = \Delta I_x$$

Dividing both sides by ΔS_x, we get

$$\frac{1}{\Delta S_x}\oint \mathbf{H} \cdot d\ell = \frac{\Delta I_x}{\Delta S_x}$$

In the limit ΔS_x approaching to zero, the right side of the above relation is equal to current density in x direction. That is,

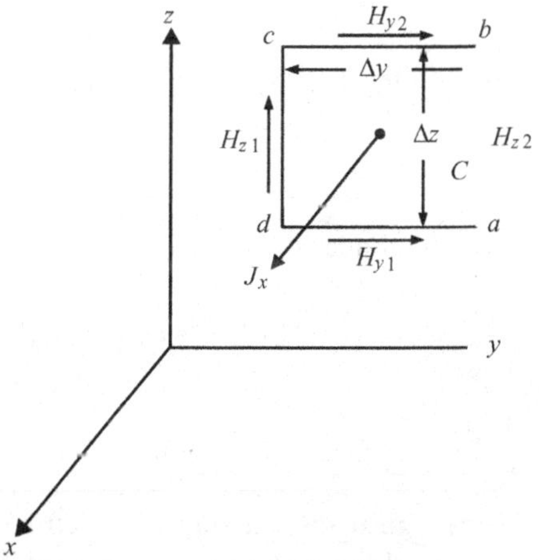

Fig. 4.21 An infinitesimal closed path *abcda* in a region of current flow

$$\lim_{\Delta S_x \to 0}\left[\frac{1}{\Delta S_x}\oint \mathbf{H} \cdot d\ell\right] = J_x \tag{1}$$

In order to evaluate the line integral, we assume H_y H_{y2}, H_{z1} and H_{z2} to be the average values of H along the four sides of the rectangle as indicated in the figure. Then, the line integral of $\mathbf{H}$ around the rectangle,

$$\oint \mathbf{H} \cdot d\ell = (\quad)\Delta z \; (\quad) y \tag{2}$$

The first term on the right side is the change of $(H_z \Delta z)$ over an infinitesimal length Δy. It is equal to the rate of change of $(H_z \Delta z)$ with respect to y multiplied by Δy. Thus,

$$(H_{z2} - H_{z1})\Delta z = \frac{\partial (H_z \Delta z)}{\partial y}\Delta y$$

As Δz is not a function of y,

$$(H_{z2} - H_{z1})\Delta z = \frac{\partial H_z}{\partial y}\Delta y\,\Delta z = \frac{\partial H_z}{\partial y}\Delta S_x \tag{3}$$

Similarly,

$$(H_{y2} - H_{y1})\Delta y = \frac{\partial H_y}{\partial z}\Delta S_x \tag{4}$$

Substituting (3) and (4) in (2), we have

$$\oint \mathbf{H}\cdot d\ell = \left(\frac{\partial H_z}{\partial y} - \frac{\partial H_y}{\partial z}\right)\Delta S_x \tag{5}$$

Dividing both sides of (5) by ΔS_x and taking the limit $\Delta S_x \to 0$, we get

$$\lim_{\Delta S_x \to 0}\left[\frac{1}{\Delta S_x}\oint \mathbf{H}\cdot d\ell\right] = \frac{\partial H_z}{\partial y} - \frac{\partial H_y}{\partial z} \tag{6}$$

Comparing (1) with (6) we see that the left-hand sides of both the equations are same. Hence, the right-hand sides must also be equal. Accordingly,

$$\frac{\partial H_z}{\partial y} - \frac{\partial H_y}{\partial z} = J_x \tag{7}$$

Now, multiplying both sides of (7) by the unit vector $\mathbf{u}_x$, we have

$$\left(\frac{\partial H_z}{\partial y} - \frac{\partial H_y}{\partial z}\right)\mathbf{u}_x = J_x\,\mathbf{u}_x \tag{8}$$

It can be similarly shown that,

$$\left(\frac{\partial H_x}{\partial z} - \frac{\partial H_z}{\partial x}\right)\mathbf{u}_y = J_y\mathbf{u}_y \tag{9}$$

And

$$\left(\frac{\partial H_y}{\partial x} - \frac{\partial H_x}{\partial y}\right)\mathbf{u}_z = J_z\,\mathbf{u}_z \tag{10}$$

We recognize that the sum of the left-hand sides of (8), (9), and (10) is equal to curl of **H**, and the sum of the right-hand sides is equal to the current density vector **J**. Hence,

$$\text{curl } \mathbf{H} = \nabla\times \mathbf{H} = \mathbf{J} \tag{11a}$$

The curl operation can be written in the determinant form. In rectangular coordinates,

$$\text{curl } \mathbf{H} = \begin{vmatrix} \mathbf{u}_x & \mathbf{u}_y & \mathbf{u}_z \\ \frac{\partial}{\partial x} & \frac{\partial}{\partial y} & \frac{\partial}{\partial z} \\ H_x & H_y & H_z \end{vmatrix} = \mathbf{J} \tag{11b}$$

Equation (11) gives the relationship between current density and magnetic field intensity at a point in a region where steady current flows. It is known as *point form* or *differential form* of Ampere's circuital law.

In a current-free region, $\mathbf{J} = 0$. Hence,

$$\nabla \times \mathbf{H} = 0 \tag{12}$$

That is, the curl operation of the magnetic field intensity in a current-free region is zero.

As a vector field of zero curl is called an irrotational field, a current-free region of a magnetic field is an irrotational field

To obtain $\nabla \times \mathbf{H}$ in cylindrical coordinates, let us consider an infinitesimal closed path on or parallel to the ϕ-z plane with side lengths $r\Delta\phi$ in ϕ direction and Δz in z direction. The line integral of $\mathbf{H}$ around the path,

$$\oint \mathbf{H} \cdot d\ell = (H_{z2} - H_{z1})\Delta z - (H_{\phi 2} - H_{\phi 1})(r\Delta\phi)$$

$$= \frac{1}{r}\frac{\partial(H_z \Delta z)}{\partial \phi}(r\Delta\phi) - \frac{\partial[H_\phi (r\Delta\phi)]}{\partial z}\Delta z$$

$$= \left(\frac{1}{r}\frac{\partial H_z}{\partial \phi} - \frac{\partial H_\phi}{\partial z}\right)(r\Delta\phi\,\Delta z)$$

$$= \left(\frac{1}{r}\frac{\partial H_z}{\partial \phi} - \frac{\partial H_\phi}{\partial z}\right)\Delta S_r \tag{13}$$

where $\Delta S_r = r\Delta\phi\Delta z$ is area of the surface element in r direction.

By dividing both sides of (13) by ΔS_r and taking the limit we will get the r component of curl of $\mathbf{H}$ in cylindrical coordinates.

$$[\nabla \times \mathbf{H}]_r = \frac{1}{r}\frac{\partial H_z}{\partial \phi} - \frac{\partial H_\phi}{\partial z}$$

After deriving the ϕ and z components and then adding the three components we will get $\nabla \times \mathbf{H}$ in cylindrical coordinates as given below.

$$\nabla \times \mathbf{H} = \left(\frac{1}{r}\frac{\partial H_z}{\partial \phi} - \frac{\partial H_\phi}{\partial z}\right)\mathbf{u}_r + \left(\frac{\partial H_r}{\partial z} - \frac{\partial H_z}{\partial r}\right)\mathbf{u}_\phi + \frac{1}{r}\left(\frac{\partial (rH_\phi)}{\partial r} - \frac{\partial H_r}{\partial \phi}\right)\mathbf{u}_z . \tag{14a}$$

$$= \begin{vmatrix} \frac{\mathbf{u}_r}{r} & \mathbf{u}_\phi & \frac{\mathbf{u}_z}{r} \\ \frac{\partial}{\partial r} & \frac{\partial}{\partial \phi} & \frac{\partial}{\partial z} \\ H_r & rH_\phi & H_z \end{vmatrix} \tag{14b}$$

In order to derive the curl of $\mathbf{H}$ in spherical coordinates, let us consider an infinitesimal closed path parallel to θ-ϕ plane and with side lengths $(r\Delta\theta)$ in θ direction and $(r\sin\theta\,\Delta\phi)$ in ϕ direction. The line integral of $\mathbf{H}$ around the path,

$$\oint \mathbf{H} \cdot d\ell = ([illegible] - [illegible])([illegible] \sin [illegible]) - (H_{\theta} - H_{\theta 1})(r\Delta\theta)$$

$$= \frac{1}{r}\frac{\partial [illegible]}{\partial \theta}(r [illegible]) - \frac{1}{r\sin\theta}\frac{\partial[H [illegible] (r\Delta\theta)]}{\partial \phi}(r\sin\theta [illegible] \phi)$$

$$= \frac{1}{r\sin\theta}\left[\frac{\partial(H_\phi \sin\theta)}{\partial\theta} - \frac{\partial H_\theta}{\partial\phi}\right]\Delta S_r \qquad (15)$$

where $\Delta S_r = r^2 \sin\theta\Delta\theta\Delta\phi$ is area of the differential surface in r direction

By dividing both sides of (15) by ΔS_r and taking the limit we will get the r component of curl of **H** in spherical coordinates.

$$[\nabla\times\mathbf{H}]_r = \frac{1}{r\sin\theta}\left[\frac{\partial(H_\phi \sin\theta)}{\partial\theta} - \frac{\partial H_\theta}{\partial\phi}\right]$$

The components of curl of **H** in θ and ϕ directions can similarly be obtained.

In spherical coordinates,

$$\nabla\times\mathbf{H} = \frac{1}{r\sin\theta}\left(\frac{\partial(H_\phi \sin\theta)}{\partial\theta} - \frac{\partial H_\theta}{\partial\phi}\right)\mathbf{u}_r + \frac{1}{r}\left(\frac{1}{\sin\theta}\frac{\partial H_r}{\partial\phi} - \frac{\partial(rH_\phi)}{\partial r}\right)\mathbf{u}_\theta$$

$$+\frac{1}{r}\left(\frac{\partial(rH_\theta)}{\partial r} - \frac{\partial H_r}{\partial\theta}\right)\mathbf{u}_\phi \qquad (16a)$$

$$= \begin{vmatrix} \dfrac{\mathbf{u}_r}{r^2\sin\theta} & \dfrac{\mathbf{u}_\theta}{r\sin\theta} & \dfrac{\mathbf{u}_\phi}{r} \\ \dfrac{\partial}{\partial r} & \dfrac{\partial}{\partial\theta} & \dfrac{\partial}{\partial\phi} \\ H_r & rH_\theta & r\sin\theta H_\phi \end{vmatrix} \qquad (16b)$$

It may be mentioned that it is easier to memorise the determinant forms of the curl of a vector function than the expanded form.

EXAMPLE 4.24

A hollow cylindrical conductor with inner radius b and outer radius c carries a uniformly distributed current of I A in z direction of cylindrical coordinates. The magnetic field intensity at a radial distance r ($b < r < c$) is

$$\mathbf{H} \quad \frac{I(r^2 - b^2)}{2\pi r(c^2 - b^2)}\mathbf{u}_\phi$$

Show that the curl of **H** is equal to the current density.

$$\text{Current density} = \frac{I}{\pi(c^2 - b^2)}\mathbf{u}_z$$

The field intensity has only one component in ϕ direction in cylindrical coordinates, and it is a function of r. Thus using (4.7-14), we have

$$\nabla\times\mathbf{H} = \frac{\mathbf{u}_z}{r}\left[\frac{\partial(rH_\phi)}{\partial r}\right] = \frac{I}{\pi(c^2 - b^2)}\mathbf{u}_z = \text{Current density}$$

EXAMPLE 4.25

The radius of a long solid conductor is a. It carries a uniformly distributed current I. Find the magnetic field intensity at $r < a$ using the relation $\nabla \times \mathbf{H} = \mathbf{J}$. Use cylindrical coordinates.

The current density, which is in z direction, $J_z = \dfrac{I}{\pi a^2}$

The field intensity is in ϕ direction and it is a function of r. Thus from (4.7-14), we have

$$\frac{1}{r}\left[\frac{\partial (rH_\phi)}{\partial r}\right] = J_z = \frac{I}{\pi a^2}$$

Or $$\frac{\partial (rH_\phi)}{\partial r} = \frac{Ir}{\pi a^2}$$

Or $$rH_\phi = \int \frac{Ir}{\pi a^2} dr = \frac{Ir^2}{2\pi a^2} + C, \text{ where } C \text{ is a constant.}$$

Therefore, the field intensity, $H_\phi = \dfrac{Ir}{2\pi a^2} + \dfrac{C}{r}$

Since $r = 0$ is included, C must be zero as otherwise the field would be infinite which is not acceptable. Thus,

$$H_\phi = \frac{Ir}{2\pi a^2}$$

4.8 Stokes' Theorem

In Section 4.6 we have expressed the magnetic circuital law in integral form (Eq. 2), and in Section 4.7 we have obtained the law in differential form (Eq. 11a). We will now use these forms of the law to derive Stokes' theorem. Let us consider an open surface S of any shape on which current is continuously distributed, as shown in Fig. 4.22. The current distribution is defined by current per unit area $\mathbf{J}$. If $\mathbf{H}$ is magnetic field intensity in the region of the surface, then

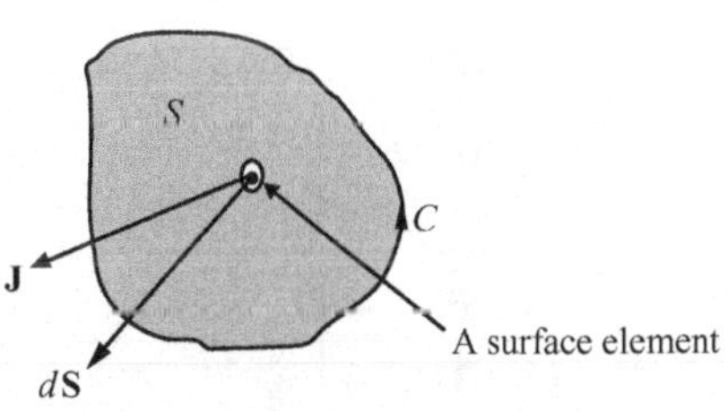

Fig.4.22 An open surface with continuous distribution of current. The current is out of the page.

$$\int_S \mathbf{J}\cdot d\mathbf{S} = \oint_C \mathbf{H}\cdot d\ell \qquad (1)$$

where the surface integration is over the surface and the line integral is around the periphery C of the surface. Also,

$$\nabla \times \mathbf{H} = \mathbf{J} \qquad (2)$$

By eliminating $\mathbf{J}$ from (1) and (2), we get

$$\int_S (\quad\quad).\ \mathbf{S} = \oint_C \mathbf{H}\cdot d\ell \qquad (3)$$

Equation (3) is valid for any vector field which is a continuous function of space variables. The relation is known as *Stokes' theorem*. The theorem states that the integral of the normal component of any vector over a surface is equal to the line integral of the same vector around the periphery of the surface.

Its proof for any vector function is a little involved.

The theorem is useful in transforming the surface integral of the curl of a vector function into a line integral of the same vector, and vice versa.

EXAMPLE 4.26

The magnetic field intensity in the region of the rectangle shown in Fig. 4.23 is given by

$\mathbf{H} \quad \mathbf{u}_{..} + 3x\mathbf{u}_y$.

Verify Stokes' theorem.

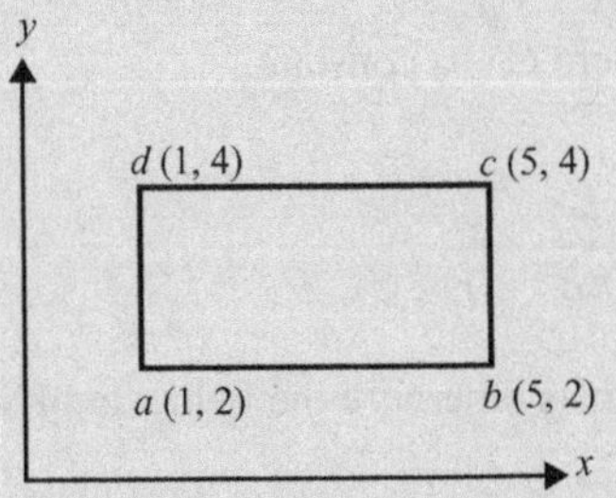

Fig. 4.23 Illustration for Example 4.26

The components of the given field intensity vector are

$H_{..} \quad y^2. \; H_y = 3x$

As the rectangle is in the x-y plane,

$d \quad dx\,dy\,\mathbf{u}_z$ and $d\ell = dx\,\mathbf{u}_x + dy\,\mathbf{u}_y$

$$\nabla \times \mathbf{H} = \begin{vmatrix} \mathbf{u}_x & \mathbf{u}_y & \mathbf{u}_z \\ \dfrac{\partial}{\partial x} & \dfrac{\partial}{\partial y} & \dfrac{\partial}{\partial z} \\ y^2 & 3x & 0 \end{vmatrix} = (3-2y)\,\mathbf{u}_z$$

The surface integral of $\nabla \times \mathbf{H}$ is

$$\int_S (\quad).d\mathbf{S} = \int_1^5 \int_2^4 (3-2y)\,dx\,dy = -24$$

Next, $\quad \mathbf{H} \cdot d\ell \quad = (y^2\,\mathbf{u}_x + 3x\,\mathbf{u}_y) \cdot (dx\,\mathbf{u}_x + dy\,\mathbf{u}_y) = y^2 dx + 3x dy$

Referring to the figure we note that from a to b, $y = 2$ and $dy = 0$; from b to c, $x = 5$ and $dx = 0$; from c to d, $y = 4$ and $dy = 0$; from d to a, $x = 1$ and $dx = 0$. Therefore, the line integral of $\mathbf{H}$ around the closed path,

$$\oint \mathbf{H} \cdot d\ell = \int \quad + \int_2^4 15 \quad + \int \quad + \int_4^2 3\,dy = -24$$

Hence, the theorem is verified.

EXAMPLE 4.27

The magnetic field intensity inside a long, straight circular wire of radius a and carrying uniformly distributed current I is given by

$$H = \frac{Ir}{2\pi a^2}\mathbf{u}_\phi,$$

in cylindrical coordinates. Show that Stokes' theorem is valid over the circular section.

Only the ϕ component of **H** is present and it is a function of r only. Using formula (4.7-14), we have

$$\nabla \times \mathbf{H} = \frac{\mathbf{u}_z}{r}\left[\frac{\partial (rH_\phi)}{\partial r}\right] = \frac{I}{\pi a^2}\mathbf{u}_z$$

The differential surface area in z direction is

$$d\_ = r\,dr\,d\phi\,d\mathbf{u}_z$$

Thus, $$\int_S (\nabla \times \mathbf{H}).d\mathbf{S} = \int_0^a \int_0^{2\pi} \frac{Ir\,drd\phi}{\pi a^2} = I \qquad (4)$$

The field intensity and differential length at $r = a$ are

$$\mathbf{H} \;\; \frac{I}{2\pi a}\mathbf{u}_\phi \text{ and } d\ell = a\,d\phi\,\mathbf{u}_\phi.$$

The line integral of **H** around the circular section of the conductor,

$$\oint \mathbf{H}\cdot d\ell = \int_0^{2\pi} \frac{I\,d\phi}{2\pi} = I \qquad (5)$$

The theorem is verified as (4) and (5) give the same result.

4.9 Divergence of B

The divergence of the electric flux density vector is defined by (6) of Section 2.13. The divergence of magnetic flux density is obtained by writing **B** in place of **D** in the above mentioned equation. Accordingly,

$$\text{div } \mathbf{B} = \nabla \cdot \mathbf{B} = \lim_{\Delta v \to 0} \frac{\oint_S \mathbf{B}\cdot d\mathbf{S}}{\Delta v} \qquad (1)$$

where the numerator of right side is the net flux out of a closed surface enclosing a volume element Δv in a magnetic field.

As magnetic field lines close on themselves, the flux entering a closed surface enclosing the volume must be equal to the flux coming out of it.

That is, net flux *out* of the surface must be zero. Hence,

$$\oint_S \mathbf{B} \;\; \mathbf{S} = 0 \qquad (2)$$

From (1) and (2), we get

$$\nabla.\mathbf{B} = 0 \tag{3}$$

Thus,

> the divergence of the magnetic flux density vector at a point in a region of magnetic field is equal to zero irrespective of whether the region has any current or not. As a vector field of zero divergence is termed as a solenoidal field, a magnetic field is solenoidal in character.
>
> Equations (2) and (3) may be considered as integral and differential forms of Gauss's law for magnetic flux density.

EXAMPLE 4.28

The magnetic flux density inside a cylindrical conductor of radius a, and carrying a uniformly distributed current of I is given by

$$\mathbf{B} \quad \frac{uIr}{2\pi a^2}\mathbf{u}_\phi$$

in cylindrical coordinates. Show that the divergence of **B** is equal to zero.

The divergence of electric flux density in cylindrical coordinates is given by (2.13-4). By substituting **B** in place of **D** in this equation, we will obtain the divergence of the magnetic flux density. Thus, the divergence **B** is given by

$$\nabla\cdot\mathbf{B} = \frac{1}{r}\frac{\partial(rB_r)}{\partial r} + \frac{1}{r}\frac{\partial B_\varphi}{\partial \varphi} + \frac{\partial B_z}{\partial z}$$

As the given flux density has only one component in ϕ direction and it is not a function of ϕ, the divergence of **B** is zero.

4.10 The Vector Magnetic Potential

In the study of static electric field we found that it is advantageous to obtain electric field intensity from a potential function. The electrostatic potential, defined as work done per unit charge, is a scalar point function of space. The potential concept is also useful for solving magnetic field problems. The vector magnetic potential provides an alternative means for calculating the magnetic flux density. It also increases the scope for analysis of magnetic field problems.

> The vector magnetic potential is not derived from any physical consideration as in the case of electrostatic potential. It is derived from the solenoidal property of magnetic flux density

defined by (3) in the preceding section.

Referring to the vector identity (2) of Section (1.9) we find that the divergence of curl of a vector function is zero. Since the divergence of **B** is zero, it can be expressed as the curl of a vector function. Denoting the vector function by **A**, we may write

$$\mathbf{B} = \nabla\times\mathbf{A} \tag{1}$$

where

A is called *vector magnetic potential* or simply *vector potential.* The unit of **A** is tesla-metre (T-m) or weber per metre (Wb/m).

Using the relation $\mathbf{B} = \mu \mathbf{H}$ in (1), we get

$$\nabla \times \mathbf{A} = \mu \mathbf{H} \tag{2}$$

Just as **H** is related to the current density vector through the differential form of the magnetic circuital law, the vector potential is also related to the current density. We will derive this relation.

Taking curl of both the sides of (2), we have

$$\nabla \times \nabla \times \mathbf{A} = \nabla \times (\mu \mathbf{H}) \tag{3}$$

For a homogeneous and isotropic medium the permeability μ is constant. Therefore, it can be written outside the curl operation of (3).

$$\nabla \times \nabla \times \mathbf{A} = \mu(\nabla \times \mathbf{H})$$

Substituting $\nabla \times \mathbf{H} = \mathbf{J}$ in the above relation, we get

$$\nabla \times \nabla \times \mathbf{A} = \mu \mathbf{J} \tag{4}$$

In rectangular coordinates it can be shown that

$$\nabla \times \nabla \times \mathbf{A} = \nabla(\nabla . \mathbf{A}) - \nabla^2 \mathbf{A} \tag{5}$$

For static magnetic field $\nabla \cdot \mathbf{A}$ is taken as zero. Hence from (4) and (5), we get

$$\nabla^2 \mathbf{A} = -\mu \mathbf{J} \tag{6}$$

Equation (6) is a second order partial differential equation relating the vectors **A** and **J**. Both the quantities may have components in all the three coordinate directions and each component may be a function of all the coordinate variables. In rectangular coordinates, let

$$\mathbf{A} = A_x \mathbf{u}_x + A_y \mathbf{u}_y + A_z \mathbf{u}_z \tag{7}$$

$$\mathbf{J} = J_x \mathbf{u}_x + J_y \mathbf{u}_y + J_z \mathbf{u}_z \tag{8}$$

Substituting (7) and (8) in (6), we get

$$\nabla^2 (A_x \mathbf{u}_x + A\mathbf{u}_y + A\mathbf{u}_z) = -(J_x \mathbf{u}_x + J_y \mathbf{u}_y + J_z \mathbf{u}_z) \tag{9}$$

Equating the like components on two sides of (9), we get

$$\nabla^2 A_x = \frac{\partial^2 A_x}{\partial x^2} + \frac{\partial^2 A_x}{\partial y^2} + \frac{\partial^2 A_x}{\partial z^2} = -\mu J_x \tag{10a}$$

$$\nabla^2 A_y = \frac{\partial^2 A_y}{\partial x^2} + \frac{\partial^2 A_y}{\partial y^2} + \frac{\partial^2 A_y}{\partial z^2} = -\mu J_y \tag{10b}$$

$$\nabla^2 A_z = \frac{\partial^2 A_z}{\partial x^2} + \frac{\partial^2 A_z}{\partial y^2} + \frac{\partial^2 A_z}{\partial z^2} = -\mu J_z \tag{10c}$$

Thus, each component of **A** in rectangular coordinates satisfies a Poisson's type second order partial differential equation. However, in other coordinate systems the components cannot be separated; the exception being the z component of **A** in cylindrical coordinates. The differential equation for A_z in cylindrical coordinates is given by

$$\frac{1}{r}\frac{\partial}{\partial r}\left(r\frac{\partial A_z}{\partial r}\right) + \frac{1}{r^2}\frac{\partial^2 A_z}{\partial \phi^2} + \frac{\partial^2 A_z}{\partial z^2} = -\mu J_z \tag{11}$$

In a current-free region the right side of (6) is zero. Hence,

$$\nabla^2 \mathbf{A} = 0 \tag{12}$$

Equation (12) is a Laplaces' type of second order partial differential equation for the vector magnetic potential.

As per Stokes' theorem, the line integral of a vector quantity around a closed path is equal to the surface integral of the curl of the vector over the surface enclosed by the closed path. Thus,

$$\oint \mathbf{A}.d\ell = \int_S (\nabla \times \mathbf{A}) \cdot d\mathbf{S}$$

But, $\nabla \times \mathbf{A} = \mathbf{B}$. Hence,

$\oint \mathbf{A}.d\ell = \int_S \mathbf{B} \cdot d\mathbf{S} = \Phi$, the magnetic flux through the surface.

4.11 Calculation of Flux Density from Vector Magnetic Potential

We obtained the magnetic flux density of some current distributions by the use of Biot–Savart law in Section 4.4. This section illustrates the procedure for calculating magnetic flux density from the vector magnetic potential of some simple current distributions. We will at first obtain the vector potential of a current element and then utilise it for the calculation of flux density of a line current, of a two-conductor transmission line, and of a current loop.

Vector Fields of a Current Element

In a region of volume distribution of steady current let us consider a point P' at x', y', z' (primed variables), as shown in Fig. 4.24. Let $\mathbf{J}'$ be the current density and dv' be the differential volume at P'. Then a differential current element at P' is given by $\mathbf{J}'dv'$. The magnetic flux density $d\mathbf{B}$ at a point P with coordinates x, y, z (unprimed variables) due to the current element is obtained by substituting $\mathbf{J}'dv'$ in place of $Id\ell$ in the Biot–Savart law given by (4.3-4).

$$d\mathbf{B} = \frac{\mu(\mathbf{J}' \times \mathbf{R})}{4\pi R^3} dv'$$

where $$\mathbf{R} = (x - x')\mathbf{u}_x + (y - y')\mathbf{u}_y + (z - z')\mathbf{u}_z$$

$$R = \sqrt{(x - x')^2 + (y - y')^2 + (z - z')^2}$$

To find the magnetic flux density at P, we have to integrate $d\mathbf{B}$ over the current distribution. The flux density at P is

$$\mathbf{B} = \mu\int_{v'} \frac{\mathbf{J}' \times \mathbf{R}}{4\pi R^3} dv' \tag{1}$$

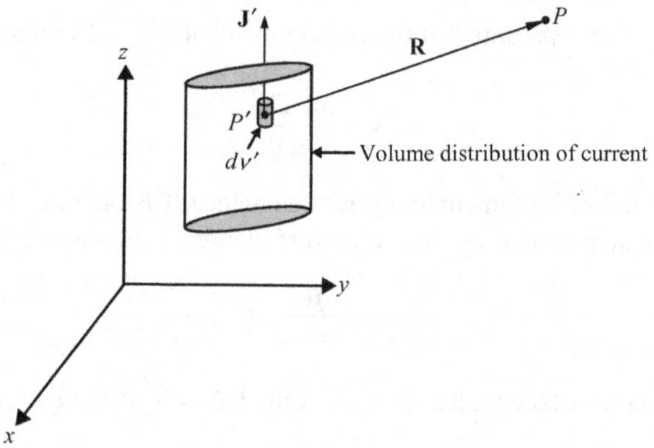

Fig. 4.24 Illustration for defining the vector magnetic potential of a current element

If we define a vector operator ∇ in terms of the unprimed variables as

$$\nabla = \mathbf{u}_x \frac{\partial}{\partial x} + \mathbf{u}_y \frac{\partial}{\partial y} + \mathbf{u}_z \frac{\partial}{\partial z},$$

then it can be shown that

$$\frac{\mathbf{J}' \times \mathbf{R}}{R^3} = \nabla\left(\frac{1}{R}\right) \times \mathbf{J}' \tag{2}$$

Using the vector identity (6) of Section (1.9), we get

$$\nabla \times \left(\frac{\mathbf{J}'}{R}\right) = \nabla\left(\frac{1}{R}\right) \times \mathbf{J}' + \frac{1}{R}(\nabla \times \mathbf{J}')$$

The second term on right side of the above relation is zero as $\mathbf{J}'$ is a function of primed variables whereas ∇ is defined in terms the unprimed variables. Thus,

$$\nabla \times \left(\frac{\mathbf{J}'}{R}\right) = \nabla\left(\frac{1}{R}\right) \times \mathbf{J}' \tag{3}$$

As the right-hand sides of (2) and (3) are same, we get the following relation from the two equations.

$$\frac{\mathbf{J}' \times \mathbf{R}}{R^3} = \nabla \times \left(\frac{\mathbf{J}'}{R}\right) \tag{4}$$

By replacing the integrand of (1) by the right side of (4), we get

$$\mathbf{B} = \frac{\mu}{4\pi} \int_{v'} \left(\nabla \times \frac{\mathbf{J}'}{R}\right) dv'$$

The ∇ operator can be brought out of the integral sign as the integration is with respect to the primed variables.

$$\mathbf{B} = \nabla \times \int_{v'} \frac{\mu \mathbf{J}' dv'}{4\pi R} \tag{5}$$

As $\mathbf{B}$ is also equal to $\nabla \times \mathbf{A}$, the integral of (5) gives the vector potential of a volume current distribution. Hence,

$$\mathbf{A} - \int_{v'} \frac{\mu \mathbf{J}' av'}{4\pi R} \tag{6}$$

The integrand of (6) is the vector potential of the current element $\mathbf{J}'dv'$. Denoting the potential function by $d\mathbf{A}$, we have

$$d\mathbf{A} = \frac{\mu \mathbf{J}'\, dv'}{4\pi R} \tag{7}$$

The vector magnetic potential of a filamentary current element $Id\boldsymbol{\ell}$ is obtained by replacing $\mathbf{J}'dv'$ by $Id\boldsymbol{\ell}$ in (7). Thus, the vector magnetic potential of a linear current element is given by

$$d\mathbf{A} = \frac{\mu I}{4\pi R} d\boldsymbol{\ell} \tag{8}$$

For current distribution over a surface with density $\mathbf{K}$ A/m, $\mathbf{J}\, dv = \mathbf{K}\, dS$. Therefore,

$$d\mathbf{A} = \frac{\mu \mathbf{K}\, dS}{4\pi R}$$

We note from (7) and (8) that

> the vector magnetic potential of a current element at a point is inversely proportional to the distance of the point from the current element and the direction of the vector potential is same as that of the current element.

Vector Fields of a Line Current

Figure 4.25 shows a filamentary line of length ℓ and carrying current I. The line is taken along the z axis of cylindrical coordinates with its mid-point at the origin. A differential current element at $P'(0, 0, z')$ on the line current is given by,

$$I\, d\boldsymbol{\ell} = I\, dz' \mathbf{u}_z$$

The vector potential at $P(r, 0, z)$ due to the current element,

$$d\mathbf{A} = \frac{\mu I dz' \mathbf{u}_z}{4\pi R} = \frac{\mu I dz' \mathbf{u}_z}{4\pi \sqrt{r^2 + (z - z')^2}} \tag{9}$$

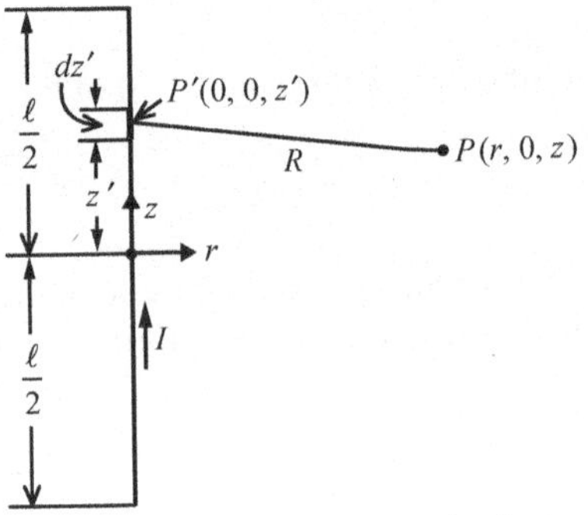

Fig. 4.25 Illustration for calculation of vector fields of a line current of finite length

The vector potential due to the line current is found by integrating (9) between the limits $-\ell/2$ and $+\ell/2$.

$$\mathbf{A} = \int_{-\ell/2}^{+\ell/2} d\mathbf{A} = \left(\frac{\mu I}{4\pi}\right) \ln \left(\frac{(z + \ell/2) + \sqrt{r^2 + (z + \ell/2)^2}}{(z - \ell/2) + \sqrt{r^2 + (z - \ell/2)^2}} \right) \mathbf{u}_z \tag{10}$$

Formula (10) shows that the vector potential has only one component A_z and it is a function of r and z. The curl of $\mathbf{A}$ in cylindrical coordinates gives the flux density. Thus using (4.7-14), we get

$$\mathbf{B} = \nabla \times \mathbf{A} = -\frac{\partial A_z}{\partial r} \mathbf{u}_\phi$$

$$= \frac{\mu I}{4\pi r} \left(\frac{z + \ell/2}{\sqrt{r^2 + (z + \ell/2)^2}} - \frac{z - \ell/2}{\sqrt{r^2 + (z - \ell/2)^2}} \right) \mathbf{u}_\phi \tag{11}$$

This expression is same as that obtained in Section 4.4

Vector Fields of a Two-Wire Parallel Transmission line

When a two-wire transmission line is used to supply electric power, the two wires will carry equal and oppositely directed currents. The wires, which are assumed to be straight and parallel, are taken on the x-z plane. They are symmetrically located on two sides of the z axis, as shown in Fig. 4.26. The length of each wire and spacing between them are taken as ℓ and s respectively. Current in Wire1 is along positive z direction and current in Wire 2 is along negative z direction. The value of current in each wire is taken as I. We will calculate the vector potential and flux density on the x-y plane.

The point at which we wish to find the field intensity is shown in the figure as $P(x, y, 0)$. The vector potential at P due to the current element dz located at $P_1(s/2, 0, z)$ on Wire 1,

$$d\mathbf{A}_1 = \frac{\mu I\, dz}{4\pi R_1'}\mathbf{u}_z \tag{12}$$

The vector potential at P due to the current element dz located at $P_2(-s/2, 0, z)$ on Wire 2,

$$d\mathbf{A}_2 = -\frac{\mu I dz}{4\pi R_2'}\mathbf{u}_z \tag{13}$$

The sum of (12) and (13), denoted by $d\mathbf{A}$, is

$$d\mathbf{A} = \frac{\mu I\mathbf{u}_z}{4\pi}\left[\frac{1}{R_1'} - \frac{1}{R_2'}\right]dz$$

$$= \frac{\mu I\mathbf{u}_z}{4\pi}\left(\frac{1}{\sqrt{R_1^2 + z^2}} - \frac{1}{\sqrt{R_2^2 + z^2}}\right)dz$$

where $R_1 = \sqrt{(x - s/2)^2 + y^2}$ and $R_2 = \sqrt{(x + s/2)^2 + y^2}$ are the distances of the mid-points of wires 1 and 2 from P.

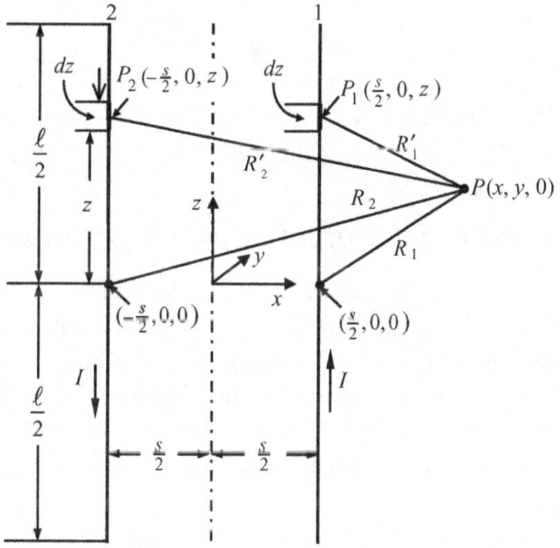

Fig. 4.26 Illustration for calculation of vector fields of a transmission line with two parallel wires

The vector potential due to the wire currents is given by

$$\mathbf{A} = \frac{\mu I\mathbf{u}_z}{4\pi}\int_{-\ell/2}^{\ell/2}\left(\frac{1}{\sqrt{R_1^2 + z^2}} - \frac{1}{\sqrt{R_2^2 + z^2}}\right)dz$$

$$= \frac{\mu I}{2\pi}\left(\ln\frac{R_a + \ell/2}{R_b + \ell/2} - \ln\frac{R_1}{R_2}\right)\mathbf{u}_z$$

In the above expression,

$$R_a = \sqrt{R_1^2 + (\ell/2)^2} \text{ and } R_b = \sqrt{R_2^2 + (\ell/2)^2}.$$

The vector potential has only one component along z direction. It is a function of x and y. The magnetic flux density,

$$\mathbf{B} = \nabla \times \mathbf{A} = \frac{\partial A_z}{\partial y}\mathbf{u}_x - \frac{\partial A_z}{\partial x}\mathbf{u}_y$$

The flux density in x direction,

$$B_x = \frac{\partial A_z}{\partial y} = \frac{2\mu I}{\pi}\left(\frac{y}{R_a(R_a+\ell/2)} - \frac{y}{R_b(R_b+\ell/2)} - \frac{y}{R_1^2} + \frac{y}{R_2^2}\right)$$

The flux density in y direction,

$$B_y = -\frac{\partial A_Z}{\partial x} = -\frac{2\mu I}{\pi}\left(\frac{x-s/2}{R_a(R_a+\ell/2)} - \frac{x+s/2}{R_b(R_b+\ell/2)} - \frac{x-s/2}{R_1^2} + \frac{x+s/2}{R_2^2}\right)$$

If the lengths of the lines are very large compared to the distances of P from their mid-points, then it can be shown that

$$\mathbf{A} \simeq \frac{\mu I}{2\pi}\ln\left(\frac{R_2}{R_1}\right)\mathbf{u}_z, \tag{14}$$

$$\mathbf{B} \simeq \frac{\mu I}{2\pi}\left(\left(\frac{y}{R_2^2} - \frac{y}{R_1^2}\right)\mathbf{u}_x - \left(\frac{x+s/2}{R_2^2} - \frac{x-s/2}{R_1^2}\right)\mathbf{u}_y\right) \tag{15}$$

Expression (15) can be derived more easily by the use of formula for the magnetic field intensity of a long line current given by (4.6-3c). The example is included here to demonstrate the procedure for finding the magnetic flux density from the vector magnetic potential.

Vector Fields of a Circular Current Loop

Let us consider a circular current loop on the x-y plane with its centre at the origin O, as shown in Fig. 4.27. The radius of the loop is a and the current I is in anticlockwise direction. We will obtain the vector magnetic potential at a point far away from the loop, at first in rectangular coordinates and then transform it into spherical coordinates. The flux density is obtained in spherical coordinates. Spherical coordinate system is not used at the outset because the direction of the current element in spherical coordinates is not constant.

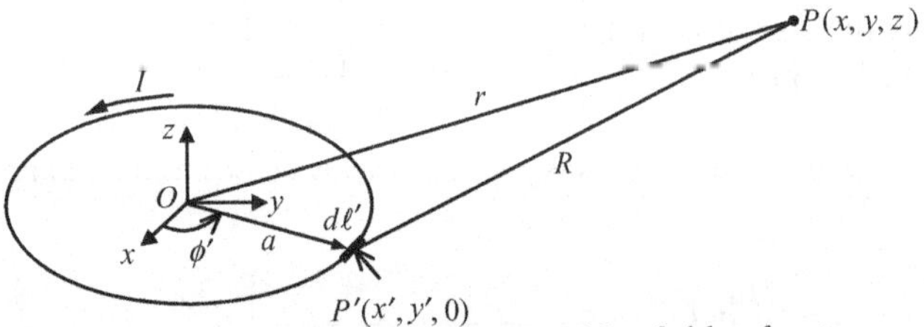

Fig. 4.27 Illustration for calculation of far-fields of a circular current loop

Let us take a differential length $d\ell'$ at $P'(x', y', 0)$ on the loop. The radial line OP' is at an angle ϕ' with the x axis. A current element at P' is defined by

$$I\,d\ell' = I(\mathbf{u}_x\,dx' + \mathbf{u}_y\,dy')$$

The vector potential at $P(\ ,y,\ )$ due to the current element,

$$d\mathbf{A} = \frac{\mu I \quad {}_x\,dx \qquad {}_y\,dy')}{4\pi R}$$

where $$\frac{1}{R} = \left[(x-x')^2 + (y-y')^2 + z^2\right]^{-\frac{1}{2}} \tag{16}$$

The components of the vector potential $d\mathbf{A}$ are given by

$$dA_x = \frac{\mu I dx'}{4\pi R}, \tag{17}$$

$$dA_y = \frac{\mu I dy'}{4\pi R} \tag{18}$$

In order to obtain A_x and A_y we have to integrate $1/R$ defined by (16) around the current loop. This is achieved by at first expressing the primed variables in terms of ϕ' and then expanding the result using the binomial theorem. Term-by-term integration of the expansion will yield the desired result. The number of terms to be integrated depends on the accuracy needed.

From the geometry of Fig. 4.27, we get

$$x' = a\cos\phi', \tag{19}$$

$$y' = a\sin\phi' \tag{20}$$

Thus $$dx' = -a\sin\phi'\, d\phi',$$

$$dy' = a\cos\phi'\, d\phi'$$

Substituting relations (19) and (20) in (16), we get

$$\frac{1}{R} = \frac{1}{r}\left(1 + \frac{a^2 - 2ax\cos\phi' - 2ay\sin\phi'}{r^2}\right)^{-\frac{1}{2}} \tag{21}$$

where $r = \sqrt{x^2 + y^2 + z^2}$ is the radial distance of P from centre of the loop. It is also the r variable of spherical coordinates. As $r \gg a$, the second term inside the bracket of (21) would be small compared to unity. Therefore, we may neglect all but the first two terms of the binomial expansion. Accordingly,

$$\frac{1}{R} \approx \frac{1}{r}\left(1 - \frac{a^2 - 2ax\cos\phi' - 2ay\sin\phi'}{2r^2}\right) \tag{22}$$

Substituting (22) in (17) and (18) and then integrating around the loop we will obtain A_x and A_y.

$$A_x = -\frac{\mu I a}{4\pi r}\int_0^{2\pi}\left(1 - \frac{a^2 - 2a\cos\phi' - 2ay\sin\phi'}{2r^2}\right)\sin\phi'\, d\phi'$$

$$= -\frac{\mu I S}{4\pi r^3}\, y$$

$$A_y = \frac{\mu I a}{4\pi r}\int_0^{2\pi}\left(1 - \frac{a^2 - 2a\cos\phi' - 2ay\sin\phi'}{2r^2}\right)\cos\phi'\, d\varphi'$$

$$= \frac{\mu I S}{4\pi r^3}\, x$$

where $S = \pi a^2$ is area of the loop.

Thus, the magnetic vector potential in rectangular coordinates is given by

$$\mathbf{A}(x, y, z) = \frac{\mu IS}{4\pi r^3}(-y\mathbf{u}_x + x\mathbf{u}_y)$$

The vector potential can be expressed in a compact form by transforming $\mathbf{A}(x, y, z)$ into spherical coordinates as follows. Let A_r, A_θ and A_ϕ be the components of the vector potential in spherical coordinates. Then, by applying the technique of vector transformation it can be shown that

$$A_r = \mathbf{u}_r \cdot \mathbf{A}(x, y, z) = 0$$

$$A_\theta = \mathbf{u}_\theta \cdot \mathbf{A}(x, y, z) = 0$$

$$A_\phi = \mathbf{u}_\phi \cdot \mathbf{A}(x, y, z) = \frac{\mu IS \sin\theta}{4\pi r^2}$$

Hence the vector potential in spherical coordinates,

$$\mathbf{A}(r, \theta) = \frac{\mu IS \sin\theta}{4\pi r^2}\mathbf{u}_\phi \tag{23}$$

The vector potential has only one component in ϕ direction and it is not a function of ϕ. The flux density is obtained by evaluating the curl of (23) in spherical coordinates. Thus, using (4.7-16) we get

$$\mathbf{B} = \nabla \times \mathbf{A} = \frac{\mathbf{u}_r}{r^2 \sin\theta}\left[\frac{\partial}{\partial\theta}(r\sin\theta A_\phi)\right] - \frac{\mathbf{u}_\theta}{r\sin\theta}\left[\frac{\partial}{\partial r}(r\sin\theta A_\phi)\right]$$

$$= \frac{\mu IS}{4\pi r^3}(2\cos\theta\,\mathbf{u}_r + \sin\theta\,\mathbf{u}_\theta) \tag{24}$$

Equation (24) shows that

> the flux density has components along r and θ directions and it is independent of ϕ. Its magnitude decreases rapidly with distance as it is inversely proportional to r^3.

Since the expression for magnetic flux density is similar to that of the electric field intensity of an electric dipole given by (2.18-12),

> a small current loop is called a *magnetic dipole*. The product of current I and area S of a small, planar current loop is its *magnetic dipole moment*. The direction of the dipole moment is along the normal to the plane of the loop. For the direction of current shown in Fig. 4.27, the direction of the dipole moment is along the z $(\theta = 0)$ axis.

The vector form of magnetic dipole moment, which is generally denoted by the symbol **m**, is given by

$$\mathbf{m} = I\mathbf{S} \tag{25}$$

EXAMPLE 4.29

A current element $Id\ell$ is located at the origin. Find the vector potential $d\mathbf{A}$ at a point $P(x, y, z)$. Obtain the magnetic field intensity at P from the potential function and verify that the result agrees with that obtained by the use of the Biot–Savart law.

If $Id\ell = 0.2\mathbf{u}_z$ A-m, find the vector magnetic potential at $x = 3$ m, $y = 5$ m, $z = -2$ m in free space.

The current element in rectangular coordinates, $Id\ell = I(dx\,\mathbf{u}_x + dy\,\mathbf{u}_y + dz\,\mathbf{u}_z)$

Distance of P from the origin, $R = \sqrt{x^2 + y^2 + z^2}$

The vector magnetic potential, $d\mathbf{A} = \dfrac{\mu I}{4\pi R} d\ell = \dfrac{\mu I(dx\,\mathbf{u}_x + dy\,\mathbf{u}_y + dz\,\mathbf{u}_z)}{4\pi\sqrt{x^2 + y^2 + z^2}}$

Magnetic field intensity, $d\mathbf{H} = \dfrac{1}{\mu}(\nabla \times d\mathbf{A}) = \nabla \times \left[\dfrac{I(dx\,\mathbf{u}_x + dy\,\mathbf{u}_y + dz\,\mathbf{u}_z)}{4\pi\sqrt{x^2 + y^2 + z^2}}\right]$

$$= \frac{I}{4\pi R^3}[(z\,dy - y\,dz)\mathbf{u}_x + (x\,dz - z\,dx)\mathbf{u}_y + (y\,dx - x\,dy)\mathbf{u}_z)]$$

This result is same as (4.2-5) that was derived using Biot–Savart law.

$$R = \sqrt{3^2 + 5^2 + 2^2} = 6.164 \text{ m}$$

$$d\mathbf{A} = \frac{(4\pi \times 10^{-7})(0.2\mathbf{u}_z)}{4\pi(6.164)} - 3.245 \times 10^{-9} \text{ T-m}$$

EXAMPLE 4.30

A small current loop is centred at the origin with its axis along the z axis of rectangular coordinates. The radius of the loop is 5 cm and it carries a current of 16 A. Find the magnetic field intensity in spherical coordinates at (25 m, 30°, 0).

$$\mathbf{H} \quad \frac{IS}{4\pi r^3}(2\cos\theta\mathbf{u}_r + \sin\theta\mathbf{u}_\theta)$$

Substituting $I = 16$ A; $S = \pi(0.05)^2 = 7.854 \times 10^{-3}$ m^3; $r = 25$ m; $\theta = 30°$

$$\mathbf{H} \quad \frac{(16)(7.854 \times 10^{-3})}{4\pi(25)^3}(2\cos 30°\,\mathbf{u}_r + \sin 30°\,\mathbf{u}_\theta)$$

$$= (6.4 \times 10^{-7})(1.732\mathbf{u}_r + 0.5\mathbf{u}_\theta) \text{ A/m}$$

4.12 The Scalar Magnetic Potential

In the preceding section we expressed magnetic flux density as the curl of a vector function by utilizing the solenoidal property of magnetic field. In this section we will derive the relation between magnetic field intensity and a scalar function using the differential form of Ampere's circuital law.

The differential form of Ampere's circuital law is

$$\nabla \times \mathbf{H} = \mathbf{J}$$

In a *current-free* region **J** = 0. Therefore,

$$\nabla \times \mathbf{H} = 0 \tag{1}$$

Referring to the vector identity (1) of Section (1.9) we find that the curl of gradient of a scalar function is always zero. Hence, **H** of (1) can be expressed as the gradient of a scalar function. Denoting the scalar function by V_m, we have

$$\mathbf{H} = -\nabla V_m \tag{2}$$

This is the relation between magnetic field intensity and the scalar function V_m. The scalar is called *scalar magnetic potential*. Its unit is ampere. A minus sign is used in (2) so that the relation is analogous to the relation $\mathbf{E} = -\nabla V$ of static electric field.

The potential difference between two points P_1 and P_2 in a static electric field is

$$V_2 - V_1 = -\int_{P_1}^{P_2} \mathbf{E} \cdot d\ell$$

Similarly, the scalar magnetic potential difference between two points in a current-free region is defined by

$$V_{m2} - V_{m1} = -\int_{P_1}^{P_2} \mathbf{H} \cdot d\ell$$

The scalar magnetic potential, however, has no physical meaning.

To know a little more about the scalar magnetic potential, let us calculate the potential difference between two points outside a line current of infinite length. The line, carrying a current I, is taken along the z axis of cylindrical coordinates. The magnetic field intensity at any point $P(r, \phi, z)$ on an arbitrary path between $P_1(r_1, \phi_1, z_1)$ and $P_2(r_2, \phi_2, z_2)$, shown in Fig. 4.28, is

$$\mathbf{H} = \frac{I}{2\pi r}\mathbf{u}_\phi$$

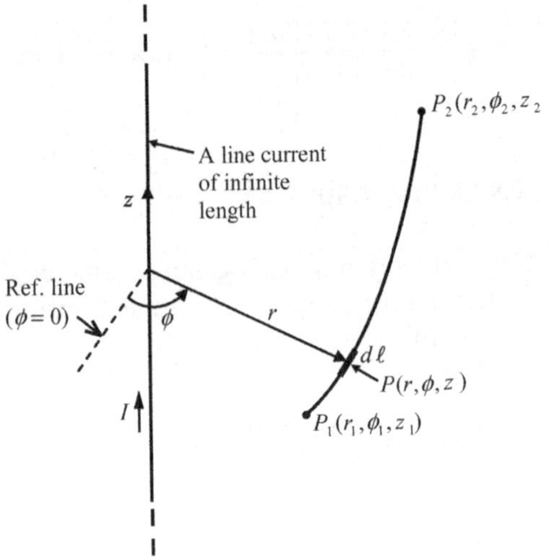

Fig. 4.28 Showing an arbitrary path P_1P_2 in the field of a line current of infinite length

A differential length at P is given by

$$d\ell = \mathbf{u_r}\, dr + \mathbf{u}_\phi\, r d\phi + \mathbf{u}_z\, dz$$

Hence, $$\mathbf{H} \cdot d\ell = \frac{I}{2\pi} d\phi$$

The potential difference between points P_1 and P_2 is

$$V_{m2} - V_{m1} = -\int_{\phi_1}^{\phi_2} \frac{Id\phi}{2} = \frac{I(\phi_1 - \phi_2)}{2\pi}$$

Thus, the potential difference between two points depends only on the angular displacements of the points. It does not depend on shape of the path of integration. Further, radial lines (ϕ = constant) are the equipotential lines. Hence equipotential lines and **H**-lines, which are in ϕ direction, intersect at right angles.

Now let us assume that the scalar potential at P_1 has some value V_{m1}. Starting from P_1 if we integrate **H** around the current in counter-clockwise direction and end at the same point, the potential at P_1 becomes

$$V_{m1} - \int_{\phi}^{\phi+2\pi} \frac{I}{2\pi} d\phi = V_{m1} - I$$

Thus, the potential at P_1 will decrease by I after each integration of **H** around the current. And if the integration **H** is carried out in clockwise direction, the potential at P_1 will increase by I after each integration. This shows that the scalar magnetic potential at P_1 is not single valued. But the electrostatic potential, which is also a scalar function, is single valued because the line integral of **E** around a closed path is equal to zero even if the path encloses electric charges.

Next, taking divergence of both sides of (2), after reversing the sides we get

$$\nabla \cdot (\nabla V_m) = -\nabla \cdot \mathbf{H}$$

$$\Rightarrow \quad \nabla^2 V_m = -\nabla \cdot \left(\frac{\mathbf{B}}{\mu}\right)$$

In a homogeneous and isotropic medium μ is constant. Hence,

$\nabla^2 V_m = -\frac{1}{\mu}(\nabla \cdot \mathbf{B}) = 0$, as divergence of **B** is zero.

This is similar to the Laplace's equation for electrostatic potential.The scalar magnetic potential is useful for plotting of magnetic fields.

4.13 Magnetic Properties of Materials

The electromagnetic force between two current elements depends, apart from other factors, on the magnetic properties of the medium in which they are located. One of the magnetic properties of a material is its magnetic susceptibility. Different materials have different susceptibility. This is because different materials interact with an external magnetic field differently at the atomic level. Another property of engineering importance is the permeability of a material that has been defined in Section 4.3. In this section we will obtain the permeability of a material in terms of its susceptibility.

The Atomic Magnetic Moment

According to the electronic theory of matter, each atom consists of a positively charged nucleus and negatively charged particles (electrons) that revolve around the nucleus.

It can be shown that an electron moving around a closed path is equivalent to a current loop. Hence an orbital electron produces a magnetic field. Magnetic field is also produced by electron spin and nuclear spin.

The field due to nuclear spin is generally neglected because it is very small. An approximate numerical value of the magnetic moment of a revolving electron can be obtained by assuming it to be orbiting around the nucleus in a circular path, as illustrated in Fig. 4.29. Let a, ω and e be the radius of the orbit, angular velocity and magnitude of electronic charge respectively. Then

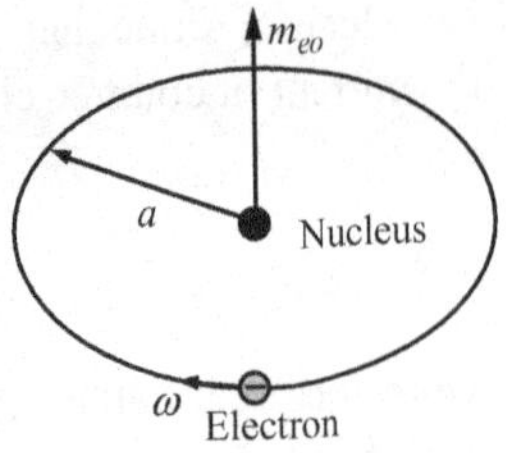

Fig. 4.29 Simplified model of an atom

Charge flow in one revolution = e

Duration of one revolution = $\frac{2\pi}{\omega}$

Magnitude of current flow, $I = \frac{e\omega}{2\pi}$ (1)

The direction of current flow is opposite to the direction of motion of the electron. Thus, an orbital electron is equivalent to a very small current loop having the same radius a. The magnetic moment of a current loop is equal to the product of current magnitude and loop area, as defined in Section 4.11. Thus, the magnitude of magnetic moment of an orbital electron

$$m_{eo} = \pi I a^2 \quad (2)$$

Eliminating I from (1) and (2), we have

$$m_{eo} = \frac{e\omega a^2}{2} \quad (3)$$

The angular momentum of an electron is equal to $m\omega a^2$, where m is mass of the electron. It is also equal to $nh/2\pi$, where h is Planck's constant and n is a positive integer. Thus for $n = 1$,

$$m\omega a^2 = \frac{h}{2\pi}$$

Or

$$\omega a^2 = \frac{h}{2\pi m} \quad (4)$$

By substituting (4) in (3), we will get

$$m_{eo} = \frac{eh}{4\pi m}$$

As $e = 1.602 \times 10^{-19}$ C, $h = 6.626\, e^{-34}$ J-s, $m = 9.109 \times 10^{-31}$ kg, the magnitude of magnetic moment of an orbital electron,

$$m_{eo} \simeq 9.3 \times 10^{-24} \text{ A-m}^2$$

The *magnetic moment*, which is a vector quantity, has its direction normal to the plane of orbit of the electron. In Fig. 4.29 the direction of rotation of the electron is shown as clockwise. Therefore, the direction of current is anticlockwise. Thus, the direction of the magnetic moment is vertically upwards as indicated in the figure.

The electrons while orbiting around the nucleus also spin about their axes. A second component of magnetic moment arises due to the electron spin. In an atom having several shells containing electrons, only electrons of the shells that are not completely full contribute to this magnetic moment. The approximate value of the spin magnetic-moment is found to be nearly equal to the orbital magnetic moment. Thus, denoting the spin magnetic moment by m_{es}, we have

$$m_{es} \approx 9.3 \times 10^{-24} \text{ A-m}^2$$

The direction of the spin magnetic moment may be the same or opposite to the externally applied field.

A third component of magnetic moment arises due to nuclear spin. Its value is found to be very small (less than 0.2 percent of the magnetic moment due to electron spin).

As its contribution to the magnetic properties of a material is very small, it will not be considered.

Thus, each atom of a material has mainly two components of magnetic moment; one associated with electron orbit and the other with electron spin. The vector sum of the two magnetic moments is the atomic magnetic moment. If the orbital and spin moments cancel out, we say that the atom has *no permanent magnetic moment.* If the two moments do not cancel, we say that the atom *has a permanent magnetic moment.* These definitions apply when a material is not under the influence of an external magnetic field.

The Magnetisation Vector M

If a magnetic material is placed in an externally applied magnetic field, its atomic dipoles align themselves in the direction of the external field. Under this condition the material is said to be in a magnetised state. The magnetic field produced by the dipoles is the induced field. The strength of the induced field is a function of the applied field.

The induced magnetic field of a large sample of a material is described by a quantity called *magnetisation* (or magnetisation density), which is defined as the atomic magnetic-moment per unit volume. It is a vector quantity and is generally denoted by **M**.

In order to obtain a simple mathematical definition of **M**, let us consider a uniformly magnetised rod having a large number of atomic dipoles uniformly distributed over its volume. Let us assume that the rod has n_v dipoles per unit volume and each dipole has a magnetic moment equal to $\mathbf{m}_e$. Then, the number of dipoles contained in a very small volume Δv is equal to $n_v \Delta v$. Assuming all dipoles have the same direction, the total moment of the dipoles within Δv is

$$\Delta \mathbf{m}_e = \mathbf{m}_e n_v \, \Delta v \tag{5}$$

The magnetisation vector, which is generally denoted by **M,** is obtained by dividing (5) by Δv, and then taking the limit. Thus,

$$\mathbf{M} = \lim_{\Delta v \to 0} \frac{\Delta \mathbf{m}_e}{\Delta v} = \mathbf{m}_e n_v \tag{6}$$

The SI unit of magnetisation is ampere per metre, which is same as that of the magnetic field intensity. It is termed as *induced* magnetic field intensity.

Magnetisation of a Uniformly Magnetised Rod: The Equivalent Solenoid

Consider that a rod of circular section of length ℓ and sectional area S is placed with its axis parallel to a steady and uniform external magnetic field, as illustrated in Fig. 4.30(*a*). The rod will get magnetised. We assume that the rod is uniformly magnetised and magnetic moments of all the dipoles are in the axial direction. Let there be n_v dipoles per unit volume of the rod, and the magnetic moment of each dipole be m_e. Each dipole moment may be considered as equivalent to an infinitesimal current loop as shown in the figure. If the current and area of each loop are I_e and S_e respectively, then its magnetic moment is equal to $I_e S_e$. Thus,

$$m_e = I_e S_e \tag{7}$$

As per (6), the value of magnetisation is given by the product of the magnetic moment of each dipole m_e given by (7) and the number of dipoles per unit volume n_v. Accordingly,

$$M \quad n_v I_e S_e \tag{8}$$

Let us now turn our attention to Fig. 4.30(b) where the distribution of dipoles over a cross section of the rod is illustrated. If there are N_S dipoles on the surface, then the magnitude of magnetic moments of all the dipoles,

$$m_S = N_S I_e S_e$$

Assuming continuous distribution of dipoles over the surface, the product of N_S and S_e is equal to the sectional area of the rod S. Hence,

$$m_S = S I_e$$

The above relation shows that the magnetic moment of all the dipoles on the surface is equivalent to a single current loop of area S and current I_e. This is illustrated in Fig. 4.30(c). If we assume that the magnetised rod consists of N_ℓ such loops, the induced magnetic moment of all the dipoles within the volume of the rod is equal to m_S multiplied by N_ℓ. Thus the magnitude of magnetisation,

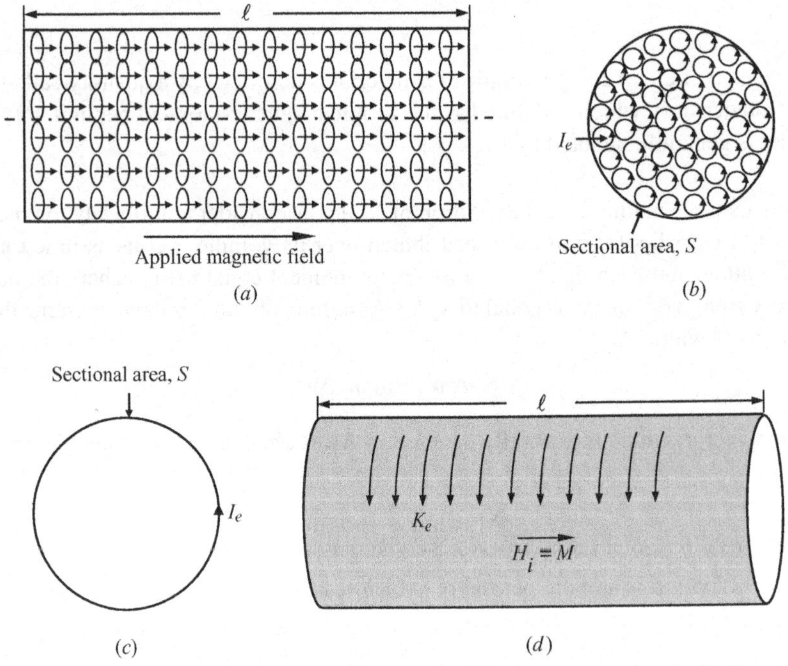

Fig. 4.30 Uniformly magnetised rod and bound current sheet: (a) Magnetised rod, (b) Distribution of dipoles over a cross section of the rod, (c) Equivalent current loop of (b), and (d) Equivalent air-core solenoid of (a).

$$M = \frac{N_\ell S I_e}{\ell S} = \frac{N_\ell I_e}{\ell} = n_\ell I_e \tag{9}$$

where n_ℓ is the number of equivalent current loops per unit length of the rod.

Comparison between (8) and (9) yields,

$$n_\ell \quad S_e n_v$$

Since spacing between the equivalent current loops is of atomic dimensions, we may assume that the loops are continuously distributed in the form of a cylindrical sheet of *bound* current I_e flowing in the circumferential direction. The equivalent current per unit axial length of the rod,

$$K_e = n_\ell I_e \text{ A/m} \tag{10}$$

As the right-hand sides of (9) and (10) are same, we have

$$M = K_e$$

Thus, we see that

a uniformly magnetised rod may be replaced by an air-core solenoid having the same dimensions as that of the magnetised rod, with an equivalent current per unit axial length equal to its magnetisation.

The equivalent solenoid is illustrated in Fig. 4.30(*d*). Since the field intensity of the equivalent solenoid is equal to K_e, the induced field intensity of the magnetised rod,

$$H_i = M \tag{11}$$

Magnetic Susceptibility and Relative Permeability

Let us consider a long air-core solenoid having a uniformly and closely spaced coil of N turns of fine wire, as shown in Fig. 4.31(*a*). If excited by a current I, the magnetic field intensity within the coil would be nearly uniform with its direction along the axis of the solenoid. The field intensity,

$$H = \frac{NI}{\ell} = nI$$

where ℓ = length of the solenoid and n = number of turns per unit length.

The flux density inside the solenoid,

$$\mathbf{B}_0 = \mu_0 \mathbf{H}$$

If the core of the solenoid is made up of some material, as illustrated in Fig.4.31(*b*), the field inside the core will consist of two components; one due to the solenoid (the applied field) and the other due to the magnetised core (the induced field).

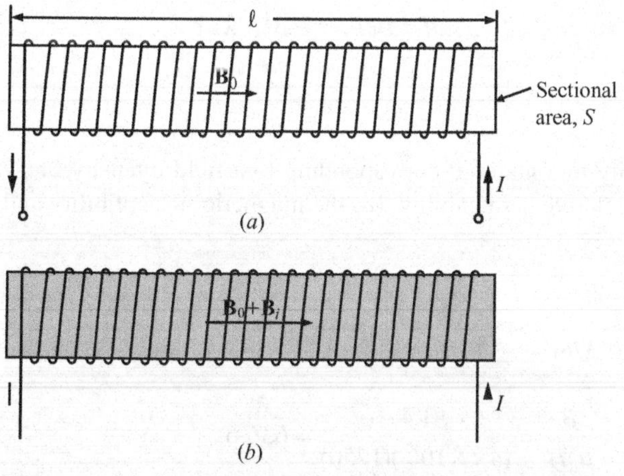

Fig. 4.31 (*a*) An air-core solenoid and (*b*) A solenoid with ferromagnetic core.

Since the induced field intensity is equal to the magnetisation M as per (11), the induced flux density,

$$\mathbf{B}_i = \mu_0 \mathbf{M}$$

The total flux density inside the core is equal to the sum of $\mathbf{B}_0$ and $\mathbf{B}_i$. Denoting the total flux density by $\mathbf{B}$, we have

$$\mathbf{B} = \mathbf{B}_0 + \mathbf{B}_i = \mu_0(\mathbf{H} + \mathbf{M}) \tag{12}$$

For isotropic and linear materials the direction of **M** is same as that of **H**, and the magnitude of **M** is proportional to that of **H**. Therefore, the ratio of **M** to **H** is a scalar quantity. The constant of proportionality is called *magnetic susceptibility*. It is a dimensionless quantity and is generally denoted by χ_m. Thus,

$$\mathbf{M} = \chi_m \mathbf{H} \tag{13}$$

Substituting (13) into (12), we get

$$\mathbf{B} = \mu_0(1 + \chi_m)\mathbf{H} \tag{14}$$

In Section 4.3 we have defined magnetic flux density in a medium as the product of magnetic field intensity and a quantity called permeability. The permeability itself was defined as free space permeability multiplied by the relative permeability of the medium. The relationship between **B** and **H** given by (4.3-3) and reproduced below, is

$$\mathbf{B} = \mu_0 \mu_r \mathbf{H} \tag{15}$$

From (14) and (15), we get

$$\mu_r = 1 + \chi_m \tag{16}$$

Thus, the relative permeability of a material is equal to one plus the magnetic susceptibility. Relation (16) is analogous to the relation between relative permittivity and electric susceptibility of a dielectric material.

The permeability of a material,

$$\mu = \mu_0 \mu_r = \mu_0(1 + \chi_m)$$

EXAMPLE 4.31

The magnetic flux density in a material corresponding to a field intensity of 1750.0 A/m is found to be 1.4 T. Calculate (a) the relative permeability, (b) the magnetic susceptibility and (c) the magnetisation of the material.

Flux density, $B = 1.4$ T

Field intensity, $H = 1750.0$ A/m

Relative permeability, $\mu_r = \dfrac{B}{\mu_0 H} = \dfrac{1.4}{(4\pi \times 10^{-7})(1750)} = 636.6$

Magnetic susceptibility, $\chi_m = \mu_r - 1 = 635.6$

Magnetisation, $M = \chi_m H = (635.6)(1750.0) = 1112.3 \times 10^3$ A/m

EXAMPLE 4.32

A magnetic material has 7.8×10^{19} atoms per mm^3. The magnetic dipole of each atom is 4.2×10^{-27} A-m^2. For a magnetic field intensity of 15 A/m, find the magnetic susceptibility and permeability of the material.

Magnetisation, $M = m_e n_v$;

$m_e = 4.2 \times 10^{-27}$ A-m^2; $n_v = (7.8 \times 10^{19})(10^9) = 7.8 \times 10^{28}$ atoms/m^3

Substituting the values, $M = (4.2 \times 10^{-27})(7.8 \times 10^{28}) = 327.6$ A/m

Magnetic susceptibility, $\chi_m = \dfrac{M}{H} = \dfrac{327.6}{15} = 21.84$

Relative permeability, $\mu_r = 1 + \chi_m = 22.84$

Permeability, $\mu = \mu_0 \mu_r = (4\pi \times 10^{-7})(22.84) = 287 \times 10^{-7}$ H/m

4.14 Classification of Magnetic Materials

Magnetic materials are classified according to the nature of their atomic magnetic moments. Materials in which the magnetic moments due to electron spin are equal and opposite to the magnetic moments due to orbital electrons, the resultant atomic dipole moment is nearly zero. Thus, these materials have no permanent magnetic moments. These materials are classified as diamagnetic materials. Some other materials have permanent but weak magnetisation. This is because the two components of magnetic moments do not cancel out. These materials are called *paramagnetic* materials. Materials in which the atoms have strong permanent magnetic moment are classified as ferromagnetic, anti-ferromagnetic and ferrimagnetic materials.

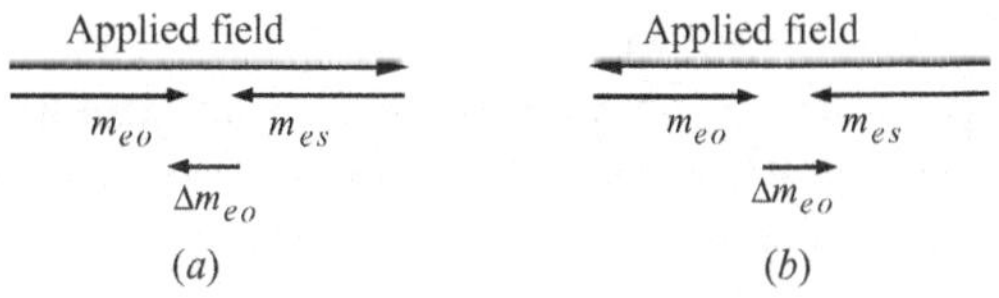

Fig. 4.32 Showing *change* in the direction of orbital magnetic moment of a diamagnetic material: (*a*) Applied field is in the same direction as the orbital moment and (*b*) Applied field is opposite to the orbital moment.

If a diamagnetic material is placed in an external time-varying magnetic field, the magnitudes of orbital magnetic moments change slightly due to change in the orbital velocity of the electrons. The change in the moment, denoted by Δm_{eo} in Fig. 4.32, is in a direction opposite to the applied field. The net result is a weak induced field that is opposite to the applied field, irrespective of whether the applied field is along or against the orbital magnetic moment. Hence,

the magnetic susceptibility of a diamagnetic material is negative and it is of a very small value. Consequently, the relative permeability is slightly less than *one*, and flux density inside the material is a little less than that due to the applied field.

Bismuth, hydrogen, copper, gold, silicon, germanium, and graphite are some of the diamagnetic materials. The magnetic susceptibility of copper is -0.95×10^{-5} at room temperature. If a diamagnetic material is brought near a pole of a bar magnet, it will be acted upon by a repulsive force.

The atoms of paramagnetic materials have weak permanent magnetic moments as the orbital and spin moments do not completely cancel out.

However, the magnetic field outside a large piece of material is found to be zero in the absence of an applied field. This is because of random orientation of the atomic dipoles within the material. If a paramagnetic material is placed in an external field, its dipoles align along the external field because of electromagnetic torque. The induced field is very weak. Hence,

the magnetic susceptibility of a paramagnetic material is very low, but it is positive.

It is of the order of 10^{-3}. Its relative permeability is only slightly more than *one*, and the flux density inside the material is a little higher than that due to the external field. Aluminium, potassium, tungsten, oxygen are some of the paramagnetic materials. If a paramagnetic material is brought near a pole of a bar magnet, it will be acted upon by an attractive force.

The atoms of ferromagnetic materials have strong permanent magnetic moments. This is mainly because of noncancellation of magnetic moments due to electron spin. The dipoles are distributed in groups throughout the material. Each group occupies a very small region. The regions are called *domains*.

The dipoles in a domain are aligned in the same direction due to inter-atomic forces, as shown in Fig. 4.33(*a*). In a virgin material the domains are randomly oriented so that the net magnetisation of macroscopic sample of the material is zero. However, the flux density inside the material becomes very high when it is placed in an external magnetic field. The relative permeability of these materials is a nonlinear function of the applied field intensity. Materials like iron, cobalt and nickel have very strong ferromagnetic properties at room temperature. The alloys of these materials also have ferromagnetic characteristics. These materials are used in various electromagnetic apparatuses such as in rotating machines, transformers, chokes, electromagnets, permanent magnets, etc.

In anti-ferromagnetic materials, the magnetic moments of neighbouring atoms are in anti-parallel, as represented in Fig. 4.33(*b*). As their magnitudes are the same, the net dipole moment of a sample is zero. These materials have very small induced magnetic effect in the presence of an external field. Some examples of anti-ferromagnetic material are cobalt chloride and nickel oxide.

In ferrimagnetic materials the neighbouring atoms are also in anti-parallel but their magnitude are not equal, as shown in Fig. 4.33(*c*). In the presence of an external magnetic field, these materials develop strong induced

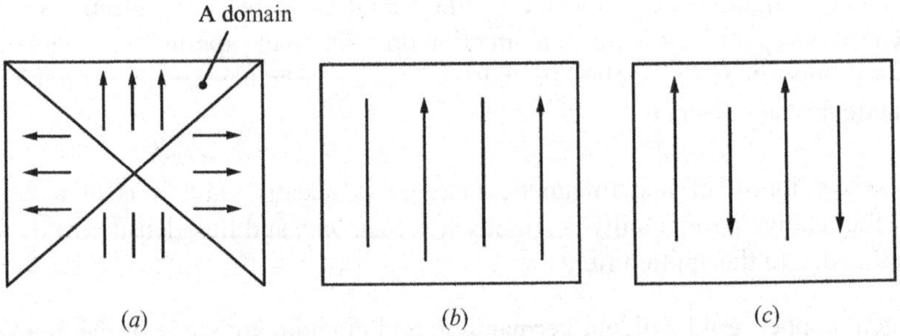

Fig. 4.33 Schematic representation of dipoles of (*a*) ferromagnetic, (*b*) anti-ferromagnetic, and (*c*) ferrimagnetic materials.

fields. Ferrites are an important group of ferrimagnetic materials. These materials are of low conductivity. Hence loss due to eddy currents, induced by time-varying external magnetic fields, is low. These are used as cores of high frequency transformers, inductors and in microwave devices.

4.15 Properties of Ferromagnetic Materials

As has been described earlier, materials which have strong atomic magnetic moments are classified as ferromagnetic materials. Further, atomic dipoles in a domain are aligned in the same direction due to strong coupling forces between adjacent dipoles. Adjacent domains are separated by very thin (about one hundred atoms) walls, known as transition walls. The size and shape of a domain depend on the size, shape and magnetic history of the material. The volume of a domain is very small; the average value is of the order of 10^{-10} m^3. A single crystallite of a ferromagnetic substance has several domains. Each domain contains a large number ($10^{17} - 10^{[illegible]}$) of atomic dipoles. In a pure ferromagnetic material that has not been magnetised earlier (virgin state), each crystallite is found to be magnetically neutral. This is because, for each domain having magnetisation in one direction there exists another domain having equal and opposite magnetisation. Further, though a single crystallite is magnetically anisotropic, a large sample of a pure magnetic material may be treated as isotropic (similar property in all directions) because of random orientation of axes of the crystallites.

On application of an external magnetic field, the domain walls of a virgin ferromagnetic material shift. This causes change in the size of the domains. The sizes of domains in which the dipoles get fully or partially oriented along the applied field increase. Consequently, the sizes of other domains decrease, and flux density B inside the material increases. The process of magnetisation is reversible for small values of applied field H. That is, if the external field is removed the material reverts to the original magnetic state. With an increase of H more dipoles align with the applied field and hence flux density inside the material increases. For moderate values of H the rate of rise of B with H is high. With further increase of H the rate of rise of B decreases and ultimately a stage is reached at which almost all the dipoles are aligned along the external field. Any further increase in H does not result in appreciable increase of B. At this stage the material is said to have reached a state of magnetic saturation. Thus, the relationship between B and H is not linear. Four stages of magnetisation are illustrated in Fig. 4.34.

The graphical relation between the applied magnetic field intensity H and flux density B inside a ferromagnetic material of uniform section is called *magnetisation characteristic* (also called *B-H* characteristic or *B-H* curve).

The shape of the *B-H* characteristic for different magnetic materials is different. One such characteristic is shown in Fig. 4.35. The characteristic is not linear (B is not proportional to H) because of magnetic saturation. Hence the permeability μ, which is defined as the ratio of B to H, and the relative permeability, which is proportional to B/H, are functions of H. The relative permeability, is given by

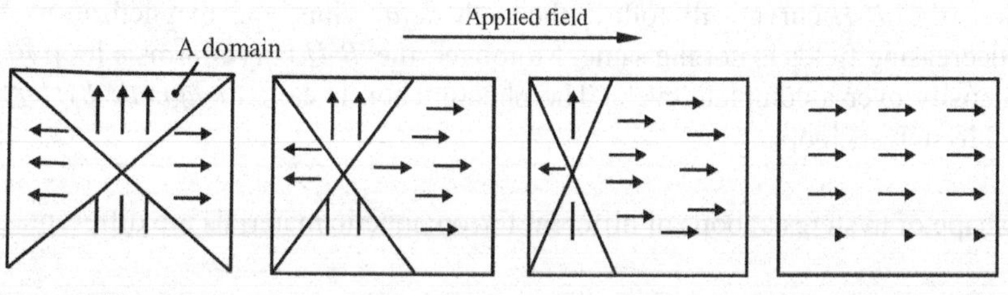

(*a*) (*b*) (c) (*d*)

Fig. 4.34 Different stages of magnetisation of a ferromagnetic substance: (*a*) No applied field, (*b*) Weak applied field, (c) Moderate applied field, and (*d*) Strong applied field.

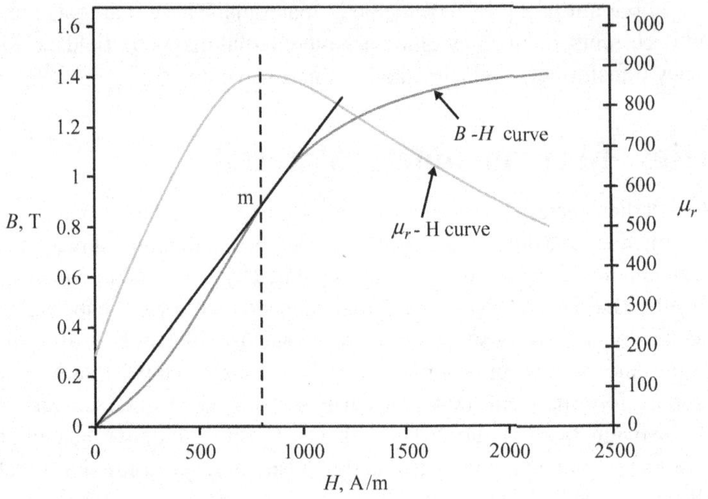

Fig. 4.35 B-H curve and μ_r-H curves of cast steel.

$$\mu_r = \frac{10^7}{4\pi}\left(\frac{B}{H}\right)$$

The variation of μ_r with H is also shown in the figure. It may be observed that μ_r is not zero at $H = 0$. This is because the slope of the curve at this point is not zero. The maximum value of relative permeability occurs at point m on the curve as B/H has maximum value at this point. The point is located by drawing a tangent to the curve from the origin, as indicated in the figure. It may however be noted that the permeability is equal to the ratio B/H. It is not given by the slope of the B-H curve at a given value of H.

The slope dB/dH at any point on the curve gives the *incremental* permeability.

The magnetisation characteristic can be determined experimentally.

Another important property of a ferromagnetic substance is the *magnetic hysteresis*.

This phenomenon is explained with reference to Fig. 4.36 where the curve *oa* represents the magnetisation characteristic of an iron sample having no initial magnetisation. That is both the flux density B and the applied field intensity H are zero initially. The applied field intensity at a is taken as H_m. Starting from H_m if the applied field is reduced to zero, then the variation of B with H instead of retracing the initial magnetisation characteristic will follow the curve *ab*. If the applied field is now reversed and then increased to H_m, the B H curve will follow the path *bcd*. By reducing the reverse field to zero and then increasing it to H_m in the forward direction, the B-H curve will follow the path *defa*. Thus, the magnetisation characteristic for increasing and decreasing fields is not the same. Moreover, the B-H curve forms a loop for variation of the applied field intensity over a complete cycle. The phenomenon is called *magnetic hysteresis* and the loop *abcdefa* is called hysteresis loop.

The size and shape of hysteresis loops of different ferromagnetic materials are different.

We find from the hysteresis loop that the flux density is not zero at $H = 0$. The flux density at $H = 0$, shown as B_r in the figure, is called *residual* flux density. The value of B_r indicates the retentive property of a magnetic material. The hysteresis loop also shows that in order to reduce the flux density to zero it is necessary

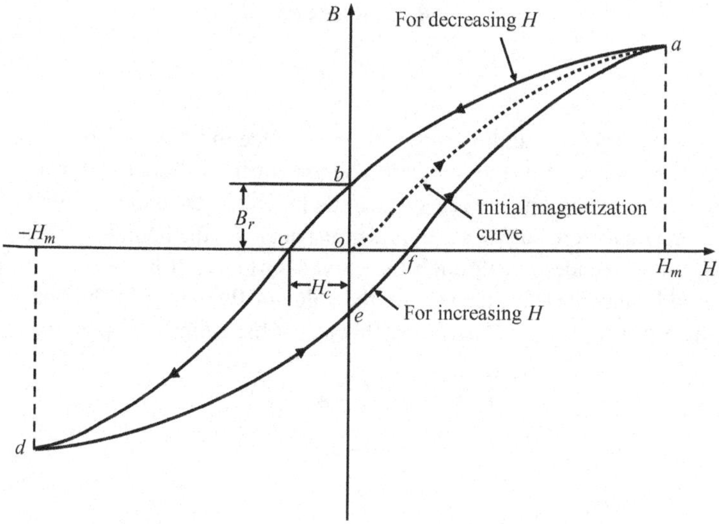

Fig. 4.36 Initial magnetisation characteristic and hysteresis loop of a ferromagnetic material

to apply a field in the reverse direction. The reverse field intensity H_c that is needed to make the flux density zero is called *coercive* force. Materials with high residual flux density are used for making permanent magnets so that they retain a high value of magnetic field. Permanent magnets should also have high coercive force so that they cannot be demagnetised easily.

4.16 Boundary Conditions

In a homogeneous medium, the flux density is a continuous function of space variables. However, at the boundary between two regions of different permeabilities the field is not continuous. The magnitude and direction of flux density on one side of the boundary are related to the corresponding quantities at the other side through their permeabilities. These relations are obtained from the relationships between normal components of flux densities and between tangential components of field intensities on two sides of the interface. The relation between normal components of flux densities is found from the consideration that the net flux out of a closed surface is zero. The relation between tangential components of field intensities is derived using the magnetic circuital law. We will at first derive the relations for a current-free boundary surface and then for a boundary having a current sheet.

Boundary with no Surface Current

Figure 4.37 shows two regions 1 and 2 separated by a plane boundary surface. The permeabilities of the two regions are taken as μ_1 and μ_2. The flux density at the boundary in Region 1 is B_1 and in Region 2 it is B_2. The fields are at angles α_1 and α_2 with respect to the normal to the interface, as shown at Fig. 4.37(*a*). The components of B_1 along the normal and tangential directions to the boundary surface, denoted by B_{n1} and B_{t1} respectively, are

$$B_{n1} \quad B_1 \cos\alpha_1 \tag{1}$$

$$B_{t1} \quad B_1 \sin\alpha_1 \tag{2}$$

The normal and tangential components of B_2, denoted by B_n B_{t2} respectively, are

$$B_{n2} = B_2 \cos\alpha_2 \tag{3}$$

$$B_{t2} = B_2 \sin\alpha_2 \tag{4}$$

To find the relationship between B_{n1} and B_{n2}, let us take a surface in the form of a closed cylinder enclosing the boundary surface, as illustrated in Fig. 4.37(*b*). The axis of the cylinder is normal to the interface. The cylinder has two end surfaces and one curved surface. Its height is Δh and area of each end surface is ΔS. If we allow Δh to approach to zero so that the cylinder just encloses the boundary, area of the curved surface would be negligibly small. Hence flux through the curved surface can be assumed to be zero. As the net magnetic flux enclosed by a closed surface is zero, the value of flux entering the bottom surface is equal to that coming out of the top surface. Thus, assuming flux *out* of the cylinder as positive, we have

$$\oint_S \mathbf{B}\cdot d\mathbf{S} = \Delta S(B_{n2} - B_{n1}) = 0.$$

Therefore,

$$B_{n1} = B_{n2} \tag{5}$$

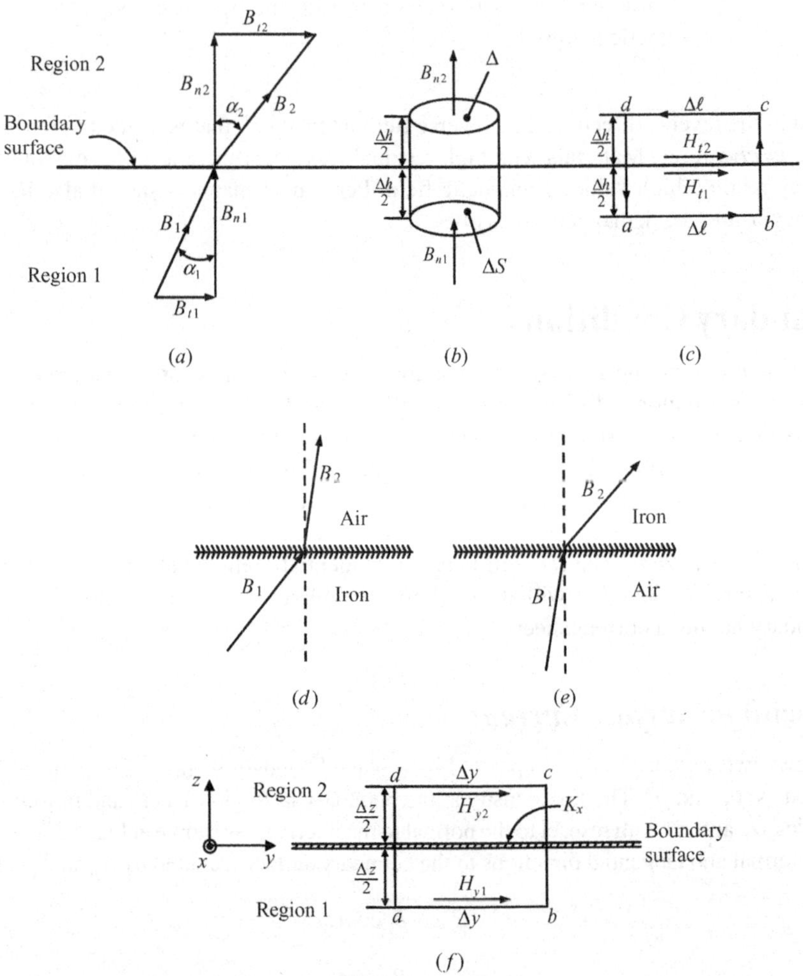

Fig. 4.37 (*a*, *b*, *c*) Illustrations for deriving boundary relations in a magnetic field, (*d*, *e*) Directions of flux densities on two sides of the boundary surface between air and iron and (*f*) Boundary having a current sheet.

Or
$$\mu_1 H_{n1} = \mu_2 H_{n2} \tag{6}$$

Relation (5) shows that the normal components of B on two sides of the boundary surface are equal. In other words, the normal component of flux density is continuous across the boundary. And, relation (6) shows that the normal components of magnetic field intensities on two sides of the boundary are not equal.

Next, we will obtain the relation between the tangential components of field intensities H_{t1} and H_{t2}. Let us take a small closed path *abcda* enclosing the interface, as shown at Fig. 4.37(*c*). Sides *ab* and *cd*, each of length $\Delta\ell$, are parallel to the interface. Sides *bc* and *da*, each of height Δh, are normal to the interface. As per the magnetic circuital law, the line integral of **H** around a closed path is equal to the current enclosed by the path. As the boundary has no current, we have

$$\oint \mathbf{H}\cdot d\boldsymbol{\ell} = 0$$

In the limit Δh tending to zero, the line integral of **H** along sides *bc* and *da* can be neglected. Accordingly,

$$\oint \mathbf{H}\cdot d\boldsymbol{\ell} = (H_{t1} - H_{t2})\Delta\ell = 0$$

Or
$$(H_{t1} - H_{t2}) = 0$$

Hence,
$$H_{t1} = H_{t2} \tag{7}$$

And
$$\frac{B_{t1}}{\mu_1} = \frac{B_{t2}}{\mu_2} \tag{8}$$

Thus,

---

if the boundary has no surface current, the tangential component of the magnetic field intensity is continuous across the boundary. However, the tangential component of magnetic flux density is discontinuous across the boundary.

Equations (5) to (8) give the boundary conditions or boundary relations for the case of a current-free boundary. We will use some of these relations to derive the relationship between B_1 and B_2. We will also obtain the relation between angles α_1 and α_2.

The magnitudes of B_1 and B_2 in terms of their normal and tangential components are given by

$$B_1 = \sqrt{B_{n1}^2 + B_{t1}^2} \tag{9}$$

$$B_2 = \sqrt{B_{n2}^2 + B_{t2}^2} \tag{10}$$

By substituting relations (5) and (8) in (9), and also in (10), we will get the flux density at the boundary of one region in terms the normal and tangential components of flux density at the boundary of the other region. These are

$$B_1 \quad B_{n}^{2} + B_{t2}^{2}(\;_1/\mu_2)^2$$

$$B_2 \quad B_{n}^{2} + B_{t1}^{2}(\;_2/\mu_1)^2$$

Next, let us substitute (1) and (3) in (5) to get

$$B_1 \cos\alpha_1 = B_2 \cos\alpha_2 \tag{11}$$

Also, substituting (2) and (4) in (8), we get

$$\mu_2 B_1 \sin\alpha_1 = \mu_1 B_2 \sin\alpha_2 \tag{12}$$

Dividing (12) by (11), we have

$$\frac{\tan\alpha_1}{\tan\alpha_2} = \frac{\mu_1}{\mu_2} \tag{13}$$

From the foregoing relations we find that the magnitude and direction of flux density on one side of the boundary surface are different from those on the other side. A case of practical importance is the condition at the iron–air boundary. If the flux densities at the air and iron sides of the boundary surface subtend angles α_a and α_i with the normal to the interface respectively, then by (13)

$$\frac{\tan\alpha_a}{\tan\alpha_i} = \frac{\mu_0}{\mu_i} = \frac{1}{\mu_r}$$

where μ_i and μ_r are absolute permeability and relative permeability of iron. Since μ_r is very large α_a is very small. Hence flux density at air side of the boundary is nearly normal to the interface, as shown at Figs. 4.37(*d* and *e*). That is, magnetic field lines enter normally into an iron surface from air and emerge normally from an iron surface into air.

Boundary having a Current Sheet

The procedure for deriving the boundary relations in the presence of a current sheet at the interface is same as in the case of a current-free boundary. As the net flux *out* of a closed surface enclosing the boundary is zero even if the closed surface encloses some current, the relation between the normal components of flux densities is same as that given by (5). However, the relation between the tangential components is not that given by (7) because in the present case the line integral of **H** around a closed path enclosing the boundary is not equal to zero.

Figure 4.37(*f*) shows a current sheet at the boundary between two regions. As per the coordinate system indicated, the boundary is at the $z = 0$ plane and the surface current is along positive *x* direction with current density of K_x amperes per unit width in *y* direction. The tangential components of magnetic field intensity on the two sides of the boundary are taken as H_{y1} and H_{y2}. Application of the magnetic circuital law around the rectangle *abcda*, in the limit Δh approaching zero, gives

$$(H_{y1} - H_{y2})\Delta y = K_x \Delta y$$

Or

$$H_{y1} - H_{y2} = K_x \tag{14}$$

In order to express relation (14) in vector form, let us multiply both sides by the unit vector $\mathbf{u}_x$ to get

$$(H_{y1} - H_{y2})\mathbf{u}_x = K_x \mathbf{u}_x \tag{15}$$

Now, let

$$\mathbf{H}_1 = H_{y1}\mathbf{u}_y + H_{z1}\mathbf{u}_z$$

and

$$\mathbf{H}_2 = H_{y2}\mathbf{u}_y + H_{z2}\mathbf{u}_z$$

It can be shown that

$$(H_{y1} - H_{y2})\mathbf{u}_x = (\mathbf{H}_1 - \mathbf{H}_2)\times\mathbf{u}_z \tag{16}$$

Since the left-hand sides of (15) and (16) are same, we have

$$(\mathbf{H}_1 - \mathbf{H}_2)\times\mathbf{u}_z = K_x\,\mathbf{u}_x = \mathbf{K} \tag{17}$$

The general form of (17) is

$$(\mathbf{H}_1 - \mathbf{H}_2)\times\mathbf{u}_{n2} = \mathbf{K} \tag{18}$$

where $\mathbf{u}_{n2}$ is a unit vector normal to the interface in Region 2. Relation (18) may be used to calculate the tangential field intensity vector at one side of the boundary from the tangential field intensity vector at the other side of the boundary, if the boundary surface has a current sheet.

Though we have considered a plane boundary surface, the relations derived are valid for any shape of the boundary as the field quantities are functions at a point in space.

EXAMPLE 4.33

Two regions 1 and 2 have a common boundary surface. The relative permeability of Region 1 is 250 and that of Region 2 is 400. The magnitude of flux density in Region 1, $B_1 = 0.4$ T and it is at an angle 30° with normal to the boundary surface. Find the magnitude of flux density B_2 in Region 2 and its angle with the normal to the interface.

Given: $\mu_{r1} = 250$; $\mu_{r2} = 400$; $B_1 = 0.4$ T; $\alpha_1 = 30°$

Component of B_1 tangential to the interface, $B_{t1} = B_1 \sin\alpha_1 = (0.4)(\sin 30°) = 0.2$ T

Component of B_1 normal to the interface, $B_{n1} = B_1 \cos\alpha_1 = (0.4)(\cos 30°) = 0.346$ T

Component of B_2 tangential to the interface, $B_{t2} = \dfrac{B_{t1}\mu_{r2}}{\mu_{r1}} = \dfrac{(0.2)(400)}{250} = 0.32$ T

Component of B_2 normal to the interface, $B_{n2} = B_{n1} = 0.346$ T

Magnitude of flux density in Region 2, $B_2 = \sqrt{B_{t2}^{\,2} + B_{n2}^{\,2}} = \sqrt{0.32^2 + 0.346^2} = 0.471$ T

Angle of B_2 with the normal, $\alpha_2 = \tan^{-1}(B_{t1}/B_{n1}) = \tan^{-1}(0.32/0.346) = 42.76°$

EXAMPLE 4.34

In a magnetic field, two regions 1 and 2 have a common boundary surface at $x = 0$ plane. The relative permeability of Region 1 $(x > 0)$, $\mu_{r1} = 250$ and that of Region 2, $\mu_{r2} = 180$. The magnetic flux density in Region 1 is given by, $\mathbf{B}_1 = \mathbf{0.6u}_x + 0.8\mathbf{u}_z$ T. Find (a) the flux density $\mathbf{B}_2$ in Region 2 and (b) the angles subtended by the flux densities with the normal to the interface.

The x component of $\mathbf{B}_1$ is normal to the interface and z component of $\mathbf{B}_1$ is tangential to the interface. Thus,

$B_n \quad B_{x1} = 0\ 6$ T

$B_t \quad B_{z1} = 0.8\ \text{T}$

(a) As per the boundary relations,

$$B_n \quad B_{n1} = 0.6\ \text{T}$$

$$\text{and} \quad B_{t2} = \mu_{r2}\frac{B_{t1}}{\mu_{r1}} = (\frac{180}{250})(0.8) = 0.567\ \text{T}$$

(b) $\alpha_1 = \tan^{-1}(B_{t1}/B_{n1}) = \tan^{-1}(0.8/0.6) = 53.1°$

$$\alpha_2 = \tan^{-1}(B_{t2}/B_{n2}) = \tan^{-1}(0.576/0.6) = 43.8°$$

EXAMPLE 4.35

The boundary surface between two regions 1 and 2 in a magnetic field is on the $z = 0$ plane. The magnetic field intensity at the interface in Medium 1 ($z < 0$) is $\mathbf{H}_1 = 25.0\mathbf{u}_x + 30.0\mathbf{u}_z$ A/m. The relative permittivity of Medium 1 is 12.0 and that of Medium 2 is 15.0. Find (a) the field intensity at the interface in Medium 2 if the interface has no surface current, and (b) if the interface has a current sheet defined by $\mathbf{K} = 6.0\mathbf{u}_y$ ampere per unit width in y direction and $\mathbf{H}_1 = 25.0\mathbf{u}_x + 30.0\mathbf{u}_z$.

Given: $H_x = 25.0$ A/m, $H_{z1} = 30.0$ A/m, $\mu_{r1} = 12.0$, $\mu_{r2} = 15.0$

Let, $\mathbf{H}_2 \quad H_{x2}\mathbf{u}_x + H_{z2}\mathbf{u}_z$

(a) The x components of the fields are tangential, and the z components of the fields are normal, to the interface. For both the cases,

$B_z \quad B_{z1}$

Or $H_{z2} = \dfrac{\mu_{r1}H_{z1}}{\mu_{r2}} = \dfrac{(12.0)(30.0)}{15.0} = 24.0$ A/m

For the current-free boundary,

$H_x \quad H_{x1} = 25.0$ A/m

Thus, the field intensity in Medium 2 for the current-free region is given by

$\mathbf{H}_2 \quad \mathbf{u}_x + 24.0\mathbf{u}_z$ A/m

(b)If the boundary contains a current sheet, we have to use formula (18). As the unit normal vector in Region 2 is $\mathbf{u}_n \quad \mathbf{u}_z$ and the current sheet on the interface has the density $\mathbf{K} = 6.0\mathbf{u}_y$, we have

$[(25.0 \quad {}_x \quad 30.0\mathbf{u}_z) - (H_{x2}\mathbf{u}_x + H_{z2}\mathbf{u}_z)] \times \mathbf{u}_z = 6.0\mathbf{u}_y$

Or, $- \quad {}_y + H_{x2}\mathbf{u}_y = 6.0\mathbf{u}_y$

$H_{x2} = 31.0$ A/m

The field intensity in Region 2 is given by

$\mathbf{H}_2 \quad \mathbf{u}_x \quad 24\ 0\mathbf{u}_z$ A/m

EXAMPLE 4.36

The unit normal vector to the interface of two regions 1 and 2 in a magnetic field away from the origin is given by $\mathbf{u}_n = 0.6\mathbf{u}_x + 0.4\mathbf{u}_y + |a|\mathbf{u}_z$. The magnetic field intensity in Region 1 containing the origin is $\mathbf{H}_1 = 150\mathbf{u}_x + 225\mathbf{u}_y - 50\mathbf{u}_z$. The relative permittivity of Region 1 is 300 and that of Region 2 is 175. Find (a) the magnitudes of field intensities that are normal and tangential to the interface in Region 1, and (b) the magnitudes of field intensities that are normal and tangential to the interface in Region 2.

As the magnitude of a unit vector is unity,

$$a = \sqrt{1 - 0.6^2 - 0.4^2} = 0.693$$

Thus, $\mathbf{u}_n = 0.6\mathbf{u}_x + 0.4\mathbf{u}_y + 0.693\mathbf{u}_z$

(a) Magnitude of field intensity at the interface in Region 1,

$$H_1 = \sqrt{150^2 + 225^2 + 50^2} = 275 \text{ A/m}$$

Magnitude of field intensity normal to the interface in Region 1,

$$H_{n1} = \mathbf{u}_n \cdot \mathbf{H}_1 = (0.6\mathbf{u}_x + 0.4\mathbf{u}_y + 0.693\mathbf{u}_z)\cdot(150\mathbf{u}_x + 225\mathbf{u}_y - 50\mathbf{u}_z)$$

$$= 145.3 \text{ A/m}$$

Magnitude of field intensity tangential to the interface in Region 1,

$$H_{t1} = \sqrt{H_1^2 - H_{n1}^2} = \sqrt{275^2 - 145.35^2} = 233.45 \text{ A/m}$$

(b) From the boundary relations,

$$H_{t2} = H_{t1} = 233.45 \text{ A/m}$$

and $H_{n2} = (\mu_{r1} H_{n1})/\mu_{r2} = (300 \times 145.3)/(175) = 249.1 \text{ A/m}$

EXAMPLE 4.37

Two regions 1 and 2 in a magnetic field have a common boundary surface defined by $S = 3x + 4y + 5z = 30$. Region 1 is in the origin side of the boundary, and the field intensity at the boundary of this region is given by $\mathbf{H}_1 = 300\mathbf{u}_y + 400\mathbf{u}_z$ A/m. The relative permittivity of Region 1 is 125 and that of Region 2 is 250. Find (a) the vector components of the field intensity that are normal and tangential to the boundary of Region 1, and (b) the field intensity vector at the boundary of Region 2.

The unit vector normal to the interface,

$$\mathbf{u}_n = \frac{\nabla S}{|\nabla S|}$$

$$\nabla = 3\mathbf{u}_x + 4\mathbf{u}_y + 5\mathbf{u}_z = 30,\ |\nabla S| = \sqrt{3^2 + 4^2 + 5^2} = 7.07$$

$\mathbf{u}_n$ [illegible] $\mathbf{u}_x +$ [illegible] $\mathbf{u}_y$ [illegible] $\mathbf{u}_z$ [illegible] $= 0.424\mathbf{u}_x$ [illegible] $\mathbf{u}_y + 0.707\mathbf{u}_z$

The scalar component of $\mathbf{H}_1$ normal to the interface, $H_{n1} = \mathbf{u}_n \cdot \mathbf{H}_1$

$=$ [illegible] $_x + 0.566\mathbf{u}_y +$ [illegible] $_z) \cdot ($ [illegible] $) = 452.6$ A/m

The vector component of $\mathbf{H}_1$ normal to the interface,

$$\mathbf{H}_n \quad \mathbf{u}_n H_{n1} = (0.424\mathbf{u}_x + 0.566\mathbf{u}_y + 0.707\mathbf{u}_z)(452.6)$$

$$= 191.9\mathbf{u}_x + 256.2\mathbf{u}_y + 320.0\mathbf{u}_z \text{ A/m}$$

The vector component of $\mathbf{H}_1$ tangential to the interface,

$$\mathbf{H}_{t1} \quad \mathbf{H}_1 - \mathbf{H}_{n1} = -191.9\mathbf{u}_x + 43.8\mathbf{u}_y + 80.0\mathbf{u}_z \text{ A/m}$$

(b) From the boundary conditions, $\mathbf{H}_{t2} = \mathbf{H}_{t1} = -191.9\mathbf{u}_x + 43.8\mathbf{u}_y + 80.0\mathbf{u}_z$ A/m

and $\mathbf{H}_n = (\mu_{r1}\mathbf{H}_{n1})/\mu_{r2} = (125/250)(191.9\mathbf{u}_x + 256.2\mathbf{u}_y + 320.0\mathbf{u}_z)$

$$= 95.95\mathbf{u}_x + 128.1\mathbf{u}_y + 160.0\mathbf{u}_z$$

Total field at the interface in Region 2,

$$\mathbf{H}_2 \quad \mathbf{H}_{t2} + \mathbf{H}_{n2} = -95.95\mathbf{u}_x + 171.9\mathbf{u}_y + 240\mathbf{u}_z$$

4.17 Magnetic Circuits

The closed path of a magnetic field line is called a magnetic circuit.

We will limit our discussion of magnetic circuits in which the field lines are either parallel or which can be assumed to be parallel. Some examples are illustrated in Fig. 4.38. The toroid shown at 4.38(*a*) has a closely and uniformly distributed exciting coil carrying current *I*. Even with a nonmagnetic core, the direction of field lines inside the core would be along the circumferential direction. The core at 4.38(*b*) is that of a ferromagnetic material of high permeability. The exciting coil is wound on a small part of the core only. Even then most of the flux will be within the core. The field lines are also nearly parallel. Some flux lines however will complete their path through air.

The component of flux within the core is called *main* flux and the component of flux through the air path is called *leakage* flux.

The leakage flux is very small if the relative permeability of the core is above one hundred.

Magnetic circuit calculations are useful in the design and analysis of electromagnetic devices.

Magnetomotive Force, Reluctance and Permeance

Let us consider a toroid of a single magnetic material of high permeability and of uniform section, as shown in Fig. 4.38(*a*). Its inner radius and outer radius are taken as *a*, and *b* respectively. The device has an exciting coil of *N* turns and it carries a direct current *I*. The core will be magnetised by the applied field. Because of toroidal symmetry, the field lines inside the core are concentric circles as shown in the figure, Moreover, the field intensity along any field line may be assumed to be constant. Thus, the field intensity at a radial distance *R*, which can be obtained using Ampere's magnetic circuital law, is

$$= \frac{NI}{2\pi R}, \quad a \quad R \le b \tag{1}$$

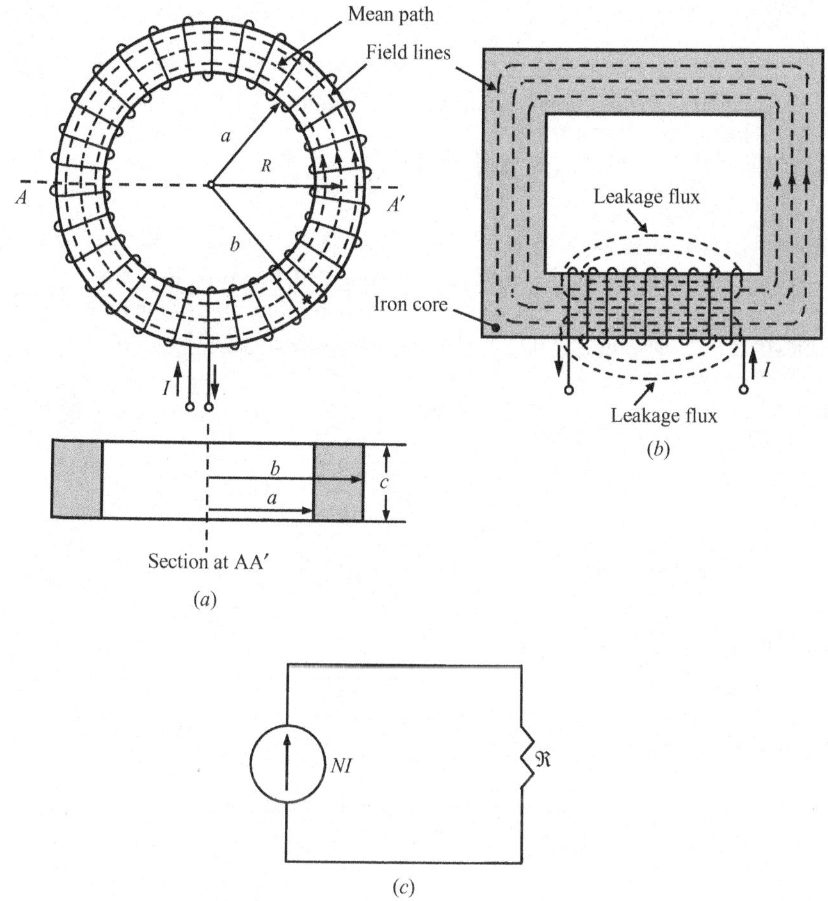

Fig. 4.38 (*a*) A toroidal core with a distributed coil, (*b*) A rectangular iron core with a concentrated coil, and (*c*) Equivalent circuit of a magnetic circuit.

We find from the above equation that H is inversely proportional to R. But the same is not true for the flux density B because of nonlinear nature of the magnetisation characteristics of the core material. Hence, the core flux cannot be obtained easily. However for approximate calculations, we may use the permeability corresponding to the average value of field intensity which can be obtained by integrating (1) from $R = a$ to R b and then dividing the result by $(b - a)$. Denoting the average field intensity by H_{av1}, we have

$$H_{av1} = \frac{NI}{2\pi(b-a)} \ln(b/a) \tag{2}$$

Alternatively, we may assume the field intensity at the mean radius of $(a + b)/2$ as the average value. Denoting it by H_{av2}, we have

$$H_{av2} = \frac{NI}{\pi(a+b)} = \frac{NI}{\ell} \tag{3}$$

where $\ell =$ (+) is length of the magnetic circuit at the mean radius.

In a problem where high accuracy is not needed we may use either (2) or (3), or even the mean of the two to find the average field intensity.

If permeability of the core material for H_{av} is μ, the average flux density,

$$B_{av} = \frac{\mu NI}{\ell}$$

The core flux Φ is equal to the sectional area of the core S multiplied by B_{av}. Thus,

$$\Phi = NI\left(\frac{\mu S}{\ell}\right) = \frac{NI}{\Re}, \quad \Re = \frac{\ell}{\mu S} \tag{4}$$

The product of number of turns and current of the exciting coil, that is NI in (4), is called applied *magnetomotive force* abbreviated as mmf. The quantity $\Re$ is called *reluctance* of the magnetic circuit. The unit of mmf is ampere (or ampere-turn) and that of reluctance is 1/henry (1/H).

Magnetomotive force, $\quad \text{mmf} = NI \text{ A}$

Reluctance, $\quad \Re = \dfrac{\ell}{\mu S} \text{ 1/H} \quad (5)$

The reciprocal of reluctance is called *permeance*.

Denoting the permeance by $\wp$, we have

$$\wp = \frac{1}{\Re} = \frac{\mu S}{\ell} \text{ H} \tag{6}$$

Thus, formula (4) can be written as

$$\text{Flux} = \frac{\text{mmf}}{\text{reluctance}} = \text{mmf} \times \text{permeance}$$

These relations are analogous to

$$\text{Current} = \frac{\text{emf}}{\text{resistance}} = \text{emf} \times \text{ conductance}$$

in an electric circuit. Moreover, formulae (5) for reluctance and formula (6) for permeance are analogous to the formulae for resistance and conductance respectively of a conductor of length ℓ and uniform sectional area S. Thus,

a physical magnetic circuit can be represented by an equivalent circuit as shown in Fig. 4.38(*c*).

The magnetic circuit quantities and the analogous electric circuit quantities are listed below.

Magnetic Circuit	Electric Circuit
Flux	Current
mmf	emf
Reluctance	Resistance
Permeance	Conductance
Permeability	Conductivity
Flux density	Current density
Magnetic field intensity	Electric field intensity

Though a static magnetic circuit is analogous to a direct current circuit, there are some basic differences between the two. For example, conductivity of a good conductor remains practically constant over a wide range of current density, whereas permeability of a ferromagnetic material is a function of flux density. In electric circuits there is continuous dissipation of energy in the resistance. In magnetic circuits excited by direct current, there is no continuous dissipation of energy. Energy is needed only to establish the magnetic field.

Series Magnetic Circuit

Figure 4.39(*a*) shows a magnetic circuit consisting of three parts where each part is of a different ferromagnetic material. The sectional areas of the parts may or may not be same. An exciting coil of N turns carrying current I is wound on Part 1 of the core. The method of solving problems on magnetic circuits made of different ferromagnetic materials (or even of a single material having different sectional areas in different parts of the core) depends on the data given. If permeabilities of different parts of the core are known, then a relation between applied mmf NI and core flux Φ can be obtained by assuming that Φ is constant throughout the magnetic circuit, and it is distributed uniformly over the sections of all the parts. The procedure for deriving the relation is outlined below.

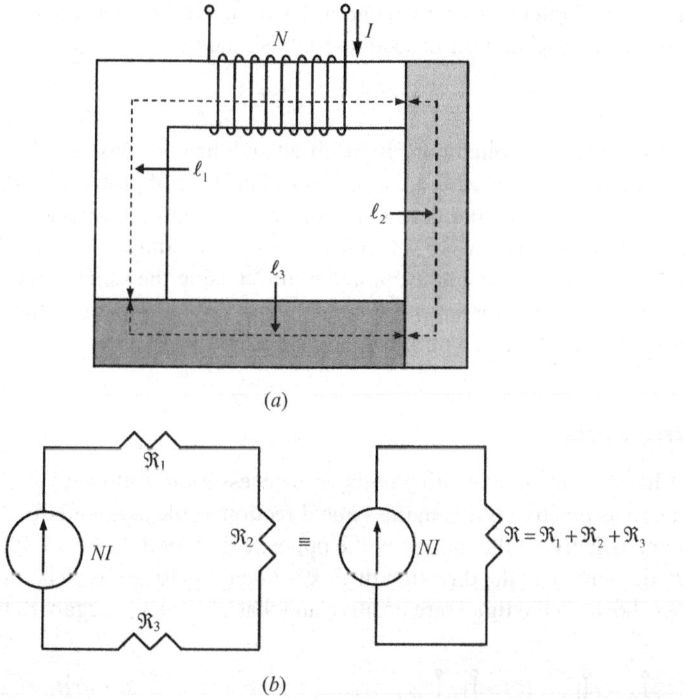

Fig. 4.39 (*a*) A series magnetic circuit consisting of three different ferromagnetic materials and (*b*) Equivalent circuit of (*a*).

The field intensities in the three parts of the core are

$$H_1 = \frac{\Phi}{\mu_1 S_1}; \quad H_2 = \frac{\Phi}{\mu_2 S_2}; \quad H_3 = \frac{\Phi}{\mu_3 S_3}.$$

where the subscripts 1, 2, 3 are meant for the three parts of the core.

Now applying the circuital law along the mean magnetic path, as indicated in the figure, we get

$$NI = H_1\ell_1 + H_2\ell_2 + H_3\ell_3$$

Or

$$NI = \Phi\left(\frac{\ell_1}{\mu_1 S_1} + \frac{\ell_2}{\mu_2 S_2} + \frac{\ell_3}{\mu_3 S_3}\right)$$

$$= \Phi(\Re_1 + \Re_2 + \Re_3) = \Phi\Re \tag{7}$$

where $\Re$ is sum of reluctances of three parts of the core.

The relation between mmf and flux given by (7) is analogous to the relationship between voltage and current of an electric circuit with three resistances in series. Thus, the equivalent circuit of the magnetic circuit is as shown in Fig. 4.39(*b*). The composite magnetic circuit of Fig. 4.39(*a*) is an example of a series magnetic circuit.

In the second category of problems, the core flux (or flux density in any part of the core) is given and the applied mmf to establish the flux is to be calculated. The magnetisation characteristics of the core materials are also available. Then from the given flux Φ and calculated values of sectional areas, we get

$$B_1 = \frac{\Phi}{S_1}; \quad B_2 = \frac{\Phi}{S_2}; \quad B_3 = \frac{\Phi}{S_3}.$$

If H_1 H_2, and H_3 are the field intensities corresponding to the above flux densities (read from the magnetisation characteristics of the materials), mmf needed to establish the flux,

$$NI = H_1\ell_1 + H_2\ell_2 + H_3\ell_3$$

In the third type of problems, the applied mmf is specified and the core flux is to be calculated. The magnetisation characteristics of the core materials are also given. This type of problem cannot be solved directly as the flux densities in various parts of the core may not be equal and the magnetisation characteristics are nonlinear. A problem in this category can be solved by the use of a semigraphical method. We at first assume a value of core flux Φ and then calculate the applied mmf NI using the same procedure as in the preceding example. The procedure is repeated for several assumed values of core flux and then the results are plotted. The core flux corresponding to the given mmf is then read from the graph.

Multiple Exciting Coils

If a magnetic circuit has two or more exciting coils, it is necessary to follow a sign convention. One way is to take the mmf of a coil as positive if it is in the same direction as the assumed direction of the flux (or path of integration of H) and negative if the mmf is in the opposite direction. In Fig. 4.40(*a*) where three exciting coils are mounted on the same leg, the direction of flux is taken as clockwise. Therefore, mmf of coils 1 and 2 having directions as shown in the figure are positive and that of Coil 3 is negative. In Fig. 4.40(*b*) the mmfs

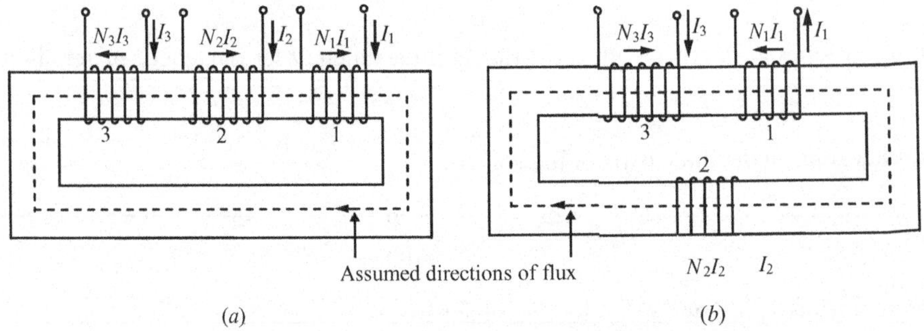

Fig. 4.40 Magnetic circuits with multiple exciting coils

of coils 1 and 2 will have to be taken as negative and mmf of Coil 3 as positive. The line integral of H around the magnetic circuit is equal to the algebraic sum of mmfs of all the coils.

Parallel Magnetic Circuits

The magnetic core shown in Fig. 4.41(*a*) consists of a central *leg* and two outer *legs*. The legs are joined at the top and bottom by *yokes* to complete the magnetic circuits. The central leg has an exciting coil of N turns carrying current I. The exciting coil may also be placed on any of the outer legs, or there may be multiple exciting coils placed on different legs. The coil mmf NI sets up magnetic flux along two paths *abcda* and *abefa* shown in the figure. As both the paths have the same mmf, the two paths are in parallel. We designate the mean magnetic path of the central leg as Path 1, of the left side yoke and leg as Path 2, and of the right side yoke and leg as Path 3. We will use subscripts 1, 2, and 3 to denote the quantities associated with the three paths. Assuming no leakage, flux Φ_1 in the central leg is equal to the sum of fluxes Φ_2 and Φ_3 in the outer legs. That is,

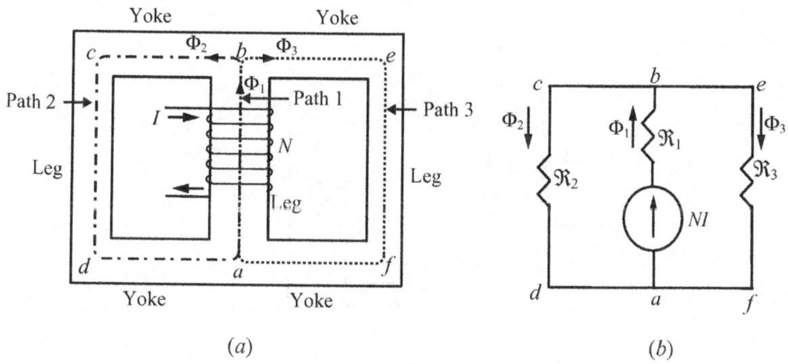

Fig. 4.41 (*a*) A parallel magnetic circuit and (*b*) Equivalent circuit of (*a*).

$$\Phi_1 = \Phi_2 + \Phi_3 \tag{8}$$

Application of the magnetic circuital law along paths 1 and 2 gives,

$$NI = H_{1l}\ell_{1l} + H_{2l}\ell_{2l} + H_{2y}\ell_{2y} \tag{9a}$$

Application of the magnetic circuital law along paths 1 and 3 gives,

$$NI = H_{1l}\ell_{1l} + H_{3l}\ell_{3l} + H_{3y}\ell_{3y} \tag{9b}$$

where the additional subscripts l and y denote quantities associated with the legs and yokes of the core.

The magnetic circuit equations involving fluxes and reluctances are

$$NI = \Phi_1\Re_{1l} + \Phi_2(\Re_{2l} + \Re_{2y}) = \Phi_1\Re_{1l} + \Phi_3(\Re_{3l} + \Re_{3y})$$

Or

$$NI = \Phi_1\Re_1 + \Phi_2\Re_2 = \Phi_1\Re_1 + \Phi_3\Re_3 \tag{10}$$

where $\Re_1$, $\Re_2$ and $\Re_3$ are reluctances of the three paths.

We can use either equations (9) or (10) to solve parallel magnetic circuit problems.

The equivalent circuit corresponding to equations (10) is shown at Fig. 4.41(*b*).

Magnetic Circuit with Air Gap

Electromechanical devices, namely, rotating and linear machines, relays, etc., consist of stationary and movable members. Air gaps are necessary between the stationary and moving members of these devices. In dc measuring instruments, the moving coil is located in the gap between two stationary ferromagnetic cores. Further, air gaps in magnetic circuits are provided so that the *B-H* characteristic becomes linear over a wide range of field intensity. Figure 4.42(*a*) shows a magnetic circuit with one air gap.

It is found that field distribution in the region of air gap of a magnetic circuit is not uniform. Moreover, the field lines spread over an area larger than that of the iron surface at the gap. This is illustrated in Fig. 4.42(*b*). The spreading of field lines at a gap is termed as *fringing* of flux.

If the gap is uniform and its length is small (less than ten percent of the smaller dimension of iron surface at the gap), we may assume uniform distribution of flux over an effective area that is larger than the iron area at the gap as illustrated at Fig. 4.42(*c*). The effective gap area is calculated from the following empirical formulae.

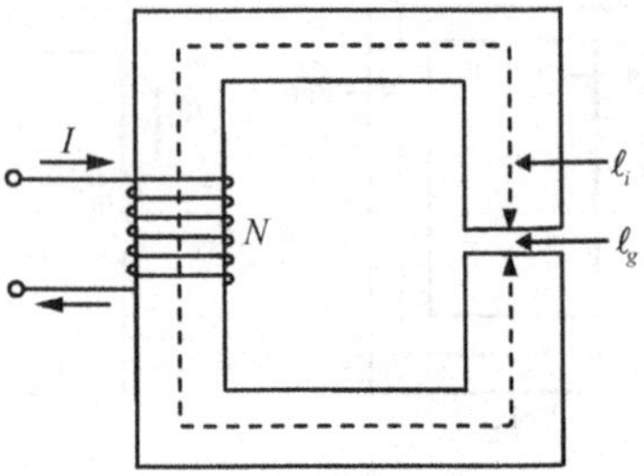

(*a*)

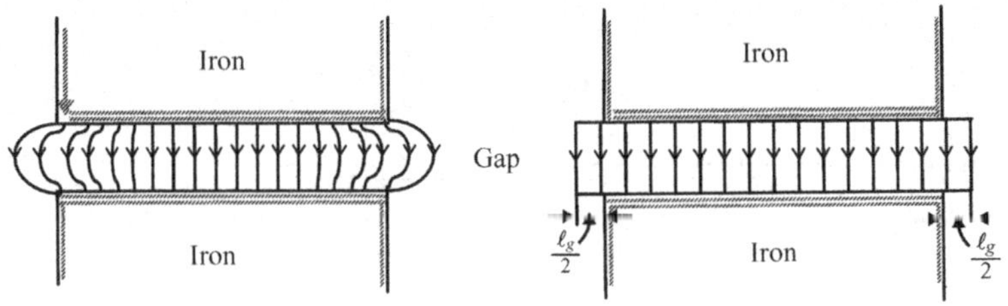

(*b*) (*c*)

Fig. 4.42 (*a*) A magnetic circuit with one air-gap, (*b*) Fringing of field lines at the gap, and (c) Assumed uniform distribution of field in the gap for rectangular section of iron surface.

Effective gap area for a core of rectangular section,

$$S_g = (w_1 + \ell_g)(w_2 + \ell_g) \tag{11}$$

where w_1 and w_2 are linear dimensions of the iron surface at the gap, and ℓ_g is the gap length.

Effective gap area for a core of circular section is

$$S_g \quad (\; S_i + \; _g)^2 \tag{12}$$

where S_i is area of iron surface at the gap.

In order to obtain the relationship between mmf and flux of a magnetic circuit with an air gap, let us assume that the iron part of the magnetic circuit shown in Fig. 4.42(*a*) is made of one material and has a uniform sectional area S_i and mean length ℓ_i. If we neglect leakage, flux throughout the magnetic circuit including the gap will have the same value. Taking the value of flux as Φ the field intensity in the iron part,

$$H_i = \frac{\Phi}{\mu_0 \mu_r S_i} \tag{13}$$

The field intensity in the gap,

$$H_g = \frac{\Phi}{\mu_0 S_g} \tag{14}$$

Application of the magnetic circuit law gives

$$NI = H_i \ell_i + H_g \ell_g \tag{15}$$

The first term on the right-hand side of (15), that is $H_i\ell_i$, is the mmf *drop* in the iron part of the magnetic circuit and the second term $H_g\ell_g$ is the mmf *drop* in the air gap. By replacing H_i and H_g of (15) by the right-hand sides of (13) and (14), we get

$$NI = \frac{\Phi \ell_i}{\mu_0 \mu_r S_i} + \frac{\Phi \ell_g}{\mu_0 S_g} = \Phi(\Re_i + \Re_g) \tag{16}$$

where $\Re_i = \dfrac{\ell_i}{\mu_0 \mu_r S_i}$ is the reluctance of iron part of the magnetic circuit.

$\Re_g = \dfrac{\ell_g}{\mu_0 S_g}$ is reluctance of air gap.

The ratio of mmf drop in the gap [the second term on the right of (16)] to the mmf drop in the iron [the first term on the right of (16)], neglecting fringing and leakage, is equal to $(\mu_r \ell_g)/\ell_i$. Thus for a core material of relative permeability of one thousand, the mmf drop in one millimetre of air gap is equal to the mmf drop in one metre of iron. Even a small gap in a magnetic circuit will increase substantially the mmf needed to establish a given flux density. Consequently, the exciting winding will be costlier. However, the *NI-B* characteristic would be linear over a wider range of *NI*. This is illustrated in Fig. 4.43.

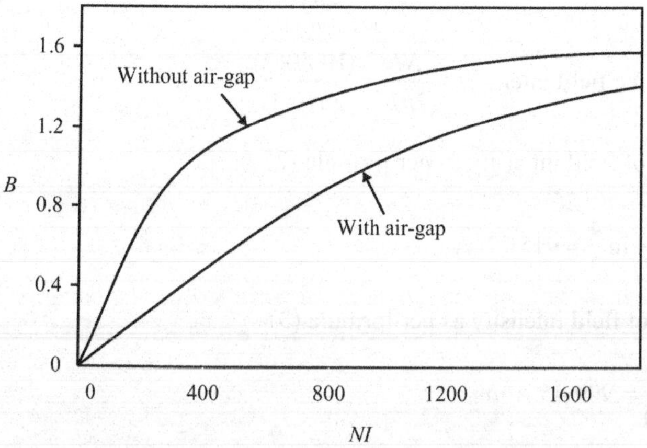

Fig. 4.43 *NI-B* characteristics of a magnetic circuit

EXAMPLE 4.38

The mean length of magnetic circuit shown in Fig. 4.38 (*b*) is 60 cm, and cross-sectional area of its core is 20 cm^2. The exciting coil has 200 turns and permeability of the core is 1.5×10^{-3} H/m. Find the coil current for a core flux of 2.4 mWb.

$$N = 200;\ \ell = 0.6 \text{ m};\ S = 2\times10^{-3} \text{ m}^2;\ \mu = 1.5\times10^{-3} \text{ H/m};\ \Phi = 2.4\times10^{-3} \text{ Wb};$$

Reluctance of the magnetic circuit, $\Re = \dfrac{\ell}{\mu S} = \dfrac{0.6}{(1.5\times10^{-3}(2\times10^{-3})} = 200\times10^{-3}\ \text{H}^{-1}$

$\text{mmf} = \Phi\Re = (2.4\times10^{-3})(200\times10^{3}) = 480$ A (amp-turns)

Coil current, $I = \dfrac{\text{mmf}}{N} = \dfrac{480}{200} = 2.4$ A

Alternatively:

Field intensity, $H = \dfrac{B}{\mu} = \dfrac{\Phi}{\mu S} = \dfrac{2.4\times10^{-3}}{(1.5\times10^{-3})(2\times10^{-3})} = 800$ A/m

$\text{mmf} = NI = H\ell = (800)(0.6) = 480$ A

Coil current, $I = 480/200 = 2.4$ A

EXAMPLE 4.39

The inner radius, outer radius and axial depth of a toroid of rectangular cross section shown in Fig. 4.38(*a*) are $a = 30$ cm, $b = 40$ cm, and $c = 15$ cm respectively. It is excited by a coil of 1000 turns carrying a steady current of 2 A. Calculate (a) maximum and minimum field intensities, (b) average field intensity using formula (2) as well as (3), (c) B_{av}, Φ, $\Re$, $\wp$ using the value of field intensity at the mean radius and taking $\mu_r = 950$. Neglect leakage.

Number of turns, $N - 1000$; Current, $I = 2$ A

(a) Maximum value of the field intensity $= \dfrac{NI}{2\pi a} = \dfrac{(1000)(2)}{2\pi(0.3)} = 1061.03$ A/m

Minimum value of the field intensity $= \dfrac{NI}{2\pi b} = \dfrac{(1000)(2)}{2\pi(0.4)} = 795.77$ A/m

(b) The average value of field intensity as per formula (2),

$$H_{av1} = \frac{(1000)(2)}{2\pi(0.4-0.3)}\ln\frac{4}{3} = 915.72 \text{ A/m}$$

The average value of field intensity as per formula (3),

$$H_{av2} = \frac{(1000)(2)}{\pi(0\,4+0\,3)} = 909.45 \text{ A/m}$$

(c) $B_{av} = \mu H_{av} = (4\pi\times10^{-7})(950)(909.45) = 1.086$ T

$$\Phi = B_{av}S = B_{av}c(b-a) = (1.086)(0.15)(0.4-0.3) = 16.29 \text{ mWb}$$

$$\mathfrak{R} = \frac{NI}{\Phi} = \frac{1000\times 2}{16.29\times 10^{-3}} = 1.228\times 10^{5} \text{ 1/H}$$

$$\wp = 1/(1.228\times 10^{5}) = 8.143\times 10^{-6} \text{ H}$$

EXAMPLE 4.40

A series magnetic circuit consists of two parts 1 and 2. The material of both the parts is made up of the same ferromagnetic material. The exciting coil has 400 turns. Calculate the exciting current needed to set up a flux of 0.3 mWb. Given: $\ell_1 = 30$ cm and $S_1 = 5$ cm^2; $\ell_1 = 10$ cm and $S_2 = 2.5$ cm^2. Calculate also the permeance of each part and the total permeance. The B-H curve for the core material is given in Fig. 4.35. Neglect leakage.

$$B_1 = \frac{\Phi}{S_1} = \frac{0\cdot 3\times 10^{-3}}{5\times 10^{-4}} = 0.6 \text{ T},\; B_2 = \frac{\Phi}{S_1} = \frac{0\cdot 3\times 10^{-3}}{2\cdot 5\times 10^{-4}} = 1.2 \text{ T}$$

The field intensities corresponding to B_1 and B_2, read from the *B-H* curve, are

$H_1 = 600$ A/m; $H_2 = 1230$ A/m.

Mmf for Part 1, $N_1I_1 = H_1\ell_1 = (600)(0.3) = 180$ A (amp-turn)

Mmf for Part 2, $N_2I_2 = H_2\ell_2 = (1230)(0.1) = 123$ A (amp-turn)

Total mmf, $NI = H_1\ell_1 + H_2\ell_2 = 303$ A (amp-turn)

Exciting current, $I = 303/400 = 0.7575$ A

Permeance of Part 1, $\wp_1 = \dfrac{\Phi}{N_1I_1} = \dfrac{0.3\times 10^{-3}}{180} = 1.667\times 10^{-6}$ H

Permeance of Part 2, $\wp_2 = \dfrac{\Phi}{N_2I_2} = \dfrac{0.3\times 10^{-3}}{123} = 2.439\times 10^{-6}$ H

Total permeance, $\wp = \dfrac{\wp_1\wp_2}{\wp_1+\wp_2} = \dfrac{(1.667)(2.439)}{1.667+2.439}(10^{-6}) = 0.99\times 10^{-6}$ H

EXAMPLE 4.41

The magnetic circuits shown in Figs. 4.40(*a* and *b*) have cores of uniform sectional area of 5 cm^2. The length of mean magnetic path of each core is 25 cm. If $N_1 = 100$, $N_2 = 50$, $N_3 = 25$, $I_1 = 0.45$ A, $I_2 = 0.4$ A, $I_3 = 0.6$ A. Find the core flux if relative permeability of the core material is 2500.

For the directions of mmfs and flux shown at Fig. 4.40(*a*), the effective mmf

$$NI = N_1I_1 + N_2I_2 - N_3I_3 = (100)(0\cdot 45) + (50)(0\cdot 4) - (25)(0\cdot 6) = 50 \text{ A}$$

The field intensity,

$$H = \frac{NI}{\ell} = \frac{50}{0\cdot 25} = 200 \text{ A}$$

$\Phi = \mu$ [illegible] $= (4\pi \times 10^{-7} \times 2500)(200)(5.0 \times 10^{-4}) = 0.3142$ mWb.

For the directions of mmfs and flux shown at Fig. 4.40(*b*) the net mmf,

$$NI = -N_1 I_1 - N_2 I_2 + N_3 I_3 = -50 \text{ A}$$

As the net mmf is negative, the flux direction is anticlockwise. The amount of flux is 0.3142 mWb.

EXAMPLE 4.42

The core of a shell type single-phase transformer is as shown in Fig. 4.41(*a*). It is made of silicon steel laminations. Find the mmf needed to establish a flux of $0 \cdot 8$ mWb in the central leg. Given: mean length of each leg = 10 cm; mean length of yoke of each parallel path = 30 cm; sectional area of central leg = 8 cm² and of those of outer legs = 4 cm². Sectional area of both the yokes = 5 cm²; relative permeability of the core material $\mu_r = 4500$.

Flux in the central leg, $\Phi_1 = 0 \cdot 8$ mWb

Since the two parallel magnetic circuits are identical, flux in the central leg is equally shared by the two parallel paths. Hence,

$$\Phi_2 \quad \Phi_3 = 0 \cdot 4 \text{ mWb}$$

Flux density in the central leg, $B_{1\ell} = \dfrac{0 \cdot 8 \times 10^{-3}}{8 \times 10^{-4}} = 1 \cdot 0$ T

Flux density in the outer legs, $B_{2\ell} = B_{3\ell} = \dfrac{0 \cdot 4 \times 10^{-3}}{4 \times 10^{-4}} = 1 \cdot 0$ T

Field intensity in the legs $= \dfrac{1.0}{(4\pi \times 10^{-7})(4500)} = 176.8$ A/m

Flux density in the yokes $= \dfrac{0{,}4 \times 10^{-3}}{5 \times 10^{-4}} = 0 \cdot 8$ T

Field intensity in the yokes $= \dfrac{0.8}{(4\pi \times 10^{-7})(4500)} = 141.5$ A/m

It is necessary to consider only one parallel path. Each parallel path has two legs each of 10 cm, and the yoke length is 30 cm. Thus, required mmf

= [illegible] $\times 0.1) + (141.5)(2 \times 0.15) = 77.81$ A (or ampere-turn)

EXAMPLE 4.43

The parallel magnetic circuit shown in Fig. 4.41(*a*) has the following dimensions. Mean length of central leg = 10 cm. Mean length of left leg and yokes = 30 cm. Mean length of right leg and yokes = 25 cm. Sectional area of central leg = 12 cm². Sectional area of left leg and yokes = 7 cm². Sectional area of right leg and yokes = 6 cm². Find flux in the legs if mmf of the exciting coil is 400 A, and relative permeability of all parts of the core is 2400.

Permeability of the core material, $\mu = (4\pi \times 10^{-7})(2400) = 3.016 \times 10^{-3}$ H/m

Reluctance of Part 1, $\Re_1 = \dfrac{\ell_1}{\mu S_1} = \dfrac{0.1}{(3.016 \times 10^{-3})(12 \times 10^{-4})} = 2.76 \times 10^4$ 1/H

Reluctance of Part 2, $\Re_2 = \dfrac{0.3}{(3{\cdot}016 \times 10^{-3})(7 \times 10^{-4})} = 14.21 \times 10^4$ 1/H

Reluctance of Part 3, $\Re_3 = \dfrac{0.25}{(3.016 \times 10^{-3})(6 \times 10^{-4})} = 13.82 \times 10^4$ 1/H

Writing equations (8) and (10) in matrix form, we have

$$\begin{bmatrix} NI \\ 0 \\ 0 \end{bmatrix} = \begin{bmatrix} \Re_1 & \Re_2 & 0 \\ 1 & -1 & -1 \\ 0 & \Re_2 & -\Re_3 \end{bmatrix} \begin{bmatrix} \Phi_1 \\ \Phi_2 \\ \Phi_3 \end{bmatrix}$$

By solving for the three fluxes, we get

$$\Phi_1 = NI\frac{\Delta_{11}}{\Delta}$$

$$\Phi_2 = NI\frac{\Delta_{12}}{\Delta}$$

$$\Phi_3 = NI\frac{\Delta_{13}}{\Delta}$$

where $\Delta_{11}, \Delta_{12}, \Delta_{13}$ are the cofactors and Δ is determinant of the square matrix.

$$\Delta_{11} = \Re_2 + \Re_3 = 28.03 \times 10^4$$

$$\Delta_{12} = \Re_3 = 13.82 \times 10^4$$

$$\Delta_{13} = \Re_2 = 14.21 \times 10^4$$

$$\Delta = \Re_1(\Re_2 + \Re_3) + \Re_2\Re_3 = 273.76 \times 10^8$$

Substituting the numerical values in the above equations, we get

$\Phi_1 = 4.096$ mWb, $\Phi_2 = 2.019$ mWb, $\Phi_3 = 2.076$ mWb.`

In the last two examples we have assumed that the permeability in all parts of the core has the same value. In practice, however, this will not be the case as the *B-H* curves of ferromagnetic materials are nonlinear and the flux densities in various parts of the core may also be different. Correct results can be obtained by using the values of permeabilities (or flux densities) corresponding to the values of *H* from the *B-H* curves of the core materials, as in Example 4.40.

EXAMPLE 4.44

The iron core of the magnetic circuit in Fig. 4.42(*a*) has a rectangular section with sides of 100 mm and 80 mm. The mean length of the flux path is 750 mm and gap length is 1.2 mm. Calculate the mmf required for a flux density of 1.2 T in the iron assuming no leakage. The relative permeability of iron is 1250. Use formula (11) for determining the effective gap area. Find also the effective gap area if the core is of circular section having the same area as in the case of the rectangular section.

Flux density in iron, $B_i = 1.2\ \text{T}$

Cross-sectional area of iron, $S_i = 8.0 \times 10^{-3}\ \text{m}^2$

Mean iron length, $\ell_i = 0.75\ \text{m}$

Gap length, $1.2 \times 10^{-3}\ \text{m}$

Effective area of the gap as per formula (11), $S_g = (100 + 1.2)(80 + 1.2) \times 10^{-6}$

$$= 8.217 \times 10^{-3}\ \text{m}^2$$

Flux density in the gap, $B_g = \dfrac{B_i S_i}{S_g} = \dfrac{(1.2)(8.0 \times 10^{-3})}{8.217 \times 10^{-3}} = 1.168\ \text{T}$

mmf required, $NI = H_i \ell_i + H_g \ell_g = \left(\dfrac{1.2}{1250\mu_0}\right)(0.75) + \left(\dfrac{1.168}{\mu_0}\right)(1.5 \times 10^{-3}) = 1967.2\ \text{A}$

If the iron is of circular section, then effective gap area as per formula (12) is given by

$$S_g = \left[\sqrt{\text{[illegible]} \times 10^{-3}} + (1.2 \times 10^{-3})\right]^2 = 8.216 \times 10^{-3}\ \text{m}^2$$

This result is nearly same as the effective gap area for the rectangular section of iron.

4.18 Inductors and Inductances

An inductor consists of a coil, generally of several turns, wound on an insulating former. The core of an inductor may be of a nonferromagnetic or a ferromagnetic material. Inductors, which come in several shapes, sizes and ratings, are widely used in electrical and electronic circuits. Wire loops, short and long solenoids, and toroidal coils are some of the examples of inductors.

The electromagnetic property of an inductor is called inductance. In order to define inductance, let us consider a coil of N turns carrying current I. The coil mmf NI will result in a flux,

$$\Phi = \wp NI \tag{1}$$

where $\wp$ is the permeance of the magnetic circuit. If we assume that all flux link with all the turns of the coil (or all the currents), the product of N and Φ, i.e., $N\Phi$, is called *flux linkage*. The ratio between the flux linkage and coil current is called *inductance*. The unit of inductance is *henry* (H). Thus, denoting the inductance by L, we have

$$L = \frac{N\Phi}{I} \tag{2a}$$

Eliminating Φ from (1) and (2*a*), we get

$$L = \wp N^2 \tag{2b}$$

We see from (2*b*) that the inductance is independent of the current. It depends only on the number of turns and permeance of the magnetic path. The permeance depends on the size, shape and material of the core. For a nonmagnetic core the permeance is independent of the mmf; hence the inductance is proportional to N^2. If the core is of a ferromagnetic substance, the permeance as well as the inductance depends on the coil mmf. Hence the inductance is not proportional N^2. Coils with iron cores have high inductance. These are called *choke* coils.

The assumption that all the flux link with all the turns (or currents) is not valid for all types of current distribution. For example, in the case of a short air core solenoid all the flux may not link with all the turns, especially if the turns are widely spaced. This is illustrated in Fig. 4.44. The total flux linkage in such cases is found by adding the flux linking each turn. Thus if a coil has N turns, then

$$N\Phi = \Phi_1 + \Phi_2 + \Phi_3 + - - - - + \Phi_N$$

A good flux plot is needed to find the total flux linkage. Inductance of standard coils can be calculated using empirical formulae. Inductance can also be determined experimentally.

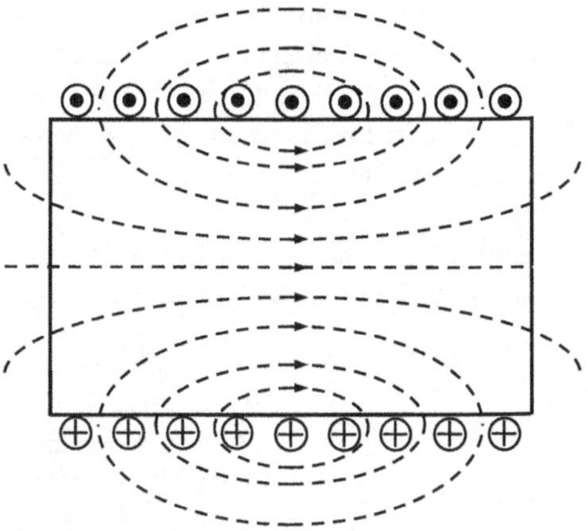

Fig. 4.44 Schematic representation of magnetic flux lines of a short solenoid with widely spaced turns

Self-Inductance

In order to define self-inductance, let us consider a single-turn coil of a nonmagnetic conductor carrying direct current. A cross-sectional view of the loop is shown in Fig. 4.45. We know that the current in the coil will produce magnetic fields inside as well as outside the conductor. The flux inside the conductor of the coil is called self-internal flux, and the flux outside the coil is called self-external flux. The sum of the two flux components is the self-flux. The two flux components are indicated in the figure. It may be noted that all the external flux link with the whole current whereas all the internal flux do not link with the total current. Hence the method of calculating internal flux linkage is different from that due to the external flux linkage.

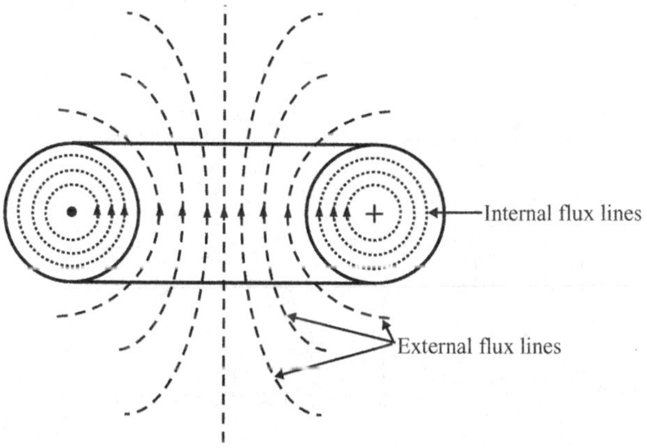

Fig. 4.45 Showing internal and external fluxes of a current loop

The inductance due to internal flux linkage is called *internal* inductance, and the inductance due to the external flux linkage is called *external* inductance. The sum of the two inductances is known as *self-inductance* of the coil.

Denoting the internal inductance by L_i, the external inductance by L_e, the self-inductance,

$$L = L_i + L_e$$

Inductances of Magnetically Coupled Coils: Mutual Inductance and Coupling Coefficient

If the configurations of two or more coils are such that the flux produced by one coil links with the other coils, the coils are said to be magnetically coupled or mutually coupled. We will define the various inductances by considering two coupled coils 1 and 2 shown in Fig. 4.46. The coils have N_1 and N_2 turns, and they have a common magnetic circuit. If Coil 1 is excited by a current I_1 as in Fig. 4.46(*a*), a part of the flux produced by the mmf N_1I_1 will link with both the coils and the remaining part will link with Coil 1 only. Similarly, if Coil 2 is excited by a current I_2, as shown in Fig. 4.46(*b*), a part of the flux produced by the mmf N_2I_2 will link with both the coils and the remaining part will link with Coil 2 only.

The flux which is common to both the coils is called *mutual* flux. The flux that links with one coil but not with the other is called *leakage flux*.

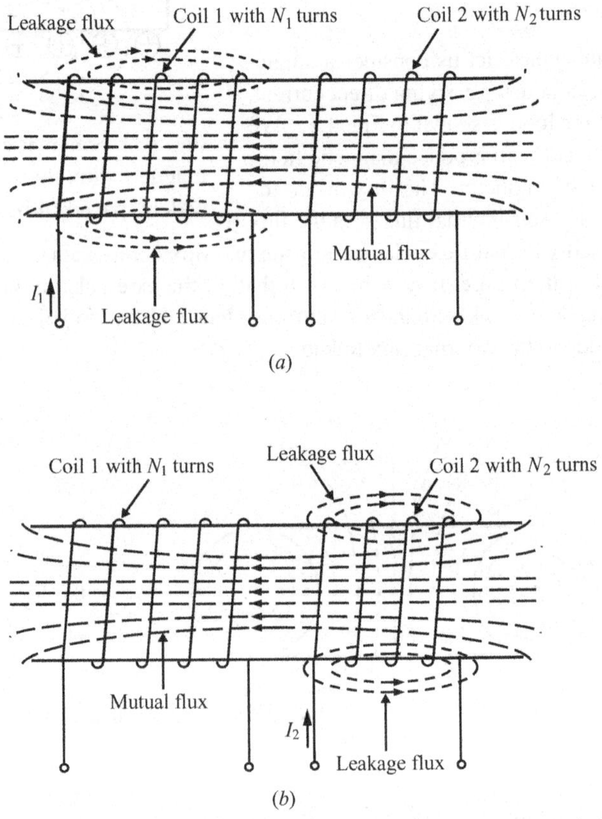

Fig. 4.46 Showing mutual flux and leakage flux of two coupled coils

The sum of the mutual flux and leakage flux is the external self-flux of a coil. If Φ_{21} is the mutual flux and $\Phi_{\ell 1}$ is the leakage flux due to the current in Coil 1, then the external self-flux Φ_{e1} that links with Coil 1 is

$$\Phi_{e1} = \Phi_{21} + \Phi_{\ell 1}$$

And if Φ_{12} is the mutual flux and $\Phi_{\ell 2}$ is the leakage flux due to current in Coil 2, then the external self-flux Φ_{e2} that links with Coil 2 is

$$\Phi_{e2} = \Phi_{12} + \Phi_{\ell 2}$$

In the foregoing discussion we have learnt about various flux components of two magnetically coupled coils. We will now use the general formula for inductance given by (2*a*) to define the inductances of the coupled coils.

The *leakage inductance* of Coil 1, $L_{\ell 1} = \dfrac{N_1\Phi_{\ell 1}}{I_1}$

The *leakage inductance* of Coil 2, $L_{\ell 2} = \dfrac{N_2 \Phi_{\ell 2}}{I_2}$

The product of Φ_{e1} and N_1 gives the external flux linkage of Coil 1 due to its own current I_1. The ratio between this flux linkage and I_1 is the *external self-inductance* of Coil 1. Denoting the self-inductance by L_{e1}, we have

$$L_{e1} = \frac{N_1 \Phi_{e1}}{I_1} \tag{3}$$

Using the same definition, the *external self-inductance* of Coil 2 is

$$L_{e2} = \frac{N_2 \Phi_{e2}}{I_2} \tag{4}$$

In addition to the self-flux linkage, each coil is also linked by the mutual flux due to the current in the other coil. The mutual flux due to the current of the first coil multiplied by the number of turns of the second coil gives the *mutual flux linkage* of the second coil. The ratio between this flux linkage and the current in the first coil is called *mutual inductance* of the second coil.

Accordingly, the mutual inductance of Coil 2, denoted by M_{21}, is

$$M_{21} = \frac{N_2 \Phi_{21}}{I_1} \tag{5}$$

For a linear magnetic circuit, Φ_{21} is proportional to the mmf of Coil 1, i.e., to $N_1 I_1$. The constant of proportionality is the permeance of the mutual flux path. Denoting the permeance by $\wp_{21}$, we have

$$\wp_{21} = \frac{\Phi_{21}}{N_1 I_1} \tag{6}$$

Eliminating Φ_{21} from (5) and (6), we get

$$M_{21} = \wp_{21} N_1 N_2 \tag{7}$$

Similarly the mutual inductance of Coil 1, denoted by M_{12}, is

$$M_{12} = \frac{N_1 \Phi_{12}}{I_2} = \wp_{12} N_1 N_2 \tag{8}$$

where

$$\wp_{12} = \frac{\Phi_{12}}{N_2 I_2} \tag{9}$$

The coupled coils shown at (*a*) and (*b*) of Fig. 4.46 have identical magnetic circuits. Further, for a linear magnetic circuit,

$$\frac{\Phi_{21}}{N_1 I_1} = \frac{\Phi_{12}}{N_2 I_2}$$

Therefore, from (6) and (9), we get

$$\wp_{12} = \wp_{21} = \wp$$

Thus, the mutual inductances given by (7) and (8) have the same value for a linear magnetic circuits. By using the symbol M for denoting the mutual inductance, we have

$$M = \wp N_1 N_2$$

The mutual inductance is proportional to the product of number of turns of the two coils. The constant of proportionality is permeance of the magnetic circuit that is common to both the coils.

If both the coils of Fig. 4.46 are excited, each coil will be linked by two flux components, namely, the external self-flux due to its own mmf and the mutual flux due to the mmf of the other coil. Accordingly, the inductance of each coil has two components. These components are external self-inductance and mutual inductance.

> If the self flux and mutual flux are in the same sense, the total inductance of Coil 1 is equal to the sum of its external self-inductance and the mutual inductance. If the two flux components are in opposite sense, the total inductance of each coil is equal to the difference between the self inductance and mutual inductance.

Thus total inductance of Coil 1 denoted by L_1, and of Coil 2 denoted by L_2, are

$$L_1 = L_{e1} \pm M$$

$$L_2 = L_{e2} \pm M$$

where L_{e1} and L_{e2} are external self-inductances of coils 1 and 2 as defined in (3) and (4) respectively.

Now, let us assume that mutual flux Φ_{21} that links with Coil 2 is $k_1\,(0 < k_1 < 1)$ times Φ_{e1}. That is, $\Phi_{21} = k_1\Phi_{e1}$. Then, the mutual inductance of the coils in terms of Φ_{e1} is

$$M = k_1 \frac{N_2\Phi_{e1}}{I_1} \tag{10}$$

Similarly, if the mutual flux Φ_{12} that links Coil 1 is $k_2\,(0 < k_2 < 1)$ times Φ_{e2}, then the mutual inductance of Coil 1 in terms of Φ_{e2} is

$$M = k_2 \frac{N_1\Phi_{e2}}{I_2} \tag{11}$$

Multiplication of (10) and (11) gives

$$M^2 = k_1 k_2 \left(\frac{N_1\Phi_{e1}}{I_1}\right)\left(\frac{N_2\Phi_{e2}}{I_2}\right) = k_1 k_2 L_{e1} L_{e2}$$

Assuming that $k_1 = k_2 = k$, the mutual inductance of the coils is given by

$$M = k\sqrt{L_{e1} L_{e2}} \tag{12}$$

Formula (12) gives the relationship between mutual inductance and external self-inductances of two magnetically coupled coils. The constant k is called coupling coefficient. The coefficient is nearly equal to unity if the coils are coupled through a magnetic medium. If the medium is air, the coupling coefficient is substantially less than unity.

If two coils of self-inductances L_{e1} and L_{e2}, and mutual inductance M are connected in series, the total inductance of the series combination,

$$L \quad L_e + L_{e2} \quad M$$

The plus sign is to be used if fluxes due to the coils have the same direction in a magnetic circuit, and the minus sign if the directions are opposite.

EXAMPLE 4.45

A closely wound coil of 150 turns produces a flux of 0.3 mWb when carrying 0.8 A. Find the external self-inductance of the coil.

$$\text{External self-inductance, } L_e = \frac{N\Phi}{I} = \frac{(150)(0.3\times10^{-3})}{0.8} = 56.25 \text{ mh}$$

EXAMPLE 4.46

The length ℓ and uniform cross-sectional area S of a rectangular magnetic circuit with iron core are 70 cm and 12.0 cm² respectively. The relative permittivity of the core material μ_r is 260. The exciting coil has 320 turns. Find the external self-inductance of the coil.

External self-inductance, $L_e = \dfrac{N^2}{\Re}$ where N is number of turns of the coil and $\Re$ is reluctance of the magnetic circuit.

$$\text{Reluctance, } \Re = \frac{\ell}{\mu S} = \frac{0.7}{(4\pi\times10^{-7})(260)(12\times10^{-4})} = 1.785\times10^{6} \text{ H}^{-1}$$

$$\text{Self-inductance, } L_e = \frac{(320)^2}{1.785\times10^{6}} = 57.37 \text{ mH}$$

EXAMPLE 4.47

Consider that the magnetic circuit of Example 4.46 has an air gap of 1.0 mm. Find external self-inductance of the coil. Neglect leakage and fringing at the gap.

$$\text{Air gap reluctance, } \Re_g = \frac{\ell_g}{\mu_0 S} = \frac{10^{-3}}{(4\pi\times10^{-7})(12\times10^{-4})} = 0.663\times10^{6} \text{ H}^{-1}$$

$$\text{Total reluctance, } \Re_i + \Re_g = (1.785 + 0.663)\times10^{6} = 2.448\times10^{-6} \text{ H}^{-1}$$

$$\text{External self-inductance, } L_{e1} = \frac{(320)^2}{2.448\times10^{6}} = 41.83 \text{ mH}$$

EXAMPLE 4.48

The length ℓ and uniform cross-sectional area S of a rectangular magnetic circuit with iron core are 120 cm and 24.0 cm² respectively. The core carries two coils 1 and 2 of number of turns $N_1 = 750$ and $N_2 = 375$ respectively. The relative permittivity of the core material is 600. Find (a) external self-inductances of the coils, and (b) the mutual inductance between the coils if the mutual flux is 92 per cent of the self-flux.

$$\text{(a) Reluctance, } \Re = \frac{\ell}{\mu S} = \frac{1.2}{(4\pi\times10^{-7})(600)(24\times10^{-4})} = [\text{illegible}]\times10^{5} \text{ H}^{-1}$$

$$\text{External self-inductance of Coil 1, } L_{e1} = \frac{(750)^2}{6.66\times10^{5}} = 0.845 \text{ H}$$

External self-inductance of Coil 2, $L_{e2} = \dfrac{(375)^2}{6.66\times10^5} = 0.211$ H

Coupling coefficient, $k = 0.92$

Mutual inductance, $M = k\sqrt{L_{e1}L_{e2}} = 0.92\sqrt{(0.845)0.211)} = 0.388$ H

EXAMPLE 4.49

The number of turns of two coupled coils 1 and 2 are $N_1 = 360$ and $N_2 = 480$ respectively. The external self-inductance of Coil 1 is $L_{e1} = 0.2$ mH, and the mutual inductance between the coils is $M = 0.24$ mH. Find the coupling coefficient.

As the self-inductances are proportional to the square of number of turns,

$$L_{e2} = \frac{N_2^{\,2} L_{e1}}{N_1^{\,2}} = \left(\frac{480}{360}\right)^2 (0.2) = 0.356 \text{ mH}$$

Coupling coefficient, $k = \dfrac{M}{\sqrt{L_1L_2}} = \dfrac{0.24}{\sqrt{(0.2)(.356)}} = 0.899$

EXAMPLE 4.50

Consider that in Example 4.49 the two coils are connected such that their mmfs around the magnetic circuit (a) are in the same direction and (b) are in the opposite direction. Find the inductances for the two types of connections.

(a) Inductance of Coil 1, $L_1 = L_{e1} + M$

Inductance of Coil 2, $L_2 = L_{e2} + M$

Total inductance, $L = L_1 + L_2 = L_{e1} + L_{e2} + 2M = 0.2 + 0.356 + 2(0.24) = 1.036$ mH

(b) Inductance of Coil 1, $L_1 = L_{e1} - M$

Inductance of Coil 2, $L_2 = L_{e2} - M$

Total inductance, $L = L_1 + L_2 = L_{e1} + L_{e2} - 2M = 0.2 + 0.356 - 2(0.24) = 0.076$ mH

4.19 Calculation of Inductances

In the preceding section we have defined various inductances in terms of flux linkages. In this section we will calculate internal inductance of cylindrical conductors, external inductance of a toroidal coil, external inductance of a long solenoid, inductance of various types of transmission lines, and mutual inductance of some simple coupled coils.

Internal Inductance of Round Conductors

Let us consider at first a long, straight and solid cylindrical conductor of radius a, as shown in Fig. 4.47(a). The conductor carries a uniformly distributed steady current I. The flux density inside the conductor at a radial distance r, as per formula (7) of Section 4.6, is given by

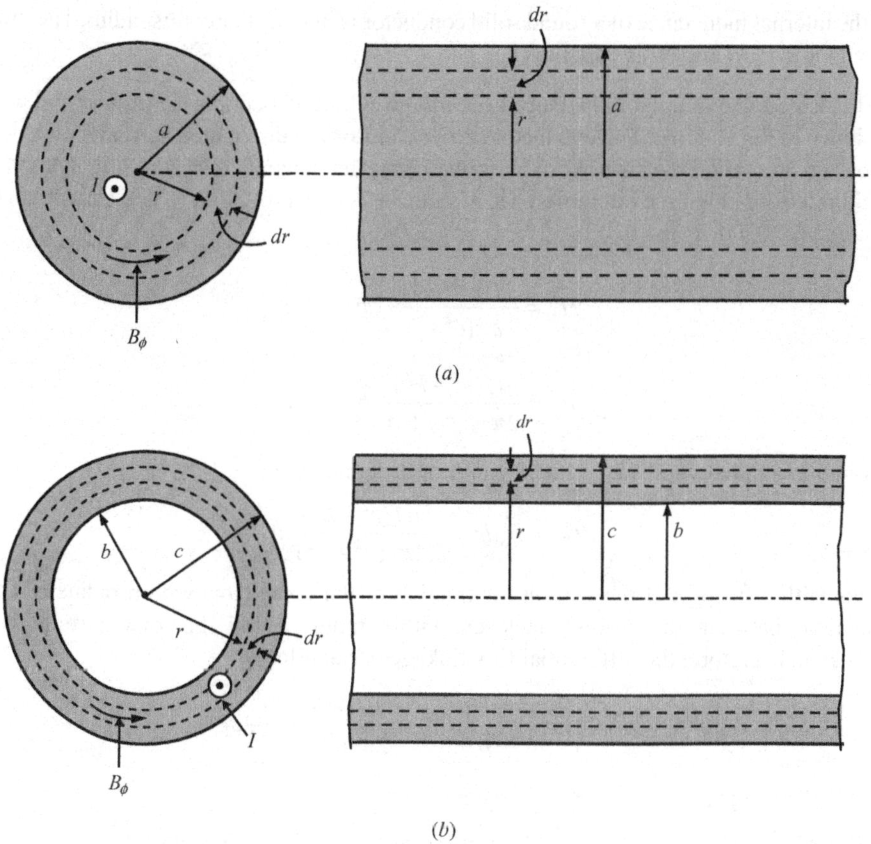

Fig. 4.47 (*a*) A solid cylindrical conductor and (*b*) A hollow cylindrical conductor.

$$B_\phi = \mu H_\phi = \frac{\mu I r}{2\pi a^2},\ 0 < r < a$$

The internal flux within a radial width dr and unit axial length is

$$d\Phi_i = B_\phi dr = \frac{\mu I r}{2\pi a^2} dr$$

This flux links with a fraction of the current (or fraction of the conductor) within radius r. The fraction is equal to the ratio of the current enclosed within radius r and the total current. It is equal to r^2/a^2. Thus, the differential flux linkage per unit length is

$$d(N\Phi_i) = \left(\frac{r^2}{a^2}\right)\left(\frac{\mu I r\, dr}{2\pi a^2}\right) = \left(\frac{\mu I r^3}{2\pi a^4} dr\right)$$

The total internal flux linkage per unit length,

$$N\Phi_i = \int_0^a d(N\Phi_i) = \frac{\mu I}{8\pi}$$

The internal inductance per unit length of the conductor,

$$L_i = \frac{N\Phi_i}{I} = \frac{\mu}{8\pi},\ \text{H/m} \tag{1}$$

We see that the internal inductance of a round solid conductor is independent of its radius. The internal inductance of the conductor of length ℓ is equal to ℓL_i.

We will next calculate the internal inductance of a *hollow* cylindrical conductor of inner radius b and outer radius c, as shown in Fig. 4.47(b). The conductor carries a uniformly distributed current I. Flux density within the conductor and at a radial distance r can be found using the magnetic circuital law. The line integral of magnetic field intensity along the circumference at radius r is equal to $2\pi r H_\phi$. The current enclosed by the path is $I(r^2 - b^2)/(c^2 - b^2)$. Thus,

$$H_\phi = \frac{I(r^2 - b^2)}{2\pi r(c^2 - b^2)}, \; b < r < c$$

and

$$B_\varphi = \frac{\mu I(r^2 - b^2)}{2\pi r(c^2 - b^2)}, \; b < r < c$$

Internal flux within a radial width dr and unit length of the conductor is

$$d\Phi_i = B_\phi dr = \frac{\mu I(r^2 - b^2)}{2\pi r(c^2 - b^2)} dr$$

This flux links with a fraction of the current (or fraction of the conductor) within radius r. The fraction is equal to the ratio between the current enclosed within radius r and the total current. It is equal to $(\quad)/(c^2 - b^2)$. Therefore, the differential flux linkage per unit length,

$$d(N\Phi_i) = \frac{r^2 - b^2}{c^2 - b^2} d\Phi_i = \frac{\mu I(r^2 - b^2)^2}{2\pi r(c^2 - b^2)^2} dr$$

Total internal flux linkage per unit length of the conductor,

$$N\Phi_i = \int_b^c d(N\Phi_i) = \frac{\mu I[c^4 + 3b^4 - 4b^2c^2 + 4b^4 \ln(c/b)]}{8\pi(c^2 - b^2)^2}$$

Thus the internal inductance of a circular hollow cylindrical conductor per unit length,

$$L_i = \frac{N\Phi_i}{I} = \frac{\mu[c^4 + 3b^4 - 4b^2c^2 + 4b^4 \ln(c/b)]}{8\pi(c^2 - b^2)^2} \text{ H/m} \tag{2}$$

The internal inductance of the conductor of length ℓ is equal to ℓL_i

External Self-Inductance of a Toroidal Coil

A toroidal coil with inner radius a and outer radius b is shown in Fig. 4.18. The coil has N number of turns which are uniformly distributed. Magnetic field intensity within the core at a radius R from the centre is

$$H = \frac{NI}{2\pi R}, \; b > R > a$$

Assuming that the permeability of the core material is independent of R, magnetic flux density,

$$B = \frac{\mu NI}{2\pi R}, \quad b > R > a$$

The flux density is along circumferential direction.

External flux within a radial width dR and axial depth c is

$$d \;\; {}_e \quad B(c\,dR) = \frac{\mu c NI\, dR}{2\pi R}$$

Hence total core flux,

$$\Phi_e = \int_a^b d\Phi_e = \frac{\mu c N I}{2\pi}\ln\frac{b}{a}$$

Assuming that all the flux links with all the turns, the external self-inductance,

$$L_e = \frac{N\Phi_e}{I} = \frac{\mu c N^2}{2\pi}\ln\frac{b}{a}$$

External Inductance of a Long Solenoid

Flux density inside a long solenoid shown in Fig. 4.19(*a*) is

$$B = \frac{\mu N I}{\ell}$$

where N is the total number of turns and ℓ is length of the solenoid.

For a circular core of radius a, the external flux,

$$\Phi_e = \frac{\pi a^2 \mu N I}{\ell}$$

Assume that all the flux links with all the turns, external inductance of the solenoid is

$$L_e = \frac{N\Phi_e}{I} = \frac{\pi a^2 \mu N^2}{\ell}$$

Inductance of a Coaxial Transmission Line

The total inductance of a coaxial transmission line having a solid inner conductor, shown in Fig. 4.17, consists of three components:

(i) Internal inductance of the inner conductor.
(ii) Inductance due to linkage of flux in the dielectric region between $r = a$ and $r = b$. The flux in this region is due to the current of the inner conductor and it link with only the inner conductor.
(iii) Inductance due to linkage of flux that is within the outer conductor.

The internal inductance of a solid inner conductor is given by (1). Thus, assuming that the conductor is non-magnetic, the first component of inductance for one meter of line length is

$$L_1 = \frac{\mu_0}{8\pi} \text{ H/m}$$

The second component of inductance L_2 is found as follows. Flux density at a radial distance r,

$$B_\phi = \frac{\mu_0 I}{2\pi r}, \quad b > r > a$$

Flux within a radial thickness dr and line length of one meter,

$$d\Phi_2 = B_\phi dr = \frac{\mu_0 I dr}{2\pi r}$$

Total flux in the region,

$$\Phi_2 = \int_a^b d\Phi_2 = \frac{\mu_0 I}{2\pi}\ln\left(\frac{b}{a}\right)$$

This flux links only with the inner conductor ($N = 1$). Hence, the second component of inductance is

$$L_2 = \frac{\Phi_2}{I} = \frac{\mu_0}{2\pi}\ln\left(\frac{b}{a}\right), \text{ H/m}$$

The third component of inductance is calculated as follows. Using formula (4.6-10), flux density at a radial distance r (r is greater than b but less than c) is

$$B_\phi = \frac{\mu_0 I(c^2 - r^2)}{2\pi(c^2 - b^2)r}, \quad c > r > b$$

Flux within a radial thickness dr and axial length of one meter,

$$d\Phi_3 = B_\phi\, dr = \frac{\mu_0 I(c^2 - r^2)\,dr}{2\pi(c^2 - b^2)r}$$

This flux links once with the current of the inner conductor ($N = 1$), and a fraction of the current of the outer conductor. The fraction is equal to $(r^2 - b^2)/(c^2 - b^2)$. Since the currents in the two conductors are in opposite directions, the differential flux linkage is given by

$$N(d\Phi_3) = \left(1 - \frac{r^2 - b^2}{c^2 - b^2}\right)d\Phi_3 = \frac{\mu_0 I(c^2 - r^2)^2\,dr}{2\pi(c^2 - b^2)^2 r}$$

The third component of inductance,

$$L_3 = \int_b^c \frac{d(N\Phi_3)}{I} = \frac{\mu_0}{8\pi}\left[\frac{4c^4 \ln(c/b)}{(c^2 - b^2)^2} - \frac{3c^2 - b^2}{c^2 - b^2}\right]$$

The total inductance of the transmission line per unit length is

$$L = L_1 + L_2 + L_3 = \frac{\mu_0}{8\pi}\left[1 + \frac{4c^4 \ln(c/b)}{(c^2 - b^2)^2} - \frac{3c^2 - b^2}{c^2 - b^2}\right] + \frac{\mu_0}{2\pi}\ln\left(\frac{b}{a}\right) \text{ H/m} \quad (3)$$

As the first term of (3) is due to the internal flux linkages, it gives the internal inductance of the transmission line per unit length. The second term which is due to the linkage of flux that is external to the conductors is the external inductance per unit length. Thus, the external inductance per unit length,

$$L_e = \frac{\mu}{2\pi}\ln(b/a) \text{ H/m} \quad (4)$$

The capacitance per unit length of a coaxial transmission line having a single dielectric between the conductors has been derived in Section 2.27. It is given by

$$C = \frac{2\pi\varepsilon}{\ln(b/a)} \text{ F/m} \quad (5)$$

Multiplying the two sides of (4) by the corresponding sides of (5), we get

$$L_e C = \mu\varepsilon \quad (6)$$

We see that the product of L_e C is not a function of either the dimensions of the conductors or the spacing between the conductors. It is a function of the constants of the medium only. This is valid for any type of two-conductor, symmetrical and lossless transmission line.

Inductance of a Transmission Line consisting of Two Parallel Cylindrical Conductors

A transmission line consisting of two parallel, solid and round conductors C_1 and C_2, each of radius a and centre to centre spacing h, is illustrated in Fig. 4.48. If the transmission line is connected to a dc source at one end and to a load at the other end, equal but oppositely directed current I will flow in the lines. The directions of current in the two wires are indicated in the figure. We assume that the conductors are nonmagnetic and these are located in free space.

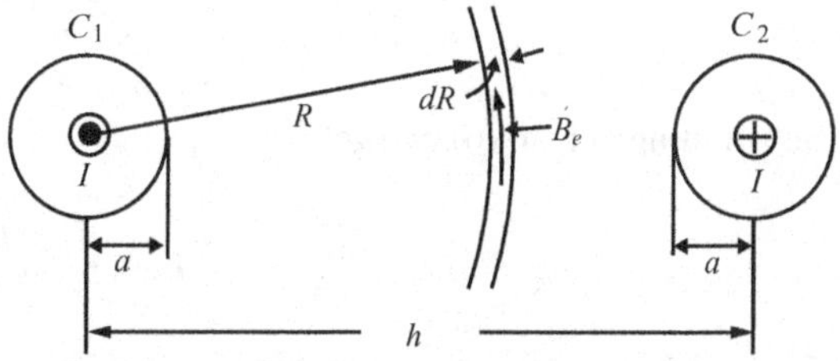

Fig. 4.48 A transmission line consisting of two parallel cylindrical conductors

The inductance of the transmission line is equal to the sum of inductances of both the conductors. Since the conductors have the same radius, it is sufficient to calculate the inductance of one conductor, say of C_1, and then multiply the result by two to get the inductance of the line. The inductance of each conductor consists of two components, namely, its internal inductance and the external inductance. If the distance between the conductors is large compared to the radius of a conductor, we can assume that its internal inductance is same as that of an isolated conductor given by (1). Accordingly, the internal inductance per meter length of one conductor, assuming it to be of a nonmagnetic material,

$$L_i = \frac{\mu_0}{8\pi} \text{ H/m} \tag{7}$$

The external inductance is derived by assuming that $h \gg a$ and the currents are concentrated along the axis of the respective conductors. Thus, the flux density at a distance R from the centre of conductor C_1 is

$$B_e = \frac{\mu_0 I}{2\pi R}, \quad a < R < h$$

Differential flux within a radial thickness dR and one meter length of the conductor,

$$d\Phi_e = B_e(dR) = \frac{\mu_0 I}{2\pi R} dR \tag{8}$$

The approximate value of external flux that links with conductor C_1 is found by integrating (8) from $R \quad a$ to $R = h$. Thus,

$$\Phi_e \approx \int_a^h d\Phi_e = \frac{\mu_0 I}{2\pi} \ln\frac{h}{a}$$

As $N = 1$, external inductance of the conductor,

$$L_e = \frac{\Phi_e}{I} = \frac{\mu_0}{2\pi} \ln\frac{h}{a} \text{ H/m} \tag{9}$$

The inductance per unit length of the transmission line is equal to *twice* the sum of (7) and (9). Thus,

$$L = \frac{\mu_0}{4\pi} + \frac{\mu_0}{\pi} \ln\frac{h}{a} \text{ H/m} \tag{10}$$

The exact formula for the external inductance can be obtained by using the expression for capacitance per unit length of the transmission line, which has been derived in Section 2.28, in relation (6). The formula for capacitance per unit length of the line located in free space,

$$C = \frac{\pi\varepsilon_0}{\ln\dfrac{h+\sqrt{h^2-4a^2}}{2a}} \text{ F/m} \tag{11}$$

Substituting (11) in (6), we get

$$L_e = \frac{\mu_0}{\pi}\ln\frac{h+\sqrt{h^2-4a^2}}{2a} \text{ H/m} \tag{12}$$

If the spacing between the conductors is large compared to the conductor radius $(h \gg a)$, the approximate formula for the external inductance per unit length is found by substituting this condition in (12). It is given by

$$L_e \approx \frac{\mu_0}{\pi}\ln\frac{h}{a} \text{ H/m} \tag{13}$$

This is same as (10)

External Inductance of a Parallel-Plane Transmission Line

A parallel-plane transmission line consists of two plane conductors separated by an insulating medium as shown in Fig. 4.49. The width and length of each conductor are w and ℓ respectively and the spacing between the conductors is taken as s. The planes carry equal and oppositely directed current I. If s is small compared to w and ℓ, we can assume that the flux density between the planes is constant. The direction of flux density is as shown in the figure.

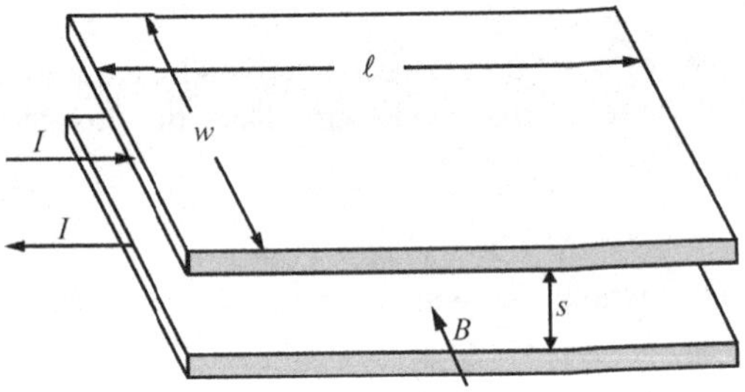

Fig. 4.49 A parallel-plane transmission line

The current per unit width,

$$K = \frac{I}{w}$$

Using formula (4.6-12), the flux density between the conductors,

$$B = \frac{\mu_0 I}{w}$$

The flux between the planes is the external flux. The external flux per unit length of the transmission line,

$$\Phi_e = sB = \frac{s\mu_0 I}{w} \tag{14}$$

Each conductor is linked by one-half of this flux. As $N = 1$, the flux linkage per unit length for each conductor is $\Phi_e/2$. Accordingly, external inductance per metre length of each conductor is $\Phi_e/(2I)$ and inductance of both the conductors is equal to Φ_e/I. Thus substituting the expression for Φ_e given by (14), the external inductance of the parallel transmission line per unit length,

$$L_e = \frac{\mu_0 s}{w} \text{ H/m}$$

Mutual Inductance between Two Coaxial Air-Cored Solenoids

Two coaxial solenoids having radii a_1 and a_2 and number of turns N_1 and N_2 are shown in Fig. 4.50. The inner solenoid, which may be used as a search coil for measurement purposes, is made of fine wire of a few turns. The mutual inductance between the solenoids is calculated assuming that the outer solenoid is long and has a uniformly distributed winding. Therefore, flux density within the solenoid is approximately given by

$$B_1 = \frac{\mu_0 N_1 I_1}{\ell_1}$$

where I_1 and ℓ_1 are the current and length of the outer solenoid.

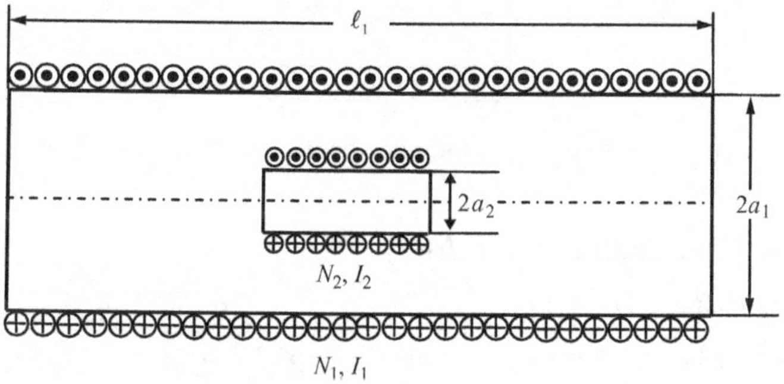

Fig. 4.50 Two concentric solenoids

The mutual flux is equal to the flux within the inner solenoid. Thus,

$$\Phi_{21} = \pi a_2^2 B_1 = \frac{\pi a_2^2 \mu_0 N_1 I_1}{\ell_1}$$

The mutual inductance, $M = \frac{N_2 \Phi_{21}}{I_1} = \frac{\pi a_2^2 \mu_0 N_1 N_2}{\ell_1}$

It may be noted that the mutual inductance does not depend on the radius of the outer solenoid.

Mutual Inductance between a Long line and a Rectangular Loop in Free Space

The arrangement of a long line carrying current I_1 and a rectangular loop of N_2 turns which are on the same plane is illustrated in Fig. 4.51. The length and width of the loop are a and b respectively. The nearest side of the loop is at a distance c from the line current.

Flux density at a distance x from the line current, as shown in the figure, is

$$B_1 = \frac{\mu_0 I_1}{2\pi x}$$

The flux density is normal to the plane of the loop. Hence flux within a differential width dx,

$$d\Phi_{21} = B_1 (a\,dx) = \frac{\mu_0 a I_1\,dx}{2\pi x}$$

The total flux through the loop,

$$\Phi_{21} = \frac{\mu_0 a I_1}{2\pi} \int_c^{c+b} \frac{dx}{x} = \frac{\mu_0 a I_1}{2\pi} \ln \frac{c+b}{c}$$

This is the mutual flux that links with Coil 2. The mutual inductance of the system,

$$M = \frac{N_2 \Phi_{21}}{I_1} = \frac{a\mu_0 N_2}{2\pi} \ln \frac{c+b}{c}$$

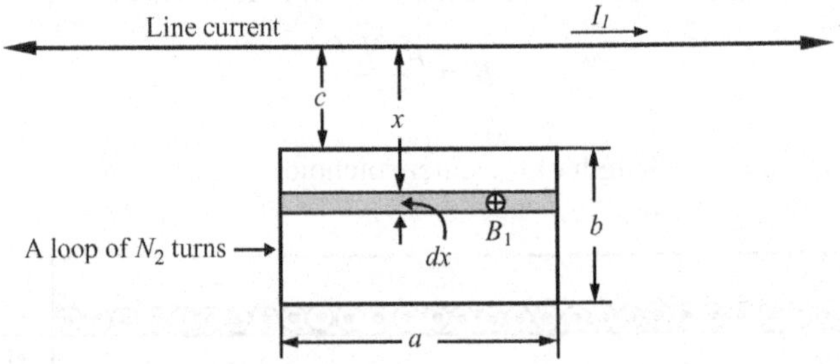

Fig. 4.51 For calculation of mutual inductance between a long wire and a rectangular loop

Mutual Inductance between a Toroidal Coil and a Long Conductor

Figure 4.52 shows a toroidal coil and a long straight cylindrical conductor of circular section. The coil of N_2 turns is wound on a core of rectangular section. The inner and outer radii of the toroid are a and b respectively. The axial length of the core is c and its permeability is μ. The conductor is along the axis of the toroid.

In order to calculate the mutual flux between the coil and the conductor, let us assume that the conductor carries a current I_1. Magnetic flux density at a radial distance R from the axis of the conductor,

$$B_1 = \frac{\mu I_1}{2\pi R}, \quad b > R > a$$

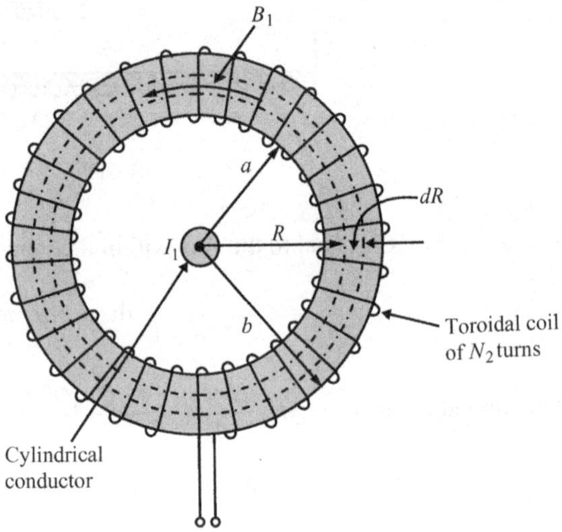

Fig. 4.52 Illustration of a toroidal coil and a cylindrical conductor for calculation of mutual inductance.

The flux density is in the circumferential direction, as indicated in the figure. Thus flux within a radial thickness dR of the core,

$$d\Phi_{21} = B_1(c\,dR) = \frac{c\mu I_1}{2\pi R} dR$$

The total flux inside the core is found by integrating the above expression from $R = a$ to $R = b$.

$$\Phi_{21} = \frac{c\mu I_1}{2\pi} \int_a^b \frac{dR}{R} = \frac{c\mu I_1}{2\pi} \ln \frac{b}{a}$$

This is the mutual flux and it links with N_2 turns. Therefore the mutual inductance,

$$M = \frac{N_2 \Phi_{21}}{I_1} = \frac{c, N_2}{2\pi} \ln \frac{b}{a}$$

EXAMPLE 4.51

The internal and external radii of a hollow nonmagnetic cylindrical conductor are $b = 4$ cm and $c = 6$ cm respectively. Find the internal inductance of the conductor of length $\ell = 10.0$ m.

$$\text{Internal inductance, } L_i = \frac{\mu_0 \ell [c^4 + 3b^4 - 4b^2c^2 + 4b^4 \ln(c/b)]}{8\pi(c^2 - b^2)^2}$$

$$= \frac{(4\pi \times 10^{-7})(10)[6^4 + (3)(4)^4 - (4)(4)^2(6)^2 + (4)(4)^4(\ln 6/4)](10^{-8})}{8\pi(6^2 - 4^2)^2(10^{-8})}$$

$$= 2.19 \times 10^{-7} \text{ H}$$

EXAMPLE 4.52

The inner radius, outer radius and axial depth of a toroid of rectangular cross section shown in Fig. 4.38(a) are $a = 30$ cm, $b = 40$ cm, and $c = 15$ cm respectively. It has a uniformly distributed coil of 1000 turns. The relative permeability of the core material is 950. Find the external inductance of the exciting coil of the toroid. Neglect leakage.

$$\text{External inductance, } L_e = \frac{\mu c N^2}{2\pi} \ln \frac{b}{a}$$

Given: $\mu_r = 950$; $a = 0.3$ m; $b = 0.4$ m; $c = 0.15$ m; $N = 1000$

$$L_e = \frac{(4\pi \times 10^{-7})(950)(0.15)(1000)^2}{2\pi} \ln \frac{40}{30} = 8.2 \text{ H}$$

EXAMPLE 4.53

Suppose that the toroid of Example 4.52 has a search coil of 125 turns in addition to the main coil. Assume perfect magnetic coupling between the coils. Find the mutual inductance between the coils.

External self-inductance of the main coil, $L_{e1} = 8.2$ H

External self-inductance of the search coil, $L_{e2} = (8.2)\left(\frac{125}{1000}\right)^2 = 0.128$ H

Coupling coefficient, $k = 1.0$

Mutual inductance between the coils, $M = k\sqrt{L_1 L_2} = \sqrt{8.2 \times 0.128} = 1.024$ H

EXAMPLE 4.54

An air-core solenoid of length 1.2 m and radius 4.0 cm has a uniformly distributed coil of 1500 turns. Find the external inductance of the solenoid.

$$\text{External inductance, } L_e = \frac{\pi \mu_0 a^2 N^2}{\ell}$$

Given: a m; m, $N = 1500$

$$L_e = \frac{\pi(4\pi\times10^{-7})(0.04)^2(1500)^2}{1.2} = 11.8\text{ mH}$$

EXAMPLE 4.55

In Example 4.54, a search coil with a nonmagnetic core is coaxially located at the centre of the solenoid. The radius of the search coil is 0.5 cm and it has 50 turns. Find the mutual inductance between the main coil and search coil.

Mutual inductance, $M = \dfrac{\pi\mu_0 a_2{}^2 N_1 N_2}{\ell}$

Given: $a_2 = 0.005$ m; $N_1 = 1500$; $N_2 = 50$; $\ell_1 = 1.2$ m

$$M = \frac{\pi(4\pi\times10^{-7})(0.005)^2(1500)(50)}{1.2} = 7.4\ \mu\text{H}$$

EXAMPLE 4.56

A coaxial transmission line has a solid inner conductor of radius 3.0 mm. The radial thickness of insulation between the conductors is 5.0 mm. Length of the line is 6.0 m. Calculate the external inductance of the line. Assume $\mu_r = 1.0$.

Radius of inner conductor, $a = 3.0$ mm

Inner radius of the outer conductor, $b = 3.0 + 5.0 = 8.0$ mm

Length of the line, $\ell = 6.0$ m

External inductance, $\ell L_e = \dfrac{\mu_0 \ell}{2\pi}\ln\dfrac{b}{a} = \dfrac{(4\pi\times10^{-7})(6.0)}{2\pi}\ln\dfrac{8.0}{3.0} = 1.177\ \mu\text{H}$

EXAMPLE 4.57

A two-conductor transmission line consists of two parallel cylindrical conductors, each of radius 1.2 mm. The centre-to-centre distance between the conductors is 16.0 mm. The length of the line is 15 m. Calculate inductance of the line assuming relative permeability of the medium between the conductors as unity.

Radius of inner conductor, $a = 1.2$ mm

Distance between conductor centres, $h = 16.0$ mm

Length of the line, $\ell = 15.0$ m

Internal inductance of both the conductors, $2\ell L_i = \dfrac{\ell\mu_0}{4\pi} = \dfrac{(15)(4\pi\times10^{-7})}{4\pi} = 1.5\ \mu\text{H}$

External inductance, $\ell L_e = \dfrac{\mu_0 \ell}{\pi}\ln\dfrac{h \quad h^2 - 4a^2}{2a}$

$$= \frac{4\pi \times 10^{-7} \times 15.0}{\pi} \ln \frac{16.0 + \sqrt{256.0 - 5.76}}{2 \times 1.2} = 15.51\ \mu\text{H}$$

Total inductance of the line = 1.5 +15.51 = 17.01 μH

The external inductance when calculated using formula (10) is found to be 15.54 μH respectively. This is nearly equal to the value obtained using the exact formula (12). If the spacing between the conductors is greater than ten times of conductor radius, the approximate formula (10) may be used.

4.20 Energy Stored in the Magnetic Field

An expression for energy stored in the static electric field of discrete point charges is obtained by finding the work done in bringing the charges from infinity to their final positions. It may be thought, by analogy, that energy in static magnetic field can also be obtained by finding the work done in bringing current elements from infinity to their desired positions.

This approach, however, is more complicated because current elements are vector quantities, and flux linking a current element changes with time resulting in an induced voltage, which must be considered while calculating the magnetic field energy.

An alternative and simple method for finding magnetic field energy is by the use of Maxwell's equations given in Section 6.18. However in this section we will obtain an expression for energy in a magnetic field by using the well known Faraday's law of electromagnetic induction given in Section 6.3. The result will be used to express magnetic field energy in terms of inductance and also to explain hysteresis loss in ferromagnetic materials.

Let us consider a long solenoid, as shown in Fig. 4.19(a). It has a uniformly distributed coil of N turns. Let the resistance of the coil be R. If we connect the coil to a dc source of voltage V current will flow into the coil and a magnetic field will be established inside the solenoid. Since the current, starting from zero, will take some time to attain its steady state value of $I = V/R$ due to inductive action of the coil, the flux linking the coil during this period changes with time and hence voltage will be induced in it. The induced voltage in the coil as per Faraday's law of electromagnetic induction is

$$\mathcal{E} = -N\frac{d\Phi}{dt} \tag{1}$$

where Φ is the flux linking the coil at time t after the coil is connected to the source. At this instant, total voltage across the coil is equal to the sum of $\mathcal{E}$ and V and current in the coil,

$$i = (V + \mathcal{E})/R \tag{2}$$

Eliminating $\mathcal{E}$ from (1) and (2), we get

Or
$$V - iR = N\frac{d\Phi}{dt} \tag{3}$$

By multiplying both sides of (3) by idt, we have

$$(\qquad)dt = (\quad)d\Phi \tag{4}$$

The first term on the left-hand side of (4) is the energy supplied by the source and the second term is the energy dissipated in the resistance of the coil in time dt. The difference between two energies is equal to the energy expended in building up the magnetic field. This energy is stored in the magnetic field. Thus, the right-hand side of (4) is the differential energy stored in the magnetic field during time dt. Denoting the differential energy by dW_m, we have

$$dW_m = (Ni)d\Phi \tag{5}$$

The flux $d\Phi$ is equal to the product of sectional area S of the solenoid and differential change of flux density dB. Also, the mmf Ni is equal to the product of length ℓ of the solenoid and the field intensity H. Therefore, relation (5) may be written as

$$dW_m = (H\ell)(S\,dB) = (H\,dB) \times \text{ Volume of the solenoid} \tag{6}$$

By dividing both sides of (6) by the volume we will get the energy transferred from the source to the magnetic field per unit volume of the field. Thus, denoting the differential energy per unit volume by dw_m, we have

$$dw_m = H\,dB \text{ J/m}^3 \tag{7}$$

Assuming that the initial flux density is zero and the final flux density is B, the total energy per unit volume,

$$w_m = \int_0^B H\,dB \text{ J/m}^3 \tag{8}$$

As H and B are functions at a point in space, equation (8) is quite general though we have used a simple example to derive it.

For a linear medium, $B = \mu H$. Hence,

$$w_m = \int_0^B \frac{B}{\mu} dB = \frac{B^2}{2\mu} \text{ J/m}^3 \tag{9}$$

The right-hand side of (9) gives magnetic energy stored per unit volume. This is called *energy density* of a magnetic field. Thus, energy density of a magnetic field,

$$w_m = \frac{B^2}{2\mu} = \frac{\mu H^2}{2} \text{ J/m}^3 \tag{10}$$

Total energy stored in a linear magnetic field is found by integrating (10) over the volume of the solenoid. Thus,

$$W_m = \frac{1}{2}\int_v \mu H^2 dv \tag{11}$$

If the medium is that of a ferromagnetic material, the relation between B and H is not linear. Therefore, direct integration of (8) is not possible. However, the magnetic energy can be computed by graphical integration.

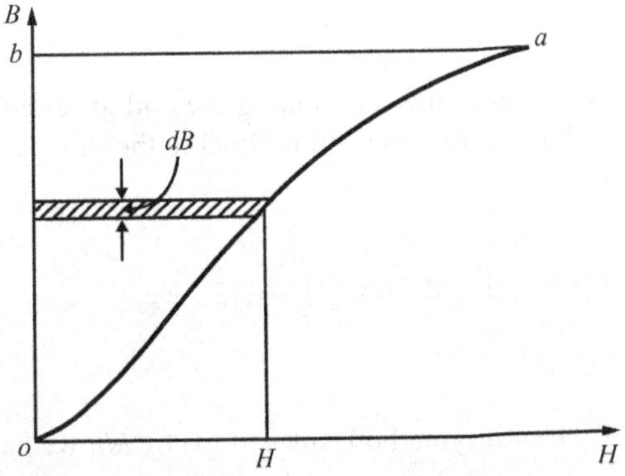

Fig. 4.53 For calculation of energy stored in a magnetic field

In Fig. 4.53, which gives the B-H characteristic of a ferromagnetic material, the hatched area shown is the differential energy density as defined by (7). Total energy stored per unit volume of the medium when flux density changes from 0 to B is proportional to the area between the B-H curve and the B axis, that is, the area enclosed by the lines *oa*, *ab* and *bo*. The area between the B-H curve and H axis is called magnetic co-energy.

Magnetic Energy in Terms of Inductance

The self-inductance of a coil of N turns in terms of flux linkage is

$$L = \frac{N\Phi}{i}$$

$$\Rightarrow \qquad Li = N\Phi$$

where Φ is the self-flux due to the coil current i

Now let a differential change of current di produce a differential change of flux $d\Phi$. Then assuming L to be constant, we get from the above relation,

$$L\,di = N(d\Phi) \tag{12}$$

By replacing $N(d\Phi)$ of (5) by the left-hand side of (12), we have

$$dW_m = Li\,di \tag{13}$$

For a change of current from 0 to I, the energy supplied to the magnetic field is obtained by integrating (13) between the limits. Thus, magnetic field energy in terms of the inductance is given by

$$W_m = \int_0^I Li\,di = \frac{1}{2}LI^2 \tag{14}$$

$$\Rightarrow \qquad L = \frac{2W_m}{I^2} \tag{15}$$

By substituting the right side of (11) in place of W_m of (15), we get

$$L = \frac{\mu}{I^2}\int_v H^2 dv \tag{16}$$

Formula (15) and (16) give alternative methods for calculation of the inductance.

If the field energy at current I_1 is W_{m1} and at current I_2 it is W_{m2}, then change in energy stored in the magnetic field when current changes from I_1 to I_2 is

$$W_{m2} - W_{m1} = \int_{I_1}^{I_2} Li\,di = \frac{1}{2}L(I_2^2 - I_1^2)$$

EXAMPLE 4.58

A long and straight coaxial transmission line has an inner solid conductor of radius, $a = 5$ mm and an outer conductor of inside radius, $b = 12$ mm. It is connected to a load which takes 25 A. Calculate magnetic energy stored in the region between the conductors for one metre length of the line. Also calculate the external inductance per unit length of the line using the calculated value of magnetic energy. Take $\mu = \mu_0$.

Radius of the inner conductor, $a = 0.005$ m

Internal radius of the outer conductor, $b = 0.012$ m

Current, $I = 25.0$ A

Assuming the axis of the line to be along the z axis of cylindrical coordinates, the magnetic field would be in ϕ direction. The field intensity at a radial distance r,

$$H_\phi = \frac{I}{2\pi r} \text{ A/m}, \quad b \geq r \geq a$$

Differential volume at radius r per unit length of the line, $dv = 2\pi r\, dr$

Magnetic energy stored per unit length, $W_m = \frac{\mu_0}{2}\int_v H_\phi^2\, dv = \frac{\mu_0}{2}\int_a^b \left(\frac{I}{2\pi r}\right)^2 (2\pi r\, dr)$

$$= \frac{\mu_0 I^2}{4\pi} \ln(b/a) = \frac{(4\pi\times 10^{-7})(25)^2}{4\pi} \ln(12/5)$$

$$= 547.2 \quad 10^{-7} \text{ J/m}$$

External inductance per unit length of the line,

$$L_e = \frac{2W_m}{I^2} = \frac{2(547.2\times 10^{-7})}{625} = 1.751\times 10^{-7} \text{ H/m}$$

Hysteresis Loss

When a ferromagnetic material is magnetised, energy has to be supplied from an external source for establishing the magnetic field. After the applied field is decreased to the original value, energy is returned by the field to the source. However, energy returned by the field is found to be less than that supplied to the field. The difference between the two energies, called hysteresis loss, is dissipated as heat. The loss of energy per unit volume of the core material is proportional to the area enclosed by the hysteresis loop, as explained below.

Change in energy stored per unit volume in a magnetic field when flux density changes from B_1 to B_2 is

$$w_m = \int_{B1}^{B2} H\, dB$$

Referring to the hysteresis loop of Fig. 4.54 we find that both H and dB are positive from a to b, whereas from b to c though H is positive dB is negative. Moreover, both H and dB are negative from c to d and from d to a the field intensity H is negative and dB is positive. Thus for a complete cycle of the loop, we have

$$w_m = \int_a^b H\, dB + \int_b^c H\,(-dB) + \int_c^d (-H)(-dB) + \int_d^a (-H)\, dB$$

$$= \int_a^b H\, dB - \int_b^c H\, dB + \int_c^d H\, dB - \int_d^a H\, dB \tag{17}$$

The first term gives the area enclosed by the lines *ab*, *be* and *ea*. The area is proportional to the energy received by the magnetic field from a source. The second term gives the area enclosed by the lines *be*, *ec* and *cb*. The area is proportional to the energy released by the field. The difference between the two areas, which is equal to the area enclosed by the lines *ab*, *bc* and *ca*, is proportional to the energy lost in the medium Similarly, the area enclosed by the lines *cd*, *da* and *ac*, which is equal to the difference between the areas represented by the third and forth integrals of (17), is also proportional to the energy lost in the medium. Thus the area enclosed by the hysteresis loop is proportional to the energy lost in one cycle per unit volume of the medium. This energy is dissipated as heat. If the cyclic variation of

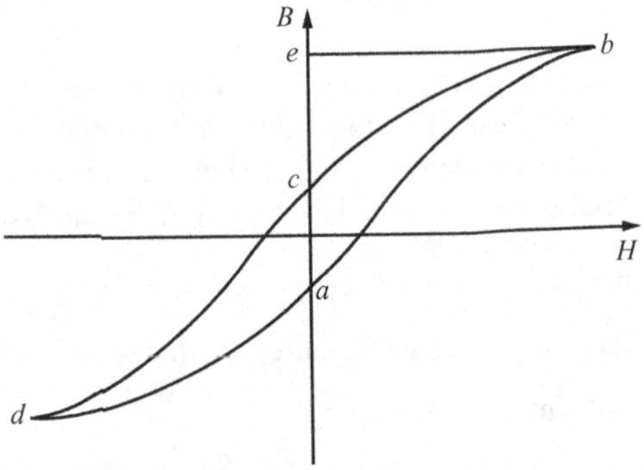

Fig. 4.54 For calculation of hysteresis loss in a ferromagnetic material

applied field is f per second, then power dissipated per unit volume of the medium is proportional to f multiplied by the loop area.

4.21 Forces between Two Iron Surfaces at a Gap

Figure 4.55 shows a part of a magnetic circuit having an air gap. The surface at the gap from which magnetic flux comes out acts as a north pole and the surface into which the flux enters acts as a south pole. Thus, a magnetic force of attraction exists between the two surfaces. A formula for the force can be derived using the technique of virtual work by assuming uniform distribution of flux in the gap area. This assumption is valid if the iron surfaces are parallel and have the same linear dimensions, and the gap length is small.

Let B_g, S_g and l_g denote the flux density in the gap, gap area and gap length respectively. The energy density in the gap,

$$w_m = \frac{B_g^{\ 2}}{2\mu_0} \text{ J/m}^3 \qquad (1)$$

As the energy density has the same value throughout the volume of the gap, energy stored in the gap is obtained by multiplying (1) by the gap volume. Thus, magnetic energy in the field between the iron surfaces,

$$W_m = \frac{B_g^{\ 2} S_g \ell_g}{2\mu_0} \text{ J} \qquad (2)$$

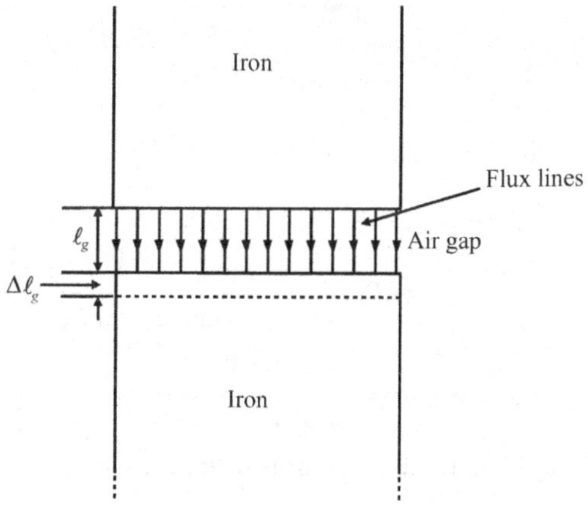

Fig. 4.55 Showing part of a magnetic circuit with an air gap for calculation of gap force

Consider now that the gap length is increased by an infinitesimal amount $\Delta\ell_g$ by an external force F that is equal in magnitude but opposite in direction to the magnetic force F_m. At the same time, the exciting current is also increased so as to keep flux density in the gap constant. Consequently, magnetic energy stored in the gap is increased because of increase in its volume. The increase of gap volume is $S_g \Delta\ell_g$, and the corresponding increase in the magnetic energy is ΔW_m. Thus, from (2), we have,

$$\Delta W_m = \frac{B_g^{\ 2} S_g \Delta\ell_g}{2\mu_0}$$

This must be equal to the energy expended in increasing the gap length. As the energy is equal to $F\Delta\ell_g$, we have

$$F\Delta\ell_g = \Delta W_m = \frac{B_g^{\ 2} S_g \Delta\ell_g}{2\mu_0}$$

$$\Rightarrow \qquad F = \frac{B_g^{\ 2} S_g}{2\mu_0}$$

Therefore, magnetic force on each iron surface, $F_m = F = \dfrac{B_g^{\ 2} S_g}{2\mu_0}$ (3)

The magnetic force on an iron surface near a gap is thus proportional to the product of square of gap density and area of the iron surface. Since there is a practical limit to the flux density because of saturation in iron parts of the magnetic circuit, the magnetic force is increased by increasing the area of iron surface at the gap. Formula (3) gives the *lifting power of a magnet.*

EXAMPLE 4.59

The magnetic circuit shown in Fig. 4.42(*a*) has a mean iron length of 0.75 m and sectional area of the core is 0.12 m². The relative permeability of iron is 1600. The exciting coil has 500 turns and it carries a current of 1.2 A. The gap length is 1.5 mm. Calculate the magnetic force on iron surfaces at the gap neglecting fringing and leakage.

Iron length, $\ell_i = 0.75$ m; Gap length, $\ell_g = 1.5$ mm; Core area, S_i = gap area, $S_g = 0.12$ m²; $\mu_r = 1600$; $N = 500$; $I = 1.2$ A

$$\text{Reluctance of iron} = \frac{\ell_i}{\mu_0\mu_r S_i} = \frac{0.75}{(4\pi\times10^{-7})(1600)(0.12)} = 3.108\times10^3 \text{ 1/H}$$

$$\text{Reluctance of gap} = \frac{\ell_g}{\mu_0 S_g} = \frac{1.5\times10^{-3}}{(4\pi\times10^{-7})(0.12)} = 9.947\times10^3 \text{ 1/H}$$

Total reluctance = 13.055×10^3 1/H

Coil mmf = $NI = 500\times1.2 = 600$ A

$$\text{Gap flux} = \text{core flux} = \frac{\text{mmf}}{\text{Reluctance}} = \frac{600}{13.055\times10^3} = 45.96\times10^{-3} \text{ Wb}$$

$$\text{Gap density, } B_g = \frac{45.96\times10^{-3}}{0.12} = 0.383 \text{ T}$$

$$\text{Magnetic force on an iron surface, } F_m = \frac{B_g^{\,2} S_g}{2\mu_0} = \frac{(0.383)^2(0.12)}{2(4\pi\times10^{-7})} = 7003.9 \text{ N}$$

4.22 Force on a Moving Charge in Stationary Electric and Magnetic Fields

Force on a stationary electric charge Q in a static electric field,

$$\mathbf{F} = Q\mathbf{E} \qquad (1)$$

where **E** is intensity of electric field at the location of Q. Now let us imagine that the charge is *moving* with a finite velocity in a static electric field, as shown in Fig. 4.56. The electric force on the charge would be same as that given by (1). In this case **E** is the intensity of the external field at the *location* of Q at a given time.

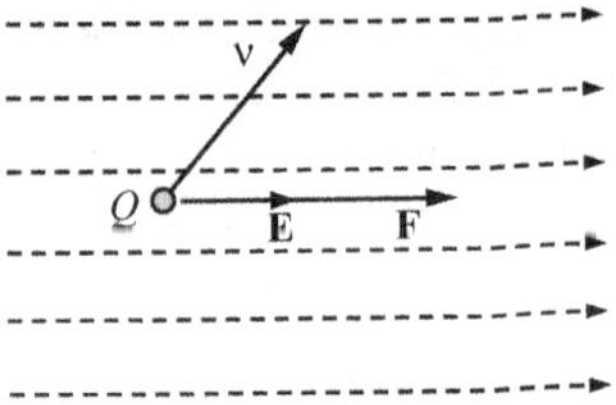

Fig. 4.56 Showing an electric charge Q moving in an electric field. The direction of force is same as that of the field. The magnitude of the force is equal to QE.

It has been found experimentally that a *stationary* charge is not acted by any force in a static magnetic field. However, a *moving* charged particle in a static magnetic field is acted upon by a force. The magnitude of the force F is proportional to the charge magnitude Q, velocity ν (Greek alphabet Nu) of the charge, flux density B at the location of the charge and sine of the angle α between the directions of ν and B. The constant of proportionality in SI unit is unity. Thus magnitude of the force on the charge,

$$F \quad Q(\; B\sin\alpha)$$

The direction of the force is normal to the plane containing the velocity vectors **v** and magnetic flux density **B** and its sense is given by the right-hand rule. The vector form of the force, therefore, is

$$\mathbf{F} = Q(\mathbf{v} \times \mathbf{B}) \qquad (2)$$

If **v** and **B** are on the plane of the paper, as illustrated in Fig. 4.57, the force is normally into the paper. Since dimensionally $Id\ell = Q\mathbf{v}$, equation (4.5-3) is equivalent to equation (2). That a moving charge is acted by a magnetic force shows that magnetic field is produced by a moving charge.

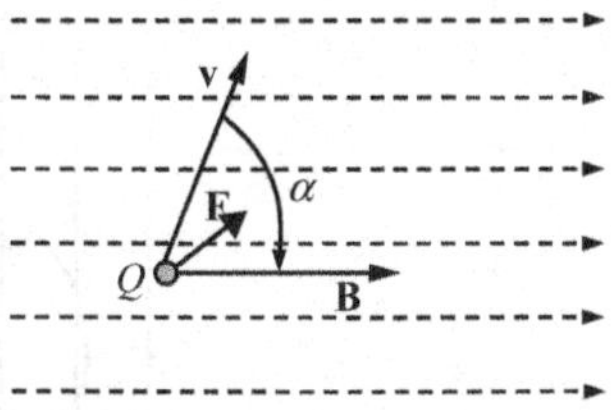

Fig. 4.57 Showing an electric charge Q moving in a magnetic field. The direction of force is normally into the plane of the paper. $F = vQB \sin\alpha$.

As the direction of the force is perpendicular to the direction of movement of the charge, transfer of energy cannot take place to or from the moving charge.

A moving charge having a finite mass will however accelerate in the direction of the electromagnetic force.

The sum of (1) and (2) is the total force on a moving charge in a region having both electric and magnetic fields. The total force,

$$\mathbf{F} = Q[\mathbf{E} + (\mathbf{v} \times \mathbf{B})] \qquad (3)$$

Equation (3) is called *Lorentz force equation*. It may be used to determine trajectory of electrons moving in vacuum in the presence of stationary electric field or magnetic field or both electric and magnetic fields.

EXAMPLE 4.60

An electron enters into a region where both electric and magnetic fields are present, at a velocity of $(200\mathbf{u}_x + 300\mathbf{u}_y)$ m/s in free space. The electric and magnetic field intensities at the location of the charge are $500\mathbf{u}_y$ V/m and $0.01\,\mathbf{u}_x$ tesla respectively. It is required to calculate the acceleration of the electron.

Charge of an electron, $Q = -1.6 \times 10^{-19}$ C

Mass of an electron, $m = 9.1 \times 10^{-31}$ kg

Force experienced by the atom, $\mathbf{F} = Q[\mathbf{E} + (\mathbf{v} \times \mathbf{B})]$

$$= -1.6 \times 10^{-19}\left[500\mathbf{u}_y + \left(200\mathbf{u}_x + 300\mathbf{u}_y\right) \times 0.01\mathbf{u}_x\right]$$

$$= \left(-800\mathbf{u}_y + 4.8\mathbf{u}_z\right)10^{-19} \text{ N}$$

$$\text{Acceleration of the electron} = \frac{\mathbf{F}}{m} = \frac{(-800\,\mathbf{u}_y + 4.8\,\mathbf{u}_z)10^{-19}}{9.1 \times 10^{-31}}$$

$$= (\ 87.9\,\mathbf{u}_y + 0.527\,\mathbf{u}_z)10^{12} \text{ m/s}^2$$

4.23 Path of a Charged Particle in a Uniform Magnetic Field

Force on a charged particle carrying charge Q and moving with a velocity **v** in a magnetic field of flux density **B** is given by (4.22-2). The force on the particle and its acceleration are perpendicular to the plane containing **v** and **B.** Maximum acceleration of the particle is obtained if the angle between **v** and **B** is 90°. Since acceleration of the particle is perpendicular to the direction of its velocity, the particle will revolve in a circular path if it enters into a region of stationary and uniform magnetic field provided its initial velocity is at right angle to **B**.

The radius of the path of the particle is found by equating the electromagnetic force with the centrifugal force acting on the particle.

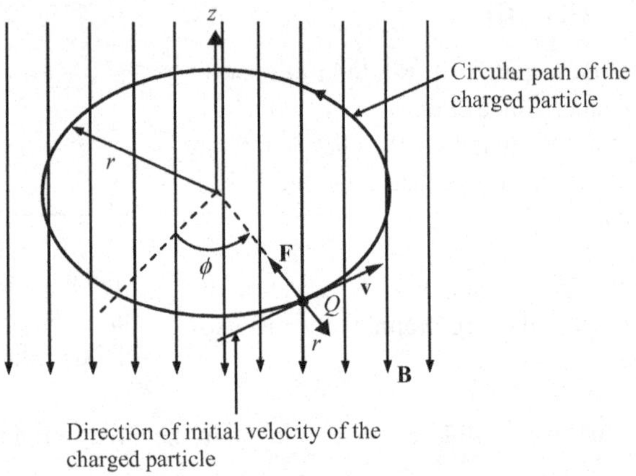

Fig. 4.58 Path of a charged particle with its initial velocity normal to a stationary and uniform magnetic field.

The magnetic field in Fig. 4.58 is assumed to be stationary and of uniform density B. The direction of the field is along negative z direction of cylindrical coordinates and the velocity of the charged particle is along positive ϕ direction. Thus, the vector forms of magnetic flux density and particle velocity are given by

$$\mathbf{B} = -B\mathbf{u}_z$$

and $\mathbf{v}$ $\cdot\cdot\mathbf{u}_\phi$

Electromagnetic force on the particle, $\mathbf{F} = Q(\mathbf{v} \times \mathbf{B}) = -QBv(\mathbf{u}_\phi \times \mathbf{u}_z) = -QBv\,\mathbf{u}_r$

If m is mass of the particle, its acceleration due to the force, $a = \dfrac{\mathbf{F}}{m} = -\dfrac{QvB}{m}\mathbf{u}_r$

Since the acceleration of the particle is perpendicular to its velocity, the particle will describe a circular path as per the laws of circular motion. The radius of the path r is determined from the consideration that under steady state condition the electromagnetic force will be balanced by the centrifugal force on the particle. Thus, we have

$$\frac{mv^2}{r} = QvB$$

Or

$$r = \frac{mv}{QB} \tag{1}$$

The radius is proportional to both mass and initial velocity of the particle. It is inversely proportional to Q as well as B.

The angular velocity of the particle, $\omega = \dfrac{v}{r} = \dfrac{QB}{m}$

Number of revolutions per second is called gyro-frequency. Denoting it by f, we have

$$f = \frac{\omega}{2\pi} = \frac{QB}{2\pi m} \tag{2}$$

Period of revolution, $T = \dfrac{1}{f}$ (3)

Formula (2) shows that the number of revolutions per second does not depend on the initial velocity of the particle.

EXAMPLE 4.61

A charged particle, when at rest but free to move, is subjected to a stationary electric field of intensity $\mathbf{E}$ $\mathbf{u}_x$ kV/m. After accelerating for 0.8 μs, the particle leaves the electric field and enters into a stationary magnetic field of flux density $\mathbf{B} = 0.25\mathbf{u}_y$ mT. Find (a) the velocity of the particle at the instant it enters the magnetic field, (b) the magnetic force on the particle and its acceleration and (c) the radius of curvature of the particle. The particle has a charge of $Q = 1.6\times10^{-19}$ C, its mass $m = 1.7\times10^{-27}$ kg.

(a) Electric force on the particle,

$$\mathbf{F}_e = Q\mathbf{E} = (1.6\times10^{-19})(2400\mathbf{u}_x) = 3.84\times10^{-16}\mathbf{u}_x \text{ N}$$

$$\text{Acceleration} = \frac{\mathbf{F}_e}{m} = \frac{3.84\times10^{-16}}{1.7\times10^{-27}} = 2.259\times10^{11} \text{ m/s}^2$$

$$\text{Velocity after } 0.8\ \mu\text{s}, v = (2.259\times10^{11})(0.8\times10^{-6}) = 1.807\times10^{5}\mathbf{u}_x \text{ m/s}$$

(b) Magnetic force,

$$\mathbf{F}_m = Q(\mathbf{v}\times\mathbf{B}) = (1.602\times10^{-19})[(1.807\times10^{5}\mathbf{u}_x)\times(0.25\times10^{-3}\mathbf{u}_y)] = 0.723\times10^{-17}\mathbf{u}_z \text{ N}$$

$$\text{Acceleration} = \frac{\mathbf{F}_m}{m} = \frac{0.723\times10^{-17}}{1.7\times10^{-27}}\mathbf{u}_z = 0.425\times10^{10}\mathbf{u}_z \text{ m/s}^2$$

(c) $$\text{Radius of curvature} = \frac{mv}{QB} = \frac{(1.7\times10^{-27})(1.807\times10^{5})}{(1.6\times10^{-19})(0.25\times10^{-3})} = 7.68 \text{ m}$$

EXAMPLE 4.62

Calculate the radius and frequency of orbit of an electron if it is injected at an initial velocity of $(300\mathbf{u}_x + 400\mathbf{u}_y)$ m/s into a magnetic field of uniform flux density of $0.08\mathbf{u}_z$ tesla.

As the velocity vector is on the *x-y* plane and the flux density vector is along *z* direction, the two vectors are at 90°.

Magnitude of velocity, $v = 500$ m/s
Magnitude of flux density, $B = 0.08$ T

$$\text{Radius of orbit, } r = \frac{mv}{QB} = \frac{(9.1\times10^{-31})(500)}{(1.6\times10^{-19})(0.08)} = 35.55\times10^{-9} \text{ m}$$

$$\text{Gyro-frequency, } f = \frac{QB}{2\pi m} = \frac{1.6\times10^{-19}\times0.08}{2\pi\times9.1\times10^{-31}} = 2.24\times10^{9} \text{ Hz}$$

4.24 The Hall Effect

The charge carriers in a current-carrying conductor or a semiconductor move at a velocity known as the drift velocity. If the material is placed in a magnetic field, the charges in motion will experience a magnetic force resulting in a slight shift in their positions. The shifting of charges from their normal positions creates an electric force. The direction of the electric force is opposite to that of the magnetic force. When the electric force becomes equal to the magnetic force, shifting of charges stops and steady state condition is reached.

The establishment of the electric force, and hence the electric field, created by the interaction of the moving charges and the magnetic field is known as the *Hall effect*.

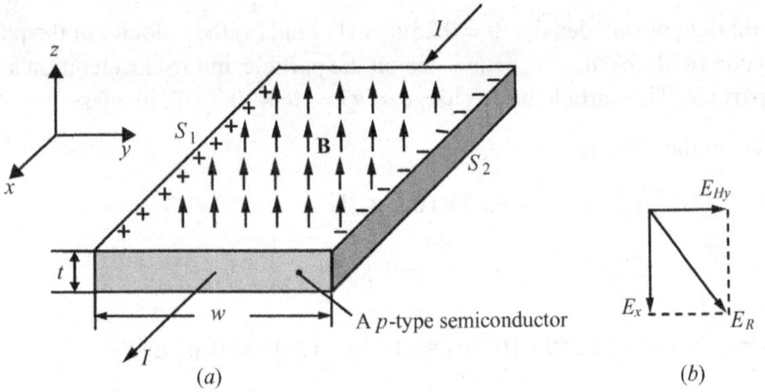

Fig. 4.59 (*a*) For explanation of Hall effect and (*b*) Electric field intensity inside the material.

To understand the Hall effect phenomenon, let us consider a rectangular block of *p*-type semiconductor, as shown in Fig. 4.59(*a*). The width and thickness of the block are *w* and *t* respectively. It carries a uniformly distributed current *I* in *x* direction. The current density, which is also in *x* direction, is given by

$$J_x = \frac{I}{wt} \tag{1}$$

The vector form of current density is

$$\mathbf{J} = J\mathbf{u}_x \tag{2}$$

The magnitude of electric field intensity due to the conduction current is found by dividing J_x of (1) by the conductivity of the material.

$$E_x = \frac{J_x}{\sigma} \tag{3}$$

The magnetic flux density *B*, assumed to be constant, is in *z* direction. That is, the flux density is normal to the plane of the sheet. The magnetic flux density vector,

$$\mathbf{B} = B\mathbf{u}_z \tag{4}$$

The drift velocity $\mathbf{v}_{dh}$ of holes as per formula (3.11-7) is

$$\mathbf{v}_{dh} = \frac{\mathbf{J}}{\rho_h}$$

where ρ_h is volume charge density due to the holes.

Substituting the right side of (2) in place of **J**, we get

$$\mathbf{v}_{dh} = \frac{J}{\rho_h}\mathbf{u}_x \tag{5}$$

The magnetic force on a hole with charge $e = 1.6\times10^{-19}$ C and moving at velocity $\mathbf{v}_{dh}$ in the magnetic field is given by

$$\mathbf{F}_{mh} = e(\mathbf{v}_{dh} \times \mathbf{B}) \tag{6}$$

Replacing **B** and v_{dh} of (6) by the right-hand sides of (4) and (5), we get

$$\mathbf{F}_{mh} = \frac{eJB}{\rho_h}(\mathbf{u}_x \times \mathbf{u}_z) = -\frac{eJB}{\rho_h}\mathbf{u}_y \tag{7}$$

We find from (7) that the magnetic force experienced by the holes is in negative *y* direction. This force will shift the hole in this direction resulting in an electric force F_H in positive *y* direction. Thus F_{mh} and F_H are in opposite directions. When the magnitudes of the two forces become equal, there will not be any further shifting of the holes. Under this condition,

$$\mathbf{F}_H = -\mathbf{F}_{mh} = \frac{eJB}{\rho_h}\mathbf{u}_y \tag{8}$$

The electric field intensity is obtained by dividing (8) by e. Thus the field intensity, called the *Hall field intensity*, is

$$\mathbf{E}_H = \frac{\mathbf{F}_H}{e} = \frac{JB}{\rho_h}\mathbf{u}_y = E_H\mathbf{u}_y \tag{9}$$

The direction of Hall field is along positive y direction, i.e., from surface S_1 towards surface S_2. The magnitude of Hall field intensity,

$$E_{Hy} = \frac{JB}{\rho_h} \tag{10}$$

The Hall field intensity given by (10) is proportional to the product of conduction current density and the magnetic flux density. The constant of proportionality is equal to the reciprocal of the volume charge density. The field will establish a potential difference between surfaces S_1 and S_2. The polarity of surface S_1 would be positive and that of S_2 would be negative. The potential difference is called *Hall voltage.*

It is equal to the product of width of the block and the Hall field intensity. Thus the Hall voltage, denoted by V_H, is

$$V_H = wE_H = \frac{wJB}{\rho_h} \tag{11}$$

The hall voltage in terms of current is found by substituting the right side of (1) in place of J in (11). Thus,

$$V_H = \frac{IB}{t\rho_h} \tag{12}$$

Equation (12) shows that the Hall voltage in terms of the current is independent of width of the sheet. The value of Hall voltage is very small.

If the p-type semiconductor of Fig. 4.59(a) is replaced by a sample of n-type semiconductor or by a good conductor, the drift velocity of the negative charge carriers (electrons) in the material would be in negative x direction. The drift velocity of electrons as per formula (3.11-6) is

$$\mathbf{v}_{de} = -\frac{J}{\rho_e}\mathbf{u}_x$$

The magnetic force on an electron,

$$\mathbf{F}_{me} = -\frac{eJB}{\rho_h}\mathbf{u}_y$$

Because of this force the electrons in the material will be shifted towards negative y direction. The Hall field intensity due to the shifting of the electrons would also be in negative y direction. This will establish a Hall voltage between the side surfaces of the material. The polarity of Hall voltage in this case will be opposite to the polarity of hall voltage in a p-type semiconductor.

The electric field intensity in the material consists of two components; one due to the conduction current density E_x given by (3) and the other is the Hall field E_{Hy} given by (10). The two fields are perpendicular to each other, as shown in Fig. 4.59(b). The resultant field E_R is equal to the vector sum of the two fields.

The Hall effect can be used to measure flux density by keeping the current constant, to measure current by keeping magnetic flux density constant, and to measure power if the flux density is made proportional to the voltage. The phenomenon can be utilised to determine charge concentration in semiconductors. Moreover, the polarity of Hall voltage indicates whether a sample of semiconductor is p-type or n-type.

EXAMPLE 4.63

A p-type semiconductor sheet of length, $\ell = 3.0$ cm; width, $w = 0.8$ cm and thickness, $t = 1.5$ mm carries a current, $I = 3.4$ mA. The direction of current flow is along the length of the sheet and voltage drop due the current, $V = 36.6$ mV. A Hall voltage, $V_H = 0.74$ mV is generated when a uniform magnetic field of density, $B = 0.6$ T is applied normal to the surface containing length and width of the slab. Calculate (a) the volume charge density, (b) number of charge carriers per unit volume, (c) conductivity of the material, and (d) total electric field intensity in the sheet.

(a) Referring to formula (12) the charge density,

$$\rho_h = \frac{IB}{tV_H} = \frac{(3.4\times10^{-3})(0.6)}{(1.5\times10^{-3})(0.74)} = 1837.8\ \text{C/m}^3$$

(b) Number of charge carriers per unit volume, $\dfrac{\rho_h}{e} = \dfrac{1837.8}{1.6\times10^{-19}} = 1.149\times10^{22}$

(c) Conductivity of the material, $\sigma = \dfrac{J}{E}$

$$J = \frac{I}{wt} = \frac{3.4\times10^{-3}}{(0.8\times10^{-2})(1.5\times10^{-3})} = 283.3\ \text{A/m}^2$$

$$E = \frac{V}{\ell} = \frac{36.6\times10^{-3}}{3.0\times10^{-2}} = 1.220\ \text{V/m}$$

$$\sigma = \frac{283.3}{1.22} = 232.2\ \text{S/m}$$

(d) Hall field intensity, $E_H = \dfrac{V_H}{w} = \dfrac{0.74\times10^{-3}}{0.8\times10^{-2}} = 0.0925\ \text{V/m}$

Total field intensity $= \sqrt{E^2 + E_H{}^2} = \sqrt{(1.22)^2 + (0.0925)^2} = 1.224\ \text{V/m}$

EXAMPLE 4.64

A thin conducting sheet of thickness 1.5 mm contains 7.5×10^{27} conduction electrons per cubic metre. A Hall voltage of 12 μV is generated across its width when the magnetic flux density normal to its surface is 1.6 T. Find the current carried by the sheet.

Thickness of the sheet, $t = 1.5\times10^{-3}$ m

Number of charge carriers per unit volume, $n = 7.5\times10^{27}$

Hall voltage, $V_H = 12.0\times10^{-6}$ V

Flux density, $B = 1.6$ T

Volume charge density, $\rho_h = ne = (7.5\times10^{27})(1.6\times10^{-19}) = 12.0\times10^{8}\ \text{C/m}^3$

We have from (12),

$$I = \frac{t\,\rho_h V_H}{B} = \frac{(\;.5\times10^{-3})(1\ldots\times\;\;)(\;.0\times10^{-6})}{1.6} = 13.5\ \text{A}$$

SUMMARY

After stating Biot–Savart law, we defined the magnetic flux density as the product of magnetic field intensity and permeability, and then we calculated flux densities due to some current distributions. Though after repeated use one feels that magnetic field intensity and magnetic flux densities are physical quantities, however these are only useful concepts. The physical quantity of course is the electromagnetic force on a current-carrying conductor in a magnetic field. The electromagnetic force on a line current is equal to the product of current, line length, flux density, and sine of the angle between flux density and current; its direction is given by the right-hand rule. We found that the net force on a current loop in a uniform magnetic field is zero but the torque on the loop has a finite value. The direction of the torque is given by the right-hand rule. We developed Ampere's magnetic circuital law in integral form and applied it for calculation of magnetic field intensities of symmetrical current distributions. We also applied the law around a closed path enclosing an infinitesimal current-carrying area and obtained the important relation between magnetic field intensity and current density ($\nabla \times \mathbf{H} = \mathbf{J}$), which is the differential form of the magnetic circuital law. Then, we utilised the two forms of the law to formulate Stokes' theorem which is useful for transformation of a surface integral into a line integral and vice versa. Then, we learnt that the divergence of magnetic flux density is zero even in a region of current flow, and used this solenoidal property to define the vector magnetic potential. Though the vector magnetic potential has no physical meaning like the electrostatic potential, it is useful for analysis and computation of magnetic field problems. A brief study of properties of magnetic materials led to the relation between magnetic susceptibility and relative permeability. We found that magnetisation characteristic of a ferromagnetic material is highly nonlinear, and learnt what a hysteresis loop is.

We applied the magnetic field laws to derive boundary conditions, for magnetic circuit calculations, for calculating the field energy, energy density, inductances, force on a charged particle in a uniform magnetic field, and the Hall voltage. The following table shows that the magnetic boundary conditions are analogous to the boundary conditions of static electric field.

Magnetic boundary conditions	Boundary conditions in static electric field
$H_{t1} = H_{t2}$	$E_{t1} = E_{t2}$
$\frac{B_{t1}}{\mu_1} = \frac{B_{t2}}{\mu_2}$	$\frac{D_{t1}}{\varepsilon_1} = \frac{D_{t2}}{\varepsilon_2}$
$B_{n1} = B_{n2}$	$D_{n1} = D_{n2}$
$\mu_1 H_{n1} = \mu_2 H_{n2}$	$\varepsilon_1 E_{n1} = \varepsilon_2 E_{n2}$
$\frac{\tan\alpha_1}{\tan\alpha_2} = \frac{\mu_1}{\mu_2}$	$\frac{\tan\alpha_1}{\tan\alpha_2} = \frac{\varepsilon_1}{\varepsilon_2}$

IMPORTANT FORMULAE

Biot–Savart law: $d\mathbf{H} = \frac{I}{4\pi R^2}(d\boldsymbol{\ell} \times \mathbf{u}_R) = \frac{I}{4\pi R^3}(d\boldsymbol{\ell} \times \mathbf{R})$

Magnetic flux crossing through a surface: $= \int_S \mathbf{B} \cdot \mathbf{S}\ \Phi$

Magnetic flux density vector on the axis of a current loop: $\mathbf{B}\ \frac{uIa^2}{2(a^2 + h^2)^{3/2}}\mathbf{u}_z$

Magnetic flux density on the axis of a short solenoid:

$$B_a(x) = \frac{\mu_0 n I}{2}\left[\frac{x+\ell/2}{\sqrt{a^2+(x+\ell/2)^2}} - \frac{x-\ell/2}{\sqrt{a^2+(x-\ell/2)^2}}\right]$$

Magnetic flux density of a line current of infinite length: $\mathbf{B} = \dfrac{\mu I}{2\pi r}\mathbf{u}_\phi$

Electromagnetic force on a current element: $d\mathbf{F} = I\,(d\boldsymbol{\ell} \times \mathbf{B}) = -I(\mathbf{B} \times d\boldsymbol{\ell})$

Force per unit length between two parallel line currents: $\dfrac{F}{\ell} = \dfrac{\mu I_1 I_2}{2\pi s}$

Electromagnetic torque on a current loop: $\mathbf{T} = (I\mathbf{S}) \times \mathbf{B}$

Ampere's magnetic circuital law: $\oint \mathbf{H}\cdot d\boldsymbol{\ell} = I$

$$\oint \mathbf{H}\cdot d\ell = \int_S \mathbf{J}\cdot d\mathbf{S}$$

Magnetic field intensity due to a current sheet: $\mathbf{H} = \dfrac{\mathbf{K} \times \mathbf{u}_n}{2}$

Curl of **H**:

In rectangular coordinates, $\nabla \times \mathbf{H} = \begin{vmatrix} \mathbf{u}_x & \mathbf{u}_y & \mathbf{u}_z \\ \dfrac{\partial}{\partial x} & \dfrac{\partial}{\partial y} & \dfrac{\partial}{\partial z} \\ H_x & H_y & H_z \end{vmatrix}$

In cylindrical coordinates, $\nabla \times \mathbf{H} = \begin{vmatrix} \dfrac{\mathbf{u}_r}{r} & \mathbf{u}_\phi & \dfrac{\mathbf{u}_z}{r} \\ \dfrac{\partial}{\partial r} & \dfrac{\partial}{\partial \phi} & \dfrac{\partial}{\partial z} \\ H_r & rH_\phi & H_z \end{vmatrix}$

In spherical coordinates, $\nabla \times \mathbf{H} = \begin{vmatrix} \dfrac{\mathbf{u}_r}{r^2 \sin\theta} & \dfrac{\mathbf{u}_\theta}{r\sin\theta} & \dfrac{\mathbf{u}_\phi}{r} \\ \dfrac{\partial}{\partial r} & \dfrac{\partial}{\partial \theta} & \dfrac{\partial}{\partial \phi} \\ H_r & rH_\theta & r\sin\theta H_\phi \end{vmatrix}$

Relation between magnetic field intensity and current density: $\nabla \times \mathbf{H} = \mathbf{J}$

Stokes' theorem: $\int_S (\nabla \times \mathbf{H}).d\mathbf{S} = \oint_C \mathbf{H}\cdot d\ell$

Divergence of magnetic flux density: $\nabla.\,\mathbf{B} = 0$

Relation between magnetic flux density and vector magnetic potential: $\mathbf{B} = \nabla \times \mathbf{A}$

Vector magnetic potential of a current element: $d\mathbf{A} = \dfrac{\mu I}{4\pi R}\,d\ell$

Magnetic field intensity of a small current loop: $\mathbf{H}$ $\dfrac{IS}{4\pi r^3}$ $\mathbf{u}_r +$ $\theta\mathbf{u}_\theta)$

Relation between relative permeability and magnetic susceptibility: μ χ_m

Boundary conditions: B_n B_{n2}

$H_{t1} = H_{t2}$ (Without surface current)

$(\mathbf{H}_1 - \mathbf{H}_2) \times \mathbf{u}_n = \mathbf{K}$ (With surface current)

Magnetic circuit:

Reluctance, $\Re = \dfrac{\ell}{\mu S}$

Permeance, $\wp = \dfrac{1}{\Re}$

$$\text{Flux} = \frac{\text{mmf}}{\text{Reluctance}} = \text{mmf} \times \text{prmeance}$$

Series magnetic circuit: $NI = \Phi\left(\dfrac{\ell_1}{\mu_1 S_1} + \dfrac{\ell_2}{\mu_2 S_2} + \dfrac{\ell_3}{\mu_3 S_3} + \cdots\right)$

Mutual Inductance: $M = k\sqrt{L_{e1} L_{e2}}$

External inductance of coaxial line per unit length: $L_e = \dfrac{\mu}{2\pi} \ln(b/a)$

External inductance of a two-conductor parallel transmission line:

$$L_e = \frac{\mu_0}{\pi} \ln \frac{h + \sqrt{h^2 - 4a^2}}{2a}$$

$$L_e \approx \frac{\mu_0}{\pi} \ln \frac{h}{a} \quad h \gg a$$

Mutual inductance between two coaxial solenoids: $M = \dfrac{\pi a_2^2 \mu_0 N_1 N_2}{\ell_1}$

Energy stored in the magnetic field: $W_m = \dfrac{1}{2} \int_v \mu H^2 dv$

Inductance in terms of energy stored in a magnetic circuit: $L = \dfrac{2W_m}{I^2}$

Radius of revolution of a charged particle in a uniform magnetic field: $r = \dfrac{mv}{QB}$

Hall voltage: $V_H = \dfrac{IB}{t\rho_h}$

MULTIPLE-CHOICE QUESTIONS

(Starred questions have two answers)

*4.1 The magnetic field intensity due to a current element at a point P is
 (a) inversely proportional to square of the distance between the current element and P
 (b) inversely proportional to the distance between the current element and P
 (c) a scalar quantity
 (d) a vector quantity

*4.2 Which of the following statements are correct?
 (a) The direction of magnetic flux density is same as that of the magnetic field intensity.
 (b) The SI unit of magnetic flux is weber.

(c) The SI unit of magnetic flux density is tesla-metre.
(d) Magnetic flux is vector quantity.

4.3 A line current of finite length is along z direction of cylindrical coordinates. The direction of magnetic field intensity
(a) is in z direction
(b) is in r direction
(c) is in ϕ direction
(d) depends on length of the line

4.4 Two parallel circular coils 1 and 2 have the same radius and have a common axis. Coil 1 has 10 turns and Coil 2 has 6 turns. They carry the same current, but in opposite directions. On the axis and at the mid-point between the coils, the magnetic field intensity of Coil 1 is 80 A/m. The magnitude of total field intensity at the mid-point is
(a) 128 A/m (b) 32 A/m (c) 160 A/m (d) 64 A/m

4.5 The Helmholtz pair consists of two identical parallel coils with a common axis. The coils carry current in the same direction. Let s be the spacing between the coils and a be the radius of each coil. The magnetic field intensity along the axis between the coils would be practically constant if
(a) $s = 2a$ (b) $s = a$ (c) $s = a/2$ (d) $s = a/4$

4.6 The direction of electromagnetic force on a differential current element $Id\ell$ located in a magnetic field of flux density **B** is
(a) in the direction of the current element
(b) in the direction of **B**
(c) tangential to the plane containing $Id\ell$ and **B**
(d) perpendicular to the plane containing $Id\ell$ and **B**

*4.7 Two long parallel conductors have currents in opposite directions. The electromagnetic forces on the conductors are
(a) attractive
(b) repulsive
(c) inversely proportional to the distance between the conductors
(d) inversely proportional to permeability of the medium in which the conductors are located.

4.8 A circular current loop of radius r is located on the r-ϕ plane with its centre at the origin of cylindrical coordinates in a uniform magnetic field. The current in the loop is in negative ϕ direction and the magnetic field is in positive z direction. The direction of electromagnetic force on a current element $rd\phi$ is
(a) in r direction towards the origin
(b) in r direction away from the origin
(c) in z direction
(d) in ϕ direction

*4.9 A square current loop is located on the y-z plane with its centre at the origin in a uniform magnetic field. The magnetic field is in x direction. Which of the following statements are correct?
(a) The direction of electromagnetic force on all the sides of the loop is same.
(b) The electromagnetic force on the loop is zero.
(c) The electromagnetic torque on the loop is zero.
(d) The electromagnetic torque on the loop has a finite value.

4.10 The radius of a circular conductor is a. It carries a uniformly distributed current of density J A/m^2. The magnetic field intensity at a radius $r < a$ is given by
(a) ( )/2 (b) rJ (c) J ¨ r (d) J r

4.11 The external and internal radii of a toroid are 45 cm and 55 cm respectively. It has a uniformly distributed winding of 1500 turns. The winding current is 1.2 A. The magnetic field intensity, in A/m, at 50 cm radius is nearest to

(a) 573 (b) 1800 (c) 636 (d) 521

4.12 Two large current sheets 1 and 2 are at $z = 0$ and $z = 2.0$ m planes respectively. Current per unit width in Sheet 1 is $-20.0\mathbf{u}_x$ A/m and that of Sheet 2 is $15.0\mathbf{u}_x$ A/m. The magnetic field intensity at $z = 1.5$ m is

(a) $-17.5\mathbf{u}_y$ A/m (b) $17.5\mathbf{u}_y$ A/m (c) $5.0\mathbf{u}_y$ A/m (d) $-5.0\mathbf{u}_y$ A/m

4.13 A closed path encloses 4 current-carrying conductors with uniform current distributions. Two conductors are of radius a, and the radius of each of the other two conductors is $3a/4$. If the current density is J in all the conductors then the line integral of the magnetic field intensity vector around the closed path is nearest to

(a) $8.6a^2J$ (b) $11.0a^2J$ (c) $8.1a^2J$ (a) $9.8a^2J$

4.14 The magnetic field intensity in a region is given by $\mathbf{H} = ax\mathbf{u}_x + by\mathbf{u}_y$. The current density at a point in the region is

(a) $a + b$ (b) a (c) zero (d) b

*4.15 Which of the following are correct?

(a) Magnetisation is defined as dipole moment per unit volume.
(b) The unit of magnetisation is tesla.
(c) The unit of magnetic susceptibility is henry/m.
(d) The magnetic susceptibility of a material is less than its relative permeability.

*4.16 Which of the following statements are correct?

(a) The magnetic susceptibility of a diamagnetic material is positive.
(b) The magnetic susceptibility of a paramagnetic material is positive.
(c) The relative permeability of a diamagnetic material is less than unity
(d) The relative permeability of a paramagnetic material is less than unity.

4.17 The magnetisation characteristic of a ferromagnetic material is the relation between

(a) the magnetic flux density and permeability
(b) magnetic flux density and magnetic field intensity
(c) magnetic field intensity and magnetisation
(d) magnetic field intensity and magnetic susceptibility

4.18 The area of the hysteresis loop of a ferromagnetic material is proportional to

(a) energy stored per unit volume of the material
(b) energy loss per unit volume of the material
(c) power loss per unit volume of the material
(d) magnetic flux in the material

*4.19 Let B, H, μ_r, M, and χ_m be the magnetic flux density, field intensity, relative permeability, magnetisation, and magnetic susceptibility of a magnetic material. Then,

(a) $\chi_m = \mu_r - 1$ (b) $H = \chi_m M$ (c) $B = \mu_0(1+\chi_m)H$ (d) $B = 1+\chi_m H$

4.20 The susceptibility of a diamagnetic material is

(a) very small and negative
(b) very small and positive
(c) nearly unity and negative
(d) nearly unity and positive

4.21 Two regions 1 and 2 in a magnetic field have a common boundary surface. The permeability of Region 2 is twice that of Region 1. The magnetic field intensity at the boundary in Region 1 subtends an angle of 30° with the normal to the boundary. The angle of the field intensity at the boundary in Region 2 is
(a) 15° (b) 60° (c) nearly 49° (d) nearly 16°

4.22 Let ℓ, S, and μ be the length, cross-sectional area, and permeability of a magnetic circuit. The permeance of the magnetic circuit is given by

(a) $\dfrac{\ell}{\mu S}$ (b) $\dfrac{\ell S}{\mu}$ (c) $\dfrac{\ell \mu}{S}$ (d) $\dfrac{\mu S}{\ell}$

*4.23 A series magnetic circuit consists of two parts of different magnetic materials. Which of the following statements are correct?
(a) The permeance of the magnetic circuit is equal to the sum of the permeances of the two parts.
(b) The magnetic field intensities in both the parts are nearly equal.
(c) The reluctance of the magnetic circuit is equal to the sum of the reluctances of the two parts.
(d) The magnetic flux in both the parts is nearly same.

*4.24 Consider that the magnetic circuit with a single ferromagnetic material is excited by a certain value of mmf. If an air gap is created in the magnetic circuit by giving a saw cut, and if the mmf is same as before, then
(a) the flux density in the iron part will increase
(b) the reluctance of the magnetic circuit will increase
(c) the permeance of the magnetic circuit will increase
(d) the core flux will decrease

*4.25 Two coils of N_1 and N_2 turns have a common magnetic circuit. If leakage is neglected, and if the external self-inductance of Coil 1 is L_1, then which of the following statements are correct.
(a) The external self-inductance of Coil 2 is $(N_2^2/N_1^2)L_1$
(b) The mutual inductance between the coils is $(\sqrt{N_2/N_1})L_1$
(c) The external self-inductance of Coil 2 is $(N_2/N_1)L_1$
(d) The mutual inductance between the coils is $(N_2/N_1)L_1$

4.26 The internal inductance of a round conductor carrying steady current is
(a) independent of conductor radius
(b) proportional to square of conductor radius
(c) inversely proportional to conductor radius
(d) directly proportional to conductor radius

4.27 In a linear magnetic field, the field intensity is increased from H_1 to H_2. The increase in the energy stored in the magnetic field is

(a) $\dfrac{\mu}{2}(H_2 - H_1)$ (b) $\dfrac{\mu}{2}(H_2 - H_1)^2$ (c) $\dfrac{\mu}{2}(H_2^2 - H_1^2)$ (d) $\mu(H_2^2 - H_1^2)$

4.28 The magnetic flux density in the air gap between two iron surfaces is B_g. The force between the iron surfaces at this flux density is F. If the flux density is reduced to $(3/4)B_g$, the decrease in the force would be
(a) $(3/4)F$ (b) $(7/16)F$ (c) $F/4$ (d) $(9/16)F$

4.29 A charge moving in an electric field is acted upon by an electromagnetic force. The direction of the force is
(a) in the direction of the magnetic field
(b) in the direction of velocity of the charge

(c) tangential to the plane containing the flux density and velocity vectors
(d) perpendicular to the plane containing the flux density and velocity vectors

4.30 A piece of semiconductor carrying direct current is located in a steady magnetic field so that the magnetic flux density is perpendicular to the direction of flow of current. The direction of Hall voltage is
(a) along the current density vector
(b) along the flux density vector
(c) perpendicular to both current density and flux density vectors
(d) depends on the relative values of current density and flux density

QUESTIONS

4.1 Biot–Savart law gives the magnetic field intensity of a current element. Describe the law.

4.2 A current element along z direction of cylindrical coordinates is centred at the origin. What is the direction of magnetic field intensity at $P(r,\phi,0)$?

4.3 A current element along z direction of rectangular coordinates is centred at the origin. The magnetic field intensity at $P(x, y, z)$ will have components in which directions?

4.4 Sketch the variation of flux density along the axis between a Helmholtz pair of current-carrying coils.

4.5 Two circular current loops 1 and 2 centred at (0, 0, 0) and (0, 0, s) in rectangular coordinates carry currents I and $1.5I$ respectively. The magnetic field intensity at $P(0,0,s/2)$ due to Loop 1 is $H_z\mathbf{u}_z$. What is the total field intensity at P if the current in Loop 2 is (a) in the same sense as in Loop 1 and (b) in the opposite sense to that of Loop 1? Justify your answer with necessary illustrations.

4.6 Describe the Helmholtz pair of coils and state its application.

4.7 Explain the law governing the electromagnetic force on a current element.

4.8 Explain with necessary illustrations that the magnetic force between two parallel wires carrying current in the same direction is attractive in nature.

4.9 A straight wire carrying direct current lies on the y axis in a uniform magnetic field. The flux density has components in x and y directions. What is the direction of force on the current carrying wire?

4.10 What is the net magnetic force on a planar current loop situated in a uniform magnetic field?

4.11 A long line current I_1 lies on the z axis of cylindrical coordinates with the current in positive z direction. A circular loop of radius r with current I_2 in ϕ direction has its centre at the origin. What is the force on the conductor? Justify your answer. Note that the field due to I_1 is not uniform.

4.12 Consider that the current loop in Question 4.10 is free to rotate. The electromagnetic torque exerted on the loop will rotate it to a position of zero torque. At this position, what is the angle between the magnetic flux density and the line that is perpendicular to the surface of the loop?

4.13 A small current loop has an area S and carries a current I. What is the magnitude, direction and unit of the dipole moment of the loop?

4.14 State and explain Ampere's magnetic circuital law.

4.15 A line current of infinite length is along the z axis of cylindrical coordinates. Write the expression for the magnetic field intensity vector at (, ,).

4.16 Ampere's magnetic circuital law cannot be used to find the field intensity of a line current of finite length. Explain why.

4.17 A long cylindrical conducting shell has its axis along the z axis. It carries a surface current $K\mathbf{u}_\phi$ A/m. What is the value of magnetic field intensity at a point on the z axis, and what is its direction?

4.18 What is the unit of curl of **H**, where **H** is the magnetic field intensity?

4.19 Give the mathematical statement of Stokes' theorem, and explain the meaning of each of its sides.

4.20 A current element is oriented along x direction. What is the direction of vector magnetic potential at a point $P(x, y, z)$ due to the current? What is the unit of vector magnetic potential?

4.21 Justify that a magnetic field is a solenoidal field.

4.22 What are the quantities on which the magnetisation of a magnetic material depends?

4.23 Distinguish between diamagnetic and paramagnetic materials.

4.24 Enumerate some of the important properties of ferromagnetic materials.

4.25 The magnetisation characteristic of a ferromagnetic material is not linear. Explain why.

4.26 Sketch the hysteresis loop of a ferromagnetic material and indicate the salient points on the loop. What does the area of the loop represent?

4.27 Which magnetic field law is used to derive the relation between tangential magnetic field intensities on two sides of the common boundary surface of two magnetic materials?

4.28 State the boundary conditions in a magnetic field.

4.29 At the interface of a ferromagnetic material and air, the field lines on the air gap side are nearly normal to the iron surface. Explain why.

4.30 A static magnetic circuit is analogous to an electric circuit. However, there are some differences. What are these differences?

4.31 With the help of an example, differentiate between internal flux and external flux due to the current in a conductor.

4.32 The magnetic field in the air gap between two iron surfaces spreads non-uniformly beyond the region of the surfaces. What is this phenomenon called? How does this affect the gap reluctance? Give a sketch of field lines in the air gap.

4.33 What is the value of internal inductance per unit length of a copper conductor of circular section?

4.34 A coil with a ferromagnetic core has a certain value of inductance. If the magnetic circuit now has an air gap, the inductance decreases. Explain why.

4.35 Identify the flux components due to currents in two mutually coupled coils.

4.36 A charged particle is moving at a certain velocity. Show that the dimension of the product of charge and velocity is the same as that of a current element.

4.37 A charged particle when injected at right angle into a stationary uniform magnetic field revolves in a circular path. What forces act on the particle? If the flux density is in z direction and the initial velocity of the particle is in the negative ϕ direction of cylindrical coordinates, what are the directions of forces?

4.38 In addition to the magnetic field in Question 4.37, if there is a stationary and uniform electric field in the positive z direction, what additional force will act on the particle? What is the direction of this force?

4.39 Name at least two applications of the Hall effect.

ANSWERS TO SELECTED QUESTIONS

4.1 Biot–Savart law gives the magnetic field intensity of a current element. Describe the law.

Answer: According to the law, the field intensity of a current element $Id\ell$ at a point P, which is at a distance R from the element, is directly proportional to $Id\ell$ and inversely proportional to R^2. It is also proportional to the sine of the angle α between the vectors $d\ell$ and **R**. The direction of the field intensity is along normal to the plane containing $d\ell$ and **R**. Its sense is in the direction of progress of a right-handed screw when it is rotated from $d\ell$ towards **R** through the smaller angle α.

4.6 Describe the Helmholtz pair of coils and state its application.

Answer: The Helmholtz pair consists of two identical, circular and parallel coils with a common axis. They carry current in the same direction. Between the coils and along the axis, the magnetic field produced by both the coils are in the same direction. If the spacing between the coils is a little more than the radius of the coils, the magnetic flux density along a large part of the axis would be nearly uniform. This is one of the devices used for the production of uniform magnetic field.

4.7 Explain the law governing the electromagnetic force on a current element.

Answer: When a current element is subjected to a magnetic field, the element is acted upon by an electromagnetic force. The magnitude of the force is directly proportional to the value of the current element and the value of magnetic flux density at the location of the element. It is also proportional to the sine of the angle between the current element vector and the flux density vector. The direction of the force is perpendicular to the plane containing the two vectors. Its sense is in the direction of progress of a right-handed screw when rotated from the current element vector towards the flux density vector.

4.14 State and explain Ampere's magnetic circuital law.

Answer: According to the law, the line integral of magnetic field intensity H taken once around a closed path is equal to the current enclosed. The chosen path may or may not be along a field line. If it is not along a field line, then as per the definition of line integral, the integration has to be evaluated by using the component of H tangential to the path.

Before the law is applied to a specific problem, it is necessary to know the nature of field variation so that a suitable closed path can be chosen. A suitable closed path is one along which the field is either tangential or normal so that the line integral of the field intensity is easily evaluated.

4.21 Justify that a magnetic field is a solenoidal field.

Answer: The divergence of a vector is equal to the surface integral of the vector over a closed surface enclosing a volume element divided by the volume of the element, in the limit the volume tending to zero. A vector field, the divergence of which is zero, is termed as a solenoidal field.

The field lines in a region of magnetic field form closed paths irrespective of whether the region has current or not. Therefore, the magnetic flux entering a closed surface is equal to the flux coming out of it. It means that the surface integral of the flux density vector over a closed surface in the region is equal to zero. Hence, the divergence of the flux density vector is zero. Accordingly, a magnetic field is a solenoidal field.

4.23 Distinguish between diamagnetic and paramagnetic materials.

Answer: In a diamagnetic material the magnetic moments due to electron spin and electron orbit nearly cancel out. Thus, these materials have no permanent dipole moments. The magnetic susceptibility of a diamagnetic material is very small and negative. Accordingly, the relative permeability of a diamagnetic material is a little less than unity.

Permanent magnetic moments exist in paramagnetic materials as the two above mentioned magnetic moments do not cancel out. However, in a large piece of the material the net magnetic moment is nearly zero because of random orientations of atomic dipoles. The magnetic susceptibility of a paramagnetic material is very small but positive. Accordingly, the relative permeability of a paramagnetic material is a little more than unity.

4.25 The magnetisation characteristic of a ferromagnetic material is not linear. Explain why.

Answer: On application of an external magnetic field to a ferromagnetic material, the atomic magnetic moments of the material align along the applied field. Hence, the flux density B in the material increases. Initially, the rate of rise of B with the applied field intensity H is high. With further increase in the external field, the rate of rise of B decreases and ultimately a stage is reached at which almost all the dipoles are aligned along the external field. Any further increase in H does not result in an appreciable increase of B. At this stage the material is said to have reached a state of magnetic saturation. Thus, the relationship between B and H is not linear.

4.28 State the boundary conditions in a magnetic field.

Answer: At the boundary between two regions in a magnetic field, the normal component of flux density is continuous across the boundary whether the boundary has a current sheet or not. However, the tangential component field intensity is continuous across the boundary only if the boundary has no current sheet. If the boundary has a current sheet, the tangential component of the field intensities on two sides of the interface would be different. The difference between the two tangential field intensities is equal to the current per unit width of the current sheet.

4.29 At the interface of a ferromagnetic material and air, the field lines on the air gap side are nearly normal to the iron surface. Explain why.

Answer: As per the magnetic boundary conditions, the normal components of flux density on two sides of the boundary are equal. Though the tangential components of field intensities on two sides of the boundary are also equal, the tangential flux density at the air side of the boundary is negligibly small as permeability of air is small compared to that of the ferromagnetic material. Hence, the field lines are nearly normal to the iron surface.

4.34 A coil with a ferromagnetic core has a certain value of inductance. If the magnetic circuit now has an air gap, the inductance decreases. Explain why.

Answer: The external inductance of a coil is inversely proportional to the reluctance of the magnetic circuit. Without the gap, the reluctance of the magnetic circuit is that of the iron core only. In the presence of an air gap, the reluctance of the magnetic circuit is equal to the sum of the reluctance of the iron and the gap reluctance. Thus, the reluctance of the magnetic circuit with the gap is higher than the reluctance of the magnetic circuit without the gap. Hence, the inductance decreases if the circuit has an air gap.

PROBLEMS

**4.1 A current element $0.08\mathbf{u}_z$ A-m is centred at the origin of cylindrical coordinates. Find the magnetic field intensity vector at P(0.2 m, 30°, 0.4 m). Give the answer in rectangular coordinates

**4.2 The mid-point of a current element $0.06\mathbf{u}_x$ A-m is at P'(0.3 m, 0.4 m, – 0.2 m). Find the magnetic field intensity vector at the origin.

**4.3 A current element described by $0.03\mathbf{u}_x + 0.04\mathbf{u}_y$ A-m is centred at P'(0.1 m, 0.2 m, 0). Calculate the magnetic field intensity vector at P(– 0.5 m, 0.7 m, – 0.4 m).

**4.4 A circular loop of radius a is located on the x-y plane with its centre at the origin. The loop current is I. Show that the magnetic field intensity at (0, 0, z) is given by

$$\mathbf{H} = \frac{Ia^2}{2(a^2 + z^2)^{3/2}}\mathbf{u}_z$$

***4.5 Two parallel circular current loops each of radius 7 cm are located in air. The centres of the loops are at P_1(0, 0, 0) and P_2(0, 0, 12.0 cm) in cylindrical coordinates. Each loop carries a current of 5 A in ϕ direction. Find the flux density vector due to the two current distributions at P(0, 0, 4.0 cm).

***4.6 An air-core solenoid of circular section having 50 turns and carrying 0.5 A is located in air. Its radius and length are 2.0 cm and 5.0 cm respectively. Find the magnetic flux density at the following points on the axis of the solenoid. (a) At the mid-point of the solenoid. (b) At the ends of the solenoid. (c) At a distance of 8.0 cm from the mid-point of the solenoid. Mention the directions of the fields.

**4.7 A Helmholtz pair consists of two identical and parallel coaxial circular coils. If the distance between the coils is equal to the radius of the coils, the magnetic field intensity on the axis between the coils is nearly constant. Verify this from the following example. Take radius of the coils $a = 10.0$ cm; current in the coils $I = 1.5$ A; spacing between the coils $s = 10.0$ cm; number of turns of each coil $N = 500$. Find the magnetic field intensity on the axis at the mid-point between the coils, and at a distance of 4.0 cm from the mid-point.

**4.8 A line current of length ℓ is oriented along the z axis of cylindrical coordinates with its centre at the origin. The current in the line is I. Show that the magnetic flux density at $(r, 0, 0)$ is

$$\mathbf{B} = \frac{\mu I \ell}{4\pi r\sqrt{r^2 + (\ell/2)^2}}\mathbf{u}_\phi,\ r > 0$$

**4.9 A straight wire of 2.0 m length has a current of 15 A. The wire, located in air, is along the positive z axis of rectangular coordinates with its mid-point at the origin. Find the magnetic flux density vector in rectangular coordinates at point P(0.3 m, 0.4, m, 0).

**4.10 Two straight wires 1 and 2, each of 1.2 m length and parallel to the z axis of rectangular coordinates, have their mid-points at P_1(0, 0.4 m, 0) and P_2(0, 0.8 m, 0). Each conductor carries a current of 5.0 A. Find the magnetic field intensity vector at P(0, 0.6 m, 0) if (a) both the currents are in positive z direction, (b) current in Wire 1 is in positive z direction and current in Wire 2 is in negative z direction.

**4.11 A straight conductor of length 2.4 m and parallel to the z axis has its mid-point at $x = 0.1$ m and $y = 0.2$ m. It carries a current of 2.0 A in the positive z direction. Find the magnetic field intensity at P(0.4 m, – 0.6 m, 0.2 m).

**4.12 A square coil of 25 turns has a mean side length of 20 cm. The coil is on the x-y plane in air with its mid-point at the origin, and carries a current of 4.0 A in clockwise direction. Find the magnetic flux density vector at $z = 15.0$ cm.

**4.13 Two long parallel lines carry equal and opposite current I. Derive the formula for the electromagnetic force per unit length on the conductors. Make necessary assumptions.

***4.14 A long straight wire lies on the z axis of cylindrical coordinates in air. It carries a current of 12.0 A in positive z direction. A second straight wire of 2.5 m length parallel to the first wire is at a distance of 10.0 cm. The current in the second wire is 8.0 A in positive z direction. Calculate the magnetic force vector on the second wire.

**4.15 A straight conductor of vector length $0.8\mathbf{u}_x + 0.5\mathbf{u}_y$ m is situated in a uniform magnetic field defined by $B = 2.0\mathbf{u}_y + 6.0\mathbf{u}_z$ T. Calculate the force on the conductor if it caries 0.2 A.

**4. 16 The flux density of a uniform magnetic field is in y direction, and its value is 0.8 T. A straight metallic bar of length 1.5 m and mass 2.4 kg is located in the field parallel to the horizontal x-y plane. What should be the value of current in the bar to neutralize its gravitational force if the bar is (a) in the x direction, and (b) at an angle of 20° to the x direction?

***4.17 Three long line currents 1, 2, 3, which are parallel to the z axis of rectangular coordinates, are located at P_1(– 0.05 m, 0), P_2(0.05 m, 0), and P_3(0, 0.866 m) in air. The value of current in Line 1 is 10.0 A in positive z direction. Lines 2 and 3 both carry 5.0 A current in negative z direction. Find the magnetic force vector on line 3 per unit length.

**4.18 A current strip with surface current per unit width of $50.0\mathbf{u}_z$ A/m is located in a uniform magnetic field with $\mathbf{B}$ $\mathbf{u}_x - 0\ 8\mathbf{u}_y$ T. The length and width of the strip are 2.0 m and 0.1 m. Calculate the magnetic force on the strip.

**4.19 A circular loop of radius a and carrying current I is located in a uniform magnetic field of density $\mathbf{B}$. The loop axis is along the z axis of rectangular coordinates and its centre is at the origin. Show that the electromagnetic torque on the loop $\mathbf{T} = IS(\mathbf{u}_z \times \mathbf{B})$ where S is area of the loop.

*4.20 A rectangular coil with 25 turns has average side lengths of 30 cm and 20 cm. The longer sides are parallel to the x axis, and the shorter sides are parallel to the y axis. The geometric centre of the coil is at (0, 0.2 m, 0). The coil carries 16 A of current in counter-clockwise direction. Find the electromagnetic torque on the coil in a uniform magnetic field defined by $0.5\mathbf{u}_x$ T.

**4.21 If the flux density in problem 4.20 is $0.5\mathbf{u}_x + 0.8\mathbf{u}_y$ T, find the torque on the coil.

***4.22 The normal to the plane of a rectangular coil of N turns is parallel to the x-y plane and is at an angle ϕ with the x axis. The surface area of the coil is S, and it carries a current I in the counter-clockwise direction when viewed from the normal to the plane. The coil is in a uniform magnetic field of flux density $\mathbf{B} = B_y\mathbf{u}_y$. Calculate the electromagnetic torque on the coil.

**4.23 A line current on the y axis carries a current of 25.0 A in the positive y direction. Find the magnetic field intensity at $x = 0.1$ m, $y = 0.2$ m, $z = 0.3$ m.

**4.24 A line current of large length and parallel to the z axis is situated at $x = 0.3$ m and $y = 0.4$ m. It carries a current of 15.0 A in the positive z direction. Find the magnetic field intensity at (0.8 m, 1.0 m, 0).

***4.25 Two long line currents 1 and 2 are along x and y axes respectively. Line 1 carries a current of 4.5 A in the positive x direction, and Line 2 carries a current of 6.0 A in the negative y direction. Find the magnetic field intensity at (0.5 m, 0.7 m, 0.8 m).

**4.26 Two line currents parallel to the z axis are positioned at $x = 0$, $y = -0.2$ m, and $x = 0$, $y = 0.2$ m. The line current on the negative side of the y axis has a current of 30.0 A in the positive z direction. The current in the other line is also 30.0 A, but in the negative z direction. Calculate the magnetic field intensity at (0.3 m, –0.1 m, 0).

**4.27 The inner and outer radii of a long hollow cylindrical conductor are b and c respectively. It carries a uniformly distributed current I. The axis of the cylinder is along the z axis of cylindrical coordinates. Show that the magnetic field intensity at $c \ge r \ge b$ is given by

$$\mathbf{H} = \frac{I(r^2 - b^2)}{2\pi r(c^2 - b^2)}\mathbf{u}_\phi$$

**4.28 The inner and outer radii of a long hollow cylindrical conductor, with its axis along the z axis of cylindrical coordinates, are 4.0 mm and 6.0 mm respectively. The current in the conductor is 150 A. Calculate the magnetic field intensity vector at $r = 3.0$ mm, 5.0 mm, and 7.0 mm.

***4.29 A toroid having a uniformly distributed winding with closely spaced fine wires carrying direct current may be treated as a current sheet wrapped around the core. The inner and outer radii of the toroid are 20.0 cm and 24.0 cm respectively. If the linear current density per metre length at the inner circumference of such a current sheet is 2000 A/m, calculate the magnetic field intensity at a radius cm $\le R \le$ 24 cm. Use this expression to calculate the average field intensity in the core. Compare this value with the value of the field at the mean radius. What is the direction of the field?

**4.30 A toroid with iron core of square cross section has a uniformly distributed winding. The inner and outer radii of the toroid are 15.0 cm and 20.0 cm respectively. The winding has $40/\pi$ turns per centimetre of the circumference at the inner radius, and the current in the winding is 4.0 A. (a) Find the maximum value, minimum value, and the average of the maximum and minimum value of magnetic field intensities inside the core. (b) Find the field intensity at the mean radius. (c) Calculate the core flux if the relative permittivity of iron is 500.

**4.31 In a region of steady current distribution, show that $\nabla \times \mathbf{H} = \mathbf{J}$ where $\mathbf{H}$ is the magnetic field intensity and $\mathbf{J}$ is current density.

***4.32 The current density in a long, solid, circular conductor of radius a is given by $\mathbf{J} = J_{z0} e^{\alpha r} \mathbf{u}_z$ A/m^2 in cylindrical coordinates. Find the magnetic field intensity at (a) $r < a$, and (b) $r > a$. Use Ampere's magnetic circuital law.

***4.33 The value of magnetic field intensity very near a current-carrying wire of finite length can approximately be calculated by the use of Ampere's magnetic circuital law. Consider that two identical, single-turn, square loops of side length 0.8 m are placed one above the other in air. The distance between the loops is 4.0 mm. Each loop carries a current of 10.0 A in the same sense. Find the magnetic force on one loop due to the current in the other loop. What is the nature of the force?

**4.34 The magnetic field intensity in a rectangular block of conductor is given by $\mathbf{H} = 50.0 e^{-3z} \mathbf{u}_y$ A/m. Find the current density at (x, y, 0.1 m).

***4.35 The magnetic field intensity inside a round conductor of radius a is given by

$$\frac{J_{zo}}{r}\left\{\frac{1}{\alpha^2} + e^{\alpha r}\left(\frac{r}{\alpha} - \frac{1}{\alpha^2}\right)\right\}\mathbf{u}_\phi$$

Find the current density at $r < a$.

*4.36 In a region of steady current distribution show that $\nabla^2 \mathbf{A} = -\mu \mathbf{J}$ where A is the vector magnetic potential and **J** is current density.

**4.37 A line current of length ℓ and carrying current I is on the z axis of cylindrical coordinates. The mid-point of the line is at the origin. The vector magnetic potential of the line current on $z = 0$ plane is given by

$$\mathbf{A} = \left(\frac{\mu I}{4\pi}\right)\ln\left(\frac{\sqrt{r^2 + (\ell/2)^2} + \ell/2}{\sqrt{r^2 + (\ell/2)^2} - \ell/2}\right)\mathbf{u}_z$$

Obtain the magnetic flux density vector using the relation $\mathbf{B} = \nabla \times \mathbf{A}$.

**4.38 The far-zone vector magnetic potential of a small current loop in spherical coordinates is approximately given by

$$\mathbf{A}(r,\theta) = \frac{\mu IS \sin\theta}{4\pi r^2}\mathbf{u}_\varphi$$

Obtain the magnetic flux density vector.

***4.39 The magnetic flux density of two line currents parallel to the z axis, derived from the vector magnetic potential, is given by (4.11-15). It can be obtained more easily by using the Ampere's magnetic circuital law. Derive the formula for the magnetic flux density using the latter method.

**4.40 A circular loop of radius 3.0 cm lies on the x-y plane with its centre at the origin in air. The current in the loop is 8.0 A in counter-clockwise direction. Find the approximate value of magnetic flux density at (2.0 m, 45°, 0) of spherical coordinates.

*4.41 Assuming that the magnetisation is proportional to the applied field intensity, obtain the relation between the magnetic flux density in a material in terms of the magnetic susceptibility.

**4.42 The interface of two regions 1 and 2 in a magnetic field is at $x = 0$ plane. The relative permeability of Region 2 ($x > 0$) is 1200, and the magnetic field intensity at the interface in this region is $\mathbf{H}_2 = 750\mathbf{u}_x + 400\mathbf{u}_y + H_{z2}\mathbf{u}_z$ A/m. The field intensity at the interface in Region 1 is $\mathbf{H}_1 = 500\mathbf{u}_x + H_{y1}\mathbf{u}_y + 300\mathbf{u}_z$ A/m. Find, μ_{r1}, H_{y1} and H_{z2}.

**4.43 The common boundary surface between the regions in a static magnetic field has a current sheet. Obtain the boundary relation.

***4.44 The $z = 0$ plane is the common boundary surface of two regions in a magnetic field. The magnetic flux density at the interface in Region 1 ($z < 0$) is $\mathbf{B}_1$ [illegible] $\mathbf{u}_x + 0.3\mathbf{u}_z$ T, and its permeability is 4.[illegible] 10^{-4} H/m. Find the angle subtended by B_1 with $\mathbf{u}_z$. Also find the flux density and its angle with $\mathbf{u}_z$ in Region 2, if Region 2 is (a) air and (b) a magnetic material of permeability 2.[illegible] 10^{-4} H/m.

***4.45 The normal to the interface between two regions 1 and 2 in a magnetic field is in y direction. The relative permeability of Region 1 is 280 and that of Region 2 is 360. The flux density in Region 2 is $0.12\mathbf{u}_x + 0.25\mathbf{u}_y + 0.08\mathbf{u}_z$ T. Find (a) the magnitude of flux density tangential to the interface in Region 1 as well as in Region 2 and (b) the angles subtended by the fields in the two regions with the normal to the interface.

**4.46 Two regions 1 and 2 in a magnetic field have a common boundary surface at $x = 0$ plane. The relative permeability of Region 1 ($x < 0$) is 1850 and that of Region 2 ($x > 0$) is 1275. The interface has a current sheet with linear current density $\mathbf{K} = 75.0\mathbf{u}_y$ ampere for one metre width in z direction. The magnetic flux density in Region 1 is $0.2\mathbf{u}_x + 0.2\mathbf{u}_z$ T. Calculate the flux density vector in Region 2.

***4.47 The unit normal vector at the plane boundary surface of two regions 1 and 2 in a magnetic field away from the origin is $\mathbf{u}_n = 0.267\mathbf{u}_x + 0.535\mathbf{u}_y + 0.802\mathbf{u}_z$. The magnetic field intensity in Region 1 containing the origin is $\mathbf{H}_1 = 675\mathbf{u}_x + 450\mathbf{u}_y$ A/m. Find (a) the angle between $\mathbf{u}_n$ and $\mathbf{H}_1$, (b) the magnetic field intensity H_2 in Region 2 and (c) the angle between $\mathbf{u}_n$ and $\mathbf{H}_2$. The relative permeability of Region 1 is 920 and that of Region 2 is 550.

**4.48 The ferromagnetic core of a magnetic circuit has two parts 1 and 2. The mean length of the flux path, the cross-sectional area and relative permeability of Part 1 are 25.0 cm, 6.0 cm^2 and 1250 respectively. The corresponding values for Path 2 are 30.0 cm, 5.0 cm^2, and 1575. Find (a) the reluctance of each part and (b) the mmf needed to set up a magnetic flux of 0.6 mWb in the core.

***4.49 If there is an air gap of 0.2 mm in Part 2 of the magnetic circuit of Problem 4.48 and the magnet is excited by a coil of 250 turns, what is the value of exciting current? Neglect leakage and fringing.

**4.50 The inside and outside radius of a toroid with iron core are 0.30 m and 0.40 m respectively. The toroid is of rectangular section with axial length 0.15 m. It has an air gap of length 2.0 mm and its effective area is 1.2 percent higher than the core area because of fringing of flux. The relative permeability of the core material is 550, and the flux density in the gap is 0.7 T. Calculate (a) reluctances of the iron part and the air gap, (b) flux in the air gap and (c) mmf needed for the iron part and air gap. Base your calculations on the average radius of the toroid.

**4.51 The primary winding of a core-type single-phase transformer with silicon steel as the core material has 250 turns. The winding is equally divided into two coils. The coils are placed on the two legs and connected in series as shown in Fig. P4.51. The sectional areas of the legs and yokes are 30.0 cm^2 and 40.0 cm^2 respectively. The length of mean flux path in each leg is 50.0 cm and in each yoke is 40.0 cm. The relative permeabilities corresponding to the flux densities in the leg and yoke are 1600 and 1800 respectively. Find the magnetising current needed to establish a flux density of 1.2 T in the legs. Indicate the direction of flux in the core.

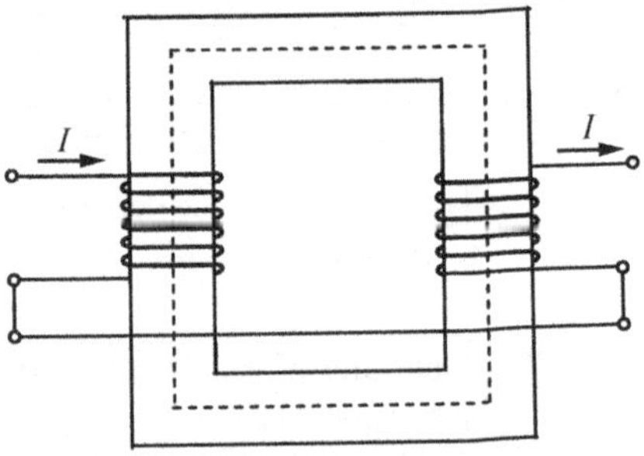

Fig. P4.51

**4.52 For some types of transformer construction, small gaps may exist at the joints of the core. Assume that the equivalent gap length and gap area in the magnetic circuit of Fig P4.51 are 0.25 mm and 40 cm^2. Find the value of the winding current to produce the same core flux as in Problem 4.51.

***4.53 For the parallel magnetic circuit shown in Fig. P4.53, the reluctances of the three iron parts are $\Re_1 = 8\times10^4\ \text{H}^{-1}$, $\Re_2 = \Re_3 = 20\times10^4\ \text{H}^{-1}$. The length and effective area of the air gap are 0.5 mm

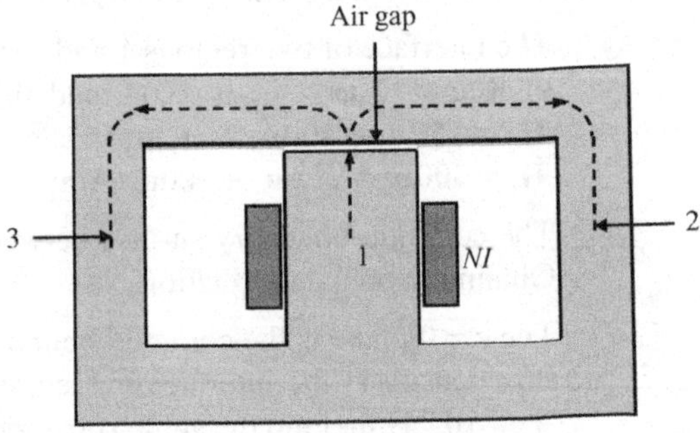

Fig. P4.53

and 15 cm^2 respectively. (a) An exciting coil on the central leg produces an mmf of $NI = 300$ ampere-turns. Find the flux in the central leg without the air gap and (b) the mmf needed to develop the same flux as in (a) with the air gap.

*4.54 A coil of 750 turns encloses a magnetic circuit of sectional area 30.0 cm^2. The flux density is 0.9 T with a coil current of 3.5 A. Calculate the value of external self-inductance of the coil.

*4.55 A long air-core solenoid has 1000 uniformly distributed turns. The length and average radius of the solenoid are 1.0 m and 2.5 cm respectively. Calculate the approximate value of its external self-inductance.

**4.56 A toroid having an iron core has a uniformly distributed winding of 1500 turns. The permeance of the magnetic circuit is 1.8×10^{-6} H. Find the value of self-inductance of the winding. If the toroid has an air gap of length 0.4 mm and effective area 5.0 cm^2, calculate the value of self-inductance.

***4.57 The exciting coil of the central leg of Fig. P4.53 has 500 turns. Calculate the values of self-inductance of the coil without and with the air gap.

*4.58 Define self-inductance of a coil in terms of flux linkage assuming that the flux linking different turns of the coil are different.

**4.59 Derive the formula for the external inductance of a uniformly distributed toroidal coil. The inner and outer radii of the toroid are a and b, and the relative permeability of the core material is μ_r.

**4.60 The outside radius of the inner conductor and the inner radius of the outside conductor of coaxial transmission line are a and b respectively. Derive the formula for the external inductance of the line per unit length.

*4.61 Obtain the mutual inductance between two magnetically coupled coils in terms of their self inductances and coupling coefficient.

*4.62 A long air-core solenoid of length 1.0 m has 1200 uniformly distributed turns. A search coil of 50 turns is located coaxially at the centre of the solenoid. The radius of the search coil is 2.0 mm. Find the value of mutual inductance between the solenoid and the search coil.

**4.63 For the arrangements of coils shown in Fig. P4.51, calculate (a) the self inductance of coil on each leg, (b) the mutual inductance between the coils if 95 per cent of flux due to one coil links with the other, (c) the total inductance of each coil, and (d) the inductance of the winding with the coils connected in series as in the figure. Use the data of Problem 4.51.

**4.64 Two long, air-core solenoids 1 and 2 have the same length and number of uniformly distributed turns. However, the radius of Solenoid 1 is a_1 and that of Solenoid 2 is $a_2\,(a_1 > a_2)$. The second solenoid is placed coaxially and symmetrically inside the first solenoid. Show that the magnetic coupling coefficient between the two solenoids is given by a_2/a_1.

**4.65 A rectangular coil of N turns is placed symmetrically between two parallel conductors of a very long transmission line in air. The arrangement is illustrated in Fig. P4.65. Derive an expression for the mutual inductance between the transmission line and the coil.

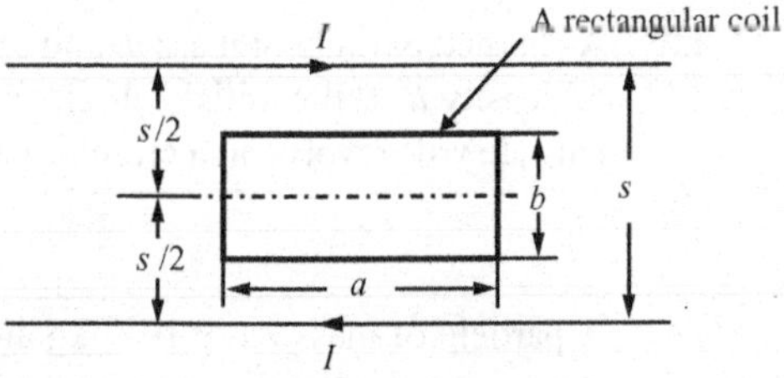

Fig. P4.65

*4.66 Electromagnetic energy stored in a magnetic circuit is given by

$$W_m = \frac{1}{2}\int_v \mu H^2 dv$$

Express the energy in terms of inductance.

**4.67 A cylindrical nonmagnetic conductor of circular section carries a uniformly distributed direct current I. The radius of the conductor is a. Using the formula for energy density w_m $BH/2$, calculate the magnetic energy stored within the conductor per unit length. Find the internal inductance of the conductor per unit length from the energy expression. For $I = 100$ A, what is the value of energy stored per unit length?

***4.68 A hollow cylindrical copper conductor of internal radius b and external radius c carries a uniformly distributed direct current I. Using the formula for energy density $w_m = BH/2$, calculate the magnetic energy stored within the conductor per unit length. Obtain the internal inductance per unit length of the conductor from the energy expression. If $b = 5.0$ mm, $c = 7.0$ mm, $I = 120$ A, what is the value of energy stored per unit length?

**4.69 The outside radius of the inner conductor and the inner radius of the outside conductor of a coaxial transmission line are a and b respectively. The current in the inner conductor is I and the permeability of the medium between the conductors is μ_0. Calculate the magnetic energy stored in the region between the conductors per unit length of the line.

**4.70 A toroid of rectangular section has an inside radius of 25 cm and an outside radius of 40 cm. Its axial length is 20 cm. It has a uniformly distributed winding of 1000 turns carrying a current of 2.0 A. The relative permeability of iron is 750. Find the magnetic energy stored in the toroid using the actual flux distribution. Neglect leakage flux.

**4.71 Using the data of the magnetic circuit described in Problem 4.48, calculate the energy stored in the core.

**4.72 The electromagnet shown in Fig. P4.72 is required to lift an iron plate of 250 kg. The area of iron surface at each gap is 30 cm^2, and the length of each gap is 1.2 mm. The reluctance of the cast steel magnet as well as the iron plate is 3.2×10^5 H^{-1}. What should be the mmf of the exciting coil?

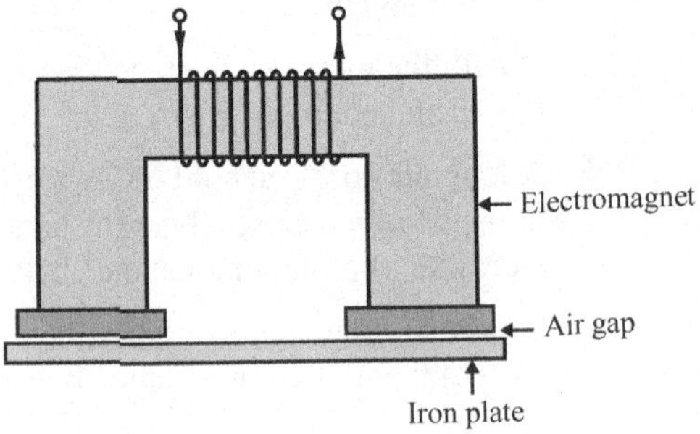

Fig. P4.72

*4.73 Two long parallel conductors carry equal current I but in opposite directions. The conductors are located in air and spacing between them is s. Derive the formula for the electromagnetic force between the conductors per unit length.

**4.74 A particle having an electric charge of 3.2×10^{-19} C and mass of 6.6×10^{-27} kg is at rest. If the particle is subjected to an electric field of intensity $1.8\mathbf{u}_x$ kV/m at $t = 0$, find the velocity and kinetic energy of the particle after 1.2 μs.

**4.75 A charged particle of mass m and charge Q is injected at a velocity v into a stationary magnetic field of density B. If the initial velocity of the particle is perpendicular to the direction of flux density, the particle will revolve in a circular path. Show that the radius of the path,

$$r = \frac{mv}{QB}$$

**4.76 A particle of mass 2.6×10^{-27} kg and charge 1.6×10^{-19} C is injected at a velocity 15×10^4 m/s into a magnetic field of flux density 0.05 T. The initial velocity of the particle is perpendicular to the flux density. Calculate (a) the magnitude of force and acceleration of the particle, (b) the radius of orbit and (c) time required for one revolution.

**4.77 A proton enters into a region that has both electric and magnetic fields. The initial velocity of the proton is 4×10^5 m/s along the ϕ direction of cylindrical coordinates. The electric field intensity of 2.5 kV/m is along positive z direction, and the magnetic flux density of 75×10^{-6} T is along the negative z direction. Find the total force on the proton. Mention the types of motion of the particle due to the electric force only, due to the magnetic force only, and due to both the forces. The charge and mass of a proton are 1.602×10^{-19} C and 1.673×10^{-27} kg.

Chapter Five

SOLUTION OF BOUNDARY-VALUE PROBLEMS

5.1 Introduction

In Chapter 2, we applied Coulomb's law and Gauss's law for calculation of electric field intensity and electrostatic potential of specified charge distributions in unbounded regions. And in chapter four, we applied Biot–Savart law and Ampere's magnetic circuit law to find the magnetic field intensity and vector magnetic potential of some specific current distributions in unbounded regions. Application of Coulomb's law and Biot–Savart law involves evolution of definite integrals. Thus, these methods are useful only if a function can be integrated. Moreover, Gauss's law and Ampere's magnetic circuit law are useful for calculation of fields of only highly symmetrical charge and current distributions.

There are many electric- and magnetic-field problems involving boundary surfaces on which a field quantity is specified. These are called boundary-value problems. Out of several techniques available for solving such problems, we will discuss only three methods. These are (1) Closed-form solution of Laplace's and Poisson's equations for potential functions; (2) The image method; (3) Graphical field mapping.

Laplace's and Poisson's equation are second order partial differential equations of a field quantity. In electric and magnetic fields, the quantity generally is a potential function. The equations, in general, are three-dimensional as the potential is a function of all the three coordinate variables. We will solve problems which can be treated as one-dimensional and two-dimensional problems. We will restrict our study to the solution of two-dimensional problems having rectangular and cylindrical symmetry. It may be noted that closed-form solutions satisfying the appropriate boundary conditions can be obtained only for a limited set of problems.

The method of images consists of replacing the effect of boundary surfaces near a charge distribution by suitable image charges. The electric field on the side of the boundary of the original charge is calculated considering both image and actual charges. The method can also be used to find the magnetic field of line currents near conducting planes. The image method is however not suitable if the boundaries are of arbitrary shape.

The technique of graphical field mapping may be used to solve two-dimensional field problems involving boundaries of any shape. The method consists of sketching the equipotential lines and field lines in the region between the boundary surfaces. Approximate values of potential gradient and capacitance between conductors can be obtained from the map of an electric field. The ohmic resistance of conductors of arbitrary shape can be found from the map of current stream lines and equipotential lines. The magnetic field map in the region of a non-uniform air gap between two highly permeable magnetic surfaces can be used to obtain the field distribution in the gap and also for calculation of gap reluctance.

5.2 The Uniqueness Theorem

As per the theorem, a potential function which is a solution of Laplace's or Poisson's equation and satisfies the boundary conditions is the only solution. In other words, the solution so obtained is unique.

One way to prove the theorem is to assume that there are two potential functions which are solutions of the equations, and then show that the two solutions are not different if they satisfy the same boundary conditions. At first let us consider the Laplace's equation given by (2.30-7). Let us assume that two potential functions V_1 and V_2 are solutions of the equation and these functions also satisfy the given boundary conditions. Laplace's equation for V_1 and V_2 are given by

$$\nabla^2 V_1 = 0 \tag{1}$$

$$\nabla^2 V_2 = 0 \tag{2}$$

Subtracting (1) from (2), we have

$$\nabla^2 (V_2 - V_1) = 0 \tag{3}$$

At a given boundary, let $V_1 = V_{b1}$ and $V_2 = V_{b2}$. Since the potential at the boundary can have only one value, V_{b1} must be equal to V_{b2}. Thus,

$$V_{b2} - V_{b1} = 0 \tag{4}$$

Now, replacing the scalar function φ by $(V_2 - V_1)$ and the vector function **A** by $\nabla(V_2 - V_1)$ in the vector identity (3) of Section 1.9, we get

$$\begin{aligned}\nabla\cdot[(V_2 - V_1)\nabla(V_2 - V_1)] &= (V_2 - V_1)[(\nabla\cdot\nabla(V_2 - V_1)] + \nabla(V_2 - V_1)\cdot\nabla(V_2 - V_1)\\ &= (V_2 - V_1)[(\nabla^2(V_2 - V_1)] + [\nabla(V_2 - V_1)]^2\end{aligned} \tag{5}$$

Integration of both sides of (5) throughout the volume enclosed by the boundary surface gives,

$$\int_v \nabla\cdot[(V_2 - V_1)\nabla(V_2 - V_1)]dv = \int_v (V_2 - V_1)[(\nabla^2(V_2 - V_1)]dv + \int_v [\nabla(V_2 - V_1)]^2 \, dv \tag{6}$$

By the use of divergence theorem, the volume integral on the left side of (6) can be transformed into a surface integral on the closed boundary surface enclosing the volume. Since at the boundary, $V_1 = V_{b1}$ and $V_2 = V_{b2}$, we have

$$\oint_S [(V_{b2} - V_{b1})\nabla(V_{b2} - V_{b1})]\cdot d\mathbf{S} = \int_v (V_2 - V_1)[(\nabla^2(V_2 - V_1)]dv + \int_v [\nabla(V_2 - V_1)]^2 \, dv \tag{7}$$

The left-hand side of (7) is zero by (4) and the first term on the right-hand side is zero by (3). Hence, we get

$$\int_v [\nabla(V_2 - V_1)]^2 \, dv = 0$$

As V_1 and V_2 are real functions, the gradient of $(V_2 - V_1)$ should also be a real function. Moreover, the square of a real function is always positive. Therefore, the integral can be zero only if the integrand is zero. Accordingly,

$$\nabla \quad - \quad {}^2 = 0$$

Or

$$\nabla(V_2 - V_1) = 0$$

Since the gradient operation is differentiation with respect to the space variables, $(V_2 - V_1)$ can have only a constant value throughout the region including the boundary surface. As at the boundary $V_{b2} - V_{b1} = 0$, the value of the constant can only be zero. Therefore,

$$V_2 - V_1 = 0$$

Or

$$V_2 = V_1$$

That is, the two potentials V_1 and V_2 which we have assumed to be different are indeed the same. Therefore, a solution of Laplace's equation which satisfies the boundary conditions is unique.

We will now consider the Poisson's equation. As before, let us assume that two potential functions V_1 and V_2 are solutions of (2.30-6) and both satisfy the same boundary conditions.

Then,

$$\nabla^2 V_1 = -\frac{\rho}{\varepsilon} \text{ and } \nabla^2 V_2 = -\frac{\rho}{\varepsilon}$$

Hence,

$$\nabla^2 (V_2 - V_1) = 0$$

As this equation is identical to (3), from here onwards the proof proceeds in the same as that for Laplace's equation.

It may be noted that for different boundary conditions the solutions of the equations would also be different.

5.3 Solutions of One-Dimensional Problems

In a given problem, if the size, shape and arrangement of boundaries are such that the field between them can be assumed to be a function of only one space variable, the problem is solved by using the one-dimensional Laplace's or Poisson's equation. The one-dimensional partial differential equation becomes an ordinary differential equation and it is solved by direct integration. The constants of integration are determined from the boundary conditions. We will obtain solutions for some simple boundary value problems. Some of the problems discussed here have been solved in earlier chapters by using other techniques. The purpose of including the same problems here is to demonstrate the procedure for solving boundary-value problems.

Parallel-Plate Capacitor with One Dielectric Layer

A parallel-plate capacitor with a single dielectric layer between the plates is illustrated in Fig. 5.1(a). The dielectric material of permittivity ε is assumed to be perfect. The electric potential of the plate at $x = 0$ plane is assumed to be V_a and that of the plate located at $x = s$ plane is taken as V_b.

If the linear dimensions of the plates are large compared to the spacing between the plates, we can assume that the potential V in the dielectric region varies with x only.

Moreover, as there is no free charge between the plates the potential function will satisfy Laplace's equation in the dielectric region. Thus, equation (2.30-8) reduces to a one-dimensional equation,

$$\frac{\partial^2 V}{\partial x^2} = 0 \tag{1}$$

Since V is a function of x only, equation (1) becomes an ordinary differential equation.

Thus,

$$\frac{d^2 V}{d x^2} = 0 \tag{2}$$

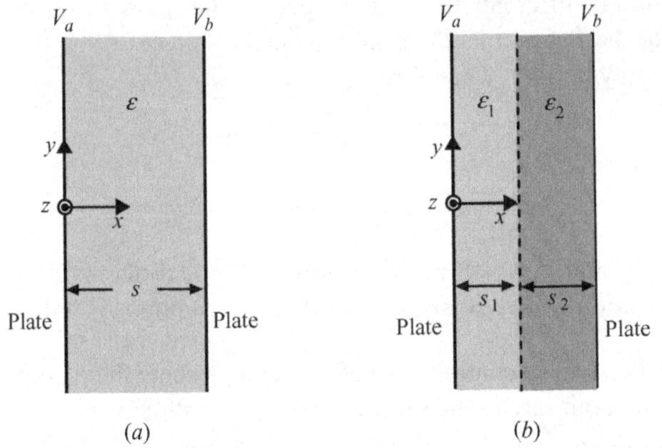

Fig. 5.1 (*a*) Parallel-plate capacitor with one dielectric layer and (*b*) Parallel-plate capacitor with two dielectric layers.

Integrating (2) twice with respect to x we get

$$V = Ax + B \tag{3}$$

where A and B are constants of integration. These are calculated using the following boundary conditions.

1. $V \quad V_a$ at $x = 0$
2. $V \quad V_b$ at $x = s$

Using the first boundary relations in (3), we get

$$V_a = 0 + B$$

The second boundary relation gives,

$$V_b = As + B$$

Hence,

$$A = \frac{V_b - V_a}{s} \tag{4}$$

and

$$B = V_a \tag{5}$$

Substituting in (3) the values of A and B given by (4) and (5), we get the solution for the potential in the dielectric region. The solution is,

$$V = \frac{(V_b - V_a)x}{s} + V_a \tag{6}$$

The electric field intensity,

$$\mathbf{E} = -\nabla V = -\frac{dV}{dx}\mathbf{u}_x = \frac{V_a - V_b}{s}\mathbf{u}_x \tag{7}$$

As the field intensity is normal to the plate surfaces, the surface charge density ρ_S on the plates is equal to the electric flux density as per the boundary condition. Hence,

$$\rho_{-} \quad \varepsilon E$$

Capacitance of the capacitor is given by

$$C = \frac{A\rho_S}{V_a - V_b} = \frac{A\varepsilon}{s} \tag{8}$$

where A is the surface area of each plate.

Parallel-Plate Capacitor with Two Dielectric Layers

A parallel-plate capacitor with two layers of perfect dielectric materials of permittivity ε_1 and ε_2 is shown in Fig. 5.1(*b*). The common boundary surface between the dielectrics is parallel to the conducting plates. The thickness of the dielectric layers are s_1 and s_2 and the electrostatic potentials of the plates are V_a and V_b, as indicated in the figure. Let V_1 be the potential at a point in the dielectric medium 1 and V_2 be the potential in the dielectric medium 2. If the dimensions of the plates are large compared to the distance between them, then we can assume that the potentials are functions of x only. Thus, Laplace's equations for the potential are,

$$\frac{\partial^2 V_1}{\partial x^2} = 0 \tag{9}$$

$$\frac{\partial^2 V_2}{\partial x^2} = 0 \tag{10}$$

The solutions of (9) and (10) are

$$V_1 = A_1 x + B_1. \tag{11}$$

$$V_2 = A_2 x + B_2. \tag{12}$$

There are four constants of integration and to determine these we need four boundary conditions. Referring to the figure, the boundary conditions for the potential functions are

1. $V_1 = V_a$ at $x = 0$
2. $V_1 = V_2$ at $x = s_1$
3. $V_2 = V_b$ at $x = s_1 + s_2$

> The fourth boundary condition is determined by assuming that the field lines are straight, and normal to the interface. Accordingly, flux densities in the two dielectric regions must be equal.

Hence, $\varepsilon_1 E_1 = \varepsilon_2 E_2$. Thus, the fourth boundary condition is

4. $\varepsilon_1 \dfrac{dV_1}{dx} = \varepsilon_2 \dfrac{dV_2}{dx}$

By using these boundary relations in (11) and (12), we get the following equations:

$$V_a = B_1$$

$$A_1 s_1 + B_1 = A_2 s_1 + B_2$$

$$V_b = A_2(s_1 + s_2) + B_2$$

$$\varepsilon_1 A_1 \quad \varepsilon_2 A_2$$

By solving these equations simultaneously, we will obtain the four constants in terms of V_a V_o. Substitution of these results in (11) and (12) will give,

$$V_1 = \frac{\varepsilon_2(V_b - V_a)x}{\varepsilon_1 s_2 + \varepsilon_2 s_1} + V_a, \; s_1 \geq x \geq 0 \tag{13}$$

$$V_2 = \frac{\varepsilon_1(V_b - V_a)x + V_b s_1(\varepsilon_2 - \varepsilon_1) + V_a \varepsilon_1(s_1 + s_2)}{\varepsilon_1 s_2 + \varepsilon_2 s_1}, \; s_1 + s_2 \geq x \geq s_1 \tag{14}$$

The field intensities in the two regions are

$$\mathbf{E}_1 = -\frac{\partial V_1}{\partial x}\mathbf{u}_x = \frac{\varepsilon_2(V_a - V_b)}{\varepsilon_1 s_2 + \varepsilon_2 s_1}\mathbf{u}_x, \; s_1 \geq x \geq 0 \tag{15}$$

And

$$\mathbf{E}_2 = -\frac{\partial V_2}{\partial x}\mathbf{u}_x = \frac{\varepsilon_1(V_a - V_b)}{\varepsilon_1 s_2 + \varepsilon_2 s_1}\mathbf{u}_x, \; s_1 + s_2 \geq x \geq s_1 \tag{16}$$

If A is the area of each plate and ρ_S is the surface charge density, capacitance

$$C = \frac{A\rho_S}{V_a - V_b} = \frac{A(\varepsilon_1 E_1)}{V_a - V_b} = \frac{A\varepsilon_1\varepsilon_2}{\varepsilon_1 s_2 + \varepsilon_2 s_1} \tag{17}$$

EXAMPLE 5.1

The distance between the plates of a parallel-plate capacitor is 8.0 mm. Find the safe operating voltage, (a) if the space between the plates is completely filled with a dielectric material for which the relative permittivity is 2.5 and dielectric strength is 15.0×10^6 V/m and (b) if the thickness of the dielectric is reduced to 7.5 mm creating an air gap of 0.5 mm. Take the dielectric strength of air as 3.0×10^6 V/m.

(a) Assuming uniform field distribution between the plates, the voltage between the plates is given by

$V \quad sE$

where s is spacing between the plates.

The breakdown voltage,

$V_{max} = sE_{max}$

where E_{max} is the dielectric strength. Thus,

$V_{max} = (8.0 \times 10^{-3})(15.0 \times 10^6) = 120.0$ kV

The operating voltage should be less than 120.0 kV.

(b) Designating the air gap as Region 1 and the dielectric layer as Region 2, we have from (15) and (16)

$$\frac{E_2}{E_1} = \frac{\varepsilon_1}{\varepsilon_2} = \frac{1}{\varepsilon_{r2}}$$

The field intensity in air is more than that in the dielectric material. Moreover, the dielectric strength of air is less than that of the dielectric. Therefore, the breakdown voltage depends on the dielectric strength of air.

The voltage between the plates is given by

$V \quad s_1E_1 \quad s_2E_2 = \;,\; {}^{-}{}_1 + s_2 \; E_1/\varepsilon_{r2})$

Substituting $s_1 = 0.5 \times 10^{-3}$ m, $s_2 = 7.5 \times 10^{-3}$ m, $E_1 = 3 \times 10^6$ V/m, $\varepsilon_{r2} = 2.5$, the breakdown voltage,

$$V_{\max} = (0.5 \times 10^{-3})(3.0 \times 10^6) + (7.5 \times 10^{-3})\left(\frac{3.0 \times 10^6}{2.5}\right) = 10.5 \text{ kV}$$

The operating voltage should be less than 10.5 kV.

Coaxial Transmission Line with One Dielectric Layer

The cross-sectional view of a coaxial transmission line having an inner conductor of radius a and an outer conductor of radius b with one layer of a dielectric material between them is shown in Fig. 5.2(a). The electrostatic potentials of the inner and outer conductors are taken as V_a and V_b respectively.

If we assume that the cable is long and straight and the dielectric is perfect then the potential in the dielectric region would be a function of r only in cylindrical coordinates and it satisfies Laplace's equation.

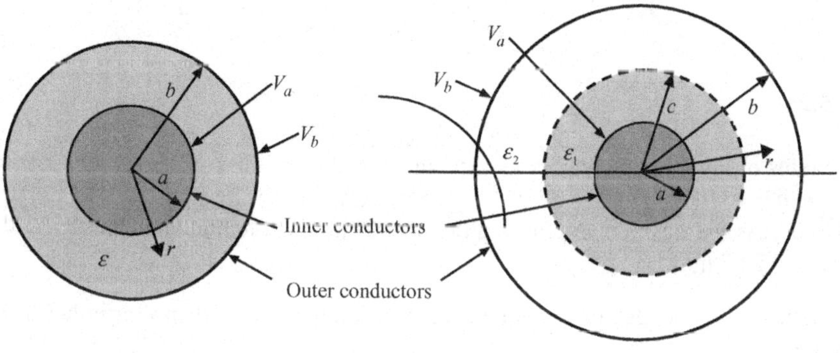

(a) (b)

Fig. 5.2 (a) Coaxial cable having one dielectric layer between conductors and (b) Coaxial cable having two dielectric layers between conductors.

Thus, from (2.30-9) we get

$$\frac{1}{r}\frac{\partial}{\partial r}\left(r\frac{\partial V}{\partial r}\right) = 0$$

Or

$$\frac{1}{r}\frac{d}{dr}\left(r\frac{dV}{dr}\right) = 0 \tag{18}$$

As $r \neq 0$, we can multiply both sides of (18) by r to obtain,

$$\frac{d}{dr}\left(r\frac{dV}{dr}\right) = 0$$

Integrating twice with respect to r, we get

$$V \quad A' \quad r \quad B \tag{19}$$

where A and B are constants of integration.

The boundary conditions are given by

1. At $r = a$, $V = V_a$
2. At $r = b$, $V = V_b$

By substituting the boundary relations in (19) and then solving for the constants we will obtain,

$$A = \frac{V_b - V_a}{\ln(b/a)} \tag{20}$$

$$B = \frac{V_a \ln b - V_b \ln a}{\ln(b/a)} \tag{21}$$

The complete solution is found by substituting in (19) the values of A and B given by (20) and (21). The solution is

$$V = \frac{V_b \ln(r/a) + V_a \ln(b/r)}{\ln(b/a)}, \quad b \geq x \geq a \tag{22}$$

The electric field intensity in the dielectric region,

$$\mathbf{E} = -\nabla V = -\frac{\partial V}{\partial r}\mathbf{u}_r = \frac{(V_a - V_b)}{\ln(b/a)}\frac{\mathbf{u}_r}{r}, \quad b \geq x \geq a \tag{23}$$

EXAMPLE 5.2

The outside radius of the inner conductor and the inside radius of the outer conductor of a coaxial transmission line are $a = 15.0$ mm and $b = 25.0$ mm respectively. The electric potential of the inner conductor $V_a = 1100$ V whereas the outer conductor is grounded. Find the potential and electric field intensity at a radial distance $r = 22.5$ mm.

As the outer conductor is grounded its potential is zero. Substituting $V_b = 0$ into formula (22), the potential at radius r is given by

$$V = \frac{V_a \ln(b/r)}{\ln(b/a)} = \frac{(1100)\ln(25.0/22.5)}{\ln(25/15)} = 226.8 \text{ V}$$

Substituting $V_b = 0$ into formula (23), the field intensity at radius r,

$$\mathbf{E} = \frac{V_a}{\ln(b/a)}\frac{\mathbf{u}_r}{r} = \frac{1100}{\ln(25/15)}\frac{\mathbf{u}_r}{22.5\times10^{-3}} = 95.71\times10^3 \text{ V/m}$$

Coaxial Transmission Line with Two Dielectric Layers

The reasons for using two or more dielectric layers between conductors of high voltage coaxial transmission lines have been given in Section 2.27. As shown in Fig. 5.2(*b*), the inner and outer conductors of a coaxial transmission line are of radii a and b and these are at potentials V_a and V_b respectively. The boundary between the two perfect dielectric layers of permittivity ε_1 and ε_2 is at a radial distance c.

If the transmission line is straight and long we can assume that the potentials V_1 and V_2 in the dielectric regions vary with r only in cylindrical coordinates.

Thus the general solutions for the potential function in the two dielectric regions are

$$V_1 = A_1 \ln r + B_1 \tag{24}$$

$$V_2 = A_2 \ln r + B_2 \tag{25}$$

where A_1, B_1, A_2 and B_2 are constants of integration.

The given and the known boundary conditions are as follows:

1. At $r = a$, $V_1 = V_a$
2. At $r = b$, $V_2 = V_b$
3. At $r = c$, $V_1 = V_2$
4. At $r = c$, $\varepsilon_1 \dfrac{dV}{dr} = \varepsilon_2 \dfrac{dV}{dr}$ because the electric flux density is continuous at the interface.

Substitution of the boundary relations into (24) and (25) yields the following equations:

$$V_a = A_1 \ln a + B_1$$

$$V_b = A_2 \ln b + B_2$$

$$A_1 \ln c + B_1 = A_2 \ln c + B_2$$

$$\varepsilon_1 A_1 = \varepsilon_2 A_2 \;\Rightarrow \varepsilon_{r1} A_1 = \varepsilon_{r2} A_2$$

Simultaneous solution of the above four equations will give the constants in terms of V_a and V_b. By substituting these results into (24) and (25) we will obtain the solution of the problem. The potential in two dielectric regions are given by

$$V_1 = \frac{\varepsilon_{r2}(V_b - V_a)\ln(r/a)}{\varepsilon_{r2}\ln(c/a) + \varepsilon_{r1}\ln(b/c)} + V_a,\; c \ge x \ge a \tag{26}$$

$$V_2 = \frac{\varepsilon_{r1}(V_b - V_a)(\ln r - \ln b)}{\varepsilon_{r2}\ln(c/a) + \varepsilon_{r1}\ln(b/c)} + V_b,\; b \ge x \ge c \tag{27}$$

The electric field intensities in the two dielectric regions are given by

$$\mathbf{E}_1 = -\frac{\partial V_1}{\partial r} = \frac{\varepsilon_{r2}(V_a - V_b)}{\varepsilon_{r2}\ln(c/a) + \varepsilon_{r1}\ln(b/c)}\frac{\mathbf{u}_r}{r} \tag{28}$$

$$\mathbf{E}_2 = -\frac{\partial V_2}{\partial r} = \frac{\varepsilon_{r1}(V_a - V_b)}{\varepsilon_{r2}\ln(c/a) + \varepsilon_{r1}\ln(b/c)}\frac{\mathbf{u}_r}{r} \tag{29}$$

EXAMPLE 5.3

Consider that the transmission line of Example 5.2 has two dielectric layers with the interface at a radial distance $c = 20.0$ mm. The relative permittivity of the dielectric layer near the inner conductor is $\varepsilon_{r1} = 4\ 8$ and near the outer conductor is $\varepsilon_{r2} = 3\ 0$. Find the potential and potential gradient at the interface.

Substituting $V_o = 0$ and r c in formula (26), the potential at the interface is given by

$$V_1(r \quad c) = \frac{-\varepsilon_{r2} V_a \ln(c/a)}{\varepsilon_{r2}\ln(c/a) + \varepsilon_{r1}\ln(b/c)} + V_a$$

$$= \frac{-(3.0)(1100)\ln(20/15)}{3.0\ln(20/15) + 4.8\ln(25/20)} + 1100 = 609.1 \text{ V}$$

The same result will be obtained if we use formula (27).

Substituting $V_b = 0$ and $r = c$ in formula (28) and (29), potential gradient on two sides of the interface are

$$\mathbf{E}_1(\quad) = \frac{\varepsilon_{r2} V_a}{\varepsilon_{r2}\ln(c/a) + \varepsilon_{r1}\ln(b/c)} \frac{\mathbf{u}_r}{c}$$

$$= \frac{(3.0)(1100)}{(3.0)\ln(20/15) + (4.8)\ln(25/20)} \frac{\mathbf{u}_r}{20.0\times10^{-3}} = 85.3\times10^3 \text{ V/m}$$

$$\mathbf{E}_2(\quad) = \frac{\varepsilon_{r1} V_a}{\varepsilon_{r2}\ln(c/a) + \varepsilon_{r1}\ln(b/c)} \frac{\mathbf{u}_r}{c}$$

$$= \frac{(4.8)(1100)}{(3.0)\ln(20/15) + (4.8)\ln(25/20)} \frac{\mathbf{u}_r}{20.0\times10^{-3}} = 136.5\times10^3 \text{ V/m}$$

Two Radial Conducting Plates

Consider two radial conducting plates located at $\phi = 0$ and $\phi = \alpha$ in cylindrical coordinates, as shown in Fig. 5.3. The plates are separated by a thin layer of insulation along the z axis. The static electric potential of the plates are taken as V_a and V_b.

If the plates are of a large size and the angle between them is small (5° to 10°), we can assume that the potential V between the plates is a function of ϕ only.

Further, if we assume that the dielectric between the plates is perfect, then the potential in the dielectric region will satisfy the Laplace's equation in cylindrical coordinates. As V does not vary with r and z, we get the following differential equation from (2.30-9).

$$\frac{1}{r^2}\frac{\partial^2 V}{\partial \phi^2} = 0$$

Or

$$\frac{d^2 V}{d\phi^2} = 0, \text{ as } r \neq 0$$

Integrating twice with respect ϕ, we get

$$V = A\phi + B \qquad (30)$$

where A and B are constants. The boundary conditions are

1. At $\phi = 0, V \quad V_a$
2. At $\phi = \alpha, V \quad V_b$

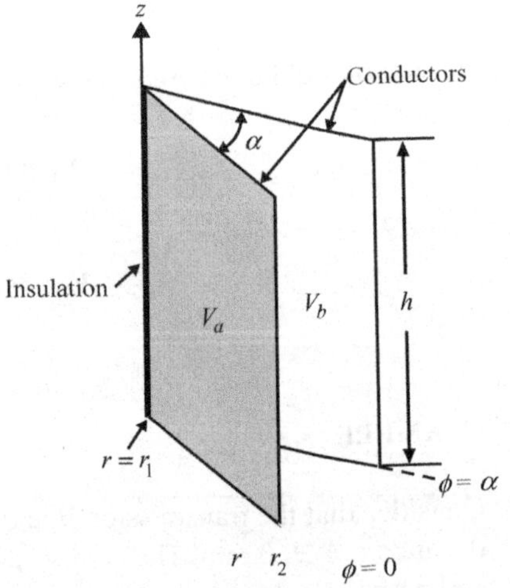

Fig. 5.3 Two radial plane conductors at an angle α

By using the boundary conditions in (30) and then solving for the constants we will obtain,

$$A = \frac{V_b - V_a}{\alpha}$$

$$B = V_a$$

Substituting these in (30) yields,

$$V = \frac{V_b - V_a}{\alpha}\phi + V_a \tag{31}$$

The electric field intensity is given by

$$\mathbf{E} = -\frac{1}{r}\frac{\partial V}{\partial \phi}\mathbf{u}_\phi = \frac{V_a - V_b}{\alpha}\frac{\mathbf{u}_\phi}{r} \tag{32}$$

We find that the field intensity is a function of r; however, it is in the ϕ direction.

Next, we will derive the capacitance between the plates. If ε is permittivity of the medium between the plates, electric flux density in the dielectric region,

$$\mathbf{D} = \varepsilon\mathbf{E} = \frac{\varepsilon(V_a - V_b)}{\alpha}\frac{\mathbf{u}_\phi}{r}$$

As the flux density is normal to the conducting plates, the surface charge density at a point on either of the plates is numerically equal to the flux density at this point. Thus,

$$\rho_S = \frac{\varepsilon(V_a - V_b)}{\alpha r}$$

The differential area in ϕ direction at the point is

$$dS_\phi = drdz$$

Total charge on the plate of height h and extending from $r = r_1$ to $r = r_2$ in the radial direction is

$$Q = \int_S \rho_S \, drdz = \int_0^h \int_{r_1}^{r_2} \frac{\varepsilon(V_a - V_b)}{\alpha r} drdz = \frac{\varepsilon h(V_a - V_b)}{\alpha} \ln\frac{r_2}{r_1}$$

Capacitance between the plates,

$$C = \frac{Q}{V_a - V_b} = \frac{\varepsilon h}{\alpha}\ln\frac{r_2}{r_1} \tag{33}$$

It may be noted that α is in *radians*.

EXAMPLE 5.4

Two radial conducting plates are at an angle 7.5°. The plates have a height of 15.0 mm in z direction and extend from 1.5 mm to 45.0 mm in r direction. The potential difference and relative permittivity of the dielectric material between the plates are 50.0 V and 4.2 respectively. Calculate the electric field intensity at a radial distance 10.0 mm and capacitance between the plates.

The field intensity as per formula (32) is

$$\mathbf{E} = \frac{V_a \quad V_b}{\alpha}\frac{\mathbf{u}_\phi}{r}$$

$V_a \quad V_o = 50\text{ V}$

$$\alpha = \frac{7\,5\pi}{180} = 0.1309\text{ rad}$$

$r = 10 \times 10^{-3}\text{ m}$

Substituting the values,

$$\mathbf{E} \quad \frac{50}{(0.1309)(10\times10^{-3})}\mathbf{u}_\phi = 3.82\times10^4\,\mathbf{u}_\phi\ \text{V/m}$$

Capacitance between the plates as per formula (33) is

$$C = \frac{\varepsilon h}{\alpha}\ln\frac{r_2}{r_1}$$

$\varepsilon_r = 4\ 2$

$h = 15 \times 10^{-3}\text{ m}$

$r_1 = 1\ 5 \times 10^{-3}\text{ m}$

$r_2 = 45 \times 10^{-3}\text{ m}$

Substituting the values,

$$C = \frac{(\ .85\times10^{-12})(4.2)(15\times10^{-3})}{0.1309}\ln\frac{45}{1.5} = 14.47\times10^{-12}\text{ F}$$

Two Concentric Spherical Conducting Shells

Two concentric spherical conducting shells separated by a single layer of a perfect dielectric material are shown in Fig. 5.4. The radius of the inner sphere is taken as *a*, and that of the outer sphere as *b*. The electrostatic potentials of inner and outer spheres are assumed to be V_a and V_b respectively.

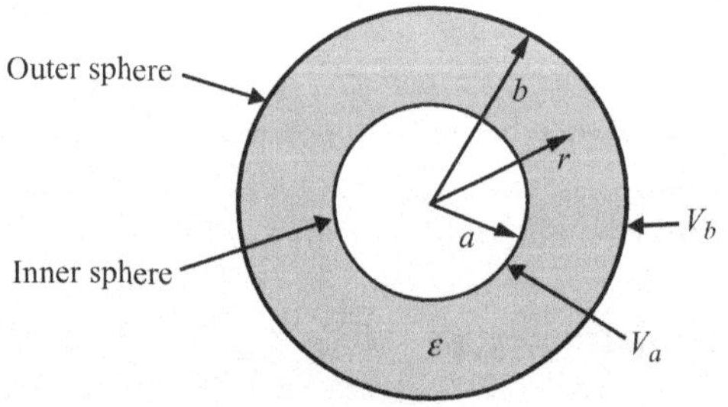

Fig. 5.4 Two concentric conducting spheres with a single dielectric layer

> Because of symmetry, the potential *V* in the dielectric region varies with *r* only in spherical coordinates.

We will obtain the variation of potential between the shells by solving the one dimensional Laplace's equation in *r*. The first term of (2.30-10) gives the required differential equation. The equation is

$$\frac{1}{r^2}\frac{\partial}{\partial r}\left(r^2\frac{\partial V}{\partial r}\right) = 0$$

Or

$$\frac{1}{r^2}\frac{d}{dr}\left(r^2\frac{dV}{dr}\right) = 0 \tag{34}$$

Since $r \neq 0$, we can multiply both sides of (34) by r^2. Thus, the equation reduces to

$$\frac{d}{dr}\left(r^2 \frac{dV}{dr}\right) = 0 \tag{35}$$

Integrating (35) twice with respect to r, we get

$$V = -\frac{A}{r} + B \tag{36}$$

where A and B are constants. The boundary conditions are

1. At $r = a$, $V = V_a$
2. At $r = b$, $V = V_b$

The constants can be determined by applying the boundary relations to (36). The results are then substituted back into (36) to get the complete solution. The solution is,

$$V = \frac{ab(V_a - V_b)}{b-a}\frac{1}{r} + \frac{bV_b - aV_a}{b-a} \tag{37}$$

The electric field intensity between the spheres,

$$\mathbf{E} = -\frac{\partial V}{\partial r} = \frac{ab(V_a - V_b)}{(b-a)r^2}\mathbf{u}_r \tag{38}$$

Two Coaxial Conducting Cones

A combination of two coaxial conducting cones having angles α_1 and α_2 and separated by a perfect dielectric material is shown in Fig. 5.5(a). The slant length of each cone is taken as a. The cones are separated at the tips by an insulating material. The scalar electric potential of the inner cone is taken as V_0 and that of the outer cone as zero. We recall that in defining a point in space in spherical coordinates one of the coordinate surfaces is a conical one. Hence, it would be convenient to solve the Laplace's equation in spherical coordinates. We assume that potential V in the dielectric region is a function θ only. Thus, referring to equation (2.30-10) the Laplace's equation in V is given by

$$\frac{1}{r^2 \sin\theta}\frac{d}{d\theta}\left(\sin\theta \frac{dV}{d\theta}\right) = 0$$

Excluding $r = 0$ and since $\sin\theta$ is not zero, the above equation reduces to

$$\frac{d}{d\theta}\left(\sin\theta \frac{dV}{d\theta}\right) = 0 \tag{39}$$

Integrating (39), we have

$$\frac{dV}{d\theta} = \frac{A}{\sin\theta} \tag{40}$$

where A is a constant

Integrating (40), we get

$$V = A\ln[(\tan(\theta/2)] + B \tag{41}$$

where B is a second constant

The boundary conditions are

1. At $\theta = \alpha_1$, $V = V_0$
2. At $\theta = \alpha_2$, $V = 0$

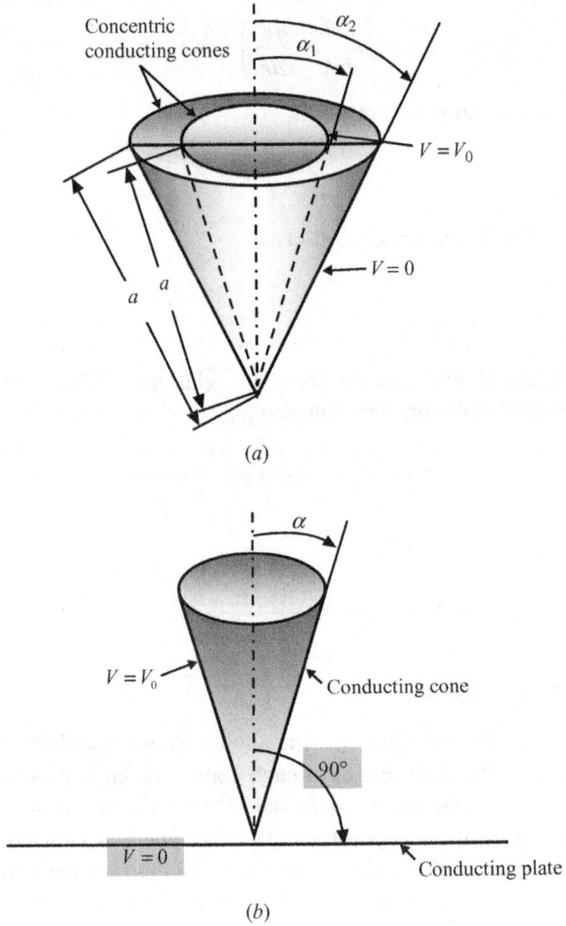

Fig. 5.5 (*a*) Two coaxial conducting cones and (*b*) A conducting cone above a conducting plate.

By applying the boundary conditions to (41), we get the following equations:

$$V_0 = A \ln[(\tan(\alpha_1/2)] + B$$

$$0 = A \ln[(\tan(\alpha_2/2)] + B$$

Solving for A and B and then substituting the results in (41), we get

$$V = V_0 \frac{\ln[\tan(\theta/2)] - \ln[\tan(\alpha_2/2)]}{\ln[\tan(\alpha_1/2)] - \ln[\tan(\alpha_2/2)]}, \; \alpha_2 \geq \theta \geq \alpha_1 \tag{42}$$

The electric field intensity is given by

$$\mathbf{E} = -\frac{1}{r}\frac{\partial V}{\partial \theta} = \frac{V_0}{\ln[\tan(\alpha_2/2)] - \ln[\tan(\alpha_1/2)]} \frac{\mathbf{u}_\theta}{r \sin\theta} = E_\theta \mathbf{u}_\theta \tag{43a}$$

$$E_\theta = \frac{V_0}{\ln[\tan(\alpha_2/2)] - \ln[\tan(\alpha_1/2)]} \frac{1}{r\sin\theta}, \; \alpha_2 \quad \theta > \alpha_1 \tag{43b}$$

The problem of a conducting cone just above the large conducting plate shown in Fig. 5.5(*b*) can be solved by proceeding in the same way as for the problem of two concentric conducting cones. The general solution of the problem is same as that given by (41). However, the boundary conditions are a little different. These are

1. At $\theta = \alpha$, $V = V_0$
2. At $\theta = 90^0$, $V = 0$

Substituting these boundary conditions into (41), we get

$$A = \frac{V_0}{\ln[\tan(\alpha/2)]},$$

$$B = 0$$

Substituting the values of A and B into (41), we get

$$V = V_0 \frac{\ln[\tan(\theta/2)]}{\ln[\tan(\alpha/2)]} \tag{44}$$

$$\mathbf{E} = -\frac{V_0}{\ln[\tan(\alpha/2)]}\frac{\mathbf{u}_\theta}{r\sin\theta} \tag{45}$$

EXAMPLE 5.5

The angles of two concentric cones are 10° and 20°. The slant length of each cone is 5.0 cm. The relative permittivity of the dielectric material and the potential difference between the cones are 4.6 and 75.0 V respectively. Calculate the charge contained in each cone and the capacitance.

The problem is illustrated in Fig. 5.5(a). The given data are

$\alpha_1 = 10°$, $\alpha_2 = 20°$, $a = 5.0\times10^{-2}$ m, $\varepsilon_r = 4.6$, $V_0 = 75.0$ V

The electric flux density is normal to the cone surfaces. Hence, the surface charge density on each cone surface is equal to the flux density at the surface. Referring to formula (43b), the surface charge density on the outside surface of the inner cone,

$$\rho_{S1} = D_\theta(\theta = \alpha_1) = \varepsilon E_\theta(\theta = \alpha_1) = \frac{\varepsilon V_0}{\ln[\tan(\alpha_2/2)] - \ln[\tan(\alpha_1/2)]}\frac{1}{r\sin\alpha_1}$$

The differential surface area in θ direction, $dS_\theta = r\sin\alpha_1\, dr\, d\phi$

Charge on the outside surface of the inner cone,

$$Q = \int_S \rho_{S1}\, dS_\theta = \int_0^a\int_0^{2\pi} \frac{\varepsilon V_0\, dr d\phi}{\ln[\tan(\alpha_2/2)] - \ln[\tan(\alpha_1/2)]}$$

$$= \frac{2\pi\varepsilon a V_0}{\ln[\tan(\alpha_2/2)] - \ln[\tan(\alpha_1/2)]}$$

Substituting the data,

$$Q_1 = \frac{2\pi(8.85\times10^{-12})(4.6)(5.0\times10^{-2})(75.0)}{\ln(\tan 10°) - \ln(\tan 5°)} = \text{[illegible]}\times10^{-9}\text{ C}$$

As the slant length a of the outer cone is same as that of the inner cone, charge on the inside surface of the outer cone

$$Q_2 = Q_1 = 1.368 \times 10^{-9}$$

Capacitance between the cones,

$$C = \frac{Q_1}{V_0} = \frac{1.368 \times 10^{-9}}{75.0} = 18.24 \times 10^{-12} \text{ F}$$

The p-n Semiconductor Junction

A simplified model of a *p-n* semiconductor junction is shown in Fig. 5.6(*a*). The device is assumed to be of rectangular section and the junction is a plane surface of negligible thickness. The junction is taken at $x = 0$. The semiconductor to the left of the junction ($x < 0$) is of *p* type and the semiconductor to the right of the junction ($x > 0$) is of *n* type.

Initially, that is soon after formation of the junction, some electrons from the *n* region diffuse to the *p* side of the junction, and some holes from the *p* region diffuse to the *n* side of the junction. Therefore, an electric field directed from *n* side towards *p* side is established near the junction. This field opposes the process of diffusion. At a certain value of the field intensity, migration of the charges completely stop resulting in a negative space charge layer in the *p* side and a positive space charge layer in the *n* side of the junction as shown in the figure.

> As the region of space charge has only a few mobile charges, the region is known as 'depletion region'. The semiconductor outside the charge layers is electrically neutral.

Our aim is to calculate the field distribution in the region of the charges.

To simplify the analysis, we assume that the value of charge on one side of the junction is same as the value of charge on the other side of the junction, and the charges are uniformly distributed with volume charge densities $+\rho_j$ and $-\rho_j$. The width of each charge layer is taken as w. The charge distributions are thus two pulses, as indicated at Fig. 5.6(*b*). We also assume that the potential in the region of the charges is a function of x only. Accordingly, the field distribution can be obtained by solving the one-dimensional Poisson's equation. Denoting the potential in the positive charge region by V^+, the differential equation to be solved is

$$\frac{\partial^2 V^+}{\partial x^2} = -\frac{\rho_j}{\varepsilon}, \; w > x > 0$$

Or

$$\frac{d^2 V^+}{dx^2} = -\frac{\rho_j}{\varepsilon}, \; w > x > 0 \qquad (46)$$

Integrating (46) twice with respect to x, we get

$$V^+ = -\frac{\rho_j}{2\varepsilon} x^2 + C_1 x + C_2 \qquad (47)$$

$$E^+ = -\frac{\partial V^+}{\partial x} = \frac{\rho_j}{\varepsilon} x - C_1 \qquad (48)$$

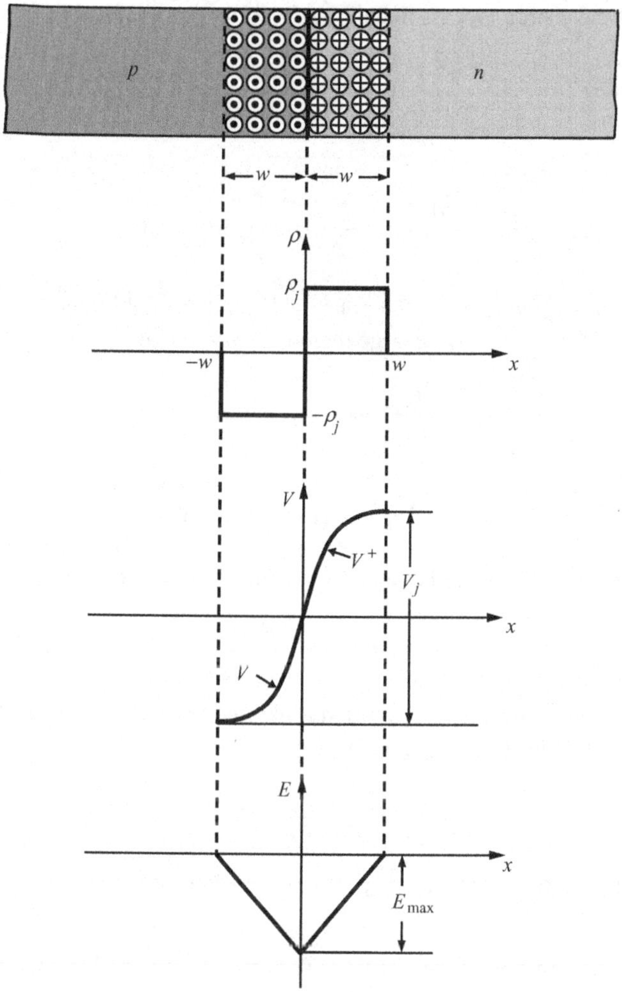

(a)

(b)

(c)

(d)

Fig. 5.6 (a) A semiconductor junction, (b) Distribution of charge in the depletion region and (c, d) Variation of potential and field intensity due the charges.

The constants are obtained from the following boundary conditions:

1. Since the positive and negative space charges are symmetrically distributed on two sides of the junction, $V^+ = 0$ at $x = 0$.
2. As field lines start from positive charge and end on negative charge, the field intensity $E^+ = 0$ at $x = w$.

Using the first boundary condition in (47), we have

$$C_2 = 0$$

Using the second boundary condition in (48), we get

$$C_1 = \frac{w\rho_j}{\varepsilon}$$

By substituting these constants in (47), we get the solution of (46).

$$V^+ = \frac{\rho_j}{2\varepsilon}(2wx - x^2),\ w > x > 0 \qquad (49)$$

The potential at the end of the charge layer is found by substituting $x = w$ in (49).

$$V^+(x = w) = \frac{\rho_j w^2}{2\varepsilon} \tag{50}$$

The electric field intensity,

$$E^+ = -\frac{\partial V^+}{\partial x} = -\frac{\rho_j}{\varepsilon}(w - x)\,,\; w > x > 0 \tag{51}$$

In the negative charge layer, expressions for the potential V^- and the field intensity E^- can be obtained by using the same procedure as for the positive charge region. The potential,

$$V^- = \frac{\rho_j}{2\varepsilon}(2wx + x^2)\,,\, -w < x < 0 \tag{52}$$

The field intensity in the region,

$$E^- = -\frac{\rho_j}{\varepsilon}(w + x)\,,\, -w < x < 0 \tag{53}$$

The potential at the end of the charge layer is found by substituting $x = -w$ in (52). Thus,

$$V^-(x = -w) = -\frac{\rho_j w^2}{2\varepsilon} \tag{54}$$

The maximum value of field intensity occurs at $x = 0$. It is obtained by substituting $x = 0$ either in (51) or (53). Denoting the maximum field intensity by E_{max}, we have

$$E_{max} = \frac{\rho_j w}{\varepsilon} \tag{55}$$

Figures 5.6(c) and 5.6(d) give the variations of potential and field intensity on two sides of the junction respectively.

Next we will calculate the capacitance of the junction. The charge contained in each layer

$$Q = \rho_j(wA)$$

where A is sectional area of the semiconductor.

The potential difference across the junction

$$V_j = V_j(x = w) - V_j(x = -w)$$

Substituting the expressions of (50) and (54) into the above relation, we get

$$V_j = \frac{\rho_j w^2}{\varepsilon} \tag{56}$$

The capacitance of the junction is

$$C = \frac{Q}{V_j} = \frac{\varepsilon A}{w} \tag{57}$$

We see that the formula for the capacitance of the junction is similar to that of a parallel-plate capacitor.

It may be noted that the results are derived with the assumption that the charges are uniformly and symmetrically distributed on two sides of the junction. The results would be different for any other form of charge distribution.

EXAMPLE 5.6

Each side of the depletion layer of a semiconductor junction has 2.8×10^{22} charge carriers per cubic metre. The relative permittivity of the material is 9.8. The junction area is 1.2×10^{-9} m^2 and the voltage across the junction is 570 mV. Calculate the value of electric field intensity at the junction and the junction capacitance assuming uniform distribution of charges.

Referring to formulae (55) and (57) we see that to solve the problem we have to first calculate the charge density and width of the charge layer. The charge density is equal to e $(=1.6\times10^{-19})$ multiplied by the number of charge carriers per unit volume.

Charge density, $\rho_j = (1.6\times10^{-19})(2.8\times10^{22}) = 4.48\times10^3$ C/m^3

The width of the charge layers from formula (56) is

$$w = \sqrt{\frac{\varepsilon V_j}{\rho_j}}$$

$$\varepsilon = (8.85\times10^{-12})(9.8) = 86.73\times10^{-12} \text{ F/m}$$

$$V_j = 570.0\times10^{-3} \text{ V}$$

Substituting the values,

$$w = \sqrt{\frac{(86.73\times10^{-12})(570\times10^{-3})}{4.48\times10^3}} = 10.5\times10^{-8} \text{ m}$$

Electric field intensity at the junction,

$$E_{\max} = \frac{w\rho_j}{\varepsilon} = \frac{(10.5\times10^{-8})(4.48\times10^3)}{86.73\times10^{-12}} = 5.424\times10^6 \text{ V/m}$$

Junction capacitance, $C_j = \dfrac{A\varepsilon}{w} = \dfrac{(1.2\times10^{-9})(86.73\times10^{-12})}{10.5\times10^{-8}} = 0.9912\times10^{-12}$ F

Round Conductor Carrying Direct Current

Figure 5.7 represents the cross section of an isolated solid cylindrical conductor of radius a. Its axis is along the z direction of cylindrical coordinates. It carries a uniformly distributed current I. The current density J_z is equal to $I/\pi a^2$. The permeability of the conductor is μ_1 and that of the space outside the conductor is μ_2. We shall calculate the magnetic fields inside and outside the conductor by solving the one-dimensional Poisson's and Laplace's equations for the vector magnetic potential.

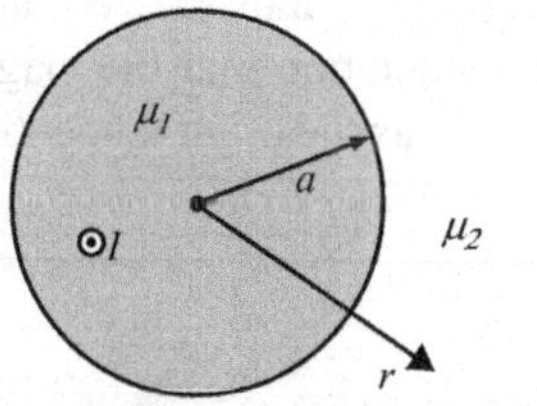

Fig. 5.7 Cross-sectional view of a circular conductor of permeability μ_1, and carrying current I. The conductor is located in a medium of permeability μ_2.

If the conductor is straight and very long, we may assume that the vector potential inside as well as outside the conductor will not vary along ϕ and z directions. Further, since the current is along z direction the vector potentials would also be in z direction.

Using the above properties in (4.10-11), the differential equation for the vector potential A_{z1} inside the conductor is given by

$$\frac{1}{r}\frac{\partial}{\partial r}\left(r\frac{\partial A_{z1}}{\partial r}\right) = -\mu_1 J_z,\ a \geq r \geq 0$$

$$\Rightarrow \quad \frac{1}{r}\frac{d}{dr}\left(r\frac{dA_{z1}}{dr}\right) = -\mu_1 J_z,\ a \geq r \geq 0 \tag{58}$$

Outside the conductor the current is zero. Hence, the differential equation for the vector potential A_{z2} outside the conductor is

$$\frac{1}{r}\frac{\partial}{\partial r}\left(r\frac{\partial A_{z2}}{\partial r}\right) = 0,\ r \geq a$$

$$\Rightarrow \quad \frac{1}{r}\frac{d}{dr}\left(r\frac{dA_{z2}}{dr}\right) = 0,\ r \geq a \tag{59}$$

Integrating (58) and (59) twice, we get

$$A_{z1} = -\frac{\mu_1 J_z r^2}{4} + C_1 \ln r + C_2$$

$$A_{z2} = C_3 \ln r + C_4$$

where C_1 C_2, C_3 and C_4 are constants.

Flux densities in the two regions can be obtained from the relation $\mathbf{B} = \nabla \times \mathbf{A}$ in cylindrical coordinates. As the vector potentials are in z direction and they do not vary with ϕ and z, we have from (4.7-14a)

$$\mathbf{B}_1 = -\frac{\partial A_{z1}}{\partial r}\mathbf{u}_\phi = \frac{\mu_1 J_z r}{2}\mathbf{u}_\phi - \frac{C_1}{r}\mathbf{u}_\phi,\ a \geq r \geq 0 \tag{60}$$

$$\mathbf{B}_2 = -\frac{\partial A_{z2}}{\partial r}\mathbf{u}_\varphi = -\frac{C_3}{r}\mathbf{u}_\varphi,\ r \geq a \tag{61}$$

In (60), C_1 must be zero as otherwise $\mathbf{B}_1$ would be infinite at $r = 0$. The constant C_3 is determined from the consideration that the magnetic field intensities on two sides of the boundary at $r = a$ must have the same value, as these are tangential to the boundary. Thus,

$$\frac{B_1(r=a)}{\mu_1} = \frac{B_2(r=a)}{\mu_2} \tag{62}$$

By substituting $r = a$ in (60) and (61) and then using these results in (62), we will obtain

$$C_3 = -\frac{\mu a^2 J_z}{2}$$

Therefore, flux densities in the two regions in terms of I are given by

$$\mathbf{B}_1 = \frac{\mu I r}{2\pi a^2}\mathbf{u}_\varphi,\ 0 < r \leq a \tag{63}$$

$$\mathbf{B}_2 \quad \frac{uI}{2\pi r}\mathbf{u}_\varphi,\ r \geq a \tag{64}$$

The constants C_2 and C_4 do not contribute to the flux density. Moreover, the vector magnetic potential has no physical meaning. Hence, these constants may be assumed to be zero.

If the conductor is of a nonmagnetic material and if it is located in air, then $\mu_1 = \mu_2 = \mu_0$.

5.4 Solutions of Two-Dimensional Laplace's Equation in Rectangular Coordinates

If the boundary conditions of a problem have to be satisfied in two spatial directions in charge-free regions, it is necessary to solve the two-dimensional Laplace's equation. The two-dimensional equation cannot be solved by direct integration as in the case of the one-dimensional equation. Here,

the potential is assumed to be the product of two potentials where each potential varies with one variable only. By such an assumption, the two-dimensional second order partial differential equation is converted into two separate one-dimensional second order ordinary differential equations.

Thus the variables are separated. Each one-dimensional equation is solved, and the total solution is given by the product of the two solutions. The constants arising out of the solutions are determined from the boundary conditions. The method of solution is called *separation of variable* method, and the solution is called *product solution*. In this section we will obtain the solution in rectangular coordinates. In the next section we will consider two-dimensional Laplace's equation in cylindrical coordinates.

In rectangular coordinates the potential V may be a function of x and y, or y and z, or z and x depending on the choice of variables we assign to the coordinate axes in a problem. Let us assume that the potential is a function of x and y. Then from (2.30-8), we have

$$\frac{\partial^2 V}{\partial x^2} + \frac{\partial^2 V}{\partial y^2} = 0 \tag{1}$$

Let us assume that V can be expressed as the product of two potentials X and Y, where X varies with x only and Y varies with y only. Then, we have

$$V = XY \tag{2}$$

By substituting (2) in (1), we get

$$\frac{\partial^2 (XY)}{\partial x^2} + \frac{\partial^2 (XY)}{\partial y^2} = 0 \tag{3}$$

Since Y in the first term of (3) does not vary with x and X in the second term does not vary with y, these can be brought out of the brackets. Also, partial derivatives can be replaced by ordinary derivatives as each term is a function of one variable only. Thus,

$$Y\frac{d^2 X}{dx^2} + X\frac{d^2 Y^2}{dy^2} = 0 \tag{4a}$$

Dividing both sides of (4a) by XY, we have

$$\frac{1}{X}\frac{d^2 X}{dx^2} + \frac{1}{Y}\frac{d^2 Y}{dy^2} = 0 \tag{4b}$$

Or

$$\frac{1}{X}\frac{d^2 X}{dx^2} = -\frac{1}{Y}\frac{d^2 Y}{dy^2} \tag{4c}$$

As the lhs of (4c) is a function of x only and the rhs is a function of y only, both the sides must be constant. This is because if the left-hand side changes with x, it cannot be equal to the right-hand side as the rhs does not contain x. By the same logic, the rhs cannot change with y.

Assuming the constant to be a positive real quantity, and denoting it by k^2, we have

$$\frac{1}{X}\frac{d^2X}{dx^2} = -\frac{1}{Y}\frac{d^2Y}{dy^2} = k^2 \tag{5}$$

Or

$$\frac{d^2X}{dx^2} = k^2 X \tag{6}$$

And

$$\frac{d^2Y}{dy^2} = -k^2 Y \tag{7}$$

Equations (6) and (7) are ordinary second order differential equations. The solution of (6) in terms of hyperbolic functions is

$$X = A\cosh kx + B\sinh kx \tag{8}$$

And the solution of (7) in terms of circular sine and cosine functions is given by

$$Y = C\cos ky + D\sin ky \tag{9}$$

In equations (8) and (9) A, B, C and D are constants. It may be verified that X and Y given by (8) and (9) satisfy equations (6) and (7) respectively. By using these solutions in (2), we will obtain the the general solution of (1). It is given by

$$V = XY = (A\cosh kx + B\sinh kx)(C\cos ky + D\sin ky) \tag{10}$$

Another useful form of solution of (1) in terms of exponential functions, which can be derived from (10), is

$$V = (A'e^{kx} + B'e^{-kx})(C\cos ky + D\sin ky) \tag{11}$$

We obtained solutions (10) and (11) assuming that the functions on the two sides of (4c) are equal to a positive constant. Now, if we assume that the constant in (5) is $-k^2$, then the product solution of (1) would be

$$V = (A\cos kx + B\sin kx)(C\cosh ky + D\sinh ky) \tag{12}$$

An exponential form of solution, which can be derived from (12), is given by

$$V = (A\cos kx + B\sin kx)(C'e^{ky} + D'e^{-ky}) \tag{13}$$

Further, if the constant k is taken as zero, it can be shown that the product solution of (1) is given by

$$V = (Ax + B)(Cy + D) \tag{14}$$

Various forms of product solutions of the two-dimensional Laplace's equation in rectangular coordinates are given by (10) through (14).

The form of solution to be used depends on the physical nature of the problem and the boundary conditions. Moreover, as the Laplace's equation is a linear equation, solutions can be superposed wherever necessary.

We find that the solutions have five constants and hence the number of boundary conditions should also be five. However in a physical problem some of the constants may be zero hence the number of constants to be determined would be less than five. The procedure for solving two dimensional boundary-value problems in rectangular coordinates is illustrated through the following examples.

EXAMPLE 5.7

Figure 5.8(a) shows the cross-sectional view of a conducting strip located at the left-hand end of two infinitely large parallel conducting plates separated by a. The strip is perpendicular to the parallel plates. The width of the strip, which is insulated from the parallel plates, is only a little less than a. The parallel plates are at zero potential and the strip is at a constant potential V_0. It is required to find the electric potential and field intensity in the region between the plates. The plates are located in air.

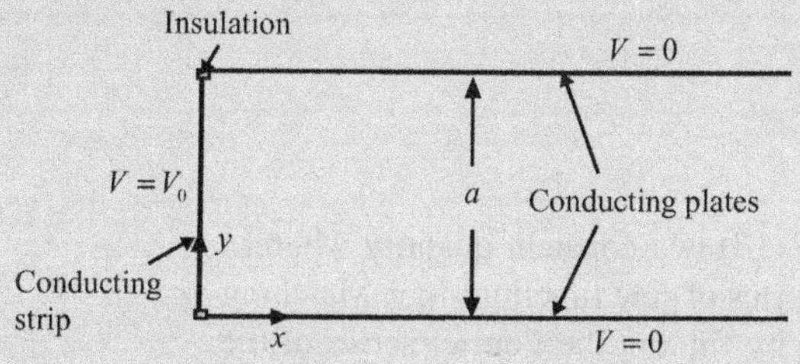

(a)

Fig. 5.8 (a) A region bounded by two parallel plates and a conducting strip normal to the plates.

The first step in solving the problem is to choose the location of the origin of the coordinate system to be used. For the problem under consideration, the origin of the rectangular coordinate system may be taken at the bottom corner or at the top corner or at the mid-point of the strip. Though the form of the solution depends on the choice of the location of the origin, the field distribution however does not depend on it. We choose the origin at the bottom corner. As per the coordinate axes shown in Fig. 5.8(a), the parallel plates are at y 0 and $y = a$ whereas the strip is at $x = 0$.

To select a suitable solution, let us list the boundary conditions. We have three boundary conditions from the given data. Another boundary condition is obtained from the consideration that the potential is zero at $x = \infty$. Thus, the boundary conditions are

1. $V = 0$ at $x = \infty$, $a \geq y \geq 0$
2. $V = 0$ at $y = 0$, for all x
3. $V = 0$ at $y = a$, for all x
4. $V = V_0$ at $x - 0$, $a > y > 0$

Since the potential is zero at $x = \infty$ as per the first boundary condition, the solution given by (11) is suitable with the constant A' equal to zero. Moreover to satisfy boundary condition 2, the constant C must also be zero. Thus, substituting $A' = C = 0$ in (11), we get

$$V \quad (B'D)e^{-kx} \sin ky \tag{15}$$

Next, using boundary condition 3 in (15), we have

$$(B'D)e^{-kx} \sin ka = 0$$

$$\Rightarrow \quad ka = n\pi$$

Or $k - \dfrac{n\pi}{a}$, where n is an integer

Substituting the value of k in (15) and replacing $(B'D)$ by another constant B_n, we get

$$V \quad B_n\, e^{-(n\pi/a)x} \sin\frac{n\pi y}{a}$$

Since the Laplace's equation is linear and the medium between the plates is air, the sum of an infinite number of terms is also a solution of the equation. Accordingly, the solution is given by

$$V \quad \sum_n B_n e^{-(n\pi/a)x} \sin\frac{n\pi y}{a} \tag{16}$$

The constant B_n is found by the use of the fourth boundary condition. Its application to (16) results in the following equation.

$$V_0 \quad \sum_n B_n \sin\frac{n\pi y}{a} \tag{17}$$

We see that the left side of (17) is a constant quantity whereas the right side is an infinite series of sine functions in y. Matching of the two sides is possible by finding the Fourier series of the positive half-cycle of a rectangular wave of amplitude V_0 and half-wavelength a, as shown in Fig. 5.8(b).

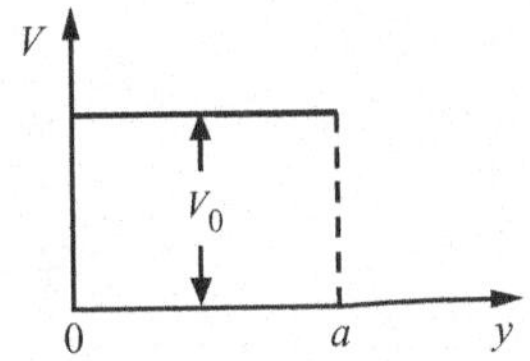

(b)

Fig. 5.8 (b) Graphical representation of potential of the strip.

The Fourier series of such a wave has been obtained in Section 1.13. By replacing ℓ of (1.13-3) by a, we will get the Fourier series of the function shown in 5.8(b). It is given by

$$V_0 = \frac{4V_0}{\pi} \sum_n \frac{1}{n} \sin\frac{n\pi y}{a}, \; n = 1, 3, 5, \ldots \tag{18}$$

Comparison between (17) and (18) gives,

$$B_n = \frac{4V_0}{n\pi}$$

By substituting this value of B_n in (16), we get

$$V = \frac{4V_0}{\pi} \sum_n \frac{1}{n} e^{-(n\pi x/a)} \sin\frac{n\pi y}{a}, \; n = 1, 3, 5, \ldots \tag{19}$$

This is the complete solution of the problem and gives the potential anywhere between the parallel plates and to the right of the strip.

The solution is in the form of a single infinite series. As the series decreases rapidly with x, only a finite number of terms may be used for approximate calculations. The sine functions of the series are called *rectangular harmonics.*

The electric field intensity in the region between the plates,

$$\mathbf{E} \quad \frac{\partial V}{\partial x}\mathbf{u}_x - \frac{\partial V}{\partial y}\mathbf{u}_x = \frac{4V_0}{a} \sum_n e^{-\frac{n\pi x}{a}} \left(\sin\frac{n\pi y}{a}\mathbf{u}_x - \cos\frac{n\pi y}{a}\mathbf{u}_y \right) \tag{20}$$

The field intensity has components in x and y directions. Each component is an infinite series and attenuates exponentially with x. At $y \quad a/2$ the field is in x direction. It is perpendicular to the surfaces of all the three plates as these are equipotential surfaces. Fig. 5.8(c) shows equipotential and field lines

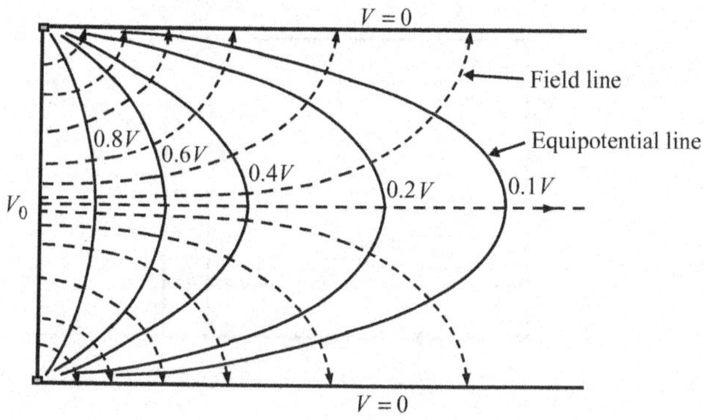

(c)

Fig. 5.8 (c) Sketch of equipotential and field lines between the plates.

To understand the use of (19) and (20), we will utilise the first three terms of each series to calculate approximately the potential in terms of V_0 and the components of field intensity in terms of a and V_0 at x $a/2$ and $y = a/4$.

$$V(\tfrac{a}{2},\tfrac{a}{4}) \approx \frac{4V_0}{\pi}\left[\left(e^{-\frac{\pi}{2}}\sin\frac{\pi}{4}\right)+\frac{1}{3}\left(e^{-\frac{3\pi}{2}}\sin\frac{3\pi}{4}\right)+\frac{1}{5}\left(e^{-\frac{5\pi}{2}}\sin\frac{5\pi}{4}\right)\right]$$

$$=\frac{4V_0}{\pi}\left[(0.2079)(0.7071)+\frac{1}{3}(0.00898)(0.7071)-\frac{1}{5}(0.00039)(0.7071)\right]$$

$$=\frac{4V_0}{\pi}[(0.1470)+(0.00212)-(0.00005)] = 0.1898V_0$$

$$E_x(\tfrac{a}{2},\tfrac{a}{4}) \approx \frac{4V_0}{a}\left[\left(e^{-\frac{\pi}{2}}\sin\frac{\pi}{4}\right)+\left(e^{-\frac{3\pi}{2}}\sin\frac{3\pi}{2}\right)+\left(e^{-\frac{5\pi}{2}}\sin\frac{5\pi}{2}\right)\right]$$

$$=\frac{4V_0}{a}[(0.1470)+(0.00635)-(0.00028)]=0.6123\left(\frac{V_0}{a}\right)$$

$$E_y(\tfrac{a}{2},\tfrac{a}{4}) \approx -\frac{4V_0}{a}\left[\left(e^{-\frac{\pi}{2}}\cos\frac{\pi}{4}\right)+\left(e^{-\frac{3\pi}{2}}\cos\frac{3\pi}{2}\right)+\left(e^{-\frac{5\pi}{2}}\cos\frac{5\pi}{2}\right)\right]$$

$$=-\frac{4V_0}{a}[(0.1470)-(0.00635)-(0.00028)] = -0.5615\left(\frac{V_0}{a}\right)$$

Slope of the electric field at the chosen point, $\tan\alpha = \dfrac{-0.5615}{0.6123}$, $\alpha = -42.5°$

EXAMPLE 5.8

Our next problem is to obtain the potential distribution in a rectangular region bounded by four plates, as illustrated in Fig. 5.9(a). Three plates are at zero potential. The fourth plate, which is separated from the other plates by a thin layer of insulation, is at a constant potential V_0. The dielectric in the region is air. We will solve the problem assuming that the potential in the region varies with x and y only.

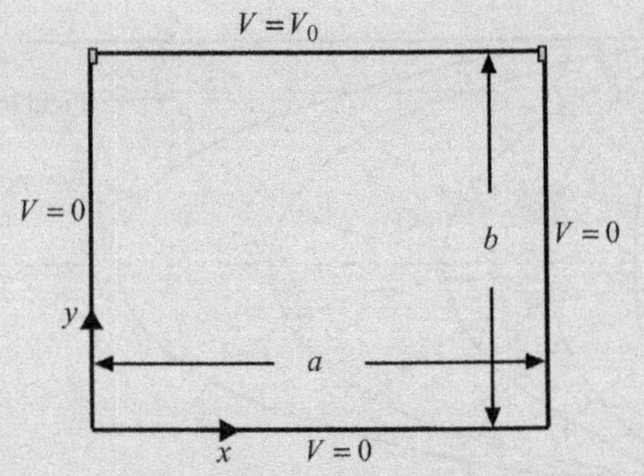

(*a*)

Fig. 5.9 (*a*) A rectangular region bounded by four plates with potentials as shown.

With coordinate directions as shown in the figure, the boundary conditions are

1. $V = 0$ at $x = 0,\ b \geq y \geq 0$
2. $V = 0$ at $y = 0,\ a \geq x \geq 0$
3. $V = 0$ at $x = a,\ b \geq y \geq 0$
4. $V = V_0$ at $y = b,\ a > x > 0$

A study of the boundary conditions shows that the product solution given by (12) is a suitable solution. So the general solution of Laplace's equation for the problem is

$$V \quad (A \quad kx + B\sin kx)(C\cosh ky + D\sinh ky) \tag{21}$$

In order to satisfy boundary conditions 1 and 2, *A* and *C* of (21) should be zero. Hence, we have

$$V \quad (BD, \quad n\, kx \sinh ky \tag{22}$$

Next, applying boundary condition 3 to (22), we get

$$(\quad)\sin{}' \quad \sinh ky = 0$$

$$\Rightarrow \quad ka = n\pi$$

Or $k = \dfrac{n\pi}{a}$, where *n* is an integer

Replacing *k* by $n\pi/a$ and (*BD*) by another constant B_n in (22),

$$V \quad B_n \sin\frac{n\pi x}{a} \sinh\frac{n\pi y}{a}$$

As Laplace's equation is linear and the medium in the rectangular region is air, the sum of infinite number of terms, *n* varying from one to infinity, is also a solution of the equation. Thus, we have

$$V \quad \sum_n B_n \sin\frac{n\pi x}{a} \sinh\frac{n\pi y}{a} \tag{23}$$

In order to find B_n of (23) we use the fourth boundary condition to (23) and obtain,

$$V_0 \quad \sum_n B_n \sinh\frac{n\ b}{a} \sin\frac{n^{-}x}{a}, \qquad a > x > 0 \tag{24}$$

We find that the left side of (24) is a constant quantity whereas the right side is an infinite series of sine functions in x. Matching of the two sides is possible by finding the Fourier series of the positive half-cycle of a rectangular wave of amplitude V_0 and half-wavelength a, as shown in Fig. 5.9(b).

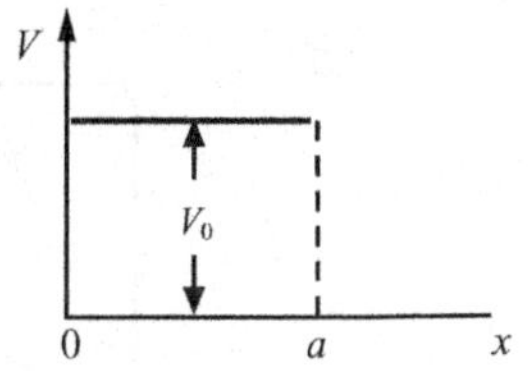

(b)

Fig. 5.9 (b) Variation of potential of the plate at $y = b$.

The Fourier series of such a wave has been obtained in Section 1.13. By replacing ℓ of (1.13-3) by a, we will get the Fourier series of the function shown in 5.9(b). It is given by

$$V_0 = \frac{4V_0}{\pi}\sum_n \frac{1}{n}\sin\frac{n\pi x}{a},\ a > x > 0,\ n = 1, 3, 5, \ldots \tag{25}$$

Equating the nth term of (24) with the nth term of (25) gives,

$$B_n = \frac{4V_0}{n\pi \sinh(n\pi b/a)}$$

The complete solution of the problem is found by substituting the expression of B_n into (23).

$$V = \sum_n \frac{4V_0 \sin(n\pi x/a)\sinh(n\pi y/a)}{n\pi \sinh(n\pi b/a)},\quad n = 1, 3, 5, \ldots \tag{26}$$

The solution of the problem is in the form of a single convergent infinite series.

Let us calculate the potential in terms of V_0 using the first three terms of (26) at $\left(\frac{a}{2}, \frac{a}{2}\right)$ and for $b/a = 1$.

$$V\left(\frac{a}{2}, \frac{a}{2}\right) \approx \frac{4V_0}{\pi}\left[\frac{\sin(\pi/2)\sinh(\pi/2)}{\sinh(\pi)} + \frac{1}{3}\frac{\sin(3\pi/2)\sinh(3\pi/2)}{\sinh(3\pi)} + \frac{1}{5}\frac{\sin(5\pi/2)\sinh(5\pi/2)}{\sinh(5\pi)}\right]$$

$$= \frac{4V_0}{\pi}\left[\frac{\sinh(\pi/2)}{\sinh(\pi)} + \frac{1}{5}\frac{\sin(5\pi/2)\sinh(5\pi/2)}{\sinh(5\pi)}\right] + \frac{1}{5}\frac{\sinh(5\pi/2)}{\sinh(5\pi)}\Bigg]$$

$$= \frac{4V_0}{\pi}[0.1994 - 0.003 + \text{very small}] = 0.25\,V_0$$

EXAMPLE 5.9

This example has the same description as that of the previous example, except that the potential of the plane at $y = b$ is not constant but is a linear function of x defined by $V = V_0(x/a)$.

Referring to Fig. 5.10(a), the boundary conditions are as given below:

1. $V = 0$ at $x = 0,\ b \geq y \geq 0$
2. $V = 0$ at $y = 0,\ a \geq x \geq 0$
3. $V = 0$ at $x = a,\ b \geq y \geq 0$
4. $V = \frac{V_0}{a}x$ at $y = b,\ a > x > 0$

The product solution of the problem satisfying the first three boundary conditions is same as that given by (23). Thus,

$$V = \sum_n B_n \sin\frac{n\pi x}{a}\sinh\frac{n\pi y}{b} \tag{27}$$

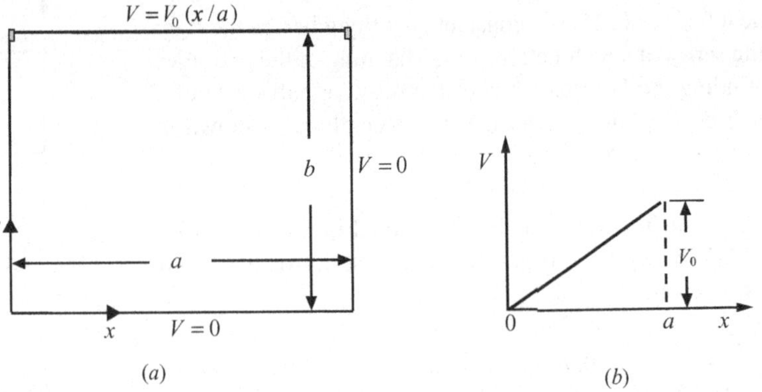

Fig. 5.10 (*a*) A rectangular region bounded by four plates at potentials as shown and (*b*) Variation of potential of the plate at $y = b$.

where n is an integer. To find B_n of (27) we use the fourth boundary condition to obtain,

$$\frac{V_0 x}{a} = \sum_n B_n \sinh\frac{n\pi b}{a} \sin\frac{n\pi x}{a}, \quad a > x > 0 \tag{28}$$

We find that the left side of (28) is a linear function of x whereas the right side is an infinite series of sine functions in x. Matching of the two sides is possible by finding the Fourier series of the positive half-cycle of a triangular wave of amplitude V_0, as shown in Fig. 5.10(*b*).

The Fourier series of such a wave has been obtained in Section 1.13. By replacing ℓ of (1.13-4) by a, we will get the Fourier series of the function shown in 5.10(*b*). It is given by

$$\frac{V_0 x}{a} = \sum_n (-1)^{n-1} \frac{2V_0}{n\pi} \sin\frac{n\pi x}{a}, \quad a > x > 0,\ n = 1, 2, 3, \ldots \tag{29}$$

By comparing (28) with (29) we obtain B_n in terms of V_0.

$$B_n = \frac{(-\ )^{n-1} 2V_0}{n^{-}\sinh(n\pi b/a)}$$

By substituting the expression of B_n in (28) we will obtain the complete solution of the problem.

$$V = \sum_n \frac{(-1)^{n-1} 2V_0}{n^{-}\sinh(n\pi b/a)} \sin\frac{n\pi x}{a} \sinh\frac{n\pi y}{a}, \quad n = 1, 2, 3, \ldots \tag{30}$$

This is an infinite series consisting of terms with both odd and even values of n. It is a convergent series.

5.5 Solution of Two-Dimensional Laplace's Equation in Cylindrical Coordinates

If the boundary surfaces of a problem are those of the cylindrical coordinate system, it is convenient to use the Laplace's equation in cylindrical coordinates to solve the problem. The three-dimensional Laplace's equation for the electrostatic potential V in cylindrical coordinates is given by formula (9) in Section (2.30). If V does not vary with one of the space variables then the three-dimensional equation gets reduced to a two-dimensional one. We shall consider two cases; (1) the potential is a function of r and ϕ only, and (2) it is a function of r and z only. The solutions are found by using the separation of variable technique.

Product Solution when the Potential varies with r and ϕ only

As the potential does not vary with z, we have from (2.30-9)

$$r^2\frac{\partial^2 V}{\partial r^2}+r\frac{\partial V}{\partial r}+\frac{\partial^2 V}{\partial \phi^2}=0 \tag{1}$$

Let us assume that V is equal to the product of two different potentials R and Φ, where the potential R is a function of only r and the potential Φ is a function of the variable only ϕ. Thus, we have

$$V=R\Phi \tag{2}$$

By substituting (1) in (2), we get

$$r^2\frac{\partial^2 (R\Phi)}{\partial r^2}+r\frac{\partial (R\Phi)}{\partial r}+\frac{\partial^2 (R\Phi)}{\partial \phi^2}=0$$

Or

$$\Phi\left(r^2\frac{\partial^2 R}{\partial r^2}+r\frac{\partial R}{\partial r}\right)+R\frac{\partial^2 \Phi}{\partial \phi^2}=0 \tag{3}$$

Dividing both sides of (3) by $R\Phi$, and then separating the variables, we get

$$\frac{1}{R}\left(r^2\frac{\partial^2 R}{\partial r^2}+r\frac{\partial R}{\partial r}\right)=-\frac{1}{\Phi}\frac{\partial^2 \Phi}{\partial \varphi^2} \tag{4}$$

Since the left-hand side of (4) does not vary with ϕ and the right-hand side does vary with r, each side can only be a constant as justified in the preceding section.

Taking this constant as k^2, we get two ordinary differential equations as given below:

$$r^2\frac{d^2 R}{dr^2}+r\frac{dR}{dr}-k^2R=0 \tag{5}$$

$$\frac{d^2\Phi}{d\varphi^2}+k^2\Phi=0 \tag{6}$$

Equation (5) is called Cauchy equation or Euler equation. A solution to the equation is found by assuming,

$$R=C'r^m \tag{7}$$

where C' is a constant and m is an integer. Substitution of (7) and its first and second derivatives into (5) yields

$$C'[m(m-1)+m-k^2]r^2=0$$

$$\Rightarrow \quad m(m-1)+m-k^2r^2=0$$

Or

$$m\text{`} \quad k^2$$

And

$$m=\pm k$$

Denoting the solution for positive k by R_1 and for negative k by R_2, we have

$$R_1 = Ar^k, \ A \text{ is a constant} \tag{8}$$

$$R_2 = Br^{-k}, \ B \text{ is another constant.} \tag{9}$$

Thus the general solution of (5), which is equal to the sum of (8) and (9), is

$$R = Ar^k + Br^{-k} \tag{10}$$

Equation (6) is a second order ordinary differential equation. Its solution in terms of circular sine and cosine functions is given by

$$\Phi = C \cos k\phi + D \sin k\phi \tag{11}$$

The product of (10) and (11) is the solution of (1). Accordingly,

$$V = (Ar^k + Br^{-k})(C \cos k\phi + D \sin k\phi), \ k \neq 0 \tag{12}$$

We will illustrate the use of (12) by considering a few examples.

EXAMPLE 5.10

The inner cylinder of a system of two coaxial cylinders is split axially into two halves, as shown in Fig. 5.11(*a*). The two halves are separated by a thin layer of insulation. The potential of one half for $\pi > \phi > 0$ is V_0 and the potential of the other half for $2\pi > \phi > \pi$ is $-V_0$. The outer cylinder is at zero potential. The radii of the inner and outer cylinders are taken as *a* and *b* respectively. The dielectric between the cylinders is air. It is required to find the potential at a point in the space between the cylinders.

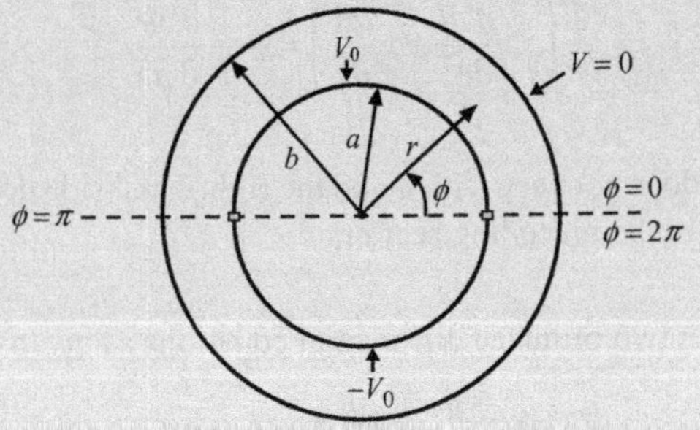

(*a*)

Fig. 5.11 (*a*) Cross section of coaxial cylinders with the inner cylinder split axially into two halves.

The boundary conditions are

1. $V = 0$ at $r = b$, for all ϕ
2. $V = V_0$ at $r = a$ and $\pi > \phi > 0$
3. $V = -V_0$ at $r = a$ and $2\pi > \phi > \pi$

Applying the first boundary condition to (12), we have

$$(\quad ^k \quad\quad ^k)(C \cos k\phi + D \sin k\phi) = 0$$

$$\Rightarrow \quad Ab^k + Bb^{-k} = 0$$

$$\text{Or} \quad B \quad Ab^{2k} \tag{13}$$

By substituting the value of *B* given by (13) into (12), we get

$$V \quad A(r \quad b^k r \quad)(C \cos k \quad D \sin k \quad), \ b \quad r \geq a \tag{14a}$$

Also $V = \sum_k A_k (r^k - b^{2k} r^{-k})(C_k \cos k\phi + D_k \sin k\phi), b \geq r \geq a$ (14b)

A graphical plot of the variation of potential on the surface of the inner conductor is shown in Fig. 5.11(b). It shows that the potential is an odd function of ϕ. Hence, C_k in the series of (14b) must be zero, as the cosine terms cannot exist. Thus, we have

$$V = \sum_k A_k D_k (r^k - b^{2k} r^{-k}) \sin k\phi \quad (15)$$

Next we substitute the second and third boundary conditions in (15) and obtain,

$$\sum_k A_k D_k (a^k - b^{2k} a^{-k}) \sin k\phi = V_0, \ \pi > \phi > 0 \quad (16a)$$

$$= -V_0, \ 2\pi > \phi > \pi \quad (16b)$$

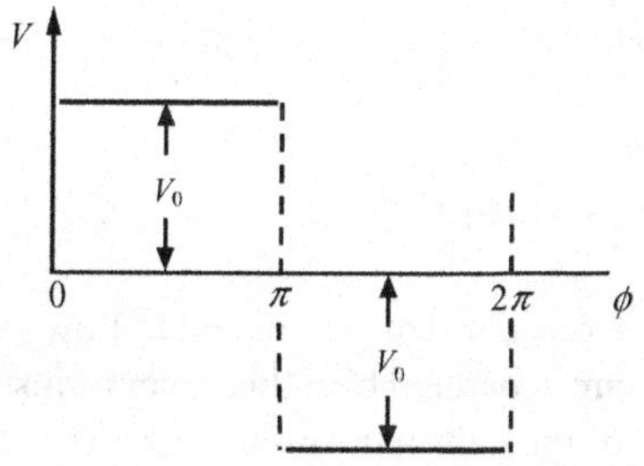

(b)

Fig. 5.11 (b) Graphical representation of potential of two halves of the inner cylinder.

To determine the constants of (16), we have to match this series on the right-hand side with the Fourier series of the wave shown in Fig. 5.11(b). As the wave is an odd function of ϕ and its negative half-cycle is identical to the positive half-cycle, the Fourier series will have only sine terms. Thus, denoting the function representing the wave by $f(\phi)$, we have

$$f(\phi) = \sum_n b_n \sin n\phi \quad (17)$$

where $f(\phi) = V_0, \ \pi > \phi > 0$

$$f(\phi) = -V_0, \ 2\pi > \phi > \pi$$

The coefficient of the nth term of (17),

$$b_n = \frac{1}{\pi} \int_0^{2\pi} f(\phi) \sin n\phi \, d\phi$$

$$= \frac{V_0}{\pi} \left(\int_0^{\pi} \sin n\phi \, d\phi - \int_{\pi}^{2\pi} \sin n\phi \, d\phi \right)$$

$$= \frac{V_0}{n\pi} [(-\cos n\phi)_0^{\pi} + (\cos n\phi)_{\pi}^{2\pi}]$$

It can be shown that

$b_n = 0$ for even values of n,

$= \dfrac{4}{n\pi}$ for odd values of n

Thus, the Fourier series of the wave shown in Fig. 5.11(b) is given by

$$\sum_n \frac{4V_0}{n\pi} \sin n\phi = V_0, \ \pi > \phi > 0 \quad (18a)$$

$$= -V_0, \ 2\pi > \phi > \pi, \ n = 1, 2, 3, \ldots \quad (18b)$$

By comparing (16) with (18), we get

$$k = n,$$

$$A_k D_k = \frac{4V_0 a^n}{n\pi(a^{2n} - b^{2n})}$$

By substituting these in (15) we will get the complete solution of the problem. It is given by

$$V = \sum_n \frac{4V_0 a^n (r^n - b^{2n} r^{-n})}{n\pi(a^{2n} - b^{2n})} \sin n\phi,\ b \ge r \ge a, n = 1, 2, 3, \ldots \tag{19}$$

The solution is described by the so called *cylindrical harmonics*.

EXAMPLE 5.11

Consider that an isolated hollow cylinder of radius a is split along its axis into two equal halves. The halves are separated by a thin layer of insulation. The potential of the half in the region $\pi > \phi > 0$ is V_0. The potential of the half in the region $2\pi > \phi > \pi$ is $-V_0$. The example is illustrated in Fig. 5.12(*a*). It is required to find the potential at any point inside the cylinder.

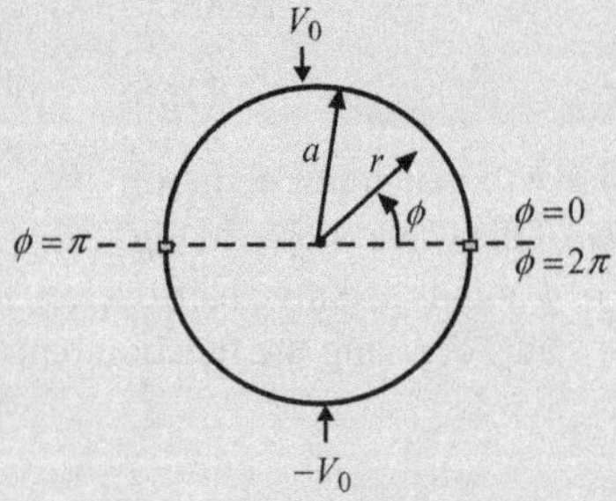

(*a*)

Fig. 5.12 (*a*) Cross section of a hollow cylinder split axially into two halves.

The given boundary conditions are

1. $V = V_0$ at $r = a$ and $0 < \phi < \pi$
2. $V = -V_0$ at $r = a$ and $\pi < \phi < 2\pi$

As $r = 0$ is included, B in (12) has to be zero as otherwise the potential would be infinite at this point. Moreover, the potential distribution at $r = a$, given in Fig. 5.12(*b*), shows that it is an odd function of ϕ. Hence the constant C of (12) must also be zero. Thus, we have

$$V \quad (AD)r^k \sin k\phi,\ r \le a \tag{20a}$$

Also $V \quad \sum_k (A_k D_k) r^k \sin k\phi,\ r \le a$ (20*b*)

Substituting the boundary condition into (20*b*), we get

$$\sum_k (\quad) a^k \sin k\phi = V_0,\ \pi > \phi > 0 \tag{21a}$$

$$= -V_0,\ 2\pi > \phi > \pi \tag{21b}$$

The constants are determined by matching the infinite series of (21) with the Fourier series of the rectangular wave shown at Fig. 5.12(*b*). The Fourier series of the wave has been derived in the previous example. It is given by

$$\sum_n \frac{4V_0}{n\pi} \sin n\phi = V_0,\ \pi > \phi > 0 \tag{22a}$$

$$= -V_0,\ 2\pi > \phi > \pi,\ n = 1, 2, 3, \ldots \tag{22b}$$

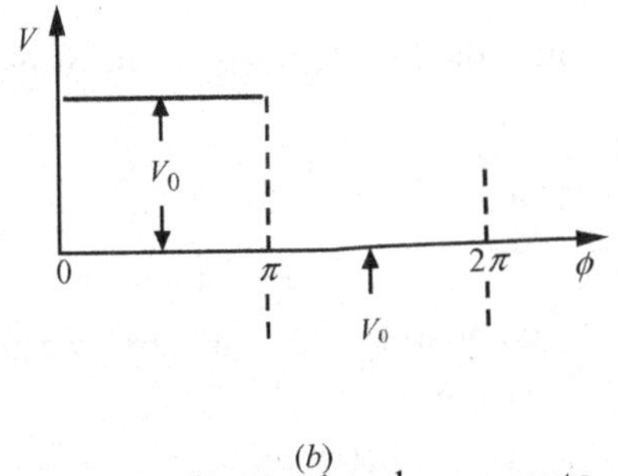

(*b*)

Fig. 5.12 (*b*) Graphical representation of potential of the two halves.

Comparison between (21) and (22) gives

$k = n$,

and $A_k D_k = \dfrac{4V_0}{n\pi a^n}$

Substituting the above relation in (20*b*) we get the complete solution for the potential inside the cylinder.

$$V = \sum_n \frac{4V_0 r^n}{n\pi a^n} \sin n\phi, \; r \le a \qquad (23)$$

The potential is described by what are known as cylindrical harmonics.

EXAMPLE 5.12

A long conducting cylinder of radius a is placed in a uniform static electric field. It is required to find the electric potential and field intensity near the surface of the cylinder.

Though the field intensity is uniform initially as shown in Fig. 5.13(*a*), the presence of the conducting cylinder will make it non-uniform near the outside surface of the cylinder. This is because the field lines will terminate on and will start from the surface of the cylinder normally. However, the field far away from the cylinder will remain uniform. We will solve the problem by assuming that the cylinder is at zero potential and its axis is aligned along the z axis, as shown in Fig. 5.13(*b*). We also assume that the far-field is in y direction with magnitude E_{y0}, the field intensity in the neighbourhood of the cylinder has components along r and ϕ only, and $E_z = 0$ everywhere. Thus, the boundary conditions are

1. $V = 0$ at $r = a$
2. $E_x = 0$ for $r \gg a$
3. $E_y = E_{y0}$ for $r \gg a$

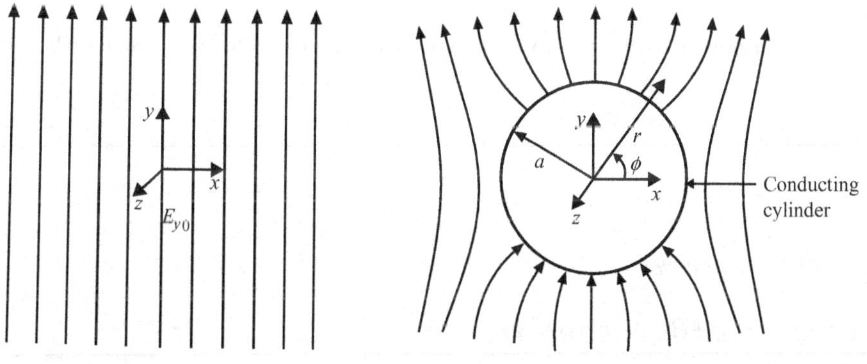

(*a*) (*b*)

Fig. 5.13 (*a*) Shows a uniform electric field in y direction. The value of the field is E_{y0} and (*b*) Electric field near a conducting cylinder placed in the field shown in (*a*).

The product solution of Laplace's equation for the problem is given by (12). Since a series type of solution is not needed, we will take $k = 1$. Accordingly, the potential outside the cylinder is given by

$$V \quad (Ar + B/r)(C\cos\phi + D\sin\phi) \qquad (24)$$

Applying the first boundary condition to (24), we get

$$0 = (\quad + \quad /a)(C\cos \;, + \;\sin\phi)$$

Or $Aa + \dfrac{B}{a} = 0$

$\Rightarrow \quad B \quad Aa^2$

By substituting the above relation into (24) and denoting the resulting potential by V_1, we get

$$V_1 \quad A(r \quad a^2/r)(C\cos\phi + D\sin\phi) \tag{25}$$

In order to use the second boundary condition we have to calculate E_x.

$$E_x \quad E_r \quad {}_r + E_\phi \mathbf{u}_\phi)\cdot \mathbf{u}_x$$

$$= E_r(\mathbf{u}_r \cdot \mathbf{u}_x) + E_\phi(\mathbf{u}_\phi \cdot \mathbf{u}_x)$$

$$= -\frac{\partial V_1}{\partial r}(\cos\phi) - \frac{1}{r}\frac{\partial V_1}{\partial \phi}(-\sin\phi)$$

$$= -[\ (\ + a^2/r^2)(C\cos\phi + D\sin\phi)](\cos\phi) \ -A[(1-a^2/r^2)(-C\sin\phi + D\cos\phi)](-\sin\phi) \tag{26}$$

As $a^2 \quad r^2$ is small compared to unity, $1 + a^2/r^2 \approx 1$. Using this approximation and application of the second boundary condition to (26) gives,

$$- \quad {}^2\ \phi + D\sin\phi\cos\phi) + \ (C\sin^2\phi - D\sin\phi\cos\phi)] = 0$$

$$\Rightarrow \quad AC = 0,$$

Or $C = 0$, as A cannot be zero in view of (25).

By substituting $C = 0$ into (25) and denoting the resulting expression by V_2, we get

$$V_2 \quad A_1(r - a^2/r)\sin\phi \tag{27}$$

where $A_1 \quad AD$

To determine the constant A_1, we have to at first find the field intensity E_y, and then apply boundary condition (3).

$$E_y \quad E_r \quad {}_r + E_\phi \mathbf{u}_\phi)\cdot \mathbf{u}_y$$

$$= E_r(\mathbf{u}_r \cdot \mathbf{u}_y) + E_\phi(\mathbf{u}_\phi \cdot \mathbf{u}_y)$$

$$= -\frac{\partial V_2}{\partial r}(\sin\phi) - \frac{1}{r}\frac{\partial V_2}{\partial \phi}(\cos\phi)$$

$$= -A_1 \quad + a^2/r^2)\sin^2\phi + (1 - a^2/r^2)\cos^2\phi] \tag{28}$$

As $a^2 \quad r^2$ is small compared to unity, application of the third boundary condition to (28) yields,

$$A_1 \quad E_{y0}$$

Substitution of the value A_1 in (27) gives the potential near the conducting cylinder. Thus, the potential near the outside surface of the cylinder,

$$V \quad E_{y0}\left(r - \frac{a^2}{r}\right)\sin\phi \tag{29}$$

The field intensity near the outside surface of the cylinder,

$$\mathbf{E} = -\frac{\partial V}{\partial r}\mathbf{u}_r - \frac{1}{r}\frac{\partial V}{\partial \phi}\mathbf{u}_\phi = \quad {}_y\left[\left(\ + \frac{a^2}{r}\right) \quad {}_r + \left(\ - \frac{a^2}{r^2}\right)\cos\phi\, \mathbf{u}_\phi\right] \tag{30}$$

Now, let us calculate the potential and field intensity in terms of E_{y0} for $a = 5.0$ cm, $r = 6.0$ cm and $\phi = \pi/4$. We will also calculate the slope of the field line at the given point.

Substituting the data into (29),

$$V = -E_{y0}\left[(6.0\times10^{-2}) - \frac{(5.0\times10^{-2})^2}{6.0\times10^{-2}}\right](0.7071) = -(1.296\times10^{-2})E_{y0}\ \text{V}$$

From (30) the r component of field intensity,

$$E_r = E_{y0}\left[\left(1+\frac{a^2}{r^2}\right)\sin\phi\right] = E_{y0}\left(1+\frac{25}{36}\right)(0.7071) = 1.198\,E_{y0}$$

The ϕ component of field intensity,

$$E_\varphi = E_{y0}\left[\left(1-\frac{a^2}{r^2}\right)\cos\phi\right] = E_{y0}\left(1-\frac{25}{36}\right)(0.7071) = 0.216\,E_{y0}$$

Slope of the field line, $\tan\alpha = \dfrac{E_y}{E_x}$

$$E_x = E_r\cos\phi - E_\phi\sin\phi = \left(1.198\cos\frac{\pi}{4} - 0.216\sin\frac{\pi}{4}\right)E_{y0} = 0.6944E_{y0}$$

$$E_y = E_r\sin\phi + E_\phi\cos\phi = \left(1.198\sin\frac{\pi}{4} + 0.216\cos\frac{\pi}{4}\right)E_{y0} = 0.9998\,E_{y0}$$

$$\tan\alpha = \frac{0.9998}{0.6944} = 1.44; \alpha = 55.2^\circ$$

Product Solution when the Potential varies with r and z but not with ϕ

If there is axial symmetry in a given problem, the potential function does not vary in ϕ direction. So Laplace's equation for this case is given by

$$\frac{\partial^2 V}{\partial r^2} + \frac{1}{r}\frac{\partial V}{\partial r} + \frac{\partial^2 V}{\partial z^2} = 0 \tag{31}$$

Now we assume that V is the product of two potential functions, namely, R that varies with r only and Z that varies with z only. Accordingly,

$$V = RZ \tag{32}$$

Substituting (32) into (31) and then separating the variables, we get

$$\frac{1}{Z}\frac{\partial^2 Z}{\partial z^2} = -\frac{1}{R}\left(\frac{\partial^2 R}{\partial r^2} + \frac{1}{r}\frac{\partial R}{\partial r}\right)$$

Since the lhs is not a function of r and the rhs is not a function z, each must be equal to a constant. Taking the constant as k^2, where k^2 is positive (k is not imaginary), we get two second order ordinary differential equations. These are

$$\frac{d^2R}{dr^2} + \frac{1}{r}\frac{dR}{dr} + k^2R = 0$$

Or
$$r^2\frac{d^2R}{dr^2}+r\frac{dR}{dr}+(kr)^2R=0 \tag{33}$$

And
$$\frac{d^2Z}{dz^2}-k^2Z=0 \tag{34}$$

Equation (33) is known as a Bessel equation. Its solution is given in Section 1.14. It is given by

$$R=AJ_0(kr) \tag{35}$$

The functions $J_0(kr)$ has been defined in Section 1.14. It is an infinite series and is called Bessel function of first kind and zero order.

The solution of (34) in terms of hyperbolic sine and cosine functions is

$$Z=C\cosh(kz)+D\sinh(kz) \tag{36}$$

The general solution of (31) is the product of (35) and (36). Thus,

$$V=[AJ_0(kr)][C\cosh(kz)+D\sinh(kz)] \tag{37}$$

This is the general solution of Laplace's equation in cylindrical coordinates if the potential V is a function of r and z, but not of ϕ, in a region where $r=0$ is included. The solution of a problem satisfying the appropriate boundary conditions may contain an infinite series of Bessel function of zero order, called Fourier–Bessel series, and also Bessel functions of higher than zero order.

5.6 The Method of Images

In order to understand the method of images for solving some types of electric and magnetic field problems, let us consider the problem of a positive point charge Q located at a height h above an infinitely large plane conductor in air. The conductor, assumed to be at zero potential, is taken on the $y=0$ plane of rectangular coordinates. The problem is illustrated in Fig. 5.14(a). The charges induced by Q in the conductor are distributed on the upper surface of the conductor in the form of surface charge. The induced charge would be negative if the point charge is positive and vice versa.

The total electric field above the conductor is equal to the sum of the field of the point charge and the field due to the induced charges.

Assuming that the point charge is positive, the electric field lines starting from the charge terminate normally on the equipotential conductor surface, as shown in the figure. No field is present below the conducting plane. The problem is to find the potential above the $y=0$ plane.

The problem cannot be solved by using Coulomb's law because the manner in which the charge is distributed on the conductor surface is not known. Gauss's law cannot be applied as the field is not uniform. One way of solving the problem is the method of images described below.

Single Point Charge above a Conducting Plane

Let us consider a system of two equal and opposite point charges located symmetrically above and below a conducting plane located at $y=0$. The positive charge Q is at height h above, and the negative charge $-Q$ is at a depth h below the conductor. The figure also shows electric field lines between the charges. A study of field maps of Fig. 5.14(a) and Fig. 5.14(b) shows that they are identical above the $y=0$ plane. Moreover, the potential at $y=0$

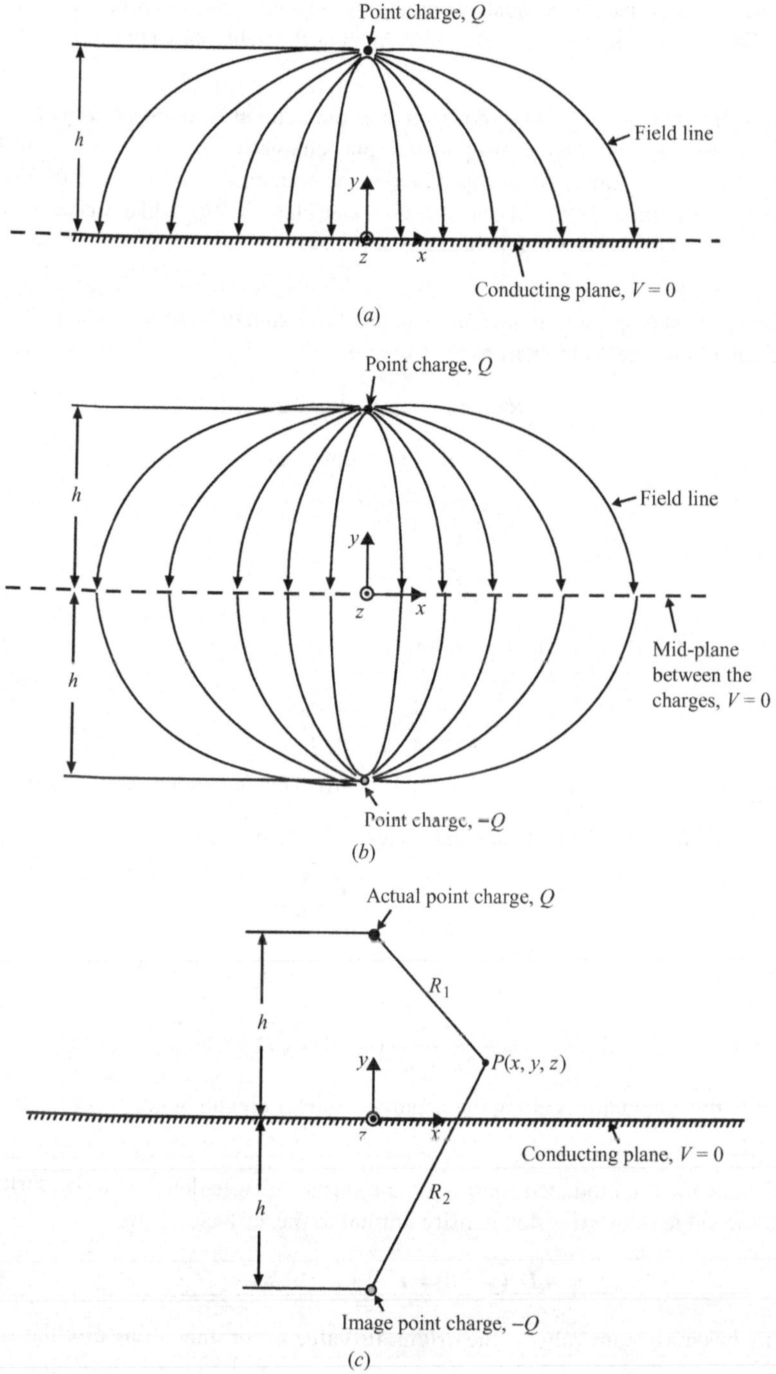

Fig. 5.14 (*a*) Field lines between a positive point charge and an infinitely large conducting plane, (*b*) Field lines between two equal and opposite point charges, and (*c*) For calculation of potential above the $y = 0$ plane due to the original charge and its image.

plane of Fig. 5.14(*b*) is zero as charges of equal magnitude and opposite sign are symmetrically located on the two sides of the plane. Thus, the two systems are analogous so far as the fields above the $y = 0$ planes are concerned.

> Therefore, a point charge located at $y = h$ above a plane conductor at zero potential can be replaced by the original charge at $y = h$ and an equal and opposite point charge located at $y = -h$, with the conductor removed. The latter charge is called the *image charge*. The potential at any point *above* the $y = 0$ plane is equal to the sum of the potential of both the actual and image charges. No field exists below the conducting plane. This method of solution is called the *method of images*.

In Fig. 5.14(*c*), let $P(x, y, z)$ be a point in the region above the conducting surface. The vector distances of the point from the actual and image point charges are

$$\mathbf{R}_1 = x\mathbf{u}_x + (y-h)\mathbf{u}_y + z\mathbf{u}_z$$

$$\mathbf{R}_2 = x\mathbf{u}_x + (y+h)\mathbf{u}_y + z\mathbf{u}_z$$

Moreover,

$$R_1 = \sqrt{x^2 + (y-h)^2 + z^2}$$

$$R_2 = \sqrt{x^2 + (y+h)^2 + z^2}$$

The electrostatic potential at P due to the two point charges is given by

$$V = \frac{Q}{4\pi\varepsilon_0}\left(\frac{1}{R_1} - \frac{1}{R_2}\right), y > 0 \tag{1}$$

As R_1 R_2 at any point on the conductor, its potential is zero. Thus, the boundary condition is satisfied.

The total field intensity at P due to the actual and image point charges,

$$\mathbf{E} = \frac{Q}{4\pi\varepsilon_0}\left(\frac{\mathbf{R}_1}{R_1^3} - \frac{\mathbf{R}_2}{R_2^3}\right), y > 0 \tag{2}$$

The field intensity at the conductor surface is found by substituting $y = 0$ in (2).

$$\mathbf{E}(y=0) = -\frac{Qh}{2\pi\varepsilon_0 R^3}\mathbf{u}_y;\ R = \sqrt{x^2 + z^2 + h^2}$$

The field intensity at the conductor is along the negative y direction and hence it is perpendicular to the conductor surface.

Next, we will calculate the total induced charge. As the surface charge density on the surface of a good conductor is equal to the value of electric flux density normal to the surface, we have

$$\rho_S = D_y(y=0) = \varepsilon_0 E_y(y=0) = -\frac{Qh}{2\pi R^3}$$

The charge density has maximum value at the origin. Its value is constant along circular paths centred at the origin.

Total charge on the surface of the conductor $= \int_0^\infty \int_0^\infty \rho_S\, dx dz$

$$= -\int_0^\infty \int_0^\infty \frac{Qh}{2\pi(x^2 + \ ^2 + h^2)^{3/2}}\, dxdz = -Q$$

That is, the value of induced charge on the conductor is equal and opposite to that of the original point charge.

EXAMPLE 5.13

A point charge of 16.0 nC is located at (1.2 m, 0, 0) above a plane conductor situated on the $x = 0$ plane in free space. The conductor is at a potential of 60.0 V. Find the potential at P(0.5 m, 0.4 m, 0.3 m).

The image charge of –16.0 nC is at (–1.2 m, 0, 0).

The distance of P from the actual charge,

$$R_1 = \sqrt{(0.5-1.2)^2 + (0.4-0)^2 + (0.3-0)^2} = 0.860 \text{ m}$$

The distance of P from the image charge,

$$R_2 = \sqrt{(0.5+1.2)^2 + (0.4-0)^2 + (0.3-0)^2} = 1.772 \text{ m}$$

Potential at P due to the charges,

$$V = \frac{Q}{4\pi\varepsilon_0}\left(\frac{1}{R_1} - \frac{1}{R_2}\right) = (9\times10^9)(16.0\times10^{-9})\left(\frac{1}{0.860} - \frac{1}{1.772}\right) = 86.18 \text{ V}$$

As the conducting plane is at 60.0 V, the total potential at P

$$= 86.18 + 60.0 = 146.\ 18 \text{ V}$$

EXAMPLE 5.14

A point charge of 8.0 μC is located above a large conducting plate. The plate is on the x-z plane and the charge is at $P(0, 5 \text{ m}, 0)$. Find the surface charge density on the plate at $P'(2 \text{ m}, 0, 3 \text{ m})$

Surface charge density, $\rho_S = \dfrac{Qh}{2\pi R^3}$

$$Q = 8\times10^{-6} \text{ C},\ h = 5 \text{ m},\ R = \sqrt{(2-0)^2 + (0-5)^2 + (3-0)^2} = 6.16 \text{ m}$$

$$\rho_S = \frac{(8\times10^{-6})(5)}{2\pi(6.16)^3} = 27.24 \text{ nC/m}^2$$

Point Charge between Two Intersecting Conducting Planes

The problem is illustrated in Fig. 5.15(a). The angle of intersection of the conducting planes is taken as π/n, where n is a positive integer. A point charge Q is located at an angle α with respect to the horizontal plane, as indicated in the figure.

Boundary conditions are satisfied with $(2n - 1)$ image charges, each of the same magnitude as that of the original charge.

Moreover, the sign of an image charge is opposite to that of the charge from which it is imaged, and the image charges are at angles $\left(\frac{2\pi}{n} - \alpha\right)$, $\left(\frac{2\pi}{n} + \alpha\right)$, $\left(\frac{4\pi}{n} - \alpha\right)$, and $\left(\frac{4\pi}{n} + \alpha\right)$ with the horizontal plane. A geometrical

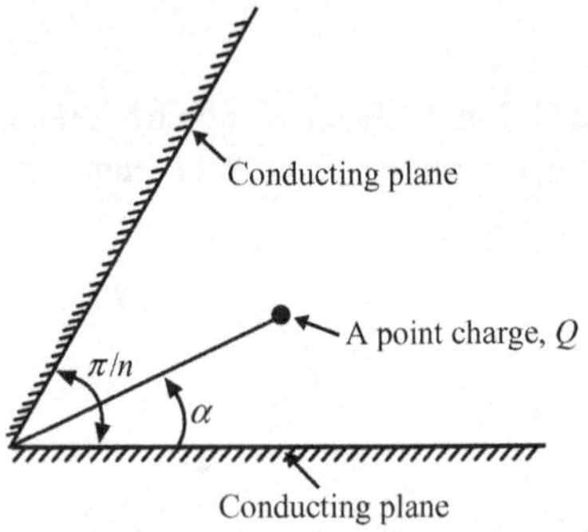

(a)

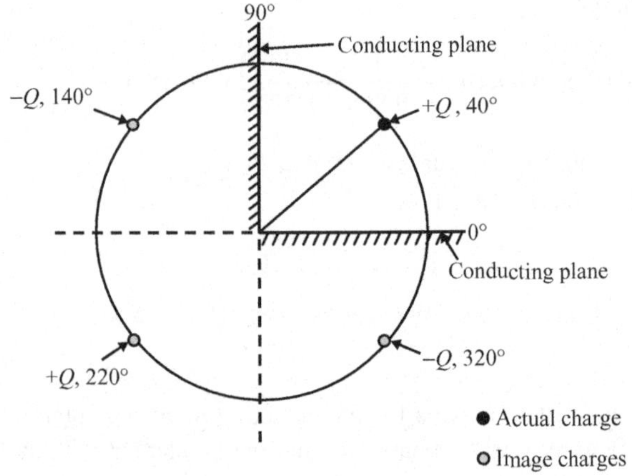

(b)

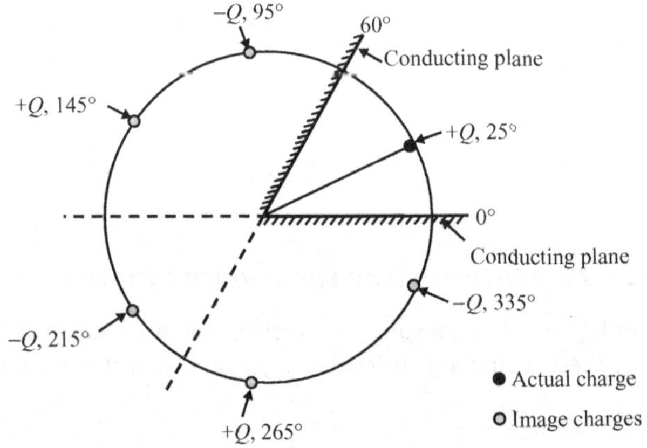

(c)

Fig. 5.15 (*a*) A point charge between two intersecting conducting planes and (*b* and *c*) Location of actual and image charges for $n = 2$ and 3.

construction (or proof) will show that the image charges lie on the circumference of a circle passing through the actual charge and centred on the line of intersection of the planes. The consecutive charges are of opposite sign. Moreover, positive and negative charges are symmetrically positioned about the planes so that the boundary condition of constant potential of the planes is satisfied. Figs. 5.15(*b* and c) show the locations of the actual charge and the image charges for $n = 2$ and $n = 3$. The potential at a point between the planes is equal to the sum of the potentials of the actual and image charges, imagining that the conducting planes are absent.

If n is not an integer, the number of image charges is not finite. In this case it is necessary to obtain images of image charges until further images are far away from the planes.

Line Charge above a Plane Conductor

Figure 5.16 illustrates a line charge of infinite length of linear charge-density of ρ_ℓ at a distance h above an infinitely large conducting plate in free space. The conducting plate is on the x-z plane, and the line charge which is parallel to the conducting plate is at $y = h$. The potential of the conductor is assumed to be zero.

Since the mid-plane between two line charges of equal and opposite linear charge density is also at zero potential as found in Section 2.18, the image charge is a line charge of density $-\rho_\ell$ positioned at $y = -h$.

The potential due to the actual and image charges at a point $P(x, y)$ above $y = 0$ plane can be obtained by using the same procedure as for two equal and opposite line charges described in Section 2.18. It is given by

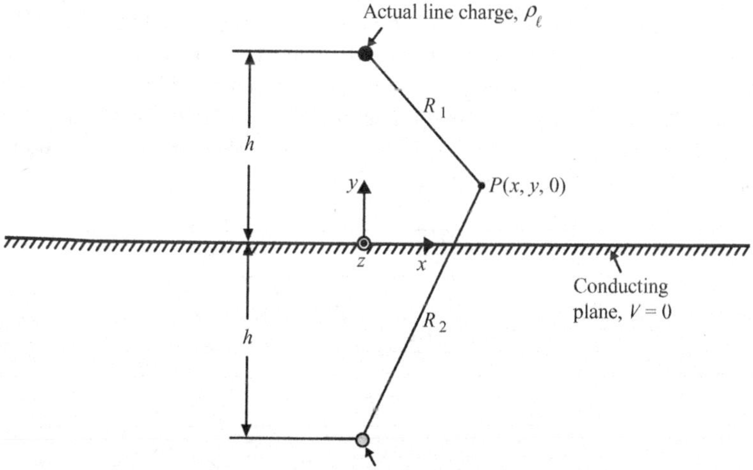

Fig. 5.16 Illustration of an infinitely long line charge above an infinitely large conducting plane at zero potential, and the image charge

$$V = \frac{\rho_\ell}{2\pi\varepsilon_0}\ln\frac{R_2}{R_1} = \frac{\rho_\ell}{2\pi\varepsilon_0}\ln\frac{\sqrt{x^2+(y+h)^2}}{\sqrt{x^2+(y-h)^2}} \tag{3}$$

The electric field intensity above the conductor,

$$\mathbf{E} = -\nabla V = \frac{\rho_\ell}{2\pi\varepsilon}\left[\frac{x}{R_1^2} - \frac{x}{R_2^2}\right]\mathbf{u}_x + \frac{\rho_\ell}{2\pi\varepsilon}\left[\frac{y-h}{R_1^2} - \frac{y+h}{R_2^2}\right]\mathbf{u}_y$$

EXAMPLE 5.15

A line charge of large length is parallel to a large conducting plate in free space. The conducting plate is on the x-z plane and the line charge is at $y = 2.4$ m. Find the potential at P(3 m, 4 m, 0). The line charge has 2.5 nC per metre of its length.

Potential at $P, V = \dfrac{\rho_\ell}{2\pi\varepsilon_0}\ln\dfrac{R_2}{R_1}$

$$\rho_\ell = 2.5\times10^{-9}\text{ C/m},\ R_1 = \sqrt{3^2+(4-2.4)^2} = 3.40\text{ m},\ R_2 = \sqrt{3^2+(4+2.4)^2} = 7.07\text{ m},$$

$$V = \frac{2.5\times10^{-9}}{2\pi(8.85\times10^{-12})}\ln\frac{7.07}{3.4} = 32.65\text{ V}$$

Line Charge Parallel to the Axis of a Long Cylindrical Conductor

A line charge of infinite length with linear charge density ρ_ℓ C/m is located in free space. The line charge is parallel to the axis of a circular cylindrical conductor of radius a and of infinite length. The axis of the cylinder is at a distance d from the line charge. The potential of the conductor has a finite value. The system is shown schematically in Fig. 5.17. We have to obtain an image charge so that the surface of the cylinder is at a constant potential (not zero).

It has been shown in Section 2.18 that the equipotential surfaces of two equal and opposite line charges are circular cylinders. Thus by analogy, the potential of a cylindrical conductor near a positive line charge would be constant if an image charge of linear charge density $-\rho_\ell$ C/m is located inside the cylinder and on the line PC, as indicated in the figure. Let us assume that the image charge is at P' and its distance from the original charge is d'. The mid-point between the charges is at O. Then, replacing s by d' in the results of the above mentioned example, we get the following relations.

$$\frac{d'k}{k^2-1} = a \tag{4}$$

And
$$\frac{d'(k^2+1)}{2(k^2-1)} = \text{OC}$$

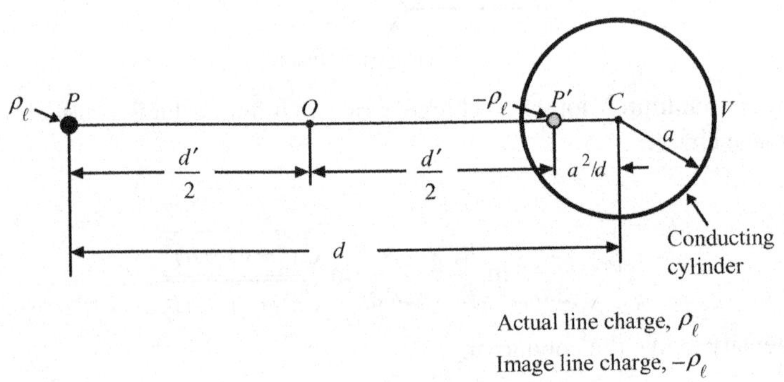

Fig. 5.17 Shows a line charge of infinite length parallel to the axis of an infinitely long cylindrical conductor and the image line charge

Thus, $$d = \text{PC} = \text{PO} + \text{OC} = \frac{d'}{2} + \frac{(k^2+1)d'}{2(k^2-1)} = ak \quad (5)$$

By eliminating k from (4) and (5), we get

$$d' = d - \frac{a^2}{d} \quad (6)$$

Equation (6) shows that the image charge is at a distance a^2/d from the axis of the cylinder. The original and image charges can be used to calculate the potential at any point outside the cylinder. The image of a line charge located parallel to the axis of a hollow conducting cylinder is a line charge located outside the cylinder.

EXAMPLE 5.16

A long line charge of linear charge density 7 nC/m is parallel to the axis of a long cylindrical conductor. The radius of the cylinder $a = 8$ cm, and the distance of its axis from the line charge $d = 30$ cm. What is the value of linear charge density of the image charge and its location so that the surface of the conducting cylinder is an equipotential one?

The linear charge density of the image line charge is –7 nC/m and its location is on the line that is perpendicular to the line charge and conductor axis. Its distance from the actual charge,

$$d' - d - \frac{a^2}{d} = 30 - \frac{8^2}{30} = 27.87 \text{ cm}$$

Point Charge Outside a Spherical Conductor

Let the centre of a spherical conductor be at a distance d from a point charge Q in free space, as illustrated in Fig. 5.18. The sphere of radius a is assumed to be at zero potential. It is required to replace the effect of induced charge on the spherical surface by an image point charge Q'. The image charge is on the radial line joining the point charge and centre of the sphere, and it is inside the sphere. The value of the image charge Q' and its distance d' from Q should be such that the sum of the potentials of the two charges on the spherical surface is zero.

Let $P(x, y)$ be a point on the surface of the sphere. The distance of P from the actual and image charges is R and R' respectively. As the potential due to Q and Q' at P must be zero, we have

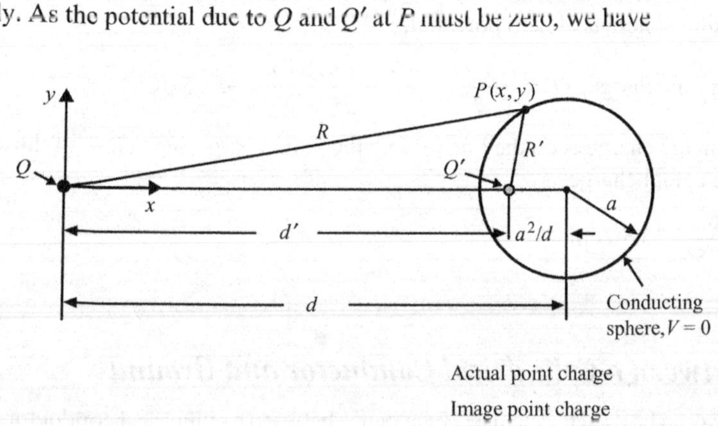

Fig. 5.18 Shows a point charge near a conducting sphere and the image point charge

$$\frac{1}{4\pi\varepsilon_0}\left(\frac{Q}{R}+\frac{Q'}{R'}\right)=0$$

$$\Rightarrow \quad \frac{R'}{R}=-\frac{Q'}{Q}$$

It has been shown in Section 2.18 that if the ratio of distances of two points from a third point has a constant value, then the locus of the third point is a circle. Since P in Fig. 5.18 is on a circle, the ratio of R' to R must be a constant. Thus, taking this constant as k, we have

$$\frac{R'}{R}=-\frac{Q'}{Q}=k \tag{7}$$

Expressing R and R' of (7) in terms of x and y, we get

$$\sqrt{(x-d')^2+y^2}=k\sqrt{x^2+y^2}$$

Squaring both sides and then rearranging, we get

$$\left(x-\frac{d'}{1-k^2}\right)^2+y^2=\left(\frac{d'k}{1-k^2}\right)^2$$

Thus, the distance of the centre of the circle from Q is,

$$d=\frac{d'}{1-k^2} \tag{8}$$

And radius of the circle,

$$a=\frac{d'k}{1-k^2} \tag{9}$$

Solving (8) and (9) for the unknown quantities k and d', we get

$$k=\frac{a}{d} \text{ and } d'=d-\frac{a^2}{d}$$

Thus, $Q'=-Q(a/d)$ and the distance of the image charge from the centre of the sphere is equal to a^2/d.

EXAMPLE 5.17

A point charge $Q = 24\ \mu\text{C}$ is located outside a spherical conductor of radius $a = 6$ cm. The distance of centre of the conductor from the point charge $d = 25$ cm. Find the value and location of the image charge so that the conducting sphere is at zero potential.

The value of image point charge, $Q'=-\dfrac{aQ}{d}=-\dfrac{(6)(24\times10^{-6})}{25}=-5.76\ \mu\text{C}$

The location of the image charge is on the line joining the point charge and centre of the spherical conductor. Its distance from the actual charge,

$$d'=d-\frac{a^2}{d}=25-\frac{6^2}{25}=23.56 \text{ cm}$$

Capacitance between a Cylindrical Conductor and Ground

The method of images can be used to calculate capacitance between a cylindrical conductor of infinite length and the ground assuming that the charge is concentrated on the axis of the conductor. Let us take a conductor of radius a with uniformly distributed charge per unit length ρ_ℓ at a height h above the ground. If we assume that the surface of the ground is flat, the image is also a cylindrical charge of infinite length of radius a with uniformly distributed charge of density $-\rho_\ell$ C/m, and located at a distance h below the ground plane. The actual and image charges are shown in Fig. 5.19. Referring to the figure, field intensity at $P(0, y, 0)$ is approximately given by

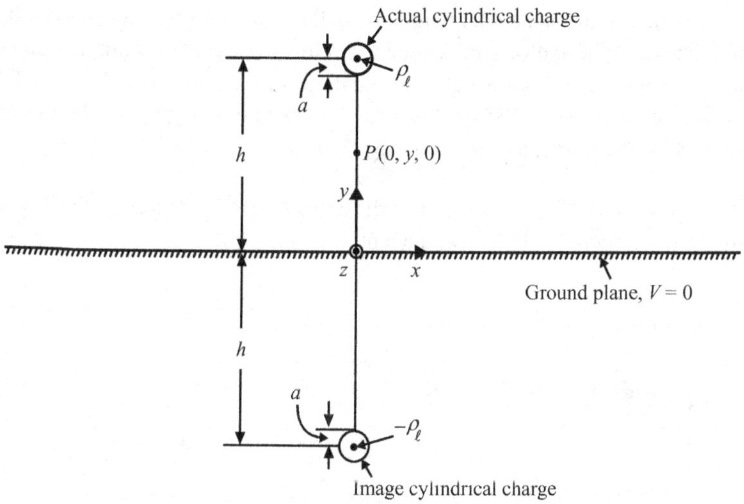

Fig. 5.19 For calculation of capacitance between a cylindrical conductor and the ground

$$\mathbf{E} = \frac{\rho_\ell}{2\pi\varepsilon_0}\left(\frac{1}{y-h} - \frac{1}{y+h}\right)\mathbf{u}_y$$

Potential of the conductor, $V = -\int_0^{h-a} \mathbf{E}\cdot d\boldsymbol{\ell} = -\frac{\rho_\ell}{2\pi\varepsilon_0}\int_0^{h-a}\left(\frac{1}{y-h} - \frac{1}{y+h}\right)\mathbf{u}_y\cdot(dy\mathbf{u}_y)$

$$= \frac{\rho_\ell}{2\pi\varepsilon_0}\ln\frac{2h-a}{a}$$

Capacitance of one metre length of the conductor, $C = \frac{\rho_\ell}{V} = \frac{2\pi\varepsilon_0}{\ln[(2h-a)/a]}$ F/m

This is twice the capacitance between two cylindrical conductors, each of radius a and separated by a distance $2h$. Note that the capacitance formula given by (2.28-10) is for a spacing h between the conductors.

EXAMPLE 5.18

A cylindrical conductor of radius a = 5.0 mm is at a height of 5.0 m above the ground plane. Find the capacitance between the conductor and ground.

$$C = \frac{2\pi\varepsilon_0}{\ln[(2h - a)/a]} = \frac{2\pi(8.85\times10^{-12})}{\ln[(10-0.005)/0.005]} = 7.32 \text{ pF}$$

Image of a Point Charge Moving Parallel to a Conducting Plane, and Image of a Line Current Parallel to a Conducting Plane

Consider that a positive point charge Q is moving at a velocity v parallel to a grounded conducting plane (or earth), as shown in Fig. 5.20(*a*). At every instant, the image charge $-Q$ is directly below the position of the real charge on the other side of the conducting surface. The distance of the image charge below the conductor is same as the height of the actual charge above the conductor. Image charges at $t = 0$ and $t = t'$ are shown in the figure. Thus, the image of a moving charge located over a conducting plane is also a moving charge of opposite sign moving at the same velocity and direction as the actual charge. It follows, by analogy, that the image of a line current I parallel to a grounded conducting plane is a line current parallel to but located on the other side of the plane, as shown in Fig. 5.20(*b*).

The magnitude of image current is same as that of the actual current. However, the direction of the image current is opposite to the direction of the real current,

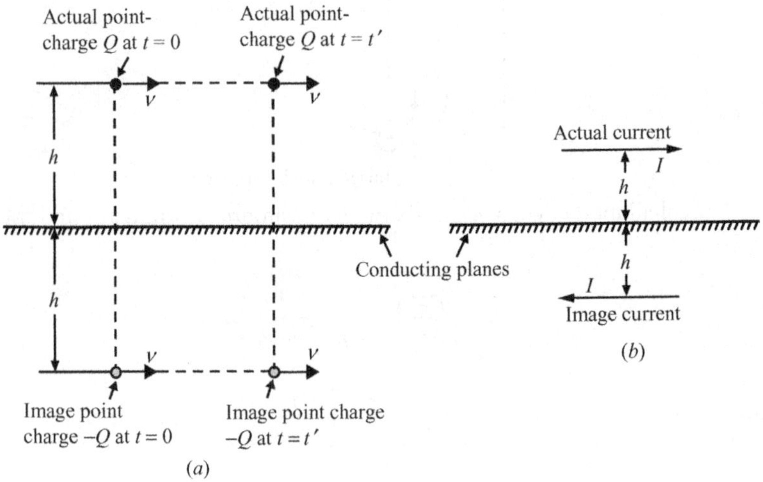

Fig. 5.20 (*a*) Images of a point charge moving parallel to a grounded conducting plane and (*b*) Image of a line current parallel to a grounded conducting plane.

because if we assume that the real current is due to flow of positive charges then the image current is due to flow of negative image charges.

5.7 Graphical Field Mapping

Two-dimensional field problems with boundaries of arbitrary shape cannot be solved easily by analytical methods. One method of solving such problems is to draw field maps. The maps, consisting of equipotential lines and field lines (stream lines), may be treated as solutions of two-dimensional boundary value problems. A good field map gives reasonably accurate results. Maps are plotted using the known properties of the field.

The properties of static fields are as follows:

(a) Static electric fields with conducting boundaries

1. Field lines start from the positively charged conductor and terminate on the negatively charged conductor.

2. As good conductors are at constant potential, the field intensity and flux density are normal to their surfaces.
3. Field lines and equipotential surfaces (shown as lines in a map) intersect at right angles.
4. The region between two consecutive field lines is a flux tube, and the flux throughout a tube has the same value.

(b) Conductors carrying direct current

1. Stream lines showing the direction of current flow and equipotential lines are orthogonal.
2. Current between stream lines is constant and continuous.
3. Current flow is tangential to a conductor insulation boundary.

(c) Static magnetic fields with iron boundaries

1. Magnetic field lines and scalar equipotential lines intersect at right angles.
2. Iron surfaces are equipotential surfaces.
3. Field intensity and flux density are normal to iron surfaces at iron–air boundaries.
4. The region between two consecutive field lines is a flux tube. The flux throughout a tube has the same value.

The Curvilinear Square

Figure 5.21 illustrates one mesh formed by two field lines *ad* and *bc*, and two equipotential surfaces represented by lines *ab* and *cd* in the map of a two-dimensional electric field. The potential of *ab* is $V + dV$ and that of *cd* is V. The distance between the mid-points of the equipotential lines and the distance between the mid-points of field lines are taken as $\Delta\ell_V$ and $\Delta\ell_\Psi$ respectively. As the field intensity is equal to the rate of change of the potential with distance, the average field intensity and flux density within the mesh are given by

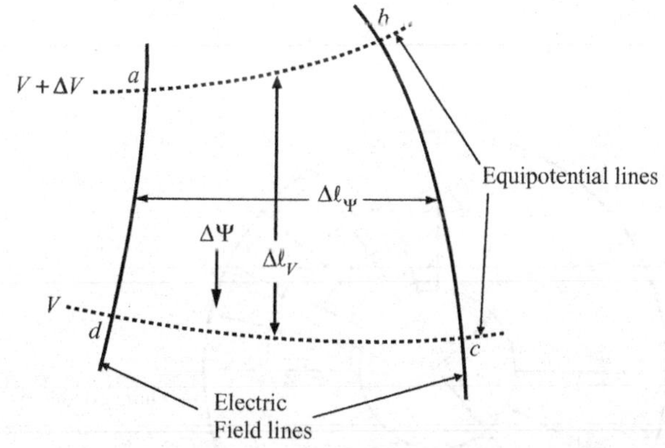

Fig. 5.21 A curvilinear square

$$E = \frac{\Delta V}{\Delta \ell_V} \tag{1}$$

and

$$D = \varepsilon \frac{\Delta V}{\Delta \ell_V} \tag{2}$$

Flux between the field lines per unit depth of the field normal to the page,

$$\Delta\Psi = D\Delta\ell_{\Psi} \tag{3}$$

From (2) and (3), we get

$$\Delta\Psi = \varepsilon\,\Delta V\,\frac{\Delta\ell_{\Psi}}{\Delta\ell_{V}} \tag{4}$$

In sketching field maps, the potential difference between consecutive equipotential lines, i.e., ΔV, is kept constant so that computation becomes easy. Further, if the ratio between $\Delta\ell_{\Psi}$ and $\Delta\ell_{V}$ is kept constant throughout the map, then all the flux tubes would have the same value of flux. The ratio is generally taken as unity. Under this condition the mesh *abcd* is called a *curvilinear square*.

Tips for Sketching Field Maps

To learn the technique of drawing a field map, it is desirable that the field map of an example with known shapes of field lines and equipotential surfaces be drawn first. A simple example is a coaxial transmission line with circular inner and outer conductors, as shown in Fig. 5.22. The electric field lines between the conductors are radial and equipotential lines are circles concentric to the inner conductor. Even for this simple example, though the orthogonal condition between field and potential lines is easily fulfilled, spacing of equipotential and field lines is not that straight forward. However, it is required to draw the map only in one quadrant because of symmetry. One way to begin is to divide a quadrant into suitable number of sectors with uniformly spaced field lines. For the example under study, there are six sectors in a quadrant, as shown in Fig. 5.22. The next step is to draw the equipotential lines with proper spacing so that all the meshes are curvilinear squares. It may be seen from the figure that the meshes near the outer conductor are not curvilinear squares. These are curvilinear rectangles. Hence, one *rectangle* is subdivided into suitable number of *squares*. This is necessary to calculate potential difference between the conductors.

The field distribution between boundaries of arbitrary shape is not uniform. Therefore, the maps have to be sketched by trial and error.

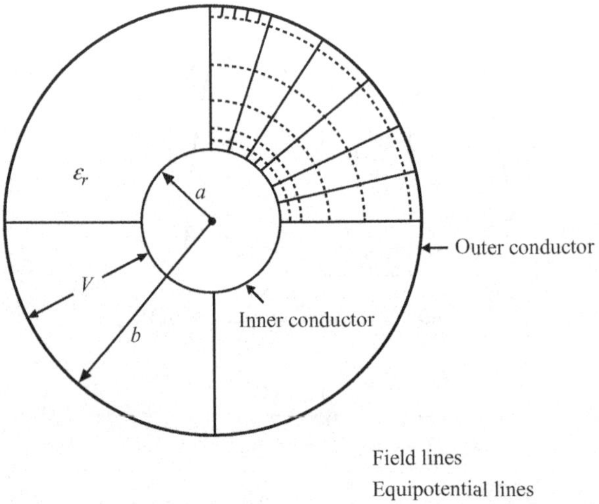

Fig. 5.22 Electric field map between inner and outer conductors of a coaxial transmission line. The map is shown for one quadrant only.

The following hints may be of some help for drawing a field map.

1. It is convenient to draw a large map preferably on a graph paper, as it is easy to check whether a mesh is a *square* or not.
2. See if there is any symmetry between boundaries. For example, the electric field map for the coaxial cable shown in Fig. 5.23 need be drawn only for one quadrant because of symmetry.
3. Start from the boundary where the field is likely to be more uniform. The map for the coaxial cable shown in Fig. 5.23 would be nearly uniform near the inner conductor, whereas the map for the current-carrying conductor shown in Fig. 5.26 would be uniform far away from the section at which the conductor size changes.
4. Decide how many equipotential lines will result in a map with neither too few meshes nor too many of them. Too few meshes will not yield results of reasonable accuracy, and too many of them will take more time to draw the map. After completion of the map any mesh can be subdivided to obtain more accurate results.
5. The shape of the equipotential contours progressively changes from the shape of one boundary surface to that of the other boundary surface.
6. Flow lines spread-out near the obtuse angles of a boundary, and close-in towards the acute angles of a boundary.
7. The meshes near a boundary may not be curvilinear *squares*. One such mesh has to be subdivided into appropriate number of *squares* for the purpose of calculation.

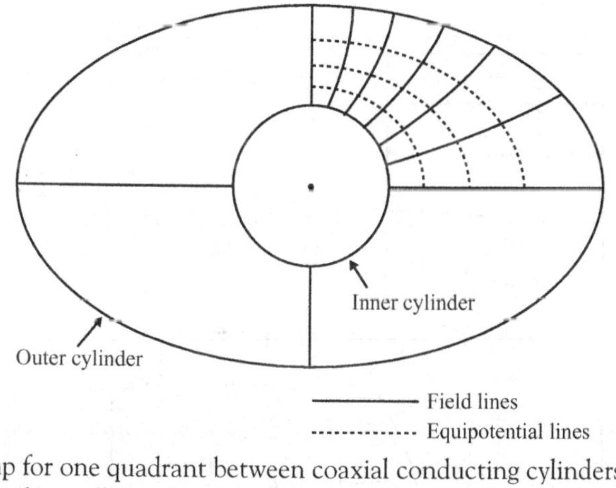

Fig. 5.23 Electric field map for one quadrant between coaxial conducting cylinders. The inner cylinder is circular and the outer cylinder is elliptical.

Application of Field Maps

In high voltage systems it is necessary to know the region where the potential gradient is high and whether its value is within the breakdown strength of insulation or not. This information can be obtained from the field map of the system. Field intensity at a point within a mesh is equal to $\Delta V/\Delta\ell_V$. As ΔV is kept constant while plotting the map, E is inversely proportional to $\Delta\ell_V$. Thus, maximum value of potential gradient occurs in a region where the value of $\Delta\ell_V$ is minimum. So, from a field map we can identify the region where E is high and its approximate value can be calculated. Potential gradient is high near the acute angle of a conducting surface, as in Fig. 5.24 and Fig. 5.25. Therefore, sharp bends are avoided in cables and windings in high voltage apparatus. If the flat conductor in Fig. 5.24 represents a charged cloud and the conical conductor is the topmost part of a lightning rod connected to earth, lightning will most likely strike at the tip of the conical surface and the energy would be diverted to earth. This is the reason why lightning will strike a man standing on an elevated (or even flat) ground. (Benjamin Franklin suggested that an earthed iron rod mounted on top of a structure can protect the structure from lightning strikes).

Another application of a field map is to find the capacitance between conductors of any shape. Let us assume that the potential difference between the conductors of the coaxial transmission line shown in Fig. 5.22 is V and the permittivity of the medium between the conductors is ε. In a field map between the conductors, let n_Ψ and n_V be the number of flux tubes and potential divisions respectively. Then electric flux in a tube per unit length normal to the page, as per formula (4), is

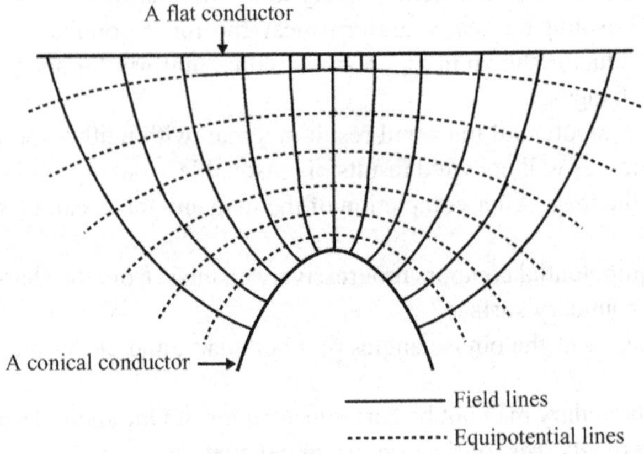

Fig. 5.24 Electric field map between a flat conductor and a conical conductor

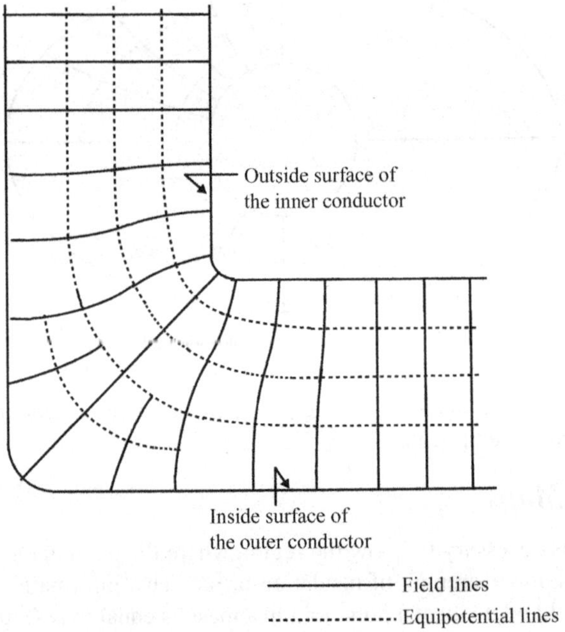

Fig. 5.25 Electric field map between the inner and outer conductors of a coaxial cable at a right angle bend

$$\Delta\Psi = \varepsilon\,\Delta V \frac{\Delta\ell_\Psi}{\Delta\ell_V} \tag{5}$$

If the field map is of curvilinear squares, then the ratio $\Delta\ell_\Psi/\Delta\ell_V = 1$. Also, $\Delta V = V/n_V$. Substituting these relations in (5), we get

$$\Delta\Psi = \frac{\varepsilon V}{n_V}$$

Total electric flux between the boundaries per unit length normal to the page,

$$\Psi = \Delta\Psi(1 \times n_\Psi) = \frac{\varepsilon V n_\Psi}{n_V}$$

By Gauss's law this is equal to the charge contained on the surface of one of the conductors per unit length. Hence, capacitance per unit length is give by

$$C = \frac{\Psi}{V} = \frac{\varepsilon n_\Psi}{n_V} \text{ F/m} \tag{6}$$

Capacitance of the line of ℓ meter length $= \ell C$.

We find from (5) that the capacitance can be calculated by estimating the number of flux tubes and number of potential divisions in a map. It may be noted that n_Ψ and n_V may or may not be integers.

The resistance of a conductor of any shape can also be determined from a field map of current stream lines and equipotential lines by following the above procedure. An example is illustrated in Fig. 5.26. It can be shown that the resistance between the ends and for a width w of the conductor (width normal to the plane of the paper) is

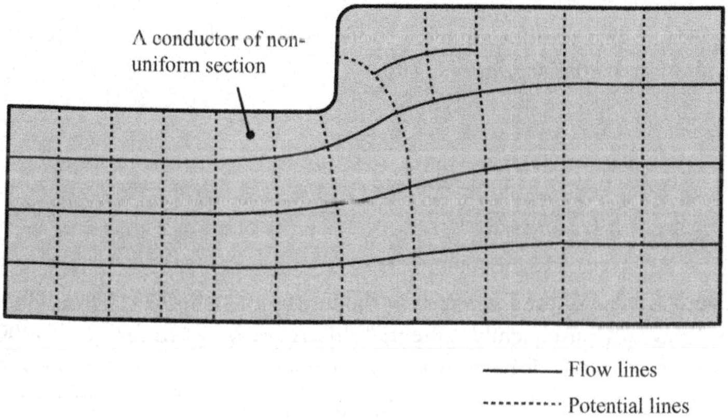

Fig. 5.26 Current flow lines and equipotential lines in a conductor of non-uniform cross section carrying direct current

$$R = \frac{n_V}{\sigma w n_I} \tag{7}$$

where σ is conductivity, n_V is number of potential divisions, and n_I is number of current tubes.

Further, the reluctance of air gap between two iron surfaces of any shape can be computed from a magnetic field map. Figure 5.27 shows the maps of magnetic field lines and equipotential lines in the non-uniform air gap between two highly permeable magnetic structures. The field distribution is similar to that of a salient pole synchronous machine at no-load without considering the effect of slots. Note that the field map shown is for one-half of a pole pitch. It can be shown that the air-gap reluctance for a depth d (normal to the plane of the page) is

$$\Re = \frac{n_V}{\mu_0 d n_\Phi} \tag{8}$$

where n_V is number of potential (scalar) divisions and n_Φ is number of flux tubes.

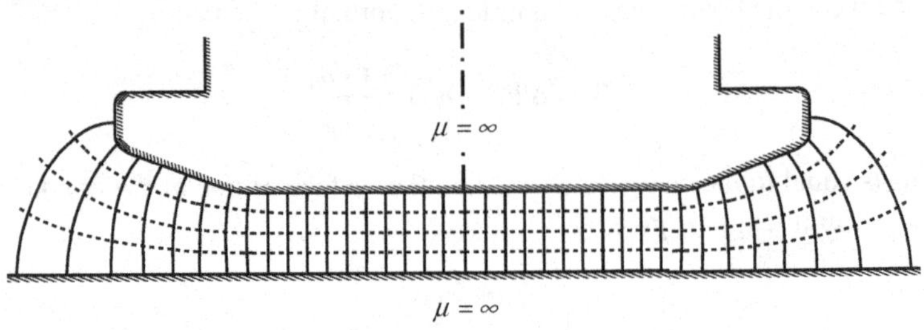

Fig. 5.27 Shows the magnetic field map in the region of a non-uniform air gap between two highly permeable magnetic structures. This is similar to the air gap field distribution of a salient pole synchronous machine at no-load neglecting the effect of slots.

EXAMPLE 5.19

We will calculate electric field intensity on the surface of the inner conductor, and capacitance per unit length of a coaxial transmission line, shown in Fig. 5.22, using the field map and also from the formulae. Let us take $a = 1.25$ cm, $b = 3.75$ cm, $\varepsilon_r = 3.6$ and $V = 10$ kV.

We find from the field map that meshes adjacent to the outer conductor are not *squares*. Hence, one mesh when subdivided resulted in 5 smaller *squares*. Thus,

$$n_V = 4 + \frac{1}{5} = 4.2$$

$$n_\Phi = 6 \quad 4 = 24$$

$$\Delta V = 10000/4.2 = 2381 \text{ V}$$

To obtain E more accurately one of the *squares* near the inner conductor has been subdivided. The subdivision resulted in 4 smaller *squares*. Consequently, potential difference between the inner conductor and the nearest equipotential surface is $\Delta V/2$, and its distance from the conductor is found to be 0.175 cm. So, field intensity

$$E = \frac{2381/2}{0.175} = 6802.9 \text{ V/cm}$$

Capacitance per meter length, $C = (3.6\varepsilon_0)\dfrac{24}{4.2} = 20.57\varepsilon_0$ F/m

From the formulae:

$$E(r \quad a) = \frac{V}{a\ln(b/a)} = \frac{10000}{(1.25)(\ln 3)} = 7281.9 \text{ V/cm}$$

$$C = \frac{2\pi\varepsilon}{\ln(b/a)} = \frac{2\pi \times 3.6\,\varepsilon_0}{\ln 3} = 20.59\,\varepsilon_0 \text{ F/m}$$

EXAMPLE 5.20

The field map of a current-carrying conductor consists of curvilinear squares. The number of equipotential divisions along the length of the conductor $n_V = 11$ and the number of current tubes along its depth $n_I = 4$. The conductor width $w = 2.4$ cm and its conductivity $\sigma =$ [illegible] $\times 10^7$ S/m. Find resistance of the conductor.

$$R = \frac{n_V}{\sigma w n_l} = \frac{11}{(3.8\times10^{7})(0.024)(4)} = 30.15\times10^{-7}\ \Omega$$

EXAMPLE 5.21

The field map of the magnetic field in the air gap between two highly permeable iron surfaces consists of curvilinear squares. The map has 4 equipotential divisions between the iron surfaces and 36 flux tubes along the width of the iron surfaces. Find reluctance of the air gap if the depth of the iron surfaces is 20 cm.

Equipotential divisions, $n_V = 4$

Flux tubes, $n_\Phi = 36$

Depth, $d = 0.2$ m

Air gap reluctance, $\Re = \dfrac{n_V}{\mu_0 d n_\Phi} = \dfrac{4}{(4\pi\times10^{-7})(0.2)(36)} = 1.77\times10^{6}\ \text{H}^{-1}$

SUMMARY

We used a vector identity and the divergence theorem to prove that the solution of a problem by the use of Laplace's or Poisson's equation that satisfies the boundary conditions is unique. We obtained analytical solutions of some one-dimensional and two-dimensional problems in rectangular and cylindrical coordinates using these equations. We found that an analytical solution is possible only if the boundary conditions can be described mathematically with respect to a coordinate system.

We described, with examples, the method of images and the graphical method for solving boundary value problems. The graphical method, though it does not give accurate results, can be used to solve problems with boundaries of any shape. We note that solution of a problem obtained either by the image method or by the graphical technique is a solution of Laplace's equation. We learnt how to calculate from a field map the capacitance between conductors separated by a dielectric, the resistance of a conductor with non-uniform sectional area, and the reluctance of a non-uniform air gap between two iron surfaces.

IMPORTANT FORMULAE

Potential between two radiating conducting plates: $V = \dfrac{V_b - V_a}{\alpha}\phi + V_a$

Potential between two spherical conducting shells: $V = \dfrac{ab(V_a - V_b)}{b-a}\dfrac{1}{r} + \dfrac{bV_b - aV_a}{b-a}$

Potential between two coaxial conducting cones:

$$V = V_0 \frac{\ln[\tan(\ /2 \ \text{[illegible]}\ _2\ 2)]}{\ln[\tan(\ _1/2\ \text{[illegible]}\ _2\ 2)]},\ \alpha_2 \ \ \theta > \alpha_1$$

Solutions of two-dimensional Laplace's equation in rectangular coordinates:

$$V = (\quad \cosh kx + B \sinh kx)(C \cos ky + D \sin ky)$$

$$V \quad (A' e^{kx} + B' e^{-kx})(C \cos ky + D \sin ky)$$

$$V \quad (A \quad kx + B \sin kx)(C \cosh ky + D \sinh ky)$$

$$V \quad (A \quad kx + B \sin kx)(C' e^{ky} + D' e^{-ky})$$

$$V \quad (Ax \quad B)(Cy + D)$$

Solution of two-dimensional Laplace's equation in cylindrical coordinates when the potential varies with r and ϕ *:*

$$V = (Ar^{k} + Br^{-k})(C \cos k\phi + D \sin k\phi),\ k \neq 0$$

MULTIPLE-CHOICE QUESTIONS

(Starred questions have two answers)

*5.1 Which of the following functions satisfy Laplace's equation?
(a) $x \sin y$ (b) $\sin x \cos y$ (c) $e^{x} \cos y$ (d) $\sinh x \sin y$

5.2 The following functions, where A is a constant, are defined in cylindrical coordinates. Which function satisfies Laplace's equation?
(a) A/r (b) $A \ln r$ (c) Ae^{-r} (d) Ar

5.3 The following functions, where C is a constant, are defined in spherical coordinates. Which function satisfies Laplace's equation?
(a) $-C/r$ (b) Cr (c) C/r^{2} (d) Cr^{2}

5.4 A potential function is given by $ax^{3} + bxy^{2}$ where a and b are constants. The function will satisfy Laplace's equation if
(a) $b = 3a$ (b) $3b = -a$ (c) $b = -3a$ (d) $3b = a$

5.5 The scalar electric potential in free space is given by $V = ax^{2}y + byz^{2}$. The expression for the charge density is
(a) $\varepsilon_0(a+b)y$ (b) $-2\varepsilon_0(a+b)y$ (c) $\varepsilon_0(ax+bz)$ (d) $\varepsilon_0(axy+byz)$

*5.6 The scalar electric potential is a function of r and ϕ in cylindrical coordinates. Which of the following expressions satisfy Laplace's equation?
(a) $Ar^{k} \cos k\phi$ (b) $Ar^{k} \cosh k\phi$ (c) $Ar^{-k} \sin k\phi$ (d) $Ar^{-k} \sinh k\phi$

5.7 A long coaxial transmission line has two dielectric layers of different permittivities between the conductors. To solve the appropriate one-dimensional Laplace's equations for the scalar potential in the dielectric regions, the number of boundary conditions needed is
(a) 2 (b) 3 (c) 4 (d) 5

5.8 A point charge $-Q$ is located above a large conducting plate. The plate is on the x-y plane and the charge is at $(0,0,z)$ in free space. The surface charge density at the origin is

(a) $\dfrac{Q}{2\pi z^2}$ (b) $\dfrac{Q}{2\pi\varepsilon_0 z^2}$ (c) $\dfrac{Q}{4\pi z^2}$ (d) $\dfrac{Q}{4\pi z^2}$

5.9 A long line charge is parallel to the axis of a long cylindrical conductor and it is located outside the conductor. The radius of the cylinder is a and its axis is at a distance d from the line charge. An image charge located on the line that is perpendicular to the line charge and the conductor axis satisfies the necessary boundary conditions. The distance of the image line charge from the real line charge is

(a) $1-\dfrac{a}{d^2}$ (b) $d-\dfrac{a^2}{d}$ (c) $d^2-\dfrac{a}{d}$ (d) $\dfrac{a^2}{d}$

5.10 A point charge $-Q$ is located outside a spherical conductor. The radius of the cylinder is a and its centre is at a distance d from the point charge. A positive image charge located on the radial line joining the point charge and centre of the conductor satisfies the necessary boundary conditions. The value of the image charge is

(a) $\dfrac{a^2Q}{d}$ (b) $\dfrac{aQ}{d^2}$ (c) $\dfrac{dQ}{a}$ (d) $\dfrac{aQ}{d}$

5.11 The capacitance per unit length of a transmission line with two parallel cylindrical conductors each of radius a and conductor spacing $2h$ is C. Then, the capacitance of a cylindrical conductor of radius a and located at a height h above the ground plane is

(a) $C/4$ (b) $C/2$ (c) $2C$ (d) C

*5.12 Which of the following statements relating to static electric field are correct?

(a) The field lines and equipotential lines intersect at right angles.
(b) The flux in a tube of flux varies along the tube.
(c) The potential gradient has maximum value near the acute angle bend of a good conductor.
(d) The angle of a field line with the normal to a conductor surface depends on the shape of the conductor

QUESTIONS

5.1 What is the order of Laplace's and Poisson's equations? Distinguish between one-dimensional, two-dimensional and three-dimensional equations. Write the Laplace's equation for the electrostatic potential V that is a function of only y and z, and that is a function of only r in cylindrical coordinates.

5.2 State the uniqueness theorem.

5.3 Two large parallel plates have equal and opposite surface charge densities on their inside surfaces. What boundary conditions are needed to obtain the solution of Laplace's equation in the dielectric region between the plates?

5.4 A coaxial transmission line has two dielectric layers between the conductors with the interface concentric with the inner conductor. In addition to the given potentials of the conductors, what other known boundary conditions are needed to find the solution of Laplace's equation in the region between the conductors?

5.5 The problem of potential distribution between two radial conducting planes may be treated as a one-dimensional problem under some assumptions. What are these assumptions? Write the appropriate Laplace's equation to be solved.

5.6 Draw a few field lines between the radiating planes shown in Fig. 5.3.

5.7 The magnetic field intensity of a cylindrical conductor carrying direct current can be found by applying Ampere's magnetic circuital law. It can also be obtained by the use of Poisson's and Laplace's equations in the vector magnetic potential. What boundary conditions are used to solve the problem by the latter method?

5.8 To obtain the solution of Laplace's equation for the potential function in the two dielectric regions between two conducting spherical shells shown in Fig. Q5.8, four boundary conditions are needed. Write the Laplace's equations to be solved, and state the four boundary conditions. The radii of the inner sphere, the outer sphere, and the interface are r_1, r_2, and r_0 respectively. The potentials of the inner and the outer spheres are V_1 and 0. The permittivities of the inner and the outer dielectric layers are ε_1 and ε_2.

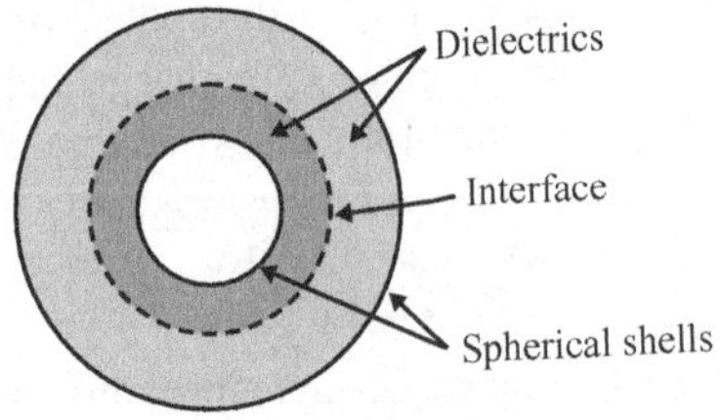

Fig. Q5.8

5.9 Explain the separation of variable method of solving Laplace's equation for the electrostatic potential in rectangular coordinates.

5.10 What is the maximum number of boundary conditions needed to obtain the complete solution of a two-dimensional field problem?

5.11 Show that the function $V = V_0 \sin nx \cosh ny$ in rectangular coordinates, and the function $V = V_0 r^{-n} \cos n\phi$ in cylindrical coordinates satisfy the two-dimensional Laplace's equations?

5.12 Show that all the four solutions given by (10) to (13) of Section 5.4 satisfy the two-dimensional Laplace's equation (5.4-1).

5.13 Both (10) and (11) of Section 5.4 are the solutions of the same Laplace's equation. Obtain A' as well as B' in terms of A and B.

5.14 The problem of a point charge above a grounded conducting plate can be solved by the image method. Explain briefly the principle of the method.

5.15 Justify that the electric field intensity below the conducting plane of Fig. 5.14(*c*) is zero.

5.16 A point charge is located between two conducting planes which are at an angle of 30°. Show the required number and positions of image charges so that the boundary conditions are satisfied.

5.17 To draw the field map between the conductors at different potentials what properties of the electric field are used?

5.18 What are the advantages of drawing a field map consisting of curvilinear squares?

5.19 A point charge is situated above a grounded conducting plane. Draw a few field lines between the charge and the plane.

5.20 Explain with the help of a field map why sharp bends in high voltage cables are avoided.

ANSWERS TO SELECTED QUESTIONS

5.2 State the uniqueness theorem.

Answer: Laplace's equation is a linear, homogeneous and second order partial differential equation. It may be one-dimensional, two-dimensional or three-dimensional. The number of constants in the general solutions depends on the dimension of the equation. In the case of boundary value problems, the constants are determined from the given and known boundary conditions. As per uniqueness theorem, a solution of Laplace's equation fulfilling the boundary condition is unique. That is, there can be only one solution of Laplace's equation satisfying the boundary conditions.

5.4 A coaxial transmission line has two dielectric layers between the conductors with the interface concentric with the inner conductor. In addition to the given potentials of the conductors, what other known boundary conditions are needed to find the solution of Laplace's equation in the region between the conductors?

Answer: Assuming that the potential in the dielectric region varies only with r of cylindrical coordinates, general solutions of one-dimensional Laplace's equation are obtained for the potential in each dielectric region. The two solutions will have four constants, and to determine the constants four boundary conditions are required. Two boundary conditions are the given potentials of the conductors. The other two boundary conditions are (1) the potentials on two sides of the interface are equal, and (2) the radial flux density, which is normal to the interface, is continuous across the boundary.

5.7 The magnetic field intensity of a cylindrical conductor carrying direct current can be found by applying Ampere's magnetic circuital law. It can also be obtained by solving Poisson's and Laplace's equations for the vector magnetic potential. What boundary conditions are used to solve the problem by the latter method?

Answer: A current-carrying conductor produces magnetic field inside as well as outside the conductor. General solutions of one-dimensional Poisson's equation for the vector magnetic potential in the region of current and Laplace's equation for the vector magnetic potential outside the conductor, both in cylindrical coordinates with r as variable, will have a total of four constants. The boundary conditions are (1) the magnetic field intensity, which is tangential to the conductor surface, is continuous across the surface, and (2) the magnetic field intensity is zero at infinity. From these two boundary conditions only two constants can be determined. As the other two constants are not coefficients of any function of r, they do not contribute to the field intensity. Moreover, the vector magnetic potential has no physical meaning. Hence, these constants may be assumed to be zero.

5.9 Explain the separation of variable method of solving Laplace's equation for the electrostatic potential in rectangular coordinates.

Answer: Consider that the electrostatic potential is a function of x and y. As per the separation of variable method of solution of two-dimensional Laplace's equation, the potential is assumed to be equal to the product of two potentials, say, X and Y where X is a function of x only and Y is a function of y only. Separation of the variables yields,

$$\frac{1}{X}\frac{d^2X}{dx^2} = -\frac{1}{Y}\frac{d^2Y}{dy^2}$$

As the lhs is a function of x only and the rhs is a function of y only, each of the sides must be a constant. This is because if the left-hand side changes with x, it cannot be equal to the right-hand side as the rhs does not contain x. By the same logic, the rhs cannot change with y. Consequently, two ordinary one-dimensional second order differential equations are obtained. The product of solutions of the two differential equations gives the general solution of Laplace's equation.

5.14 The problem of a point charge above a grounded conducting plate can be solved by the image method. Explain briefly the principle of the method.

Answer: The total electric field above the plate is equal to the sum of the field of the point charge and the field due to the induced surface charges on the plate. The field due to the induced charge cannot be found by Coulomb's law because the manner of charge distribution on the plate surface is not known. Gauss's law cannot be applied because the field is not uniform.

As per the method of images, the field above the plate is equal to the sum of the field of the actual charge and an image charge with the plate removed. The image charge, which is also a point charge, is equal in magnitude but opposite in sign to the actual charge. The image charge is at the same distance directly below the plate as the distance of the actual charge above the plate.

PROBLEMS

**5.1 Two large circular parallel plates 1 and 2 are located at $x = 0$ and $x = 1.0$ cm. The potential of Plate 1 is 300 V, and that of Plate 2 is 500 V. Derive an expression for the potential at a point between the plates by solving the one-dimensional Laplace's equation, assuming that the potential varies with x only. If the radius of each plate is 20.0 cm, and the relative permittivity of the single dielectric between the plates is 4.0, find the charge on each plate.

***5.2 Figure P5.2 shows a capacitor with three parallel plates. The thickness of the intermediate plate is very small. Obtain solutions of Laplace's equation for the potentials V_1 V_2 in the two dielectric regions.

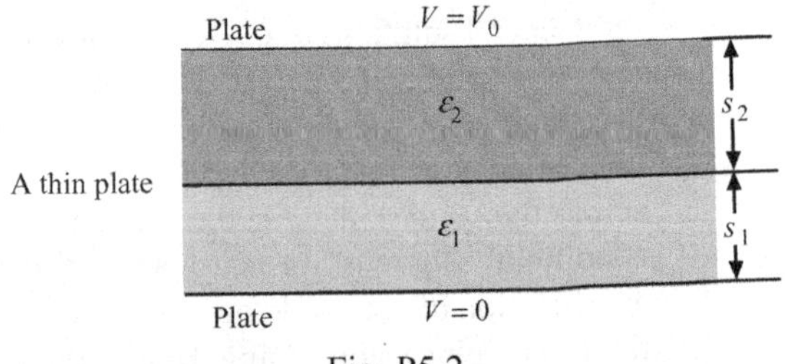

Fig. P5.2

***5.3 The outside radius of the inner conductor and the inside radius of the outer conductor of a coaxial transmission line are a and b respectively. There are two dielectric layers between the conductors with the interface at radius c. The permittivity of the dielectric layer near the inner conductor is ε_1 and that of the other dielectric is ε_2. The potential of the inner conductor is V_0 and that of the outer conductor is zero. Obtain the solutions of Laplace's equation for the potential in the two dielectric regions.

**5.4 Two radial conducting Plates 1 and 2 are at $\phi = \alpha_1$ and $\phi = \alpha_2$ in cylindrical coordinates. Plate 1 is at potential V_1 and Plate 2 is at potential V_2. Obtain a solution of Laplace's equation for the potential function between the plates. Assume that the potential varies with ϕ only.

**5.5 Using the data in Example 5.5 for the coaxial cones of Fig. 5.5(*a*), find the electric potential and electric field intensity at $\theta = 15°$ and $r = 1.5$ cm.

**P5.6 The coaxial conducting cones shown in Fig. 5.5(a) have angles α_1 and α_2. The inner cone is at potential V_1 and the outer cone is at potential V_2. Solve the Laplace's equation in spherical coordinates for the potential in the region between the cones, assuming that the potential varies with θ only.

If $\alpha_1 = 10°$, $\alpha_2 = 20°$, $V_1 = 200$ V, and $V_2 = -200$ V, find the angle of the conical surface over which the potential is zero.

***5.7 Charge is distributed in the form of a solid sphere of radius a with volume charge density ρ in free space. Obtain solutions for the potential functions using Poisson's and Laplace's equations. Find also the electric field intensity inside as well as outside the charge distribution.

**5.8 Example 3.7 is illustrated in Fig. 3.8. Solve the problem using Laplace's equation for the potential function within the conductor. Use the solution to find the expression for the current density.

***5.9 Solve Example 5.7 by taking the origin of the rectangular coordinates at the mid-point of the conducting strip in Fig. 5.8(a).

**5.10 A point charge of 0.5 μC is situated over a large plane conductor in free space. The conductor is on the $y = 0$ plane and its potential is zero. The point charge is located at (0, 2.0 m, 0). Find (a) the electric potential and field intensity at P(1.0 m, 1.0 m, 0) and (b) the charge density on the conductor surface at P'(1.0 m, 0, 2.0 m).

*5.11 The solutions for the electric potential and field intensity between a conducting cone and a conducting plate shown in Fig. 5.5(b) are given by (44) and (45) of Section 5.3. Give a sketch of the field map showing equipotential and field lines between the conductors.

*5.12 Figure P5.12 shows a conducting cylinder placed between two large parallel conducting plates with uniform charge distributions of surface charge densities ρ_S and $-\rho_S$ on their inside surfaces. Sketch a few electric field lines between the plates.

*5.13 Figure P5.13 shows two conducting spherical shells of equal radius. The spheres are maintained at equal and opposite potentials. Sketch a few electric field lines between the plates. Also show the surface of zero potential.

**5.14 Figure P5.14 illustrates two symmetrically placed parallel conducting sheets of different widths, as in a strip transmission line. Sketch the electric field map between the plates.

**5.15 Derive formula (7) and (8) of Section 5.7.

*5.16 The field map of current stream lines and equipotential lines of an aluminium conductor of arbitrary shape consists of curvilinear squares. There are 15 equipotential divisions between the ends and 6 current divisions across the thickness of the conductor. The width of the conductor is 2.5 cm and its resistivity is 2.9×10^{-8} Ω-m. Find the resistance between the ends of the conductor.

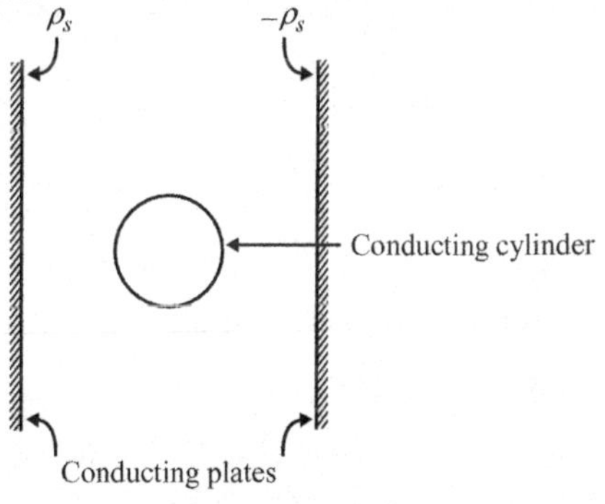

Fig. P5. 12

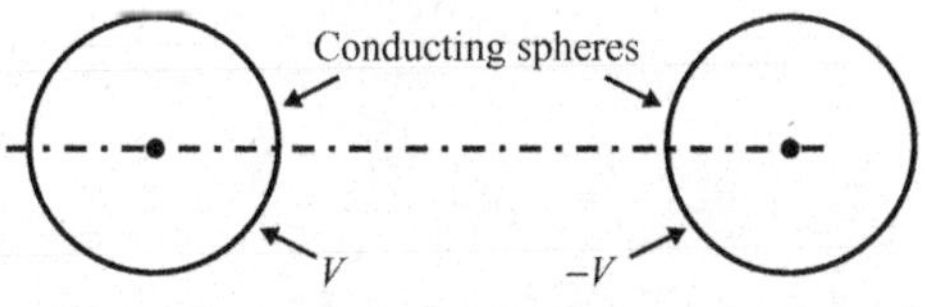

Fig. P5. 13

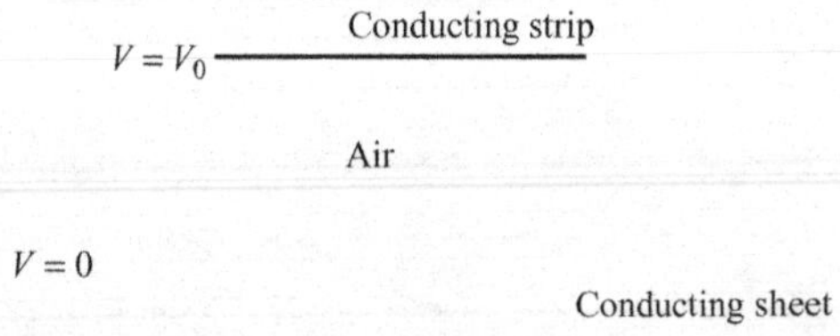

Fig. P5. 14

**5.17 The magnetic field map in the air gap of a salient pole electromagnetic machine consists of curvilinear squares. There are 12 divisions of flux stream lines between the pole axis and the neutral axis, and 6 equipotential divisions between the iron surfaces. The length of the pole is 30 cm. Find (a) the reluctance of the air gap between two neutral axes, (b) Show that the flux density, say at the armature surface, is inversely proportional to the width between the stream lines. The table below gives serially the widths between consecutive stream lines from the pole axis to the neutral axis. The width at the pole axis is taken as one unit. Plot the graph showing the variation of flux density (field form) from the pole axis to the neutral axis. Take the flux density at the pole axis as one unit.

Serial number	1	2	3	4	5	6	7	8	9	10	11	12
Relative width	1.0	1.0	1.0	1.02	1.04	1.1	1.2	1.5	2.0	3.0	5.0	12.0

Chapter Six

TIME-VARYING ELECTRIC AND MAGNETIC FIELDS

6.1 Introduction

The electric field quantities arising out of a stationary charge distribution of constant magnitude are also stationary in space and are independent of time. As the field quantities are proportional to the value of the charge, the fields will vary with time if the value of the charge varies with time. Moving charges also produce time-varying electric fields. By the same reasoning, the magnetic field quantities of time-varying currents are also functions of time. An important property of time-varying electric and magnetic fields is that the two fields coexist. This is because time-varying electric fields produce time-varying magnetic fields and vice-versa. The combination of electric and magnetic fields is termed as *electromagnetic* field.

In a homogenous and linear medium, most of the static field laws including boundary condition are valid for time-varying fields without or with some modification. However, the static field laws only are not sufficient for the study of time-varying fields. Some important additions for the time-varying fields are Faraday's law of electromagnetic induction, the concept of displacement current, the retarded potential, Maxwell's equations, Poynting's theorem and wave equations.

The formulae for capacitance between conductors and inductance of electric circuits derived using the static field laws can be used for the time-varying case at low and medium frequency ranges. But, the ac resistance of a conductor is higher than its dc value because of the skin effect phenomenon.

Time-varying electric and magnetic fields have numerous applications. The principle of rotating electrical machines, transformers, induction heating and many other electrical devices are based on Faraday's law of electromagnetic induction. Electric power at low and power frequencies can be transmitted efficiently through transmission lines only. However at high frequencies, electric power can be transmitted efficiently through air (free-space) as well as through transmission lines. The power is carried by electromagnetic waves of time-varying electric and magnetic fields. This property of high frequency fields is utilised in terrestrial and satellite communications, in radio and television broadcasting, in radars, in remote control devices, etc.

The main symbols used for denoting static-field quantities will also be used to denote time-varying field quantities. However, additional notations will be used wherever necessary.

6.2 The Continuity Equation

Consider that time-varying positive charges are continuously distributed in a region of volume with charge density ρ, as shown in Fig. 6.1. The total charge in the region is taken as $Q(t)$. At any instant, let the current flowing out of the region be I_2 and current flowing into the region be I_1. We assume that the currents are distributed over the closed surface S enclosing the volume. The net current flowing *out* of the region is

$$I = I_2 - I_1$$

Electric charge cannot be created or destroyed because of its conservation property. Hence, charge within the volume will decrease if I_2 is greater than I_1 and will increase if I_2 is less than I_1. If we assume that I is positive, charge inside the region will decrease and the rate of *decrease* of charge would be equal to the current. Hence,

$$I = -\frac{\partial Q(t)}{\partial t} \tag{1}$$

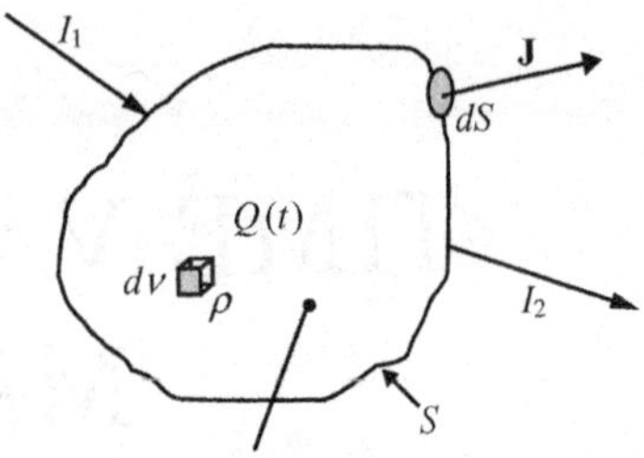

Fig. 6.1 For derivation of the continuity equation

If we assume that the net current is distributed over the surface S and the current density vector at a point on the surface as **J**, then

$$I = \oint_S \mathbf{J}\cdot d\mathbf{S} \tag{2}$$

Also, the total charge in the region is equal to the integration of the charge density over the volume. That is,

$$Q(t) = \int_v \rho\, dv \tag{3}$$

By substituting (2) and (3) in (1), we get

$$\oint_S \mathbf{J}\cdot d\mathbf{S} = -\frac{\partial}{\partial t}\int_v \rho\, dv \tag{4}$$

The lhs of (4) may be transformed into a volume integral using the divergence theorem. Thus,

$$\oint_S \mathbf{J}\cdot d\mathbf{S} = \int_v (\nabla\cdot\mathbf{J})\, dv \tag{5}$$

As the integral on the rhs (4) involves space variables, we can shift the partial differentiation with time from outside to inside of the integral sign and get

$$\frac{\partial}{\partial t}\int_v \rho\, dv = \int_v \frac{\partial}{\partial t}(\rho\, dv) \tag{6}$$

By substituting (5) and (6) in (4), we have

$$\int_v (\nabla\cdot\mathbf{J})\, dv = -\int_v \frac{\partial \rho}{\partial t} dv \tag{7}$$

As both sides of (7) are volume integrals, we get the following relation:

$$\nabla\cdot\mathbf{J} = -\frac{\partial \rho}{\partial t} \tag{8}$$

Equation (8) gives the relation between current density and the volume charge density at a point. It is called *Continuity Equation*.

We note that the lhs of (8) gives the rate of change of current density with space variables whereas the rhs gives the negative of the rate of change of the charge density with time. The relation has been derived from the conservative property of charge. Moreover, as per the definition of the divergence of a vector,

$$\nabla\cdot\mathbf{J} = \frac{\oint_S \mathbf{J}\cdot \mathbf{S}}{\lim_{\Delta \to \infty} \Delta v} \tag{9}$$

where the numerator of rhs of (9) is the current out of the closed surface enclosing a very small volume. Thus, the rhs is the current out of a small volume per unit volume at any instant. And as per (8), this is equal to the time rate of decrease of volume charge density at the same instant.

6.3 Faraday's Law of Electromagnetic Induction

Let us consider a stationary coil of N turns linked by a magnetic field. If the flux *linking* the coil changes with time t, an electromotive force (emf) or voltage will be induced between its open ends. Electric current will flow if the coil is connected to a resistor (load). The magnitude of the induced voltage in a turn is proportional to the time rate of change of magnetic flux linking it. In SI unit the constant of proportionality is one. The total emf induced at any instant in a coil is equal to the sum of emfs in all its turns. If Φ_1 $\Phi_2, \Phi_3, \ldots \Phi_N$ are magnetic fluxes that link with different turns of a coil, total induced voltage $\mathcal{E}$ in the coil at any instant is

$$\mathcal{E} = \frac{d}{dt}(\Phi_1 + \Phi_2 + \Phi_3 + \cdots + \Phi_N)$$

If all the turns are linked by the same flux Φ then emf induced in the coil is given by

$$\mathcal{E} = \frac{d(N\Phi)}{dt} = N\frac{d\Phi}{dt}, \text{ assuming that } N \text{ does not change with time.}$$

The above law was established by Michael Faraday and Joseph Henry independently. The law is called Faraday's law of electromagnetic induction. The direction of induced emf (current) is given by Lenz's law. As per the law, the direction of induced current is determined from the consideration that flux produced by the current opposes the change of flux linking the coil.

We note that the flux produced by the induced current opposes the *change* of flux, *not the flux* itself.

To interpret the law let us consider a stationary coil linked by a time-varying flux due to an external source, as shown in Fig. 6.2(*a*). Let us assume that at a given instant the upward flux is increasing with time. That is the rate of change of flux with time, $d\Phi/dt$, is positive. By Lenz's law the flux due to the induced current should be downward and increasing with time. Hence, direction of the induced current is as shown in the figure. When the upward flux is decreasing with time at a given instant ($d\Phi/dt$ is negative), then flux due to the induced current should be upwards and increasing with time. Hence, direction of the induced current would be as shown in Fig. 6.2(*b*). Thus, by combining Faraday's and Lenz's laws, we get

$$\mathcal{E} = -N\frac{d\Phi}{dt} \tag{1}$$

The change in flux linking a coil can be brought about in three ways. These are (a) both the coil and the flux are stationary but flux linking the coil changes with time; (b) due to relative motion between a coil and a steady flux. The relative motion is caused when either the coil or the flux is moving and the other one is stationary, or when both the coil and the flux are moving but at different speeds; (c) a combination of (a) and (b), that is, due to relative motion between a time-varying flux and a coil.

The working principle of a number of devices such as transformers, rotating electric machines, linear induction motors, loop antennas, etc., are based on Faraday's law.

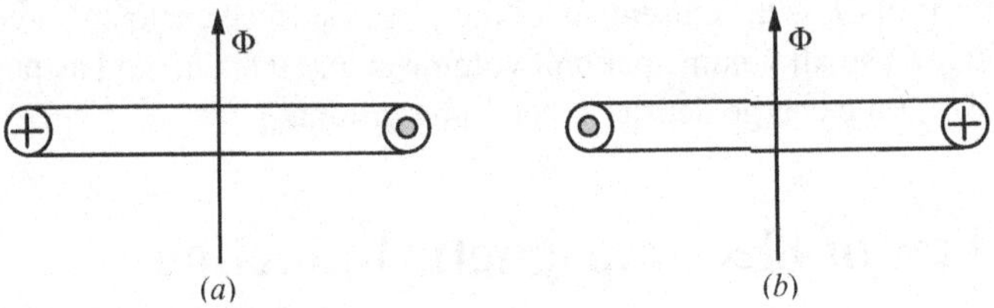

Fig. 6.2 Direction of induced currents in a planar coil due to time-varying magnetic flux. (*a*) Upward flux increasing with time and (*b*) Upward flux decreasing with time.

6.4 The Transformer Voltage

The voltage induced in a coil by a time-varying magnetic flux is called *transformer* voltage, because such a voltage is induced in the transformer windings due to flux produced by the primary winding. We will use (6.3-1) to calculate the transformer voltage in a rectangular coil of N turns, as shown in Fig. 6.3. The average length and width of the coil are taken as a and b respectively. Let us assume that the flux is uniformly distributed over the coil surface with flux density B. The flux density vector makes an angle α with the normal to the plane of the coil. The component of flux density that is normal to the coil surface,

$$B_n = B\cos\alpha$$

Area of the coil, $S = ab$

Flux Φ linking the coil is equal to the product of B_n and S. Thus,

$$\Phi = SB\cos\alpha$$

Voltage induced in the coil,

$$\mathcal{E} = -N\frac{d\Phi}{dt} = -SN\cos\alpha\frac{dB}{dt} \quad (1)$$

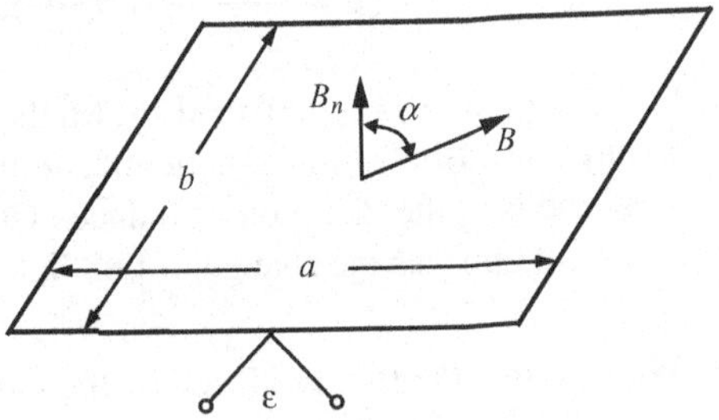

Fig. 6.3 Showing a rectangular coil of N turns in a time-varying uniform magnetic field of flux density B. The field is assumed to be uniform.

We will use (1) to calculate the emf induced in a stationary coil of N turns for some simple variation of flux density with time.

1. If $B = B_0 e^{-gt}$ where B_0 is flux density at $t = 0$, $\mathcal{E} = gNSB_0\cos\alpha\, e^{-gt}$.
2. If $B = B_0(1 - kt)$, $\mathcal{E} = kNSB_0\cos\alpha$.
3. If $B(t) = B_0\sin\omega t$, $\mathcal{E}(t) = -NSB_0\omega\cos\alpha\cos\omega t$

$$= -(2\pi fNSB_0\cos\alpha)\sin(\omega t - 90°) = -\mathcal{E}_0\sin(\omega t - 90°)$$

where $= 2\pi fNSB_0\cos\alpha$ is the maximum value of the induced voltage.

We see that for sinusoidal variation of flux with time, the induced voltage is also sinusoidal. But, it lags the flux by 90° in time.

EXAMPLE 6.1

A rectangular coil of 10 turns of side lengths 20 cm and 15 cm is situated in a uniform magnetic field. The flux density vector is at an angle $\alpha = 30°$ with the normal to the plane of the coil. The flux density varies sinusoidally with time with maximum flux density $B_0 = 0.01$ T. The frequency is 1000 Hz. Calculate the maximum value of voltage induced in the coil.

Area of the coil, $S = (20\times10^{-2})(15\times10^{-2}) = 300\times10^{-4}$ m^2

Frequency, $f = 1000$ Hz

Number of turns, $N = 10$

Maximum value of induced voltage,

$$= 2\pi f N S B_0 \cos\alpha$$

$$= 2\pi(1000)(10)(300\times10^{-4})(0.01)(\cos 30°) = 16.32 \text{ V}$$

6.5 Integral and Differential Forms of Faraday's Law

In the preceding section we have derived an expression for the voltage induced in a coil of rectangular shape assuming uniform distribution of flux over the coil area. In this section, we will use Faraday's law to calculate the induced emf in a coil of any shape and for non-uniform spatial distribution of the flux linking the coil. This will result in two important relations between the flux density and the induced electric-field intensity. One of the relations is in integral form and the other is in differential form.

Let us consider a planar loop of one turn linked by a time-varying magnetic flux, as shown in Fig. 6.4. At any point on the plane of the loop, let us consider an element of area $d\mathbf{S}$ and let flux density at the point be $\mathbf{B}$. Then, flux through the differential area,

$$d\Phi = \mathbf{B}\cdot d\mathbf{S} \quad (1)$$

Total flux linking the loop is found by integrating (1) over the area of the loop.

$$\Phi = \int_S \mathbf{B}\cdot d\mathbf{S}$$

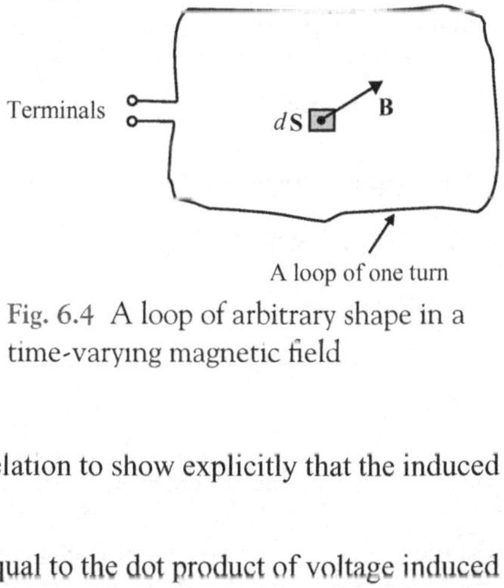

Fig. 6.4 A loop of arbitrary shape in a time-varying magnetic field

Voltage induced between the terminals of the loop is given by

$$\mathcal{E} = -\frac{d\Phi}{dt} = -\int_S \frac{\partial}{\partial t}(\mathbf{B}\cdot d\mathbf{S}) = -\frac{\partial}{\partial t}\int_S \mathbf{B}\cdot d\mathbf{S} \quad (2)$$

Partial derivative with respect to time has been used in the above relation to show explicitly that the induced emf is due to a time-varying flux.

The induced voltage $d\mathcal{E}$ across a differential length of the loop is equal to the dot product of voltage induced per unit length, i.e., the induced electric field intensity $\mathbf{E}$ (V/m), and the differential vector length $d\boldsymbol{\ell}$. Thus,

$$d\mathcal{E} = \mathbf{E}\cdot d\boldsymbol{\ell} \quad (3)$$

The total induced voltage is obtained by integrating (3) around the loop. Thus,

$$\mathcal{E} = \oint \mathbf{E}\cdot d\boldsymbol{\ell} \quad (4)$$

By equating the right-hand sides of (2) and (4), we get

$$\oint \mathbf{E}\cdot d\boldsymbol{\ell} = -\frac{\partial}{\partial t}\int_S \mathbf{B}\cdot d\mathbf{S} \quad (5)$$

Equation (5) is the *integral form of Faraday's law* where the line integral on the left-hand side of (5) is around the loop whereas the surface integral is over the area bounded by the loop. The loop may be over a conductor or in free space.

It may be noted that though the same symbol **E** is used for the electric field intensity due to static electric charges and also for the induced field intensity due to a time-varying magnetic field, the two fields are different qualitatively.

The line integral of field intensity due to static charges around a closed path is zero. However, the line integral of the field intensity due to a time-varying magnetic field around the loop is not zero. It is equal to the induced voltage. Moreover, the potential difference between two points in a static electric field does not depend on the shape of the path along which $\mathbf{E}\cdot d\boldsymbol{\ell}$ is integrated. But, the induced voltage between two points in a time-varying magnetic field depends on shape of the path along which $\mathbf{E}\cdot d\boldsymbol{\ell}$ is integrated, as the value of the flux now depends on the loop area.

The differential form of Faraday's law can be obtained by transforming the lhs of (5) to a surface integral using Stokes' theorem. Accordingly,

$$\int_S (\nabla\times\mathbf{E})\cdot d\mathbf{S} = -\frac{\partial}{\partial t}\int_S \mathbf{B}\cdot d\mathbf{S}$$

Since both the sides are surface integrals, we have

$$\nabla\times\mathbf{E} = -\frac{\partial\mathbf{B}}{\partial t} \qquad (6)$$

Equation (6) is the *differential form of Faraday's law*. We will find in a later section that it is one of Maxwell's equations.

EXAMPLE 6.2

The magnetic field intensity of an electromagnetic field in free space is given by

$\mathbf{H}(\;)\qquad e^{-5x}\sin(\pi t\times10^7)\mathbf{u}_z$ A/m

Find the electric field intensity. Assume the constant of integration to be zero.

$$\nabla\times\mathbf{E}(\;) = -\mu_0\frac{\partial\mathbf{H}(t)}{\partial t} = -(4\pi\times10^{-7})(\pi\times10^7)e^{-5x}\cos(\pi t\times10^7)\mathbf{u}_z$$

$$= -39.48e^{-5x}\cos(\pi t\times10^7)\mathbf{u}_z$$

As the rhs is in z direction and it is a function of x only, hence only the z component of $\nabla\times\mathbf{E}(t)$ that varies with x need be considered. Thus,

$$\frac{\partial E_y\,t)}{\partial x} = -39.48e^{-5x}\cos(\pi t\times10^7)$$

$$E_y\,t) = -39.48\cos(\pi t\times10^7)\int e^{-5x}dx = 7.9e^{-5x}\cos(\pi t\times10^7)\ \text{V/m}$$

6.6 Transformer Voltage in terms of Self-Inductance and Mutual Inductance

We know that a current-carrying coil produces a magnetic field, the value of which depends on the coil mmf and reluctance of the flux path. The flux multiplied by the number of turns of the coil is the self-flux linkage. The self-inductance of the coil has been defined in terms of the self-flux linkage in Section 4.18. If the current in a coil varies with time, the flux would also vary with time. The time-varying flux in turn will induce a voltage in the coil. As the induced voltage and the self-inductance are both functions of the same flux, a relation between the induced voltage and self-inductance can be derived. Also, voltage induced in a coil due the time-varying flux produced by the current in another coil can be expressed in terms of the mutual inductance.

Let us consider a coil of N turns carrying a time-varying current I, as shown in Fig. 6.5(a). The mmf produced by the coil establishes a time-varying flux Φ. This flux induces a voltage in the coil. Assuming that the flux links all the turns, the induced voltage is

$$\mathcal{E} = -N\frac{d\Phi}{dt} = -\frac{d(N\Phi)}{dt} \tag{1}$$

The self-inductance of the coil,

$$L = \frac{N\Phi}{I}$$

Hence

$$LI = N\Phi \tag{2}$$

Eliminating $N\Phi$ from (1) and (2), and assuming that L is not a function of time, we get

$$\mathcal{E} = -L\frac{dI}{dt} \tag{3}$$

Formula (3) gives transformer voltage in a coil in terms of its self-inductance.

Let us now consider two magnetically coupled coils 1 and 2 having number of turns N_1 and N_2 respectively, as shown in Fig. 6.5(b). Let flux linking Coil 2 due to current I_1 in Coil 1 be Φ_{21}. Then voltage induced in Coil 2 by this flux,

$$\mathcal{E}_{21} = -N_2\left(\frac{d\Phi_{21}}{dt}\right) = -\frac{d(N_2\Phi_{21})}{dt} \tag{4}$$

Similarly, if flux linking Coil 1 due to current I_2 in Coil 2 is Φ_{12}, voltage induced in Coil 1 due to this flux is

$$\mathcal{E}_{12} = -N_1\left(\frac{d\Phi_{12}}{dt}\right) = -\frac{d(N_1\Phi_{12})}{dt} \tag{5}$$

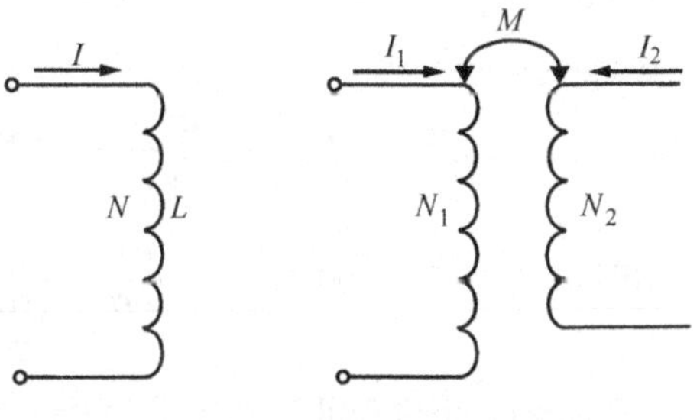

(a) (b)

Fig. 6.5 (a) A coil of N turns and self-inductance L and (b) Two magnetically coupled coils of mutual inductance M between them.

The mutual inductance between the coils is given by

$$M = \frac{N_1\Phi_{12}}{I_2} = \frac{N_2\Phi_{21}}{I_1}$$

Substituting these relation in (4) and (5), we get the transformer voltage in terms of the mutual inductance. These are

$$\mathcal{E}_{12} = -M\frac{dI_2}{dt} \tag{6}$$

$$_{21} = -M\frac{dI_1}{dt} \tag{7}$$

Equations (3), (6) and (7) are used in the analysis of electrical and electronic networks, to develop equivalent circuits of various types of electromagnetic devices such as transformers, to derive transmission line equations, etc.

EXAMPLE 6.3

The number of turns of two magnetically coupled coils 1 and 2 are 2500 and 1250 respectively. Coil 1 is connected to a 230 V, 50 Hz ac supply and Coil 2 is connected to a high resistance voltmeter. The measured value of current in Coil 1 is 1.2 A and voltage across Coil 2 is 105 V. Calculate (a) the mutual inductance between the coils, (b) the self-inductance of each coil, and (c) the coupling coefficient. Neglect resistance of Coil 1.

Let the instantaneous current in Coil 1, $I_1(t) = I_{10} \sin \omega t$

Then, ${}_{21} = -M \dfrac{dI_1(t)}{dt} = -\omega M I_{10} \cos \omega t = -\mathcal{E}_{210} \cos \omega t$

where, ${}_{210} = \omega M I_{10}$ is the maximum value of voltage induced in Coil 2.

Thus, $M = \mathcal{E}_{210}/(\omega I_{10}) = \mathcal{E}_{21}/(2\pi f I_1)$, where $\mathcal{E}_{21}$ and I_1 are rms values.

Number of turns of Coil 1, $N_1 = 2500$

Number of turns of Coil 2, $N_2 = 1250$

Current in Coil 1, $I_1 = 1.2$ A (rms)

Voltage across Coil 2, $\mathcal{E}_{21} = 105$ V (rms)

Frequency, $f = 50$ Hz

$$\text{Mutual inductance, } M = \frac{105}{(100\pi)(1.2)} = 0.279 \text{ H}$$

$$\text{Self-inductance of Coil 1, } L_1 = \frac{E_1}{2\pi f I_1} = \frac{230}{(100\pi)(1.2)} = 0.61 \text{ H}$$

$$\text{Self-inductance of Coil 2, } L_2 = \frac{I_1 N_2^2}{N_1^2} = 0.61\left(\frac{1250}{2500}\right)^2 = 0.1525 \text{ H}$$

$$\text{Coupling coefficient} = \frac{M}{\sqrt{L_1 L_2}} = \frac{0.279}{\sqrt{(0.61)(0.1525)}} = 0.915$$

6.7 The Speed Voltage

When a conductor moves in a magnetic field it cuts or sweeps the flux during its motion. A voltage is induced in the conductor because of relative motion between the conductor and flux. Alternatively, if the conductor is stationary and the flux moves relative to the conductor or in general if both conductor and flux move at different speeds, then also a voltage is induced in the conductor. This type of induced voltage is called *speed or motional voltage*.

The speed voltage is proportional to the rate at which flux is swept by the conductor.

An equation for the speed voltage is derived by considering a straight conductor of length ℓ moving at a constant velocity v in a stationary magnetic field of uniform density B. As shown in Fig. 6.6, the conductor is along x direction. The velocity is in y directions and the flux density is in z directions. Thus, the conductor, velocity of the conductor, and the flux density are mutually perpendicular to one another. We will first calculate the flux swept by the conductor in a differential time dt and then divide the flux by dt to obtain the induced voltage. Let the displacement of the conductor in time dt be dy. Then,

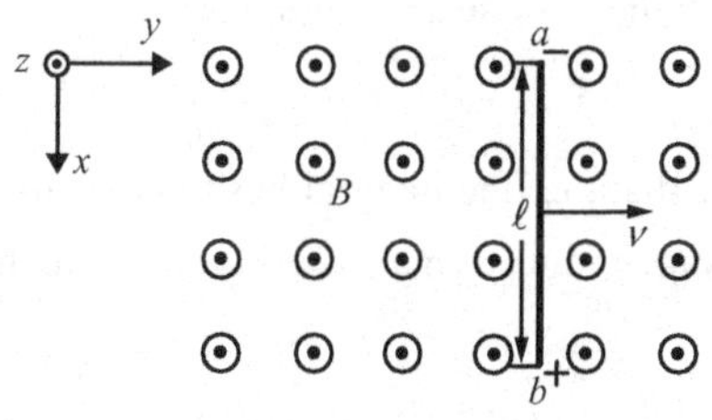

Fig. 6.6 A straight conductor *ab* of length ℓ moving at a velocity v in a uniform magnetic field of flux density B

$$dy = v\,dt$$

Area swept by the conductor in time dt,

$$dS = \ell\,dy = \ell(v\,dt)$$

Differential flux $d\Phi$ cut by the conductor during this period is equal to the product of area swept by the conductor and flux density. Accordingly,

$$d\Phi = B\,dS = B(v\ell dt)$$

The magnitude of the induced voltage, as per Faraday's law, is

$$\mathcal{E} = \frac{d\Phi}{dt} = vB\ell \tag{1}$$

The direction of the speed voltage is obtained using the right-hand rule.

According to the rule, the direction of induced voltage is normal to the plane containing the velocity vectors **v** and the flux density vector **B**, and its sense is along the advance of a right-handed screw when it is rotated from **v** towards **B** through the smaller angle. Accordingly, the direction of induced voltage is in positive x direction, i.e., from a to b. The polarities of the voltage are as shown in the figure. The moving conductor is a voltage source and current in it will flow from a to b if it is connected to an electric load such as a resistor. Interaction between the induced current and the flux results in an electromagnetic force in the negative y direction. This force opposes the motion of the conductor which is the cause of the induced voltage. Thus, Lenz's law is satisfied.

Note that v in formula (1) denote velocity of the conductor *relative to the flux*. Accordingly, the direction of induced voltage would still be from a to b if the conductor is stationary and the flux moves in the negative y direction. Moreover, the induced voltage given by (1) is for the case when **v**, **B** and $\boldsymbol{\ell}$ are mutually perpendicular to one another. Any departure from this condition will reduce the voltage. If **v** is at an angle α with **B**, the component of **v** perpendicular to **B** is equal to v multiplied by sin α. Hence the induced voltage is

$$\mathcal{E} = v\,(B\sin\alpha)\,\ell = |\,\mathbf{v}\times\mathbf{B}|\,\ell$$

where $|\mathbf{v}\times\mathbf{B}|$ is the scalar value of the cross product.

Moreover, if the direction of cross product of **v** and **B** is at an angle β with respect to ℓ, the induced voltage would be

$$\mathcal{E} = \left(|\mathbf{v}\times\mathbf{B}|\right)(\ell\cos\beta) = (\mathbf{v}\times\mathbf{B})\cdot\boldsymbol{\ell} \tag{2}$$

The speed voltage given by (2) is for the case of uniform distribution of the magnetic flux and for a straight conductor moving at a constant velocity. If the conductor is of arbitrary shape, then voltage induced in a differential length is integrated along the conductor between its ends. Voltage induced in a differential length is obtained by substituting $d\boldsymbol{\ell}$ in place of $\boldsymbol{\ell}$ in (2). Thus, we have

$$d\quad=(\qquad)\cdot d\boldsymbol{\ell}$$

The voltage between b and a is,

$$\mathcal{E}_{ba} = \int_a^b (\mathbf{v}\times\mathbf{B})\cdot d\boldsymbol{\ell} \tag{3}$$

Note that the line integral has to be evaluated along the direction of $(\mathbf{v}\times\mathbf{B})$.

Voltage induced in a closed loop of one turn is

$$\mathcal{E} = \oint (\mathbf{v}\times\mathbf{B})\cdot d\boldsymbol{\ell} \tag{4}$$

EXAMPLE 6.4

One end of a straight conductor is at the origin and the other end is at (0, 2.5 m, 0) in rectangular coordinates. The conductor moves at a constant velocity $\mathbf{v} = 5.0\mathbf{u}_z$ m/s in a uniform magnetic field $\mathbf{B} = 0.3\mathbf{u}_x + 0.4\mathbf{u}_z$ T. Calculate the voltage induced in the conductor.

As the conductor is on the y axis, $d\boldsymbol{\ell} = dy\,\mathbf{u}_y$

Voltage induced in the differential length,

$d \;\; = (\mathbf{v} \;\; \mathbf{B})\cdot d\boldsymbol{\ell} = (5.0\mathbf{u}_z)\times(0.3\mathbf{u}_x + 0.4\mathbf{u}_z)\cdot dy\mathbf{u}_y = 1.5dy$ V

Voltage induced in the conductor, $\mathcal{E} = \int_0^{2.5} 1.5\,dy = 3.75$ V

EXAMPLE 6.5

If the magnetic flux density in the preceding example varies harmonically with time defined by $\mathbf{B}\;\; 6\sin 1000t)\mathbf{u}_x$, calculate the voltage induced in the conductor.

As velocity of the conductor, the flux density and direction of orientation of the conductor are perpendicular to each other, voltage induced at any instant,

$(t) = vB\,t)\ell = (5.0)(0.6\sin 1000t)(2.5) = 7.5\sin 1000t$ V

The induced voltage is in time-phase with the flux density, and its frequency is same as that of the flux density.

EXAMPLE 6.6

A conducting sheet of width w is travelling at a constant speed v in a steady magnetic field of flux density B, as illustrated in Fig. 6.7. Assuming that B is perpendicular to the plane of the strip and also to v, calculate the voltage induced between the edges.

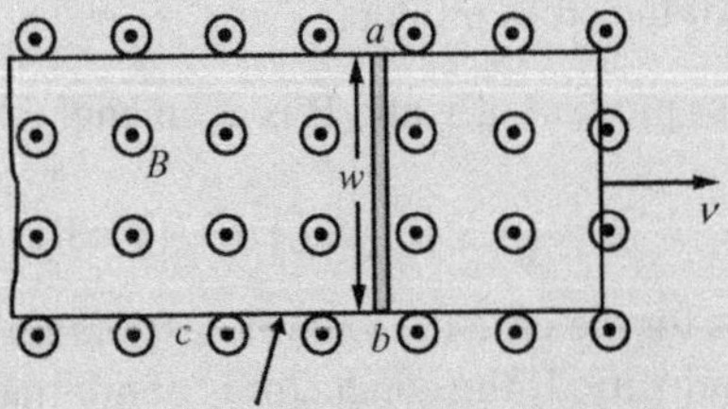

A conducting sheet moving at a constant velocity v in a uniform magnetic field of density B

Fig. 6.7 Illustration for Example 6.6

Let us imagine that the strip is made up of a large number of very small strips. Such a strip *ab*, shown shaded in the figure, may be treated as a straight conductor of length w. Hence voltage induced between the edges is same as that given by (1). Thus,

$$= Bwv$$

It may be noted that the induced voltage between point b and point c on the edge of the strip is zero.

If $v = 12$ cm/s, $B = 0.75$ T, $w = 16$ cm, then

$$= (0.75)(0.16)(0.12) = 0.0144 \text{ V}$$

EXAMPLE 6.7

Figure 6.8 shows a conducting bar of length ℓ pivoted at one end. It revolves at a constant angular speed ω_s in a steady magnetic flux of uniform density B. The flux density is normal to the plane of revolution. Find the voltage induced between the ends of the bar.

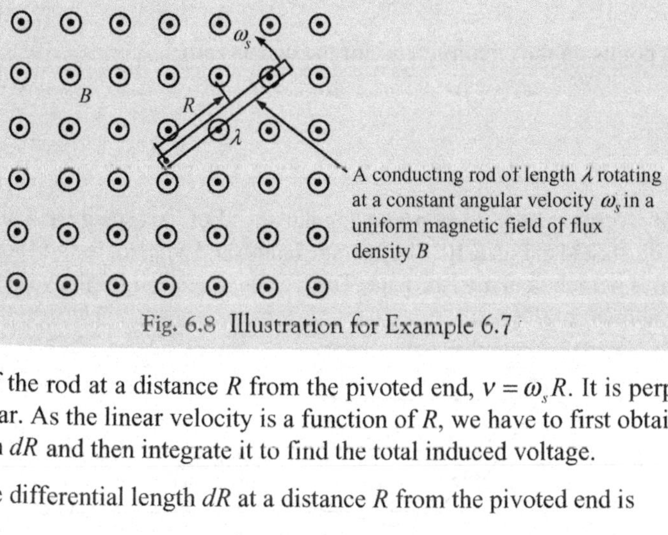

Fig. 6.8 Illustration for Example 6.7

The linear velocity of the rod at a distance R from the pivoted end, $v = \omega_s R$. It is perpendicular to both the flux density and the bar. As the linear velocity is a function of R, we have to first obtain the induced voltage in a differential length dR and then integrate it to find the total induced voltage.

Induced voltage in the differential length dR at a distance R from the pivoted end is

$$d \quad = B(\omega_s R)dR$$

Total induced voltage

$$- \omega_s B \int_0^{\ell} R\, dR = \frac{1}{2}\omega_s B\ell^2$$

If length of the rod is 68 cm and it revolves at an angular speed of 60.0 radians per second in a uniform magnetic field of flux density of 0.6 T, then

$$= \frac{1}{2}(60)(0.6)(0.68)^2 = 8.32 \text{ V}$$

EXAMPLE 6.8

A circular conducting disc of radius a is rotating at a constant angular speed ω_s in a uniform magnetic field of flux of density B. The flux density is normal to the plane of the disc. Calculate voltage induced between the centre and rim of the disc. The problem is illustrated in Fig. 6.9.

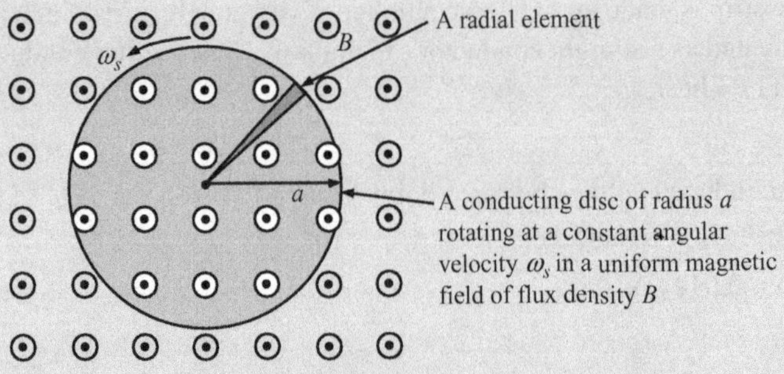

Fig. 6.9 Illustration for Example 6.8

If we imagine that the disc is made up of a large number of radial elements, one of which is shown in the figure, voltage induced between the axis and rim can be calculated in the same way as for the revolving bar. Thus, voltage induced between the centre and rim of the disc is given by

$$\mathcal{E} = \frac{1}{2}\omega_s B a^2$$

Voltage between two points on the circumference of the disc is zero.

EXAMPLE 6.9

A rectangular coil of N turns rotates at a constant angular speed ω_s in a steady and uniform magnetic field of flux density B, as illustrated in Fig. 6.10. The average length and width of the coil are a and b respectively. The plane of the coil is perpendicular to the page. Derive an expression for the voltage induced across the open terminals (not shown) of the coil.

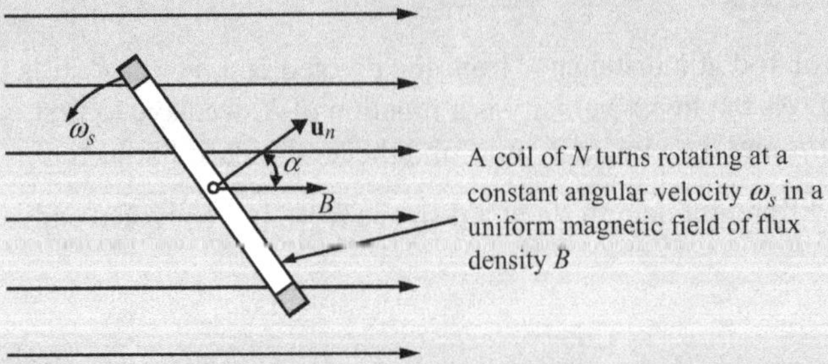

Fig. 6.10 Illustration for Example 6.9

Voltage is induced in the coil because the flux linking it changes during its rotation. At any time t, let the unit vector $\mathbf{u}_n$ normal to the plane of the coil be at an angle α with the flux density. If we assume that $\alpha = 0$ at $t = 0$, then at time t,

$\alpha \quad \omega_s t$

The component of flux density normal to plane of the coil $= \mathbf{B}\cdot\mathbf{u}_n = B\cos\alpha = B\cos\omega_s t$

Flux linking the coil at time t, $\Phi(t) = abB\cos\omega_s t = SB\cos\omega_s t$ (5)

where $S = ab$ is area of the coil.

Voltage induced in the coil,

$$(t) = -N\frac{d \quad t)}{dt} = \omega_s SNB \quad {}_s t = \quad {}_s SNB \quad {}_s t - \quad {}^\circ) \tag{6}$$

The maximum value of induced voltage, $\mathcal{E}_0 = \omega_s SNB = 2\pi f_s SNB$

where f_s is the rotational frequency.

The rms value of induced voltage, $\mathcal{E} = \mathcal{E}_0/\sqrt{2} = \sqrt{2}\pi f_s SNB$

We find that though the magnetic field is stationary and time-invariant, the flux linking the coil varies with time. Hence, the induced voltage also varies with time. It is clear from relations (5) and (6) that the induced voltage lags the *flux linking* the coil by 90° in time. It means that the induced voltage is *zero* when *maximum flux links* the coil and vice versa. The same result would be obtained if the coil is stationary and the field rotates at a constant angular velocity ω_s in the opposite direction.

This method of voltage generation is used in rotating electromagnetic machines. In practical machines the flux density is generally perpendicular to the coil sides. In alternating current generators the air-gap field distribution is nearly sinusoidal.

EXAMPLE 6.10

A rectangular coil of length 15.0 cm and width 10.0 cm has 60 turns. It rotates at a constant speed of 10.0 revolutions per second (rps) in a uniform magnetic field of 0.8 T. Calculate the maximum voltage induced in the coil and frequency of the voltage.

Area of the coil, $S = (15.0\times10^{-2})(10.0\times10^{-2}) = 15.0\times10^{-3}\ \text{m}^2$

Rotational speed of the coil, $n_s = 10.0$ rps

Angular velocity of the coil, $\omega_s = 2\pi n_s = 20\pi$ rad/s

Flux density, $B = 0.8$ T

Maximum value of induced voltage,

$$_0 = \omega_s SNB = (20\pi)(15.0\times10^{-3})(60)(0.8) = 45.23\ \text{V}$$

Frequency of the voltage, $f_s = \dfrac{\omega_s}{2\pi} = 10.0$ Hz

Note that the rotational frequency is equal to the speed of the coil in revolution per second.

6.8 Combined Transformer and Speed Voltages

In the case of a coil moving in a magnetic field that changes with time, the voltage induced in the coil has two components; (1) a speed voltage due to relative motion between the flux and the coil, and (2) a transformer voltage due to time variation of the flux. Thus, combining (6.7-4) and (6.5-2) we get the total induced voltage as

$$= \int (\times)\cdot d\ell - \frac{\partial}{\partial t}\int_S \mathbf{B}\cdot d\mathbf{S} \qquad (1)$$

where the line integral has to be evaluated along the direction of **v** **B**.

EXAMPLE 6.11

The example is illustrated in Fig. 6.11. The conducting bar of length ℓ is parallel to the x axis and it moves between two conducting guides in y direction at constant velocity v in a uniform magnetic field. The flux density is in z direction and it varies sinusoidally with time given by $B(t) = B_0 \sin \omega t$. A high resistance voltmeter is connected between the rails at $y = 0$ to complete the loop. Find the voltage induced in the coil.

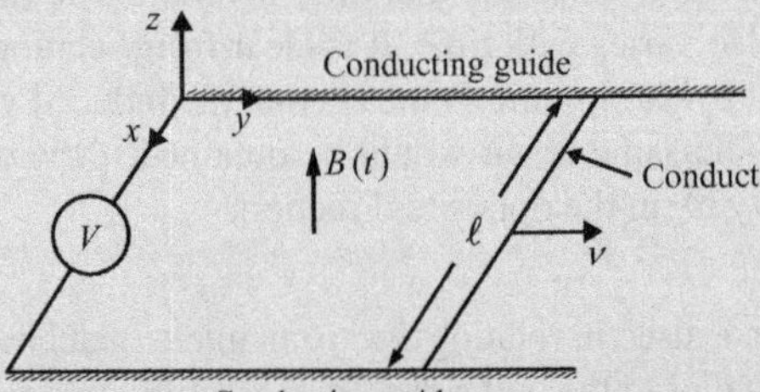

Fig. 6.11 Illustration for Example 6.11

At time $t = 0$, let the bar be at $y = 0$. After time t_1 the bar would be at $y = vt_1$ resulting in a one-turn loop on the $z = 0$ plane, as shown in the figure. The flux density therefore is normal to the plane of the loop. Speed voltage would be induced in the moving bar and transformer voltage would be induced in the loop due the time-changing flux. The total voltage induced in the loop can be obtained by using (1); or by calculating the speed and transformer voltages separately and then adding the two voltages.

At time $t = t_1$, the position of the bar, $y = vt_1$

The vector forms of various quantities are

$\mathbf{v}$ $\mathbf{u}_y$, $\mathbf{B}(t) = B(t)\mathbf{u}_z$, $d\boldsymbol{\ell} = dx\,\mathbf{u}_x$, $d\mathbf{S} = dS\,\mathbf{u}_z = (\ell dy)\,d\mathbf{u}_z$

Thus, $[\mathbf{v} \times \mathbf{B}(t)] \cdot d\boldsymbol{\ell} = (vB_0 \sin \omega t\,\mathbf{u}_x) \cdot (dx\,\mathbf{u}_x) = vB_0 \sin \omega t\, dx$,

and $\mathbf{B}$ $d\mathbf{S} = (B_0 \sin \omega t)(\ell dy)$

Voltage induced in the loop,

$$= vB_0 \sin \omega t \int_0^{\ell} dx - \ell B_0 \frac{\partial(\sin \omega t)}{\partial t} \int_0^{y} dy$$

$$= v\ell B_0 (\sin \omega t - \omega t_1 \cos \omega t)$$

$$= v\ell B_0 [\sin \omega t + \omega t_1 \sin(\omega t - 90°)]$$

The speed voltage given by the first term is in phase with the flux density whereas the transformer voltage given by the second term lags the flux density by 90° in time.

Let $\ell = $ 25 m; $v = 12.0$ m/s; $S = 0.3$ m^2; $B_0 = 0.16$ T; frequency, $f = 10.0$ Hz

The calculated values are

$B(t) = 0.$ $\sin 20\pi t$ T

$\Phi(\;)$ $(\;)(0.16 \sin 20\pi t) = 0.048 \sin 20\pi t$ Wb

Speed voltage, $\mathcal{E}_s(t) = v\ell B(t) = (12.0)(0.25)(0.16 \sin 20\pi t) = 0.48 \sin 20\pi t$ V

Transformer voltage, $\mathcal{E}_t(t) = -\dfrac{\partial \Phi(t)}{\partial t} = -(\;.048)(20\pi) \cos 20\pi t = -$ 3 02 $20\pi t$

Total voltage, $= 0$ 48 20 $3.02 \cos 20\pi t = 3$ 06 20 81)

The total induced voltage has a maximum value of 3.06 V and it lags the flux density by 81°.

EXAMPLE 6.12

The description of the example is same as that of Example 6.9. However, in this example the flux density varies harmonically with time described by $B(t) = B_0 \sin \omega t$. It is required to calculate the voltage induced in the coil.

At any time t, let the unit vector $\mathbf{u}_n$ normal to the plane of the coil be at an angle α with the flux density. If we assume that $\alpha = 0$ at $t = 0$, then at time t, $\alpha = \omega_s t$

The component of flux density normal to plane of the coil $= \mathbf{B}(t) \cdot \mathbf{u}_n = B_0 \sin \omega t \cos \omega_s t$

Flux linking the coil at time t, $\Phi(t) = abB_0 \sin \omega t \cos \omega_s t$

Voltage induced in the coil, $\mathcal{E}(t) = -N\dfrac{d\Phi(t)}{dt} = abNB_0(\omega_s \sin \omega t \sin \omega_s t - \omega \cos \omega t \cos \omega_s t)$

The first term on rhs is the speed voltage and the second term is the transformer voltage.

6.9 Power Loss due to Eddy Current in a Thin Iron Sheet

We have learnt in Section 4.20 that when a ferromagnetic material is subjected to an alternating magnetic field, power loss takes place in the material due to magnetic hysteresis. The loss is called hysteresis loss. Another form of loss, called eddy-current loss, also occurs in any conducting material when it is placed in a time-changing magnetic field. To understand the cause of this loss, let us consider an iron slab of rectangular section subjected to an alternating magnetic field produced by a coil carrying alternating current, as shown in Fig. 6.12. The flux density is normal to the cross-section of the slab. Let us imagine that the slab is made up of a large number of concentric rectangular loops. One such loop is as indicated in the figure. Flux enclosed by the loop induces a voltage in the loop, and hence induced current flows in the loop as it is short-circuited.

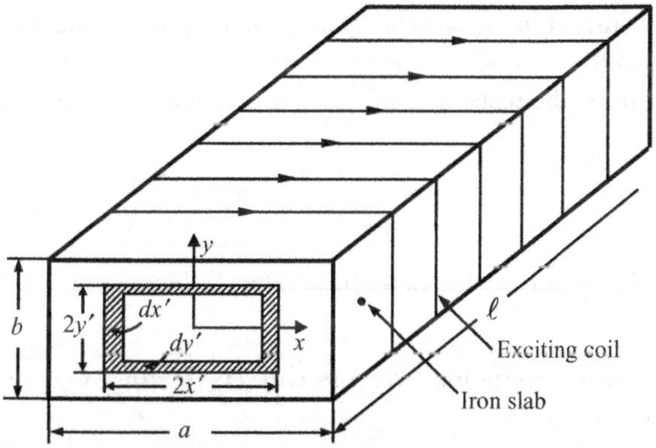

Fig. 6.12 Illustration for calculation of eddy-current loss in an iron slab

The induced current, called *eddy current*, produces its own magnetic field. The direction of this field is opposite to the applied field as per Lenz's law. Consequently, distribution of the resultant flux over the section of the sheet is not uniform.

Eddy currents cause power loss. The loss is called *eddy-current loss*. The loss is converted to heat and contributes to the temperature rise of the material and the surrounding medium. Since the flux in the iron cores of many electromagnetic devices vary with time, estimation of eddy-current loss is needed to predetermine efficiency and temperature rise. A rigorous analysis for the estimation of eddy-current loss in iron cores of rectangular section involves solution of two-dimensional second order partial differential equations satisfying the necessary boundary conditions. However, the problem can be reduced to a one-dimensional one with judicious assumptions. The problem is further simplified by assuming that the flux is uniformly distributed, i.e., by neglecting the demagnetising effect of the eddy currents. This assumption is valid only if the thickness of the slab is small compared to the width. Moreover, the periphery of an iron core may be taken as a flux line.

In order to derive formula for the eddy-current loss, with the above mentioned assumptions, let us consider an iron sheet of rectangular section of width a and thickness b. The origin of the rectangular coordinates is taken at the centre of the rectangular section of the sheet. The flux density is normal to the section of the core and it varies sinusoidally with time defined by

$$B(t) = B_0 \sin \omega t$$

where B_0 is the maximum value of flux density.

Let us consider a rectangular loop with side lengths $2x'$ and $2y'$, as shown in Fig. 6.12. The width of the loop of side length $2x'$ is dy' and that of side length $2y'$ is dx'.

Flux enclosed by the loop, $\Phi(t) = 4x'y'B_0 \sin \omega t$

Magnitude of voltage induced in the loop, $\mathcal{E}(t) = 4x'y'\omega B_0 \cos \omega t$ (1)

Current in the loop is equal to the ratio of the induced voltage and resistance of the loop. Noting that the area in the formula for the resistance of a conductor is the conductor area that is normal to the direction of current flow, the resistance of the loop for an axial length ℓ of the sheet is given by

$$dR = \frac{4}{\sigma \ell}\left(\frac{x'}{dy'} + \frac{y'}{dx'}\right) = \frac{4}{\sigma \ell}\left(\frac{x'dx' + y'dy'}{dx'dy'}\right)$$

Eddy current in the loop, $dI(t) = \mathcal{E}/dR$

Eddy-current loss, $dP_e(t) = \mathcal{E}\, dI(t) = \mathcal{E}^2/dR = \mathcal{E}^2 \dfrac{\sigma \ell}{4} \dfrac{dx'dy'}{xdx' + ydy'}$ (2)

In order to obtain the total loss we have to integrate $dP_e(t)$ over the section of the sheet. A study of the expression shows that the integration of the function may be possible but it will not be easy. To simplify the problem, let us assume

$$\frac{y'}{x'} = \frac{b}{a}$$

$\Rightarrow$
$$y' = \frac{b}{a} x' \quad (3)$$

Or
$$dy' = \frac{b}{a} dx' \quad (4)$$

By substituting (1), (3) and (4) in (2), we get

$$dP_e(t) = \frac{4\sigma b \quad B_0^2 \cos^2 \omega t}{a(a \quad b)} x\, dx$$

Total eddy-current loss, $P_e(t) = \int_0^{a/2} dP_e(t) = \dfrac{\sigma \ell a^3 b^3 \omega^2 B_0^{\,2} \cos^2 \omega t}{16(a^2 + b^2)}$

Time-average of eddy-current loss, $P_e = \dfrac{\sigma \ell a^3 b^3 \omega^2 B_0^{\,2}}{32(a^2 + b^2)}$

Replacing ω by $2\pi f$, power loss per unit volume $= \dfrac{\pi^2 \sigma a^2 b^2 f^2 B_0^{\,2}}{8(a^2 + b^2)}$

If $b/a \ll 1$, power loss per unit volume $\approx \dfrac{\pi^2 \sigma b^2 f^2 B_0^{\,2}}{8}$

Thus, the eddy-current loss per unit volume in a thin iron sheet is proportional to square of the thickness of the sheet, square of the frequency, and square of the maximum flux density. It is also proportional to conductivity of the sheet. The cores of transformers and other electromagnetic devices carrying time-varying flux are therefore constructed with thin (0.35 to 0.45 mm) silicon (less than 5 % by weight) steel sheets to reduce eddy-current loss. At high frequency ferrites are used as cores because of their low conductivity.

EXAMPLE 6.13

The leg of a transformer is made up of 200 silicon steel sheets. The thickness b, width a, and length ℓ of each sheet are 0.5 mm, 120 mm, and 600 mm respectively. The conductivity of silicon steel $\sigma = 2.4 \times 10^6$ S/m. The maximum value of flux density $B_0 = 1.4$ T and its frequency $f = 50$ Hz. Find the eddy-current loss in the leg.

Eddy-current loss per unit volume of a sheet $= \dfrac{\pi^2 \sigma b^2 f^2 B_0^2}{8} \dfrac{\pi^2 (2.4\times 10^6)(0.0005)^2 (50)^2 (1.4)^2}{8} = 3627.1\ \text{W/m}^3$

Eddy-current loss in the leg = (200) (0.0005) (0.12) (0.6) (3627.1) = 26.2 W

6.10 The Convection Current

We have derived equations for current density and current due to flow of charges in conductors and semiconductors in Chapter 3. Current in a material medium is called conduction current. The current in ampere through an area is equal to the amount of charge in coulombs crossing the area in one second. The same definition can be applied to calculate current due to moving charges in free space.

Let us imagine that a block of charge is moving with a velocity $\mathbf{v}$ in free space. At any point in the block, let the volume charge density be ρ. Also, let us assume that ΔS is an infinitesimal surface area at the point, and its direction is same as that of the velocity at the point. Then, charge crossing the area in time Δt is given by

$$\Delta Q \quad (\rho \mathrm{v} \Delta S)\Delta t$$

Magnitude of current through the surface element, $\Delta I = \dfrac{\Delta Q}{\Delta t} = \rho \mathrm{v} \Delta S$

Current density at the point, $J = \lim_{\Delta S \to 0} \dfrac{\Delta I}{\Delta S} = \rho v$

As the direction of current density is same as that of the velocity of the moving charge, the vector form of the current density is

$$\mathbf{J} = \rho \mathbf{v}$$

Current I associated with the moving charge is obtained by integrating the normal component of current density over a cross-sectional surface of the block. Thus,

$$I = \int_S \mathbf{J} \cdot d\mathbf{S} = \int_S \rho(\mathbf{v} \cdot d\mathbf{S})$$

The current is called convection current.

> The convection current density is equal to the product of the volume charge density and the velocity.

6.11 The Displacement Current

The differential form of Ampere's circuital law for static magnetic field, derived in Chapter 4, is

$$\nabla \times \mathbf{H} = \mathbf{J} \tag{1}$$

where **J** is the time-invariant (steady) current density. By taking divergence of both sides of (1), we get

$$\nabla \cdot (\nabla \times \mathbf{H}) = \nabla \cdot \mathbf{J} \tag{2}$$

As the divergence of curl of a vector function is always zero, the lhs of (2) is equal to zero. Hence, the rhs must also be zero. That is,

$$\nabla \cdot \mathbf{J} = 0 \tag{3}$$

Though equation (3) satisfies the continuity condition for steady current, it does not agree with the continuity equation given by (6.2-8). So, equation (1) was modified by Maxwell (James Clerk Maxwell) by adding the term $\partial \quad \partial t$ to the rhs of the equation. The modified equation is

$$\nabla \times \mathbf{H} = \mathbf{J} + \frac{\partial \mathbf{D}}{\partial t} \tag{4}$$

In order to check that (4) satisfies the continuity equation, let us take the divergence of both sides of (4) and obtain,

$$\nabla \cdot \nabla \times \mathbf{H} = \nabla \cdot \mathbf{J} + \nabla \cdot \frac{\partial \mathbf{D}}{\partial t} \tag{5}$$

As the lhs of (5) is zero, the rhs must also be zero. Accordingly,

$$\nabla \cdot \mathbf{J} + \nabla \cdot \frac{\partial \mathbf{D}}{\partial t} = 0$$

Or

$$\nabla \cdot \mathbf{J} = -\nabla \cdot \frac{\partial \mathbf{D}}{\partial t} = -\frac{\partial}{\partial t}(\nabla \cdot \mathbf{D}) \tag{6}$$

As per the point form of Gauss's law, derived for static electric fields in Section 2, we have

$$\nabla \cdot \mathbf{D} = \rho \tag{7}$$

This is the relation between flux density D and volume charge density ρ in static electric field. In the case of time-varying fields, the quantities represent instantaneous values. Moreover, the divergence of $\mathbf{D}$ is also equal to the electric flux Ψ out of a closed surface as in the case of static electric field. For time-varying fields $\mathbf{D}$ and Ψ are instantaneous values.

By substituting (7) into (6), we get

$$\nabla \cdot \mathbf{J} = -\frac{\partial \rho}{\partial t}$$

This, we know, is the equation of continuity. Thus, the equation of continuity is satisfied because of the second term on the rhs of (4). Equation (4) may be considered as the general form of Ampere's circuital law that is valid both for static and time-varying fields. For static fields the second term on the right side is zero.

The first term on the rhs of (4) is the conduction (or convection) current density. Hence, the second term should also be a current density. This current density is equal to $\partial \mathbf{D}/\partial t$ and it is a vector quantity. It is called *displacement current density*.

Denoting the displacement current density by the symbol $\mathbf{J}_d$, we have

$$\text{Displacement current density, } \mathbf{J}_d = \frac{\partial \mathbf{D}}{\partial t} \tag{8}$$

Displacement current I_d crossing a given surface is obtained by integrating the normal component of $\mathbf{J}_d$ over the surface. Thus,

$$I_d = \int_S \mathbf{J}_d \cdot d\mathbf{S}$$

By substituting (8) in (4) we get the following relation:

$$\nabla \times \mathbf{H} = \mathbf{J} + \mathbf{J}_d \tag{9}$$

Now by integrating both sides of (9) over a surface that is enclosed by a closed path, we get

$$\int_S (\nabla \times \mathbf{H}) \cdot d\mathbf{S} = \int_S (\mathbf{J} + \mathbf{J}_d) \cdot d\mathbf{S} \tag{10}$$

As per Stokes' theorem, the lhs of (10) is equal to the line integral of $\mathbf{H}$ around the closed path. The rhs is equal to the sum of conduction and displacement currents. Hence, we have

$$\oint \mathbf{H} \cdot d\ell = I + I_d \tag{11}$$

Equation (11) may be considered as the integral form of Ampere's magnetic circuital law for time-varying fields.

EXAMPLE 6.14

Calculate the ratio of conduction current density to displacement current density in (a) a good conductor for which the conductivity is 3.8×10^7 S/m and relative permittivity is 1.0 and (b) a good dielectric for which the conductivity is 1.2×10^{-9} S/m and relative permittivity is 5.0. The frequency of field intensity that varies harmonically with time is 10^9 Hz.

Let the field intensity in the material, $E(t) = E \quad t$

Conduction current density in the material,

$J(t) = \sigma E(t) = \sigma E \sin \omega t = J \sin \omega t \; ; \; J = \sigma E$

Displacement current density in the material, $J_d(t) = \dfrac{\partial D(t)}{\partial t}$

Assuming the relative permittivity to be constant,

$J_d(t) = \varepsilon \dfrac{\partial E(t)}{\partial t} = \omega \varepsilon E \sin \omega t = J_d \sin \omega t \; ; \; J_d = \omega \varepsilon E$

The ratio between the conduction current density and the displacement current density,

$$\frac{J}{J_d} = \frac{\sigma}{\omega \varepsilon} \tag{12}$$

(a) $\sigma = 3.8 \times 10^7$ S/m; $\omega = 2\pi \times 10^9$ rad/s; $\varepsilon = 8.84 \times 10^{-12}$ F/m.

Substituting these values in (12), $\dfrac{J}{J_d} = \dfrac{3.8 \times 10^7}{(2\pi \times 10^9)(8.84 \times 10^{-12})} = 68.42 \times 10^7$

(b) $\sigma = 1.2 \times 10^{-9}$ S/m; $\omega = 2\pi \times 10^9$ rad/s; $\varepsilon = (5.0)(8.84 \times 10^{-12}) = 44.2 \times 10^{-12}$ F/m

Substituting these values in (12), $\dfrac{J}{J_d} = \dfrac{1.2 \times 10^{-9}}{(2\pi \times 10^9)(44.2 \times 10^{-12})} = 4.32 \times 10^{-9}$

It may be noted that the displacement current density in a good conductor is very small compared to the conduction current density; whereas in a good dielectric the conduction current density is very small compared to the displacement current density.

6.12 Continuity of Current in a Capacitor

We know that when a capacitor is connected to a dc source no current flows except during a brief charging period. However, if a capacitor is connected to an ac source, as shown in Fig. 6.13, the ammeter shows a reading indicating that current flows continuously in the circuit even though the capacitor plates are separated by a dielectric material. It can be shown that continuity of current on two sides of the capacitor is maintained by the displacement current between the plates.

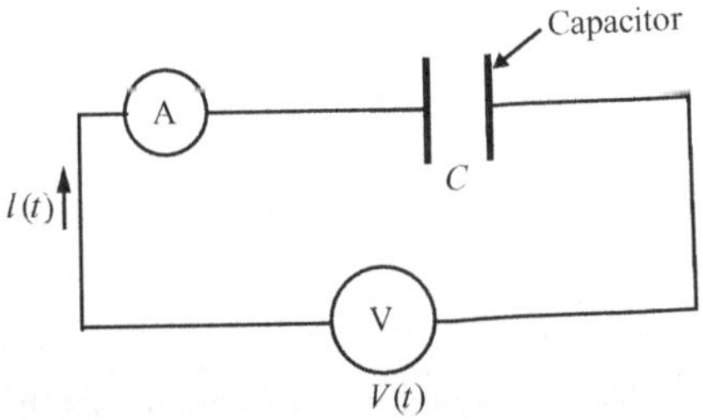

Fig. 6.13 A capacitor connected to an ac source

Let voltage across the capacitor, $V(t) = V_0 \sin \omega t$

Magnitude of charge on each plate, $Q(t) = CV(t) = CV_0 \sin \omega t$,

where C is capacitance of the capacitor.

Conduction current in the circuit, $I(t) = \dfrac{dQ(t)}{dt} = \omega C V_0 \cos \omega t$ (1)

Let us assume that the capacitor shown in the figure is a lossless parallel plate capacitor. Spacing between the plates, area of each plate and permittivity of the medium between the plates are taken as s, A and ε respectively. Assuming A to be large compared to s, electric field intensity between the plates is

$$E(t) = \frac{V(t)}{s} = \frac{V_0 \sin \omega t}{s}$$

Electric flux density, $D(t) = \varepsilon E(t) = \dfrac{\varepsilon V_0 \sin \omega t}{s}$

Displacement current density, $J_d(t) = \dfrac{\partial D(t)}{\partial t} = \dfrac{\omega \varepsilon V_0 \cos \omega t}{s}$

The displacement current between the plates is equal to the product of surface area of a plate and the displacement current density. Thus, displacement current,

$$I_d(t) = A\frac{\partial D(t)}{\partial t} = \omega\left(\frac{\varepsilon A}{s}\right)V_0 \cos \omega t$$

As $\varepsilon A/s$ is equal to the capacitance of the capacitor,

$$I_d(t) = \omega C V_0 \cos \omega t \tag{2}$$

We find from (1) and (2) that the displacement current is equal to the conduction current. This shows that current flow through the capacitor is maintained by the displacement current.

6.13 Maxwell's Equations in Differential Forms

A set of two curl equations and two divergence equations constitute Maxwell's equations in differential form. These are

$$\nabla \times \mathbf{E} = -\frac{\partial \mathbf{B}}{\partial t} \tag{1}$$

$$\nabla \times \mathbf{H} = \mathbf{J} + \frac{\partial \mathbf{D}}{\partial t} \tag{2}$$

$$\nabla \cdot \mathbf{B} = 0 \tag{3}$$

$$\nabla \cdot \mathbf{D} = \rho \tag{4}$$

Equation (1) is the differential form Faraday's law and equation (2) is the differential form of Ampere's circuital law. The divergence equations are same as those for static fields. The divergence of **B** and **D** for time-changing fields may be interpreted as the *instantaneous* values of magnetic and electric flux respectively per unit volume. These four equations and some additional equations, derived in earlier chapters and reproduced below, are the fundamental equations of electromagnetics. The additional equations are

$$\mathbf{B} = \mu \mathbf{H} = \mu_0 \mu_r \mathbf{H}$$

$$\mathbf{D} = \varepsilon \mathbf{E} = \varepsilon_0 \varepsilon_r \mathbf{E}$$

$\mathbf{J} = \sigma \mathbf{E}$ (For conduction current)

$\mathbf{J} = \rho \mathbf{v}$ (For convection current)

$\nabla \cdot \mathbf{J} = -\dfrac{\partial \rho}{\partial t}$ (Continuity equation)

$\mathbf{F}$ [illegible] $\mathbf{E} + \mathbf{V} \times \mathbf{B})$ (Lorentz force equation)

Equations (1) and (2) clearly show that a time-varying magnetic field is supported by a time-varying electric field and vice versa. That is, time-changing electric and magnetic fields are interdependent. Maxwell's equations in free space are obtained by substituting $\mathbf{J} = 0$ in (2) and $\rho = 0$ in (4). Maxwell's equations for static electric and magnetic field are found by replacing the terms $\partial\mathbf{B}/\partial t$ in (1) and $\partial\mathbf{D}/\partial t$ in (2) by zero. Thus, Maxwell's equations for static electric and magnetic field are given by

$$\nabla \times \mathbf{E} = 0$$

$$\nabla \times \mathbf{H} = \mathbf{J}$$

$$\nabla \cdot \mathbf{B} = 0$$

$$\nabla \cdot \mathbf{D} = \rho$$

In the case of static fields, the electric field is independent of the magnetic field and vice-versa.

It can be shown that equations (3) and (4) are not independent as these can be derived from (1) and (2) respectively as follows. By taking divergence of both sides of (1), we get

$$\nabla \cdot (\nabla \times \mathbf{E}) = -\frac{\partial}{\partial t}(\nabla \cdot \mathbf{B})$$

The lhs is zero because the divergence of curl of a vector is always equal to zero. Hence, the rhs should also be zero. If the rate of change of **B** with time is finite, then

$$\nabla \cdot \mathbf{B} = 0$$

Next, taking the divergence of both sides of (2), we get

$$\nabla \cdot (\nabla \times \mathbf{H}) = \nabla \cdot \mathbf{J} + \frac{\partial}{\partial t}(\nabla \cdot \mathbf{D})$$

As the lhs is equal to zero, the rhs must also be zero. Thus,

$$\nabla \cdot \mathbf{J} + \frac{\partial}{\partial t}(\nabla \cdot \mathbf{D}) = 0 \tag{5}$$

By replacing $\nabla \cdot \mathbf{J}$ of (5) by the rhs of the continuity equation given by (6.2-8), we get

$$-\frac{\partial \rho}{\partial t} + \frac{\partial}{\partial t}(\nabla \cdot \mathbf{D}) = \frac{\partial}{\partial t}(\nabla \cdot \mathbf{D} - \rho) = 0$$

Or
$$\nabla \cdot \mathbf{D} = \rho$$

Though the divergence equations are not independent, these are equally important.

EXAMPLE 6.15

The electric field intensity in a perfect dielectric medium is given by

$$\mathbf{E} = E_{x0} e^{j(\omega t - \beta z)} \mathbf{u}_x \text{ V/m} \tag{6}$$

(a) Use this electric field intensity in Maxwell's equation (1) and find the magnetic field intensity.
(b) Substitute the magnetic field intensity found in (a) into Maxwell's equation (2), and then calculate the electric field intensity. What should be the value of β so that this electric field intensity is same as that given by (6)?

(a) From equation (1), we have

$$\mu \frac{\partial \mathbf{H}}{\partial t} = -\nabla \times \mathbf{E}$$

As **E** given by (6) is a function of z only, and it has only one component in x direction,

$$\mu\frac{\partial \mathbf{H}}{\partial t} = -\nabla\times\mathbf{E} = -\frac{\partial E_x}{\partial z}\mathbf{u}_y = j\beta E_{x0}e^{j(\omega t-\beta z)}\mathbf{u}_y$$

Or $$\mathbf{H} = \int j\frac{\beta}{\mu}E_{x0}\,e^{j(\omega t-\beta z)}\,dt\,\mathbf{u}_y = \frac{\beta}{\omega\mu}E_{x0}\,e^{j(\omega t-\beta z)}\mathbf{u}_y \tag{7}$$

The constant of integration is not considered as it is not useful in time-varying fields.

(b) From Maxwell's equation given by (2), we have

$$\frac{\partial \mathbf{D}}{\partial t} = \nabla\times\mathbf{H}$$

As the magnetic field intensity of (7) has only one component in y direction, and it is a function of z only,

$$\varepsilon\frac{\partial \mathbf{E}}{\partial t} = \nabla\times\mathbf{H} = -\frac{\partial H_y}{\partial z} = \frac{j\beta^2}{\omega\mu}E_{x0}\,e^{j(\omega t-\beta z)}\mathbf{u}_y$$

Or $$\mathbf{E} = \int \frac{j\beta^2}{\omega\mu\varepsilon}E_{x0}\,e^{j(\omega t-\beta z)}\,dt\,\mathbf{u}_y = \frac{\beta^2}{\omega^2\mu\varepsilon}E_{x0}\,e^{j(\omega t-\beta z)}\,\mathbf{u}_y \tag{8}$$

For $\beta = \omega\sqrt{\mu\varepsilon}$ the field intensity given by (8) would be same as that given by (6).

6.14 Maxwell's Equations in Integral Form

The integral forms of Maxwell's equations can be obtained from those of the differential forms given in the preceding section. By integrating both sides of (6.13-1) over an open surface, we will obtain

$$\int_S(\nabla\times\mathbf{E})\cdot d\mathbf{S} = -\int_S\left(\frac{\partial \mathbf{B}}{\partial t}\right)\cdot d\mathbf{S} \tag{1}$$

As per Stokes' theorem, the lhs is equal to the line integral of **E** around the closed path that encloses the surface. Thus, shifting the partial derivative with time to outside of the integral sign, we have

$$\oint \mathbf{E}\cdot d\ell = -\frac{\partial}{\partial t}\int_S \mathbf{B}\cdot d\mathbf{S} \tag{2}$$

Next, integrating both sides of (6.13-2) over an open surface and then applying Stokes' theorem to the surface integral of $\nabla\times\mathbf{E}$, we get

$$\oint \mathbf{H}\cdot d\ell = \int_S \mathbf{J}\cdot d\mathbf{S} + \frac{\partial}{\partial t}\int_S \mathbf{D}\cdot d\mathbf{S} \tag{3}$$

Now, by integrating both sides of (6.13-3) and (6.13-4) over a given volume and then applying the divergence theorem to the volume integrals of $\nabla\cdot\mathbf{B}$ and $\nabla\cdot\mathbf{D}$ we will obtain,

$$\oint_S \mathbf{B}\cdot d\mathbf{S} = 0 \tag{4}$$

$$\oint \mathbf{D}\cdot \quad = \int_v \,, \; dv \tag{5}$$

Equations (2) to (5) are the integral forms of Maxwell's equations. It may be mentioned that these equations have been obtained earlier. Equation (2) is Faraday's law of electromagnetic induction; equation (3) is the general form of Ampere's magnetic circuital law; equation (4) shows that magnetic flux entering a closed surface is equal to the flux coming out of the surface at any instant – a known property of the magnetic field, and equation (5) is the general form of Gauss's law where **D** and ρ are instantaneous values.

6.15 Maxwell's Equations for Sinusoidal Variation of Fields with Time

The differential and integral forms of Maxwell's equations given in the last two sections are for any type of variation of the fields with time. In this section we will obtain Maxwell's equations for sinusoidal (cosinusoidal) time variation of the field quantities. We will use the complex exponential function $e^{j\omega t}$ to represent variation of the fields with time.

> This form of representation of sinusoidal functions, introduced in Section 1.12, has some advantages. One advantage is that differentiation or integration of a function with time does not change the form of the function. Differentiating a function with time is equivalent to multiplying the function by $j\omega$, and integrating a function with time is equivalent to dividing the function by $j\omega$. No change of sign is involved in both the cases. Another advantage is that as $e^{j\omega t}$ appears on both sides of an equation it gets cancelled out.

Hence writing an equation becomes simpler. At the end of an analysis, either the real part or the imaginary part of a phasor solution can be used to obtain the solution in real time.

The field quantities will be denoted using the same symbols as for static fields. However, the meaning of the symbols here would be a little different. The electric and magnetic field intensities are vectors as well as phasors. Phasor quantities will be denoted by placing a bar over the main symbols. For example, the electric field intensity is denoted as

$$\bar{\mathbf{E}}(x, y, z, t) = \bar{\mathbf{E}}(x, y, z)e^{j\omega t} = \bar{\mathbf{E}}e^{j\omega t},$$

where **E** is a function of the space variables only. It is a phasor quantity and represents the amplitude. Similarly, a scalar function like charge density, which is a function of space variables and varies sinusoidally with time, is denoted by $\bar{\rho}$.

Maxwell's equations for time-harmonic case can be derived from differential forms of Maxwell's equations in Section 6.13. These are given by

$$(\nabla \times \bar{\mathbf{E}})e^{j\omega t} = -j\omega \bar{\mathbf{B}} e^{j\omega t}$$

$$(\nabla \times \bar{\mathbf{H}})e^{j\omega t} = (\bar{\mathbf{J}} + j\omega \bar{\mathbf{D}})e^{j\omega t}$$

$$(\nabla \cdot \bar{\mathbf{B}})e^{j\omega t} = 0$$

$$(\nabla \cdot \bar{\mathbf{D}})e^{j\omega t} = \bar{\rho}\, e^{j\omega t}$$

As the exponential factor on two sides of the equations cancel out, we get

$$\nabla \times \bar{\mathbf{E}} = -j\omega \bar{\mathbf{B}} \tag{1}$$

$$\nabla \times \bar{\mathbf{H}} = \;\bar{}\; + \quad \bar{\mathbf{D}}) \tag{2}$$

$$\nabla \cdot \mathbf{B} = 0 \tag{3}$$

$$\nabla \cdot \mathbf{D} = \rho \tag{4}$$

Equations (1) to (4) are Maxwell's equations for time-harmonic case when the complex exponential $e^{j\ \ t}$ is used to represent time-variation of field quantities.

6.16 Boundary Conditions

The boundary relations for time-varying electric and magnetic fields have the same meaning and significance as in the case of static fields. The relations for time-varying fields are derived using the integral forms of Maxwell's equations.

Relationship between tangential components of electric field intensities on two sides of a boundary surface is obtained from the equation,

$$\oint \mathbf{E}\cdot d\ell = -\frac{\partial}{\partial t}\int_S \mathbf{B}\cdot d\mathbf{S} \tag{1}$$

Let us apply relation (1) to an infinitely small rectangular region *abcda* enclosing the boundary surface of two regions 1 and 2, as shown in Fig. 6.14(*a*). The side lengths of the rectangle are $\Delta\ell$ and Δh. In the limit Δh tending to zero, line integrals from *b* to *c* and from *d* to *a* may be assumed to be zero. Hence,

$$\lim_{\Delta h\to 0}\oint \mathbf{E}\cdot d\ell = (E_{t1} - E_{t2})\Delta\ell \tag{2}$$

The rhs of (1) gives the rate of change of magnetic flux within the rectangle at any instant. As in the limit Δh approaching to zero the area of the rectangle also tends to zero, the rhs of (1) may be assumed to be zero provided that the rate of change of flux density with time is finite. Therefore,

$$\lim_{\Delta h\to 0}\frac{\partial}{\partial t}\int_S \mathbf{B}\cdot d\mathbf{S} = 0 \tag{3}$$

Substituting (2) and (3) in (1), we get

$$(E_{t1} - E_{t2})\Delta\ell = 0$$

Or

$$E_{t2} = E_{t1}. \tag{4}$$

The relation between normal components of electric flux densities on two sides of the boundary surface can be found applying the equation,

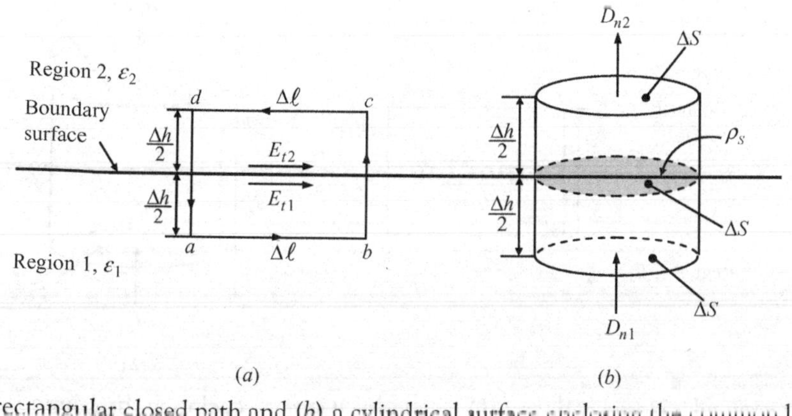

Fig. 6.14 (*a*) A rectangular closed path and (*b*) a cylindrical surface enclosing the common boundary surface of two regions of permittivity ε_1 and ε_2.

$$\oint_S \mathbf{D}\cdot d\mathbf{S} = \int_v \rho\, dv \tag{5}$$

to a region of volume in the form of an infinitely small cylinder, as illustrated at Fig. 6.14(*b*). The left side of (5) gives the electric flux out of the closed surface. In the limit Δh tending to zero, flux through the curved surface of the cylinder may be taken as zero. Hence, we have

$$\lim_{\Delta h\to 0} \oint_S \mathbf{D}\cdot d\mathbf{S} = (D_{n2} - D_{n1})\Delta S \tag{6}$$

where D_{n1} and D_{n2} are normal components of electric flux densities on two sides of the interface.

The right side of (5) gives the electric charge enclosed by the cylinder. If the boundary surface has a surface charge of density of ρ_S C/m^2, then in the limit Δh approaching zero the charge enclosed is equal to the product of ρ_s and ΔS. Thus,

$$\lim_{\Delta h\to 0} \int_v \rho\, dv = \rho_s \Delta S \tag{7}$$

Equating (6) with (7), we get

$$D_{n2} - D_{n1} = \rho_s \tag{8}$$

If the interface has no surface charge, then $\rho_S = 0$, hence

$$D_{n2} = D_{n1} \tag{9}$$

Relationship between tangential components of magnetic field intensities on two sides of a boundary surface is derived by applying the following equation to an infinitely small rectangular region *abcda* enclosing the boundary surface of two regions 1 and 2, as shown in Fig. 6.15(*a*).

$$\oint \mathbf{H}\cdot d\ell = \int_S \mathbf{J}\cdot d\mathbf{S} + \frac{\partial}{\partial t}\int_S \mathbf{D}\cdot d\mathbf{S} \tag{10}$$

In the limit Δh tending to zero, the contribution to the line integral of **H** from b to c and from d to a would be negligibly small and hence can be neglected. Thus, we have

$$\lim_{\Delta h\to 0} \oint \mathbf{H}\cdot d\ell = (H_{t1} - H_{t2})\Delta\ell \tag{11}$$

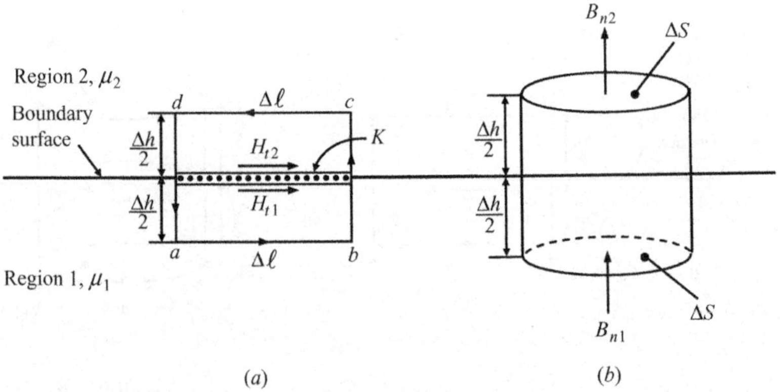

Fig. 6.15 (*a*) A rectangular closed path and (*b*) a cylindrical surface enclosing the common boundary surface of two regions of permeability μ_1 and μ_2.

This is equal to the surface integral of **J** plus the surface integral of $\partial\mathbf{D}/\partial t$ over the area enclosed by the rectangle *abcda*. If the boundary contains a sheet of conduction current of K ampere per metre, then in the limit Δh tending to zero, we have

$$\lim_{\Delta h\to 0}\int_S \mathbf{J}\cdot d\mathbf{S} = K\Delta\ell \tag{12}$$

Moreover in the limiting case, the area of the rectangle approaches zero. So, the second surface integral on the rhs of (10) may be taken as zero provided that the rate of change of **D** with time is finite. Accordingly,

$$\lim_{\Delta h\to 0}\frac{\partial}{\partial t}\int_S \mathbf{D}\cdot d\mathbf{S} = 0 \tag{13}$$

By substituting (11), (12) and (13) in (10), we get the following relation:

$$H_{t1} - H_{t2} = K \tag{14}$$

If the boundary has no conduction current, then $K = 0$. Hence,

$$H_{t1} - H_{t2} = 0$$

$$\Rightarrow \qquad H_{t2} = H_{t1} \tag{15}$$

For deriving the relationship between the normal components of magnetic flux densities on two sides of a boundary surface, we apply the equation,

$$\oint_S \mathbf{B}\cdot d\mathbf{S} = 0 \tag{16}$$

to an infinitesimal volume in the form of a cylinder enclosing the interface, as illustrated in Fig. 6.15(*b*). As per equation (16), the net magnetic flux out of any closed surface is equal to zero. When applied to the cylindrical region in the limiting case of Δh tending to zero, the magnetic flux out of the curved surface can be neglected. Thus, we have

$$\lim_{\Delta h\to 0}\oint_S \mathbf{B}\cdot d\mathbf{S} = (B_{n1} - B_{n2})\Delta S = 0$$

Or

$$B_{n2} = B_{n1} \tag{17}$$

Equations (4), (8), (9), (14), (15) and (17) are the boundary conditions. It may be noted that these relations are same as those for static fields.

6.17 Potentials for Time-Varying Fields: The Retarded Potential

We found in Chapter 2 that it is advantageous, especially in boundary-value problems, to obtain the electric field intensity **E** from the scalar potential function V. The relation between **E** and V for static fields is given by

$$\mathbf{E} = -\nabla V \tag{1}$$

The potential function can be calculated either from the differential equation,

$$\nabla^2 V = -\frac{\rho}{\varepsilon} \tag{2}$$

or from the integral equation,

$$V = \int_v \frac{\rho\, dv}{4\pi\varepsilon R} \tag{3}$$

Moreover, we obtained in Chapter 4 the relationship between the magnetic flux density **B** and the vector magnetic potential **A** due to steady electric current. The relation is

$$\mathbf{B} = \nabla \times \mathbf{A} \tag{4}$$

The potential function can be found either from the differential equation,

$$\nabla^2 \mathbf{A} = -\mu \mathbf{J} \tag{5}$$

or from the integral equation,

$$\mathbf{A} = \int_v \frac{\mu \mathbf{J}\, dv}{4\pi R} \tag{6}$$

In the same way, potential functions for time-varying fields are useful for calculating electric and magnetic field intensities. Equations relating the field intensities with the potential functions are derived in this section using Maxwell's equations. The differential forms of Maxwell's equations are reproduced below replacing **D** by $\varepsilon\mathbf{E}$ and **H** by $\mathbf{B}/\mu$ in equation (2) of Section 6.13.

$$\nabla \times \mathbf{E} = -\frac{\partial \mathbf{B}}{\partial t} \tag{7}$$

$$\nabla \times \mathbf{B} = \mu \mathbf{J} + \mu\varepsilon \frac{\partial \mathbf{E}}{\partial t} \tag{8}$$

$$\nabla \cdot \mathbf{B} = 0 \tag{9}$$

$$\nabla \cdot \mathbf{E} = \rho/\varepsilon \tag{10}$$

We know that the divergence of curl of a vector is always equal to zero. Hence, the magnetic flux density **B** in (9) can be expressed as the curl of a vector function. This vector function is the vector magnetic potential **A,** as in the case of static magnetic field due to steady current. Accordingly,

$$\mathbf{B} = \nabla \times \mathbf{A} \tag{11}$$

This relation is identical to that obtained for static magnetic field.

Next, by substituting the rhs of (11) in place of **B** of equation (7), we get

$$\nabla \times \mathbf{E} = -\frac{\partial}{\partial t}(\nabla \times \mathbf{A})$$

Or

$$\nabla \times \left(\mathbf{E} + \frac{\partial \mathbf{A}}{\partial t} \right) = 0$$

As the curl of gradient of a scalar function is zero, the sum of the two terms inside the bracket can be expressed as the gradient of a scalar function. The obvious choice of a scalar function is the scalar potential V due to time-varying charges. Thus, using a minus sign before the gradient operation, we have

$$\mathbf{E} + \frac{\partial \mathbf{A}}{\partial t} = -\nabla V$$

Or

$$\mathbf{E} = -\nabla V - \frac{\partial \mathbf{A}}{\partial t} \tag{12}$$

It may be noted that **E** is a function of both V and **A** for time-varying fields. The second term on the rhs of (12) is zero if the fields do not vary with time. Hence, formula (12) is in agreement with the relation between **E** and V for static electric fields given by (1).

We will now use the other two Maxwell's equations to derive differential equations for V and **A**. Substitution of the rhs of (12) in place of **E** of (10) yields,

$$-\nabla \cdot \left(\nabla V + \frac{\partial \mathbf{A}}{\partial t} \right) = \frac{\rho}{\varepsilon}$$

Or

$$\nabla^2 V + \frac{\partial}{\partial t}(\nabla \cdot \mathbf{A}) = -\frac{\rho}{\varepsilon} \tag{13}$$

Next, replacing **B** and **E** of (8) by the right-hand sides of (11) and (12) respectively, we get

$$\nabla \times \nabla \times \mathbf{A} = \mu \mathbf{J} - \mu\varepsilon \frac{\partial}{\partial t}(\nabla V) - \mu\varepsilon \frac{\partial^2 \mathbf{A}}{\partial t^2}$$

Or

$$\nabla(\nabla \cdot \mathbf{A}) - \nabla^2 \mathbf{A} = \mu \mathbf{J} - \nabla\left(\mu\varepsilon \frac{\partial V}{\partial t} \right) - \mu\varepsilon \frac{\partial^2 \mathbf{A}}{\partial t^2}, \tag{14}$$

as $\nabla \times \nabla \times \mathbf{A} = \nabla(\nabla \cdot \mathbf{A}) - \nabla^2 \mathbf{A}$ at least in rectangular coordinates.

We see that both (13) and (14) contain the scalar potential V as well as the vector potential **A**. But our aim is to derive the differential equation for V independent of **A** and vice versa.

This is accomplished by defining,

$$\nabla \cdot \mathbf{A} = -\mu\varepsilon \frac{\partial V}{\partial t} \tag{15}$$

Relation (15) is the *Lorentz condition*. Since the vector potential **A** is described by its curl as well as by its divergence, it satisfies the *Helmholtz theorem* which states that a vector function is completely specified by its curl and divergence.

The desired differential equations are obtained by substituting (15) into (13) and (14). The equations are given by

$$\nabla^2 V - \mu\varepsilon \frac{\partial^2 V}{\partial t^2} = -\frac{\rho}{\varepsilon} \tag{16}$$

$$\nabla^2 \mathbf{A} - \mu\varepsilon \frac{\partial^2 \mathbf{A}}{\partial t^2} = -\mu \mathbf{J} \tag{17}$$

In charge-free and current-free regions, as in a perfect dielectric or in free space, the right-hand sides of (16) and (17) are zero. Hence, we have

$$\nabla^2 V = \mu\varepsilon \frac{\partial^2 V}{\partial t^2} \tag{18}$$

$$\nabla^2 \mathbf{A} = \mu\varepsilon \frac{\partial^2 \mathbf{A}}{\partial t^2} \tag{19}$$

Equations (18) and (19) are called *wave equations*. It is shown in Chapter 7 that the solutions of the wave equations are travelling waves and the velocity of the waves, $v = 1/\sqrt{\mu\varepsilon}$.

Solution of the differential equations is one method of finding the potential functions. Another method is by the use of the integral equations similar to (3) and (6). In order to obtain the integral equations for time-varying field, let us consider that a source element, either ρdv or $\mathbf{J} dv$, is located at P'. Point P at which the potential is to be found is at a distance R from P', as shown in Fig. 6.16. If the charge or the current magnitude at P' changes with time, then the potential at P will also change with time. But, the change in the value of potential will occur after a finite time. The delay is caused because time-changing electromagnetic fields travel at a finite velocity, as

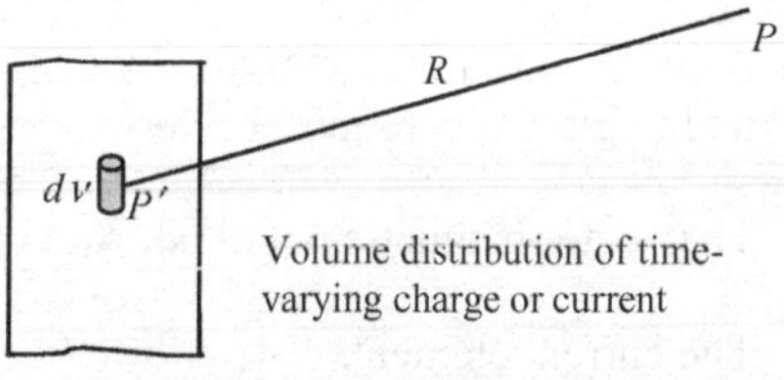

Fig. 6.16 Illustration for calculation of retarded potential

mentioned earlier. Let us assume that v is velocity of propagation of the field. Then, any change in the value of a source element at P' with time will reach P after R/v seconds. Thus, to calculate the potential at P at time t we have to use the value of charge density or current density at time, $t' = t - R/v$. Therefore, formulae (3) and (6), which are for static fields, have to be suitably modified for the time-varying case. The modified formulae are

$$V(t) = \int_v \frac{\rho(t')}{4\pi\varepsilon R} dv \tag{20}$$

$$\mathbf{A}(t) = \int_v \frac{\mu \mathbf{J}(t')}{4\pi R} dv \tag{21}$$

The interpretation of equations (20) and (21) is that in order to calculate the potential at P at a given time t, we have to use the value of the charge or the current element at time, $t' = t - R/v$.

The potentials are called *retarded* potentials because of the delay between the cause and effect. The potential functions are useful in the study of electromagnetic radiation problems.

EXAMPLE 6.16

To arrive at the differential equations (16) and (17), the Lorentz condition given by (15) is used. Use this relation to calculate the vector magnetic potential if the scalar electric potential is

$$V \quad V_0 \cos(\omega t - \beta z), \text{ where } \beta = \omega\sqrt{\mu\varepsilon} \tag{22}$$

Substituting (22) into (15),

$$\nabla \cdot \mathbf{A} = -\mu\varepsilon \frac{\partial V}{\partial t} = \omega\mu\varepsilon V_0 \sin(\omega t - \beta z)$$

As the rhs is a function of z only, the vector magnetic potential can have only the z component. Hence,

$$\frac{\partial A_z}{\partial z} = \omega\mu\varepsilon V_0 \sin(\omega t - \beta z)$$

Integrating,

$A_z \quad \frac{\omega\mu\varepsilon}{\beta} V_0 \sin(\omega t - \beta z) = \sqrt{\mu\varepsilon}\, V_0 \cos(\omega t - \beta z)$, assuming the constant of integration to be zero.

EXAMPLE 6.17

A current element of length $\ell = 2.0$ cm is located in air. It is oriented along the z axis of rectangular coordinates with its mid-point at the origin. It carries a current $I = 5\cos(2\pi \times 10^6)t$ A. Find the vector magnetic potential at P(200 m, 0, 250 m) and $t = 1.0\ \mu s$

As the length of the current element is small compared to the distance at which the potential is to be found, no integration is required. Thus,

$$\text{The vector magnetic potential, } \mathbf{A}(\ ) = \frac{\mu_0 \mathbf{J}(t')dv}{4\pi R} \tag{23}$$

The current element, $\mathbf{J}(\ )dv = I(\ )d\ell$

$= [\ \cos(\qquad)(\qquad /v)](2. \qquad)\mathbf{u}_z$

$= 0.1\cos(2\pi\times10^6)(t - R/v)\mathbf{u}_z$ A-m

Distance of P from the origin, $R = \sqrt{20^2 + 25^2} = 32.02$ m

Velocity of the field, $v = 3\times10^8$ m/s

$\frac{R}{v} = \frac{32.02}{3\times10^8}$ s

$t = 10^{-6}$ s

Substituting the values in (23),

$$\mathbf{A}(\ = 1\ \mu s) = \frac{(4\pi\times10^{-7})[0.1\cos(2\pi\times10^6)(10^{-6} - 10.67\times10^{-8})]}{4\pi(32.02)}\mathbf{u}_z$$

$$= 2.45\times10^{-10}\,\mathbf{u}_z \text{ Wb/m (or T-m)}$$

6.18 Flow of Power in Electromagnetic Fields: The Poynting Vector

In an electric circuit consisting of an active source, a transmission line, and a load, electric power flows from the source to the load. Power flow through the line is proportional to the product of voltage and current. Voltage V between the transmission lines and current I through it establish electric and magnetic fields. The electric field intensity is a function of V and the magnetic field intensity is a function of I. This suggests that power flow from source to load can be expressed in terms of the field quantities. An expression for power flow in electromagnetic fields is derived in terms of field quantities using Maxwell's equations, as described below.

Maxwell's two curl equations are

$$\nabla\times\mathbf{E} = -\frac{\partial\mathbf{B}}{\partial t} \tag{1}$$

$$\nabla\times\mathbf{H} = \mathbf{J} + \frac{\partial\mathbf{D}}{\partial t} \tag{2}$$

For any two vectors, it can be shown that

$$-\nabla\cdot(\mathbf{E}\times\mathbf{H}) = \mathbf{E}\cdot(\nabla\times\mathbf{H}) - \mathbf{H}\cdot(\nabla\times\mathbf{E}) \tag{3}$$

Substitution of (1) and (2) in (3) gives,

$$-\nabla\cdot(\mathbf{E}\times\mathbf{H}) = \mathbf{E}\cdot\mathbf{J} + \mathbf{E}\cdot\frac{\partial\mathbf{D}}{\partial t} + \mathbf{H}\cdot\frac{\partial\mathbf{B}}{\partial t} \tag{4}$$

By integrating both sides of (4) over a region of volume with no active sources and moving charges, we get

$$-\int_v \nabla\cdot(\mathbf{E}\times\mathbf{H})dv = \int_v\left(\mathbf{E}\cdot\mathbf{J} + \mathbf{E}\cdot\frac{\partial\mathbf{D}}{\partial t} + \mathbf{H}\cdot\frac{\partial\mathbf{B}}{\partial t}\right)dv$$

The lhs may be transformed into a closed surface integral over the surface enclosing the volume by the use of divergence theorem. Accordingly,

$$-\oint_s(\mathbf{E}\times\mathbf{H})\cdot d\mathbf{S} = \int_v\left(\mathbf{E}\cdot\mathbf{J} + \mathbf{E}\cdot\frac{\partial\mathbf{D}}{\partial t} + \mathbf{H}\cdot\frac{\partial\mathbf{B}}{\partial t}\right)dv \tag{5}$$

This is the mathematical statement of Poynting's theorem. To interpret the theorem, let us assume that the permeability μ, the permittivity ε and the conductivity σ of the region are constants. Under this assumption, it can be shown that

$$\mathbf{E}\cdot\frac{\partial \mathbf{D}}{\partial t}=\frac{1}{2}\frac{\partial}{\partial t}(\mathbf{E}\cdot\mathbf{D})=\frac{\partial}{\partial t}\left(\frac{\varepsilon E^2}{2}\right) \tag{6}$$

$$\mathbf{H}\cdot\frac{\partial \mathbf{B}}{\partial t}=\frac{1}{2}\frac{\partial}{\partial t}(\mathbf{H}\cdot\mathbf{B})=\frac{\partial}{\partial t}\left(\frac{\mu H^2}{2}\right) \tag{7}$$

Substituting (6) and (7) in (5), we get

$$-\oint_S(\mathbf{E}\times\mathbf{H})\cdot d\mathbf{S}=\int_v(\mathbf{E}\cdot\mathbf{J})\,dv+\int_v\frac{\partial}{\partial t}\left(\frac{\varepsilon E^2}{2}\right)dv+\int_v\frac{\partial}{\partial t}\left(\frac{\mu H^2}{2}\right)dv \tag{8}$$

The first term of rhs of (8) gives power dissipation due to the conduction current, as $\mathbf{E}\cdot\mathbf{J}=J^2/\sigma$ is the ohmic loss per unit volume. The function within the bracket of the second term is the energy stored per unit volume of the electric field. And the quantity within the bracket of the third term is the energy stored per unit volume of the magnetic field. Hence, the second and third terms are the instantaneous increase in power stored in the electric and magnetic fields respectively. The sum of the three powers is supplied through the closed surface. Therefore, the lhs of (8) gives power flow *into* the region at any instant. Thus, electromagnetic power flow *out* of the region can be calculated by integrating $\mathbf{E}\times\mathbf{H}$ over a closed surface enclosing the volume. Denoting the power flow out of a region by P, we have

$$P=\oint_S(\mathbf{E}\times\mathbf{H})\cdot d\mathbf{S}=\oint_S\mathbf{P}_w\cdot d\mathbf{S} \tag{9}$$

where

$$\mathbf{P}_w=\mathbf{E}\times\mathbf{H} \tag{10}$$

Relation (9) is known as *Poynting's theorem* and relation (10) is called *Poynting vector.*

The unit of Poynting vector is W/m^2. Its direction, and hence the direction of power flow, is normal to the plane containing **E** and **H**.

Even though it is not possible to establish that the Poynting vector gives power per unit area at every point in a region of an electromagnetic field, its integration over a surface however does give the power flowing out of the surface.

Poynting vector is particularly useful for calculating power flow in electromagnetic waves.

Equation (8) has been derived assuming that the region does not contain moving charges and active sources. If the region has moving charges and active sources, then two additional terms will have to be added to the rhs of (8); one for the power needed to accelerate the moving charge and the other for the power supplied by the source which has to be taken as negative.

6.19 Power Flow in a Coaxial Transmission Line

Let us consider a long and straight coaxial transmission line with a solid inner conductor of radius a and an outer conductor of inside radius b and outside radius c. The axis of the line is taken along the z axis of cylindrical coordinates. One end of the line is connected to a dc source and the other end to a load resistance, as shown in Fig. 6.17. At first let us assume that the conductivities of the conductors are infinitely large so that the electric field intensity vector inside the conductors is zero for finite values of current densities. Accordingly, the Poynting vector and hence the power in the region of the conductors is zero. If the voltage between the conductors is V and current in each conductor is I, then we know that power transmitted by the line,

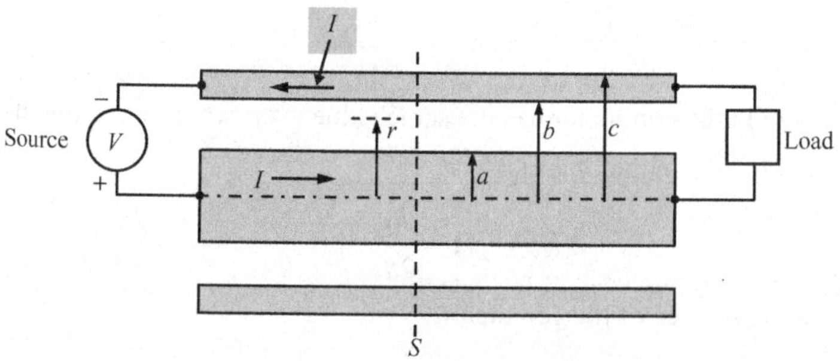

Fig. 6.17 Illustration of a coaxial transmission line for calculation of power flow using Poynting vector

$$P = VI \tag{1}$$

In order to calculate power carried by the line using the Poynting vector, let us consider a cross-sectional surface S at any point in the line, as indicated in the figure. At a radial distance r, the electric and magnetic field intensities on this surface are

$$\mathbf{E} = \frac{V}{\ln(b/a)}\frac{\mathbf{u}_r}{r}, \; b > r > a$$

and

$$\mathbf{H} = \frac{I}{2\pi}\frac{\mathbf{u}_\phi}{r}, \; b > r > a$$

The Poynting vector at radius r,

$$\mathbf{P}_w = \mathbf{E} \times \mathbf{H}$$

$$= \frac{VI}{2\pi \ln(b/a) r^2}(\mathbf{u}_r \times \mathbf{u}_\phi) = \frac{VI\,\mathbf{u}_z}{2\pi \ln(b/a) r^2}, \; b > r > a \tag{2}$$

A differential area on the surface has only one component in z direction. It is given by

$$d\mathbf{S} = (r\,dr\,d\phi)\mathbf{u}_z \tag{3}$$

The dot product of $\mathbf{P}_w$ and $d\mathbf{S}$ given by (2) and (3) respectively is

$$\mathbf{P}_w \cdot d\mathbf{S} = \frac{VI}{2\pi \ln(b/a)}\frac{dr\,d\phi}{r}$$

Power flow out of the surface,

$$P = \frac{VI}{2\pi \ln(b/a)}\int_a^b \int_0^{2\pi} \frac{dr\,d\phi}{r} = VI$$

We see that this result is same as given by (1).

Though we have derived the power flow in a coaxial transmission line by assuming that its conductors have infinite conductivity, in practice all good conductors have finite conductivity and ohmic loss occurs when current flows in a conductor. It is found that integration of the normal component of Poynting vector over the surface of a conductor gives the ohmic loss. As an example, let us consider the inner conductor of the coaxial line carrying current in the positive z direction. Current density in the conductor is given by

$$\mathbf{J} = \frac{I\mathbf{u}_z}{\pi a^2}$$

The electric field intensity arising out of the current flow,

$$\mathbf{E} = \frac{\mathbf{J}}{\sigma} = \frac{I\mathbf{u}_z}{\pi\sigma a^2}$$

The electric field intensity in the conductor has the same value everywhere including its surface.

Magnetic field intensity on the conductor surface is

$$\mathbf{H} = \frac{I\mathbf{u}_\phi}{2\pi a}$$

The Poynting vector on the surface of the conductor,

$$\mathbf{P}_w = \mathbf{E} \times \mathbf{H} = \frac{I^2}{2\pi^2 a^3 \sigma}(\mathbf{u}_z \times \mathbf{u}_\phi) = -\frac{I^2\mathbf{u}_r}{2\pi^2 a^3 \sigma} \tag{4}$$

A differential area on the surface of the conductor has only one component in r direction. It is given by

$$d\mathbf{S} = (ad\phi dz)\mathbf{u}_r \tag{5}$$

The dot product of $\mathbf{P}_w$ given by (4) and $d\mathbf{S}$ given by (5) is

$$\mathbf{P}_w \cdot d\mathbf{S} = -\frac{I^2 d\phi dz}{2\pi^2 a^2 \sigma} \tag{6}$$

Power flow *into* the surface of the conductor per unit length is found by integrating the negative of (6) over the conductor surface.

$$-\oint_S \mathbf{P}_w \cdot d\mathbf{S} = -\int_S \mathbf{P}_w \cdot d\mathbf{S} = \int_0^{2\pi}\int_0^1 \frac{I^2 d\phi dz}{2\pi^2 a^2 \sigma} = I^2\left(\frac{1}{\pi a^2 \sigma}\right) = I^2 R \tag{7}$$

This is the ohmic loss in the conductor per unit length, as the function within the brackets of (7) is resistance of the conductor per unit length.

This simple example shows that as per Poynting's theorem power flows from the source to load in the dielectric region between the conductors. Power within the conductors is only the ohmic loss.

6.20 Poynting Vector for Fields Varying Sinusoidally with Time

The Poynting vector defined in Section 6.18 is for any type of variation of fields with time. In this section we will obtain the Poynting vector when variation of fields with time is sinusoidal. At first we will derive the instantaneous Poynting vector and then its complex form.

Instantaneous Poynting Vector

To define the instantaneous Poynting vector, let us assume the electric field is in x direction and the magnetic field is in y direction as given below.

The electric field intensity,

$$\mathbf{E}(t) = E_{x0}\cos(\omega t + \psi_e)\mathbf{u}_x \tag{1}$$

The magnetic field intensity,

$$\mathbf{H}(t) = H_{y0}\cos(\omega t + \psi_h)\mathbf{u}_y \tag{2}$$

where E_{x0} and H_{y0} are the amplitudes.

The instantaneous Poynting vector is defined as the cross product of instantaneous values of electric and magnetic field intensity vectors. Thus, denoting the instantaneous Poynting vector by $\mathbf{P}_w(t)$, we have

$$\mathbf{P}_w(t) = \mathbf{E}(t) \times \mathbf{H}(t) \tag{3}$$

Substituting (1) and (2) in (3), we get

$$\mathbf{P}_w(t) = E_{x0} H_{y0} \cos(\omega t + \psi_e) \cos(\omega t + \psi_h)\mathbf{u}_z$$

$$= \frac{1}{2} E_{x0} H_{y0} [\cos\psi + \cos(2\omega t + \psi_e + \psi_h)]\mathbf{u}_z \tag{4}$$

where $\psi = \psi_e - \psi_h$, the time-phase difference between the two fields.

Thus, the instantaneous Poynting vector of (4) has two components. The first component is independent of time and gives average power flow per unit area. The second component varies at twice the frequency. As the time average of the second term over a complete cycle is zero, it has no contribution to active power. As the Poynting vector is in z direction, power flow per unit area is also in z direction. Note that the fields are in x and y direction.

Complex Poynting Vector

We learnt in Section 1.12 the complex form of representation of sinusoidally time-varying voltage and currents. In the same way, sinusoidally time-varying field quantities can also be represented in complex form.

The complex power in electric circuits is given by $\frac{1}{2}\bar{V}_0 \bar{I}_0{}^*$. In the same way, the complex form of Poynting vector having the unit of power per unit area is defined as $\frac{1}{2}(\bar{\mathbf{E}} \times \bar{\mathbf{H}}^*)$ where $\bar{\mathbf{H}}^*$ is conjugate of the $\bar{\mathbf{H}}$ phasor. Denoting the complex Poynting vector by $\bar{\mathbf{P}}_w$, we have

$$\bar{\mathbf{P}}_w = \frac{1}{2}(\bar{\mathbf{E}} \times \bar{\mathbf{H}}^*) \tag{5}$$

The field quantities in (5) are vectors as well as phasors. As vectors, these are represented by their components along coordinate directions. As each component is a phasor quantity, the components are represented by their real and imaginary parts. Let us assume that the fields have components in x and y directions. Then,

$$\bar{\mathbf{E}} = \bar{E}_x \mathbf{u}_x + \bar{E}_y \mathbf{u}_y \tag{6}$$

$$\bar{\mathbf{H}} = \bar{H}_x \mathbf{u}_x + \bar{H}_y \mathbf{u}_y \tag{7}$$

The conjugate of (7) is

$$\bar{\mathbf{H}}^* = \bar{H}_x{}^* \mathbf{u}_x + \bar{H}_y{}^* \mathbf{u}_y \tag{8}$$

By substituting (6) and (8) in (5), we get

$$\bar{\mathbf{P}}_w = \frac{1}{2}(\bar{E}_x \mathbf{u}_x + \bar{E}_y \mathbf{u}_y) \times (\bar{H}_x{}^* \mathbf{u}_x + \bar{H}_y{}^* \mathbf{u}_y)$$

$$= \frac{1}{2}(\bar{E}_x \bar{H}_y{}^* - \bar{E}_y \bar{H}_x{}^*)\mathbf{u}_z \tag{9}$$

In (6) and (7) the field intensities are complex quantities, each having a real part and an imaginary part. Let the complex forms of various components in terms of the real and imaginary parts be given by

$$\bar{E}_x = E_{xa} + jE_{xi} \tag{10}$$

$$E_y \quad E_{ya} + jE_{yi} \tag{11}$$

$$H_x \quad H_{xa} + jH_{xi} \tag{12}$$

$$H_y \quad H_{ya} + jH_{yi} \tag{13}$$

The additional subscripts a and i are for the real and imaginary parts of the fields.

By substituting (10) and (11) and conjugates of (12) and (13) into (9), we get

$$\bar{\mathbf{P}}_w = \frac{1}{2}[(E_{xa} + jE_{xi})\,(H_{ya} - jH_{yi}) - (E_{ya} + jE_{yi})\,(H_{xa} - jH_{xi})]\mathbf{u}_z$$

$$= \frac{1}{2}[(E_{xa}H_{ya} + E_{xi}H_{yi}) - (E_{ya}H_{xa} + E_{yi}H_{xi})]\mathbf{u}_z$$

$$+ \frac{j}{2}[(E_{xi}H_{ya} - E_{xa}H_{yi}) - (E_{yi}H_{xa} - E_{ya}H_{xi})]\mathbf{u}_z \tag{14}$$

The first term within the square bracket of (14) is the time-average of active or real power per unit area and the second term within the square bracket is the reactive power per unit area. We find that the field components, which are perpendicular to each other in space, contribute to the power. The real power is due to the components of the fields which are in time-phase, and the reactive power is due to the field components which are in time-quadrature. It may be noted that the field intensities have components in x and y directions whereas the power flow is in z direction. Thus, the real part of the complex Poynting vector gives the average power per unit area and the imaginary part gives the reactive power per unit area. The direction of flow of power is normal to the plane containing the field intensity vectors.

The real part of the complex Poynting vector, $\mathbf{P}_{wa} = \text{Re}\,\frac{1}{2}(\bar{\mathbf{E}} \times \bar{\mathbf{H}}^*)$ (15)

The imaginary part of the complex Poynting vector, $\mathbf{P}_{wi} = \text{Im}\,\frac{1}{2}(\bar{\mathbf{E}} \times \bar{\mathbf{H}}^*)$ (16)

Average power flow P_{av} *out* of a given surface is obtained by integrating the normal component $\mathbf{P}_{wa}$ over the surface. Thus,

$$P_{av} = \int_S \text{Re}\,\frac{1}{2}(\bar{\mathbf{E}} \times \bar{\mathbf{H}}^*) \cdot d\mathbf{S} \tag{17}$$

The results of this section are useful for computation of power flow in electromagnetic waves.

EXAMPLE 6.18

Let the electric field intensity in a lossless medium be given by

$$\mathbf{E}(\;)\quad{}_x \cos(\omega t - \beta z)\mathbf{u}_x \tag{18}$$

where $E_{x0} = 150$ V/m, $\omega = 2\pi \times 10^8$ rad/s, $\beta = 4\pi/3$ rad/m

Find (a) the instantaneous Poynting vector, (b) the complex Poynting vector, using the complex forms and phasor forms of the field intensities, and (c) the average power flow through an area $\mathbf{S} = 5.0\mathbf{u}_z$ m^2.

It is easier to make the calculations with the symbols and then use the values of various quantities to get the results.

In order to find the Poynting vector, we have to at first calculate the magnetic field intensity. The magnetic field intensity can be found either from (6.13-1) or (6.13-2). We will use (6.13-1) given by

$$\nabla \times \mathbf{E}(\;) = -\mu_0 \frac{\partial \mathbf{H}(t)}{\partial t}$$

As the electric field intensity has only one component in x direction and it is a function of only z, we have

$$\nabla \times \mathbf{E}\;\; = \frac{\partial E_{\;}(\;)}{\partial z}\mathbf{u}_y = \beta E_{x\;} \sin(\;\; t - \beta z)\mathbf{u}_y$$

Thus, $$\frac{\partial \mathbf{H}(t)}{\partial t} = -\frac{\beta E_{x0}}{\mu_0}\sin(\omega t - \beta z)\mathbf{u}_y$$

Or $$\mathbf{H}(t) = \frac{\beta E_{x0}}{\omega\mu_0}\cos(\omega t - \beta z)\mathbf{u}_y \tag{19}$$

(a) The instantaneous Pointing vector,

$$\mathbf{P}_w(t) = \mathbf{E}(t)\times\mathbf{H}(t) = \frac{\beta E_{x0}^{\ 2}}{\omega\mu_0}\cos^2(\omega t - \beta z)\mathbf{u}_z$$

$$= \frac{\beta E_{x0}^{\ 2}}{2\omega\mu_0}[1+\cos 2(\omega t - \beta z)]\mathbf{u}_z$$

By substituting the values,

$$\mathbf{P}_w(t) = \frac{(4\pi/3)(150)^2}{2(2\pi\times 10^8)(4\pi\times 10^{-7})}[1+\cos 2[(2\pi\times 10^8)t - (4\pi/3)z]\mathbf{u}_z$$

$(59.68)[1+\cos 2[(2\pi\times 10^8)t-(4\pi/3)z]\mathbf{u}_z$ W/m²

As the time average of the second term is zero, the average power per unit area is given by the first term.

(b) The complex forms of the field intensities, which can be written from (18) and (19), are

$$\bar{\mathbf{E}} = E_{x0}(\cos\beta z - j\sin\beta z)\mathbf{u}_x$$

$$\bar{\mathbf{H}} = \frac{\beta E_{x0}}{\omega\mu_0}(\cos\beta z - j\sin\beta z)\mathbf{u}_y$$

$$\bar{\mathbf{H}}^* = \frac{\beta E_{x0}}{\omega\mu_0}(\cos\beta z + j\sin\beta z)\mathbf{u}_y$$

$$\bar{\mathbf{P}}_w = \frac{1}{2}(\bar{\mathbf{E}}\times\bar{\mathbf{H}}^*) = \frac{\beta E_{x0}^{\ 2}}{2\omega\mu_0}(1+j0)\mathbf{u}_z = 59.68\,\mathbf{u}_z \text{ W/m}^2$$

The phasor forms of the field intensities, which can be written from (18) and (19) are

$$\bar{\mathbf{E}} = E_{x0}e^{-j\beta z}\ \mathbf{u}_x$$

$$\bar{\mathbf{H}} = \frac{\beta E_{x0}}{\omega\mu_0}e^{-j\beta z}\ \mathbf{u}_y$$

$$\bar{\mathbf{H}}^* = \frac{\beta E_{x0}}{\omega\mu_0}e^{j\beta z}\ \mathbf{u}_y$$

The complex Poynting vector, $\bar{\mathbf{P}}_w = \frac{1}{2}(\bar{\mathbf{E}}\times\bar{\mathbf{H}}^*) = \frac{\beta E_{x0}^{\ 2}}{2\omega\mu_0}\mathbf{u}_z = 59.68\,\mathbf{u}_z$ W/m²

The real part of the complex Poynting vector,

$\mathbf{P}_{wa}$ [illegible] $\mathbf{u}_z$

(c) Average power flow through the area, $P_{av} = \mathbf{P}_{wa}\cdot\mathbf{S} = ($ [illegible] $.68\mathbf{u}_z)\cdot($ [illegible] $\mathbf{u}_z) = 298.4$ W

6.21 Distribution of Current in a Good Solid-Conductor of Circular Section: The Skin Effect

When time-varying current flows in a good conductor, the internal magnetic field induces currents within the conductor. The induced currents are in opposite sense to the main current as per Lenz's law. This results in a non-uniform distribution of current within the conductor.

The current density is maximum at the surface of a round conductor and minimum at its centre.

This phenomenon is called *skin effect.* Aluminium cables with steel reinforcements (ACSR) used in ac transmission lines have central steel wires for mechanical strength. Current in the steel wire is negligibly small not only because its permeability is high but also because it is at the centre of the cable. For high frequency applications where mechanical strength is also a consideration, hollow conductors are more economical than solid conductors.

Expressions showing variation of current density over the circular cross section of a cylindrical conductor can be derived using Maxwell's equations in cylindrical coordinates, assuming complex exponential form of variation of current with time. A simple treatment of the problem is possible by assuming the conductor to be straight and long and that flow of current is along the axial direction. If the axis of the conductor is taken along z direction, the current density as well as the electric field intensity will be in z direction, and the magnetic field intensity would be in ϕ direction. Further, the fields will vary with r only. Also, for a good conductor the displacement current density is very small compared to the conduction current density. Accordingly, the displacement the current-density term may be taken as zero. Complex exponential form of variation of fields with time will be assumed. With the above assumptions, the electric and magnetic field intensities in cylindrical coordinates are given by

$$\bar{\mathbf{E}} = \bar{E}_z \mathbf{u}_z$$

$$\bar{\mathbf{H}} = \bar{H}_\phi \mathbf{u}_\phi$$

As the electric field intensity has only one component in z direction and it does not vary with ϕ and z, its curl in cylindrical coordinates is given by

$$\nabla \times \mathbf{E} = -\frac{\partial E_z}{\partial r}\mathbf{u}_\phi \tag{1}$$

By substituting (1) in (6.15-1), we get

$$\frac{\partial \bar{E}_z}{\partial r} = j\omega \bar{B}_\phi = j\omega\mu \bar{H}_\phi \tag{2}$$

Moreover, the magnetic field intensity has only one component in ϕ direction and it does not vary with ϕ and z. Hence, the curl of $\mathbf{H}$ in cylindrical coordinates is

$$\nabla \times \bar{\mathbf{H}} = \frac{\partial \bar{H}_\phi}{\partial r} + \frac{1}{r}\bar{H}_\phi \tag{3}$$

By substituting (3) in (6.15-2), we get

$$\frac{\partial \bar{H}_\phi}{\partial r} + \frac{1}{r}\bar{H}_\phi = \bar{J}_z = \sigma \bar{E}_z \tag{4}$$

By solving (2) and (4) for E_z and using ordinary derivatives as the fields are function of a single variable, we get the following differential equation.

$$\frac{d^2 E_z}{dr^2} + \frac{1}{r}\frac{dE_z}{dr} = j \quad E_z$$

Or

$$\frac{d^2 E_z}{dr^2} + \frac{1}{r}\frac{dE_z}{dr} + \tau^2 E_z = 0 \tag{5}$$

where $$\tau^2 = -j\omega\mu\sigma$$

$$\tau = \sqrt{-j\omega\mu\sigma} = \left(\sqrt{2\pi f\mu\sigma}\right)\left(\cos\frac{\pi}{2} - j\sin\frac{\pi}{2}\right)^{\frac{1}{2}}$$

$$= \left(\sqrt{2\pi f\mu\sigma}\right)\left(\cos\frac{\pi}{4} - j\sin\frac{\pi}{4}\right), \text{ [using formula of De Moivre]}$$

$$= \frac{1-j}{\delta}$$

where $$\delta = \frac{1}{\sqrt{\pi f\mu\sigma}} \tag{6}$$

The parameter δ is called *depth of penetration* or *skin depth*. Its unit is metre.

Equation (5) is a Bessel equation of zero order. As $r = 0$ is included in the problem, the solution of the equation is

$$\bar{E}_z = AJ_0(\tau r) \tag{7}$$

where $J_0(\tau r)$ is the Bessel function of zero order and A is a constant. To determine the constant, let us take the electric field intensity on the surface of the conductor of radius a as $\bar{E}_z(a)$. The Bessel function at $r \quad a$ is $J_\upsilon(\tau a)$. Substituting these in (7), we have

$$A = \frac{\bar{E}_z(a)}{J_0(\tau a)} \tag{8}$$

The solution of (5) in terms of $\bar{E}_z(a)$ is obtained by substituting (8) in (7). Using the symbol $\bar{E}_z(r)$ instead of $\bar{E}_a$ to show explicitly that the field intensity is a function of r, the solution is

$$\bar{E}_z(r) = \bar{E}_z(a)\frac{J_0(\tau r)}{J_0(\tau a)} \tag{9}$$

The current density, $$\bar{J}_z(r) = \sigma\bar{E}_z(r) = \bar{J}_z(a)\frac{J_0(\tau r)}{J_0(\tau a)} \tag{10a}$$

Or $$\frac{\bar{J}_z(r)}{\bar{J}_z(a)} = \frac{J_0(\tau r)}{J_0(\tau a)} \tag{10b}$$

where $\bar{J}_a(a) = \sigma\bar{E}_z(a)$ is the current density on the surface of the conductor of radius a. It may be noted that the symbol $\bar{J}_z$ denotes the current density phasor whereas the symbol J_0 stands for the Bessel function. Both J_υ $r)$ and $J_0(\tau a)$ are complex functions, and each is an infinite series. Denoting the real parts of the functions by ber (Bessel-real) and the imaginary parts by bei (Bessel-imaginary), equation (10*b*) may be written as

$$\bar{J}_z(r) = \bar{J}_z(a)\frac{\text{ber}(kr) + j\,\text{bei}(kr)}{\text{ber}(ka) + j\,\text{bei}(ka)} \tag{11a}$$

Or $$\frac{\bar{J}_z(r)}{\bar{J}_z(a)} = \frac{\text{ber}(kr) + j\,\text{bei}(kr)}{\text{ber}(ka) + j\,\text{bei}(ka)} \tag{11b}$$

where $k = \quad \omega\mu\sigma = \quad \delta$

The ber and bei functions are defined by the following series,

$$\text{ber}(kr) = 1 - \frac{(\quad /2)^4}{(\quad !)^2} + \frac{(\quad /2)^8}{(\quad !)^2} - \cdots \tag{12}$$

$$\mathrm{bei}(kr) = \frac{(kr/2)^2}{(1!)^2} - \frac{(kr/2)^6}{(3!)^2} + \frac{(kr/2)^{10}}{(5!)^2} - \cdots \tag{13}$$

It is clear from (11) that the magnitude and phase angle of the current density inside the conductor is a function of r. The ratio between the magnitudes of current density at r and the current density on the conductor surface is

$$\frac{|\bar{J}_z(r)|}{|\bar{J}_z(a)|} = \frac{|\mathrm{ber}(kr) + j\,\mathrm{bei}(kr)|}{|\mathrm{ber}(ka) + j\,\mathrm{bei}(ka)|} \tag{14}$$

For numerical calculations, the values of ber and bei functions can be found from numerical tables of these functions. Alternatively, a finite number of terms of the series given by (12) and (13) may be used for approximate calculations. Figure 6.18 shows the variations of current density from the surface of the conductor to its centre for three values of a/δ. The graphs show that the degree of non-uniformity in the distribution of current over the section of the conductor increases with an increase in this ratio.

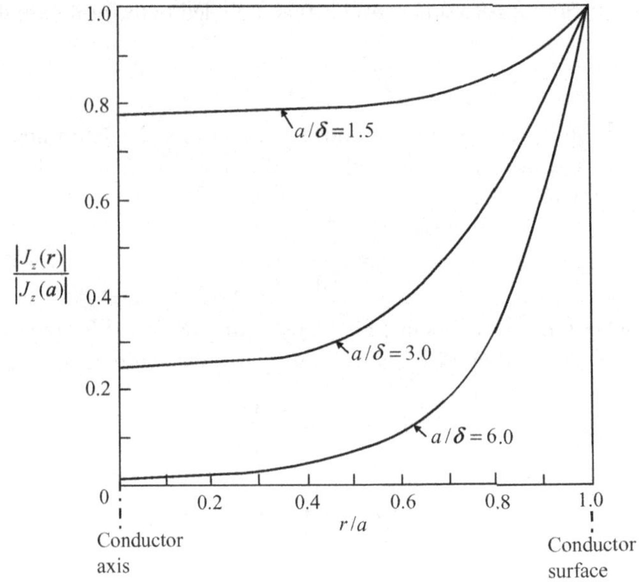

Fig. 6.18 Showing variation of magnitude of current density from the surface to the axis of a cylindrical conductor

EXAMPLE 6.19

Calculate the current density at the centre of a circular conductor in terms of the current density on its surface assuming $a/\delta = 3.0$, and using the first four terms of the infinite series of ber and bei functions.

We have, $\dfrac{ka}{2} = \dfrac{a}{\sqrt{2}\,\delta} = 2.121$

$$\mathrm{ber}(2.\quad) = 1 - \frac{(2.121)^4}{(2!)^2} + \frac{(2.121)^8}{(4!)^2} - \frac{(2.121)^{12}}{(6!)^2} = -3.366$$

$$\mathrm{bei}(2.\quad) = \frac{(\ .121)^2}{(\ !)^2} - \frac{(\ .121)^6}{(\ !)^2} + \frac{(\ .121)^2}{(\ !)^2} - \frac{(\ .121)^6}{(\ !)^2} = 2.098$$

$\mathrm{ber}(r = 0) = 1$; $\mathrm{bei}(r = 0) = 0$

$$\frac{J_z(r=0)|}{J_z(r=a)|} = \frac{1}{\sqrt{(-3.366)^2 + (2.098)^2}} = 0.252$$

SUMMARY

Using the property of conservation of electric charges, we obtained the continuity equation, and then described Faraday's law of electromagnetic induction and Lenz's law. We distinguished between voltage induced in a coil due to time-varying flux (transformer voltage) and voltage induced due to the relative motion between a conductor and magnetic field (speed voltage). We established that if the direction of speed voltage is determined by the use of the right-hand rule, the direction satisfies Lenz's law. Moreover, the induced voltage in a coil due to sinusoidally time-varying flux lags the flux by 90° in time, and the voltage induced in a coil rotating at constant speed in a uniform steady magnetic field varies sinusoidally with time.

We derived an expression for eddy-current loss in a thin iron plate and found that the loss is proportional to square of thickness of the plate, square of the frequency and square of the maximum flux density.

After defining displacement current density, we stated Maxwell's equations in differential form, in integral form and for sinusoidally time varying fields. We applied Maxwell's equations to derive boundary conditions and to define potential functions for time-varying fields. The boundary conditions for time-varying fields are identical to those for static fields. We also applied Maxwell's equations for calculation of power flow leading to the definition of the Poynting vector and its instantaneous and complex forms. Poynting vector is useful for calculation of power flow in electromagnetic fields. Finally, we used Maxwell's equations in cylindrical coordinates to study the distribution of current over the cross section of a circular conductor and learnt that the current distribution is not uniform. The current density has maximum value at the conductor surface and minimum value at its centre. Because of this phenomenon, known as the skin effect, the ac resistance of a conductor is higher than its dc value (see Appendix B).

IMPORTANT FORMULAE

Continuity equation: $\nabla \cdot \mathbf{J} = -\frac{\partial \rho}{\partial t}$

Transformer voltage: $\mathcal{E} = -SN \cos\alpha \frac{dB}{dt}$

Integral form of Faraday's law: $\oint \mathbf{E} \cdot d\ell = -\frac{\partial}{\partial t}\int_S \mathbf{B} \cdot d\mathbf{S}$

Differential form of Faraday's law: $\nabla \times \mathbf{E} = -\frac{\partial \mathbf{B}}{\partial t}$

Speed voltage: $\mathcal{E} = \oint (\mathbf{v} \times \mathbf{B}) \cdot d\ell$

Combined transformer and speed voltage: $\mathcal{E} = \oint (\mathbf{v} \times \mathbf{B}) \cdot d\ell - \frac{\partial}{\partial t}\int_S \mathbf{B} \cdot d\mathbf{S}$

Eddy-current loss in a thin iron sheet per unit volume $\approx \frac{\pi^2 \sigma \upsilon^2 f^2 B_0^{\ 2}}{8}$

Convection current density: $\mathbf{J}$, $\mathbf{v}$

Displacement current density: $\mathbf{J}_d = \dfrac{\partial \mathbf{D}}{\partial t}$

Maxwell's equations in differential form:

$$\nabla \times \mathbf{E} = -\frac{\partial \mathbf{B}}{\partial t},\ \nabla \times \mathbf{H} = \mathbf{J} + \frac{\partial \mathbf{D}}{\partial t},\ \nabla \cdot \mathbf{B} = 0,\ \nabla \cdot \mathbf{D} = \rho$$

Maxwell's equations for harmonic variation of fields with time:

$$\nabla \times \mathbf{E} = -_j\omega \bar{\mathbf{B}},\ \nabla \times \bar{\mathbf{H}} = (\bar{\mathbf{J}} + j\omega \bar{\mathbf{D}}),\ \nabla \cdot \bar{\mathbf{B}} = 0,\ \nabla \cdot \bar{\mathbf{D}} = \bar{\rho}$$

Poynting's theorem: $P = \oint_S (\mathbf{E} \times \mathbf{H}) \cdot d\mathbf{S}$

Poynting vector: $\mathbf{P}_w = \mathbf{E} \times \mathbf{H}$

Instantaneous Poynting vector: $\mathbf{P}_w(t) = \mathbf{E}(t) \times \mathbf{H}(t)$

Complex Poynting vector: $\bar{\mathbf{P}}_w = \dfrac{1}{2}(\bar{\mathbf{E}} \times \bar{\mathbf{H}}^*)$

MULTIPLE-CHOICE QUESTIONS

(Starred questions have two answers)

6.1 Which of the following is the continuity equation?
(a) $\nabla \times \mathbf{E} = \mathbf{J}$ (b) $\nabla \cdot \mathbf{D} = \rho$ (c) $\nabla \cdot \mathrm{J} = -\partial \rho / \partial t$ (d) $\nabla \times \mathbf{E} = -\partial \mathbf{B} / \partial t$

6.2 Voltage induced in a stationary planar rectangular coil by a uniform sinusoidally time-varying magnetic field is $V_0 \sin \omega t$ when the coil axis is along the direction of flux density. If the coil is tilted to a position where its plane is at an angle of 30° with direction of flux density, then voltage induced in the coil will be
(a) $0.866V_0 \sin \omega t$ (b) $0.5V_0 \sin \omega t$ (c) $V_0 \sin(\omega t + 30)$ (d) $V_0 \sin(\omega t - 30)$

6.3 The amplitude of voltage induced in a stationary planar circular coil of radius a and number of turns N, located in a uniform time-varying magnetic field, is V_0. Given that the orientations of both the coils with respect to the magnetic field are same, the amplitude of voltage induced in a circular coil of radius $a/2$ and number of turns $2N$ when placed in the same magnetic field is
(a) $2V_0$ (b) V_0 (c) $0.5V_0$ (d) $0.25V_0$

6.4 A stationary single-turn circular conducting loop is located in a uniform magnetic field with its axis along the direction of flux density. The radius of the loop is a and the flux density is $B_0 \sin \omega t$. The electric field intensity along the loop is
(a) $a\omega B_0 \sin \omega t$ (b) $-0.5a\omega B_0 \cos \omega t$ (c) $0.5a\omega B_0 \sin \omega t$ (d) $0.5a\omega B_0 \cos \omega t$

6.5 A conducting bar moves in a uniform magnetic field at a constant velocity between two conducting rails of negligible resistance. The flux density, velocity, and distance between the rails are 0.25 T, 12 m/s and 40 cm respectively. The flux density is at an angle of 60° with normal to the plane containing the rails. A high resistance voltmeter is connected between the ends of the rails. The reading of the voltmeter is
(a) 0.6 V (b) 1.2 V (c) 1.04 V (d) 1.9 V

6.6 A conducting bar moves in a uniform magnetic field between two conducting rails of negligible resistance at a constant velocity. The flux density, which is normal to the plane containing the rails, is 0.4 T. The velocity of the bar and its length are 8 m/s and 0.6 m respectively. The angle between the bar and its direction of motion is 60°. A high resistance voltmeter is connected between the ends of the rails. The reading of the voltmeter is
(a) 0.96 V (b) 1.92 V (c) 1.66 V (d) 1.4 V

6.7 A planar coil rotates at a constant angular velocity in a uniform magnetic field. The maximum voltage induced in the coil is 12 V. What is the angle between the direction of flux density and normal to the plane of the coil when the induced voltage is 6 V?
(a) 75° (b) 60° (c) 45° (d) 30°

6.8 Two parallel-plate capacitors 1 and 2 having the same plate area are connected in parallel to an ac source. The spacing between the plates, and the permittivity of the dielectric of Capacitor 1 are s, and ε respectively. The corresponding values of Capacitor 2 are $s/2$, and 2ε. If the displacement current in Capacitor 1 is I_d, then the total displacement current of the parallel combination of the capacitors is
(a) $2I_d$ (b) $3I_d$ (c) $4I_d$ (d) $5I_d$

6.9 The eddy-current loss per unit volume in a thin iron sheet of thickness b, subjected to a sinusoidally time-varying magnetic field of frequency f and maximum flux density B_0, is proportional to
(a) $b^2 f^2 B_0^2$ (b) $b^2 f B_0^2$ (c) $b^2 f^2 B_0$ (d) $b f^2 B_0^2$

*6.10 Which of the following Maxwell's equations will remain unchanged in a perfect dielectric medium?
(a) $\nabla \times \mathbf{E} = -\dfrac{\partial \mathbf{B}}{\partial t}$ (b) $\nabla \times \mathbf{H} = \mathbf{J} + \dfrac{\partial \mathbf{D}}{\partial t}$ (c) $\nabla \cdot \mathbf{B} = 0$ (d) $\nabla \cdot \mathbf{D} = \rho$

6.11 The unit of Poynting vector is
(a) watt per metre square (b) joule per metre square
(c) joule per metre cube (d) watt per metre cube

6.12 The instantaneous electric and magnetic field intensities in an electromagnetic field are $\mathbf{H}(t) = H_x(t)\mathbf{u}_x + H_y(t)\mathbf{u}_y$ and $\mathbf{E}(t) = E_x(t)\mathbf{u}_x - E_y(t)\mathbf{u}_y$. Which of the following expressions gives the instantaneous Poynting vector?
(a) $[E_x(t)H_y(t) - E_y(t)H_x(t)]\mathbf{u}_z$ (b) $[E_x(t)H_y(t) + E_y(t)H_x(t)]\mathbf{u}_z$
(c) $[E_x(t)H_x(t) - E_y(t)H_y(t)]\mathbf{u}_z$ (d) $[E_x(t)H_x(t) + E_y(t)H_y(t)]\mathbf{u}_z$

6.13 The complex electric and magnetic field intensities in an electromagnetic field in terms of real and imaginary parts are $\bar{\mathbf{E}} = (E_a + jE_i)\mathbf{u}_x$ and $\bar{\mathbf{H}} = (H_a + jH_i)\mathbf{u}_y$ where the quantities are amplitudes. Which of the following expressions gives the real part of the complex Poynting vector?
(a) $\dfrac{1}{2}(E_a H_a - E_i H_i)\mathbf{u}_z$ (b) $\dfrac{1}{2}(E_a H_i - E_i H_a)\mathbf{u}_z$
(c) $\dfrac{1}{2}(E_a H_i + E_i H_a)\mathbf{u}_z$ (d) $\dfrac{1}{2}(E_a H_a + E_i H_i)\mathbf{u}_z$

6.14 Calculation of power in a coaxial transmission line using the Poynting vector shows that useful power flows
(a) through the inner conductor
(b) through the outer conductor
(c) through both inner and outer conductors
(d) between the conductors

QUESTIONS

6.1 The continuity equation for time-varying fields gives the relation between conduction-current density and volume charge density in the differential form. Interpret the equation.

6.2 A planar conducting loop with open terminals is located on the $x = 0$ plane. It encloses a time-varying magnetic field directed along x direction. When the terminals are short circuited, induced current will flow through the loop. What are the directions of the induced current when the flux is increasing and when the flux is decreasing with time? Justify with proper explanation.

6.3 Describe how transformer voltage is induced in a coil. What are the factors on which the voltage depends? What should be the angle between the magnetic flux density and normal to the plane of the coil so that the induced emf has the maximum value?

6.4 What is the time-phase difference between a sinusoidally time-varying flux and the voltage induced in a coil by the flux? Which is the leading quantity? Draw the waveforms of the flux and the induced voltage.

6.5 Justify that the polarities of speed voltage induced in a conductor by a magnetic field, determined by the use of the right-hand rule, satisfies the Lenz's law.

6.6 Give the differential and integral forms of Faraday's law of electromagnetic induction due to a time-varying flux.

6.7 The voltage induced in a coil rotating in a stationary uniform magnetic field varies with time even when the field does not vary with time. Explain. What is the time-phase difference between the flux linkage and induced voltage?

6.8 Discuss the phenomena of eddy currents and eddy-current loss in a conductor subjected to an alternating field.

6.9 Eddy-current loss reduces the efficiency of electromagnetic equipments. What measures are taken to reduce the loss?

6.10 Distinguish between conduction, convection and displacement current densities. On what factors each of these currents depends?

6.11 Except for a brief charging period no current flows in a capacitor connected to a dc source. However, if a capacitor is connected to an ac source current flows continuously. Explain.

6.12 Justify that the displacement current is zero in a perfect conductor and the conduction current is zero in a perfect dielectric.

6.13 If the dielectric medium of a capacitor is not perfect (conductivity is not zero) then the dielectric will carry both conduction and displacement currents when the capacitor is connected to an ac source. Prove that the ratio of the conduction current density to the displacement current density is equal to σ_u '$\omega\varepsilon$, where σ_d is conductivity of the dielectric material.

6.14 Show that in order to satisfy both the Maxwell's curl equations for time-varying electric and magnetic fields, the fields must also vary in space.

6.15 Obtain giving reasons thc differential forms of Maxwell's equations applicable to free space.

6.16 Write the differential forms of Maxwell's equations assuming that the variation of electric and magnetic fields with respect to time are represented by the complex exponential function $e^{j\ t}$

6.17 Consider that the magnetic flux density in a region of perfect dielectric is given by

$$\mathbf{B} \qquad \mathbf{u}_x$$

Find the electric field intensity in the region. What is the time-phase difference and space-phase difference between the two field intensities?

6.18 Derive the relation between the normal components of time-varying magnetic field intensities on two sides of the common boundary of two regions using the appropriate Maxwell's equation. Assume that the boundary does not contain any surface current.

6.19 The electric field intensity in a perfect dielectric medium is given by

$$\overline{\mathbf{E}} = E_0 \sin \beta z e^{j\omega t} \mathbf{u}_y$$

Obtain the complex Poynting vector and interpret the result.

6.20 The electric field intensity in a perfect dielectric medium is given by

$$\overline{\mathbf{E}} = E_0 e^{j(\omega t - \beta z)} \overline{\mathbf{u}}_x$$

Find the complex Poynting vector and interpret the result.

6.21 Figure Q6.1 shows a coaxial transmission line connected to a source at one end and a load at the other end. Neglecting ohmic and dielectric losses, what is the net power out of each closed surface S_1, S_2 and S_3? Justify your answers.

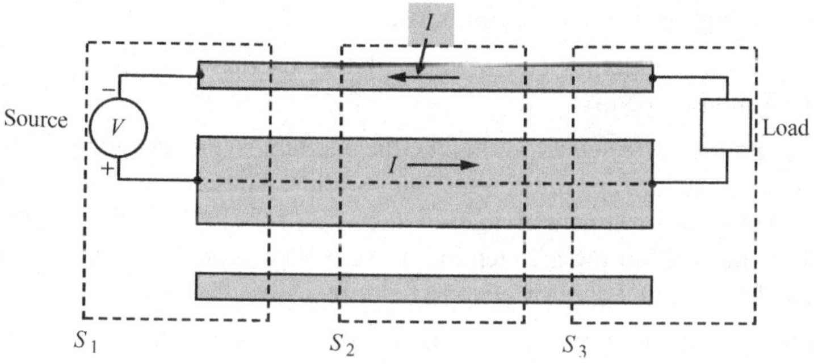

Fig. Q6.1

6.22 Explain the skin effect phenomenon in a conductor carrying alternating current.

ANSWERS TO SELECTED QUESTIONS

6.1 The continuity equation for time-varying fields gives the relation between conduction-current density and volume charge density in the differential form. Interpret the equation.

Answer: The continuity equation for a time-varying field is given by

$$\nabla \cdot \mathbf{J} = -\frac{\partial \rho}{\partial t}$$

where **J** and ρ are the instantaneous current density and volume charge density respectively. The divergence of **J** is equal to current flow at any instant out of a closed surface enclosing a volume element Δv divided by Δv, in the limit $\Delta \quad \to 0$. The right-hand side of the continuity equation gives the time rate of decrease of volume charge density at the same instant and at a point in the region under consideration. Thus, the continuity equation satisfies the continuity condition of current flow.

6.7 The voltage induced in a coil rotating in a stationary uniform magnetic field varies with time even when the field does not vary with time. Explain. What is the time-phase difference between the flux linkage and induced voltage?

Answer: When a coil rotates at a constant speed in a steady and uniform magnetic field, flux linkage changes with time t. It is a cosine function of t, if t is taken as zero when the normal to the plane of the coil is along the flux direction. Hence, the induced voltage in the coil, which is equal to the negative of rate of change of flux linkage with time, is a sine function of time. Since the sine function lags the cosine function by 90°, the induced voltage also lags the flux linkage by 90°. This also means that the induced voltage is zero when the flux linkage has the maximum value and vice-versa.

6.9 Eddy-current loss reduces the efficiency of electromagnetic equipments. What measures are taken to reduce the loss?

Answer: The eddy-current loss in a thin iron sheet subjected to a sinusoidally time-varying magnetic field is approximately proportional to its conductivity and square of its thickness. It is also proportional to square of the frequency and square of the amplitude of the flux density. Since the values of frequency and flux density are determined from other considerations, the magnetic cores of equipments operating at power and low frequencies are constructed with thin silicon steel sheets. The conductivity of silicon steel is less than that of normal steel.

As ferrites have low value of conductivity, these are used as cores in equipments operating at high frequency so that eddy-current loss is minimised.

6.10 Distinguish between conduction, convection and displacement current densities. On what factors do each of these currents depend?

Answer: Currents due to free electrons in metallic conductors and due to positive and negative charge carriers in semiconductors are the conduction currents. The conduction current density is equal to current through a surface element divided by the area of the surface element normal to the direction of current flow, in the limit the area tending to zero. The conduction current density is equal to the product of conductivity and applied electric field intensity.

The current density due to a moving charge in free space is the convection current density. The convection current density at a point is equal to the product of volume charge density at the point and velocity.

Displacement current is generated in a dielectric material by a time-varying applied electric field. The displacement current density is equal to the time rate of change of electric flux density in the dielectric.

6.11 Answer: Except for a brief charging period no current flows in a capacitor connected to a dc source. However, if a capacitor is connected to an ac source current flows continuously. Explain.

When a capacitor is connected to a dc source no current flows through it because the region between the plates is an insulator. However, current flows continuously through a capacitor when it is connected to an ac source. Continuity of current between the plates is maintained by the displacement current, the value of which is equal to the conduction current in the circuit. The displacement current is generated by the time-varying applied electric field.

When the capacitor is connected to a dc source, current flows in the circuit until the potential difference between the plates becomes equal to the applied voltage. As during this period the field varies with time, displacement current is generated which maintains the current flow.

6.22 Explain the skin effect phenomenon in a conductor carrying alternating current.

Answer: When time-varying current flows in a good conductor, the internal magnetic field induces currents within the conductor. The induced currents are in opposite sense to the main current as per Lenz's law. This results in a non-uniform distribution of current within the conductor. The current density is maximum at the surface of a round conductor and minimum at its centre. The degree of non-uniformity in the distribution of current over the conductor section increases with increase of the ratio a/δ where a is radius of the conductor and δ is the skin depth. This phenomenon is called the *skin effect*.

PROBLEMS

*6.1 A circular coil of 25 turns and mean radius 2.5 cm is on the $z = 0$ plane of cylindrical coordinates in a uniform magnetic field. The flux density is given by $\mathbf{B} = (0.06\mathbf{u}_y + 0.08\mathbf{u}_z)\sin 5000t$ T. Find the value of root-mean-square (rms) voltage induced in the coil.

**6.2 A planar coil of 20 turns has an area of 30 cm^2. The unit vector normal to the plane of the coil is given by $0.6\mathbf{u}_x + 0.8\mathbf{u}_y$. The coil is linked by an alternating magnetic field of flux density $\mathbf{B} = (0.4\mathbf{u}_y + 0.5\mathbf{u}_z)\cos(100\pi)t$ T. Find the voltage induced in the coil.

*6.3 Derive the integral form of Faraday's law considering a conducting loop of one turn.

**6.4 A circular coil of N turns and mean radius 2.5 cm is on the $z = 0$ plane of cylindrical coordinates. The coil is linked by an alternating magnetic field of flux density $\mathbf{B} = 0.08\sin 2500t\,\mathbf{u}_z$ T. Calculate the electric field intensity along the wire of the coil using the integral form of Faraday's law.

*6.5 A coil of N turns is carrying a time-varying current. Obtain the induced voltage in the coil due to its self-flux in terms of the self-inductance.

**6.6 The number of turns of two magnetically coupled coils 1 and 2 are N_1 and N_2 and currents in them are I_1 and I_2. Obtain the voltage induced in Coil 2 due the current in Coil 1 in terms of the mutual inductance.

***6.7 A search coil of 25 turns has an effective area of 2.4 cm^2. It is placed coaxially at the centre of an air-core solenoid having a uniformly and closely distributed winding of 2000 turns. The length of the solenoid is 25 cm. Find the rms value of induced voltage in the search coil if the current in the solenoid $I = 0.5\sin(2500\pi)t$ A.

*6.8 A conductor oriented along y direction has a length of 15 cm. It moves in a magnetic field with a velocity of 25 m/s in x direction. The magnetic flux density $\mathbf{B} = 0.6\sin x\,\mathbf{u}_z$ T. Find the value of voltage induced in the conductor.

**6.9 A conductor of 30 cm length is oriented along z direction of cylindrical coordinates. It revolves about the z axis at a speed $\omega_s = 100\pi$ rad/s in a magnetic field. The radius of revolution $r = 20$ cm and the magnetic flux density $\mathbf{B} = -0.5\cos\phi\,\mathbf{u}_r$ T, where $\phi = \omega_s t$. Calculate the voltage induced in the conductor. What is the frequency of the induced voltage?

*6.10 Let $\mathbf{B} = -0.6e^{-y}\mathbf{u}_z$ in Problem P6.8. The other data are the same. Find the value of induced voltage.

**6.11 A rectangular coil of N turns and area S rotates at constant angular velocity ω_s in a uniform, stationary and steady magnetic field of flux density B. Calculate the voltage induced in the coil. What is the phase difference between the flux linkage and induced voltage?

**6.12 A rectangular coil has 25 insulated turns. The mean length and width of the coil are 30 cm and 20 cm respectively. It rotates at a speed of 12.5 revolutions per second in a uniform magnetic field of density 0.7 T. Find the value of voltage induced in the coil when the normal to the plane of the coil is (a) at 45° and (b) at 90° with the direction of the flux density.

*6.13 The magnetic field intensity in a region of electromagnetic field is given by $\mathbf{H} = 0.4\sin(2\pi ft - \beta z)\mathbf{u}_y$ A/m^2, where $f = 5.0$ MHz and $\beta = 0.2$ rad/m. Find the displacement current density vector at any time t and at any position z.

*6.14 The conductivity and relative permittivity of the dielectric material in a capacitor are 5.0×10^{-12} S/m and 3.8 respectively. Find the ratio of conduction current density to displacement current density in the dielectric medium at 50Hz.

**6.15 A parallel-plate capacitor is connected to an ac source. Show that the displacement current between the plates is equal to the conduction current in the circuit.

**6.16 A coaxial transmission line has an inner conductor of radius 1.6 mm. The relative permittivity of the dielectric material between the conductors is 2.5. The electric field intensity between the conductors and at a radial distance 2.0 mm is given by $\mathbf{E} = 4000\sin(2\pi\times10^{7})t\,\mathbf{u}_r$ V/m. Find the amplitude of displacement current density on the surface of the inner conductor.

**6.17 The electric field intensity in free-space is given by

$$\mathbf{E} = 15.0\cos(10^{8}t - 0.4z)\mathbf{u}_y.$$

Find the expression for the magnetic field associated with the electric field. Assume the constant of integration to be zero

**6.18 The plain boundary surface between two dielectric regions 1 and 2 has no surface charge. The relative permittivity of Region 1 is 2.5 and that of Region 2 is 4.0. The electric field intensity that is normal to the boundary in Region 1 is given by $80\sin\omega t$ V/m, and the electric flux density that is tangential to the boundary in Region 2 is $300\varepsilon_0 \sin\omega t$ C/m². Find the amplitudes of electric field intensity in the two regions.

**6.19 The vector magnetic potential at a point in free space,

$$\mathbf{A}(t) = (5\times10^{-6})\cos[(2\pi\times10^{8})t - (2\pi/3)z]\mathbf{u}_z \text{ T-m}$$

Find the scalar electric potential at the point using the Lorentz identity $\nabla\cdot\mathbf{A} = -\mu\varepsilon\dfrac{\partial V}{\partial t}$.

**6.20 A current element $Id\ell = 0.5\sin(2\pi\times10^{7})t\,\mathbf{u}_z$ is located in air with its mid-point at the origin of rectangular coordinates. Find the value of vector magnetic potential at time t and at a point $P(r = 300 \text{ m}, \theta = 60°, \phi = 0)$.

**6.21 The magnetic and electric field intensities in an electromagnetic field are given by $\bar{\mathbf{H}} = H_x\mathbf{u}_x + H_y\mathbf{u}_y$ and $\bar{\mathbf{E}} = E_x\mathbf{u}_x - E_y\mathbf{u}_y$ where the quantities are amplitudes and each has a real part and an imaginary part. Obtain the expression for power flow per unit area using the complex form of Poynting vector and interpret the result.

***6.22 A coaxial transmission line has an inner conductor of outside radius 2.0 mm and an outer conductor of inside radius 6.0 mm. The instantaneous electric and magnetic field intensities at a radial distance r (2.0 mm $\leq r \leq$ 6.0 mm) are given by

$$\mathbf{E}(t) = \frac{50}{r}\sin\left(\omega t - \frac{\pi}{8}\right)\mathbf{u}_r \text{ V/m}$$

$$\mathbf{H}(t) = \frac{2}{r}\sin\left(\omega t - \frac{\pi}{6}\right)\mathbf{u}_\phi \text{ A/m}$$

(a) Represent the fields in complex form, (b) Find the complex Poynting vector and (c) Calculate the power flow in the loss-free dielectric region between the conductors.

Chapter Seven

THE UNIFORM PLANE ELECTROMAGNETIC WAVE

7.1 Introduction

An important property of an electromagnetic field is that electric and magnetic fields are interdependent. This may be verified from Maxwell's equations. Maxwell's equations can be used to obtain partial differential equations separately for **E** and **H**. By substituting the solution of either of these equations into the appropriate Maxwell's equation, relation between the two fields can be derived. The form of differential equations, their solutions and relation between the fields depend on whether the displacement current is considered or neglected. For example, Equation (6.21-5) has been derived for a region of good conductor neglecting the displacement current density. The differential equation is of order two with respect to the space variable r and of order one with respect to time. If displacement current density is considered, the differential equations for **E** and **H** would be of different form. Solution of these equations shows that the electromagnetic fields can propagate in the form of waves. The velocity of propagation depends on constants of the medium in which the waves travel. Further, the direction of Poynting vector of the fields is found to be same as the direction of propagation of the waves. It means that electric energy can be transmitted through electromagnetic fields from a source to distant places. The fields also undergo reflection and refraction at the common boundary of two regions of different electric properties.

7.2 The Wave Equation

In order to derive the wave equations for the electromagnetic fields, we will use Maxwell's equations in differential form given in Section 6.13. In a homogeneous and charge-free medium we have $\mathbf{B} = \mu\mathbf{H}$, $\mathbf{D} = \varepsilon\mathbf{E}$, and $\rho = 0$. Using these relations in Maxwell's equations, we get

$$\nabla \times \mathbf{E} = -\mu \frac{\partial \mathbf{H}}{\partial t} \tag{1}$$

$$\nabla \times \mathbf{H} = \sigma\mathbf{E} + \varepsilon \frac{\partial \mathbf{E}}{\partial t} \tag{2}$$

$$\nabla \cdot \mathbf{H} = 0 \tag{3}$$

$$\nabla \cdot \mathbf{E} = 0 \tag{4}$$

We find that both (1) and (2) contain the electric field intensity **E** as well as the magnetic field intensity **H**. By solving the two equations simultaneously we will get two partial differential equations—one for the electric field intensity and the other for the magnetic field intensity. This is achieved as follows. Taking the curl of both sides of (1), we have

$$\nabla \times \nabla \times \mathbf{E} = -\nabla \times \left(\mu \frac{\partial \mathbf{H}}{\partial t} \right) = -\mu \frac{\partial}{\partial t} (\nabla \times \mathbf{H})$$

As per the vector identity (5) of Section (1.9),

$$\nabla \times \nabla \times \mathbf{E} = \nabla(\nabla \cdot \mathbf{E}) - \nabla^2\mathbf{E}.$$

Thus,
$$\nabla(\nabla \cdot \mathbf{E}) - \nabla^2\mathbf{E} = -\mu\frac{\partial}{\partial t}(\nabla \times \mathbf{H}) \tag{5}$$

Replacing $\nabla \times \mathbf{H}$ of (5) by the rhs of (2), and since $\nabla \cdot \mathbf{E} = 0$ as per (4), we obtain

$$\nabla^2\mathbf{E} = \mu\sigma\frac{\partial \mathbf{E}}{\partial t} + \mu\varepsilon\frac{\partial^2\mathbf{E}}{\partial t^2} \tag{6}$$

Equation (6) has been derived by taking the curl of both sides of (1) and then eliminating **H**. Alternatively, if we take the curl of both sides of (2) and then eliminate **E**, then we will get

$$\nabla^2\mathbf{H} = \mu\sigma\frac{\partial \mathbf{H}}{\partial t} + \mu\varepsilon\frac{\partial^2\mathbf{H}}{\partial t^2} \tag{7}$$

This equation is of same form as (6).

Equations (6) and (7) are second order partial differential equations for the vector fields **E** and **H**. These are the *vector wave equations* for the electromagnetic field in a charge-free homogeneous medium.

Wave Equations in a Loss-Free Dielectric Medium

If the medium is loss-less ($\sigma = 0$), then the wave equations are

$$\nabla^2\mathbf{E} = \mu\varepsilon\frac{\partial^2\mathbf{E}}{\partial t^2} \tag{8}$$

$$\nabla^2\mathbf{H} = \mu\varepsilon\frac{\partial^2\mathbf{H}}{\partial t^2} \tag{9}$$

The wave equations in free space are obtained by replacing μ by μ_0 and ε by ε_0 in the above equations.

The field intensities may have components along all the three coordinate directions and each component may be a function of all the space variables x, y and z. The electric field intensity in terms of its components is

$$\mathbf{E} = E_x\mathbf{u}_x + E_y\mathbf{u}_y + E_z\mathbf{u}_z \tag{10}$$

By substituting (10) in (8), we obtain

$$\nabla^2(E_x\mathbf{u}_x + E_y\mathbf{u}_y + E_z\mathbf{u}_z) = \mu\varepsilon\frac{\partial^2}{\partial t^2}(E_x\mathbf{u}_x + E_y\mathbf{u}_y + E_z\mathbf{u}_z)$$

Equating each component of the rhs with the corresponding component of the lhs, we get

$$\nabla^2 E_x = \mu\varepsilon\frac{\partial^2 E_x}{\partial t^2} \tag{11}$$

$$\nabla^{`}\ {}_{y} = \mu\varepsilon\frac{\partial^2 E_y}{\partial t^2} \tag{12}$$

$$\nabla^{`}\ {}_{z} = \mu\varepsilon\frac{\partial^2 E_z}{\partial t^2} \tag{13}$$

Thus, the differential equation of each component of **E** (and also of **H**) is also a wave equation. As the components are scalars, the equations are the *scalar wave equations*. Using the expanded form of the ∇^2 operator in (11), we have

$$\frac{\partial^2 E_x}{\partial x^2} + \frac{\partial^2 E_x}{\partial y^2} + \frac{\partial^2 E_x}{\partial z^2} = \mu\varepsilon \frac{\partial^2 E_x}{\partial t^2} \tag{14}$$

Equation (14) is the three-dimensional wave equation in E_x. The equations for the other components of the electric field intensity and for the components of magnetic field intensity are similar to (14).

Wave Equations for Sinusoidal Variation of Fields with Time

Wave equations for sinusoidal variation of fields with time using the complex exponential $e^{j\omega t}$ form of representation can be derived either from Maxwell's equations in Section 6.15 or by simply substituting $\partial/\partial t = j\omega$ and $\partial^2/\partial t^2 = -\omega^2$ in (6) and (7) and using the phasor symbols. The equations are given by

$$\nabla^2 \bar{\mathbf{E}} = j\omega\mu\sigma\bar{\mathbf{E}} - \omega^2\mu\varepsilon\bar{\mathbf{E}} \tag{15}$$

$$\nabla^2 \bar{\mathbf{H}} = j\omega\mu\sigma\bar{\mathbf{H}} - \omega^2\mu\varepsilon\bar{\mathbf{H}} \tag{16}$$

The equations in a loss-free medium are obtained by substituting $\sigma = 0$ in (15) and (16). Thus,

$$\nabla^2 \bar{\mathbf{E}} = -\omega^2\mu\varepsilon\bar{\mathbf{E}} \tag{17}$$

$$\nabla^2 \bar{\mathbf{H}} = -\omega^2\mu\varepsilon\bar{\mathbf{H}} \tag{18}$$

Equations (17) and (18) are known as *Helmholtz vector wave equations.*

The equations for each component of $\bar{\mathbf{E}}$ and $\bar{\mathbf{H}}$ can be derived using the same procedure as described. For example, the three dimensional wave equation for the electric field intensity in x direction is

$$\frac{\partial^2 \bar{E}_x}{\partial x^2} + \frac{\partial^2 \bar{E}_x}{\partial y^2} + \frac{\partial^2 \bar{E}_x}{\partial z^2} = -\omega^2\mu\varepsilon\bar{E}_x \tag{19}$$

This may be considered as the *phasor wave equation* in $\bar{E}_x$.

7.3 Solution of the One-Dimensional Wave Equation

Most of the properties of an electromagnetic field satisfying the wave equation can be derived by solving the one-dimensional wave equation in a perfect dielectric medium. It is obtained by considering only one component of **E** or **H** and assuming that it is a function of only one space variable.

We assume that the electric field intensity has only the x component E_x and it is a function of z and t.

The wave equation for the field intensity may be obtained from (19) of the preceding section by setting variations with x and y to zero. Thus, the wave equation is

$$\frac{\partial^2 E_x}{\partial z^2} = \mu\varepsilon \frac{\partial^2 E_x}{\partial t^2} \tag{1}$$

One way of solving (1) is by the method of separation of variables described in Section 5.4. As per the method, the electric field intensity E_x is assumed to be the product of two field intensities $E_x(z)$ and $E_x(t)$, where $E_x(z)$ is a function of z only, and $E_x(t)$ is a function of t only. That is,

$$E_x = E_x(z)\, E_x(t) \tag{2}$$

Substitution of (2) in (1) yields

$$E_x(t)\frac{\partial^2 E_x(z)}{\partial z^2} = \mu\varepsilon E_x(z)\frac{\partial^2 E_x(t)}{\partial t^2}$$

Dividing both sides by $\mu\varepsilon E_x(z)\, E_x(t)$, we obtain

$$\frac{1}{\mu\varepsilon E_x(z)}\frac{\partial^2 E_x(z)}{\partial z^2} = \frac{1}{E_x(t)}\frac{\partial^2 E_x(t)}{\partial t^2} \tag{3}$$

As the expression on the lhs is a function of z only and that on the rhs is a function of t only, each must be equal to a constant. Because, if the lhs changes with z it cannot be equal to the rhs, which is not a function of z. By the same argument, the expression on rhs cannot change with t.

Taking the constant as $-k^2$, where k is a real quantity, we will get two second order ordinary differential equations from (3). These are

$$\frac{d^2 E_x(t)}{dt^2} = -k^2 E_x(t) \tag{4}$$

$$\frac{d^2 E_x(z)}{dz^2} = -k^2 \mu\varepsilon E_x(z) \tag{5}$$

The solutions of (4) and (5) in complex exponential form are given by

$$E_x(t) = Ae^{-jkt} + Be^{jkt} \tag{6}$$

$$E_x(z) = Ce^{-j(k\sqrt{\mu\varepsilon})z} + De^{j(k\sqrt{\mu\varepsilon})z} \tag{7}$$

where A, B, C and D are constants.

The complete solution of (1) is the product of (6) and (7). Thus,

$$E_x = \left(Ae^{-jkt} + Be^{jkt}\right)\left(Ce^{-j(k\sqrt{\mu\varepsilon})z} + De^{j(k\sqrt{\mu\varepsilon})z}\right)$$

$$= C_1 e^{-jk(t+\sqrt{\mu\varepsilon}z)} + C_2 e^{-jk(t-\sqrt{\mu\varepsilon}z)} + C_3 e^{jk(t-\sqrt{\mu\varepsilon}z)} + C_4 e^{jk(t+\sqrt{\mu\varepsilon}z)} \tag{8}$$

This is the general solution of (1) when both sides of (3) are equal to a negative constant.

It can be shown that each term of (8) and any function with $(t-\sqrt{\mu\varepsilon z})$ as its variable is a solution of (1). The solution to be used depends on the nature of the problem. In the next section we will find that a function with $(t-\sqrt{\mu\varepsilon z})$ as the variable represents a travelling wave.

EXAMPLE 7.1

Show that the electric field,

$E = C_1 \cos\left(t-\sqrt{\mu\varepsilon}z\right) + C_2\cos\left(t+\sqrt{\mu\varepsilon}z\right)$

satisfies the appropriate wave equation.

As E is a function of z, the appropriate wave equation is

$$\frac{\partial^2 E}{\partial z^2} = \mu\varepsilon\frac{\partial^2 E}{\partial t^2}$$

Differentiating E twice with t, we get

$$\frac{\partial^2 E}{\partial t^2} = -C_1 \cos\left(t - \sqrt{\mu\varepsilon z}\right) - C_2 \cos\left(t + \sqrt{\mu\varepsilon z}\right)$$

Differentiating E twice with z, we get

$$\frac{\partial^2 E}{\partial z^2} = -\mu\varepsilon\left[C_1 \cos\left(t - \sqrt{\mu\varepsilon z}\right) + C_2 \cos\left(t + \sqrt{\mu\varepsilon z}\right)\right] = \mu\varepsilon \frac{\partial^2 E}{\partial t^2}$$

7.4 Solution of Wave Equation for Sinusoidally Time-Varying Fields

Since in practice solutions of wave equations for fields varying sinusoidally with time are mostly used, we will solve the phasor form of one-dimensional wave equation and analyse the solution in some detail. As in the preceding section, we will consider only the x component of the electric field intensity and assume that it is a function of z and t only. By using the complex exponential $e^{j\omega t}$ to represent variation of the fields with time, the x component of electric field intensity may be written as

$$\bar{E}_x = \bar{E}_x(z)e^{j\omega t} \tag{1}$$

where $\bar{E}_x(z)$ is a function of z only. The wave equation for the field intensity is

$$\frac{\partial^2 \bar{E}_x}{\partial z^2} = \mu\varepsilon \frac{\partial^2 \bar{E}_x}{\partial t^2} \tag{2}$$

By substituting (1) in (2), we get

$$\frac{\partial^2 [\bar{E}_x(z)e^{j\omega t}]}{\partial z^2} = \mu\varepsilon \frac{\partial^2 [\bar{E}_x(z)e^{j\omega t}]}{\partial t^2}$$

Or

$$\frac{\partial^2 \bar{E}_x(z)}{\partial z^2} = -\omega^2 \mu\varepsilon \bar{E}_x(z) = -\beta^2 \bar{E}_x(z) \tag{3}$$

where

$$\beta = \omega\sqrt{\mu\varepsilon} = 2\pi f\sqrt{\mu\varepsilon} \tag{4}$$

The solution of (3) in exponential form is

$$\bar{E}_x(z) = C_1 e^{-j\beta z} + C_2 e^{j\beta z} \tag{5}$$

The arbitrary constants C_1 and C_2 are assumed to be real, though, in general, these may be complex quantities. By inserting the factor $e^{j\omega t}$ in (5) we get the solution in terms of z and t. Thus,

$$\bar{E}_x = C_1 e^{j(\omega t - \beta z)} + C_2 e^{j(\omega t + \beta z)} \tag{6}$$

The real-time solution of the wave equation is given by either the real part or the imaginary part (without the factor j) of (6). We will use the real part to study the properties of the solution. It is given by

$$E_x = C_1 \cos(\omega t - \beta z) + C_2 \cos(\omega t + \beta z) \tag{7}$$

The Positively Travelling Wave: Phase Velocity, Phase Constant and Wavelength

We find that the solution given by (7) consists of two cosine functions. One is a cosine function of $(\omega t - \beta z)$ and the other is a cosine function of $(\omega t + \beta z)$. Each function is also a solution of the equation. The characteristic of the field depends on the values of constants C_1 and C_2. Let us at first assume that C_2 is zero. Accordingly,

$$E_x = C_1 \cos(\omega t - \beta z) \tag{8}$$

The field intensity defined by (8) has a maximum value of C_1 and it occurs when $\omega t = \beta z$. It means that the position of the maximum value along z direction depends on t. If we consider three instants of time, namely, $t = 0$, $t = t_1$ and $t = t_2$, then the corresponding values of z at which E_x is maximum are as follows: $z = 0$, $z = z_1 = t_1/\sqrt{\mu\varepsilon}$, and $z = z_2 = t_2/\sqrt{\mu\varepsilon}$. The plots of E_x versus z for the three instants are shown in Fig. 7.1. A study of the plots shows that the field has progressed by a distance equal to $(z_2 - z_1)$ metres along positive z direction in time $(t_2 - t_1)$ seconds, without change of form. Thus relative to the stationary origin, the field is a travelling wave. The velocity of the wave is

$$v = \frac{z_2 - z_1}{t_2 - t_1} = \frac{1}{\sqrt{\mu\varepsilon}} \text{ m/s} \tag{9}$$

In a single medium the field propagates at a constant velocity. The velocity of the waves is called *phase velocity*. The phase velocity in free space,

$$v_0 = \frac{1}{\sqrt{\mu_0 \varepsilon_0}} = 3\times10^8 \text{ m/s} \tag{10a}$$

This is same as the velocity of light.

The phase velocity of a wave in a material medium of relative permeability μ_r and relative permittivity ε_r in terms of v_0 is

$$v = \frac{v_0}{\sqrt{\mu_r \varepsilon_r}} \tag{10b}$$

This is less than the velocity of the wave in free space.

The parameter β defined by (4) is called *phase constant* as it is a measure of the phase shift of a wave per unit distance in the direction of its travel. The velocity given by (10) is termed as phase velocity because a phase point P of the wave, shown in Fig. 7.1, travels at this velocity or because the wave travels with a fixed phase constant in a given medium and at a given frequency.

The travelling wave described by (8) repeats itself after every 2π radian during its progress in the positive z direction. Hence, the phase difference between two consecutive cycles is equal to 2π. Thus, if λ denotes the wavelength then

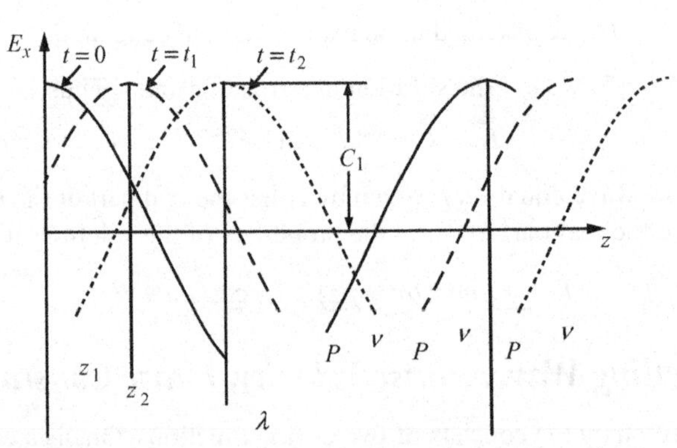

Fig. 7.1 Representation of a travelling wave in a loss-free medium

$$\beta\lambda = 2\pi$$

Or

$$\lambda = \frac{2\pi}{\beta} \tag{11}$$

The relation between phase velocity and phase constant is

$$\beta = \omega\sqrt{\mu\varepsilon} = \frac{\omega}{v} \tag{12}$$

The Negatively Travelling Wave

If in (7) C_1 is taken as zero, it can similarly be shown as in the case of the positively travelling wave that the field defined by the function $C_2 \cos(\omega t + \beta z)$ is a cosine wave travelling along the negative z direction. The amplitude of the wave is C_2. The phase constant, phase velocity and wavelength are same as those of the wave travelling along positive z direction.

The Standing Wave

Next, if in (7) $C_1 = C_2 = C$ (say), then

$$E_x = C[\cos(\omega t - \beta z) + \cos(\omega t + \beta z)] = 2C \cos \omega t \cos \beta z \tag{13}$$

The field described by (13) is a cosine wave along the spatial direction z.

As the maximum value of the wave occurs at fixed values of z, the wave is a stationary wave or a standing wave.

The amplitude of the wave is $2C \cos \omega t$ with wavelength equal to $2\pi/\beta$. A plot of the wave is shown in Fig. 7.2.

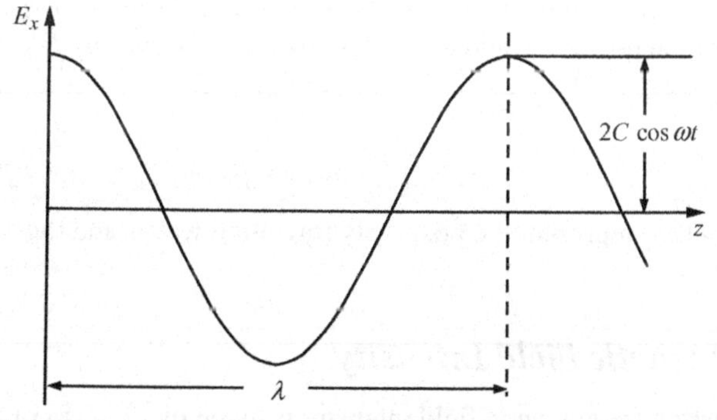

Fig. 7.2 Representation of a standing cosine wave in a loss-free medium. The amplitude of the wave varies harmonically with time.

Combined Travelling and Standing Waves

In (7), if $C_1 \neq C_2$ and assuming C_1 to be greater than C_2, it can be shown that

$$E_x = (C_1 - C_2) \cos(\omega t - \beta z) + 2C_2 \cos \omega t \cos \beta z \tag{14}$$

The first term on rhs of (14) is a wave travelling in positive z direction with amplitude $(C_1 - C_2)$, and the second term is a standing wave with amplitude $2C_2 \cos \omega t$. Thus, the field is a combination of a travelling wave and a standing wave.

Amplitude of the Waves

In equation (7), C_1 is the amplitude of the positively travelling wave and C_2 is the amplitude of the negatively travelling wave. Replacing C_1 by E^+_{x0} and C_2 by E^-_{x0}, the solution of the wave equation in phasor form is given by

$$\bar{E}_x(z) = E^+_{x0}e^{-j\beta z} + E^-_{x0}e^{j\beta z} \tag{15}$$

The real-time solution of the wave equation is the real part of (15). Thus denoting the solution by E_x, we have

$$E_x = E^+_{x0}\cos(\omega t - \beta z) + E^-_{x0}\cos(\omega t + \beta z) \tag{16}$$

Denoting the first and second terms of (16) by E^+_x and E^-_x respectively, the equations for the positively and negatively travelling waves are

$$E^+_x = E^+_{x0}\cos(\omega t - \beta z) \tag{17}$$

$$E^-_x = E^-_{x0}\cos(\omega t + \beta z) \tag{18}$$

Note that the fields described by (17) and (18) are in x direction, whereas these are functions of z and t.

Next, let us consider that the electric field intensity is in y direction and it varies in space with z only. For variation of the field with time represented by $e^{j\omega t}$, the phasor form of the field is described by

$$\bar{E}_y = \bar{E}_y(z)e^{j\omega t}$$

where $E_y(z)$ is a function of z only.

The wave equation for the field is

$$\frac{\partial^2 \bar{E}_y(z)}{\partial z^2} = -\omega^2 \mu\varepsilon \bar{E}_y(z) \tag{19}$$

The phasor and real-time solutions of (19) are of the same form as given by (15) and (16) respectively. Thus denoting the amplitudes of the positively and negatively travelling waves by E^+_{y0} and E^-_{y0}, the solutions are

$$\bar{E}_y(z) = E^+_{y0}e^{-j\beta z} + E^-_{y0}e^{j\beta z} \tag{20}$$

and

$$E_y = E^+_{y0}\cos(\omega t - \beta z) + E^-_{y0}\cos(\omega t + \beta z) \tag{21}$$

The first terms of (20) and (21) represent the positively travelling waves and the second terms give the negatively travelling waves.

Solutions for the Magnetic Field Intensity

The Helmholtz wave equation for magnetic field intensity is given by (7.2-18) of Section 7.2. Assuming that the x- and y-directed magnetic field intensities are functions only of z, we can obtain the one-dimensional wave equations and their solutions by proceeding in similar lines as for the electric field intensities. The solutions involving phasors of the magnetic fields are

$$\bar{H}_x(z) = H^+_{x0}e^{-j\beta z} + H^-_{x0}e^{j\beta z} \tag{22}$$

$$\bar{H}_y(z) = H^+_{y0}e^{-j\beta z} + H^-_{y0}e^{j\beta z} \tag{23}$$

The real-time solutions are given by

H_x [illegible] H^+_{x0} [illegible] t [illegible] $z) + H_{x0}$ [illegible] $t +$ [illegible] $z)$ (24)

H_y [illegible] H^+_{y0} [illegible] t [illegible] $z) + H_{y0}$ [illegible] $t +$ [illegible] $z)$ (25)

Relation between E and H

The relationship between electric and magnetic field intensities can be derived by substituting (16) and (21) in one of Maxwell's curl equations. We will use the curl equation.

$$\mu \frac{\partial \mathbf{H}}{\partial t} = -\nabla \times \mathbf{E} \tag{26}$$

At first let us assume that the electric field intensity is in the x direction. So, the field-intensity vector is given by

$$\mathbf{E} = E_x \mathbf{u}_x$$

Since E_x does not vary with x and y, hence

$$\nabla \times \mathbf{E} = -\frac{\partial E_x}{\partial z}\mathbf{u}_y \tag{27}$$

Substituting (27) in (26), we obtain

$$\mu \frac{\partial \mathbf{H}}{\partial t} = -\frac{\partial E_x}{\partial z}\mathbf{u}_y \tag{28}$$

As the rhs of (28) is in y direction, the magnetic field intensity can only be in y direction. Thus, the magnetic-field vector

$$\mathbf{H} = H_y \mathbf{u}_y$$

By substituting in (28), we get

$$\mu \frac{\partial H_y}{\partial t} = -\frac{\partial E_x}{\partial z} \tag{29}$$

Substitution of (16) in (29) gives

$$\mu \frac{\partial H_y}{\partial t} = -\frac{\partial}{\partial z}[E_{x0}^{+}\cos(\omega t - \beta z) + E_{x0}^{-}\cos(\omega t + \beta z)]$$

$$= -\beta[E_{x0}^{+}\sin(\omega t - \beta z) - E_{x0}^{-}\sin(\omega t + \beta z)]$$

Or,

$$H_y = -\frac{\beta}{\mu}\int [E_{x0}^{+}\sin(\omega t - \beta z) - E_{x0}^{-}\sin(\omega t + \beta z)]dt$$

$$= \frac{1}{\eta}[E_{x0}^{+}\cos(\omega t - \beta z) - E_{x0}^{-}\cos(\omega t + \beta z)] \tag{30}$$

where

$$\eta = \sqrt{\frac{\mu}{\varepsilon}} \tag{31}$$

The constant of integration is taken as zero because any function that is independent of time cannot be a travelling wave.

The first and second terms of (30) are the waves of magnetic field intensity travelling along positive and negative z directions. The two fields are given by

$$H_y^{+} = H_{y0}^{+}\cos(\omega t - \beta z) = \frac{E_{x0}^{+}}{\eta}\cos(\omega t - \beta z) \tag{32}$$

$$H_y^{-} = H_{y0}^{-}\cos(\omega t + \beta z) = \frac{E_{x0}^{-}}{\eta}\cos(\omega t + \beta z) \tag{33}$$

Thus, the relationship between the x-directed electric field and the y-directed magnetic field travelling along positive and negative z directions are

$$\frac{E_x^+}{H_y^+} = \frac{E_{x0}^+}{H_{y0}^+} = \eta \qquad (34)$$

$$\frac{E_x^-}{H_y^-} = \frac{E_{x0}^-}{H_{y0}^-} = -\eta \qquad (35)$$

Next, assuming that the electric field intensity is in y direction and using the same procedure as above it can be shown that

$$\frac{E_y^+}{H_x^+} = \frac{E_{y0}^+}{H_{x0}^+} = -\eta \qquad (36)$$

$$\frac{E_y^-}{H_x^-} = \frac{E_{y0}^-}{H_{x0}^-} = \eta \qquad (37)$$

The parameter η defined by (31) is called *intrinsic impedance* or *intrinsic wave impedance* of the medium in which the waves propagate. The unit of η is ohm. The intrinsic impedance of free space is

$$\eta_0 = \sqrt{\frac{\mu_0}{\varepsilon_0}} = 120\pi \cong 377\ \Omega$$

We see that the waves of the x-directed electric field intensity are associated with the waves of y-directed magnetic field intensity. Also, the waves of y-directed electric field intensity are associated with the waves of x-directed magnetic field intensity. The values of phase velocity, phase constant, and wavelength of the magnetic fields are same as those of the associated electric fields. The waves are called *electromagnetic waves.* One set of waves is shown in Fig. 7.3.

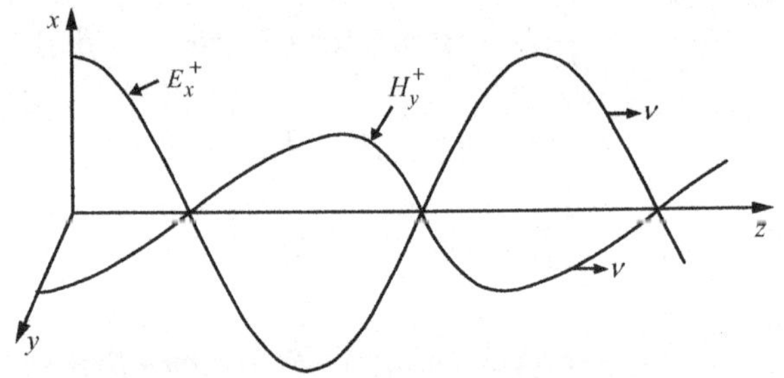

Fig. 7.3 Representation of electric and magnetic field waves travelling along positive z direction. The electric field is in x direction and the magnetic field is in y direction.

Some other properties of electromagnetic waves are described below.

1. The distribution of the fields at any instant of time over planes parallel to the x-y plane is uniform, as shown in Fig. 7.4. Therefore the waves are called *uniform plane waves*.

2. If one field has both x and y components, the other field will also have both the components. The ratio between the magnitudes of the resultant electric and the resultant magnetic field intensities is equal to η. To derive the relation, let the magnetic field intensity for the positively travelling wave be given by

$$\mathbf{H}^+ = H_x^+\mathbf{u}_x + H_y^+\mathbf{u}_y$$

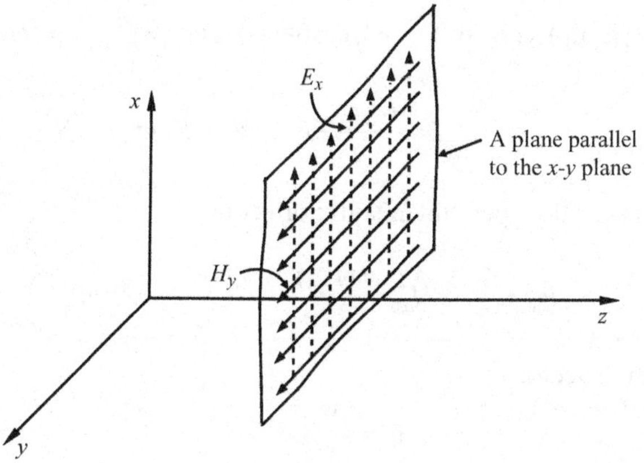

Fig. 7.4 Distribution of electric and magnetic fields of a uniform plane electromagnetic wave

The magnitude of the magnetic field,

$$|\mathbf{H}^+| = \sqrt{(H_x^+)^2 + (H_y^+)^2}$$

The electric field intensity associated with the magnetic field is

$$\mathbf{E}^+ = E_x^+\mathbf{u}_x + E_y^+\mathbf{u}_y$$

The magnitude of the electric field,

$$|\mathbf{E}^+| = \sqrt{(E_x^+)^2 + (E_y^+)^2}$$

Using relations (34) and (36),

$$|\mathbf{E}^+| = \eta\sqrt{(H_y^+)^2 + (-H_x^+)^2} = \eta|\mathbf{H}^+|$$

Or
$$\frac{|\mathbf{E}^+|}{|\mathbf{H}^+|} = \eta$$

This relation is same as the relation between the electric field intensity having a single component and the corresponding magnetic field intensity.

3. The electric and magnetic fields are in space quadrature even if they have components in both x and y directions. This is proved as follows. For the positively travelling fields we have

$$\mathbf{H}^+ = H_x^+\mathbf{u}_x + H_y^+\mathbf{u}_y$$

and
$$\mathbf{E}^+ = E_x^+\mathbf{u}_x + E_y^+\mathbf{u}_y = \eta(H_y^+\mathbf{u}_x - H_x^+\mathbf{u}_y)$$

Taking the dot product of $\mathbf{E}^+$ and $\mathbf{H}^+$, we have

$$\mathbf{E}^+ \cdot \mathbf{H}^+ = \eta(H_y^+H_x^+ - H_x^+H_y^+) = 0$$

This shows that the two fields are perpendicular to each other.

4. The power flow per unit area is equal to the average value of the instantaneous or the real part of the complex Poynting vector. Considering the positively travelling wave with instantaneous fields E_x^+ and H_y^+, the Poynting vector for any z and t,

$$\mathbf{P}_w(z,t) = (E_x^+\mathbf{u}_x)\times(H_y^+\mathbf{u}_y) = [E_{x0}^+\cos(\omega t-\beta z)\mathbf{u}_x]\times[H_{y0}^+\cos(\omega t-\beta z)\mathbf{u}_y]$$

$$= \frac{E_{x0}^+H_{y0}^+}{2}\cos^2(\omega t-\beta z)\mathbf{u}_z = \frac{E_{x0}^+H_{y0}^+}{2}[1+\cos 2(\omega t-\beta z)]\mathbf{u}_z$$

Thus, the average power flow per unit area in z direction is

$$P_{av} = \frac{1}{2}E_{x0}^+H_{y0}^+ = \frac{(E_{x0}^+)^2}{2\eta} = \frac{\eta(H_{y0}^+)^2}{2}\ \text{W/m}^2$$

The complex Poynting vector,

$$\bar{\mathbf{P}}_w = \frac{1}{2}(\bar{\mathbf{E}}\times\bar{\mathbf{H}}^*)$$

For the positively travelling wave with x-directed electric field and y-directed magnetic field, we have

$$\bar{\mathbf{E}} = \bar{E}_x^+\mathbf{u}_x = E_{x0}^+e^{-j\beta z}\mathbf{u}_x$$

$$\bar{\mathbf{H}} = \bar{H}_y^+\mathbf{u}_y = H_{y0}^+e^{-j\beta z}\mathbf{u}_y$$

$$\bar{\mathbf{H}}^* = H_{y0}^+e^{j\beta z}\mathbf{u}_y$$

So,
$$\bar{\mathbf{P}}_w = \frac{1}{2}[(E_{x0}^+e^{-j\beta z}\mathbf{u}_x)\times(H_{y0}^+e^{j\beta z}\mathbf{u}_y)] = \frac{E_{x0}^+H_{y0}^+}{2}\mathbf{u}_z$$

We find that the complex Poynting vector has no imaginary part. This is because we have assumed E_{x0}^+ and H_{y0}^+ to be real quantities. Thus, the average power flow per unit area in z direction is given by

$$P_{av} = \frac{E_{x0}^+H_{y0}^+}{2} = \frac{E_{x0}^{+\,2}}{2\eta} = \frac{\eta H_{y0}^{+\,2}}{2}$$

which is same as that obtained from the instantaneous Poynting vector.

5. The time average of energy stored per unit volume (energy density) in the electric field is equal to that stored in the magnetic field. This is proved as follows. Energy densities in the static electric and static magnetic fields are given by $\varepsilon E^2/2$ and $\mu H^2/2$ respectively. The formulae for the energy densities for sinusoidally time-varying electromagnetic fields are obtained by using rms values of the field intensities in these expressions.

$$\text{Energy density in the magnetic field} = \frac{1}{2}\left(\frac{\mu H_{y0}^+}{\sqrt{2}}\right)^2 = \frac{\mu(H_{y0}^+)^2}{4}$$

$$\text{Energy density in the electric field} = \frac{1}{2}\left(\frac{\varepsilon E_{y0}^+}{\sqrt{2}}\right)^2 = \frac{\varepsilon(E_{x0}^+)^2}{4} = \frac{\varepsilon(\eta H_{y0}^+)^2}{4} = \frac{\mu(H_{y0}^+)^2}{4}$$

Thus, the two energy densities are equal.

6. In a charge-free region,

$$\nabla\cdot\mathbf{E} = \frac{\partial E_x}{\partial x} + \frac{\partial E_y}{\partial y} + \frac{\partial E_z}{\partial z} = 0$$

The first two terms are equal to zero as both E_x and E_y have been assumed to be independent of x and y. So, the last term must also be equal to zero. That is,

$$\frac{\partial E_z}{\partial z} = 0$$

Also,

$$\frac{\partial^2 E_z}{\partial z^2} = 0$$

As this is not a wave equation, its solution will not represent a wave. Therefore, E_z may be assumed to be zero. Also, as $\nabla \cdot \mathbf{B} = 0$, it can be shown in the same way that $H_z = 0$. So, a uniform plane wave travelling in z direction will not contain components of the fields in the direction of wave propagation.

7. As the electric and magnetic field intensities of a uniform plane wave are on a plane that is transverse (perpendicular) to the direction of propagation of the wave, it is called *transverse electromagnetic* (TEM) wave.

EXAMPLE 7.2

The electric field intensity of a uniform plane wave in a perfect dielectric medium is given by $E = E_0 \cos[(\pi\times 10^8)t + 1.6z]$. If the relative permeability of the medium is unity, find its relative permittivity.

Angular frequency, $\omega = \pi \times 10^8$ rad/s

Phase constant, $\beta = 1.6$ rad/m

$$\beta \quad \omega\sqrt{\mu_0\varepsilon_0\varepsilon_r} = \frac{\pi\times 10^8}{3\times 10^8}\sqrt{\varepsilon_r} = 1.6$$

$\varepsilon_r = 2.332$

EXAMPLE 7.3

The phase constant of a uniform plane electromagnetic wave travelling in a perfect dielectric medium is 10 radians per metre. Calculate the phase velocity, wave length, frequency and intrinsic impedance of the field. The relative permittivity and relative permeability of the medium are 4.8 and 1.0 respectively.

Phase constant, $\beta = 10.0$ rad/m

Phase velocity, $v = 1/\sqrt{\mu_0(4.8\varepsilon_0)} = (3\times10^8)/\sqrt{4.8} = 1.369\times10^8$ m/s

Wavelength, $\lambda = 2\pi/\beta = 2\pi/10 = \pi/5$ m

Frequency, $f = v/\lambda = (1.369\times10^8)/(\pi/5) = 2.179\times10^8$ Hz

Angular frequency, $\omega = 2\pi f = 2\pi(2.179) = 13.69\times10^8$ rad/s

Intrinsic impedance, $\eta = \sqrt{\mu_0/(4.8\varepsilon_0)} = 120\pi/\sqrt{4.8} = 172.07\ \Omega$

EXAMPLE 7.4

A uniform plane electromagnetic wave at 10^7 Hz is travelling along positive z direction in free space. The electric field intensity of the wave is in positive y direction and its amplitude is 5.0 V/m. Obtain expressions for the electric and magnetic field intensities both in phasor and real-time forms.

Frequency, $f = 10^7$ Hz

Amplitude of the electric field, $E_0 = 5.0$ V/m

The phase constant,

$$\beta \quad \omega \sqrt{\mu_{.} \varepsilon_0} = 2\pi(10.0\times10^6)/(3\times10^8) = 0.2094 \text{ rad/m}$$

The intrinsic impedance, $\eta_0 = \sqrt{\mu_0/\varepsilon_0} = 120\pi$ Ohm

The expressions for the electric field are

Phasor form: $\bar{\mathbf{E}}(z) = E^+_{y0} e^{-j\beta z}\mathbf{u}_y = 5.0e^{-j0.2094z}\,\mathbf{u}_y$ V/m

In real time: $\mathbf{E} = E^+_{y0}\cos(\omega t - \beta z)\mathbf{u}_y = 5.0\cos[(2\pi\times10^7)t - 0.2094z]\mathbf{u}_y$ V/m

As the electric field is in y direction and the wave propagates along positive z direction,

$$H_x^{'} \qquad E^+_{y0}/\eta_0 = -5.0/120\pi = -1/24\pi$$

Thus, the expressions for the magnetic field are

Phasor form: $\bar{\mathbf{H}}(z) = -(1/24\pi)e^{-j0.2094z}\,\mathbf{u}_x$ A/m

In real time: $\mathbf{H} = -(1/24\pi)\cos[(2\pi\times 10^7)t - 0.2094z]\mathbf{u}_x$ A/m

EXAMPLE 7.5

The amplitude and phase constant of magnetic field intensity in a uniform plane electromagnetic wave travelling along negative z direction in a perfect dielectric medium are 0.5 V/m and 8 rad/m respectively. For the medium, $\varepsilon_r = 5.0$ and $\mu_r = 1.0$. Obtain expressions for the magnetic field intensity and electric field intensity in real time and in phasor form.

Amplitude of magnetic field intensity, $H_0 = 0.5$ V/m

Phase constant, $\beta = 8$ rad/m

$$\text{Radian frequency, } \omega = \frac{\beta}{\sqrt{\mu_0\varepsilon_0\varepsilon_r}} = \frac{(8)(3\times10^8)}{\sqrt{5}} = 10.73\times10^8 \text{ rad/s}$$

$$\text{Intrinsic impedance, } \eta = \frac{\eta_0}{\sqrt{\varepsilon_r}} = \frac{377}{\sqrt{5}} = 168.6\ \Omega$$

Amplitude of electric field intensity, $E_0 = \eta H_0 = (168.6)(0.5) = 84.3$ V/m

Expressions of the fields in real time are

$E = 84.3\cos(10.73t\times10^8 + 8z)$ V/m

$H = 0.5\cos(10.73t\times10^8 + 8z)$ A/m

Expressions of the fields in phasor form are

$E \qquad e^{j\ z}$ V/m

$H \qquad e^{j\ z}$ A/m

EXAMPLE 7.6

The amplitude of electric field intensity of a uniform plane wave travelling along positive z direction in air is 25.0 V/m. Calculate the power flowing through a plane surface of area 1.2 square metres if the unit vector normal to the surface is

$\mathbf{u}_n = (5/13)\mathbf{u}_x + (12/13)\mathbf{u}_z$.

Amplitude of the electric field, $E_0^+ = 25.0$ V/m

Intrinsic impedance of air, $\eta_0 = 120\pi\,\Omega$

The vector surface, $\mathbf{S} = 1.2\mathbf{u}_n = (6/13)\mathbf{u}_x + (14.4/13)\mathbf{u}_z$ m^2

The electric and magnetic fields of a uniform plane wave have the same phase constant and are in time phase. Therefore, the complex Poynting vector has no imaginary part. Also, its direction is same as the direction of propagation of the wave. Thus,

$$\mathbf{P}_{wa} = \frac{E_0^{+2}}{2\eta_0}\mathbf{u}_z = \frac{25^2}{2(120\pi)}\mathbf{u}_z = 0.829\,\mathbf{u}_z \text{ W/m}^2$$

Power through the given surface $= \mathbf{P}_{wa} \cdot \mathbf{S}$

$$= (0.829\mathbf{u}_z) \cdot [(6/13)\mathbf{u}_x + (14.4/13)\mathbf{u}_z] = 0.918 \text{ W}$$

EXAMPLE 7.7

A uniform plane wave propagates along positive z direction in free space. The electric field intensity in the wave has components both in x and y directions. The amplitude of the field in x direction is 250 V/m and along y direction the amplitude is 300 V/m. The frequency of the wave is 50 MHz. Obtain the phasor expressions of electric and magnetic field intensities.

Amplitudes of electric field intensity in x and y directions are

$E_x' = 250$ V/m, $E_{y0}' = 300$ V/m

Phase constant, $\beta = \omega\sqrt{\mu_0\varepsilon_0} = 2\pi(50\times10^6)/(3\times10^8) = 1.047$ rad/m

Intrinsic impedance, $\eta_0 = 377\,\Omega$

For the positively travelling wave,

$H_x' = E_{y0}^+/\eta_0 - -300/377 = -0.798$ A/m

$H_y' = E_{x0}^+/\eta_0 = 250/377 = 0.663$ A/m

Phasor expressions of the fields are

$\mathbf{E}\ \ (250\mathbf{u}_x + 300\mathbf{u}_y)e^{-1.047z}$ V/m

$\mathbf{H}\ \ (-0.798\mathbf{u}_x + 0.663\mathbf{u}_y)e^{-1.047z}$ A/m

(b) The complex Poynting vector, $\overline{\mathbf{P}}_w = \frac{1}{2}(\overline{\mathbf{E}}\times\overline{\mathbf{H}}^*)$

$$= \frac{1}{2}[(\text{[illegible]}\mathbf{u}_x + 300\mathbf{u}_y)\ ^{-1\,0472}]\times[(-0.798\mathbf{u}_x + 0.603\mathbf{u}_y)\ ^{1\,0472}]$$

$$- 195.8\mathbf{u}_z \text{ W/m}^2$$

7.5 Plane Waves in Good Dielectrics and Good Conductors

In the last section we have studied the properties of uniform plane electromagnetic waves in a medium of zero conductivity. However, all the materials including the dielectrics have finite conductivity. Hence, displacement current as well as conduction current flows in the material under the influence of a time-varying external field. Conduction current causes ohmic loss. So, the amplitudes of the fields get attenuated while travelling in a material medium.

The conductivities of different materials are different. The value of conductivity of a good dielectric material (not perfect) is quite small, whereas the conductivity of a good conducting material is very large. Accordingly, the conduction current density in a good dielectric is small compared with the displacement current density. In the case of good conductors, the displacement current density is much less than the conduction current density.

We will derive the wave equation in a material medium without neglecting its conductivity σ. Sinusoidal variation of fields with time will be assumed. We will solve the one-dimensional wave equation for the general case and define various parameters associated with the waves. We will also obtain approximate formulae for various parameters of waves in good dielectrics and good conductors assuming that the value of loss tangent is small for good dielectrics, and it is large for good conductors.

The Wave Equations: General Case

The two curl equations by Maxwell for sinusoidal variation of fields with time, represented by the complex exponential function $e^{j\omega t}$ are

$$\nabla \times \bar{\mathbf{E}} = -j\omega\mu\bar{\mathbf{H}} \tag{1}$$

$$\nabla \times \bar{\mathbf{H}} = (\sigma + j\omega\varepsilon)\bar{\mathbf{E}} \tag{2}$$

In these equations, $\bar{\mathbf{E}}$ and $\bar{\mathbf{H}}$ are functions of space variables only.

By taking the curl of both sides of (1), we get

$$\nabla \times \nabla \times \bar{\mathbf{E}} = -j\omega\mu(\nabla \times \bar{\mathbf{H}}) \tag{3}$$

Replacing $\nabla \times \bar{\mathbf{H}}$ of (3) by the rhs of (2) and using the vector identity $\nabla \times \nabla \times \bar{\mathbf{E}} = \nabla(\nabla \cdot \mathbf{E}) - \nabla^2\mathbf{E}$, the following relation is obtained.

$$\nabla(\nabla \cdot \bar{\mathbf{E}}) - \nabla^2\bar{\mathbf{E}} = -j\omega\mu(\sigma + j\omega\varepsilon)\bar{\mathbf{E}} \tag{4}$$

In a conducting medium, we may assume $\nabla \cdot \mathbf{E} = 0$. Substituting this in (4), we have

$$\nabla^2\bar{\mathbf{E}} = j\omega\mu(\sigma + j\omega\varepsilon)\bar{\mathbf{E}} = (\bar{\gamma})^2\bar{\mathbf{E}} \tag{5}$$

where

$$\bar{\gamma} = \sqrt{j\omega\mu(\sigma + j\omega\varepsilon)} \tag{6}$$

It can similarly be shown that

$$\nabla^2\bar{\mathbf{H}} = (\bar{\gamma})^2\bar{\mathbf{H}} \tag{7}$$

Equations (5) and (7) are the wave equations in $\bar{\mathbf{E}}$ and $\bar{\mathbf{H}}$. As has been explained before, the field intensities are vectors as well as phasors. As vectors, both the fields may have components in all the three coordinate directions, and as phasors each component may have a real part and an imaginary part.

Propagation Constant The parameter γ defined by (6) is called *propagation constant*. It is a complex quantity and its unit is m^{-1}. Denoting the real and imaginary parts of the propagation constant by α and β respectively, we have

$$\bar{\gamma} = \alpha + j\beta = \sqrt{j\omega\mu(\sigma + j\omega\varepsilon)} \tag{8}$$

Squaring both sides of (8) and then separating the real and imaginary parts, two independent equations will be obtained. Simultaneous solution of these equations will yield

$$\alpha = \omega\sqrt{\frac{\mu\varepsilon}{2}\left(-1+\sqrt{1+\left(\frac{\sigma}{\omega\varepsilon}\right)^2}\right)} \tag{9}$$

$$\beta = \omega\sqrt{\frac{\mu\varepsilon}{2}\left(1+\sqrt{1+\left(\frac{\sigma}{\omega\varepsilon}\right)^2}\right)} \tag{10}$$

The unit of α is m^{-1} (or neper per metre) and that of β is m^{-1} (or radians per metre).

The One-Dimensional Wave Equation and its Solution: General Case

By assuming that the electric field intensity is in x direction and it is a function only of z, we can obtain the following one-dimensional wave equation from (5).

$$\frac{\partial^2 \bar{E}_x(z)}{\partial z^2} = \bar{\gamma}^2 \bar{E}_x(z) \tag{11}$$

The solution of (11) is

$$\bar{E}_x(z) = \bar{E}_{x0}^{+} e^{-\bar{\gamma}z} + \bar{E}_{x0}^{-} e^{\bar{\gamma}z} = \bar{E}_{x0}^{+} e^{-(\alpha+j\beta)z} + \bar{E}_{x0}^{-} e^{(\alpha+j\beta)z} \tag{12}$$

The complete solution is obtained by multiplying (12) with $e^{j\omega t}$. Thus,

$$\bar{E}_x = \bar{E}_{x0}^{+} e^{-\alpha z} e^{(j\omega t-\beta z)} + \bar{E}_{x0}^{-} e^{\alpha z} e^{(j\omega t+\beta z)} \tag{13}$$

The real part of (13) is given by

$$E_x = E_{x0}^{+} e^{-\alpha z} \cos(\omega t - \beta z) + E_{x0}^{-} e^{\alpha z} \cos(\omega t + \beta z) \tag{14}$$

Attenuation Constant and Phase Constant The first term of (14) represents a wave travelling in the positive z direction. It is a cosine function of $(\omega t - \beta z)$ and contains the exponential factor $e^{-\alpha z}$. As the exponent of the factor is negative, the amplitude of the wave decreases as the wave progresses along the positive z direction. The rate of decrease of the amplitude depends on α. The parameter α described by (9) is called *attenuation constant*. And the parameter β defined by (10) has the same meaning as in the case of electromagnetic waves in a lossless medium. It is called *phase constant*. Comparison between formula (4) of Section 7.4 and formula (10) of this section shows that the value of the phase constant of a plane wave is higher in a medium of nonzero conductivity than in a medium of zero conductivity. In a loss-free medium the attenuation constant is zero.

Phase Velocity and Wavelength The phase velocity and wavelength of the plane wave are

$$v = \frac{\omega}{\beta}$$

$$\lambda = \frac{2\pi}{\beta}$$

where β is the phase constant defined by (10).

The second term of (14) represents a wave travelling along negative z direction. As z in this case is negative, the amplitude of this wave also decreases as the wave moves along negative z direction. The values of various parameters of the negatively travelling wave are same as those of the positively travelling wave.

Solution for the Magnetic Field Intensity We found in the preceding section that the waves of the x-directed electric field are associated with the y-directed magnetic field. The one-dimensional wave equation involving the y-directed magnetic field is

$$\frac{\partial^2 \bar{H}_y(z)}{\partial z^2} = \bar{\gamma}^2 \bar{H}_y(z) \tag{15}$$

The solution of (15) is

$$\bar{H}_y^+(z) = \bar{H}_{y0}^+ \, e^{-(\alpha + j\beta)z} + \bar{H}_{y0}^- \, e^{(\alpha + j\beta)z} \tag{16}$$

Also,

$$\bar{H}_y = \bar{H}_{y0}^+ \, e^{-\alpha z} \, e^{(j\omega t - \beta z)} + \bar{H}_{y0}^- \, e^{\alpha z} \, e^{(j\omega t + \beta z)} \tag{17}$$

Relation between $\bar{\mathbf{E}}$ and $\bar{\mathbf{H}}$: The Complex Intrinsic Impedance In order to obtain the relation between the electric and magnetic field intensities, we will use the solution given by (13), and one of Maxwell's curl equations for harmonic variation of the fields. The vector form of electric field intensity is

$$\bar{\mathbf{E}} = \bar{E}_x \mathbf{u}_x \tag{18}$$

By substituting (18) in (2), we get

$$\nabla \times \bar{\mathbf{H}} = (\sigma + j\omega\varepsilon)\bar{E}_x \, \mathbf{u}_x \tag{19}$$

As the electric field is in x direction, we have to use only the x component of $\nabla \times \bar{\mathbf{H}}$. Moreover, the magnetic field intensity can vary only with z as the electric field intensity is a function of z only. Thus,

$$\nabla \times \bar{\mathbf{H}} = -\frac{\partial \bar{H}_y}{\partial z} \mathbf{u}_x \tag{20}$$

By equating the right-hand sides of (19) and (20), we have

$$\frac{\partial \bar{H}_y}{\partial z} = -(\sigma + j\omega\varepsilon)\bar{E}_x$$

Or

$$\bar{H}_y = -(\sigma + j\omega\varepsilon)\int \bar{E}_x \, dz \tag{21}$$

By substituting (13) in (21) and integrating, we get the following relation.

$$\bar{H}_y = \frac{\sigma + j\omega\varepsilon}{\bar{\gamma}}[\bar{E}_{x0}^+ e^{(j\omega t - \bar{\gamma} z)} - \bar{E}_{x0}^- e^{(j\omega t + \bar{\gamma} z)}]$$

$$= \frac{1}{\bar{\eta}}[\bar{E}_{x0}^+ e^{(j\omega t - \bar{\gamma} z)} - \bar{E}_{x0}^- e^{(j\omega t + \bar{\gamma} z)}] \tag{22}$$

where

$$\bar{\eta} = \frac{\bar{\gamma}}{\sigma + j\omega\varepsilon} \tag{23}$$

Substituting (8) in (23) and simplifying, we have

$$\eta = \frac{j\omega\mu}{\sigma + j\omega\varepsilon} \tag{24}$$

The first and second terms of (22) represent the positively travelling magnetic field $\bar{H}_y^+$ and the negatively travelling magnetic field $\bar{H}_y^-$ respectively. Thus, referring to (13) and (22), the relation between the positively travelling electric and magnetic fields is

$$\bar{H}_y^+ = \bar{H}_{y0}^+ e^{(j\omega t - \bar{\gamma} z)} = \frac{1}{\bar{\eta}} \bar{E}_{x0}^+ e^{(j\omega t - \bar{\gamma} z)} = \frac{\bar{E}_x^+}{\bar{\eta}}$$

Or

$$\frac{\bar{E}_x^+}{\bar{H}_y^+} = \frac{\bar{E}_{x0}^+}{\bar{H}_{y0}^+} = \bar{\eta} \tag{25}$$

And the relation between the negatively travelling electric and magnetic fields is

$$\frac{\bar{E}_x^-}{\bar{H}_y^-} = \frac{\bar{E}_{x0}^-}{\bar{H}_{y0}^-} = -\bar{\eta} \tag{26}$$

The parameter $\bar{\eta}$ defined by (24) is the *intrinsic impedance* (*or intrinsic wave impedance*). It is a complex quantity with a real part and an imaginary part. The two parts are analogous to the resistance and reactance of an electric circuit. As the magnetic and electric field intensities are related through the complex intrinsic impedance the two fields are out of phase in time. The fields are however in space phase.

The Loss Tangent

The first term on the rhs of (2) is the conduction current density **J**, and the second term is the displacement current density $\mathbf{J}_d$. That is,

$$\mathbf{J} = \sigma \mathbf{E}$$

$$\mathbf{J}_d = j\omega\varepsilon \mathbf{E}$$

The conduction current density is in time phase with the electric field intensity. However, the presence of the factor j in the relation between $\mathbf{J}_d$ and **E** shows that the displacement current density leads the electric field intensity by 90°. Hence, the displacement current density leads the conduction current density by 90°.

If the conduction current density is taken along the real axis of a complex plane, then the displacement current density will be along the imaginary axis as shown in Fig. 7.5. Such a representation of sinusoidally time-varying quantities is called *phasor diagram*. The resultant current density J_R is the phasor sum of the two current densities. The time-phase difference between J_d and J_R is denoted by δ_d. Accordingly,

$$\tan \delta_d = \frac{J}{J_d} = \frac{\sigma}{\omega\varepsilon} \tag{27}$$

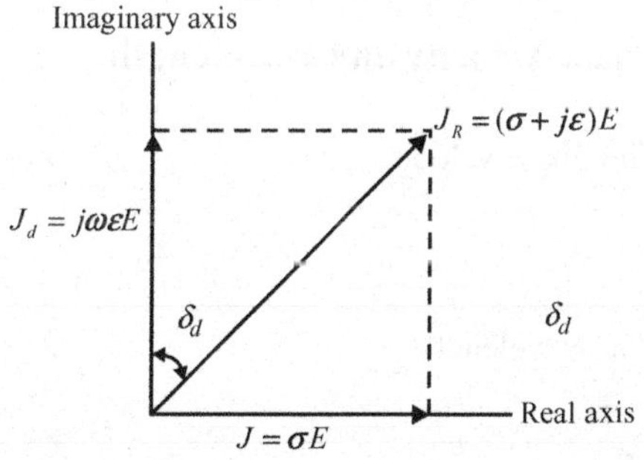

Fig. 7.5 Phasor diagram of current densities. The displacement current density leads the conduction current density by 90°.

The value of $\tan \delta_d$ is a measure of the magnitude of conduction current density relative to that of the displacement current density. A low value of $\tan \delta_d$ for a material means that the material is of low-loss type. The attenuation constant of the material is also small. The ratio $\sigma/\omega\varepsilon$ is called *loss tangent* or *dissipation factor*.

As both σ and ε vary with frequency, the loss tangent is not inversely proportional to the frequency. For example, the loss tangent of dry sand at 20 °C is nearly 0.2 at 50 Hz and 0.02 at 10^6 Hz.

Plane Waves in Good Dielectric Materials

> If the loss tangent $\sigma/\omega\varepsilon$ of a dielectric material is less than 0.01 then it may be considered as a good dielectric substance.

Under this condition, we will obtain approximate formulae for the parameters associated with waves travelling in good dielectrics from the general formulae.

Attenuation Constant and Phase Constant The attenuation constant given by (9) and the phase constant given by (10) contain the function $\sqrt{1+(\sigma/\omega\varepsilon)^2}$. As for good dielectrics the function $\sigma/\omega\varepsilon$ is very small, we may use only the first two terms of the binomial expansion of the function. Accordingly,

$$\sqrt{1+\frac{\sigma^2}{\omega^2\varepsilon^2}} \cong 1+\frac{\sigma^2}{2\omega^2\varepsilon^2}$$

Substituting this in (9) and (10) we will get approximate formulae for the attenuation constant α and phase constant β.

$$\alpha \cong \omega\sqrt{\frac{\mu\varepsilon}{2}\left(-1+1+\frac{\sigma^2}{2\omega^2\varepsilon^2}\right)} = \frac{\sigma}{2}\sqrt{\frac{\mu}{\varepsilon}} \tag{28}$$

$$\beta \cong \omega\sqrt{\frac{\mu\varepsilon}{2}\left(1+1+\frac{\sigma^2}{2\omega^2\varepsilon^2}\right)} = \omega\sqrt{\mu\varepsilon}\sqrt{1+\frac{\sigma^2}{4\omega^2\varepsilon^2}}$$

$$\cong \omega\sqrt{\mu\varepsilon}\left(1+\frac{\sigma^2}{8\omega^2\varepsilon^2}\right) \tag{29}$$

The attenuation constant of electromagnetic waves in a good dielectric material is quite small, and the phase constant is only slightly higher than the value in a perfect dielectric. Note that the approximate formula for the attenuation constant does not include frequency of the field. It means that above the frequency at which the loss tangent is less than 0.01, the attenuation constant is independent of frequency.

Phase Velocity and Wavelength

The phase velocity,

$$v = \frac{\omega}{\beta} \cong \frac{1}{\sqrt{\mu\varepsilon}\left(1+\frac{\sigma^2}{8\omega^2\varepsilon^2}\right)} \tag{30}$$

The wavelength,

$$\lambda = \frac{2\pi}{\beta} \cong \frac{2\pi}{\omega\sqrt{\mu\varepsilon}\left(1+\frac{\sigma^2}{8\omega^2\varepsilon^2}\right)} \tag{31}$$

The Intrinsic Impedance Referring to (23), the intrinsic impedance for the general case,

$$\bar{\eta} = \sqrt{\frac{j\omega\mu}{\sigma + j\omega\varepsilon}} = \sqrt{\frac{\mu}{\varepsilon}}\left(1-j\frac{\sigma}{\omega\varepsilon}\right)^{-\frac{1}{2}}$$

Using only the first two terms of the binomial expansion of $(1-j\sigma/\omega\varepsilon)^{-1/2}$, the approximate formula for the complex intrinsic impedance,

$$\eta \cong \frac{\mu}{\varepsilon}\left(\cdot + j\frac{\sigma}{2\omega\varepsilon}\right) = \eta \quad \psi + j\sin\psi \quad \eta e^{j\psi} \tag{32}$$

The magnitude of the complex intrinsic impedance,

$$\eta = \sqrt{\frac{\mu}{\varepsilon}\left[1+\left(\frac{\sigma}{2\omega\varepsilon}\right)^2\right]}$$

The phase angle of the complex intrinsic impedance,

$$\psi = \tan^{-1}(\sigma/2\omega\varepsilon)$$

Average Power To calculate the average power carried by the positively travelling fields, let us assume that $\bar{E}_{x0}^{+}$ in (12) is a real quantity. Then, the phasor forms of electric and magnetic field intensities are

$$\bar{E}_x^{+}(z) = (E_{x0}^{+}e^{-\alpha z})e^{-j\beta z}$$

$$\bar{H}_y^{+}(z) = \frac{\bar{E}_x^{+}(z)}{\bar{\eta}} = \frac{E_{x0}^{+}e^{-\alpha z}e^{-j\beta z}}{\eta e^{j\psi}} = \left(\frac{E_{x0}^{+}e^{-\alpha z}}{\eta}\right)e^{-j(\beta z+\psi)}$$

The amplitude of the electric field is $E_{x0}^{+}e^{-\alpha z}$ and that of the magnetic field is $E_{x0}^{+}e^{-\alpha z}/\eta$. The time-phase difference between the two fields is ψ. The average power per unit area carried by the fields

$$P_{av}(z) = \frac{E_{x0}^{+}H_{y0}^{+}}{2}e^{-2\alpha z}\cos\psi = \frac{(E_{x0}^{+})^2}{2\eta}e^{-2\alpha z}\cos\psi$$

$$= \frac{\eta(H_{y0}^{+})^2}{2}e^{-2\alpha z}\cos\psi \text{ W/m}^2 \quad (33)$$

Average power flow per unit area at $z = 0$, $P_{av}(z=0) = \dfrac{(E_{x0}^{+})^2}{2\eta}\cos\psi$ W/m^2 (34)

Power loss per unit area of the medium over a distance z is equal to the difference between the powers given by (34) and (33).

Expressions (28) to (32) give the approximate formulae for various parameters of a plane wave travelling in an imperfect dielectric medium. However,

the exact formulae for various parameters are not complicated. Hence, these can also be used for numerical calculations even for good dielectrics and good conductors. Further, both σ and ε vary with frequency. So, values of the constants at the working frequency have to be used for calculating the parameters.

A plot of electric field intensity of a positively travelling wave at $t = 0$ is shown in Fig. 7.6

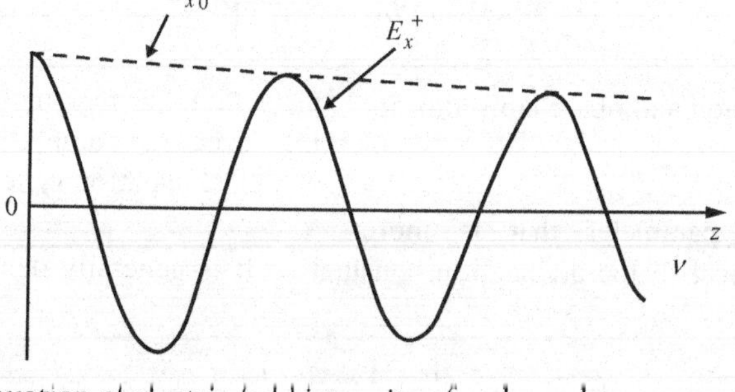

Fig. 7.6 Illustration of attenuation of electric field intensity of a plane electromagnetic wave in a good dielectric medium. The symbol α denotes attenuation constant.

Plane Waves in Good Conductors

If the ratio $\sigma/\omega\varepsilon$ of a material is greater than 100, then it may be considered as a good conductor.

In order to study the properties of electromagnetic waves travelling in a good conductor, let us consider a flat conductor with its surface on the *x-y* plane and its depth in *z* direction, as illustrated in Fig. 7.7(*a*). The linear dimensions of the conductor are assumed to be large. We will assume that the *x*-directed electric field and the *y*-directed magnetic field travel along positive *z* direction.

The approximate formulae for various parameters of the waves are derived in the following.

The Propagation Constant The general expression for the propagation constant is

$$\bar{\gamma} = \sqrt{j\omega\mu(\sigma + j\omega\varepsilon)} = \sqrt{-\omega^2\mu\varepsilon\left(1 - j\frac{\sigma}{\omega\varepsilon}\right)} \tag{35}$$

For a good conductor the value of $\sigma/\omega\varepsilon$ in the working frequency range is very high. Hence,

$$1 - j\frac{\sigma}{\omega\varepsilon} \approx -j\frac{\sigma}{\omega\varepsilon} \tag{36}$$

Substituting (36) into (35),

$$\bar{\gamma} = \sqrt{j}\,\sqrt{\omega\mu\sigma}$$

Since

$$j\left(\frac{\pi}{2} + j\sin\frac{\pi}{2}\right)^{\frac{1}{2}} = \cos\frac{\pi}{4} + j\sin\frac{\pi}{4} = e^{j\pi/4} = \frac{1+j}{\sqrt{2}},$$

$$\gamma = (\ +\ )\sqrt{\frac{\omega\mu\sigma}{2}} = (1+j)\sqrt{\pi f\mu\sigma} = \alpha + j\beta \tag{37a}$$

The propagation constant given by (37*a*) is a complex quantity. Its real part α is the attenuation constant and the imaginary part β is the phase constant. Both the parameters have the same value defined by

$$\alpha = \beta = \sqrt{\pi f\mu\sigma} \tag{37b}$$

The values of the attenuation and phase constants are very large.

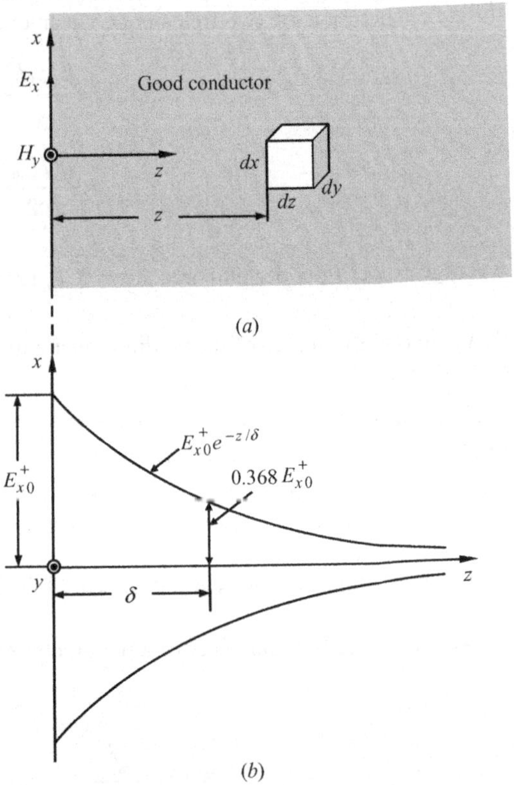

Fig. 7.7. (*a*) A region of good conductor and (*b*) Variation of amplitude of electric field intensity of a plane electromagnetic wave in a good conductor.

The Skin Depth A parameter that is more informative is the reciprocal of the attenuation constant α. It is generally denoted by δ. Thus, using the expression for α given by (37*b*),

$$\delta = \frac{1}{\alpha} = \frac{1}{\beta} = \frac{1}{\pi'\mu\sigma} \tag{38}$$

The unit of δ is metre.

The phasor form of the x-directed electric field intensity travelling along positive z direction is given by the first term of (12). Using the parameter δ in place of α and β in this equation and assuming that $\bar{E}_{x0}^{+}$ is a real quantity, the field intensity is given by

$$\bar{E}_x^+(z) = E_{x0}^+ e^{-z/\delta} e^{-jz/\delta} \tag{39}$$

The field in real-time form is

$$E_x^+ = E_{x0}^+ e^{-z/\delta} \cos(\omega t - z/\delta) \tag{40}$$

This is a cosine function of $(\omega t - z/\delta)$ and its amplitude is $E_{x0}^+ e^{-z/\delta}$. The amplitude has maximum value at $z = 0$. The amplitude decreases exponentially with increase of z. The amplitude at $z = \delta$ is

$$E_{x0}^+ e^{-z/\delta}(z = \delta) = E_{x0}^+/e = 0.368 E_{x0}^+$$

The parameter δ defined by (38) is called *skin depth* or *depth of penetration.*

The parameter is inversely proportional to the square root of the products of *frequency, permeability* and *conductivity*. As δ is very small for a good conductor, the amplitudes of the fields travelling in a good conductor decrease rapidly with z, as shown in Fig. 7.7(*b*). It may be noted that the depth of penetration is not the depth at which the value of the field becomes zero. It is the depth at which the amplitude of the field is $1/e$ times the value at $z = 0$.

The Intrinsic Impedance The general expression for the intrinsic impedance,

$$\bar{\eta} = \sqrt{\frac{j\omega\mu}{\sigma + j\omega\varepsilon}} = \sqrt{\frac{j\omega\mu}{j\omega\varepsilon(1 - j\frac{\sigma}{\omega\varepsilon})}}$$

Using the approximate relation of (36) it can be shown that the complex intrinsic impedance,

$$\bar{\eta} \cong \sqrt{j}\sqrt{\frac{\omega\mu}{\sigma}} = \sqrt{\frac{\omega\mu}{\sigma}}\left(\cos\frac{\pi}{2} + j\sin\frac{\pi}{2}\right)^{1/2}$$

$$= \sqrt{\frac{\omega\mu}{\sigma}}\left(\cos\frac{\pi}{4} + j\sin\frac{\pi}{4}\right) = \sqrt{\frac{\pi f\mu}{\sigma}}(1 + j) \tag{41a}$$

By eliminating $\sqrt{\pi f\mu}$ from (38) and (41*a*), the intrinsic impedance can be expressed in terms of the skin depth. Accordingly,

$$\bar{\eta} == \frac{\sqrt{2}e^{j\pi/4}}{\sigma\delta} = \frac{1}{\sigma\delta} + j\frac{1}{\sigma\delta} = \frac{\alpha}{\sigma}(1 + j), \text{ as } \frac{1}{\delta} = \alpha \tag{41b}$$

The intrinsic impedance is a complex quantity with equal real and imaginary parts. The real part is called *surface resistance* or *skin-effect* resistance and it is denoted by R_s. The resistance is per unit length (in x direction) and per unit width (in y direction). The imaginary part is called *internal reactance*. The internal reactance is equal to ω times the *internal inductance* L_i.

The surface resistance, $R_s = \sqrt{\frac{\pi f \mu}{\sigma}} = \frac{1}{\sigma\delta}$

The internal reactance, $X_i = \sqrt{\frac{\pi f \mu}{\sigma}} = \frac{1}{\sigma\delta}$

The internal inductance, $L_i = \frac{1}{\omega\sigma\delta}$

The magnitude of the intrinsic impedance, $\eta = \frac{\sqrt{2}}{\sigma\delta} = \sqrt{\frac{2\pi f \mu}{\sigma}}$

The last expression is obtained by substituting the expression (38) for δ. For a good conductor, we can take $\mu = \mu_0$.

Phase Velocity and Wavelength

The phase velocity, $v = \frac{\omega}{\beta}$

Substituting the expression of β given by (37*b*), $v = \sqrt{\frac{4\omega}{\mu\sigma}}$

The wavelength, $\lambda = \frac{2\pi}{\beta} = 2\pi\sqrt{\pi f \mu\sigma}$

Average Power The phasor form of magnetic field travelling along positive *z* direction is

$$\bar{H}_y^+(z) = \frac{\bar{E}_x^+(z)}{\bar{\eta}} \tag{42}$$

Substituting (39) and (41) in (42), we get

$$\bar{H}_y^+(z) = \frac{E_{x0}^+}{\eta e^{j\pi/4}} = \frac{E_{x0}^+}{\eta} e^{-z/\delta}\, e^{-j(z/\delta + \pi/4)} \tag{43}$$

The magnetic field intensity in real time is

$$H_y^+ = \frac{E_{x0}^+}{\eta} e^{-z/\delta} \cos(\omega t - \beta z - \pi/4) \tag{44}$$

The amplitude of magnetic field intensity is $(E_{x0}^+/\eta)e^{-z/\delta}$. Comparison between (40) and (44) shows that the magnetic field lags the electric field by 45° in time.

The average power flow into the conductor per unit surface area can be found from the real part of the complex Poynting vector or from the instantaneous Poynting vector at $z = 0$. It is given by

$$P_{av} = \frac{1}{2} E_{x0}^+ H_{y0}^+ \cos 45° = \frac{(E_{x0}^+)^2}{2\sqrt{2}\,\eta} = \frac{\delta\sigma(E_{x0}^+)^2}{4} \text{ W/m}^2 \tag{45}$$

It can be shown that this power is equal to the ohmic loss in the conductor if the depth of the conductor is large compared to the skin depth. Let us consider a conductor block of length *dx*, width *dy* and depth *dz* at a distance *z* from the surface of the conductor (see inset in Fig. 7.7(*a*)).

Amplitude of current density at *z* $J_{x0}^+ = \sigma E_{x0}^{'} e^{-z/\delta}$

Amplitude of current through the block, $dI_0 = \sigma E_{x0}^{'} e^{-z/\delta}(dydz)$

The dc resistance of the block (in the direction of current density), $dR = \dfrac{dx}{\sigma\, dy dz}$

Ohmic loss in the block, $dP_\Omega = \dfrac{1}{2}(dI_0)^2\, dR = \dfrac{1}{2}\sigma (E_{x0}^+)^2 e^{-2z/\delta}(dxdydz)$

Ohmic loss per unit length, unit width and depth z_1 of the conductor,

$$P_\Omega = \int_0^1\int_0^1\int_0^{z_1} dP_\Omega = \frac{1}{4}\sigma\delta (E_{x0}^+)^2\left(1 - e^{-2z_1/\delta}\right)$$

If $z_1 \gg \delta$, the second term would be negligibly small and hence may be neglected. Thus,

$$P_\Omega = \frac{\sigma\delta (E_{x0}^+)^2}{4}\ \text{W/m}^2$$

This is same as (45).

EXAMPLE 7.8

A uniform plane wave at frequency $f = 20$ MHz is travelling along positive z direction in a medium having $\sigma = 4.5 \times 10^{-5}$ S/m, $\varepsilon_r = 8.0$ and $\mu_r = 1.0$. Calculate the attenuation constant, phase constant, propagation constant, intrinsic impedance, and phase velocity of the wave.

The loss tangent, $\dfrac{\sigma}{\omega\varepsilon} = \dfrac{(4.5\times 10^{-5})}{(40\pi\times 10^6)(8\times 8.85\times 10^{-12})} = 5.06\times 10^{-3}$

As the loss tangent is less than 0.01, the medium may be considered to be a good dielectric.

Attenuation constant, $\alpha \sim \dfrac{\sigma}{2}\sqrt{\dfrac{\mu}{\varepsilon}} = \dfrac{4.5\times 10^{-5}}{2}\sqrt{\dfrac{4\pi\times 10^{-7}}{8\times 8.85\times 10^{-12}}} = 0.003$ 1/m

Phase constant,

$$\beta\ \ \omega\sqrt{\mu\varepsilon}\left(1 + \frac{\sigma^2}{8\omega^2\varepsilon^2}\right) = (40\pi\times 10^6)\sqrt{\mu_0(8\varepsilon_0)}\left[1 + \frac{(5.06\times 10^{-3})^2}{8}\right] = 1.185\ \text{1/m}$$

Propagation constant, $\bar{\gamma} = \alpha + j\beta = 0.003 + j1.158$ 1/m

Intrinsic impedance,

$$\eta \approx \sqrt{\frac{\mu}{\varepsilon}}\left(1 + j\frac{\sigma}{2\omega\varepsilon}\right) - \sqrt{\frac{\mu_0}{8\varepsilon_0}}\left(1 + j\frac{5.06\times 10^{-3}}{2}\right) - 133.3 + j0.337 = 133.3e^{j0.0025}\ \Omega$$

Phase velocity, $v = \dfrac{\omega}{\beta} = \dfrac{40\pi\times 10^6}{1.185} = 1.06\times 10^8$ m/s

EXAMPLE 7.9

The conductivity σ, relative permittivity ε_r, and relative permeability μ_r of a medium are 1.6×10^{-3} S/m, 11.8, and 1.0 respectively. Find the propagation constant, attenuation constant, phase constant and intrinsic impedance of the medium at a frequency of 160 MHz.

The loss tangent, $\dfrac{\sigma}{\omega\varepsilon} = \dfrac{1.6\times 10^{-3}}{(2\pi\times 160\times 10^6)(11.8\times 8.85\times 10^{-12})} = 0.0152$

As the loss tangent is greater than 0.01, we have to calculate the parameters using the actual formulae.

Propagation constant, $\bar{\gamma} = \sqrt{j\omega\mu(\sigma + j\omega\varepsilon)}$

$$j \qquad j(\pi \times 160 \times 10^{6})(4\pi \times 10^{-7}) = 1263.3\angle 90^\circ$$

$$\sigma + j\omega\varepsilon = 1.6 \times 10^{-3} + j(2\pi \times 160 \times 10^{6})(11.8)(8.85 \times 10^{-12})$$

$$= (1.6 + j104.98) \times 10^{-3} = 0.105 \angle 89.1^\circ$$

$$\gamma = \sqrt{(\quad .3\angle 90^\circ)(0.105\angle 89.1^\circ)} = 11.51\angle 89.6^\circ = 0.08 + j11.51 \text{ m}^{-1}$$

Attenuation constant, $\alpha = 0.08$ neper/m

Phase constant, $\beta = 11.51$ rad/m

Intrinsic impedance, $\bar{\eta} = \sqrt{\dfrac{j\omega\mu}{\sigma + j\omega\varepsilon}} = \sqrt{\dfrac{1263.3\angle 90^\circ}{0.105\angle 89.1^\circ}}$

$$= 109.69 \angle 0.45^\circ = 109.68 + j0.86 \ \Omega$$

EXAMPLE 7.10

At frequency $f = 150$ MHz, the propagation constant $\bar{\gamma}$ and intrinsic impedance $\bar{\eta}$ of a medium are $0.01 + j12$ m^{-1} and $110 + j1.2\ \Omega$ respectively. The amplitude of magnetic field intensity $\bar{H}_{y0}$ of a plane electromagnetic wave travelling along positive z direction in the medium at $z = t = 0$ is $0.9\angle 0$ A/m. (a) Write expressions for the instantaneous electric field intensity and instantaneous magnetic field intensity. (b) Calculate power flow per unit area along the direction of propagation of the wave (average Poynting vector) at $z = 0$ and $z = 10$ m.

Attenuation constant, $\alpha = 0.01$ neper/m

Phase constant, $\beta = 12$ rad/m

(a) The instantaneous magnetic field intensity, $H_y(t) = \bar{H}_{y0} e^{-\alpha z} \cos(\omega t - \beta z)$

$= 0.9\, e^{-0.01z} \cos[(300\pi \times 10^{6})t - 12z]$ A/m

The electric field intensity is in x direction. Its amplitude,

$E_x = \bar{\eta} H_{y0} = (110 + j1.2)(0.9\angle 0^\circ) = 99.0\angle 0.625^\circ$ V/m

Phase difference between the fields, $\psi = 0.625^\circ$

The instantaneous electric field intensity, $E_x(t) = E_{x0} e^{-\alpha z} \cos(\omega t - \beta z + \psi)$

$= 99.0\, e^{-0.01z} \cos[(300\pi \times 10^{6})t - 12z + 0.625^0]$ A/m

(b) Power flow per unit area at $z = 0$, $P_{wa}(z = 0) = \dfrac{1}{2} E_{x0} H_{y0} \cos\psi$

$= \dfrac{1}{2}(99.0)(0.9)(1) = 44.55$ W/m^2

Power flow per unit area at $z = 10$ m, $P_{wa}(z \quad) = \dfrac{1}{2} E_x H_{y0} e^{-2\alpha z} \cos\psi$

$= (44.55)(e^{-0.2}) = 36.47$ W/m^2

EXAMPLE 7.11

The conductivity, relative permittivity, and relative permeability of sea water are 4.0 S/m, 80.0, and 1.0 respectively. (a) Find the frequency below which sea water can be considered as a good conductor. (b) Calculate the skin depth of sea water at 0.5 MHz.

If the loss tangent of a medium is equal to or greater 100, then it can be considered as a good conductor. Let f be the frequency at which the loss tangent of sea water is equal 100. Then,

(a) Loss tangent, $\dfrac{\sigma}{\omega\varepsilon} = \dfrac{4}{(2\pi f)(80)(8.85\times 10^{-12})} = 100$

$$f = 8.99 \text{ MHz}$$

Thus, sea water can be treated as a good conductor below this frequency.

(b) At the given frequency, the loss tangent of sea water is higher than 100. Hence we can use the skin depth formula of a good conductor.

$$\text{Skin depth, } \delta = \frac{1}{\sqrt{\pi f \mu \sigma}} = \frac{1}{\sqrt{(\pi\times 0.5\times 10^6)(4\pi\times 10^{-7})(4)}} = 0.356 \text{ m}$$

EXAMPLE 7.12

Consider that a plane wave at 1.5×10^6 Hz is travelling along positive z direction in a good conductor for which $\sigma = 3.8 \times 10^7$ S/m, $\mu = \mu_0$. Calculate the attenuation constant, phase constant, propagation constant, skin depth, phase velocity, and intrinsic impedance of the wave.

$$\text{Attenuation constant, } \alpha = \sqrt{\pi f \mu \sigma} = \sqrt{\pi(1.5\times 10^6)(4\pi\times 10^{-7})(3.8\times 10^7)} = 15\times 10^3 \text{ 1/m}$$

$$\text{Phase constant, } \beta = \sqrt{\pi f \mu \sigma} = 15\times 10^3 \text{ 1/m}$$

$$\text{Propagation constant, } \bar{\gamma} = \alpha + j\beta = 15\times 10^3 (1+j)\, 1/\text{m}$$

$$\text{Skin depth, } \delta = \frac{1}{\alpha} = \frac{1}{15\times 10^3} = 66.6\times 10^{-6} \text{ m}$$

$$\text{Phase velocity, } v = \frac{\omega}{\beta} = \frac{2\pi(1.5\times 10^6)}{15\times 10^3} = 682.3 \text{ m/s}$$

$$\text{Intrinsic impedance, } \bar{\eta} = \frac{1+j}{\sigma\delta} = \frac{1+j}{(3.8\times 10^7)(66.6\times 10^{-6})} = 3.95\times 10^{-4}(1+j)$$

$$= 5.588\times 10^{-3} \angle 45^\circ \ \Omega$$

EXAMPLE 7.13

The electric and magnetic field intensities of a uniform plane wave in a good conductor of conductivity 3.5×10^7 S/m and unity relative permeability are in x and y directions. The frequency of the wave is 15 MHz, and amplitude of the electric field intensity at the surface ($z = 0$) of the conductor is $0.02 \angle 0$ V/m. (a) Find the surface resistance and internal inductance of the conductor. (b) Calculate the amplitude of the magnetic field intensity at the conductor surface.

What is the phase difference between the two fields?

(a) Surface resistance, $R_s = \sqrt{\dfrac{\pi f \mu_0}{\sigma}} = \sqrt{\dfrac{\pi(15\times10^6)(4\pi\times10^{-7})}{3.5\times10^7}} = 13.0\times10^{-4}\ \Omega$

Internal inductive reactance, $X_i = \sqrt{\dfrac{\pi f \mu_0}{\sigma}} = 13.0\times10^{-4}\ \Omega$

Internal inductance, $L_i = \dfrac{X_i}{\omega} = \dfrac{13.0\times10^{-4}}{2\pi(15\times10^6)} = 0.138\times10^{-10}$ H

(b) Intrinsic impedance, $\bar{\eta} = \sqrt{\dfrac{\pi f \mu_0}{\sigma}}(1+j) = (13.0\times10^{-4})(1+j)\ \Omega$

Amplitude of magnetic field intensity,

$$H_{y0} = \frac{\bar{E}_{x0}}{\bar{\eta}} = \frac{0.02\angle 0}{(13.0\times10^{-4})(1+j)} = 10.88\angle -45°\ \text{A/m}$$

The magnetic field lags the electric field by 45°.

EXAMPLE 7.14

The frequency of a plane electromagnetic wave is 25 kHz, and the amplitude of its electric field intensity at the surface of a thick good conductor is 0.06 ∠0 V/m. The conductivity of the conductor is 1.1 × 10^7 S/m, and its relative permeability is unity. (a) Find the amplitude of the electric field at a distance of 0.5 mm from the conductor surface. (b) Calculate average power flow into the conductor (or power loss) per unit surface area.

(a) Attenuation constant,

$$\alpha \quad \sqrt{\pi f \mu \sigma} = \sqrt{\pi(25\times10^3)(4\pi\times10^{-7})(1.1\times10^7)} = 1.042\times10^3\ \text{neper/m}$$

Amplitude of the field at a distance z from the surface,

$$E_0(z) = E_0(z=0)e^{-\alpha z} = 0.06e^{-(1.042\times10^3)(0.5\times10^{-3})} = 0.0356\ \text{V/m}$$

(b) Intrinsic impedance,

$$\eta = \sqrt{\frac{\pi f \mu}{\sigma}}(1+j) = \sqrt{\frac{\pi(25\times10^3)(4\pi\times10^{-7})}{1.1\times10^7}}(1+j) = (9.47\times10^{-5})(1+j)\ \Omega$$

Amplitude of magnetic field intensity, $\bar{H}_0 = \dfrac{\bar{E}_0}{\bar{\eta}} = \dfrac{0.06\angle 0}{(9.47\times10^{-5})(1+j)} = 448.0\angle -45°$ A/m

Power flow per unit surface area of the conductor,

$$P_{wa} = \frac{1}{2}E_0 H_0 \cos 45° = \frac{1}{2}(0.06)(448.0)(0.707) = 9.50\ \text{W/m}^2$$

7.6 Polarisation of Uniform Plane Waves

To understand the meaning of wave polarisation, let us consider a uniform plane electromagnetic wave travelling along positive z direction in a lossless medium. The wave, in general, has electric fields both in x and

y directions. We assume that the fields vary harmonically with time and the time phase difference between the fields is ψ. Thus, taking the x-directed field as reference, the two fields are described by

$$E_x = E_{x0} \cos(\omega t - \beta z) \tag{1}$$

and

$$E_y = E_{y0} \cos(\omega t - \beta z - \psi) \tag{2}$$

Note that the superscript '+' for denoting a positively travelling field is not used here for the sake of simplicity. Since the direction of the electric field intensity on any z plane is independent of z, we may take $z = 0$ to describe wave polarisation. Thus substituting $z = 0$ in (1) and (2), we get

$$E_x = E_{x0} \cos \omega t \tag{3}$$

$$E_y = E_{y0} \cos(\omega t - \psi) \tag{4}$$

The total electric field is equal to the vector sum of (3) and (4). That is,

$$\mathbf{E} = E_x\mathbf{u}_x + E_y\mathbf{u}_y = \mathbf{u}_x E_{x0} \cos \omega t + \mathbf{u}_y E_{y0} \cos(\omega t - \psi) \tag{5}$$

The vector diagram of the electric field is shown in Fig. 7.8. The direction of propagation of the wave is normally out of the page.

The orientation of the electric field intensity **E** depends on the values of E_{x0}, E_{y0}, and ψ.

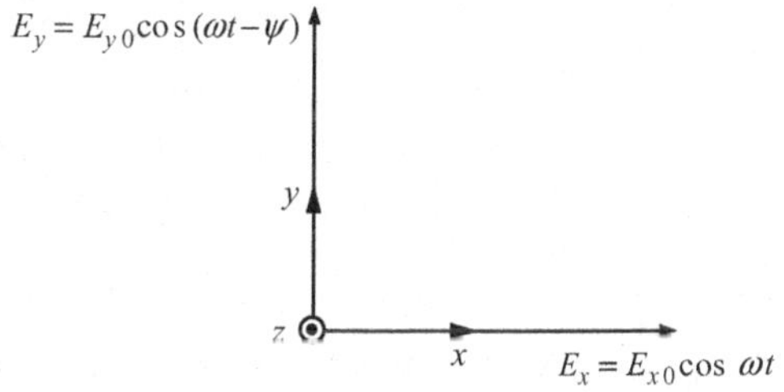

Fig. 7.8 Sinusoidally time-varying electric fields in x and y directions

The term *polarisation* is used to describe the orientation of the electric field intensity in a plane electromagnetic wave.

Linear Polarisation

Let us assume that the two field components of (5) are in time phase so that $\psi = 0$. Thus, the total field,

$$\mathbf{E} = E_x\mathbf{u}_x + E_y\mathbf{u}_y = (\mathbf{u}_x E_{x0} + \mathbf{u}_y E_{y0}) \cos \omega t \tag{6}$$

If E_y in (6) is zero, then the field intensity would always be in x direction, and if E_x is zero, the field intensity would be in y direction at all time. However when both E_x and E_y are present, then the magnitude of the resultant field E_R and its angle with the x axis Θ are given by

$$E_R = \sqrt{E_{x0}^2 + E_{y0}^2} \cos \omega t \tag{7}$$

$$\Theta = \tan^{-1}(E_{y0}/E_{x0}), \tag{8}$$

Relations (7) and (8) show that the magnitude of E_R is a function of time and the angle Θ is independent of time. The value Θ will remain constant if the ratio E_{x0}/E_{y0} is kept constant.

The orientations of the electric fields for the three cases are illustrated in Fig. 7.9. We find that the field intensities in all the three cases are along fixed directions.

An electromagnetic wave in which the *direction* of the electric field intensity always remains constant is called a *linearly polarised wave*. Moreover, because the field intensity describes a plane as it progresses along the z direction, it is also called a *plane polarised wave*. The polarisation is termed as *vertical* if the electric field is on a vertical plane and *horizontal* if it is on a horizontal plane.

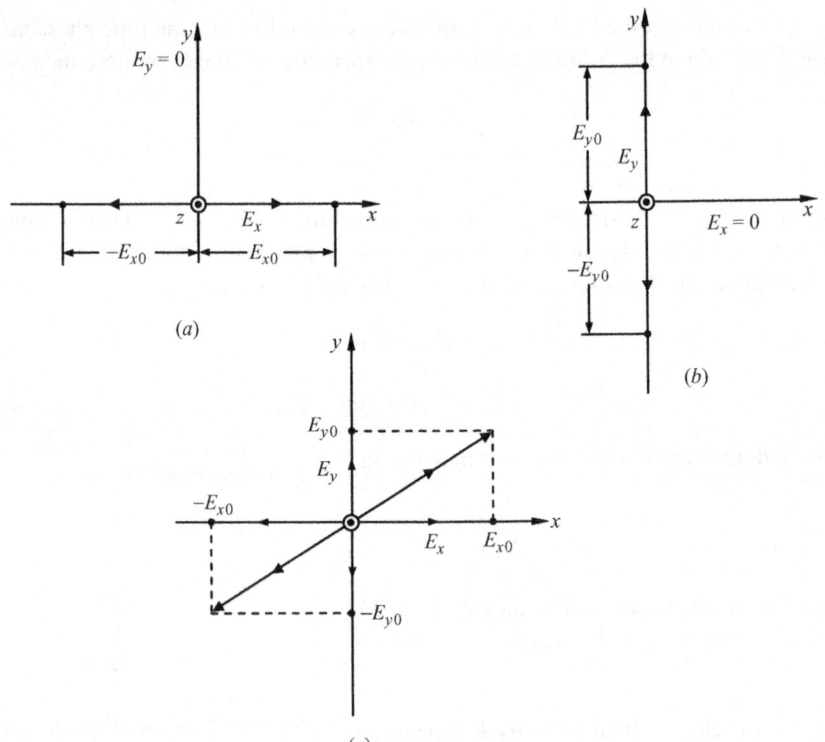

Fig. 7.9 Illustrations of linear polarisation of electromagnetic waves. (*a*) Polarisation along *x* direction, (*b*) Polarisation along *y* direction, and (c) Polarisation when both E_x and E_y are present.

Circular Polarisation

Next, let us assume that fields described by (3) and (4) have the same amplitude E_0 and the phase angle $\psi = \pm \pi/2$. Thus,

$$E_x = E_0 \cos \omega t \tag{9}$$

$$E_y = \pm E_0 \sin \omega t \tag{10}$$

The vector sum of (9) and (10) is

$$\mathbf{E} = E_0(\mathbf{u}_x \cos \omega t \pm \mathbf{u}_y \sin \omega t)$$

The magnitude of the resultant field is independent of t and it has a constant value E_0. And if Θ is the angle of the resultant field with x direction, then

$$\tan \Theta = \pm \tan \omega t$$

$$\Rightarrow \qquad \Theta = \pm \omega t$$

Differentiating both sides with t, we get

$$\frac{d\Theta}{dt} = \pm \omega$$

Thus, the total electric field of constant magnitude rotates at an angular velocity equal to the angular frequency of time variation of the fields. The tip of the field vector describes a circle on a given z plane. The direction of rotation of the resultant field is from the leading field towards the lagging field. If $\psi = \pi/2$, then E_x is the

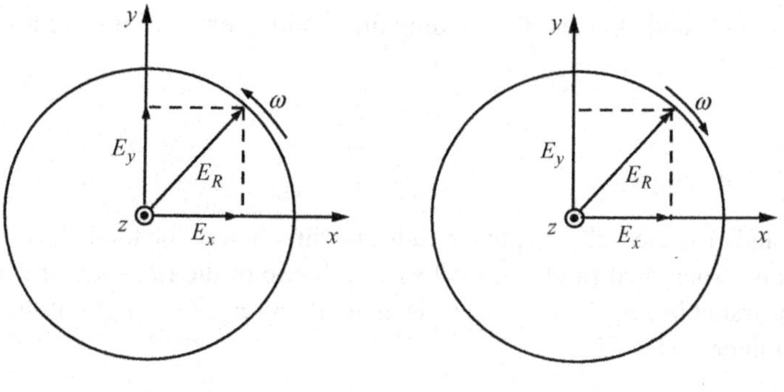

(a) (b)

Fig. 7.10 Showing direction of rotation of the resultant vector E_R of circularly-polarised electromagnetic waves. (a) E_x leads E_y by 90° in time and (b) E_y leads E_x by 90° in time. The direction of propagation of the wave is normally out of the page.

leading field; hence the direction of rotation of the resultant field is from E_x towards E_y, as in Fig. 7.10(a). And if $\psi = -\pi/2$, the direction of rotation of the resultant field is from E_y towards E_x, as in Fig. 7.10(b).

An electromagnetic wave where the electric field intensity of constant magnitude rotates at a constant angular velocity is termed as a *circularly polarised wave.*

If the direction of rotation of the resultant field is same as that of a right-handed screw advancing along the progress of the wave as in Fig. 7.10(a), the wave is called *right circularly* polarised wave. And, if the direction of rotation of the resultant field is same as that of a left-handed screw advancing along the progress of the wave, as in Fig. 7.10(b), the wave is called *left-circularly* polarised wave.

Elliptical Polarisation

Now, let us assume that amplitudes of the fields described by (3) and (4) are not equal ($E_{x0} \neq E_{y0}$) and the time angle $\psi = \pi/2$. So, the x and y components of electric field intensities are given by

$$E_x = E_{x0} \cos \omega t$$

Or

$$\frac{E_x}{E_{x0}} = \cos \omega t \tag{11}$$

$$E_y = E_{y0} \sin \omega t$$

Or

$$\frac{E_y}{E_{y0}} = \sin \omega t \tag{12}$$

The total field, which is equal to the vector sum of E_x and E_y, is

$$\mathbf{E} = \mathbf{u}_x E_{x0} \cos \omega t + \mathbf{u}_y E_{y0} \sin \omega t$$

Magnitude of the total field is

$$E_R = \sqrt{E_{xo}^2 \cos^2 \omega t + E_{y0}^2 \sin^2 \omega t} \tag{13}$$

And if Θ is angle of inclination of E_R with the x axis, then

$$\tan \Theta = \frac{E_y}{E_x} = \frac{E_{y0}}{E_{xo}} \tan \omega t \tag{14}$$

Squaring both sides of (11) and (12) and then adding the resulting expressions, we get

$$\frac{E_x^2}{E_{x0}^2}+\frac{E_y^2}{E_{y0}^2}=1 \tag{15}$$

This equation is that of an ellipse.

Relations (13), (14) and (15) show that the magnitude and direction of the total electric field are functions of time. Also, the field is a rotational field and the locus of the tip of the field vector describes an ellipse on a given z plane, as illustrated in Fig. 7.11(a). If E_{x0} is greater than E_{y0}, the length of the major axis is equal to $2E_{x0}$ and that of the minor axis is $2E_{y0}$.

An electromagnetic wave in which the locus of the end of the electric field intensity vector is an ellipse is called an *elliptically polarised* wave.

If the phase difference between E_x and E_y is other than $\pi/2$ and $E_{x0} \neq E_{y0}$, it can be shown that the wave is still elliptically polarised. Referring to (3) and (4), we have

$$\cos \omega t = E_x/E_{x0} \tag{16}$$

$$\sin \omega t = \sqrt{1-E_x^2/E_{x0}^2} \tag{17}$$

and $$E_y = E_{y0}\cos(\omega t-\psi) = E_{y0}(\cos \omega t \cos \psi + \sin \omega t \sin \psi) \tag{18}$$

By substituting (16) and (17) in (18), we will get

$$\left[\sqrt{1-(E_x^2/E_{x0}^2)}\right]\sin\psi = E_y/E_{y0} - (E_x/E_{x0})\cos\psi \tag{19}$$

Squaring both the sides of (19) and then simplifying, we get

$$\frac{E_x^2}{E_{x0}^2}+\frac{E_y^2}{E_{y0}^2}-\frac{2E_xE_y}{E_{x0}E_{y0}}\cos\psi=\sin^2\psi$$

This also is the equation of an ellipse. However, the axes of the ellipse in this case are inclined to the coordinate directions. The angle of tilt of the major axis is the angle of inclination of the maximum value of E_R with respect to a coordinate axis. The formula for the angle is derived as follows.

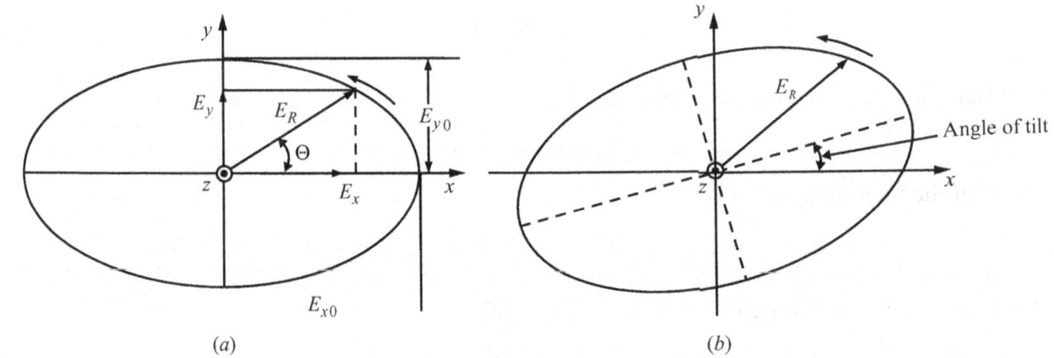

Fig. 7.11 Illustration of elliptically polarised electromagnetic waves. (a) Major axis is along x direction and (b) Major axis is tilted with respect to x direction.

The resultant field intensity at any instant of time is

$$E_R = \sqrt{E_{x0}^{\ 2} \cos^2 \omega t + E_{y0}^{\ 2} \cos^2(\omega t - \psi)} \tag{20}$$

$$\tan \Theta = \frac{E_{y0} \cos(\omega t - \psi)}{E_{x0} \cos \omega t} \tag{21}$$

where Θ is angle of inclination of E_R with the x axis at any instant. The time-angle at which E_R has maximum value can be determined by differentiating (20) with respect to t and then equating the result to zero. This process yields the following relation.

$$\tan \omega t' = \frac{E_{y0}^{\ 2} \sin 2\psi}{E_{x0}^{\ 2} + E_{y0}^{\ 2} \sin 2\psi}$$

where t' is the value of t at which E_R is maximum.

The angle $\omega t'$ can be calculated for given values of E_{x0}, E_{y0} and ψ. Substituting this in (20) and (21), maximum value of E_R and its angle with the x axis can be determined. The ellipse of Fig. 7.11(*b*) has been obtained taking $E_{x0} = 6$ units, $E_{y0} = 4$ units and $\psi = 60°$. The maximum value of E_R and angle of tilt for this example are found to be 6.45 units and 25.3° respectively.

7.7 Fields at the Surface of a Perfect Conductor

Materials having very high conductivity may be considered as perfect conductors. The skin depth, given by Formula (7.5-38), for these materials is negligibly small and may be assumed to be zero. Hence all the fields, applied as well as induced, are confined to the surface of a perfect conductor. With these assumptions, the fields at the surface of a perfect conductor can be derived from the boundary relations derived in Section 6.16. These are

$$E_{t1} - E_{t2} = 0$$

$$D_{n2} - D_{n1} = \rho_S$$

$$H_{t1} - H_{t2} = K$$

$$B_{n2} - B_{n1} = 0$$

Let us take Region 2 to be a perfect conductor. Then all the field quantities with the second Subscript 2 in the above formulae would be zero. Also, the tangential component of electric field intensity as well as the normal component of magnetic flux density at the surface of a perfect conductor is zero. Further, the induced surface-charge density ρ_S is equal to the normal component electric flux density, and the induced surface-current density K is equal to the tangential component of magnetic field intensity. Thus, dropping the second Subscript 1, fields at the surface of a perfect conductor are

1. $E_t = 0 \Rightarrow \mathbf{u}_n \times \mathbf{E} = 0$
2. $D_n = \rho_S \Rightarrow \mathbf{u}_n \cdot \mathbf{D} = \rho_S$
3. $H_t = K \Rightarrow \mathbf{u}_n \times \mathbf{H} = \mathbf{K}$
4. $B_n = 0 \Rightarrow \mathbf{u}_n \cdot \mathbf{B} = 0$

where $\mathbf{u}_n$ is a unit vector normal to the conductor surface. These are the boundary conditions at the surface of a perfect conductor for time-varying fields.

Silver and copper can be considered as perfect conductors; especially at high frequencies.

7.8 Reflection of a Uniform Plane Wave Incident Normally on a Perfect Conductor

Consider that a uniform plane electromagnetic wave, propagating along positive z direction in a perfect dielectric medium of permeability μ and permittivity ε, is incident normally on the surface of a perfect flat conductor. The conductor surface is taken at $z = 0$ plane, as shown in Fig. 7.12(a). Since the depth of penetration for a perfect conductor is zero, the wave will not travel beyond the $z = 0$ plane. Hence, there will be a normally reflected wave from the conductor surface. The directions of propagation of incident and reflected waves are shown in Fig. 7.12(a). The relationship between the fields of incident and reflected waves and the magnitude of current induced on the conductor surface can be obtained using the boundary relations given in Section 7.7.

Fields of Incident and Reflected Waves

Let us assume that the electric field intensity of the incident wave is in positive x direction (x-polarised) and it varies harmonically with time with an angular frequency ω. Thus, the cosine form of variation of the field with z and t is

$$E_x^+ = E_{x0}^+ \cos(\omega t - \beta z) \tag{1}$$

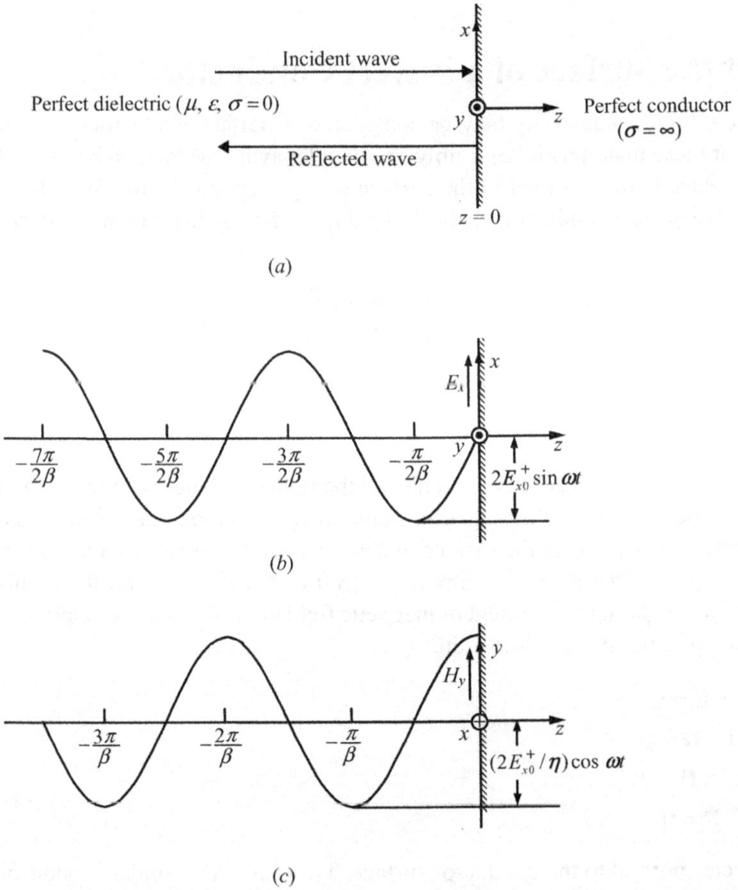

Fig. 7.12 Reflection of electromagnetic wave from the surface of a perfect conductor. (a) Rays of incident and reflected waves, (b) Standing wave of total electric field, and (c) Standing wave of total magnetic field.

where $\beta = \omega\sqrt{\mu\varepsilon}$ is phase constant of the wave.

The direction of propagation of the reflected wave is along the negative z direction. Its frequency and phase constant are same as those of the incident wave. In order that boundary conditions are satisfied for all time, the reflected electric field should also be a cosine function. Thus,

$$E_x^- = E_{x0}^- \cos(\omega t + \beta z) \tag{2}$$

Total Electric and Magnetic Fields

The total electric field intensity to the left of the $z = 0$ plane is equal to the sum of (1) and (2). Denoting the total field by E_x, we have

$$E_x = E_x^+ + E_x^- = E_{x0}^+ \cos(\omega t - \beta z) + E_{x0}^- \cos(\omega t + \beta z) \tag{3}$$

Since E_x is tangential to the conductor surface, it is equal to zero at $z = 0$ as per Boundary Condition 1 of the preceding section. Using this condition in (3), we get

$$E_{x0}^- = -E_{x0}^+$$

By substituting the above relation in (2) and (3), the electric field of reflected wave and the total field are obtained in terms of E_{x0}^+. The reflected electric field is

$$E_x^- = -E_{x0}^+ \cos(\omega t + \beta z) \tag{4}$$

Thus,

the electric field gets totally reflected with change of sign.

The total instantaneous electric field for any value of z in the dielectric region is equal to the sum of the incident and reflected waves. Denoting the total electric field be E_x, we get

$$E_x = E_{x0}^+[\cos(\omega t - \beta z) - \cos(\omega t + \beta z)] = 2E_0^+ \sin\omega t \sin\beta z \tag{5}$$

The function on rhs of (5) represents a standing (stationary) wave that varies sinusoidally with βz. The amplitude of the field is $2E_{x0}^+ \sin\omega t$ and it occurs at $\beta z = -\pi/2, -3\pi/2, -5\pi/2\ldots.$ A plot of the function is given in Fig. 7.12(*b*).

The magnetic field intensity of the positively travelling incident wave,

$$H_y^+ = \frac{E_x^+}{\eta} = \frac{E_{x0}^+}{\eta}\cos(\omega t - \beta z) \tag{6}$$

The magnetic field intensity of the negatively travelling reflected wave,

$$H_y^- = -\frac{E_x^-}{\eta} = -\frac{E_{x0}^-}{\eta}\cos(\omega t + \beta z) = \frac{E_{x0}^+}{\eta}\cos(\omega t + \beta z) \tag{7}$$

From (6) and (7), we get

$$H_y \quad H_{y0}^+$$

Thus,

the magnetic field gets totally reflected without change of sign.

The total instantaneous magnetic field intensity at any z in the dielectric region is equal to the sum of (6) and (7). Denoting the total field by H_y, we have

$$H_y = H_y^+ + H_y^- = \frac{2E_0^+}{\eta}\cos\omega t\cos\beta z$$

$$= \frac{2E_0^+}{\eta}\sin(\omega t + \pi/2)\sin(\beta z + \pi/2) \quad (8)$$

The total magnetic field intensity is also a standing wave with amplitude equal to $(2E_0/\eta)\cos\omega t$. Maximum values of the field occur at $\beta z = 0, -\pi, -2\pi \ldots$. A plot of the field is shown in Fig. 7.12(c).

Surface Current Density

The conductor will have current on its surface induced by the time-varying field. The surface current density is found using Boundary Condition 3 of the preceding section. The total magnetic field intensity is

$$\mathbf{H} = H_y\mathbf{u}_y$$

The unit vector normal to the conductor surface,

$$\mathbf{u}_n = -\mathbf{u}_z$$

The surface current density,

$$\mathbf{K} = \mathbf{u}_n \times \mathbf{H} = (-\mathbf{u}_z \times \mathbf{u}_y)H_y(z=0) = H_y(z=0)\,\mathbf{u}_x \quad (9)$$

From (8) and (9) we get

$$\mathbf{K} = \frac{2E_{x0}^+}{\eta}\cos\omega t\,\mathbf{u}_x \text{ A/m} \quad (10)$$

The current density is in x direction (same as that of the electric field intensity) and its value per unit width in y direction is equal to the total magnetic field on the surface of the conductor.

The reflection properties of a uniform plane wave incident normally on a perfect conductor are as follows.

1. As mentioned earlier, the electric field intensity is totally reflected with change of sign, whereas the magnetic field intensity is totally reflected without change of sign.
2. Comparison between (5) and (8) shows that the total electric and magnetic fields, which are standing waves, are perpendicular to each other in space. The phase difference between the two fields is $\pi/2$ radians in time as well as in space.
3. The surface current induced in the conductor is numerically equal to the magnetic field intensity tangential to the surface.
4. The electric and magnetic fields of the incident wave given by (1) and (6) are in time phase. The fields of the reflected wave given by (4) and (7) are also in time phase. Thus, the average power per unit area carried by the incident and reflected waves are

Incident wave: $P_{av}^{+} = \dfrac{E_{x0}^{+} H_{y0}^{+}}{2} = \dfrac{E_{x0}^{+\,2}}{2\eta}$

Reflected wave: $P_{av}^{-} = \dfrac{E_{x0}^{-} H_{y0}^{-}}{2} = \dfrac{1}{2}(-E_{x0}^{+})\dfrac{E_{x0}^{+}}{\eta} = -\dfrac{E_{x0}^{+\,2}}{2\eta}$

That is, the value of average power per unit area of the reflected wave in the reverse direction is equal to that of the incident wave in the forward direction. Hence, the net average power at any z plane in the dielectric medium is zero.

EXAMPLE 7.15

An electromagnetic wave travelling along positive z direction in air strikes a perfect conducting surface located at $z = 0$. The electric field intensity of the incident wave is in x direction and has a maximum value of 25 V/m at $z = t = 0$. Obtain equations of the electric fields of the incident and reflected waves if the first zero of the total field occurs 5.0 m in front of the conducting surface. Also, find the linear current density on the conductor surface. Take $\mu = \mu_0$ and $\varepsilon = \varepsilon_0$ for the conductor.

The equation of the electric field of the incident wave is of the form

$\mathbf{E}^{+} = 25\cos(\omega t - \beta z)\,\mathbf{u}_x$

As the total electric field is zero at the conductor surface and the first zero occurs 5.0 m away from the conductor surface, the wavelength of the fields $\lambda = 10.0$ m.

Frequency, $f = \dfrac{v_0}{\lambda} = \dfrac{3\times 10^8}{10.0} = 30.0$ MHz

Phase constant, $\beta = 2\pi f\sqrt{\mu_0\varepsilon_0} = (2\pi \times 30\times 10^6)/(3\times 10^8) = 0.2\pi\ \text{m}^{-1}$

Intrinsic impedance, $\eta_0 = 377\ \Omega$

The electric field of the reflected wave is in negative x direction and its amplitude is 25 V/m. The equation for the reflected electric field,

$\mathbf{E}^{-} = -25\cos[(60\pi\times 10^6)t + 0.2\pi z)\,\mathbf{u}_x$ V/m

The surface current density, $\mathbf{K} = (2E_{x0}^{+}/\eta_0)\cos\omega t\,\mathbf{u}_x = (50/377)\cos(60\pi\times 10^6)t\,\mathbf{u}_x$

$= 0.318\cos(60\pi\times 10^6)t\,\mathbf{u}_x$ A/m

EXAMPLE 7.16

A uniform plane wave travelling along positive z direction in air is incident normally on a perfect conducting surface located at $z = 0$ plane. The electric field intensity of the incident wave, $\mathbf{E}_0^{+} = (200\mathbf{u}_x + 150\mathbf{u}_y)e^{-j\beta z}$. Obtain (a) the electric field intensity of the reflected wave, (b) the magnetic field intensity of the incident wave, (c) the magnetic field intensity of the reflected wave, and (d) current density on the conductor surface.

(a) The x and y components of electric field of the incident wave are

$E_{x0}^{+} = 200$ V/m, $\quad E_{y0}^{+} = 150$ V/m

The electric field gets totally reflected with change of sign from a perfect conductor. Thus, the electric field intensity of the reflected wave travelling along negative z direction,

$$\mathbf{E}_{\upsilon} = -(200\mathbf{u}_x + 150\mathbf{u}_y)e^{j\beta z}$$

(b) The x and y components of magnetic field in the incident wave are

$$H_x^{\cdot} = -E_{y0}^{+}/\eta_0 = -150/377 = -0.398 \text{ A/m}$$

$$H_y^{\cdot} = E_{x0}^{+}/\eta_0 = 200/377 = 0.531 \text{ A/m}$$

Magnetic field intensity of the incident wave,

$$\mathbf{H}_{\upsilon}^{\cdot} = (-0.398\mathbf{u}_x + 0.531\mathbf{u}_y)e^{-j\beta z} \text{ A/m}$$

(c) The magnetic field gets totally reflected without change of sign from a perfect conductor. Thus, the magnetic field intensity of the reflected wave travelling along negative z direction,

$$\mathbf{H}_{\upsilon} = (-0.398\mathbf{u}_x + 0.531\mathbf{u}_y)e^{j\beta z} \text{ A/m}$$

(d) Total magnetic field at the conductor surface ($z = 0$),

$$\mathbf{H}_{\upsilon} = \mathbf{H}_0^{+} + \mathbf{H}_0^{-} = -0.796\mathbf{u}_x + 1.062\mathbf{u}_y \text{ A/m}$$

As per the boundary condition, the surface current density on a perfect conductor surface is given by

$$\mathbf{K} \quad \mathbf{u}_n \times \mathbf{H}_0 = (-\mathbf{u}_z)(-0.796\mathbf{u}_x + 1.062\mathbf{u}_y) = 1.062\mathbf{u}_x + 0.796\mathbf{u}_y \text{ A/m}$$

7.9 Reflection and Refraction of Plane Waves Incident Normally at the Interface of Two Dielectric Regions

Figure 7.13 shows two dielectric regions 1 and 2 with a common boundary surface at $z = 0$ plane. Let a uniform plane electromagnetic wave travelling along positive z direction in Region 1 be incident normally at the boundary surface. This is the *incident wave*. Because of the discontinuity, there will be reflection of the wave into Region 1 and transmission of the wave into Region 2. The wave that is reflected to Region 1 is called *reflected wave*, and the wave that crosses the interface is termed as *transmitted wave*. Both transmitted and reflected waves must be present in order to satisfy the boundary conditions. The waves are indicated by rays in the figure. A part of the energy of the incident wave is transmitted to Region 2 and the remaining part is reflected back to Region 1. The division of power between the incident and reflected waves depends on the magnitudes of field intensities of the respective waves, and the magnitudes of the fields are governed by the boundary relations.

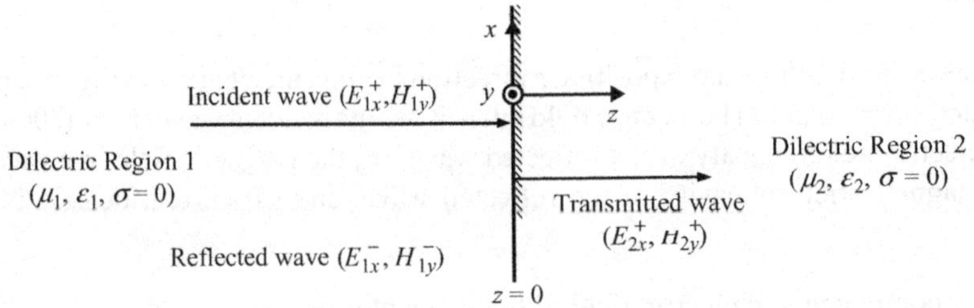

Fig. 7.13 Showing directions of the incident, reflected and transmitted waves at the common boundary surface of two dielectric regions

We will study the phenomenon of wave reflection and transmission by considering that the electric field intensity of the incident wave is in x direction and it varies harmonically with time. We will define the reflection and transmission coefficients of the fields and calculate the reflected and transmitted powers by assuming that the dielectric regions have no losses and there is no discontinuity in Region 2. The constants of the regions are taken as μ_1, ε_1 and μ_2, ε_2. Fields in the two regions will be denoted by additional subscripts 1 and 2.

Incident, Reflected, and Transmitted Waves

Let us assume that the electric field intensity of the incident wave travelling along positive z direction is given by

$$E_{1x}^{+} = E_{1x0}^{+} \cos(\omega t - \beta_1 z) \tag{1}$$

As the electric field given by (1) is in x direction, the electric fields of reflected and transmitted waves will also be in x direction. Further, the reflected and transmitted waves travel along negative and positive z directions respectively. So, the electric fields of the two waves are given by

$$E_{1x}^{-} = E_{1x0}^{+} \cos(\omega t + \beta_1 z) \tag{2}$$

and

$$E_{2x}^{+} = E_{2x0}^{+} \cos(\omega t - \beta_2 z) \tag{3}$$

where $\beta_1 = \omega\sqrt{\mu_1\varepsilon_1}$ and $\beta_2 = \omega\sqrt{\mu_2\varepsilon_2}$ are the phase constants.

The magnetic fields in terms of the respective electric fields associated with the three waves are

$$H_{1y}^{+} = \frac{E_{1x}^{+}}{\eta_1} = \frac{E_{1x0}^{+}}{\eta_1} \cos(\omega t - \beta_1 z) \tag{4}$$

$$H_{1y}^{-} = -\frac{E_{1x}^{-}}{\eta_1} = -\frac{E_{1x0}^{-}}{\eta_1} \cos(\omega t + \beta_1 z) \tag{5}$$

$$H_{2y}^{+} = \frac{E_{2x}^{+}}{\eta_2} = \frac{E_{2x0}^{+}}{\eta_2} \cos(\omega t - \beta_2 z) \tag{6}$$

where $\eta_1 = \sqrt{\mu_1/\varepsilon_1}$ and $\eta_2 = \sqrt{\mu_2/\varepsilon_2}$ are the intrinsic impedances of the two dielectric regions.

Reflection and Transmission Coefficients

The total electric field in Region 1 is equal to the sum of electric fields of the incident and reflected waves. Similarly, the total magnetic field in Region 1 is equal to the sum of the magnetic fields of incident and reflected waves. Region 2 has only the transmitted wave. Also, all the fields are tangential to the interface.

We will use the continuity condition of tangential electric field intensity and tangential magnetic field intensity across the boundary surface to relate the fields of the reflected and the transmitted waves in terms of the corresponding fields of the incident wave. The ratio of electric field intensity of the reflected wave and the electric field intensity of the incident wave is called reflection coefficient. And the ratio of electric field intensity of the transmitted wave and the electric field intensity of the incident wave is called transmission coefficient.

From the continuity condition of tangential electric field at the interface, we have

$$E_{1x}^{+}(z=0)+E_{1x}^{-}(z=0)=E_{2x}^{+}(z=0) \tag{7}$$

By substituting $z = 0$ in (1), (2) and (3), and then using in (7), we get

$$(E_{1x0}^{+}+E_{1x0}^{-})\cos\omega t = E_{2x0}^{+}\cos\omega t$$

Or

$$E_{1x0}^{+}+E_{1x0}^{-}=E_{2x0}^{+} \tag{8}$$

If we assume that the interface has no surface current, then the magnetic field intensity on one side of the boundary must be equal to that on the other side. Accordingly,

$$H_{1y}^{+}(z=0)+H_{1y}^{-}(z=0)=H_{2y}^{+}(z=0) \tag{9}$$

Thus using (4), (5) and (6) in relation (9), we have

$$\frac{E_{1x0}^{+}}{\eta_1}-\frac{E_{1x0}^{-}}{\eta_1}=\frac{E_{2x0}^{+}}{\eta_2} \tag{10}$$

Equations (8) and (10) can be solved simultaneously to obtain E_{1x0}^{-} and E_{2x0}^{+} in terms of E_{1x0}^{+}. The ratio between E_{1x0}^{-} and E_{1x0}^{+} is the *reflection coefficient* for the electric field intensity at $z = 0$. We will denote the reflection coefficient by the Greek alphabet Γ (Gamma).

The reflection coefficient,

$$\Gamma=\frac{E_{1x0}^{-}}{E_{1x0}^{+}}=\frac{\eta_2-\eta_1}{\eta_2+\eta_1} \tag{11a}$$

If the regions are nonmagnetic, then $\mu_1=\mu_2=\mu_0$. Hence,

$$\Gamma=\frac{\sqrt{\varepsilon_{r1}}-\sqrt{\varepsilon_{r2}}}{\sqrt{\varepsilon_{r1}}+\sqrt{\varepsilon_{r2}}} \tag{11b}$$

The ratio between E_{2x0}^{+} and E_{1x0}^{+} is the *transmission coefficient* for electric field intensity at $z = 0$. Thus, denoting the transmission coefficient by the lowercase Greek alphabet τ (Tau), the transmission coefficient for the electric field intensity,

$$\tau=\frac{E_{2x0}^{+}}{E_{1x0}^{+}}=\frac{2\eta_2}{\eta_2+\eta_1}=1+\Gamma \tag{12}$$

The reflection and transmission coefficient for the magnetic field intensity can be found by using the relation E_x H_y in (11) and (12). The transmission and reflection coefficients for the magnetic field are

$$\frac{H_{1y0}^{-}}{H_{1y0}^{+}}=\frac{\eta_1-\eta_2}{\eta_1+\eta_2}=-\Gamma \tag{13}$$

and

$$\frac{H_{2y0}^{+}}{H_{1y0}^{+}}=\frac{2\eta_1}{\eta_1+\eta_2}=1-\Gamma \tag{14}$$

The sign of the reflection coefficients depend on the relative values of η_1 and η_2. The sign of the transmission coefficients however can only be positive.

Reflection Coefficient at Any Point in Region 1

The reflection coefficient at the interface in Region 1 is equal to the ratio of electric field intensities of the reflected and incident waves. In the same way, the reflection coefficient at any point z in Region 1 is defined as the ratio of electric field intensities of the reflected and incident waves at z. We will use the phasor forms of the field intensities to calculate the reflection coefficient.

The phasor forms of the electric field intensity of the incident wave travelling along positive z direction, and the electric field intensity of the reflected wave travelling along negative z direction in Region 1 are given by

$$\bar{E}_{1x}^{+}(z) = E_{1x0}^{+}\, e^{-j\beta z}$$

and

$$\bar{E}_{1x}^{-}(z) = E_{1x0}^{-}\, e^{j\beta z}$$

where the amplitudes E_{1x0}^{+} and E_{1x0}^{-} are assumed to be real quantities. Denoting the reflection coefficient, which in general is a complex quantity, by $\bar{\Gamma}_1(z)$, we have

$$\bar{\Gamma}_1(z) = \frac{E_{1x}^{-}(z)}{E_{1x}^{+}(z)} = \frac{E_{1x0}^{-}\, e^{j\beta z}}{E_{1x0}^{+} e^{-j\beta z}} = \frac{E_{1x0}^{-}}{E_{1x0}^{+}} e^{2j\beta z} = \Gamma_1(0) e^{2j\beta z}$$

where $\Gamma_1(0)$ is the reflection coefficient at $z = 0$ in Region 1 given by (11).

Power Relations

As both the dielectric regions are lossless, the electric and magnetic field intensities of each wave are in time phase. Hence, the values of average power per unit area of the three waves are as follows.

For the incident wave, $P_{1,av}^{+} = \dfrac{(E_{1x0}^{+})^2}{2\eta_1}$

For the reflected wave, $P_{1,av}^{-} = \dfrac{(\Gamma E_{1x0}^{+})^2}{2\eta_1} = \Gamma^2 P_{1,av}^{+}$

For the transmitted wave, $P_{2,av}^{+} = \dfrac{(E_{2x0}^{+})^2}{2\eta_2} = (1-\Gamma^2) P_{1,av}^{+}$

As there are no losses, the power per unit area of the reflected wave plus the power per unit area of the transmitted wave is equal to the power per unit area of the incident wave.

Relation (11) and (13) show that if $\eta_1 > \eta_2$, the electric field is reflected with change of sign, whereas the magnetic field intensity is reflected without change of sign. Reverse is the case if $\eta_1 < \eta_2$. From relations (12) and (14) we find that the signs of transmitted electric field and magnetic field intensities are same as those of the incident wave irrespective of whether η_1 is greater or less than η_2.

If Region 2 is a perfect conductor, the electric field of the reflected wave is equal and opposite to the electric field of the incident wave. The reflection coefficient for this case is minus one.

If the dielectric in any of the region is not perfect or if the dielectrics of both the regions are not perfect, as is the case in practice, the intrinsic impedances would be complex quantities. So, the fields of the reflected and transmitted waves would be out of phase in time relative to the fields of the incident waves. The electric and magnetic fields of each wave will also not be in time phase as the intrinsic impedance would be a complex quantity. The amplitude of the fields will attenuate, and the energy in the waves will decrease with the progress of the waves in their respective directions.

EXAMPLE 7.17

The reflection coefficient for the electric field intensity at the interface of two nonmagnetic regions is 0.15. The relative permittivity of Region 1 is 4.5. If the incident wave is in Region 1, find the relative permittivity of Region 2.

We have, $\mu_1 = \mu_2 = \mu_0$

Reflection coefficient, $\Gamma = \dfrac{\eta_2 - \eta_1}{\eta_2 + \eta_1} = 0.15$

Or $(.15)(\eta_2 + \eta_1) = \eta_2 - \eta_1,$

$1\ 15 _{\prime\prime} \quad 0\ 85\eta_2,$

$\dfrac{\eta_1}{\eta_2} = \dfrac{\varepsilon_{r2}}{\varepsilon_{r1}} = \dfrac{0.85}{1.15} = 0.739,$

$\varepsilon_{r2} = (0.739)^2(4.5) = 2.46.$

EXAMPLE 7.18

The transmission coefficient for the electric field between two dielectric regions 1 and 2 when the incident wave is in Region 1 is 0.8. The constants for Region 1 are $\varepsilon_{r1} = 4, \mu_{r1} = 1, \sigma_1 = 0$. Find the relative permeability of Region 2 if $\varepsilon_{r2} = 16, \sigma_2 = 0$.

Transmission coefficient, $\tau = \dfrac{2\eta_2}{\eta_2 + \eta_1}$

$\eta_{.} \quad \sqrt{\mu_0/4\varepsilon_0} = 0.5\eta_0\ \Omega$

$\eta_{\_} \quad \sqrt{(\mu_0\mu_{r2})/(16\varepsilon_0)} = (0.25\eta_0)\sqrt{\mu_{r2}}\ \Omega$

Substituting the values, $0.8 = \dfrac{0.5\sqrt{\mu_{r2}}}{0.25\sqrt{\mu_{r2}} + 0.5}$

By solving, we get $\mu_{r2} = 1.778$

EXAMPLE 7.19

Two nonmagnetic and perfect dielectric regions 1 and 2 have a common plane boundary surface. The relative permittivity of Region 1 is 6.25. A uniform plane wave from Region 1 is incident normally on the interface. If 90 per cent power of the incident wave is transmitted to Region 2, find the relative permittivity of Region 2.

Reflection coefficient, $\Gamma = \dfrac{\eta_{\_} - \eta_1}{\eta_{\_} + \eta_1} = \dfrac{\varepsilon_{r1} - \varepsilon_{2}}{\varepsilon_{r1} + \varepsilon_{2}}$, as $\mu_{.} = \mu_2 = \mu_0$

Ratio of transmitted power to incident power is equal to $1 - \Gamma^2$ where Γ is the reflection coefficient. Thus,

$1 - \Gamma^2 = 0.9$

And $\Gamma = \pm 0.316$

Considering the positive value of Γ,

$$0.316 = \frac{\sqrt{6.25} - \sqrt{\varepsilon_{r2}}}{\sqrt{6.25} + \sqrt{\varepsilon_{r2}}}$$

By solving, we get $\varepsilon_{r2} = 1.69$

Next, considering the negative value of Γ, we get $\varepsilon_{r2} = 23.14$

EXAMPLE 7.20

A uniform plane wave travelling along positive z direction in air strikes normally on the surface of a dielectric with $\mu = \mu_0$ and $\varepsilon = 6.25\varepsilon_0$. The amplitude of electric field of the incident wave is 10 V/m. Calculate the amplitudes of electric field intensities associated with the reflected and transmitted waves, assuming that the dielectric extends to infinity. Also, calculate the power per unit area carried by each wave.

Amplitude of electric field in the incident wave, $E^+_{1x0} = 10.0$ V/m

Intrinsic impedance of air, $\eta_1 = \eta_0 = 377\ \Omega$

Intrinsic impedance of the dielectric, $\eta_2 = \sqrt{\frac{\mu_0}{6.25\varepsilon_0}} = \frac{377}{2.5} = 150.8\ \Omega$

Reflection coefficient, $\Gamma = \frac{\eta_2 - \eta_1}{\eta_2 + \eta_1} = \frac{150.8 - 377}{150.8 + 377} = -0.4286$

Transmission coefficient $= 1 + \Gamma = 1 - 0.4286 = 0.5714$

Amplitude of electric field in the reflected wave, $E^-_{1x0} = |\Gamma| E^+_{1x0} = 4.286$ V/m

Amplitude of electric field in the transmitted wave, $E^+_{2x0} = |1 + \Gamma| E^+_{1x0} = 5.714$ V/m

Powers per unit area carried by the waves are

Incident wave: $P^+_{1,av} = \frac{(E^+_{1x0})^2}{2\eta_1} = \frac{10^2}{2 \times 377} = 0.1326$ W/m^2

Reflected wave: $P^-_{1,av} = \frac{(E^-_{1x0})^2}{2\eta_1} = \frac{4.286^2}{2 \times 377} = 0.0244$ W/m^2

Transmitted wave: $P^+_{2,av} = \frac{(E^+_{2x0})^2}{2\eta_2} = \frac{5.714^2}{2 \times 150.8} = 0.1082$ W/m^2

EXAMPLE 7.21

A uniform plane electromagnetic wave is incident normally at the interface of two nonmagnetic, perfect dielectric regions 1 and 2. The incident wave is in Region 1 and the amplitude of its electric field intensity is 24 V/m. The intrinsic impedance of Region 2 is 188 Ω and the transmission coefficient for the magnetic field is 0.75. Find average power per unit area in the incident, reflected, and transmitted waves.

Intrinsic impedance of Region 2, $\eta_2 = 188\ \Omega$

Amplitude of electric field of the incident wave, $E_{10}^{+} = 24\ \text{V/m}$

$$\text{Transmission coefficient for the magnetic field} = \frac{2\eta_1}{\eta_1 + \eta_2} = \frac{2\eta_1}{\eta_1 + 188} = 0.75$$

By solving, we get $\eta_1 = 112.8\ \Omega$

$$\text{Reflection coefficient for the electric field, } \Gamma = \frac{\eta_2 - \eta_1}{\eta_2 + \eta_1} = \frac{188 - 112.8}{188 + 112.8} = 0.25$$

Average power per unit area in the incident wave,

$$P_{1,av}^{+} = \frac{E_{10}^{+2}}{2\eta_1} = \frac{24^2}{(2)(112.8)} = 2.553\ \text{W/m}^2$$

Average power per unit area in the reflected wave,

$$P_{1,av}^{-} = \Gamma^2 P_{1,av}^{+} = (0.25)^2 (2.553) = 0.16\ \text{W/m}^2$$

Average power per unit area in the transmitted wave,

$$P_{2,av}^{+} = P_{1,av}^{+} - P_{1,av}^{-} = 2.393\ \text{W/m}^2$$

EXAMPLE 7.22

A perfect dielectric slab of relative permittivity 5.8 is located in air with its plane surfaces at $z = 0$ and at $z = 5$ cm. A uniform plane wave incident normally on the interface at $z = 0$, comes out from the other surface of the dielectric. If amplitude of the x-directed electric field intensity of the incident wave is 15 V/m, find the amplitude of the field transmitted to air at $z = 5$ cm. Assume the relative permeability of the dielectric to be unity.

Amplitude of electric field of the incident wave, $E_{1x0} = 15\ \text{V/m}$

Relative permittivity of the dielectric slab, $\varepsilon_r = 5.8$.

Let η and ε_r be the intrinsic impedance and relative permittivity of the dielectric slab.

$$\text{Transmission coefficient at } z = 0,\ \tau_1 = \frac{2\eta}{\eta + \eta_0} = \frac{2\sqrt{1}}{\sqrt{\varepsilon_{r2}} + \sqrt{1}} = \frac{2\sqrt{1}}{\sqrt{5.8} + \sqrt{1}} = 0.587$$

Amplitude of the field transmitted to the dielectric at $z = 0$,

$$E_{2x0} = \tau_1 E_{1x0} = (0.587)(15) = 8.81\ \text{V/m}$$

As the dielectric is perfect, the amplitude of the field does not change between the two interfaces. Thus, amplitude of the electric field incident on the interface at $z = 5$ cm is 8.81 V/m.

$$\text{Transmission coefficient at } z = 5\text{ cm},\ \tau_2 = \frac{2\eta_0}{\eta + \eta_0} = \frac{2\sqrt{\varepsilon_r}}{\sqrt{\varepsilon_r} + \sqrt{1}} = \frac{2\sqrt{5.8}}{\sqrt{5.8} + \sqrt{1}} = 1.41$$

Amplitude of the field transmitted to air at $z = 5$ cm,

$$E_{3x0} = \tau_2 E_{2x0} = (1.41)(8.81) = 12.42\ \text{V/m}$$

7.10 The Standing Wave Ratio

In the preceding two sections we found that when a plane electromagnetic wave is incident normally on an interface, a reflected wave exists in the region from where the incident wave has originated. The total electric field intensity in the region is the sum of the electric fields of the incident and reflected waves. At some point of time during the progress of the waves in their respective direction of propagation, the electric fields of the two waves are in space phase. The amplitude of the total electric field at this instant is equal to the sum of the amplitudes of the incident and reflected waves. Thus, the amplitude of the total field has the maximum value. After each wave progresses by one-quarter of a wavelength from this position, the fields of the waves would be in anti-phase. The amplitude of total electric field at this instant is the difference between the amplitudes of electric fields of the two waves. Thus, the magnitude of the amplitude of the total field at this position has the minimum value. The amplitude of the total magnetic field intensity also varies between a maximum value and a minimum value in the region.

The ratio of the maximum amplitude to the minimum amplitude of the total electric field is called *standing wave ratio.*

The ratio between the maximum amplitude and the minimum amplitude of the total magnetic field also gives the standing wave ratio. The standing wave ratio is a measurable quantity and its value provides information about the quality of transmission of electromagnetic waves.

In order to calculate the standing wave ratio, let us take the common boundary of two loss-free dielectric regions at the $z = 0$ plane, as shown in Fig. 7.14(a). Let a uniform plane electromagnetic wave originating in the region to the left of the interface and travelling along the positive z direction incident normally on the interface. There will be a normally reflected wave from the interface, and it will travel along the negative z direction. Let the electric field of the incident wave be in x direction and suppose it varies harmonically with time. Then, the electric field of the reflected wave is also in x direction and it varies harmonically with time. The magnetic fields of the incident and reflected waves, which are in positive and negative y direction respectively, also vary harmonically with time.

Let the expressions for the electric fields of the incident and reflected waves in real time be given by

$$E_x^+ = E_{x0}^+ \cos(\omega t - \beta z) \tag{1}$$

$$E_x^- = E_{x0}^- \cos(\omega t + \beta z) \tag{2}$$

The total electric field E_x at any instant and at any z in the region is equal to the sum of (1) and (2). Thus,

$$\begin{aligned} E_x &= E_{x0}^+ \cos(\omega t - \beta z) + E_{x0}^- \cos(\omega t + \beta z) \\ &= E_{x0}^+ (\cos\omega t \cos\beta z + \sin\omega t \sin\beta z) + E_{x0}^- (\cos\omega t \cos\beta z - \sin\omega t \sin\beta z) \\ &= [(E_{x0}^+ + E_{x0}^-)\cos\beta z]\cos\omega t + [(E_{x0}^+ - E_{x0}^-)\sin\beta z]\sin\omega t \\ &= A\cos\omega t + B\sin\omega t \end{aligned} \tag{3}$$

where $A = (E_{x0}^+ + E_{x0}^-)\cos\beta z$, $\mathrm{B} = (E_{x0}^+ - E_{x0}^-)\sin\beta z$

Equation (3) can be written as a single function by taking $B/A = \tan\xi$ so that

$$A \quad A^2 + B^2 \; \cos\xi$$

$$B \quad A^2 + B^2 \; \sin\xi$$

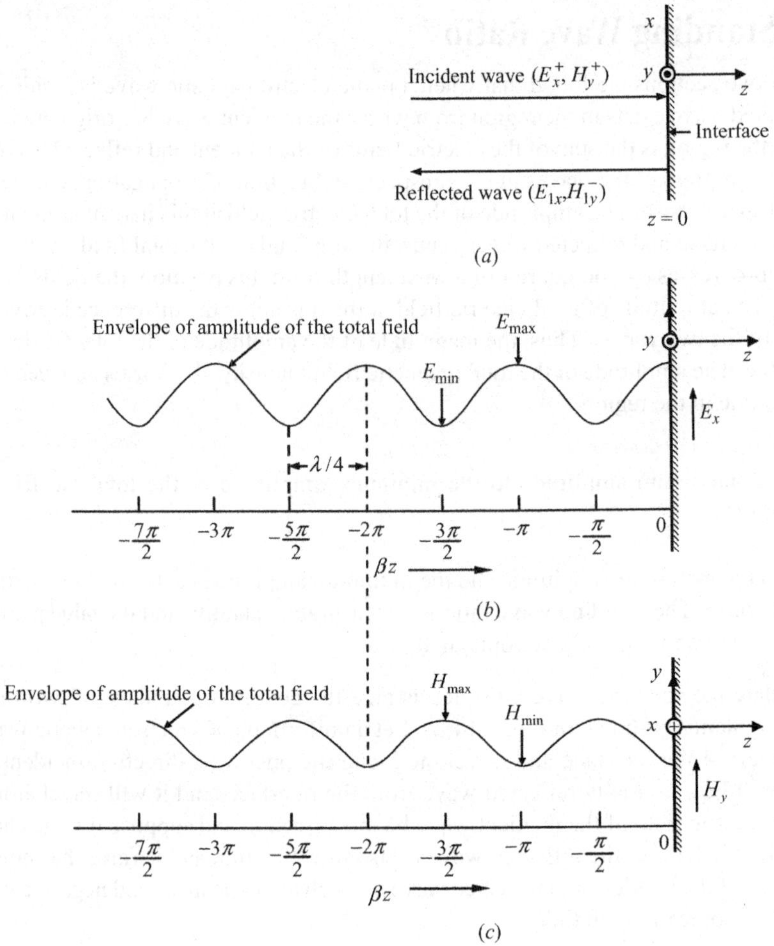

Fig. 7.14 (*a*) Showing directions of incident and reflected waves at the interface of two dielectric regions, (*b*) variation of total electric field intensity on the positive side of *x*, and (*c*) variation of amplitude of the total magnetic field intensity at any *z* on the positive side of *y*. The envelopes are standing waves.

Using these relations in (3),

$$E_x = \sqrt{A^2 + B^2}\ (\cos \omega t \cos \xi + \sin \omega t \sin \xi)$$

$$= \sqrt{A^2 + B^2}\ \cos(\omega t - \xi)$$

$$= \sqrt{(E_{x0}^+ + E_{x0}^-)^2 \cos^2 \beta z + (E_{x0}^+ - E_{x0}^-)^2 \sin^2 \beta z}\ \cos(\omega t - \xi)$$

$$= \left[\sqrt{(E_{x0}^+)^2 + (E_{x0}^-)^2 + 2E_{x0}^+ E_{x0}^- \cos 2\beta z}\right] \cos(\omega t - \xi) \tag{4}$$

The function within the square brackets of (4) is the amplitude of the total electric field intensity. It is given by

$$E_x \ [illegible]\ E_{x0}^+ [illegible]^2 + (E_x [illegible]\ E_{x0} E_{x0} [illegible]\ z \tag{5}$$

The amplitude is a function of z. The value of the amplitude is maximum where $\cos 2\beta z = +1$, and it has the minimum value where $\cos 2\beta z = -1$. Accordingly, the maximum value of amplitude of the total electric field occurs at $\beta z = 0, -\pi, -2\pi, \ldots$, and the minimum value of the amplitude occurs at $\beta z = -\pi/2, -3\pi/2, -5\pi/$. Using these conditions in (5), the maximum and minimum values of amplitude of the electric field are given by

$$E_{\max} = |E_{x0}^+| + |E_{x0}^-| = E_{x0}^+(1+|\Gamma|)$$

$$E_{\min} = |E_{x0}^+| - |E_{x0}^-| = E_{x0}^+(1-|\Gamma|)$$

where $|\Gamma|$ is magnitude of the reflection coefficient at $z = 0$.

We will next calculate the maximum and minimum amplitudes of the total magnetic field in the region. The total magnetic field,

$$H_y = H_y^+ + H_y^- = \frac{1}{\eta}(E_x^+ - E_x^-) \quad (6)$$

Substitution of (1) and (2) in (6) gives

$$H_y = \frac{1}{\eta}[E_{x0}^+ \cos(\omega t - \beta z) - E_{x0}^- \cos(\omega t + \beta z)]$$

$$= \frac{1}{\eta}\{[E_{x0}^+ - E_{x0}^-)\cos\beta z]\cos\omega t + [(E_{x0}^+ + E_{x0}^-)\cos\beta z]\cos\omega t]\}$$

Using the procedure as described, it can be shown that

$$H_y = \frac{1}{\eta}\left[\sqrt{(E_{x0}^+)^2 + (E_{x0}^-)^2 - 2E_{x0}^+E_{x0}^-\cos 2\beta z}\,\right]\cos(\omega t - \varphi') \quad (7)$$

where $\varphi' = \tan^{-1}\dfrac{(E_{x0}^+ + E_{x0}^-)\sin\beta z}{(E_{x0}^+ - E_{x0}^-)\cos\beta z}$

The function within the square brackets of (7) is the amplitude of the total magnetic field intensity at any z. The amplitude,

$$H_{y0} = \frac{1}{\eta}\left[\sqrt{(E_{x0}^+)^2 + (E_{x0}^-)^2 - 2E_{x0}^+E_{x0}^-\cos 2\beta z}\,\right] \quad (8)$$

The amplitude of the magnetic field intensity has minimum value at $\beta z = 0, \pi, 2\pi, \ldots$, and the amplitude has maximum value at $\beta z = \pi/2, 3\pi/2, 5\pi/2\ldots$. Using these conditions in (8), the maximum and minimum values of the total magnetic field intensity are

$$H_{\max} = \frac{E_{x0}^+}{\eta}(1+|\Gamma|)$$

$$H_{\min} = \frac{E_{x0}^+}{\eta}(1-|\Gamma|)$$

With progress of time, the amplitudes of the total electric field and the total magnetic field vary between their respective maximum and minimum values. Figure 7.14(b) shows variation of amplitudes of the total electric field intensity with βz for the positive side of the x axis, and Fig 7.14(c) shows the variation of amplitudes of the total magnetic field intensity with βz for the positive side of the y axis. Envelopes of the amplitudes of the total fields are generally not sinusoidal, and these are stationary in space. We find that maximum amplitude of the total electric field and minimum amplitude of the total magnetic field occur at the same space angle. Also, minimum amplitude of the total electric field and maximum amplitude of the total magnetic field occur at the same space angle. Moreover if λ is wavelength of the fields, the distance between the points of maximum value of the amplitude and the nearest minimum value of the amplitude, either of the total electric field or of the total magnetic field, is $\lambda/4$.

The standing wave ratio, which we will denote by the symbol S_R, is defined as the ratio of E_{max} to E_{min} or H_{max} to H_{min}. That is,

$$S_R = \frac{E_{\max}}{E_{\min}} = \frac{H_{\max}}{H_{\min}} = \frac{1+|\Gamma|}{1-|\Gamma|} \tag{9}$$

The standing wave ratio is determined experimentally by measuring E_{max} and E_{min} using a probe connected to a measuring instrument. Absolute values of E_{max} and E_{min} are not required as the standing wave ratio depends on relative magnitude of these quantities. The magnitude of reflection coefficient lies between 0 and 1, hence the standing wave ratio may vary from 1 to ∞.

The reflection coefficient is a measure of efficiency of transmission of electromagnetic power through an interface. For example, if the measured value of $S_R = 2.0$. the magnitude of reflection coefficient is 1/3. Thus, the power transmitted per unit area through the interface is 8/9 times of the incident power per unit area.

We can calculate the reflection coefficient in terms of the standing wave ratio from (9). It is given by

$$|\Gamma| = \frac{S_R - 1}{S_R + 1} \tag{10}$$

EXAMPLE 7.23

A uniform plane wave travelling in free space is incident normally on (a) a perfect conductor and (b) a perfect dielectric for which $\varepsilon_r = 5.0$ and $\mu_r = 1.0$. Find the standing wave ratio in each case.

(a) The wave gets totally reflected from a perfect conducting surface. Hence the magnitude of reflection coefficient $|\Gamma|$ is 1.0.

Standing wave ratio, $S_R = \frac{1+|\Gamma|}{1-|\Gamma|} = \infty$

The standing wave ratio is infinity because the minimum value of the total electric field is zero.

(b) Intrinsic impedance of the dielectric region, $\eta_2 = \sqrt{\frac{\mu_0}{5\varepsilon_0}} = \frac{377}{\sqrt{5}} = 168.6\ \Omega$

Reflection coefficient, $\Gamma = \frac{\eta_2 - \eta_1}{\eta_2 + \eta_1} = \frac{168.6 - 377}{168.6 + 377} = -0.382$

Standing wave ratio, $S_R = \frac{1+|\Gamma|}{1-|\Gamma|} = \frac{1+0.382}{1-0.382} = 2.236$

EXAMPLE 7.24

The instantaneous electric field intensity of the incident wave and instantaneous magnetic field intensity of the reflected wave, which are uniform plane waves in a region are given by

E [illegible] $t) = 16\cos(\omega t - \beta z)$ V/m

and H [illegible] $(t) =$ [illegible] $\cos($[illegible]t [illegible] $z)$ A/m

For the region, ε [illegible] $\mu_r = 1$, and $\sigma = 0$. Find the standing wave ratio.

Amplitude of the incident electric field intensity, $E_0^+ = 16.0$ V/m

Amplitude of the reflected magnetic field intensity, $|H_0^-| = 0.04$ A/m

Intrinsic impedance of the region, $\eta = \sqrt{\dfrac{\mu_0}{3.2\varepsilon_0}} = \dfrac{377}{\sqrt{3.2}} = 210.7\ \Omega$

Amplitude of the reflected electric field intensity, $|E_0^-| = |\eta H_0^-| = (210.7)(0.04) = 8.43$

Maximum amplitude of total electric field, $E_{max} = |E_0^+| + |E_0^-| = 16.0 + 8.43 = 24.43$ V/m

Minimum amplitude of total electric field, $E_{min} = |E_0^+| - |E_0^-| = 16.0 - 8.43 = 7.57$ V/m

Standing wave ratio, $S_R = E_{max}/E_{min} = 24.43/7.57 = 3.23$

EXAMPLE 7.25

A uniform plane wave travelling in free space is incident normally on a perfect, nonmagnetic, plane dielectric surface. In free space, the maximum and minimum amplitudes of total electric field intensities are found to be 27.4 V/m and 12.3 V/m respectively. Find relative permittivity of the dielectric region.

Given: $E_{max} = 27.4$ V/m, $E_{min} = 12.3$ V/m

Standing wave ratio, $S_R = \dfrac{E_{max}}{E_{min}} = \dfrac{27.4}{12.3} = 2.228$

Magnitude of reflection coefficient, $|\Gamma| = \dfrac{S_R - 1}{S_R + 1} = \dfrac{2.228 - 1}{2.228 + 1} = 0.380$

Reflection coefficient, $\Gamma = \dfrac{\eta_2 - \eta_1}{\eta_2 + \eta_1}$

As $\eta_1 > \eta_2$, the reflection coefficient is negative. Thus,

$$\frac{\eta_2 - 377}{\eta_2 + 377} = -0.380$$

This gives $\eta_2 = 169.38$

Or $377/\sqrt{\varepsilon_{r2}} = 169.38$

Hence, $\varepsilon_{r2} = 4.954$

EXAMPLE 7.26

The standing wave ratio S_R in a dielectric medium where an incident wave as well as a reflected wave is present is found to be 3.8. Calculate the reflected power per unit area in terms of the incident power per unit area.

The magnitude of reflection coefficient, $|\Gamma| = \dfrac{S_R - 1}{S_R + 1} = \dfrac{3.8 - 1}{3.8 + 1} = 0.5833$

Reflected power per unit area, P_{av} ... $P_{av}^+ = (0.5833)^2 = 0.3402 P_{av}^+$ W/m^2

7.11 The Wave Impedance

The *intrinsic* impedance is defined as the ratio of the sinusoidally time-varying electric field intensity to the magnetic field intensity of a *single* plane electromagnetic wave. In a region where both the incident and reflected waves are present,

> the ratio of the *total* electric field intensity to the *total* magnetic field intensity of plane waves is defined as the *wave impedance.*

In Region 1 of Fig. 7.13, the total or resultant electric field intensity is equal to the sum of electric fields of incident and reflected waves. Similarly, the total magnetic field intensity in the region is the sum of magnetic fields of incident and reflected waves. Thus, the wave impedance (also called field impedance) at any z in this region is obtained by dividing the total electric field intensity by the total magnetic field intensity. Since Region 2 has only one wave, the wave impedance in this region is same as its intrinsic impedance.

Now suppose that both the regions shown in Fig. 7.13 are lossless; the incident wave is x-polarised, and the fields vary harmonically with time. Then the phasor forms of incident and reflected electric fields in Region 1 are

$$\bar{E}_{1x}^{+}(z) = E_{1x0}^{+} e^{-j\beta_1 z} \tag{1}$$

$$\bar{E}_{1x}^{-}(z) = E_{1x0}^{-} e^{j\beta_1 z} \tag{2}$$

where the amplitudes E_{1x0}^{+} and E_{1x0}^{-} are assumed to be real quantities. The total electric field in this region is the sum of (1) and (2). Thus, the expression for total electric field intensity,

$$\bar{E}_{1x}(z) = \bar{E}_{1x}^{+}(z) + \bar{E}_{1x}^{-}(z) = E_{1x0}^{+} e^{-j\beta_1 z} + E_{1x0}^{-} e^{j\beta_1 z}$$

$$= E_{1x0}^{+} e^{-j\beta_1 z}\left(1 + \frac{E_{1x0}^{-} e^{j\beta_1 z}}{E_{1x0}^{+} e^{-j\beta_1 z}}\right)$$

$$- E_{1x0}^{+} e^{-j\beta_1 z}[1 + \Gamma_1(0) e^{j2\beta_1 z}]$$

$$= E_{x01}^{+} e^{-j\beta_1 z}[1 + \bar{\Gamma}_1(z)] \tag{3}$$

In equation (3), the parameter $\Gamma_1(0) = E_{1x}^{+}(0)/E_{1x}^{-}(0) = E_{1x0}^{+}/E_{1x0}^{-}$ is the reflection coefficient at $z = 0$ at the Region 1 side of the boundary and the parameter $\bar{\Gamma}_1(z) = \Gamma_1(0) e^{j2\beta_1 z}$ is the reflection coefficient at any z, as defined in Section 7.9.

The total magnetic field intensity in Region 1 is

$$\bar{H}_{1y}(z) = \bar{H}_{1y}^{+}(z) + \bar{H}_{1y}^{-}(z) = \frac{1}{\eta_1}[\bar{E}_{1x}^{+}(z) - \bar{E}_{1x}^{-}(z)]$$

$$= \frac{E_{1x0}^{-}}{\eta_1} e^{-j\beta_1 z}(1 - \Gamma_1(0) e^{j2\beta_1 z})$$

$$= \frac{E_{1x0}^{-}}{\eta_1} e^{-j\beta_1 z}[1 - \bar{\Gamma}_1(z)] \tag{4}$$

The wave impedance in Region 1 is equal to the ratio of (3) and (4). Thus, denoting the wave impedance by the symbol $\bar{Z}_1(z)$, we have

$$\bar{Z}_1(z) = \frac{\bar{E}_{1x}(z)}{\bar{H}_{1y}(z)} = \eta_1 \frac{1+\Gamma_1(0)e^{j2\beta_1 z}}{1-\Gamma_1(0)e^{j2\beta_1 z}} = \eta_1 \frac{1+\bar{\Gamma}_1(z)}{1-\bar{\Gamma}_1(z)} \tag{5}$$

We find from relation (5) that the wave impedance is a function of z. It is a complex quantity and its unit is ohm.

As there is only a single wave in Region 2, the wave impedance in this region is equal to the intrinsic impedance and it is independent of z. Thus,

$$\bar{Z}_2(z) = \eta_2$$

If there is no surface charge on both sides of the interface, the tangential electric field intensity on one side of the boundary is equal to that on the other side. Similarly, if there is no surface current on both sides of the interface, the tangential magnetic field intensity on one side of the boundary is equal to that on the other side. Since the electric fields are in the x direction and the magnetic fields are in the y direction, both the fields are tangential to the interface at $z = 0$. Accordingly,

$$\bar{E}_{1x}(0) = \bar{E}_{2x}(0) \tag{6}$$

$$\bar{H}_{1y}(0) = \bar{H}_{2y}(0) \tag{7}$$

Dividing each side of (6) by the corresponding side of (7), we get

$$\frac{\bar{E}_{1x}(0)}{\bar{H}_{1y}(0)} = \frac{\bar{E}_{2x}(0)}{\bar{H}_{2y}(0)}$$

The left side of the above relation is the wave impedance at the interface in Region 1, and the right side is the wave impedance at the interface in Region 2. Hence,

$$\bar{Z}_1(0) = \bar{Z}_2(0) \tag{8}$$

That is,

the wave impedances on two sides of the interface are equal. This property of the wave impedance is useful for studying the reflection and transmission phenomena of plane waves involving more than one boundary surface.

For the problem under consideration, the wave impedance in Region 2, including the boundary surface, has a constant value of η_2. Therefore,

$$\bar{Z}_1(0) = \bar{Z}_2(0) = \eta_2 \tag{9}$$

where η_2 is a real quantity.

Now, by substituting $z = 0$ in (5) we will get $Z_1(0)$ in terms of $\Gamma_1(0)$.

$$Z_1(0) = \eta_1 \frac{1+\Gamma_1(0)}{1-\Gamma_1(0)} \tag{10}$$

Equation (10) when solved for $\Gamma_1(0)$ gives

$$\Gamma_1(0) = \frac{Z_1(0)-\eta_1}{Z_1(0)+\eta_1} \tag{11}$$

As per relation (9), $Z_1(0) = \eta_2$. Hence, the reflection coefficient at the boundary in Region 1 is given by

$$\Gamma_1(0) = \frac{\eta_2 - \eta_1}{\eta_2 + \eta_1}$$

This agrees with (11) of Section 7.9.

EXAMPLE 7.27

A uniform plane wave travelling in air is incident normally on a perfect dielectric medium. The standing wave ratio in air is found to be 2.0. Find the wave impedance at the point(s) where the amplitude of the total magnetic field intensity has (a) the maximum value and (b) the minimum value.

Let $|E_{10}^{+}|$ and $|E_{10}^{-}|$ be the amplitudes of the incident and the reflected electric fields. Then, the standing wave ratio,

$$S_R = \frac{|E_{10}^{+}| + |E_{10}^{-}|}{|E_{10}^{+}| - |E_{10}^{-}|}$$

(a) The point at which maximum amplitude of the total electric field occurs, at that point the amplitude of the total magnetic field has the minimum value. If $|H_{10}^{+}|$ and $|H_{10}^{-}|$ are the amplitudes of incident and reflected magnetic fields, then the minimum amplitude of the total magnetic field is equal to $|H_{10}^{+}| - |H_{10}^{-}|$. Therefore, the wave impedance at the point,

$$Z_{max} = \frac{|E_{10}^{+}| + |E_{10}^{-}|}{|H_{10}^{+}| - |H_{10}^{-}|} = \eta_0 \frac{|E_{10}^{+}| + |E_{10}^{-}|}{|E_{10}^{+}| - |E_{10}^{-}|} = \eta_0 S_R = (377)(2) = 754\ \Omega$$

(b) It can similarly be shown that the wave impedance at which the minimum amplitude of the total electric field occurs,

$$Z_{min} = \eta_0 / S_R = 377/2 = 188.5\ \Omega$$

EXAMPLE 7.28

A uniform plane electromagnetic wave travelling along positive z direction in air (Region 1) is incident normally at the plane surface of a lossless dielectric region (Region 2). The interface is at the $z = 0$ plane. The intrinsic impedance of the dielectric region is 188.5 ohm. (a) Calculate the reflection coefficient and wave impedance at $z = -\lambda/4$. (b) Show that the wave impedances on two sides of the interface have the same value.

(a) In air, both incident and reflected waves are present. Wave impedance at any z in air,

$$Z_1(z) = \eta_1 \frac{1 + \Gamma_1(0) e^{j2\beta_1 z}}{1 - \Gamma_1(0) e^{j2\beta_1 z}}$$

The intrinsic impedance in air, $\eta_1 = \eta_0 = 377\ \Omega$. The reflection coefficient at $z = 0$ at the air side of the boundary,

$$\Gamma_1(0) = \frac{\eta_2 - \eta_0}{\eta_2 + \eta_0} = \frac{188.5 - 377}{188.5 + 377} = -\frac{1}{3}$$

At z [illegible] $\beta z = 2($ [illegible] $)(-\lambda$ [illegible] $) = -\pi$.

Thus, the wave impedance at $z = -\lambda/4$ is

$$\bar{Z}_1(z=-\lambda/4) = 377\frac{1-(1/3)e^{-j\pi}}{1+(1/3)e^{-j\pi}} = 377\frac{1-(1/3)(\cos\pi - j\sin\pi)}{1+(1/3)(\cos\pi - j\sin\pi)} = 754 + j0\ \Omega$$

(b) The wave impedances at the air side of the interface is

$$\bar{Z}_1(z=0) = 377\frac{1-(1/3)}{1+(1/3)} = 188.5 + j0\ \Omega$$

The reflection coefficient in Region 2 is zero as there is no wave reflection in this region. Hence, the wave impedance everywhere in this region, including the boundary surface, is 188.5. Thus, the wave impedances on two sides of the interface have the same value.

7.12 Oblique Incidence of a Plane Wave at the Boundary between Two Regions

In this section we will define some terminologies used in the study of reflection and transmission of waves resulting from oblique incidence of a uniform plane wave at the boundary between two regions of different electrical properties. We will derive expressions for electric and magnetic fields of the wave travelling along a specific direction in terms of rectangular coordinate variables. We will assume that the regions are loss-free, the interface is a plane surface, and the fields vary harmonically with time.

Angle of Incidence and Plane of Incidence

Let the x-y plane be the boundary surface between two regions as shown in Fig. 7.15. The directed line AO represents the direction of the incident wave and the line ON is normal to the boundary surface. The angle θ_i between OA and ON is the *angle of incidence*. The *plane of incidence* is the plane containing the ray AO and the line ON that is normal to the boundary surface. As AO and ON are on the x-z plane, this is the plane of incidence.

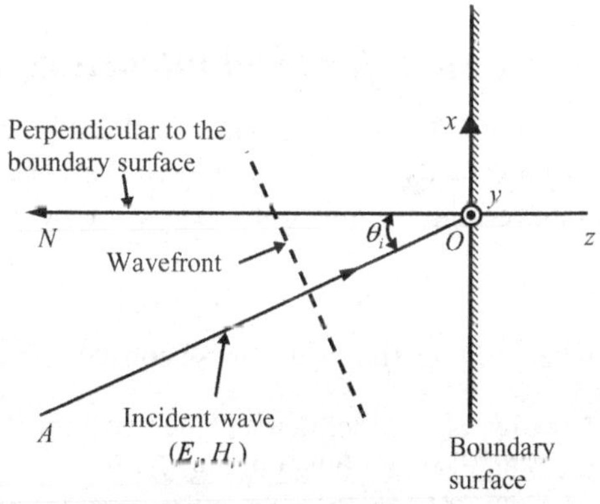

Fig. 7.15 Illustration of oblique incidence of a plane wave at the boundary surface of two regions

The Wavefront

A surface over which the field intensity of a uniform plane wave has the same phase angle is termed as a wavefront. For example, the phase angle of electric field intensity of a plane wave defined by $E = E_0\cos(\omega t - \beta z)$ has the same value over a constant z plane. Hence, in this case any plane that is parallel to the x-y plane is a wavefront of this wave. The wavefront is perpendicular to the direction of propagation of the wave. A wavefront of a wave travelling along AO direction is shown in Fig. 7.15.

Perpendicular and Parallel Polarisations

In the case of uniform plane waves, the electric and magnetic fields are transverse to the direction of propagation. Figure 7.16 shows two types of field orientations. In each case the direction of wave propagation is same as that of the Poynting vector so that flow of electromagnetic energy is along the direction of propagation of the wave. In Fig. 7.16(a), the electric field is perpendicular and the magnetic field is parallel

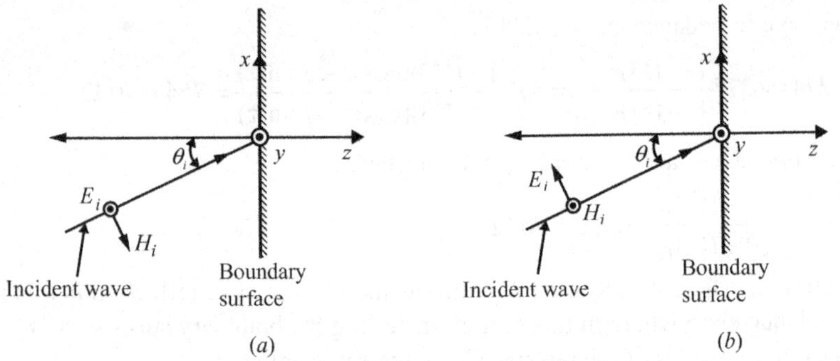

Fig. 7.16 Showing two types of orientation of fields of the incident wave. (*a*) Perpendicular polarisation and (*b*) Parallel polarisation.

to the plane of incidence. This orientation of fields is called *perpendicular polarisation*. In Fig. 7.16(*b*), the electric field is parallel and the magnetic field is perpendicular to the plane of incidence. This orientation of fields is called *parallel polarisation*. In the case of perpendicular polarisation, the electric field intensity is in y direction and hence is tangential to the boundary surface, but the magnetic field intensity has components in x and z directions, as per the coordinate system used. The x component of the magnetic field is tangential and the z component is normal to the boundary. For parallel polarisation, the magnetic field intensity is tangential to the interface, whereas the electric field intensity has both tangential and normal components.

If the electric field is at an arbitrary angle to the plane of incidence, it can be resolved into components perpendicular and parallel to the plane of incidence. Thus, the wave will have both perpendicular and parallel polarisations.

Expressions for Field Intensities, Phase Velocities and Wavelengths

In Section 7.4 we have obtained the solution of the one-dimensional wave equation for sinusoidally time-varying fields with z as the space variable. The solutions are for uniform plane waves and the fields are transverse to the direction of travel of the waves. The solution for the electric field intensity in real time is

$$E = E_0 \sin(\omega t - \beta z) \tag{1}$$

where $\beta \;\; \omega\sqrt{\mu\varepsilon}$ is the phase constant, and E_0 is amplitude of the field.

Now let us consider that a plane wave travelling along AO direction in Region 1 is incident at an angle θ_i on the boundary, as shown in Fig. 7.17(*a*). At a given instant of time, let the point of intersection of AO and the wave front be at P, and the distance of P from the origin be R. The expression for the electric field of the incident wave, for both types of polarisation, is obtained by substituting R in place of z in (1). Thus, the expression for the electric field intensity of the incident wave is

$$E_i = E_{i0} \cos(\omega t - \beta_1 R) \tag{2}$$

where $\beta \;\; \omega\sqrt{\mu_1\varepsilon_1}$ is the phase constant of the wave in Region 1.

The phase velocity and wavelength of the wave along the direction of propagation AO are

$$v_1 = \frac{\omega}{\beta_1} = \frac{1}{\mu\,\varepsilon_1}$$

and

$$\lambda_1 = \frac{2\pi}{\beta_1} = \frac{v_1}{f}$$

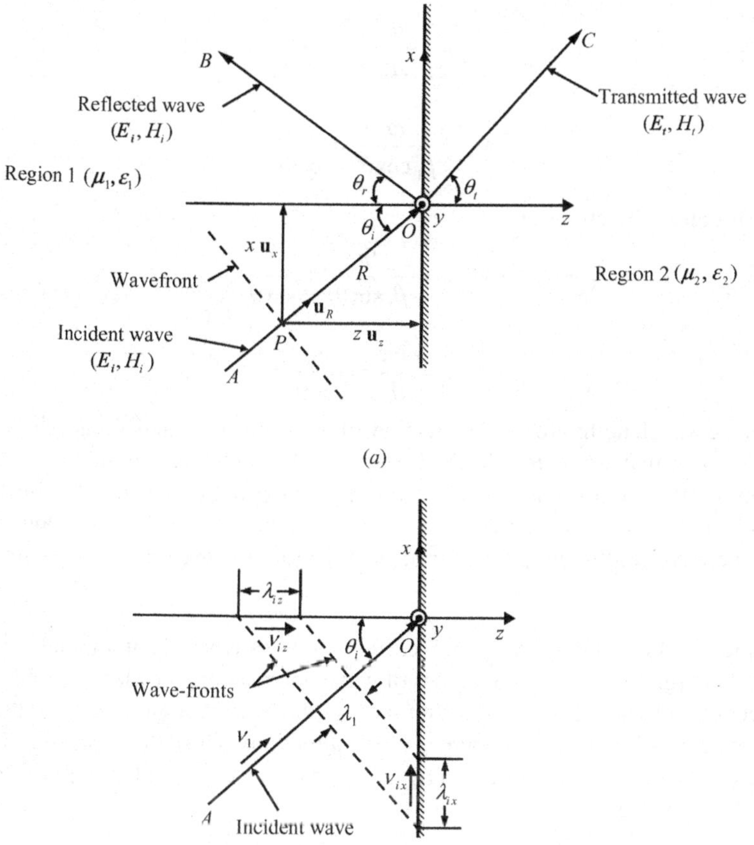

Fig. 7.17 (*a*) Vector diagram for expressing the electric field intensity of a plane wave in terms of the coordinate variables and (*b*) Shows two wavefronts separated by one wavelength, and wavelengths along the coordinate directions.

We will next obtain the expression for the electric field intensity described by (2) in terms of the coordinate variables. This is done by expressing R in terms of x and z. By referring to Fig. 7.17(*a*) we find that the vector distance of O from P is given by

$$\mathbf{R} = x\mathbf{u}_x + z\mathbf{u}_z$$

The scalar value of **R** is

$$R = \mathbf{R}\cdot\mathbf{u}_R = x(\mathbf{u}_x\cdot\mathbf{u}_R) + z(\mathbf{u}_z\cdot\mathbf{u}_R) = x\sin\theta_i + z\cos\theta_i \tag{3}$$

By using (3) in (2) we will get the expression for the electric field intensity of the incident wave.

$$E_i = E_{i0}\cos[\omega t - (\beta_1\sin\theta_i)x - (\beta_1\cos\theta_i)z] \tag{4}$$

The function $(\beta_1\sin\theta_i)x$ in (4) gives the phase angle of E_i in x direction. Similarly, the function $(\beta_1\cos\theta_i)z$ gives the phase angle of the field in z direction. Hence, we can consider $(\beta_1\sin\theta_i)$ and $(\beta_1\cos\theta_i)$ to be the phase constants of the wave associated with x and z directions respectively. Thus, the phase velocity and the wavelength have components along both x and z directions. The phase velocity v_x in x direction is derived by assuming z to be constant, and the phase velocity v_z along z direction is obtained by assuming x to be constant. These are

$$v_x = \frac{\omega}{\beta_1 \sin\theta_i} = \frac{v_1}{\sin\theta_i}$$

$$v_z = \frac{\omega}{\beta_1 \cos\theta_i} = \frac{v_1}{\cos\theta_i}$$

The wavelengths in x and z directions are

$$\lambda_x = \frac{2\pi}{\beta_1 \sin\theta_i} = \frac{\lambda_1}{\sin\theta_i}$$

$$\lambda_z = \frac{2\pi}{\beta_{iz}} = \frac{\lambda_1}{\cos\theta_i}$$

The velocities and the wavelengths along AO direction, along x direction, and along z direction are illustrated in Fig. 7.17(b). We note that when $\theta_i = 0$, the phase velocity and wavelength along x direction become infinity. And when $\theta_i = 90°$, both v_z and λ_z become infinity. For θ_i between 0 and 90°, the phase velocity and the wavelength in both the coordinate directions are more than the actual values of v_1 and λ_1. This means that the velocities and the wavelengths along the coordinate directions in the case of oblique incidence are not real quantities.

Next, let us consider a reflected plane wave travelling along OB which is at an angle θ_r with the negative z axis, as illustrated in Fig. 7.17(a). For both types of wave polarisation, the electric field intensity is perpendicular to the direction of travel of the wave. With the coordinate directions as shown, the phase velocity of the wave has components along positive x and negative z directions. Thus, the expression for the electric field intensity E_r of the wave can be written from (4) by using $-z$ in place of z and θ_r in place θ_i. The electric field intensity of the reflected wave,

$$E_r = E_{r0} \cos[\omega t - (\beta_1 \sin\theta_r)x + (\beta_1 \cos\theta_r)z] \tag{5}$$

Finally, let us consider that a transmitted plane wave is travelling along OC direction in Region 2, as shown in Fig. 7.17(a). The angle of OC with the positive z axis is θ_t. The velocity of the wave has components along positive x and positive y directions. Hence, the equation for electric field intensity E_t of the transmitted wave can be obtained from (4) by using appropriate subscripts with the main symbols of the quantities associated with the field. Thus, the electric field intensity is described by

$$E_t = E_{t0} \cos[\omega t - (\beta_2 \sin\theta_t)x - (\beta_2 \cos\theta_t)z] \tag{6}$$

where $\beta\_ \ \omega\sqrt{\mu_2 \varepsilon_2}$ is the phase constant of the wave.

The directions of the magnetic field intensities relative to the electric field intensities of all the three waves are determined from the consideration that the Poynting vector and hence the direction of energy flow is along the direction of propagation of the respective waves.

The relation between the electric and magnetic field intensities of the waves are

$$H_i = \frac{E_i}{\eta_1}$$

$$H_r = \frac{E_r}{\eta_1}$$

$$H_t = \frac{E_t}{\eta_2}$$

where $\eta_1 = \sqrt{\mu_1/\varepsilon_1}$ is the intrinsic impedance of Region 1, and $\eta_2 = \sqrt{\mu_2/\varepsilon_2}$ is the intrinsic impedance of Region 2.

Note that the subscript r is also used to denote the radial components of a vector in cylindrical and spherical coordinates. And, the subscript t is also used to denote tangential components of field intensities at the boundary of two regions.

EXAMPLE 7.29

A uniform plane wave travelling along positive z direction in air is incident at an angle of 30° with the normal to the plane surface of a dielectric medium. The x-z plane is the plane of incidence. Amplitude of electric field intensity of the wave is 20 V/m and frequency of the wave is 10 MHz. (a) Obtain expressions for instantaneous electric and magnetic field intensities of the wave. (b) Find phase velocities of the wave along x and z directions.

Instantaneous electric field intensity, $E_i = E_0 \cos(\omega t - \beta x \sin\theta_i - \beta z \cos\theta_i)$

$E_0 = 20$ V/m, $\omega = 20\pi \times 10^6$ rad/s,

$\beta \quad \omega\sqrt{\mu_0\varepsilon_0} = (20\pi \times 10^6)/(3\times 10^8) = 0.209$ rad/m

$\sin\theta_i = 0.5,\ \cos\theta_i = 0.866,\ \eta_0 = 377$

(a) Substituting the values, $E_i = 20\cos[(20\pi\times 10^6)t - 0.1045x - 0.1801z]$ V/m

Instantaneous magnetic field intensity,

$H_i = E_i/\eta_0 = 0.0503\cos[(20\pi\times 10^6)t - 0.1045x - 0.1801z]$ A/m

(b) Phase velocity in x direction $= \dfrac{v_0}{\sin\theta_i} = \dfrac{3\times 10^8}{0.5} = 6\times 10^8$ m/s

Phase velocity in z direction $= \dfrac{v_0}{\cos\theta_i} = \dfrac{3\times 10^8}{0.866} = 3.464\times 10^8$ m/s

Note that both the velocities are more than that of light. So, these are only apparent velocities.

7.13 Oblique Incidence of a Plane Wave on a Flat Perfect Conductor

When a plane electromagnetic wave originating in a dielectric medium is incident at an angle with the normal to the surface of a conductor, the wave will get reflected from the interface. The reflected wave will also be at an angle with the normal to the boundary surface. The polarisation of the reflected wave will be same as that of the incident wave. If the conductivity of the conductor is infinitely large, there will not be any transmitted wave as the depth of penetration is zero. Thus, the power per unit area of the reflected wave will be equal to the power per unit area carried by the incident wave. In Fig. 7.18, the plane boundary surface common to a perfect dielectric region and a perfect conductor is taken on the x-y plane.

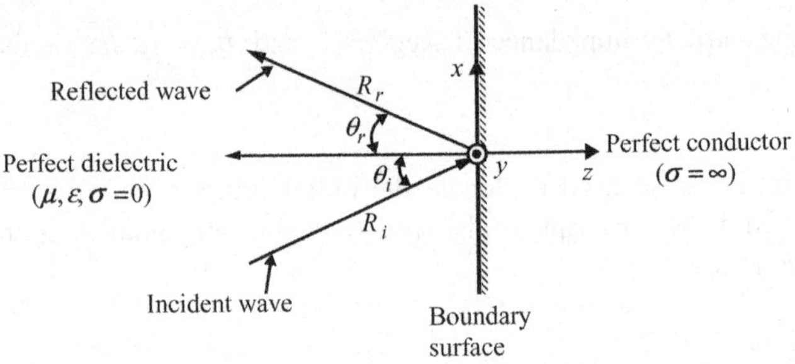

Fig. 7.18 Oblique incidence of a plane wave on the surface of a flat perfect conductor

The directions of the electric and magnetic fields of a wave are determined from the consideration that the Poynting vector is in the same direction as the direction of travel of the wave.

The direction of the incident wave, represented by the directed line R_i, is at an angle θ_i with the z axis. The direction of the reflected wave, shown by the directed line R_r, makes an angle θ_r with the negative z axis. The relation between θ_i and θ_r, and relation between field magnitudes of incident and reflected waves are derived by the use of the boundary conditions. As the direction of propagation of incident and reflected waves are same as those of AO and OB of Fig. 7.17(a), the equation for electric field intensities of the incident and reflected waves are of the same form as given by (4) and (5) respectively of the preceding section.

Perpendicular Polarisation

The orientation of fields of incident and reflected waves with perpendicular polarisation are shown in Fig. 7.19. The electric field intensities of the incident and reflected waves are along the y direction. The corresponding magnetic field intensities have components in both x and z directions. As mentioned before, the quantities associated with the incident wave are denoted using the subscript i, and the subscript r will be used to denote quantities associated with the reflected wave. Expressions for the electric field intensity E_i of the incident wave and E_r of the reflected wave are

$$E_i = E_{i0} \cos[\omega t - (\beta \sin \theta_i)x - (\beta \cos \theta_i)z] \tag{1}$$

$$E_r = E_{r0} \cos[\omega t - (\beta \sin \theta_r)x + (\beta \cos \theta_r)z] \tag{2}$$

where the phase constant $\beta = \omega\sqrt{\mu\varepsilon}$

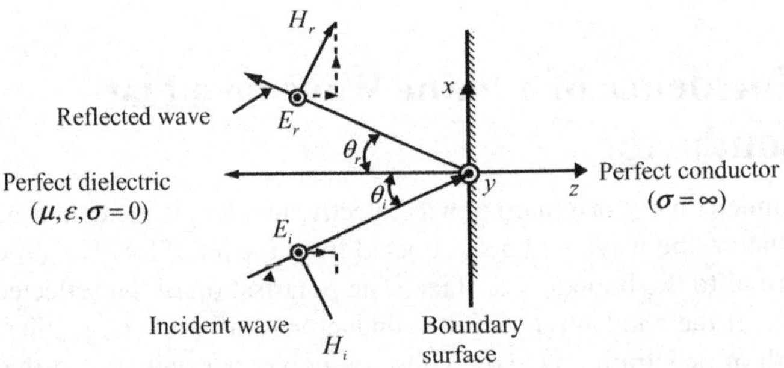

Fig. 7.19 Oblique incidence of a plane wave on the surface of a flat perfect conductor for perpendicular polarisation

The total electric field in the dielectric region is equal to the sum of (1) and (2), as both are in y direction. Denoting the total field by E_y, we have

$$E_y = E_i + E_r = E_{i0}\cos[\omega t - (\beta\sin\theta_i)x - (\beta\cos\theta_i)z]$$

$$+E_{r0}\cos[\omega t - (\beta\sin\theta_r)x + (\beta\cos\theta_r)z] \quad (3)$$

As E_y is tangential to the surface of the conductor, it must be equal to zero at $z = 0$. Thus, substituting $z = 0$ into (3), we have

$$E_y(z=0) = E_{i0}\cos[\omega t - (\beta\sin\theta_i)x] + E_{r0}\cos[\omega t - (\beta\sin\theta_r)x] = 0$$

This equation will be satisfied for all x if

$$\theta_r = \theta_i \quad (4)$$

and

$$E_{r0} = -E_{i0} \quad (5)$$

Thus, the angle of the reflected wave is equal to the angle of the incident wave, and the amplitude of electric field intensity of the reflected wave is equal and opposite to the amplitude of incident wave.

By substituting relations (4) and (5) into (3) we will get the expression for the total field in the dielectric region. It is given by

$$E_y = E_{i0}\cos[\omega t - (\beta\sin\theta_i)x - (\beta\cos\theta_i)z] - E_{i0}\cos[\omega t - (\beta\sin\theta_i)x + (\beta\cos\theta_i)z]$$

$$= 2E_{i0}\sin[\omega t - (\beta\sin\theta_i)x)]\sin(\beta\cos\theta_i)z \quad (6)$$

Referring to Fig. 7.19 we find that the magnetic field intensity of the incident wave has components in negative x and positive z directions, whereas the reflected wave has components in positive x and positive z directions. The components of the magnetic field of the incident wave are

$$H_{ix} = -H_i\cos\theta_i = -\frac{E_i}{\eta}\cos\theta_i = -\frac{E_{y0}}{\eta}\cos\theta_i\cos[\omega t - (\beta\sin\theta_i)x - (\beta\cos\theta_i)z]$$

$$H_{iz} = H_i\sin\theta_i = \frac{E_i}{\eta}\sin\theta_i = \frac{E_{i0}}{\eta}\sin\theta_i\cos[\omega t - (\beta\sin\theta_i)x - (\beta\cos\theta_i)z]$$

The components of the magnetic field of the reflected wave are

$$H_{rx} = H_r\cos\theta_r = \frac{E_r}{\eta}\cos\theta_i = -\frac{E_{i0}}{\eta}\cos\theta_i\cos[\omega t - (\beta\sin\theta_i)x + (\beta\cos\theta_i)z]$$

$$H_{rz} = H_r\sin\theta_r = \frac{E_r}{\eta}\sin\theta_i = -\frac{E_{i0}}{\eta}\sin\theta_i\cos[\omega t - (\beta\sin\theta_i)x + (\beta\cos\theta_i)z]$$

The total magnetic field intensity in x direction,

$$H_x = H_{ix} + H_{rx} = -\frac{E_{i0}\cos\theta_i}{\eta}\{\cos[\omega t - (\beta\sin\theta_i)x - (\beta\cos\theta_i)z] + \cos[\omega t - (\beta\sin\theta_i)x + (\beta\cos\theta_i)z]\}]$$

$$= -\frac{2E_{i0}\cos\theta_i}{\eta}\cos[\omega t - (\beta\sin\theta_i)x]\cos(\beta\cos\theta_i)z \quad (7)$$

Similarly, it can be shown that the total magnetic field intensity in z direction,

$$H_z = \frac{2E_{i0}\sin\theta_i}{\eta}\sin[\omega t - (\beta\sin\theta_i)x]\sin(\beta\cos\theta_i)z \quad (8)$$

Over a given x plane, the electric field intensity E_y described by (6) and the magnetic field H_x described by (7) are out of phase by 90° in time because E_y is a sine function and H_x is a cosine function of time. The factors $\sin(\beta \cos\ \ )z$ and $\cos(\beta \cos\theta_i)z$ in the expressions for electric and magnetic fields show that the fields are standing waves along z direction. Hence, the average power flow in z direction is zero. But, H_z given by (8) and E_y are in time phase as both the quantities are sine functions of time on a given x plane. Moreover, the factors $\omega\ \ (\ \sin\theta_i)x]$ in these expressions show that the fields are travelling waves along x direction. The fields have the same phase velocity along x direction. The fields are also in time phase and space phase. Thus,

the standing waves travel as a unit along x direction or parallel to the interface.

Parallel Polarisation

In this case the electric fields of incident and reflected waves are parallel, and the magnetic fields of the waves are perpendicular to the x-z plane, as illustrated in Fig. 7.20. The electric field of the incident wave has components in positive x and negative z directions. The electric field of the reflected wave has components in negative x and z directions. The magnetic fields of incident and reflected waves are in the positive y direction. The expressions for the incident and reflected electric fields are same as those given by (1) and (2).

The components of the incident electric field in x and z directions are given by

$$E_{ix} = E_i \cos\theta_i$$

$$E_{iz} = -E_i \sin\theta_i$$

The components of the reflected electric field in x and z directions are given by

$$E_{rx} = -E_r \cos\theta_r$$

$$E_{rz} = -E_r \sin\theta_r$$

The total electric field intensity in x direction in the dielectric region is equal to the sum of the x-directed electric fields of the incident and reflected waves. It is given by

$$\begin{aligned} E_x &= E_i \cos\theta_i - E_r \cos\theta_r \\ &= E_{i0} \cos\theta_i \cos[\omega t - (\beta \sin\theta_i)x - (\beta \cos\theta_i)z] \\ &\quad - E_{r0} \cos\theta_r \cos[\omega t - (\beta \sin\theta_r)x + (\beta \cos\theta_r)z] \end{aligned} \tag{9}$$

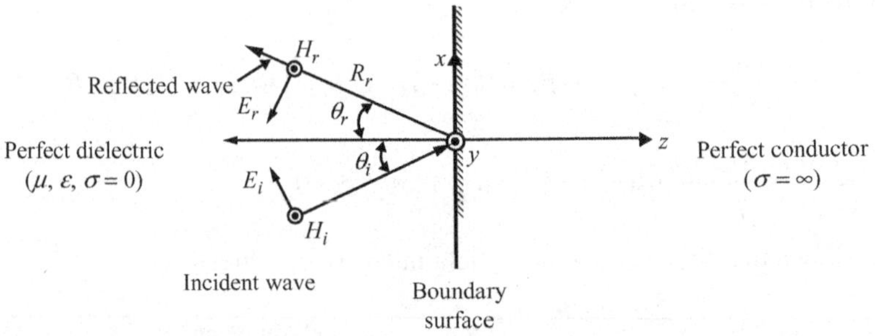

Fig. 7.20 Oblique incidence of a plane wave on the surface of a flat perfect conductor for parallel polarisation

Total electric field in z direction is

$$E_z = -E_i \sin\theta_i - E_r \sin\theta_r$$
$$= -E_{i0} \sin\theta_i \cos[\omega t - (\beta \sin\theta_i)x - (\beta\cos\theta_i)z]$$
$$-E_{r0} \sin\theta_r \cos[\omega t - (\beta\sin\theta_r)x + (\beta\cos\theta_r)z] \qquad (10)$$

The magnetic field intensities of the incident and the reflected waves are in y direction. Therefore, the total magnetic field is

$$H_y = H_r + H_i = \frac{1}{\eta}(E_r + E_i) = \frac{1}{\eta}\{E_{i0}\cos[\omega t - (\beta\sin\theta_i)x - (\beta\cos\theta_i)z]$$
$$+E_{r0}\cos[\omega t - (\beta\sin\theta_r)x + (\beta\cos\theta_r)z]\} \qquad (11)$$

The relationship between the electric fields of the incident wave and the reflected wave, and the relation between the incident angle and reflected angle can by derived from the consideration that the electric field intensity E_x, which is tangential to the conducting surface, must be equal to zero at $z = 0$. That is,

$$E_x(z=0) = E_{i0}\cos\theta_i \cos[\omega t - (\beta\sin\theta_i)x] - E_{r0}\cos\theta_r \cos[\omega t - (\beta\sin\theta_r)x] = 0$$

This equation will be satisfied for all x if

$$\theta_r = \theta_i \qquad (12)$$

and

$$E_{r0} = E_{i0} \qquad (13)$$

Thus, the angle at which reflection of the wave occurs is equal to the angle of incidence, as in the case of perpendicular polarisation. Moreover, the amplitude of electric field of the reflected wave is equal to that of the incident wave. By substituting relations (12) and (13) into expressions (9), (10) and (11), we get

$$E_x = 2E_{i0}\cos\theta_i \sin[\omega t - (\beta\sin\theta_i)x]\sin(\beta\cos\theta_i)z \qquad (14)$$

$$E_z = -2E_{i0}\sin\theta_i \cos[\omega t - (\beta\sin\theta_i)x]\cos(\beta\cos\theta_i)z \qquad (15)$$

$$H_y = \frac{2E_{i0}}{\eta}\cos[\omega t - (\beta\sin\theta_i)x]\cos(\beta\cos\theta_i)z \qquad (16)$$

On a given x plane, the time-phase difference between the total electric field intensity E_x given by (14) and the magnetic field intensity H_y given by (16) is equal to 90°. However, E_z described by (15) and H_y are in time phase on a given x plane. All the fields have the character of standing waves along z direction and travelling waves along x direction, as in the case of parallel polarisation. Figure 7.21 illustrates the standing wave pattern of the electric field E_x for $x = 0$ and when $\sin \omega t$ is not zero. The standing wave travels as a whole along x direction with velocity v_x.

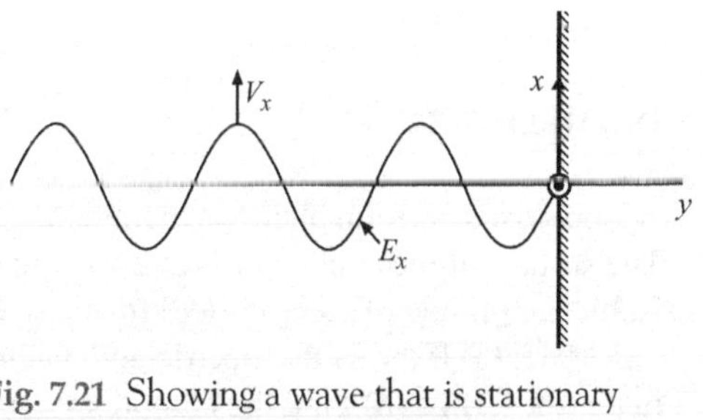

Fig. 7.21 Showing a wave that is stationary along z direction and travelling along x direction

EXAMPLE 7.30

The boundary surface between air and a flat perfect conductor is on the $z = 0$ plane. A 60 MHz uniform plane wave from free space is incident at an angle of 45° with the normal to the boundary surface. The x-z plane is the plane of incidence, the wave is horizontally polarised, and amplitude of the electric field intensity of the wave is 50 V/m. Obtain expressions for the components of electric and magnetic fields of (a) the incident wave and (b) the reflected wave.

We have, $\theta_r = \theta_i = 45°$, $\sin\theta_i = \cos\theta_i = 0.707$

Phase constant, $\beta = \omega\sqrt{\mu_0\varepsilon_0} = \dfrac{120\pi \times 10^6}{3\times10^8} = 1.259$ rad/m

Intrinsic impedance, $\eta_0 = 377\ \Omega$

For parallel polarisation, $E_{r0} = E_{i0} = 50$ V/m

(a) Instantaneous electric field intensity of the incident wave,

$$E_i = E_{i0}\cos(\omega t - \beta x \sin\theta_i - \beta z\cos\theta_i)$$
$$= 50\cos[(120\pi\times10^6)t - 0.889x - 0.889z] \text{ V/m}$$

The electric field has components along positive x direction and negative z direction. The field components are

$$E_{ix} = E_i\cos\theta_i = 35.35\cos[(120\pi\times10^6)t - 0.889x - 0.889z] \text{ V/m}$$

$$E_{iz} = -E_i\sin\theta_i = -35.35\cos[(120\pi\times10^6)t - 0.889x - 0.889z] \text{ V/m}$$

The magnetic field of the incident wave is in y direction.

$$H_i = H_{iy} = E_i/\eta_0 = 0.133\cos[(120\pi\times10^6)t - 0.889x - 0.889z] \text{ A/m}$$

(b) The electric field of the reflected wave,

$$E_r = E_{r0}\cos(\omega t - \beta x\sin\theta_r + \beta z\cos\theta_r)$$
$$= 50\cos[(120\pi\times10^6)t - 0.889x + 0.889z] \text{ V/m}$$

The electric field has components along negative x and z directions. The field components are

$$E_{rx} = -E_r\cos\theta_r = -35.35\cos[(120\pi\times10^6)t - 0.889x + 0.889z] \text{ V/m}$$

$$E_{rz} = -E_r\sin\theta_r = -35.35\cos[(120\pi\times10^6)t - 0.889x + 0.889z] \text{ V/m}$$

The magnetic field of the incident wave is in y direction.

$$H_r = H_{ry} = E_r/\eta_0 = 0.133\cos[(120\pi\times10^6)t - 0.889x + 0.889z] \text{ A/m}$$

EXAMPLE 7.31

A y-polarised uniform plane wave travelling in free space is incident obliquely on a perfect plane conductor. The surface of the conductor is at $z = 0$ and the x-z plane is the plane of incidence. The angle of incidence is 30°, amplitude of electric field intensity of the incident wave is 15 V/m, and frequency of the wave is 125 MHz. (a) Express the electric and magnetic field intensity vectors of the incident and reflected waves in phasor forms. (b) Find the complex Poynting vectors of the incident wave, reflected wave, and sum of the complex Poynting vectors of the incident and reflected waves.

The directions of the fields are illustrated in Fig. 7.19. The electric field of the incident wave is in y direction and the magnetic field has components in x and z directions.

$E_{i0} = 15$ V/m, $\theta_r = \theta_i = 30°$, $\sin\theta_i = 0.5$, $\cos\theta_i = 0.866$

$\beta = 2\pi(125\times10^6)\sqrt{\mu_0\varepsilon_0} = 2.62$ rad/m

$\eta_0 = 377\ \Omega$

As the polarisation of the wave is perpendicular, $E_{r0} = -E_{i0} = -15$ V/m

(a) The phasor form of electric field intensity vector of the incident wave is

$$\bar{\mathbf{E}}_i = \bar{E}_i \mathbf{u}_y = E_{i0} e^{-j\beta(x\sin\theta_i + z\cos\theta_i)} \mathbf{u}_y = 15 e^{-j(1.31x+2.27z)} \mathbf{u}_y \text{ V/m}$$

The x and z components of the magnetic field intensity phasors of the incident wave are

$$\bar{H}_{ix} = -\bar{H}_i \cos\theta_i = -H_{i0}\cos\theta_i e^{-j\beta(x\sin\theta_i + z\cos\theta_i)}$$

$$\bar{H}_{iz} = \bar{H}_i \sin\theta_i = H_{i0}\sin\theta_i e^{-j\beta(x\sin\theta_i + z\cos\theta_i)}$$

The magnetic field intensity vector of the incident wave,

$$\bar{\mathbf{H}}_i = \bar{H}_{ix}\mathbf{u}_x + \bar{H}_{iz}\mathbf{u}_z = \bar{H}_i(-\cos\theta_i \mathbf{u}_x + \sin\theta_i \mathbf{u}_z)$$

$$= H_{i0}(-\cos\theta_i \mathbf{u}_x + \sin\theta_i \mathbf{u}_z) e^{-j\beta(x\sin\theta_i + z\cos\theta_i)} \mathbf{u}_x$$

$$= (-0.0345\mathbf{u}_x + 0.0199\mathbf{u}_z) e^{-j(1.31x+2.27z)} \text{ A/m}$$

The phasor form of electric field intensity vector of the reflected wave is

$$\bar{\mathbf{E}}_r = \bar{E}_r \mathbf{u}_y = E_{r0} e^{-j\beta(x\sin\theta_r - z\cos\theta_r)} \mathbf{u}_y = -15 e^{-j(1.31x-2.27z)} \mathbf{u}_y \text{ V/m}$$

The x and z components of the magnetic field intensity phasors of the reflected wave are

$$\bar{H}_{rx} = \bar{H}_r \cos\theta_r = H_{r0}\cos\theta_i e^{-j\beta(x\sin\theta_r - z\cos\theta_r)}$$

$$\bar{H}_{rz} = \bar{H}_r \sin\theta_r = H_{r0}\sin\theta_r e^{-j\beta(x\sin\theta_r + z\cos\theta_r)}$$

The magnetic field intensity vector of the reflected wave,

$$\bar{\mathbf{H}}_r = \bar{H}_{rx}\mathbf{u}_x + \bar{H}_{rz}\mathbf{u}_z = \bar{H}_r(\cos\theta_r \mathbf{u}_x + \sin\theta_r \mathbf{u}_z)$$

$$= H_{r0}(\cos\theta_r \mathbf{u}_x + \sin\theta_r \mathbf{u}_z) e^{-j\beta(x\sin\theta_r - z\cos\theta_r)}$$

$$= -(0.0345\mathbf{u}_x + 0.0199\mathbf{u}_z) e^{-j(1.31x-2.27z)} \text{ A/m}$$

(b) The complex Poynting vector for the incident wave is

$$\frac{1}{2}(\bar{\mathbf{E}}_i \times \bar{\mathbf{H}}_i^*) = \frac{1}{2}[(15 e^{-j(1.31x+2.27z)} \mathbf{u}_y) \times \{(-0.0345\mathbf{u}_x + 0.0199\mathbf{u}_z) e^{j(1.31x+2.27z)}\}]$$

$$= 0.149\mathbf{u}_x + 0.259\mathbf{u}_z \text{ W/m}^2$$

The complex Poynting vector for the reflected wave is

$$\frac{1}{2}(\bar{\mathbf{E}}_r \times \bar{\mathbf{H}}_r^*) = \frac{1}{2}[(-15 e^{-j(1.31x-2.27z)} \mathbf{u}_y) \times \{(-0.0345\mathbf{u}_x - 0.0199\mathbf{u}_z) e^{j(1.31x-2.27z)}\}]$$

$$= 0.149\mathbf{u}_x - 0.259\mathbf{u}_z \text{ W/m}^2$$

Sum of the two Poynting vectors $= 0.298\mathbf{u}_x$ W/m^2

The Poynting vectors have no imaginary parts and the average power along z direction is zero

7.14 Oblique Incidence of Plane Wave at the Boundary between Two Perfect Dielectric Regions

Figure 7.22 depicts two perfect dielectric regions 1 and 2 with a common boundary surface. The permeability and permittivity of Region 1 are taken as μ_1 and ε_1 and those of Region 2 as μ_2 and ε_2. With the coordinate axes as shown in the figure, the boundary surface is on the x-y plane.

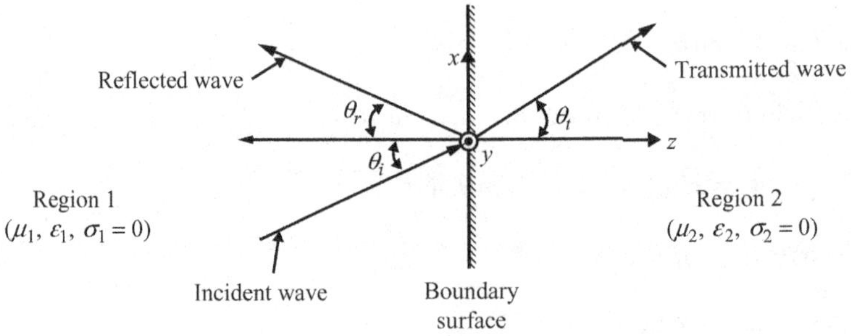

Fig. 7.22 Illustrations of incident, reflected and transmitted waves at the interface between two dielectric regions

Let a uniform plane wave travelling in Region 1 be incident on the boundary at an angle θ_i with normal to the interface. A part of power incident on the boundary will be transmitted to Region 2 and the remaining part will be reflected to Region 1. Let us take the angles of reflected and transmitted (refracted) waves with the z axis as θ_r and θ_t respectively. The orientations of the fields are chosen so that the directions of Poynting vectors correspond to the direction of propagation of the waves. We see that the directions of propagation of incident, reflected and transmitted waves are same as those of OA, OB and OC of Fig. 7.17(a). Hence, the expressions for the electric field intensities of incident, reflected and transmitted waves for both the types of polarisations are same as those given by (4), (5) and (6) respectively of Section 7.12. The expressions for electric field intensities of the three waves are

$$E_i = E_{i0} \cos[\omega t - (\beta_1 \sin\theta_i)x - (\beta_1 \cos\theta_i)z]$$

$$E_r = E_{r0} \cos[\omega t - (\beta_1 \sin\theta_r)x + (\beta_1 \cos\theta_r)z]$$

$$E_t = E_{t0} \cos[\omega t - (\beta_2 \sin\theta_t)x - \beta_2 \cos\theta_t)z]$$

The magnetic fields intensities, in terms of the electric the electric field intensities of respective waves are

$$H_i = E_i/\eta_1$$

$$H_r = E_r/\eta_1$$

$$H_t = E_t/\eta_2$$

The phase constants of the two regions are

$$\beta_1 = \omega\sqrt{\mu_1\varepsilon_1} \text{ and } \beta_2 = \omega\sqrt{\mu_2\varepsilon_2}.$$

The intrinsic impedances of the two regions are

$$\eta \quad \mu_1/\varepsilon_1 \text{ and } \eta \quad \mu_2/\varepsilon_2 .$$

The subscripts 1 and 2 denote quantities in regions 1 and 2 respectively.

Formulae relating the quantities of reflected as well as of transmitted waves with those of the incident wave are derived using the boundary conditions.

The boundary conditions are

1. At $z = 0$, the tangential component of electric field intensity of the incident wave plus the tangential component of electric field intensity of the reflected wave is equal to the tangential component of electric field intensity of the transmitted wave.
2. At $z = 0$, the tangential component of magnetic field intensity of the incident wave plus the tangential component of magnetic field intensity of the reflected wave is equal to the tangential component of magnetic field intensity of the transmitted wave.

Perpendicular Polarisation

By referring to Fig. 7.23, we find that the electric fields of all the waves are in y direction. Hence, these are tangential to the interface. The magnetic fields of all the waves have components along both x and z directions. The x components of magnetic fields are tangential, whereas the z components of the magnetic fields are perpendicular, to the interface.

The total electric field intensity in Region 1 is equal to the sum of the electric field intensity of the incident and reflected waves. Similarly, the total magnetic field in the region is equal to the sum of the magnetic field intensities of the incident and reflected waves. As both the electric field intensities of the two waves are in y direction, the total electric field,

$$E_{1y} = E_i + E_r$$

$$= E_{i0}\cos[\omega t - (\beta_1 \sin\theta_i)x - (\beta_1 \cos\theta_i)z] + E_{r0}\cos[\omega t - (\beta_1 \sin\theta_r)x + (\beta_1 \cos\theta_r)z] \qquad (1)$$

The x component of total magnetic field,

$$H_{1x} = -H_i \cos\theta_i + H_r \cos\theta_r$$

$$= -\frac{1}{\eta_1}\{E_{i0}\cos\theta_i \cos[\omega t - (\beta_1 \sin\theta_i)x - (\beta_1 \cos\theta_i)z]$$

$$-E_{r0}\cos\theta_r \cos[\omega t - (\beta_1 \sin\theta_r)x + (\beta_1 \cos\theta_r)z]\} \qquad (2)$$

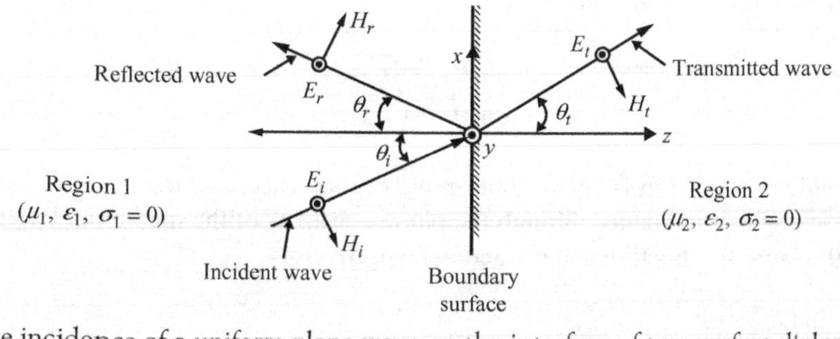

Fig. 7.23 Oblique incidence of a uniform plane wave on the interface of two perfect dielectric regions for perpendicular polarisation

The z component of total magnetic field,

$$H_{1z} = H_i \sin\theta_i + H_r \sin\theta_r$$

$$= \frac{1}{\eta_1}\{E_{i0} \sin\theta_i \cos[\omega t - (\beta_1 \sin\theta_i)x - (\beta_1 \cos\theta_i)z]$$

$$+ E_{r0} \sin\theta_r \cos[\omega t - (\beta_1 \sin\theta_r)x + (\beta_1 \cos\theta_r)z]\} \tag{3}$$

The expressions for electric and magnetic fields along the coordinate direction in Region 2 are,

$$E_{2y} = E_t = E_{t0} \cos[\omega t - (\beta_2 \sin\theta_t)x - (\beta_2 \cos\theta_t)z] \tag{4}$$

$$H_{2x} = -H_t \cos\theta_t = -\frac{E_{t0}}{\eta_2} \cos\theta_t \cos[\omega t - (\beta_2 \sin\theta_t)x - (\beta_2 \cos\theta_t)z] \tag{5}$$

$$H_{2z} = H_t \sin\theta_t = \frac{E_{t0}}{\eta_2} \sin\theta_t \cos[\omega t - (\beta_2 \sin\theta_t)x - (\beta_2 \cos\theta_t)z] \tag{6}$$

Now, we will make use of the boundary conditions to derive formulae for reflection and transmission coefficients. According to the first boundary condition,

$$E_{1y}(z=0) = E_{2y}(z=0)$$

Thus, substituting $z = 0$ in (1) and (4) and then equating the resulting expressions, we get

$$E_{i0} \cos[\omega t - (\beta_1 \sin\theta_i)x] + E_{r0} \cos[\omega t - (\beta_1 \sin\theta_r)x]$$

$$= E_{t0} \cos[\omega t - (\beta_2 \sin\theta_t)x] \tag{7}$$

For this equation to be satisfied for all x at any time t, all the terms must be in phase along x direction. This is possible if

$$\beta_1 \sin\theta_i = \beta_1 \sin\theta_r = \beta_2 \sin\theta_t \tag{8}$$

These relations show that all the waves have the same phase constant and phase velocity along x direction. The equalities of (8) yield the following relations.

$$\theta_r = \theta_i \tag{9}$$

and

$$\frac{\sin\theta_t}{\sin\theta_i} = \frac{\beta_1}{\beta_2} \tag{10a}$$

For nonmagnetic media, the relation is given by

$$\frac{\sin\theta_t}{\sin\theta_i} = \sqrt{\frac{\varepsilon_{r1}}{\varepsilon_{r2}}} \tag{10b}$$

We find that the angle of reflection is equal to the angle of incidence, and the sine of angle of refraction is related to the sine of angle of incidence through the phase constants of the mediums. Relations (9) and (10) are known as Snell's laws of reflection and refraction respectively.

Substitution of (8) in (7) gives,

$$E_{i} \quad E_{r0} = E_{t0} \tag{11}$$

Now, as per the second boundary condition we have,

$$H_{ix}(z=0) = H_{tx}(z=0)$$

By substituting $z = 0$ in (2) and (5) and utilising relations (8) and (9), we get the following equation.

$$\frac{1}{\eta_1}(E_{i0} - E_{r0})\cos\theta_i = \frac{1}{\eta_2} E_{t0}\cos\theta_t \tag{12}$$

Solving (11) and (12) simultaneously we will obtain E_{r0} and E_{t0} in terms of E_{i0}. The ratio of the reflected electric field to the incident electric field is the *Reflection coefficient*, and the ratio between the transmitted electric field and the incident electric field is termed as *transmission coefficient*. Using notations $\Gamma_\perp$ and $\tau_\perp$ for the reflection and transmission coefficients for the perpendicularly polarised wave respectively, we have

$$\Gamma_\perp = \frac{E_{r0}}{E_{i0}} = \frac{\eta_2\cos\theta_i - \eta_1\cos\theta_t}{\eta_2\cos\theta_i + \eta_1\cos\theta_t} \tag{13}$$

$$\tau_\perp = \frac{E_{t0}}{E_{i0}} = \frac{2\eta_2\cos\theta_i}{\eta_2\cos\theta_i + \eta_1\cos\theta_t} = 1 + \Gamma_\perp \tag{14}$$

If the dielectric mediums are nonmagnetic, we may assume that $\mu_1 \approx \mu_2 \approx \mu_0$. Also, $\varepsilon_1 = \varepsilon_0\varepsilon_{r1}$ and $\varepsilon_2 = \varepsilon_0\varepsilon_{r2}$. Hence, the reflection and transmission coefficients of (13) and (14) are

$$\Gamma_\perp = \frac{\sqrt{\varepsilon_{r1}}\cos\theta_i - \sqrt{\varepsilon_{r2}}\cos\theta_t}{\sqrt{\varepsilon_{r1}}\cos\theta_i + \sqrt{\varepsilon_{r2}}\cos\theta_t} \tag{15}$$

$$\tau_\perp = \frac{2\sqrt{\varepsilon_{r1}}\cos\theta_i}{\sqrt{\varepsilon_{r1}}\cos\theta_i + \sqrt{\varepsilon_{r2}}\cos\theta_t} = 1 + \Gamma_\perp \tag{16}$$

where

$$\cos\theta_t = \sqrt{1-\sin^2\theta_t} = \sqrt{1-(\beta_1/\beta_2)^2\sin^2\theta_i} = \sqrt{1-(\varepsilon_{r1}/\varepsilon_{r2})\sin^2\theta_i} \tag{17}$$

For cos θ_t in (17) to be a real quantity, the function under the radical sign should also be a real quantity. Accordingly, $\varepsilon_{r1}\sin^2\theta_i$ should be less than ε_{r1}.

We can express the total electric and total magnetic field intensities in Region 1 in terms of the amplitude of the electric field intensity of the incident wave by substituting $E_{r0} = \Gamma_\perp E_{i0}$ in (1), (2) and (3). The fields in Region 2 can also be expressed in terms of the amplitude of the electric field intensity of the incident wave by using the relation $E_{t0} = \tau_\perp E_{i0}$ in (4), (5) and (6).

Parallel Polarisation

In the case of parallel polarisation, the electric field intensities of the waves are parallel to the plane of incidence and the magnetic field intensities are perpendicular to the plane of incidence, as shown in Fig. 7.24. Thus, the electric fields have components along x and z directions, whereas the magnetic fields are in y direction only. The x components of electric fields and the y components of the magnetic fields are tangential to the interface.

The expressions for the total electric and magnetic fields along the coordinate directions in Region 1 are

$$E_{1x} \quad E_i \cos\theta_i - E_r\cos\theta_r$$

$$= E_i \; [illegible] \; t - (\beta_1 \sin\theta_i)x - [illegible] \; z]$$

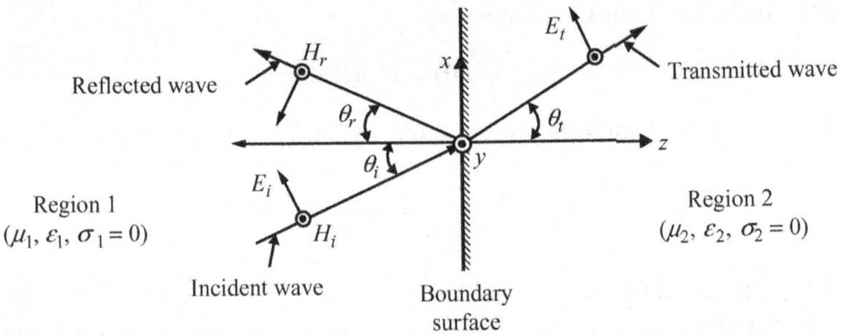

Fig. 7.24 Oblique incidence of a uniform plane wave on the interface of two perfect dielectric regions for parallel polarisation

$$-E_{r0} \cos\theta_r \cos[\omega t - (\beta_1 \sin\theta_r)x + (\beta_1 \cos\theta_r)z] \tag{18}$$

$$E_{1z} = -E_i \sin\theta_i - E_r \sin\theta_r$$

$$= -E_{i0} \sin\theta_i \cos[\omega t - (\beta_1 \sin\theta_i)x - (\beta_1 \cos\theta_i)z]$$

$$-E_{r0} \sin\theta_r \cos[\omega t - (\beta_1 \sin\theta_r)x + (\beta_1 \cos\theta_r)z] \tag{19}$$

$$H_{1y} = \frac{1}{\eta_1}(E_i + E_r) = \frac{1}{\eta_1}\{E_{i0} \cos[\omega t - (\beta_1 \sin\theta_i)x - (\beta_1 \cos\theta_i)z]$$

$$+E_{r0} \cos[\omega t - (\beta_1 \sin\theta_r)x + (\beta_1 \cos\theta_r)z]\} \tag{20}$$

The expressions for fields along coordinate directions in Region 2 are

$$E_{2x} = E_{t0} \cos\theta_t \cos[\omega t - (\beta_2 \sin\theta_t)x - (\beta_2 \cos\theta_t)z] \tag{21}$$

$$E_{2z} = -E_{t0} \sin\theta_t \cos[\omega t - (\beta_2 \sin\theta_t)x - (\beta_2 \cos\theta_t)z] \tag{22}$$

$$H_{2y} = \frac{1}{\eta_2}[E_{t0} \cos[\omega t - (\beta_2 \sin\theta_t)x - (\beta_2 \cos\theta_t)z] \tag{23}$$

As per the second boundary condition, we have

$$H_{1y}(z = 0) = H_{2y}(z = 0)$$

By substituting $z = 0$ in (20) and (23) and then equating the resulting expressions we will get

$$\frac{1}{\eta_1}\{E_{i0} \cos[\omega t - (\beta_1 \sin\theta_i)x] + E_{r0} \cos[\omega t - (\beta_1 \sin\theta_r)x]\}$$

$$= \frac{E_{t0}}{\eta_2} \cos[\omega t - (\beta_2 \sin\theta_t)x] \tag{24}$$

For this equation to be satisfied for all x at any time t, all the terms must be in phase along x direction. This is possible if

$$\beta \quad \theta \quad \beta \quad \theta_i = \beta \quad \theta_t \tag{25}$$

The first equality of (25) yields

$$\theta_r = \theta_i \tag{26}$$

And from the second equality we get

$$\frac{\sin\theta_t}{\sin\theta_i} = \frac{\beta_1}{\beta_2} \tag{27}$$

Relations (26) and (27) are identical to (9) and (10) respectively.

By using (26) and (27) in (24) gives

$$\frac{E_{i0} + E_{r0}}{\eta_1} = \frac{E_{t0}}{\eta_2} \tag{28}$$

Now, as per the first boundary condition,

$$E_{x1}(z=0) = E_{x2}(z=0)$$

Or $\quad E_{i0}\cos\theta_i \cos[\omega t - (\beta_1 \sin\theta_i)x] - E_{r0}\cos\theta_r \cos[\omega t + (\beta_1 \sin\theta_r)x]$

$$= E_{t0}\cos\theta_t \cos[\omega t - (\beta_2 \sin\theta_t)x] \tag{29}$$

Using relations (25) and (26) in (29), we get

$$(E_{i0} - E_{r0})\cos\theta_i = E_{t0}\cos\theta_t \tag{30}$$

By solving (28) and (30) simultaneously we will get E_{r0} and E_{t0} in terms of E_{i0}. If we define the ratio of the reflected electric field to the incident electric field as *reflection coefficient*, and the ratio between the transmitted electric field and the incident electric field as *transmission coefficient*, then using notations $\Gamma_{\parallel}$ and $\tau_{\parallel}$ for the reflection and transmission coefficients for the parallel polarised wave respectively, we have

$$\Gamma_{\parallel} = \frac{E_{r0}}{E_{i0}} = \frac{\eta_1\cos\theta_i - \eta_2\cos\theta_t}{\eta_1\cos\theta_i + \eta_2\cos\theta_t} \tag{31}$$

$$\tau_{\parallel} = \frac{E_{t0}}{E_{i0}} = \frac{2\eta_2\cos\theta_i}{\eta_1\cos\theta_i + \eta_2\cos\theta_t} \tag{32}$$

If we assume that the permeabilities of the regions are that of free space, then (31) and (32) will be

$$\Gamma_{\parallel} = \frac{\sqrt{\varepsilon_{r2}}\cos\theta_i - \sqrt{\varepsilon_{r1}}\cos\theta_t}{\sqrt{\varepsilon_{r2}}\cos\theta_i + \sqrt{\varepsilon_{r1}}\cos\theta_t} \tag{33}$$

$$\tau_{\parallel} = \frac{2\sqrt{\varepsilon_{r1}}\cos\theta_i}{\sqrt{\varepsilon_{r2}}\cos\theta_i + \sqrt{\varepsilon_{r1}}\cos\theta_t} \tag{34}$$

where $\cos\theta_t = \sqrt{1-(\varepsilon_{r1}/\varepsilon_{r2})\sin^2\theta_i}$

Note that for the reflection and transmission coefficient as defined, $\tau_{\parallel} \neq 1 + \Gamma_{\parallel}$.

The reflection and transmission coefficients derived in this section are useful for calculating the fields and power carried by reflected and transmission waves. Two special cases of reflection coefficient are discussed in the next two sections.

EXAMPLE 7.32

The common boundary between a nonmagnetic perfect dielectric medium and free space is a plane surface. A uniform plane wave with perpendicular polarisation from the dielectric medium is incident at an angle of 35° with the normal to the boundary surface. The relative permittivity of the dielectric is 1.8. Find the reflection and transmission coefficients for the electric field.

Let us take the dielectric medium as Region 1 and free space as Region 2.

$$\theta_r \quad \theta_i = 35°, \ \varepsilon_{r1} = 1.8, \ \varepsilon_{r2} = 1.0$$

$$\frac{\sin\theta_t}{\sin\theta_i} = \frac{\beta_1}{\beta_2} = \frac{\sqrt{\varepsilon_{r1}}}{\sqrt{\varepsilon_{r2}}} = \sqrt{1.8} = 1.342$$

$$\sin \hat{}_t \quad {}^{\cdot}.342 \sin 35° = 0.769$$

$$\cos \hat{}_t \quad .639, \ \cos\theta_i = 0.819$$

Reflection coefficient,

$$\Gamma_\perp = \frac{\varepsilon_r \cos\theta_i - \sqrt{\varepsilon_{r2}} \cos\theta_t}{\varepsilon_r \cos\theta_i + \sqrt{\varepsilon_{r2}} \cos\theta_t} = \frac{(\sqrt{1.8})(0.819) - (\sqrt{1})(0.639)}{(\sqrt{1.8})(0.819) + (\sqrt{1})(0.639)} = 0.264$$

Transmission coefficient, $\tau_\perp = 1 + \Gamma_\perp = 1.264$

EXAMPLE 7.33

Two nonmagnetic perfect dielectric regions have a plane common boundary surface. A uniform plane wave of parallel polarisation is incident at an angle θ_i with the normal to the interface. The relative permittivities of the regions are 3.5 and 4.8. If the incident wave is from the region of lower permittivity and angle of refraction is 25°, find the transmission and reflection coefficients for the electric field.

Let us take the dielectric medium of lower permittivity as Region 1 and dielectric medium of higher permittivity as Region 2.

$$\theta_i \qquad \varepsilon_{r1} = 3.5, \ \varepsilon_{r2} = 4.8$$

$$\frac{\sin\theta_i}{\sin\theta_t} = \frac{\beta_2}{\beta_1} = \frac{\sqrt{\varepsilon_{r2}}}{\sqrt{\varepsilon_{r1}}} = \frac{\sqrt{4.8}}{\sqrt{3.5}} = 1.171$$

$$\sin \hat{}_i \quad {}^{\cdot}.171 \sin 25° = 0.495$$

$$\cos \hat{}_i \quad .869, \ \cos\theta_t = 0.906$$

Reflection coefficient,

$$\Gamma_\parallel = \frac{\varepsilon_r \cos\theta_i - \sqrt{\varepsilon_{r1}} \cos\theta_t}{\varepsilon_r \cos\theta_i + \sqrt{\varepsilon_{r1}} \cos\theta_t} = \frac{(\sqrt{4.8})(0.869) - (\sqrt{3.5})(0.906)}{(\sqrt{4.8})(0.869) + (\sqrt{3.5})(0.906)} = 0.058$$

Transmission coefficient,

$$\tau_\parallel = \frac{2\sqrt{\varepsilon_{r1}} \cos\theta_i}{\varepsilon_{r1} \cos\theta_i + \sqrt{\varepsilon_{r2}} \cos\theta_t} = \frac{2(\sqrt{3.5})(0.869)}{(\sqrt{4.8})(0.869) + (\sqrt{3.5})(0.906)} = 0.904$$

EXAMPLE 7.34

Using that data of Example 7.33, calculate Poynting vectors for the incident, reflected, and transmitted waves. Amplitude of electric field intensity of the incident wave is 75 V/m.

$\varepsilon_r = 3.5,\ \varepsilon_{r2} = 4.8$

$\sin\theta_i = \sin\theta_r = 0.495, \cos\theta_i = \cos\theta_r = 0.869$

$\sin\theta_t = 0.423,\ \cos\theta_t = 0.906$

$\eta\ \ 377/\sqrt{3.5} = 201.51\ \Omega,\ \eta_2 = 377/\sqrt{4.8} = 172.08\ \Omega$

$\Gamma_{\|} = 0.058,\ \tau_{\|} = 0.904$

Refer to Fig. 7.24.

As the media are loss-free, the electric and magnetic field intensities of a wave are in time phase. Hence, the average value of Poynting vector of a wave,

$$\mathbf{P}_{wa} = \frac{E_0^2}{2\eta}\mathbf{u}$$

where **u** is a unit vector in the direction of propagation of the wave. We will denote the unit vectors in the direction of travel of the incident, reflected and transmitted waves by $\mathbf{u}_i$, $\mathbf{u}_r$ and $\mathbf{u}_t$ respectively.

Average Poynting vector of the incident wave $= \frac{E_{i0}^2}{2\eta_1}\mathbf{u}_i = \frac{75^2}{2\times 201.51}\mathbf{u}_i = 13.96\mathbf{u}_i$ W/m²

The unit vector, $\mathbf{u}_i = \sin\theta_i\mathbf{u}_x + \cos\theta_i\mathbf{u}_z = 0.495\mathbf{u}_x + 0.869\mathbf{u}_z$

Average Poynting vector of the incident wave in terms of the components

$= (13.96)(0.495\mathbf{u}_x + 0.869\mathbf{u}_z) = 6.91\mathbf{u}_x + 12.13\mathbf{u}_z$ W/m²

Average Poynting vector of the reflected wave

$= \frac{E_{r0}^2}{2\ _1}\mathbf{u}_r = \frac{(\Gamma_{\|}E_{i0})^2}{2\eta_1}\mathbf{u}_r = (0.058)^2(13.96)\mathbf{u}_r = 0.047\mathbf{u}_r$ W/m²

The unit vector, $\mathbf{u}_r = \sin\theta_r\mathbf{u}_x\ \ \cos\theta_r\mathbf{u}_z = 0.495\mathbf{u}_x - 0.869\mathbf{u}_z$

Average Poynting vector of the reflected wave in terms of the components

$= (\ .047)(0.495\mathbf{u}_x - 0.869\mathbf{u}_z) = 0.023\mathbf{u}_x - 0.041\mathbf{u}_z$ W/m²

Average Poynting vector of the transmitted wave

$\frac{E_{t0}^2}{2\ _2}\mathbf{u}_t = \frac{(\tau_{\|}E_{i0})^2}{2\eta_2}\mathbf{u}_t = \frac{(0.904)^2(75)^2}{2\times 172.08}\mathbf{u}_t = 13.36\mathbf{u}_t$ W/m²

The unit vector, $\mathbf{u}_t = \sin\theta_t\mathbf{u}_x + \cos\theta_t\mathbf{u}_z = 0.423\mathbf{u}_x + 0.906\mathbf{u}_z$

Average Poynting vector of the reflected wave in terms of the components

$= (13.36)(0.423\mathbf{u}_x + 0.906\mathbf{u}_z) = 5.651\mathbf{u}_x + 12.104\mathbf{u}_z$ W/m²

The results show that power per unit area along z direction in Region 2 is equal to the net power per unit area along z direction in Region 1. However, powers per unit area along x direction, as given by the Poynting vector, on two sides of the interface are not equal. A likely reason for the discrepancy is that the normal component (z component) of electric field intensity is not continuous across the boundary surface.

7.15 Critical Angle of Incidence and Total Reflection

When a plane wave travelling in one dielectric region is incident obliquely on the surface of another dielectric region having a common boundary, wave reflection and transmission occurs. One of the parameters of wave reflection is the reflection coefficient described in the last section. For nonmagnetic and perfect dielectric regions, the magnitudes of the reflection coefficients, for both types of polarisation, are functions of the incident angle and the relative permittivities of the regions.

For a given value of the ratio between the relative permittivities of the regions, there exists an angle of incidence at which the reflection coefficient becomes unity. The angle is called *critical angle of incidence.*

It is generally denoted by θ_c. For values of incident angle equal to or higher than the critical angle, *total reflection* of the incident wave takes place.

Considering the reflection phenomenon for perpendicular polarisation, the reflection coefficient is given by

$$\Gamma_\perp = \frac{\sqrt{\varepsilon_{r1}}\cos\theta_i - \sqrt{\varepsilon_{r2}}\cos\theta_t}{\sqrt{\varepsilon_{r1}}\cos\theta_i + \sqrt{\varepsilon_{r2}}\cos\theta_t} \tag{1}$$

Substituting $\theta_i = \theta_c$ and $\Gamma_\perp = 1$ into (1), we obtain

$$1 = \frac{\sqrt{\varepsilon_{r1}}\cos\theta_c - \sqrt{\varepsilon_{r2}}\cos\theta_t}{\sqrt{\varepsilon_{r1}}\cos\theta_c + \sqrt{\varepsilon_{r2}}\cos\theta_t}$$

Or

$$\sqrt{\varepsilon_{r1}}\cos\theta_c + \sqrt{\varepsilon_{r2}}\cos\theta_t = \sqrt{\varepsilon_{r1}}\cos\theta_c - \sqrt{\varepsilon_{r2}}\cos\theta_t$$

$$\Rightarrow\ 2\sqrt{\varepsilon_{r2}}\cos\theta_t = 0$$

As ε_{r2} has a finite value, hence

$$\cos\theta_t = 0$$

Or

$$\theta_t = 90°$$

As $\cos\theta_t = \sqrt{1-\sin^2\theta_t}$ and by Snell's law of reflection $\sin\theta_t = (\sqrt{\varepsilon_{r1}/\varepsilon_{r2}})\sin\theta_c$, we have

$$\cos\theta_t = \sqrt{1-(\varepsilon_{r1}/\varepsilon_{r2})\sin^2\theta_c} = 0$$

Or

$$\sin\theta_c = \sqrt{\frac{\varepsilon_{r2}}{\varepsilon_{r1}}} \tag{2a}$$

and

$$\theta_c = \sin^{-1}(\varepsilon_{r2}/\varepsilon_{r1}) \tag{2b}$$

Formula (2) gives the relation between the critical angle of incidence and the relative permittivities of the regions. It can be shown that the relation is same for parallel polarisation of the incident wave.

As the value of sin θ_c cannot be greater than one, a real value of critical angle of incidence exists only if $\varepsilon_{r1} > \varepsilon_{r2}$. That is, the permittivity of the region from where the incident wave originates should be greater than that of the other region.

The transmission coefficient at the critical angle of incidence, obtained by substituting cos $\theta_t = 0$ into (7.14-17), is equal to two. It means that fields exist in Region 2 but they travel only along x direction, that is parallel to the interface because $\theta_t = 90°$. Such waves are called *surface waves.*

The relation between the angle of the transmitted wave and critical angle of incidence is obtained by using (2) in (7.14-17). The relation is

$$\cos\theta_t = \sqrt{1-(\sin\theta_i/\sin\theta_c)^2} \tag{3}$$

Expression (3) shows that if the angle of incidence is higher than the critical angle, then cos θ_t and hence the phase constant along z direction will become imaginary. In order to understand the implication of the imaginary phase constant, let us write the electric field of the transmitted wave in phasor form. For $\theta_i < \theta_c$, cos θ_t is a positive real quantity and hence

$$\bar{E}_t = E_{t0}e^{-j\beta_2(x\sin\theta_t + z\cos\theta_t)} \tag{4}$$

The phase velocity of the wave along x and z directions are v/sin θ_t and v/cos θ_t respectively.

And for angles of incidence higher than the critical angle, the function under the square-root sign of (3) becomes negative. So we may write,

$$\cos\theta_t = \pm j\sqrt{(\sin\theta_i/\sin\theta_c)^2 - 1},\ \theta_i > \theta_c \tag{5}$$

Choosing the negative sign,

$$\beta_2\cos\theta_t = -j\beta_2\sqrt{(\sin\theta_i/\sin\theta_c)^2 - 1} = -j\alpha\,(\text{Say}),\quad \theta_i > \theta_c \tag{6}$$

where α is a real quantity.

Using (6) in (4), the transmitted electric field is given by

$$\bar{E}_t = (E_{t0}e^{-\alpha z})e^{-j\beta_2 x\sin\theta_t} \tag{7}$$

The function within the brackets of (7) is the amplitude of the field at any z. Thus, the amplitudes of electric and magnetic fields of the transmitted wave decay exponentially along z direction and the decaying fields travel along positive x direction, that is, parallel to the boundary surface as surface waves. By using the negative value of (5) in (7.14-15) and (7.14-33) we will get the reflection coefficient for both types of polarisation. These are given by

$$\bar{\Gamma}_\perp = \frac{\sqrt{\varepsilon_{r1}}\cos\theta_i + j\sqrt{\varepsilon_{r2}[(\sin\theta_i/\sin\theta_c)^2 - 1]}}{\sqrt{\varepsilon_{r1}}\cos\theta_i - j\sqrt{\varepsilon_{r2}[(\sin\theta_i/\sin\theta_c)^2 - 1]}}$$

$$\bar{\Gamma}_\parallel = \frac{\sqrt{\varepsilon_{r2}}\cos\theta_i + j\sqrt{\varepsilon_{r1}[(\sin\theta_i/\sin\theta_c)^2 - 1]}}{\varepsilon_r \qquad \theta_i - j\ \ \varepsilon_r \qquad \theta_i/\sin\theta_c) \qquad]}$$

We see that the reflection coefficients are complex quantities. However, their magnitudes are equal to unity as the numerators and denominators have same real parts and same imaginary parts.

Thus, we conclude that if the permittivity of Region 1 (the region from where the incident wave originates) is higher than that of Region 2 and the angle of incidence is equal to or greater than the critical angle of incidence, then the magnitude of the reflection coefficient is unity. This phenomenon is called *total reflection* of the incident wave. However, fields exist in the second region, and they propagate parallel to the boundary surface.

The positive root of (5) is not chosen because the amplitude of the field will increase with z, which is not permissible from physical considerations.

EXAMPLE 7.35

The common boundary surface of a perfect dielectric region and air is a plane surface. The constants of the dielectric region are $\mu = \mu_0$ and $\varepsilon = 5\varepsilon_0$. (a) Find the critical angle of incidence. (b) Show that the magnitude of the reflection coefficient is unity even if the angle of incidence is more than the critical angle of incidence.

(a) A critical angle of incidence exists only if the incident wave is in the region of higher permittivity. As the permittivity of the dielectric is more than that of air, the incident wave must be in the dielectric region. Hence, $\varepsilon_{r1} = 5.0$ and $\varepsilon_{r2} = 1.0$. If θ_c is the critical angle of incidence, then

$$\sin{\hat{}}_{c} = \sqrt{\frac{\varepsilon_{r2}}{\varepsilon_{r1}}} = \sqrt{\frac{1}{5}} = 0.447$$

$$\theta_c = 26.6°$$

(b) Let the angle of incidence be 30°. Then as per Snell's law,

$$\frac{\sin\theta_t}{\sin\theta_i} = \sqrt{\frac{\varepsilon_{r1}}{\varepsilon_{r2}}} = \sqrt{5}$$

$$\sin{\hat{}}_{t} \quad \sqrt{5}\sin 30° - 1.118$$

$$\cos{\hat{}}_{t} \quad \sqrt{1 - 1.118^2} = j0.5$$

Considering the reflection coefficient of a perpendicularly polarised wave given by (7.14-15), we have

$$\Gamma_{\perp} = \frac{\sqrt{5}\cos 30° - j0.5}{\sqrt{5}\cos 30° + j0.5} = \frac{1.936 - j0.5}{1.936 + j0.5}$$

$$\Gamma_{\perp} = \frac{|\ .936 - j0.5|}{|\ .936 + j0.5|} = 1$$

7.16 Brewster Angle

In the last section we learnt that the value of the reflection coefficient becomes unity if the angle of an obliquely incident wave is equal to or greater than the critical angle of incidence. The necessary condition is that the permittivity of the region of the incident wave is greater than the permittivity of the region on the other side of the interface. In this section we will find that the reflection coefficient becomes zero at a specific angle of incidence governed by the permittivities of the regions on two sides of the interface.

It is found that if the regions are nonmagnetic, the reflection coefficient can become zero only for parallel polarisation of the wave. However, there is no restriction regarding the relative values of the permittivities of the regions. The angle of incidence at which the reflection coefficient becomes zero is called *Brewster angle or Polarising angle.*

Consider that at some angle of incidence $\theta_i = \theta_p$ the reflection coefficient of a plane wave with parallel polarisation becomes zero. Then, substituting θ_p in place of θ_i in the formula for the reflection coefficient given by (7.14-33) and equating the numerator to zero, we get

$$\sqrt{\varepsilon_{r2}}\cos\theta_p = \sqrt{\varepsilon_{r1}}\cos\theta_t \tag{1}$$

As per Snell's law of refraction,

$$\beta_1 \sin\theta_p = \beta_2 \sin\theta_t$$

Assuming that the two regions are nonmagnetic, Snell's law of refraction is given by

$$\sqrt{\varepsilon_{r1}}\sin\theta_p = \sqrt{\varepsilon_{r2}}\sin\theta_t \tag{2}$$

Solving (1) and (2) simultaneously for sin θ_p, we get

$$\sin\theta_p = \sqrt{\frac{\varepsilon_{r2}}{\varepsilon_{r1}+\varepsilon_{r2}}} \tag{3}$$

Also,

$$\cos\theta_p = \sqrt{\frac{\varepsilon_{r1}}{\varepsilon_{r1}+\varepsilon_{r2}}} \tag{4}$$

And

$$\tan\theta_p = \sqrt{\frac{\varepsilon_{r2}}{\varepsilon_{r1}}},$$

$$\theta_p = \tan^{-1}\sqrt{\varepsilon_{r2}/\varepsilon_{r1}} \tag{5}$$

Thus for angle of incidence equal to θ_p as given by (5), the reflection coefficient for parallel polarisation will be zero. The angle θ_p is called *Brewster angle* or *Polarising angle.* It may be noted that a real value of Brewster angle exists whether ε_{r2} is greater or less than ε_{r1}. The value of θ_p is more than 45 degrees if $\varepsilon_{r2} > \varepsilon_{r1}$ and less than 45 degrees if $\varepsilon_{r2} < \varepsilon_{r1}$.

The solution for cos θ_t is found by substituting (4) into (1). This gives,

$$\cos\theta_t = \sqrt{\frac{\varepsilon_{r2}}{\varepsilon_{r1}+\varepsilon_{r2}}} = \sin\theta_p = \cos(90-\theta_p)$$

Thus, the angle of refraction is equal to 90° minus the Brewster angle.

Now let us examine whether it is possible to make the reflection coefficient zero for a perpendicularly polarised wave. For zero reflection, the numerator of (7.14-15) must be zero. Accordingly,

$$\frac{\cos\theta_i}{\cos\theta_t} = \frac{\varepsilon_{r2}}{\varepsilon_{r1}} \tag{6}$$

From Snell's law of refraction we have,

$$\frac{\sin\theta_i}{\sin\theta_t} = \sqrt{\frac{\varepsilon_{r2}}{\varepsilon_{r1}}} \tag{7}$$

Dividing (7) by (6), we get

$$\tan\theta_t = \tan\theta_i$$

$$\theta_i = \theta_t$$

The above relation means that the permittivity of the two mediums must be equal. That is, no discontinuity exists. This is a trivial condition. Hence, Brewster angle does not exist in the case of perpendicular polarisation if the regions are nonmagnetic.

EXAMPLE 7.36

A parallel-polarised electromagnetic wave is incident obliquely at the interface of two perfect dielectric regions 1 and 2. The constants of the Region 1 are $\mu_1 = \mu_0$ and $\varepsilon_1 = 4\varepsilon_0$. The constants of Region 2 are $\mu_{\_}$ μ_0 and $\varepsilon_2 = 6.25\varepsilon_0$. Find the Brewster angle if the incident wave is, (a) in Region 1, and (b) in Region 2.

We have, $\varepsilon_{r1} = 4.0$ and $\varepsilon_{r2} = 6.25$

(a) Let the Brewster angle be θ_{p1}. Then,

$$\tan\theta_{p1} = \sqrt{\frac{\varepsilon_{r2}}{\varepsilon_{r1}}} = \sqrt{\frac{6.25}{4}} = 1.25$$

$$\theta_{p1} = 51.34°$$

(b) Let the Brewster angle be θ_{p2}. Then,

$$\tan\theta_{p2} = \sqrt{\frac{\varepsilon_{r1}}{\varepsilon_{r2}}} = \sqrt{\frac{4}{6.25}} = 0.8$$

$$\theta_{p2} = 38.66°$$

Note that the sum of θ_{p1} and θ_{p2} is equal to 90°.

SUMMARY

After developing the Helmholtz vector wave equations, we obtained the one-dimensional wave equation for the electric field intensity in an unbounded ideal dielectric medium in rectangular coordinates, and solved it. The solution, in general, has two parts; one solution is a positively travelling wave and the other is a negatively travelling wave. We also derived the solution for the magnetic field intensity and found that the electric and magnetic field intensities are at 90° in space, they are in time phase, and the waves are uniform plane waves. We defined various parameters associated with the fields and learnt that the phase constant and wavelength are functions of ω, μ, and ε, whereas the phase velocity and intrinsic impedance are functions of μ and

ε only. We also calculated average power per unit area carried by a uniform plane wave using the instantaneous and complex Poynting vectors.

We studied wave propagation in an imperfect dielectric medium and found that amplitudes of electric and magnetic field intensities decrease as they progress in the medium, and the fields are not in time phase. We found from the analysis of wave propagation in a good conductor that the fields attenuate rapidly, the amplitude of the field at a depth, called the skin depth, is equal to $1/e$ times of its value at the conductor surface, and the time-phase angle between electric and magnetic fields is 45°.

We described wave polarisation and noted that the meaning of wave polarisation is different from dielectric polarisation.

We analysed wave reflection from a perfect conductor and from the interface of two perfect dielectrics both for normal and oblique incidence, and defined reflection and transmission coefficients. In the case of normal incidence on a perfect conductor, the electric field gets totally reflected with change of sign, and the magnetic field gets totally reflected without change of sign. The magnitude and sign of amplitudes of the reflected fields in the case of normal incidence on the boundary surface of two dielectrics depend on the intrinsic impedances of the regions. We derived expressions for total electric and magnetic field intensities for the case of normal incidence, and defined standing wave ratio, which is the ratio between the maximum amplitude and the minimum amplitude of electric/magnetic field intensities.

From the analysis of oblique incidence of a uniform plane wave on a dielectric interface, we obtained Snell's laws of reflection and refraction and defined critical angle of incidence and the Brewster angle. We learnt that a real value of critical angle of incidence exists only if the permittivity of the region from where the incident wave originates is greater than that of the other region. We also found that that no Brewster angle exists for perpendicular polarisation.

IMPORTANT FORMULAE

One-dimensional wave equations:

General form, $\dfrac{\partial^2 E_x}{\partial z^2} = \mu\varepsilon \dfrac{\partial^2 E_x}{\partial t^2}$

Phasor form, $\dfrac{\partial^2 \bar{E}_x(z)}{\partial z^2} = -\omega^2 \mu\varepsilon \bar{E}_x(z)$

Parameters of uniform plane wave in an ideal medium:

Phase constant, $\beta = \omega\sqrt{\mu\varepsilon} = 2\pi f\sqrt{\mu\varepsilon}$

Phase velocity, $v = \dfrac{1}{\sqrt{\mu\varepsilon}} = \dfrac{\omega}{\beta}$

Wavelength, $\lambda = \dfrac{2\pi}{\beta}$

Intrinsic wave impedance, $\lambda = \dfrac{2\pi}{\beta}$

Average power per unit area, $P_{av} \; \dfrac{1}{2} E_{x}^{+} H_{y0}^{+} = \dfrac{(E_{x0}^{+})^2}{2\eta} = \dfrac{\eta(H_{y0}^{+})^2}{2}$

Plane waves in a general medium:

Propagation constant, $\bar{\gamma} = \sqrt{j\omega\mu(\sigma + j\omega\varepsilon)}$

Attenuation constant, $\alpha = \omega\sqrt{\frac{\mu\varepsilon}{2}(-1+\sqrt{1+\left(\frac{\sigma}{\omega\varepsilon}\right)^2}}$

Phase constant, $\beta = \omega\sqrt{\frac{\mu\varepsilon}{2}\left(1+\sqrt{1+\frac{\sigma}{\omega\varepsilon}}\right)^2}$

Intrinsic impedance, $\bar{\eta} = \sqrt{\frac{j\omega\mu}{\sigma + j\omega\varepsilon}}$

Loss tangent, $\tan\delta_d = \frac{J}{J_d} = \frac{\sigma}{\omega\varepsilon}$

Plane waves in a good dielectric medium:

Attenuation constant, $\alpha \cong \frac{\sigma}{2}\sqrt{\frac{\mu}{\varepsilon}}$

Phase constant, $\beta \cong \omega\sqrt{\mu\varepsilon}\left(1+\frac{\sigma^2}{8\omega^2\varepsilon^2}\right)$

Intrinsic impedance, $\bar{\eta} \cong \sqrt{\frac{\mu}{\varepsilon}}\left(1+j\frac{\sigma}{2\omega\varepsilon}\right)$

Plane waves in a flat good conductor:

Attenuation constant, $\alpha = \sqrt{\pi f\mu\sigma}$

Phase constant, $\beta = \sqrt{\pi f\mu\sigma}$

Skin depth, $\delta = \frac{1}{\alpha} = \frac{1}{\sqrt{\pi f\mu\sigma}}$

Intrinsic impedance, $\bar{\eta} = \sqrt{\frac{\pi f\mu}{\sigma}}(1+j)$

Power flow per unit surface area, $P_{av} = \frac{1}{2}E_{x0}^{+}H_{y0}^{+}\cos 45° = \frac{(E_{x0}^{+})^2}{2\sqrt{2}\,\eta}$

Normal incidence of wave on a perfect conductor:

Relation between incident and reflected fields, $E_{x0}^{-} = -E_{x0}^{+},\ H_{y0}^{-} = H_{y0}^{+}$

Sum of incident and reflected electric fields, $E_x = 2E_0^{+}\sin\omega t\sin\beta z$

Sum of incident and reflected magnetic fields, $H_y = \frac{2E_0^{+}}{\eta}\cos\omega t$ [illegible] z

Normal incidence of wave on the interface of two perfect dielectric regions:

Reflection coefficient, $\Gamma = \dfrac{E_{1x0}^-}{E_{1x0}^+} = \dfrac{\eta_2 - \eta_1}{\eta_2 + \eta_1}$

Transmission coefficient, $\Gamma = \dfrac{E_{1x0}^-}{E_{1x0}^+} = \dfrac{\eta_2 - \eta_1}{\eta_2 + \eta_1}$

Reflected power, $P_{1,av}^- = \Gamma^2 P_{1,av}^+$

Transmitted power, $P_{2,av}^+ = \dfrac{(E_{2x0}^+)^2}{2\eta_2} = (1 - \Gamma^2) P_{1,av}^+$

Standing wave ratio: $S_R = \dfrac{|E_{x0}^+| + |E_{x0}^-|}{|E_{x0}^+| - |E_{x0}^-|} = \dfrac{1 + |\Gamma|}{1 - |\Gamma|}$

Reflection coefficient in terms of standing wave ratio: $|\Gamma| = \dfrac{S_R - 1}{S_R + 1}$

Wave impedance: $\bar{Z}_1(z) = \eta_1 \dfrac{1 + \Gamma_1(0) e^{j2\beta_1 z}}{1 - \Gamma_1(0) e^{j2\beta_1 z}}$

Oblique incidence of wave on a perfect conductor: $\theta_r = \theta_i$, $E_{r0} = -E_{i0}$

Oblique incidence of wave on the interface of two perfect dielectrics:

Snell's law of reflection, $\theta_r = \theta_i$

Snell's law of refraction, $\dfrac{\sin\theta_t}{\sin\theta_i} = \dfrac{\beta_1}{\beta_2}$

Critical angle of incidence: $\theta_c = \sin^{-1}(\varepsilon_{r2}/\varepsilon_{r1})$

Brewster angle: $\theta_p = \tan^{-1}\sqrt{\varepsilon_{r2}/\varepsilon_{r1}}$

MULTIPLE-CHOICE QUESTIONS

(Starred questions have two answers)

*7.1 Which of the following are the proper wave equations?

(a) $\dfrac{\partial^2 H_x}{\partial x^2} = \mu\varepsilon \dfrac{\partial H_x}{\partial t}$ (b) $\dfrac{\partial^2 H_x}{\partial x^2} = \dfrac{\mu}{\varepsilon} \dfrac{\partial^2 H_x}{\partial t^2}$

(c) $\dfrac{\partial^2 E_y}{\partial y^2} = \mu\varepsilon \dfrac{\partial^2 E_y}{\partial t^2}$ (d) $\dfrac{\partial^2 E_y}{\partial x^2} = \mu\varepsilon \dfrac{\partial^2 E_y}{\partial t^2}$

*7.2 For sinusoidally time-varying fields represented by the complex exponential function $e^{j\omega t}$, which of the following equations are the wave equations?

(a) $\dfrac{\partial^2 E_x}{\partial x^2} = -\omega^2 \mu\varepsilon E_x$ (b) $\dfrac{\partial^2 E_y}{\partial y^2} = -\omega\mu\varepsilon E_y$

(c) $\dfrac{\partial^2 \bar{H}_y}{\partial y^2} = -\omega^2 \mu\varepsilon \bar{H}_y$ (d) $\dfrac{\partial^2 \bar{H}_x}{\partial x^2} = -\omega\mu\varepsilon \bar{H}_x$

7.3 $E_x = E_{x0} \cos(\omega t - \beta z)$ is a solution of the one-dimensional wave equation for sinusoidally time-varying fields. The phase velocity of the wave is
(a) β/ω (b) ω/β (c) $\omega\beta$ (d) $\sqrt{\omega/\beta}$

7.4 The intrinsic wave impedance of a loss-free medium of permeability μ and permittivity ε is given by
(a) $\sqrt{\mu/\varepsilon}$ (b) $\sqrt{\varepsilon/\mu}$ (c) $\sqrt{\mu\varepsilon}$ (d) μ/ε

7.5 A 5 MHz uniform plane wave travels in a perfect dielectric medium. The relative permeability and relative permittivity of the medium are 1.0 and 4.0 respectively. The wavelength of the wave in the medium is
(a) 60 m (b) 60π m (c) 30π m (d) 30 m

7.6 The instantaneous magnetic field intensity of a uniform plane wave travelling in perfect dielectric medium is $H_y = 0.5\cos(\omega t + \beta z)$ A/m. The relative permeability and relative permittivity of the medium are 1.0 and 2.25 respectively. The instantaneous electric field intensity in V/m of the wave is given by
(a) $125.7\cos(\omega t + \beta z)$ (b) $-125.7\cos(\omega t + \beta z)$
(c) $167.6\cos(\omega t + \beta z)$ (d) $-167.6\cos(\omega t + \beta z)$

*7.7 A perfect dielectric medium has a uniform plane wave. Which of the following statements for the wave are correct?
(a) The electric and magnetic fields are in the same direction.
(b) The electric and magnetic fields are perpendicular to each other.
(c) The electric and magnetic fields are in phase.
(d) The phase difference between the electric and magnetic fields is 90°.

*7.8 The amplitudes of electric and magnetic fields of a uniform plane wave in a perfect dielectric medium are E_0 and H_0. The intrinsic impedance of the medium is η. Average power flow per unit area of the wave is given by
(a) ηH_0^2 (b) $\eta H_0^2/2$ (c) E_0^2/η (d) $E_0^2/2\eta$

7.9 A uniform plane wave is also called a Transverse Electro-Magnetic (TEM) wave because
(a) direction of electric field is perpendicular to, and direction of magnetic field is along the direction of travel of the wave
(b) direction of magnetic field is perpendicular to, and direction of electri field is along the direction of travel of the wave
(c) directions of both the fields are in the direction of travel of the wave
(d) directions of both the fields are perpendicular to the direction of travel of the wave

7.10 A dielectric medium has losses, but no free charges. The propagation constant of the medium is given by
(a) j $j\omega\varepsilon)$ (b) $\omega\mu(\sigma$ $\omega\varepsilon)$
(c) ( )( ) (d) ( )( )

7.11 The amplitude of electric field intensity on the surface of a good conductor is E_0. The amplitude of the field at a depth equal to the skin depth is

(a) zero (b) $0.368E_0$ (c) $0.5E_0$ (d) $0.736E_0$

7.12 The amplitude of magnetic field intensity of a plane wave on the surface of a good conductor, the depth of which is large compared to the skin depth, is H_0 and the intrinsic impedance is η. The power loss per unit surface area of the conductor is

(a) ηH_0^2 (b) $\eta H_0^2/\sqrt{2}$ (c) $\eta H_0^2/2$ (d) $\eta H_0^2/(2\sqrt{2})$

7.13 The conductivity and permittivity of a medium are $\pi \times 10^{-4}$ S/m and 5×10^{-11} F/m respectively. The loss tangent of the medium at a frequency of 100 MHz is

(a) 0.1 (b) 0.05 (c) 0.01 (d) 0.005

7.14 A uniform plane wave is said to be linearly polarised

(a) if magnitude of the electric field intensity of the wave remains constant, and direction of the field need not be constant
(b) if direction of the electric field intensity of the wave remains constant, and magnitude of the field may or may not be constant
(c) only if both magnitude and direction of electric field of the wave remain constant
(d) if the magnetic field intensity of the wave is a linear function of its electric field intensity

*7.15 A perfect conductor is located in a time-varying field. If $\mathbf{u}_n$ is a unit vector normal to the conductor surface, the boundary conditions on the surface of the conductor are

(a) $\mathbf{u}_n \times \mathbf{E} = 0$ (b) $\mathbf{u}_n \times \mathbf{D} = \rho_S$ (c) $\mathbf{u}_n \times \mathbf{H} = 0$ (d) $\mathbf{u}_n \cdot \mathbf{B} = 0$

*7.16 When a uniform plane wave from air is incident normally on the surface of a perfect conductor, wave reflection takes place. Which of the following statements regarding the reflected wave are correct?

(a) The electric field intensity gets reflected with change of sign, and the magnetic field intensity gets reflected without change of sign
(b) The electric field intensity gets reflected without change of sign, and the magnetic field intensity gets reflected with change of sign.
(c) The magnetic and electric fields are in the same phase.
(d) The magnetic and electric fields differ in phase by 90°.

*7.17 A uniform plane wave is incident normally on the interface of two perfect nonmagnetic dielectric media 1 and 2. The relative permittivity of Medium 1 is 4.0 and that of Medium 2 is 9.0. If the incident wave is in Medium 2, then

(a) the reflection coefficient is –0.2
(b) the reflection coefficient is 0.2
(c) the reflected power is 0.2 times the incident power
(d) the reflected power is 0.04 times the incident power

*7.18 A uniform plane wave of wavelength λ is incident normally at the interface of two perfect dielectric regions. The magnitude of reflection coefficient in the region where both incident and reflected waves are present is 0.25. In this region,

(a) standing wave ratio is 4.0
(b) standing wave ratio is 1.67
(c) distance between the points of maximum amplitude and minimum amplitude of the total voltage is $\lambda/4$
(d) distance between the points of maximum amplitude and minimum amplitude of the total voltage is $\lambda/2$

*7.19 The interface of air and a perfect conductor is on the $z = 0$ plane. A uniform plane wave from air is incident obliquely at the interface. The total (incident plus reflected) electric field intensity in air
(a) is a standing wave along z direction
(b) is a travelling wave along z direction
(c) is a standing wave along the direction parallel to the interface
(d) is a travelling wave along the direction parallel to the interface

*7.20 A uniform plane wave is incident obliquely at the interface of two perfect dielectric regions. The boundary conditions needed to obtain the relation between electric field intensity of the reflected wave and electric field intensity of the incident wave are
(a) the normal components of magnetic field intensity on two sides of the boundary are equal
(b) the tangential components of magnetic field intensity on two sides of the boundary are equal
(c) the normal components of electric field intensity on two sides of the boundary are equal
(d) the tangential components of electric field intensity on two sides of the boundary are equal

*7.21 The angle of incidence at which the reflection coefficient at the common boundary between two perfect dielectric regions becomes unity is called critical angle of incidence. For nonmagnetic regions, the magnitude of the reflection coefficient can become unity
(a) only if the incident wave is in the region of lower permittivity
(b) only if the incident wave is in the region of higher permittivity
(c) even when the angle of incidence is higher than the critical angle
(d) even when the angle of incidence is lower than the critical angle

*7.22 The angle of incidence at which the reflection coefficient at the common boundary between two perfect dielectric regions becomes zero is called Brewster angle. For nonmagnetic regions, a real value of Brewster angle exists
(a) only if the wave polarisation is parallel to the plane of incidence
(b) only if the wave polarisation is perpendicular to the plane of incidence
(c) only if the incident wave is in the region of higher permittivity
(d) irrespective of whether the incident wave is in the region of lower permittivity or higher permittivity

QUESTIONS

7.1 Mention the assumptions under which the one-dimensional wave equation $\partial^2 E_x/\partial z^2 = (\mu\varepsilon)(\partial^2 E_x/\partial t^2)$ for the electric field intensity E_x is obtained.

7.2 Show that $Ce^{jk(t-\sqrt{\mu\varepsilon}z)}$, where k is a real quantity, is a solution of the appropriate wave equation.

7.3 The magnetic field intensity in x direction varies with z, and also varies harmonically with time at ω rad/s. Write the one-dimensional wave equation of the field in a perfect dielectric medium. Show that the function $C\sin(\omega t - \sqrt{\mu\varepsilon z})$ is a solution of the wave equation.

7.4 Explain with necessary illustrations that the function $C\cos(\omega t - \beta z)$, where $\beta = \omega\sqrt{\mu\varepsilon}$, is a travelling wave that moves at a velocity v ' ' $\mu\varepsilon$ in the positive z direction.

7.5 The parameter β and the velocity v of a uniform plane wave are called *phase* constant and *phase* velocity. What is the significance of the adjective 'phase' in each case?

7.6 Distinguish between a travelling wave and a standing wave.

7.7 What is the ratio between the electric field intensity and the magnetic field intensity of a single plane electromagnetic wave called? What is its unit and how it is arrived at?

7.8 Why is the electric field described by $E_{x0}^{+}\cos(\omega t - \beta z)$ called a uniform plane wave?

7.9 Consider that the electric and magnetic field intensities in a uniform plane electromagnetic wave have components along both x and y directions. The total electric field is equal to the vector sum of the x and y components, and the total magnetic field is the vector sum of the component fields. What is the angle between the total electric field and total magnetic field in space, and what is the value of the ratio between the two quantities?

7.10 The amplitude of magnetic field intensity in a uniform plane electromagnetic wave travelling in a loss-free dielectric medium of intrinsic impedance η is H_{y0}. How much power is carried by the wave per unit area in the direction of propagation?

7.11 The propagation constant of a plane electromagnetic wave travelling in an imperfect dielectric medium is a complex quantity. What are the real and imaginary parts called, and what are their units?

7.12 Consider that a uniform plane wave is travelling in an imperfect dielectric medium. The average power flow per unit area at $z = 0$ is P_{av}. What is the power flow per unit area at $z = z'$ if the attenuation constant is α and the phase angle of the intrinsic impedance is 10°?

7.13 The loss tangent is defined as $\tan\delta = \sigma/(\omega\varepsilon)$ where the symbols have their usual meanings. Is the loss tangent inversely proportional to the frequency at all frequencies? Explain.

7.14 For efficient transmission of electromagnetic power through a dielectric medium, the loss tangent $\sigma/\omega\varepsilon$ of the medium should be low. Explain why.

7.15 Assuming the conductivity and permittivity of a dielectric medium to be independent of frequency, at what frequency the conduction current density in the medium becomes equal to the displacement current density? What is the value of loss tangent at this frequency?

7.16 Below what value of loss tangent a dielectric material can be considered as 'good'? What is the approximate formula of attenuation constant of a good dielectric?

7.17 Show that the attenuation constant of a plane wave in a good conductor is approximately equal to the phase constant.

7.18 The intrinsic impedance of a good conductor is a complex quantity with a real part and an imaginary part. What are these parts called, and what are the approximate formulae of these parts.

7.19 What is the approximate value of time-phase difference between the electric and magnetic field intensities of a uniform plane wave travelling in a good conductor? Which is the leading phase?

7.20 The amplitude of field intensities of an electromagnetic wave travelling in a good conductor decreases rapidly with distance, and the value of skin depth provides information regarding the rapidity of decrease of the amplitudes. What are the factors on which the skin depth depends? What is the amplitude of a field at a depth equal to the skin depth compared to its amplitude on the surface of the conductor?

7.21 The meaning of wave polarisation is different from dielectric polarisation and magnetic polarisation. Explain.

7.22 Describe the linear polarisation of a plane electromagnetic wave.

7.23 Describe the circular polarisation of a plane electromagnetic wave.

7.24 A plane electromagnetic wave incident normally on the surface of a perfect conductor gets totally reflected. Explain why the electric field intensity gets reflected with change of sign, whereas the magnetic field intensity gets reflected with the same sign.

7.25 A plane wave incident normally on the surface of a perfect conductor gets totally reflected into the region from where the incident wave has originated. The total electric and magnetic fields in the region are equal to sum of the respective incident and reflected fields. Describe the properties of the total fields.

7.26 When an electromagnetic wave is incident on the common boundary surface of two dielectric regions, reflection and transmission of waves occur. The amplitudes of field intensities of reflected and transmitted waves are expressed in terms of the fields of the incident wave by the use of boundary relations. State and explain the boundary relations.

7.27 Let the reflection coefficient at the common boundary surface of two perfect dielectric regions be Γ. If the power per unit area in the normally incident wave along its direction of propagation is P_1^+, what are the values of power per unit area in the reflected and transmitted waves along their respective direction of propagation?

7.28 The amplitudes of electric field intensities in a region where both incident and reflected waves are present are E_{x0}^+ and E_{x0}^-. Assuming the amplitudes to be real, what is the standing wave ratio?

7.29 The standing wave ratio is a measurable quantity. Its value provides information regarding the efficiency of transmission of electromagnetic power. Explain how.

7.30 A lossless dielectric region has both incident and reflected waves. The phasor forms of the incident and reflected waves are $E_{x0}^+ e^{-j\beta z}$ and $E_{x0}^- e^{j\beta z}$. Define reflection coefficient and wave impedance at any point in the region.

7.31 In the case of oblique incidence of a plane wave at an interface, the wave polarisation may be parallel, or perpendicular, or may have both types of polarisation. Define parallel and perpendicular polarisations of the wave with respect to the plane of incidence.

7.32 If the angle of incidence of a plane wave at the boundary surface of two perfect dielectric regions is equal to or higher than the critical angle of incidence, total reflection of the wave occurs. What is the necessary condition for a real critical angle to exist?

7.33 The angle of incidence of a plane wave at the common boundary surface of two lossless and non-magnetic dielectric regions is equal to the Brewster angle. What is the value of the reflection coefficient?

ANSWERS TO SELECTED QUESTIONS

7.1 Mention the assumptions under which the one-dimensional wave equation $\partial^2 E_x/\partial z^2 = (\mu\varepsilon)(\partial^2 E_x/\partial t^2)$ for the electric field intensity E_x is obtained.

Answer: The wave equation is obtained under the following assumptions.

1. The region of the field has no free charges and no losses.
2. The region is homogeneous and linear.
3. The electric field does not vary with the space variables x and y.

7.5 The parameter β and the velocity v of a uniform plane are called *phase* constant and *phase* velocity. What is the significance of the adjective 'phase' in each case?

Answer: The parameter β is called *phase constant* as it is a measure of the phase shift of a wave per unit distance in the direction of its travel. The velocity is termed as phase velocity because a phase point of the wave travels at this velocity or because the wave travels with a fixed phase constant in a given medium and at a given frequency.

7.6 Distinguish between a travelling wave and a standing wave.

Answer: One form of solution of the one-dimensional wave equation in E_x is

$$E_x = E_{x0} \cos(\omega t - \beta z)$$

where E_{x0} is amplitude and $\beta = \omega\sqrt{\mu\varepsilon}$ is the phase constant of the cosine wave. We find from the above expression that the maximum value of the field E_{x0} occurs when $z = \omega t/\beta = t/\sqrt{\mu\varepsilon}$. Accordingly, at $t = 0$ the position of maximum value of the field is at $z = 0$. Let P be the point on the peak of the wave at $t = 0$. At some later time $t = t_1$, the position of P will be at $z_1 = t_1/\sqrt{\mu\varepsilon}$. It means that after time t_1 the wave representing the field has moved by a distance z_1 from its position at $t = 0$. Hence, the cosine function is a travelling wave. The velocity of the wave is $z_1/t_1 = 1/\sqrt{\mu\varepsilon}$

Another form of solution of the wave equation is

$$E_x = E_{x0} \cos \omega t \cos \beta z$$

The field is a cosine wave along the spatial direction z. As the maximum value of the wave occurs at fixed values of z, the wave is a stationary wave or a standing wave. The amplitude of the wave is $E_{x0} \cos \omega t$

7.8 Why is the electric field described by $E^{+}_{x0} \cos(\omega t - \beta z)$ called a uniform plane wave?

Answer: The field is a travelling wave along positive z direction. The distribution of the field at any instant of time t is uniform over planes parallel to the x-y plane (constant z planes). Hence the waves are called uniform plane waves.

7.14 For efficient transmission of electromagnetic power through a dielectric medium, the loss tangent $\sigma/\omega\varepsilon$ of the medium should be low. Explain why.

Answer: If the numerator and denominator of loss tangent are multiplied by the electric field intensity, then the numerator will be equal to the conduction current density and the denominator will be the displacement current density. As the ohmic loss in the medium is proportional to the square of the conduction current density, a low value of conduction current density and hence a low value of loss tangent is desirable for efficient transmission of electromagnetic power.

7.19 What is the approximate value of time-phase difference between the electric and magnetic field intensities of a uniform plane wave travelling in a good conductor? Which is the leading phase?

Answer: The electric field intensity of a plane wave is equal to the product of the intrinsic field intensity and its magnetic field intensity. The intrinsic impedance of a good conductor is a complex quantity with the real part approximately equal to the positive imaginary part. Thus, the phase angle of the intrinsic impedance is $+\pi/4$ radian. Hence, the phase difference between the two fields is also $+\pi/4$ radian. The electric field leads the magnetic field.

7.23 Describe the circular polarisation of a plane electromagnetic wave.

Answer: The polarisation of a plane wave is described by the orientation of its electric field intensity. A circular polarised wave results from two components of electric fields that are perpendicular to each other and of same frequency. The magnitudes of the field components are equal and the time-phase difference between them is 90°. The resultant of the two components is of constant magnitude, and it rotates at a constant angular velocity equal to the angular frequency of the component fields. Thus, the tip of the resultant field describes a circle on the plane of the fields. The direction of rotation of the resultant field is from the leading field component toward the lagging field component.

7.26 When an electromagnetic wave is incident on the common boundary surface of two dielectric regions, reflection and transmission of waves occur. The amplitudes of field intensities of reflected and transmitted waves are expressed in terms of the fields of the incident wave by the use of boundary relations. State and explain the boundary relations.

Answer: The number of boundary relations at the common boundary surface between two dielectrics is four. Out of the four, only two relations are necessary to express the fields of the reflected and transmitted wave in terms of the corresponding fields of the incident wave. These are (1) the tangential electric field intensities on two sides of the boundary surface are equal, and (2) the tangential magnetic field intensities on two sides of the boundary surface are equal. On the side of the boundary surface where both the incident and reflected waves are present, the tangential fields (electric and magnetic) at the boundary are equal to the sum of the tangential fields of the two waves. Whereas on the other side of the boundary surface, the tangential fields are those of the transmitted wave, assuming that in this region only the transmitted wave is present.

7.29 The standing wave ratio is a measurable quantity. Its value provides information regarding the efficiency of transmission of electromagnetic power. Explain how.

Answer: Consider that in the region where both incident and transmitted waves are present, amplitude of the incident electric field is E_{i0} and amplitude of the reflected electric field is E_{r0}. Then, the standing wave ratio is defined as $(|E_{i0}|+|E_{r0}|)/(|E_{i0}|-|E_{r0}|)$. A high value of standing wave ratio means that the amplitude of the reflected field is high. As power per unit area of the wave is proportional to the square of electric field intensity, the value of reflected power is high and hence the value of transmitted power is low. Reverse is the case if the value of standing wave ratio is small. Thus, the standing wave ratio provides information about the efficiency of transmission of electromagnetic power.

7.31 In the case of oblique incidence of a plane wave at an interface, the wave polarisation may be parallel, or perpendicular, or may have both types of polarisation. Define parallel and perpendicular polarisations of the wave with respect to the plane of incidence.

Answer: The plane of incidence of a uniform plane wave is the plane containing a line normal to the interface of two dielectric media and a ray representing the direction of travel of the incident wave. If the incident electric field intensity is parallel to the plane of incidence, then the polarisation is parallel. And if the direction of the incident field intensity is perpendicular to the plane of incidence, the polarisation of the wave is said to be perpendicular. If the electric field of the incident wave is at an angle with the plane of incidence, the field is resolved into two components; one parallel to and the other perpendicular to the plane of incidence. In this case, the wave has both types of polarisation.

PROBLEMS

**7.1 The x-directed electric field intensity of a uniform plane wave travelling in a perfect dielectric medium is given by

$$E_x = E_{x0} \cos(\omega t - \beta z)$$

where $\beta = \omega\sqrt{\mu\varepsilon}$ is the phase constant. Obtain the expression for the magnetic field intensity of the wave, and the formula for the intrinsic impedance of the medium.

**7.2 The frequency of a uniform plane electromagnetic wave travelling in a perfect dielectric medium is 50 MHz. The relative permeability and permittivity of the medium are 1.0 and 4.0 respectively. (a) Calculate the phase velocity, phase constant and wavelength of the wave. (b) Find the intrinsic impedance of the wave.

*7.3 The x-directed electric field intensity of an electromagnetic wave travelling along positive z direction in air is 120 V/m at $t = 0$ and $z = 0$. Find the values of electric and magnetic field intensities of the wave at $z = 10$ m and $t = 0.5$ μs if the frequency is 25 MHz. What is the direction of the magnetic field?

**7.4 The instantaneous magnetic field intensity of a uniform plane electromagnetic wave travelling in air is given by $\mathbf{H}(t) = (0.08\mathbf{u}_x + 0.06\mathbf{u}_y) \cos(\omega t - \beta z)$ A/m. Calculate (a) the instantaneous electric field intensity, (b) the amplitudes of total electric and magnetic field intensities and (c) the angle between the total electric field and total magnetic field.

*7.5 The magnetic field intensity of a uniform plane wave in a perfect dielectric medium is given by

$$H_x = H_{x0} \cos(\omega t - \beta z)$$

Obtain the expression for instantaneous power per unit area of the wave and interpret the result.

**7.6 The amplitude of the y-directed electric field intensity propagating along the positive z direction in a perfect dielectric medium is 60 V/m. The relative permittivity and permeability of the medium are 3.6 and 1.0 respectively. The frequency is 15 MHz. (a) Express the electric and magnetic field intensities in real time and phasor forms. (b) Calculate the instantaneous and complex Poynting vectors.

***7.7 The propagation constant associated with a plane wave travelling in a dielectric medium having finite conductivity is given by

$$\bar{\gamma} = \sqrt{j\omega\mu(\sigma + j\omega\varepsilon)}$$

where the symbols have their usual meaning. The real part of $\bar{\gamma}$ is the attenuation constant α and the imaginary part β is the phase constant. Derive the expressions for α and β.

**7.8 The propagation constant of a plane wave in an imperfect dielectric medium is given by

$$\bar{\gamma} = \sqrt{j\omega\mu(\sigma + j\omega\varepsilon)}$$

If $\mu_r = 1.0$, $\varepsilon_r = 4.5$, $\sigma = 5 \times 10^{-5}$ S/m and $f = 25$ MHz, find the attenuation and phase constants.

**7.9 The x-directed electric field intensity phasor of a plane wave in a general medium is given by

$$E_x(z) = E_{x0}e^{-\bar{\gamma}z}$$

where $\bar{\gamma} = \sqrt{j\omega\mu(\sigma + j\omega\varepsilon)}$ is the propagation constant. Use Maxwell's equation $\nabla \times \mathbf{H} = (\sigma + j\omega\ )\mathbf{E}$, to obtain the expressions for the magnetic field intensity of the wave and intrinsic impedance of the medium.

**7.10 The intrinsic impedance of an imperfect dielectric medium is given by

$$\eta = \frac{j\omega\mu}{(\sigma + j\omega\varepsilon)}$$

Calculate the impedance using the data of Problem 7.8

**7.11 The attenuation constant of a plane electromagnetic wave in a general medium is

$$\alpha = \omega\sqrt{\frac{\mu\varepsilon}{2}(-1+\sqrt{1+(\frac{\sigma}{\omega\varepsilon})^2}$$

Show that for a good dielectric medium,

$$\alpha \approx \frac{\sigma}{2}\sqrt{\frac{\mu}{\varepsilon}}.$$

**7.12 If the loss tangent of a material is less than 0.01, then the material may be treated as a good dielectric. Find the value of frequency above which clay can be treated as a good dielectric. The conductivity and relative permittivity of clay are 5×10^{-3} S/m and 14.0 respectively. What are the approximate values of attenuation and phase constants of clay if the loss tangent is 0.01?

***7.13 If the loss tangent of a material is more than 100, then the material may be treated as a good conductor. Find the value of frequency below which sea water can be treated as a good conductor. The conductivity and relative permittivity of sea water are 5.0 S/m and 80.0 respectively. Calculate the values of (a) attenuation constant, (b) phase constant, (c) propagation constant, (d) skin depth, (e) phase velocity, and (f) wavelength if the loss tangent is 100?

**7.14 The intrinsic impedance of a medium in general is given by

$$\bar{\eta} = \sqrt{\frac{j\omega\mu}{(\sigma + j\omega\varepsilon)}}$$

Obtain approximate expressions for the intrinsic impedance, surface resistance and internal inductance of a good conductor.

**7.15 The conductivity of a conductor is 58 MS/m. Calculate the attenuation constant, phase constant, propagation constant, skin depth, and intrinsic impedance of a uniform plane wave of at a frequency of 0.5 MHz.

**7.16 A uniform plane wave travelling in air along the positive z direction is incident normally on a perfect conductor located at $z = 0$. The magnitude of the y-directed electric field at $t = 0$ and $z = 0$ is 25 V/m. The frequency of the incident wave is 15 MHz. Obtain expressions for the total electric field intensity and total magnetic field intensity in air.

***7.17 A uniform plane wave travelling in air along positive z direction is incident normally on the surface of a good conductor located at $z = 0$, as shown in Fig. P7.17. The conductivity of the conductor is 3.5×10^4 S/m, amplitude of electric field intensity on the surface of the conductor is 12 V/m, frequency of the wave is 150 kHz, and the relative permeability of the conductor is 1.0. As the conductor is not perfect, the field will travel into the conductor. In the conducting region, find (a) real time expressions of the magnetic and electric field intensities, (b) velocity of the wave, and (c) ratio between power per unit area of the wave at the skin depth and at $z = 0$.

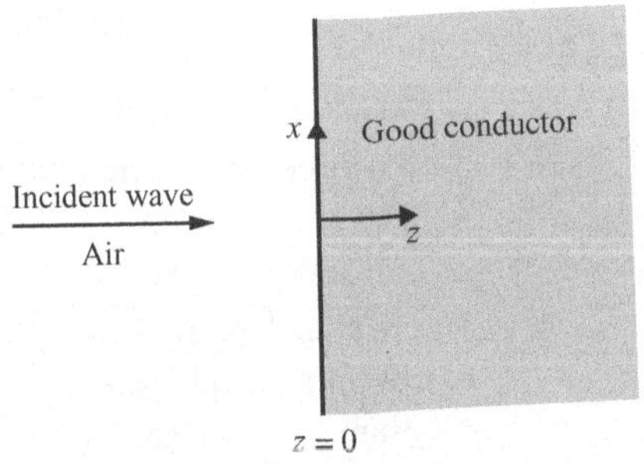

Fig. P7.17

**7.18 At z = 0, the electric field intensity of a plane wave travelling along z direction is given by

$$\mathbf{E} = E_0(\cos\omega t\,\mathbf{u}_x + \sin\omega t\,\mathbf{u}_y)$$

Prove that the wave polarisation is circular.

***7.19 For normal incidence of a uniform plane electromagnetic wave at the interface between two perfect dielectric regions, derive the formula for reflection coefficient in terms of the intrinsic impedances of the regions.

**7.20 A uniform plane wave travelling in air along the positive z direction is incident normally on the plane surface of a nonmagnetic perfect dielectric region covering the $z \geq 0$ region. The amplitude and frequency of the x-directed electric field intensity of the incident waves are 12 V/m and 1.6 MHz respectively. The relative permittivity of the dielectric is 3.6. (a) Calculate the reflection and transmission coefficients. (b) Write the expressions of electric field vectors of the incident and reflected waves in phasor form. (c) Write the expressions of the electric and magnetic field vectors of transmitted wave in phasor forms. (d) Find the transmitted power per unit area.

*7.21 The reflection coefficient at the boundary surface between two perfect dielectric regions is Γ. Prove that the average power per unit area transmitted across the boundary is $1 - \Gamma^2$ times the power per unit area of the incident wave.

*7.22 In a loss-free region where both incident and reflected waves are present, the amplitude of total electric field intensity at any point in the region is given by

$$E_{x0} = \sqrt{(E_{x0}^+)^2 + (E_{x0}^-)^2 + 2E_{x0}^+ E_{x0}^- \cos 2\beta z}$$

where E_{x0}^+ and E_{x0}^- are the amplitudes of the incident and reflected field intensities respectively, and z is the direction of travel of the incident wave. If S_R is the standing wave ratio, show that the magnitude of the reflection coefficient,

$$|\Gamma| = \frac{S_R - 1}{S_R + 1}$$

**7.23 A plane electromagnetic wave travelling along positive z direction in air gets partially reflected from a perfect dielectric surface located at $z = 0$. The amplitudes of electric field intensities of the incident and reflected waves are 48 V/m and 12 V/m respectively. (a) Find the standing wave ratio. (b) If the distance between the maximum and the nearest minimum field intensity of the total field is 1.5 m, find the frequency of the incident wave.

***7.24 The boundary surface between two perfect nonmagnetic regions 1 and 2 is located at the $z = 0$ plane. The relative permittivity of Region 1 is 2.25 and that of Region 2 is 4.0. The amplitude of electric field intensity of the incident wave originating in Region 1 is 16 V/m. (a) Show that the wave impedances on two sides of the boundary have the same value. (b) Find the wave impedance at $z = -\lambda/4$.

***7.25 The common boundary surface of a perfect nonmagnetic dielectric region and a perfect nonmagnetic conducting region is at $z = 0$. The x-z plane is the plane of incidence of a y-polarised incident wave. The angle of incidence is 30°, the amplitude of the electric field intensity of the incident wave is 25 V/m, and the relative permittivity of dielectric medium is 3.2. Find the amplitudes of the x and z components of the magnetic field intensities of (a) the incident wave and (b) the reflected wave.

**7.26 The common boundary surface of two perfect nonmagnetic dielectric regions 1 and 2 is at the $z = 0$ plane. The x-z plane is the plane of incidence, and the polarisation of the incident wave is perpendicular. The angle of incidence is 45°, and the amplitude of electric field intensity of the incident wave is 50 V/m. The relative permittivity of Region 1 is 2.5 and that of Region 2 is 3.6. Find the amplitude and mention the direction of the electric field intensity of the transmitted wave.

**7.27 The common boundary surface of two perfect nonmagnetic dielectric regions 1 and 2 is at the $z = 0$ plane. The x-z plane is the plane of incidence, and the polarisation of the incident wave is parallel. The angle of incidence is 45°, and the amplitude of electric field intensity of the incident wave is 50 V/m. The relative permittivity of Region 1 is 2.5 and that of Region 2 is 3.6. Find the amplitudes of the x and z components of electric field intensity of the transmitted wave.

*7.28 The reflection coefficient for oblique incidence of a perpendicularly polarised uniform plane wave at the interface of two nonmagnetic dielectric regions is given by

$$\Gamma_{\perp} = \frac{\sqrt{\varepsilon_{r1}}\cos\theta_i - \sqrt{\varepsilon_{r2}}\cos\theta_t}{\sqrt{\varepsilon_{r1}}\cos\theta_i + \sqrt{\varepsilon_{r2}}\cos\theta_t}$$

Show that a *real* value of critical angle of incidence exists only if the incident wave is in the region of higher permittivity.

**7.29 The interface of two perfect nonmagnetic dielectric regions is a plane surface. The relative permittivities of regions 1 and 2 are 3.0 and 4.0 respectively. (a) Show that no critical angle of incidence exists if the incident wave is in Region 1. (b) Find the critical angle of incidence if the incident wave is in Region 2. (c) Find the magnitude of reflection coefficient if the angle of incidence is greater than the critical angle by 5°.

**7.30 The permittivities of two nonmagnetic perfect dielectric media 1 and 2 are ε_1 and ε_2 respectively. (a) Find the ratio of permittivities of the two media if the Brewster angle is 30° and the incident wave is in Region 1. (b) For the ratio of permittivities found in (a), find the Brewster angle when the incident wave is in Region 2.

Chapter Eight

TWO-CONDUCTOR TRANSMISSION LINES

8.1 Introduction

Transmission lines are used to transmit electric power from an active source (input or sending end) to an electric load (output or receiving or load end). A two-conductor transmission line consists of two good conductors, which are separated by air or by an insulating material. The service connection used for supplying electric power from a 50 Hz or 60 Hz electrical distribution system to domestic or other types of loads is a simple example of a two-conductor transmission line. A cable which connects a television to a receiving antenna is also a two-conductor transmission line. In this case the frequency of the signals is in megahertz range. Telephone lines for transmission of speech; shielded cables used for interconnecting two electronic equipments; cables used for transmitting radio frequency (RF) signals from transmitters to antennas; high-voltage dc transmission lines are a few other examples of two-conductor transmission lines. Small lengths of low-loss lines are used as circuit elements such as inductance, capacitance and resonant circuits at very high frequencies.

A two-conductor transmission line is represented by the resistance and external inductance of both the lines, and capacitance and conductance between the lines. The conductance is reciprocal of insulation resistance between the lines. The representation of a transmission line by its constants is its circuit model. The mathematical model of a transmission line is derived from its circuit model for analytical study.

A transmission line may be classified as *short* or *medium* or *long* depending on its length relative to the wavelength of the signal it carries. A transmission line of a few kilometres may be treated as a short line at power frequency (50 Hz), whereas a line of a few metres has to be treated as a long line if the frequency of signal is in the megahertz range. Voltage and current in a long transmission line travel from the source to load in the form of waves. The velocity of travel depends mainly on the inductance and capacitance per unit length of the line. Lumped parameters can be used for circuit representation of short and medium transmission lines. However, distributed parameters are used for long transmission lines. The equations for voltage and current of a long transmission line are second order partial differential equations. The solutions of these equations show that voltage and current travel as waves along the line.

The time variation of signals carried by a transmission line may be nonsinusoidal such as pulses, coded messages, audio signals, etc. Many transmission lines also carry electric power and signals that vary sinusoidally with time. The brief study of transmission lines presented here is restricted to steady state analysis of symmetrical two-conductor long transmission lines carrying signals which vary sinusoidally with time

8.2 Types of Two-Conductor Transmission Lines

Some types of two-conductor transmission lines are shown in Figs. 8.1 to 8.6. The parallel-conductor transmission line consists of two parallel cylindrical conductors, either solid or stranded, separated by a dielectric medium or air. Overhead transmission lines with two bare conductors are used for transmission of dc power at high voltages and in single-phase ac distribution systems. Insulated twin-conductors with weatherproof covering find application as service lines for the supply of single-phase power to domestic and other small-scale consumers.

The two-conductor parallel transmission lines shown in Fig. 8.1(*b, c, d*) are embedded in an insulating material (generally vinyl or foam). The 'tubular' type and the 'window' type of lines have lower dielectric loss compared to the loss in the 'flat' type of line. However, the 'flat' type is cheaper.

Parallel transmission lines are used as leads between a receiving antenna and television, and for general purposes. These are suitable at frequencies up to 500 MHz. These are not suitable at higher frequencies because power loss takes place due to radiation at bends and joints thus affecting the efficiency of transmission.

The electric and magnetic fields of *open* two-conductor parallel lines extend far beyond their physical location.

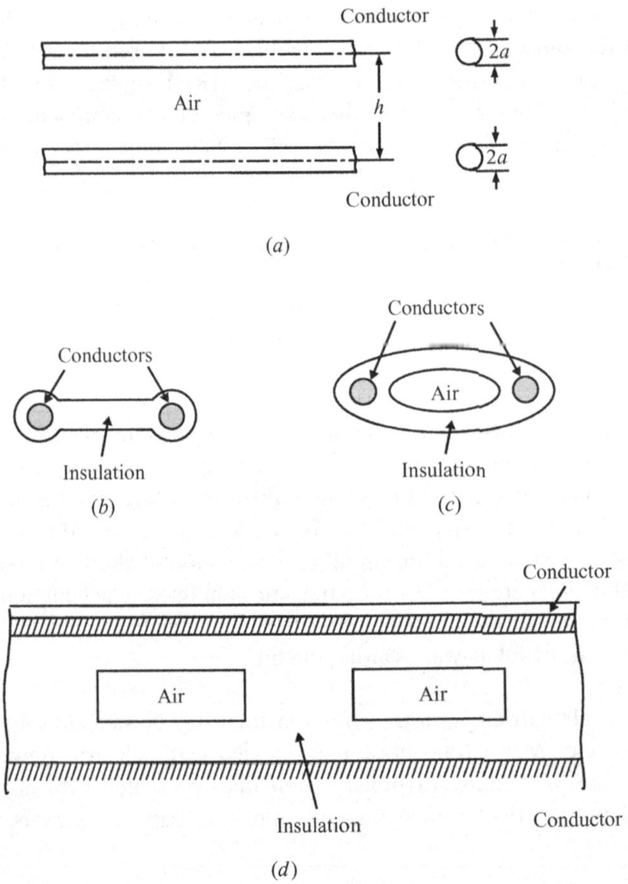

Fig. 8.1 Parallel-wire transmission lines. (a) Overhead line, (b) flat type, (c) Tubular Type, and (d) Window Type.

If these carry radio frequency signals, then there will be interference to and from neighbouring open lines because of inductive and capacitive coupling. In order to reduce such interference, these lines are enclosed in metal sheaths, as shown in Fig. 8.2. The effectiveness of shielding depends on the thickness of the sheath in comparison to its skin depth. Since the electric fields of the lines covered by dielectric materials are mostly confined to the dielectric region, partial shielding is provided by the dielectric cover.

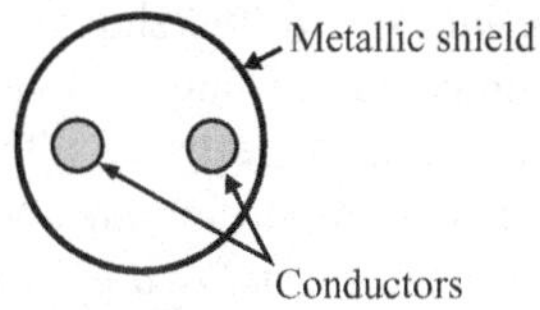

Fig. 8.2 A shielded transmission line

In telephone systems, multiple pairs of lines with insulated conductors are laid under the ground. Each pair is shielded to reduce interference, and all the pairs are covered with insulation to form a cable.

The *coaxial* transmission line (cables), shown in Fig. 8.3, are used to transmit electric power and to send RF signals at various frequencies. The electric and magnetic fields in a coaxial cable are confined to the region between the conductors.

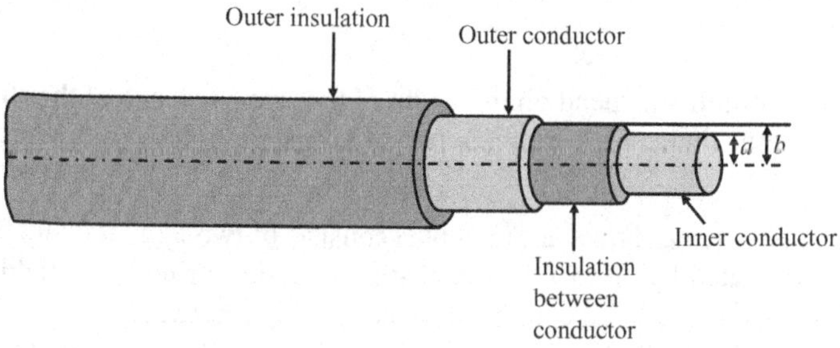

Fig. 8.3 Illustration of a coaxial transmission line

The outer conductor acts as a shield. Thus, practically there is no radiation from a coaxial cable.

Therefore, coaxial cables are more efficient compared to the open (unshielded) lines.

The efficiency of transmission is not affected even if a coaxial cable is twisted or bent up to certain extent. Another property of a coaxial cable is that the shielding effect of the outer conductor greatly reduces interference with and from signals of open lines located outside the cable, even when the cable is tied to metallic structures. Coaxial cables of very low dielectric loss can be used to transmit signals at frequencies up to five gigahertz.

A *stripline* (also called symmetrical stripline) consists of a central conducting strip and two outer conducting planes, as illustrated in Fig. 8.4. The strip is parallel to, and symmetrically located between the outer conductors. The region between the strip and the outer conductors is filled with a low-loss homogenous dielectric substance.

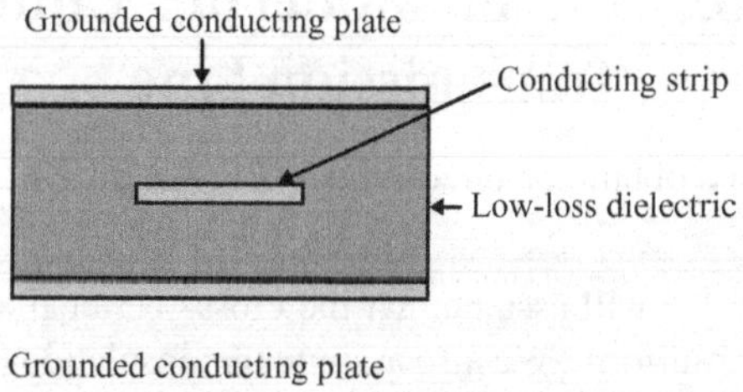

Fig. 8.4 A strip transmission line

The constants of a strip line depend on the width of the strip, distance of the strip from the outer planes, and permittivity of the dielectric medium surrounding the strip.

The *microstrip* line shown at Fig. 8.5 consists of a thin conducting strip and a conducting ground plane mounted on two sides of a dielectric substrate. The electric-field lines between the planes are partly in air and partly in the dielectric. So, the field lines are discontinuous at the interface. The microstrip line is a inhomogeneous transmission line. Microstrip lines if enclosed in a metallic shield to reduce radiation from its sides are termed as *enclosed microstrip* lines.

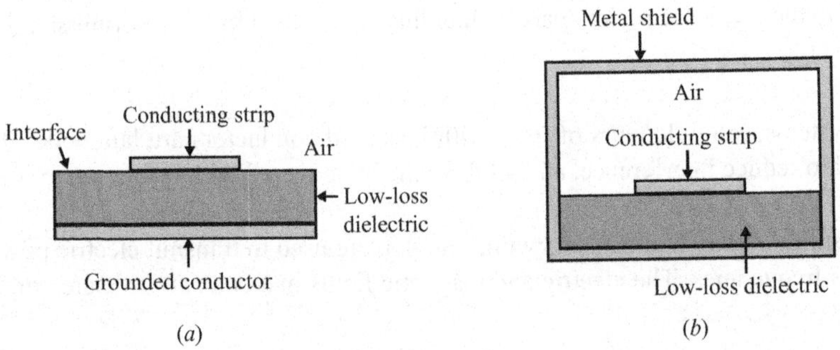

Fig. 8.5 Illustrations of a microstrip transmission lines. (*a*) Unshielded and (*b*) shielded.

The constants of a microstrip line depend on the width of the strip, distance of the strip from the ground plane and permittivity of the dielectric. Striplines and microstrip lines are used in microwave systems.

The parallel-plate transmission line shown at Fig. 8.6(*a*) consists of two large flat and thin conducting parallel plates of equal size separated by air or by an insulating material. Fringing of field at the edges may be neglected if width of the plates w is large compared to the spacing s between them $(w \gg s)$. If the width of the conducting strips is small, as illustrated in Fig. 8.6(*b*), fringing of the fields has to be considered in computing the constants. Parallel-plane lines are rarely used as transmission lines in practice. However, its properties are utilised in the analysis of striplines and microstrip lines.

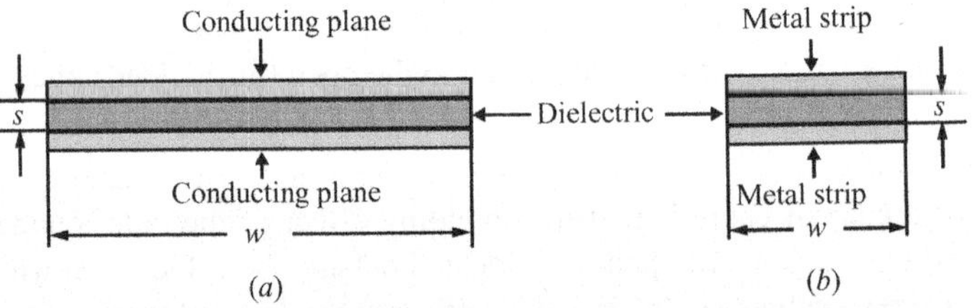

Fig. 8.6 Parallel-plane transmission lines

8.3 Circuit Model of a Uniform Two-Conductor Transmission Line

To obtain the circuit model of a long two-conductor transmission line,

we will assume that the cross-sectional area of the conductors and spacing between the conductors are uniform. We also assume that insulation (dielectric) between the conductors is homogenous. A transmission line fulfilling these conditions is considered to be a uniform transmission line. Thus, parallel-wire, coaxial, and parallel-plane transmission lines can be treated as uniform lines.

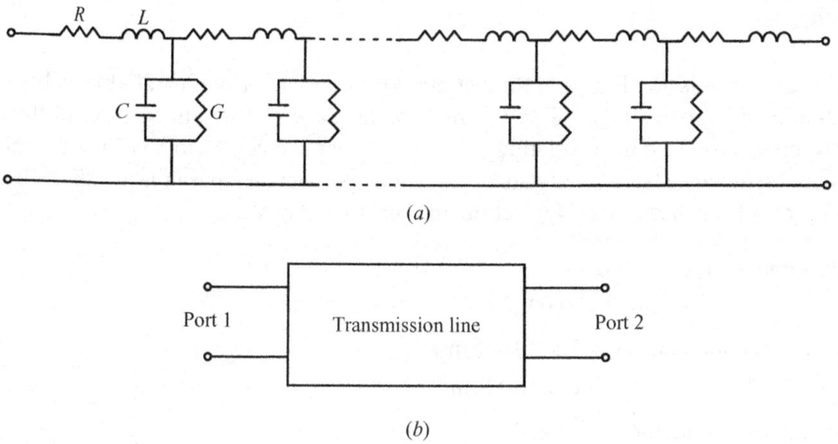

Fig. 8.7 (*a*) A circuit model of a two-conductor transmission line and (*b*) Two-port equivalent of (*a*).

Generally, both the conductors are of the same material.

A transmission line has four basic constants. These are series resistance per unit length *R*, inductance per unit length *L*, capacitance per unit length *C*, and shunt conductance per unit length *G*. The internal inductance, which is small compared to the external inductance, may be neglected. A long transmission line may be regarded as made up of a large number of very small lengths, all connected in series. Accordingly, the constants of a uniform transmission line are assumed to be distributed uniformly throughout its length, as shown in Fig. 8.7(*a*). This is the circuit model of the line and it is equivalent to a two-port network, as illustrated in Fig. 8.7(*b*). The ports are interchangeable as the network is symmetrical.

The formulae for the constants of parallel-wire, coaxial, and parallel-plane transmission lines are given in the following table. The resistance formulae are for both the conductors and represent values at high frequencies. Note that at high frequencies the skin depth in good conductors is small.

If the radius of a round conductor is greater than ten times the skin depth ($a > 10\delta$), it can be treated as a flat conductor with width equal to the circumference of the conductor and thickness equal to the skin depth.

The value of resistance of a flat conductor of unit length and width w is equal to its surface resistance divided by w. The surface resistance per unit length and unit width of a flat conductor is nearly equal to $1/(\sigma\delta)$. The dimensions of the three transmission lines are shown in Fig 8.1(*a*), Fig 8.3 and Fig. 8.6(*a*). The symbol σ_c and σ_d appearing in the formulae denote conductivities of the conductors and the dielectric respectively.

Transmission line constants per metre length

Line types	Resistance (*R*)	Inductance (*L*)	Capacitance (*C*)	Conductance (*G*)
Parallel wire	$\dfrac{1}{\pi a\delta\sigma_c}$	$\dfrac{\mu}{\pi}\ln\dfrac{h+\sqrt{h^2-4a^2}}{2a}$	$\dfrac{\pi\varepsilon}{\ln\left[\left(h+\sqrt{h^2-4a^2}\right)/2a\right]}$	$\dfrac{\pi\sigma_d}{\ln(h/a)}$
Coaxial	$\dfrac{1}{2\pi\delta\sigma_c}\left(\dfrac{1}{a}+\dfrac{1}{b}\right)$	$\dfrac{\mu}{2\pi}\ln\left(\dfrac{b}{a}\right)$	$\dfrac{2\pi\varepsilon}{\ln(b/a)}$	$\dfrac{2\pi\sigma_d}{\ln(b/a)}$
Parallel-plate	$\dfrac{2}{w\delta\sigma_v}$	$\dfrac{\mu s}{w}$	$\dfrac{\varepsilon w}{s}$	$\dfrac{w\sigma_d}{s}$

EXAMPLE 8.1

A two-conductor transmission line operating at 500 kHz consists of two parallel round conductors each of radius 2.0 mm and conductivity 3.8 × 10^7 S/m. The distance between the centres of the conductors is 20.0 mm. The conductivity of the insulating material between the conductors is 5.0 × 10^{-8} S/m. Calculate the inductance, capacitance, series resistance, and shunt conductance per unit length of the transmission line, assuming $\mu_r = 1.0$ and $\varepsilon_r = 4.2$. Neglect the internal inductance.

Radius of each conductor, $a = 0.002$ m

Distance between the conductors, $h = 0.02$ m

Conductivity of the conductors, $\sigma_c = 3.8 \times 10^7$ S/m

Conductivity of the dielectric, $\sigma_d = 5.0 \times 10^{-8}$ S/m

Skin depth for a good conductor,

$$\delta = \frac{1}{\sqrt{\pi f \mu \sigma_c}} = \frac{1}{\sqrt{\pi(500\times10^3)(4\pi\times10^{-7})(3.8\times10^7)}} = 0.115\times10^{-3}\text{ m}$$

Series resistance per unit length of the line, $R = \dfrac{1}{\pi a \delta \sigma_c}$

$$= \frac{1}{\pi(0.002)(0.115\times10^{-3})(3.8\times10^7)} = 0.138\ \Omega/\text{m}$$

Inductance per unit length of the line, $L = \dfrac{\mu}{\pi}\ln\dfrac{h+\sqrt{h^2-4a^2}}{2a}$

$$= \left(\frac{4\pi\times10^{-7}}{\pi}\right)\ln\frac{0.02+\sqrt{0.02^2-4(0.002)^2}}{2(0.002)} = 9.21\times10^{-7}\text{ H/m}$$

Capacitance per unit length of the line, $C = \dfrac{\pi\varepsilon}{\ln\left[\left(h+\sqrt{h^2-4a^2}\right)/2a\right]}$

$$= \frac{\pi(4.2)(8.84\times10^{-12})}{\ln\left[0.02+\sqrt{0.02^2-4(0.002)^2}\right]-\ln(2\times0.002)} = 50.66\times10^{-12}\text{ F/m}$$

Shunt conductance per unit length of the line,

$$G = \frac{\pi\sigma_d}{\ln(h/a)} = \frac{\pi(5\times10^{-8})}{\ln(0.02/0.002)} = 6.82\times10^{-8}\text{ S/m}$$

EXAMPLE 8.2

The radii of the inner and outer conductors of a coaxial transmission line are 2.0 mm and 4.0 mm respectively. The conductivity of the conductors is 5.8 × 10^7 S/m. The relative permittivity, relative permeability, and conductivity of the insulation between the conductors are 3.5, 1.0, and 6.0 × 10^{-8} S/m respectively. Find series resistance, inductance, capacitance, and shunt conductance per unit length of the line at a frequency of 150 kHz.

Radius of inner conductor, $a = 0.002$ m

Radius of outer conductor, $b = 0.004$ m

Conductivity of the conductors, $\sigma_c = 5.8 \times 10^7$ S/m

Conductivity of the dielectric, $\sigma_d = 6.0 \times 10^{-8}$ S/m

Skin depth, $\delta = \dfrac{1}{\sqrt{\pi f \mu \sigma_c}} = \dfrac{1}{\sqrt{\pi(150\times10^3)(4\pi\times10^{-7})(5.8\times10^7)}} = 0.170\times10^{-3}$ m

Series resistance per unit length of the line, $R = \dfrac{1}{2\pi\delta\sigma_c}\left(\dfrac{1}{a}+\dfrac{1}{b}\right)$

$$= \frac{1}{(2\pi)(0.170\times10^{-3})(5.8\times10^7)}\left(\frac{1}{0.002}+\frac{1}{0.004}\right) = 0.012\ \Omega/\text{m}$$

Inductance per unit length of the line,

$$L = \frac{\mu_0}{2\pi}\ln\frac{b}{a} = \frac{4\pi\times10^{-7}}{2\pi}\ln(4/2) = 1.386\times10^{-7}\ \text{H/m}$$

Capacitance per unit length of the line,

$$C = \frac{2\pi\varepsilon}{\ln(b/a)} = \frac{(2\pi)(3.5\times8.85\times10^{-12})}{\ln(4/2)} = 280.78\times10^{-12}\ \text{F/m}$$

Shunt conductance per unit length of the line,

$$G = \frac{\pi\sigma_d}{\ln(b/a)} = \frac{\pi(6.0\times10^{-8})}{\ln(4/2)} = 54.39\times10^{-8}\ \text{S/m}$$

8.4 Uniform Transmission Line Neglecting Losses

Some of the basic properties of a uniform two-conductor transmission line, such as variation of voltage and current along the line, relation between voltage and current, reflection and refraction phenomena at a discontinuity, etc., can be studied by analysing an ideal transmission line.

A transmission line that has negligible ohmic and dielectric losses may be treated as an ideal line. The series resistance and shunt conductance of such a line are assumed to be zero.

Accordingly, the equivalent circuit of an ideal transmission line is as shown in Fig. 8.8.

If a time-changing voltage (including sudden application of a dc voltage) is applied to the input terminals of the line, time-changing current will flow through the first inductor. However, voltage across the first capacitor cannot build up immediately because of delay caused by the induced voltage across the inductor. Similarly, the charging of the second capacitor is delayed further due to the presence of the second inductor, and so on. Thus, the voltage and current through the transmission line travel in the form of waves. The velocity of travel depends on the constants of the line. Partial differential equations relating voltage and current of the line are derived by

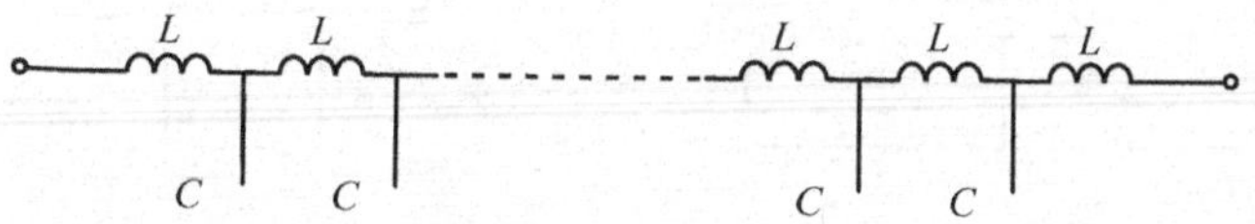

Fig. 8.8 Circuit model of a lossless two-conductor transmission line

considering a small element of the line. The equations are designated as *Transmission Line Equations*. The transmission line equations are of first order. Simultaneous solution of the transmission line equations yields second order partial differential equations. The second order equations are known as *wave equations*.

Transmission Line Equations and Wave Equations

In order to derive the transmission line equations, let us consider an infinitesimal length Δz at any position z of the line. If L and C are inductance and capacitance per metre length of the line, the constants of a line of length Δz are equal to $L\Delta z$ and $C\Delta z$. If we wish to represent the line element by a symmetrical network, there are two ways of doing so. The two configurations are shown in Fig. 8.9. In the model shown at Fig. 8.9(*a*), the inductance is divided into two halves and the capacitance is connected at the mid-point. In the second model shown at Fig. 8.9(b), the capacitance is divided into two halves and the inductor is connected between them. We will use the model shown at Fig. 8.9(*a*). Let us assume that the voltage and current at z are V and I respectively. The voltage at $z + \Delta z$ would be different from V because of the induced voltage across the inductor. The current at $z + \Delta z$ would also be different from I because of the charging current in the capacitor. At $z + \Delta z$, let the voltage and current be $V - \Delta V$ and $I - \Delta I$. The voltage difference ΔV between the two ends, which is equal to the sum of the voltage drops in the two halves of the inductors, is given by

$$(V-\Delta V)-V=-\Delta V=\left(\frac{1}{2}L\Delta z\right)\left[\frac{\partial I}{\partial t}+\frac{\partial(I-\Delta I)}{\partial t}\right] \tag{1}$$

Partial derivatives have been used because the induced voltages in the inductors are due to the rate of change of current with time only. The change of voltage ΔV in (1) is also equal to the rate of change of V with z, i.e., $\partial V/\partial z$, multiplied byΔz. That is,

$$\Delta V=\frac{\partial V}{\partial z}\Delta z \tag{2}$$

From (1) and (2), we get

$$\frac{\partial V}{\partial z}\Delta z=-L\frac{\partial I}{\partial t}\Delta z+\frac{1}{2}L\frac{\partial(\Delta I)}{\partial t}\Delta z \tag{3}$$

In the limit Δz tending to zero, ΔI also approaches zero and hence we can neglect the second term on rhs of (3). Thus, we get

$$\frac{\partial V}{\partial z}=-L\frac{\partial I}{\partial t} \tag{4}$$

The difference between the current at the ends of Fig. 8.9(*a*) is equal to the charging current in the capacitor. Accordingly,

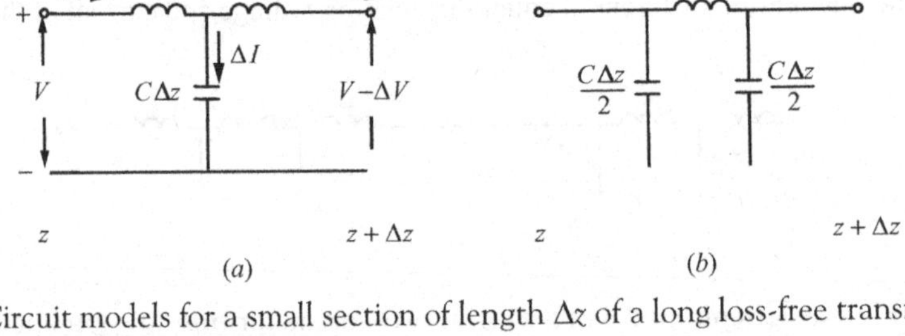

Fig. 8.9 Circuit models for a small section of length Δz of a long loss-free transmission line

$$(I - \Delta I) - I = -\Delta I = (C\Delta z)\frac{\partial V}{\partial t}$$

Or

$$\frac{\partial I}{\partial z}\Delta z = -(C\Delta z)\frac{\partial V}{\partial t}$$

$$\Rightarrow \quad \frac{\partial I}{\partial z} = -C\frac{\partial V}{\partial t} \tag{5}$$

We find from equations (4) and (5) that the rate of change of voltage along the line is proportional to the time rate of change of current and vice-versa. These are the *transmission line equations* of an ideal line. These are also known as *telegraph equations*. Both equations contain V and I. In order to separate the two quantities, let us partially differentiate (4) with respect to z to get

$$\frac{\partial^2 V}{\partial z^2} = -L\frac{\partial}{\partial z}\left(\frac{\partial I}{\partial t}\right) = -L\frac{\partial}{\partial t}\left(\frac{\partial I}{\partial z}\right) \tag{6}$$

By substituting (5) in (6), we obtain

$$\frac{\partial^2 V}{\partial z^2} = LC\frac{\partial^2 V}{\partial t^2} \tag{7}$$

It can similarly be shown that

$$\frac{\partial^2 I}{\partial z^2} = LC\frac{\partial^2 I}{\partial t^2} \tag{8}$$

Equations (7) and (8) are second order partial differential equations in V and I which are analogous to the wave equation in a loss-free medium derived in Chapter 7. So, these are the wave equations in voltage and current of an ideal transmission line.

General Solutions of Wave Equations: Wave Velocity

The general form of solutions of (7) and (8) are given by

$$V = f_1(vt - z) + f_2(vt + z)$$

$$I = f_3(vt - z) + f_4(vt + z)$$

where

$$v = 1/\sqrt{LC} \tag{9a}$$

That these solutions satisfy the wave equations can be verified by using them in the respective wave equations. Considering the solution for V, we have

$$\frac{\partial^2 V}{\partial t^2} = v^2[f_1''(vt - z) + f_2''(vt + z)]$$

where the double primes denote partial differentiation of the functions twice. Next partially differentiating the function twice with z, we get

$$\frac{\partial^2 V}{\partial z^2} = [f_1''(vt - z) + f_2''(vt + z)] = \frac{1}{v^2}\frac{\partial^2 V}{\partial t^2} = LC\frac{\partial^2 V}{\partial t^2}$$

which is the wave equation in V.

The functions with argument $vt - z$ are waves travelling along positive z direction and the functions with argument $vt + z$ are waves travelling along negative z direction. The velocity of the waves can be found by considering that a point of a function, say of $vt - z$, remains stationary relative to an observer. The condition for this is that $vt - z$ must also remain constant relative to the observer. Hence,

$$vt - z = \text{a constant.}$$

By differentiating with t, we have

$$\frac{dz}{dt} = v$$

As dz/dt has the dimension of velocity, v gives velocity of the observer. Hence, this is also velocity of the wave. Thus from (9*a*),

$$\text{velocity of the wave, } v = 1/\sqrt{LC} \tag{9b}$$

This is analogous to the velocity of propagation of a uniform plane wave in a loss-free medium.

Solution of the Wave Equation for Sinusoidally Time-Varying Voltage

In communication systems a variety of time-changing signals are sent through transmission lines. We will limit our study to transmission lines carrying sinusoidally time-varying signals. Sinusoidally time-varying voltages and currents when represented by the use of the complex exponential $e^{j\omega t}$ are given by

$$\bar{V} = \bar{V}(z)e^{j\omega t} \tag{10}$$

$$\bar{I} = \bar{I}(z)e^{j\omega t} \tag{11}$$

where the phasors $\bar{V}(z)$ and $\bar{I}(z)$ are functions of z only.

The equations for the sinusoidally time-varying case are obtained by substituting both (10) and (11) in (4) as well as (5). The transmission line equations are given by

$$\frac{\partial \bar{V}(z)}{\partial z} = -j\omega L\bar{I}(z) \tag{12}$$

$$\frac{\partial \bar{I}(z)}{\partial z} = -j\omega C\bar{V}(z) \tag{13}$$

The wave equations in V and I are found by using (10) in (7) and (11) in (8). The wave equations for voltage and current are found to be

$$\frac{\partial^2 \bar{V}(z)}{\partial z^2} = -\omega^2 LC\bar{V}(z) \tag{14}$$

$$\frac{\partial^2 \bar{I}(z)}{\partial z^2} = -\omega^2 LC\bar{I}(z) \tag{15}$$

The second order partial differential equations (14) and (15) are functions of only one variable in z. Hence they can be treated as ordinary differential equations. As equations (14) and (15) are of identical form, their solutions will also be of the same form. Therefore, solution of only one equation is necessary to study its properties. We will use the solution of (14). The exponential forms of solution of (14) is

$$V\,z) = V_{\upsilon}' e^{-j\;z\;\;p)} + V_{\upsilon} e^{+j\;z\;\;n)} \tag{16}$$

where $$\beta \quad \omega \quad LC \tag{17}$$

The complex form of the solution is obtained by introducing the factor $e^{j\omega t}$ into (16). Accordingly,

$$\bar{V} = V_0^+ e^{j(\omega t - \beta z + \psi_p)} + V_0^- e^{j(\omega t + \beta z + \psi_n)} \tag{18}$$

The first term of (18) is the voltage wave travelling along positive z direction, whereas the second term is the voltage wave travelling along negative z direction. The amplitude and phase angle of the positively travelling voltage wave are V_0^+ and ψ_p, and the amplitude and phase angle of the negatively travelling wave are V_0^- and ψ_n. Thus, denoting the positively and negatively travelling voltage waves by $\bar{V}^+$ and $\bar{V}^-$ respectively, we have

$$\bar{V}^+ = V_0^+ e^{j(\omega t - \beta z + \psi_p)} = (V_0^+ e^{j\psi_p})\, e^{j(\omega t - \beta z)} = \bar{V}_0^+ e^{j(\omega t - \beta z)} \tag{19}$$

$$\bar{V}^- = V_0^- e^{j(\omega t + \beta z + \psi_n)} = (V_0^- e^{j\psi_n}) e^{j(\omega t + \beta z)} = \bar{V}_0^- e^{j(\omega t + \beta z)} \tag{20}$$

where $\bar{V}_0^+ = V_0^+ e^{j\psi_p} = V_0^+ (\cos\psi_p + j\sin\psi_p)$,

$$\bar{V}_0^- = V_0^- e^{j\psi_n} = V_0^- (\cos\psi_n + j\sin\psi_n)$$

are complex quantities.

The real parts of (19) and (20) are:

$$V^+ = V_0^+ \cos(\omega t - \beta z + \psi_p) \tag{21}$$

$$V^- = V_0^- \cos(\omega t + \beta z + \psi_n) \tag{22}$$

Phase Constant, Phase Velocity, Wavelength

The function βz gives the phase of the voltages for any value of z relative to their phases at $z = 0$. The parameter β, defined by (17), is called *phase constant.*

Phase constant, $\beta = \omega\sqrt{LC}$

As defined by (9), the velocity of the wave, $v = \dfrac{1}{\sqrt{LC}} = \dfrac{\omega}{\beta}$

The velocity of wave propagation is more specifically called *phase velocity.*

As the waves repeat after every change of βz by 2π radians, the wavelength of the wave,

$$\lambda = \frac{2\pi}{\beta} = \frac{2\pi}{\omega\sqrt{LC}} = \frac{v}{f} \tag{23}$$

The electric and magnetic field intensities produced by the voltage and current waves also travel in the z direction and at the same phase velocity v, as $LC = \mu\varepsilon$.

As the field due to the voltage and current are transverse to the direction of propagation, the waves are known as *transverse* electromagnetic waves, or TEM waves in abbreviated form. The waves are uniform plane waves for a uniform transmission line.

Relation between Voltage and Current: The Characteristic Impedance

One method of deriving the relation between the voltage and current in the wave travelling along positive z direction is by substituting the expression of V^+ given by (21) in (5) and then solving the equation. Thus, denoting the instantaneous current associated with the wave by I^+, we have

$$\frac{\partial I^+}{\partial z} = -C\frac{\partial V^+}{\partial t} = -CV_0^+\frac{\partial}{\partial t}[\cos(\omega t - \beta z + \psi_p)] = \omega C V_0^+ \sin(\omega t - \beta z + \psi_p)$$

$$I^+ = \omega C V_0^+ \int \sin(\omega t - \beta z + \psi_p)\,dz = \frac{\omega C}{\beta}[V_0^+ \cos(\omega t - \beta z + \psi_p)]$$

The expression within the square bracket represents the positively travelling voltage wave. Hence, we have

$$I^+ = \frac{\omega C}{\beta} V^+$$

Or

$$\frac{V^+}{I^+} = \frac{\beta}{\omega C} = \frac{\omega\sqrt{LC}}{\omega C} = \sqrt{\frac{L}{C}} \tag{24}$$

As the lhs of this relation in (24) is equal to the ratio between voltage and current, the rhs must be an impedance parameter. The impedance is called *characteristic impedance* and its unit is ohm.

It is generally denoted by Z_0. The characteristic impedance of a two-conductor, uniform and ideal transmission line therefore is,

$$Z_0 = \sqrt{\frac{L}{C}} \tag{25}$$

We find that the characteristic impedance of a lossless transmission line is a function of its inductance and capacitance per unit length. The characteristic impedance for a given line configuration is independent of the line length. It is a real quantity and hence purely resistive. It is also called *characteristic resistance*. The voltage and current waves of an ideal line are in phase. From (24) and (25), we get

$$\frac{\bar{V}^+}{\bar{I}^+} = Z_0 \tag{26}$$

Also

$$\frac{\bar{V}_0^+}{\bar{I}_0^+} = \frac{V_0^+}{I_0^+} = Z_0 \tag{27}$$

The relation between voltage and current of the negatively travelling wave can be derived in the same way as for the positively travelling wave. It can be shown that

$$-\frac{\bar{V}^-}{\bar{I}^-} = -\frac{V_0^-}{I_0^-} = Z_0 \tag{28}$$

It may be noted that Z_0 is voltage-to-current ratio of a *single* wave.

EXAMPLE 8.3

A two-conductor transmission line is excited by a 50 MHz source. The inductance and capacitance per metre length of the line are 300 nH and 120 pF respectively. Calculate the characteristic impedance of the transmission line. Calculate also the phase constant, phase velocity and wavelength of the voltage wave.

Frequency, $f = 50 \times 10^6$ Hz

Inductance, $L = 300 \times 10^{-9}$ H/m

Capacitance, $C = 120 \times 10^{-12}$ F/m

Characteristic impedance, $Z_0 = \sqrt{\dfrac{L}{C}} = \sqrt{\dfrac{300\times10^{-9}}{120\times10^{-12}}} = 50.0\ \Omega$

Phase constant, $\beta = \omega\sqrt{LC} = 2\pi(50\times10^6)\sqrt{(300\times10^{-9})(120\times10^{-12})} = 1.885$ rad/m

Phase velocity, $v = \dfrac{1}{\sqrt{LC}} = \dfrac{1}{\sqrt{(300\times10^{-9})(120\times10^{-12})}} = 1.667\times10^8$ m/s

Wavelength, $\lambda = \dfrac{v}{f} = \dfrac{1.667\times10^8}{50\times10^6} = 3.334$ m

---

EXAMPLE 8.4

The inductance and capacitance per metre length of a two-conductor transmission line that is assumed to be lossless are 150 nH and 50 pF respectively. The voltage at the input end of the line is given by $5\ 0\cos(20\pi\times10^6)t$ V. Obtain expressions for the voltage and current waves produced by the source in real time and in phasor forms, assuming that the line is in positive z direction and the input point is at $z = 0$.

Inductance, $L = 150.0\times10^{-9}$ H/m

Capacitance, $C = 50.0\times10^{-12}$ F/m

Radian frequency, $\omega = 20\pi\times10^6$ rad/s

Amplitude of the voltage, $V_0^+ = 5.0$ V

Phase constant, $\beta = \omega\sqrt{LC} = (20\pi\times10^6)\sqrt{(150\times10^{-9})(50\times10^{-12})} = 1.721$ rad/m

Characteristic impedance, $Z_0 = \sqrt{\dfrac{L}{C}} = \sqrt{\dfrac{150\times10^{-9}}{50\times10^{-12}}} = 54.77\ \Omega$

The expression of the voltage wave travelling along positive z direction is

$V^+ = V_0^+\cos(\omega t - \beta z) = 5.0\cos[(20\pi\times10^6)t - 1.721z]$ V

The expression of the current wave travelling along positive z direction is

$$I^+ = \frac{V_0^+}{Z_0}\cos(\omega t - \beta z) = \frac{5.0}{54.77}\cos[(20\pi\times10^6)t - 1.721z]$$

$$= 0.0913\cos[(20\pi\times10^6)t - 1.721z]\ \text{A}$$

The voltage and current in phasor forms are

$V^+ = 5.0e^{-1.721z}$ V

$I^+ = 0.0913e^{-1.721z}$ A

EXAMPLE 8.5

The instantaneous voltage and instantaneous current of a loss-free two-conductor transmission line are given by

V $\cos[(5\pi \times 10^6)t - 1.4z]$

I $\mathrm{os}[(5\pi \times 10^6)t - 1.4z]$

Find the inductance and capacitance for a line length of 50 m.

Amplitude of the voltage, $V_0 = 50$ V

Amplitude of the current, $I_0 = 0.5$ A

Angular frequency, $\omega = 5\pi \times 10^6$ rad/s

Phase constant, $\beta = 1.4$

Characteristic impedance, $Z_0 = \sqrt{\dfrac{L}{C}} = \dfrac{V_0}{I_0} = 100\ \Omega$

Hence, $\dfrac{L}{C} = 10^4$

Phase constant, $\beta = \omega\sqrt{LC} = (5\pi \times 10^6)\sqrt{LC} = 1.4$

Hence, $LC = \left(\dfrac{1.4}{5\pi \times 10^6}\right)^2 = 79.44 \times 10^{-16}$

Solving for L and C, we get $L = 8.91 \times 10^{-6}$ H/m, $C = 8.91 \times 10^{-10}$ F/m

For a line length of 50 m

inductance $= 50L = 445.5 \times 10^{-6}$ H

capacitance $= 50C = 445.5 \times 10^{-10}$ F

8.5 Wave Reflection at a Discontinuity in an Ideal Transmission Line

To study the reflection phenomenon at the discontinuity of a transmission system let us consider a transmission line of length ℓ, as shown in Fig. 8.10. The transmission line, assumed to be loss-free, is connected to a source at one end (input end) and to a load at the other end (load end). We will take $z = 0$ at the load end instead of at the input end because the form of expressions for reflection and transmission coefficients will be simpler. The positive direction of z is still from the source towards the load. Thus, the terminals of the source are at $z = -\ell$, as indicated in the figure. The waves travelling from the source towards the load are in positive z direction, and those travelling from the load towards the source are in the negative z direction. Let us assume that voltage and current in the line are zero before the switch is closed ($t = 0-$). After the switch is closed ($t = 0+$), the voltage and current waves will travel towards the load with phase velocity v. After time $t = \ell / v$, the waves will reach the load end. These are the incident waves.

If impedance of the load is not equal to the characteristic impedance of the line, both in magnitude and in phase, then the transmission line is said to be discontinuous at the load.

In general, wave reflection as well as wave transmission will occur at the discontinuity. Consequently, a part of the power of the incident wave will be reflected back to the line and the remaining part will be transmitted to the load. The reflected wave will travel from the load towards the source. Wave reflection and transmission will also occur at the junction of two or more transmission lines of different characteristic impedances.

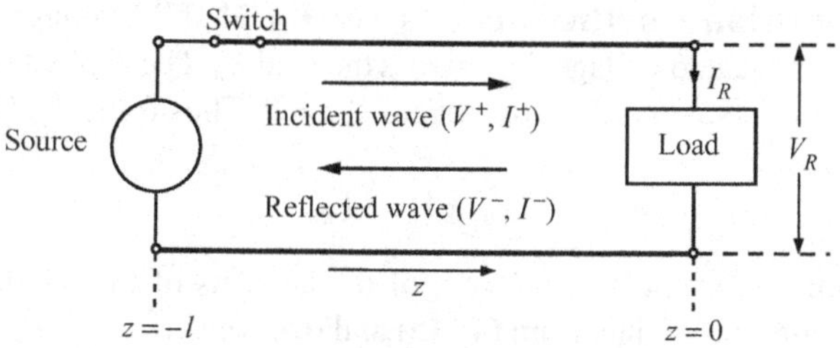

Fig. 8.10 Illustration of a two-conductor transmission lines for calculation of reflection coefficient

---

The voltage reflection coefficient is defined as the ratio between the reflected and incident voltages, and the voltage transmission coefficient is defined as the ratio between the transmitted voltage and incident voltage at the load terminals.

The fraction of incident power reflected back to the line is a function of the reflection coefficient. We will derive the coefficient assuming at first that the load is purely resistive and then generalise it for any type of load. The coefficients are derived from the terminal conditions at the load point using Kirchhoff's voltage and current laws.

Transmission Line Terminated with a Pure Resistance

The characteristic impedance of an ideal transmission line is purely resistive. If the line is terminated even with a purely resistive load, the resistance of which is different from the characteristic impedance of the line, wave reflection takes place at the load discontinuity. The values of the reflected voltage and voltage across the load are found by the use of Kirchhoff's voltage and current laws. With the positive direction as shown in Fig. 8.10, the complex exponential form of voltages of the incident and reflected waves are given by

$$\bar{V}^+ = V_0^+ e^{j(\omega t - \beta z)} \tag{1}$$

$$\bar{V}^- = V_0^- e^{j(\omega t + \beta z)} \tag{2}$$

where V_0^+, V_0^- and V_{0R} are real quantities.

The voltage across the load resistance,

$$\bar{V}_R = V_{R0} e^{j\omega t} \tag{3}$$

The current $\bar{I}^+$ in the incident wave and I^- in the reflected wave are,

$$\bar{I}^+ = \frac{V_0^+}{Z_0} e^{j(\omega t - \beta z)} \tag{4}$$

$$\bar{I}^- = -\frac{V_0^-}{Z_0} e^{j(\omega t + \beta z)} \tag{5}$$

The current in the load of resistance R_R is

$$I_R = \frac{V_{R0}}{R_R} e^{j\ \ t} \tag{6}$$

Reflection and Transmission Coefficients As per Kirchhoff's voltage law, the applied voltage between the load terminals is equal to voltage drop across the load V_R. The applied voltage at the load terminals is equal to the sum of the incident and reflected voltages at $z = 0$. Thus from (1), (2) and (3), we get

$$V_0^+ + V_0^- = V_{R0} \tag{7}$$

Moreover, as per Kirchhoff's current law, the sum of the currents of the incident and reflected waves at $z = 0$ is equal to the loads current I_R. Thus from (4), (5) and (6), we get

$$\frac{1}{Z_0}(V_0^+ - V_0^-) = \frac{V_{R0}}{R_R} \tag{8}$$

By solving equations (7) and (8) simultaneously we will obtain the reflected and load voltages in terms of the incident voltage. The solution for the reflected voltage is

$$\frac{V_0^-}{V_0^+} = \frac{R_R - Z_0}{R_R + Z_0}$$

The ratio of reflected voltage to the incident voltage is called *voltage reflection coefficient* ('reflection coefficient' will also mean 'voltage reflection coefficient').

The reflection coefficient for an ideal transmission line connected to a pure resistance is a real quantity. Denoting the voltage reflection coefficient by Γ, we have

$$\Gamma = \frac{V_0^-}{V_0^+} = \frac{R_R - Z_0}{R_R + Z_0} \tag{9}$$

The load voltage in terms of the incident voltage is given by

$$\frac{V_{R0}}{V_0^+} = \frac{2R_R}{R_R + Z_0}$$

The ratio between the load voltage and incident voltage is the *voltage transmission coefficient*. It is equal to one plus the voltage reflection coefficient.

Denoting the voltage transmission coefficient by τ, we have

$$\tau = \frac{V_{R0}}{V_0^+} = \frac{2R_R}{R_R + Z_0} = 1 + \Gamma \tag{10}$$

The ratio between the amplitudes of currents of the reflected and incident waves, and the ratio between the load current and current of the incident wave are

$$\frac{I_0^-}{I_0^+} = \frac{(-V_0^-/Z_0)}{(V_0^+/Z_0)} = -\frac{R_R - Z_0}{R_R + Z_0} = -\Gamma \tag{11}$$

$$\frac{I_{R0}}{I_0^+} = \frac{(V_{R0}/R_R)}{(V_0^+/Z_0)} = \frac{2Z_0}{R_R + Z_0} = 1 - \Gamma \tag{12}$$

Relations (11) and (12) respectively give the *current reflection coefficient* and *current transmission coefficient*.

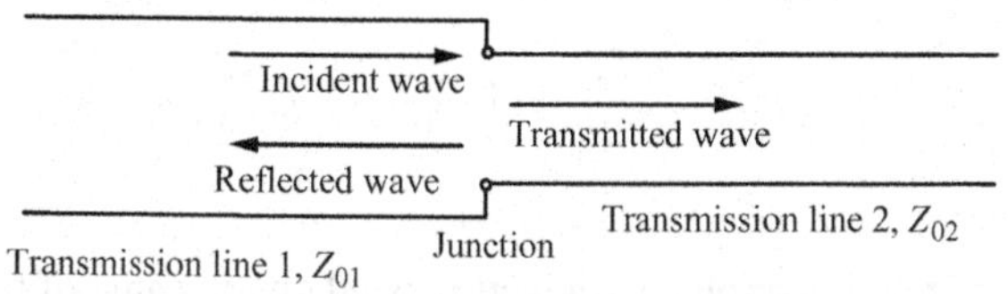

Fig. 8.11 Series connection of two transmission lines of different characteristic impedances

The reflection coefficient at the junction between two lossless transmission lines of different characteristic impedances Z_{01} and Z_{02} shown in Fig. 8.11 is derived by treating the characteristic impedance of the second line as a load at the end of the first line. Assuming that there is only a single transmitted wave in Line 2, the reflection coefficient at the junction is

$$\Gamma = \frac{Z_{01} - Z_{02}}{Z_{01} + Z_{02}}$$

If $Z_{01} \neq Z_{02}$, part of the power incident at the junction will get reflected to Line 1.

The voltage and current reflection coefficients of an open-circuited transmission line, which can be found by substituting $R_R = \infty$, in (9) and (11), are equal to +1 and −1 respectively. So $V_0^- = V_0^+$ and $I_0^- = -I_0^+$. Thus, the voltage wave gets totally reflected without change of sign, whereas the current wave gets totally reflected with change of sign. For a transmission line terminated with a short circuit ($R_R = 0$), the voltage reflection coefficient is −1, and the current reflection coefficient is +1. Hence, $V_0^- = -V_0^+$ and $I_0^- = I_0^+$. Thus, the voltage wave gets totally reflected with change of sign, whereas the current wave gets totally reflected without change of sign, If resistance of the load is equal to the characteristic impedance ($R_R = Z_0$), the reflection coefficient is equal to zero.

Another important parameter associated with the reflection phenomenon is the voltage reflection coefficient at any position $-z$ of the transmission line. It is defined as the ratio between the reflected voltage and incident voltage at $-z$. Thus, denoting the reflection coefficient by $\bar{\Gamma}(-z)$, we have

$$\bar{\Gamma}(-z) = \frac{\bar{V}^-(-z)}{\bar{V}^+(-z)} = \frac{V_0^- e^{j(\omega t - \beta z)}}{V_0^+ e^{j(\omega t + \beta z)}} = \Gamma e^{-j2\beta z}$$

where Γ is the reflection coefficient at the load end defined by (9).

Power Relations The voltage and current in a wave of an ideal transmission line connected to a pure resistance are in phase at any position of the line. The power flow is also same throughout the line. The value of average power carried by the incident wave is

$$P_{av}^+ = \frac{V_0^+ I_0^+}{2Z} = \frac{(V_0^+)^2}{2Z_0}$$

The value of average power carried by the reflected wave is

$$P_{av}^+ = \frac{V_0^+ I_0^+}{2Z} = \frac{(V_0^-)^2}{2Z_0}$$

Thus,

$$\frac{P_{av}^-}{P_{av}^+} = \frac{(V_0^-)^2}{(V^+)^2} = \Gamma^2$$

Power transmitted to the load is the difference between incident and reflected powers. Power flow into the load is

$$P_{R,av} \quad P_{av}^+ - P_{av}^-$$

Hence,
$$\frac{P_{R,av}}{P_{av}^{+}} = 1 - \frac{P_{av}^{-}}{P_{av}^{+}} = 1 - \Gamma^2$$

The magnitude and sign of the reflection coefficients depend on the value of the load resistance R_R relative to the value of the characteristic resistance Z_0. If $R_R = Z_0$, the reflection coefficient is zero; therefore, no power is reflected. That is, all the incident power is absorbed by the load. This is the case of a perfectly matched system. If R_R is greater than Z_0, the voltage is reflected without change of sign, whereas the current is reflected with change of sign. Opposite will be the case if R_R is less than Z_0. In both cases, a part of the incident power is reflected. These are mismatched systems.

For example, if the characteristic resistance of a loss-free transmission line is 150 Ω and it is connected to a purely resistive load of 150 Ω, the reflected power would be zero. However if two resistive loads (equipments), each of 150 Ω, are connected in parallel to the line, the combined resistance of the equipments is 75 Ω. Therefore not only a part of the incident power is reflected, each of the equipments will get only one half of the transmitted power. If the resistances of the two equipments are not equal, then the equipment having lower resistance will get larger share of the transmitted power.

For open-circuited and short-circuited of transmission lines, the magnitude of reflection coefficient is unity. Hence, the reflected power is equal to the incident power.

Transmission Line Terminated with a Complex Impedance

Let us next consider that an ideal transmission line is terminated with a complex load impedance consisting of a resistance part R_R and a reactance part X_R. The expression for the complex impedance is

$$\bar{Z}_R = R_R \pm jX_R$$

The positive sign in the impedance function is to be used if the reactance is inductive, and the negative sign is to be used if the reactance is capacitive. Because of the reactance component in the impedance, the load current will not be in phase with the load voltage. Consequently, amplitude of the reflected voltage will not be in phase with the amplitude of the incident voltage at $z = 0$. The procedure for deriving the reflection coefficient is same as for the resistive load. The voltage relation at the load terminals is

$$\bar{V}_0^{+} + \bar{V}_0^{-} = \bar{V}_{R0} \qquad (13)$$

the current relation at the load terminals is given by

$$\frac{1}{Z_0}(\bar{V}_0^{+} - \bar{V}_0^{-}) = \frac{\bar{V}_{R0}}{\bar{Z}_R} \qquad (14)$$

Solution of (13) and (14) simultaneously gives,

$$\frac{\bar{V}_0^{-}}{\bar{V}_0^{+}} = \frac{\bar{Z}_R - Z_0}{\bar{Z}_R + Z_0} \qquad (15)$$

Relation (15) gives the reflection coefficient at the load end of a loss-free transmission line. The reflection coefficient is a complex quantity having a real part Γ_a and an imaginary part Γ_i. Thus,

$$\bar{\Gamma} = \frac{Z_R - Z_0}{Z_R + Z_0} = \Gamma_a + j\Gamma_i = \Gamma e^{j\psi_r} = \Gamma \angle \psi_r$$

where $\psi_r = \tan^{-1}(\Gamma_i/\Gamma_a)$ is phase angle of the reflection coefficient.

The transmission coefficient, $\bar{\tau} = \dfrac{\bar{V}_{R0}^+}{\bar{V}_0^+} = \dfrac{2\bar{Z}_R}{\bar{Z}_R + \bar{Z}_0} = 1 + \bar{\Gamma}$

Since the characteristic impedance of an ideal line is purely resistive, voltage and current of the incident wave are in time phase. Also, voltage and current of the reflected wave are in phase. Thus, power carried by the incident and reflected waves are

$$P_{av}^+ = \frac{|\bar{V}_0^+||\bar{I}_0^+|}{2} = \frac{|\bar{V}_0^+|^2}{2Z_0}$$

$$P_{av}^- = \frac{|\bar{V}_0^-||\bar{I}_0^-|}{2} = \frac{|\bar{V}_0^-|^2}{2Z_0} = |\bar{\Gamma}|^2 P_{av}^+$$

Power absorbed by the load is equal to the difference between the incident power and reflected power.

Reflection Coefficient at Any Point

The reflection coefficient at any position $z = -z'$ of the line,

$$\bar{\Gamma}(-z') = \frac{\bar{V}^-(-z')}{\bar{V}^+(-z')} = \frac{\bar{V}_0^- e^{j(\omega t - \beta z')}}{\bar{V}_0^+ e^{j(\omega t + \beta z')}} = \bar{\Gamma}(0)e^{-j2\beta z'} = |\bar{\Gamma}(0)|e^{j(\psi_r - 2\beta z')} \tag{16}$$

where $\bar{\Gamma}(0)$ is reflection coefficient at $z = 0$. Thus, the reflection coefficient at $z = -z'$ has the same magnitude as the reflection coefficient at $z = 0$, but its phase angle differs by $-2\beta z'$ radian.

EXAMPLE 8.6

The characteristic impedance of an ideal two-conductor transmission line is 300 Ω. It is terminated with a load resistance of 200 Ω. At the load terminals, find (a) the voltage and current reflection coefficients and (b) the voltage and current transmission coefficients.

Characteristic impedance, $Z_0 = 300\ \Omega$

Load resistance, $R_R = 200\ \Omega$

(a) Voltage reflection coefficient, $\Gamma = \dfrac{R_R - Z_0}{R_R + Z_0} = \dfrac{200 - 300}{200 + 300} = -0.2$

Current reflection coefficient $= -\Gamma = +0.2$

(b) Voltage transmission coefficient, $1 + \Gamma = 0.8$

Current transmission coefficient, $1 - \Gamma = 1.2$

EXAMPLE 8.7

The characteristic impedance of an ideal two-conductor transmission line is 75 Ω. It is terminated with a load of resistance 150 Ω and inductive reactance of 20 Ω. Find the voltage reflection and voltage transmission coefficients at the load terminals.

Characteristic impedance, $Z_{\_}$ $j0\ \Omega$

Load resistance, $\bar{Z}_R = 150 + j20\ \Omega$

Voltage reflection coefficient,

$$\Gamma = \frac{Z_R - Z_0}{Z_R + Z_0} = \frac{(150 + j20) - 75}{(150 + j20) + 75} = \frac{77.62\angle 14.93^\circ}{225.89\angle 5.08^\circ} = 0.344\angle 9.85^\circ$$

Voltage transmission coefficient

$$= \;+ \quad = 1 + 0.344(\cos 9.85^\circ + j\sin 9.85^\circ) = 1.339 + j0.058$$

EXAMPLE 8.8

(a) The characteristic impedance of an ideal two-conductor transmission line is 75 Ω. It is terminated with a load of resistance 150 Ω and capacitive reactance of 20 Ω. Find the voltage reflection and voltage transmission coefficients at the load terminals.
(b) An ideal two-conductor transmission line of characteristic impedance 175 Ω is connected to a load of resistance 200 Ω. Find power of the incident wave and power transmitted to the load, if the amplitude of voltage of the incident wave is 24 V.

(a) Characteristic impedance, $\bar{Z}_0 = 75 + j0\ \Omega$

Load resistance, $\bar{Z}_R = 150 - j20\ \Omega$

Voltage reflection coefficient,

$$\Gamma = \frac{\bar{Z}_R - \bar{Z}_0}{\bar{Z}_R + \bar{Z}_0} = \frac{(150 - j20) - 75}{(150 - j20) + 75} = \frac{77.62\angle -14.93^\circ}{225.89\angle -5.08^\circ} = 0.344\angle -9.85^\circ$$

Voltage transmission coefficient

$$= \;+ \bar{\Gamma} = 1 + 0.344(\cos 9.85^\circ - j\sin 9.85^\circ) = 1.339 - j0.058$$

(b) Characteristic impedance, $Z_0 = 175\ \Omega$

Load resistance, $R_R = 300\ \Omega$

Amplitude of incident voltage, $V_0^+ = 24\ \text{V}$

$$\text{Power of the incident wave, } P_{av}^+ = \frac{(V_0^+)^2}{2Z_0} = \frac{24^2}{350} = 1.646\ \text{W}$$

$$\text{Transmission coefficient} = \tau = \frac{2R_R}{R_R + Z_0} = \frac{600}{475} = 1.263$$

Load voltage, $V_{R0} = \tau V_0^+ = (1.263)(24) = 30.312\ \text{V}$

$$\text{Power transmitted to the load, } P_{R,av} = \frac{V_{R0}^{\ 2}}{2R_R} = \frac{(\quad .312)^2}{600} = 1.531\ \text{W}$$

EXAMPLE 8.9

A coaxial transmission line 1 is joined with another coaxial transmission line 2. The conductor sizes and relative permeabilities of the lines are same. But, the relative permittivity of insulation of Line 2 is 1.6 times that of Line 1. Find the reflection coefficient at the junction if the incident voltage is in Line 1.

The inductance per unit length of the lines are same. But, as the capacitance per unit length is proportional to the relative permittivity, capacitance per unit length of Line 2 is 1.6 times that of Line 1. Let L_1 and C_1 be the inductance and capacitance per unit length of Line 1.

Characteristic impedance of Line 1, $Z_{01} = \sqrt{\dfrac{L_1}{C_1}}$

Characteristic impedance of Line 2, $Z_{01} = \sqrt{\dfrac{L_1}{1.6C_1}} = 0.791$

Reflection coefficient, $\Gamma = \dfrac{Z_{02} - Z_{01}}{Z_{02} + Z_{01}} = \dfrac{0.791 - 1}{0.791 + 1} = -0.117$

EXAMPLE 8.10

A two-conductor transmission line of characteristic impedance 150 Ω is terminated with a load of unknown impedance. If the reflection coefficient at the load end is 0.3 – *j*0.1 Ω, find the load impedance.

Characteristic impedance, $\bar{Z}_0 = 150 + j0\ \Omega$

Reflection coefficient, $\bar{\Gamma} = 0.3 - j0.1$

Let the load impedance, $\bar{Z}_R = R_R + jX_R$, then

$$\Gamma = \frac{R_R + jX_R - 150}{R_R + jX_R + 150} = 0.3 - j0.1$$

Or $R_R + jX_R - 150 = (0.3 - j0.1)(R_R + jX_R + 150)$

By equating the real parts of the two sides and imaginary parts of the two sides, we get

$$0.7R_R - 0.1X_R = 195$$

and $0.1R_R + 0.7X_R = -15$

Solving the two equations simultaneously, we get

R_R $\qquad$ $X_R = -60\ \Omega$

Thus the load impedance, Z $\qquad$ $j60\ \Omega$

EXAMPLE 8.11

A lossless two-conductor transmission line of characteristic impedance 150 Ω is terminated with an impedance of $150 - j300$ Ω. Calculate (a) the voltage reflection coefficient at the load terminals, (b) the voltage transmission coefficient and (c) power carried by the incident wave, reflected wave, and power transferred to the load. The amplitude of the incident voltage is $10.0 \angle 0$ V.

Characteristic impedance, $\bar{Z}_0 = 150 + j0\ \Omega$

Load impedance, $\bar{Z}_R = 150 - j300\ \Omega$

Amplitude of incident voltage, $\bar{V}_0^+ = 10 + j0$ V

(a) The voltage reflection coefficient, $\bar{\Gamma} = \dfrac{\bar{Z}_R - \bar{Z}_0}{\bar{Z}_R + \bar{Z}_0} = \dfrac{(150 - j300) - 150}{(150 - j300) + 150} = \dfrac{-j}{1-j}$

$$= \frac{1.0\angle -90^\circ}{\sqrt{2}\angle -45^\circ} = \frac{1}{\sqrt{2}}\angle -45^\circ$$

(b) The voltage transmission coefficient

$$= 1 + \bar{\Gamma} = 1 - \frac{j}{1-j} = \frac{1-j2}{1-j} = \frac{\sqrt{5}\angle -63.4^\circ}{\sqrt{2}\angle -45^\circ} = 1.58\angle -18.4^\circ$$

(c) Power carried by the incident wave, $P_{av}^+ = \dfrac{|\bar{V}_0^+|^2}{2|\bar{Z}_0|} = \dfrac{100}{300} = \dfrac{1}{3}$ W

Power carried by the reflected wave, $P_{av}^- = |\bar{\Gamma}|^2 P_{av}^+ = \dfrac{1}{6}$ W

Power transferred to the load, $P_{av}^+ - P_{av}^- = \dfrac{1}{6}$ W

EXAMPLE 8.12

The input point of an ideal two-conductor transmission line operating at 10 MHz is at $z = -30$ m, and the load point is at $z = 0$. The inductance and capacitance per unit length of the line are 12 μH and 533 pF respectively. The line is terminated with a load resistance of 275 Ω. Find the magnitude and phase angle of reflection coefficient at $z = -10$ m

Inductance per unit length, $L = 12 \times 10^{-6}$ H/m

Capacitance per unit length, $C = 533 \times 10^{-12}$ F/m

Frequency, $f = 10$ MHz

Load resistance, $R_R = 275\ \Omega$

Characteristic impedance, $Z_0 = \sqrt{\dfrac{L}{C}} = \sqrt{\dfrac{12\times 10^{-6}}{533\times 10^{-12}}} = 150.0\ \Omega$

Phase constant, $\beta \quad \omega \ LC = (\ \pi \times \quad \times \quad \times \quad) = 5\ 03$ rad/m

Voltage reflection coefficient at $z = 0$, $\bar{\Gamma}(0) = \dfrac{R_R - Z_0}{R_R + Z_0} = \dfrac{275-150}{275+150} = 0.294\angle 0$

Voltage reflection coefficient at $Z = -z'$

$$\Gamma(-z') = \bar{\Gamma}(0)e^{-j2\beta z'} = 0.294e^{-j(2)(5.03)(10)} = 0.294e^{-j100.6}$$

The magnitude of the reflection coefficient at $z = -10$ m is same as that of the reflection coefficient at the load point. But, it lags the reflection coefficient at the load point by 100.6 radians.

8.6 Variation of Voltage and Current along a Transmission Line with Discontinuity

A transmission line terminated with impedance that is different from the characteristic impedance of the line will have both incident and reflected waves. The total voltage at any position of a transmission line is equal to the sum of the voltages of the incident and reflected waves. And the total current is equal to the sum of the currents of the incident and reflected waves. The amplitude of the total voltage along the line varies between a maximum value and a minimum value. Similarly, the amplitude of the total current also varies between maximum and minimum values. It is found that the point at which the amplitude of the total voltage has maximum value, the current has the minimum value and vice versa. Moreover, envelope of the amplitude of the total voltage as well as that of total current is a standing wave. The ratio between the maximum amplitude and the minimum amplitude either of the total voltage or of the total current is called *standing wave ratio.*

Expressions for the Total Voltage and Total Current

By taking the voltage of the incident wave in Fig. 8.12(a) as reference, we have

$$\bar{V}^+ = V_0^+ e^{j(\omega t - \beta z)} \tag{1}$$

The amplitude of the voltage V_0^+ is a real quantity. The reflected voltage is equal to the product of incident voltage and the reflection coefficient. The reflection coefficient in general is a complex quantity. The reflected voltage is given by

$$\bar{V}^- = \bar{\Gamma} V_0^+ e^{j(\omega t + \beta z)} = |\bar{\Gamma}| V_0^+ e^{j(\omega t + \beta z + \psi_r)} = V_0^- e^{j(\omega t + \beta z + \psi_r)} \tag{2}$$

where $V_0^- = |\bar{\Gamma}| V_0^+$ is the real amplitude of the reflected voltage and ψ_r is phase angle of the voltage reflection coefficient.

The real parts of (1) and (2) are

$$V^+ = V_0^+ \cos(\omega t - \beta z) \tag{3}$$

$$V^- = V_0^- \cos(\omega t + \beta z + \psi_r) \tag{4}$$

The total voltage V at any position of the transmission line is equal to the sum of (3) and (4).

$$V = V^+ + V^- = V_0^+ \cos(\omega t - \beta z) + V_0^- \cos(\omega t + \beta z + \psi_r) \tag{5}$$

$$= (V_0^+ - V_0^-)\cos(\omega t - \beta z) + V_0^-[\cos(\omega t - \beta z) + \cos(\omega t + \beta z + \psi_r)]$$

Applying the formula $\cos a + \cos b = 2\cos\frac{1}{2}(a+b)\cos\frac{1}{2}(a-b)$ to the second term, we get

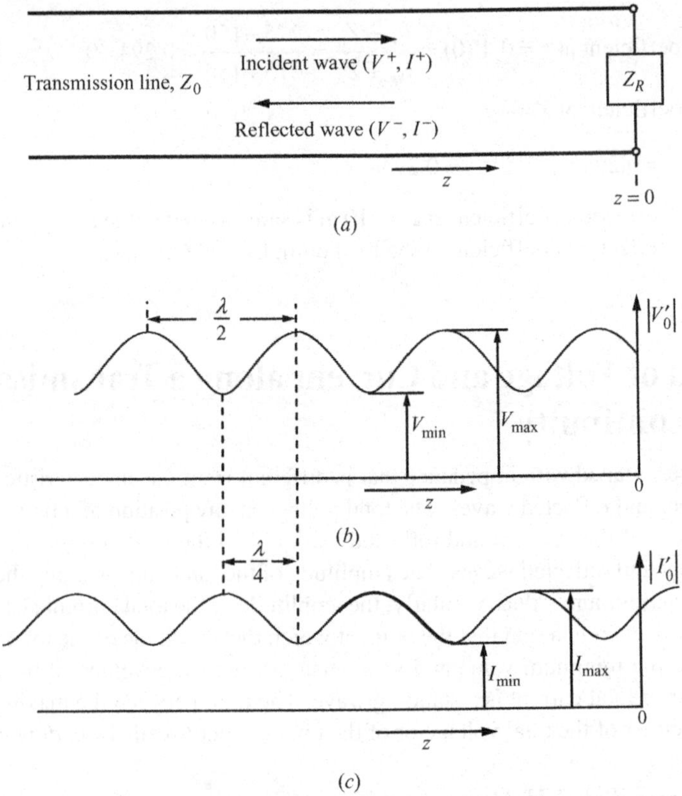

Fig. 8.12 (*a*) Ideal transmission line connected to an unmatched load, (*b*) variation of amplitude of total voltage, and (*c*) variation of amplitude of total current.

$$V = (V_0^+ - V_0^-)\cos(\omega t - \beta z) + 2V_0^- \cos\left(\omega t + \tfrac{\psi_r}{2}\right)\cos\left(\beta z + \tfrac{\psi_r}{2}\right) \tag{6}$$

The total current at any position of the transmission line,

$$I = I^+ + I^- = \frac{1}{Z_0}(V^+ - V^-)$$

$$= \frac{1}{Z_0}[V_0^+ \cos(\omega t - \beta z) - V_0^- \cos(\omega t + \beta z + \psi_r)]$$

$$= \frac{1}{Z_0}(V_0^+ - V_0^-)\cos(\omega t - \beta z) + \frac{V_0^-}{Z_0}[\cos(\omega t - \beta z) - \cos(\omega t + \beta z + \psi_r)]$$

$$= \frac{1}{Z_0}(V_0^+ - V_0^-)\cos(\omega t - \beta z) + \frac{2V_0^-}{Z_0}\sin\left(\omega t + \tfrac{\psi_r}{2}\right)\sin\left(\beta z + \tfrac{\psi_r}{2}\right) \tag{7}$$

For an ideal transmission line terminated with resistance R_R, the value of which is higher than the characteristic impedance Z_0 of the line ($R_R > Z_0$), the magnitude of the reflection coefficient at the load end is less than unity and its phase angle is zero. If $R_R < Z_0$, the magnitude of the reflection coefficient at the load end is also less than unity but its phase angle is 180°. Thus, the values of voltage as well as current in the reflected wave for both the cases) are less than the values of voltage and current respectively of the incident wave. The total voltage and total current are partly travelling waves and partly standing waves. The travelling waves are represented by the first terms, and the standing waves are represented by the second terms, of (6) and (7).

Standing Wave Ratio

In the course of travel of the incident and reflected waves in their respective directions, the voltages of the waves come into phase at some positions of the transmission line. The amplitude of the total voltage at this position is the sum of the amplitudes of voltages of the incident and reflected waves; hence the amplitude of the total voltage has maximum value. After each wave progresses by a quarter wave length, the amplitudes of the voltages are in anti-phase. Thus, the amplitude of the total voltage is the difference between the amplitudes of the two voltages. At this position the amplitude of the total voltage has minimum value.

The ratio between the maximum and minimum amplitudes of the total voltage is the *standing wave ratio*. The ratio between the maximum and minimum amplitudes of total current also gives the standing wave ratio.

Equation (6) gives one way of describing the total voltage in an ideal transmission line terminated with an unmatched load. Another form of the total voltage is derived by applying the trigonometric identity

$$\cos(a \mp b) = \cos a \cos b \pm \sin a \sin b$$

to (5). Thus, we have

$$V = V_0^+(\cos\omega t \cos\beta z + \sin\omega t \sin\beta z) + V_0^-[\cos\omega t \cos(\beta z + \psi_r) - \sin\omega t \sin(\beta z + \psi_r)]$$

$$= [V_0^+ \cos\beta z + V_0^- \cos(\beta z + \psi_r)]\cos\omega t + [V_0^+ \sin\beta z - V_0^- \sin(\beta z + \psi_r)]\sin\omega t \tag{8}$$

Taking $A = V_0^+ \cos\beta z + V_0^- \cos(\beta z + \psi_r)$ and $B = V_0^+ \sin\beta z - V_0^- \sin(\beta z + \psi_r)$,

$$V = A\cos\omega t + B\sin\omega t \tag{9}$$

Now let, $B/A = \tan\xi$ so that

$$A = \sqrt{A^2 + B^2}\cos\xi$$

$$B = \sqrt{A^2 + B^2}\sin\xi$$

Using these relations in (9), we get

$$V = \sqrt{A^2 + B^2}\,(\cos\beta z \cos\xi + \sin\beta z \sin\xi)$$

$$= \sqrt{A^2 + B^2}\cos(\beta z - \xi) \tag{10}$$

Substituting the expressions of A and B in (10) and then simplifying, we have

$$V = \sqrt{(V_0^+)^2 + (V_0^-)^2 + 2V_0^+ V_0^- \cos(2\beta z + \psi_r)}\,\cos(\omega t - \xi) \tag{11}$$

$$= V_0' \cos(\omega t - \xi)$$

where
$$V_0' = \sqrt{(V_0^+)^2 + (V_0^-)^2 + 2V_0^+ V_0^- \cos(2\beta z + \psi_r)} \tag{12}$$

is amplitude of the total voltage.

Both (6) and (11) represent the sum of the incident and reflected voltages. As has been described earlier, the voltage as per (6) is the combination of a travelling wave and a standing wave. The amplitude of the travelling-wave component has a constant value equal to the difference between the amplitudes of the incident and reflected voltages. The amplitude of the standing wave is a function of time. Whereas the total voltage as per (11) represents a travelling wave because the phase angle ξ is a function of z. The velocity of the wave, however, is not constant as the phase angle is not a linear function of z. Moreover, the amplitude of the voltage, defined by (12), varies with z, but not with time. Thus, the envelope of the amplitude is a standing wave along z direction. The amplitude varies between a maximum value and a minimum value. The maximum value occurs at locations along the line where

$$\cos(2\beta z + \psi_r) = 1$$

Or
$$2\beta z + \psi_r = -2n\pi, n = 0, 1, 2, 3, \dots \tag{13}$$

The minus sign is chosen because the transmission line is along the negative side of z. The magnitude of maximum amplitude is

$$V_{\max} = V_0^+ + V_0^- = V_0^+(1+|\Gamma|) \tag{14}$$

The minimum value of the amplitude occurs at locations along the line where

$$\cos(2\beta z + \psi_r) = -1$$

Or
$$2\beta z + \psi_r = -(2n+1)\pi, n = 0, 1, 2, 3, \dots \tag{15}$$

The value of minimum amplitude of the total voltage is given by

$$V_{\min} = V_0^+ - V_0^- = V_0^+(1-|\Gamma|) \tag{16}$$

Using the procedure as described, it can be shown that the amplitude of the total current,

$$I_0' = \frac{1}{Z_0}\sqrt{(V_0^+)^2 + (V_0^-)^2 - 2V_0^+V_0^- \cos(2\beta z + \psi_r)}$$

The amplitude of the current has maximum value at locations along the line where

$$\cos(2\beta z + \psi_r) = -1$$

Or
$$2\beta z + \psi_r = -(2n+1)\pi, n = 0, 1, 2, 3, \dots \tag{17}$$

The value of maximum amplitude of the total current,

$$I_{\max} = \frac{1}{Z_0}(V_0^+ + V_0^-) = \frac{V_0^+}{Z_0}(1+|\Gamma|) \tag{18}$$

The amplitude of the current has minimum value at locations along the line where

$$\cos(2\beta z + \psi_r) = 1$$

Or
$$2\beta z + \psi_r = -2n\pi, n = 0, 1, 2, 3, \dots \tag{19}$$

$$I_{\min} = \frac{1}{Z_0}(V_0^+ - V_0^-) = \frac{V_0^+}{Z_0}(1-|\Gamma|) \tag{20}$$

Figure 8.12(b) and (c) show the variation of the magnitudes of the voltage and current amplitudes. The waves are generally not sinusoidal. We see that maximum amplitude of the voltage and minimum amplitude of the current occur at the same points of the transmission line. Also, the minimum amplitude of the voltage and maximum value of the current occur at the same locations of the line. The distance between the positions of maximum voltage amplitude and maximum current amplitude is equal to one-quarter of a wavelength. The distance between two successive voltage maxima or two voltage minima is equal to one-half of a wavelength.

The standing wave ratio,

$$S_R = \frac{V_{\max}}{V_{\min}} = \frac{I_{\max}}{I_{\min}} = \frac{1+\Gamma}{1-\Gamma} \tag{21}$$

Equation (21) when solved for $|\Gamma|$ gives

$$|\bar{\Gamma}| = \frac{S_R - 1}{S_R + 1} \tag{22}$$

For an open-circuited transmission line $V_0^- = V_0^+$ and $\psi_r = 0$. Substituting these values in (6) and (7) we obtain the total voltage V_{oc} and total current I_{oc} of an open-circuited transmission line. These are

$$V_{oc} = (2V_0^+ \cos \omega t) \cos \beta z \tag{23}$$

$$I_{oc} = \left(\frac{2V_0^+}{Z_0} \sin \omega t \right) \sin \beta z \tag{24}$$

In the case of a transmission line terminated with a short-circuit, $V_0^- = V_0^+$ and $\psi_r = 180°$. Substituting these values in (6) and (7) we obtain the total current V_{sc} and total current I_{sc} of a short-circuited transmission line. These are

$$V_{sc} = (2V_0^+ \sin \omega t) \sin \beta z \tag{25}$$

$$I_{sc} = \left(\frac{2V_0^+}{Z_0} \cos \omega t \right) \cos \beta z \tag{26}$$

Equations (23) to (26) show that the total voltage and total current of open-circuited and short-circuited transmission lines are standing waves. The standing waves along the line are either sine or cosine waves. The amplitude of the waves are either sine or cosine function of time.

If the magnitude of reflection coefficient is zero, then the value of standing ratio is unity. And if the magnitude of the reflection coefficient is unity, as in the case of a short-circuited or open-circuited line, then the value of standing wave ratio is infinity. Thus, the standing wave ratio may vary from unity to infinity depending on the terminal condition. A low value of standing wave ratio means the ratio of reflected power to incident power is low.

The standing wave ratio can be determined experimentally. Since slope of the voltage envelope is higher at V_{min} than at V_{max}, the positions of minimum voltage along the line can be determined more accurately experimentally than the positions of maximum voltage. The reflection coefficient is calculated from the standing wave ratio using formula (22).

We note that the procedure for deriving the standing wave ratio of a transmission line is similar to that used in the case of plane waves.

EXAMPLE 8.13

An ideal two-conductor transmission line of characteristic impedance Z_0 is terminated with a load resistance R_R at $z = 0$. The expressions for the incident voltage V^+ and reflected voltage V^- are given by

$$V^+ = 120 \cos(\omega t - \beta z) \text{ V}$$

and $V^- = -40 \cos(\omega t - \beta z)$ V

(a) Find the standing wave ratio. (b) Obtain R_R in terms of Z_0.

Amplitude of the incident voltage, $|V_0^+| = 120$ V

Amplitude of the reflected voltage, $V_0^- = 40$ V

(a) Standing wave ratio, $S_R = \frac{V_0}{V_0'} \quad \frac{V_0^-}{V_0^-} = \frac{120+40}{124-40} = 2\ 0$

(b) Incident voltage at $z = 0$, $V^+(0) = 120\cos\omega t$ V

Reflected voltage at $z = 0$, $V^-(0) = -40\cos\omega t$ V

Reflection coefficient, $\Gamma = \frac{R_R - Z_0}{R_R + Z_0} = \frac{V^-(0)}{V^+(0)} = \frac{-40}{120} = -\frac{1}{3}$

By solving for R_R, we get $R_R = Z_0/2$

EXAMPLE 8.14

The characteristic impedance and phase constant of a two-conductor transmission line are 100 Ω and 2.5 rad/m respectively. The load impedance is $150 - j25$ Ω. Find (a) distance between two consecutive minima of the total voltage, (b) the standing wave ratio, (c) distance of the nearest maximum voltage from the load, and (d) distance of the nearest minimum voltage from the load.

Characteristic impedance, $\bar{Z}_0 = 100 + j0\ \Omega$

Phase constant, $\beta = 2.5$ rad/m

Load impedance, $Z_R = 150 - j25\ \Omega$

The reflection coefficient,

$$\bar{\Gamma} = \frac{\bar{Z}_R - \bar{Z}_0}{\bar{Z}_R + \bar{Z}_0} = \frac{150 - j25 - 100}{150 - j25 + 100} = \frac{55.9\angle -26.6^\circ}{251.2\angle -5.7^\circ} = 0.22\angle -20.9^\circ$$

$\Gamma = 0\ 22,\ \psi_r = -20.9^\circ = -0.116\pi$ rad

(a) Wavelength, $\lambda = \frac{2\pi}{\beta} = \frac{2\pi}{2.5} = 2.513$ m

Distance between two successive voltage minima $= \frac{\lambda}{2} = 1.251$ m

(b) Standing wave ratio $S_R = \frac{1 + |\bar{\Gamma}|}{1 - |\bar{\Gamma}|} = \frac{1 + 0.22}{1 - 0.22} = 1.564$

(c) Let z_{max} be the distance of the first maximum of total voltage from the load end. Thus substituting $n = 1$ (for $n = 0$ the distance would be positive)? in (13), we get

$$2_{r}\,_{\max} + \psi_r = -2\pi$$

$$\text{Or } z_{\max} = -\frac{2\pi + \psi_r}{2\beta} = -\frac{\pi(2 - 0.116)}{2(2.5)} = -1.184 \text{ m}$$

This is also the position of maximum voltage and minimum current.

(d) Let z_{min} be the distance of the first minimum of total voltage from the load end. Thus substituting $n = 0$ in (15), we get

$$2\beta\,_{\min} + \psi_r = -\pi$$

$$\text{Or } z_{\min} = -\frac{\pi + \psi_r}{2\beta} = -\frac{\pi(\quad .116)}{2(\ .5)} = -0.555 \text{ m}$$

This is also the position of minimum voltage and maximum current.

8.7 Transmission Line Impedances

The ratio between the voltage and current of a single travelling wave in a two-conductor uniform transmission line is the characteristic impedance of the line. If a transmission line has both incident (forward) and reflected (reverse) waves, the total voltage at any position of the line is equal to the sum of the voltages of the forward and reflected waves. And the total current is equal to the sum of the currents of the forward and reflected waves.

The impedance at any position of the line is defined as the ratio of total voltage to total current at that position.

We will use the phasor forms of voltage and current to define various impedances of a lossless line. Referring to Fig. 8.10, the total voltage at $z = -z'$ of the transmission line is

$$\bar{V}(-z') = \bar{V}^+ + \bar{V}^- = \bar{V}_0^+ e^{j\beta z'} + \bar{V}_0^- e^{-j\beta z'} = \bar{V}_0^+ (e^{j\beta z'} + \bar{\Gamma} e^{-j\beta z'}) \quad (1)$$

The total current at $z = -z'$ is given by

$$\bar{I}(-z') = I^+ + I^- = \frac{\bar{V}^+ - \bar{V}^-}{Z_0} = \frac{\bar{V}_0^+}{Z_0}(e^{j\beta z'} - \bar{\Gamma} e^{-j\beta z'})$$

$$= \bar{I}_0^+ (e^{j\beta z'} - \bar{\Gamma} e^{-j\beta z'}) \quad (2)$$

where $\bar{V}_0^+$ and $\bar{I}_0^+$ are amplitude of voltage and current of the incident wave at the load terminals, and

$$\bar{\Gamma} = \frac{\bar{Z}_R - Z_0}{\bar{Z}_R + Z_0} \quad (3)$$

is reflection coefficient at the load terminals ($z = 0$) and $\bar{Z}_R$ is impedance of the load.

The line impedance at $z = -z'$ is equal to the ratio of (1) and (2). The impedance in general is a complex quantity even for a lossless line. Thus, denoting the impedance by $\bar{Z}(-z')$, we have

$$\bar{Z}(-z') = \frac{\bar{V}(-z')}{\bar{I}(-z')} = Z_0 \frac{e^{j\beta z'} + \bar{\Gamma} e^{-j\beta z'}}{e^{j\beta z'} - \bar{\Gamma} e^{-j\beta z'}} = Z_0 \frac{1 + \bar{\Gamma}(0) e^{-j2\beta z'}}{1 - \bar{\Gamma}(0) e^{-j2\beta z'}} \quad (4)$$

By substituting (3) into (4) and simplifying, we get

$$\bar{Z}(-z') = Z_0 \frac{\bar{Z}_R (e^{j\beta z'} + e^{-j\beta z'}) + Z_0 (e^{j\beta z'} - e^{-j\beta z'})}{Z_0 (e^{j\beta z'} + e^{-j\beta z'}) + \bar{Z}_R (e^{j\beta z'} - e^{-j\beta z'})}$$

$$= Z_0 \frac{\bar{Z}_R \cos \beta z' + j Z_0 \sin \beta z'}{Z_0 \cos \beta z' + j \bar{Z}_R \sin \beta z'} \quad (5)$$

Maximum and Minimum Impedances

We will now define some other impedance functions. The impedance at locations of the line where the total voltage has maximum or minimum value is found to be a real quantity. As the maximum voltage and minimum current occurs at the same point, the impedance at these points has the maximum value. The maximum impedance,

$$Z_{max} = \frac{V_{max}}{I_{min}} = \frac{V_{max}}{V_{min}\ Z_0} = \frac{V_0^+(\ +\ )}{(V\ Z_0)(\ -\ )} = S_R Z_0$$

where $S_R = (1+|\overline{\Gamma}|)/(1-|\overline{\Gamma}|)$ is the standing wave ratio.

The line impedance has the minimum value at points of minimum voltage. The minimum impedance,

$$Z_{\min} = \frac{V_{\min}}{I_{\max}} = \frac{V_0^+(1-|\overline{\Gamma}|)}{(V_0^+/Z_0)(1+|\overline{\Gamma}|)} = \frac{Z_0}{S_R}$$

The characteristic impedance in terms of $Z_{\max}$ and $Z_{\min}$ is given by

$$Z_0 = \sqrt{Z_{\max} Z_{\min}}$$

The Input Impedance

An important impedance function of a transmission line is the input impedance. It is given by the ratio of total voltage to total current at the input end. Referring to Fig. 8.10 we find that the input point is at $z = -\ell$. Accordingly, the expression for the input impedance Z_i is obtained by substituting ℓ in place of z' in (4). It is given by

$$\overline{Z}_{in} = \overline{Z}(-\ell) = \frac{\overline{V}(-\ell)}{\overline{I}(-\ell)} = Z_0 \frac{1+\overline{\Gamma}e^{-j2\beta\ell}}{1-\overline{\Gamma}e^{-j2\beta\ell}} = Z_0 \frac{\overline{Z}_R \cos\beta\ell + jZ_0 \sin\beta\ell}{Z_0 \cos\beta\ell + j\overline{Z}_R \sin\beta\ell}$$

$$= Z_0 \frac{\overline{Z}_R \cos 2\pi(\ell/\lambda) + jZ_0 \sin 2\pi(\ell/\lambda)}{Z_0 \cos 2\pi(\ell/\lambda) + j\overline{Z}_R \sin 2\pi(\ell/\lambda)} \quad (6)$$

The input impedance is a function of wavelength, load impedance and characteristics impedance

Two special cases of practical importance are the input impedances of a short-circuited line and that of an open-circuited line. The input impedance of a short-circuited line is obtained by substituting $\overline{Z}_R = 0$ in (6). The impedance,

$$\overline{Z}_{sc} = jZ_0 \tan\beta\ell = jZ_0 \tan 2\pi(\ell/\lambda) \quad (7)$$

Because of the presence of the factor j, the input impedance of a short-circuited lossless transmission line is a pure reactance. The reactance $X_{sc} = Z_0 \tan 2\pi(\ell/\lambda)$. The reactance may be either inductive or capacitive depending on the value of ℓ/λ. The reactance varies from $-\infty$ to $+\infty$.

A short-circuited line is called a *stub*.

A low-loss stub of appropriate length is used as a circuit element (L or C) for the purpose of impedance matching at high frequencies (above 100 MHz).

The input impedance of an open-circuited line is obtained by substituting $\overline{Z}_R = \infty$ in (6). The impedance,

$$\overline{Z}_{oc} = -jZ_0 \cot\beta\ell = -jZ_0 \cot 2\pi(\ell/\lambda) \quad (8)$$

The input impedance of an open-circuited lossless transmission line is also a pure reactance. The reactance X_{oc} $Z_0 \cot 2\pi(\ell/\lambda)$. The reactance may be either inductive or capacitive depending on the value of ℓ/λ. Its reactance varies from $-\infty$ to $+\infty$.

We find from (7) and (8) that for a transmission line of a given length the product of short-circuit input impedance and open-circuit input impedance is equal to Z_0^2. Hence,

$$Z_0 \quad Z_{sc} Z_{oc}$$

EXAMPLE 8.15

An ideal two-conductor transmission line of characteristic impedance 150 Ω is terminated with a load of resistance 100 Ω. The input point and the load point of the transmission line are at $z = -30$ m and $z = 0$ respectively. The phase constant is 2.4 rad/m. Find (a) the line impedance at $z = -12.0$ m, and (b) the maximum and minimum values of line impedance.

Characteristic impedance, $Z_0 = 150\ \Omega$

Load resistance, $R_R = 100\ \Omega$

Phase constant, $\beta = 2.4$ rad/m

(a) Line impedance, $Z(-z') = Z_0 \dfrac{1+\bar{\Gamma}e^{-j2\beta z'}}{1-\bar{\Gamma}e^{-j2\beta z'}} = Z_0 \dfrac{1+\bar{\Gamma}(\cos 2\beta z' - j\sin 2\beta z')}{1-\bar{\Gamma}(\cos 2\beta z' - j\sin 2\beta z')}$

$$\bar{\Gamma} = \frac{R_R - Z_0}{R_R + Z_0} = \frac{100-150}{100+150} = -0.2$$

$$z' = 12 \text{ m}$$

$$2\beta z' = (2)(2.4)(12) = 57.6 \text{ rad} = 3300.2^\circ$$

$$Z(-z') - (150)\frac{1-(0.2)(0.497-j0.868)}{1+(0.2)(0.497-j0.868)}$$

$$= (150)\frac{0.91\angle 10.94^\circ}{1.11\angle -8.99^\circ} = 122.97\angle 19.93^\circ\ \Omega$$

(b) Maximum value of line impedance, $Z_{max} = \dfrac{V_0^+\left(1+|\bar{\Gamma}|\right)}{(V_0^+/Z_0)\left(1-|\bar{\Gamma}|\right)} = (150)\dfrac{1+0.2}{1-0.2} = 225\ \Omega$

Minimum value of line impedance, $Z_{min} = \dfrac{V_0^+\left(1-|\bar{\Gamma}|\right)}{(V_0^+/Z_0)\left(1+|\bar{\Gamma}|\right)} = (150)\dfrac{1-0.2}{1+0.2} = 100\ \Omega$

EXAMPLE 8.16

A two-conductor transmission line, assumed to be lossless, has inductance of 120 nH/m and capacitance of 50 pF/m. The line length is 15 m and the load impedance is $300 + j0\ \Omega$. Calculate the input impedance of the line at 100 MHz.

Inductance, $L = 120 \times 10^{-9}$ F/m

Capacitance, $C = 50 \times 10^{-12}$ H/m

Load impedance, $\bar{Z}_R = 300 + j0\ \Omega$

Length of transmission line, $\ell = 15.0$ m

Input impedance, $Z_{in}\quad Z_0 \dfrac{Z \quad\quad jZ_0 \sin\beta\ell}{Z \quad\quad jZ_R \sin\beta\ell}$

Characteristic impedance, $Z_0 = \dfrac{L}{C} = \dfrac{120\times 10^{-9}}{50\times 10^{-12}} = 49\ \Omega$

Phase constant, $\beta = \omega\sqrt{LC} = 2\pi(100\times10^{6})\sqrt{(120\times10^{-9})(50\times10^{-12})} = 1.54 \text{ rad/m}$

β rad, $\sin\beta\ell = -0.895$, $\cos\beta\ell = -0.446$

Substituting the values,

$$Z_{in} = (\ )\frac{(300)(-0446) + j(49)(-0.895)}{(49)(-0.446) + j(300)(-0.895)} = (49)\frac{-133.8 - j43.9}{-21.85 - j268.5}$$

$$= (\ )\frac{140.81\angle 18.2^\circ}{269.4\angle 85.3^\circ} = 25.61\angle -67.1^\circ = 9.97 - j23.6\ \Omega$$

EXAMPLE 8.17

The characteristic impedance and phase constant of a stub are same as those of the transmission line in Example 8.16. What minimum length ℓ of the stub will have input impedance of (a) $j25\ \Omega$, and (b) $-j30\ \Omega$?

$Z_0 = 49\ \Omega$, $\beta = 1.54$ rad/m

The input impedance of a stub $= jZ_0 \tan\beta\ell = jX_{sc}$

(a) $\tan\beta\ell = \dfrac{X_{sc}}{Z_0} = \dfrac{25}{49} = 0.51$

Hence, $\beta\ell = 0.472$ rad

$$\ell = \frac{0.472}{1.54} = 0.306 \text{ m}$$

(b) $\tan\beta\ell = \dfrac{X_{sc}}{Z_0} = \dfrac{-30}{49} = -0.612$

Hence, $\beta\ell = -0.612$ rad

As ℓ cannot be a negative quantity,

$\beta \ = -0.612 + \pi = 2.593$ rad

$$\ell = \frac{2.593}{1.54} = 1.684 \text{ m}$$

EXAMPLE 8.18

A transmitter of internal resistance $\bar{Z}_g = 20 + j0\ \Omega$ is connected to a load of $\bar{Z}_R = 100 + j0\ \Omega$ through a two-conductor transmission line. The length ℓ of the transmission line is 0.8λ m and its characteristic impedance Z_0 is $50 + j0\ \Omega$. The system is shown in Fig. 8.13(a). Calculate (a) the input impedance and (b) the average power taken by the load if amplitude of the transmitter voltage V_{g0} is $60 + j0$ V.

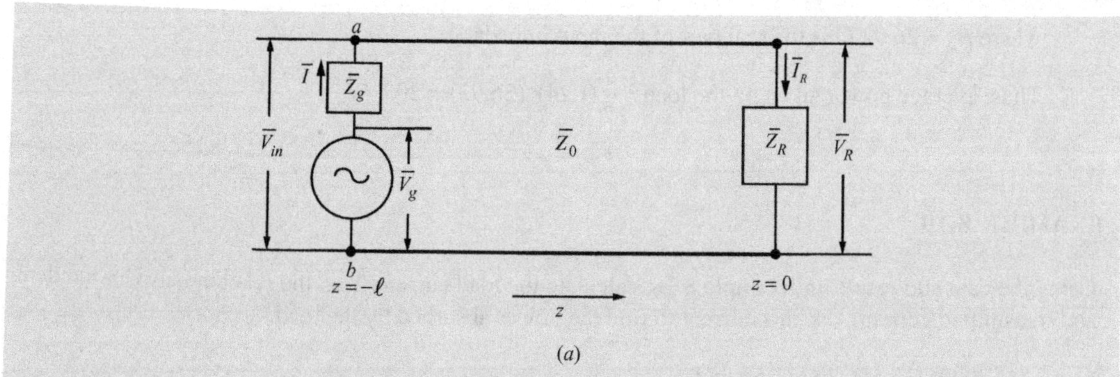

Fig. 8.13 (*a*) Illustration for Example 8.18

The equivalent circuit of the system is shown in Fig. 8.13(*b*), where $\bar{Z}_{in}$ is input impedance of the transmission line to the right of *ab*.

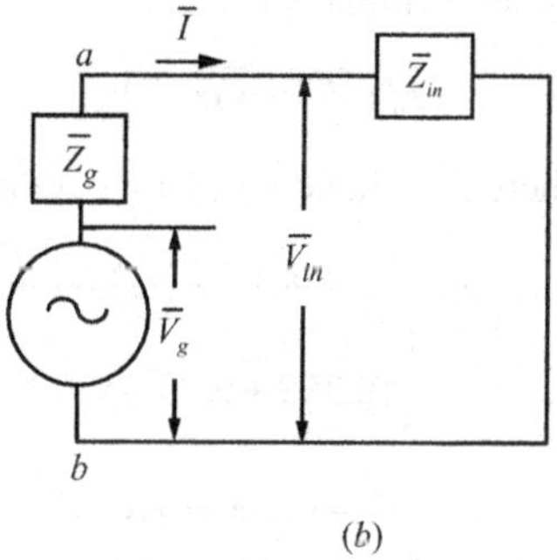

Fig. 8.13 (*b*) Illustration for Example 8.18

(a) $\bar{Z}_{in} = Z_0 \dfrac{1+\bar{\Gamma}e^{-j2\beta\ell}}{1-\bar{\Gamma}e^{-j2\beta\ell}}$

where $\bar{\Gamma}$ is the reflection coefficient at the load end.

$$\bar{\Gamma} = \frac{Z_R - Z_0}{Z_R + Z_0} = \frac{100-50}{100+50} = \frac{1}{3}$$

$$2\beta\ell = 2\left(\frac{2\pi}{\lambda}\right)(0.8\lambda) = 3.2\pi \text{ rad}$$

$$\bar{\Gamma}e^{-j2\beta\ell} = \frac{1}{3}(\cos 3.2\pi - j\sin 3.2\pi) = -0.27 + j0.196$$

$$1+\bar{\Gamma}e^{-j2\beta\ell} = 0.73 + j0.196 = 0.756\angle 15.0^\circ$$

$$1-\bar{\Gamma}e^{-j2\beta\ell} = 1.27 - j0.196 = 1.285\angle -8.7^\circ$$

$$\bar{Z}_{in} = (50)\frac{0.756\angle 15.0^\circ}{1.285\angle -8.7^\circ} = 29.4\angle 23.7^\circ = 26.92 + j11.82\ \Omega$$

(b) Total Impedance, $\bar{Z} = \bar{Z}_g + \bar{Z}_{in} = 46.92 + j11.82 = 48.4\angle 14.1^\circ\ \Omega$

Amplitude of current at *ab*, $\bar{I}_0 = \dfrac{\bar{V}_{g0}}{\bar{Z}} = \dfrac{60}{48.4\angle 14.1^\circ} = 1.24\angle -14.1^\circ$ A

This is also the input current. As the transmission line is loss-free, average power taken by the load is same as the average power delivered at the input terminals. Average power delivered at the input terminals,

$$P_{av} \quad \frac{1}{2} \quad I_0^{\ 2} \quad R_{in}$$

where $R_{in} = 26.92\ \Omega$ is the real part of the input impedance.

$$\text{Thus, average power taken by the load} = \frac{1}{2}(1.24)^2(26.92) = 20.7\ \text{W}$$

EXAMPLE 8.19

Using the data and results in Example 8.18, calculate the load current from the relation between incident and transmitted current. Use this current to find the power absorbed by the load.

$$Z_{..} \quad j0\ \Omega;\ \bar{\Gamma} = 1/3;\ \beta\ell = 1.6\pi\ \text{rad}$$

The current $\bar{I}_0$ is also the current at the input point. Thus,

$$I_{\upsilon} \quad I_0 = 1.24\angle -14.1^\circ$$

As per formula (2),

$$I_{\upsilon} \quad I_0^+(e^{j\beta\ell} - \bar{\Gamma}e^{-j\beta\ell})$$

where I_0^+ is incident current at the load terminals.

$$e^{j} \quad e^{j\ \ \pi} = \cos 1.6\pi + j\sin 1.6\pi = 0.309 - j0.951$$

$$\Gamma e^{\ \ j\beta\ell} = \frac{1}{3}(0.309 + 0.951) = 0.103 + j0.317$$

$$e^{j} \quad e^{\ \ j\beta l} = 0.206 - j1.268 = 1.284\angle -80.8^\circ$$

Substituting the values, we get

$$I_0^+ = \frac{1\ 24\angle -14.1^\circ}{1.284\angle -80.8^\circ} = 0.966\angle 66.7^\circ\ \text{A}$$

$$\text{Load current, } \bar{I}_{0R} = (1 - \bar{\Gamma})\bar{I}_0^+ = \frac{2}{3}(0.966\angle 66.6^\circ) = 0.644\angle 66.6^\circ$$

$$\text{Power absorbed by the load} = \frac{1}{2}\left|\bar{I}_{0R}\right|^2 R_R = \frac{1}{2}(0.644)^2(100) = 20.7\ \text{W}$$

This is same as the value obtained in Example 8.18.

8.8 Transmission Line having Losses

All types of practical transmission lines dissipate some amount of the input power due to losses in the resistance of the conductors, and also in the dielectric medium. The loss in the conductor resistance (ohmic loss) is equal to the product of square of the current in the conductor and its effective resistance. The dielectric loss is equal to the product of square of voltage between the conductors and conductance of an imperfect dielectric material. Hence in the circuit representation of a transmission line, resistance of the conductors is shown as a series element and the conductance of the dielectric material is shown as a shunt element. The circuit model of a long transmission line having losses is as shown in Fig. 8.7(*a*). The treatment of a transmission line having losses is similar to that of a lossless transmission line. Sinusoidal variation of voltage and current with time is assumed.

The Wave Equations and Propagation Constant

Let R, L, C, and G denote resistance, external inductance, capacitance, and conductance respectively for one metre length of a uniform transmission line. Then, the series impedance $\bar{Z}$ and shunt admittance $\bar{Y}$ per unit length of the line are given by

$$\bar{Z} = R + j\omega L$$

$$\bar{Y} = G + j\omega C$$

Considering a very small length Δz of the line as shown in Fig. 8.14, the series impedance of the line element is equal to $\bar{Z}\Delta z$ and its shunt admittance is equal to $\bar{Y}\Delta z$. We find that this network is similar to the network shown in Fig. 8.9(*a*). Hence, the transmission line equations and the wave equations of a two-conductor transmission line having losses may be obtained from those of the lossless line, derived in Section 8.4, by substituting $\bar{Z}$ in place of $j\omega L$ and $\bar{Y}$ in place of $j\omega C$. Accordingly, the transmission line equations using phasor notations are given by

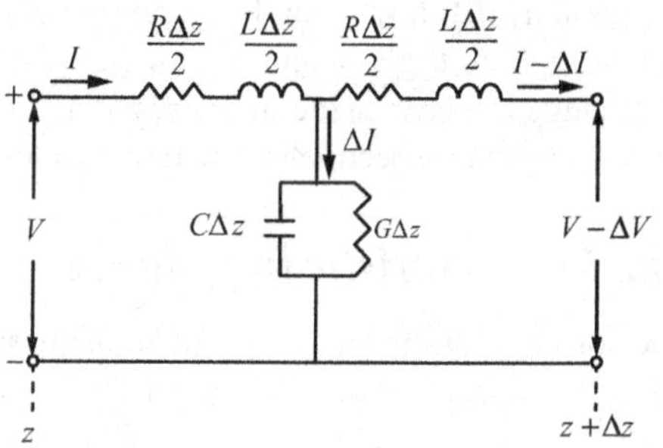

Fig. 8.14 Circuit models for a small section of length Δz of a long transmission line with losses

$$\frac{\partial \bar{V}(z)}{\partial z} = -\bar{Z}\,\bar{I}(z) \qquad (1)$$

$$\frac{\partial \bar{I}(z)}{\partial z} = -\bar{Y}\,\bar{V}(z) \qquad (2)$$

The wave equations for voltage and current of the line are

$$\frac{\partial^2 \bar{V}(z)}{\partial z^2} = (\bar{Z}\bar{Y})\bar{V}(z) = (\bar{\gamma})^2\,\bar{V}(z) \qquad (3)$$

$$\frac{\partial^2 \bar{I}(z)}{\partial z^2} = (\bar{\gamma})^2\,\bar{I}(z) \qquad (4)$$

where

$$\bar{\gamma} = \sqrt{\bar{Z}\bar{Y}} = \sqrt{(R + j\omega L)(G + j\omega C)} \qquad (5)$$

The parameter $\bar{\gamma}$ defined by (5) is called *propagation constant*. It is a complex quantity and its unit is 1/m (also written as m^{-1}). The real and imaginary parts of the propagation constants are denoted by α and β respectively. Thus,

$$\bar{\gamma} = \alpha + j\beta$$

The exponential form of solution of (3) is given by

$$\bar{V}(z) = \bar{V}_0^+ e^{-\bar{\gamma}z} + \bar{V}_0^- e^{\bar{\gamma}z} \qquad (6)$$

Introducing the exponential factor $e^{j\omega t}$ and substituting $\alpha + j\beta$ in place of $\bar{\gamma}$ in (6), we have

$$\bar{V} = \bar{V}_0^+ e^{-\alpha z} e^{j(\omega t - \beta z)} + \bar{V}_0^- e^{\alpha z} e^{j(\omega t + \beta z)} \qquad (7)$$

The first and second terms of (7) represent voltages travelling along positive and negative z directions respectively. Thus, the equations for the two voltages are

$$V^+ = V_0^+ e^{-\alpha z} e^{j(\omega t - \beta z)} \qquad (8)$$

$$\bar{V}^- = \bar{V}_0^- e^{\alpha z} e^{j(\omega t+\beta z)} \tag{9}$$

If we take the amplitude of positively travelling voltage as reference, then

$$\bar{V}_0^+ = V_0^+ + j0 = V_0^+$$

Then the real part of (8) is

$$V^+ = V_0^+ e^{-\alpha z} \cos(\omega t - \beta z) \tag{10}$$

Expression (10) shows that the amplitude of the voltage at any position z along the transmission line is $V_0^+ e^{-\alpha z}$ and the amplitude gets attenuated as the voltage wave progresses along positive z direction. Moreover, βz gives the change in phase angle of the wave at any z. The parameters α and β are called *attenuation constant* and *phase constant* respectively. The unit of attenuation constant is neper/m, and that of phase constant is rad/m.

The Characteristic Impedance

The solution for the current in the transmission line is found by substituting (6) into (1) and then solving the resulting equation. The solution for the current is given by

$$\bar{I}(z) = \frac{1}{\bar{Z}_0}(\bar{V}_0^+ e^{-\bar{\gamma} z} - \bar{V}_0^- e^{\bar{\gamma} z}) \tag{11}$$

where

$$\bar{Z}_0 = \sqrt{\frac{R + j\omega L}{G + j\omega C}} = |\bar{Z}_0| e^{j\psi_0} = |\bar{Z}_0| \angle \psi_0 \tag{12}$$

The parameter $\bar{Z}_0$ is the *characteristic impedance* of the transmission line. It is a phasor, with magnitude $|\bar{Z}_0|$ and phase angle $\angle \psi_0$.

Introducing the exponential factor $e^{j\omega t}$ and substituting $\alpha + j\beta$ in place of $\bar{\gamma}$ in (11), currents associated with voltages travelling along positive and negative z directions are

$$\bar{I}^+ = \frac{V_0^+}{\bar{Z}_0} e^{-\alpha z} e^{j(\omega t - \beta z)} = \frac{V_0^+}{|\bar{Z}_0|} e^{-\alpha z} e^{j(\omega t - \beta z - \psi_0)}$$

$$\bar{I}^- = -\frac{\bar{V}_0^-}{\bar{Z}_0} e^{\alpha z} e^{j(\omega t + \beta z)} = -\frac{V_0^-}{|\bar{Z}_0|} e^{\alpha z} e^{j(\omega t + \beta z - \psi_0)}$$

Reflection Coefficient and Standing Wave Ratio

Expressions for reflection coefficient and standing wave ratio can be obtained by following the same procedure as in the case of the ideal transmission line. Let a transmission line of length ℓ and characteristic impedance Z_0 be connected to a source at one end and to a load at the other end, as shown in Fig. 8.10. The load is at $z = 0$ and the source is at $z = -\ell$. The load impedance consists of resistance R_R and reactance $\pm jX_R$. The complex form of the impedance is

$$\bar{Z}_R = R_R \pm jX_R$$

If the load impedance is not equal to the characteristic impedance both in magnitude and phase, wave reflection will occur at the load end. The reflection coefficient at the load end is given by the ratio of reflected voltage to the incident voltage at $z = 0$. The voltage reflection coefficient $\bar{\Gamma}$ at the load end is

$$\bar{\Gamma} = \frac{\bar{V}_0^-}{V_0^+} = \frac{\bar{Z}_R - \bar{Z}_0}{\bar{Z}_R + \bar{Z}_0} = \Gamma e^{j\psi_r} = \Gamma \angle \psi_r$$

Thus,

$$\bar{V}^- = (\bar{\Gamma} V_0^+) e^{\alpha z} e^{j(\omega t + \beta z)} = (\Gamma V_0^+) e^{\alpha z} e^{j(\omega t + \beta z + \psi_r)}$$

By taking the incident voltage as reference, the expression for the reflected voltage in real time is

$$V^- = \left(|\bar{\Gamma}|V_0^+\right)e^{\alpha z}\cos(\omega t + \beta z + \psi_r) = V_0^- e^{\alpha z}\cos(\omega t + \beta z + \psi_r) \tag{13}$$

The instantaneous currents in the incident and reflected waves are:

$$I^+ = \frac{V_0^+}{|\bar{Z}_0|}e^{-\alpha z}\cos(\omega t - \beta z - \psi_0) = I_0^+ e^{-\alpha z}\cos(\omega t - \beta z - \psi_0) \tag{14}$$

$$I^- = -\frac{|\bar{\Gamma}|V_0^+}{|\bar{Z}_0|}e^{+\alpha z}\cos(\omega t + \beta z + \psi_r - \psi_0)$$

$$= -I_0^- e^{+\alpha z}\cos(\omega t + \beta z + \psi_r - \psi_0) \tag{15}$$

The standing wave ratio in terms of the magnitude of voltage reflection coefficient is given by

$$S_R = \frac{1 + |\bar{\Gamma}|}{1 - |\bar{\Gamma}|}$$

Power Relations and Power Loss

Referring to (10) and (14) we find that the phase difference between the voltage and current of the positively travelling wave is ψ_0. And from (13) and (15) we find that the phase difference between the voltage and current in the reflected wave is $\pi + \psi_0$. The average value of power in the incident wave at the load end ($z = 0$) is

$$P_{av}^+ = \frac{V_0^+ I_0^+}{2}\cos\psi_0 = \frac{(V_0^+)^2}{2|\bar{Z}_0|}\cos\psi_0$$

Average value of reflected power at $z = 0$ is

$$P_{av}^- = \frac{V_0^- I_0^-}{2}\cos\psi_0 = \frac{(V_0^-)^2}{2|\bar{Z}_0|}\cos\psi_0 = \frac{|\bar{\Gamma}|^2 (V_0^+)^2}{2|\bar{Z}_0|}\cos\psi_0 = |\bar{\Gamma}|^2 P_{av}^+$$

The reflected power is in negative z direction. Power flow into the load is the difference between the values of incident and reflected powers.

Power loss in the conductor resistance per unit length of the transmission line is equal to $I_0^2 R/2$, where I_0 is amplitude of current and R is value of the skin-effect resistance per unit length of the line. The approximate formulae for apparent resistances per unit length of three types of transmission line are given in the table in Section 8.3. Power loss in the dielectric medium between the conductors per unit length of the transmission line is equal to $V_0^2 G/2$, where V_0 is amplitude of the voltage and G is conductance of the dielectric per unit length. Thus considering the voltage and current at the load end ($z = 0$) for the sake of simplicity, the total power loss per unit length is given by

$$P_\Omega = \frac{I_0^2 R}{2} + \frac{V_0^2 G}{2} = \frac{V_0^2}{2}\left(\frac{R}{Z_0^2} + G\right);$$

where Z_0 is magnitude of the characteristic impedance of the line.

To obtain expression for the wave impedance at a distance $-z$ from the load end of Fig. 8.10, we will use the phasor forms of voltage and current. The expression for the total voltage and total current at any point are obtained by substituting $-z$ in place of z in (6) and (11). The total voltage is

$$V(-z) = V_0^+ e^{+\gamma z} + V_0^- e^{-\gamma z} = V_0^+ (e^{\gamma z} \text{[illegible]} e^{-\gamma z})$$

The total current is

$$\bar{I}(-z) = \frac{1}{\bar{Z}_0}(\bar{V}_0^+ e^{+\bar{\gamma}z} - \bar{V}_0^- e^{-\bar{\gamma}z}) = \frac{\bar{V}_0^+}{\bar{Z}_0}(e^{+\bar{\gamma}z} - \bar{\Gamma} e^{-\bar{\gamma}z})$$

The wave impedance $Z(-z)$ is equal to the ratio between $\bar{V}(-z)$ and $\bar{I}(-z)$.

$$\bar{Z}(-z) = \frac{\bar{V}(-z)}{\bar{I}(-z)} = \bar{Z}_0 \frac{e^{\bar{\gamma}z} + \bar{\Gamma} e^{-\bar{\gamma}z}}{e^{\bar{\gamma}z} - \bar{\Gamma} e^{-\bar{\gamma}z}} \tag{16}$$

The formula for the input impedance is found by replacing z by ℓ in (16). The input impedance,

$$\bar{Z}_{in} = \bar{Z}_0 \frac{1 + \bar{\Gamma} e^{-2\bar{\gamma}\ell}}{1 - \bar{\Gamma} e^{-2\bar{\gamma}\ell}} \tag{17}$$

EXAMPLE 8.20

A two-conductor transmission line has the following constants.

Series resistance, $R = 0.5\ \Omega$/m

Inductance, $L = 50 \times 10^{-9}$ H/m

Capacitance, $C = 0.02 \times 10^{-9}$ F/m

Calculate characteristic impedance, attenuation constant and phase constant at 25 MHz. Neglect the shunt conductance.

Characteristic impedance, $\bar{Z}_0 = \sqrt{\dfrac{R + j\omega L}{G + j\omega C}}$

$R \quad j\ldots L \quad 0.5 + j2\pi(25 \times 10^6)(50 \times 10^{-9}) = 0.5 + j7.85 = 7.87\angle 86.4°\ \Omega$/m

$G \quad j\ C \quad 0 + j2\pi(25 \times 10^6)(0.02 \times 10^{-9}) = (3.14 \times 10^{-3})\angle 90°$ S/m

$$Z_0 = \sqrt{\frac{7.87\angle 86.4°}{(3\ 14 \times 10^{-3})\angle 90°}} = \sqrt{2506.4\angle -3.6°} = 50.1\angle -1.8° = 50.1 - j1.57\ \Omega/\text{m}$$

The propagation constant,

$$\gamma \quad \sqrt{[illegible]\ j\omega L)(G + j\omega C)} = \sqrt{(7.87\angle 86.4°)(3.14 \times 10^{-3}\angle 90°)}$$

$$= \sqrt{[illegible]\angle 176.4} = 0.157\angle 88.2° = 0.0049 + j0.157\ \text{m}^{-1}$$

Attenuation constant, $\alpha = 0.0049$ neper/m

Phase constant, $\beta = 0.157$ rad/m

EXAMPLE 8.21

A two-conductor transmission line of length 6.5 m is terminated with a load of impedance $\bar{Z}_R = 40 + j25\ \Omega$. The transmission line has the following parameters:

Characteristic impedance, Z [illegible] $j2\ \Omega$

Propagation constant, $\gamma =$ [illegible] $+ j0\ 6\ \text{m}^{-1}$

Calculate the input impedance.

Phase angles are expressed in radians.

Input impedance, $\bar{Z}_{in} = \bar{Z}_0 \dfrac{1+\bar{\Gamma}e^{-2\bar{\gamma}\ell}}{1-\bar{\Gamma}e^{-2\bar{\gamma}\ell}}$

$$Z_{\_} = 50 - j2 = 50.04\angle{-0.04}\ \Omega$$

$$\Gamma = \frac{\bar{Z}_R - \bar{Z}_0}{\bar{Z}_R + \bar{Z}_0} = \frac{(40+j25)-(50-j2)}{(40+j25)+(50-j2)} = \frac{28.79\angle 1.93}{92.89\angle 0.25} = 0.31\angle 1.68 = 0.31e^{j1.68}$$

$$\Gamma e^{-2\bar{\gamma}\ell} = [0.31e^{j1.68}][e^{-2(0.004+j0.6)(6.5)}] = 0.31e^{-0.052}e^{-j6.12} = 0.287 + j0.041$$

$$1\ \bar{\Gamma}e^{-2\bar{\gamma}\ell} = 1.287 + j0.041 = 1.28\angle 0.032$$

$$1\ \bar{\Gamma}e^{-2\bar{\gamma}\ell} = 0.713 - j0.041 = 0.714\angle{-0.057}$$

$$Z_{in} = (50.04\angle{-0.04})\frac{1.28\angle 0.032}{0.714\angle{-0.057}} = 89.7\angle 0.049 = 89.59 + 4.39\ \Omega$$

8.9 Matching of Transmission Lines with Load

If the characteristic impedance of a transmission line is equal to the load impedance, both in magnitude and phase, then there will not be any reflection of power from the load end. The input impedance in this case is also equal to the characteristic impedance of the line. In the case of a mismatch between characteristic impedance of the line and the load impedance, wave reflection will occur. As a result, part of the input power will be reflected, thereby reducing the efficiency of the transmission system. To overcome the problem, partially or fully, a matching device is incorporated near the load-end of the transmission line.

Matching of characteristic impedance of a transmission line with load impedance maximises power transmitted to the load, reduces losses, and minimises signal distortion.

Some of the matching techniques are briefly described below.

Impedance Matching by the Use of a Transformer

A transformer of suitable turn-ratio, with its primary winding connected to the transmission line and the secondary winding connected to the load, is used as a matching device in the audio-frequency range. This is illustrated in Fig. 8.15(*a*). Notations N_1 and N_2 in the figure denote number of turns of primary and secondary windings whereas R_R denotes the load resistance. The characteristic impedance of the transmission line, which is assumed to be of low-loss type, is taken as Z_0. If we assume that the transformer is ideal (losses, magnetising current and leakage fluxes are negligibly small), the volt-ampere rating of both the windings would be same. Thus, we have

$$V_2 I_2 = V_1 I_1$$

Or

$$\frac{V_2}{V_1} = \frac{I_1}{I_2} \tag{1}$$

where I_1 and I_2 are primary and secondary currents. Moreover, ratio between the terminal voltages V_2 and V_1 of primary and secondary windings can be taken as equal to the ratio of N_2 and N_1. That is,

$$\frac{V_2}{V_1} = \frac{N_2}{N_1} \tag{2}$$

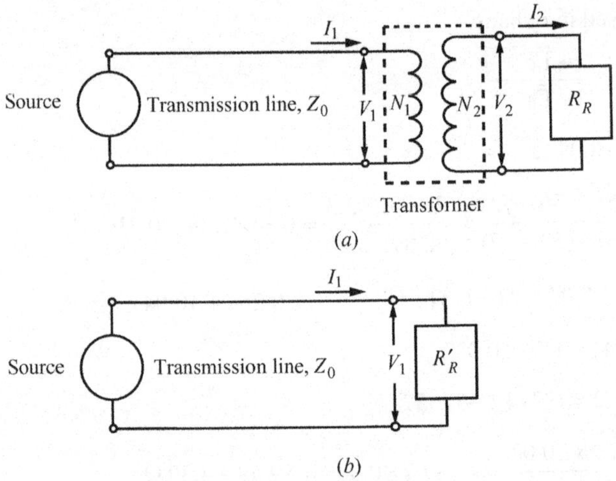

Fig. 8.15 Illustration of impedance matching of a transmission line with load by the use of a transformer

From relations (1) and (2), we get

$$I_2 = \frac{I_1 N_1}{N_2} \tag{3}$$

Assuming the voltages and currents to be rms values, average power taken by the load

$$P_R = I_2^2 R_R \tag{4}$$

By replacing I_2 of (4) by the right-hand side of (3) we will get the power in terms of the line current I_1. It is given by

$$P_R = I_1^2 \left(\frac{N_1^2}{N_2^2} R_R \right) = I_1^2 R_R' \tag{5}$$

where $$R_R' = \frac{N_1^2}{N_2^2} R_R$$

The resistance R_R' is called *equivalent* load resistance referred to the primary side of the transformer. Relation (5) shows that power taken by the load is equal to power taken by the resistance R_R' through which current I_1 flows. Thus, the scheme is equivalent to a transmission line terminated with a resistance R_R', as shown in Fig. 8.15(*b*). Thus, matching is possible if R_R' is made equal to the characteristic impedance Z_0 of the transmission line. That is,

$$R_R' = \frac{N_1^2}{N_2^2} R_R = Z_0 \tag{6}$$

Relation (6) determines the turn ratio of the transformer. It is given by

$$\frac{N_2}{N_1} = \sqrt{\frac{R_R}{Z_0}} \tag{7}$$

For loads having resistance as well as reactance, a transformer may be used for matching the magnitude of load impedance with the magnitude of line impedance. The turn ratio of a matching transformer in this case is given by

$$\frac{N_2}{N_1} = \frac{Z_R}{Z_0} \tag{8}$$

By matching the magnitudes of impedances the power transfer is maximised. But, it is less than the maximum possible power transfer. For maximum power transfer the two impedances must be complex conjugates of each other.

Impedance Matching by the Use of a Quarter-Wave Line

Another impedance matching technique, suitable at high frequencies, makes use of the impedance transformation property of a transmission line of length equal to one-quarter of a wavelength at the operating frequency. For purely resistive loads, a low-loss (assumed to be lossless) quarter-wave line of appropriate characteristic impedance is connected in series between the load terminals and the main transmission line.

If the load consists of both resistance and reactance, then the load-side end of the quarter-wave line is connected at the position of the voltage maximum (or voltage minimum) nearest to the load. At these positions the line impedance is purely resistive.

The condition for matching is that the characteristic impedance of the main line is equal to the input impedance of the quarter-wave line at the point of connection. The input impedance of the quarter-wave line is a function of its characteristic impedance and the load impedance. We will derive the formulae for the characteristic impedance of a matching quarter-wave line assuming the main line to be lossless.

Let us consider a quarter-wave transmission line of characteristic impedance Z_{0q} and wavelength λ_q at signal frequency. One end of the quarter-wave line is connected to the main transmission line at *ab* and the other end is terminated with a load of resistance R_R, as illustrated in Fig. 8.16(*a*). The characteristic impedance of the main line is taken as Z_{0m}. Referring to formula (6) of Section 8.7, the input impedance of a lossless transmission line of characteristic impedance Z_0, phase constant β and load resistance R_R is

$$\bar{Z}_{in} = Z_0 \frac{R_R \cos 2\pi(\ell/\lambda) + jZ_0 \sin 2\pi(\ell/\lambda)}{Z_0 \cos 2\pi(\ell/\lambda) + jR_R \sin 2\pi(\ell/\lambda)} \tag{9}$$

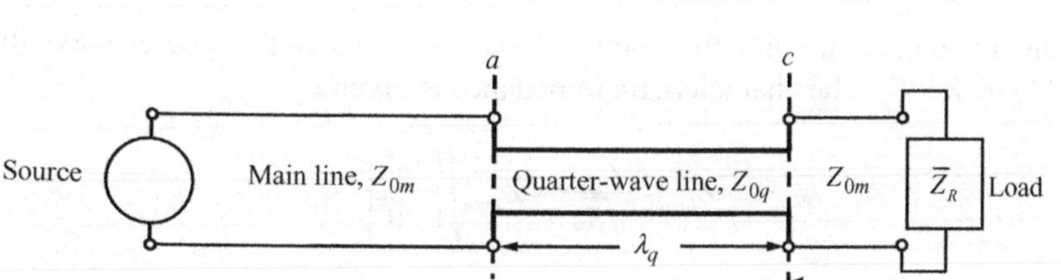

(*a*)

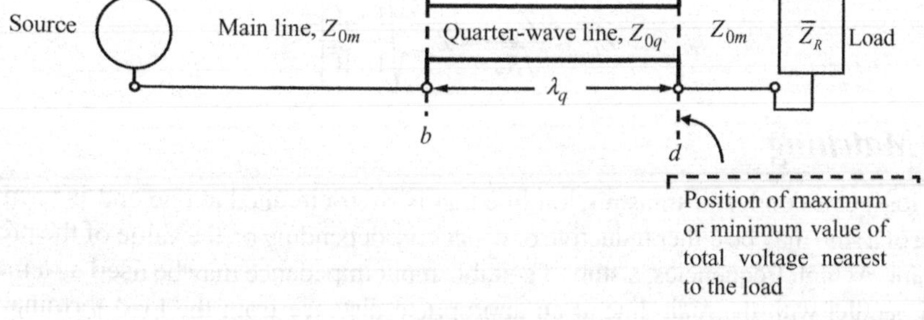

(*b*)

Fig. 8.16 Impedance matching schemes by the use of quarter-wave transmission line

Substituting $\ell/\lambda_q = 1/4$ and replacing Z_0 by Z_{0q} in (9), the input impedance Z_{inq} of the quarter-wave line at *ab* is

$$Z_{inq} = \frac{Z_{0q}^{\ 2}}{R_R} \tag{10}$$

In order that there is no reflection at *ab*, Z_{inq} must be equal to the characteristic resistance of the main line Z_{0m}. Using this condition in (10), we get

$$Z_{0q} = \sqrt{Z_{0m} R_R} \tag{11}$$

Thus, a lossless quarter-wave line of characteristic impedance equal to the rhs of (11) transforms the load resistance to the characteristic impedance of the main line. The matching line is therefore called a *quarter-wave transformer.*

Relation (11) is also satisfied if the length of the matching line is an odd multiple of a quarter-wavelength. As the wavelength is a function of frequency, matching is possible only at one frequency. However, the technique can still be used over a limited frequency range if the standing wave ratio at *ab* is within 1.5.

If the load consists of resistance and reactance then matching cannot be achieved by connecting a quarter-wave line directly across the load terminals. However, by connecting the load-end of the line where the amplitude of total voltage is maximum or minimum, matching is possible because the line impedance at these positions is purely resistive.

The line impedances at the positions of voltage maximum and voltage minimum are

$$Z_{max} = S_R Z_{0m}$$

and

$$Z_{min} = Z_{0m}/S_R$$

where S_R is the standing wave ratio at the load end. This suggests that matching can be achieved by the use of a low-loss quarter-wave transmission line by connecting it as in Fig. 8.16(*b*) where the load-side end of the quarter-wave line at *cd* coincides with the position of either the first maximum or first minimum of total voltage from the load end. The resistance $S_R Z_{0m}$ or Z_{0m}/S_R may be treated as a load resistance connected to the quarter-wave line at *cd*. Thus, if *cd* is the position of voltage maximum, the characteristic impedance of the quarter-wave line is obtained by replacing R_R of (11) by $S_R Z_{0m}$. The characteristic impedance is given by

$$Z_{0q} = Z_{0m}\sqrt{S_R} = Z_{0m}\sqrt{\frac{1+|\bar{\Gamma}|}{1-|\bar{\Gamma}|}} \tag{12}$$

If *cd* is the position of voltage minimum, the characteristic impedance of the quarter-wave line is obtained by replacing R_R of (11) by Z_{0m}/S_R. The characteristic impedance is given by

$$Z_{0q} = \frac{Z_{0m}}{\sqrt{S_R}} = Z_{0m}\sqrt{\frac{1-|\bar{\Gamma}|}{1+|\bar{\Gamma}|}} \tag{13}$$

Single-Stub Matching

A lossless (low loss) two-conductor transmission line that is short-circuited at one end is termed as a stub. The input impedance of a stub may be either inductive or capacitive depending on the value of the product of its length and phase constant. At high frequencies, a stub of suitable input impedance may be used as a matching device by connecting it in parallel with the main line at an appropriate distance from the load terminals. The scheme is shown in Fig. 8.17. The stub of length ℓ_s, characteristic impedance Z_{0s}, and phase constant β_s is connected in

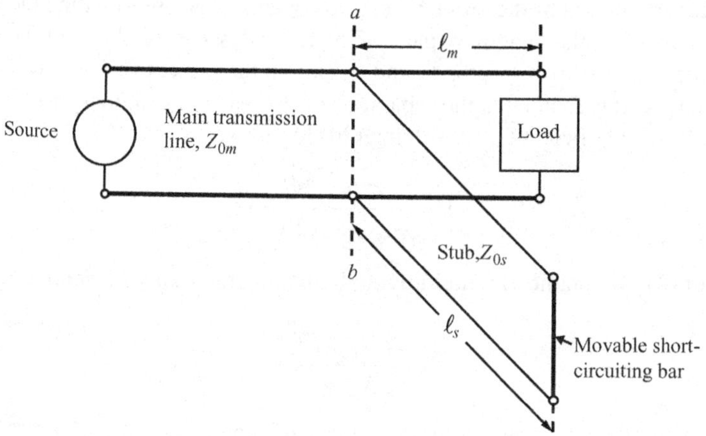

Fig. 8.17 Illustration for single-stub matching of transmission line with load

parallel to the main line at *ab*. The distance of *ab* from the load terminals is ℓ_m. The characteristic impedance of the main line, which is assumed to be purely resistive, is Z_{0m}. It may be seen from the figure that that there are two parallel branch lines to the right of *ab*. One is a part of the main line connected to the load and the other is the stub. In order that there is no reflection at *ab*, the total impedance of the two parallel branches should be equal to the characteristic impedance of the main line. The impedance of the main line to the right of *ab* is its input impedance seen from *ab* towards the load end. And the impedance of the stub is its input impedance at *ab*. The matching problem is to calculate the stub length ℓ_s and distance of its location ℓ_m from the load terminals.

The input impedance of the main line viewed from *ab* towards the load is

$$\bar{Z}_{in} = Z_{0m}\frac{1+\bar{\Gamma}e^{-j2\beta_m\ell_m}}{1-\bar{\Gamma}e^{-j2\beta_m\ell_m}} \tag{14}$$

where β_m is the phase constant of the main line and $\bar{\Gamma}$ is the reflection coefficient at the load terminals. It is given by

$$\bar{\Gamma} = \left|\bar{\Gamma}\right|e^{j\psi_r} \tag{15}$$

The input impedance of the stub at *ab* is

$$\bar{Z}_{sc} = jZ_{0s}\tan\beta_s\ell_s = jX_{sc} \tag{16}$$

The reactance of the stub,

$$X_{sc} = Z_{0s}\tan\beta_s\ell_s \tag{17}$$

For no wave reflection at *ab*, the parallel combination of $\bar{Z}_{in}$ and $\bar{Z}_{sc}$ should be equal to the characteristic impedance of the main line Z_{0m}. Accordingly,

$$Z_{0m} = \frac{\bar{Z}_{in}\bar{Z}_{sc}}{\bar{Z}_{in}+\bar{Z}_{sc}} \tag{18}$$

Or

$$\bar{Z}_{sc} = \frac{\bar{Z}_{in}\bar{Z}_{0m}}{\bar{Z}_{in}-\bar{Z}_{0m}} \tag{19}$$

By substituting (14) and (16) in (19), we get

$$jX_{sc} = Z_{0m}\frac{1+Z_{vm}\;^{-}e^{-j2\beta_m\ell_m}}{2\Gamma e^{-j2\beta_m\ell_m}} \tag{20}$$

This apparently gives the solution of the problem, as two equations, one by equating the real parts of the two sides and the other by equating the imaginary parts of the two sides, can be obtained. The solution of the two equations will give the two unknown lengths ℓ_m and ℓ_s. But, in the present form the real and imaginary parts cannot be separated easily. It is found that the separation of the real and imaginary parts becomes easy if (20) is solved for the reflection coefficient $\bar{\Gamma}$. By solving (20) for $\bar{\Gamma}$ we will get

$$\bar{\Gamma} = \frac{Z_{0m} e^{j(2\beta_m \ell_m)}}{-Z_{0m} + j(2X_{sc})} \tag{21}$$

A convenient form of (21) is obtained by multiplying both numerator and the denominator of the rhs by the operator $-j$. Thus,

$$\bar{\Gamma} = \frac{-jZ_{0m} e^{j(2\beta_m \ell_m)}}{2X_{sc} + jZ_{0m}}$$

Writing the reflection coefficient by its magnitude and phase angle, and since $-j = e^{-j(\pi/2)}$, we have

$$|\bar{\Gamma}| e^{j\psi_r} = \frac{Z_{0m} e^{j(2\beta_m \ell_m - \pi/2)}}{\left(\sqrt{4X_{sc}^2 + Z_{0m}^2}\right) e^{j\varphi}} = \frac{Z_{0m} e^{j(2\beta_m \ell_m - \pi/2 - \varphi)}}{\sqrt{4X_{sc}^2 + Z_{0m}^2}} \tag{22}$$

where
$$\varphi = \tan^{-1}\left(\frac{Z_{0m}}{2X_{sc}}\right) \tag{23}$$

The value of the rhs of (22) will not change even if it is multiplied by $e^{j(2n\pi)}$; $n = 0, \pm 1, \pm 2, \ldots$,. So,

$$|\bar{\Gamma}| e^{j\psi_r} = \frac{Z_{0m} e^{j(2\beta_m \ell_m - \pi/2 - \varphi + 2n\pi)}}{\sqrt{4X_{sc}^2 + Z_{0m}^2}} \tag{24}$$

Equating magnitudes of two sides of (24) gives

$$|\bar{\Gamma}| = \frac{Z_{0m}}{\sqrt{4X_{sc}^2 + Z_{0m}^2}}$$

Or
$$X_{sc} = \pm \frac{Z_{0m}\sqrt{1 - |\bar{\Gamma}|^2}}{2|\bar{\Gamma}|} \tag{25}$$

From (17) and (25), we have

$$\tan \beta_s \ell_s = \tan 2\pi(\ell_s/\lambda_s) = \pm \frac{X_{sc}}{Z_{0s}}$$

Thus,
$$\ell_s = \frac{\lambda_s}{2\pi} \tan^{-1}\left(\pm \frac{X_{sc}}{Z_{0s}}\right) \tag{26}$$

where λ_s is the wavelength in the stub. This gives the solution for the length of the stub ℓ_s

Now, by equating the angles of two sides of (24), we get

$$2\beta_m \ell_m = \psi_r + \pi/2 + \varphi - 2n\pi$$

As β_m π/λ_m where λ_m is the wavelength in the main line at the operating frequency, we have

$$\frac{\ell_m}{\lambda_m} = \frac{\psi_r \quad \pi/2 + \varphi \quad n\pi}{4\pi}$$

Or
$$\ell_m = \frac{\lambda_m}{4\pi}(\psi_r + \pi/2 + \varphi - n\pi) \tag{27}$$

This gives the solution for ℓ_m.

In (27), ψ_r is the phase angle of the reflection coefficient and φ is given by (23). We find from (25) that the value of input reactance of the stub X_{sc} is positive as well as negative. Therefore, the stub length ℓ_s, and the distance of the stub ℓ_m from the load point will have two sets of values; one set for positive X_{sc} and another set for negative X_{sc}.

The following procedure may be used for solving numerical problems.

1. Calculate the reflection coefficient from the formula
$\bar{\Gamma} = \dfrac{\bar{Z}_R - \bar{Z}_0}{\bar{Z}_R + \bar{Z}_0} = |\bar{\Gamma}| e^{j\psi_r}$; $|\bar{\Gamma}|$ and ψ_r are the magnitude and phase angle of the reflection coefficient.
2. Calculate input reactance of the stub X_{sc} from (25).
3. Obtain the stub length from (26) both for positive and negative values of X_{sc}.
4. Calculate φ from (23) for both positive and negative values of X_{sc}
5. Find the position of the stub ℓ_m from (27) for the two values of φ obtained in 4.

EXAMPLE 8.22

A two-conductor lossless transmission line of characteristic impedance 300 Ω is terminated with a load resistance of 450 Ω. The line length is 3 times the wavelength at the operating frequency. The load is matched by a suitable two-winding transformer connected at the load end. The transmission line is fed by a source of voltage with amplitude of $120\angle 0$ V and internal resistance 50 Ω. Find (a) the voltage and current of the transformer windings and (b) the ratio of power absorbed by the load before and after using the matching transformer. Assume perfect matching and an ideal transformer.

Characteristic impedance, $Z_0 = 300\ \Omega$

Load resistance, $R_R = 450\ \Omega$

Line length, $\ell = 3\lambda$

Source voltage = 120 V

Source resistance = 50 Ω

(a) Ratio of secondary winding turns to primary winding turns of the transformer for perfect matching,

$$\frac{N_2}{N_1} = \sqrt{\frac{R_R}{Z_0}} = \sqrt{\frac{450}{300}} = 1.225$$

The equivalent load-resistance $R_R' = (N_1/N_2)^2 (R_R) = 300\ \ \Omega$

As the equivalent load-resistance is same as the characteristic impedance, the input impedance at the feeding point is equal to the characteristic impedance of 300 Ω. Thus, amplitude of current at the feeding point,

$$I_{in} = \frac{120}{300 + 50} = 0.343\ \text{A}$$

Voltage at the feeding point, $V_{in} = 120 - (0.343)(50) = 102.85$ V

Let V_R' be the amplitude of voltage at the primary terminals of the transformer. Then, as there is no reflection,

$$V_R' \ [illegible]\ V_{in} e^{-j} \ [illegible]\ e^{j6\pi} = [illegible] + j0\ \text{V}$$

This is the amplitude of voltage V_{10} across the primary winding.

Amplitude of current in the primary winding, $I_{10} = V_{10}/R_R' = 102.85/300 = 0.343$ A

Amplitude of voltage across the secondary winding, $V_{20} = V_{10}(N_2/N_1) = 126.0$ V

Amplitude of current in the secondary winding, $I_{20} = I_{10}(N_1/N_2) = 0.280$ A

(b) Reflection coefficient before using the transformer, $\Gamma = \dfrac{R_R - Z_0}{R_R + Z_0} = \dfrac{450-300}{450+300} = 0.2$

The reflection coefficient with the matching transformer is zero; hence power received by the load is equal to the incident power. As the transmission line is lossless, power absorbed by the load is also equal to the input power. Before using the matching transformer, power received by the load is $1-|\Gamma|^2 = 1-0.2^2 = 0.96$ of the input power. Thus, the ratio is 0.96.

EXAMPLE 8.23

A two-conductor transmission line of characteristic impedance $50 + j0\ \Omega$ is terminated with a resistor of $70\ \Omega$. A low-loss quarter-wave coaxial transmission line is to be used between the load and the main line for the purpose of matching. Find the ratio of radius of the outer conductor to that of the inner conductor of the quarter-wave line if the relative permittivity of the nonmagnetic dielectric between its conductors is 2.2.

Characteristic impedance of the main line, $Z_{0m} = 50\ \Omega$
Terminal resistance, $R_R = 70\ \Omega$
Relative permittivity of the dielectric in the quarter-wave line, $\varepsilon_r = 2.2$
Characteristic impedance of the quarter-wave line for no reflection,

$$Z_{0q} = \sqrt{Z_{0m}R_R} = \sqrt{(50)(70)} = 59.16\ \Omega$$

If a and b are the radii of the inner and outer conductors of the quarter-wave line, its characteristic impedance,

$$Z_{0q} = \sqrt{\frac{L}{C}} = \sqrt{\frac{\mu}{\varepsilon}}\frac{\ln(b/a)}{2\pi} = \frac{377}{\sqrt{2.2}}\frac{\ln(b/a)}{2\pi}$$

Substituting the values,

$$59.16 = \frac{377}{\sqrt{2.2}}\frac{\ln(b/a)}{2\pi} = 40.45\ln(b/a)$$

$$\text{Or } \ln\frac{b}{a} = \frac{59.16}{40.45} = 1.463$$

And, $b/a = 4.317$

EXAMPLE 8.24

A two-conductor transmission line of characteristic impedance $Z_{0m} = 150\ \Omega$ is terminated with a load of impedance Z $j300\ \Omega$. A single stub is to be used for the purpose of matching. Obtain the length of the stub in terms of wavelength in the stub. Also, find the distance from the load end at which the stub is to be connected in terms of the wavelength in the main line. The characteristic impedance of the stub $Z_{0s} = 100\ \Omega$.

Reflection coefficient at the load end,

$$\Gamma = \frac{\bar{Z}_R - Z_0}{\bar{Z}_R + Z_0} = \frac{150 - j300 - 150}{150 - j300 + 150} = \frac{300\angle -\pi/2}{300\sqrt{2}\angle -\pi/4} = 0.707\angle -\pi/4$$

$\Gamma = 0.707;\ \psi_r = -\pi/4$ rad

$$X_{sc} = \pm\frac{Z_{0m}\sqrt{1-|\bar{\Gamma}|^2}}{2|\bar{\Gamma}|} = \pm\frac{150\sqrt{1-(0.707)^2}}{2(0.707)} = \pm 75\ \Omega$$

$$\tan 2\pi(\ell_s/\lambda_s) = \frac{X_s}{Z_{0s}} = \pm\frac{75}{100} = \pm 0.75$$

For the positive value,

$$\frac{2\pi\ell_s}{\lambda_s} = \tan^{-1}(0.75) = 0.644 \text{ radian}$$

$$\frac{\ell_s}{\lambda_s} = 0.102$$

For the negative value,

$$\frac{2\pi\ell_s}{\lambda_s} = \tan^{-1}(-0.75) = -0.644 \text{ radian}$$

As ℓ_s cannot be negative,

$$\frac{2\pi\ell_s}{\lambda_s} = \pi - 0.644$$

$$\frac{\ell_s}{\lambda_s} = 0.398$$

The ratio between ℓ_m, the distance of position of the stub from the load, and the wavelength λ_m of the main line is given by

$$\frac{\ell_m}{\lambda_m} = \frac{\psi_r + \pi/2 + \varphi - 2n\pi}{4\pi}$$

where $\varphi = \tan^{-1}\left(\frac{Z_{0m}}{2X_s}\right)$ radian

For positive X_{sc},

$$\varphi = \tan^{-1}\left(\frac{150}{2(75)}\right) = \frac{\pi}{4} \text{ radian}$$

$$\frac{\ell_m}{\lambda_m} = \frac{-(\pi/4) + \pi/2 + \pi/4 - 2n\pi}{4\pi} = \frac{1}{8}, \quad \text{for } n = 0$$

For negative X_{sc},

$$\varphi = \tan^{-1}\left(\frac{150}{2(-\ \ )}\right)$$

As the angle is in the second quadrant, $\varphi = 3\pi/4$

$$\frac{\ell_m}{\lambda_m} = \frac{-(\pi/4) + \pi/2 + 3\pi/4 - 2n\pi}{4\pi} = \frac{1}{4}, \quad \text{for } n = 0$$

Hence, either $\ell_s = 0.102\lambda_s$ and $\ell_m = 0.125\lambda_m$ or $\ell_s = 0.398\lambda_s$ and $\ell_m = 0.25\lambda_m$

SUMMARY

At first we described briefly various types of two-conductor transmission lines, and then developed the circuit model of a long uniform transmission line. From the circuit model, we derived the transmission line equations and wave equations for voltage and current. The wave equations and their solutions are similar in form to those of the fields of a uniform plane wave discussed in Chapter 7. We calculated various parameters of a uniform line and noted that the phase constant and wavelength are functions of ω, L, and C, and the phase velocity and characteristic impedance are functions of L and C only.

We studied the reflection phenomenon at discontinuities and obtained expressions for the reflection and transmission coefficients. We discussed two special cases, namely, line terminated with an open circuit and with a short circuit. In the case an open-circuited line, the incident voltage and current get totally reflected without and with a change of sign respectively, whereas for a short-circuited line, the incident voltage and current get totally reflected, respectively with and without a change of sign. We also calculated average powers in the reflected and transmitted waves in terms of the incident power. The transmitted power is found to be $1 - \Gamma^2$ times the incident power, where Γ is the reflection coefficient. If a line is terminated with an unmatched impedance, then the total voltage/current at any point in the line is equal to sum of the voltages/currents of the incident and reflected waves. We derived expressions for the total voltage and total current and defined the standing wave ratio. The standing wave ratio is equal to the ratio of amplitudes of maximum voltage (or current) to the minimum voltage (or current). We obtained expressions for the input impedance and found that the input impedance of a lossless transmission line terminated with an open circuit or with a shirt circuit is purely reactive. We also discussed matching of transmission lines with load.

The following table gives analogy between a uniform transmission line and a uniform plane wave.

Transmission Line	**Plane Wave**
Voltage	Electric field intensity
Current	Magnetic field intensity
Characteristic impedance, $Z_0 = \sqrt{L/C}$	Intrinsic impedance, $\eta = \sqrt{\mu/\varepsilon}$
Phase constant, $\beta = \omega\sqrt{LC}$	Phase constant, $\beta = \omega\sqrt{\mu\varepsilon}$
Propagation constant, $\bar{\gamma} = \sqrt{(R + j\omega L)(G + j\omega C)}$	Propagation constant, $\bar{\gamma} = \sqrt{(j\omega\mu)(\sigma + j\omega\varepsilon)}$
Phase velocity, $v = 1/\sqrt{LC}$	Phase velocity, $v = 1/\sqrt{\mu\varepsilon}$
Wavelength, $\lambda = 2\pi/\omega\sqrt{LC}$	Wavelength, $\lambda = 2\pi/\omega\sqrt{\mu\varepsilon}$
Reflection coefficient, $\Gamma = \dfrac{R_R - Z_0}{R_R + Z_0}$	Reflection coefficient, $\Gamma = \dfrac{\eta_2 - \eta_1}{\eta_2 + \eta_1}$
Standing wave ratio, $S_R = \dfrac{1+\Gamma}{1-\Gamma}$	Standing wave ratio, $S_R = \dfrac{1+\Gamma}{1-\Gamma}$

IMPORTANT FORMULAE

Ideal Transmission Line

Generalised transmission line equations, $\dfrac{\partial V}{\partial z} = -L\dfrac{\partial I}{\partial t}, \dfrac{\partial I}{\partial z} = -C\dfrac{\partial V}{\partial t}$

Generalised wave equations, $\dfrac{\partial^2 V}{\partial z^2} = LC\dfrac{\partial^2 V}{\partial t^2}, \dfrac{\partial^2 I}{\partial z^2} = LC\dfrac{\partial^2 I}{\partial t^2}$

Sinusoidal variation with time:

Wave equations, $\dfrac{\partial^2 \bar{V}(z)}{\partial z^2} = -\omega^2 LC\bar{V}(z), \dfrac{\partial^2 \bar{I}(z)}{\partial z^2} = -\omega^2 LC\bar{I}(z)$

Phase constant, $\beta = \omega\sqrt{LC}$

Phase velocity, $v = \dfrac{1}{\sqrt{LC}} = \dfrac{\omega}{\beta}$

Wavelength, $\lambda = \dfrac{2\pi}{\beta} = \dfrac{v}{f}$

Characteristic impedance, $Z_0 = \sqrt{\dfrac{L}{C}}$

Voltage reflection coefficient (purely resistive load), $\Gamma = \dfrac{R_R - Z_0}{R_R + Z_0}$

Voltage transmission coefficient (purely resistive load), $\tau = \dfrac{2R_R}{R_R + Z_0} = 1 + \Gamma$

Voltage reflection coefficient (complex load), $\bar{\Gamma} = \dfrac{\bar{Z}_R - Z_0}{\bar{Z}_R + Z_0}$

Voltage transmission coefficient (complex load), $\bar{\tau} = \dfrac{\bar{V}_{R0}}{\bar{V}_0^+} = 1 + \bar{\Gamma}$

Standing wave ratio, $S_R = \dfrac{V_{\max}}{V_{\min}} = \dfrac{I_{\max}}{I_{\min}} = \dfrac{1+|\bar{\Gamma}|}{1-|\bar{\Gamma}|}$

Reflection coefficient in terms of standing wave ratio, $|\bar{\Gamma}| = \dfrac{S_R - 1}{S_R + 1}$

Maximum impedance, $Z_{\max} = \dfrac{V_{\max}}{I_{\min}} = S_R Z_0$

Minimum impedance, $Z_{\min} = \dfrac{V_{\min}}{I_{\max}} = \dfrac{Z_0}{S_R}$

Input impedance, $\bar{Z}_{in} = Z_0 \dfrac{\bar{Z}_R \cos\beta\ell + jZ_0 \sin\beta\ell}{Z_0 \cos\beta\ell + j\bar{Z}_R \sin\beta\ell}$

Input impedance of a stub, $\bar{Z} = jZ_0 \tan 2\pi(\ell/\lambda)$

Line with Losses

Sinusoidal variation with time:

Propagation constant, γ ′ ω)($+ \omega$)

Characteristic impedance, $\bar{Z}_0 = \sqrt{\dfrac{R + j\omega L}{G + j\omega C}}$

Reflection coefficient, $\bar{\Gamma} = \dfrac{\bar{Z}_R - \bar{Z}_0}{\bar{Z}_R + \bar{Z}_0}$

Average power of incident wave, $P_{av}^+ = \dfrac{V_0^+ I_0^+}{2} \cos\psi_0$

Average power of reflected wave, $P_{av}^- = |\bar{\Gamma}|^2 P_{av}^+$

Single stub matching:

Stub length, $\ell_s = \dfrac{\lambda_s}{2\pi} \tan^{-1}\left(\pm \dfrac{X_{sc}}{Z_{0s}}\right)$

Position of the stub, $\ell_m = \dfrac{\lambda_m}{4\pi}(\psi_r + \pi/2 + \varphi - 2n\pi)$

MULTIPLE-CHOICE QUESTIONS

(Starred questions have two answers)

*8.1 A two-conductor transmission line is said to be a uniform line if
(a) the spacing between the conductors is same throughout the line
(b) the spacing between the conductors is large compared to the radius of the conductors
(c) the insulation between the conductors is homogeneous
(d) it has no losses

8.2 An ideal transmission line is one in which
(a) ohmic loss is zero but dielectric loss may not be zero
(b) dielectric loss is zero but ohmic loss may exist
(c) both ohmic and dielectric losses are small
(d) there are no losses

*8.3 The inductance and capacitance per unit length of an ideal two-conductor transmission line are L and C respectively. For the transmission line operating at frequency f, which of the following formulae are correct?
(a) Phase constant $= f\sqrt{LC}$
(b) Wavelength $= \dfrac{1}{f\sqrt{LC}}$
(c) Characteristic impedance $= \sqrt{\dfrac{L}{C}}$
(d) Velocity of the wave $= \sqrt{LC}$

8.4 If the incident voltage wave in a long loss-free transmission line is given by $V = V_0^+ \cos(2\pi ft - \beta z)$ where β is the phase constant, then velocity of the wave is

(a) $\dfrac{2\pi f}{\beta}$ (b) $\dfrac{\pi f}{\beta}$ (c) $\dfrac{f}{\beta}$ (d) $\dfrac{f}{2\beta}$

*8.5 Wave reflection takes place if a transmission line is terminated with an unmatched load. Which of the following conditions at the load terminals are used to derive the reflection coefficient?
(a) The sum of the incident voltage and reflected voltage is equal to the load voltage.
(b) The sum of the reflected voltage and load voltage is equal to the incident voltage.
(c) The sum of the reflected current and load current is equal to the incident current.
(d) The sum of the reflected current and incident current is equal to the load current.

8.6 An ideal long two-conductor transmission line is terminated with a purely resistive load. If average power of the incident wave is P_{av}^{+} and the reflection coefficient is 0.25, the average power transmitted to the load is
(a) 0.0625 P_{av}^{+} (b) 0.25 P_{av}^{+} (c) 0.75 P_{av}^{+} (d) 0.9375 P_{av}^{+}

8.7 The signal frequency in an ideal two-conductor transmission line is 5.0 MHz. If the inductance and capacitance per unit length of the line are 1.0 μH and 16.0 pF, the wavelength of the signal is
(a) 25.0 m (b) 50.0 m (c) 75.0 m (d) 100 m

*8.8 An ideal transmission line is terminated with a purely resistive load. If the characteristic impedance of the line is 1.5 times the load resistance, then
(a) the current reflection coefficient is –0.2
(b) the current reflection coefficient is 0.2
(c) the current transmission coefficient is 1.2
(d) the current transmission coefficient is 0.6

8.9 A long two-conductor transmission line is terminated with a complex impedance. If the average power of the incident wave is P_{av}^{+} and the reflection coefficient is $0.4 + j0.3$, then average power of the reflected wave is
(a) 0.25 P_{av}^{+} (b) 0.3 P_{av}^{+} (c) 0.4 P_{av}^{+} (d) 0.5 P_{av}^{+}

8.10 An ideal transmission line is short circuited at one end. The voltage reflection coefficient at the short-circuited end is
(a) 1.0 (b) 0.5 (c) 0.0 (d) –1.0

8.11 The amplitude of incident and reflected voltages of a transmission line are 50.0 V and 12.5 V respectively. The standing wave ratio is
(a) –4.0 (b) 4.0 (c) 1.667 (d) 0.6

*8.12 If the maximum impedance and minimum impedance of an ideal two-conductor transmission line are 180 Ω and 20 Ω respectively, then
(a) the characteristic impedance of the line is 100 Ω
(b) the characteristic impedance of the line is 60 Ω
(c) the standing wave ratio is 3
(d) the standing wave ratio is 9

8.13 An ideal two-conductor transmission line is terminated with a short circuit. The length of the line is 0.25 times the wavelength at the signal frequency. If the characteristic impedance of the line is 75.0 Ω, the magnitude of input impedance is equal to
(a) 75.0 Ω (b) infinity (c) $75\sqrt{3}$ Ω (d) $75/\sqrt{3}$ Ω

8.14 The input impedance of an open-circuited ideal two-conductor transmission line is 260.0 Ω and the input impedance when the line is short-circuited is 87.0 Ω. The characteristic impedance of the line is nearest to
(a) 110 Ω (b) 120 Ω (c) 150 Ω (d) 173 Ω

8.15 The series impedance and shunt admittance per unit length of a long transmission line are $8.0 \angle 85° \ \Omega/m$ and $0.32 \angle 90°$ S/m respectively. The propagation constant per metre of the line is nearest to
(a) $1.6 + j0.07 \ m^{-1}$ (b) $1.6 - j0.07 \ m^{-1}$ (c) $0.07 + j1.60 \ m^{-1}$ (d) $0.07 - j1.60 \ m^{-1}$

8.16 The series impedance and shunt admittance per unit length of a long transmission line are $12.0 \angle 82°$ Ω/m and $3.0 \times 10^{-4} \angle 88°$ S/m respectively. The characteristic impedance per unit length of the line is nearest to
(a) $199.7 - j10.5 \ \Omega/m$ (b) $199.7 + j10.5 \ \Omega/m$ (c) $398 - j42 \ \Omega/m$ (d) $398 + j42 \ \Omega/m$

QUESTIONS

8.1 Various types of transmission lines are used for transmission of electrical power. Describe three types of transmission lines.

8.2 A coaxial transmission line consists of two concentric cylindrical conductors separated by an insulating medium. Another type of two-conductor transmission line has two parallel wires embedded in a dielectric medium. Describe the advantages of the coaxial line over the parallel-wire line for transmission of high frequency signal.

8.3 A two-conductor transmission line can be designated as uniform transmission line under some assumptions. What are these assumptions?

8.4 Two important constants of a two-conductor transmission line are its capacitance and external inductance. What are the factors on which the values of these constants depend?

8.5 The circuit model of a short transmission line consists of lumped parameters, whereas that of a long transmission line consists of distributed parameters. Explain why.

8.6 The transmission line equations of a long transmission line are derived by considering a very short length of a long transmission line. Derive the equations.

8.7 The transmission line equations of a long uniform transmission line are given by

$$\frac{\partial V}{\partial z} = -L\frac{\partial I}{\partial t} \text{ and } \frac{\partial I}{\partial z} = -C\frac{\partial V}{\partial t}$$

Derive the *wave equations* of the line from these relations.

8.8 The phase velocity of voltage and current waves in a uniform and long transmission line are given by $\sqrt{1/(LC)}$ where L is inductance and C is capacitance per unit length of the line. Show, using the example of a coaxial transmission line, that $\sqrt{1/(LC)} = \sqrt{1/\mu\varepsilon}$ if L is the external inductance per unit length of the line; μ and ε are permeability and permittivity of the dielectric medium between the lines.

8.9 The phase constant of a uniform and long transmission line is given by $\omega\sqrt{LC}$; the symbols have their usual meanings. Show, using the example of a coaxial transmission line, that $\omega\sqrt{LC} = \omega\sqrt{\mu\varepsilon}$ if the internal inductances are not included in L. μ and ε are permeability and permittivity of the dielectric medium between the lines.

8.10 The capacitance and external inductance per unit length of a coaxial transmission line depend on the ratio of the radii of the outer and inner conductors. Moreover, these constants also depend on the relative permittivity and relative permeability of the insulation between the conductors. Discuss how these factors affect the characteristic impedance of the line.

8.11 For a given value of conductor radius of a transmission line with two parallel conductors, the characteristic impedance of the line depends on the distance h between the conductors. Explain how the value of characteristic impedance of the line varies if h is varied.

8.12 Show that the characteristic impedance of a coaxial transmission line, $Z_0 = \sqrt{\frac{\mu}{\varepsilon}}\left[\frac{\ln(b/a)}{2\pi}\right]$ if internal inductances are neglected.

8.13 Show that for a lossless two-conductor transmission line, $\omega L = \beta Z_0$ and $\omega C = \beta / Z_0$. The symbols have their normal meaning.

8.14 Wave reflection takes place if a transmission line is terminated with impedance that is different from the characteristic impedance of the line. In order to find the reflection and transmission coefficients it is necessary to use the terminal conditions. What are the terminal conditions?

8.15 The ratio of the reflected current to the incident current at the discontinuity of a transmission line is called current-reflection coefficient. Show that the current-reflection coefficient is equal to $-\Gamma$ where Γ is the voltage-reflection coefficient at the discontinuity.

8.16 If the load resistance is not equal to the characteristic impedance of a lossless transmission line, a part of the incident power is reflected back to the line. If Γ is the voltage-reflection coefficient, establish that the reflected power is equal to Γ^2 times the incident power.

8.17 What are the magnitude and sign of reflection coefficients of a loss-free transmission line (a) terminated with open-circuit, (b) terminated with short-circuit and (c) terminated with a resistance equal to one-half of the characteristic resistance of the line?

8.18 For a lossless transmission line what are the values of standing wave ratio if the line (a) is terminated with open circuit, (b) is terminated with short circuit and (c) is terminated with a resistance equal to the characteristic resistance of the line?

8.19 The amplitudes of voltages of incident and reflected waves in a transmission line are V_0^+ an V_0^-. What is the value of the standing wave ratio? Express the standing wave ratio in terms of the voltage reflection coefficient at the load end.

8.20 The total voltage in a transmission line terminated with unmatched load impedance is equal to sum of the voltages of the incident and reflected waves. The standing wave ratio S_R of the line is equal to ratio of the maximum amplitude of the total voltage to the minimum amplitude of the total voltage. Show that the magnitude of voltage-reflection coefficient at the load end,

$$|\Gamma| = (S_R - 1)/(S_R + 1)$$

8.21 Consider that a transmission line has both incident and reflected waves. Define impedance at a point of the transmission line viewed from the point towards the load end. Express the impedance in terms of the characteristic impedance Z_0 of the line and voltage-reflection coefficient Γ at the load end. Assume that the line is lossless and the load is purely resistive.

8.22 A transmission line has both incident and reflected waves. The standing wave ratio and characteristic impedance of the line are S_R and Z_0 respectively. Show that the maximum value of line impedance is equal to S_R multiplied by Z_0.

8.23 Define *input impedance* of a transmission line having both incident and reflected waves. What are the quantities on which the value of the input impedance depends?

8.24 A lossless two-conductor transmission line of characteristic impedance Z_0 is terminated with unmatched impedance Z_R. What is the input impedance if the line length (a) is equal to the wavelength and (b) is equal to ¾ of the wavelength.

8.25 A low loss transmission line that is short-circuited at one end is known as a stub. The input impedance of a stub is proportional to its characteristic impedance. It is also a function of ℓ/λ where ℓ is length of the stub and λ is wavelength at the signal frequency. Discuss how the input impedance of a stub varies with ℓ/λ. Draw a graph showing the variation of the input impedance with ℓ/λ.

8.26 In the circuit representation of a transmission line, the line resistance is shown as a *series* element, whereas the conductance of the dielectric is shown as a *shunt* element. Explain why.

8.27 Write the complex forms of series impedance and shunt admittance of a transmission line having both ohmic and dielectric losses. State what the symbols represent, and give their units.

8.28 The propagation constant of a transmission line having losses has a real part and an imaginary part. The real part is called attenuation constant. State whether the attenuation constant should be low or high. Justify your answer.

8.29 What are the benefits of matching the characteristic impedance of a transmission line with the load impedance?

8.30 Describe the single-stub matching of a transmission line.

ANSWERS TO SELECTED QUESTIONS

8.2 A coaxial transmission line consists of two concentric cylindrical conductors separated by an insulating medium. Another type of two-conductor transmission line has two parallel wires embedded in a dielectric medium. Describe the advantages of the coaxial line over the parallel-wire line for transmission of high frequency signal.

Answer: The transmission lines consisting of two parallel conductors embedded in a dielectric medium, though they provide shielding to a certain degree, are not suitable for transmission of very high frequency signals as radiation takes place at joints and bends causing interference. These are cheaper than coaxial transmission lines.

In a coaxial transmission line the outer conductor acts as a good shield. Hence, radiation from a coaxial line is very small. The transmission efficiency is not affected even if it is bent or twisted. It has low dielectric loss and is suitable for transmission of signal up to five gigahertz. Coaxial lines are comparatively costlier.

8.3 A two-conductor transmission line can be designated as uniform transmission line under certain assumptions. What are these assumptions?

Answer: A two-conductor transmission line can be considered as uniform if (1) the spacing between the conductors is same throughout the line, (2) the radius of the conductors throughout the line is same, and (3) the insulation between the conductors is homogeneous.

8.5 The circuit model of a short transmission line consists of lumped parameters whereas that of a long transmission line consists of distributed parameters. Explain why.

Answer: A transmission line is classified as *short* or *medium* or *long* depending on its length relative to the wavelength of the signal it carries. In a short line the capacitance and hence the charging current or shunt current is negligibly small. Hence, current throughout the line is nearly same. Therefore, it makes no difference whether the ohmic resistance and inductance of the line are shown as distributed parameters or lumped parameters.

In a long transmission line, the value of current throughout the line is not same because of the shunt current. This is analogous to the current distribution in a circuit with distributed parameters. As line resistance and inductance are series elements, and capacitance and insulation conductance are shunt elements, its equivalent circuit also has the same configuration. Use of a circuit model with distributed parameters to represent a long transmission line makes it possible to explain the wave nature of voltage and current in the line.

8.10 The capacitance and external inductance per unit length of a coaxial transmission line depend on the ratio of the radii of the outer and inner conductors. Moreover, these constants also depend on the relative permittivity and relative permeability of the insulation between the conductors. Discuss how these factors affect the characteristic impedance of the line.

Answer: The characteristic impedance of an ideal transmission line, $Z_0 = \sqrt{L/C}$. Inductance and capacitance per unit length of a coaxial line, are given by $L = (\mu/2\pi)\ln b/a$ and $C = (2\pi\varepsilon)/\ln b\ a$ respectively where b and a are the radii of the outer and inner conductors. Thus, the characteristic impedance, $Z_0 = \sqrt{\mu}(\ln b/a)/\left(2\pi\sqrt{\varepsilon}\right)$. As the relative permeability of the dielectric between the conductors is generally unity, the characteristic impedance increases with an increase in the ratio b/a, and decreases with an increase in the relative permittivity of the dielectric.

8.14 Wave reflection takes place if a transmission line is terminated with impedance that is different from the characteristic impedance of the line. In order to find the reflection and transmission coefficients it is necessary to use the terminal conditions. What are the terminal conditions?

Answer: Kirchhoff's voltage and current laws determine the conditions at the load terminal. The terminal conditions are (a) sum of the incident voltage and reflected voltage is equal to the load voltage, and (b) sum of the incident current and reflected current is equal to the load current.

8.18 For a lossless transmission line what are the values of standing wave ratio if the line (a) is terminated with open circuit, (b) is terminated with short circuit and (c) is terminated with a resistance equal to the characteristic resistance of the line?

Answer: (a) For a short-circuited transmission line, the voltage reflection coefficient Γ at the point of short circuit is -1.0. Hence, the standing wave ratio $S_R = (1+|\Gamma|)/(1-|\Gamma|)$ is infinity. (b) For an open-circuited transmission line, the voltage reflection coefficient Γ at the point of open circuit is 1.0. Hence, the standing wave ratio $S_R = (1+|\Gamma|)/(1-|\Gamma|)$ is infinity. (c) When the transmission line is terminated with a resistance equal to the characteristic resistance of the line, the reflection coefficient is zero. Hence, the standing wave ratio is unity.

8.25 A low loss transmission line that is short-circuited at one end is known as a stub. The input impedance of a stub is proportional to its characteristic impedance. It is also a function of ℓ/λ where ℓ is length of the stub and l is wavelength at the signal frequency. Discuss how the input impedance of a stub varies with ℓ/λ. Draw a graph showing the variation of the input impedance with ℓ/λ.

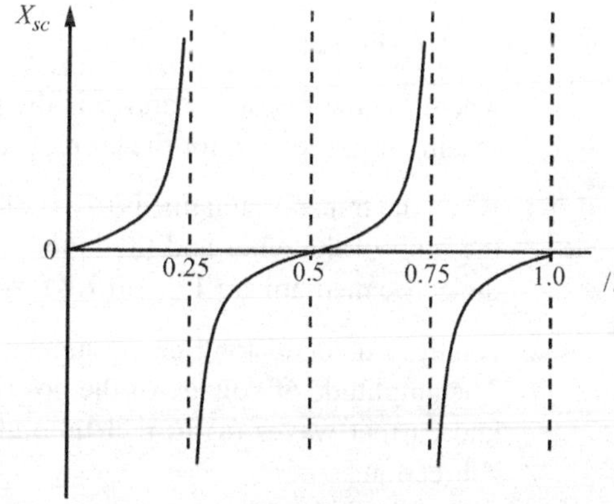

Fig. Q8.25

Answer: The input impedance of a short-circuited transmission line is given by

Z_{sc} $jZ_0 \tan 2$ (/ ˙)

Due to the presence of the factor j, the input impedance of a short-circuited, low loss transmission line is a pure reactance, X_{sc} $Z_0 \tan 2$ (/ ˙). The reactance may be either inductive or capacitive depending on the value of $\ell/$˙. The reactance varies from $-\infty$ to $+\infty$ crossing the zero value at $\ell/\lambda =$ 0.5, 1.0,..., not considering the trivial case of $\ell/\lambda = 0$. Fig. Q8.25 shows variation of the reactance with ℓ/λ.

8.28 The propagation constant of a transmission line having losses has a real part and an imaginary part. The real part is called attenuation constant. State whether the attenuation constant should be low or high. Justify your answer.

Answer: The real part of the propagation constant, called attenuation constant, arises due to ohmic loss in the conductors and dielectric loss in the insulation between the conductors. Consequently, the amplitudes of voltage and current waves decrease exponentially during progress of the waves in a line. It can be shown that the rate of decrease of voltage and current amplitudes along the line is proportional to the attenuation constant. And the rate of decrease of power flow is proportional to twice of the attenuation constant. Moreover, the rate of decrease of power is equal to power loss per unit length. Hence, a low value of attenuation constant indicates a low value of power loss and high value of transmission efficiency.

8.29 What are the benefits of matching the characteristic impedance of a transmission line with the load impedance?

Answer: In the case of a mismatched transmission system, a part of incident power gets reflected from the load terminals. Hence, power received by the load is less than the incident power. In a matched transmission system, the transmitted power is maximised. Moreover, in a perfectly matched system, the input impedance is equal to the characteristic impedance.

PROBLEMS

*8.1 The radii of the inner and outer conductors of a coaxial transmission line are 2 mm and 10 mm respectively, and their conductivity is 5.8×10^7 S/m. Find approximately the skin-effect resistance of the line per unit length at 50 MHz.

***8.2 Develop the transmission line equation for a long, ideal, two-conductor transmission line.

**8.3 A voltage wave travelling along positive z direction in a uniform, ideal, two-conductor transmission line is described by

$$V^+ = V_0^+ \cos(2\pi ft - \beta z)$$

where f is the frequency and β is the phase constant. Using one of the transmission line equations, obtain the expression for the current wave in the line and hence define the characteristic impedance.

*8.4 A coaxial transmission line is to have a characteristic impedance of 75 Ω. Calculate the ratio between the radii of the outer and the inner conductors if the relative permeability and permittivity of the dielectric medium are 1.0 and 2.7 respectively.

*8.5 A long two-conductor transmission line has an inductance of 0.8 μH/m and a capacitance of 50 pF/m. The amplitude of voltage of the positively travelling wave is 75 V. Obtain expressions for voltage and current waves (a) in real time and (b) in phasor form at 100 MHz by taking the line along z direction.

***8.6 An ideal two-conductor transmission line of characteristic impedance Z_0 is terminated with a load of resistance R_R. Derive the formula for voltage reflection coefficient at the load end.

**8.7 An ideal two-conductor transmission line 1 of characteristic impedance Z_{01} is connected with another ideal transmission line 2 of characteristic impedance Z_{02}. If the incident wave is in Line 1, find the average power transmitted to Line 2 in terms of the average incident power.

**8.8 A long two-conductor transmission line of characteristic impedance 150 Ω is connected to a 200 Ω resistor. Find the power transmitted to the load if amplitude of the voltage is 60 V. Neglect losses.

***8.9 The inductance and capacitance per unit length of a long two-conductor transmission line are 0.25 μH and 100 pF. The line is terminated by a 150 Ω resistor in series with an 8 pF capacitor. Calculate the voltage reflection coefficient at the load end at 250 MHz. Neglect losses.

**8.10 The load point of a long two-conductor transmission line is at $z = 0$ and the input point is at $z = -\ell$. The reflection coefficient at the load end is $\bar{\Gamma}$. Obtain the reflection coefficient at $z = -z'$.

***8.11 Find the voltage reflection coefficient if the 8 pF capacitor is in parallel with the 150 Ω resistor in Problem 8.9.

***8.12 A loss-free, long two-conductor transmission line of characteristic impedance of 100 Ω is terminated with a 150 Ω resistor in series with an inductor that has a reactance of 50 Ω at the supply frequency. Find the power transmitted to the load if amplitude of the voltage is 120 V. Neglect losses.

**8.13 A section of an ideal two-conductor transmission line has both incident and reflected voltage waves. The sum of the incident and reflected voltages is given by $V = \sqrt{(V_0^+)^2 + (V_0^-)^2 + 2V_0^+V_0^- \cos 2\beta z} \cos(\omega t - \text{[illegible]})$ where V_0^+ and V_0^- are real amplitudes of the incident and reflected waves, β is the phase constant. Obtain the standing wave ratio of the line in terms of the reflection coefficient Γ at the load terminals.

***8.14 A loss-free, long, two-conductor transmission line of characteristic impedance 300 Ω is terminated by a 200 Ω resistor in series with a capacitor that has a reactance of 75 Ω at the supply frequency. (a) Find the standing wave ratio. (b) Calculate the wavelength and phase constant if the distance between two successive voltage minima is 4.0 m.

**8.15 A long two-conductor transmission line is terminated by an unmatched resistor of 200 Ω. The values of maximum and minimum voltages along the line are 12.5 V and 7.2 V. The distance between a maximum voltage point and the nearest minimum voltage point is 2.4 m. (a) Find the characteristic impedance of the line. (b) If the signal frequency is 50 MHz, what are the inductance and capacitance per unit length of the line. Neglect the line loss.

***8.16 An ideal two-conductor transmission line of characteristic impedance Z_0 and phase constant β is terminated with a pure resistance R_R located at $z = 0$. The input point is at $z = -\ell$. If the reflection coefficient at the load end is Γ, obtain the input impedance of the line.

*8.17 Show that the input impedance of an ideal two-conductor transmission line short circuited at one end is

$$Z_{sc} = jZ_0 \tan 2\pi(\ell/\lambda)$$

where Z_0, ℓ, and λ are the characteristic impedance, length, and wavelength respectively.

**8.18 The ratio of the outer radius to the inner radius of a coaxial transmission line is 5.0, and the relative permittivity of the nonmagnetic dielectric between the conductors is 2.2. The line is terminated with a 150 Ω resistor. Find (a) the impedance of the line where minimum and maximum currents occur and (b) the input impedance for a line length equal to twice the wavelength.

***8.19 A two-conductor transmission line has the following constants.

Resistance, $R = 0.5$ Ω/m

Inductance, $L = 0.3\ \mu$H/m

Capacitance, $C = 150$ pF/m

Conductance, $G = 0.8$ ns/m

Find the attenuation constant, the phase constant, and the characteristic impedance of the line at a frequency of 25 MHz.

***8.20 A two-conductor transmission line of characteristic impedance $150.0 - j2.5\ \Omega$ is terminated with an impedance of $250.0 - j80.0\ \Omega$. Find (a) the reflection coefficient at the load end, and (b) the standing wave ratio.

*8.21 The characteristic impedance of a 2.0 m long two-conductor transmission line is $125 + j0\ \Omega$. The line is connected to a load of resistance of 250 Ω. The phase velocity in the line at the operating frequency of 30 MHz is 2.4×10^8 m/s. Find the input impedance of the line.

***8.22 A 150 Ω two-conductor transmission line is connected to a load of resistance of 220 Ω. The length of the line is 2.4 times the wavelength in the line at the operating frequency. The power source connected to the input terminals of the line has a peak voltage of $120 \angle 0°$V, and internal resistance of 20 Ω. Find (a) amplitude of the source current; (b) power consumed by the load; (c) amplitude of the load current.

***8.23 The series resistance, external inductance, capacitance, and shunt conductance per unit length of a two-conductor transmission line are R, L, C, and G respectively. Show that the characteristic impedance of the line,

$$\bar{Z}_0 = \sqrt{\frac{R + j\omega L}{G + j\omega C}}.$$

***8.24 A two-conductor transmission line of characteristic impedance $\bar{Z}_0 = 150 - j5\ \Omega$ is terminated with load impedance $\bar{Z}_R = 120 - j60\ \Omega$. The line length is 1.2 times the wavelength in the line at the operating frequency. Find the impedance at the input terminals of the line.

***8.25 A lossless two-conductor transmission line having a characteristic impedance of 300 Ω is terminated with a load resistance of 250 Ω. The line length is 2.75 times the wavelength at the operating frequency. The line is fed by a source of voltage amplitude $V_{g0} = 150 \angle 0$ and internal resistance of 60 Ω. Find (a) amplitudes of load voltage and load current, and (b) power delivered to the load as percentage of power supplied by the source.

***8.26 A lossless two-conductor transmission line having characteristic impedance of $150 - j8\ \Omega$ is connected to a load of impedance of $200 - j50\ \Omega$. A suitable two-winding transformer is used to match the magnitude of the characteristic impedance with that of the load. (a) Find the ratio of number of turns of the secondary winding to that of the primary winding. (b) Calculate the ratio between the reflected power and the incident power, without and with the matching transformer.

***8.27 A two-conductor lossless transmission line of characteristic impedance 75 Ω is terminated with a load impedance of $125 - j30\ \Omega$. A low-loss quarter-wave line is used for the purpose of matching the line with the load. Find the characteristic impedance of the quarter-wave line if its load-end is connected where the total voltage in the main line has (a) maximum value and (b) minimum value.

***8.28 If matching of the transmission line with load in Problem 8.27 is to be achieved by a single short-circuited stub having the same characteristic impedance as the main line, find (a) the length of the stub and (b) the distance of its location from the load end in terms of the wavelengths.

Chapter Nine

WAVEGUIDES

9.1 Introduction

In the preceding chapter, we learnt that voltage and current in long transmission lines travel in the form of waves. As voltage between the lines produces electric field intensity and current through each line produces magnetic field intensity, the fields also travel in the form of waves and in the same direction as the voltage and current waves. For example, the radial electric field and the circumferential magnetic field between the conductors of a coaxial transmission line travel along the line with the conductors acting as boundaries or guides. The directions of the fields are perpendicular (transverse) to the direction of propagation. Thus, a coaxial transmission line is a form of waveguide.

At high frequencies, from about 1 GHz to 100 GHz, electromagnetic energy can be transmitted inside hollow conducting cylinders. These structures are the waveguides and the electromagnetic waves are called guided waves. Waveguides are generally of uniform rectangular or circular section, as illustrated in Fig. 9.1. Waveguides can transmit large amount of electromagnetic energy from one point to another point efficiently.

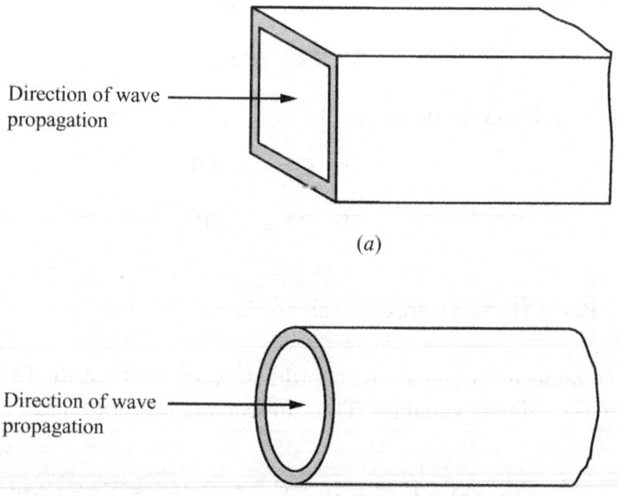

Fig. 9.1 Hollow metallic waveguides of (*a*) Rectangular Section and (*b*) Cylindrical Section.

Electromagnetic energy can also be transmitted through a dielectric slab enclosed by another dielectric medium of lower permittivity, as illustrated in Fig. 9.2. The principle of total internal reflection is used to guide the waves in the desired direction.

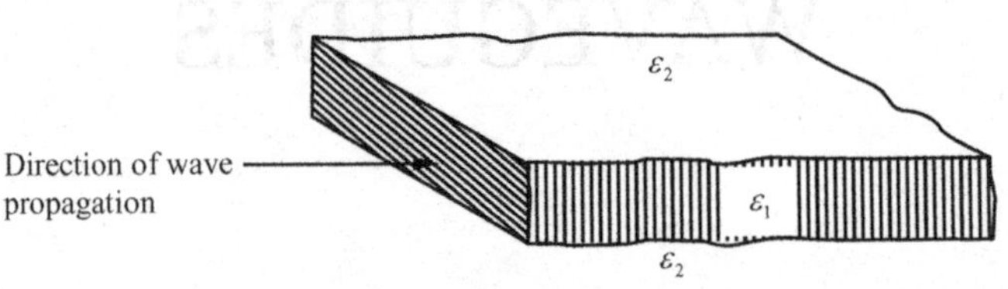

Fig. 9.2 A dielectric-sheet waveguide of permittivity ε_1 surrounded by another dielectric medium of permittivity ε_2 ; $\varepsilon_1 > \varepsilon_2$

We will analyse at first the phenomenon of wave propagation between two large parallel conducting plates, and then inside a hollow conducting cylinder of uniform rectangular section. The study is limited to fields, which vary sinusoidally with time.

9.2 Formulation of Field Equations

In this section, we utilise the Helmholtz wave equations and Maxwell's curl equations to derive the field equations to study the propagation of guided waves.

We assume that the dielectric medium in which the waves travel is lossless. That is, conductivity of the dielectric medium is zero. We also assume that the conductivity of the metallic guides is infinitely large so that the skin depth is zero and hence no losses occur in the guide walls.

The wave equations in **E** and **H** for a loss-free dielectric medium, assuming harmonic variation of fields with time represented by $e^{j\omega t}$, are given by

$$\nabla^2\bar{\mathbf{E}} = -\omega^2\mu\varepsilon\bar{\mathbf{E}} \tag{1}$$

$$\nabla^2\bar{\mathbf{H}} = -\omega^2\mu\varepsilon\bar{\mathbf{H}} \tag{2}$$

The phasor forms of Maxwell's curl-equations are

$$\nabla\times\bar{\mathbf{E}} = -j\omega\mu\bar{\mathbf{H}} \tag{3}$$

$$\nabla\times\bar{\mathbf{H}} = j\omega\varepsilon\bar{\mathbf{E}} \tag{4}$$

The field intensities $\bar{\mathbf{E}}$ and $\bar{\mathbf{H}}$ are vectors as well as phasors.

The fields may have components along all the coordinate axes. Further, the field components in general are functions of all the three coordinate variables. Thus, in rectangular coordinates we may write the electric field intensity as

$$\bar{\mathbf{E}}(x\ y\ z) = \bar{E}_x(x\ y\ z)\mathbf{u}_x + \bar{E}_y(x\ y\ z)\mathbf{u}_y + \bar{E}_z(x\ y\ z)\mathbf{u}_z \tag{5}$$

By substituting (5) in (1) and equating the x component of the two sides, we get

$$\nabla^{\hat{}}\ {}_x \qquad = - \quad E_x(x\ y\ z)$$

Or
$$\frac{\partial^2 \bar{E}_x(x,y,z)}{\partial x^2} + \frac{\partial^2 \bar{E}_x(x,y,z)}{\partial y^2} + \frac{\partial^2 \bar{E}_x(x,y,z)}{\partial z^2} = -\omega^2 \mu\varepsilon \bar{E}_x(x,y,z) \tag{6}$$

If we assume that the electric field travels along positive z direction with propagation constant γ, variation of the field in z direction may be expressed by the exponential function $e^{-\gamma z}$. So,

$$\bar{E}_x(x,y,z) = \bar{E}_x(x,y)e^{-\gamma z} = \bar{E}_x e^{-\gamma z} \tag{7}$$

where $\bar{E}_x(x,y)$ is replaced by $\bar{E}_x$ to simplify writing the equations. By substituting (7) in (6), and as $e^{-\gamma z}$ of the two sides cancels out, we get

$$\frac{\partial^2 \bar{E}_x}{\partial x^2} + \frac{\partial^2 \bar{E}_x}{\partial y^2} = -(\omega^2 \mu\varepsilon + \gamma^2)E_x = -k^2 \bar{E}_x; \tag{8}$$

$$k^2 = \gamma^2 + \omega^2 \mu\varepsilon \tag{9}$$

The wave equations for other field intensities are

$$\frac{\partial^2 \bar{E}_y}{\partial x^2} + \frac{\partial^2 \bar{E}_y}{\partial y^2} = -k^2 \bar{E}_y \tag{10}$$

$$\frac{\partial^2 \bar{E}_z}{\partial x^2} + \frac{\partial^2 \bar{E}_z}{\partial y^2} = -k^2 \bar{E}_z \tag{11}$$

$$\frac{\partial^2 \bar{H}_x}{\partial x^2} + \frac{\partial^2 \bar{H}_x}{\partial y^2} = -k^2 \bar{H}_x \tag{12}$$

$$\frac{\partial^2 \bar{H}_y}{\partial x^2} + \frac{\partial^2 \bar{H}_y}{\partial y^2} = -k^2 \bar{H}_y \tag{13}$$

$$\frac{\partial^2 \bar{H}_z}{\partial x^2} + \frac{\partial^2 \bar{H}_z}{\partial y^2} = -k^2 \bar{H}_z \tag{14}$$

Next, we will derive relations between the electric field intensity and magnetic field intensity involving their components by expanding the curl operations. With variation of fields with respect to z described by $e^{-\gamma z}$, we get from (3),

$$\left(\frac{\partial \bar{E}_z}{\partial y} + \gamma \bar{E}_y\right)\mathbf{u}_x - \left(\frac{\partial \bar{E}_z}{\partial x} + \gamma \bar{E}_x\right)\mathbf{u}_y + \left(\frac{\partial \bar{E}_y}{\partial x} - \frac{\partial \bar{E}_x}{\partial y}\right)\mathbf{u}_z = -j\omega\mu(\bar{H}_x\mathbf{u}_x + \bar{H}_y\mathbf{u}_y + \bar{H}_z\mathbf{u}_z)$$

By equating the like components of the two sides, we have

$$\frac{\partial \bar{E}_z}{\partial y} + \gamma \bar{E}_y = -j\omega\mu \bar{H}_x \tag{15a}$$

$$\frac{\partial \bar{E}_z}{\partial x} + \text{[illegible]} = j\omega u \bar{H}_y \tag{15b}$$

$$\frac{\partial E_y}{\partial x} - \frac{\partial E_x}{\partial y} = -j \quad H_z \tag{15c}$$

In a similar way, we can derive the following relations from (4).

$$\frac{\partial \bar{H}_z}{\partial y} + \gamma \bar{H}_y = j\omega\varepsilon \bar{E}_x \tag{16a}$$

$$\frac{\partial \bar{H}_z}{\partial x} + \gamma \bar{H}_x = -j\omega\varepsilon \bar{E}_y \tag{16b}$$

$$\frac{\partial \bar{H}_y}{\partial x} - \frac{\partial \bar{H}_x}{\partial y} = j\omega\varepsilon \bar{E}_z \tag{16c}$$

These are the basic equations of field intensities in rectangular coordinates. Solutions of these equations satisfying the appropriate boundary conditions determine the properties of waves in a waveguide having rectangular symmetry.

It may be noted that the field intensities are functions of x and y only and variations with t and z are as per the functions $e^{j\omega t}$ and $e^{-\gamma z}$. Moreover, the superscript '+' that has been used in earlier chapters to denote positively travelling waves, is not necessary here as the direction of propagation of all the fields is along positive z direction.

9.3 Wave Types

We have derived the field equations in the preceding section assuming that the field intensities have components along all the coordinate axes and the waves travel along z direction. If the axis of waveguide is oriented in z direction, then the field components will be either normal or tangential to the inner surfaces of the guide walls. Such an orientation of the waveguide relative to the field components simplifies the application of boundary conditions. A further simplification is achieved in classifying the waves by assuming that either H_z is zero or E_z is zero or both H_z and E_z are zero. Accordingly, we have the following three types of waves.

Transverse Magnetic Waves

Waves of this type do not have magnetic field intensity but have electric field intensity along the direction of propagation. That is, $\bar{H}_z = 0$ and $\bar{E}_z \neq 0$. As the other two components of magnetic field intensities $\bar{H}_x$ and $\bar{H}_y$ are perpendicular to the direction of wave propagation, these waves are known as *Transverse Magnetic Waves* or TM waves. The waves are also known as E waves.

Transverse Electric Waves

Waves of this type do not contain electric field intensity but have magnetic field intensity along the direction of propagation. That is, $\bar{E}_z = 0$ and $\bar{H}_z \neq 0$. As the electric field intensity is perpendicular to the direction of wave propagation, the waves are called *Transverse Electric Waves* or TE waves. The waves are also known as H waves.

Transverse Electromagnetic Waves

These waves do not have either electric-field intensity or magnetic-field intensity in the direction of propagation. That is, E_z $H_z = 0$. As both electric field intensity and magnetic field intensity are perpendicular to the direction of propagation, these waves are called *Transverse Electromagnetic Waves* or TEM waves. These waves are similar to the electromagnetic waves produced by voltage and current of long and uniform two-conductor transmission lines. Hence, the waves are also known as *transmission line waves*.

9.4 Electromagnetic Waves between Two Conducting Plates

A simple wave-guiding system consists of two large parallel conducting plates separated by a low-loss dielectric medium (normally air), as illustrated in Fig 9.3. The lower plate is at $x = 0$ and the upper plate is at $x = a$. The system, called parallel-plate waveguide, is identical to a parallel-plate transmission line.

We make the following assumptions to simplify the analysis.

1. The plates are infinitely large along y and z directions.
2. The fields do not vary with y.
3. The fields travel along positive z direction. Hence, the field equations derived in Section 9.2 are applicable.
4. The conducting plates and the dielectric medium between the plates are lossless. Therefore as per the boundary conditions (See Section 7.7), the electric field intensity that is tangential, and the magnetic field intensity that is normal, to the plate surfaces are equal to zero. Moreover, the amplitudes of the fields do not vary with z.

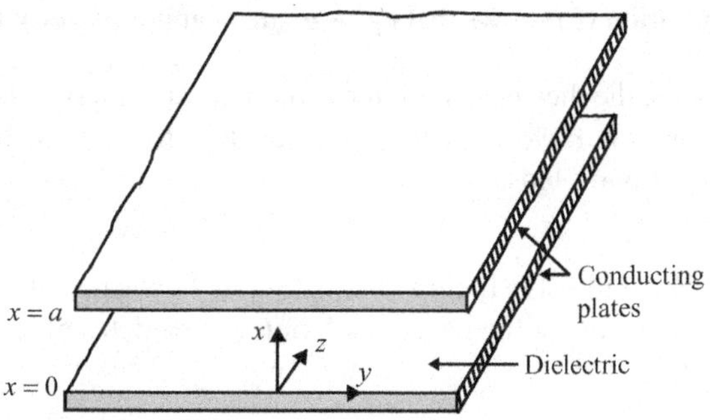

Fig. 9.3 A waveguide consisting of two conducting parallel-plates

We will study the properties of wave propagation between the plates for the three classes of waves described in the previous section. Though the system considered is an ideal one, the results obtained from its study are useful for analysing practical wave guiding systems.

TM Waves between Parallel Plates

In the case of TM waves, magnetic field intensity along the direction of propagation is zero. As we have chosen z direction as the direction of travel of the fields, hence $\bar{H}_z = 0$. Moreover as per the assumptions, the fields between the plates do not vary with y. Using these conditions in (9.2-15) and (9.2-16), we get

$$\bar{E}_y = -j\frac{\omega\mu}{\gamma}\bar{H}_x \tag{1}$$

$$\frac{\partial \bar{E}_z}{\partial x} + \gamma \bar{E}_x = j\omega\mu \bar{H}_y \tag{2}$$

$$\bar{H}_y = j\frac{\omega\varepsilon}{\gamma}\bar{E}_x \tag{3}$$

$$\bar{H}_x = -j\frac{\omega\varepsilon}{\gamma}\bar{E}_y \tag{4}$$

We do not need equations (9.2-15c) and (9.2-16c). By eliminating $\bar{H}_y$ from (2) and (3) and then solving for E_x, we get

$$E_x = -\frac{\gamma}{\gamma^2 + \omega^2\mu\varepsilon}\frac{\partial E_z}{\partial x} = -\frac{\gamma}{k^2}\frac{\partial E_z}{\partial x} \tag{5}$$

Now, by substituting the rhs of (5) in place of E_x in (3), we obtain

$$\bar{H}_y = -j\frac{\omega\varepsilon}{k^2}\frac{\partial \bar{E}_z}{\partial x} \tag{6}$$

And solving (1) and (4) simultaneously, we get the following relations.

$$\bar{E}_y(\gamma^2 + \omega^2\mu\varepsilon) = 0$$

$$\bar{H}_x(\gamma^2 + \omega^2\mu\varepsilon) = 0$$

Relation (5) shows that $(\gamma^2 + \omega^2\mu\varepsilon)$ cannot be zero, hence, both $\bar{E}_y$ and $\bar{H}_x$ must be zero.

Thus, the dielectric region has three field components in $\bar{E}_x, \bar{E}_z$ and $\bar{H}_y$. All the components satisfy the wave equation. If we solve the wave equation for any one of the field intensities, we can obtain expressions for the other two fields.

We will solve the wave equation in $\bar{E}_z$ because it is tangential to the surfaces of the plates and hence the constants of the general solution can be determined easily.

The wave equation in $\bar{E}_z$ is obtained by setting $\partial^2\bar{E}_z/\partial y^2 = 0$ in (9.2-11). Accordingly,

$$\frac{\partial^2\bar{E}_z}{\partial x^2} = -k^2\bar{E}_z \tag{7}$$

The general solution of (7) in terms of circular sine and cosine functions is

$$\bar{E}_z = C_1\cos kx + C_2\sin kx \tag{8}$$

where C_1 and C_2 are constants. The constants are determined using boundary conditions for a perfect conductor. As the value of electric field intensity that is tangential to a perfect conducting surface is zero, the boundary conditions, referring to Fig. 9.3, are

1. At $x = 0$, $\bar{E}_z = 0$
2. At $x = a$, $\bar{E}_z = 0$

Application of the first boundary conditions to (8) yields

$$C_1 = 0$$

The second boundary condition is satisfied if

$$C_2\sin ka = 0 \tag{9}$$

Or $\qquad k = m\pi/a, \quad m = 1,2,3,\ldots$

Though (9) is also satisfied for $m = 0$, it is not admissible because $\bar{E}_z$ would become zero everywhere. Substituting $C_1 = 0$ and $k = m\pi/a$ in (8), we have

$$\bar{E}_z = C_2\sin\frac{m\pi}{a}x \tag{10}$$

This is the solution of (7) satisfying the boundary conditions. The constant C_2 is the amplitude of the electric field in z direction. Using the notation E_{z0} for the amplitude,

$$E_z \quad E_{z0}\sin\frac{m^{-}x}{a} \tag{11}$$

We now substitute $k = m\pi/a$ into (9.2-9) to obtain the propagation constant. Denoting the propagation constant by γ_m, we have

$$\gamma_m = \sqrt{(m\pi/a)^2 - \omega^2 \mu\varepsilon} \tag{12}$$

The complete solution for the electric field intensity phasor $\bar{E}_z$ is found by multiplying the rhs of (11) by $e^{-\gamma_m z}$. The solution is

$$\bar{E}_z = E_{z0} \sin\frac{m\pi x}{a} e^{-\gamma_m z} \tag{13}$$

By substituting the partial derivative of (13) with respect to x into (5) and (6) we get the solutions for E_x and $\bar{H}_y$. These are given by

$$\bar{E}_x = -(\frac{\gamma_m a}{m\pi} E_{z0})\cos\frac{m\pi x}{a} e^{-\gamma_m z} \tag{14}$$

$$\bar{H}_y = -j(\frac{\omega\varepsilon a}{m\pi} E_{z0})\cos\frac{m\pi x}{a} e^{-\gamma_m z} \tag{15}$$

The foregoing expressions show that variation of the fields along x direction is either sinusoidal or cosinusoidal. The number of half cycles of the fields between the conducting plates is equal to m. Each value of m defines a specific field configuration or mode of the guide.

The TM wave having m number of half cycles between the plates is named as TM_m wave or TM_m mode. The propagation properties of each mode depend on the propagation constant given by (12). The propagation constant may be zero, a real quantity, or an imaginary quantity depending on the frequency of the fields, constants of the dielectric medium, and spacing between the plates. The frequency at which the propagation constant is zero is called *cut-off frequency* or *critical frequency* and is denoted by $\omega_{c,m}$ (or $f_{c,m}$).

Thus at the cut-off frequency,

$$\omega_{c,m}^2 \mu\varepsilon = (m\pi/a)^2 \tag{16a}$$

Or

$$\omega_{c,m} = \frac{m\pi}{a\sqrt{\mu\varepsilon}} \tag{16b}$$

and

$$f_{c,m} = \frac{m}{2a\sqrt{\mu\varepsilon}} \tag{16c}$$

It will be more useful if the propagation constant is expressed in terms of the cut-off frequency by substituting (16*a*) in (12). Accordingly,

$$\gamma_m = \sqrt{\mu\varepsilon}\sqrt{\omega_{c,m}^2 - \omega^2} \tag{17}$$

At the cut-off frequency ($\omega = \omega_{c,m}$), the propagation constant becomes zero, hence the waves do not propagate. The waves simply bounce back and forth between the plates.

At frequencies less than the cut-off frequency, the function under the radical sign of (17) is positive. Therefore, the propagation constant is a positive real quantity.

That is,

$$\gamma_{...} \quad \mu\varepsilon \quad \omega_{c,m}^2 - \omega^2 \quad \alpha_m + j0, \quad \omega \quad \omega_{c,m}$$

where α_m is the attenuation constant for the mth mode. It means that at frequencies below the cut-off frequency the amplitudes of the fields attenuate exponentially with z. The phase angle however does not change as β is zero. The fields do not propagate. The attenuation of the fields is not due to any losses.

Properties of waves above cut-off frequency

> At frequencies above the cut-off frequency, the function under the radical sign of (17) is negative. Hence, the propagation constant becomes imaginary. The waves propagate without any change in form or magnitude.

The propagation constant for this case is

$$\gamma_m = j\sqrt{\mu\varepsilon}\sqrt{\omega^2 - \omega_{c,m}{}^2} = j\beta_m \qquad \omega > \omega_{c,m}$$

where $$\beta_m = \sqrt{\mu\varepsilon}\sqrt{\omega^2 - \omega_{c,m}{}^2} = \omega\sqrt{\mu\varepsilon}\sqrt{1-(\omega_{c,m}/\omega)^2} = \beta\sqrt{1-(\omega_{c,m}/\omega)^2} \tag{18}$$

The parameter β_m is the phase constant of the mth mode. We know that $\beta = \omega\sqrt{\mu\varepsilon}$ is phase constant of a uniform plane wave propagating in a homogeneous loss-free unbounded region. Therefore, in a given dielectric medium, the phase constant of waves between parallel plates is less than the phase constant of a plane wave.

By substituting $j\beta_m$ in place of γ_m in (13), (14) and (15) we get the following expressions for the field intensities.

$$\bar{E}_z = E_{z0}\sin\frac{m\pi x}{a}e^{-j\beta_m z} \tag{19}$$

$$\bar{E}_x = -j\left(\frac{a\beta_m}{m\pi}E_{z0}\right)\cos\frac{m\pi x}{a}e^{-j\beta_m z} = -jE_{x0}\cos\frac{m\pi x}{a}e^{-j\beta_m z} \tag{20}$$

$$\bar{H}_y = -j\left(\frac{a\omega\varepsilon}{m\pi}E_{z0}\right)\cos\frac{m\pi x}{a}e^{-j\beta_m z} = -jH_{y0}\cos\frac{m\pi x}{a}e^{-j\beta_m z} \tag{21}$$

where E_{x0} and H_{y0} are the functions inside the brackets and represent amplitudes of the respective fields in terms of E_{z0}.

Some other parameters of TM waves are the intrinsic wave impedance, phase velocity, and wavelength. All these parameters are functions of the phase constant β_m. The intrinsic wave impedance is equal to the ratio between E_x given by (20) and $\bar{H}_y$ defined by (21). Denoting the impedance of the mth mode by $\eta_{TM,\,m}$, we have

$$\eta_{TM,m} = \frac{\bar{E}_x}{\bar{H}_y} = \frac{E_{x0}}{H_{y0}} = \frac{\beta_m}{\omega\varepsilon} = \eta\sqrt{1-(\omega_{c,m}/\omega)^2} = \eta\sqrt{1-(f_{c,m}/f)^2} \tag{22}$$

$$\eta = \sqrt{\mu/\varepsilon}$$

The phase velocity,

$$v_{p,m} = \frac{\omega}{\beta_m} = \frac{v}{\sqrt{1-(\omega_{c,m}/\omega)^2}}; \tag{23}$$

$$v = \frac{1}{\mu\varepsilon}$$

The phase shift of the fields along z direction is equal to $\beta_m z$. Since a phase shift of 2π radians is equal to one wavelength, the wavelength along the waveguide for the mth mode,

$$\lambda_m = \frac{2\pi}{\beta_m} = \frac{\lambda}{\sqrt{1-(\omega_{c,m}/\omega)^2}}; \tag{24}$$

$$\lambda = v/f$$

The parameters η, v and λ are the intrinsic impedance, phase velocity, and wavelength respectively of a uniform plane wave propagating in a homogeneous loss-free unbounded region.

We will now calculate the average power flow between the plates by the use of the complex Poynting vector. Noting that the electric field has components in x and z directions and the magnetic field intensity has only one component in y direction, the complex Poynting vector is

$$\bar{\mathbf{P}}_w = \frac{1}{2}(\bar{\mathbf{E}} \times \bar{\mathbf{H}}^*) = \frac{1}{2}(\bar{E}_x\mathbf{u}_x + \bar{E}_z\mathbf{u}_z) \times (\bar{H}_y{}^*\mathbf{u}_y) = \frac{1}{2}(\bar{E}_x\bar{H}_y{}^*\mathbf{u}_z - \bar{E}_z\bar{H}_y{}^*\mathbf{u}_x)$$

Substituting expressions (19), (20) and the conjugate of (21) in the above expression, we get

$$\mathbf{P}_w = \frac{1}{2}\left(-jE_{x0}\cos\frac{m\pi x}{a}e^{-j\beta_m z}\right)\left(jH_{y0}\cos\frac{m\pi x}{a}e^{j\beta_m z}\right)\mathbf{u}_z$$

$$-\frac{1}{2}\left(E_{z0}\sin\frac{m\pi x}{a}e^{-j\beta_m z}\right)\left(jH_{y0}\cos\frac{m\pi x}{a}e^{j\beta_m z}\right)\mathbf{u}_x$$

$$= \frac{1}{2}E_{x0}H_{y0}\cos^2\frac{m\pi x}{a}\mathbf{u}_z - \frac{j}{4}E_{z0}H_{y0}\sin\frac{2m\pi x}{a}\mathbf{u}_x$$

The real part of the Poynting vector,

$$\mathbf{P}_{wa} = \frac{1}{2}E_{x0}H_{y0}\cos^2\frac{m\pi x}{u}\mathbf{u}_z \text{ W/m}^2$$

Average power flow between the plates for a width b (in y direction) of the guide is

$$P_{av} = \int_S \mathbf{P}_{wa}\cdot d\mathbf{S} = \frac{E_{x0}H_{y0}}{2}\int_0^a\left(\cos^2\frac{m\pi x}{a}\mathbf{u}_z\right).(b\,dx\mathbf{u}_z)$$

$$= \frac{abE_{x0}H_{y0}}{4} \text{ W}$$

As $H_{y0} = E_{x0}/\eta_{\text{TM},m}$, the average power flow in terms of E_{x0} is

$$P_{av} = \frac{abE_{x0}{}^2}{4\eta_{\text{TM,m}}} \tag{25}$$

As the real part of Poynting vector is along z direction, the direction of energy flow is same as the direction of propagation of the waves. However, the velocity of energy flow is not the same as the phase velocity. This is evident from the fact that at $f = f_c$ though the phase velocity is infinite, the energy velocity is zero because the waves do not propagate. Moreover, the phase velocity is greater than velocity of light at frequencies higher than the cut-off frequency, and the velocity of electromagnetic waves cannot be greater than the velocity of light.

The phase velocity gives only *the rate of change of phase* of the waves with distance along the direction of travel of the waves; it is not equal to the velocity of energy flow.

The phase velocity of a plane wave in a lossless unbounded dielectric medium is not a function of frequency and the velocity of energy flow is the same as the phase velocity. Such a system is a *nondispersive system*. If the phase velocity in a transmission system is a function of frequency, then it is a *dispersive system* even if there are no losses.

In a dispersive system, the waveform of a nonsinusoidal signal changes in shape during its propagation. This is because the Fourier series of the signal consists of sinusoids of different frequencies and in a dispersive medium the phase velocity of different sinusoids are different. Therefore, the phase differences between different sinusoids change with time and they do not add up to the original signal as they travel. Consequently, a nonsinusoidal signal carrying information or data transmitted at the sending end will not reproduce itself at the receiving end. However, if the signal travels as a single group spread over a small range of frequency, as in the case of modulated signals, the velocity of the group is equal to the *reciprocal* of the rate of change of the phase constant with frequency at the mean frequency of the group. This velocity is called *group velocity*. Energy carried by the signal travels at group velocity. Denoting the group velocity by $v_{g,m}$, we have

$$v_{g,m} = \frac{1}{d\beta_m/d\omega}$$

By differentiating the phase constant defined in (18) with respect to ω and then taking the reciprocal of the resulting expression we will obtain the formula for group velocity.

$$\frac{d\beta_m}{d\omega} = \frac{d}{d\omega}\left(\sqrt{\mu\varepsilon}\sqrt{\omega^2 - \omega_{c,m}^{\;2}}\right) = \frac{\sqrt{\mu\varepsilon}}{\sqrt{1-(\omega_{c,m}/\omega)^2}}$$

And $$v_{g,m} = \frac{\sqrt{1-(\omega_{c,m}/\omega)^2}}{\sqrt{\mu\varepsilon}} = v\sqrt{1-(\omega_{c,m}/\omega)^2} \tag{26}$$

This is the velocity of energy flow. We note that the group velocity is less than the phase velocity.

Equations (19) to (21) give the phasor forms of the field intensities. In order to obtain the formulae in real time, we multiply each expression by $e^{j\omega t}$ and then take its real part. Thus,

$$\overline{E}_x = -jE_{x0}\cos\frac{m\pi x}{a}e^{j(\omega t-\beta_m z)} = E_{x0}\cos\frac{m\pi x}{a}e^{j(\omega t-\beta_m z-\pi/2)}, \text{ as } (-j) = e^{-j\pi/2}$$

The real part of $\overline{E}_x$, denoted by E_x, is

$$E_x = E_{x0}\cos\frac{m\pi x}{a}\sin(\omega t - \beta_m z) \tag{27}$$

The expressions for the other two field intensities are

$$E_z = E_{z0}\sin\frac{m\pi x}{a}\cos(\omega t - \beta_m z) \tag{28}$$

$$H_y = H_{y0}\cos\frac{m\pi x}{a}\sin(\omega t - \beta_m z) \tag{29}$$

Figure 9.4 shows a sketch of electric field lines for $m = 1$, $t = 0$, and when E_{z0} is positive. The electric field lines terminate normally at the surface of the plates. The electric field induces surface charges with density $\rho_s = \varepsilon E_n$.

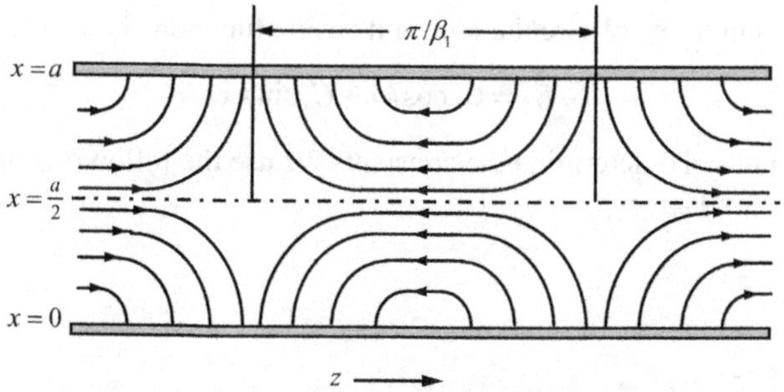

Fig. 9.4 Electric field lines of TM_1 wave mode between two conducting parallel-plates

The charges are negative in regions where the field lines enter the plates and positive where the field lines emerge from the plates. Over one-half of a wavelength, induced charges on both the plates have the same sign (positive or negative) at a given instant of time.

TE Waves between Parallel Plates

As per the definition, TE waves do not contain electric field intensity in the direction of propagation. As we have chosen z direction as the direction of travel of the fields, hence $\bar{E}_z = 0$. Moreover, the fields do not vary along y direction. Applying these restrictions to (9.2-15) and (9.2-16), we get the following relations.

$$\bar{E}_y = -j\frac{\omega\mu}{\gamma}\bar{H}_x \tag{30}$$

$$\bar{E}_x = j\frac{\omega\mu}{\gamma}\bar{H}_y \tag{31}$$

$$\bar{H}_y = j\frac{\omega\varepsilon}{\gamma}\bar{E}_x \tag{32}$$

$$\frac{\partial \bar{H}_z}{\partial x} + \gamma \bar{H}_x = -j\omega\varepsilon \bar{E}_y \tag{33}$$

Eliminating $\bar{H}_x$ from (30) and (33) and then solving for $\bar{E}_y$, we get

$$\bar{E}_y = \frac{j\omega\mu}{k^2}\frac{\partial \bar{H}_z}{\partial x} \tag{34}$$

where

$$k^2 = \gamma^2 + \omega^2\mu\varepsilon$$

By solving (31) and (32) simultaneously, and since $\gamma^2 + \omega^2\mu\varepsilon$ cannot be zero, we have

$$\bar{E}_x = 0, \bar{H}_y = 0$$

Thus, the dielectric region has three field components in $\bar{E}_y, \bar{H}_x$ and $\bar{H}_z$. All the components satisfy the wave equation. If we solve the wave equation for any one of the field intensities, we can find the solutions for the other two components. We will solve the wave equation involving $\bar{E}_y$ because it is tangential to the plate surfaces and hence the constants arising out of the general solution can be determined easily. The wave equation in E_y is obtained by setting $\partial^2/\partial y^2 = 0$ in equation (9.2-10). Accordingly,

$$\frac{\partial^2 E_y}{\partial x^2} = -k^2 E_y \tag{35}$$

The general solution of (3) in terms of circular sine and cosine functions is

$$\bar{E}_y = C_3 \cos kx + C_4 \sin kx \qquad (36)$$

where C_3 and C_4 are constants. To determine the constants we use the following known boundary conditions in the surface of a perfect conductor.

1. At $x = 0,\ \bar{E}_y = 0$
2. At $x = a,\ \bar{E}_y = 0$

Application of the first boundary condition to (36) gives

$$C_3 = 0$$

Application of the second boundary condition yields

$$C_4 \sin ka = 0$$

The above equation is satisfied if k is an integral multiple of π/a. Denoting the integer by m, we get

$$k = m\pi/a, \quad m = 1, 2, 3, \ldots$$

Even though $m = 0$ also satisfies the boundary condition, it is not considered because then $\bar{E}_y$ will become zero everywhere.

Substituting $C_3 = 0$ and $k = m\pi/a$ into (36), the solution of (35) satisfying the boundary conditions is

$$\bar{E}_y = C_4 \sin \frac{m\pi x}{a} \qquad (37)$$

Equation (37) shows that the number of half cycles of the field between the conducting plates is equal to m. Each value of m defines a specific field configuration or mode of the guide. The TE wave having m number of half cycles between the plates is the TE_m wave or TE_m mode.

The propagation constant of the mth mode,

$$\gamma_m = \sqrt{k^2 - \omega^2 \mu\varepsilon} = \sqrt{(m\pi/a)^2 - \omega^2 \mu\varepsilon}$$

We find that

the formula for the propagation constant of the TE_m mode is the same as that of the TM_m mode. Therefore, the formulae for the parameters $\omega_{c,m}, \alpha_m, \beta_m, \lambda_m, v_{p,m}, v_{g,m}$ of TE waves are also same as those of TM waves.

The propagation constant above the cut-off frequency, $\gamma_m = j\beta_m$. Thus, inserting the factor $e^{-j\beta_m z}$ into (37) and denoting the amplitude of the electric field by E_{y0}, we get

$$\bar{E}_y = E_{y0} \sin \frac{m\pi x}{a} e^{-j\beta_m z} \qquad (38)$$

Now substituting (38) into (30) and replacing γ by $j\beta$, we get the solution for H_x.

$$H_x = \left(\frac{\beta}{\omega\mu} E_{y0}\right) \sin \frac{m\pi x}{a} e^{-j\beta_m z} = H_{x0} \sin \frac{m\pi x}{a} e^{-j\beta_m z} \qquad (39)$$

Next, by substituting (38) into (34) and then integrating we get the solution for $\bar{H}_z$. Taking the constant of integration as zero, we have

$$\bar{H}_z = j\left(\frac{m\pi}{a\omega\mu}E_{y0}\right)\cos\frac{m\pi x}{a}e^{-j\beta_m z} = jH_{z0}\cos\frac{m\pi x}{a}e^{-j\beta_m z} \tag{40}$$

In (39) and (40) H_{x0} and H_{z0} represent the functions inside the brackets. These are the amplitudes of respective field intensities in terms of E_{y0}.

The intrinsic wave impedance,

$$\eta_{\text{TE},m} = -\frac{\bar{E}_y}{\bar{H}_x} = -\frac{E_{y0}}{H_{x0}} = \frac{\omega\mu}{\beta_m} = \frac{\eta}{\sqrt{1-(\omega_{c,m}/\omega)^2}} = \frac{\eta}{\sqrt{1-(f_{c,m}/f)^2}} \tag{41}$$

$$\eta = \sqrt{\frac{\mu}{\varepsilon}}$$

Note that the formulae for intrinsic wave impedance of TM wave given by (22) and of TE wave given by (41) are different.

The expressions for the instantaneous field intensities are given by

$$E_y = E_{y0}\sin\frac{m\pi x}{a}\cos(\omega t - \beta_m z)$$

$$H_x = -H_{x0}\sin\frac{m\pi x}{a}\cos(\omega t - \beta_m z)$$

$$H_z = -H_{z0}\cos\frac{m\pi x}{a}\sin(\omega t - \beta_m z)$$

Average power flow over a width b (in y direction),

$$P_{av} = \text{Re}\frac{1}{2}\int_0^a\left(E_{y0}\sin\frac{m\pi x}{a}e^{-j\beta_m z}\mathbf{u}_y\right)\times\left(-H_{x0}\sin\frac{m\pi x}{a}e^{-j\beta_m z}\mathbf{u}_x\right)^*.(b\,dx\mathbf{u}_z)$$

$$= \frac{abE_{y0}H_{x0}}{4} = \frac{ab\eta_{\text{TE},m}H_{x0}^2}{4} = \frac{abE_{y0}^2}{4\eta_{\text{TE},m}}$$

Figure 9.5 shows a sketch of the magnetic field lines for $m = 1$, at $t = 0$, and when E_{y0} is positive. The field lines are closed curves enclosing displacement currents in the region between the plates.

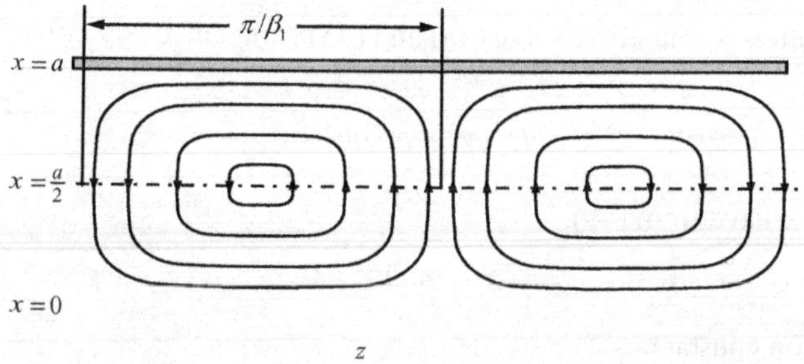

Fig. 9.5 Magnetic field lines of TE_1 wave mode between two conducting parallel-plates

TEM Waves between Parallel Plates

The electric field intensity and the magnetic field intensity along the direction of wave travel for TEM waves are zero. As we have assumed that waves travel along z direction, hence both $\bar{E}_z$ and $\bar{H}_z$ are equal to zero. We have also assumed that the fields between the parallel plates do not vary with y. By using these conditions in (9.2-15) and (9.2-16), we get the following relations.

$$\gamma \bar{E}_y = -j\omega\mu \bar{H}_x \tag{42}$$

$$\gamma \bar{E}_x = j\omega\mu \bar{H}_y \tag{43}$$

$$\frac{\partial \bar{E}_y}{\partial x} = 0 \tag{44}$$

$$\gamma \bar{H}_y = j\omega\varepsilon \bar{E}_x \tag{45}$$

$$\gamma \bar{H}_x = -j\omega\varepsilon \bar{E}_y \tag{46}$$

$$\frac{\partial \bar{H}_y}{\partial x} = 0 \tag{47}$$

As a first step, we may solve either (44) or (47) and then use the solution to find the other components. We will solve (44) because it is easier to determine the constant of integration from the boundary conditions. The integration of (44) results in a constant value for $\bar{E}_y$. The constant must have the same value everywhere between the plates including the plate surfaces. As the field $\bar{E}_y$ is tangential to the perfect conducting surfaces, it is equal to zero at the surfaces as per the boundary condition. Hence, $\bar{E}_y$ must be zero everywhere. Using this condition in (42) gives $\bar{H}_x = 0$. Thus, the region between the plates has only one component of electric field intensity $\bar{E}_x$ and one component of magnetic field intensity $\bar{H}_y$. Equation (43) as well as (45) gives relation between $\bar{E}_x$ and $\bar{H}_y$. The field $\bar{H}_y$ also satisfies the first order differential equation (47) and its solution is,

$$\bar{H}_y = \text{A constant}$$

The constant can only be the amplitude of the field. Denoting the amplitude by H_{y0} and introducing the factor $e^{-\gamma z}$, the solution is given by

$$\bar{H}_y = H_{y0} e^{-\gamma z} \tag{48}$$

Substituting (48) into (43), we get the solution for $\bar{E}_x$.

$$\bar{E}_x = j\frac{\omega\mu}{\gamma} \bar{H}_{y0} e^{-\gamma z} \tag{49}$$

To calculate the propagation constant, we solve (43) and (45) simultaneously either for $\bar{E}_x$ or for $\bar{H}_y$. Solving for H_y, we get

$$\bar{H}_y(\gamma^2 + \omega^2\mu\varepsilon) = 0 \tag{50}$$

As H_y cannot be zero, we have from (49),

$$\gamma^2 \quad \omega^2\mu\varepsilon = 0$$

This gives the propagation constant,

$$\gamma \quad \omega \quad \mu\varepsilon = j\beta.$$

where $\qquad \beta = \omega\sqrt{\mu\varepsilon}$, is the phase constant of TEM waves.

The phase velocity, $\qquad v = \dfrac{\omega}{\beta} = \dfrac{1}{\sqrt{\mu\varepsilon}}$

The wavelength, $\qquad \lambda = \dfrac{\beta}{2\pi} = \dfrac{f}{v}$

The intrinsic wave impedance, $\qquad \eta_{\text{TEM}} = \dfrac{\bar{E}_x}{\bar{H}_y} = \dfrac{\omega\mu}{\beta} = \sqrt{\dfrac{\mu}{\varepsilon}} = \eta \qquad (51)$

We see that

---

all the parameters of TEM waves between two conducting plates are same as those of the plane waves. Moreover, the field distribution between the plates at any z is uniform as in the case of static fields. In addition, there is no cut-off frequency for TEM waves.

Replacing γ by $j\beta$ in (48) and (49), we get

$$\bar{H}_y = H_{y0}e^{-j\beta z} \qquad (52)$$

$$\bar{E}_x = \left(\frac{\omega\mu}{\beta}\bar{H}_{y0}\right)e^{-j\beta z} = E_{x0}e^{-j\beta z} = \eta\bar{H}_{y0}e^{-j\beta z} \qquad (53)$$

The average power flow between the plates may be calculated using the complex Poynting vector. The complex Poynting vector,

$$\mathbf{P}_w = \frac{1}{2}(\bar{\mathbf{E}}\times\bar{\mathbf{H}}^*) = \frac{1}{2}(E_{x0}e^{-j\beta z})\times(H_{y0}e^{-j\beta z})^* = \frac{E_{x0}H_{y0}}{2}\ \text{W/m}^2$$

The Poynting vector is a real quantity and it is independent of the coordinate variables. Thus, average power flow between the plates over a width b (in y direction) is

$$P_{av} = \frac{abE_{x0}H_{y0}}{2} = \frac{ab\eta H_{y0}^2}{2}$$

The velocity of power flow is same as the phase velocity because the phase velocity is independent of frequency.

To obtain the solutions of the fields at any time t, we multiply (48) and (49) by $e^{j\omega t}$ and then use either the real or the imaginary parts. The solutions using the real parts are

$$H_y = H_{y0}\cos(\omega t - \beta z)$$

$$E_x = E_{x0}\cos(\omega t - \beta z) = \eta H_{y0}\cos(\omega t - \beta z)$$

EXAMPLE 9.1

The relative permeability and relative permittivity of the dielectric between the perfect conducting plates of a parallel-plate waveguide are 1.0 and 2.5 respectively. Frequency of signal in the waveguide $f = 3.0$ GHz. Find spacing between the plates if the signal frequency is 1.25 times the cut-off frequency for TM_1 mode.

Signal frequency, 3.0×10^9 Hz; $\quad = 1, \quad _r = 1.0 \quad _r = 2.5$

Cut-off frequency, $\qquad f_{c,m} = \dfrac{m}{2a\ \ \mu\varepsilon}$

Using the given values, $$f_{c,1} = \frac{3\times10^8}{2a\sqrt{(1.0)(2.5)}} = \frac{0.949\times10^8}{a} = \frac{3.0\times10^9}{1.25}$$

Thus, $$a = \frac{(1.25)(0.949\times10^8)}{3.0\times10^9} = 0.04 \text{ m}$$

EXAMPLE 9.2

The spacing between the perfect conducting plates of a parallel-plate waveguide $a = 5.0$ cm. The relative permeability and relative permittivity of the dielectric between the plates are 1.0 and 1.8 respectively. (a) Find the cut-off frequency. (b) Find the intrinsic wave impedance for TM_1 and TE_1 modes if the working frequency is 30 per cent higher than the cut-off frequency.

(a) Cut-off frequency, $$f_{c,1} = \frac{1}{2a\sqrt{\mu\varepsilon}} = \frac{3\times10^8}{(2)(0.05)\sqrt{1.8}} = 22.36\times10^8 \text{ Hz}$$

(b) $$\eta = \sqrt{\frac{\mu}{\varepsilon}} = \frac{120\pi}{\sqrt{1.8}} = 281\ \Omega;\ f_{c,1}/f = 1/1.3$$

Intrinsic wave impedance for the TM_1 mode,

$$\eta_{[illegible]} \quad \eta\sqrt{1-(f_{c,1}/f)^2} = (281)\sqrt{1-(1/1.3)^2} = 179.63\ \Omega$$

Intrinsic wave impedance for the TM_1 mode,

$$\eta_{TE,1} = \frac{\eta}{\sqrt{1-(f_{c,1}/f)^2}} = \frac{281}{\sqrt{1-(1/1.3)^2}} = 439.78\ \Omega$$

EXAMPLE 9.3

A waveguide consists of two conducting parallel plates with a spacing of 3 cm in air. Calculate the value of cut-off frequency of the TE_1 wave. Also, find the phase constant, phase velocity, and group velocity at a source frequency equal to twice the critical frequency.

Distance between the plates, $a = 0.03$ m

Permeability, $\mu = \mu_0 = 4\pi\times10^{-7}$ H/m

Permittivity, $\varepsilon = \varepsilon_0 = 8.85\times10^{-12}$ F/m

Mode, $m = 1$

Cut-off frequency, $$f_{c,1} = \frac{m}{2a\sqrt{\mu\varepsilon}} = \frac{3\times10^8}{2(0.03)} = 5\times10^9 \text{ Hz}$$

Source frequency, f [illegible] $f_{c,1} =$ [illegible] $\times10^9$ Hz

Radian frequency, ω [illegible] π [illegible] $= 62.83\times10^9$ [illegible] /s

Phase constant, β [illegible] $\omega\ \mu\varepsilon\ 1-(\omega_{c,1}/\omega)^2 = \frac{62.83\times10^9}{3\ \ 10^8}$ 1 [illegible] $(1/2)^2 = 181.4$ rad/m

Phase velocity, $v_{p,1} = \dfrac{\omega}{\beta_1} = \dfrac{62.83 \times 10^9}{181.4} = 3.46 \times 10^8$ m/s

Note that the phase velocity is greater than the velocity of light.

Group velocity, $v_{g,m} = \dfrac{\sqrt{1-(\omega_{c,m}/\omega)^2}}{\sqrt{\mu\varepsilon}} = (3 \times 10^8)\sqrt{1-(1/2)^2} = 2.6 \times 10^8$ m/s

The group velocity is less than the velocity of light even though the medium between the plates is air.

EXAMPLE 9.4

The amplitude of voltage between the plates of a lossless parallel-plate waveguide located in air is 200 V. Calculate the average power flow per unit width of the waveguide at source frequencies of 5 GHz and 7.5 GHz if the cut-off frequency is 3.0 GHz in the TM_1 mode.

Voltage between the plates, $V_0 = 200$ V

Cut-off frequency, $f_{c,1} = 3 \times 10^9$ Hz

Permeability, $\mu = \mu_0 = 4\pi \times 10^{-7}$ H/m

Permittivity, $\varepsilon = \varepsilon_0 = 8.85 \times 10^{-12}$ F/m

Mode, $m = 1$

Average power flow between the plates, $P_{av} = \dfrac{abE_0^2}{4\eta_{\text{TM},1}}$

Distance between the plates, $a = \dfrac{m(=1)}{2f_{c,1}\sqrt{\mu\varepsilon}} = \dfrac{3.0 \times 10^8}{2(3.0 \times 10^9)} = 0.05$ m

Width, $b = 1.0$ m

Electric field intensity, $E_0 = \dfrac{V_0}{a} = \dfrac{200.0}{0.05} = 4000.0$ V/m

Intrinsic impedance, $\eta_{\text{TM},1} = \eta\sqrt{1-(f_{c,1}/f)^2}$, where $\eta = \eta_0 = 120\pi\ \Omega$

At $f = 5$ GHz,

$$\eta_{\text{TM},1} = 120\pi\sqrt{1-(3/5)^2} = 301.6\ \Omega$$

$$P_{av} = \frac{(0.05)(1)(4000)^2}{(4)(301.6)} = 663.1 \text{ W}$$

At $f = 7.5$ GHz,

$$\eta_{\text{TM},1} = 120\pi\sqrt{1-(3/7.5)^2} = 345.5\ \Omega$$

$$P_{av} = \frac{(0.05)(1)(4000)^2}{(4)(345.5)} = 579.0 \text{ W}$$

9.5 The Rectangular Waveguide

A rectangular waveguide consists of a hollow conducting cylinder of rectangular cross section, as shown in Fig. 9.6. The rectangular section of the waveguide is parallel to the x-y plane and its axis is parallel to the z axis, as shown in the figure. The dielectric region within the waveguide is of height a in x direction and width b in y direction. We assume a to be larger than b. We will study the propagation of electromagnetic waves in the waveguide with the following simplifying assumptions.

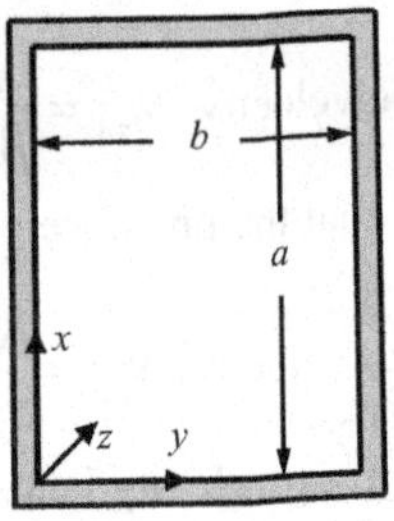

Fig. 9.6 Cross-sectional view of a hollow rectangular metallic waveguide

1. There is no discontinuity along the length of the waveguide; hence, there is no reflection of waves. Accordingly, the waveguide has a single set of travelling waves.
2. Conductivity of the waveguide walls is infinitely large so that fields do not penetrate into the walls. Therefore, no losses occur in them. In addition, the electric field intensity that is tangential and the magnetic field intensity that is normal to the waveguide walls are equal to zero as per the boundary conditions on the surface of a perfect conductor.
3. Conductivity of the dielectric medium is infinitely small; hence, it is loss-free.
4. Assumptions 2 and 3 also mean that amplitudes of the fields do not vary along the direction of propagation above the cut-off frequency.
5. The fields travel along positive z direction.
6. The variations of fields with z and t are in accordance with exponential functions $e^{-\gamma z}$ and $e^{j\omega t}$ respectively, where γ, generally a complex quantity, is the propagation constant and ω is the angular frequency of the fields. Thus, the wave equations and relations between field intensities given in Section 9.2 will be the basis for analysing the phenomenon of wave propagation inside the waveguide.

We will find that many properties of waves in a rectangular waveguide are similar to those of a parallel-plate waveguide studied in the last section. However, there are certain differences as well. For example

in a parallel-plate waveguide, all the wave types (TM, TE and TEM) can exist whereas rectangular guides can transmit power in TM and TE modes only. Moreover, fields in the dielectric region of a rectangular waveguide vary with both x and y.

TM Waves in Rectangular Waveguides

The magnetic field intensity along the direction of propagation is zero in the case of TM waves. That is, $H_z = 0$. Thus, relations (9.2-15*a*) and (9.2-15*b*) remain unchanged, whereas the first terms on left sides of (9.2-16*a*) and (9.2-16*b*) become zero. Hence, we get the following relations for the TM mode in a rectangular waveguide.

$$\frac{\partial \bar{E}_z}{\partial y} + \gamma \bar{E}_y = -j\omega\mu \bar{H}_x \tag{1}$$

$$\frac{\partial \bar{E}_z}{\partial x} + \gamma \bar{E}_x = j\omega\mu \bar{H}_y \tag{2}$$

$$\bar{H}_y = j\frac{\omega\varepsilon}{\gamma}\bar{E}_x \tag{3}$$

$$\bar{H}_x = -j\frac{\omega\varepsilon}{\gamma}\bar{E}_y \tag{4}$$

We will not require equations (9.2-15*c*) and (9.2-16*c*).

We find from the above expressions that there are three components of electric field intensity E_z, E_y, and E_z. We can express E_x in terms of E_z by eliminating H_y from (2) and (3), and also express E_y in terms of E_z by eliminating H_x from (1) and (4). This process will result in the following relations.

$$\bar{E}_x = -\frac{\gamma}{\gamma^2 + \omega^2 \mu\varepsilon} \frac{\partial \bar{E}_z}{\partial x} = -\frac{\gamma}{k^2} \frac{\partial E_z}{\partial x} \tag{5}$$

$$\bar{E}_y = -\frac{\gamma}{k^2} \frac{\partial E_z}{\partial y} \tag{6}$$

The foregoing relations show that the dielectric region of the waveguide has five field components. Each of these components satisfies the two-dimensional wave equation given in Section 9.2. We will solve the wave equation for $\bar{E}_z$ and then use this solution to find the solution for other field components. The wave equation in $\bar{E}_z$ is

$$\frac{\partial^2 \bar{E}_z}{\partial x^2} + \frac{\partial^2 \bar{E}_z}{\partial y^2} = -k^2 \bar{E}_z \tag{7}$$

Equation (7) is a two-dimensional partial differential equation of second order. To solve the equation we assume $\bar{E}_z$ to be the product of two fields. One field is a function of x only and the other is a function of y only. Let the field that varies with x be $\bar{X}$ and the field that varies with y be $\bar{Y}$. Then,

$$\bar{E}_z = \bar{X}\,\bar{Y} \tag{8}$$

By substituting (8) in (7), we get

$$\bar{Y}\frac{\partial^2 \bar{X}}{\partial x^2} + \bar{X}\frac{\partial^2 \bar{Y}}{\partial y^2} = -k^2 \bar{X}\,\bar{Y}$$

Dividing both sides by $\bar{X}\,\bar{Y}$, we have

$$\frac{1}{\bar{X}}\frac{\partial^2 \bar{X}}{\partial x^2} + \frac{1}{\bar{Y}}\frac{\partial^2 \bar{Y}}{\partial y^2} = -k^2 \tag{9}$$

$$\Rightarrow \frac{1}{\bar{X}}\frac{\partial^2 \bar{X}}{\partial x^2} = -k^2 - \frac{1}{\bar{Y}}\frac{\partial^2 \bar{Y}}{\partial y^2} \tag{10}$$

Since the lhs of (10) is a function of x only and the rhs is a function of y only, each side can only be the same constant. This is because if the lhs varies with x, it cannot be equal to the rhs as the rhs is not a function of x, and if the rhs varies with y it cannot be equal to the lhs as the lhs is not a function of y.

Moreover, we know from the analysis of the parallel-plate guide that the constant must be a negative quantity so that the solutions will be in terms of circular sine and cosine functions and the boundary conditions can be determined easily. Thus, taking the constant as $-g^2$ and using ordinary derivatives, as each side is a function of only one variable, we have

$$\frac{d^2 \bar{X}}{dx^2} = -g^2 \bar{X}$$

and

$$\frac{d^2 \bar{Y}}{dx^2} = -(k^2 - g^2)\bar{Y} = -h^2 \bar{Y}$$

where

$$h^2 = k^2 - g^2$$

As $k^2 = \gamma^2 + \omega^2 \mu\varepsilon$, we have

$$\gamma^2 + \omega^2 \mu\varepsilon = g^2 + h^2 \tag{11}$$

The general solutions of the differential equations are

$$X \quad C_1 \quad gx + C \quad gx \tag{12}$$

$$\bar{Y} = C_3 \cos hy + C_4 \sin hy \tag{13}$$

Substitution of (12) and (13) into (8) yields

$$\bar{E}_z = (C_1 \cos gx + C_2 \sin gx)(C_3 \cos hy + C_4 \sin hy) \tag{14}$$

To calculate the constants we use of the boundary conditions for a perfect conductor. The boundary condition is that no tangential electric field intensity can exist on the surface of a perfect conductor. Accordingly, the electric field $\bar{E}_z$ must be zero at the surfaces of all the walls of the waveguide as it is tangential to these surfaces. Thus referring to Fig. 9.6, we have

1. $E_z = 0$ at $x = 0$
2. $E_z = 0$ at $x = a$
3. $E_z = 0$ at $y = 0$
4. $E_z = 0$ at $y = b$

Using the first two boundary conditions into (14), we get

$$C_1 = 0 \text{ and } g = m\pi/a, \qquad m = 1, 2, 3, \ldots$$

The third and fourth boundary conditions are satisfied if

$$C_3 = 0 \text{ and } h = n\pi/b, \qquad n = 1, 2, 3, \ldots$$

By substituting the above values into (14), we get

$$\bar{E}_z = (C_2 C_4) \sin \frac{m\pi x}{a} \sin \frac{n\pi y}{b} = E_{z0} \sin \frac{m\pi x}{a} \sin \frac{n\pi y}{b} \tag{15}$$

where $C_2 C_4 = E_{z0}$ is the amplitude of the field.

We note that the boundary conditions are satisfied even if $m = 0$ or $n = 0$ or if both m and n are equal to zero. However, these are not admissible because for these cases $\bar{E}_z$ will become zero everywhere. And other fields will also become zero everywhere as these are functions of $\bar{E}_z$.

Multiplication of (15) by $e^{-\gamma z}$ gives the complete solution of the phasor wave equation in $\bar{E}_z$. Accordingly, we have

$$\bar{E}_z = E_{z0} \sin \frac{m\pi x}{a} \sin \frac{n\pi y}{b} e^{-\gamma z} \tag{16}$$

In order to obtain the formula for the propagation constant, we substitute, $g = m\pi/a$ and $h = n\pi/b$ into (11). Denoting the propagation constant by γ_{mn}, we have

$$\gamma_{mn} = \sqrt{(m\pi/a)^2 + (n\pi/b)^2 - \omega^2 \mu\varepsilon} \tag{17}$$

Formula (17) shows that the propagation constant for a given waveguide depends on the frequency of the fields, the constants of the dielectric medium, and dimensions of the waveguide. The frequency at which the propagation constant becomes zero is defined as the cut-off or critical frequency. Thus equating (17) to zero and denoting the cut-off frequency by $\omega_{c,\,mn}$, we have

$$\omega_{c,mn} = \frac{1}{\sqrt{\mu\varepsilon}}\sqrt{\left(\frac{m\pi}{a}\right)^2 + \left(\frac{n\pi}{b}\right)^2} \tag{18a}$$

And

$$f_{c,mn} = \frac{1}{2\sqrt{\mu\varepsilon}}\sqrt{\left(\frac{m}{a}\right)^2 + \left(\frac{n}{b}\right)^2} \tag{18b}$$

It will be convenient if the propagation constant is expressed in terms of the cut-off frequency. This is achieved by substituting (18*a*) in (17). Thus,

$$\gamma_{mn} = \sqrt{\mu\varepsilon}\sqrt{\omega_{c,mn}^{\;2} - \omega^2} \tag{19}$$

As in the case of a parallel-plate waveguide, the propagation constant is a real quantity if the frequency of the fields is less than the cut-off frequency. Hence, we may write

$$\gamma_{mn} = \sqrt{\mu\varepsilon}\sqrt{\omega_{c,mn}^{\;2} - \omega^2} = \alpha_{mn} + j0, \quad \omega < \omega_{c,mn} \tag{20}$$

Substitution of (20) in (16) gives

$$\bar{E}_z = (E_{z0}\, e^{-\alpha_{mn}})\sin\frac{m\pi x}{a}\sin\frac{n\pi y}{b}$$

The function within the bracket is the amplitude of the field at any z. Thus, the amplitude of the field attenuates exponentially with z. Moreover as the phase constant is zero, there is no change of phase and hence the fields do not travel below the cut-off frequency.

The parameter α_{mn} is the attenuation constant. Note that the attenuation is not because of any kind of losses in the waveguide.

Properties of TM Waves above *Cut-off Frequency*: At frequencies greater than the cut-off frequency, the function under the radical sign of (19) is a negative quantity and hence the propagation constant becomes imaginary. We may write the propagation constant as

$$\gamma_{mn} = j\sqrt{\mu\varepsilon}\sqrt{\omega^2 - \omega_{c,mn}^{\;2}}$$

$$= j\omega\sqrt{\mu\varepsilon}\sqrt{1-(\omega_{c,mn}/\omega)^2} = j\beta_{mn}, \quad \omega > \omega_{c,mn} \tag{21}$$

where

$$\beta_{mn} = \omega\sqrt{\mu\varepsilon}\sqrt{1-(\omega_{c,mn}/\omega)^2} = \beta\sqrt{1-(\omega_{c,mn}/\omega)^2}\,;\ \beta = \omega\sqrt{\mu\varepsilon} \tag{22}$$

Substituting $\gamma = j\beta_{mn}$ into (16), the solution of the wave equation in $\bar{E}_z$ is

$$\bar{E}_z = E_{z0}\sin\frac{m\pi x}{a}\sin\frac{n\pi y}{b}e^{-j\beta z} \tag{23}$$

To find the solutions for $\bar{E}_x$ and $\bar{E}_y$, we substitute partial derivatives of (23) with respect to x into (5), and with respect to y into (6) respectively. Thus, we have

$$E_{..} \quad j\left(\frac{m\pi\beta_{mn}}{ak^2}E_{z0}\right)\cos\frac{m\pi x}{a}\sin\frac{n\pi y}{b}e^{-j\ z}$$

$$= -jE_{x0}\cos\frac{m\ x}{a}\sin\frac{n\ y}{b}e^{-j\ z} \tag{24}$$

$$\bar{E}_y = -j\left(\frac{n\pi\beta_{mn}}{bk^2}E_{z0}\right)\sin\frac{m\pi x}{a}\cos\frac{n\pi y}{b}e^{-j\beta_{mn}z}$$

$$= -jE_{y0}\sin\frac{m\pi x}{a}\cos\frac{n\pi y}{b}e^{-j\beta_{mn}z} \tag{25}$$

By substituting (25) into (4) and (24) into (3), and as $\gamma = j\beta_{mn}$, we will get the solutions for $\bar{H}_x$ and $\bar{H}_y$. These are given by

$$\bar{H}_x = j\left(\frac{n\pi\omega\varepsilon}{bk^2}E_{z0}\right)\sin\frac{m\pi x}{a}\cos\frac{n\pi y}{b}e^{-j\beta_{mn}z}$$

$$= jH_{x0}\sin\frac{m\pi x}{a}\cos\frac{n\pi y}{b}e^{-j\beta_{mn}z} \tag{26}$$

$$\bar{H}_y = -j\left(\frac{m\pi\omega\varepsilon}{ak^2}E_{z0}\right)\cos\frac{m\pi x}{a}\sin\frac{n\pi y}{b}e^{-j\beta_{mn}z}$$

$$= -jH_{y0}\cos\frac{m\pi x}{a}\sin\frac{n\pi y}{b}e^{-j\beta_{mn}z} \tag{27}$$

In the above expressions E_{x0}, E_{y0}, H_{x0} and H_{y0} denote the functions inside the brackets. These are the amplitudes of the respective fields. Moreover,

$$k^2 = g^2 + h^2 = (m\pi/a)^2 + (n\pi/b)^2.$$

The ratio between amplitudes of electric field intensity in y direction given by (25) and amplitude of electric field intensity in x direction given by (24) is

$$\frac{E_{y0}}{E_{x0}} = \frac{na}{mb}$$

It can be verified that all the field components satisfy the boundary conditions. The distribution of the fields along x and y directions are either sine or cosine functions of space variables in these directions. The number of half cycles between the guide walls is m in x direction and n in y direction. The field configuration over the x-y plane depends on the values of m and n; each set of m and n gives a specific field configuration or mode.

The TM wave having m half-cycles along x direction and n half-cycles along y direction is called TM_{mn} wave or TM_{mn} mode. The boundary conditions are satisfied for $m = 0$ or $n = 0$ as mentioned earlier. However, these are not admissible because in these cases $\bar{E}_z$ will become zero everywhere. Thus, the lowest order of TM wave is the TM_{11} mode. The fields travel along z direction with phase constant β_{mn} defined by (22). The phase constant is less than that of a plane wave in an unbounded dielectric region having the same permeability and permittivity as in the waveguide.

The formulae for various parameters of the waveguide are

Intrinsic wave impedance,

$$\eta_{TM,mn} = \frac{\bar{E}_x}{\bar{H}_y} = -\frac{\bar{E}_y}{\bar{H}_x} = \frac{\beta_{mn}}{\omega\varepsilon} = \eta\sqrt{1-(\omega_{c,mn}/\omega)^2}\,;\ \eta = \sqrt{\mu/\varepsilon}$$

Phase velocity, $$v_{p\,mn} = \frac{\omega}{\beta_{mn}} = \frac{v}{1-(\omega_{c,mn}/\omega)^2}\,;\ v = \frac{1}{\mu\varepsilon}$$

Wavelength along the waveguide, $\lambda_{mn} = \dfrac{2\pi}{\beta_{mn}} = \dfrac{\lambda}{\sqrt{1-(\omega_{c,mn}/\omega)^2}}; \lambda = v/f$

Group velocity, $v_{g,mn} = \dfrac{1}{(d\beta_{mn}/d\omega)} = \dfrac{\sqrt{1-(\omega_{c,mn}/\omega)^2}}{\sqrt{\mu\varepsilon}} = v\sqrt{1-(\omega_{c,mn}/\omega)^2}$

The parameters β, η, v and λ have the same meaning as for a uniform plane wave travelling in an unbounded lossless dielectric medium.

Next, we will calculate power flow in the dielectric region of the wave guide using the complex Poynting vector. The electric field intensity has all the three components, whereas the magnetic field intensity has components in x and y directions. The complex Poynting vector,

$$\bar{\mathbf{P}}_w = \frac{1}{2}(\bar{\mathbf{E}} \times \bar{\mathbf{H}}^*)$$

$$= \frac{1}{2}(\bar{E}_x\mathbf{u}_x + \bar{E}_y\mathbf{u}_y + \bar{E}_z\mathbf{u}_z) \times (\bar{H}_x\mathbf{u}_x + \bar{H}_y\mathbf{u}_y)^*$$

$$= \frac{1}{2}[(\bar{E}_x\bar{H}_y^* - \bar{E}_y\bar{H}_x^*)\mathbf{u}_z + \bar{E}_z\bar{H}_x^*\,\mathbf{u}_y - \bar{E}_z\bar{H}_y^*\,\mathbf{u}_x] \tag{28}$$

By substituting the expressions of electric field intensities and the conjugate of magnetic field intensities in (28) we will find that the z component of the Poynting vector is real, whereas the x and y components are imaginary. The real part of the complex Poynting vector,

$$\mathbf{P}_{wa} = \frac{1}{2}(\bar{E}_x\bar{H}_y^* - \bar{E}_y\bar{H}_x^*)\mathbf{u}_z$$

$$= \frac{E_{x0}H_{y0}}{2}\cos^2\frac{m\pi x}{a}\sin^2\frac{n\pi y}{b}\mathbf{u}_z + \frac{E_{y0}H_{x0}}{2}\sin^2\frac{m\pi x}{a}\cos^2\frac{n\pi y}{b}\mathbf{u}_z$$

The average power flow in the dielectric region of the guide,

$$P_{av} = \int_S \mathbf{P}_{wa}\cdot d\mathbf{S} = \int_{y=0}^{b}\int_{x=0}^{a}\Big(\frac{E_{x0}H_{y0}}{2}\cos^2\frac{m\pi x}{a}\sin^2\frac{n\pi y}{b}$$

$$+\frac{E_{y0}H_{x0}}{2}\sin^2\frac{m\pi x}{a}\cos^2\frac{n\pi y}{b}\mathbf{u}_z\Big).(dx\,dy)$$

$$= \frac{ab}{8}(E_{x0}H_{y0} + E_{y0}H_{x0})$$

Since $\dfrac{E_{x0}}{H_{y0}} = \dfrac{E_{y0}}{H_{x0}} = \eta_{\text{TM},mn}$ and $\dfrac{E_{y0}}{E_{x0}} = \dfrac{na}{mb}$, the average power flow in terms of E_{x0} is

$$P_{av} = \frac{abE_{x0}^2}{8\eta_{\text{TM},mn}}\left(\frac{m^2b^2 + n^2a^2}{m^2b^2}\right) \tag{29}$$

In order to show variation of the fields with time, we have to multiply each of (23) to (27) by $e^{j\omega t}$ and then use either the real or the imaginary part. Using the real parts, the expressions of the fields at any instant are

$$E_z \quad E_{z0}\sin\frac{m\pi x}{a}\sin\frac{n\pi y}{b}\cos(\omega t \quad \beta_{mn}z)$$

$$E_x = E_{x0} \cos\frac{m\pi x}{a} \sin\frac{n\pi y}{b} \sin(\omega t - \beta_{mn} z)$$

$$E_y = E_{y0} \sin\frac{m\pi x}{a} \cos\frac{n\pi y}{b} \sin(\omega t - \beta_{mn} z)$$

$$H_x = -H_{x0} \sin\frac{m\pi x}{a} \cos\frac{n\pi y}{b} \sin(\omega t - \beta_{mn} z)$$

$$H_y = H_{y0} \cos\frac{m\pi x}{a} \sin\frac{n\pi y}{b} \sin(\omega t - \beta_{mn} z)$$

Figure 9.7 shows field maps of the TM_{11} mode at $t = 0$, and for positive E_{z0}. In Fig. 9.7(*a*), the electric field pattern is on the $y = b/2$ plane, and the electric field pattern shown at Fig. 9.7(*b*) is on the $x = a/2$ plane. Fig. 9.7(*c*) shows electric and magnetic field lines on a plane parallel to the *x*-*y* plane.

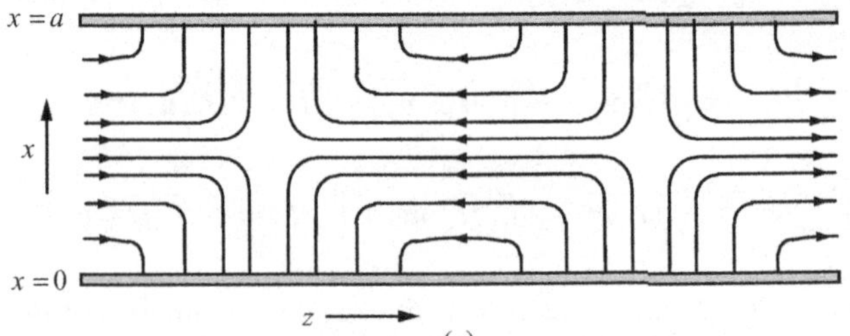

(*a*)

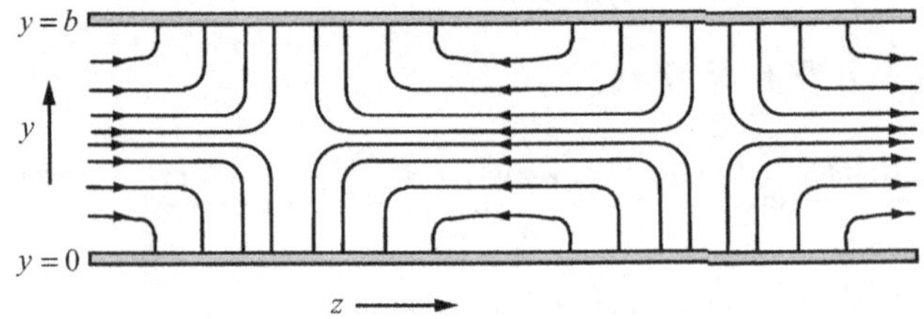

(*b*)

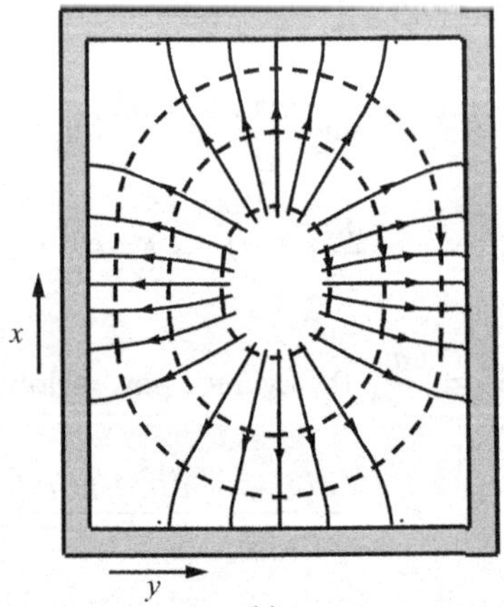

(*c*)

Fig. 9.7 Field patterns for TM_{11} mode in a rectangular waveguide. (*a*) Electric-field lines on the mid-plane parallel to the *x*-*z* plane. (*b*) Electric-field lines on the mid-plane parallel to the *y*-*z* plane. (*c*) Electric-field lines and magnetic field lines on a plane parallel to the *x*-*y* plane; solid lines are for the electric field and dashed lines are for the magnetic field.

TE Waves in Rectangular Waveguides

For TE waves, the electric field intensity along the direction of propagation, that is $\bar{E}_z$, is zero. Hence, relations (9.2-16*a*) and (9.2-16*b*) remain unchanged, whereas the first terms on the left sides of (9.2-15*a*) and (9.2-15*b*) become zero. The remaining two equations are not required. Accordingly, we have the following equations relating the electric and magnetic field intensities for TE waves.

$$\bar{E}_y = -j\frac{\omega\mu}{\gamma}\bar{H}_x \tag{30}$$

$$\bar{E}_x = j\frac{\omega\mu}{\gamma}\bar{H}_y \tag{31}$$

$$\frac{\partial \bar{H}_z}{\partial y} + \gamma \bar{H}_y = j\omega\varepsilon \bar{E}_x \tag{32}$$

$$\frac{\partial \bar{H}_z}{\partial x} + \gamma \bar{H}_x = -j\omega\varepsilon \bar{E}_y \tag{33}$$

By eliminating $\bar{H}_y$ from (31) and (32) and $\bar{H}_x$ from (30) and (31) we can express $\bar{E}_x$ and $\bar{E}_y$ in terms of H_z. The relations are

$$\bar{E}_x = -j\frac{\omega\mu}{k^2}\frac{\partial \bar{H}_z}{\partial y} \tag{34}$$

$$\bar{E}_y = j\frac{\omega\mu}{k^2}\frac{\partial \bar{H}_z}{\partial x} \tag{35}$$

We will obtain the general solution of the wave equation in $\bar{H}_z$ first and then use the solution in (34) and (35) to get the general solution for $\bar{E}_x$ and $\bar{E}_y$ respectively. Application of boundary conditions to $\bar{E}_x$ and $\bar{E}_y$ will give the constants of the general solution.

The wave equation in $\bar{H}_z$ is

$$\frac{\partial^2 \bar{H}_z}{\partial x^2} + \frac{\partial^2 \bar{H}_z}{\partial y^2} = -k^2 \bar{H}_z \tag{36}$$

The general solution of (36) in terms of cosine and sine functions is given by

$$\bar{H}_z = (C_1 \cos gx + C_2 \sin gx)(C_3 \cos hy + C_4 \sin hy) \tag{37}$$

where the constants g and h satisfy the following relation.

$$g^2 + h^2 = k^2 = \gamma^2 + \omega^2\mu\varepsilon \tag{38}$$

By substituting the partial derivatives of (37) with respect to y in (34) and with respect to x in (35) we will get the general solution for $\bar{E}_x$ and $\bar{E}_y$. The solutions are

$$\bar{E}_x = -j\frac{\omega\mu h}{k^2}(C_1 \cos gx + C_2 \sin gx)(-C_3 \sin hy + C_4 \cos hy) \tag{39}$$

$$\bar{E}_y = j\frac{\omega\mu g}{k^2}(-C_1 \sin gx + C_2 \cos gx)(C_3 \cos hy + C_4 \sin hy) \tag{40}$$

The electric field E_x is tangential to walls of the waveguide at $y = 0$ and $y = b$, whereas E_y is tangential to the guide walls at $x = 0$ and $x = a$. Thus, the boundary conditions are

1. $E_x = 0$ at $y = 0$
2. $E_x = 0$ at $y = b$
3. $E_y = 0$ at $x = 0$
4. $E_y = 0$ at $x = a$

Application of the first two boundary conditions to (39) yields

$$C_4 = 0,$$

$$h = \frac{n\pi}{b}; \quad n = 1, 2, 3, \ldots \tag{41a}$$

The third and fourth boundary conditions when applied to (40) give

$$C_2 = 0$$

$$g = \frac{m\pi}{a}; \quad m = 1, 2, 3, \ldots \tag{41b}$$

Substituting (41a) and (41b) into (37) and denoting the product of C_1 and C_3 by H_{z0}, we get

$$\bar{H}_z = H_{z0} \cos\frac{m\pi x}{a} \cos\frac{n\pi y}{b} \tag{42}$$

To obtain the propagation constant we substitute $g = (m\pi/a)$ and $h = (n\pi/b)$ in (38). Denoting the propagation constant by γ_{mn}, we have

$$\gamma_{mn} = \sqrt{(m\pi/a)^2 + (n\pi/b)^2 - \omega^2\mu\varepsilon}$$

This is identical to the formula for propagation constant of TM waves. Hence, the formulae for the cut-off frequency and the attenuation constant below cut-off frequency are the same as those of TM waves. The phase constant, phase velocity, wavelength along the waveguide, and group velocity above the cut-off frequency are also same as those of TM waves.

The solution for $\bar{H}_z$, showing its variation with z above the cut-off frequency, is obtained by multiplying (42) by $e^{-j\beta_{mn}z}$.

$$\bar{H}_z = H_{z0} \cos\frac{m\pi x}{a} \cos\frac{n\pi y}{b} e^{-j\beta_{mn}z} \tag{43}$$

where the phase constant β_{mn} is given in (22).

In order to get solutions for $\bar{E}_x$ and $\bar{E}_y$, we substitute the partial derivatives of (43) with respect to y and x into (34) and (35) respectively. The solutions are

$$\bar{E}_x = j\left(\frac{n\pi\omega\mu}{bk^2} H_{z0}\right) \cos\frac{m\pi x}{a} \sin\frac{n\pi y}{b} e^{-j\beta_{mn}z}$$

$$= jE_{x0} \cos\frac{m\pi x}{a} \sin\frac{n\pi y}{b} e^{-j\beta_{mn}z} \tag{44}$$

$$\bar{E}_y = -j\left(\frac{m\pi\omega\mu}{ak^2} H_{z0}\right) \sin\frac{m\pi x}{a} \cos\frac{n\pi y}{b} e^{-j\beta_{mn}z}$$

$$= -jE_{y0} \sin\frac{m\pi x}{a} \cos\frac{n\pi y}{b} e^{-j\beta_{mn}z} \tag{45}$$

Now, we use (45) in (30) and (44) in (31) to get the solutions for H_x and H_y. Thus noting that $\gamma_{mn} = j\beta_{mn}$, we have

$$\bar{H}_x = j\left(\frac{m\pi\beta_{mn}}{ak^2}H_{z0}\right)\sin\frac{m\pi x}{a}\cos\frac{n\pi y}{b}e^{-j\beta_{mn}z}$$

$$= jH_{x0}\sin\frac{m\pi x}{a}\cos\frac{n\pi y}{b}e^{-j\beta_{mn}z} \tag{46}$$

$$\bar{H}_y = j\left(\frac{n\pi\beta_{mn}}{bk^2}H_{z0}\right)\cos\frac{m\pi x}{a}\sin\frac{n\pi y}{b}e^{-j\beta_{mn}z}$$

$$= jH_{y0}\cos\frac{m\pi x}{a}\sin\frac{n\pi y}{b}e^{-j\beta_{mn}z} \tag{47}$$

where E_{x0}, E_{y0}, H_{x0} and H_{y0} denote the quantities inside the brackets.

The TE wave having m half-cycles in x direction and n half-cycles in y direction between the guide walls is designated as TE_{mn} wave.

The ratio between amplitudes of electric field intensity in y direction given by (25) and amplitude of electric field intensity in x direction given by (26) is given by

$$\frac{E_{y0}}{E_{x0}} = \frac{mb}{na}$$

The intrinsic wave impedance,

$$\eta_{\text{TE},mn} = \frac{\bar{E}_x}{\bar{H}_y} = -\frac{\bar{E}_y}{\bar{H}_x} = \frac{\omega\mu}{\beta_{mn}} = \frac{\eta}{\sqrt{1-(\omega_{c,mn}/\omega)^2}}$$

Note that

the formulae for the intrinsic wave impedance of TM and TE waves are different.

Let us now calculate the average power flow using the complex Poynting vector. As the electric field has components in x and y directions only, and the magnetic field has components in all the three coordinate directions, the complex Poynting vector,

$$\bar{\mathbf{P}}_w = \frac{1}{2}(\bar{E}_x\mathbf{u}_x + \bar{E}_y\mathbf{u}_y)\times(\bar{H}_x\mathbf{u}_x + \bar{H}_y\mathbf{u}_y + \bar{H}_z\mathbf{u}_z)^*$$

$$= \frac{1}{2}[(\bar{E}_x\bar{H}_y^* - \bar{E}_y\bar{H}_x^*)\mathbf{u}_z - \bar{E}_x\bar{H}_z^*\mathbf{u}_y + \bar{E}_y\bar{H}_z^*\mathbf{u}_x] \tag{48}$$

The presence of the operator j in (44) and (45) and its absence in (43) means that $\bar{H}_z$ is in time-quadrature to both $\bar{E}_x$ and $\bar{H}_y$. Therefore, the x and y components of (48) are imaginary. The real part of the complex Poynting vector,

$$\bar{\mathbf{P}}_{wa} = \frac{1}{2}(\bar{E}_x\bar{H}_y^* - \bar{E}_y\bar{H}_x^*)\mathbf{u}_z$$

Average transmitted-power in the dielectric region of the waveguide,

$$P_{av} = \int_0^a\int_0^b \frac{1}{2}(E_xH_y^* - E_yH_x^*)\,dxdy \tag{49}$$

Substituting expressions (44) and (45), and conjugates of (46) and (47) in (49) and then evaluating the definite integrals, we get

$$P_{av} = \frac{ab}{8}(E_{x0}H_{y0} + E_{y0}H_{x0})$$

As H_y $E_{x0} = H_{x0}/E_{y0} = \beta_{mn}/\omega\mu = 1/\eta_{\text{TE},mn}$ and $E_{y0}/E_{x0} = (mb)/(na)$, the average power in terms of E_{x0} is

$$P_{av} = \frac{ab}{8\eta_{\text{TE},mn}}(E_{x0}^2 + E_{y0}^2) = \frac{abE_{x0}^2}{8\eta_{\text{TE},mn}}\left(\frac{m^2b^2 + n^2a^2}{n^2a^2}\right) \tag{50}$$

Properties of TE$_{10}$ Wave

The boundary conditions in solutions (43) to (47) are satisfied if either $m = 0$ or $n = 0$. However both m and n cannot be zero, as in this case all the fields (except H_z) would become zero and no wave propagation is possible. Thus, the lowest order of TE waves in a rectangular waveguide is either the TE$_{10}$ mode or the TE$_{01}$ mode. The cut-off frequencies of the two modes are

$$f_{c,10} = \frac{1}{2a\sqrt{\mu\varepsilon}} = \frac{v}{2a}; \; v = \frac{1}{\sqrt{\mu\varepsilon}} \tag{51}$$

$$f_{c,01} = \frac{1}{2b\sqrt{\mu\varepsilon}} = \frac{v}{2b} \tag{52}$$

Relations (51) and (52) show that the cut-off frequency of TE$_{10}$ wave will be less than that of TE$_{01}$ wave if the height of the waveguide is greater than its width ($a > b$). Moreover, TM$_{11}$ mode is the lowest order for TM waves. Hence, the TE$_{10}$ wave has the lowest cut-off frequency.

> A wave mode that has the lowest cut-off frequency is called the *dominant mode*. Thus, the TE$_{10}$ mode is the dominant mode in a rectangular waveguide.

As a wave-mode with lowest cut-off frequency is generally used for practical studies, it is important to know the properties of the TE$_{10}$ wave. Some of the properties are outlined below.

The cut-off wavelength,

$$\lambda_{c,10} = \frac{v}{f_{c,10}} = 2a$$

That is, the cut-off wavelength is twice the longer side of the rectangular section of the waveguide.

It is useful for the purpose of comparison to express the cut-off frequency of the TM$_{mn}$ (or TE$_{mn}$) waves in terms of the cut-off frequency of the TE$_{10}$ wave. The cut-off frequencies of TM$_{mn}$ and TE$_{10}$ are defined in (18*b*) and (51) respectively. The ratio of the two cut-off frequencies is

$$\frac{f_{c,mn}}{f_{c,10}} = \sqrt{m^2 + \left(\frac{a}{b}n\right)^2}$$

The relative values of cut-off frequencies of some lower-order TM and TE waves are indicated in Fig. 9.8. We find from the figure that for values of a/b equal to or more than two, the *second lowest* cut-off frequency is twice of $f_{c,10}$. If the source frequency is kept between $f_{c,10}$ and twice of $f_{c,10}$, transmission of signal will be only in the dominant mode. In practice, the longer side of the rectangular section of a waveguide is chosen so that $f_{c,10}$ is nearly thirty percent less than the source frequency.

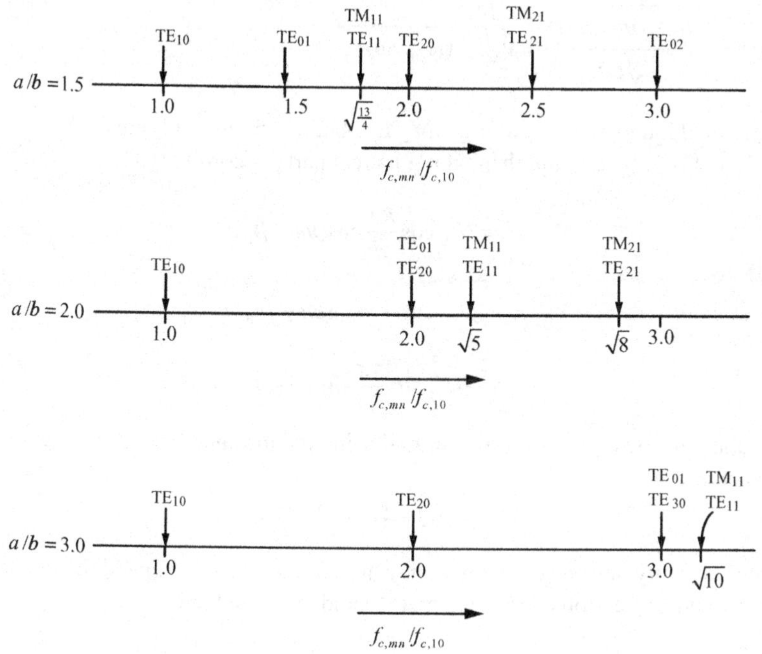

Fig. 9.8 Cut-off frequencies of some low order wave modes in terms of the cut-off frequency of TE_{10} mode

The solutions for the fields of TE_{10} wave may be obtained by substituting $m = 1$ and $n = 0$ in the solutions given by (43) to (47). Thus,

$$\bar{H}_z = H_{z0} \cos\frac{\pi x}{a} e^{-j\beta_{10}z} \tag{53}$$

$$\bar{E}_y = -j\left(\frac{\pi\omega\mu}{ak^2} H_{z0}\right)\sin\frac{\pi x}{a} e^{-j\beta_{10}z} = -jE_{y0} \sin\frac{\pi x}{a} e^{-j\beta_{10}z} \tag{54}$$

$$\bar{H}_x = j\left(\frac{\pi\beta_{10}}{ak^2} H_{z0}\right)\sin\frac{\pi x}{a} e^{-j\beta_{10}z} = jH_{x0} \sin\frac{\pi x}{a} e^{-j\beta_{10}z} \tag{55}$$

$$\bar{E}_x = 0, \bar{H}_y = 0$$

In the above expressions,

$$k = \pi/a,$$

$\beta_{10} = \omega\sqrt{\mu\varepsilon}\sqrt{1-(\omega_{c,10}/\omega)^2}$ is the phase constant.

The formulae for some other parameters are as follows.

The intrinsic wave impedance, $\eta_{TE,10} = \frac{\omega\mu}{\beta_{10}} = \frac{\eta}{\sqrt{1-(\omega_{c,10}/\omega)^2}}$

Phase velocity, $v_{p,10} = \frac{\omega}{\beta_{10}} = \frac{v}{1-(\omega_{c,10}/\omega)^2}$

Wavelength, $\lambda_{10} = \frac{2\pi}{\beta_{10}} = \frac{\lambda}{1-(\omega_{c,10}/\omega)^2}$

Group velocity, $v_{g,10} = \dfrac{\sqrt{1-(\omega_{c,10}/\omega)^2}}{\sqrt{\mu\varepsilon}} = v\sqrt{1-(\omega_{c,10}/\omega)^2}$

Expressions for the field intensities in real time may be obtained by multiplying each of the phasor solutions given by (53), (54) and (55) by $e^{j\omega t}$ and then taking its real part. Accordingly,

$$H_z = H_{z0}\cos\frac{\pi x}{a}\cos(\omega t - \beta_{10}z) \tag{56}$$

$$E_y = E_{y0}\sin\frac{\pi x}{a}\sin(\omega t - \beta_{10}z) \tag{57}$$

$$H_x = -H_{x0}\sin\frac{\pi x}{a}\sin(\omega t - \beta_{10}z) \tag{58}$$

We will next calculate the average power transmitted using the instantaneous Poynting vector. The instantaneous Poynting vector,

$$\mathbf{P}_w(t) = \mathbf{E}(t)\times\mathbf{H}(t)$$

As the electric field intensity has one component in y direction given by (57) and the magnetic field intensity has components in x and z directions described by (58) and (56), we have

$$\mathbf{P}_w(t) = (E_y\mathbf{u}_y)\times(H_x\mathbf{u}_x + H_z\mathbf{u}_z) = -E_yH_x\mathbf{u}_z + E_yH_z\mathbf{u}_x$$

By substituting the expressions of the fields, we get

$$\mathbf{P}_w(t) = E_{y0}H_{x0}\sin^2\frac{\pi x}{a}\sin^2(\omega t - \beta_{10}z)\mathbf{u}_z$$
$$+\frac{1}{4}E_{y0}H_{z0}\sin\frac{2\pi x}{a}\sin 2(\omega t - \beta_{10}z)\mathbf{u}_x$$

We know that the time-average of $\sin^2(\omega t - \beta_{10}z)$ over one time period is equal to ½ and that of $\sin 2(\omega t - \beta_{10}z)$ is zero. Therefore, the average power flow through the cross section of the waveguide at any z and for width b (in y direction) is

$$P_{av} = \frac{1}{2}\int_0^a \left(E_{y0}H_{x0}\sin^2\frac{\pi x}{a}\mathbf{u}_z\right)\cdot(b\,dx\mathbf{u}_z)$$
$$= \frac{abE_{y0}H_{x0}}{4} = \frac{abE_{y0}^2}{4\eta_{\mathrm{TE},10}} = \frac{ab\eta_{\mathrm{TE},10}H_{x0}^2}{4} \tag{59}$$

Figure 9.9 shows field distributions of the TE_{10} mode. The magnetic-field lines on the x-z plane are closed loops enclosing the displacement current. On the transverse plane, the y-directed electric field intensity is zero at $x = 0$ and $x = a$, and has the maximum value at $x = a/2$. The non-uniform spacing of the field lines indicates this. As the x-directed magnetic field intensity does not vary with y, the field lines are shown as uniformly spaced parallel lines.

The main features of TE_{10} mode are as follows:

1. There are only three field components and they do not vary with y.
2. The cut-off frequency does not depend on the length of the shorter side of the rectangular section. The length of the shorter side is chosen from the consideration of power density and wall losses.
3. There is only one component of electric field intensity. Thus, the polarisation is fixed and it is convenient to excite the guide at the input end and extract power at the output end.
4. Field distributions are simple.

5. The electric field intensity E_y establishes surface charges on the inside surfaces of the waveguide at $y = 0$ and $y = a$. The magnitude and sign of the charges are governed by the boundary condition $\rho_S = \mathbf{u}_n \cdot \mathbf{D}$ where $\mathbf{u}_n$ is a unit vector normal (outward) to the surfaces and ρ_S is the surface charge density. On the surface at $y=0$, $\mathbf{u}_n = \mathbf{u}_y$ and $\mathbf{D} = \mathbf{u}_y(\varepsilon E_y)$; hence, $\rho_S = \varepsilon E_y$. And at $y=b$, $\mathbf{u}_n = -\mathbf{u}_y$ and $\mathbf{D} = -\mathbf{u}_y(\varepsilon E_y)$; hence, $\rho_S = -\varepsilon E_y$. No charge is induced on the walls at $x = 0$ and $x = a$, as E_y is zero on these surfaces.
6. The magnetic field intensities H_x and H_z set up surface currents on the walls of the waveguide. To calculate the surface current densities we make use of the boundary relation $\mathbf{K} = \mathbf{u}_n \times \mathbf{H}$ where the symbol $\mathbf{K}$ denotes surface current density. The magnetic field H_x sets up surface currents on the walls of the waveguide at $y = 0$ and $y = b$. On the surface at $y = 0$, $\mathbf{u}_n = \mathbf{u}_y$ and $\mathbf{H} = \mathbf{u}_x H_x$; hence, $\mathbf{K} = \mathbf{u}_y \times (\mathbf{u}_x H_x) = -H_x \mathbf{u}_z$. And at $y = b$, $\mathbf{K} = (-\mathbf{u}_y) \times (\mathbf{u}_x H_x) = H_x \mathbf{u}_z$. No current is induced at $x = 0$ and $x = a$ by H_x as it is zero on these surfaces. Thus, the magnetic field H_x induces surface currents in the longitudinal direction. The magnetic field H_z produces surface currents on all the walls of the waveguide. On the surface at $x = 0$, $\mathbf{u}_n = \mathbf{u}_x$ and $\mathbf{H} = \mathbf{u}_z H_z$; hence, the surface current density $\mathbf{K} = \mathbf{u}_x \times (\mathbf{u}_z H_z) = -H_z \mathbf{u}_y$. We can similarly show that $\mathbf{K} = H_z \mathbf{u}_y$ on the surface at $x = a$, $\mathbf{K} = H_z \mathbf{u}_x$ on the surface at $y = 0$, and $\mathbf{K} = -H_z \mathbf{u}_x$ on the surface at $y = b$. We find that the surface current due to H_z is clockwise when viewed towards the z direction.
7. The TE_{10} mode in rectangular waveguides is similar in many respects to the TE_1 mode between two parallel conducting plates.

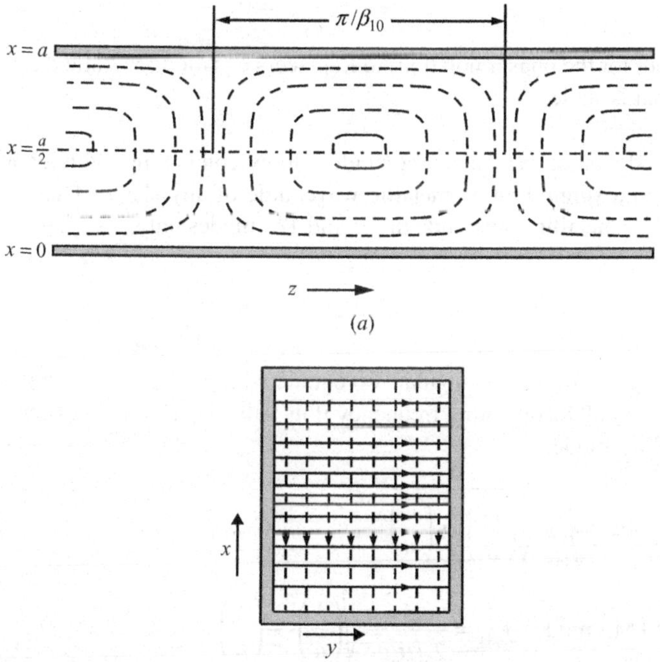

Fig. 9.9 Field patterns for TE_{10} mode in a rectangular waveguide. (*a*) Magnetic-field lines parallel to the *x*-*z* plane and (*b*) Electric-field lines and magnetic field lines on a plane parallel to the *x*-*y* plane. Solid lines are for the electric field and dashed lines are for the magnetic field.

Non-existence of TEM Wave

The study of wave propagation in TM and TE modes in a rectangular waveguide shows that electric energy can be transmitted through a hollow conducting cylinder in these modes. Energy transfer in these modes is possible even if the section of the hollow metallic cylinder is circular or of any other shape. It is however found that TEM waves cannot exist inside a *single* hollow metallic cylinder as explained below.

We know that electromagnetic waves not having electric field intensity and magnetic field intensity in the direction of propagation are called TEM waves. Thus, if a wave propagates in z direction, $E_z = H_z = 0$. The equations relating the components of the electric field intensity with the components of the magnetic field intensity are given in Section 9.2. Substituting $E_z = 0$ in (9.2-15a) and $H_z = 0$ in (9.2-16b), we have

$$\gamma \bar{E}_y = -j\omega\mu \bar{H}_x$$

$$\gamma \bar{H}_x = -j\omega\varepsilon \bar{E}_y$$

Solving the two equations simultaneously, we obtain

$$\bar{E}_y(\gamma^2 + \omega^2\mu\varepsilon) = 0$$

$$\bar{H}_x(\gamma^2 + \omega^2\mu\varepsilon) = 0$$

If we assume that the fields $\bar{E}_y$ and $\bar{H}_x$ are not zero, then

$$\gamma^2 + \omega^2\mu\varepsilon = 0$$

Now, substituting the above relation in (9.2-10), we get

$$\frac{\partial^2 \bar{E}_y}{\partial x^2} + \frac{\partial^2 \bar{E}_y}{\partial y^2} = 0$$

This is identical (except for the phasor notation) to Laplace's equation for static electric field and its solution will result in a field that is static.

As static electric field cannot exist inside a single hollow conducting cylinder without sources, TEM waves cannot exist in a *single* hollow metallic waveguide of any shape. Thus, electric energy can be transmitted in a hollow metallic waveguide in TE and TM modes only.

EXAMPLE 9.5

The height and width of a hollow rectangular waveguide with perfect conducting walls are $a = 5$ cm and $b = 3$ cm respectively. Find the operating frequency if its value is midway between the cut-off frequencies of TM_{11} mode and TM_{21} mode.

$$\text{Cut-off frequency, } f_{c,mn} = \frac{1}{2\sqrt{\mu\varepsilon}}\sqrt{\left(\frac{m}{a}\right)^2 + \left(\frac{n}{b}\right)^2}$$

$$\text{Cut-off frequency for TM}_{11}\text{ mode, } f_{c,11} = \frac{1}{2\sqrt{\mu_0\varepsilon_0}}\sqrt{\left(\frac{1}{a}\right)^2 + \left(\frac{1}{b}\right)^2}$$

$$= (1.5\times10^8)\sqrt{\left(\frac{1}{0.05}\right)^2 + \left(\frac{1}{0.03}\right)^2} = 58.3\times10^8 \text{ Hz}$$

$$\text{Cut-off frequency for TM}_{21}\text{ mode, } f_{c,21} = \frac{1}{2\sqrt{\mu_0\varepsilon_0}}\sqrt{\left(\frac{2}{a}\right)^2 + \left(\frac{1}{b}\right)^2}$$

$$= (1.5\times10^8)\sqrt{\left(\frac{2}{0.05}\right)^2 + \left(\frac{1}{0.03}\right)^2} = 78.1\times10^8 \text{ Hz}$$

$$\text{Operating frequency, } f = \frac{58.3 + 78.1}{2} = 68.2 \times 10^8 \text{ Hz}$$

EXAMPLE 9.6

A hollow rectangular waveguide with perfect conducting walls is operating in the TE_{10} mode. The signal frequency is 1.4 times the cut-off frequency. The side lengths of cross section of the waveguide along x and y directions are $a = 7.5$ cm and $b = 4.0$ cm respectively. If the amplitude of magnetic field intensity $H_{x0} = 15.0$ A/m, find the average power flow in the waveguide.

$$\text{Average power, } P_{av} = \frac{ab\eta_{TE,10} H_{x0}^{\ 2}}{4}$$

Intrinsic wave impedance,

$$\eta_{TE,10} = \frac{\eta(=\eta_0)}{\sqrt{1-(f_{c,10}/f)^2}} = \frac{377}{\sqrt{1-(1/1.4)^2}} = 538.7\ \Omega$$

$$P_{av} = \frac{(0.075)(0.04)(538.7)(15)^2}{4} = 90.9 \text{ W}$$

EXAMPLE 9.7

The height and width of a hollow rectangular waveguide with perfect conducting walls are 2.4 cm and 1.2 cm respectively. Calculate the cut-off frequency of the lowest order TM mode. Find also the phase constant, phase velocity, intrinsic wave impedance and energy velocity at a source frequency equal to 1.6 times the cut-off frequency.

Height of the waveguide, $a = 0.024$ m

Width of the waveguide, $b = 0.012$ m

Permeability, $\mu = \mu_0 = 4\pi \times 10^{-7}$ H/m

Permittivity, $\varepsilon = \varepsilon_0 = 8.85 \times 10^{-12}$ F/m

For the lowest order TM mode, $m = 1$ and $n = 1$

$$\text{Cut-off frequency, } f_{c,11} = \frac{1}{2\sqrt{\mu\varepsilon}}\sqrt{\left(\frac{m}{a}\right)^2 + \left(\frac{n}{b}\right)^2}$$

$$= \frac{3 \times 10^8}{2}[(1/0.024)^2 + (1/0.012)^2]^{1/2} = 139.75 \times 10^8 \text{ Hz}$$

Source frequency, $f = 1.6(139.75 \times 10^9) = 223.6 \times 10^8$ Hz

$$\text{Phase constant, } \beta_{11} = 2\pi f \sqrt{\mu\varepsilon}\sqrt{1-(f_{c,11}/f)^2}$$

$$= \frac{2\pi(223.6 \times 10^8)}{3 \times 10^8} \quad 1-(1/1.6)^2 = 365.6 \quad \text{/m}$$

$$\text{Phase velocity, } v_{p,11} = \frac{\omega}{\beta_{11}} = \frac{2\pi(\ \ .6\ \ 10^8)}{356.6} = 3. \quad \text{/s}$$

Group velocity, $v_{g,11} = \dfrac{\sqrt{1-(f_{c,11}/f)^2}}{\sqrt{\mu\varepsilon}} = (3\times10^8)\sqrt{1-(1/1.6)^2} = 2.34\times10^8$ m/s

Intrinsic wave impedance, $\eta_{TM,11} = \dfrac{\beta_{11}}{\omega\varepsilon} = \dfrac{365.6}{2\pi(223.6\times10^8)(8.85\times10^{-12})} = 294.4\ \Omega$

EXAMPLE 9.8

The inside dimensions of a hollow rectangular waveguide, with air as dielectric, are $a = 2.5$ cm and $b = 1.6$ cm in x and y directions respectively. It transmits electromagnetic energy at a source frequency 1.5 times the cut-off frequency in the dominant TE mode. The amplitude of electric field intensity in y direction is 7500 V/m. Calculate the average power transmitted.

The TE_{10} mode is the dominant mode.

Height of the waveguide, $a = 0.025$ m

Width of the waveguide, $b = 0.016$ m

Permeability, $\mu = \mu_0 = 4\pi\times10^{-7}$ H/m

Permittivity, $\varepsilon = \varepsilon_0 = 8.85\times10^{-12}$ F/m

Amplitude of electric field intensity in y direction, $E_{y0} = 7500$ V/m

The cut-off frequency, $f_{c,10} = \dfrac{1}{2a\sqrt{\mu\varepsilon}} = \dfrac{3\times10^8}{2(0.025)} = 60\times10^8$ Hz

Source frequency, $f = 1.5f_{c,10} = (1.5)(6\times10^9) = 90\times10^8$ Hz

Phase constant, $\beta_{10} = 2\pi f\sqrt{\mu\varepsilon}\sqrt{1-(f_{c,10}/f)^2} = \dfrac{2\pi(90\times10^8)}{3\times10^8}\sqrt{1-(1/1.5)^2} = 140.5$ rad/m

The intrinsic wave impedance, $\eta_{TE,10} = \dfrac{\omega\mu}{\beta_{10=}} = \dfrac{2\pi(90\times10^8)(4\pi\times10^{-7})}{140.5} = 505.8\ \Omega$

Average power, $P_{av} = \dfrac{abE_{y0}^2}{4\eta_{TE,10}} = \dfrac{(0.025)(0.016)(7500)^2}{4(505.8)} = 11.12$ W

A rectangular waveguide is suitable for transmission of electromagnetic energy at frequencies higher than 1 GHz in TE and TM modes. At much lower frequencies, the dimensions of the waveguide would be too large and hence use of rectangular waveguide is uneconomical. Two-conductor transmission lines are more suitable at lower frequencies.

SUMMARY

We formulated the field equations from Helmholtz wave equations and Maxwell's curl equations in rectangular coordinates by assuming that the fields travel along positive z direction. We classified the waves into three types, namely, Transverse Magnetic (TM) waves, Transverse Electric (TE) waves, and Transverse Electromagnetic (TEM) waves.

Wave propagation between two conducting parallel plates separated by an ideal dielectric medium was analysed, and it was found that wave propagation is possible only if the frequency of the fields is above the cut-off frequency. The cut-off frequency for a given wave mode depends on the distance between the plates, and permeability and permittivity of the medium separating the plates. Various parameters above the cut-off frequency were defined, and power flow between the plates was calculated.

We analysed propagation of electromagnetic waves in a rectangular waveguide and learnt that waves can travel in a specific mode only if the frequency of the fields is above the cut-off frequency for the mode. The cut-off frequency for a particular mode depends on the inside dimensions of the waveguide, and permeability and permittivity of the medium through which waves travel. We calculated various parameters associated with the fields, and also calculated power flow through the waveguide. We found that TE_{10} wave mode has the lowest cut-off frequency, and this mode is designated as the dominant mode. We enumerated various properties of the TE_{10} mode and learnt that the properties are similar in many respect to the TE_1 mode of the parallel-plate waveguide. Finally, we established that TEM waves cannot exist inside a hollow metallic waveguide.

IMPORTANT FORMULAE

Two-dimensional wave equation in E_x, $\dfrac{\partial^2 \bar{E}_x}{\partial x^2} + \dfrac{\partial^2 \bar{E}_x}{\partial y^2} = -(\omega^2 \mu\varepsilon + \gamma^2) E_x$

TM and TE waves between parallel plates:

Propagation constant, $\gamma_m = \sqrt{(m\pi/a)^2 - \omega^2 \mu\varepsilon}$

Cut-off frequency, $\omega_{c,m} = \dfrac{m\pi}{a\sqrt{\mu\varepsilon}}$, $f_{c,m} = \dfrac{m}{2a\sqrt{\mu\varepsilon}}$

Propagation constant in terms of cut-off frequency, $\gamma_m = \sqrt{\mu\varepsilon}\sqrt{\omega_{c,m}{}^2 - \omega^2}$

Phase constant, $\beta_m = \sqrt{\mu\varepsilon}\sqrt{\omega^2 - \omega_{c,m}{}^2} = \omega\sqrt{\mu\varepsilon}\sqrt{1-(\omega_{c,m}/\omega)^2}$

Phase velocity, $v_{p,m} = \dfrac{v}{\sqrt{1-(\omega_{c,m}/\omega)^2}}$, $v = \dfrac{1}{\sqrt{\mu\varepsilon}}$

Group velocity, $v_{g,m} = v\sqrt{1-(\omega_{c,m}/\omega)^2}$

TM waves between parallel plates:

Intrinsic wave impedance, $\eta_{TM,m} = \eta\sqrt{1-(\omega_{c,m}/\omega)^2}$, $\eta = \sqrt{\mu/\varepsilon}$

Average power, $\dfrac{abE_{x0}{}^2}{4\eta_{TM,m}}$

TE waves between parallel plates:

Intrinsic wave impedance, $\eta_{TE,m} = \dfrac{\eta}{\sqrt{1-(\omega_{c,m}'\omega)^2}}$

Average power, $\dfrac{abE_{y0}{}^2}{4\eta_{TE,m}}$

TM and TE waves in rectangular waveguide:

Propagation constant, $\gamma_{mn} = \sqrt{(m\pi/a)^2 + (n\pi/b)^2 - \omega^2 \mu\varepsilon}$

Cut-off frequency, $\omega_{c,mn} = \dfrac{1}{\sqrt{\mu\varepsilon}}\sqrt{\left(\dfrac{m\pi}{a}\right)^2 + \left(\dfrac{n\pi}{b}\right)^2}$, $f_{c,mn} = \dfrac{1}{2\sqrt{\mu\varepsilon}}\sqrt{\left(\dfrac{m}{a}\right)^2 + \left(\dfrac{n}{b}\right)^2}$

Propagation constant in terms of cut-off frequency, $\gamma_{mn} = \sqrt{\mu\varepsilon}\sqrt{\omega_{c,mn}{}^2 - \omega^2}$

Phase constant, $\beta_{mn} = \omega\sqrt{\mu\varepsilon}\sqrt{1-(\omega_{c,mn}/\omega)^2}$

Phase velocity, $v_{p,mn} = \dfrac{v}{\sqrt{1-(\omega_{c,mn}/\omega)^2}}$, $v = \dfrac{1}{\sqrt{\mu\varepsilon}}$

Group velocity, $v_{g,mn} = v\sqrt{1-(\omega_{c,mn}/\omega)^2}$

TM waves in rectangular waveguide:

Intrinsic wave impedance, $\eta_{\text{TM},mn} = \eta\sqrt{1-(\omega_{c,mn}/\omega)^2}$, $\eta = \sqrt{\mu/\varepsilon}$

Relation between electric fields, $\dfrac{E_{y0}}{E_{x0}} = \dfrac{na}{mb}$

Average power, $P_{av} = \dfrac{abE_{x0}{}^2}{8\eta_{\text{TM},mn}}\left(\dfrac{m^2b^2 + n^2a^2}{m^2b^2}\right)$

TE waves in rectangular waveguide:

Intrinsic wave impedance, $\eta_{\text{TE},mn} = \dfrac{\eta}{\sqrt{1-(\omega_{c,mn}/\omega)^2}}$, $\eta = \sqrt{\mu/\varepsilon}$

Relation between electric fields, $\dfrac{E_{y0}}{E_{x0}} = \dfrac{mb}{na}$

Average power, $P_{av} = \dfrac{abE_{x0}{}^2}{8\eta_{\text{TE},mn}}\left(\dfrac{m^2b^2 + n^2a^2}{n^2a^2}\right)$

MULTIPLE-CHOICE QUESTIONS

(Starred questions have two answers)

*9.1 Which of the following statements are correct?

(a) Magnetic field intensity along the direction of propagation of a transverse magnetic wave is zero.

(b) Electric field intensity along the direction of propagation of a transverse magnetic wave is zero.

(c) Both electric field intensity and magnetic field intensity along the direction of propagation of a transverse electromagnetic wave are zero.

(d) Magnetic field intensity along the direction of propagation of a transverse electric wave is zero.

9.2 If the signal frequency in a perfect waveguide is greater than the cut-off frequency, then
(a) the propagation constant is zero
(b) the propagation constant is imaginary
(c) the propagation constant is a real positive quantity
(d) the propagation constant is a real negative quantity

9.3 The spacing between perfectly conducting plates of a parallel-plate waveguide is 5.0 cm. If the region between the plates is air, the cut-off frequency for TM_2 mode is
(a) 0.6 GHz (b) 3.0 GHz (c) 6.0 GHz (d) 12.0 GHz

*9.4 If $\eta = \sqrt{\mu/\varepsilon}$, then the *intrinsic wave impedance* above the cut-off frequency in a perfect rectangular waveguide is
(a) less than η for TM mode
(b) greater than η for TM mode
(c) less than η for TE mode
(d) greater than η for TE mode

*9.5 If $v = 1/\sqrt{\mu\varepsilon}$, then the *phase velocity* above the cut-off frequency in a perfect rectangular waveguide is
(a) less than v for TM mode
(b) greater than v for TM mode
(c) less than v for TE mode
(d) greater than v for TE mode

*9.6 If $v = 1/\sqrt{\mu\varepsilon}$, then the *group velocity* above the cut-off frequency in a perfect rectangular waveguide is
(a) less than v for TM mode
(b) greater than v for TM mode
(c) less than v for TE mode
(d) greater than v for TE mode

9.7 The side lengths of the cross section of a rectangular waveguide are 7.5 cm and 5.0 cm respectively. The cut-off frequency for the TE_{10} mode is
(a) 2.0 GHz (b) 3.0 GHz (c) 4.0 GHz (d) 6.0 GHz

*9.8 Which of the following statements for a perfect rectangular waveguide are correct?
(a) The cut-off frequency of the TM_{mn} mode is equal to that of TE_{mn} wave mode.
(b) The cut-off frequency of the TM_{mn} mode is higher than that of TE_{mn} mode.
(c) The cut-off frequency of the TM_{mn} mode is less than that of TE_{mn} mode.
(d) The cut-off frequency of the TE_{10} mode does not depend on the shorter side length of the cross section.

9.9 A parallel-plate waveguide consists of two perfect conducting plates. The intrinsic wave impedance of the waveguide for TM_1 mode is $60\pi\ \Omega$ and for TE_1 mode is $240\pi\ \Omega$. If the relative permeability of the dielectric between the plates is 1.0, the relative permittivity of the dielectric is
(a) 2.5 (b) 2.0 (c) 1.5 (d) 1.0

QUESTIONS

9.1 Consider that an electromagnetic wave is travelling along positive z direction in a perfect dielectric medium. The x-directed electric field intensity in the wave varies with x and y in space, and harmonically with time represented by $e^{j\omega t}$. The propagation constant is γ. Write the two-dimensional wave equation for the field.

9.2 What simplifying assumptions are made to study wave propagation between two parallel conducting plates?

9.3 An electromagnetic wave is travelling along positive z direction of rectangular coordinates. Which is the transverse plane relative to the direction of propagation?

9.4 Define Transverse Magnetic Waves, Transverse Electric Waves, and Transverse Electromagnetic Waves.

9.5 In the propagation range of frequency, the electric field intensity E_z between perfect conducting plates located at $x = 0$ and $x = a$, of a parallel plate waveguide is given by

$$\bar{E}_z = E_{z0} \sin(m\pi x/a) e^{-j\beta z}$$

Interpret the solution and explain what is meant by *mode* of the field.

9.6 Two large lossless parallel conducting plates are located at $x = 0$ and $x = a$ in air. The propagation constant of TM waves travelling between the plates along positive z direction is γ. Assuming that the fields do not vary with y, and variation with time is represented by $e^{j\omega t}$, the following relations are obtained from Maxwell's equations.

$$\bar{E}_y = -j\frac{\omega\mu}{\gamma}\bar{H}_x$$

$$\frac{\partial \bar{E}_z}{\partial x} + \gamma \bar{E}_x = j\omega\mu \bar{H}_y$$

$$\bar{H}_y = j\frac{\omega\varepsilon}{\gamma}\bar{E}_x$$

$$\bar{H}_x = -j\frac{\omega\varepsilon}{\gamma}\bar{E}_y$$

Show that the wave does not contain electric field in y direction, and magnetic field in x direction.

9.7 An ideal parallel-plate waveguide has conducting plates located at $x = 0$ and $x = a$. The TM wave travelling along positive z direction between the plates has electric field in x and z directions and magnetic field in y direction. What are the values of the fields on the plate surfaces?

9.8 The propagation constant of TM_m or TE_m wave mode in a parallel-plate waveguide is given by

$$\gamma_m = \sqrt{(m\pi/a)^2 - \omega^2\mu\varepsilon}$$

where a is spacing between the plates and ω is the signal frequency. If the signal frequency is higher than the critical frequency, then $\gamma_m = j\beta_m$. Obtain the expression for the phase constant β in terms of the cut-off frequency.

9.9 An ideal parallel-plate waveguide has conducting plates located at $x = 0$ and $x = a$. The TM_m wave in the waveguide, with propagation constant γ, contains electric fields in x and z directions, and magnetic field in y direction. The solution for the electric field intensity phasor $\bar{E}_z$ in z direction is given by

$$\bar{E}_z = E_{z0} \sin \frac{m\pi x}{a}$$

For satisfying the boundary conditions, What should be the values of m? Why m cannot be zero?

9.10 The conducting plates of an ideal parallel-plate waveguide are located at $x = 0$ and $x = a$ in air. The propagation constant of the TE wave travelling along positive z direction is γ. Assuming that the fields do not vary with y, and variation with time is represented by $e^{j\omega t}$, the following relations are obtained from Maxwell's equations.

$$\bar{E}_y = -j\frac{\omega\mu}{\gamma}\bar{H}_x$$

$$\bar{E}_x = j\frac{\omega\mu}{\gamma}\bar{H}_y$$

$$\bar{H}_y = j\frac{\omega\varepsilon}{\gamma}\bar{E}_x$$

$$\frac{\partial \bar{H}_z}{\partial x} + \gamma \bar{H}_x = -j\omega\varepsilon \bar{E}_y$$

Show that the wave does not contain electric field in x direction, and magnetic field in y direction.

9.11 The propagation constant of both TM_m and TE_m waves in an ideal parallel-plate waveguide with conducting planes located at $x = 0$ and $x = a$ is given by

$$\gamma_m = \sqrt{(m\pi/a)^2 - \omega^2\mu\varepsilon}$$

Explain why the waves cannot travel if the signal frequency ω is equal to or less than the cut-off frequency.

9.12 The various parameters of TE and TM waves in an ideal parallel-plate waveguide are propagation constant, phase constant, phase velocity, wavelength, and intrinsic wave impedance. Except for one parameter, the formulae for all other parameters are same for both types of waves. Which is this parameter? Give its formulae for the two types of waves.

9.13 The TEM wave travelling along z direction in an ideal parallel-plate waveguide has electric field intensity in x direction and magnetic field intensity in y direction. The relations between the electric field phasor $\bar{E}_x$ and the magnetic field phasor $\bar{H}_y$ are

$$\gamma \bar{E}_x = j\omega\mu \bar{H}_y$$

$$\gamma \bar{H}_y = j\omega\varepsilon \bar{E}_x$$

Show that the propagation constant of the wave is same as the phase constant of uniform plane waves in unbounded regions, provided the relative permeability and relative permittivity through which the waves travel in both the cases are same.

9.14 The TEM wave travelling along z direction in an ideal parallel-plate waveguide has electric field intensity in x direction and magnetic field intensity in y direction. The relations between the two fields are given by

$$\bar{H}_y = H_{y0}e^{-j\beta z} \text{ and } \gamma\bar{E}_x = j\omega\mu H_y$$

Obtain the solution for the electric field intensity. Define and interpret the intrinsic wave impedance.

9.15 The two-dimensional wave equation for the electric field intensity phasor $\bar{E}_z$ of TM waves in an ideal rectangular waveguide is

$$\frac{\partial^2 \bar{E}_z}{\partial x^2} + \frac{\partial^2 \bar{E}_z}{\partial y^2} = -k^2\bar{E}_z;\, k^2 = \gamma^2 + \omega^2\mu\varepsilon$$

The 'separation of variable' method of solution of the equation yields the following relation at one of the steps.

$$\frac{1}{\bar{X}}\frac{\partial^2 \bar{X}}{\partial x^2} = -k^2 - \frac{1}{\bar{Y}}\frac{\partial^2 \bar{Y}}{\partial y^2}$$

where $\bar{E}_z = \bar{X}\,\bar{Y}$. Explain why both sides of the above relation are taken as a constant.

9.16 The height and width of an ideal rectangular waveguide with perfect conducting walls are a and b in x and y directions respectively. The electric field of the TM wave travelling along z direction in the waveguide has components in all the three coordinate directions, and the magnetic field intensity has components in x and y directions. State the boundary conditions to be used to obtain solutions for the fields.

9.17 The height and width of an ideal rectangular waveguide with conducting walls are a and b in x and y directions respectively. The propagation constant for both TE_{mn} and TM_{mn} modes is given by

$$\gamma_{mn} = \sqrt{(m\pi/a)^2 + (n\pi/b)^2 - \omega^2\mu\varepsilon}$$

Obtain the expression for the cut-off frequency in hertz.

9.18 Explain why TM_{01} and TM_{10} wave modes cannot exist in a hollow waveguide with perfect conducting walls.

9.19 Distinguish between a nondispersive transmission system and a dispersive transmission system.

9.20 The propagation constant of TM_{mn} or TE_{mn} wave mode in a rectangular waveguide is given by

$$\gamma_{mn} = \sqrt{(m\pi/a)^2 + (n\pi/b)^2 - \omega^2\mu\varepsilon}$$

Find the expression for the phase constant above the cut-off frequency and interpret it.

9.21 Show that the cut-off frequency of TE_{10} wave does not depend on the length of the shorter side of the rectangular section of a rectangular waveguide.

9.22 The side lengths of a hollow rectangular waveguide with perfect conducting walls are a and b, and a is greater than b. Justify that TE_{10} mode is the dominant mode.

9.23 Explain why TEM waves cannot exist in a single hollow metallic waveguide.

ANSWERS TO SELECTED QUESTIONS

9.4 Define Transverse Magnetic Waves, Transverse Electric Waves, and Transverse Electromagnetic Waves.

Answer: Transverse Magnetic Waves do not contain magnetic field intensity but have electric field intensity along the direction of propagation. As the magnetic field intensity is perpendicular to the direction of wave propagation, these waves are known as Transverse Magnetic Waves or TM waves.

Transverse Electric Waves do not contain electric field intensity but have magnetic field intensity along the direction of propagation. As the electric field intensity is perpendicular to the direction of wave propagation, the waves are called Transverse Electric Waves or TE waves.

Transverse Electromagnetic Waves do not have either electric-field intensity or magnetic-field intensity in the direction of propagation. As both electric and magnetic field intensities are perpendicular to the direction of wave propagation, these waves are called Transverse Electromagnetic Waves or TEM waves.

9.5 In the propagation range of frequencies, the electric field intensity $\bar{E}_z$ between perfect conducting plates, located at $x = 0$ and $x = a$, of a parallel-plate waveguide is given by

$$\bar{E}_z = E_{z0} \sin(m\pi x/a) e^{-j\beta z}$$

Interpret the solution and explain what is meant by *mode* of the field.

Answer: One of the boundary condition at the surface of a perfect conductor for time-varying electromagnetic fields is that the tangential electric field intensity on the surface is zero. As $\bar{E}_z$ is tangential to the surface of plates, the boundary conditions are satisfied for $m = 1, 2, 3, \ldots$ at $x = 0$ and $x = a$. Though $m = 0$ also satisfies the boundary conditions, this value is not admissible because $\bar{E}_z$ would become zero everywhere.

The number of half cycles of the fields between the conducting plates is equal to m. Each value of m defines a specific field configuration or mode of the waveguide. The TM wave having m number of half cycles between the plates is named as TM_m wave or TM_m mode.

9.7 An ideal parallel-plate waveguide has conducting plates located at $x = 0$ and $x = a$. The TM wave travelling along positive z direction between the plates has electric field intensity along x and z directions and magnetic field in y direction. What are the values of the fields on the plate surfaces?

Answer: Three out of four boundary conditions for the time-varying fields at the surface of a perfect conductor are as follows. (1) The electric flux density normal to the surface is equal to surface charge density. (2) The electric field intensity tangential to the surface is zero. (3) The magnetic field intensity tangential to the surface is equal to the surface current density. As the x-directed electric field intensity is normal to the surfaces, its value is $(\mathbf{u}_x \cdot E_x \mathbf{u}_x) = \rho_S/\varepsilon$ at $x = 0$ and $(-\mathbf{u}_x \cdot E_x \mathbf{u}_x) = -\rho_S/\varepsilon$ at $x = a$. The electric field intensity in z direction is tangential to the surfaces. Hence, it is zero at $x = 0$ and at $x = a$. The magnetic field intensity in y direction is tangential the plate surfaces. Its value is $(\mathbf{u}_x \times H_y \mathbf{u}_y) = K_z \mathbf{u}_z$ at $x = 0$ and $(-\mathbf{u}_x \times H_y \mathbf{u}_y) = -K_z \mathbf{u}_z$ at $x = a$.

9.11 The propagation constant of both TM_m and TE_m waves in an ideal parallel plate waveguide with perfect conducting plates located at $x = 0$ and $x = a$ is given by

$$\gamma_m = \sqrt{(m\pi/a)^2 - \omega^2 \mu\varepsilon}$$

Explain why the waves cannot travel if the signal frequency ω is less than or equal to the cut-off frequency.

Answer: Consider that the x-directed electric field travels along positive z direction in a parallel-plate waveguide and the propagation constant is γ_m where

$$\gamma_m = \sqrt{\mu\varepsilon}\sqrt{\omega_{c,m}^2 - \omega^2}$$

Let the field be a function of x. Then assuming that harmonic variation of field with time is represented by $e^{j\omega t}$, the expression for the field may be written as $E_x(x)e^{-\gamma z}e^{j\omega t}$. At the cut-off frequency ($\omega = \omega_{c,m}$), the propagation constant becomes zero, hence the field is a standing wave. The waves simply bounce back and forth between the plates.

At frequencies less than the cut-off frequency, the propagation constant is a real positive quantity and the phase constant is zero. It means that below the cut-off frequency, the amplitudes of the field attenuate exponentially with z and it does not travel.

9.18 Explain why TM_{01} and TM_{10} wave modes cannot exist in a hollow waveguide with perfect conducting walls.

Answer: Let $\bar{E}_z$ be the electric field intensity phasor travelling along z direction in a hollow rectangular waveguide with perfect conducting walls. In the propagation range of frequencies, the solution for the field is given by

$$\bar{E}_z = E_{z0} \sin\frac{m\pi x}{a} \sin\frac{n\pi y}{b} e^{-j\beta z}$$

where a and b are side lengths of the waveguide. The other fields are functions of $\bar{E}_z$. For TM_{01} mode, $m = 0$ and for TM_{10} mode, $n = 0$. Though the boundary conditions are satisfied even if $m = 0$ or $n = 0$ or if both m and n are equal to zero, these are not admissible because in such a case $\bar{E}_z$ and hence all the fields will become zero everywhere. Therefore, these wave modes cannot exist.

9.19 Distinguish between a nondispersive transmission system and a dispersive transmission system.

Answer: The phase velocity of a plane wave in a lossless unbounded dielectric medium is not a function of frequency and the energy velocity is same as the phase velocity. Such a system is a nondispersive system. If the phase velocity in a transmission system is a function of frequency, the system is a dispersive system even if there are no losses. In a dispersive system, the waveform of a nonsinusoidal signal changes in shape during its propagation. This is because the Fourier series of the signal consists of sinusoids of different frequencies and in a dispersive medium the phase velocity of different sinusoids are different.

9.22 The side lengths of a hollow rectangular waveguide with perfect conducting walls are a and b, and a is greater than b. Justify that TE_{10} mode is the dominant mode.

Answer: The wave mode for which the cut-off frequency has the lowest value is termed as the dominant mode. The cut-off frequency of a rectangular waveguide with perfect conducting walls is given by

$$f_{c,mn} = \frac{1}{2\sqrt{\mu\varepsilon}} \sqrt{\left(\frac{m}{a}\right)^2 + \left(\frac{n}{b}\right)^2}$$

As TM_{10} and TM_{01} modes cannot exist, the TM_{11} mode is of the lowest order for TM waves. However, both TE_{10} and TE_{01} can exist for TE waves. The cut-off frequency for these two modes is less than that of TM_{11} mode. Moreover, the cut-off frequency of TE_{10} mode is less than that of the TE_{01} mode as a is greater than b. Hence, TE_{10} mode is the dominant mode.

9.23 Explain why TEM waves cannot exist in a single hollow metallic waveguide.

Answer: The electric field intensity and magnetic field intensity in a TEM wave are perpendicular to the direction of propagation of the wave. Thus, if the direction of propagation in a hollow rectangular waveguide is along z direction, the fields are along x and y direction. The fields satisfy Laplace's equation. For example, assuming that the electric field intensity is in y direction,

$$\frac{\partial^2 \bar{E}_y}{\partial x^2} + \frac{\partial^2 \bar{E}_y}{\partial y^2} = 0$$

This is identical (except for the phasor notation) to the Laplace's equation for static electric field and its solution will result in a field that is static. As static electric field cannot exist inside a single hollow conducting cylinder without sources, TEM waves cannot exist in a *single* hollow metallic waveguide of any shape. Thus, electric energy can be transmitted in a hollow metallic waveguide in TE and TM modes only.

PROBLEMS

**9.1 The electric and magnetic field intensities in an electromagnetic wave, travelling along positive z direction, have components in x and y directions. The fields vary harmonically with time represented by the complex exponential function $e^{j\omega t}$. Obtain equations relating the components of electric field intensity with the components of the magnetic field intensity from Maxell's equation $\nabla \times \bar{\mathbf{E}} = -j\omega\mu\mathbf{H}$, where $\bar{\mathbf{E}}$ and $\bar{\mathbf{H}}$ are vectors as well as phasors. The propagation constant is γ.

**9.2 An ideal parallel-plate waveguide has conducting plates located at $x = 0$ and $x = a$. The wave equation for the z-directed electric field intensity phasor $\bar{E}_z$ of the TM wave, travelling along positive z direction between the plates, is given by

$$\frac{\partial^2 \bar{E}_z}{\partial x^2} = -(\gamma^2 + \omega^2 \mu\varepsilon)\bar{E}_z$$

where the symbols have their normal meaning. Write the general solution of the equation and then obtain the complete solution using appropriate boundary conditions and interpret the result.

*9.3 The perfectly conducting plates of a parallel-plate waveguide, located in air, are separated by 7.5 cm. Find the cut-off frequencies for the TE_1 and TM_2 wave modes.

*9.4 The conducting plates, assumed to be perfect, of a parallel-plate waveguide are located in free space. The distance between the plates is 10 cm. Find the operating frequency if it has to be 20 per cent higher than the cut-off frequency for the TE_2 wave mode.

**9.5 The conducting plates of a parallel-plate waveguide are located in free space. The distance between the plates is 6.0 cm and the operating frequency is 7.5 GHz. Find the modes in which wave propagation is possible. Assume the plates to be perfect.

***9.6 The perfectly conducting plates of a parallel-plate waveguide, located in air, are separated by 6.0 cm. The operating frequency is 30 percent less than the critical frequency corresponding to TM_2 wave mode. Find (a) the phase constant; (b) wavelength along the waveguide; (c) the intrinsic wave impedance.

*9.7 The intrinsic wave impedance of a parallel-plate waveguide in TM mode is 250 Ω with air as medium between the plates. What is the value of the impedance if the space between the plates is completely filled with a nonmagnetic dielectric of permittivity $\varepsilon = 4\varepsilon_0$?

**9.8 The conducting plates of a parallel-plate waveguide are located at $x = 0$ and $x = 5.0$ cm in air. The amplitude of electric field intensity of the electromagnetic wave travelling along positive z direction in TE_1 mode is 50.0 V/m. The operating frequency is 25 per cent higher than the cut-off frequency. Obtain real time expressions of electric field intensity in y direction and magnetic field intensity in x direction.

***9.9 The space between the plates of a parallel-plate waveguide is completely filled with a nonmagnetic dielectric material of permittivity $6\varepsilon_0$. Find (a) the phase velocity; (b) the group velocity; (c) intrinsic wave impedance in TE mode; (d) intrinsic wave impedance in TM mode if the operating frequency is 1.5 times the cut-off frequency.

***9.10 The perfect conducting plates of a parallel-plate waveguide are at $x = 0$ and $x = a$. The TE_m wave between the plates has electric field intensity along y direction, and magnetic field intensity in x-direction. The phasor expression for the electric field is given by

$$E_y \quad E_{y0} \sin \frac{m\pi x}{a} \, e^{-j\beta z}$$

where β_m is phase constant for the mth wave mode. (a) Derive expressions for magnetic field intensity phasors along x and z directions, and for the intrinsic wave impedance. (b) Obtain the expression for power flow between the plates over a width b in y direction. The medium between the plates is air.

**9.11 The amplitude of electric field intensity in the plane transverse to the direction of wave propagation in a parallel-plate waveguide is 200 V/m. The distance between the plates is 5.0 cm and the ratio of cut-off frequency to operating frequency is 0.714. The relative permittivity of the nonmagnetic dielectric medium between the plates is 2.25. Find power transmitted per unit width of the waveguide in (a) TM wave mode; (b) TE wave mode.

**9.12 The internal dimensions of a hollow rectangular waveguide are 16.0 cm and 9.0 cm. Calculate the cut-off frequency for TE_{01} and TM_{11} wave modes.

**9.13 The height and width of the interior of a rectangular metallic waveguide are 7.2 cm and 3.4 cm respectively. Calculate the frequencies below which electromagnetic waves in TM_{12} and TE_{21} modes cannot propagate.

**9.14 The phase constant of both TM_{mn} and TE_{mn} wave modes in rectangular waveguide is given by

$$\beta_{mn} = \omega\sqrt{\mu\varepsilon}\sqrt{1-(\omega_{c,mn}/\omega)^2}\,;\ \omega > \omega_{c,mn}$$

where $\omega_{c,\,mn}$ is the cut-off frequency and ω is the operating frequency. Find the expression for the group velocity.

***9.15 The internal dimensions of a hollow rectangular waveguide are 15.0 cm and 10.0 cm. Find (a) the phase constant; (b) the phase velocity; (c) wavelength along the waveguide; and (d) the group velocity of TE_{10} mode if the operating frequency is 1.6 times of the critical frequency.

**9.16 An air-filled rectangular waveguide is required to propagate in TE_{10} mode above a frequency of 1.5 GHz. (a) Find the longer side of the rectangular section of the waveguide. (b) Consider that the ratio of longer to shorter sides of the rectangular section of the waveguide is 2.0. Investigate and find the frequency range within which only the TE_{10} mode will propagate.

**9.17 The inside dimensions of a hollow metallic rectangular waveguide are 7.5 cm and 5.0 cm. (a) find the frequency above which power transmission through the waveguide in TE_{11} mode would be possible. (b) Calculate the intrinsic wave impedance of TM_{11} and TE_{11} wave modes if the operating frequency is 1.5 times the cut-off frequency.

**9.18 The axis of a hollow rectangular waveguide is along z direction. Its height in x direction is 8.0 cm and width along y direction is 5.0 cm. The amplitude of electric field intensity in x direction, $E_{x0} = 250$ V/m in TM_{11} mode, and the operating frequency is 5.0 GHz. Find E_{y0}, H_{x0}, and H_{y0}.

**9.19 For the waveguide described in Problem 9.18, calculate E_{y0}, H_{x0}, and H_{y0} in the TE_{11} mode.

**9.20 A hollow rectangular waveguide has its axis along z direction. It's internal dimensions in x and y directions are 10 cm and 7 cm respectively. The amplitude of electric field intensity in x direction, E_{x0} = 500 V/m, and the operating frequency is 30 per cent higher than the cut-off frequency. Find the power transmitted in TM_{11} and TE_{11} modes.

*9.21 For the waveguide described in Problem 9.20, find the power transmitted in TE_{10} mode.

Chapter Ten

RADIATION FROM ANTENNAS

10.1 Introduction

In the wireless communication system illustrated in Fig. 10.1, a radio-frequency transmitter is connected to a transmitting antenna through a transmission system. The transmitting antenna radiates energy into free space in the form of electromagnetic waves. A receiving antenna, located suitably, can pick up some of the radiant energy and pass it on to a receiver through a transmission line. Thus, an antenna is a transition device that acts as a link between a transmission system and free-space. The function of a transmitting antenna is to radiate electromagnetic energy efficiently into free-space in the desired direction.

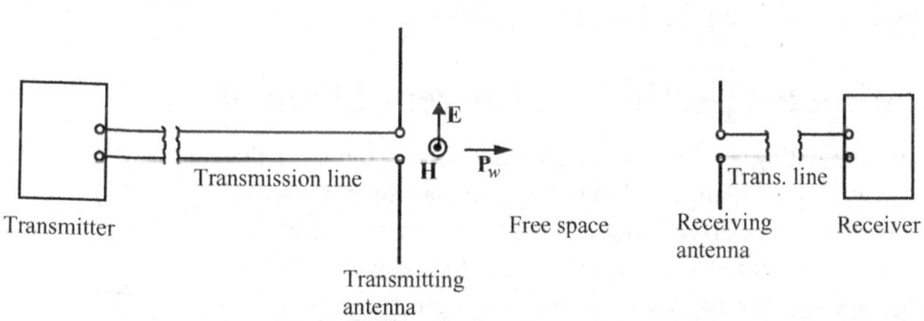

Fig. 10.1 Illustration of a wireless communication system

In order to understand the principle of radiation, let us consider an open-circuited two-conductor parallel transmission line connected to a transmitter, as shown in Fig. 10.2(*a*). If the transmission line is uniform with spacing between conductors small compared to the wavelength, then electromagnetic field outside the lines is quite small because of near cancellation of fields of equal and oppositely directed standing wave currents. However, if the lines at the open-circuited ends are opened out at p_1 and p_2 as shown in Fig. 10.2(*b*), the currents in the vertical parts of the open-circuited lines will be in the same direction. Therefore, the fields due to the currents in the vertical parts of the line add up. The fields, which satisfy Maxwell's equations, travel as waves to great distances in free space. This is an example of a linear antenna that acts as a source for radiating electromagnetic energy into space. A conducting loop connected to a transmitter through a transmission system, as illustrated in Fig. 10.2(*c*), also acts as a source for launching electromagnetic waves into space. The loop, called loop antenna, may consist of several turns. Transmitting and receiving antenna systems are designed for specific applications.

The free-space radiating properties of a Hertzian dipole, a half-wave dipole, and a monopole are described in this chapter. Descriptions of a simple folded dipole, antenna arrays including the Yagi–Uda antenna, and receiving antennas are also given.

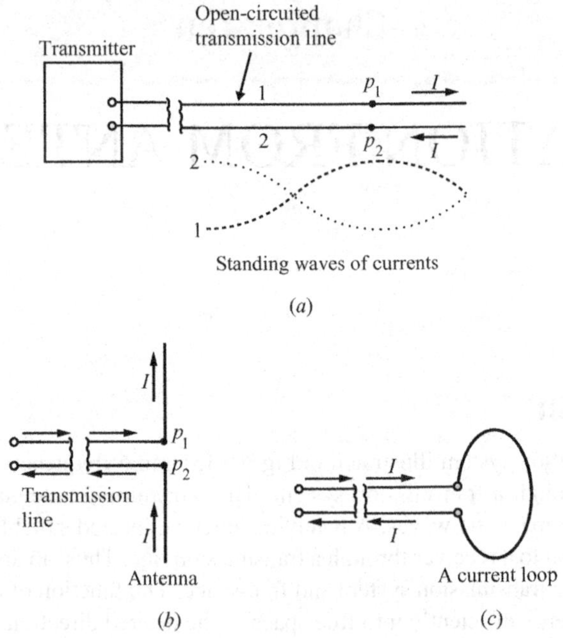

Fig. 10.2 (*a*) An open-circuited transmission line connected to a transmitter, (*b*) A linear antenna is formed by folding out the conductors at p_1 and p_2, and (c) A loop antenna.

10.2 Radiation Properties of a Current Element

One difficulty in calculating the fields even of a linear antenna is that distribution of currents and charges along the antenna are not known. Therefore, field equations are derived with an assumed current distribution. The simplest current distribution is a current element $I(d\ell)$ where I is the sinusoidally time-varying current in a thin wire of length $d\ell$, as shown in Fig. 10.3(*a*). The wire length is assumed to be very small compared to the wavelength $\lambda\,(d\ell \ll \lambda)$. The current distribution over the wire length is assumed to be uniform. For the direction of current shown, there is accumulation of positive charge at the upper end of the wire, and as per the law of conservation of charges, an equal amount of negative charge gets accumulated at the lower end. The charges also vary sinusoidally with time. Thus,

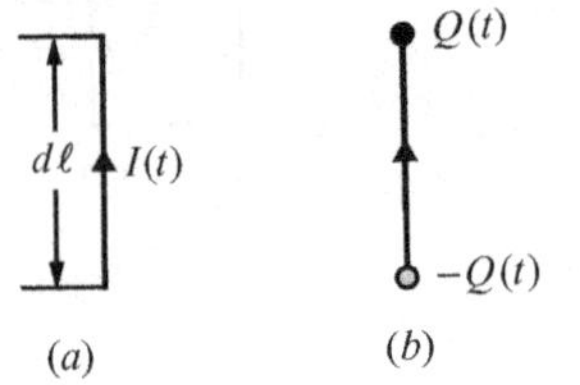

Fig. 10.3 (*a*) Illustration of a time-varying current element and (*b*) Electric dipole equivalent to the current element.

the hypothetical current element is equivalent to an electric dipole with time-varying charges, as illustrated in Fig. 10.3(*b*). The current element is known as *Hertzian Dipole*.

If the current is described by $I = I_0 \sin \omega t$, the magnitude of charge at each end of the dipole,

$$Q(t) = \int I\,dt = -(I_0 \cos \omega t)/\omega.$$

In this section, we will derive expressions for electric- and magnetic-field intensities due to a time-varying current element located in free space. We will use these expressions to calculate power radiated by the dipole, and various other parameters related to electromagnetic radiation. The basis of the analysis is the integral equation for the retarded vector magnetic potential given by (6.17-21). The equation is

$$\mathbf{A}(\ ) = \int_v \frac{\mu\,\mathbf{J}(\qquad /v_0)}{4\pi R}\,dv \tag{1}$$

In the above equation, **A**(t) is the vector magnetic potential at a point at time t due the current at time $(t - R/v_0)$ in a differential volume dv. The function R/v_0 is the time difference between the cause (the current in the element) and the effect (the potential at a point).

When Equation (1) is applied to a linear current element, the differential volume is equal to the product of the sectional area of the element and its length $d\ell$. Also current in the element is equal to the product of current density and the sectional area. Therefore,

$$\mathbf{J}(t - R/v_0) = I(t - R/v_0)dvd\boldsymbol{\ell} \tag{2}$$

Substituting (2) in (1), we get

$$\mathbf{A}(t) = \int \frac{\mu_0 I(t - R/v_0)}{4\pi R} d\boldsymbol{\ell} \tag{3}$$

Since we are considering a current element, the integrand of (3) itself is the potential of the current element. Thus, the retarded vector magnetic potential at a distance R from the current element is

$$\mathbf{A}(t) = \frac{\mu_0 I(t - R/v_0)}{4\pi R} d\boldsymbol{\ell} \tag{4}$$

Now, let us assume that the current element is located at the origin of spherical coordinates and it is oriented along positive z direction of rectangular coordinates, as shown in Fig 10.4. We also assume that the variation of current with time is sinusoidal and its amplitude is I_0. Thus using the spherical coordinate variable r instead of R and the complex exponential function $e^{j\omega t}$ to represent variation with time, we have

$$I(t - R/v_0) = I_0 e^{j\omega(t - r/v_0)} \tag{5}$$

Using (5) in (4), we get

$$\mathbf{A}(t) = \frac{\mu_0 I_0}{4\pi r} e^{-j\omega r/v_0} e^{j\omega t} d\boldsymbol{\ell} = \bar{\mathbf{A}} e^{j\omega t}$$

where $$\bar{\mathbf{A}} = \frac{\mu_0 I_0}{4\pi r} e^{-j\omega r/v_0} d\boldsymbol{\ell} \tag{6}$$

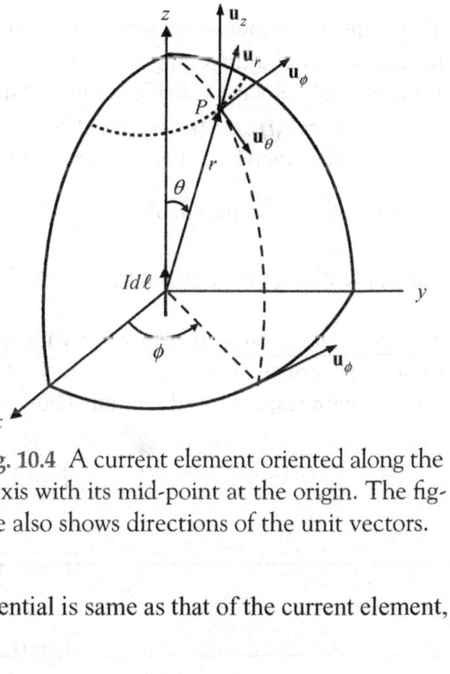

Fig. 10.4 A current element oriented along the z axis with its mid-point at the origin. The figure also shows directions of the unit vectors.

Expression (6) is the potential phasor. As the direction of the potential is same as that of the current element, we have

$$\bar{\mathbf{A}} = \bar{A}_z \mathbf{u}_z$$

The differential length, $d\boldsymbol{\ell} = d\ell\, \mathbf{u}_z$

Substituting these relations in (6),

$$\bar{A}_z = \frac{\mu_0 (I_0 d\ell)}{4\pi r} e^{-j\omega r/v_0} \tag{7}$$

Note that the subscript '0' with I is meant to show that the current is amplitude of the alternating current, whereas v_0 and μ_0 denote free-space values of velocity and permeability.

In arriving at the integral formula for the vector magnetic potential described by (1), we assumed v_0 to be the velocity of the potential field. The question now is: what is the value of this velocity?

We shall get the answer by considering the differential equation for the vector magnetic potential given by 6.17. The equation for free space is

$$\nabla^2 \mathbf{A} = \mu_0 \varepsilon_0 \frac{\partial^2 \mathbf{A}}{\partial t^2} \tag{8}$$

The vector magnetic potential due to the current element under consideration is $\mathbf{A} = \bar{\mathbf{A}} = \bar{A}_z \mathbf{u}_z$. Moreover, $\partial\hat{} \;\; \partial^2 = -\omega^2 \mu\varepsilon\bar{\mathbf{A}}$. Hence Equation (8) becomes

$$\nabla^2 \bar{A}_z = -\omega^2 \mu_0 \varepsilon_0 \bar{A}_z \tag{9}$$

Equation (7) shows that $\bar{A}_z$ is a function of only a single variable r in spherical coordinates. Thus, substituting A_z in place of V in the first term of the Laplacian given by (2.30-10), we obtain the lhs of (9). Accordingly, the differential equation for $\bar{A}_z$ is

$$\frac{1}{r^2}\frac{\partial}{\partial r}\left(r^2 \frac{\partial \bar{A}_z}{\partial r} \right) = -\omega^2 \mu_0 \varepsilon_0 \bar{A}_z \tag{10}$$

This is the one-dimensional wave equation in spherical coordinates. The solution of this equation should have the properties of a travelling wave. One of the properties of the wave is that its velocity is equal to $1/\sqrt{\mu_0 \varepsilon_0}$. It means that velocity v_0 in (7) is same as the velocity of electromagnetic waves. In addition, it can be shown that with $v = 1/\sqrt{\mu_0 \varepsilon_0}$, the potential function of (7) satisfies the differential equation (10); hence, the vector potential described by (7) is a solution of (10).

Continuing our discussion about

the potential function given by (4), we find that it has spherical symmetry.

Therefore, it is convenient to work with spherical coordinates. This requires calculation of the components of A_z in spherical coordinates. Figure 10.4 shows that r, θ and ϕ directions are at angles θ, $(90 + \theta)$ and 90 degrees with respect to z direction. Thus, the components of $\bar{A}_z$ along r, θ and ϕ directions are

$$\bar{A}_r = \bar{A}_z \cos\theta = \frac{\mu_0 (I_0 d\ell)\cos\theta}{4\pi r} e^{-j\omega r/v_0} \tag{11}$$

$$\bar{A}_\theta = -\bar{A}_z \sin\theta = -\frac{\mu_0 (I_0 d\ell)\sin\theta}{4\pi r} e^{-j\omega r/v_0} \tag{12}$$

$$\bar{A}_\phi = 0$$

In order to calculate the magnetic-field intensity from the vector potential we use the relation,

$$\mathbf{H} = \frac{1}{\mu_0}(\nabla \times \mathbf{A})$$

The curl of a vector function in spherical coordinates is given in Section 4.7. As the vector magnetic potential has only two components in A_r A_θ and these do not vary with ϕ, we have

$$\mathbf{H} \quad \frac{1}{\mu_0}\left[\frac{\partial (rA_\cdot)}{\partial} \quad \frac{\partial A}{\partial \theta} \right] \mathbf{u}_\phi$$

By substituting the expressions of $\bar{A}_r$ and $\bar{A}_\theta$ given in (11) and (12) and then evaluating the partial derivatives we will get $\bar{\mathbf{H}}$.

$$\bar{\mathbf{H}} = \frac{I_0 d\ell \sin\theta}{4\pi}\left(j\frac{\omega}{v_0 r} + \frac{1}{r^2}\right)e^{-j\omega r/v_0}\mathbf{u}_\phi = \bar{H}_\phi \mathbf{u}_\phi$$

where

$$\bar{H}_\phi = \frac{I_0 d\ell \sin\theta}{4\pi}\left(\frac{1}{r^2} + j\frac{\omega}{v_0 r}\right)e^{-j\omega r/v_0} \tag{13}$$

is the magnetic field intensity phasor in ϕ direction.

The electric field associated with the magnetic field may be obtained from Maxwell's equations for time harmonic fields. We will make use of the following equation.

$$\nabla \times \bar{\mathbf{H}} = \bar{\mathbf{J}} + j\omega\varepsilon\bar{\mathbf{E}} \tag{14}$$

In free space, $\mathbf{J} = 0$ and $\varepsilon = \varepsilon_0$. Substituting these relations and by reversing the sides, we have

$$j\omega\varepsilon_0 \bar{\mathbf{E}} = \nabla \times \bar{\mathbf{H}}$$

As the magnetic field intensity given by (13) has only one component in ϕ direction and it is independent of ϕ in spherical coordinates, we have

$$j\omega\varepsilon_0 \bar{\mathbf{E}} = \nabla \times \bar{\mathbf{H}} = \frac{1}{r\sin\theta}\frac{\partial(\bar{H}_\phi \sin\theta)}{\partial\theta}\mathbf{u}_r - \frac{1}{r}\frac{\partial(r\bar{H}_\phi)}{\partial r}\mathbf{u}_\theta \tag{15}$$

Substituting the expression of $\bar{H}_\phi$ into (15) and then evaluating the partial derivatives, we will get the solution for $j\omega\varepsilon_0 \bar{\mathbf{E}}$. Dividing the solution by $j\omega\varepsilon_0$ will give $\bar{\mathbf{E}}$. Thus,

$$\bar{\mathbf{E}} = \frac{I_0 d\ell \cos\theta}{2\pi\varepsilon_0}\left(\frac{1}{v_0 r^2} + j\frac{1}{\omega r^3}\right)e^{-j\omega r/v_0}\mathbf{u}_r + \frac{I_0 d\ell \sin\theta}{4\pi\varepsilon_0}\left(\frac{1}{v_0 r^2} + j\frac{\omega}{v_0^2 r} - j\frac{1}{\omega r^3}\right)e^{-j\omega r/v_0}\mathbf{u}_\theta$$

The electric field has two components in r and θ directions.

The components of electric and magnetic field intensities of the dipole in phasor notation are given below.

$$\bar{E}_r = \frac{I_0 d\ell \cos\theta}{2\pi\varepsilon_0}\left(\frac{1}{v_0 r^2} - j\frac{1}{\omega r^3}\right)e^{j\omega r/v_0} \tag{16}$$

$$\bar{E}_\theta = \frac{I_0 d\ell \sin\theta}{4\pi\varepsilon_0}\left(\frac{1}{v_0 r^2} + j\frac{\omega}{v_0^2 r} - j\frac{1}{\omega r^3}\right)e^{-j\omega r/v_0} \tag{17}$$

$$\bar{H}_\phi = \frac{I_0 d\ell \sin\theta}{4\pi}\left(\frac{1}{r^2} + j\frac{\omega}{v_0 r}\right)e^{-j\omega r/v_0} \tag{18}$$

The complex Poynting vector,

$$\mathbf{P}_w = \frac{1}{2}\mathbf{E}\times\mathbf{H}^* = \frac{1}{2}[(E_r\mathbf{u}_r + E_\theta\mathbf{u}_\theta)\times(H_\phi{}^*\mathbf{u}_\phi)]$$

Substituting the expressions of E_r and E_θ given in (16) and (17), and conjugate of H_ϕ given in (18), we will get the following expression for the complex Poynting vector.

$$\overline{\mathbf{P}}_w = \frac{(I_0 d\ell)^2}{32\pi^2\varepsilon_0}\left[\left(\frac{\omega^2}{v_0^3 r^2} - j\frac{1}{\omega r^5}\right)\sin^2\theta\,\mathbf{u}_r + j\left(\frac{\omega}{v_0^2 r^3} + \frac{1}{\omega r^5}\right)\sin 2\theta\,\mathbf{u}_\theta\right]$$

The complex Poynting vector has components in r and θ directions. The component in θ direction is totally imaginary and hence there is no active power flow along this direction. The r component has a real part and an imaginary part. The imaginary part gives reactive power per unit area.

The real part, which gives active power flow per unit area, is inversely proportional to r^2.

Thus, a small piece of wire carrying time-varying current radiates electromagnetic energy along r direction. The energy is due to the terms containing the factor $1/r$ in the expressions of θ-directed electric field intensity and ϕ-directed magnetic field intensity.

The fields are given by

$$\overline{E}_\theta = j\frac{\omega I_0 d\ell \sin\theta}{4\pi\varepsilon_0 v_0^2 r}e^{-j\omega r/v_0} \tag{19}$$

$$\overline{H}_\phi = j\frac{\omega I_0 d\ell \sin\theta}{4\pi v_0 r}e^{-j\omega r/v_0} \tag{20}$$

These are known as *radiation* fields.

Properties of Near and Far Fields

Now let us examine the nature of electric and magnetic fields defined in (16), (17) and (18). The expression of the electric field E_r contains terms which are proportional to $1/r^2$ and $1/r^3$, and E_θ has terms containing $1/r$, $1/r^2$ and $1/r^3$. In regions very close to the current element, the terms containing $1/r$ and $1/r^2$ may be neglected as these would be quite small compared to the term involving $1/r^3$. Also for small r, the ratio r/v_0 is small hence $e^{-j\omega r/v_0}$ is nearly equal to unity. Thus, the total electric field very near the current element is approximately given by

$$\overline{\mathbf{E}} = -j\frac{I_0 d\ell}{4\pi\varepsilon_0 \omega r^3}(2\cos\theta\,\mathbf{u}_r + \sin\theta\,\mathbf{u}_\theta) \tag{21}$$

The electric field at any instant,

$$\mathbf{E}(t) = \mathrm{Re}\left[-j\frac{I_0 d\ell}{4\pi\varepsilon_0 r^3}(2\cos\theta\,\mathbf{u}_r + \sin\theta\,\mathbf{u}_\theta)e^{j\omega t}\right]$$

$$= \frac{I_0 d\ell \sin\omega t}{4\pi\varepsilon_0 \omega r^3}(2\cos\theta\,\mathbf{u}_r + \sin\theta\,\mathbf{u}_\theta) \tag{22}$$

By comparing (22) with the far electric-field of an electrostatic dipole derived in Section 2.18, we find that both the fields have identical variation with r and θ. However instead of being static, the charges at the ends of a Hertzian dipole vary with time with magnitudes ($\cos\omega t/$) and ($\cos\omega t/$).

The expression for the magnetic field intensity given by (18) has two terms; one is proportional to $1/r$ and the other is proportional to $1/r^2$. In regions very near the current element, the term containing $1/r^2$ dominates; therefore the term involving $1/r$ can be neglected. Accordingly, the magnetic field intensity very near the current element at any instant is

$$\mathbf{H}(t) = \frac{I_0 d\ell \cos \omega t}{4\pi r^2} \sin\theta\, \mathbf{u}_\phi = \frac{I_0 d\ell \sin(\omega t + \pi/2)}{4\pi r^2} \sin\theta\, \mathbf{u}_\phi$$

This is Biot–Savart law for a differential alternating-current element. As $\mathbf{E}(t)$ is sine function of time and $\mathbf{H}(t)$ is cosine function of time, the time average of Poynting vector involving the two fields is zero. Moreover, $\mathbf{H}(t)$ leads $\mathbf{E}(t)$ by $\pi/2$ radians in time.

Thus, the power near the current element is purely reactive as in an ideal capacitor.

Far away from the current element, the terms with factors $1/r^2$ and $1/r^3$ in the expressions (16), (17) and (18) may be neglected as their values become negligibly small. Accordingly, the distant fields of a Hertzian dipole are as given below

$$\bar{E}_\theta = j\frac{\omega I_0 d\ell \sin\theta}{4\pi\varepsilon_0 v_0^2 r} e^{-j\omega r/v_0} \tag{23}$$

$$\bar{H}_\phi = j\frac{\omega I_0 d\ell \sin\theta}{4\pi v_0 r} e^{-j\omega r/v_0} \tag{24}$$

$$\bar{E}_r = 0$$

We find from (23) and (24) that the expressions of the distant electric and magnetic fields are identical to those of the radiation fields described by (19) and (20). Thus, the distant fields are also *radiation* fields. The fields described by (23) and (24) are in phasor form. To obtain the field in real time, we multiply (23) and (24) by $e^{j\omega t}$ and use their real parts. Accordingly,

$$E_\theta = -\frac{\omega I_0 d\ell \sin\theta}{4\pi\varepsilon_0 v_0^2 r} \sin[\omega t - (\omega/v_0)r] \tag{25}$$

$$\bar{H}_\phi = -\frac{\omega I_0 d\ell \sin\theta}{4\pi v_0 r} \sin[\omega t - (\omega/v_0)r] \tag{26}$$

The radiation fields defined by (25) and (26) have the following properties.

1. The fields are inversely proportional to r and directly proportional to $\sin\theta$. Thus, the maximum values of the fields occur at $\theta = \pi/2$. The fields do not vary with ϕ.
2. The fields E_θ and H_ϕ are in time phase. They are travelling waves, and travel along r direction with velocity $v_0 = 1/\sqrt{\mu_0\varepsilon_0}$. The velocity is same as the velocity of light.
3. The phase constant, $\beta_0 = \omega/v_0 = 2\pi f\sqrt{\mu_0\varepsilon_0}$
4. Over a spherical surface of radius r, centred at the location of the dipole, the phase-factor given by $(\omega/v_0)r = \beta_0 r$ is constant. However, amplitudes of the fields over the surface are not constant. The amplitudes are proportional to $\sin\theta$.
5. Both E_θ and H_ϕ are perpendicular to the direction of propagation. Therefore, the waves are TEM waves.
6. The intrinsic-wave impedance, $\eta_0 = E_\theta/H_\phi = \sqrt{\mu_0/\varepsilon_0} = 120\pi\ \Omega$.
7. The wavelength, $\lambda_0 = v_0/f = (2\pi)/\beta_0$.

We find that the properties of the field are same as those of uniform plane waves in free space. However, there is an important difference as well. The amplitudes of fields of uniform plane waves do not vary as they progress along the direction of propagation whereas the amplitudes of fields of a Hertzian dipole decrease with r.

More compact forms of expressions for the fields are obtained by eliminating ω, v_0 and ε_0 from the expressions in (25) and (26) in favour of η_0, β_0 and λ_0. Thus, we have

$$E_\theta = -\frac{\eta_0 I_0 \sin\theta}{2r}\left(\frac{d\ell}{\lambda_0}\right)\sin(\omega t - \beta_0 r) \qquad (27)$$

$$H_\phi = -\frac{I_0 \sin\theta}{2r}\left(\frac{d\ell}{\lambda_0}\right)\sin(\omega t - \beta_0 r) \qquad (28)$$

We note that the amplitudes of the distant fields are directly proportional to $d\ell/\lambda_0$ and inversely proportional to r. This is valid only if $d\ell/\lambda_0$ is very small and the current distribution is uniform.

The ratio $d\ell/\lambda_0$ is the *electrical length* of the dipole.

We will now calculate the instantaneous Poynting vector, time-average Poynting vector and real power radiated by the current element using the expressions of the distant fields. The instantaneous Poynting vector,

$$\mathbf{P}_\mathbf{w}(t) = (E_\theta \mathbf{u}_\theta)\times(H_\phi \mathbf{u}_\phi) = E_\theta H_\phi \mathbf{u}_r$$

Substituting the expressions given in (27) and (28), we get

$$\mathbf{P}_\mathbf{w}(t) = \frac{\eta_0 I_0^2}{4r^2}\left(\frac{d\ell}{\lambda_0}\right)^2 \sin^2\theta \sin^2(\omega t - \beta_0 r)\mathbf{u}_r$$

As the time-average of $\sin^2(\omega t - \beta_0 r)$ over one time period is equal to ½, the time-average of the instantaneous Poynting vector is

$$\mathbf{P}_{wa} = \frac{\eta_0 I_0^2}{8}\left(\frac{d\ell}{\lambda_0}\right)^2 \frac{\sin^2\theta}{r^2}\mathbf{u}_r \qquad (29)$$

The average Poynting vector is directly proportional to $\sin^2\theta$ and inversely proportional to r^2.

To calculate the total power radiated by the current element, we integrate (29) over the surface of a sphere centred at the origin, and of any radius r.

$$P_{av} = P_{rad} = \oint_S \overline{\mathbf{P}}_{wa} \cdot d\mathbf{S}$$

As the area of a differential surface element on the spherical surface $d\mathbf{S} = r^2 \sin\theta\, d\theta\, d\phi\, \mathbf{u}_r$, we have

$$P_{rad} = \frac{\eta_0 I_0^2}{8}\left(\frac{d\ell}{\lambda_0}\right)^2 \int_{\phi=0}^{2\pi}\int_{\theta=0}^{\pi} \sin^3\theta\, d\theta\, d\phi$$

$$= \frac{\pi\eta_0 I_0^2}{4}\left(\frac{d\ell}{\lambda_0}\right)^2 \left[-\frac{1}{3}\cos\theta\left(\sin^2\theta + 2\right)\right]_0^\pi$$

$$= \frac{\pi\eta_0 I_0^2}{3}\left(\frac{d\ell}{\lambda_0}\right)^2 \qquad (30)$$

As $\eta_0 = 120\pi$,

$$P_{rad} = 40\pi^2 I_0^2 \left(\frac{d\ell}{\lambda_0}\right)^2 \tag{31}$$

Relations (29) and (30) show that the time-average of the Poynting vector, and the total radiated power of a current element are proportional to $(d\ell/\lambda_0)^2$.

The power radiated by a Hertzian dipole is quite small. For example, if $I_0 = 1.5$A and $d\ell/\lambda_0 = 0.01$ power radiated by the dipole is 0.089 W. Hence, a Hertzian dipole is not a useful radiator. However, the results obtained are useful for analysing practical linear antennas of finite length.

Radiation Resistance

The formula for power radiated into free space by a Hertzian dipole, given in (31), is

$$P_{rad} = 40\pi^2 I_0^2 \left(\frac{d\ell}{\lambda}\right)^2 = \frac{1}{2} I_0^2 \left(\frac{80\pi^2 d\ell^2}{\lambda^2}\right) \tag{32}$$

We know that power dissipated in a resistor of resistance R carrying sinusoidal current of amplitude I_0 is equal to $\frac{1}{2} I_0^2 R$.

Therefore, we may consider that the quantity within the brackets of (32) as a resistance. This resistance however is not the ohmic resistance of the dipole wire as it does not contain the conductivity factor. This is an equivalent resistance and it is called *radiation resistance*.

Thus, the radiation resistance R_{rad} of the dipole located in free space is given by

$$R_{rad} = 80\pi^2 \left(\frac{d\ell}{\lambda}\right)^2 \ \Omega$$

Power radiated in terms of the radiation resistance,

$$P_{rad} = \frac{1}{2} I_0^2 R_{rad}$$

The free-space value of radiation resistance of a Hertzian dipole of length $d\ell = 0.01\lambda$ is equal to 0.079 Ω.

Radiation Intensity

The *radiation intensity* is defined as the power radiated per unit solid angle by an antenna in a given direction.

In order to calculate the radiation intensity, let us consider a sphere of radius r with centre at the mid-point of a vertical Hertzian dipole. This is illustrated in Fig. 10.5. The solid angle subtended at the centre by a differential area dS on the spherical surface,

$$d \quad _s \quad dS \quad r^2 \text{ steradian}$$

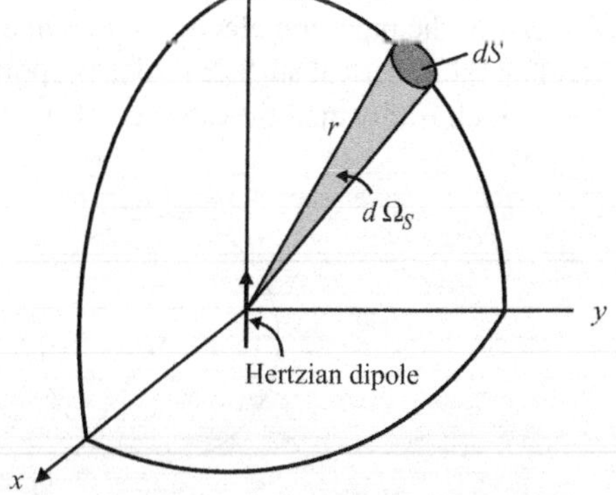

Fig. 10.5 Showing a surface element dS on a spherical surface and the solid angle $d\Omega_s$ subtended by the surface element at the centre of the sphere

Time-average power radiated through the differential area dP_{av} is equal to the product of the average Poynting vector and dS. Thus,

$$dP_{av} = P_{wa} dS = r^2 P_{wa} d\Omega_S$$

Or

$$\frac{dP_{av}}{d\Omega_S} = r^2 P_{wa}$$

This gives the active power radiated per unit solid angle. And as per the definition, the lhs is the radiation intensity. In general, the radiation intensity is a function of both θ and ϕ. Thus denoting the radiation intensity by $R_I(\theta, \phi)$, we have

$$R_I(\theta,\phi) = r^2 P_{wa} \quad \text{W/steradian} \tag{33}$$

To find the radiation intensity of a Hertzian dipole, we substitute the magnitude of the average Poynting vector given in (29) into (33). We may denote the radiation intensity by $R_I(\theta)$, as it is a function of θ only. The radiation intensity of a Hertzian dipole,

$$R_I(\theta) = \frac{\eta_0 I_0^2 \sin^2\theta}{8}\left(\frac{d\ell}{\lambda_0}\right)^2 \tag{34}$$

Radiation Patterns

The radiation intensity described by (34) is proportional to $\sin^2\theta$. It is not a function of r. It is zero along $\theta = 0$ and $\theta = \pi$ directions, and has maximum value along $\theta = \pi/2$ and $\theta = 3\pi/2$ directions. A polar plot of radiation intensity on a vertical plane containing the dipole is shown in Fig. 10.6. The plot is called *Radiation Intensity Pattern*. The pattern consists of two identical lobes. The length of a line drawn from the origin to the curve along θ direction is proportional to the radiation intensity along that direction.

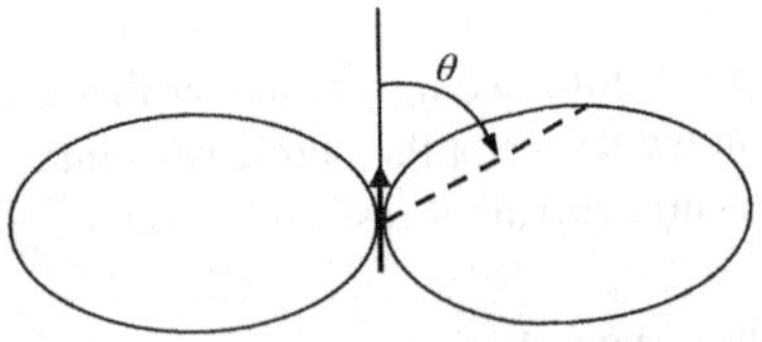

Fig. 10.6 Polar plot of radiation intensity of vertically oriented Hertzian dipole on a vertical plane containing the dipole

Next, let us consider the radiation electric-field E_θ due to a vertically oriented Hertzian dipole. The formula for the electric field intensity is given in (19). The field is tangential to the surface of a sphere with its centre at the mid-point of the dipole. On a spherical surface of radius r, the magnitude of the field is proportional to $\sin\theta$. Thus, the field has a constant maximum value on the $\theta = \pi/2$ plane. Therefore, the polar plot of the electric field on the $\theta = \pi/2$ plane is a horizontal circle, as shown in Fig. 10.7(a). The plot is called *Horizontal Pattern* of the radiation electric field. On a vertical plane containing the dipole, the magnitude of the electric field on the spherical surface is also proportional to $\sin\theta$. It can be shown that the polar plot of the field intensity, which is obtained by varying θ from 0 to 2π, consists of two circles. This is shown in Fig. 10.7(b). This

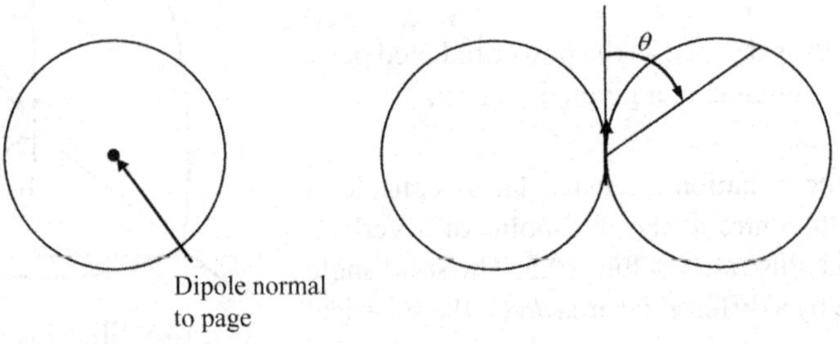

Fig. 10.7 Polar plot of radiation electric-field intensity of the Hertzian dipole. (a) Horizontal pattern and (b) Vertical pattern.

pattern is called *Vertical Pattern* of the radiation electric field. The value of the field on vertical planes passing through the dipoles is zero along $\theta = 0$ and $\theta = \pi$ directions. It has maximum value along $\theta = \pi/2$ and $\theta = 3\pi/2$ directions. The vertical pattern in three-dimensional space is the rotation of the pattern about the dipole axis.

Directive Gain, Directivity and Power Gain

We will at first define the *directive gain* of an antenna. The total power radiated by an antenna,

$$P_{rad} = \oint_S \mathbf{P}_{wa} \cdot d\mathbf{S}$$

where the rhs is the surface integral of time-average of the Poynting vector over a closed surface enclosing the antenna. If we assume that the power is uniformly radiated in all directions, as in the case of an isotropic radiator, the *average radiation intensity*,

$$R_{I,av} = \frac{P_{rad}}{4\pi} \text{ W/steradian}$$

The *directive gain* of an antenna is the ratio of its radiation intensity along any direction to the average radiation intensity.

In general, the directive gain is a function of both θ and ϕ. Thus, denoting the directive gain by $g_d(\theta, \phi)$, we have

$$g_d(\theta, \phi) = \frac{R_I(\theta, \phi)}{R_{I,av}} = \frac{R_I(\theta, \phi)}{(P_{rad}/4\pi)} \tag{35}$$

The directive gain is a nondimensional quantity.

Another radiation property of an antenna is its *directivity*. Along a particular direction, the radiation intensity has its maximum value, say $R_{I(\max)}$. Hence, the directive gain along this direction has the maximum value.

The maximum value of directive gain is called *directivity*.

Thus denoting the directivity by g_d, we have

$$g_d = \frac{R_{I(\max)}}{R_{I,av}}$$

To find the directive gain and directivity of the Hertzian dipole, we substitute (31) and (34) into (35). The directive gain of the current element, which is a function of θ only, is found to be

$$g_d(\theta) = 1.5 \sin^2 \theta \tag{36}$$

The maximum value of directive gain, which occurs at $\theta = \pi/2$, is equal to 1.5. Thus, the directivity of the Hertzian dipole is 1.5.

Lastly, we will calculate the power gain of an antenna. The total power supplied (input power) P_{input} to an antenna is the sum of radiated power and the ohmic loss. That is,

$$P_{input} = P_{rad} + \text{ohmic loss} = \frac{I_0^2}{2}(R_{rad} + R_\Omega)$$

where R_Ω is the ohmic resistance of the antenna wire. If the skin depth δ is small compared to the wire radius a ($\delta < 0.1a$), we may assume that the wire is equivalent to a flat conductor of width equal to the circumference $2\pi a$ of the wire, and thickness δ. Accordingly, the approximate resistance of the antenna wire,

$$R_\Omega = \frac{d\ell}{\sigma_c (2\pi a)\delta}$$

where σ_c is conductivity of the antenna wire.

Antenna efficiency = output power/input power = $P_{rad}/P_{input} = R_{rad}/(R_{rad} + R_\Omega)$

Now, let us assume that an antenna is loss-free so that all the power received by it is radiated. The average radiation intensity of this ideal antenna is equal to $P_{input}/4\pi$. The power gain $g_p(\theta, \phi)$ is defined as the ratio between the radiation intensity of a practical antenna $R_I(\theta, \phi)$ and the average radiation intensity of the antenna assuming it to be loss-free. Accordingly,

$$g_p(\theta,\phi) = \frac{R_I(\theta,\phi)}{(P_{input}/4\pi)} \tag{37}$$

Eliminating $R_I(\theta, \phi)$ from (35) and (37), we have

$$\frac{g_p(\theta,\phi)}{g_d(\theta,\phi)} = \frac{P_{rad}}{P_{input}} = \frac{R_{rad}}{R_{rad} + R_\Omega} = \text{Antenna efficiency}$$

Thus, the power gain of an antenna = directive gain × antenna efficiency.

EXAMPLE 10.1

A vertically oriented Hertzian dipole has a uniformly distributed sinusoidally time-varying current of amplitude 25 A at 60 MHz. It is located in free space with its mid-point at the origin of spherical coordinates. The length of the dipole is 3 cm. Find (a) the radiation electric- and magnetic-field intensities at (150 m, 90°, 0°), (b) total radiated power, and (c) maximum and average values of radiation intensity.

Amplitude of current, $I_0 = 25$ A

Length of the antenna, $d\ell = 0.03$ m

Frequency, $f = 60 \times 10^6$ Hz

Radial distance, $r = 150$ m

Angle, $\theta = 90°$

Wavelength, $\lambda_0 = \frac{v_0}{f} = \frac{3\times10^8}{60\times10^6} = 5$ m

Electrical length of the wire, $d\ell/\lambda_0 = 0.03/5 = 0.006$

(a) Amplitude of θ-directed electric field intensity, $E_{\theta 0} = \frac{\eta_0 I_0 \sin\theta}{2r}\left(\frac{d\ell}{\lambda_0}\right)$

At the given point, $$E_{\theta 0} = \frac{(377)(25)(1)}{2(150)}(0.006) = 0.189 \text{ V/m}$$

Amplitude of ϕ-directed magnetic field intensity, $H_\varphi = E_{\theta 0}/\eta_0 = 0.189/377 = 0.501$ mA/m

(b) Total radiated power, $P_{rad} = \frac{\pi\eta_0 I_0^2}{3}\left(\frac{d\ell}{\lambda_0}\right)^2 = (40\pi^2)(25)^2(0.006)^2 = 8.88$ W

(c) Radiation intensity, $R_I(\theta) = \dfrac{\eta_0 I_0^2 \sin^2\theta}{8}\left(\dfrac{d\ell}{\lambda_0}\right)^2$

The maximum value of radiation intensity occurs along $\theta = \pi/2$ direction. Thus maximum value of radiation intensity,

$$R_I(\max) = \frac{(377)(25)^2}{8}(0.006)^2 = 1.06 \text{ W/steradian}$$

Average value of radiation intensity, $R_{I,av} = P_{rad}/(4\pi) = 8.88/(4\pi) = 0.707$ W/steradian

EXAMPLE 10.2

The radius and conductivity of the antenna wire of Example 10.1 are 1.5 mm and 57 MS/m respectively. Calculate radiation efficiency of the antenna.

Radius of the wire, $a = 1.5 \times 10^{-3}$ m

Conductivity, $\sigma_c = 57 \times 10^6$ S/m

Antenna efficiency $= \dfrac{R_{rad}}{R_{rad} + R_\Omega}$

$$R_{rad} = 80\pi^2 (d\ell/\lambda_0)^2 = 80\pi^2 (0.006)^2 = 0.0284 \ \Omega$$

Approximate formula for the ohmic resistance, $R_\Omega = \dfrac{d\ell}{\sigma_c(2\pi a)\delta}$

Skin depth, $\delta = \dfrac{1}{\sqrt{\pi f \mu_0 \sigma_c}} = \dfrac{1}{\sqrt{\pi(60\times10^6)(4\pi\times10^{-7})(57\times10^6)}} = 8.6\times10^{-6}$ m

$$R_\Omega = \frac{0.03}{(57\times10^6)(2\pi\times1.5\times10^{-3})(8.6\times10^{-6})} = 0.0065 \ \Omega$$

$$\text{Efficiency} = \frac{0.0284}{0.0284 + 0.0065} = 0.814$$

10.3 Radiation Properties of a Half-Wave Dipole

A half-wave dipole is a common type of linear antenna. The antenna consists of two straight conducting wires each of one-quarter of a wavelength. With arrangement of the two wires as shown in Fig. 10.8, the total length of the antenna is one-half of a wavelength. The antenna is known as centre-fed half-wave dipole. The radiation properties of the antenna considering only far fields in free space, due to an assumed current distribution along the wires, are briefly described in this section.

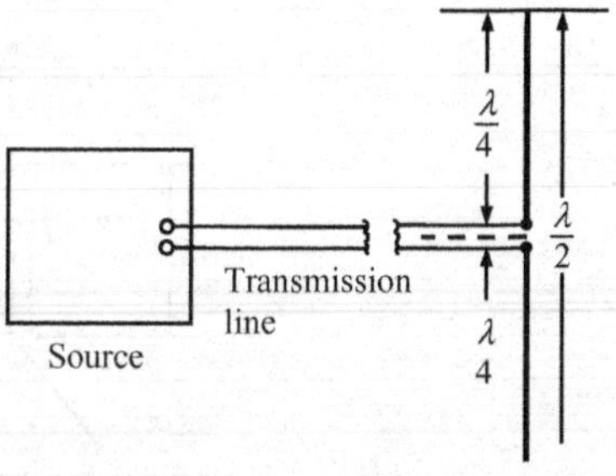

Fig. 10.8 A centre-fed vertically-oriented half-wave dipole

In the preceding section, we derived the fields due to a differential current element assuming the current distribution along the

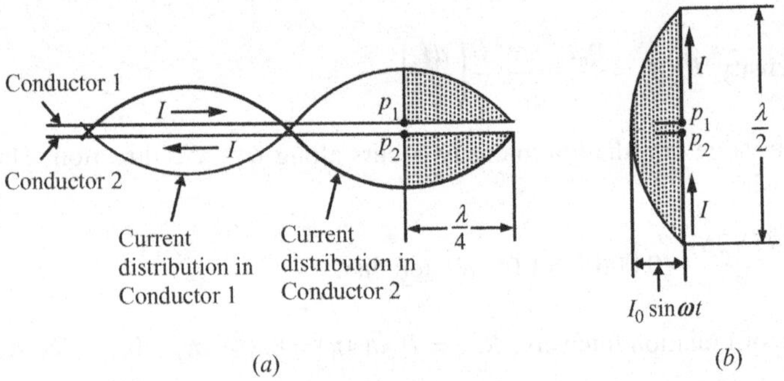

Fig. 10.9 (*a*) Current standing waves in the two conductors of an open-circuited transmission line connected to an ac source and (*b*) Assumed current distribution along a half-wave dipole.

element to be uniform. However for a linear antenna of finite length, this assumption is not valid as current at its open ends must be zero. For such an antenna, the current distribution is generally taken as sinusoidal with zero current at the open ends. The assumption of sinusoidal current distribution is justified if we imagine that the antenna is simply an extension of an open-circuited two-conductor transmission line, shown in Fig. 10.9(*a*). The currents along the two conductors of the transmission line are sinusoidal standing waves. The distributions of currents in the two conductors are in anti-phase, as shown in the figure. If the line conductors are folded at a quarter-wave distance from the open end, a half-wave dipole is formed. Thus, we may assume that the current distributions along two halves of the dipole are same as those of the transmission line conductors before they were folded. This is illustrated in Fig. 10.9(*b*).

The equation for current distribution of an open-circuited transmission line, taking its end as reference, is given in Section 8.5. If we take the antenna along the z axis of rectangular coordinates with its mid-point at $z = 0$ as shown in Fig. 10.10, then equations for current distribution along the two halves of the antenna wire are given by

$$I_1 = I_0 \sin\omega t \sin\beta_0(\lambda_0/4 - z), \quad \lambda_0/4 > z > 0$$

$$I_2 = I_0 \sin\omega t \sin\beta_0(\lambda_0/4 + z), \quad -\lambda_0/4 < z < 0$$

where I_0 is the maximum value of the current wave, and β_0 is the phase constant. As $\beta_0 = 2\pi/\lambda_0$, we have

$$I_1 = I_2 = I_0 \sin\omega t \cos\beta_0 z \tag{1}$$

Referring to Fig. 10.10, the point P at which we wish to calculate the fields is at a distance r from the origin of spherical coordinates. The line OP makes an angle θ with the z axis. Because of symmetry of the fields with ϕ, the point is taken at $\phi = 0$ for convenience. As only the radiation fields are important, we assume that r is very large compared to the wavelength. We also note that the fields of the antenna do not vary with ϕ. Now,

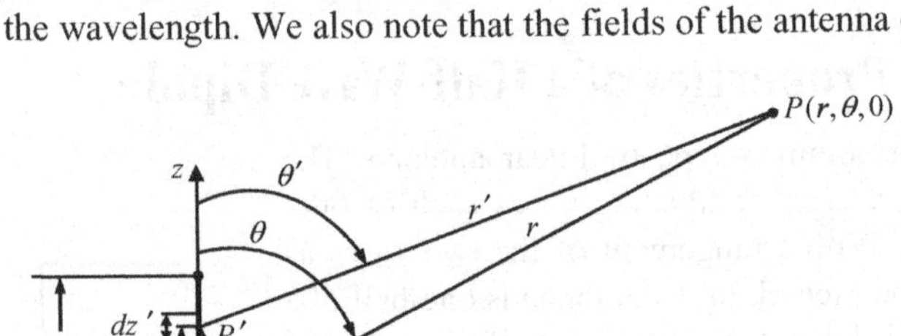

Fig. 10.10 Illustration for calculation of far-fields of a half-wave dipole

let us consider a differential length dz' of the wire at a point P' which is at a distance z' from the origin. The distance of P from P' is r' and $P'P$ makes an angle θ' with the z axis. Substituting z' in place of z in (1), the current at P' is $I_0 \sin \omega t \cos \beta_0 z'$. Assuming that current in dz' is same as at P', the differential current element is equivalent to a Hertzian dipole. Thus replacing I_0 by $I_0 \cos \beta_0 z'$, θ by θ', r by r', and $d\ell$ by dz' in (27) of the last section, we will get the expression for the electric field due to the current element. It is given by

$$dE_\theta = -\frac{\eta_0 I_0 \sin\theta'}{2\lambda_0 r'} \sin(\omega t - \beta_0 r') \cos \beta_0 z' \, dz' \tag{2}$$

As P is at a great distance from the antenna,

$$\theta' \approx \theta$$

$$r' \approx r - z' \cos\theta$$

Substituting the above approximate relations in (2), we get

$$dE_\theta = -\frac{\eta_0 I_0 \sin\theta}{2\lambda_0 (r - z'\cos\theta)} \sin[\omega t - \beta_0 (r - z'\cos\theta)] \cos \beta_0 z' \, dz' \tag{3}$$

As the maximum value of z' is $\lambda_0/4$ and r is very large compared to λ_0, we may take $(r - z'\cos\theta)$ in the denominator of (3) as equal to r. We cannot however use this approximation for the phase factor $\beta_0 (r - z'\cos\theta)$ as its variation with z' is appreciable.

Thus, $$dE_\theta = -\frac{\eta_0 I_0 \sin\theta}{2\lambda_0 r} \sin(\omega t - \beta_0 r + \beta_0 z' \cos\theta) \cos \beta_0 z' \, dz'$$

$$= -\frac{\eta_0 I_0 \sin\theta}{4\lambda_0 r} \{\sin[\omega t - \beta_0 r + \beta_0 z'(1 + \cos\theta)] + \sin[\omega t - \beta_0 r - \beta_0 z'(1 - \cos\theta)\} \, dz' \tag{4}$$

The total electric field at P is found by integrating (4) from $z' = -\lambda_0/4$ to $z' = \lambda_0/4$.

$$E_\theta = \int_{-\lambda/4}^{+\lambda/4} dE_\theta$$

$$= \frac{\eta_0 I_0 \sin\theta}{4\lambda_0 r} \left\{ \frac{\cos[\omega t - \beta_0 r + \beta_0 z'(1 + \cos\theta)]}{\beta_0 (1 + \cos\theta)} - \frac{\cos[\omega t - \beta_0 r - \beta_0 z'(1 - \cos\theta)]}{\beta_0 (1 - \cos\theta)} \right\}_{-\lambda_0/4}^{+\lambda_0/4}$$

$$= -\frac{\eta_0 I_0}{2\pi r} \frac{\cos[(\pi/2)\cos\theta]}{\sin\theta} \sin(\omega t - \beta_0 r) \tag{5}$$

The magnetic-field intensity is along ϕ direction, and it is equal to E_θ/η_0. Therefore,

$$H_\phi = -\frac{I_0}{2\pi r} \frac{\cos[(\pi/2)\cos\theta]}{\sin\theta} \sin(\omega t - \beta_0 r) \tag{6}$$

The instantaneous Poynting vector,

$$\mathbf{P}_w(t) = (E_\theta \mathbf{u}_\theta) \times (H_\phi \mathbf{u}_\phi) = E_\theta H_\phi \mathbf{u}_r \tag{7}$$

Substituting the expressions of (5) and (6) into (7), we get

$$\mathbf{P}_w \quad \frac{\eta_0 I_0^2}{4\pi^2 r^2} \frac{\cos^2[(\pi/2)\cos\theta]}{\sin^2\theta}$$ [illegible] $\mathbf{u}_r$

As the time-average of $\sin^2(\omega t - \beta r)$ is equal to ½, the time-average of Poynting vector is given by

$$\mathbf{P}_{wa} = \frac{\eta_0 I_0^2}{8\pi^2 r^2} \frac{\cos^2[(\pi/2)\cos\theta]}{\sin^2\theta}\mathbf{u}_r = P_{wa}\mathbf{u}_r \tag{8}$$

Power radiated by the dipole can be calculated by integrating the above expression over the surface of a sphere of radius r centred at the origin. Thus,

$$P_{rad} = \oint_s \mathbf{P}_{wa} \cdot d\mathbf{S}$$

As a differential surface element on the spherical surface $d\mathbf{S} = r^2 \sin\theta\, d\theta\, d\phi\, \mathbf{u}_r$, we have

$$P_{rad} = \frac{\eta_0 I_0^2}{8\pi^2} \int_{\theta=0}^{\pi} \int_{\phi=0}^{2\pi} \frac{\cos^2[(\pi/2)\cos\theta]}{\sin\theta} d\theta\, d\phi$$

$$= \frac{\eta_0 I_0^2}{4\pi} \int_{\theta=0}^{\pi} \frac{\cos^2[(\pi/2)\cos\theta]}{\sin\theta} d\theta \tag{9}$$

The integration in (9) involves change of variables twice, and results in another integral known as the Fresnel integral. The values of Fresnel sine and cosine integrals are available in mathematical tables. The value of the integral in (9) is found to be 1.128. Using this value in (9), we get

$$P_{rad} = \frac{1.218\,\eta_0 I_0^2}{4\pi}$$

Since the free-space value of intrinsic wave impedance is equal to 120π, *power radiated by a half-wave dipole* into free space is given by

$$P_{rad}\,(\text{free space}) = 36.54 I_0^2 = \frac{I_0^2}{2}(73.08) \tag{10}$$

The quantity within the bracket of (10) is the *radiation resistance*. Thus,

the free-space value of radiation resistance of a half-wave dipole is equal to 73.08 Ω.

We will calculate the radiation intensity of a half-wave dipole using formula (33) of the preceding section. The formula is,

$$R_I(\theta, \phi) = r^2 P_{wa}$$

The *radiation intensity* is obtained by substituting the expression for P_{wa} given in (8). As $\eta_0 = 120\pi$, and P_{wa} is a function of θ only, the *radiation intensity* of a half-wave dipole in free space is

$$R_I(\theta) = \frac{15 I_0^2 \cos^2[(\pi/2)\cos\theta]}{\pi \sin^2\theta} \quad \text{W/steradian} \tag{11}$$

The general formula for the directive gain given by (35) of the previous section is

$$g_d(\theta, \phi) = \frac{4\pi R_I(\theta, \phi)}{P_{rad}}$$

Substituting the expressions for radiated power given in (10) and radiation intensity given in (11), the *directive gain* of a half-wave dipole is given by

$$g_d(\ \ ) = 1.642 \frac{\cos^{\ }[(\ \ /\ \ )\cos\theta]}{\sin^2\theta}$$

It can be shown that the maximum directive gain occurs at $\theta = \pi/2$. Thus, the directivity of a half-wave dipole is equal to 1.642.

This is higher than the directivity of a Hertzian dipole.

Radiation Patterns

The expression for electric-field intensity given by (5) shows that for given values of r and t the field is a function of θ only. The polar plot of the radiation electric field on any plane containing the dipole is a pattern of the radiation electric field. This is shown in Fig. 10.11(a). The pattern for radiation intensity, which is a polar plot of the function given by (11), is illustrated in Fig. 10.11(b). As the electric field and radiation intensity are independent of ϕ, the shapes of radiation patterns on any vertical plane containing the dipole are same as those shown in Fig. 10.11. Thus, rotation of the patterns about the dipole axis gives the patterns in three-dimensional space. The radiation patterns on the horizontal plane ($\theta = \pi/2$) are circles.

The radiation electri-field and the radiation intensity of a half-wave dipole are more directional than the corresponding quantities of a Hertzian dipole.

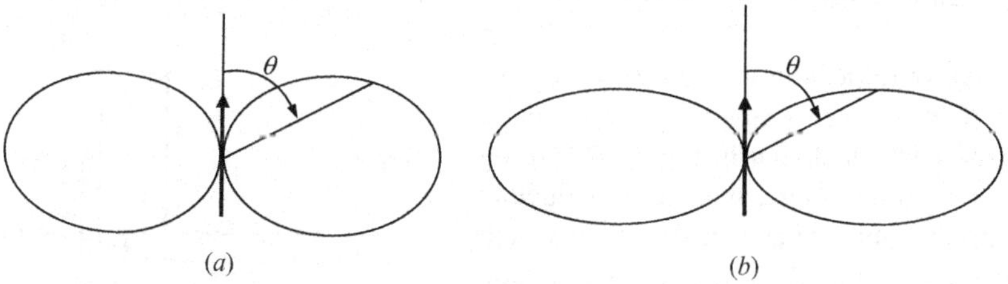

Fig. 10.11 Polar plots of (a) electric field intensity and (b) radiation intensity on a vertical plane containing a vertically oriented half-wave dipole.

EXAMPLE 10.3

A vertically oriented half-wave linear antenna is located in free space. It is excited by a sinusoidally time-varying current at 100 MHz. The antenna wire made of copper has a radius of 2 mm and its conductivity is 57 MS/m. Find the radiation efficiency of the antenna.

The antenna is same as that described in this section.

Frequency, $f = 10^8$ Hz

Wire radius, $a = 0.002$ m

Conductivity, $\sigma_c = 57 \times 10^6$ S/m

Wavelength, $\lambda_0 = v_0/f = 3$ m

Antenna length, $\ell = \lambda_0/2 = 1.5$ m

$$\text{Radiation efficiency} = \frac{R_{rad}}{R_{rad} + R_\Omega}$$

Radiation resistance, $R_{rad} = 73.08\ \Omega$

$$\text{Ohmic resistance, } R_\Omega = \frac{\ell}{\sigma_c([illegible]\pi a)\delta}$$

$$\text{Skin depth, } \delta = \frac{1}{\pi[illegible]\mu_0\sigma_c} = \frac{1}{\pi[illegible]\pi\times[illegible]\times 10^6)} = [illegible]\times 10^{-6}$$

$$R_{\Omega} = \frac{1.5}{(57\times10^{6})(2\pi\times0.002)(6.67\times10^{-6})} = 0.314\ \Omega$$

$$\text{Efficiency} = \frac{73.08}{73.08+0.314} \cong 1$$

The Folded Half-Wave Dipole

The radiation resistance of a dipole antenna with a single conductor of one half-wavelength is approximately 73 Ω. As the transmission line current feeding the antenna is same as the antenna current or radiation current, the input impedance of the antenna is also 73 Ω. If such an antenna is fed by a balanced two-conductor transmission line the characteristic impedance of which is much higher than 73 Ω, then there will be impedance mismatch at the input point. Hence, it is necessary to incorporate a matching device to minimise power reflection.

A half-wave folded dipole antenna which has much higher input impedance than that of a single dipole can directly be fed by a suitable transmission line.

One type of half-wave folded dipole consists of two parallel conductors (elements) each of one half-wavelength and shorted at the ends as shown in Fig. 10.12. In the case of closely spaced and equal diameter conductors, the radiating currents (antenna currents) in the two elements are equal in magnitude and they are also in the same sense. Thus, for radiating current of amplitude I_0 in each element, the effective radiating current of the folded dipole has amplitude $2I_0$. As the power radiated is proportional to square of the current, the power radiated by the folded dipole is approximately equal to $\frac{1}{2}(2I_0)^2(73) = \frac{1}{2}(I_0)^2(4\times73)$. It means that, the radiation resistance of the folded dipole is *four* times that of a single-element dipole, and it is nearly equal to 300 Ω. This is also the input resistance of a folded dipole with two elements.

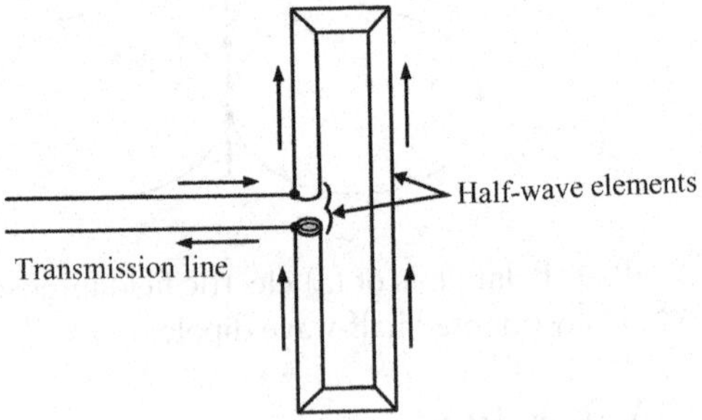

Fig. 10.12 A two-element folded dipole

This property enables direct connection of the folded dipole to a 300 Ω, balanced, two-wire transmission line. The match may not be perfect; even then the reflected power would be quite small.

Other advantages of a folded dipole are that its radiation resistance does not vary much over a vide range of frequency, and it has an increased bandwidth which is important for FM and TV reception.

The radiation resistance and hence the input resistance of a folded dipole with three parallel elements is *nine* times that of a one-element dipole.

A folded dipole made of conducting tubes has the advantages of low eddy-current loss, good mechanical strength, and low cost.

10.4 The Monopole

The monopole is a linear antenna positioned above the earth or a grounded conducting plane. The antenna may be perpendicular or horizontal or at an angle with respect to the ground surface. As the monopole and its image form a dipole, the field distributions of a monopole above the ground surface are identical to those of the corresponding dipole. A brief description of the vertically oriented monopole antenna is given here.

Consider that a positive point charge Q is moving vertically upwards above the ground plane (or above a large flat good conductor), as illustrated in Fig. 10.13(a). The velocity of the charge is v. At any given instant of time, the distance of the image charge $-Q$ below the conducting plane is equal to the height of the actual charge above the ground plane. Thus, the image charge also moves at velocity v in the downward perpendicular direction as shown in the figure. We know that the direction of current flow is same as the direction of flow of positive charges, and it is opposite to the direction of flow of negative charges. Accordingly, the upward motion of positive charges above the ground is equivalent to current flow in the upward direction, and the downward flow of negative image charges below the ground is analogous to current flow in the upward direction.

It means that the image of a vertically directed line current above the ground plane is also a line current positioned along the same line as the original current.

This is illustrated in Fig. 10.13(b).

The length and direction of the image line-current are same as those of the actual current.

Let us now consider a straight wire of length ℓ positioned vertically above the ground plane. One of the output terminals of a radio frequency transmitter is connected to the bottom of the wire and the other terminal is connected to earth, as shown in Fig. 10.13(c). If current in the wire is I at a given instant, then the image current at the same instant is also I. The direction and length of the image current are same as those of the actual current. Thus we see that the combination of the actual current and the image current form a dipole antenna. The current-carrying conductor above the ground is the *monopole antenna*. The field, electric as well as magnetic, above the ground plane is sum of the fields of the actual and the image monopoles. However, the total power radiated by a monopole is one-half of the power radiated by a corresponding dipole because the limits of θ in the definite integral (30) is from 0 to $\pi/2$ as the monopole radiates power above the ground plane only. Therefore, the radiation resistance of a monopole is one-half of the radiation resistance of a corresponding dipole.

The radiation resistance of a monopole of one-quarter wavelength is 36.54 Ω.

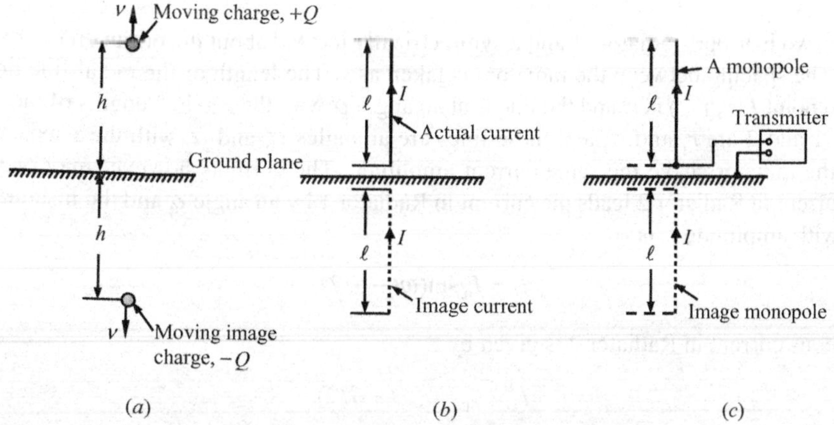

Fig. 10.13 (a) Image of a moving point charge, (b) Image of a vertically oriented linear current, and (c) Image of a monopole.

10.5 Antenna Arrays

An antenna array is an arrangement of several similar antennas all oriented in the same direction. The term *similar* means that all the elements of an array have the same radiation pattern.

The simplest array is the one-dimensional array where all the elements are arranged in a straight line.

Each element of an array contributes to the field at a point in space. The *total or resultant* electric field at a remote point is equal to the phasor sum of the radiation electric fields of all the elements. The resultant electric field consists of two factors. One of the factors is the field of any individual element and the other is a nondimensional factor, called *array factor*, which arises because of the array. Consequently, the pattern of the resultant field, known as the *resultant pattern*, has two pattern components. One of the components is the pattern of the radiation electric field of the element. This pattern is called *unit pattern* or *element pattern*. The other pattern due to the array factor is called *array pattern* or *group pattern*. The array pattern of non-isotropic radiators is found to be similar to the pattern of an identical array of isotropic radiators. Thus, the resultant pattern of an array of non-isotropic radiators can be obtained by multiplying the unit pattern of the array with the pattern of an identical array of isotropic radiators.

We will examine in this section the field patterns of an array of two isotropic radiators and an array of two Hertzian dipoles. It is found that the field patterns of an antenna array are superior to those of a single element.

Array of Two Isotropic Radiators

An isotropic radiator is one that radiates electrical energy uniformly in all directions. For example, as the radiation patterns of a vertical Hertzian dipole or a vertical half-wave dipole on the horizontal plane are circles, these radiators may be considered as isotropic radiator on this plane. A point source is also an isotropic radiator. The pattern of radiation electric field intensity of a point source is a sphere in three-dimensional space. It follows that the pattern on any plane containing a point source is a circle with the source at the centre.

Though in practice a point source is not realisable, the pattern of an array of point sources is useful in obtaining the patterns of nondirectional radiators.

Let us consider two isotropic radiators 1 and 2 symmetrically located about the origin on the x axis, as shown in Fig. 10.14. The spacing between the radiators is taken as s. The length of the radial line drawn from the origin to a field point $P(x, y, z)$ is r, and the line is at an angle α with the x axis. Lengths of radial lines drawn from radiators 1 and 2 are r_1 and r_2, and these lines are at angles α_1 and α_2 with the x axis. It is generally assumed that the radiators have the same current amplitude. The currents however may or may not be in phase. If the current in Radiator 2 leads the current in Radiator 1 by an angle δ, and the instantaneous current in Radiator 1 with amplitude I_0 is

$$I_1 = I_0 \sin(\omega t - \delta/2),$$

then instantaneous current in Radiator 2 is given by

$$I_2 = I_0 \sin(\omega t + \delta/2).$$

We will calculate the resultant radiation electric field in free space due to the currents in the two radiators by taking the origin (mid-point between the radiators in general) as reference, and also by taking the location of

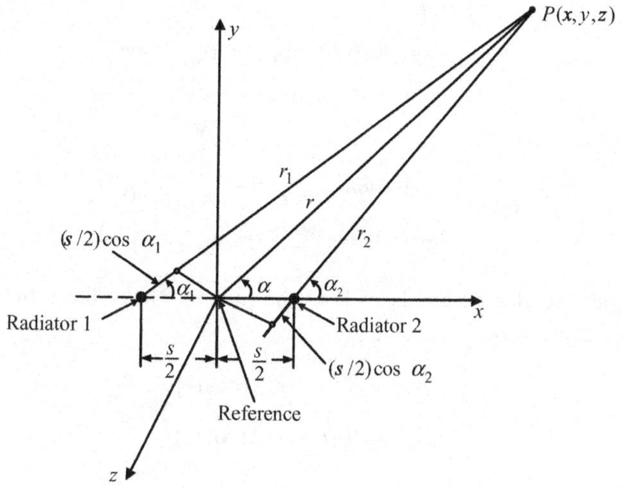

Fig. 10.14 Shows an array of two isotropic radiators and the field point $P(x, y, z)$. Reference is at the mid-point between the radiators.

Radiator 1 as the reference. We will then compare the two results. Let the amplitude of radiation electric field intensity of Radiator 1 at P be $E_0(r_1)$, and that of Radiator 2 be $E_0(r_2)$. Then, the respective fields at P due to the two radiators at time t are given by

$$E_1 = E_0(r_1)\sin(\omega t - \beta_0 r_1 - \delta/2) \tag{1}$$

$$E_2 = E_0(r_2)\sin(\omega t - \beta_0 r_2 + \delta/2) \tag{2}$$

where $\beta_0 = 2\pi/\lambda_0$ is phase constant of free space, λ_0 is the wavelength in free space, and $\beta_0 r_1$ and $\beta_0 r_2$ are the phase factors.

The phasor forms of the fields are given by,

$$\bar{E}_1 = E_0(r_1)e^{-j(\beta_0 r_1 + \delta/2)} \tag{3}$$

$$\bar{E}_2 = E_0(r_2)e^{-j(\beta_0 r_2 - \delta/2)} \tag{4}$$

If the distance of P from the origin is sufficiently large compared to the spacing between the radiators $(s \ll r)$, we can assume that $\alpha_1 = \alpha_2 = \alpha$, and

$$r_1 = r + (s/2)\cos\alpha_1 \approx r + (s/2)\cos\alpha$$

$$r_2 = r - (s/2)\cos\alpha_2 \approx r - (s/2)\cos\alpha$$

So far as distances are concerned, we may also assume that

$$r + (s/2)\cos\alpha \approx r,$$

and

$$r - (s/2)\cos\alpha \approx r.$$

However, we cannot use the above approximations for the phase factors because these change appreciably with s and α.

For example, if $r = 100\lambda_0$ and $s = 0.4\lambda_0$, then

$$r_1 = 100.2r \text{ for } \alpha = 0,$$
$$r_1 = 99.8r \text{ for } \alpha = 180°.$$

Whereas the function,

$$e^{-j\beta_0\{r+(s/2)\cos\alpha)\}} = e^{-j0.4\pi} \text{ for } \alpha = 0$$
$$e^{-j\beta_0\{r+(s/2)\cos\alpha)\}} = e^{j0.4\pi} \text{ for } \alpha = 180°$$

Thus, we can assume that the electric fields of both the radiators have the same amplitude at P, that is, $E_\upsilon\ r_1) = E_\upsilon(r_2) = E_0$. Moreover,

$$\beta_0 r_1 \approx \beta_0[r + (s/2)\cos\alpha]$$

and
$$\beta_0 r_2 \approx \beta_0[r - (s/2)\cos\alpha]$$

By using these expressions in (3) and (4), we get

$$\bar{E}_1 = E_0 e^{-j\beta_0 r} e^{-j\{\beta_0(s/2)\cos\alpha + \delta/2\}} \tag{5}$$

$$\bar{E}_1 = E_0 e^{-j\beta_0 r} e^{j\{\beta_0(s/2)\cos\alpha + \delta/2\}} \tag{6}$$

The resultant field at P is

$$\bar{E}_R = \bar{E}_1 + \bar{E}_2 = E_0 e^{-j\beta_0 r}\left[e^{-j\{\beta_0(s/2)\cos\alpha+\delta/2\}} + e^{j\{\beta_0(s/2)\cos\alpha+\delta/2\}}\right]$$

$$= 2E_0 e^{-j\beta_0 r}\left[\cos\left(\frac{\beta_0 s\cos\alpha}{2} + \frac{\delta}{2}\right)\right]$$

The amplitude of the resultant field is given by

$$|E_{R0}| = 2E_0\left|e^{-j\beta_0 r}\right|\left|\cos\left(\frac{\beta_0 s\cos\alpha}{2} + \frac{\delta}{2}\right)\right|$$

$$= 2E_0\left|\cos\left(\frac{\beta_0 s\cos\alpha}{2} + \frac{\delta}{2}\right)\right| = 2E_0\left|\cos\left(\frac{\pi s\cos\alpha}{\lambda_0} + \frac{\delta}{2}\right)\right| \tag{7a}$$

As x is equal to r cos α and it is also equal to $r\sin\theta\cos\phi$ where θ and ϕ are angles of spherical coordinates, we have cos α = sin θ cos ϕ. Using this relation in (7*a*), we have

$$|E_{R0}| = 2E_0\left|\cos\left(\frac{\pi s\sin\theta\cos\phi}{\lambda_0} + \frac{\delta}{2}\right)\right| \tag{7b}$$

The amplitude of the radiation field of an array of two isotropic radiators, given by (7), depends on the spacing s and current phase angle δ. It is a function of the angle α between the x axis and the radial line drawn from the origin to the field point. The polar plot of the expression inside the square brackets of (7*a*) as a function of α gives the pattern of the array for specified values of s and δ. We will obtain the patterns for two pairs of s and δ, and examine their properties.

1. Let $s = \lambda_0$ $\delta = 0$. The pattern for this pair is defined by the function,

$$\cos\left(\frac{\pi\cos\alpha}{\lambda_0} + \frac{\delta}{2}\right) = \cos\left(\frac{\pi}{2}\cos\alpha\right)$$

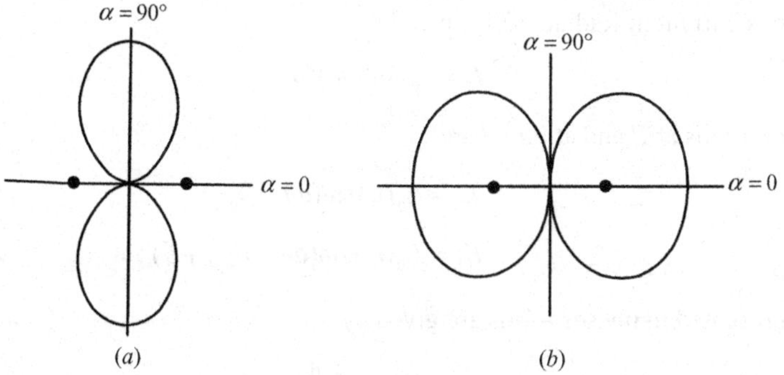

Fig. 10.15 Patterns of an array of two isotropic radiators. (a) $s = \lambda_0/2$, $\delta = 0$ and (b) $s = \lambda_0/2$, $\delta = -180°$.

The value of the function is zero at $\alpha = 90°$, 0.707 at $\alpha = \pi/3$, and unity at $\alpha = \pi/2$. The pattern for variation of α from 0 to 2π is shown in Fig. 10.15(a). The amplitude of the resultant field has the maximum value of $2E_0$ on the plane perpendicular to the array axis.

2. Let $s = \lambda_0/2$ and $\delta = -\pi$. The pattern for this pair is defined by the function,

$$\cos\left(\frac{\pi s \cos\alpha}{\lambda_0} + \frac{\delta}{2}\right) = \sin\left(\frac{\pi}{2}\cos\alpha\right)$$

The value of the function is unity at $\alpha = 0$, is 0.707 at $\alpha = \pi/3$, and is zero at $\alpha = \pi/2$. The pattern for variation of α from 0 to 2π is shown in Fig. 10.15(b). In this case also the amplitude of the resultant field has the maximum value of $2E_0$, but it occurs along the axis of the array.

We obtained the pattern of an array of two isotropic radiators by taking the reference at the mid-point between the radiators. Let us now take the reference at Radiator 1, as indicated in Fig. 10.16, and calculate the resultant radiation electric field at $P(x, y, z)$. As before, both the radiators have the same current amplitude and current in Radiator 2 leads the current in Radiator 1 by an angle δ. If the instantaneous current in Radiator 1 is taken as

$$I_1 = I_0 \sin\omega t,$$

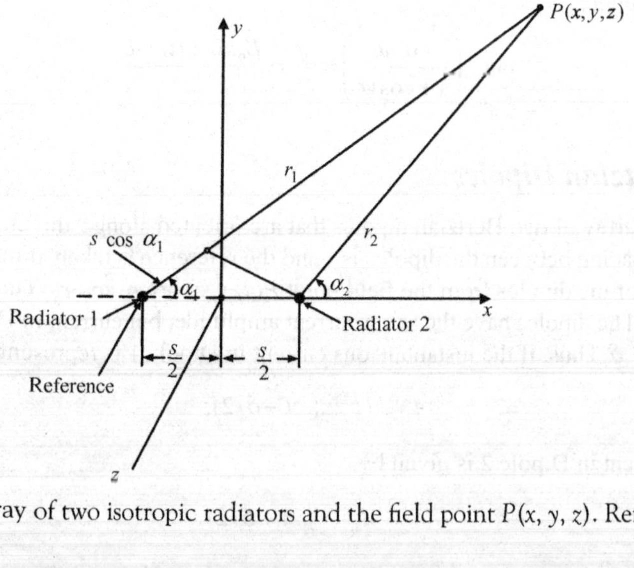

Fig. 10.16 Shows an array of two isotropic radiators and the field point $P(x, y, z)$. Reference is at Radiator 1.

then the instantaneous current in Radiator 2 is given by

$$I_2 = I_0 \sin(\omega t + \delta)$$

The radiation electric fields at P and at time t are

$$E_1 = E_0(r_1)\sin(\omega t - \beta_0 r_1)$$

$$E_2 = E_0(r_2)\sin(\omega t - \beta_0 r_2 + \delta)$$

The fields when represented in phasor forms are given by

$$\bar{E}_1 = E_0(r_1)e^{-j\beta_0 r_1}$$

$$\bar{E}_2 = E_0(r_2)e^{-j(\beta_0 r_2 - \delta)}$$

As we have assumed $s \ll r$,

$$r_2 \approx r_1;\ \alpha_2 \approx \alpha_1;\ E_0(r_2) \approx E_0(r_1) = E_0;\ \beta_0 r_2 \approx \beta_0(r_1 - s\cos\alpha_1).$$

Thus,

$$\bar{E}_1 = E_0 e^{-j\beta_0 r_1}$$

$$\bar{E}_2 = E_0 e^{-j\beta_0 r_1} e^{j(\beta_0 s\cos\alpha_1 + \delta)} = E_0 e^{-j\beta_0 r_1} e^{j\psi}$$

where

$$\psi = \beta_0 s\cos\alpha_1 + \delta$$

The resultant electric field at P is

$$\bar{E}_{R0} = \bar{E}_1 + \bar{E}_2 = E_0 e^{-j\beta_0 r}(1 + e^{j\psi})$$

The amplitude of the resultant electric field is given by

$$E_{R0} = E_0\left|e^{-j\beta_0 r_1}\right|\left|1 + e^{j\psi}\right|$$

$$= E_0\left|e^{-j\beta_0 r_1}\right|\left|1 + \cos\psi + j\sin\psi\right| = 2E_0\left|\cos\frac{\psi}{2}\right| = 2E_0\left|\cos\left(\frac{\beta_0 s\cos\alpha}{2} + \frac{\delta}{2}\right)\right|$$

This expression is identical to that obtained by taking the reference at midway between the radiators. However, its phase differs by an angle,

$$\tan^{-1}\left(\frac{\sin\psi}{1 + \cos\psi}\right) = \frac{\psi}{2} = \frac{\beta_0 s\cos\alpha_1 + \delta}{2}$$

Array of Two Hertzian Dipoles

Figure 10.17 shows an array of two Hertzian dipoles that are oriented along z direction with their mid-points along the x axis. The spacing between the dipoles is s and the reference is taken at the mid-point between the dipoles. The distances of the dipoles from the field point $P(x, y, z)$ are r_1 and r_2. The angles α, α_1 and α_2 are as shown in the figure. The dipoles have the same current amplitude, but current in Dipole 2 leads the current in Dipole 1 by an angle δ. Thus, if the instantaneous current in Dipole 1 is represented by

$$I_1 = I_0 \sin(\omega t - \delta/2),$$

then instantaneous current in Dipole 2 is given by

$$I_2 = I_0 \sin(\omega t + \delta/2)$$

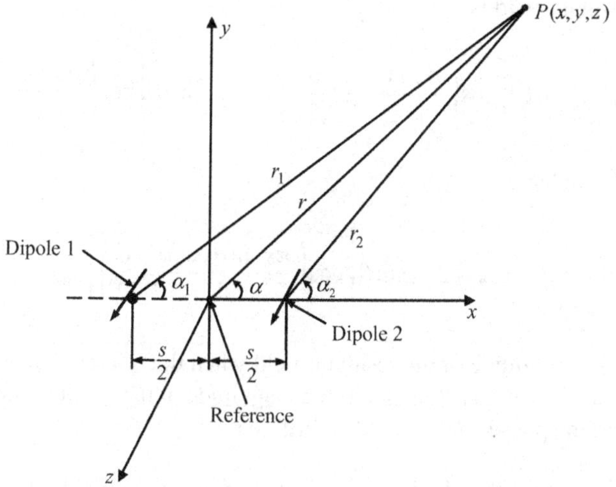

Fig. 10.17 Shows an array of two Hertzian dipoles and the field point $P(x, y, z)$. Reference is at the mid-point between the radiators.

If the distance of P from the location of the dipoles is very large compared to the distance between the dipoles, then only the radiation electric fields have to be considered. Denoting the radiation electric field intensities of the dipoles as E_1 and E_2 at point P and time t, we have from (27) of Section 10.2,

$$E_1 = -\frac{\eta_0 I_0 \sin\theta}{2r_1}\left(\frac{d\ell}{\lambda_0}\right)\sin(\omega t - \beta_0 r_1 - \delta/2) \tag{8}$$

$$E_2 = -\frac{\eta_0 I_0 \sin\theta}{2r_2}\left(\frac{d\ell}{\lambda_0}\right)\sin(\omega t - \beta_0 r_2 + \delta/2) \tag{9}$$

The field intensity phasors, which can be written from (8) and (9), are given by

$$\bar{E}_1 = \frac{\eta_0 I_0 \sin\theta}{2r_1}\left(\frac{d\ell}{\lambda_0}\right)e^{-j(\beta_0 r_1 + \delta/2)} \tag{10}$$

$$\bar{E}_2 = \frac{\eta_0 I_0 \sin\theta}{2r_2}\left(\frac{d\ell}{\lambda_0}\right)e^{-j(\beta_0 r_2 - \delta/2)} \tag{11}$$

As $s \ll r$, we can assume that both α_1 and α_2 are nearly equal to α. Accordingly, $r_1 \approx r$; $r_2 \approx r$; $\beta_0 r_1 = \beta_{\cdot}[\ + (s/2)\cos\alpha]$; $\beta_0 r_2 = \beta_0[r - (s/2)\cos\alpha]$. Using these approximations in (10) and (11), we have

$$\bar{E}_1 = E_0 \sin\theta\, e^{-j\beta_0 r} e^{-j\{\beta_0 (s/2)\cos\alpha + \delta/2\}}$$

$$\bar{E}_2 = E_0 \sin\theta\, e^{-j\beta_0 r} e^{j\{\beta_0 (s/2)\cos\alpha + \delta/2\}}$$

where
$$E_0 = \frac{\eta_0 I_0}{2r}\left(\frac{d\ell}{\lambda_0}\right)$$

The resultant electric field at P,

$$E_{R\cdot}\ \ E_1 + E_{\cdot}\ \ E_0 [illegible]\, e^{-j\beta_0 r}\left[e^{-j [illegible] s/2 [illegible]} + e^{j [illegible] s/2)\cos\alpha + \delta/ [illegible]\}}\right]$$

$$= [illegible]\, e^{-j\beta_0 r}\left[[illegible] \left(\frac{\beta_0 s}{2} [illegible]\, \alpha + \frac{\delta}{2}\right)\right]$$

The amplitude of the resultant field is

$$E_{R0} = [2E_0 \sin\theta]\left[\cos\left(\frac{\beta_0 s \cos\alpha}{2} + \frac{\delta}{2}\right)\right] = [2E_0 \sin\theta]\left[\cos\left(\frac{\pi s \cos\alpha}{\lambda_0} + \frac{\delta}{2}\right)\right] \quad (12a)$$

Since cos ¨ = $\sin\theta \cos\phi$, we have

$$E_{R0} = 2E_0 \sin\theta\left[\cos\left(\frac{\pi s \sin\theta \cos\phi}{\lambda_0} + \frac{\delta}{2}\right)\right] \quad (12b)$$

Expression (12*a*) gives the amplitude of the resultant field when the reference point is midway between the dipoles. If the reference is taken at, say Radiator 1, the magnitude of the resultant field would be same as that given by (12*a*). However, the phase angle would be different.

The function inside the first square brackets of (12*a*) is the radiation electric field intensity of one dipole. The pattern of this function is defined by sin θ. This pattern is called

unit pattern or primary pattern.

The function inside the second square brackets, arising out of the array, is called

array factor.

The pattern of the array factor, defined by cos α is known as

array pattern or secondary pattern.

The product of the unit pattern and array pattern gives the *resultant pattern* of an array of two Hertzian dipoles.

Principle of Pattern Multiplication By comparing Expression (12*a*) with Expression (7*a*) we find that

the array pattern of two Hertzian dipoles is identical to the pattern of equal number of isotropic radiators, provided the spacing s and phase difference between currents δ are same in both the cases. Thus, the resultant pattern of a dipole array is equal to the product of the unit pattern and the pattern of an identical array of isotropic radiators. This is the principle of *pattern multiplication*.

As an example, let us consider an array of two Hertzian dipoles with $s = \lambda_0/2$ and $\delta = 0$. The vertical pattern of a single dipole is as shown in Fig. 10.7(*b*), and it is reproduced as Fig. 10.18(*a*). The pattern of two isotropic radiators, that is the array pattern, for $s = \lambda_0/2$ and $\delta = 0$ is as in Fig. 10.18(*b*). The resultant pattern of the dipole array on a vertical plane midway between the dipoles is found by multiplying the two previously mentioned patterns. The resultant pattern is shown in Fig. 10.18(*c*). Rotation of the resultant pattern about the z axis gives the pattern in three-dimensional space. Moreover, the patterns of an array of two Hertzian dipoles for $s = \lambda_0/2$ and $\delta = -\pi$ on the horizontal plane ($\theta = \pi/2$, $\phi = \alpha$) are as shown in Fig. 10.19.

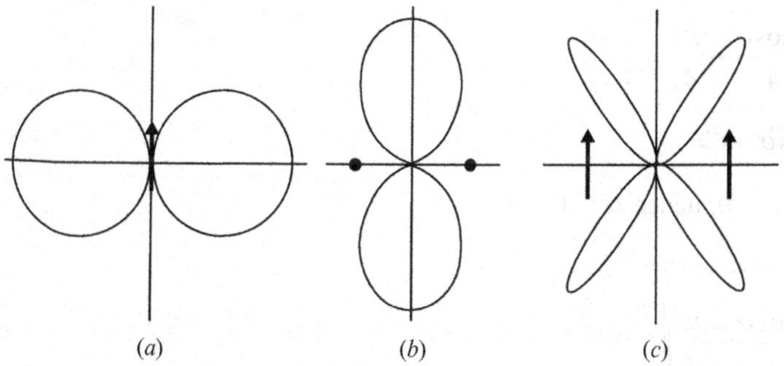

Fig. 10.18. (*a*) Unit pattern on a vertical plane of one Hertzian dipole (*b*) Array pattern of two Hertzian dipoles for s = $\lambda_0/2$ and $\delta = 0$ and (c) Resultant pattern on a vertical plane of the array.

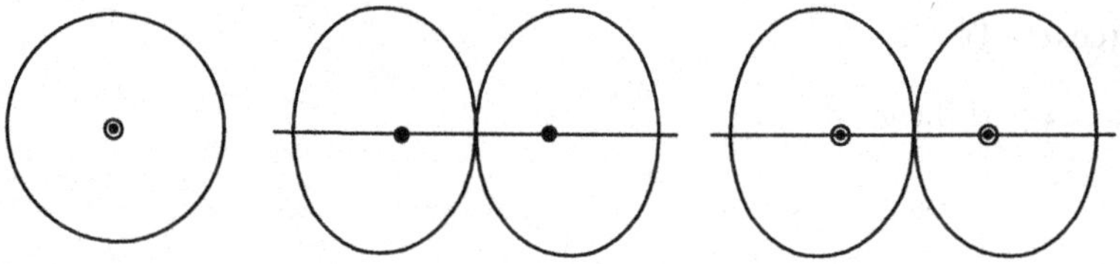

Fig. 10.19. (*a*) Unit pattern on the horizontal plane of one Hertzian dipole, (*b*) Array pattern of two Hertzian dipoles for s = $\lambda_0/2$ and $\delta = -\pi$, and (c) Resultant pattern on the horizontal plane of the array.

EXAMPLE 10.4

The distance between two isotropic radiators 1 and 2 located in free space is one-fourth of the wavelength. Both the radiators have the same current amplitude but current in Radiator 2 lags the current in Radiator 1 by $\pi/2$ radians. Find the angle with the array axis at which the resultant radiation electric field intensity of the array (a) has the highest value, (b) is 0.707 times the highest value and (c) is zero.

Substituting $s = \lambda_0/4$ and $\delta = -\pi/2$ in (7*a*),

$$E_{RO} = 2E_0 \left|\cos\left(\frac{\pi\cos\alpha}{4} - \frac{\pi}{4}\right)\right|$$

(a) The condition for the highest value of E_{RO} is

$$\left|\cos\left(\frac{\pi\cos\alpha}{4} - \frac{\pi}{4}\right)\right| = 1$$

Or $$\left|\frac{\pi\cos\alpha}{4} - \frac{\pi}{4}\right| = 0$$

Or $\cos\alpha = 1$; hence $\alpha = 0$

(b) For this case,

$$\cos\left(\frac{\pi\cos\alpha}{4} - \frac{\pi}{4}\right) = 0.707$$

$$\text{Or} \quad \left|\frac{\pi\cos\alpha}{4} - \frac{\pi}{4}\right| = \frac{\pi}{4}$$

$$\text{Or} \quad |\cos\alpha - 1| = 1$$

$$\text{Or} \quad \cos\alpha = 0;\ \text{hence}\ \alpha = \pm\,\pi/2$$

(c) For this case,

$$\cos\left(\frac{\pi\cos\alpha}{4} - \frac{\pi}{4}\right)\Bigg| = 0$$

$$\text{Or} \quad \left|\left(\frac{\pi\cos\alpha}{4} - \frac{\pi}{4}\right)\right| = \frac{\pi}{2}$$

$$\text{Or} \quad |(\cos\alpha - 1)| = 2$$

$$\text{Or} \quad \cos\alpha = -1;\ \text{hence}\ \alpha = \pi$$

EXAMPLE 10.5

Two Hertzian dipoles, each of length $d\ell$, are located in free space with their mid-points along the x axis. The distance between the dipoles is one-half of a wavelength. The dipoles have the same current amplitude, and the currents are in phase. Find the ratio of amplitudes of radiation electric field intensities of the array at field points P_1(50 m, 90°, 0) and P_2(100 m, 60°, 30°).

Substituting $s = \lambda_0/2$ and $\delta = 0$ in (12a),

$$E_{KU} \quad E_0 \sin\theta\left(\cos\frac{\pi\sin\theta\cos\phi}{2}\right)$$

$$\text{where} \quad E_0 = \frac{\eta_0 I_0}{2r}\left(\frac{d\ell}{\lambda_0}\right)$$

$$\text{The ratio of the amplitudes at the two points} = \frac{100\ \sin 90°(0.5\pi\sin 90°\cos 0°)}{50\ \sin 60°(0.5\pi\sin 60°\cos 30°)}$$

$$= \frac{16}{3\sqrt{3}} = 3.08$$

Note that the ratio is independent of distance between the dipoles.

10.6 Uniform Linear Array of *n* Isotropic Radiators

A linear array of antennas is an orderly arrangement of elements along a line with *equal spacing s* between consecutive elements, and in which all the elements have the *same current amplitude*. When the phase differences between the currents of neighbouring elements of a linear array have the same value, the array is a *uniform linear array*. Thus taking the current amplitude as I_0, the current phase angle between neighbouring elements as δ, and current in Element 1 as reference, the instantaneous currents in various elements of a uniform linear array are

$$I_1 = I_0 \sin \omega t$$

$$I_2 = I_0 \sin(\omega t + \delta)$$

$$I_3 = I_0 \sin(\omega t + 2\delta)$$

$$\cdots\cdots\cdots\cdots\cdots$$

$$\cdots\cdots\cdots\cdots\cdots$$

$$I_n = I_0 \sin[\omega t + (n-1)\delta]$$

We found from the analysis of an array of two Hertzian dipoles that the resultant pattern is given by the product of the unit pattern and the array pattern. The array pattern is same as the pattern of an identical array of isotropic radiators. Thus to obtain the resultant pattern of an antenna array consisting of n elements, it is necessary to find the pattern of only one element and the pattern of an identical array of n isotropic radiators. We learnt in Sections 10.2 and 10.3 the basic principle involved in finding the unit pattern of a linear antenna. We will now obtain the pattern of a uniform linear array of isotropic radiators.

Figure 10.20 shows an array of n isotropic radiators located along the x axis, which is also the array axis. The spacing between two consecutive radiators is s, and the instantaneous currents in them are as defined earlier. We assume that the field point is far away from the array so that we need to consider only the radiation fields. We also assume that the distance of the field point from all the radiators is r and the lines joining the radiators with the field point are parallel lines. Thus, all the lines subtend the same angle with the array axis. We will take the field point on the x-y plane for easy understanding. On this plane we have $\theta = \pi/2$ and hence $\alpha = \phi$.

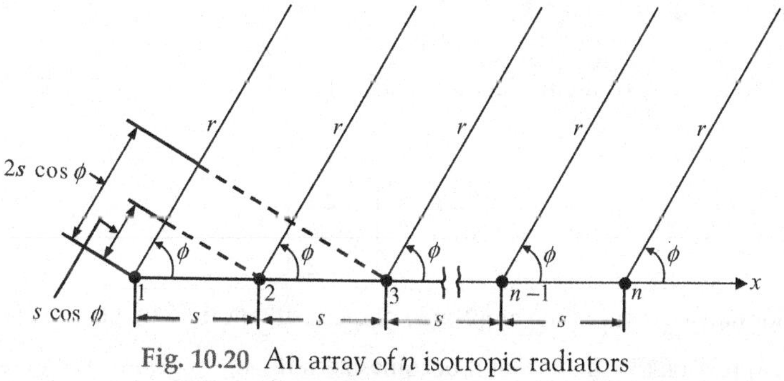

Fig. 10.20 An array of n isotropic radiators

Thus by taking the position of Radiator 1 as reference and considering retardation so far as the phase factors are concerned, the phasor forms of electric field intensities at the field point due to the radiators are given by

$$\bar{E}_1 = E_0 e^{-j\beta_0 r}$$

$$\bar{E}_2 = E_0 e^{-j\beta_0 r} e^{j(\beta_0 s\cos\phi + \delta)}$$

$$\bar{E}_3 = E_0 e^{-j\beta_0 r} e^{j2(\beta_0 s\cos\phi + \delta)}$$

$$\cdots\cdots\cdots\cdots\cdots$$

$$\cdots\cdots\cdots\cdots\cdots$$

$$\bar{E}_n = E_0 e^{-j\beta_0 r} e^{j(n-1)(\beta_0 s\cos\phi + \delta)}$$

where E_0 is amplitude of the electric field intensity of a single radiator. The resultant electric field intensity E_R is equal to the phasor sum of fields of all the sources. Thus,

$$\bar{E}_R = E_0 e^{-j\beta_0 r}[1 + e^{j\psi} + e^{2j\psi} + \cdots + e^{j(n-1)\psi}] \tag{1}$$

where
$$\psi = \beta_0 s \cos\phi + \delta \tag{2}$$

The function inside the square brackets of (1) is in the form of a geometric series. Using the summation rule for such a series, we have

$$\bar{E}_R = E_0 e^{-j\beta_0 r} \frac{1 - e^{-jn\psi}}{1 - e^{-j\psi}}$$

The amplitude of the resultant field,

$$|E_{r0}| = E_0 \left|e^{-j\beta_0 r}\right| \frac{\left|1 - e^{jn\psi}\right|}{\left|1 - e^{j\psi}\right|} = E_0 \left|\frac{\sin\frac{n\psi}{2}}{\sin\frac{\psi}{2}}\right| \tag{3}$$

Expressions (2) and (3) show that the amplitude of the resultant field depends on the number of radiators n in the array, spacing s, and current phase angle δ between consecutive radiators, and the angular position ϕ of the field point. The condition for the highest value of the amplitude of the resultant field, which can be obtained by differentiating (3) with respect to ψ and then equating the resulting function to zero, is given by

$$\psi = \beta_0 s \cos\phi + \delta = 0 \tag{4}$$

Though the expression given by (3) becomes indeterminate at $\psi = 0$, in the limit $\psi \to 0$ the value of the function is unity. Thus the highest amplitude of the resultant field,

$$E_{h0} = nE_0,$$

The ratio of the amplitude of the resultant electric field intensity for any ψ and the highest amplitude is given by

$$\left|\frac{E_{R0}}{E_{h0}}\right| = \frac{1}{n} \left|\frac{\sin\frac{n\psi}{2}}{\sin\frac{\psi}{2}}\right| \tag{5}$$

Relation (5) is the normalised form of the amplitude of the resultant field in terms of ψ.

An examination of expression (3) shows that it becomes zero if

$$\frac{n\psi}{2} = \pm k\pi, \quad k = 1, 2, 3, \ldots$$

Or
$$\psi = \pm\frac{2k\pi}{n}, \tag{6}$$

Broadside Array

The condition for the amplitude of the resultant field to have the highest value is given by (4). If currents in all the radiators are in phase, then the current phase angle $\delta = 0$. Hence,

$$\psi = \beta_0 s \cos\phi = 0$$

Or
$$\phi \quad \pi/ \ .$$

It means that the highest amplitude of the resultant field occurs on the plane that is perpendicular to the array axis. This type of the array is designated as *broadside array*.

Moreover, the values of ψ at which the resultant field becomes zero is given by (6). For variation of ψ from $-\pi$ to $+\pi$, the values of ψ at which the null points occur depends on the number of radiators. For example, if $n = 6$, the null points occur at $\psi = \pm\pi/3$ for $k = 1$, at $\psi = \pm 2\pi/3$ for $k = 2$, and $\psi = \pm\pi$ for $k = 3$. The graph between two null points is called a lobe. Thus, there are altogether 5 lobes. At $\psi = 0$, the resultant field has the highest value of $6E_0$, and the lobe having this value is the principal lobe. Other 4 lobes are called secondary lobes. For large values of n, the maximum value of a secondary lobe occurs nearly at the mid-point between the nulls. A plot of magnitude of the expression inside the modulus sign of (3) as a function of ψ for $n = 6$ and $s = \lambda_0/2$ is shown in Fig. 10.21. Rotation of the pattern about the x axis gives the pattern in three-dimensional space.

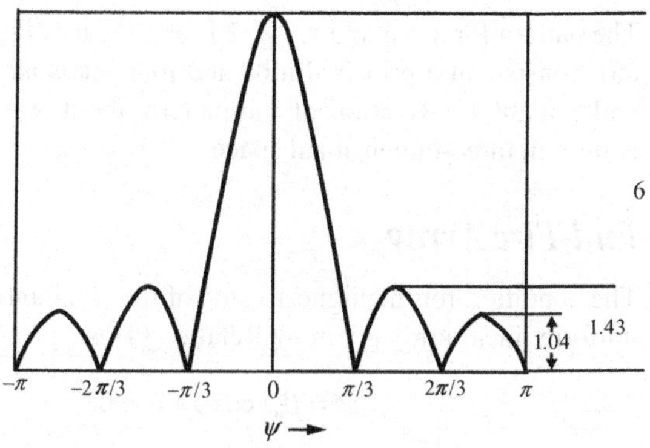

Fig. 10.21 Plot of the function $|(\sin n\psi/2)/(\sin \psi/2)|$ for six isotropic radiators with $\lambda_0/2$ spacing. For the broadside array, $\psi = \beta_0 s \cos\phi$ and for the end-fire array, $\psi = \beta_0 s(\cos\phi - 1)$.

To obtain the polar plot or pattern of the resultant electric field intensity of the array, it is necessary to express the resultant of (3) in terms of ϕ. For the broadside array under consideration, $\delta = 0$; hence from relation (2) we have

$$\psi = \beta_0 s\cos\phi = \frac{2\pi s\cos\phi}{\lambda_0} \tag{7}$$

By substituting (7) into (3), we get

$$|E_{RO}| = E_0 \left|\frac{\sin[(n\pi s\cos\phi)/\lambda_0]}{\sin[(\pi s\cos\phi)/\lambda_0]}\right| \tag{8}$$

The polar plot of the expression in (8) gives the electric field intensity pattern of the array for n radiators. Rotation of the pattern about the array axis gives the pattern in three-dimensional space.

An important parameter of the polar plot is the width of the principal lobe. It is equal to the width measured between the null points of the lobe. If width of the lobe is taken as $2\phi_w$ then the null points occur at $\pm\phi_w$ from the point of the highest value of the lobe. As the highest value occurs at $\phi = \pi/2$, one of the nulls is at $\phi = (\pi/2) - \phi_w$. Also, as per (6) the value of the principal lobe is zero at $\psi = 2\pi/n$, and for the broadside array $\psi \quad \beta_0 s\cos\phi$. Thus, we have

$$\beta_0 s\cos\left(\frac{\pi}{2} - \phi_w\right) = \frac{2\pi}{n}$$

Or

$$\sin\phi_w = \frac{2\pi}{n\beta_0 s} = \frac{\lambda_0}{ns} \tag{9}$$

We can calculate ϕ_w from (9) for known values of n and s. The width of the principle lobe is twice of this value. The width of the principal lobe decreases with an increase in the number of radiators and spacing between the radiators. Thus, the directivity of the array increases with an increase in the number of radiators, and with increase in spacing between the radiators. The spacing is generally kept within half of a wavelength to avoid unwanted lobes.

The pattern for $n = 6$ and $s = \lambda_0/2$ is as shown in Fig. 10.22. The pattern consists of a principal lobe and four secondary lobes. The lobe width is 38.9°. Rotation of the pattern about $\phi = 0$ axis gives the pattern in three-dimensional space.

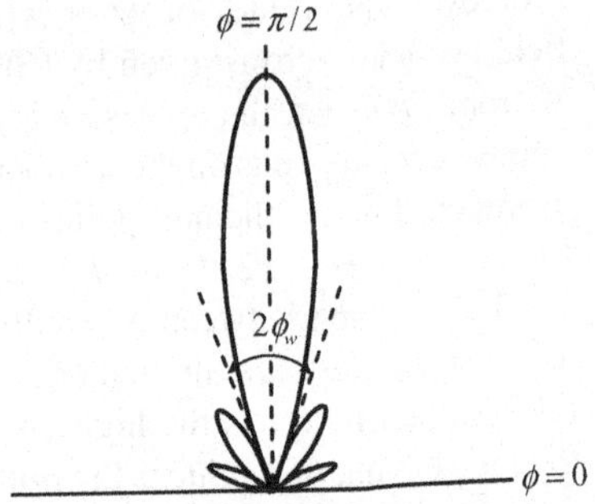

Fig. 10.22 Electric field intensity pattern (polar plot) of broadside array of six isotropic radiators with $\lambda_0/2$ spacing

End-Fire Array

The condition for the highest value of the resultant electric field of a uniform linear array given by Relation (4) is

$$\psi = \beta_0 s \cos\phi + \delta = 0$$

Or $$\cos\phi = -\frac{\delta}{\beta_0 s},$$

Where δ and s are the phase difference of currents and spacing respectively between adjacent radiators. The parameter β_0 is the phase constant of free space. If δ is made equal to $-\beta_0 s$ so that it compensates for the retardation effect along the array axis, then $\cos\phi = 1$ and $\phi = 0$. Thus, the highest value of the resultant field intensity occurs along the array axis. Such an array is called an *end-fire array*. Accordingly, we have for this type of array,

$$\psi = \beta_0 s \cos\phi + \delta = \beta_0 s(\cos\phi - 1) = (2\pi s/\lambda_0)(\cos\phi - 1) \tag{10}$$

It may be noted that the expression for the amplitude of the resultant field in terms ψ, given by (5), is the same for both the broadside array and the end-fire array. However, for the broadside array $\psi = \beta_0 s\cos\phi$ and for the end-fire array $\psi = \beta_0 s(\cos\phi - 1)$.

In order to obtain the polar plot or pattern of the resultant electric field of an end-fire array, we substitute Relation (10) into (3) and obtain,

$$|E_{RO}| = E_0 \left| \frac{\sin\dfrac{n\beta_0 s(\cos\phi - 1)}{2}}{\sin\dfrac{\beta_0 s(\cos\phi - 1)}{2}} \right| = E_0 \left| \frac{\sin\dfrac{n\pi s(\cos\phi - 1)}{\lambda_0}}{\sin\dfrac{\pi s(\cos\phi - 1)}{\lambda_0}} \right| \tag{11}$$

The polar plot of Expression (11) as a function of ϕ gives the pattern of an end-fire array with n isotropic radiators. The maximum value of the pattern occurs at $\phi = 0$, and as per (6) the nulls of the principal lobe occur at

$$\psi = \beta_0 s(\cos\phi - 1) = \pm\frac{2\pi}{n}$$

Since the function within the brackets is negative for values of ϕ greater than zero, we have

$$\beta_0 s(\cos\phi - 1) = -\frac{2\pi}{n} \tag{12}$$

Now, let width of the principal lobe between its maximum value and one of its null points be ϕ_w. Then as the maximum value is at $\phi = 0$, we have from (12)

$$\beta \quad \phi_w - 1) = -\frac{2\pi}{n}$$

Or
$$(\cos\phi_w - 1) = -\frac{2\pi}{n\beta_0 s} = -\frac{\lambda_0}{ns}$$

Or
$$\sin^2\frac{\phi_w}{2} = \frac{\lambda_0}{2ns}$$

Or
$$\sin\frac{\phi_w}{2} = \pm\sqrt{\frac{\lambda_0}{2ns}} \qquad (13)$$

By solving (13) for given values of n and s we can calculate the lobe width. If the array has 6 radiators with spacing equal to $\lambda_0/2$, the lobe width is found to be 96.4°. Thus, the lobe width of an end-fire array is greater than that of an identical broadside array. The pattern is shown in Fig. 10.23. Rotation of the pattern about the $\phi = 0$ axis gives the pattern in three-dimensional space.

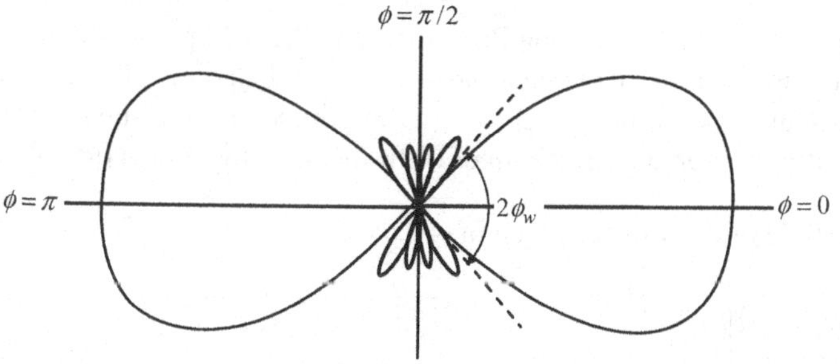

Fig. 10.23 Electric field intensity pattern (polar plot) of end-fire array of six isotropic radiators with $\lambda_0/2$ spacing

EXAMPLE 10.6

A uniform linear broadside array of 6 isotropic radiators is located in free space. The distance between neighboring radiators is one-half of a wavelength. The point at which the field is to be calculated is remote from the array. The line joining Radiator 1 and the field point is at an angle of 80° with the array axis. The amplitude of the radiation electric field due to one radiator at the point is 0.2 V/m. Find (a) the amplitude of resultant electric field intensity, (b) amplitude of the resultant magnetic field intensity and (c) width of the principal lobe of the polar plot.

The resultant electric field of the array as given in (8) is

$$E_{R0} = E_0\frac{\sin[(n\pi s\cos\phi)/\lambda_0]}{\sin[(\pi s\cos\phi)/\lambda_0]}$$

(a) Substituting $E_0 = 0.2$ V/m; $n = 6$; $s = \lambda_0/2$; $\phi = 80°$, we have

$$E_{R0} = 0.2\frac{\sin(3\pi\cos 80°)}{\sin\left(\frac{\pi}{2}\cos 80°\right)} = 0.741 \text{ V/m}$$

(b) $H_R = E_{R0}/\eta_0 = 0.741/377 = 1.966\ \mu$A/m

(c) From (9), $\sin\phi_w = \dfrac{\lambda_0}{ns}$

where ϕ_w is one half of the lobe width. Substituting the values, we have

$$\sin\phi_w = \frac{1}{3}$$

$$\phi_w = 19.47°$$

Width of the lobe $= 2\phi_w = 38.94°$

EXAMPLE 10.7

A uniform linear end-fire array having 6 isotropic radiators is located in free space. The distance between consecutive radiators is one-half of a wavelength. The point at which the field is to be calculated is remote from the array. The angle of the line joining Radiator 1 and the field point with the array axis is 20°. The amplitude of the radiation electric field due to one radiator at the point is 0.2 V/m. Calculate at the point (a) amplitude of resultant electric field intensity, (b) amplitude of the resultant magnetic field intensity, (c) value of the Poynting vector and (d) Width of the principal lobe of the polar plot.

The resultant electric field of the array as given in (11) is

$$E_{RO} \quad E_0 \frac{\sin\dfrac{n\pi s(\cos\phi - 1)}{\lambda_0}}{\sin\dfrac{\pi s(\cos\phi - 1)}{\lambda_0}}\Bigg|$$

(a) Substituting $E_0 = 0.2$ V/m; $n = 6$; $s = \lambda_0/2$; $\phi = 20°$, we have

$$E_{RO}\Big| = 0.2\left|\frac{\sin[3\pi(\cos 20° - 1)]}{\sin[0.5\pi(\cos 20° - 1)]}\right| = 1.138\text{ V/m}$$

(b) $H_{RO} = E_{RO}/\eta_0 = 1.338/377 = 3.377\ \mu$A/m

(c) value of the Poynting vector,

$$P_w = (E_{RO}H_{RO})/2 = (1.138)(3.377 \times 10^{-3})/2 = 1.992\text{ mW/m}^2$$

(d) Let the width of the principal lobe be $2\phi_w$. Then from (13), we have

$$\sin\frac{\phi_w}{2} = \sqrt{\frac{\lambda_0}{2ns}} = \sqrt{\frac{1}{6}}$$

Or $\dfrac{\phi_w}{2} = 24.1°$

Width of the lobe $= 2\phi_w = 96.4°$

By comparing the results of Examples 10.6 and 10.7 we find the lobe width of the end-fire array is greater than that of the broadside array.

10.7 Yagi–Uda Antenna

The disadvantage of an antenna array consisting of discrete elements is the necessity of feed currents with controlled phase angles. The problem is overcome by the Yagi–Uda antenna. The antenna is an array of parallel linear conducting elements, usually of tubes to reduce eddy-current loss and to provide mechanical strength, mounted on a conducting movable arm, as shown in Fig. 10.24. Only one element, which may be a simple half-wave dipole or a half-wave folded dipole, is excited by a generator through a suitable transmission line. This is the driven element.

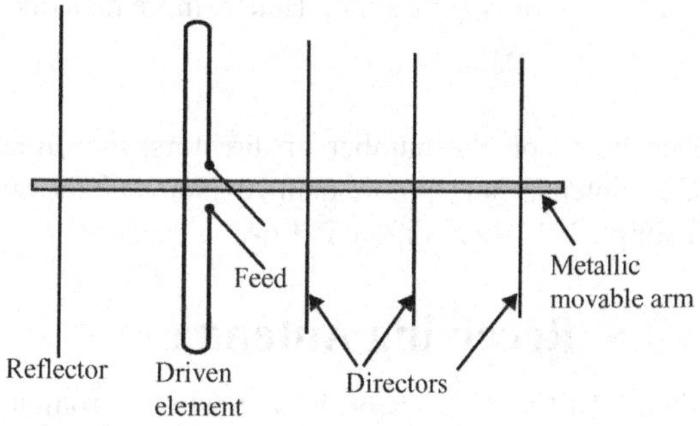

Fig. 10.24 Yagi–Uda antenna array having three directors

The other elements, the so called parasitic elements, are excited by currents induced by the driven element. One element of the array, the length of which is nearly ten percent longer than the driven element, is called reflector.

The other elements of the array the lengths of which are less than that of the driven element are called directors.

The reflector is placed on one side of the driven element and the directors are placed on the other side. The director that is nearest to the driven element is nearly ten percent shorter than the driven element. The other directors are of progressively shorter lengths.

Working Principle

The impedance of a linear dipole depends on its length and diameter. The impedance of a dipole, the length of which is nearly equal to half wavelength, is purely resistive. If the length of a dipole is more than a half wavelength, the impedance is inductive and if the length is less than a half wavelength, the impedance is capacitive. Figure 14.7, page 548 of reference 11 gives the values of resistance and reactance for a monopole; the values for a dipole are twice that of a monopole.

As the reflector is longer than a half wavelength, its impedance is inductive; hence, its current lags the voltage induced in it. Moreover, the induced voltage in the reflector is opposite to the feed voltage of the driven element; and the feed voltage and current are in phase as the reactance of the driven element is nearly zero. Therefore, the feed current lags the reflector current. Now considering the directors, their impedances are capacitive as their lengths are shorter than half of a wavelength. Hence by the same reasoning, the induced currents of the directors progressively lag the feed current. The values of phase angles depend on the lengths and diameters of the elements, and the spacing between the consecutive elements. Thus, the Yagi–Uda array, starting from the reflector, is a phased array with progressively lagging phase shift. Therefore, an array with vertically oriented elements has the horizontal field pattern similar to that of an end-fire array.

Most of the power radiated by the antenna is in the direction from the driven element towards the directors. The field pattern is as shown in Fig. 10.25.

Even with one reflector, power radiated in the direction of the reflector is quite small. Hence, more than one reflector is not necessary.

Depending on the number of directors, the directivity of a Yagi antenna can be more than 10. Note that the directivity of a simple half-wave dipole is 1.64.

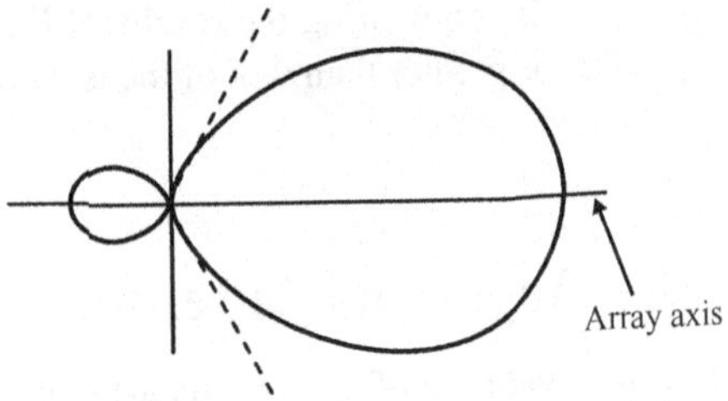

Fig. 10.25 Field pattern of principal lobes of a Yagi–Uda antenaa. The shapes are representative only.

10.8 Receiving Antenna

Figure 10.1 schematically shows a wireless communication system. The transmitting antenna radiates electromagnetic power into free space and a suitably oriented receiving antenna picks up the signal and conveys it to a receiver through a transmission system. If the receiving antenna is far away from the transmitting antenna, the electromagnetic waves arriving at the antenna are nearly uniform plane waves. In a linear antenna voltage is induced by the electric field and in a loop antenna voltage is induced by the magnetic field of the incoming wave. The magnitude of induced voltage depends on the orientation of the antenna axis with respect to the fields.

A part of electromagnetic power received by a receiving antenna is also radiated. The radiation patterns of a receiving antenna are similar to those of an identical transmitting antenna.

In this respect the receiving and transmitting antennas are reciprocal provided the medium is bilateral. Free space is of course a bilateral medium.

To maximise power received at the receiver or the load, not only the antennas should be properly oriented, there should also be proper matching of impedances at the antenna terminals and at the receiver terminals.

Linear Antenna

Let us consider a vertically oriented linear antenna of length ℓ made of a solid or a tubular conductor located in free space, as shown in Fig. 10.26. The antenna is immersed in the electromagnetic field of a remote transmitting antenna. Let the incident wave be a uniform plane wave and its polarisation be parallel to the plane of incidence. The plane of incidence is a plane containing the dipole. The angle of incidence is α as shown in the figure. As the electric field intensity E is at an angle of 90° – α with the antenna axis, its component along the antenna axis is equal to $E \sin \alpha$.

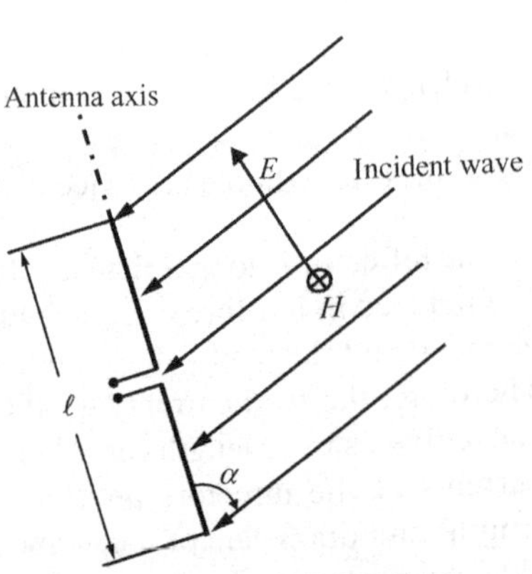

Fig. 10.26 A linear dipole antenna

If length of the antenna is small compared to the wavelength of the incoming signal, we can assume that the electric field intensity is constant over its length. Under this assumption, the voltage induced between the terminals is simply equal to the product of $E \sin \alpha$ and the antenna length ℓ. For an antenna, the length of which is not small in comparison with wavelength of the signal, the voltage induced in a differential length has to be integrated along the antenna to get the induced voltage. Thus, the magnitude of the induced voltage is given by

$$\mathcal{E}_0 = \sin\alpha \int E\, d\ell$$

The value of induced voltage would be maximum if α is 90°, that is, if the electric field is parallel to the antenna.

If the wave polarisation is at an angle with the plane of incidence, then only the component that is parallel to the plane of incidence is effective. Note that the parallel component itself, in general, is at an angle of 90° – α, as described above.

Loop Antenna

A simple planar loop antenna is generally of circular shape with two output terminals. The loop may be of one turn or may have several turns. A loop antenna assumed to be in free space is shown in Fig. 10.27. If positioned with appropriate orientation in an incoming electromagnetic field, a voltage is induced in the loop by the magnetic field of the wave.

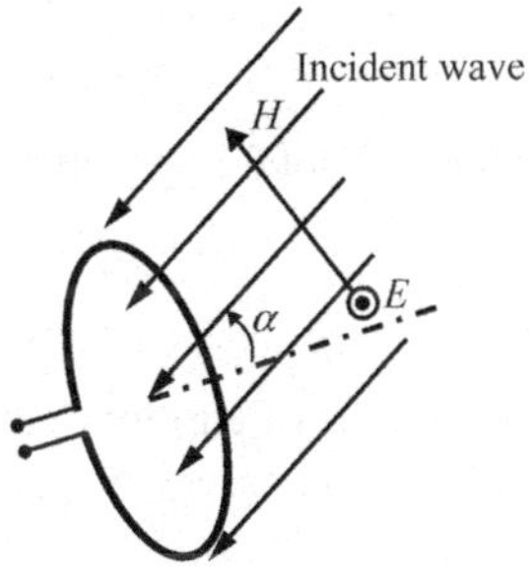

Fig. 10.27 A loop antenna

Let us assume that the fields at the location of the antenna are that of a uniform plane electromagnetic wave and the wave polarisation is perpendicular to the plane of incidence. The plane of incidence is the plane containing the antenna axis and a ray representing the direction of travel of the wave. The direction of travel of the incident wave is at an angle α with loop axis, as shown in the figure. Thus, the magnetic field is at an angle 90° – α with the loop axis and hence the component of H along the antenna axis is equal to $H \sin \alpha$, Then, magnetic flux linking the loop,

$$\Phi = \mu_0 \sin\alpha \int_S H dS$$

where dS is differential area of the loop along the antenna axis. The time-varying magnetic flux will induce a voltage in the coil as per Faraday's law of electromagnetic induction. The magnitude of the induced voltage between the terminals,

$$\mathcal{E}_0 = \frac{\partial \Phi}{\partial t} = (\mu_0 \sin\alpha)\frac{\partial}{\partial t}\int_S H dS$$

If the circumference of the loop is small compared to the signal wavelength, then we may assume the magnetic field intensity to be constant over the loop surface. Accordingly, for a loop of N turns and average surface area S, magnitude of the voltage induced is given by

$$\mathcal{E}_0 = \mu_0 NS \sin\alpha \frac{\partial H}{\partial t}$$

To get the maximum induced voltage the orientation of the loop should be such that its axis is parallel to the magnetic field of the incoming wave.

Effective Area

An important parameter associated with a receiving antenna is the *effective area* or *effective aperture*.

It is defined as the ratio of power delivered by the antenna to a matched load and power density (power per unit area) due to a suitably polarised electromagnetic wave.

A load (receiver) connected to a receiving antenna is said to be matched if its impedance is the complex conjugate of the antenna impedance. Under matched condition, the impedance of the antenna-load system is purely resistive, and the load receives maximum power. The power received depends on the voltage induced in the antenna and its radiation resistance. The power density is given by the average value of the Poynting vector.

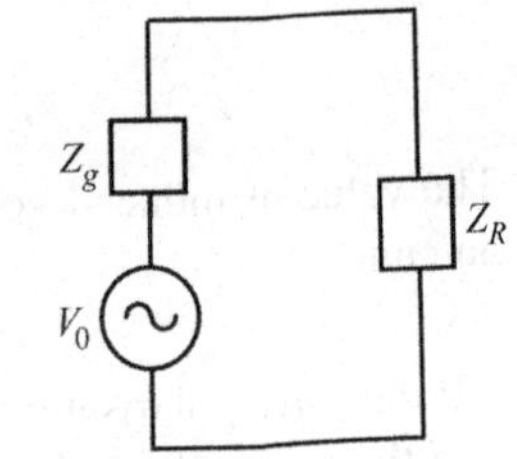

Fig. 10.28 Circuit representation of a receiving antenna and load

The equivalent circuit of a linear antenna connected to a load is as given in Fig. 10.28 where the receiving antenna is shown as a source. The amplitude of the open-circuit voltage of the antenna is V_0 and its impedance $\bar{Z}_g = R_g \pm jX_g$. As the matched load impedance is equal to the complex conjugate of the antenna impedance, the load impedance is given by

$$\bar{Z}_R = R_g \mp jX_g$$

Thus, the total impedance of the circuit,

$$\bar{Z}_g + \bar{Z}_R = 2R_g$$

Amplitude of current in the circuit, $I_0 = \dfrac{V_0}{2R_g}$

Power delivered to the load, $P_R = \dfrac{1}{2} I_0^{\,2} R_g = \dfrac{1}{2}\left(\dfrac{V_0}{2R_g}\right)^2 R_g = \dfrac{V_0^{\,2}}{8R_g}$ (1)

If the length ℓ of a linear antenna is small compared to the wavelength of the signal, as of a Hertzian dipole, we can assume the electric field intensity along the antenna to be constant. Thus, amplitude of open-circuit voltage across the antenna terminals,

$$V_0 = \ell E_0 \tag{2}$$

where E_0 is the amplitude of the electric field intensity.

The resistance of an antenna is equal to the sum of its ohmic resistance and radiation resistance. As the ohmic resistance is generally very small compared to the radiation resistance, we can assume the antenna resistance to be equal to the radiation resistance R_{rad}. Thus, substituting (2) into (1) and replacing R_g by R_{rad},

$$P_R = \frac{\ell^2 E_0^{\,2}}{8R_{rad}}$$

Next, we will calculate the power density. In a uniform plane wave, the electric and magnetic fields are perpendicular to each other; they are also in time phase. Thus average power density associated with the wave,

$$P_{wa} = \frac{1}{2} E_0 H_0 = \frac{E_0^{\,2}}{2\eta_0} = \frac{E_0^{\,2}}{240\pi}$$

The effective area is the ratio of P_R and P_{wa}. Denoting the effective area by A_e, we have

$$A_e = \frac{P_R}{P_{wa}} = \frac{30\pi \ell^2}{R_{rad}} \tag{3}$$

For a short dipole antenna,

$$R_{rad} = 80\pi^2 \quad \lambda_0)^2 \tag{4}$$

By substituting (4) into (3), we get

$$A_e = \frac{3\lambda_0^{\ 2}}{8\pi}$$

As the directivity g_d of the antenna under consideration is 1.5, we have

$$A_e = \frac{\lambda_0^{\ 2}}{4\pi} g_d \tag{5}$$

We find from (5) that the effective area is proportional to the square of the wavelength of the signal. It is also proportional to directivity of the antenna. The formula gives the maximum value of the effective area. Though Expression (5) is derived taking the example of a Hertzian dipole, it is valid for any type of antenna.

Friis Formula

Another important relation for calculation of power delivered to a matched load by a receiving antenna is in terms of power radiated by a transmitting antenna. The relation is called *Friis transmission formula*. The formula is derived under the following conditions.

1. The receiving antenna is connected to a matched receiver. Hence, the effective area is given by (5).
2. Orientation of the receiving antenna with respect to wave polarisation is such that maximum power is received by it.

To derive the formula, let us consider that a receiving antenna fulfilling the above conditions is at a distance r from the transmitting antenna. Power delivered to the receiver P_r is equal to the product of power density P_{wa} and effective area of the receiving antenna A_{er}. Thus,

$$P_r = P_{wa} A_{er} \tag{6}$$

The power density in terms of power radiated by the transmitting antenna P_t is obtained as follows. The directive gain of the transmitting antenna as per equation (10.2-35) is

$$g_{dt}(\theta, \phi) = \frac{R_I(\theta, \phi)}{(P_t/4\pi)} \tag{7}$$

where $R_I(\theta, \phi)$ is the radiation intensity.

Using (10.2-33) in (7), we have

$$g_{dt}(\theta,\phi) = \frac{4\pi r^2 P_{wa}}{P_t}$$

Or

$$P_{wa} = \frac{P_t g_{dt}(\theta,\phi)}{4\pi r^2}$$

For given values of P_t and r, the power density is maximised if the directive gain has the maximum value. As the maximum value of the directive gain is directivity g_{dt} of the transmitting antenna, the maximum value of the power density is given by

$$P_{wa} = \frac{P g_{dt}}{4\pi r^2} \tag{8}$$

Substitution of (8) into (6) gives,

$$P_r = \frac{P_t g_{dt} A_{er}}{4\pi r^2}$$

Or

$$\frac{P_r}{P_t} = \frac{g_{dt} A_{er}}{4\pi r^2} \tag{9}$$

Relation (9) is known as *Friis transmission formula.*

It gives the maximum value of power received under the ideal conditions stated. If these conditions are not met, power received would be less than the maximum possible value.

By using Relation (5), which is also valid for transmitting antenna, in (9) we get the formula in terms of the effective areas as well as in terms of directivities. Thus,

$$\frac{P_r}{P_t} = \frac{A_{et} A_{er}}{r^2 \lambda_0^2} = \frac{\lambda_0^2 g_{dt} g_{dr}}{(4\pi r)^2} \tag{9a}$$

EXAMPLE 10.8

A linear antenna of length 20 cm is located in the field of a uniform plane wave. The amplitude of electric field intensity of the parallel polarised wave is 5.0 mV/m. The direction of the wave travel is at an angle of 30° with the antenna axis. Assuming that the electric field is constant along the antenna, find amplitude of the induced voltage.

Induced voltage, $\mathcal{E}_0 = \ell E_0 \sin\alpha$

Substituting the given values, $\mathcal{E}_0 = (0.2)(5\times10^{-3})(\sin 30^\circ) = 0.5$ mV

EXAMPLE 10.9

A linear antenna of 50 cm length is located in the field of a uniform plane wave. The direction of electric field intensity is at an angle of 20° with the plane of incidence, and the direction of wave travel is perpendicular to the antenna axis. Find amplitude of the induced voltage assuming the electric field to be constant along the antenna. The amplitude of the electric field intensity is 4.0 mV/m.

Induced voltage, $\mathcal{E}_0 = \ell E_0 \sin\alpha$

Only the field component that is parallel to the plane of incidence is effective. As $\alpha = 90^\circ$,

$_0 =$ [illegible] $.004\cos 20^\circ) = 1.88$ mV

EXAMPLE 10.10

A loop antenna, the circumference of which is small compared to the signal wavelength, is placed in the field of a perpendicularly polarised wave. The direction of travel of the wave is at an angle of 75° with the loop axis, and the amplitude of the magnetic field intensity is 25 μA/m at 50 MHz. The radius of the loop is 20 cm and it is located in air. Find the amplitude of voltage induced in the loop.

$$_0 = \mu_0 S \sin\alpha \frac{\partial H}{\partial t}$$

$$H \quad (25\times10^{-6})\sin(2\pi\times50\times10^{6})t$$

$$\frac{dH}{dt} = (2500\pi)\cos(2\pi\times50\times10^{6})t$$

Amplitude of induced voltage, $\mathcal{E}_0 = (4\pi\times10^{-7})(\pi\times0.2^2)(\sin 75°)(2500\pi) = 1.198$ mV.

EXAMPLE 10.11

Calculate the maximum effective area of a Hertzian dipole in free space and at signal frequency of 60 MHz.

$$A_e = \frac{\lambda_0^{\,2}}{4\pi} g_d$$

$$g_d = 1.5; \quad \lambda_0 = \frac{3\times10^8}{60\times10^6} = 5 \text{ m}$$

$$A_e = \frac{(5)^2(1.5)}{4\pi} = 2.98 \text{ m}^2$$

EXAMPLE 10.12

One linear receiving and one linear transmitting antenna, the lengths of which are small compared to the signal wavelength, are in free space. The antennas are parallel to each other and they are perpendicular to the line joining their mid-points. Find the ratio of the received to the transmitted power at a frequency of 25 MHz. Distance between the antennas is 500 m.

The orientation of the antennas is for optimum reception of signal. Hence Friis formula given by (9*a*) is to be used. As the antenna lengths are small compared to the wavelength, the directivity of each antenna is 1.5.

$$\frac{P_r}{P_t} = \frac{\lambda_0^{\,2} g_{dt} g_{dr}}{(4\pi r)^2}$$

$$\lambda_0 = \frac{3\times10^8}{25\times10^6} = 12.0 \text{ m}; \quad r = 500 \text{ m}$$

$$\frac{P_r}{P_t} = \frac{(12)^2(1.5)(1.5)}{(4\pi\times500)^2} = 8.2\times10^{-6}$$

SUMMARY

An antenna connected to a source acts as load to the source and at the same time it is also a source that radiates power into space. Though a short current element with uniform distribution of current, known as Hertzian dipole, is a hypothetical antenna, the study of its radiation property is helpful for analysis of practical antennas.

From the expression of the vector magnetic potential of the Hertzian dipole, we derived the electric and magnetic field intensities and calculated the complex Poynting vector. We learnt that there is no radiation of active power in the θ direction, and active power radiated along r direction is due to the far fields, called radiation fields, only. We defined various radiation parameters and obtained radiation patterns. We noted that the radiation resistance is different from the ohmic resistance.

We analysed a half-wave dipole antenna assuming sinusoidal current distribution along the length of the antenna and for sinusoidal variation of current with time. We obtained expressions for various radiation parameters, calculated the instantaneous Poynting vector, and found the radiation resistance to be 73.08 ohm. We also noted that radiation electric field and radiation intensity of a half-wave dipole antenna are more directional than those of a Hertzian dipole.

We learnt that the image of a linear vertical current located above a conducting plane is also a linear current, and the combination of the actual current and image current form a dipole antenna. The radiation properties of a linear current above a conducting plane, termed as monopole, are similar to those of a dipole. The radiation resistance of quarter-wave monopole is 36.54 ohm.

From the description of a folded dipole with two half-wave parallel elements we found that its radiation resistance is about 300 ohm and hence it can directly be connected to a two-wire transmission line.

We obtained the resultant field of an array of similar antennas and found that the multiplication of the unit pattern and the array pattern gives the resultant pattern, and the array pattern is same as the pattern of an identical array of isotropic radiators. We also defined a broadside array and an end-fire array and learnt that for a broadside array the radiation takes place in the perpendicular direction to the array axis, and in the case of an end-fire array the radiation is along the array axis. One advantage of a Yagi-Uda antenna array is that only the driven element is fed; the reflector and the directors are excited by induced currents.

We found that maximum voltage is induced in a linear receiving antenna if the electric field of the signal is parallel to it; but in the case of a loop antenna, maximum voltage is induced if the magnetic field is parallel to the loop axis.

IMPORTANT FORMULAE

Hertzian Dipole:

Far-zone electric field, $E_\theta = -\dfrac{\eta_0 I_0 \sin\theta}{2r}\left(\dfrac{d\ell}{\lambda_0}\right)\sin(\omega t - \beta_0 r)$

Average Poynting vector, $\mathbf{P}_{wa} = \dfrac{\eta_0 I_0^2}{8}\left(\dfrac{d\ell}{\lambda_0}\right)^2 \dfrac{\sin^2\theta}{r^2}\mathbf{u}_r$

Total power radiated, $P_{rad} = 40\pi^2 I_0^2\left(\dfrac{d\ell}{\lambda_0}\right)^2$

Radiation resistance, $R_{rad} = 80\pi^2\left(\dfrac{d\ell}{\lambda}\right)^2$

Radiation intensity, $R_I(\theta) = \dfrac{\eta_0 I_0^2 \sin^2\theta}{8}\left(\dfrac{d\ell}{\lambda_0}\right)^2$

Directive gain, $g_d(\theta) = 1.5\sin^2\theta$

Half-wave dipole with sinusoidal current distribution:

Far-zone electric field, $E_\theta = -\dfrac{\eta_0 I_0}{2\pi r}\dfrac{\cos[(\pi/2)\cos\theta]}{\sin\theta}\sin(\omega t - \beta_0 r)$

Average Poynting vector, $\mathbf{P}_{wa} = \dfrac{\eta_0 I_0^2}{8\pi^2 r^2}\dfrac{\cos^2[(\pi/2)\cos\theta]}{\sin^2\theta}\mathbf{u}_r$

Total power radiated, $P_{rad} = \dfrac{1.218\,\eta_0 I_0^2}{4\pi}$

Radiation intensity, $R_I(\theta) = \dfrac{15 I_0^2\cos^2[(\pi/2)\cos\theta]}{\pi\sin^2\theta}$

Directive gain, $g_d(\theta) = 1.642\dfrac{\cos^2[(\pi/2)\cos\theta]}{\sin^2\theta}$

Array of Two Isotropic Radiators:

Resultant radiation electric field, $|E_{R0}| = 2E_0\left|\cos\left(\dfrac{\pi s\sin\theta\cos\phi}{\lambda_0} + \dfrac{\delta}{2}\right)\right|$

Array of Two Hertzian Dipoles, $|E_{R0}| = 2E_0\left|\cos\left(\dfrac{\pi s\sin\theta\cos\phi}{\lambda_0} + \dfrac{\delta}{2}\right)\right|;\ E_0 = \dfrac{\eta_0 I_0}{2r}\left(\dfrac{d\ell}{\lambda_0}\right)$

Uniform Linear Array of n Isotropic Radiators:

Resultant radiation electric field, $|E_{R0}| = E_0\left|\dfrac{\sin\dfrac{n\psi}{2}}{\sin\dfrac{\psi}{2}}\right|;\ \psi = \beta_0 s\cos\phi + \delta$

The ratio of the amplitude of the resultant and the highest electric field intensities,

$$\left|\frac{E_{R0}}{E_{h0}}\right| = \frac{1}{n}\left|\frac{\sin\dfrac{n\psi}{2}}{\sin\dfrac{\psi}{2}}\right|$$

Broadside array, $|E_{R0}| = E_0\left|\dfrac{\sin[(n\pi s\cos\phi)/\lambda_0]}{\sin[(\pi s\cos\phi)/\lambda_0]}\right|$

Effective area, $A \quad \dfrac{\lambda^2}{4\pi} g_d$

Friis formula, $A \quad \dfrac{\lambda^2}{4\pi} g_d$

MULTIPLE-CHOICE QUESTIONS

(Starred questions have two answers)

10.1 The assumed current distribution along the length of a Hertzian dipole is
(a) uniform
(b) triangular with maximum value at the mid-point
(c) linear with maximum value at one end and zero at the other end
(d) sinusoidal with maximum value at one end and zero at the other end

10.2 The vector magnetic potential of a Hertzian dipole that is oriented in z direction has spherical symmetry; hence it is advantageous to work with spherical coordinates. In spherical coordinates, the vector magnetic potential has components
(a) in r and ϕ directions
(b) in ϕ and θ directions
(c) in r and θ directions
(d) only in r direction

10.3 The far fields of a vertically oriented Hertzian dipole has
(a) electric field in r direction and magnetic field in ϕ direction
(b) electric field in r direction and magnetic field in θ direction
(c) electric field in θ direction and magnetic field in r direction
(d) electric field in θ direction and magnetic field in ϕ direction

*10.4 Which of the following statements are valid for the radiation fields of a small vertically oriented dipole antenna?
(a) They are inversely proportional to r and directly proportional to sin θ.
(b) They are independent of θ.
(c) They are in time phase.
(d) They are inversely proportional to r^2 and directly proportional to sin θ.

10.5 The radiation electric and magnetic fields of a vertically oriented half-wave dipole are in θ and ϕ directions respectively. The direction of power flow is
(a) along θ direction
(b) along both θ and ϕ directions
(c) along ϕ direction
(d) along r direction

*10.6 The radiation intensity of an antenna is defined as
(a) power radiated per unit area
(b) power radiated per unit solid angle
(c) the product of r^2 and the Poynting vector
(d) field energy per unit volume

10.7 The directive gain of an antenna is
(a) nondimensional
(b) has the unit of power

(c) has the unit of energy
(d) has the unit of power per unit area

10.8 Maximum radiation due to an array of two isotropic radiators with in-phase current occurs
(a) along the array axis
(b) perpendicular to the direction of the array axis
(c) at 45° to the array axis

10.9 The spacing between two isotropic radiators is s, and the free-space phase constant is β_0. If the phase angle between currents of the radiators is $-\beta_0 s$, maximum radiation occurs
(a) along the array axis
(b) perpendicular to the direction of the array axis
(c) at 45° to the array axis

*10.10 At a given wavelength and distance between a receiving antenna and a transmitting antenna, the maximum ratio of received power to the transmitting power is
(a) proportional only to the effective area of the receiving antenna
(b) proportional only to the effective area of the transmitting antenna
(c) proportional to the product of effective areas of both the antennas
(d) proportional to the product of the directivity of the antennas

QUESTIONS

10.1 The currents in the wires of a parallel wire transmission line connected to an ac source at one end and terminated with open circuit at the other end are standing waves. Explain why no electromagnetic radiation is possible from the open ends. However if the wires are opened out from a suitable point from the open end, electromagnetic radiation occurs. Explain.

10.2 A small current element of length $d\ell$ and carrying current $I_0 \sin \omega t$ is equal to an electric dipole. What are the magnitude and direction of the dipole moment?

10.3 A current loop of radius a, and carrying current $I_0 \sin \omega t$ is equivalent to a magnetic dipole. What are the magnitude and direction of the dipole moment?

10.4 Assume that a vertical current element $I_0 d\ell \sin \omega t$ is oriented along z direction of rectangular coordinates, with its centre at the origin. Give the expressions for the instantaneous and phasor forms of vector magnetic potential at a distance r in spherical coordinates.

10.5 The expressions of electric and magnetic field intensities of a Hertzian dipole in spherical coordinates contain terms that are inversely proportional to various powers of r. The terms that are inversely proportional to r are called Radiation Fields. Explain why.

10.6 Distinguish between near and far fields of a Hertzian dipole. State their properties.

10.7 Show that the intrinsic wave impedance and other parameters associated with radiation fields of a vertical ac current element are same as those of uniform plane waves.

10.8 Explain the difference between the ohmic resistance and radiation resistance of a linear antenna.

10.9 Show that the radiation intensity of a vertical ac current element, with its mid-point at the origin of spherical coordinates is equal to r^2P_{wa} where P_{wa} is the time-average of Poynting vector involving the radiation fields.

10.10 The amplitude of radiation electric field intensity of a vertical Hertzian dipole in spherical coordinates is given by

$$\bar{E}_\theta = j\frac{\omega I_0 d\ell \sin\theta}{4\pi\varepsilon_0 v_0^2 r}$$

For a given r, show that the polar plot of the function on a vertical plane containing the dipole, obtained by varying θ from 0 to 2π, results in two circles.

10.11 Explain the terms Directive Gain and Directivity of an antenna.

10.12 A half-wave dipole is fed at its centre by a sinusoidal source. Give a sketch of current distribution along the dipole.

10.13 The radiation intensity of a vertical half-wave dipole is given by

$$R_I(\theta) = \frac{15I_0^2 \cos^2[(\pi/2)\cos\theta]}{\pi \sin^2\theta}$$

in spherical coordinates. Show that the intensity is zero in $\theta = 0$ and $\theta = \pi$ directions and has maximum value in $\theta = \pi/2$ and $\theta = 3\pi/2$ directions.

10.14 Sketch the pattern of radiation intensity of a half-wave dipole on a vertical plane containing the dipole, and state its salient properties.

10.15 Describe the technique used to find the fields of a monopole antenna oriented vertically above the ground plane.

10.16 Explain the working principle of a half-wave folded dipole antenna with two parallel conductors.

10.17 Explain why the input impedance of a half-wave folded dipole antenna with two parallel conductors is *four* times that of a half-wave dipole antenna with one conductor.

10.18 What are the advantages of a folded dipole antenna over a simple dipole antenna?

10.19 State the radiation properties of an isotropic radiator.

10.20 What assumptions are made for deriving the resultant field of an antenna array?

10.21 The magnitude of the resultant electric field at a remote point from an array of two isotropic radiators is given by

$$|E_{R0}| = 2E_0\left|\cos\left(\frac{\pi s \sin\theta\cos\phi}{\lambda_0} + \frac{\delta}{2}\right)\right|$$

Sketch the pattern of the function inside the modulus sign on the horizontal plane ($\theta = \pi/2$) for the following cases.

(a) $s = \lambda_0/2$; $\delta = 0$. (b) $s = \lambda_0/2$; $\delta = -\pi$

10.22 The amplitude of the resultant radiation electric field of an array of two Hertzian dipoles is given by

$$E_{RO} = 2E_0 \sin\theta \left[\cos\left(\frac{\pi s \sin\theta \cos\phi}{\lambda_0} + \frac{\delta}{2} \right) \right]$$

Define *unit pattern*, *array pattern*, and *resultant pattern* of the array. Sketch the patterns for the following cases.

(a) $s = \lambda_0/2$; $\delta = 0$; $\phi = 0$. (b) $s = \lambda_0/2$; $\delta = 0$; $\theta = \pi/2$

10.23 Explain what is meant by a *linear array* and a *uniform linear array*. Support your answers with sketches and current equations.

10.24 The amplitude of the resultant electric field intensity of a uniform linear array of n isotropic radiators is given by

$$|E_{RO}| = E_0 \left| \frac{\sin\frac{n\psi}{2}}{\sin\frac{\psi}{2}} \right| \text{ where } \psi = \beta_0 s \cos\phi + \delta.$$

State the conditions under which the array is a *broadside array* and the condition under which it is an *end-fire array*. Justify the terminologies.

10.25 The amplitude of the resultant electric field intensity of a uniform lincar array of n isotropic radiators is given by

$$|E_{RO}| = E_0 \left| \frac{\sin\frac{n\psi}{2}}{\sin\frac{\psi}{2}} \right|$$

Obtain a sketch the expression with the modulus sign as a function of ψ for $n = 4$.

10.26 In the expression inside the modulus sign of test question 10.25, $\psi = \beta_0 s \cos\phi + \delta$. Plot of magnitude of the expression as a function of ϕ gives the pattern of the field. Sketch the pattern of the *broadside array* and of the *end-fire array* for $s = \lambda_0/2$, $n = 4$.

10.27 An important parameter of antenna array is the lobe width. Explain its meaning with sketches for the broadside array and end-fire array.

10.28 What are the quantities on which the lobe width of a uniform linear antenna array depends? How do they affect the lobe width and hence the directional properties of the array?

10.29 Explain the constructional features of a Yagi–Uda antenna and state the terminologies used to describe its elements.

10.30 In a Yagi–Uda antenna only the driven element is fed by a generator; other elements are excited by currents induced by the driven element. Explain how phasing of the array is achieved.

10.31 In a Yagi–Uda antenna more than one directors are used to obtain the required value of directivity. However, multiple reflectors are not necessary? Why is it so?

10.32 A linear antenna is located in the field of a uniform plane wave. Under what condition the voltage induced in the antenna will have (a) the maximum value and (b) the minimum value.

10.33 A small loop antenna is located in the field of a uniform plane wave. Under what condition the voltage induced in the antenna will have (a) the maximum value and (b) the minimum value.

10.34 Power associated with a receiving antenna depends on its effective area and power density. Name the quantities on which the effective area depends and state how they affect the area.

10.35 The Friis transmission formula gives the power delivered to a matched load with ideal orientation of the antenna in the field of a transmitting antenna. Name the quantities on which the received power depends, and how they affect the power.

ANSWERS TO SELECTED QUESTIONS

10.6 Distinguish between near and far fields of a Hertzian dipole. State their properties.

Answer: The expressions for electric and magnetic fields have terms which are proportional to $1/r$, $1/r^2$ and $1/r^3$. Very close to the antenna, the terms containing $1/r$ and $1/r^2$ are neglected as these are small compared to the terms containing $1/r^3$. The magnetic field in this region leads the electric fields by a time angle of 90° and hence power associated with these fields is purely reactive as in a capacitor. Far away from the dipole, the terms with factors $1/r^2$ and $1/r^3$ are neglected. The far-zone magnetic and electric fields, which are inversely proportional to r, are in time phase and their directions are perpendicular to each other. The product of expressions of the fields give power radiated per unit area.

10.8 Explain the difference between the ohmic resistance and radiation resistance of a linear antenna.

Answer: The ohmic resistance of a linear antenna depends, apart from other factors, on conductivity of the material. It is equal to power dissipated by the antenna in the form of heat divided by square of the rms current. On the other hand, the radiation resistance is defined as power radiated by the antenna divided by the rms value of antenna current. The radiation resistance is not a function of conductivity. The value of ohmic resistance is very small compared to the radiation resistance.

10.18 What are the advantages of a half-wave folded dipole antenna with two parallel elements over a simple half-wave dipole antenna?

Answer: The input resistance of a dipole antenna with a single element of a half wavelength is approximately 73 Ω. If such an antenna is fed by a balanced two-conductor transmission line the characteristic impedance of which is much higher than 73 Ω, then there will be impedance mismatch at the input point. Hence, it is necessary to incorporate a matching device to minimise power reflection. A half-wave folded dipole antenna which has much higher input impedance than that of a single dipole can directly be fed by a suitable transmission line. The other advantages are input impedance of a folded dipole does not vary much over a wide range of frequencies and it has increased bandwidth which is important for FM and television receptions.

10.29 In a Yagi–Uda antenna only the driven element is fed by a generator; other elements are excited by currents induced by the driven element. Explain how phasing of the array is achieved.

Answer: In a Yagi–Uda antenna, the impedance of the driven element which is a little less than a half wavelength is nearly resistive. As length of the reflector is longer than half of a wavelength, its impedance is inductive. The impedances of directors are capacitive as their lengths are less than one-half of a wavelength. Therefore, induced currents in the parasitic elements differ in phase in relation to the phase angle of current in the driven element. The phase angle also depends on spacing between the elements. By using appropriate lengths and spacing between the elements the required phasing is obtained.

PROBLEMS

*10.1 Find the equivalent electric dipole moment of a current element $0.5\sin(15\pi\times10^6)t$ A-m.

**10.2 The amplitude of alternating current in a straight wire of length 5 cm is 30 A. The current-element is oriented along the z axis of rectangular coordinates with its mid-point at the origin. Find the amplitude of vector magnetic potential due to the current element at a distance of 0.75 m from the origin. What is the direction of the potential?

***10.3 The vector potential of a Hertzian dipole in spherical coordinates is given by

$$\bar{\mathbf{A}} = \frac{\mu_0(I_0 d\ell)}{4\pi r}(\cos\theta\,\mathbf{u}_r - \sin\theta\,\mathbf{u}\theta)e^{-j\omega r/v_0}$$

Obtain expressions for the magnetic and electric field intensities. Use the relation

$$\nabla\times\bar{\mathbf{A}} = \frac{1}{r}\left[\frac{\partial(r\bar{A}_\theta)}{\partial r} - \frac{\partial\bar{A}_r}{\partial\theta}\right]\mathbf{u}_\phi$$

**10.4 A vertical current element of $0.4\sin(5\pi\times10^6)t$ A-m in air is oriented along the z axis of rectangular coordinates with its mid-point at the origin. Find the amplitudes of the radiation electric- and magnetic-field intensities in spherical coordinates at $r = 500$ m and $\theta = 90°$.

**10.5 A vertical Hertzian dipole of length 2.5 cm in air is oriented along the vertical z axis of rectangular coordinates with its mid-point at the origin. The dipole carries a uniformly distributed current of 30 A at 120 MHz. Find (a) power radiated by the dipole, (b) the radiation resistance of the dipole, (c) the radiation intensity along $\theta = 45°$ with the z axis and (d) the directive gain along $\theta = 45°$ with the z axis.

**10.6 A small length of straight wire is oriented along the z axis of rectangular coordinates with its mid-point at the origin in free space. Find power radiated by the current element between $\theta = 45°$ and $\theta = 135°$, measured from the z axis, as a fraction of the total radiated power.

**10.7 The amplitude of alternating current in a vertical half-wave dipole is 4.0 A. The dipole is located in air, and lies along the z axis of rectangular coordinates with its mid-point at the origin. Find the amplitudes of radiation electric- and magnetic-field intensities in spherical coordinates at

(a) $r = 750$ m and $\theta = 45°$; (b) $r = 750$ m and $\theta = 90°$.

**10.8 The expression for the instantaneous radiation electric field intensity of a dipole of small length and uniform current distribution is

$$E_\theta = -\frac{\eta_0 I_0 \sin\theta}{2r}\left(\frac{d\ell}{\lambda_0}\right)\sin(\omega t - \beta_0 r)$$

Find the expression for average value of Poynting vector.

***10.9 The average Poynting vector of a dipole of small length and uniform current distribution is

$$\mathbf{P}_{wa} = \frac{\eta_0 I_0^2}{8}\left(\frac{d\ell}{\lambda_0}\right)^2 \frac{\sin^2\theta}{r^2}\mathbf{u}_r$$

Find the total power radiated by the antenna.

**10.10 Show that the radiation intensity of an antenna is equal to $r^2 P_{wa}$ where P_{wa} is the average Poynting vector.

**10.11 The amplitude of current in a vertical half-wave dipole antenna is 5.0 A. The dipole is located in air along the z axis of rectangular coordinates with its mid-point at the origin. Find the radiation intensity in spherical coordinates at (a) (r, 45°, 0); (b) (r, 90°, 0).

**10.12 A vertical half-wave dipole antenna is located in air. It lies along the z axis of rectangular coordinates with its mid-point at the origin. The amplitude of radiation electric-field intensity at P(600 m, θ, 0) in spherical coordinates is 0.5 V/m. Find the radiation intensity at P.

**10.13 A vertical half-wave dipole antenna, located in air, is oriented along the z axis of rectangular coordinates with its mid-point at the origin. Find the directive gain of the antenna in terms of its directivity along $\theta = 60°$ where θ is in spherical coordinates.

***10.14 A vertical half-wave dipole antenna is located in air. It is oriented along the z axis of rectangular coordinates with its mid-point at the origin. The amplitude of current in the antenna is 5.0 A at 500 MHz. The radius of the antenna is 2.0 mm and its conductivity is 57 MS/m. Calculate (a) the power radiated by the antenna, (b) Approximate value of its resistance considering the skin effect and (c) efficiency of the antenna.

**10.15 Two isotropic radiators are located at $x = -s/2$ and $x = s/2$ in free space. The amplitudes of in-phase currents of both the radiators are same. Find the resultant electric field intensity at a distance from the origin and angle ϕ measured from the x axis.

**10.16 Two isotropic radiators 1 and 2 are at $x = -\lambda_0/8$ and $x = \lambda_0/8$ in free space. The amplitude of currents in both the radiators is the same. But the phase angle of current in Radiator 1 is $+\pi/4$ rad and that of Radiator 2 is $-\pi/4$ rad. The frequency of current is 100 MHz. Each radiator has an amplitude of radiation electric field-intensity of 0.15 V/m at P(100 m, 100 m, 0). Taking the reference at the origin, obtain expressions for (a) the instantaneous electric field of each radiator at P, (b) the instantaneous resultant electric field at P and (c) Find the amplitude of the resultant field at P in terms of the highest amplitude of the resultant field in the plane under consideration.

***10.17 Repeat Problem 10.16 with the reference point at Raditor 1. Take the phase angle of current in Radiator 1 as zero and that of Radiator 2 as $-\pi/2$. In addition, find the phase difference between the resultant fields of problems 10.16 and 10.17.

**10.18 Two Hertzian dipoles 1 and 2 are located in free space. They are oriented along the z direction and are symmetrically placed about the origin with their mid-points on the x axis. The distance between the dipoles is λ_0 and their currents are defined by

$$I_1 = I_2 = I_0 \sin \omega t$$

Calculate the amplitude of the resultant radiation electric-field of the array at (r, 90°, 45°) in terms of the highest amplitude on $\theta = 90°$ plane.

***10.19 A Hertzian dipole oriented along z direction is in front of a perfect conducting plane that is parallel to the y-z plane. The distance of the dipole from the conducting plane is $\lambda_0/4$. Find the array factor on the x-y plane in front of the conducting plane.

**10.20 A broadside array of 8 half-wave dipoles oriented in z direction are situated along the x axis. The spacing between consecutive dipoles is $\lambda_0/2$. Calculate at (r, 90°, 70°) (a) the amplitude of the resultant radiation electric field intensity as the ratio of the highest value of the field on the x-y plane and (b) the Poynting vector as the ratio of maximum value of the Poynting vector on the x-y plane.

**10.21 An end-fire array of 5 half-wave dipoles oriented in z direction are positioned along the x axis. The spacing between consecutive dipoles is $\lambda_0/4$. On the x-y plane, calculate the amplitude of the resultant radiation magnetic field intensity and Poynting vector in terms of their highest values.

***10.22 A uniform linear array of 8 half-wave dipoles oriented in z direction is located along the x axis. The spacing and current phase-angle between consecutive dipoles are $\lambda_0/2$ and $-\pi/2$ respectively. On the x-y plane, find the angle with the x axis along which the principal lobe has the maximum value.

***10.23 A uniform linear array consists of 10 isotropic radiators. The spacing between consecutive radiators is $\lambda_0/2$. Calculate the width of the principal lobe (a) if the currents in all the radiators are in phase and (b) the currents progressively differ in phase by $-\pi$ radians. Interpret the results from the point of view of the directional properties.

**10.24 A linear antenna of length 40 cm is on the y-z plane and it is at an angle of 30° with the z axis. At the location of the antenna, the electric field intensity of a uniform plane wave with parallel polarisation is 12.0 mV/m. The y-z plane is the plane of incidence, and the direction of wave travel is 45° measured from the y axis. Find the amplitude of voltage induced in the antenna.

**10.25 If the electric field of problem 10.24 is at an angle of 25° with the plane of incidence, find the amplitude of induced voltage in the antenna.

**10.26 A circular coil of 10 turns and 15.0 cm radius has its axis at 30° with the y axis in free space. The plane containing the loop and the y axis is the plane of incidence of a perpendicularly polarised electromagnetic wave, and the direction of wave travel is 105° measured from the y axis. Amplitude of electric field intensity at the location of the coil is 0.2 V/m at 20 MHz. Find amplitude of voltage induced in the coil.

**10.27 The maximum effective aperture of a receiving antenna appropriately oriented with respect to wave polarisation and connected to a matched load is given by

$$A_e = \frac{\lambda_0^2}{4\pi} g_d$$

where g_d is directivity. Derive the formula.

*10.28 Find the maximum effective area of a linear antenna of 2 m length at 75 MHz.

**10.29 The Friis transmission formula in terms of the directivities of the receiving and transmitting antennas is given by

$$\frac{P_r}{P_t} = \frac{\lambda_0^{\ 2} g_{dt} g_{dr}}{(4\pi r)^2}$$

Show that it is equal to $(A_{et}A_{er})/(r^2\lambda_0^{\ 2})$ where A_{et} and A_{er} are effective apertures.

*10.30 A half-wave transmitting antenna and a receiving antenna of small length are parallel to each other, and they are perpendicular to the line joining their mid-points. The distance between the antennas is 200 m in free space. Find the ratio of received to transmitted power at 50 MHz.

**10.31 A small dipole antenna is oriented with respect to a transmitting half-wave dipole to receive maximum power. If current in the transmitting antenna is 10.0 A (rms) at 60 MHz and the distance between the antennas is 10 km, find the power received. The antennas are located in air.

APPENDIX A

MUTUAL INDUCTANCE (NEUMANN'S FORM)

In Section 4.18 we established that the mutual inductance between two magnetically coupled coils 1 and 2, M_{12}, and between coils 2 and 1, M_{21}, are equal. The fact that the two mutual inductances are equal can also be proved as follows. Let us consider two filamentary loops 1 and 2 carrying currents I_1 and I_2, as shown in Fig. A.1. The vector magnetic potential at point P within the surface enclosed by Loop 2 due to a current element $I_1 d\ell_1$ at a point P_1 of Loop 1 is

$$d\mathbf{A}_{21} = \frac{\mu I_1}{4\pi R} d\ell_1 \tag{1}$$

The vector potential at P due to the total current in Loop 1 is found by integrating (1) around loop 1. Thus,

$$\mathbf{A}_{21} = \oint_{C_1} \frac{\mu I_1}{4\pi R} d\ell_1 \tag{2}$$

Flux density at P is given by

$$\mathbf{B}_{21} = \nabla \times \mathbf{A}_{21}$$

Flux crossing a differential area $d\mathbf{S}_2$ at P is

$$d\Phi_{21} = \mathbf{B}_{21} \cdot d\mathbf{S}_2$$

Therefore, flux linking Loop 2 due to current in Loop 1 is

$$\Phi_{21} = \int_{S_2} \mathbf{B}_{21} \cdot d\mathbf{S}_2 = \int_{S_2} (\nabla \times \mathbf{A}_{21}) \cdot d\mathbf{S}_2 \tag{3}$$

Now we transform the surface integral of (3) to a line integral by the application of Stoke's theorem. Accordingly,

$$\Phi_{21} = \oint_{C_2} \mathrm{A}_{21} \cdot d\ell_2 \tag{4}$$

Next, we substitute (2) into (4), and also replace R by R_{12} as the line integral is around Loop 2.

Thus,

$$\Phi_{21} = \oint_{C_2} \oint_{C_1} \frac{\mu I_1}{4\pi R_{12}} (d\ell_1 \cdot d\ell_2) \tag{5}$$

By dividing (5) by I_1 ($N_2 = 1$) we get the mutual inductance

$$M_{21} = \oint_{C_2} \oint_{C_1} \frac{\mu}{4\pi R_{12}} (d\ell_1 \cdot d\ell_2) \tag{6}$$

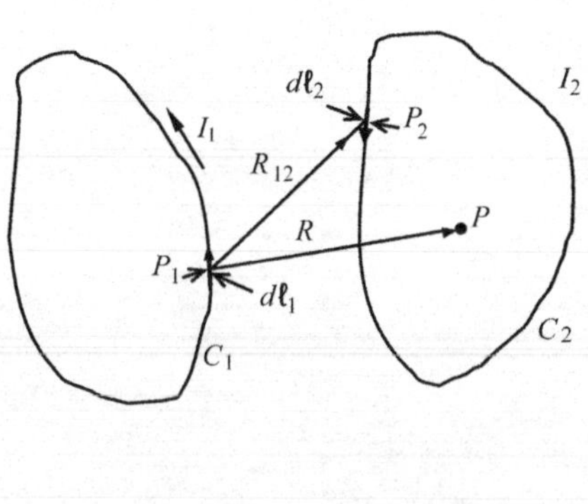

Fig. A.1 Two current loops

Similarly, it can be shown that

$$M_{12} = \oint_{C_1} \oint_{C_2} \frac{\mu}{4\pi R_{12}} (d\boldsymbol{\ell}_2 \cdot d\boldsymbol{\ell}_1) \tag{7}$$

Since $R_{21} = R_{12}$, and as the order of the line integrals and dot products of differential lengths are interchangeable, the right-hand sides of (6) and (7) are equal. Therefore,

$$M_{21} = M_{12}$$

The expression for mutual inductance given by (6) or (7) is called *Neumann's form* for the mutual inductance between two magnetically coupled loops.

APPENDIX B

INTERNAL IMPEDANCE OF ROUND CONDUCTORS

We found in Section 6.21 that current distribution over the cross section of a conductor carrying alternating current is not uniform. Because of non-uniform current distribution, the resistance and internal inductance of conductors are different from their dc values. We will at first calculate the internal impedance of a round conductor with high-frequency approximations, and then obtain a general formula for the internal impedance. The internal impedance per unit length of a good conductor, by definition, is equal to the ratio between the electric field intensity on the conductor surface and the total current carried by the conductor.

B.1 Internal Impedance of a Round Conductor at High Frequency

To calculate the resistance and internal inductance of a round conductor of radius a at high frequency, we will use Equation (2) and Expression (9) of Section 6.21. These, with some modification, are given below.

$$\frac{d\bar{E}_z(r)}{dr} = j\omega\mu\bar{H}_\phi(r) \tag{1}$$

$$\bar{E}_z(r) = \bar{E}_z(a)\frac{J_0(\tau r)}{J_0(\tau a)} \tag{2}$$

where
$\bar{E}_z(r)$ = electric field intensity at radial distance r from the centre,

$\bar{E}_z(a)$ = electric field intensity on the conductor surface,

$\bar{H}_\phi(r)$ = magnetic field intensity at radial distance r,

J_0 = Bessel function of order zero, and

$$\tau^2 = -j\omega\mu\sigma \tag{3}$$

The internal impedance per unit length $\bar{Z}_i$ of the conductor is equal to the ratio of electric field intensity on the surface of the conductor $\bar{E}_z(a)$ and current $\bar{I}$ carried by it. The current carried by the conductor is related to the magnetic field intensity on its surface $\bar{H}_\phi(a)$ through Ampere's magnetic circuital law. Accordingly,

$$I = 2\pi a\bar{H}_\phi(a)$$

The internal impedance,

$$\bar{Z}_i = \frac{\bar{E}_z(a)}{\bar{I}} = \frac{\bar{E}_z(a)}{2\pi a\bar{H}_\phi(a)} \tag{4}$$

The next step is to use Relation (1) and (2) to obtain $\bar{H}_\phi(a)$ in terms of $\bar{E}_z(a)$. From (1),

$$H_\varphi(r) = \frac{1}{j\omega\mu}\frac{\partial \bar{E}_z(r)}{\partial r} \tag{5}$$

Now,

$$\frac{\partial E_z(r)}{\partial r} = \tau\frac{\partial E_z(r)}{\partial(\tau r)} \tag{6}$$

Differentiating (2) with respect to τr and then substituting the result into (6), we get

$$\frac{\partial \bar{E}_z(r)}{\partial r} = \tau \bar{E}_z(a)\frac{J_0'(\tau r)}{J_0(\tau a)} \tag{7}$$

where $$J_0'(\tau r) = \frac{d[J_0(\tau r)]}{d(\tau r)}$$

We first substitute Relation (7) in (5) and then use a in place of r to obtain

$$\bar{H}_\phi(a) = \frac{\tau \bar{E}_z(a)}{j\omega\mu}\frac{J_0'(\tau a)}{J_0(\tau a)} \tag{8}$$

Substitution of (8) into (4) gives,

$$\bar{Z}_i = \frac{j\omega\mu}{2\pi a\tau}\frac{J_0(\tau a)}{J_0'(\tau a)} \tag{9}$$

It can be shown that $J_0'(\tau a) = -J_1(\tau a)$ where $J_1(\tau a)$ is Bessel function of order one. Hence,

$$\bar{Z}_i = -\frac{j\omega\mu}{2\pi a\tau}\frac{J_0(\tau a)}{J_1(\tau a)} \tag{10}$$

At high frequencies, the value of τ would be very large. Hence, we can use the asymptotic forms of $J_0(\tau a)$ and $J_1(\tau a)$ in (10). The asymptotic forms are given by

$$J_0(\tau a) \approx \sqrt{\frac{2}{\pi\tau a}}\, e^{j\left(\tau a - \frac{\pi}{4}\right)} \quad \text{(High-frequency approximation)}$$

$$J_1(\tau a) \approx \sqrt{\frac{2}{\pi\tau a}}\, e^{j\left(\tau a - \frac{\pi}{4} - \frac{\pi}{2}\right)} \quad \text{(High-frequency approximation)}$$

By substituting the above relations in (10), we get

$$\bar{Z}_i = \frac{\omega\mu}{2\pi a\tau} \tag{11}$$

By multiplying the numerator and the denominator by τ and then using (3), we obtain

$$\bar{Z}_i = \frac{j\tau}{2\pi a\sigma} \tag{12}$$

Now from (3), $$\tau = \sqrt{-j\omega\mu\sigma} = \frac{1-j}{\delta} \tag{13}$$

where $$\delta = \frac{1}{\sqrt{\pi f \mu\sigma}}$$

Using (13) in (12),

$$Z_i = \frac{(1+j)}{2\pi a\delta\sigma} \tag{14}$$

The real part of the internal impedance given by (14) is the approximate resistance per unit length of the conductor at high frequency. The resistance is termed as *skin-effect resistance* or *surface resistance* per unit length. The imaginary part of $\bar{Z}_i$ gives the approximate formula for the internal inductive reactance per unit length. Thus, denoting the resistance by R, and the reactance by ωL_i where L_i is *internal inductance* per unit length, we have

$$R = \frac{1}{2\pi a\delta\sigma}\ \Omega/\text{m} \tag{15}$$

$$\omega L_i = \frac{1}{2\pi a\delta\sigma}\ \Omega/\text{m} \tag{16}$$

As the reactance is equal the resistance, the internal-impedance angle is 45°.

We know that the dc resistance per unit length of a conductor with uniform cross-sectional area S and conductivity σ is equal to $1/(\sigma S)$. Thus, we find from (15) that the skin-effect resistance of a round conductor is equivalent to the dc resistance of a flat conductor of width equal to the circumference of the round conductor and thickness equal to δ. The parameter δ is called *depth of penetration* or *skin depth*.

B.2 Internal Impedance of a Round Conductor at any Frequency

To calculate the internal impedance at any frequency, it is convenient to use the solution for current density given by (11*a*) of Section 6.21. We multiply it by σ to get the expression for electric field intensity $\bar{E}_z(r)$, and then follow the procedure outlined in B.1 to obtain current I in terms of the electric field intensity $\bar{E}_z(a)$ on the conductor surface. The ratio between $\bar{E}_z(a)$ and I will give the internal impedance at any frequency.

The electric field intensity,

$$\bar{E}_z = \bar{E}_z(a)\frac{\text{ber}(kr) + j\text{bei}(kr)}{\text{ber}(ka) + j\text{bei}(ka)};\ \bar{E}_z(a) = \sigma\bar{J}_z(a) \tag{1}$$

Current carried by the conductor is given by

$$\bar{I} = \frac{2\pi ak\bar{E}_z(a)}{j\omega\mu}\frac{\text{ber}'(ka) + j\text{bei}'(ka)}{\text{ber}(ka) + j\text{bei}(ka)} \tag{2}$$

where

$$\text{ber}(kr) = 1 - \frac{(kr/2)^4}{(2!)^2} + \frac{(kr/2)^8}{(4!)^2} - \cdots$$

$$\text{bei}(kr) = \frac{(kr/2)^2}{(1!)^2} - \frac{(kr/2)^6}{(3!)^2} + \frac{(kr/2)^{10}}{(5!)^2} - \cdots$$

$$k = \sqrt{\omega\mu\sigma} = \sqrt{2}/\delta$$

$$\text{ber}'(kr) = \frac{d}{d(kr)}[\text{ber}(kr)]$$

$$\text{bei}'(kr) = \frac{d}{d(kr)}[\text{bei}(kr)]$$

Dividing (1) by (2), we get the internal impedance. Thus,

$$Z_i = \frac{E(z)}{I} = \frac{j\omega\mu}{2\pi ak}\frac{\text{ber}(ka) + j\ [illegible]\ ka)}{\text{ber}'(ka) + j\ [illegible]\ ka)} \tag{3}$$

After rationalising (3), the real and imaginary parts will give the skin-effect resistance per unit length and the internal reactance per unit length of the conductor. These are given by

$$R = \frac{\omega\mu}{2\pi ak} \frac{\text{ber}(ka)\text{bei}'(ka) - \text{bei}(ka)\text{ber}'(ka)}{[\text{ber}'(ka)]^2 + [\text{bei}'(ka)]^2} \ \Omega/\text{m} \tag{4}$$

$$\omega L_i = \frac{\omega\mu}{2\pi ak} \frac{\text{ber}(ka)\text{ber}'(ka) + \text{bei}(ka)\text{bei}'(ka)}{[\text{ber}'(ka)]^2 + [\text{bei}'(ka)]^2} \ \Omega/\text{m} \tag{5}$$

APPENDIX C

THE SMITH CHART

The reflection coefficient defined in Section 8.5 is a function of load impedance and characteristic impedance of the line. In general, the load impedance is a complex quantity. Hence, the reflection coefficient is also a complex function even if the transmission line is lossless. By expressing the load impedance in terms of the reflection coefficient and then equating the real and imaginary parts of one side to those of the other side, two separate equations are obtained. These equations represent circles with the real and imaginary parts of reflection coefficient as variables, and normalized resistance and normalized reactance as parameters. The circle diagram and additional information given on it constitute the Smith chart. The chart provides an easy means to solve simple transmission line problems with a fair degree of accuracy.

Mathematical Basis and Description of the Chart

Let us consider a loss-free transmission line of length ℓ and characteristic impedance Z_0. The line is connected to a load at the receiving end and to a generator at the input end. The reflection coefficient *at the load end* is given by

$$\bar{\Gamma} = \frac{\bar{Z}_R - Z_0}{\bar{Z}_R + Z_0} = \frac{(R_R \pm jX_R) - Z_0}{(R_R \pm jX_R) + Z_0} \tag{1}$$

where $\bar{Z}_R = R_R \pm jX_R$ is the complex load impedance. The plus sign before jX_R is for inductive loads and the minus sign is for capacitive loads. By dividing the numerator and the denominator of the right side of (1) by Z_0, we get

$$\bar{\Gamma} = \frac{\bar{Z}_R/Z_0 - 1}{\bar{Z}_R/Z_0 + 1} = \frac{(R_R \pm jX_R)/Z_0 - 1}{(R_R \pm jX_R)/Z_0 + 1}$$

The ratios $\bar{Z}_R/Z_0$, R_R/Z_0 and X_R/Z_0 are the *normalised load impedance, normalised load resistance*, and *normalised load reactance* respectively. Denoting the normalised load impedance by $\bar{z}_R$, the normalised load resistance by r, and the normalised load reactance x, we have

$$\bar{z}_R = \frac{\bar{Z}_R}{Z_0}, \; r = \frac{R_R}{Z_0}, \; x = \frac{X_R}{Z_0}$$

Thus, the reflection coefficient in terms of the normalised parameters is given by

$$\bar{\Gamma} = \frac{\bar{z}_R - 1}{\bar{z}_R + 1} = \frac{(r \pm jx) - 1}{(r \pm jx) + 1} \tag{2}$$

Solving (2) for the normalised load impedance, we get

$$\bar{z}_R = r \pm jx = \frac{1 + \bar{\Gamma}}{1 - \bar{\Gamma}} \tag{3}$$

The phasor form of reflection coefficient in terms of its real part Γ_a and imaginary part Γ_i is

$$\Gamma \quad \Gamma_a + i\Gamma_i \tag{4}$$

Substitution of (4) into (3) gives,

$$r \pm jx = \frac{(1+\Gamma_a)+j\Gamma_i}{(1-\Gamma_a)-j\Gamma_i} = \frac{(1-\Gamma_a^2-\Gamma_i^2)+j2\Gamma_i}{(1-\Gamma_a)^2+\Gamma_i^2} \tag{5}$$

By equating the real parts of the two sides of (5), and then rearranging the terms we have

$$\Gamma_a^2(1+r) - 2r\Gamma_a + \Gamma_i^2(1+r) = 1-r \tag{6}$$

Dividing both sides (6) by 1+*r*, we obtain the following relations.

$$\Gamma_a^2 - \frac{2r\Gamma_a}{1+r} + \Gamma_i^2 = \frac{1-r}{1+r}$$

Or
$$\left(\Gamma_a - \frac{r}{1+r}\right)^2 + \Gamma_i^2 = \left(\frac{1}{1+r}\right)^2 \tag{7}$$

Equation (7) represents a family of circles in the complex plane with Γ_a along positive real axis and Γ_i along positive imaginary axis as variables, and r as the parameter. The radius of a circle is equal to $1/(1+r)$, and its centre is at $[r/(1+r), 0]$. The radius for $r = 0$ is unity and the centre of the circle is at the origin. The radius for $r = \infty$ is zero and its centre is at (1, 0). Thus, all circles for values of r between 0 and ∞ lie within the unit circle, and the centres of the circles are on the positive real axis. For any value of r, the sum of radius of the circle and distance of its centre from the origin is equal to 1. Thus, all the circles pass through the point (1, 0) in the complex plane. A few constant r circles are shown in Fig. C.1. The origin of the complex plane is at *O*.

At $r = 0$, the reflection coefficient, $\bar{\Gamma} = \dfrac{(\pm jx)-1}{(\pm jx)+1}$

The magnitude of reflection coefficient, $|\bar{\Gamma}| = \dfrac{|\pm jx - 1|}{|\pm jx + 1|} = 1$

Thus, the unit circle for $r = 0$ also represents unity magnitude of the reflection coefficient.

Next, by equating the imaginary parts of both sides of (5) we get

$$(\Gamma_a - 1)^2 + \Gamma_i^2 \mp \frac{2\Gamma_i}{x} = 0$$

Or
$$(\Gamma_a - 1)^2 + \left(\Gamma_i \mp \frac{1}{x}\right)^2 = \left(\frac{1}{x}\right)^2 \tag{8}$$

Equation (8) also represents a family of circles with Γ_a along positive real axis and Γ_i along positive imaginary axis as variables, and x as the parameter. The radius of a circle is equal to $1/x$, and its centre is located at $(1, \pm 1/x)$. For $x = 0$, the radius of the circle is infinite and its centre is at $(1, \pm\infty)$. The circle of infinite radius is the real axis. For $x = \infty$, the radius of the circle is zero and its centre is located at (1,0), as indicated in the figure. The circles for positive values of x (inductive load) are above the real axis. The circles for negative values of x (capacitive load) lie below the real axis. Parts of the circles for certain values of x and lying within the unit circle are illustrated in Fig. C.1.

The phasor form of reflection coefficient at a distance $z = -z'$ from the load end is given by

$$\Gamma \quad \Gamma e^{-j2\beta z'} = \Gamma e^{j(\quad)}$$ (Refer to Section 8.5)

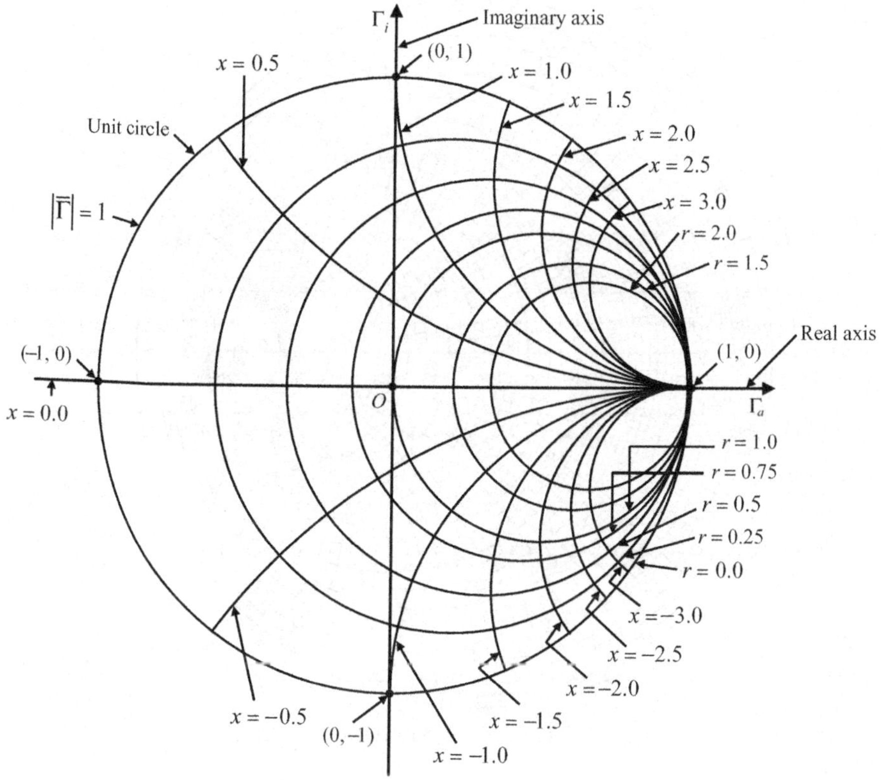

Fig. C.1 Description of a partial Smith chart

where $|\bar{\Gamma}|$ and ψ_r are magnitude and phase angle of reflection coefficient at the load end. The above relation shows that the magnitude of reflection coefficient remains constant throughout a lossless line. However, its phase angle changes. The reflection coefficient at some other point $z = -z''$ is

$$\bar{\Gamma}(-z'') = |\bar{\Gamma}| e^{j(\psi_r - 2\beta z'')} = |\bar{\Gamma}| e^{j(\psi_r - 2\beta z')} e^{-j2\beta(z''-z')} = \bar{\Gamma}(-z') e^{-j2\beta(z''-z')}$$

So, the change in phase angle of reflection coefficient between the two positions is equal to $-2\beta(z'' - z')$. If z'' is greater than z', the shift in the position along the line is towards the generator end. The change of phase angle is in the clockwise direction. And if $z'' - z'$ is equal to one-half of a wavelength ($\lambda/2$), the change in phase angle is 2π radians. That is, the reflection coefficient phasor goes through one revolution in the clockwise direction. If z'' is less than z', the shift in position along the line is towards the load. If the shift is equal to one-half of a wavelength, the reflection coefficient phasor goes through one revolution in the counter-clockwise direction.

Figure C.1 is a partial Smith Chart. In a more complete Smith chart, shown in Fig. C. 2, constant r circles and constant x circles are spaced very closely. The chart also has three circular scales. One scale is marked with degrees. This scale is useful to find the phase angle of reflection coefficient. The other two scales are marked in fractions of a wavelength; one scale gives the distance in terms of wavelength from the load towards the source (generator) and the other gives the distance from the source towards the load. It may be mentioned that location of the zero point on the wavelength scale is arbitrary. That is, the point $\lambda = 0$ does not indicate the load end. Distance (in terms of wave length) between two points along a transmission line can be calculated using the scale.

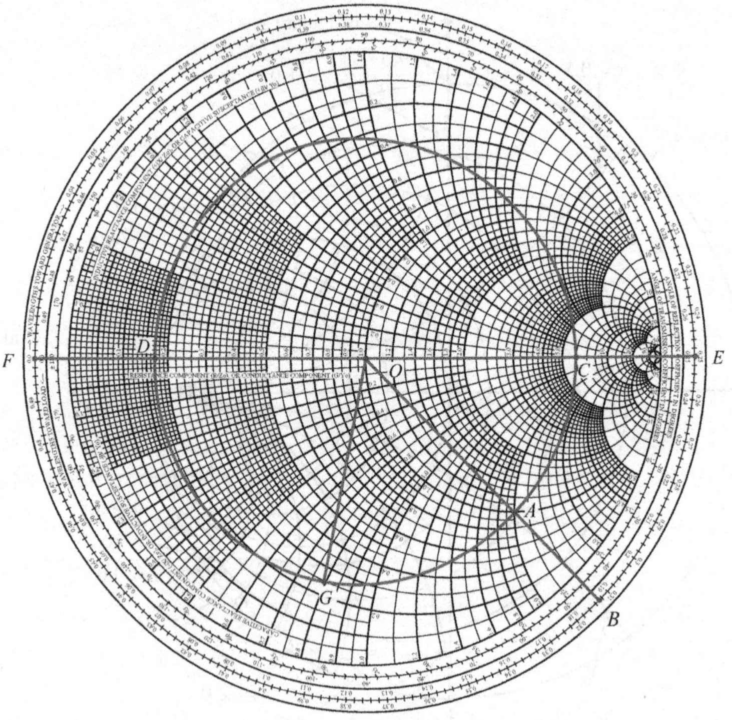

Fig. C.2 A Smith chart

Solution of Some Transmission Line Problems

To understand the procedure for calculating some of the quantities related to reflection phenomenon in a lossless transmission line using the Smith chart, let us consider a line of characteristic impedance 150 Ω connected to a load of 150 – *j*300 Ω. The length of the line is 116 m and its phase constant is 0.6 rad/m. The amplitude of the positively travelling voltage wave is 10.0 + *j*0 V.

We have from the given data,

$$Z_0 = 150\ \Omega,\ \overline{Z}_R = 150 - J300\ \Omega,\ \ell = 116\text{ m},\ \beta = 0.6\text{ rad/m},\ V_0^+ = 10.0\text{ V}$$

Normalised load impedance, $\overline{z}_R = \dfrac{150 - J300}{150} = 1.0 - j2.0$

Normalised load resistance, $r = 1.0$

Normalised load reactance, $x = -2.0$

Wavelength, $\lambda = \dfrac{2\pi}{\beta} = \dfrac{2\pi}{0.6} = 10.472$ m

The following quantities are calculated using the Smith chart.

Reflection Coefficient The load point at $r = 1.0$ and $x = -2.0$ is marked as A in Fig. C.2. The length and position of *OA* give magnitude and phase angle of the reflection coefficient at the load end. The magnitude of reflection coefficient is equal to the length of *OA* divided by radius of the unit circle. The phase angle of the reflection coefficient is equal to the angle subtended by *OA* with the real axis. The magnitude and phase angle of the reflection coefficient at the load end are found to be 0.708 and $-45°$ respectively. The circumference of the circle drawn with *OA* as radius is the locus of the reflection coefficient phasor. This is the constant Γ circle. Extension of *OA* up to the wavelength scale marked as *towards the generator* will intersect the scale at *B*. Point *B* gives the load position. The reading at B is 0.312. Thus, the position of the load is at 0.312λ.

Standing Wave Ratio

$$\text{The standing wave ratio, } S = \frac{1+|\bar{\Gamma}|}{1-|\bar{\Gamma}|} = \frac{1+0.708}{1-0.708} = 5.849$$

Since *S* has a constant value, the constant $|\bar{\Gamma}|$ circle also represents the constant *S* circle. The scales are however different.

Maximum and Minimum Amplitudes of the Total Voltage and Related Quantities The maximum amplitude of total (incident plus reflected) voltage,

$$V_{max} = V_0^+\left(1+|\bar{\Gamma}|\right) = 10.0(1+0.708) = 17.08 \text{ V}$$

The minimum amplitude of total voltage,

$$V_{min} = V_0^+\left(1-|\bar{\Gamma}|\right) = 10.0(1-0.708) = 2.92 = 2.92 \text{ V}$$

The line impedances at the positions where V_{max} and V_{min} occur are purely resistive. Therefore, the points of intersection of the constant $|\bar{\Gamma}|$ circle with the real axis (*x* is zero along the real axis) give these impedances. The points are marked in the chart as *C* and *D*. Since the impedance at V_{max} is higher than at V_{min}, maximum amplitude of the total voltage occurs at *C* and minimum amplitude of total voltage occurs at *D*. The phase angle of reflection coefficient at V_{max} is zero because *OC* is along the positive real axis, and the phase angle at V_{min} is 180° because *OD* is along the negative real axis. The value of *r* at *C* gives the maximum normalised line impedance and the value of *r* at *D* gives the minimum normalised line impedance. The normalised values of the impedances are found to be 5.9 and 0.17. Hence, the maximum and minimum values of the line impedance are

$$Z_{max} = 5.9Z_0 = (5.9)(150) = 885.0\ \Omega$$

$$Z_{min} = 0.17Z_0 = (0.17)(150) = 25.5\ \Omega$$

The distance of the point of occurrence of V_{max} from the load end is obtained by extending the line *OC* up to the circle that gives the wavelength scale *towards the generator*. The point is shown as *E* in the chart. The arc length from the load point *B* to point *E* in the clockwise direction (towards the generator) multiplied by λ gives the distance of the V_{max} point. The circumference of the circle is 0.5λ and the arc length *EB* in the clockwise direction is equal to $(0.312 - 0.25)\lambda$. Accordingly, denoting the distance by d_{max}, we have

$$d_{max} = 0.5\lambda - (0.312 - 0.25)\lambda = 0.438\lambda = (0.438)(10.472) = 4.587 \text{ m}$$

The distance from the load end of the point at which V_{min} occurs is found by extending the line *OD* up to the circle that gives the wavelength scale *towards the generator*. The point is shown as *F* in the chart. The arc length from *B* to *F* in the clockwise direction multiplied by λ gives the distance. Thus denoting the distance by d_{min}, we have

$$d_{min} = (\ [illegible].5 - 0.[illegible]\ )\lambda = 0.[illegible] \quad (0.[illegible])([illegible].472) \quad [illegible].969 \text{ m}$$

Impedance and Reflection Coefficient at the Input End The normalised load impedance, given by 3, in terms of the reflection coefficient at the load end is

$$\bar{z}_R = \frac{1+\bar{\Gamma}}{1-\bar{\Gamma}} \tag{9}$$

The formula for normalised input impedance,

$$\bar{z}_{in} = \frac{1+\bar{\Gamma}e^{-j2\beta\ell}}{1-\bar{\Gamma}e^{-j2\beta\ell}} \quad \text{(See Section 8.7)} \tag{10}$$

Comparison between (9) and (10) shows that the phasor OA has to be rotated by an angle equal to $2\beta\ell$ in the clockwise direction to locate the input position. As $2\beta\ell = (2)(0.6)(116) = 139.2$ rad $= 7975.6° = (22 \times 360)° + 55.6°$, OA has to be rotated by an angle of 55.6° in clockwise direction. This position is shown in the figure as OG. At G, $r = 0.29$ and $x = -0.82$. Hence the input impedance,

$$\bar{Z}_{in} = (150)(0.29 - j0.83) = 43.5 - j123.0\ \Omega$$

The angle of OG with the positive real axis is $-100.6°$. The reflection coefficient at the input end,

$$\bar{\Gamma}_{in} = 0.707\angle -100.6°$$

The line impedance and reflection coefficient at any position $z = -z'$ is found by rotating OA through an angle equal to $-2\beta z'$ radians.

APPENDIX D

MATERIAL CONSTANTS

Values of conductivity, relative permittivity, dielectric strength, and relative permeability of a few materials are given respectively in Table D.1, Table D.2, Table D.3, and Table D.4. The relative permittivities of dielectric materials are functions of frequency, and their dielectric strengths depend on temperature and aging. The conductivity of a material depends on temperature. The permeabilities of ferromagnetic materials depend on flux density and magnetic history. The permeabilities of ferrites depend on frequency; it increases with the increase in frequency but not proportionately. The values of material constants listed may be considered as representative at normal temperature and at medium frequency.

Table D.1

Conductivities

Material	σ(S/m)	Material	σ(S/m)
Aluminium	3.8×10^{7}	Mica	1×10^{-13}
Bakelite	1×10^{-9}	Mineral oil	1×10^{-14}
Brass	1.6×10^{7}	Nichrome	1×10^{6}
Carbon steel	6×10^{6}	Nickel	1.5×10^{7}
Cobalt	1.7×10^{7}	Phosphor bronze	1×10^{7}
Copper	5.8×10^{7}	Porcelain	1×10^{-10}
Earth (dry)	1×10^{-5}	Silver	6.1×10^{7}
Ferrite	100	Solder	7×10^{6}
Glass	1×10^{-12}	Stainless steel	1.1×10^{6}
Gold	4.1×10^{7}	Tin	8.7×10^{6}
Graphite	1×10^{5}	Tungsten	1.8×10^{7}
Intrinsic germanium	2.2	Water (distilled)	2×10^{-4}
Intrinsic silicon	1.6×10^{-3}	Water (fresh)	1×10^{-3}
Iron	1×10^{7}	Water (sea)	4
Lead	4.8×10^{6}	Zinc	1.7×10^{7}
Mercury	1×10^{6}		

Table D.2

Relative Permittivities

Material	μ_r	Material	μ_r
Air	1.0006	Polyvinyl chloride (PVC)	3.4
Amber	2.7	Polyvinyl	3
Bakelite	4.3	Porcelain	6.0
Ebonite	2.7	Quartz	5.0
Epoxy resin	3.7	Rubber (hard)	3.0
Germanium	16	Silica (SiO_2)	3.8
Glass	5.5	Silicon	11.8
Mica	5.7	Sodium chloride	5.9
Mineral oil	2.2	Soil (dry)	2.8
Nylon	4.5	Teflon	2.2
Paper (dry)	2	Water (distilled)	80
Paper (impregnated)	3	Water (sea)	80
Polyethylene	2.3	Wood (dry)	4
Polystyrene	2.6		

Table D.3

Dielectric Strength

Material	Dielectric strength (MV/m)	Material	Dielectric strength(MV/m)
Air at NPT	3	Paper (dry)	4 – 10
Amber	90	Polyethylene	19 – 22
Bakelite	20 – 25	Polystyrene	20
Ebonite	10 – 40	Porcelain	9 – 20
Glass (plate)	5 – 12	Rubber (hard)	10 – 25
Impregnated paper	15	Quartz	35
Mica	150 – 200		
Mineral oil	10 – 15	Teflon	60

Table D.4

Relative Permeability

Material	μ_r
Diamagnetic	
Bismuth	0.99983
Copper	0.999994
Germanium	0.999992
Gold	0.999964
Graphite	0.99988
Lead	0.999983
Silver	0.99998
Paramagnetic	
Aluminium	1.000022
Ferric oxide	1.0014
Ferrous chloride	1.0037
Ferrous sulphate	1.0028
Nickel sulphate	1.0012
Ferromagnetic	
Cast iron	350 at $B = 0.2$ T 300 at $B = 0.4$ T
Cast steel	1400 at $B = 0.8$ T 1100 at $B = 1.0$ T
Cobalt	250
Ferrite (Nickel Zinc)	600 at $f = 1.0$ MHz
Iron (0.2 % impurity)	5000
Mild steel	1220 at $B = 1.0$ T 950 at $B = 1.2$ T
Nickel (1 % impurity)	600
Silicon steel	4600 at $B = 1.0$ T 3100 at $B = 1.2$ T

APPENDIX E

LIST OF SYMBOLS

Symbols used in the text have been defined wherever introduced. However, the following list will serve as a reference. Vector quantities are denoted by straight bold letters, and their magnitudes are denoted by italic symbols or by the use of the modulus sign. Scalar quantities are denoted by italic letters. Moreover, phasor quantities are designated with a bar ($\bar{}$) at the top of the symbol. Because of large number of quantities involved, some letters are used to represent more than one quantity. For example, the letter r denotes the radial direction in cylindrical coordinates as well as in spherical coordinates. Some authors use the Greek letter ρ for the radial direction in cylindrical coordinates. In the text, ρ represents the volume charge density. Though in general S is used to designate the surface area, the plate area of a parallel plate capacitor is denoted by A. Otherwise, the capacitance formula would contain the upper case S for area and the lower case s for spacing between the plates, which may create confusion. The first letter of a unit that is named after a scientist is of lower case; however its abbreviation is an uppercase letter. For example, the unit of charge is 'coulomb' (named after Charles Augustus de Coulomb) whereas its abbreviation is 'C'.

Table E.1

Main Symbols

Symbol	Quantity	Unit (Abbreviation)
A	Vector magnetic potential	tesla-metre (T-m)
A	Area	m^2
B	Magnetic flux density	tesla (T)
C	Capacitance	farad (F)
D	Electric flux density	coulomb/m^2 (C/m^2)
E	Electric field intensity	volt/m (V/m)
	Electromotive force	volt (V)
F	Force	newton (N)
f	Frequency	hertz (Hz)
g	Gain	
G	Conductance	siemen (S)
G	Shunt conductance per unit length of transmission line	siemen/m (S/m)
H	Magnetic field intensity	ampere/m (A/m)
I	Current	ampere (A)
J	Current density	ampere/m^2 (A/m^2)
K	Current per unit length of a current sheet	ampere/m (A/m)
L	Inductance	henry (H)
ℓ	Length	metre (m)

(Contd.)

Table E.1 (*Contd.*)

M	Magnetisation	ampere/m (A/m)
m	Magnetic dipole moment	ampere-m^2 (A-m^2)
m	Mass	kilogram (kg)
P	Polarisation (of dielectric)	coulomb/m^2 (C/m^2)
P	Power	watt (W)
$\wp$	Permeance	henry (H)
Q	Electric charge	coulomb (C)
R	Displacement	metre (m)
R, R_Ω	Ohmic resistance	ohm (Ω)
R	Series resistance per unit length of transmission line	ohm (Ω)
R_{rad}	Radiation resistance	ohm (Ω)
R_I	Radiation intensity	watts/steradian (W/steradian)
S	Surface area	m^2
s	Spacing	m
T	Torque	Newton-m (N-m)
t	Time	second (s)
V	Potential	volt (V)
v	Velocity (vector)	m/s
v	Volume	m^3
W	Work, Energy	joule (J)
X	Reactance	ohm (Ω)
Z	Impedance	ohm (Ω)
Z	Series impedance per unit length of transmission line	ohm (Ω)
Z_0	Characteristic impedance	ohm (Ω)
α	Angle	degree or radian (°) or (rad)
α (Alpha)	Attenuation constant	neper/m (Np/m)
β (Beta)	Phase constant	radian/m(rad/m)
Γ (Gamma)	Reflection coefficient	
γ (Gamma)	Propagation constant	m^{-1}
δ (Delta)	Skin depth	m
ε (Epsilon)	Permittivity	farad/m (F/m)
η (Eta)	Intrinsic (wave) impedance	ohm (Ω)
λ (Lambda)	Wavelength	m
μ (Mu)	Permeability	henry/m (H/m)
v (Nu)	Velocity (magnitude)	m/s
ρ (Rho)	Volume charge density	coulomb/m^3 (C/m^3)
ρ_ℓ	Linear charge density	coulomb/m (C/m)
ρ_S	Surface charge density	coulomb/m^2 (C/m^2)
Ψ (Psi)	Electric flux	coulomb (C)

(*Contd.*)

Table E.1 (*Contd.*)

σ (Sigma)	Conductivity	siemen/m (S/m)
τ (Tau)	Transmission coefficient	
Φ (Phi)	Magnetic flux	weber (Wb)
Ω (Omega)	Abbreviation for *ohm*, the unit of resistance	
Coordinate Directions (Variables)		
x, y, z	Rectangular coordinates	
r, ϕ, z	Cylindrical coordinates	
r, θ, ϕ	Spherical coordinates	

Table E.2

Meanings of Prefixes Used with Units

Prefix	Abbreviation	Unit to be multiplied by
pico	p	10^{-12}
nano	n	10^{-9}
micro	μ	10^{-6}
milli	m	10^{-3}
centi	c	10^{-2}
kilo	k	10^{3}
mega	M	10^{6}
giga	G	10^{9}

APPENDIX F

REFERENCES

1. Bewley, L. W.: "Two-Dimensional Fields in Electrical Engineering," Dover Publications, INC., 1963.
2. Binns, K. J., and Lawerenson, P. J.: "Analysis and Computation of Electric and Magnetic Field Problems," The Macmillan Company, 1963.
3. Dekker, A. J.: "Electrical Engineering Materials," Prentice-Hall of India Limited, 1992.
4. Edminister, J. A.: "Electromagnetics," Schaum's outlines, Second Edition, Tata McGraw-Hill Book Company Limited, Special Indian Edition, 2006
5. Edminister, J. A.: "Theory and Problems of Electromagnetics," Schaum's Outline Series, McGraw-Hill Book Company, Asian Student Edition, First impression March 1986.
6. Golding, E. W.: "Electrical Measurements and Measuring Instruments," Fourth Edition, Sir Isaac Pitman & Sons, 1954.
7. Hague, B.: "The Principles of Electromagnetism Applied to Electrical Machines," Dover Publications, INC., 1962.
8. Hayt, W. H. Jr.: "Engineering Electromagnetics," Fourth Edition, McGraw-Hill Publishing Company Limited, 1981.
9. Jackson, J. D.: "Classical Electrodynamics," Second Edition, Wiley Eastern Limited, 1975.
10. Johnk, C. T. A.: "Engineering Electromagnetic Fields and Waves," John Wiley and Sons, 1988.
11. Jordan, E. C., and Balmain, K. G.: "Electromagnetic Waves and Radiating Systems," Second Edition, Prentice-Hall of India Private Limited, 1976.
12. Kraus, J. D.: "Electromagnetics," Fourth Edition, McGraw-Hill, Inc., 1992.
13. Kreyszig, E.: "Advanced Engineering Mathematics," New Age International (P) Limited, 1996.
14. Laud, B. B.: "Electromagnetics," Second Edition, Wiley Eastern Limited, 1952.
15. Mendiratta, R. G., Sawhney, B. K.: "Fundamentals of Electricity and Magnetism," Affiliated East-West Press Pvt. Limited, 1987.
16. Pearce, B. O., Foster, R. M.: "A Short Table of Integrals," Fourth Edition, Oxford & IBH Publishing CO., 1966.
17. Ramo, S., Whinnery, J. R., and Duzer, T. V.: "Fields and Waves in Communication Electronics," Third Edition, John Wiley & Sons, 2002.
18. Rao, N. N.: "Elements of Engineering Electromagnetics," Sixth Edition, Pearson Education, 2006.
19. Rao, N. N.: "Elements of Engineering Electromagnetics," Third Edition, Prentice-Hall of India Private Limited, 1992.
20. Reddick, H. W., and Miller F. H.: "Advanced Mathematics for Engineers," Asia Publishing House, 1962.
21. Say, M. G.: "Alternating Current Machines," Fifth Edition, ELBS with Longman.
22. Sinha, U.: "Electromagnetic Theory and its Applications," Satya Prakasan, Sixth Edition, 1988.
23. Smith, S. G.: "Basic Theory of Classical Electromagnetism," Cambridge University Press, 1997.

APPENDIX G

ANSWERS

Chapter 1

Answers to Multiple-choice Questions

1.1 (b). **1.2** (a). **1.3** (d). **1.4** (c). **1.5** (b). **1.6** (d). **1.7** (a) and (c)
1.8 (a) and (d). **1.9** (d). **1.10** (a). **1.11** (b). **1.12** (b). **1.13** (c). **1.14** (a).
1.15 (c).

Answers to Problems

1.1 14 m.

1.2 0.5385 m.

1.3 0.4 m.

1.4 (a) (3.61 m, –56.3°, –1 m); (b) (3.74 m, 105.5°, –56.3°).

1.6 3.61 m, 56.26°, 30°.

1.7 (a) 4.366 m, (b)1.942 m; (c) 0.803 m.

1.8 $0\ 5\mathbf{u}_x - 2.0\mathbf{u}_y + 1.0\mathbf{u}_z$ m; $-0.218\mathbf{u}_x + 0.873\mathbf{u}_y - 0.436\mathbf{u}_z$.

1.9 $0.464\mathbf{u}_x + 0.804\mathbf{u}_y + 0.371\mathbf{u}_z$.

1.10 $1\ 5\mathbf{u}_x + 2.5\mathbf{u}_y + 4\mathbf{u}_z$ m.

1.11 $2\ 69\ \ _x + 1.11\mathbf{u}_y + 0.5\mathbf{u}_z$ m.

1.12 $0\ 5\mathbf{u}_x + 0.866\mathbf{u}_y$.

1.13 $(\mathbf{u}_x\ \ \mathbf{u}_y)/\sqrt{2}$.

1.14 $15, 80\mathbf{u}_x - 30\mathbf{u}_y - 40\mathbf{u}_z, 81°$.

1.15 1.69.

1.16 0, the vectors are parallel.

1.17 $5\ \ _x\ \ 11\mathbf{u}_y + 10\mathbf{u}_z$.

1.18 $0.371\mathbf{u}_x + 0.557\mathbf{u}_y + 0.743\mathbf{u}_z$.

1.19 $0.743\mathbf{u}_x + 0.557\mathbf{u}_y - 0.371\mathbf{u}_z$.

1.20 41.32°.

1.21 2/3.

1.22 $-4\ \ _x - 16\mathbf{u}_y - \mathbf{u}_z$.

1.25 $\dfrac{\mathbf{u}_r}{4} - \dfrac{\mathbf{u}_\phi}{4} + \dfrac{\mathbf{u}_z}{3}$.

1.26 51.6°.

1.27 $F_x = \dfrac{x^2 - y^2}{(x^2 + y^2)^{3/2}}$, $F_y = \dfrac{2xy}{(x^2 + y^2)^{3/2}}$, $F_z = \dfrac{y}{xz}$.

1.28 0.353.

1.29 6.9 units of area.

1.30 (a) 10.61 V and 1.414 A; (b) 0.866; (c) 13 W; (d) $6.494 + j3.75\ \Omega$. $13 + 15\cos(2\omega t - \pi/6)$ W; $13 + j7.5$ W.

1.31 $18.46 + j7.68\ \Omega$; 18.46 Ω; 7.68 Ω; 24.4 mH; 960 W.

Chapter 2

Answers to Multiple-choice Questions

2.1 (a). **2.2** (c). **2.3** (b) and (d). **2.4** (a) and (c). **2.5** (d). **2.6** (b).
2.7 (a). **2.8** (c). **2.9** (c). **2.10** (d). **2.11** (d). **2.12** (b) and (c).
2.13 (a) and (d). **2.14** (b) and (d). **2.15** (a) and (b). **2.16** (b). **2.17** (a). **2.18** (a).
2.19 (c). **2.20** (b) and (c). **2.21** (d). **2.22** (c) and (d). **2.23** (b) and (c).
2.24 (b) and (c). **2.25** (a) and (c). **2.26** (d). **2.27** (b). **2.28** (b) and (d).
2.29 (a) and (b). **2.30** (a). **2.31** (a). **2.32** (c).

Answers to Problems

2.1 $-40.91\mathbf{u}_x - 24.55\mathbf{u}_y$ N.

2.2 $-16.07\mathbf{u}_x - 6.49\mathbf{u}_y + 8.41\mathbf{u}_z$ N.

2.3 1.921 N, –51.6°.

2.4 $(14.464 \times 10^{-6})(0.534\mathbf{u}_x - 0.802\mathbf{u}_y - 0.267\mathbf{u}_z)$ N.

2.5 $x_2 = 1.05$ m, $y_2 = -0.9$ m.

2.6 $(-8.05\mathbf{u}_x + 1.98\mathbf{u}_y - 6.54\mathbf{u}_z) \times 10^{-3}$ N.

2.7 $(-2.364\mathbf{u}_x + 5.360\mathbf{u}_z) \times 10^{-3}$ N.

2.8 $-2.06\mathbf{u}_x - 5.76\mathbf{u}_y$ V/m.

2.9 (4/3 m, 8/3 m, 14/3 m).

2.12 $0.9\mathbf{u}_r$ V/m.

2.13 (a) 0.432 $\mathbf{u}_y$ V/m; (b) $0.259\mathbf{u}_x + 0.346\mathbf{u}_y$ V/m.

2.14 $0.692\mathbf{u}_x + 1.038\mathbf{u}_y$ V/m.

2.15 $(52.9\mathbf{u}_x + 211.8\mathbf{u}_z) \times 10^{-3}$ V/m.

2.17 $0.154\mathbf{u}_z$ V/m.

2.18 $48.6\mathbf{u}_x - 24.3\mathbf{u}_y$ V/m.

2.19 $E_x = -\dfrac{\rho_\ell}{4\pi\varepsilon_0}$ [illegible] a [illegible] b [illegible]. $E_y = \dfrac{\rho_\ell}{4\pi\varepsilon_0 b}\left([illegible] + \dfrac{a}{a [illegible]\ b^2} \right)$, – [illegible] $_x + 1.866\mathbf{u}_y$ V/[illegible].

2.20 $1028.6\mathbf{u}_y$ V/m.

2.21 28.8 $\mathbf{u}_x$ kV/m; $3.172\mathbf{u}_x$ [illegible] $5\ 75\mathbf{u}_y$ kV/[illegible].

2.23 (a) 141.1$\mathbf{u}_z$ V/m; (b) 705.9 $\mathbf{u}_z$; V/m; (c) –141.1 $\mathbf{u}_z$ V/m.

2.25 [illegible]2 [illegible]$_x$ $+11.00\mathbf{u}_y + 18.32\mathbf{u}_z$ V/m.

2.26 y^2 [illegible] $1.5x^2$ + constant.

2.27 2.547 nC.

2.28 (a) (619.18 + 24.79yz) $\mathbf{u}_x$ nC/m²; (b) 5.736 μC.

2.29 12 μC.

2.30 15.15 mC.

2.32 5 mC.

2.33 554 μC.

2.34 130.26 pC.

2.35 (a) 0; (b) 33.93 kV/m; (c) 37.7 nC/m.

2.36 79.58 μC/m³, 30 kV/m, 22.5 kV/m.

2.37 5256 V/m, 4046 V/m.

2.38 88.3 kV/m, 59.7 kV/m.

2.39 366.5 kV/m, 1.629 μC located at the centre of the sphere.

2.40 15.429 kV/m, 2.859 kV/m.

2.41 2.826 kV/m, 1.158 kV/m.

2.42 73.03 kV/m, 22.67 kV/m.

2.43 88 μC/m³.

2.44 0.5 μC/m³.

2.46 The value of each of the integrals is $5\pi^2 r_0^4$.

2.47 (a) [illegible] 48 μJ; (b) – 64 μJ; (c) – 144 μJ.

2.48 (a) 460; (b) 385; (c) Not conservative.

2.50 9.5 mJ; Conservative.

2.51 – 600 μJ.

2.52 23.34 V.

2.53 25.59 kV; 20.61 kV; –4.98 kV.

2.54 2025 V.

2.55 1247.7 V.

2.57 0; –3.77 kV.

2.59 176.22 V, [illegible] $_x$ [illegible] $_y$ [illegible] $_z$ V/m.

2.61 2 [illegible] 5 V, [illegible] $_r$ [illegible] $_z$ V/m.

2.63 –[illegible] $_r$ $+ 0.354\mathbf{u}_\theta + 0.866$ [illegible]$_\varphi$ [illegible]/m.

2.65 (a) 1.5×10^{-10} C-m; (b) 0.027 V, $10.8\mathbf{u}_r + 9.35\mathbf{u}_\theta$ mV/m; (c) 0.0276 V.

2.66 0.0686 V, $14.52\mathbf{u}_r + 4.54\mathbf{u}_\theta$ mV/m.

2.67 0.0686 V, $14.63\mathbf{u}_r + 4.57\mathbf{u}_\theta$ mV/m.

2.68 102.2 mJ.

2.69 1564.5 nJ.

2.70 (a) $(\pi\rho^2 a^4)/(16\varepsilon_0)$; (b) 3.47 μJ/m.

2.71 $(2\pi\rho_S{}^2 a^3)/\varepsilon_0$, 177.4 μJ.

2.72 442.7 nC/m^2.

2.73 11.17 nC/m^2.

2.75 (a) 3.0; (b) 2.0; (c) 70.8 nC/m^2.

2.76 185.9 nC/m^2.

2.78 1.121 μC/m^2, 22.1°.

2.79 4517.7μJ/m^3, 6149.1μJ/m^3.

2.80 $342.4\mathbf{u}_x + 223.2\mathbf{u}_y$ V/m.

2.81 $V_1 = \dfrac{VC_2}{C_1 + C_2}, V_2 = \dfrac{VC_1}{C_1 + C_2}$.

2.82 0.711.

2.83 1.143.

2.84 7.87 V.

2.85 62.1 pF.

2.86 (a) 2.656 nF, 4.427 nF, 1.659 nF; (b) 64 kV; (c) 3.398 J.

2.87 (a) 60 kV; (b) 4 kV; (c) 9 kV.

2.89 (a) 229. 8 pF/m, 389.9 pF/m, 144.5 pF/m; (b) 8.9 V.

2.90 40.06 pF.

2.91 12.65 μN, 50.6 nJ.

Chapter 3

Answers to Multiple-choice Questions

3.1 (c). **3.2** (b). **3.3** (d). **3.4** (a). **3.5** (a). **3.6** (d). **3.7** (c) and (d).
3.8 (b) and (d).

Answers to Problems

3.1 $t_2 = 65.79 + 1.25t_1$.

3.2 0.005687 per degree centigrade at 0°C.

3.3 786.8 °C.

3.4 $8.06 \times 10^{-4}\ \Omega$.

3.5 249 A.

3.6 280.7 A.

3.8 281.5 W, 168.1 W.

3.11 $5\ 0\mathbf{u}_x + 6.105\mathbf{u}_y + 3.053\mathbf{u}_z$ A/mm^2.

3.12 $3\ 77\quad_x + 3.381\mathbf{u}_y + 5.638\mathbf{u}_z$ A/mm^2.

Chapter 4

Answers to Multiple-choice Questions

4.1 (a) and (d). **4.2** (a) and (b). **4.3** (c). **4.4** (b). **4.5** (b). **4.6** (d).
4.7 (b) and (c). **4.8** (a). **4.9** (b) and (c). **4.10** (a). **4.11** (a). **4.12** (b).
4.13 (d). **4.14** (c). **4.15** (a) and (d). **4.16** (b) and (c). **4.17** (b). **4.18** (b).
4.19 (a) and (c). **4.20** (a). **4.21** (c). **4.22** (d). **4.23** (c) and (d).
4.24 (b) and (d). **4.25** (a) and (d). **4.26** (a). **4.27** (c). **4.28** (b). **4.29** (d).
4.30 (c).

Answers to Problems

4.1 $(\quad .712\mathbf{u}_x + 1.232\mathbf{u}_y) \times 10^{-2}$ A/m.

4.2 $(\quad .133\mathbf{u}_y - 12.227\mathbf{u}_z) \times 10^{-3}$ A/m.

4.3 $(\quad .884\mathbf{u}_x + 1.413\mathbf{u}_y + 4.593\mathbf{u}_z) \times 10^{-3}$ A/m.

4.5 $42.191 \times 10^{-6}\,\mathbf{u}_z$ T.

4.6 (a) 4906.4×10^{-7} T; (b) 2916.9×10^{-7} T; (c) 133.8×10^{-7} T.

4.7 5366.6 A/m; 5234.4 A/m. The field is along the axial direction.

4.9 $(\qquad .93\mathbf{u}_x + 32.20\mathbf{u}_y) \times 10^{-7})$ T.

4.10 (a) 0; (b) $-7.548\mathbf{u}_x$ A/m.

4.11 $0.2818\mathbf{u}_x + 0.1057\mathbf{u}_y$ A/m.

4.12 $-1599.2 \times 10^{-7}\mathbf{u}_z$ T.

4.14 $-4.8 \times 10^{-4}\mathbf{u}_r$ N.

4.15 $0.6\mathbf{u}_x - 0.96\ \mathbf{u}_y + 0.32\ \mathbf{u}_z$ N.

4.16 (a) 19.62 A, (b) 20.87 A.

4.17 $(75.0\ \mathbf{u}_x + 43.3\ \mathbf{u}_y) \times 10^{-7}$ N/m.

4.18 $8.0\ \mathbf{u}_x + 5.0\ \mathbf{u}_y$ N.

4.20 $12.0\ \mathbf{u}_y$ N-m.

4.21 $-19.2\mathbf{u}_x + 12.0\ \mathbf{u}_y$ N-m.

4.22 $NISB_y \cos\phi\, \mathbf{u}_z$.

4.23 $11.94\mathbf{u}_x - 3.98\ \mathbf{u}_z$ A/m.

4.24 $-2.35\mathbf{u}_x + 1.96\ \mathbf{u}_y$ A/m.

4.25 $-0.858\mathbf{u}_x - 0.507\ \mathbf{u}_y + 0.908\ \mathbf{u}_z$ A/m.

4.26 $-12.73\mathbf{u}_x + 6.36\ \mathbf{u}_y$ A/m.

4.28 0; $2148.3\mathbf{u}_\phi$ A/m; $3410.5\ \mathbf{u}_\phi$ A/m.

4.29 (400/*R*) A/m; Average $H = 1823.2$ A/m; *H* at the mean radius = 1818.2 A/m. The value of *H* at the mean radius is nearly equal to the average value because the radial depth of the toroid is not large compared to its radius. The field is in circumferential direction.

4.30 (a) 5093.0 A/m, 3819.7 A/m, 4456.4 A/m; (b) 4365.4A/m; (c) 6.904 mWb.

4.32 (a) $\frac{J_{z0}}{r}\left\{\frac{1}{\alpha^2}+e^{\alpha r}\left(\frac{r}{\alpha}-\frac{1}{\alpha^2}\right)\right\}\mathbf{u}_\phi$ A/m; (b) $\frac{J_{z0}}{r}\left\{\frac{1}{\alpha^2}+e^{\alpha a}\left(\frac{a}{\alpha}-\frac{1}{\alpha^2}\right)\right\}\mathbf{u}_\phi$ A/m.

4.33 16.0 mN; Attractive.

4.34 $111.12\mathbf{u}_x$ A/m².

4.35 $J_{z0}e^{\alpha r}\mathbf{u}_z$.

4.37 $\mathbf{B} = \dfrac{\mu I \ell\, \mathbf{u}_\phi}{(4\pi r)\sqrt{r^2+(\ell/2)^2}}$.

4.40 $(40.0\mathbf{u}_r + 20.0\mathbf{u}_\theta) \times 10^{-11}$ T.

4.42 1800; 400 A/m; 300 A/m.

4.44 59.0°; (a) $5 \times 10^{-4}\mathbf{u}_x + 0.3\ \mathbf{u}_z$ T, 0.095°; (b) $0.25\mathbf{u}_x + 0.3\mathbf{u}_z$ T, 39.8°.

4.45 (a) 0.1122 T, 0.1442 T; (b) 24.2°; 30.0°.

4.46 $0.20\mathbf{u}_x + 0.0176\mathbf{u}_z$ T.

4.47 (a) 58.7°; (b) $750.6\mathbf{u}_x + 601.6\mathbf{u}_y + 227.1\ \mathbf{u}_z$ T; (c) 44.5°.

4.48 (a) $265.3 \times 10^3 H^{-1}$; $303.2 \times 10^3 H^{-1}$; (b) 341.1 A (ampere-turns).

4.49 2.13 A.

4.50 (a) $212.1 \times 10^3 H^{-1}$, $104.9 \times 10^3 H^{-1}$; (b) 10.625 mWb; (c) 2253.8 ampere-turn; 1114.6 ampere-turn.

4.51 3.66 A; Clockwise.

4.52 4.38 A.

4.53 1.667 mWb; 741.8 ampere-turns.

4.54 0.5786 H.

4.55 2.468 mH.

4.56 4.05 H; 1.89 H.

4.57 1.389 H; 0.562 H.

4.62 0.947 μH.

4.63 (a) 61.46 mH; (b) 58.39 mH; (c) 119.85 mH; (d) 239.7 mH.

4.65 $(a \quad N/\pi)\ln[(s+b)/(s-b)]$.

4.67 $(\quad)/(16\pi); \mu_0/(8\pi)$; 0.25 mJ/m.

4.68 $(\quad)\{c^4+3b^4-4b^2c^2+4b^4\ln(c/b)\}/\{(16\pi(c^2-b^2)^2\}$;

$\mu \; c^4+3b^4-4b^2c^2+4b^4\ln(c/b)\}/\{8\pi(c^2-b^2)^2\}$; 0.136 mJ/m.

4.70 28.2 J.

4.71 102.3 mJ.

4.72 2910 amp-turns.

4.74 $10.472\times10^4\mathbf{u}_x$ m/s; 361.89×10^{-19} J.

4.76 (a) 1.2×10^{-15} N, 46.0×10^{10} m/s^2; (b) 487.5×10^{-4} m; (c) 2.042 μs.

4.77 $4.005\times10^{-16}\mathbf{u}_z - 4.806\times10^{-18}\mathbf{u}_r$ N; Linear, Circular, and Helical.

Chapter 5

Answers to Multiple-choice Questions

5.1 (c) and (d). **5.2** (b). **5.3** (a). **5.4** (c). **5.5** (b). **5.6** (a) and (c). **5.7** (c).
5.8 (a). **5.9** (b). **5.10** (d). **5.11** (c). **5.12** (a) and (c).

Answers to Problems

5.1 $V \quad 20000x+300$ V; $Q_1=-89.01$ nC; $Q_2=89.01$ nC.

5.2 $V_1=\dfrac{V_0\varepsilon_2 x}{\varepsilon_1 s_2+\varepsilon_2 s_1}; V_2=\dfrac{V_0\{\varepsilon_1 x+s_1(\varepsilon_2-\varepsilon_1)\}}{\varepsilon_1 s_2+\varepsilon_2 s_1}$.

5.4 $V=\dfrac{(V_1-V_2)(\phi-\alpha_1)}{\alpha_1 \quad \alpha_2}+V_1$.

5.5 $V=31.27$ V; $E=27.57\mathbf{u}_\theta$ kV/m.

5.6 $V \quad V_1-V_2)\dfrac{\ln\left(\tan\dfrac{\theta}{2}\right)-\ln\left(\tan\dfrac{\alpha_1}{2}\right)}{\ln\left(\tan\dfrac{\alpha_1}{2}\right)-\ln\left(\tan\dfrac{\alpha_2}{2}\right)}+V_1; 14.16^\circ$.

5.7 $V_1 \quad \dfrac{\rho}{6\varepsilon_0}(3a^2-r^2), r<a; V_2=\dfrac{\rho a^3}{3\varepsilon_0 r}, r>a; \mathbf{E}_1=\dfrac{\rho r}{3\varepsilon_0}\mathbf{u}_r, r<a; \mathbf{E}_2=\dfrac{\rho a^3}{3\varepsilon_0 r^2}\mathbf{u}_r, r>a$

5.9 $V=\dfrac{4V_0}{\pi}\left(e^{-(\pi x/a)}\cos\dfrac{\pi y}{a}-\dfrac{1}{3}e^{-(3\pi x/a)}\cos\dfrac{3\pi y}{a}+\dfrac{1}{5}e^{-(5\pi x/a)}\cos\dfrac{5\pi y}{a}-\cdots\right.$.

5.10 (a) 1.76 kV, $(1.449\mathbf{u}_x-2.02\mathbf{u}_y)\times10^3$ V/m; (b) -5.9 nC/m^2.

5.16 $2.9\times10^{-6}\,\Omega$.

5.17 (a) 663.1×10^3 H^{-1}.

Chapter 6

Answers to Multiple-choice Questions

6.1 (c). **6.2** (b). **6.3** (c). **6.4** (b). **6.5** (a). **6.6** (c). **6.7** (d). **6.8** (d).
6.9 (a). **6.10** (a) and (c). **6.11** (a). **6.12** (b). **6.13** (c). **6.14** (d).

Answers to Problems

6.1 13.88 V.

6.2 6.03 sin $(100\pi)t$ V.

6.4 –2.5 cos $2500t\,\mathbf{u}_\phi$ V/m.

6.7 0.168 V.

6.8 2.25 sin x V.

6.9 9.43 cos$(100\pi)t$ V; 50 Hz.

6.10 2.09 V.

6.12 58.3 V; 82.47 V.

6.13 $\mathbf{J}_d = 0.08\cos\{(\pi \times 10^7)t - 0.2z)\}\mathbf{u}_x$ A/m².

6.14 473.3×10^{-6}.

6.16 6.95 A/m².

6.17 $-(3.54 \times 10^{-4})\cos(10^8 t - 0.4z)\mathbf{u}_x$ A/m.

6.18 109.66 V/m; 90 14 V/m.

6.19 $1499.7\cos[(2\pi\times 10^8)t - (2\pi/3)z]\mathbf{u}_z$ V.

6.20 $(1.667 \times 10^{-10})\sin[(2\pi\times 10^7)(t - 10^{-6})]\mathbf{u}_z$ T-m.

6.22 (a) $\frac{1}{r}(46.19 - j19.13)\mathbf{u}_r$ V/m; $\frac{1}{r}(1.73 - j)\mathbf{u}_\phi$ A/m; (b) $\frac{1}{r^2}(49.52 + j6.52)\mathbf{u}_z$ W/m²; (c) 341.8 W.

Chapter 7

Answers to Multiple-choice Questions

7.1 (c) and (d). **7.2** (a) and (c). **7.3** (b). **7.4** (a). **7.5** (d). **7.6** (b).
7.7 (b) and (c). **7.8** (b) and (d). **7.9** (d). **7.10** (a). **7.11** (b). **7.12** (d).
7.13 (c). **7.14** (b). **7.15** (a) and (d). **7.16** (a) and (c). **7.17** (b) and (d).
7.18 (b) and (c). **7.19** (a) and (d). **7.20** (b) and (d). **7.21** (b) and (c). **7.22** (a) and (d).

Answers to Problems

7.2 (a) 1.5×10^8 m/s; 2.09 rad/m, 3.0 m; (b) 188.5 Ω.

7.3 –60.0 V/m; –0.159 A/m in y direction.

7.4 (a) **E** [illegible] $\mathbf{u}_x$ [illegible] $\mathbf{u}$)cos($t - \beta z$) [illegible]/m; (b) 37.7 V/m; 0.1 A/m; (c) 90°

7.6 (a) 60cos[2 [illegible] 15 10 [illegible] 0.597 [illegible] y [illegible] $60e^{0.597z}\mathbf{u}_z$ V/ [illegible]

[illegible] $\times 10$ [illegible] $-0.597z$] [illegible]; –0.302 [illegible] 0597 [illegible] /m;

(b) $9\,06[1+\cos 2\{(2\pi\times 15\times 10^6)t-0.597z\}]\mathbf{u}_z$ W/m^2.

7.8 0.0044 neper/m ; 1.11 rad/m.

7.10 $177.62 + j0.062\ \Omega$.

7.12 0.251 neper/m; 50.34 rad/m.

7.13 11.24 MHz; 14.89 neper/m; 14.89 rad/m; $14.89(1 + j)$ m^{-1}; 6.71 cm; $2.98(1 + j)\ \Omega$; 4.74×10^6 m/s; 42.2 cm.

7.15 10.7×10^3 neper/m; 10.7×10^3 rad/m; $10.7 \times 10^3\ (1 + j)$ m^{-1}; 0.0935 mm; $1.84 \times 10^{-4}(1 + j)\ \Omega$.

7.16 $50\sin(30\pi\times 10^6 t)\sin(0.314z)\mathbf{u}_y$ V/m; $-0.133\cos(30\pi\times 10^6 t)\cos(0.314z)\mathbf{u}_x$ A/m.

7.17 (a) $\mathbf{E} = 12e^{-144z}\sin[2\pi(150\times 10^3)t - 144z]\mathbf{u}_x$ V/m,

$\mathbf{H} = (2.04\times 10^3)e^{-144z}\sin[2\pi(150\times 10^3)t - 144z - \pi/4]\mathbf{u}_y$ A/m;

(b) 6.54 m/s; (c) 0.135.

7.20 (a) -0.3097, 0.6903; (b) $12e^{-j0.0335z}\,\mathbf{u}_x$ V/m; $-3.716e^{-j0.0335z}\,\mathbf{u}_x$ V/m; (c) $8.284e^{-j0.0636z}\,\mathbf{u}_x$ V/m, $0.0417e^{-j0.0636z}\,\mathbf{u}_y$ A/m; (d) 0.1727 W/m^2 .

7.23 (a) 5/3; (b) 50 MHz .

7.24 (a) 188.5 Ω and 188.5 Ω; (b)188.5 Ω.

7.25 (a) -0.1027 A/m, 0.0593 A/m; (b) -0.1027 A/m, -0.0593 A/m.

7.26 42.2 V/m.

7.27 34.50 V/m; –25.15 V/m.

7.29 60°.

7.30 (a) $\varepsilon_1/\varepsilon_2 = 3$; (b) 60°.

Chapter 8

Answers to Multiple-choice Questions

8.1 (a) and (c). **8.2** (d). **8.3** (b) and (c). **8.4** (a). **8.5** (a) and (d). **8.6** (d).
8.7 (b). **8.8** (b) and (c). **8.9** (a). **8.10** (d). **8.11** (c). **8.12** (b) and (c).
8.13 (b). **8.14** (c). **8.15** (c). **8.16** (a).

Answers to Problems

8.1 0.176 Ω/m.

8.4 7.768.

8.5 (a) $75\cos\{(2\pi\times 10^8)t - 3.974z\}$ V, $0.593\cos\{(2\pi\times 10^8)t - 3.974z\}$ A;

(b) $75e^{-j3.974z}$ V; $0.593e^{-j3.974z}$ A.

8.8 11.76 W.

8.9 $0.594\ \angle -16.8°$.

8.11 $0.621\ \angle -68.5°$.

8.12 66.5 W.

8.14 (a) 1.656; (b) 8.0 m; $\pi/4$ rad/m.

8.15 (a) 115.2 Ω; (b) 23.98×10^{-8} H; 18.07×10^{-12} F.

8.18 (a) 150.1 Ω, 28.23 Ω; (b)150 Ω.

8.19 0.005 neper/m; 1.05 rad/m, $44.71 - j0.47$ Ω.

8.20 $0.3098 \angle -26.1°$; 1.9.

8.21 62.5 Ω.

8.22 (a) 0.642 A; (b) 32.46 W; (c) 0.543 A.

8.24 $115.7 + j56.1$ Ω.

8.25 (a) 107.15 V, 0.429 A; (b) 99.17 %.

8.26 (a) 1.17; (b) 0.034, 0.009.

8.27 (a) 101 Ω; (b) 55.6 Ω.

8.28 $\ell_s/\lambda_s = 0.164$ or 0.336; $\ell_m/\lambda_m = 0.117$ or 0.321.

Chapter 9

Answers to Multiple-choice Questions

9.1 (a) and (c). **9.2** (b). **9.3** (c). **9.4** (a) and (d). **9.5** (b) and (d). **9.6** (a) and (c).
9.7 (a). **9.8** (a) and (d). **9.9** (d).

Answers to Problems

9.3 2.0 GHz; 4.0 GHz.

9.4 3.6 GHz.

9.5 1 and 2.

9.6 (a) 51.3 rad/m; (b) 0.123 m; (c) 263.9 Ω.

9.7 125 Ω.

9.8 $50(\sin 20\pi x) \cos [(7.5\pi \times 10^9)t - 47.1z]$ V/m; $-0.08(\sin 20\pi x) \cos [(7.5\pi \times 10^9)t - 47.1z]$ A/m.

9.9 (a) 1.643×10^8 m/s; (b) 0.913×10^8 m/s; (c) 206.5 Ω; (d) 114.7 Ω.

9.11 (a) 3.32 W/m; (b) 1.63 W/m.

9.12 1.667 GHz; 1.912 GHz.

9.13 9.07 GHz; 6.68 GHz.

9.15 (a) 26.17 rad/m; (b) 3.84×10^8 m/s; (c) 24.0 cm; (d) 2.34×10^8 m/s.

9.16 (a) 10.0 cm; (b) 1.5 GHz to 3.0 GHz.

9.17 (a) 3.606 GHz; (b) 281.0 Ω, 505.8 Ω.

9.18 400 V/m; 1.502 A/m; 0.938 A/m.

9.19 156.25 V/m; 0.293 A/m; 0.469 A/m.

9.20 2.76 W; 0.552 W.

9.21 0.363 W.

Chapter 10

Answers to Multiple-choice Questions

10.1 (a). **10.2** (c). **10.3** (d). **10.4** (a) and (c). **10.5** (d). **10.6** (b) and (c).
10.7 (a). **10.8** (b). **10.9** (a). **10.10** (c) and (d).

Answers to Problems

10.1 $-(1.06 \times 10^{-8}) \cos (15\pi \times 10^6)t$ C-m.

10.2 2.0×10^{-7} T-m in z direction.

10.4 0.04 V/m; 10.61×10^{-5} A/m.

10.5 (a) 35.53 W; (b) 0.079 Ω; (c) 2.12 W/steradian; (d) 0.75.

10.6 88.4 %.

10.7 (a) 0.201 V/m, 0.533 mA/m; (b) 0.32 V/m, 0.849 mA/m.

10.11 (a) 47.06 W/steradian; (b) 119.4 W/steradian.

10.12 119.38 W/steradian.

10.13 $(2/3)g_d(\max)$.

10.14 (a) 913.5 W; (b) 0.056 Ω; (c) nearly 100 %.

10.16 (a) $E_1 = 0.15\sin\{[(200\pi\times10^6)t - 296.1]-[0.555-\pi/4]\}$ V/m,

$E_2 = 0.15\sin\{[(200\pi\times10^6)t - 296.1]+[0.555-\pi/4]\}$ V/m;

(b) $E_R = 0.292\sin[(200\pi\times10^6)t - 296.1]$ V/m; (c) 0.973.

10.17 (a) $E_1 = 0.15\sin[(200\pi\times10^6)t - 296.1]$ V/m,

$E_2 = 0.15\sin\{[(200\pi\times10^6)t - 296.1]-0.46\}$ V/m;

(b) $E_R = 0.292\sin\{[(200\pi\times10^6)t - 296.1]-0.23\}$ V/m;

(c) 0.973. The phase difference is 0.23 rad.

10.18 0.605.

10.19 $-\sin[(\pi \cos \phi)/2]$.

10.20 (a) 0.223; (b) 0.0439.

10.21 0.956; 0.914.

10.22 60°.

10.23 23.07°; 73.74°.

10.24 4.64 mV.

10.25 4.21 mV.

10.26 57.2 mV.

10.28 2.09 m^2.

10.30 14.03×10^{-6}.

10.31 28.5 μW.

INDEX

A

B

C

D

E

F

G

H

I

J

K

L

M

N

O

P

R

S

T

V

W

Y

Made in the USA
Monee, IL
11 October 2026

65356106R10404